Cattle Plague

A History

Cattle Plague
A History

C. A. Spinage

Kluwer Academic / Plenum Publishers
New York, Boston, Dordrecht, London, Moscow

Library of Congress Cataloging-in-Publication Data

Spinage, C. A. (Clive A.)
Cattle plague : a history / C.A. Spinage.
p. cm.
Includes bibliographical references (p.).
ISBN 0-306-47789-0
1. Rinderpest--History. I. Title.

SF966.S65 2003
636.2'0894575--dc21

2003047708

ISBN 0-306-47789-0

233 Spring Street, New York, NY 10013

http//www.wkap.nl/

10 9 8 7 6 5 4 3 2 1

A C.I.P. record for this book is available from the Library of Congress.

Printed in the United States of America

Frontispiece. The memorial tablet on "Plague Cottages" (photo C. A. Spinage).

FOREWORD

As one who has spent 50 years fighting the murrain known as 'Rinderpest" it gives me pleasure to draw attention to "A Veritable Rinderpest Encyclopaedia' and to a clade of Morbilliviruses descended aeons ago from Rinderpest virus. The related terrestrial viruses are peste des petits ruminants, measles, and canine distemper. In 1988 aquatic viruses suddenly emerged and killed many phocine and cetacean mammals.

All the related clade viruses exhibit a characteristic behaviour. If they infect a naive host population they provoke epidemics and pandemics that kill thousands. If the host population is large, epidemics may become persistent endemics restricted to the young with waning maternal immunity. All the clade viruses are excellent immunogens producing life-long immunities in fully susceptible hosts.

Rinderpest, for centuries, was the most dreaded ox plague known. Effective control measures were edicts drawn up in 1715 by Giovanni Maria Lancisi, the Pope's personal physician. Quarantine and isolation were enforced. Suspected sick cattle were killed. The efforts of the Papal State succeeded but elsewhere the belief that rinderpest could be cured.

The principal professor of medicine in Padua was Bernardino Ramazzi and he attempted to keep cattle alive by scarification in the belief that the pustules would protect as if it was smallpox. The majority died but a few survived. Others like Professor Pieter Camper stated that in Holland inoculation with infected fluids had both failed and succeeded.

When European countries closed their borders to cattle the Tsar was stimulated to build several Experiment Stations in 1860 to elucidate the cause of rinderpest and its nature. Professor E. Semmer led the Dorpat Group and they discovered in 1893 that if serum collected from recovered cattle was inoculated into a susceptible one it was protected against rinderpest. Shortly, thereafter the Great African Rinderpest Pandemic hit Southern Africa and the prophylactic role of rinderpest – antiserum was confirmed.

Antiserum alone was in vogue until the 1920s but its protection was short-lived. Aid personnel from Europe developed the antiserum-virus simultaneous method of immunisation which conferred a life-long resistance.

The continental pandemic in Southern Africa was a classic example of a morbillivirus slaughter of a naive population; rinderpest virus burnt itself out. In most African states north of 10 degrees south the virus persisted endemically for years. For example, French Equatorial Africa was infected in 1893 to 1913.

In the first World War rinderpest spread from Asiatic Russia into Poland and then into south-east Europe. The remnants of the epidemic were eliminated in Greece in 1921. In 1930 Europe was free of the virus except in eastern Turkey.

The virus was endemic also in several countries in Asia including China which was the source of epidemics in Indonesia and Japan. Japan's response was to develop an inactivated tissue vaccine which was a novel glycerinated organ pulp. It was developed by Dr. C. Kakizaki who published in Japanese in 1902. By 1925 every rinderpest laboratory was producing belatedly an inactivated vaccine.

In the first World War rinderpest spread from Asiatic Russia into Poland and then into south-east Europe. The remnants of the epidemic were eliminated in Greece in 1921.

A spectacular illegal outbreak occurred in 1920 when a cargo of infected zebus were sent from India to Brazil via Antwerp. The disease broke out in Belgium and Brazil. This episode is the only known rinderpest in the Americas. An important sequel was the foundation of the Office International des Epizooties (O.I.E) in Paris as a world clearing-house on animal diseases data.

A second spectacular outbreak occurred in Australia in 1923. The origin was never established but suspicion fell upon pigs that were live provisions bought in Singapore. The surviving pigs were sold to a butcher in Freemantle. The outbreak was stamped out within days by slaughtering 3,000 cattle, pigs, sheep and goats. Australia remains free of rinderpest. Meantime in India Dr. J. T. Edwards obviated the major risk in the anti-serum – virus simultaneous method of immunisation of transmitting other pathogens in the virulent virus inoculum. He passaged a bovine strain of rinderpest virus serially through goats to "fix" it and, fortuitously, produced a stable goat-adapted virus virulent for goats and yet attenuated for cattle. Anti-serum simultaneously was not necessary. Moreover, one dose of goat-adapted vaccine inoculated into a yearling calf protects it from rinderpest for life. In the 1940s Dr. R. Daubney predicted rinderpest would be eradicated through mass vaccination of national herds.

Goat-adapted rinderpest vaccine was far from being the ideal. Vaccinated animals develop fevers that last for 4–5 days. The vaccine rarely kills but it immuno-suppresses activating other latent infections.

Dr. Junji Nakamura visited the Indian Veterinary Research Institute in the 1920s and was impressed by Edwards's ideas on passaging a virus serially in a foreign host. When he returned to Korea he passaged his local strain of rinderpest serially in rabbits, goats being few in Korea. His Nakamura III strain was considered "fixed" after 600 passages and it proved to be more attenuated than Edwards's goat vaccine. This lapinised vaccine was used first in Mongolia and later in China. Nakamura forwarded a sample of his strain to the Kenya Veterinary Department in 1948 which was passed onto the new East African Veterinary Research Organization. Two years later I was producing lapinised vaccine. In Kenya, the vaccine was used only to immunize pure-bred and grade cattle. Stockowners in Africa elsewhere clamoured for the new vaccine because of its better attenuation.

Tentative steps at mass vaccination to control rinderpest were taken in China in the late 1940s and in equatorial Africa in the 1950s. India initiated a National Eradication Scheme in 1954 using goat adapted vaccine. Over the first few years the success was remarkable; outbreaks fell from thousands per year to a few hundred per year. In the first 10 years, 117 million cattle and buffaloes were vaccinated out of a population of 132 million; as a result two-thirds of the country was freed from rinderpest.

In 1952 an Inter-African Bureau for Epizootic Diseases was created by the Commission for Technical Diseases in Africa South of the Sahara specifically to study rinderpest. Dr. W. G. Beaton was the first Director and he initiated Joint Project 15 to rid Africa of rinderpest. JP 15 operated in the field between 1962 and 1976 in 22 African countries. In 1962, 17 had active rinderpest whereas in 1976 only two countries reported the disease. The goat-adapted vaccine was replaced by Plowright's new cell-cultured virus vaccine in the 1960s.

The gains of JP15 were tragically undone. A virulent virus focus was missed in southern Sudan. It spread south through Uganda in cattle looted by the victorious Tanzania troops. The looted cattle were infected and died en route. Many carcases were abandoned in the Tanzania Ngorongoro National Park. Meantime, a hidden focus in the Niger suddenly erupted and spread across the sahel. One third of the cattle of the Fulani nation died. In Nigeria alone two million fell sick and half a million perished.

In 1987 a fresh Pan-Africa Rinderpest Campaign (PARK) began field operations simultaneously in 34 countries. Mass vaccination was limited to two years and thereafter all vaccines were withheld.

In the 1970s FAO, together with the International Atomic Energy Agency [IAEA], checked veterinary diagnostic laboratories in allegedly high risk countries in Africa and Asia. The objective was to identify the true prevalence of rinderpest. From 1980 FAO recruited three consultants to prepare regional rinderpest campaigns by assessing real disease prevalences, diagnostic facilities, vaccine production plants, vaccine quality controls, available manpower, equipment, and estimates of budgets for the eradication of rinderpest. Their detailed report entitled "Global Eradication of Rinderpest" was presented, discussed and approved at a FAO Expert Consultation in Rome in October 1992.

The Expert Consultation in Rome issued a statement that global eradication of rinderpest was achievable by the year 2010. The 106th FAO Council authorised the Director-General to establish a priority programme the Emergency System for Transboundary Animal and Plant Pests and Diseases to be known by the acronym EMPRES. The initial focus was the Global Rinderpest Eradication Programme [GREP]. The Secretariat collaborates with other international organisations such as OIE, IAEA, European Union and the Inter-African Bureau and the Inter-African Bureau for Animal Resources [IBAR]. The important link is OIE. It oversees the zoosanitary approach necessary to achieve eradication using a practical moving pathway which every country has to follow culminating in a declaration of freedom from rinderpest virus.

GREP was severely tested when launched at the end of 1994 by two novel epidemics; the first was caused by a rare lineage 11 strain of the virus that is lethal in wildlife but avirulent in domesticated cattle. Lineage 11 goes back to the 1930s and is the one rinderpest strain that might delay the global rinderpest eradication programme. The source was attributed to a family group of refugee-owned cattle that crossed Kenya through three National Parks to northern Tanzania from Somalia. Wildlife mortality estimates exceeded 29,000 Cape buffaloes, 3,000 eland and 7,000 giraffes. The refugee cattle in contact with diseased wildlife were infected and caused in late 1996 mild clinical disease in Masai cattle just south of the Nairobi National Park. The Kenya Veterinary authorities warned their Tanzanian counterparts. They declared an emergency which was provided by FAO, UNDP and the European Union through PARK. The border was breeched in 1997. In August and October 1.6 million cattle were vaccinated or revaccinated.

The second novel epidemic was in naive yaks and cattle in Pakistan in 1994. Newly calved buffaloes from the Sind and Punjab Provinces were purchased to stock a new milk colony for the first time in Gilgit. Losses were very high particularly among the yaks and their crosses. The numbers at risk were 500,000 of which 40,000 died. Survivors were sold in Afghanistan but they infected the local cattle.

Minor outbreaks occurred in 1994. Illegal cross-border movements of calves placed cattle at risk in Iran and Turkey.

In 1995 rustled cattle were infected in Turkana, Kenya. There were 480 cattle in 13 herds but only 20 sickened of which 15 died.

Turkey had a minor outbreak in 1996, in 24 young cattle which were slaughtered.

A serious outbreak in 1998 occurred in a village herd of 164 cattle in the Amur Region of Russia. It was totally unexpected because the Russian Federation was considered to be free of the disease. The source is still unknown. There were 70 clinical cases of which

43 died and 4 were killed for samples. The remaining cattle were re-vaccinated and remained healthy.

Rinderpest was not reported in 1999 but in 2000 the Manzoor Milk Colony in Karachi, Pakistan purchased 65 milk buffaloes up-country of which 3 were infected and one died. In 2001, 3,564 samples from cattle and buffaloes reared in the suspect areas of Pakistan were tested by two different serological methods; all were negative in February 2001.

GREP's target is to stop routine vaccination programmes in 2002 and to declare global provisional freedom from rinderpest in 2003.

Past endemic foci have been plotted. Most vulnerable are Nepal, Afghanistan, Tajikistan, Kyrgyzstan, Uzbekistan, Turkmenistan, Kazakstan, Iran, Iraq, Yemen, United Arab Emirates, Oman, Saudi Arabia, Turkey, Georgia, Armenia and Azerbaijan.

Gordon Scott O.B.E.
Ph.D., F.R.C.V.S.

ACKNOWLEDGMENTS

I am most grateful to Norman Comben MRCVS who, over many years, has supplied me with rare books, pamphlets, ephemera, and photocopies, and whose idea it was that I should write this history. I also wish to thank James Brooks for help with references, William Beinart for Afrikaans translations, and especially Professor Gordon Scott OBE FRCVS for reading a draft of the manuscript and making helpful comments, and Gil Child for his information on Tanzania. During the years, many others have contributed knowingly or unknowingly, and to those, my thanks are equally extended.

I also wish to thank the following for permission to use material: the Hertfordshire Record Office for quotation of Thomas Bates's original letter; the Royal Veterinary College for Figure 2 of Chapter 11; ARC-Onderstepoort Veterinary Institute for Figures 4 and 5 of Chapter 23 and Figure 1 of Chapter 24; Atlas Van Stolk, Rotterdam for Figure 4 of Chapter 6.

CONTENTS

LIST OF FIGURES

LIST OF TABLES

Cattle Plague

A History

Part I
INTRODUCTION AND BACKGROUND

1

THE NATURE OF CATTLE PLAGUE OR RINDERPEST

1. INTRODUCTION

"And there was a mortality upon the animals in all the island of Britain," wrote a Welsh chronicler of the year AD 810. Cattle plague or rinderpest, the highly infectious and most lethal of bovine diseases, described as the most dreaded above all of animal contagions, had repeatedly ravaged Europe since time immemorial. Perhaps first brought west from central Asia by the Scythians in the 9th to 8th century BC, the irruption of the Huns from the east of Europe in the 4th century AD could have originated a chain of epizootics which was to be continued by the armies of Charlemagne in the 9th century, leading to the introduction of rinderpest into Britain with the cross-channel exploits of the Danes, or the Saxons themselves. Almost a thousand years later, in the 18th century, rinderpest is estimated to have carried off more than 200 million head of cattle in Europe, exclusive of Siberia and Tartary. Germany alone lost 28 million head between 1711 and 1865, three in every four animals dying. Following its introduction into Britain in 1745, the losses in 1745–57 were estimated at in excess of half a million head. Its introduction in 1865 with a dozen oxen led to the death, including those which were slaughtered, of 278,943 animals, some estimates putting the loss as high as 420,000, representing 7% of the national herd, according to some, affecting livestock farming and the meat trade for the next 25 years. But it was pointed out in 1877 that this loss was over 13,000 head less per year than the average loss each year in the 1870s from various other diseases (especially pleuropneumonia and foot-and-mouth disease (FMD)) (Brown, 1877). In 1931, one worker wrote: "The study of the [rinderpest] plague is at the moment followed throughout the whole world." Belgians, Dutch, English, French, Italians, Japanese, and Persian veterinairians were occupied in studying it (Delpy, 1931; Figure 1). The years 1979–84 saw an upsurge in Africa lead to losses of an estimated 1 million head, a stark reminder of its continuing presence and devastating potential in the world.

One of the most virulent cattle diseases known, rinderpest is acute, fever-producing, and usually fatal, ranking second only to FMD in contagiousness. It was described in Hungary in 1865 as a "febrile contagious malady, peculiar to all ruminants, as cattle, sheep, goats, and deer, which does not attack the same animal more than once in its life, which is not developed with us [i.e., in Hungary] originally, but is always introduced from some neighboring country on the east. When once introduced the malady spreads by means of a very active [intensive] contagious matter, of which ruminants of every race, age, sex, constitution, etc. are susceptible, either directly or indirectly..." (Egan, 1865). An apt description, yet elements of it were to be disputed in Britain for several years later. Ostertag (1916) refers to it being known "centuries ago" in Germany as "Great Gall" (*Grossgalle*) due to the characteristic distension of the gall bladder, its most constant sign. But the name seems to have been unknown by the 18th century.

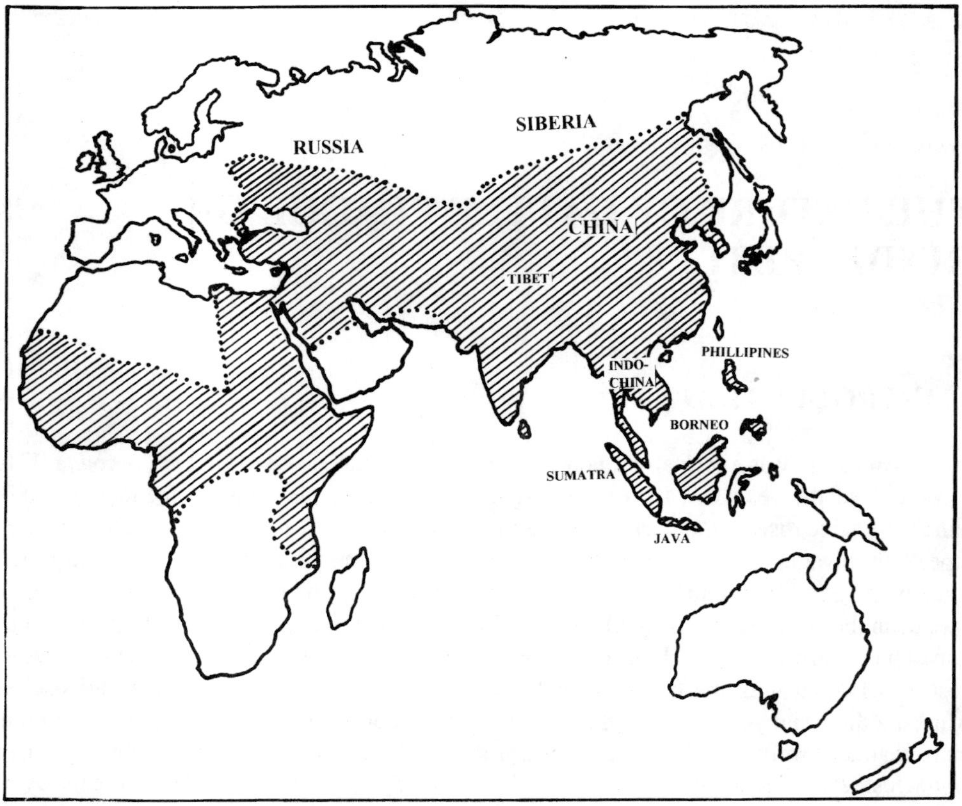

Figure 1. The worldwide incidence of rinderpest in 1931 (after Said, 1931).

At the end of the 19th century, it cut a devastating swathe of death the length and breadth of Africa. Killing countless numbers of cattle and wild ungulates, it plunged pastoralists into starvation and caused at least one war, changing the ecology of vast areas of the African continent and setting in train policies still observed today affecting the survival of wildlife. As the last great manifestation of this disease, rinderpest is usually associated today with Africa; but it had repeatedly ravaged Europe with just as destructive an effect.

On the positive side, the study of rinderpest from the beginning of the 18th century led to a more rapid and greater understanding of human epidemic diseases and their control, its outbreak in Europe demonstrating the immense importance of preventive and suppressive measures in dealing with contagious diseases, contributing more than any other disease to sanitary policing and the progress of veterinary medicine. In the 19th century, it led to legislation for the more humane treatment of livestock; as well as for public health, stopping the sale of diseased meat for human consumption, and raising standards of cleanliness in dairies and slaughterhouses.

In the 1860s outbreak in Britain, some considered that the disease resulted from the mode of shipment of cattle from the Continent. Subjected to sickening conditions, cattle were crammed together for several days and nights, ill fed, and supplied with insufficient water, in the holds of ships which were rarely, if ever, cleansed. Vivid accounts were given of the appalling stench and heat, and the steaming sweat which arose from the holds. Transport from the docks by the railways was no better; cattle trucks allegedly never cleansed from year's end to year's end: "... after conveying one lot of diseased cattle, the mass of filth bespattered on the sides, and lying thick at the bottom of the trucks, is charged with poison which readily infects any animals coming in contact with it." wrote John Gamgee to *The Times* (Gamgee, 1863a). But he was fully aware that "The filthier these places are [markets, farmyards and town cowsheds], the greater undoubtedly the facilities for infection; but no amount of filth ever produced a specific outbreak of Lung Disease or Rinderpest. We concentrate the poison in foul places, but we do not create it." (Gamgee, 1866).

2. CLINICAL EFFECTS OF RINDERPEST[1]

The virus enters the body through the mucosa of the upper respiratory tract, multiplying after several hours in the pharyngeal and submaxillary lymph nodes and tonsils, targeting epithelial cells of the alimentary, respiratory and urinogenital tracts; provoking fever, stomatitis, rhinitis, and extensive gastroenteritis, producing a loathsome stench (Taylor *et al.*, 1965). Seldom occurring free in the plasma, the virus has a selective affinity for lymphocytes and the epithelium of the mouth and alimentary tract. Destroying the superficial epithelium, the entire alimentary tract is affected by erosion of the mucous membranes. Attaching firmly to polymorphonuclear neutrophils, it destroys the lymphoid tissue, particularly lymphocytes, which suppresses the normal defense mechanism of the host, allowing other latent infections (such as coccidiosis, theileriasis, babesiosis, piroplasmosis, and trypanosomiasis in African animals) to express themselves. Initially, lymphocyte numbers rise to combat the infection but, after 4 days, decline to some 18% of normal at termination of the disease; leukocytes decline to some 28% of normal. An almost total destruction of lymphocytes in spleen and lymph nodes in some animals, especially calves, causes rapid death without the characteristic gross lesions occurring (Maurer *et al.*, 1955).

Acute infection is characterized by an unusually rapid course, and signs of illness may be noted within 3–5 days of contact with the virus, although it may take 9 days. Inoculation can cause symptoms in 2–3 days. In 1774, Dr. Weise in Germany inoculated 100 cows, of which 51 died on the 11th day. Previously, the animals had been fed on hay, and on the second day after the inoculation, they were given oat straw. Autopsies revealed only hay in the rumen and reticulum, from which it was deduced that the omasum was already affected by the second day (Weiß, 1783).

Loss of appetite accompanied by fever is followed by salivation and discharges from the eyes and nose. This is succeeded by buccal ulceration, diarrhea, dehydration, and death in 6–12 days after the onset of the fever. In survivors, diarrhea stops within a week, and recovery then begins, taking several weeks to complete. Pease (1894) in India considered the post-mortem lesions more characteristic than the symptoms.

Robert Dossie, an eminent 18th-century British agriculturist, stated that the disease had various appearances, symptoms, and degrees of mortality, at different times and in different places. In the United Provinces of the Netherlands, profuse diarrhea was common and skin eruptions more rare. In Britain, diarrhea seldom occurred and was mostly a fatal symptom, but skin eruptions were frequent (Dossie, 1771). Ribbe (1814) noted *occasionally* survivors exhibited a cutaneous eruption, which was not like smallpox but rather like the well-known scab of invalids who died of bubonic plague. D'Arboval (1816) was to write that, in his opinion, the development of skin pustules was only an epiphenomenon, occasionally and yet somewhat rarely observed in the course of the disease. In the 1814–15 epizootic in France, he

[1]Plowright (1968) has provided a detailed technical history of the experimental studies of the properties of the virus.

was unable to find a single case of pustular eruption. In India in 1870, infected animals in the same herds exhibited both the "external" eruptive form marked by pustules and the "internal" form (Hallen *et al.*, 1871). Pease (1894) stated there were numerous cases in India of the eruption of pustules or papules and that cases marked by a considerable skin eruption were less likely to terminate fatally.[2]

In Denmark in 1759, it was far more acute than had ever been known in England or Holland, or during some previous outbreaks in Denmark, with animals dying in 2–3 days and even sooner. Young calves and exotic species may be subjected to peracute reaction, death taking place within 2–3 days after the onset of fever, but this is not common. Faust (1797) tabulated susceptibility as follows:

- Animals from 6 months to 2 years of age survived the best.
- Pregnant cows aborted and died easily.
- Fat rather than lean cattle died.
- Strong, vigorous animals, were more affected than weaker animals (perhaps they were simply worked harder).
- Animals on their first green feed of the season, or on pasture, died easily.

Rapid reaction time in experimental wild animals may be due to stress on the part of the animal, and the conditions of captivity prior to the experiment could be important, as perhaps in the case of the sambar *Cervus unicolor* in Indo-China, which reacted 2 days after infection with virulent bovine virus (Jacotot, 1927). But in Kenya, reedbuck *Redunca redunca* and bush pig *Potamochoerus porcus* took 7 days to react after experimental infection (Daubney, 1943). Domestic pigs appear to react more quickly than cattle.

Barker (1745) seems the first to have observed that it could cause blindness, a relatively rarely reported effect among cattle. Contrary to Plowright (1982), it is reported. A single observer reported it among cattle in the Assam Division of India in 1868: "It has been noticed in some cases of recovery that the animals are blind, or nearly so, and in some instances total disorganization of the structures of the eye has taken place" (Purves, 1868). An Indian villager also reported in Allahabad Division that some cattle had their eyes affected, the eyes becoming white; while another in the Punjab reported that eruptions were generally found inside the thighs and inside the eyelids, in a few cases the animals becoming blind (Hallen *et al.*, 1871).

It was apparently not uncommon in some outbreaks in German East Africa in 1905–15 (Ostertag, 1916), and blindness was seen in eland *Taurotragus oryx* in outbreaks in Tanganyika in 1944 (Thomas and Reid, 1944); and giraffe *Giraffa camelopardalis reticulata* in Kenya in 1960 (Grimwood, 1961). In the 1993–7 outbreak in Kenya, in addition to corneal opacity, keratoconjunctivitis, corneal ulceration, uveitis, and cataract were recorded in lesser kudu *Tragelaphus imberbis* and African buffalo *Syncerus caffer* (Kock *et al.*, 1999). Lesser kudu in Kenya in 1993–7 also exhibited tenosynovitis, or swelling of the knee and hock joints.

[2]The pustules, originally thought to be rinderpest exanthema, are caused by exacerbations of latent cowpox as a result of massive destruction of lymphocytes. The activated latent pathogens emerge and become obvious 7–10 days after entry of the rinderpest virus, and their detection is considered to indicate a good prognosis of survival, in contrast to fulminating victims that usually die in 3–12 days. In addition to cowpox, the eruptions can also be caused by dermatophilosis in zebu cattle in Africa and Plains cattle of India, that, and cowpox eruptions, being wrongly labelled "skin rinderpest."

Much more common may be subacute reactions which most animals survive, usually experienced where the disease is enzootic. Incubation may take 15 days, and one or more of the classic signs may be absent so that often rinderpest is not diagnosed. Studies in Russia in the 1850s and 1860s, where subacute reactions were more prevalent than further west, suggested an incubation period of 9 days, so it was argued that only a 10-day quarantine period was necessary. But Professor Jessen, director of the Dorpat Veterinary Institute, stated that experience had taught him that rinderpest might appear in the steppes in such mild form that no veterinarian would be able to recognize it (Jessen, 1852). As with all diseases, the constitution of the host is important in suppressing its symptoms; and it was noted by Joseph Ravitsch, Professor of Pathology in the veterinary division of the Imperial Medico-Surgical Academy at St. Petersburg, that Russian steppe cattle could leave the steppes apparently perfectly healthy, and after 3 months from arrival at their destination, the disease might appear, induced by the long marches and privations. Schern and coworkers (1919) confirmed in Anatolian and Podolian cattle in Asia Minor that the period of incubation often extended to 10 days and could be 15 days.

After it reached South Africa in 1895, Verney (1898) recorded that he could diagnose the disease fairly well from following English textbooks, but what struck him was the remarkable dissimilarity in the post-mortem appearances; the rumen, reticulum, and omasum appeared normal. In no case had he found any well-marked change in these organs.

The 1865–7 plague in Britain led to one great clinical discovery, the use of the clinical thermometer as a means of determining early infection by a rise in body temperature. Although Professor Chauveau had apparently come to the same conclusion simultaneously in France, credit for the discovery generally goes to John Gamgee. Gamgee inoculated two heifers with the virus, which were then studied by his physician brother, Dr. Arthur Gamgee, who suggested the importance of examining temperature changes. Arthur Gamgee also attempted to discover whether any changes took place in the milk, blood, and urine.

John Gamgee pursued the temperature experiments, taking care to disinfect the thermometer between use on different animals. He was able to determine from the rectal temperature which animals in a herd of 18 already had the infection. Although outward symptoms were barely discernible, infected animals had a temperature of about 0.55–2.22°C higher than uninfected. In another case, he found that every animal in a herd of 42 had a high temperature, and 37 died within 7 days. When the temperature dropped, and the pulse rate increased, death was imminent "... it is evident that the thermometer affords unerring as well as early evidence of an animal sickening, recovering, or about to die. No other indication is so unmistakeable and satisfactory ..." (Gamgee, 1866). This conclusion was challenged by an anonymous writer in *The Veterinarian*: "Increase in the temperature of the body has no appreciable value as an evidence of the existence of cattle plague, because in simple fever arising from an inflamed udder ... and in many other diseases, the internal heat is found to be much higher than in a state of health." Indicating that, after rising to a peak, the temperature may fall sharply in cases of rinderpest, the writer continued: "These facts, and they have been verified in our own investigations frequently, do not militate against the use of the thermometer as an infallible indicator of the amount of fever in the system, but they entirely negative the idea that the rise of temperature is a positive evidence of plague" (Anon., 1866a).

But the abrupt rise in temperature has been confirmed as the most striking feature of both experimental and natural infections. Known as the prodromal phase, it rises from a normal temperature of 38.61°C to approximately 40.55°C within 24 hr, reaching a peak within another 48 hr, and then falling to near normal less rapidly than it rose, usually remaining above normal until the animal becomes prostrate from weakness. It may then become subnormal until death takes place (Maurer *et al.*, 1955).

3. ETIOLOGY OF RINDERPEST

Formerly lumped together with other acute diseases, rinderpest was distinguished as the "contagious typhus" of cattle sometime prior to 1721 by Kanold in Germany. After the European outbreaks of 1865, it became more generally known by its German name of rinderpest. As the term "plague" or "cattle plague" is sometimes used for other epizootics, the German name is preferable to this "sinister and gloomy denomination from the middle ages," as Bourgignon (1865) put it. Many other names were given to it in Germany such as Därmeseuche (gut sickness), fliegendes Feuer (flying fire), Gilbe or jaundice, Grossgalle, Herzweich (heart soft), Hungarian or stomach sickness, Lungenfäule, Löserdurre, Magenseuche, Pockenpest, Ruhr (dysentery), faule Ruhr, Uebergallung, and smallpox.[3]

Tentative conclusions concerning its nature and epizootiology were reached at the very beginning of the 18th century by two Italian physicians, Bernardo Ramazzini (1633–1714), Professor Primarius of the medical faculty of the University of Padua; and Giovanni Maria Lancisi (1654–1720), personal physician to the popes.

Another Italian physician, Antonio Vallisnieri, anticipated phagocytosis in 1711 when he stated his belief that rinderpest was due to a "multitude of little worms" which were always found in the blood of cattle affected by the disease. The fatal consequences were not so much due to these "strangers" as to the "hostility" which occurred between them and the "small worms usually residing in the body of cattle." The latter took up arms to repulse the "invaders," but the invaders were victorious. An Austrian physician, Marko Anton Plenciz, also believed that such contagious diseases were caused by little worms, noting that if one examined microscopically matter from the nasal sores of cattle suffering from rinderpest, one always found innumerable worms (Plenciz, 1762).

Its infectiousness was first established experimentally in 1754 by Dutch physicians inserting setons dipped in morbid material under the skin of the dewlap of 17 cows which all took the disease in a few days (Nozeman *et al.*, 1755). In the same year, Sir William St. Quintin of Scampston in Yorkshire, was seemingly the first to use inoculation for rinderpest in England, claiming to have inoculated eight calves, seven of which took the infection and recovered, and were afterwards turned out into a herd of infected cattle without ill effect (Anon., 1754a).

In 1866, John Burdon-Sanderson, Professor Superintendent of the Brown Animal Sanatory Institution in London, attempted to discover the "virulent principle" of rinderpest, using a method of dilution and decantation which had been used independently by Chauveau in France for identifying smallpox and cowpox agents. Sanderson attempted to define whether the agent was "colloid" or "crystalline," whether particulate and not dissolved in the serum. His results suggested to him that the infectious agent was associated with the "colloid" rather than the "crystalline" constituents of infected blood. Coincident with the first elevation of temperature and long before there was any outward appearance of ill health, the blood, he found, was infective. This, he claimed, was "the most important pathological discovery yet made in cattle-plague: it is pregnant with consequences in medical doctrine; for though the existence of a similar fact has been long suspected in several human diseases, it has never been proved in any" (Sanderson, 1866).

It was noted by the 1865 Royal Commission in Britain that the poison in a minute portion of discharge from the eyes or mouth of an infected animal, placed in the blood of a healthy animal, increased so fast that in less than 48 hr the whole of the animal's blood was

[3] A contemporary annotation in an Elizabethan hand to a first edition of Mascall (1587) against the subtitle *The Cattle Plague or Murren* reads "Alias the mountaine evil," a term I have not been able to trace elsewhere. It perhaps derives from 16th-century outbreaks of anthrax in the Swiss Alps.

sufficiently infected, and every small particle of blood had enough to infect another animal, thus accounting for the rapid spread of the plague (Bernard, 1866a). But it was not until the 1870s, although the exact cause of the the disease was not known other than that the agent was a "micrococcus," that its symptoms could be distinguished with some certainty from similar diseases such as aphthous fever (FMD), contagious bovine pleuropneumonia, malignant catarrh, anthrax, and other fatal infections.

In 1883, Eugene Semmer, a German by birth who held the chair in the Epizootiology Department of the Institute for Experimental Medicine at St. Petersburg, reviewed the literature of various attempts to incriminate bacteria as the cause of rinderpest. He referred to other Russian workers such as Kolesnikov and Mechnikov, both of whom had reported bacteria to be the cause. Semmer pointed out that his own deductions in 1871–4, together with those of Archangelski, asserting from the discovery of micrococci in the blood that certain bacteria were responsible, were erroneous (also Klebs in 1872), as was a report by himself blaming streptococci (Semmer and Archangelski, 1883). But 10 years later, Semmer reported that the virus did not pass through a bacterial filter, and Kolle and Turner came to the same conclusion in South Africa in 1898. Unbeknown to these workers, this was due to the firm manner in which the virus locks on to the polymorphonuclear neutrophil white blood cells and is not liberated by simple methods calculated to release a virus which is merely adsorbed to the cells (Daubney, 1928).

Schein (1922) expressed his belief that the virus was located in the leukocytes, and not in the red blood corpuscles as many workers thought, and 3 years later, Hornby correctly deduced that the virus must be attached to the leukocytes (Hornby, 1926). In 1899, Nicolle and Adil Bey had been able to infect susceptible cattle with aqueous peritoneal washings after passage through special Berkefeld filters, making it the second disease, after FMD in 1898, to be recorded as a filterable virus[4] (Nicolle and Adil Bey, 1899).

4. CLASSIFICATION AND STRUCTURE

For want of a better term, in 1865, rinderpest was classed as a "zymotic" or fermentative disease, a category introduced by Farr in 1842. Many veterinary authorities in Germany maintained for more than 50 years to the mid-19th century that rinderpest was the precise counterpart of abdominal typhus or enteric fever of man (Ravitsch, 1864), an opinion so widely supported at the beginning of the 1865 outbreak in Britain due to reports of veterinarians and others who were not familiar with medical cases that it was almost a matter of popular belief. British troops in the Crimea in 1854–5 also suffered from epidemics of enteric fever, as well as typhus, at the same time cattle were suffering an epizootic of rinderpest. But Britain's leading authority on fevers, Dr. Charles Murchison of the Middlesex Hospital, was unable to discover any analogy with the lesions of human enteric fever, stating categorically, "The statement that the 'cattle-plague' is pathologically analagous to human enteric fever has been proved to be devoid of foundation ..." (Murchison, 1873). Others believed it to be identical with typhus, based on a supposed analogy between the etiology: the facts both were very contagious and, like typhus, rinderpest followed in the train of armies. In Britain, there was an epidemic of typhus fever from 1862 to 1869, coincident with the major rinderpest epizootic, but summer droughts may well have exacerbated the former, just as they can the latter,

[4]No longer in use, "filterable virus" referred to an organism small enough to pass through a porcelain filter which would hold back bacteria. Meaning poison in Latin, the word virus for a disease organism had been in use since the 16th century.

although for different reasons. Dr. James Tucker, in a report to the Lord Lieutenant of Ireland, stated, "The 'purple gum,' the black, saltless blood, and some other symptoms of the African typhus, may be recognized in the Rinderpest" (Tucker, 1866). In France, it was called the "*typhus contagieux*."

Its real classification is in the in the Order Mononegavirales of negative-sense, single-stranded, RNA viruses in the Family Paramyxoviridae, Subfamily Paramyxovirinae, Genus *Morbillivirus*. The genus is a clade of six viral species descended from rinderpest virus. The other related species are *Peste des petits ruminants* (PPR), a disease of sheep and goats, Cetacean morbillivirus, canine distemper, and Phocid distemper (Figure 2a,b). They differ in their natural host ranges but otherwise all look alike, have similar physicochemical properties, share antigens, and express identical epizootiological patterns characterized by devastating dieback epizootics and panzootics when they infect susceptible unchallenged host populations. The sequelae are either the disease is burnt out or infection becomes enzootic if there are sufficient fertile. All are related antigenically, sharing the same antigens which evoke the antibodies produced by the host's immune system. Dogs which feed on rinderpest-infected meat can sometimes be immunized against canine distemper, at least briefly, because of cross-reaction but, when inoculated with rinderpest virus, show no clinical response, although their blood is infective to cattle (Polding *et al.*, 1959). Nonetheless, there was probably no connection between rinderpest in Kenya in 1906 and a large outbreak of canine distemper at the same time which killed jackals "by the hundreds." Although this was the first record of canine distemper in Kenya, several other outbreaks had been recorded by 1918 (Percival, 1918).

Believed to represent the original or archaevirus of the group, some consider that rinderpest may have evolved with ancestral Bovidae in the Pleistocene. Measles virus was the first descendant, emerging about 5,000–7,000 years ago when humans began to associate together in riverine communities and domesticate cattle. But close human–animal contact first began with sheep, kept in large numbers on the Eurasian steppe, prior to the great herds of cattle which seem to be relatively recent, within the last 500 years or so, and perhaps it was in sheep that the rinderpest virus first developed before passing to cattle. Canine distemper virus followed measles, and lastly, the PPR virus split off within the the past 3,000–7,000 years or so (McCullogh *et al.*, 1986; Scott, personal communication). A virus, Hh1, isolated from both sick and healthy hedgehogs in the United Kingdom, was probably a morbillivirus, but all stocks of it have been lost.

Polymerase chain reaction (PCR) molecular analysis, in which the DNA product can be amplified and sequenced so that genomes can be compared, has enabled grouping of recent strains of the rinderpest virus into three distinct lineages, one Asian and two African, which allows the possible origins of outbreaks to be traced, although the lineages do not appear to persist for long lengths of time, branches dying out and being replaced by others. Historically, five lineages could be distinguished, but only two were represented in outbreaks after 1983 (Table 1) (Chamberlain *et al.*, 1993; Wamwayi *et al.*, 1995). Recent (1980s and 1990s) Kenyan and Sudanese isolates are closely related to a virus previously found in Egypt, and this may represent a mild strain of the disease circulating in eastern and northeast Africa in the 1990s. The strains circulating in the Middle East did not have their origins in Africa but appear to have been derived from India (Figure 3a).

Similarly, in PPR, four distinct lineages have been identified, two African, one Asian, and one Near East, of which only three (one African, one Asian, and one Near East), have been detected since the beginning of the 1990s (Shaila *et al.*, 1996; Figure 3b). Goats are the primary host in West Africa and the Middle East, with sheep predominantly so in India, while natural infection has been found in such species as the nilgai *Boselaphus tragocamelus* in India. In South Asia, there was a long-held belief that rinderpest was maintained in sheep and

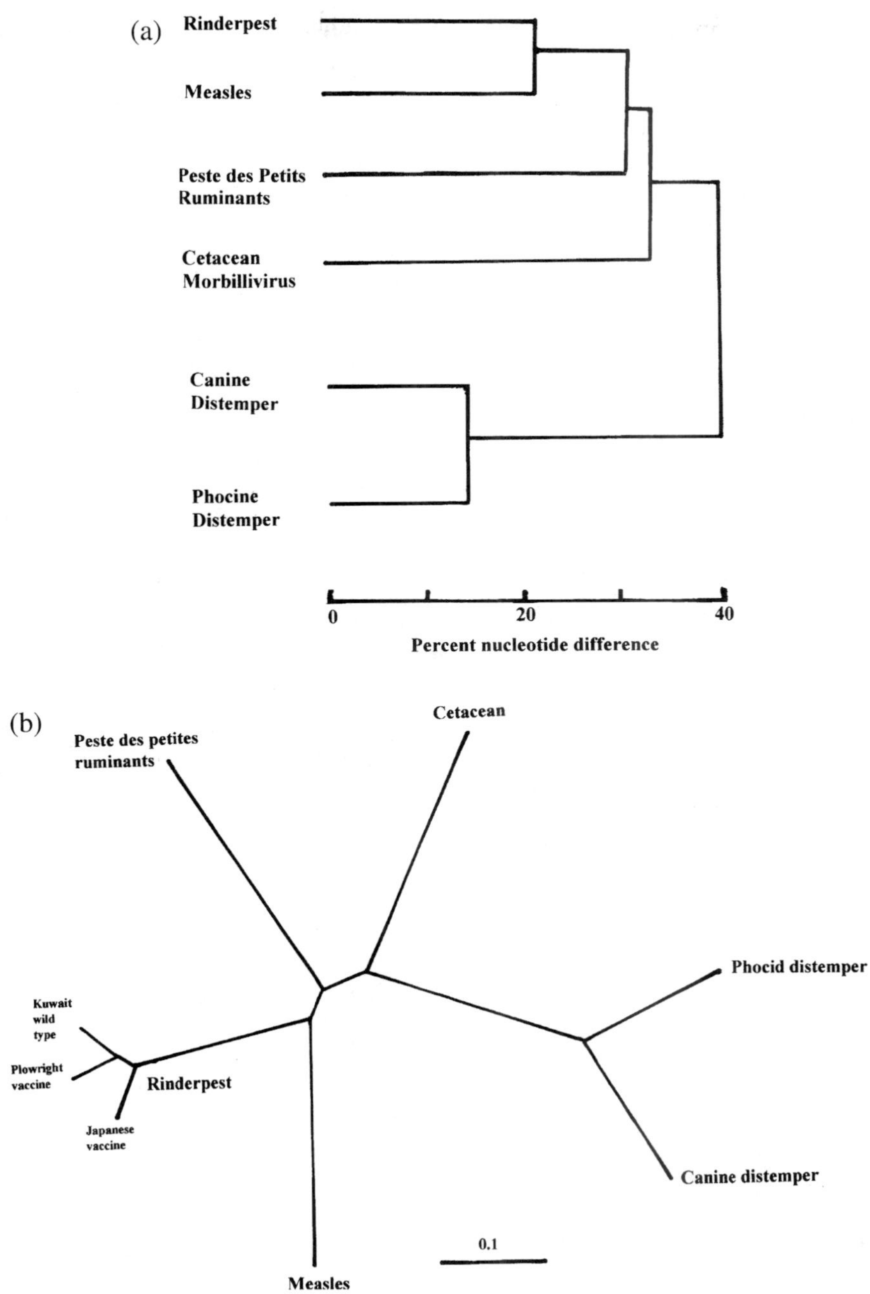

Figure 2. (a) The crurrent morbillivirus clade (after Barrett, 1993). (b) The phylogentic relationship of the morbilliviruses (after Barrett, 2001). The scale represents nucleotide changes per position.

goats, but this is now being reconsidered with the identification for the first time in 1990 of the majority of disease outbreaks attributed to rinderpest as being caused by PPR. Nevertheless, although PPR was diagnosed in sheep and goats in three states in northern India in 1994, rinderpest was still being found in sheep and goats in southern India in 1995, after rinderpest had almost disappeared. The 1990s have witnessed a major spread of PPR through

Table 1. Isolation of rinderpest virus lineages.

Lineage	Year	Country	Host
Asia	1979	Oman	Cattle
	1981	Saudi Arabia	Cattle
		Yemen	Cattle
	1982	Kuwait	Cattle
	1983	Pakistan	Cattle
	1985	Iraq	Cattle
	1987	Sri lanka	Cattle
	1989	India	Bison
		Iran	Cattle
		Russia	Cattle
	1991	Turkey	Cattle
	1994	Turkey	Cattle
		Iran	Cattle
		Pakistan	Cattle
	1995	Yemen	Cattle
Africa 1	1958	Kenya (Marsabit)	Cattle
	1983	Nigeria	Buffalo
	1984	Egypt	Cattle
	1986	Kenya (W. Pokot)	Cattle
		Kenya (Kiambu)	Cattle
	1987	Kenya (Marsabit)	Cattle
	1992	Sudan	Cattle
	1993	Sudan	Cattle
		Ethiopia (Bedelle)	Cattle
	1994	Ethiopia (Wollo)	Cattle
	1995	Ethiopia (Wollo)	Cattle
Africa 2	1958	Nigeria	Cattle
	1961	Kenya	Giraffe
	1966	Kenya (Tala)	Cattle
	1983	Nigera (Sokoto)	Cattle
	1994	Kenya	Buffalo
	1995	Kenya	Lesser kudu
	1996	Kenya	Buffalo
			Eland

Asia into the Indian Subcontinent. Apparently enzootic in the Sahel region of Africa from Senegal to western Ethiopia in pastoralist livestock, sometime prior to 1985 it extended its range in Africa from the Sudan and Ethiopia to Kenya and Uganda. As rinderpest antigens confer immunity to PPR, these extensions may be due to lack of the hosts' exposure to rinderpest, making animals susceptible to this alternative disease.

The common shape of the rinderpest virus is a spheroid 100–300 nm in diameter, containing tightly coiled serrated nucleocapsids. Less common are filaments up to 1 μm in length with regularly coiled serrated nucleocapsids. Both forms are encased in protein envelopes bristling with minute projections (Scott *et al.*, 1986). In 1995, the entire genome sequence of the virus was described, 15,881 bases in length, similar to that of the measles virus and slightly longer than that of canine distemper (Baron and Barrett, 1995). Then, in 1996, Baron and Barrett succeeded in rescuing virus from DNA copies, previously considered impossible with negative strand viruses, although this had been done with positive strand genomes for several decades. This reverse genetic technology means that more precise questions concerning the molecular determinants of pathogenicity can be addressed and may also be exploited to

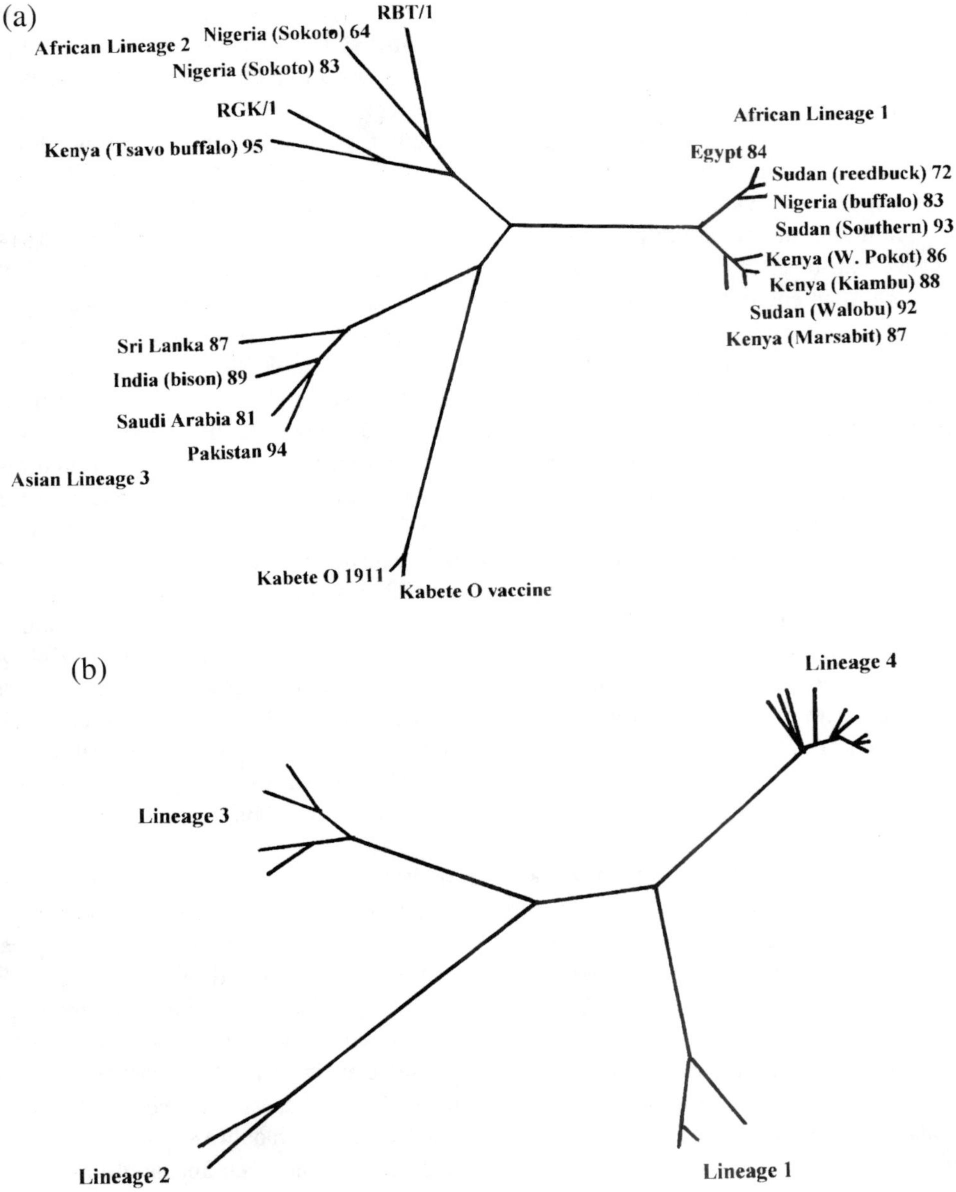

Figure 3. (a) The phylogenetic relationships of the Asian and African rinderpest virus strains (after Barrett, 1996). (b) The phylogenetic relationships of PPR (after Barrett, 1996).

produce more precisely characterized vaccines. If extra coding sequences can be introduced into the rinderpest genome, it would be possible to produce an antigenically marked vaccine to enable vaccination to be distinguished from natural infection (Barrett, 1996), of obvious interest in determining the efficacity of eradication programmes.

The virus is not robust, surviving outside the host only at low or high relative humidities and readily destroyed when the humidity is between 50 and 60%. It is sensitive to heat, light, and ultrasonic waves, as well as very acid or alkaline conditions, and rapidly destroyed when dried by evaporation. In the 18th and early 19th centuries, its durability was greatly exaggerated, although Abildgaard in Denmark showed that it was destroyed by heat (Abildgaard, 1795), while Adami had shown that it was quickly destroyed in the air (Adami, 1781). Alleged infections after long periods presumably stemmed from current contamination. Weiß (1783), for example, claimed that a thread of wool soaked in virulent matter and kept dried for 6 years in a glass bottle still retained its virulence when dampened with steam, while van Oertzen claimed that dried matter was not virulent, and infective matter maintained in humid or cold conditions remained infective for only 14 days. Zimmerman (1814) reported that, in the air, it lost its infectiousness after 10 days but remained infective for a long time in an enclosed space. It was destroyed by heat in a few moments, but the degree of heat was not known. Since then, numerous researchers have shown that drying leads to rapid sterilization. Koch (1897a) demonstrated that, when dried at 31°C, the virus was avirulent after 4 days; others claimed 2 days. Theiler (1897a) claimed that virulent blood smears exposed to the sun were avirulent after 2 hr. Braddon (1902) showed that excretions exposed to the sun were sterilized in less than 5 hr, and Shilston (1917) demonstrated that the sun rapidly sterilized pastures, although at the lower temperatures of high altitudes, faeces were still infective after 54 hr. Generally, because of fermentation, faeces lose their infectivity more rapidly than, for example, urine. In Kenya, Jackson and Cabot (1930) safely placed cattle in sheds which had housed infective animals 72 hr before, and paddocks or pastures were safe to use after 24 hr. Clearly these figures would vary according to the season and its temperature and humidity. Friedberger and Fröhner (1886–7) recorded that virulent blood kept aseptically lost its virulence after 14 days, but nasal mucous protected from air could remain infective for 6 weeks to 9 months. But they also noted that buried cadavers could remain infective for 9 months, which seems unlikely. Reynal (1873) inoculated cattle with material from an animal 58 days after its death without result, while in 1874, Visear D'Arras in France exhumed some rinderpest cadavers after 10–12 months' burial and exposed cattle to them also without result.

Zimmerman (1814) reported that in breath or evaporation on a hot summer day, it was not infective beyond 1.5 m, and in winter 6 m. Beyond this, it had to be in a large concentrated mass of infective matter. The virus has limited survival outside the host, especially in tropical conditions where high ambient temperatures and rapid putrefaction of dead infected animals soon destroy it. Van Saceghem (1923) found that if he scarified an animal lightly on the thigh and applied infective matter to the area, keeping the animal inside, death occurred. But kept in the open, nothing happened, which lack of response he attributed to the destructive action of the sun on the virus. Littlewood (1905) reported in Egypt that animals which contracted the disease in the hot summer months and were kept in the open with only a slight shelter recovered in large numbers; also, those contracting the disease in the open appeared to suffer a milder form than those which were stabled. But the two factors which most affect survival of the virus, heat and light, are in the first case diminished, and in the second absent, during the hours of darkness, and so are only effective for some 50% of the time.

As long as putrefaction has not begun, meat and other products can remain infective for a relatively long time, and there are numerous allegations of infection being transmitted for short distances by infected meat. Shilston (1917) found at 1200 m altitude in India that meat preserved without alteration remained infective between 3 and 6 days, and bone marrow for 5–9 days, even up to 15 days. Nencki (1900) claimed that infected meat in 10% saline solution remained infective for 1–3 months; but infective blood is rendered harmless by the addition of 10% salt by volume.

All other things being equal, the virus is not affected by altitude. Pease (1894) reported that it was highly virulent to yaks at 3,600–3,900 m in the Punjab.

5. TRANSMISSION

The ease with which the disease is transmitted depends upon the strain of the virus. Infected animals are infectious before the end of the incubation period of the disease although still apparently healthy, and they remain infectious into the start of the convalesence (Scott, 1955). Some infected animals do not sicken but are nevertheless infectious to others. The idea of carriers, animals which remain unaffected in the midst of the contagion, was recognized by "A Physician" in 1749 in his *Considerations Concerning the Distemper Which Still Spreads Itself among the Horned Cattle in This Kingdom*, and in 1865 postulated by medical men in Dalmatia. But such carrier states are transient, and it is considered that development of a persistent carrier state, if it ever occurs, is rare.

It had long been observed that a damp atmosphere favored propagation, and transmission is believed to be by droplet, either in the breath of an infected animal or in its secretions and excretions, but close contact is required lest the virus is destroyed in passage. Réfik-Bey and Réfik-Bey (1899) concluded that transmission was by direct contact, and the roles of soil and water seemed to be nil, while insects also did not appear to be involved. Wind-borne transmission at night cannot be ruled out, especially in the humid conditions encountered at the end of the dry season in countries like Africa. It has been reported that it can drift by the wind some distance: "experience having shown that a space of considerably more than a hundred yards affords no protection ..." (Bernard, 1865). But in the latter part of 1865, Sanderson claimed to have demonstrated for the Commission that infection could be transmitted only a few yards in air, and greater distances were the result of some indirect method of contagion (Sanderson, 1865). Pease (1894) was positive that the disease was generally propagated by ingestion, and spreading in the air was only secondary. Koch found that it did not spread among animals kept 6 m apart, nor if separate utensils and attendants were used.

At the beginning of the 18th century, Carlo Francesco Cogrossi, a distinguished Italian medical teacher, asserted that the first attacks of the disease had been observed mainly in the respiratory and digestive systems, quoting Lancisi as holding the correct opinion that the poisonous fermentation was introduced, more than in other ways, through the fauces and nostrils (Cogrossi, 1714); over two centuries before Hornby (1926) confirmed that it was most easily transmitted through the nostrils and concluded that the commonest manner by which the disease was spread was probably by inhalation of infective excretions. But Chauveau (1871) considered that the main route of infection was by ingestion, for, whereas the virus could be maintained in excretions in the open air for several weeks, there was less of the infective virus in expired air than in other sources. The infective distance in the air he considered to be only a few meters. Hornby found that it was difficult to cause infection by mouth, an experimental animal being unaffected by eating contaminated food and drinking contaminated water.

It has been known for over a century that the virus can retain its infectiveness for a long time if kept from contact with the atmosphere and high temperature. Sometimes, it has been successfully inoculated after being kept for months, even more than a year. In 19th-century studies in Russia, carefully protected mucous discharges had kept their infectivity for periods of 7 and 11 months. But when, in Belgium in 1868, it reappeared in two localities 8 months after the last cases had been observed, this was probably due to a smouldering infection picked up by calves losing their immunity. Cases of the disease appearing after "infected" hay (which had lain in a stable for a year) had been given (in another case, it was 5 months' old) were undoubtedly attributable to the same cause: animals with subacute reactions passing infection to nonimmunes, rather than infection being in the hay. Remains of infected animals which had been interred were claimed to retain their infectivity for 3 months, but this does not fit with what we know of susceptibility of the virus to putrefaction. In recovered animals, the virus is excreted in milk for up to 45 days (Curasson, 1932), but milk was usually condemned only from sick animals, and not always even then.

Dr. Bernhard Faust was firmly of the conviction that the disease only originated in Germany through infection, by transmission of the infectious "pest-poison" from a sick to a healthy animal (Faust, 1797), but these early German works generally passed unnoticed. Eugène Renault, Director of the Alfort Imperial Veterinary School in France, asserting that it spread solely by the contagious property of affected animals, their products and excretions, and the remains of dead animals, was stating nothing that had not been said already. This "contagious property" could be carried to other animals, people, or things which had touched, or only been near to, sick animals, or their excretions or remains (Renault, 1860). Fleming (1875) stated that it had long been recognized that the dung of infected animals was a potent agent in diffusing the disease and appeared to keep its infective properties for several months. While the former would be true, virus being shed in the dung, it is unlikely to survive for long in that environment.

Although large amounts of the virus are passed in the urine and faces, rarely can transmission be attributed to contaminated bedding, fodder, or water, although urine, which does not become infective until about the third day of fever, can maintain its infectiveness for 9 days. The Commission concluded in its Third Report (Bernard, 1866b): "The supply of impure water also appears to increase the fatality, but we have not yet been able to satisfy ourselves that the virus is actually taken in with the drinking water, although such a mode of communication is not improbable."

Instances in which the disease was propagated by the flesh of diseased animals were numerous. Dufot (1773) alleged that a dog infected calves at Morecourt, near Paris, by digging up and eating an infected carcass and then drinking from the calves' pail. In the 1870–71 Franco-German war in France and the appearance of the disease in many localities, several further outbreaks were traced to the use of infected flesh as food. A case was reported of two oxen contracting the disease after carrying the flesh of diseased animals.

In evidence to the House of Commons Select Committee in June 1877, James Beart Simonds, professor of cattle pathology at the Royal Veterinary College London and technical adviser to the Royal Agricultural Society of England (RASE), a body to which the government looked for advice, stated that the disease was undoubtedly capable of being communicated by dead meat, and he had known of several such cases. He quoted an incidence of a calf slaughtered unknowingly with rinderpest, and a farmer from outside the village buying a part for his dogs, taking it home, and hanging it near his cows. As a result, the disease was communicated to his cows, which all died (Simonds, 1877). In Germany, it was considered that spreading the disease by dead meat was second only to that by live cattle, peasantry in the villages slaughtering the first diseased animals and selling the meat. Water used in washing the meat was put into the washtub from which the cattle were very often fed.

Fresh hides were known as another source of spreading infection and thought to be a particular cause in India. However, the Marquis de Courtivron, in July 1745, claimed to demonstrate that fresh skins removed from animals which had died of the disease did not transmit it when healthy cows were covered with them. He took the skins of two cows which had died of 200 out of 400 which succumbed within 1 month at Aizerey, wrapped them in straw and a tanned skin, and took them to his estate where he covered one young and one old cow with them, shut in a shed with the windows stopped up. Five to six days later, the cows showed signs of distress in the heat, so he removed the skins. On August 1, he put them back among his other cattle, and none took the disease but, "not regarding the experiment I have made as a reason to doubt the popular opinion, it may perhaps be expedient to order that skins may be kept in an infected village in a particular place and may be rendered saleable after a certain period has elapsed" (Courtivron, 1748). In 1776, Vicq d'Azyr tested the Marquis de Courtivron's results by rubbing six healthy beasts with infected skins. Only one became infected, and he suspected that this was from a different source.

But then in 1769, Professor Pieter Camper, the distinguished Dutch physician and anatomist, allegedly demonstrated that fresh skins were infective for up to 6 days after the death of an animal. He claimed to have infected healthy animals by inserting into them strips of skin from animals which had been dead for 6 days from the disease. Vicq d'Azyr caused the death of two animals by introducing into sores pieces of skin taken from pits where animals which had died of the disease had been for 3 months. Both Camper's and Vicq d'Azyr's animals probably died from septicaemia rather than rinderpest. Camper also injected four calves with grease from animals which had died of rinderpest 1, 2, 4, and 6 days, after their deaths. All of the calves showed infection after 6 days and eventually died.

Renault (1860) argued at great length that tanned, salted, or hard dried skins were not infective. Only fresh skins carried infection. Likewise, rendered fat was not infective, and laws prohibiting commerce in these articles originating from countries where the disease was present were unnecessary. Salted skins shipped by sea, for example, were at least 2 months old by the time they were unloaded. Nevertheless, if the disease was present in a border state, all imports should be forbidden, because even dry skins might pick up the "contagious property" from the atmosphere, or from contact with infected animals or things. In evidence to the Commission in 1865, Simonds stated that he thought there was no chance whatsoever of the disease being imported with hides, unless it existed in Holland or in some other countries from which fresh hides might be imported, but not with Russian hides (Simonds, 1865). Kraiewsky in 1885 claimed to have shown that "undried" skins could still be infective after 3 weeks, which is perhaps why, almost 30 years later, reporting on the rinderpest raging in East Africa, Lugard (1892) noted with concern that "the hides of these diseased animals are being exported by tens of thousands, and it is quite within the bounds of possibility that we may find, when too late, that we have imported the germ of this fatal disease into England." Pease (1894), however, was convinced that the skin of an animal dead of the rinderpest had no contagious property when tanned or dried: "No record of the introduction into a country by the medium of tanned or dried skins has ever been published…." Nocard and Leclainche (1903) showed that skins lost any infective property after 2 weeks at temperatures between 12.5 and 25°C in humid conditions. In dry conditions, infectivity disappeared in 2 hr. In 1921, Curasson showed that a few hours under the tropical sun rendered skins uninfective. Schein and Jacotot (1925) found that 24 hours' immersion in salt water eliminated infectivity in fresh skins.

In France, it was demonstrated in the 18th century that when clothes worn by persons attending sick cattle were placed on healthy animals, three in six of the animals contracted the disease. The discovery that rinderpest could be transmitted by infected clothes meant that, in Britain, attendants upon infected animals were obliged to have their hair tied up in bags lest it transmitted infection. It was also considered that infection could be carried on shoes. A more extreme theory suggested by a group of doctors at La Haye in 1769 was that the contagion could be spread by the vapor of candles.

Adam Neale, a physician, introduced the idea of insects effecting the mechanical transmission of disease in 1831, suggesting that both human plague and rinderpest were transmitted by this means. In Galicia in 1846, where the disease was well known, it was believed that it could be transmitted in spite of quarantine by flies passing from one animal to another, communicating it to the eyes and tender parts of the sound animal. Dr Masch, Veterinary Professor at the Agricultural College at Altenburg in Hungary, had already stated the year before that flies which sought their food at "the margin of those apertures of the body that are provided with but a very delicate skin," as well as those which "plunge their suckers into the skin to suck blood," might be responsible for spreading the plague (Masch, 1865). Gamgee (1866) was firmly of the opinion that "flies" (presumably he was thinking of the blow fly *Calliphora* spp., and the housefly *Musca domestica*) could transmit it: "One way in which the Cattle Plague may be carried from diseased cattle or from the slaughter-house is unquestionably by flies, which, after resting on the carcase or offal of sick animals, fly about, rest again on the animal, especially on any wounded parts, and thus produce a direct inoculation.... Unaccountable outbreaks may sometimes be due to this cause." Hallen in India wrote, "and it is probable that even flies may convey the contagium from one subject to another" (Hallen *et al.*, 1871).

Although this was early thinking on the disease-carrying capabilities of the fly, a role in rinderpest transmission has not been demonstrated. Russian regulations concerning rinderpest suggested that biting flies could be implicated but were confusing anthrax with rinderpest:

> If a fly, after having been sitting on a diseased or dead animal, should bite or sting a man, he should immediately rub the place with warm olive or hempseed oil, preferably the former. Should the swelling not go down a plaster of flour, treacle and bruised onions, should be applied, for the sake of creating a blister; and when this latter has gone down, the place should be rubbed with hempseed oil. In the meanwhile he should drink freely a mixture of vinegar and warm water, and dress warm to keep up a perspiration. (Still, 1866)

A greater likelihood of mechanical transmission is provided by ophthalmotropic moths which are attracted to the eyes of animals for their moisture requirements. In the Sudan, Reid (1954) found that moths of the genus *Arcyophora* were attracted to the discharges from the eyes of cattle suffering from rinderpest. In 1993–7 in Kenya, the conjunctiva of the lesser kudu appeared to be the point of entry of the virus, possibly occasioned by mechanical transmission of dipterous insects of which there was an epizootic of "head fly" (Kock *et al.*, 1999), presumably *Lyperosia* or *Siphona* species. Once the animal begins to weep, such diptera are attracted to the site and pass from one animal to another. However, Hornby (1926) was unable to transmit the virus by swabbing the conjuctiva of an ox with lachrymal discharge of a beast dying of acute infection, but another, swabbed with infected liver extract, did react.

Pigeons, rats, dogs, and hares have all been implicated in spreading the disease in Europe. Simonds (1866) believed that it was spread by domestic pigeons, particularly in Yorkshire. Vallisnieri wrote to Lancisi in 1715 that dogs carried the infection from one area to another. But a study in India reached the conclusion that its spread by bloodsucking insects, leeches, vultures, and other carrion feeders was minimal (Minett, 1954), although Boynton (1917a) in the Philippines found that the virus could remain active for 25 days in the body of the large water leech *Hirudo boyntoni* Wharton, fed upon an infected animal, but the leech did not transmit it. Under natural conditions, only one rare transmission in this manner might be sufficient to trigger an epizootic. Hornby (1926), working at the veterinary laboratory Mpwapa in Tanganyika, claimed that he had proved experimentally that the blood-sucking tsetse fly *Glossina morsitans* could transmit the infection. In his first experiment, feeding three captive flies on an infected animal and then allowing them to finish their feed on a susceptible ox, no reaction in the latter was apparent. He then fed two flies separately, the susceptible ox reacting after 4 days and subsequently dying. However, unless the fly was disturbed while feeding, it would not feed on another animal but retire after engorgement to digest its meal. Bliss (1922) suggested that it might be transmitted by ticks as infection had been produced by injection of blood expressed from cattle ticks taken from sick cattle. Ixodid ticks feed on three different hosts, but it is not known if the virus can survive the blood-meal digestion process.

In Africa, a country populated with large numbers of potentially susceptible artiodactyls, the wildebeest *Connochaetes gnu* and African buffalo have been shown to be the most important transmitters, due to a potentially high degree of susceptibility and close contact with cattle on grazing lands and at watering points (Wilde, 1953). The importance of other species lies more in their capacity to act as contributors to the disease reservoir.

6. FACTORS INFLUENCING OUTBREAKS

As Reid (1949) stated in 1948, the virus is most likely maintained in game animals where concentrations of susceptible species are large, thus being able to maintain the virus for longer than small groups so that herd susceptibility can be raised by natural increase before the virus has "burned itself out." Where the disease has not existed for 2 or 3 years, a natural increase of potential hosts will have outweighed the residual immunity of those animals which have survived the most recent attacks. Anderson and May (1979) have pointed out that disease outbreaks in host populations at high densities may be triggered by stress due to overcrowding and malnutrition rather than enhanced transmission with high density populations, and we see among high-density populations of susceptible African game species that epizootics are often triggered by drought, when the two conditions of stress and malnutrition are met. Talbot and Talbot (1961), attempting to explain an annual die-off of wildebeest calves in the Serengeti, postulated that prolonged drought, extended migration, predator activity, and the fresh flush of grass induced by rain in the calving season could all contribute to a "stress factor," increasing calf mortality; subsequently confirmed to be due to rinderpest. When rinderpest struck South Africa in 1896, the country had witnessed what was said to have been the worst drought in living memory.

The sensitivity of the virus to its ambient environment means that the course of epizootics can be influenced by climatic factors. In Britain in 1865: "The last [1865] summer was the warmest and longest ever known in the memory of living men," followed by "the wettest winter ever known," just as the summer preceding the 1745 outbreak in Britain was "unprecedentedly hot" (Constable, 1866a). However Professor Gerlach, Director of the Royal Veterinary College in Hanover, claimed that rinderpest was worse in cold and wet weather, and better in warm and dry (Gerlach, 1865); but Ernes (1865) was of the opinion that it spread as fast in a cold season as in a hot season. In the Philippine Islands, in the first decade of the 20th century, it nearly disappeared in the dry season, only to spring up again in widespread epizootics soon after the onset of the rains (Ruediger, 1909). In India, it was claimed that heavy rainfall checked the disease, but if not sustained, the disease broke out again. In extreme heat, it apparently assumed a milder type (Farrell, 1870). Stähle (1981) has shown from detailed records of 18th-century outbreaks in the German dukedom of Bremen and Verden, that of 160 outbreaks, 65.1% occurred between September and December; and of 147 outbreaks, 79.7% ended between October and March.

The reason for these differences has probably as much to do with management as it has with physical factors. In cold, wet weather, the animals may be herded into shelter, bringing them into closer contact with one another and spreading infection more rapidly, as well as being more susceptible in the winter to secondary infection by pneumonia. In excessively dry weather, the same effect is produced by animals crowding together at water points; but in normal dry weather, they tend to spread out and have less intimate contact with each other. Stähle (1981) concluded that the seasonal incidence in Bremen and Verden reflected the disease spreading when cattle were pastured after the harvest, driven, and marketed, and was less when they were spread out on the pastures.

Dossie (1771) asserted that it was an epizootical disease, though contagious: "That is, it does not ... ever reign, but when certain unfavorable circumstances of season have created a predisposition in the cattle to receive the infection; and thence rendered them temporarily susceptible of it." He considered that outbreaks began when cattle were weakened by unseasonable weather, thus rendering them temporarily susceptible to infection: "But the irregularities of the late seasons; and, particularly the great alternations of heat and cold, the continuance of wet weather, and the frequency of easterly winds ... have consequently rendered the cattle peculiarly susceptible of this infection ...." He pointed out the season before that the year 1711 had been so bad as to destroy many of the sheep in England and was followed by the plague spreading throughout Europe. Again in 1741, the outbreak followed a season of intense frosts which lasted from December to April. In Eastern countries, he noted, the disease tended to abate in winter. In Britain, however, such difference did not exist, which Dossie attributed to "the less damp state of the air, and the keeping the cattle more out in the fields."

In mid-18th-century outbreaks in Friesland, mortality was heaviest in the autumn, with the chance of infection being increased when the cattle were housed. It also appears that infection spread mostly in the pasturelands where there was a higher cattle density and more intensive cattle trade, while in the arable regions, the risk of infection was less, but mortality rates were higher when outbreaks did occur. In the arable areas, cattle were kept mainly for producing manure and were probably less well cared for than those of the pasture regions which were the owner's chief capital, but it perhaps reflected the use of draught animals worked when sick.

In the Russian steppes, where the disease was enzootic, temperature was considered to have very little influence. But when steppe cattle were being moved, and there was snow on the ground, the disease did not spread so much, simply because the cattle confined themselves to the road, from which they otherwise deviated in search of food (Ernes, 1865). However, the Russian Commission concluded that the disease was milder in summer and winter than it was in spring and autumn (Anon., 1865a).

Since the beginning of the 20th century in the West African sahelian zone, the dry season was regularly accompanied by a recrudescence of rinderpest because migration of cattle herds was more active at that time, they were weakened by poor food and reduced access to water, and they were parasitized in the preceding wet season. Further south, where conditions were less extreme, the cattle were better able to resist infection.

7. IMMUNITY

Cogrossi recorded an example of immunity in cattle in 1711: "I have just been told that in Montodine, a large village in this territory, there is in the midst of the infected animals a cow which has not suffered at all, and that all her sons, though they are in other infected stalls, have been immune from all illness" (Cogrossi, 1714). This was the first recognition in any species of the phenomenon of maternally derived immunity, although Cogrossi postulated that if his theory of contagion by invisible animalcules was correct, this could be explained easily by the fact that "these poisonous insects find in oxen alone food adapted to their subsistence and that in certain oxen they do not flourish for the same reason that in certain men the prurient worms of the itch cannot subsist."

Ramazzini, perhaps basing his thoughts upon Cogrossi's observations, asked whether it was safe to make use of oxen which were believed to be healthy:

> Even although the ox appears to be vigorous and lively before it is led to the butcher's stall and ... no pathological signs are to be noted in the viscera ... we cannot be completely certain that that ox did not harbour contagious morbific matter which is capable of transmission. For it is conceivable that there are those among our oxen which will not have experienced an overwhelming dose of infection and they will pass as being sound and healthy. (Ramazzini, 1711)

Brocklesby (1746a) observed that cattle might develop immunities so that one man's herd would remain whole in the midst of the disease, which, like the plague, moved west, that is, from London. This statement on immunity preceded by far that of Girard and Dupuy (1816), who reported that an infection conferred immunity. Reinders (1774) and Stolte (1777) in Holland had also observed that calves from recovered cows were resistant to further infection. The lifelong immunity conferred by recovery was emphasized by Budd in 1865:

> ... oxen which have once passed through the disease are for the remainder of life proof against it. Not only do they never take it again in the natural way, but experience has shown that they cannot even be inoculated with it. The disease may have occurred in the sucking calf, and the subject of experiment may be the same calf grown to the stature of an ox ... (Budd, 1865a, 1865b)

Galtier believed that the immunity of calves delivered by dams infected during gestation was inherited. This was disputed by Hornby (1926) in Tanganyika, who correctly concluded that immunity was acquired only during an animal's lifetime. In populations subjected to the virus, most adult cattle are immune, and the calves are protected by maternal antibodies in the antibody-rich colostrum which they ingest within the first few hours of birth. The amount of gamma-globulin in the milk falls off rapidly after parturition, and a calf's capacity to absorb the large molecules through the gut wall wanes after a few hours. Protection conferred by the maternal antibodies lasts for varying lengths of time according to species. In cattle, it is effective for 4–8 months with a half-life of 48 days, 7% of calves in tests still having maternal antibodies at 8 months, and none after 8 months; but in wildebeest it lasts only 3–5 months. After this period has elapsed, the calves become susceptible, so that if the host population is large enough, infection can, theoretically, constantly smoulder. In the classical course of the enzootic disease in cattle, once born, a calf has immunity for up to 8 months. It then contracts the virus and is infected and infective for a period of approximately 1 month. But the only susceptible animals are the calves themselves, the adults being immune, for if they had not attained immunity, they would not be able to pass on antibodies. In a strictly seasonal breeder, by the time the next calf is produced, the previously infected animals are no longer infectious, added to which the new calf will receive immunity for periods of up to several months. Thus, 11 months could pass, assuming a sequence of infection as in the cow, which could occur in a species such as the African buffalo without a means for infection to be transmitted. Also, in continuous breeders, infection can be achieved only by cross-infection, because any subsequent calf will be immune for periods of up to 10 months or so.

Under natural conditions, increases in the size of the calf population in good years will also mean a larger susceptible population. This may be compounded by the fact that epizootics take hold under drought conditions, when the potential hosts are weakened and crowd together in close contact at watering places. If the susceptible calf population falls below a certain critical level, the virus dies out and must be reintroduced from another population. If the time span is long enough, the population will have lost its immunity, and reaction to infection may be peracute. However, because of the rapidity with which they multiply, viruses tend to change, and a new infection may not have the same degree of morbidity upon a population. Among wild artiodactyls, different waves of infection may affect different species, with host susceptibility changing with time. Maasai herdsmen of East Africa, whether cognitively or through some spurious homoeopathic belief, take advantage of this (Scott, 1970), deliberately mingling their cattle with animals to get "vaccinated," driving them through herds of sick buffalo or eland in the expectation that, if the disease were only of moderate severity in such normally highly susceptible species, it was unlikely to be other than mild in their own cattle. Kock *et al.* (1999) suggested that in East Africa, the Lineage II virus has a cycle of infection approximately every decade. If this is true, for they were unable to test the hypothesis, it would seem to be most likely a reflection of the periodicity of droughts, but Anon (2002a) suggests a cycle of appearance of about 5 years.

The "plains" cattle of India have a high resistance to the disease, and Edwards (1930) considered there to be an evident periodicity with 3-year periods of abatement, when outbreaks were few, mortality was low, and the disease tended to become readily extinguished spontaneously. This alternated with 3-year periods of recrudescence marked by numerous outbreaks, high mortality, and a tendency for rapid spread. This was attributed to the fact that after a 3-year period, the numbers of highly susceptible stock became exhausted, and propagation of the infection could no longer take place until a succeeding susceptible generation of cattle became available. In Egypt, an outbreak in 1903–4 was followed by another peak 3 years later, followed by periods of 5, 5, and 6 years (Askar, 1924; Table 2 and Figure 4).

In Southeast Asia, two patterns of infection were apparent among wild animals (Vittoz, 1954). Epizootics occurred in cycles of every 3–5 years as the young became susceptible to infection among banteng *Bos banteng* in Burma, Thailand, Laos, Vietnam, and Cambodia; kouprey *Bos sauveli* in North Cambodia; gaur *Bos gaurus* in Annam, Burma, Thailand, Laos, Cambodia, Vietnam, and Malaysia; gayal *B. gaurus* × *Bos domesticus* in India, Pakistan, and

Table 2. Recorded rinderpest mortality in Egypt 1904–24.

Year	Number of Cattle Dying	Year	Number of Cattle Dying	Year	Number of Cattle Dying
1904	146,748	1911	1,447	1918	
1905	689	1912	8,355	1919	873
1906	470	1913	2,314	1920	1,784
1907	4,874	1914	1,348	1921	1,244
1908		1915	358	1922	128
1909	1,998	1916	18	1923	188
1910	2,500	1917	4,794	1924	>1,500

After Askar (1924).

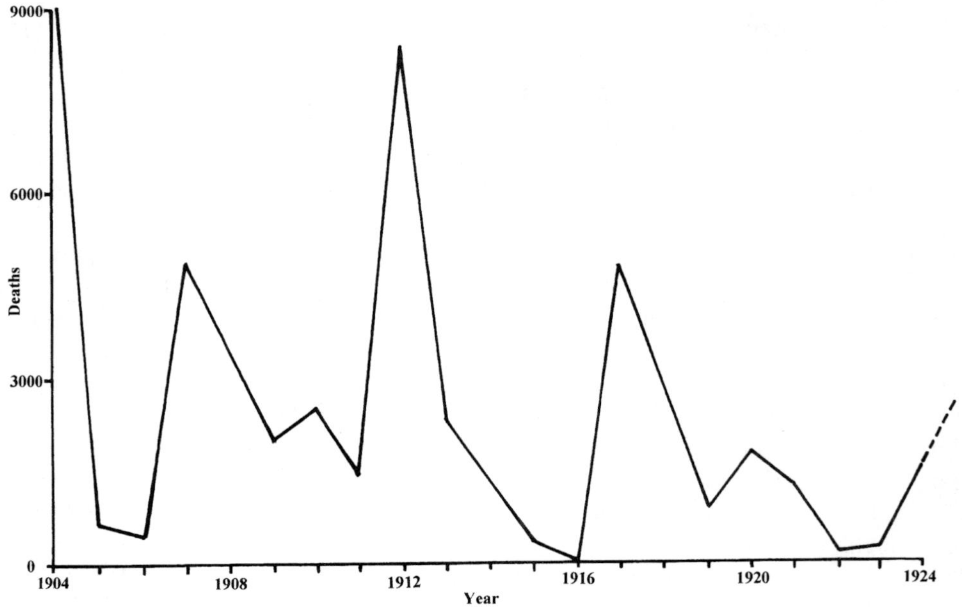

Figure 4. Rinderpest deaths in Egypt 1904–24.

east Burma; and the wild buffalo *Bubalus bubalis* throughout; as well as thamin *Cervus eldi*, hog deer *Axis porcinus* and wild boar *Sus scrofa cristatus*. Survivors became immune and perpetuated the disease, particularly wild buffalo.

Among wild species living close to human settlement such as sambar *Cervus aristotelis*, chital *Axis axis*, muntjac *Muntiacus muntjak* and mouse deer *Tragulus* spp.; outbreaks occurred at the same time as among the domestic herds. However, some species such as sambar, chital, muntjac, and certain pigs living close to villages could contract a subacute infection which made them, during certain periods, carriers.

In countries where the disease has long been enzootic, cattle possess a high innate resistance that slows the spread of the disease, lessens the clinical response, and ensures survival of the majority of infected animals. Among the native European cattle of countries west of the Russian Steppe, 90% mortality was not an uncommon average, whereas among Steppe oxen, known as "grey" cattle from their uniform grey color, mortality was often as low as 25%, sometimes even 10%. These cattle, of the Kirghiz and Kalmuck breeds, were less liable to contract the disease and also passed through it more easily. Schern and coworkers (1919) found that over 20% of Anatolian and Podolian cattle in Asia Minor proved immune against heavy artificial infection. Hungary and Romania (former Bessarabia and Moldavia) were known in the 19th century to have rinderpest more or less present continuously in a benign form. Professor Jessen, of whom Gamgee wrote that no man had greater experience on the subject, stated that steppe cattle could be so mildly affected that no veterinarian, however great their experience and ability, could recognize it. Among Hungarian cattle, the recovery rate was often 50%, but mortality could rise as high as 65%. Whereas Russian cattle experienced less severe and far less malignant reaction to the disease, Devon cattle imported into Russia and bred there for several generations were highly susceptible and rapidly exterminated by it.

To this extent, some maintained that it was unknown in the Steppe country of Russia, but generated in the long journeys performed by the cattle before reaching the frontiers of Poland and other countries. In recognition of this, noting that "too quick driving and bad feeding or watering may produce the 'ksiegosusz' (*pestis bovum*)," Polish quarantine regulations of 1857 made it obligatory to prevent steppe cattle being driven more than 25 versts (15 km) a day on average, and to be fed and watered at least twice in every 24 hr.

Curasson (1932) considered that the geographical distribution of resistant grey cattle, the grey steppe race and the grey race of Roumelia in Turkey, matched the geographical region of the Indo-European civilization. Stretching from Cambodia to Italy, it covered Asia Minor, Egypt, southern Russia, Hungary, Podolia, Roumania, Serbia, Bulgaria, Macedonia, Greece, and Venetia.

Where countries have long been free of rinderpest, the introduction of an apparently avirulent strain from an enzootic area often results in an explosive epizootic with high mortality. In enzootic areas, outbreaks tend to be restricted to immature and young adult stock, but in previously disease-free areas, outbreaks involve animals of all ages. In some epizootics, mortality was greatest at the outset, becoming less with time, but this was not always the case. Sometimes, outbreaks were comparatively mild at first but gained in virulence, maintaining their virulence for at least 2 years. The virus is not attenuated to any sensible degree by passaging it repeatedly through cattle. Early experimenters in Russia alleged that the virus still maintained full infectivity after 15 passages (Anon., 1865a). In an outbreak in the Kharkov district of Russia in 1852, it was noted that the disease did not spread uniformly:

> In some districts few animals only were attacked, and then it ceased completely; in others, it was at first very violent, then it became milder, and lastly disappeared. But it must not be supposed that the disease left for want of animals; on the contrary, in many villages it broke out only on separate farms, and left others unharmed... An immediate contact between all cattle of the village continued after the appearance of the disease as before... (Solotowsky, 1856)

In Austria in 1863, recovery rates were 5.7, 7.8, and 16.8%; depending on the district. At the same time, the recovery of sheep in the same districts was 30.5, 39.5, and 27.7%. In Poland in the same year, only 20.8% of sheep recovered. But much of this variation may be due to the manner in which the stock was kept. In 1865, the rate of mortality was considerably lower in Holland than in Britain. In Table 3 and Figure 5, the increase in cases in Holland compared with Britain for the same period shows the rate of infection in Britain was more rapid.

In 1769 in the Province of Holland for the single month of September, recovery from natural infection averaged 17.9% in the range 13.7–24.1%, a total of 21,083 dying. In North Holland or West Friesland, from April to August, recovery averaged 24%, with deaths totalling 32,907 (Anon., 1769).

In Britain, in 1865–6 of the 279,023 cattle reported affected, 40,165 recovered (14.4%). In Holland during the same period 51,562 recovered out of 156,592 affected (32.9%),

Table 3. Reported cases of rinderpest in Britain and Holland in 1865.

	Britain		Holland		
Date	Total	*a*	Total	*b*	*a*/*b*
Oct 7	11,300	1	4,084	1	1
Nov 4	20,897	2	6,744	2	1.12
Dec 2	39,714	4	11,348	3	1.26
Dec 30	73,549	7	21,437	5	1.24
Mean					1.21

a = ratio of n2/n1 …; *b* = ratio of n2/n1 ….

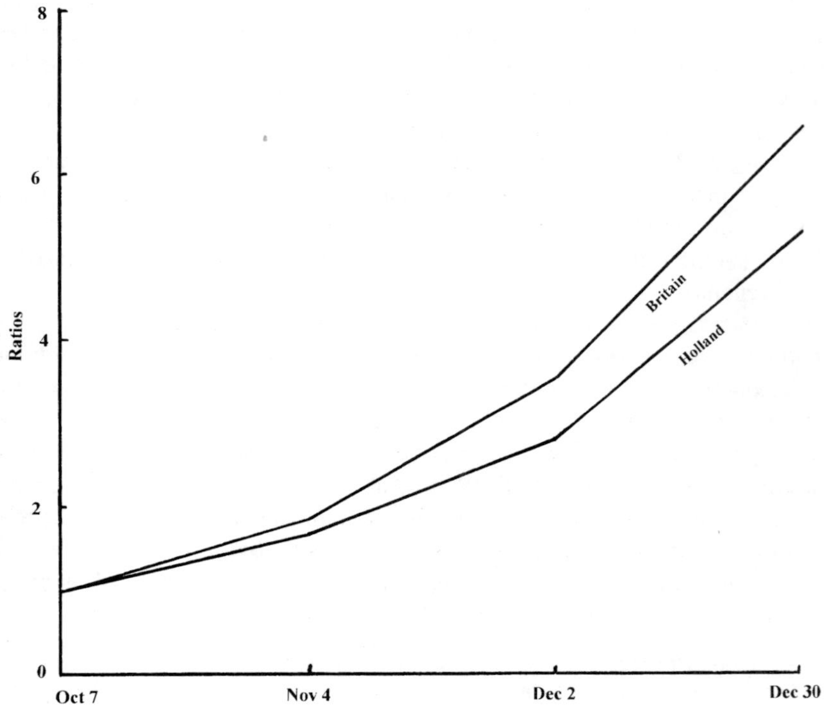

Figure 5. Reported cases of rinderpest in Britain and Holland as ratios.

although half of those suspected of infection were slaughtered. Gamgee (1866) considered that the saving of two or three animals of every 10 in an infected herd was as good an average as could be reasonably expected.

In Britain, it was found in one affected herd that all could die in 5 days, while a similar herd was not killed out in a month, and several animals might not suffer at all; but cases of animals suffering only subacute reactions were extremely rare although sufficient to show that one animal might be very susceptible and another less so.

Maurer *et al.* (1955) reported using 16 different strains in experimental trials, all immunologically identical but varying widely in virulence and hence in the severity of the lesions produced. Most strains produce less severe lesions in calves compared with mature cattle, but individual animals tend to show wide variations in resistance to the virus. There are suggestions that varying degrees of resistance to different diseases in some domestic species of animal may be attributed to differences in the blood, such as the composition of the hemoglobin (absence of hemoglobin B) and the level of potassium in the red blood cells. However, despite there being only one serotype, indications are that there are variations in the pathogenic nature of different isolates of the virus (Chamberlain *et al.*, 1993).

Among African wild ungulates, it has long been suggested (e.g., Cornell, 1934) that the passage of a fully virulent rinderpest isolate through antelope, giraffe, buffalo, or wild pig at first decreases its pathogenicity for cattle and wild animals, but further passages in susceptible cattle calves allow recovery of full virulence presumably by the selection of genotypes pathogenic for cattle. The first demonstration that a strain from a wild animal was of stable and low pathogenicity was made by Robson *et al.* (1959), passing an eland isolate nine times serially in cattle without increasing its virulence. A buffalo strain was passaged 11 times serially in cattle with only a slight increase in its low pathogenicity (Plowright, 1963a).

There is a number of observations to the effect that recovery does not always confer lasting immunity. Bliss (1922) claimed to have observed undoubted second attacks at intervals of $1\frac{1}{2}$ to 3 years after the first with all the typical symptoms present with marked severity.

8. STOCHASTIC MODELS

Youatt, professor at the Royal Veterinary College, wrote in 1834 that it was the character of epizootics to gradually wear themselves out: "They are frequent, and malignant, and fatal enough at first; but in process of time they become more rare and more tractable, and at length they disappear ..." (Youatt, 1834).

William Farr, Britain's first medical statistician, was the first to attempt to describe epidemic diseases quantitatively. Using the smallpox epidemic of 1840, he fitted numbers of deaths to the logarithmic curve of normal distribution enunciated in 1733 by DeMoivre. Farr never stated what formula he had used, referring to it simply as a "law," but in 1866 used it to predict the end of the 1865–6 rinderpest outbreak in Britain with some accuracy. Robert Lowe, one of the Cattle Plague Commissioners, had claimed in the House of Commons that if the disease was not under control by the middle of April 1866, there would be a "calamity beyond all calculation," and the thousands which were dying would grow into tens of thousands "for there is no reason why the same terrible law of increase which has prevailed hitherto should not prevail henceforth."

Farr took Lowe to task in a letter to the *Daily News* (February 17, 1866), calculating that contrary to Lowe's prediction, the outbreak would soon reach its maximum and then decline just as rapidly as it had increased. He referred to epidemic human diseases, citing particularly cholera, as always subsiding. But he was careful to qualify his remarks with the words "I will not be so presumptuous [as to predict], and will admit that the law which has "hitherto prevailed" may be modified by the introduction of unknown elements."

Farr's letter was virtually ignored, except by the *British Medical Journal*, which disputed his assumptions—disease outbreaks required measures to prevent them and it was this which resulted in their decline. The *British Medical Journal* added, "When the plague first appeared, thanks to the stupid and puffed-up ignorance of *The Times* ... the disease was generally set down as a local London cowshed nuisance, got, as "S.G.O." sagely suggested, by "Moonbeam" out of "Dungheap" (Anon., 1866*b*).

Table 4. Farr's prediction of the course of the 1865–66 rinderpest outbreak in Britain.

Period Ending	Reported Infections	Calculated Infections
4 Nov 1865	9,597	9,597
2 Dec	18,817	18,817
30 Dec	33,835	33,835
27 Jan 1866	47,287	47,191
24 Feb	57,004	43,182
24 Mar	27,958	21,927
21 Apr	15,856	5,226
19 May	14,734	494
16 Jun	~5,000	16

Table 5. The course in Yorkshire with Farr's countrywide prediction (1/20).

Period Ending	Infected	Slaughtered	Predicted
6 Jan	2,028	28	
13 Jan	1,508	48	
20 Jan	1,314	40	
27 Jan	2,034	36	2,364
3 Feb	1,426	23	
10 Feb	1,455	17	
17 Feb	1,836	51	
24 Feb	1,369	277	2,850
3 Mar	1,193	739	
10 Mar	811	882	
17 Mar	999	972	
24 Mar	683	659	1,398
31 Mar	534	517	
7 Apr	338	330	

Table 6. Progress of infection in 1866 after the Act requiring slaughter.

Period Ending	Infected	Died	Recovered	Percentage Recovered
3 Mar	7,310	3,271	2,102	28.8
10 Mar	6,518	1,185	1,151	17.7
17 Mar	6,261	779	1,014	16.2
24 Mar	4,704	318	543	11.5
31 Mar	3,956	154	394	10
7 Apr	3,361	131	267	7.9
14 Apr	2,582	108	354	13.7
21 Apr	2,823	93	210	7.4

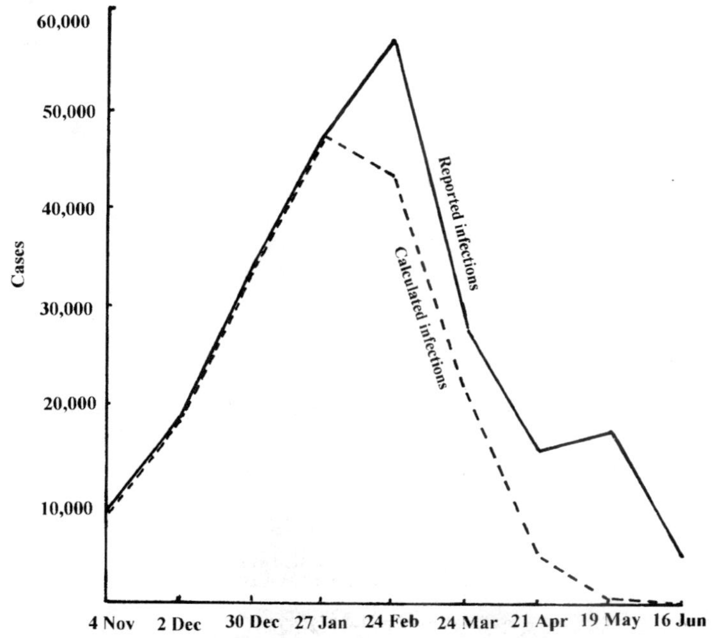

Figure 6. The course of the 1865–66 rinderpest in Britain as predicted by Farr.

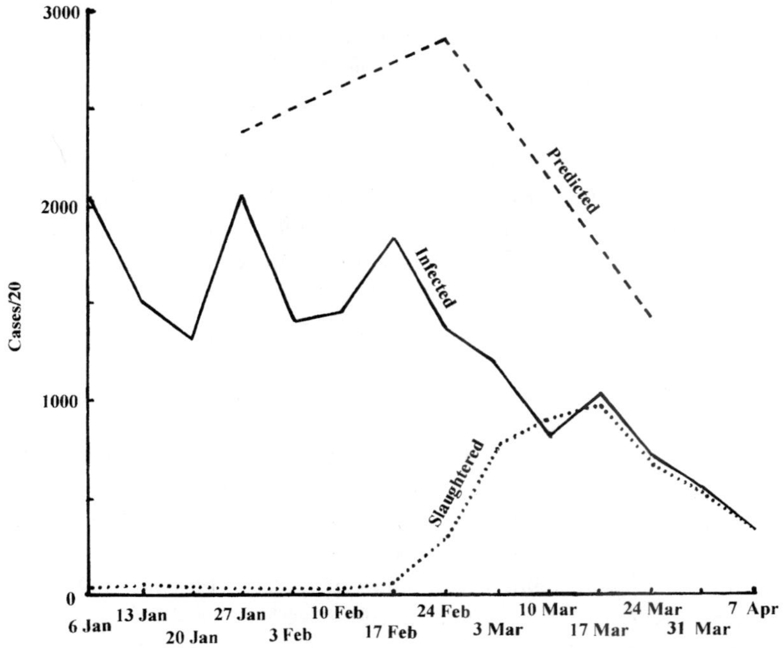

Figure 7. Course of the outbreak in Yorkshire in 1866 compared with Farr's prediction, and the numbers slaughtered (cases divided by 20).

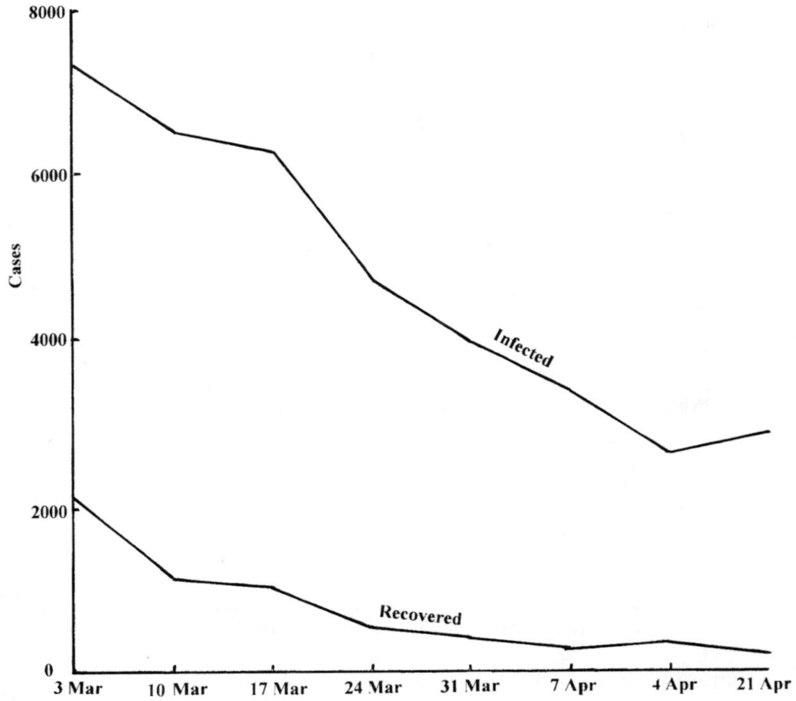

Figure 8. The progress of rinderpest in Britain in 1866 after the Act requiring slaughter.

On March 20, Farr followed up his letter with one to the *Journal of Social Science*, repeating his *Daily News* letter in full and claiming that the epizootic had apparently attained its maximum and was now declining as the "law" had led him to believe it would. Although not declining as rapidly as his equation predicted, it did decline according to the shape of the predicted curve. However the decline followed the passing of the Act on February 20, empowering slaughter of infected and suspect beasts, and the figures for Yorkshire, where the conditions were strenuously applied, suggest that it was this which caused the decline (Tables 4–6 and Figures 6–8).

In a letter to *The Lancet* of August 26, 1865, Murchison, who contributed a section "On the Relation of the Cattle Plague to Human Diseases" to the Commission's report, implied that rinderpest died out in the 18th century in Britain by passing into a mild form of disease, as with vaccinated animals. Gamgee (1866) stressed that this was contrary to all experience of the disease: "It killed the last animals in Egypt in 1863 as rapidly as the first. It is as deadly now whenever it attacks an animal as it was last July, and we shall not see it subsiding *per se* into that benignant disease which must be regarded as rather favorable than otherwise to the life of men and animals, and never calculated to annihilate the horned or other stock of any part of the globe."

A century later, Rossiter (1996) wrote "What the 1990s outbreaks [cattle plague in Pakistan; severe outbreaks in West Asia; mild disease in cattle in East Africa; typical severe diseases in hypersusceptible East African wildlife; mild disease in sheep and goats] do show is that the complete range of clinical syndromes of rinderpest are still present in the world today," echoing Faust (1797) two centuries before, who wrote that rinderpest was always essentially the same in all lands, in all climates, and in every year, noting that Kanold (1713) had already stated it was the same everywhere. Differences in expression of the disease were due to the age of the animal, nutrition, health, climate, or maintenance and care.

2

SPECIES AFFECTED AND GEOGRAPHICAL DISTRIBUTION

1. SPECIES AFFECTED

1.1. Aberrant Hosts?

The concept of host specificity was first enunciated *c.* 400 BC allegedly by Hippocrates: "... some one may ask, why these diseases do not affect all animals in the same manner, but confine themselves to some particular sort? This is the reason; because in their bodies, their nature, and their food, they differ from each other ...." During the Great Plague of London, the physician Hodges wrote in 1672: "And the Conjecture that sickness amongst Cattle is transferable to the humane Species, hath not yet appeared on any good Foundation ... I cannot be induced to believe that the Pestilence amongst Cattle from a private Cause, can ever obtain any Dominion over Mankind" (Hodges, 1720). Another English physician, Richard Mead, in 1720 wrote that he was well aware "that there are plagues among animals, which do not indifferently affect all kinds of them, some being confined to a particular species like the disease of the black cattle here, a few years since, which neither proved infectious to other brutes, nor to men." (Mead, 1720). Layard (1757) was of the opinion that rinderpest affected only horned cattle, but although a disease of all cloven-hoofed ungulates, not all species may be affected by particular strains.

Domestic buffaloes and cattle are the most commonly affected species, but from time to time, infection has been alleged in a number of aberrant hosts, although suggestions of a number of non-artiodactyls being affected, such as the elephant, are considered by Scott (1970) as unbelievable. Nevertheless, a number has contracted infection experimentally.

It was reported in Italy in 1711 that ducks and hens died after scratching in the excrement of infected oxen (Anon., 1711), and a chicken was reported in the *Journal of Venice* as dying soon after scratching in the dung of an infected cow, while Salchow (1755) reported that poultry were affected; but all these cases were more likely due to bacteria. In India, fowls were found with an "eruptive febrile infection" coincident with rinderpest in a village, but inoculation of their blood into cattle failed to induce any disease (Hallen *et al.*, 1871). Much later, Kylasamier (1931) claimed to have killed fowls by injecting with the virus, allegedly confirmed by Rajgopalan (1937). In South Africa, it was reported that guinea fowl (which pecked in animal droppings for maggots) had died of the 1896 rinderpest, and a number were reported dying near Mabuli allegedly after eating rinderpest-infected meat. Arnold Theiler, the South African Republic's veterinary surgeon, stated that they had died in the Republic all along the line where the disease was present (Theiler, 1897b). In Chad, Africans asserted that the ostrich was affected. In 1898, it was reported by Africans that along the banks of the Pongola River, crocodile which took the bodies of mountain reedbuck dead of the rinderpest were also dying of the disease, but the deaths were unconfirmed (Van Oordt, 1898a). Lugard reported in 1896:

> It is strange that the natives here [near Lake Ngami in northern Bechuanaland Protectorate] insist that the crocodiles and hippopotami are dying of it. I cannot believe it true, but we came upon a dead hippopotamus; an unusual thing, since the people are such keen hunters, and its hide so valuable here ... Now here, on the Botletle River, they say crocodiles and hippopotami, and also, I have heard, elephants, donkeys, and dogs are affected. Animals are here said to go mad before death, and become very dangerous. (Kirk, 1896)

This suggests perhaps concurrent outbreaks of anthrax and rabies.

In the German outbreak of 1720, it was thought that cats might spread the infection from one stall to another, especially at night. In India, dogs, jackals, wolves, cats, tigers, and horses were all blamed for transmitting the disease, but there was no proof of this. Carrion animals, however, were considered an undoubted means of spreading it.

The 19th-century Austrian border-control regulations, recognizing the occurrence of rinderpest in species other than cattle and the possibility of mechanical transmission by non-affected species, stated:

> Should the Plague break out in districts close to the frontier, all sorts of sheep, goats and swine are not to be admitted, and particular attention is to be paid to dogs and fowls which frequently carry about the disease, and it is desirable to check as much as possible their wandering about; this especially applies to dogs.

In December 1865, a donkey was inoculated at the Royal Veterinary College, London, with material from the nostrils of an infected cow, but showed no effects. Guinea pigs were infected by Semmer in 1893, who produced an attenuated virus by passage through them. Although infection in the guinea pig was inapparent, its blood was infective to cattle. In 1925, Edwards was able to obtain an attenuated virus by passage through rabbits, a species which had been found by several workers to react in the same manner as the guinea pig, having an inapparent infection, although its blood was virulent to cattle. In recent experiments with laboratory-bred rabbits, only 40% survived injection with virulent lapinized rinderpest virus (Patil *et al.*, 1997). In 1929, Inoue and coworkers found the Mongolian ground squirrel *Citellus citellus ramosus* experienced diarrhoea followed by death in a few days when exposed to fresh virus, but after the fifth passage, the symptoms were almost undetectable. This led to the suggestion that the ground squirrel might be involved in maintaining the disease in enzootic areas, but this is generally discounted. Nakamura and coworkers (1957) infected Syrian hamsters *Mesocricetus auratus* with lapinized avianized virus, Scott and Witcomb (1958) doing the same with a bovine strain, while Scott and Brown (1958) found that serial passaging in the hamster lowered the virulence for cattle.

The hippopotamus *Hippopotamus amphibius* in Uganda has been found to carry neutralizing antibodies, indicating exposure to rinderpest some 30 years before, described by Plowright *et al.* (1964) as an outstanding example of the stability of such antibodies in species other than man. But in the case of neutralizing antibodies which reacted with rinderpest virus recovered from wild spotted hyenas *Crocuta crocuta*, there is a known antigenic overlap between the viruses of canine distemper and rinderpest, so that distinction is difficult to verify.

Dr. Gilchrist, in a letter to the *Glasgow Citizen* in 1857, stated that the elephant and the camel were subject to murrain. Gamgee asked "What murrain? Most probably not the contagious typhus" (Gamgee, 1857a). Experiments on elephant conducted in India in 1906–7 were inconclusive. A cow elephant "in perfect health" was subcutaneously inoculated with 500 cc of virulent defibrinated blood from an infected ox. Apart from a diffused swelling at the site, there was no rise in temperature and no untoward symptoms in the next 23 days. As Evans (1910) pointed out, although he considered the elephant insusceptible, a single elephant proved nothing. It might have already contracted rinderpest and become immune to it.

1.2. The Camel

The camel has proved to be a controversial species but is one which could have played a role in introducing the disease into sub-Saharan Africa. Gilchrist (1848) stated that it had been satisfactorily determined that the camel was subject to a similar disease to "the murrain of Europe"—"... indeed it has prevailed to a considerable extent amongst the camels of the Public Cattle establishment during the current year (1848)"—and described the symptoms of affected camels and post-mortem appearances. In Egypt in 1861, an immense number of camels allegedly succumbed to rinderpest. Verdernikoff (1893) in Russia considered that the camel showed the same signs as cattle, with an average 5 days' incubation followed by a rise of temperature to 42°C, the breathing rate increasing to 50 per min, and an accelerated pulse rate of 90 beats per min. Vesicles formed on the buccal membranes, and red spots developed which ulcerated. There was intense conjunctivitis, sometimes with keratinization. Constipation was followed by diarrhea, irregular breathing, and a cough. The temperature fell near the end, and average mortality was 95%. Its susceptibility was also allegedly demonstrated by Pease *c.* 1894, who stated: "The camel also suffers very severely from the disease, it being very fatal in him; numerous instances of the disease in this animal have been cited by various authorities, and some have come under my own personal observation" (Pease, 1894). But camels were reported as unaffected in an outbreak in 1898 in the Turkish province of Aïdan. Koch was unable to infect a dromedary experimentally, but Tartakovskii (1899) in Turkestan inoculated two dromedaries and four bactrian camels, the two presenting an insignificant reaction and the four symptoms of rinderpest, and one dying. He was highly critical of all previously published work on susceptibility of the camel, claiming that the observations were suspect.

Lingard (1905a) reported that experiments conducted on five camels of the United Provinces, India, showed them to be susceptible to inoculated bovine rinderpest and vice versa. When inoculated with virulent hill bovine blood, incubation lasted from 5 to 7 days and the course of the disease 15 to 18 days, always terminating in recovery. Vesicles formed in the mouth, but there was rarely diarrhea: "But little can be diagnosed by a casual observer as to the real condition of the animals or even with regard to the fact that they were the subjects of a serious epizootic malady." When camels were inoculated with virulent blood from other camels exhibiting the disease, the infection was milder than when primarily inoculated with virulent bovine serum. When hill bulls were inoculated with infected camel blood, they contracted a virulent form of rinderpest indistinguishable from that contracted after inoculation from a bovine source. Leese (1927), who, during 6 years, was unable to find a case of natural infection, among camels in India, after observing an experimental infection, criticized Lingard's findings, claiming that the results were far from the classical symptoms of rinderpest. But other studies supported Lingard's conclusions. Thus, Cross (1917), noting that from time to time, racing camels died of rinderpest in India, inoculated three subjects each of 1, 6, and 12 years of age. The two youngest showed typical rinderpest symptoms but recovered, and the old animal showed only a slight diarrhoea and a loss of appetite, slowly recovering. "The only conclusions to be drawn from the above are that camels develop severe lesions of rinderpest and are more susceptible than some cattle to rinderpest" (Cross, 1918). These results, he claimed, confirmed the observations of a Russian worker, Petrowsky, on Steppe camels. However, in Turkey, camels living in permanent contact with cattle subject to several serious outbreaks did not themselves contract it. Inoculation of 13 camels and seven young produced no reaction, and their blood was non-infective to cattle (Anon., 1923). But Haji (1932) described an outbreak in India among camels with a mortality of 20–40%, and Dhillon (1959) reported 60 of 128 camels dying of the disease in India. But in neither of these cases was it confirmed by serology, only assumed from clinical symptoms.

Curasson (1932) pointed out that when referring to the camel, workers rarely stated which camel was referred to, whether the single-humped *Camelus dromedarius*, or two-humped *Camelus bactrianus*, but there did not seem to be any difference between them in their reactions to rinderpest.

Africans in West Africa claimed that the camel was infected in 1892, and it was reported as affected with mild symptoms in Eritrea in 1913 (Conti, 1913). In the outbreak in northern Kenya in 1960, Scott and MacDonald (1962) examined sera from 60 camels from outbreak areas but found no trace of rinderpest antibodies. Yet, in 1967, Singh and Ata (1967) did demonstrate antibodies in 8–10% ($n = 194$) of camels slaughtered in the Cairo abattoir from rinderpest enzootic areas, which included animals from the Sudan (9.7%). These workers found that camels were susceptible to the virus experimentally, developing high levels of neutralizing antibody following inoculation with virulent virus, but developed inapparent infections. However, an infected camel failed to transmit the disease to a susceptible calf, but that is not to say that infection could not be transmitted given the right circumstances. The course of the disease might be different under the stress which camels are frequently exposed to, such as long desert journeys without water. The only clinical sign of infection was recorded as a transient fever which easily could be missed. The camel must therefore be regarded as a possible source of transmission. Hamers-Casterman *et al.* (1993) demonstrated that the camel had an "amazing" immune system, producing immunoglobulins unique among mammals. It is this unique capacity which may explain their variable response to rinderpest (Jassim and Naji, 2001).

1.3. Pigs

Pigs have been infected experimentally through eating infected raw meat. Whether the infection was contracted from nosing the meat beforehand, or from ingestion, is not clear (Plowright, 1968 attributes it to ingestion). European domestic pigs experience inapparent infections. In 1865 in Britain, it was reported from Rotherham in Yorkshire on two farms that pigs were constantly fed on the refuse food, covered with saliva, of infected cattle, and mixed freely with them but did not suffer in any way, and a "strong healthy pig" inoculated with matter from an infected cow, at the Royal Veterinary College, London, failed to display any effects. But *The Times* of December 12, 1865 reported that an infected pig spread the disease to a number of cattle. A domestic pig fed on the entrails of a cow dead of the disease was reported to have died in the outbreak in Germany in 1744, when wild boar *S. scrofa* were reported to have died in the woods (Bucard-Mauchart, 1745). Salchow (1755), possibly repeating Bucard-Mauchart, reported that pigs die if they eat the guts or dung of infected animals.

In India, evidence of infection in pigs was considered strong, but inoculation failed to produce a reaction in them (Hallen *et al.*, 1871). Pease (1894) noted that the wild pig contracted it. In Asia, the domestic sway-back pig was affected, and in Sri Lanka, wild pigs are reported to have died from eating carcasses of infected cattle (Mahamooth, 1943). It was not until 1993 that there was the first confirmed report in European pigs in India, when an outbreak among Large White Yorkshires in Tamil Nadu resulted in the deaths of 40 of 56 infected. Theiler (1897a) in South Africa, and Angeloff (1917) in Bulgaria, failed to obtain any reaction in pigs, although Theiler claimed that the passaged blood used was infective to cattle.

Leblanc reported rinderpest in peccaries in 1886 (Carré and Fraimbault, 1898), and Penning (1894) claimed to have infected wild boar (Carré and Fraimbault, 1898). Pease (1894) stated that it had been observed in the wild pig of Sumatra and transmitted experimentally in them. Carré and Fraimbault (1898) demonstrated, in Indo-China in 1896, transmissibility from pig to pig by contact and by blood inoculation, also from bull to pig and pig to bull by inoculation. Pluning observed the disease in pigs in Sumatra, and in Southeast Asia, it has been reported in wild pigs (Vittoz, 1954). In the Philippines, it had been noted for many years prior to 1916 that, simultaneously with the appearance of rinderpest in cattle and water buffalo, in certain localities pigs also became infected. On the island of Rombon, where the disease had previously never existed and there had been no cattle imports for some time, an outbreak caused serious losses among cattle following an import of pigs from an island where rinderpest was prevalent. Pigs died first in considerable numbers before cattle and water buffalo began to succumb. Pigs exposed to sick cattle in the laboratory in 1914 developed symptoms similar to the disease in cattle. Experiments by Boynton showed that cattle, water buffaloes, and pigs varied but slightly in susceptibility, and the disease could be transmitted practically as readily from one type of animal to the other as among each kind (Boynton, 1916). In Africa, the wild *Suidae* are affected: warthog *Phacochoerus aethiopicus*, bushpig and giant forest hog *Hylochoerus meinertzhageni*.

2. RUMINANT ARTIODACTYLA

2.1. Deer and Other Ruminants

Deer were reported to have died in the woods in Germany in an epizootic in 1744 (Bucard-Mauchart, 1745; Salchow, 1755). In the 1770s, Camper inoculated a stag with virulent blood without apparent effect but also failed to induce infection in sheep and goats. He also obtained a reindeer from Norway, but it died before he could experiment with it (Camper, 1783). Bristowe, in December 1865, examined a deer at the Royal Albert Veterinary College, London which had died of rinderpest but also was infected heavily with tuberculosis. The origin of the deer is not stated (Bristowe, 1866). Pease (1894) reported rinderpest in India in the banteng, gaur, buffalo, muntjac, goral *Naemorhedus goral*, "gazelle" and "stag." Renshaw (1907) noted that a banteng calf which had died of rinderpest was post-mortemed in Calcutta. In Southeast Asia, it has been reported also in banteng, gaur, buffalo, and smaller antelopes and deer (Vittoz, 1954). In 1923, Japanese workers demonstrated that sika deer *Cervus sika* were susceptible, noting that the period of incubation in natural infection did not exceed 10 days, while in the laboratory, it was 3–6 days. Serum was found to be curative when administered at the onset of fever and repeated 4 days later, the virus disappearing from the circulation within 6 days of recovery (Ono and Kondo, 1923). Jacotot (1927) obtained a very benign experimental infection in the sambar in Indo-China. Dieckerhoff (1890) attributed the decline of the European bison *Bison bonasus* to rinderpest.

The *Daily News* of October 20, 1865 reported two bison and three Italian cattle affected at the London Zoo, contradicted the following day by the Superintendant, A. D. Bartlett. He stated that there were no bison in the zoo, and of the three Italian cattle, one bull had died of inflammation of the lungs, and another was slaughtered for showing symptoms of the same. All other ruminants were in perfect health, and only a baboon showed signs of sickness. In June the following year, in a letter to *The Veterinarian*, W. Egerton, reporting on a case of infection in eland, stated he believed it to be the first case in England of an antelope being affected: "... last week one of the elands at Tatton Park was seized with some of the usual symptoms of the cattle disease now so prevalent in that part of Cheshire. The first were running at the eyes and nose, great soreness of the mouth, refusal of food, followed by rapid prostation of strength, and death in a few days ... the female has been attacked with similar symptoms, but there are some hopes of her recovery." The district cattle inspector pronounced it a decided case of rinderpest, apparently contracted from infected cattle placed in an adjoining paddock, separated by a brook and a wall.

Oddly, Gamgee (1866) stated that domesticated buffaloes used for draught, although taking the infection, did not experience very severe symptoms, and the disease was not deadly: "from the contrast in the rate of mortality offered by herds of buffaloes as compared with oxen, it is sometimes scarcely credible that the same type of Steppe Murrain attacks both." But buffaloes are particularly susceptible, whether wild or domestic, and there was great destruction amongst domestic buffaloes in the 1860 outbreak in Hong Kong, but not apparently in Egypt, and it may have been the latter to which Gamgee referred. Todd (1912) reported that the Egyptian buffalo, although sometimes dying of the disease, often escaped it altogether in outbreaks or only contracted a benign form. Its reaction was complicated by an enzootic form of hemorrhagic septicemia, which was often the real cause of death. Experiments suggested during an epizootic that they might contract the disease in a very mild form and, while apparently healthy, might in reality be highly infectious and play a considerable role in maintaining and spreading infection.

2.2. Goats

The rinderpest-like disease of sheep and goats in northern equatorial Africa, not recorded from eastern Africa with the exceptions of Ethiopia and Sudan, is the PPR. PPR antibodies were found in small ruminants in Oman, where, prior to 1981, rinderpest was unknown in living memory, while unvaccinated cattle had higher antibody titres to PPR than to rinderpest virus. In the 1990s, PPR circulated with devastating effect in goats and sheep across the Asian continent from Arabia through Pakistan and India, to Nepal and Bangladesh. Zonchello (1917) reported that local people in Eritrea asserted that goats did not contract rinderpest naturally in the Sahel district (a strip of land bordering the Red Sea) and believed that cattle living among the herds of goats did not contract the disease in the scene of an outbreak, the author corroborating this with an observation of his own. Goats and sheep often die from what is attributed to rinderpest in India, but elsewhere are affected only sporadically. In some outbreaks, they remain unaffected, while in others, a large percentage might contract the disease.

In Germany, in 1744, a goat stabled with a cow which died of rinderpest also succumbed (Bucard-Mauchart, 1745). In the outbreak in France the following year, goats and sheep contracted a disease similar to, if not identical with, that of the cattle, as they did in other outbreaks (Engelman, 1763; Sauvages, 1746). But it was not until 1835 that there was the first confirmed observation of alleged rinderpest in a goat. A Norfolk farmer gave evidence to the Commission in 1865 that his children's pet goat died of rinderpest contracted from his oxen. The goat died within 3 days of being reported sick.

In October 1861, rinderpest broke out in Cankowitz in Bohemia after the passage of a number of Polish cattle which rested there. A goat was placed in the stall where some of the cattle had been and fell sick on November 21, dying on November 25 with symptoms similar to sheep affected with rinderpest. When, in April 1863, the disease broke out in Carniola, it also affected goats.

Scott and Brown (1961) confirmed the spread of the virus from cattle to goats under field conditions in East Africa, and in the 1980s, antibodies were found in goats and sheep in southern Tanzania. Later experiments conducted in Tchad also transmitted the virus from goats to cattle and vice versa (Bidjeh *et al.*, 1997).

2.3. Sheep

Fracastoro (1546) wrote of the plague sweeping away not only "the wretched cattle, but also nearly the whole of the unhappy flocks of sheep," which probably prompted Gamgee (1865) to allege that people had known "from time immemorial" that sheep would communicate the disease.

Several observers in the 18th and 19th century, namely Sauvages, Jessen, Sergejew, and Paschkewitsch, had noted the susceptibility of sheep and goats, and one writer in 1747 asserted that the disease could be spread by the wool of sheep. But Gasparin, in an essay on diseases of sheep, noted that "miasmatic illness" was very rare among them, and although one could cite cases of sheep having contracted "gastro-pulmonary catarrhs" from common grazing or by being placed with infected cattle, it was not certain whether they could transmit it to other sheep. However, if a large number fell sick at the same time, it was probable that the corrupted air could be charged with morbific particles, but examples were lacking. Nothing gave us the right to establish a contagion from afar in sheep (Gasparin, 1821). In 1865, the *Medical Times* (no. 796) erroneously reported that Professor Simonds had made the first discovery that sheep were capable of taking the infection of rinderpest, and no such discovery had hitherto been made on the Continent, for, in 1857, Dr. Kreutzer had first described the symptoms in sheep inoculated with virus from cattle. Then in 1859, Dr. Maresch, District Veterinary Surgeon in Bohemia, first noted the similarity of a disease in sheep with that of rinderpest in December of that year at Wysoka, where 18 of 108 sheep, kept close to a cattle shed where the infection was present, were afflicted with analogous symptoms, and all died. Later in December, the disease broke out at Mseno. A butcher placed affected cattle with some sheep, 10 of which sheep were taken to Sedletz where they were distributed among five properties. On all properties, the disease broke out, and of 80 sheep affected, 44 died, and 21 were slaughtered. On four properties where the sheep had been kept in close proximity to cattle, rinderpest broke out. In November 1861 at Chlumetz, Dr. Maresch found 21 of 52 sheep kept with infected cattle affected by it. Where sheep were kept separate from infected cows, they remained free of the disease, and Maresch showed that sheep most frequently contracted it when constantly penned with infected cattle but did not do so if only mixing with cattle in the open (Maresch, 1863).

At the end of 1861, Dr. Galambos, Professor of the Veterinary School of Pesth, reported a similar case in Hungary; and in Carniola, next to Croatia, where the disease was introduced by the passage of some Polish cattle in April 1863, there was clear evidence of the disease being transmitted to sheep, which spread it to other sheep, and finally back to cattle again. In Poland, 4,000 sheep are reported to have died in 1863 from the disease, in one district 74% of those contracting it. Seifman, Professor of Veterinary Surgery in the University of Warsaw, estimated that about 20% of infected animals recovered (Seifman, 1865).

In 1863, Professor Röll of the Veterinary Institute in Vienna, who, in his veterinary handbook published in 1856 had stated that only the ox and buffalo could be infected with rinderpest (Röll, 1856), studied the disease in sheep and conducted experiments demonstrating its transmission from sheep to cows and showed that passage of the virus through sheep did not render it less virulent.

In Sicily, 20,000 sheep and goats died of rinderpest in 1863. Two months after the eruption in Britain in 1865, at least two outbreaks among sheep had occurred. The first was reported among a flock of 120 lambs bought at Thetford in June. Two died 10 days after purchase, and 36 days later, 46 had died or had been slaughtered, and 27 were very ill. Ten days after its last contact with one of the sheep, a cow showed symptoms, and shortly after, eight were dead. In October 1865, eight sheep were experimented on at the Royal Veterinary College. Divided into lots of two each, one lot was placed in a shed which had not been disinfected and previously occupied by infected cattle. The sheep contracted the disease after 13 days, and one died. Another lot was placed in a shed with diseased cattle, and both died. Two were inoculated with material from sheep which had allegedly died of the disease, and both died. The final lot was inoculated with material taken from the eye of an infected bullock, and both became affected on the sixth day, one dying.

Overcrowding (and therefore poor health) accentuated the disease, probably accounting for the reversal of mortalities in November 1866 when 90 cows died from the disease compared with 208 sheep; but the variations in the symptoms in sheep were just as great as those amongst cattle. Various inoculation experiments showed that sometimes sheep took the disease and at others showed no apparent effects. Successful experiments were carried out in England and at the Inoculation Institute at Cherson, in Russia, where infected sheep and goats were found to be just as infective to cattle as if the virus had been transferred from cattle themselves. Maresch in Prague, Galambos in Hungary, Simonds in England, Leblanc and Bouley in France, Gerlach in Germany, and observers in Belgium, Holland, and Scotland had fully established that other ruminants could contract rinderpest naturally.

In 1860 when the disease was imported into the Island of Chusan with cattle from Hong Kong for provisioning the expedition sent to capture the island prior to the British attack on north China, a few months later there was great mortality among sheep and goats over the whole island with symptoms attributable to rinderpest.

In 1973, there was an alarming increase in rinderpest amongst sheep and goats in Kamataka and Tamil Nadu, India. Whereas there was some spread to cattle, it mainly affected sheep, successive waves of a severe form over several months in Tamil Nadu having little involvement of cattle. Although this lack of spread to cattle might be partly accounted for by the vaccination programme, antibody tests showed that only some 31–68% of cattle were immune. Goats inoculated with an isolate once passaged in sheep showed a reaction with fever and diarrhoea, but all survived. But after three passages in sheep, mortality in goats was total. In sheep, its course was rapid, possibly leading to low levels of replication of the virus and thus low levels of virus excretion; but although highly pathogenic to sheep, the strain was less so to buffalo and grade zebu cattle. Serial passage of the virus in sheep created an enhanced pathogenicity with a shortened period of incubation and an extension of the infective phase (Ramani *et al.*, 1974).

3. MAN

It had long been a subject of discussion whether the plague affecting man was the same as that affecting cattle (and other animals). Joubert (1567) noted that the Catalan physician Arnaldo de Vilanova (1240–1311) stated that the plague of man never attacked cattle, and that of cattle never attacked man. Marsilio Ficino (1433–99) claimed that it often passed from man to pigs, and Joubert noted that a number of "serious" writers stated that the plague which came to man was signified by the death of other animals. But, Joubert argued, if it was carried in the air, birds would die first; if it came from the earth, four-footed beasts would die first, especially sheep because their heads were closest to the ground. But this would not affect man unless the contagion was very abundant (because of the height of his head above the ground). This would mean there was no difference between the plague of man and animals, but often stud farms and herds of cattle, sometimes sheep alone, were affected without men being so. He concluded that the plague could be common to man and animals, but man was more susceptible, with animals not showing it.

Although, in the 1711 outbreak in Italy, Lancisi reported many cases of misfortune of people eating the flesh of animals which had died of the disease, the misfortunes resulted from other, bacterial, infections of the meat, for people are not known to be affected by eating the diseased meat: "... recently two peasants who chose to eat the flesh of a dead ox which was infected suffered very cruel diarrhoea" (Anon., 1711). Erring on the side of caution, a German edict of 1732 warned that in treating boils and ulcers of infected cattle, care should be taken that no pus touched the human skin. If so it was to be washed with lye and soap, followed by warm water and vinegar. Although there was no evidence that the disease was contagious to humans, the precaution, it was argued, would do no harm (Dorwart, 1959). Caspar (1777) reported that records from Breslau in Silesia noted that in 1723 people were affected by a general sickness from eating meat of infected cows, but if one slaughtered those which started to belch, the meat could be eaten.

In the region of Dijon in 1744, it is alleged that a number of people who bought rinderpest-infected meat from a butcher selling it in contravention of a law issued at Dijon died (Anon., 1744); but this may have been a case of anthrax. In 1745, it was reported that "a very sufficient experiment was made in our army in Flanders last campaign in favour of this," the flesh of cattle which had died of the infection having been fed to condemned soldiers without any ill effect on them. A flayer told Dr. Mortimer he had made a hearty meal of some steaks he had cut from a diseased carcass and had not suffered from it (Mortimer, 1745a). Brocklesby noted that the flesh of infected cattle could be safely eaten if it remained untainted for 3 or 4 days unsalted (Brocklesby, 1746a). In 1746, Bucard-Mauchart, writing of an outbreak in Swabia stated: "All the men who were during this time engaged in feeding and curing the sick cattle, in burying, removing, skinning and cutting up their bodies, and who breathed heavy clouds of effluvia, remained in perfect health ... ." While Hannæo (1746) advised that rinderpest-infected meat should not be eaten. He quoted instances of people dying from eating a pig which had died of disease, but this was probably anthrax; and in 1677, people at the University of Leipzig died from eating the meat of sick cows served up in the refectory.

The people of Puglia in southern Italy ate the flesh of animals which had died of rinderpest for 3 months in 1837 without any ill effects; likewise during the 1860 outbreak in Hong Kong, Chinese ate the meat of the animals which had died of rinderpest, including those carcasses which had been towed out to sea and were later washed up on the beaches. In Bohemia in 1863, as fast as infected animals were slaughtered and buried, so the people dug them up and ate them with impunity. In the British outbreak in 1865–6, there was a regular but illegal traffic in diseased flesh but no records of any ailment resulting from its consumption. Liebig (1843) considered that as the poison was neutral or alkaline, stomach acid destroyed its power of contagion.

During the Franco-German war of 1870–71 in Paris after the siege, the population had nothing but diseased cattle to eat, and no harm allegedly came of it. It was claimed that people were generally healthier from the excess of meat consumed.

In South Africa's Transvaal in 1896, there was a reported outbreak of food poisoning among Africans who had eaten cattle which had died of rinderpest, and Lugard reported from the northern Bechuanaland Protectorate in September 1896, "Beddoe tells me the natives are dying of eating too much 'rinderpest' beef. I do not know if this is true; all the natives eat it, and convert it into Biltong for future use" (Kirk, 1896).

Sanderson asserted that the disease could not be communicated in any form to man (Sanderson, 1866b), while John Gamgee (1866) stated that he had inoculated himself frequently with rinderpest, "more times than I venture to name," and had not suffered in any way or experienced any eruptions. Much consternation had been caused by an inquest into the death of a young veterinary surgeon, Robert Plumbly of Sudbury, who had shot and then dissected a cow suffering from rinderpest, and himself died 4 days later. His death originally being attributed to apoplexy, the coffin was opened and the body re-examined 2 days later, when it was found to be already bloated and putrid and the tissues utterly disintegrated. The jury "were so overcome by the fearful spectacle that they were permitted by the coroner to retire to partake of stimulants before they could further proceed with the inquiry." The veterinary surgeon had of course conducted his postmortem examination with bare hands, and it transpired at the inquest that he was suffering from a boil or boils on his right arm. It seems most likely that he was either already suffering from anthrax or contracted septicaemia at the post-mortem. The verdict of the coroner, based upon the evidence of a surgeon, Mr. Maurice Mason, was that the deceased had died from the effects of a virus or poison absorbed into his system while carrying out a post-mortem of a cow which had died (it had been shot) "from a certain disease called or known as the cattle plague."

Another case of a veterinary surgeon contracting infection while dissecting a rinderpest-infected carcass was that of Mr. Hancock of Uxbridge. He was punctured on the back of the hand by his assistant who was skinning the animal, but not sufficiently so to draw blood. A vesicle developed, and Mr. Hancock suffered a fever. The vesicle had all the appearance of a vaccine vesicle (Mr. Hancock had been vaccinated in infancy), and it seems most likely that it was caused by a cowpox infection rather than rinderpest, the only differences in the vesicle being its unusual size and duration (Quain, 1866).

The veterinary surgeon Farrell reported in India in 1870 that one of his assistants in making a post-mortem examination accidentally cut his finger: "He suffered fever a few days afterwards, I can't say how many, and his hand and arm swelled and pimples appeared upon them; he lingered on for about a month and eventually recovered. Ulcers formed on the arm; they were small and about quarter of an inch in diameter" (Hallen *et al.*, 1871).

In June 1831 in India, the Medical Board issued a circular to the effect that public confidence had been so shaken in the powers of [human] vaccination, owing to its failing in many instances to protect against severe and even fatal attacks of variola, that it was desirable to institute a series of investigations and experiments for the purposes of regenerating the virus. As a result, in 1832, Dr. Macpherson, Superintendent of Vaccination at Moorshedabad, tried to induce variola in cows by inoculating them with smallpox virus and covering them with contaminated blankets. Unable to obtain any satisfactory results from his experiments, and hearing that cows in Bengal were subject to a disease which usually made its appearance about the end of August or beginning of September *matah* (= rinderpest), he clearly thought it to be cowpox. The Indians used the same name for the disease in cattle as for human smallpox, although they firmly believed it was not transferrable to humans. "I consequently determined on again attempting to regenerate the vaccine virus from the original source" (Macpherson, 1842). Accordingly on the twelfth day, he removed crusts, which formed on the udders on the ninth and tenth days of several sick cows and used them to inoculate 11 Indian children. Six showed no reaction, the remainder varying degrees of reaction with one, a 5-month-old child who suffered much fever for 4 days, producing a vesicle which assumed all the characteristics of the vaccine. In subsequent vaccinations, the symptomatic fever was particularly severe. Two of the children thus vaccinated were inoculated with smallpox and both found to be protected; five children thus vaccinated were exposed to smallpox infection without result, and several European children were successfully vaccinated. From the vesicles so produced, lymph was taken and sent all over India for vaccination. One doctor questioned why Macpherson had used scabs in the first place instead of lymph, but rinderpest failed to produce a well-defined vesicle or pustule. When the children were inoculated initially with the crusts, the effects were more severe than those caused by the usual vaccine virus, and this was looked upon as an advantage. Its effectiveness, however, was not recorded for, in Calcutta, the lymph soon became mixed up with that hitherto in use, and there was no follow-up of cases. When rinderpest broke out again severely in Moorshedabad in April 1833, with great mortality among cattle, Macpherson tried the experiments again, but failed to obtain pustules. One child inoculated from a pustule on the udder of a cow showed slight inflammation of the arm on the fourth and fifth days, but there was no other result.

In the same year, the civil surgeon of Sylhet, Dacca Division, Dr. Furnell, attempted to repeat Macpherson's experiments. Four children were inoculated with matter from a cow recovering from rinderpest, without effect. In Furnell's absence, a Mr. Brown renewed the trial. Inoculating several children, the first four directly from a sick cow, and many others from these. A vaccine vesicle appeared on the eighth day in all cases. One European child suffered severe fever on the eighth day, succeeded by an eruption resembling smallpox which ran through the same course as natural smallpox. One of Mr. Brown's own children reacted similarly, but its mouth became so sore that no nourishment could be taken and it died. Another European child also suffered severely. None of the Indian children showed any apparent symptoms. The experiments were not continued further because of the unfavorable results (Hallen *et al.*, 1871).

Plowright (1968) notes that rinderpest-neutralizing antibody was present in the serum of nearly all human adults examined and appeared in children convalescing from measles, this undoubtedly being due to a similarity in the viruses sharing a common ancestor, but of course it could protect people against possible infection.

4. GEOGRAPHICAL DISTRIBUTION

By 1877, sufficient information was known that Simonds was able to state: "Cattle plague belongs to Russia and it not infrequently makes its way from Russia into Eastern Prussia... into all those countries which border immediately upon Russia... they are quite free from cattle plague themselves, except when it is introduced by cattle coming from the valleys of the Dnieper, the Dneister, the Bug, and the Lower Danube." (Simonds, 1877).

When Fleming wrote in 1875 (Fleming, 1875), rinderpest was known, apart from Russia and India, to be present in Mongolia, China, Cochin-China, Burma, Hindustan, Persia, Tibet, and Ceylon (allegedly introduced into the latter country in 1830). Pease (1894) noted that all the countries bordering China were subject to never-ceasing ravages, and its decreased virulence in those regions suggested that it had been present there for a very long time. Unknown in America, Australia, New Zealand, and Africa, except for Egypt, in central and western Europe it only appeared as an imported contagious disease. Eliminated in Russia in 1896, it remained only in a few of Russia's asiatic regions but, by the turn of the century, was enzootic in China, India, Egypt, Africa south of the Sahara, and the Philippine Islands. Japan was infected several times with cattle imported from the mainland but eventually prevented it by stopping cattle imports from Korea and Manchuria.

In 1749, a correspondent in the *Gentleman's Magazine* noted that rum punch, administered in quantities of 1 to $3\frac{1}{2}$ pints, according to the size of the animal, was a successful cure for murrain or cattle plague in Barbados, but this presumably referred to some other disease (Anon., 1749a). In the same journal, it was reported in 1751 that the distemper in Charleston, South Carolina which had broken out in 1742, was declining. Great losses had been experienced, and the animals generally died within 6 hr, so this was not rinderpest (Anon., 1751a). Probably the tick-borne protozoal disease redwater or Piroplasmosis, it was reported appearing in New England in 1738.

In June 1866, "a new and terrible type of disease" was reported from Sydney, Australia, probably imported from India, and a letter from Panama in *The Standard* of July 24th stated that hundreds of cattle were dying daily from rinderpest along the route of the Aspinall Railway, which probably prompted the Earl of Cathcart to state in an address to the RASE in March 1866, "we have sent the cattle disease to America and Australia." In the summer of 1868, a virulent plague did break out in the United States, at first thought to be rinderpest. Recognized as indigenous to Texas, it spread rapidly to New York and even to Canada. Showing many similarities with rinderpest, it was in fact Texas redwater fever. Whole herds were swept away by it, and the State of New York was thrown into "a sort of panic." Infected cattle were slaughtered in Illinois, and States prohibited the import or passage of cattle from other States. Canada banned all imports from America. The import into New York was entirely cut off, reducing the supply of meat by a half. Commentators drew attention to the lack of opposition by New Yorkers to this measure, unlike that which had ensued in London in 1865 at the threat of reduced supplies.

Bourguignon's (1865) reference to South America being affected in 1835 may refer to another disease, but in 1920 a shipload of zebu cattle from India *en route* infected cattle in Belgium and then, although the origin was never traced, apparently introduced rinderpest across the Atlantic to Martinique and Brazil, where it extended in the latter country to within an 80-km radius of São Paulo. Occurring first among work oxen, it then spread to the dairy herds, with wild deer species also dying. It was brought under control within 3 months by the imposition of quarantines, prohibition of stock movement, and slaughter of all sick animals and those in contact with them. Of 30,000 cattle, 600 died, and 1,000 were slaughtered (Roberts, 1921). Owners were offered compensation for slaughtered animals (it is unlikely they were ever paid), roughly one quarter to one fifth of the value of the animal. If they refused to have their healthy animals slaughtered, they were paid half rates for each one which died. The last official report of a case of the disease was on May 23, 1921, and restrictions were lifted in August of that year.

The cattle from India had originally been refused transit at Marseilles and were taken on to Antwerp, where they were put into quarantine sheds. This proving to be an unsuspected route of infection, France called an international conference to discuss rinderpest and other contagious diseases, as a result of which the *Office International des Epizooties* (OIE) was created with its headquarters in Paris.[1]

In November 1923, an outbreak occurred in western Australia on the outskirts of Fremantle, eradicated before the end of December by the slaughter of all animals around the centres of infection and the quarantining of an area within a radius of 50 km around Fremantle. Fifteen hundred cattle, 1,000 pigs, 24 sheep, and 317 goats died or were slaughtered. Its origin was never convincingly identified, but during October, three ships docked at Fremantle with live ship's stores. One, the *SS Charon*, arrived from Singapore via Derby and Carnarvon in Western Australia. At Derby, she picked up 297 cattle, 13 of which died before the ship reached Fremantle; and at Carnarvon, she collected 936 sheep. Before leaving Singapore, she embarked some live pigs as stores, and three of them were still alive on board when the ship entered Derby. The Chief Veterinary Officer considered that the cattle and sheep on board the *SS Charon* were the most likely source of the outbreak, having been infected by the Singapore pigs. The first cases occurred in a dairy herd in the Beaconsfield area, and within a short time, a number of herds became affected, with a second center of infection at Belmont. Eradication was aided by the somewhat sparse cattle population in the area and the existence of a cattle-free desert around Fremantle.

[1]Blystone (2001) quotes the "BBC News" as stating that the OIE was set up in 1924 after the distribution of rinderpest-infected cattle from Germany used as payment for reparations after the World War I. Such a source and action seem most improbable in view of Germany's strict legislation and control methods.

The shipment of captive wild animals has accounted for only two known instances in zoos, the first being in 1865, when two gazelles were exported from a London dealer to the Paris *Jardin d'Acclimatation*. In all, 34 animals died or were slaughtered, including yaks, springbok, Cuvier's and other gazelles, an auroch bull, a zebu bull, muntjac and Brazilian deer, and four peccaries. Only sheep and camels were not affected (Saint-Hilaire, 1865). The second was in 1950 in the Rome Zoo, consequent upon the importation of bushbuck *Tragulus scriptus*, lesser kudu *Tragulus imberbis*, and gerenuk *Litocranius walleri*, from Somalia, which was controlled by slaughtering all of the ruminants and pigs (Cilli *et al.*, 1951). The Italian authorities exercised more care the next time when, in 1954, two African buffalo exported to Italy that had been in quarantine in Kenya exhibited the disease. One died on the voyage, and the other was diagnosed as having rinderpest when the ship docked at Trieste. It was considered that cattle carried on the same ship from Mogadishu to Suez were the source of the infection (Boldrini, 1954).

In the autumn of 1931, it overran several Provinces in Persia, but it was to be 1969 before it gained access to the Near East again, killing large numbers of cattle and buffaloes in Iran and spreading westwards as far as Jordan, with outbreaks in eastern Turkey the following year. Vaccination programmes brought it under control, and by 1970, the disease was restricted to parts of tropical Africa, parts of Asia, particularly India and Nepal; and Southeast Asia, where it was believed to be enzootic in wild ruminants in the forests and mountains. In the years prior to 1983, the situation deteriorated, not only in Africa but also again in the Near East, resurging in 1979, when the disease spread to Lebanon, Syria, and the Gulf countries as a result of increased movement of slaughter stock from the enzootic zones of Africa. In February 1982, it broke out again in Iran near the Turkish border, and later that year in Syria. An intensive vaccination campaign in Turkey prevented its further spread. It was also present in India, Nepal, and Kampuchea. In 1987, Sri Lanka became infected after 40 years of freedom, the disease becoming enzootic in the north and east but considered controlled by 1998.

At the end of 1998, it was believed that rinderpest outside Africa remained present only in Pakistan, Afghanistan, and the Arabian Peninsula, with an an area of uncertainty in the "Kurdish triangle" of Turkey/Iraq/Iran. Yemen experienced sporadic outbreaks up to 1996, and a persisting enzootic infection was suspected, while the status of Saudi Arabia was uncertain. But in August, an outbreak was reported in the Russian Federation in Shimanovsk District of Amur Region, close to the border with China, which raised a suspicion there could still exist a hidden focus somewhere in Asiatic Russia, Mongolia, or China, although Mongolia was thought to be free of infection since 1993 and China since 1955 (Anon., 1998). In July 1998, an outbreak of PPR was reported in northern Iraq among smuggled animals designed for eventual sale in Turkey. By the end of 2002, rinderpest was believed to remain in only one site in the world, among the Somali cattle herds in southern Somalia and northeast Kenya. Table 1 lists the natural host range for several species and countries.

Table 1. Natural host range (excluding Africa).

Species	Common Name	Country
Antilopidae		
Antilope	Blackbuck	India
Bovidae		
Bibos	Gaur	India
Bison	European bison	Europe
Bos	Domestic ox	Europe
	Kouprey	Cambodia
	Yak	Central Asia
Boselaphus	Nilghai	India
Bubalus	Domestic buffalo	India
Tetracerus	Four-horned antelope	India
Camelidae		
Camelus	Camel	Egypt, India
Capridae		
Ammotragus	Arui	North Africa
Capra	Goat	Palaearctic, Asia, India
Naemorhedus	Goral	Asia
Ovis	Sheep	Palaearctic, Asia, India
Cervidae		
Axis	Hog deer	Asia
	Spotted deer	India
Cervus	Red deer	Palaearctic, Asia, India, North Africa
	Sambar	Asia, India
Muntiacidae		
Muntiacus	Barking deer	Asia, India
Odocoileidae		
Capreolus	Roe deer	Palaearctic
Mazama	Brocket	South America
Suidae		
Sus	Domestic pig	Ubiquitous
Tayassuidae		
Tayassu	Peccary	Southern and South America

After Scott (1971).

3

THE SEAT OF INFECTION

1. ORIGINS OF RINDERPEST OUTBREAKS

Conventional wisdom favored the belief that diseases which cause death of the host are ill-adapted, evolving diseases, a view now challenged (Hudson *et al.*, 2002). A disease such as rinderpest, which causes over 90% mortality in nonimmune artiodactyls and yet has persisted perhaps from the Pleistocene, can hardly be considered unsuccessful. The hosts in which mortality occurs may be adventitious, the disease existing in balance with some other host which forms a reservoir, so that it is neither evolving toward a balance with its susceptible hosts nor illadapted to its enzootic host.

A recurrent theme in epidemiology is that a disease remains of localized occurrence until carried from its focus by mass movements of man, usually through wars. This theory has been applied to the spread of rinderpest just as it was to the great European epidemics of bubonic plague in 1613–66. In the latter, German soldiers were believed to have carried the plague into Italy in 1629, and later that year, French troops were believed to have carried it into Piedmont. But there had been excessive rains and humidity in northern Italy in 1629, followed by a plague of rats the next year, conditions naturally favoring spread of bubonic plague.

Theory holds a disease of viral or bacterial nature originates by a mutation creating an organism tolerated by the host with which it has evolved (or ancestral host). Circumscribing the focus of such a disease a zone of immune hosts extends to a zone of susceptible potential hosts; susceptible because they have had no contact with the disease. This balance remains until disturbed by transfer of infected hosts from the locus of infection to the susceptible population. There is some evidence for such theory in the sleeping sickness foci of the Congo, foci apparently of some antiquity. But we must be aware that this is a convenient explanation of epidemics, shifting the cause to man's activities and predisposing that foci of diseases do not erupt into epidemics due to the natural reactions of potential hosts.

One can only speculate on the possible original home of rinderpest. Sergejew and Semmer, quoted by Friedberger and Fröhner (1886–7), considered that it may have been the black earth regions of southern Russia near the Black Sea and on the lower Don, where there is a deep humus and luxuriant vegetation. Others, they concede, consider an origin outside of European Russia in the Asiatic steppes as more probable. We know that there was an epizootic cattle disease in the region of Inner Mongolia in 88 BC (Di Cosmo, 1994). It seems likely to have evolved with domestication of cattle on the Great Hungarian Plain and then been taken eastwards through Asia by the wars and movements of nomads which characterized the steppes, but only when nomad herds had reached a certain critical density might we expect a virus strain to have become enzootic among them.

The Russian name for rinderpest is *tchouma* or *tchouma reina*, which indicates an Asiatic origin as the word *tchouma* was used by the Mongols and nomad Tartars of Central

Asia to denote a malevolent deity, something of the nature of a vampire. The Osmanli Turks, the Persians, and the Afghans called it *taoun*, a word believed to be derived from *tchouma*.

Considered enzootic in the Asiatic steppes and parts of India, early epizootics could have exercised a marked influence on the migrations of pastoralists away from centers of permanent infection, but thus spreading the disease. It is possible that the Indo-Aryans, in establishing themselves in Central Asia, were fleeing rinderpest and other diseases.

The vast steppes of Central Asia, before nomads peopled them with stock, were probably the home of countless herds of wild migratory bison *B. bonasus* seasonally roaming the pastures in vast herds like the bison of North America, together with the wild horse *Equus przewalskii* and the saiga antelope *Saiga tatarica mongolica*, while the aurochs or wild ox *Bos primigenius* inhabited the fringing wooded steppe. The steppe region is a distinctive ecological system with an almost continuous cover of open grassland in a broad belt from north of the Black Sea in the west to the Manchurian prairies in the east. A moisture-deficient zone of average rainfall 250–500 mm, it is characterized by a continental and semi-arid climate with Arctic air masses causing prolonged winters with temperatures below freezing for 6 months, most severe in Mongolia. Pacific air masses in the east cause warm, moist summers in Manchuria. In the Neolithic Age prior to 900 BC, the climate was warm and humid but gave way to one cooler and drier.

Somerville's Geography described the steppes in the 19th century thus:

> Hundreds of leagues may be traversed east from the Dneiper without variation of scene-a dead level of thin but luxuriant pasture ... Horses and cattle beyond numbers give some animation to the scene so long as the steppes are green ... the scorching summer's sun is as severe in these wild regions as the winter's cold ... In some seasons the drought is excessive; the air is filled with dust in impalpable powder; the springs become dry, and cattle perish in thousands ... (Somerville, 1848)

The true steppe is a treeless pasture, a grassy plain between mountains, admirably adapted to the breeding of cattle, sheep, and goats, with the sub-Alpine valleys of the Altai, or Golden Mountains, providing exceptionally fine grazing grounds. Although the grazing grounds of Mongolia and Kirghiz are more extensive, the heart of the steppe country has always been the pastures along the northern edge of the T'ien Shan and the southern edge of the Altai. From June to August, it is a green carpet of grass and flowers promoted by heavy rains. In September, the cold is already severe, in October, snowstorms sweep across the land, by November, the watercourses are frozen, and snowfalls are frequent until the following May, accompanied by winds of great force. As far as the eye can see, the steppe is but a flat wilderness, broken by occasional ravines and stony hills devoid of trees.

Cattle were first domesticated using the aurochs *B. primigenius* soon after 6500 BC, bones from this era appearing in Thessaly and Greek Macedonia pre-6200 BC, and domestication reached the Carpathian Basin before the end of the 6th millennium BC. The breeding of domestic animals, as opposed to simply capturing and restraining wild animals, probably began in the 4th millennium BC at the earliest. The steppe cattle of Central Asia and eastern Europe, the Hungarian grey cattle, are of the *orthoceros* type, probably a variant of *B. primigenius*. Domestic cattle did not reach the forest belt of eastern Europe until 1500–500 BC (Bökönyi, 1974). From 1300 BC onward, bovine motifs on artefacts in Inner Asia are rare, the popular motifs being horse, deer, goat, and ram, decorating bronze handles of knives and other artefacts; and after the 4th century BC camels, when they first became known to the Chinese, accompanying the caravans of the northern traders. A rare bovine motif harness ornament dates from the 6th century BC, and a belt ornament from the 3rd to 2nd century BC (So and Bunker, 1995), but the rarity of bovine motifs, and yak appear to be nonexistent, is surprising considering that stock plays such an intimate part in the lives of nomads. It could suggest that cattle were unfamiliar to the inhabitants of northern China in the Bronze Age, and therefore they were not interested in bovine-motif trade goods, choosing instead those of animals with which they were familiar. According to Schwabe (1978), however, domesticated cattle were already present in China by the 4th millennium BC. One of the oldest Chinese deities was a buffalo with tiger's teeth, another was the King of Oxen. They worshipped a god of cattle-breeding and Shen Tung, a later bovine god. This bovine culture was replaced by a horse culture in the 1st millenium BC. The Chou dynasty (1050–247 BC) recognized animal healers, and one, Cho Li T'en Kuan, advocated recording livestock deaths, the Chou government apparently compensating farmers for losses.

Until the beginning of the 9th century BC the steppe zone appears to have been a peaceful region of settled agricultural communities with a warm, dry climate, favoring cattle, there being evidence of a sedentary cattle-keeping culture from the first half of the 2nd millennium BC. But from the end of the Neolithic, the cooler and drier climate saw cattle decline in relation to horses and sheep. The Mongolian steppe region north of China underwent a process of desiccation, with increased lower temperatures, less rain, and a shorter growing season. This changed the way of life of the Neolithic sedentary farmers in the region to herding and hunting, with full-scale pastoralism developing. Not later than the 8th century BC in south Siberia, and to the east and west of the steppe zone, it stimulated the rearing of large herds of horses and, in places, herds of cattle and flocks of sheep. North of China, tribes were predominantly employed in sheep rearing, and the change to the use of horses and the multiplication of herds and flocks led to a search for new grazing and water with rapid exploitation of the uninhabited steppe. The mixed economy with seasonal agricultural settlements came to an end as a result of nomadic invasions, and there followed a sharper division between cattle breeders and cultivators.

2. THE HISTORY OF INNER ASIA

At the end of the 9th century BC the Chinese emperor attacked the nomadic Hsiung-nu, who were raiding his western borders, setting in train a wide-scale emigration, for, forced back, they displaced the Massagetae, who occupied grazing grounds to the north of the Oxus river along the middle Jaxartes and south of the Aral Sea. The Massagetae then turned on the Scythians, initiating an exodus to southern Russia and Iran, but some remained behind in western Siberia and the Altai. At the end of the 8th century BC, the Scythians began to drive out the Cimmerians living to the north of the Caucuses and the Sea of Azov, into western Asia. It is suggested that in about 800 BC, the northern regions suffered severe drought for 150 years (Prüsek, 1966), leading to the arrival of Scythian tribes in the Altai. But possibly the center of radiation was Ordos and adjacent areas, where, as the climatic situation deteriorated and soils became impoverished, the people changed to a nomadic way of life. There was mixed agriculture and nomadism in Xinjiang, northwest China, in 780 BC. In Inner Mongolia and Siberia from 722 to 481 BC, agricultural sites reveal remains of pigs and a large increase in bones of cattle, horses, and sheep. In 800–700 BC, mounted tribes arrived in the northwest and northeast of China and probably introduced the horse into China in a vastly expanded trade.

Between the 8th and 2nd century BC, the fragile Mongoloid type with an economy based on sheep-rearing disappeared, replaced by horse and cattle breeders of an old Europoid type. In western parts of Siberia, the transition from settled agriculture to nomadism was first effected by old Europoid stock. The west was marked by the arrival into southern Russia of the Cimmerians, closely followed by the Scythians, and their penetration into various regions of Inner Asia creating a unified "Scythian" culture from the Black Sea steppes to Ordos and Mongolia, and at about the same time in the east in China by invasions of tribes such as the

Hsien-yün, Pei Jung, and Ti. But the change from settled farming communities to mounted warrior nomads was a slow process, animal husbandry becoming paramount in certain areas of the steppe region in the second half of the 1st millennium BC.

From the end of the 7th to about the 3rd century BC, Scythian nomads occupied the steppes of the area north of the Black Sea from the Don in the east to the Danube in the west. Sinor (1990) postulates that they were a nomadic or semi-nomadic people originating in the Volga-Ural steppes. One section was called *Georgi*, meaning "esteeming livestock." Virgil tells us in *The Georgics* (31 BC) that they were nomads who drank milk curdled with blood obtained from the opened vein in a cow, as do the Maasai nomads of East Africa today. They were skilled horsemen with a lifestyle seemingly similar to the Mongols nearly 3,000 years later, quickly overcoming all resistance and creating a great empire. All the non-Scythian tribes except two lived in the forest–steppe zone of eastern Europe which, from 650 to 475 BC, was completely dominated by the Scythians. Their eastern neighbors were the Sarmatians, who developed in the 6th to 4th century BC and, in the 2nd to 4th century AD, led a nomadic life north of the Black Sea and along the Volga and north of the Caucuses.

In 530 BC, Cyrus, the Achaemenid Persian ruler, died fighting the Massagetae, who lived somewhere east of the Caspian Sea. In 517 BC, another Achaemenid ruler invaded Central Asia. Both campaigns precipitated the displacement of certain tribal groups across the steppes as far as the northwest border of China.

The transition to nomadism in Inner Mongolia developed from about 480–222 BC. South of Lake Baikal, agricultural settlements of the Hsiung-nu, and others in northern Mongolia and Siberia, suggest that although animal husbandry was important, agriculture, hunting, and fishing were essential auxiliaries.

In the 4th century BC, powerful warring tribes displaced by the campaign of Alexander the Great in 329–327 BC arrived in the region northwest of China, their emphasis on riding suggesting a shift from semi-sedentary pastoralism to full-scale transhumance.

The Hsiung-nu appear suddenly before the 3rd century BC from the northwest and formed the first great nomadic empire in the north from the borders of North Korea in the east to Xinjiang in the west and Lake Baikal in the north; raising horses, cattle, sheep, and camels. The Ting-ling inhabited an area from Lake Baikal to slightly beyond the Yenisei River. East of the Hsiung-nu, the Tung-hu or eastern barbarians comprised mainly Wu-huan and Hsien-pi, their territories toward the end of the 3rd century BC extending from the southern part of Inner Mongolia to southern Manchuria. In about 200 BC, the Hsiung-nu defeated the Tung-hu, and the Wu-huan and Tsien-pi then moved from Inner Mongolia to the upper Liao river in Manchuria but, subject to the Hsiung-nu, sent annual tribute of oxen, horses, sheep, and skins of sable. The Hsiung-nu then conquered westwards subjecting the sedentary peoples around the Tarim basin. The favourite Hsiung-nu pastures were in the Ch'i-lien and Yen-chih mountains, and the Yüeh-chih, who lived between the Ch'i-lien mountains and Tun huang, were forced out by the Hsiung-nu in the 2nd century BC westwards to the Oxus valley. In 128 BC, they were still a race of nomadic herdsmen. Over a period of 30 years, they migrated to Ta Hsia through the domains of other nomadic tribes, forcing some, loosely known as "Scythians," to move into areas of the Parthians, and were finally contained in Seistan about 100 BC.

In 104 BC, the army of Han Wu Ti marched out of northern China to conquer the area to the west, accompanied by 100,000 oxen, 30,000 horses, asses, and mules, and camels "by the ten-thousands."

In 88 BC, many of the Hsiung-nu cattle died, and the people fell ill and could not harvest (Di Cosmo, 1994), suggesting a lack of oxen for ploughing and the first possible indication of rinderpest. In the north of China, agriculture coexisted with the Hsiung-nu nomads, as in the Scythian period in Central Asia.

In 72 BC, the Han and Wu-sun captured an alleged 750,000 horses, oxen, sheep, donkeys, and camels from the Hsiung-nu, who, the following year, allegedly lost half of their livestock in further raids. In AD 50, the Han court sent a gift to the Hsiung-nu of, among other things, 36,000 cattle and sheep. In AD 91, the northern Hsiung-nu empire in Outer Mongolia and the western regions collapsed. The Hsien-pi then became the rising power, migrating into vacated territories in Outer Mongolia and taking over all the former Hsiung-nu lands (Figure 1).

These movements of people and stock for the first 2,000 years beginning in the Neolithic were restricted to east of the 45° meridian, the furthest westward displacement being that of the Yuëh-chih who settled south of the Caspian Sea about 100 BC, in turn displacing Scythians to the Pontic steppe. The exception was the forces of Alexander the Great in 327 BC, which returned to Egypt after having traveled west as far as about the 70° meridian and could well have passsed through enzootic rinderpest areas carrying the disease back to Egypt. It was not until the 4th century AD that there were large-scale movements of people and animals west of the Caspian Sea region with the irruption of the Huns who were, as far as is known, from that region, originating beyond the Sea of Azov to the east on the south Russian steppe, or further east. Aggressive mounted pastoralists owning herds of cattle and horses, they first attacked the Alan nomads along the Don in the early 370s then, together with the Alans, attacked the eastern Goths (Ostrogoths) inhabiting the Don to the Dniester in the south Russian steppe and the Black Sea to the Pripet Marshes.

The Huns had no defined leader until about 395, when they raided across the Caucuses into Armenia, Syria, Palestine, and northern Mesopotamia. The cause of this raid may have been famine, as it was only a plundering raid, and they apparently took back herds of cattle and sheep but, having unsucessfully attacked the Persians during their return, were forced to abandon most of their plunder. To feed the 10,000-strong Hun army at the sack of Rome in 409, sheep and oxen were brought from Dalmatia, the origin of rinderpest-infected cattle 1,300 years later. The Huns probably continued to occupy the Pontic steppe to the turn of the 4th century when at the end of 436, they entered into a war against the Visigoths in the south of France until they were, overcome at Toulouse. Attila first appears as the overall ruler of the Huns in about 448. Based in Pannonia, his army reached the Rhine near Mainz and crossed to Gaul in 451, but the Huns ceased to play an effective role after 455. The movement of the Huns toward Europe introduced new peoples into the Ponto-Caspian steppes, but Türkic speakers from Inner Mongolia were probably already present before the Huns crossed the Volga. By the middle of the 5th century, the Visigoths were in Gaul and Spain, the Vandals had reached North Africa, and the Ostrogoths were in Italy. These mounted barbarians undoubtedly looted cattle and moved them from place to place. It is generally considered that the Huns were responsible for introducing rinderpest into Europe in 376, spreading it from an enzootic centre in the Caspian Sea area, but Europe had probably witnessed outbreaks for centuries before this (Table 1 in Chapter 5), and the attacks of the Huns were restricted to not far west of the 30° meridian at this time but may have caused refugees to flee still further west.

Sometime after 450, drought, famine, and starvation among the population of the Wei capital in northern China were accompanied by an epizootic devastating the herds of cattle, making it necessary to employ camels in unfamiliar tasks such as drawing carts and even ploughing (*Wei shu* quoted by Schafer, 1950). It may have been drought which, in about AD 463, sparked off a series of nomadic migrations in Inner Asia, the Uighurs being driven to the Pontic steppe, western Siberia, and Kazakh.

In 546, the Türks of the Altai, the first nomads to build a formidable empire in Asia, overcame the Juan-juan and the Hephthalites, people inhabiting pastures north of the Gobi. Some Juan-juan fled west beyond the Urals and settled in the plains of Hungary north of the Black Sea, where they became known as Avars. In the next 50 years, the Türks mastered the

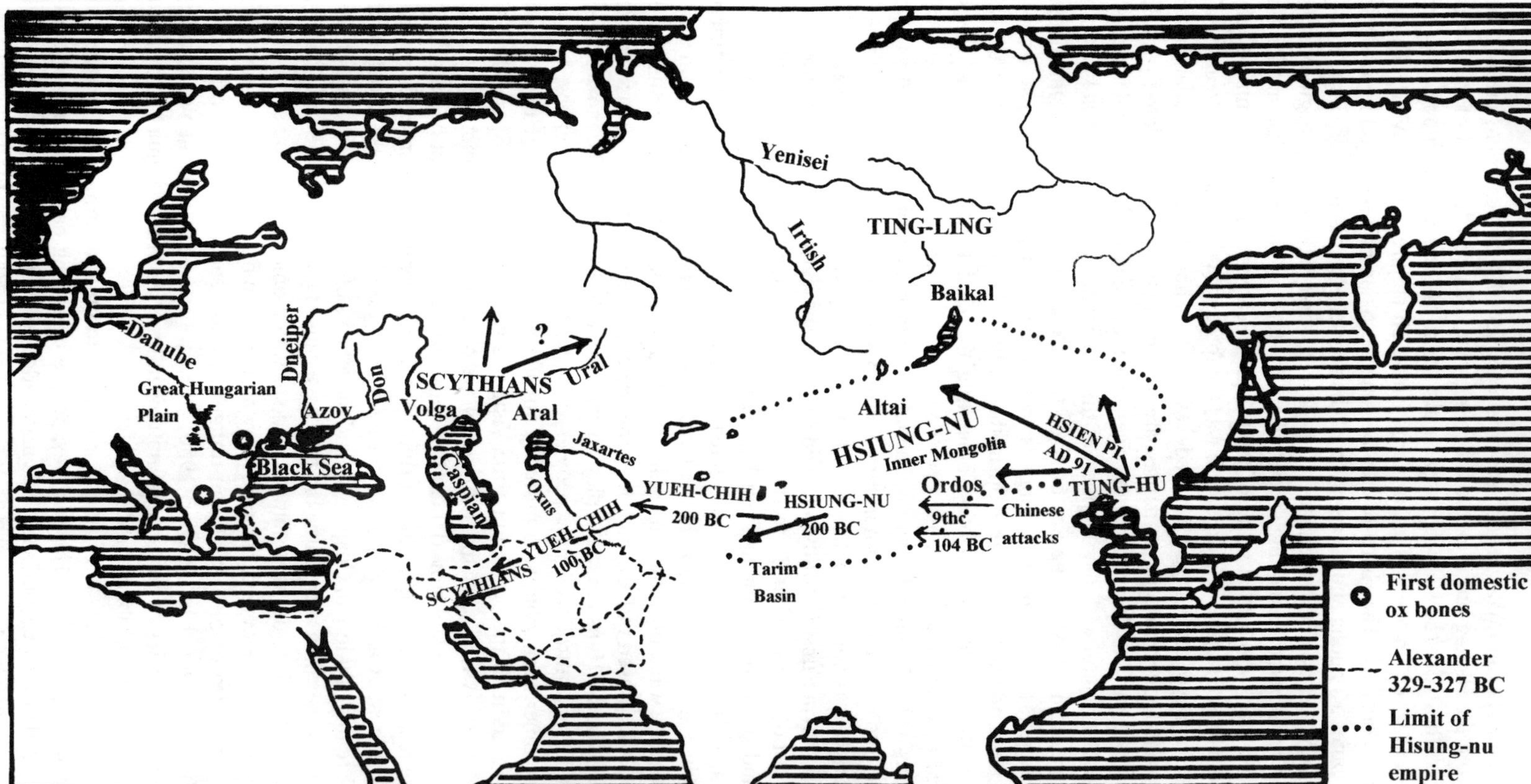

Figure 1. Inner Asian steppe migrations *c*.1000 BC to AD 91.

Eurasian steppe, reaching as far west as the Crimea, but did not maintain their hold over it. There followed a period of 450 years, during which time no one tribe or confederacy attained supremacy over the steppes.

The Avars had conquered the middle Danube regions by 568. Keepers of cattle, horses, and caprines, pigs being rare, they were noted in war for traveling with a multitude of animals which they used for food and also took into battle to look impressive. A migration from Great Bulgaria settled in the Avar home country, of which Hungary was the centre, in 670–80. In 791–9, the Franks under Charlemagne fought the Avars, annihilating them, Charlemagne using grey steppe oxen as baggage animals. Until the middle of the 9th century the Hungarians of the Pontic steppe, who had transformed from forest peoples, were seminomadic cattle breeders but retained their traditional animals when they became sedentary. Cattle were the most frequent domestic animals on the Great Hungarian Plain from the Neolithic to the end of the Middle Ages, the Hungarian white or grey cattle being developed in the late Middle Ages. In the mid-9th century, the Khazars, who led a nomadic life from spring to autumn, ruled the Volga delta and north Caucasian steppes east almost to the Ural River and west to the Don-Donets region. During the 10th to 11th centuries, tribes such as the Pechenegs, Oghuz, and Cumans surged back and forth. In 1017–18, there was a large-scale nomadic movement in the Karakhanid realm, and the Quin migration brought the Kipchak-Cumans to the south Russian steppes and a mass migration of the Oghuz into the Near and Middle East (Figure 2).

The second great nomad empire of Mongolia was that of the Uighurs from 744 to 840 whose headquarters were on the High Orkon River. In 839, a severe winter killed off much of the Uighur stock leading to a collapse of power, but only sheep and horses are mentioned when they moved their northwest tribes south under threat from the Karluks. Before 744, they were typical nomads with an economy dominated by the sheep and horse, cattle and camels playing lesser roles. To them, sheep were the most valuable because of their wool, while horses and oxen provided sources of meat and milk as well as draught. During battles in 757, the Uighurs' daily army ration included 20 cattle and 200 sheep.

The Kitans originated in about the 4th century in Manchuria near the upper course of the Liao River and its tributary the Laoha Muren, surviving as a nation until the 14th century. The river plains of this region with their abundant grass teemed with game, chiefly deer, but also wild boar, tiger, and bears. The Kitans were occupied in cattle-raising, hunting, and some agriculture. The Jurchen lived in land adjoining the Gulf of Liaotung between Korea and China, a forest people regarded by the Kitan as their vassals. Oxen played a great role in Jurchen folklore, and there is a reference to a marriage ceremony being solemnized by the gift of a dark ox. Although the ox was the principal domestic animal, the Jurchen were not nomads, also raising horses, sheep, and pigs, but not camels. Their products were horses, cattle, furs, and falcons.

In about 860, the Kirghiz, a primitive race of Siberian Türks from the Yenesei valley, migrated south, the majority settling north of the Tarim River in an area which became a centre of Uighur power until the time of Chingis Khan.

In 1126–8, the Jurchen conquered the greater part of northern China, but sometime after 1211, the Mongols invaded Manchuria and northern China. Raiding south of the Great Wall, they began a struggle with China which lasted for 23 years. Chingis Khan assembled an army in 1219 allegedly of nearly 200,000 in the upper Irtish valley and conquered Bukhara the following year, followed by Samarkand, using grey steppe oxen for baggage transport. The Mongols advanced to Azerbaijan, whose lush pastures attracted every nomad invader, and thrust into Georgia moving on to the steppes north of the Caucuses. Advancing up the Volga, they rounded the northern shores of the Caspian Sea in 1222. Chingis then overcame Gurganj and crossed the Oxus into Khurasan. Another Mongol patrol penetrated as far as Multan, but no pursuit on Indian territory was attempted.

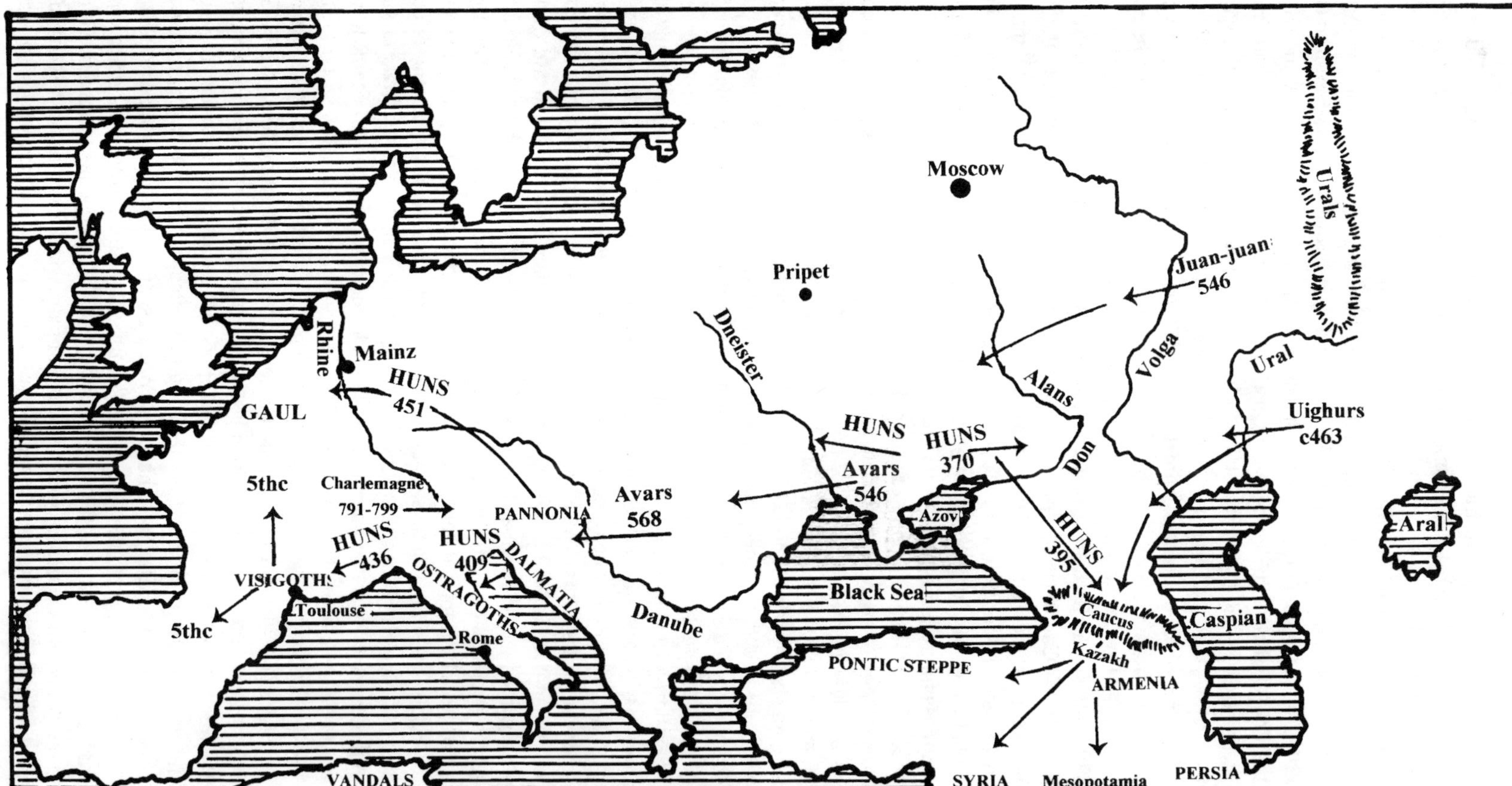

Figure 2. Huns AD 370 to Charlemagne AD 799.

In 1225, Jalal al-Din, the Khwarizm-Shah of Persia, attacked Azerbaijan, and the following year, Georgia. Fighting through the Armenian highlands, he reached the upper western Euphrates before being routed by the Mongols in 1230. Chingis then bestowed upon his younger son Batu the area beyond the Volga "as far as the soil has been trodden by hooves of Mongol horses." But the overlordship was only nominal, and Batu hence had to subdue it as a matter of honor. In the spring of 1236, an alleged army of 150,000 men marched out from Siberia on a journey which was to take them to the Adriatic Sea. First, they conquered the Bulghar cattle-owning nomads of the middle Volga, then the Kipchaks of the lower Volga north of the Caspian Sea, an alleged 40,000 tents fleeing to Hungary. In December 1240, their attack on Kiev in the middle Dnieper was described as "clouds of Tartars" approaching, and the noise of the carts, the bellowing of the camels and cattle, the neighing of the horses, and the savage war cries of the attackers made such a noise that the people within the walls of the city could not hear one another speak. But having reached Hungary, perhaps with the object of obtaining pasture for their horses on the rich Hungarian prairie as a base for attacking further west, the offensive against Europe was broken off in 1242 with the death of Ogedei, one of Chingis's sons who had succeeded him (Figure 3).

Further offensives were retaliatory. In 1258–9, Poland was invaded and the land ravaged as far west as Oppeln before the forces withdrew. Then in 1381, the Mongols attacked Russia, but the Mongol leader Toktamish antagonized Timur, who then attacked him, reaching Yeletz on the upper Don. From thence, Timur ravaged the lands of the Circassians and Alans, the people of Sarai fleeing before his army. His ravages left a desolate hinterland, and the trade which once flowed through Constantinople across the Black Sea and the Russian steppes toward Transoxiana and the Far East was reduced to a trickle. A Mongol army approached Moscow for the last time in 1480 but withdrew without battle. In 1502, the Khan of the Crimea captured Sarai, and the Mongols finally disappeared from the scene.

This diversion into history has been in some detail because Inner Asia appears to have been the enzootic center of rinderpest among the grey steppe oxen, an area subject to continual vast movements of cattle whether accompanying armies for transport or food, or being looted by them. Bradley (1721) suggested that infectious distempers were caused by pestiferous insects produced in Tartary and wafted westwards by the wind, thus being the first to consider Tartary the center of orgin. Fleming (1871) considered from the start of the 13th century until the year 1241 the Mongol invasions from Asia through Russia to Poland could have introduced many epizootics into the western hemisphere. This is more likely to have begun with the irruption of the Huns carrying it from the area of the Caspian Sea, for, in the early years of conquest, the Mongol raiders were probably few in number, mounted, and highly mobile. They did not trek with cumbersome armies, neither did they depend upon herds of cattle for food. Their food was mostly yogurt, or fermented mare's milk, and millet. These swiftly moving horsemen would hardly have waited for trains of milk cows to follow them, and they more likely depended for milk upon cattle they seized in their advance. But their incursions could have caused widespread emigrations of refugees and their herds. By the year 1240, the description of the Mongol attack on Kiev suggests that their behavior had changed radically, now moving in great, cumbersome hordes, with cattle in their train. As aggressive expansionism or migration is usually associated with a need to avoid disease or find new pasture, it is tempting to think that the irruption of the Mongols stemmed from a need to find disease-free grazing for their immense herds. But from what little we know of events, whereas such a reason may have been the trigger for the original irruption, by the time of Chingis Khan, the aim was simply that of conquest, for the Mongol leadership considered itself charged by God with the task of world conquest.

Pease (1894) considered that the literature pointed to an origin for rinderpest in the extreme east of China. Domestic cattle first appeared in northern China in the middle of

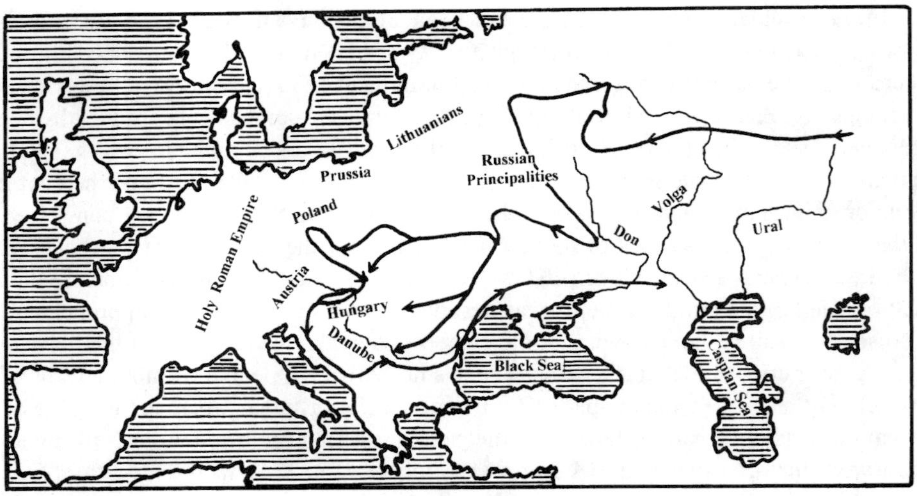

Figure 3. The Mongol invasion 1237–42.

the 3rd millennium BC, but the vast herds of cattle in Manchuria which throve on the fertile plains are believed to be of recent date, introduced by Chinese immigrants or prisoners of war (Epstein, 1969), although, as we have seen, the Kitans appear to have been raising cattle in the region in the 4th century. These were known as "yellow-cattle" regardless of color (white cattle are rare), their origin being brown, brindled, black, or yellow Mongolian cattle. The Manchus who conquered China in the 17th century AD, although excellent horsemen were ignorant of the art of milking and were essentially not pastoral nomads. They are believed to have been descendants of the Jurchen who had inhabited northeast China from before 200 BC and who, by AD 1115, controlled it until annihilated by the Mongols in 1234, the survivors remaining in the northeast until they rose to power again in the 17th century. Milking of cows was not introduced into the northeast of Manchuria (Heilungkiang) until 1945.

3. ATTACKS ON THE IMPORTATION THEORY

Some (e.g., Scott, 1970) hold that wars were the primary disseminator of rinderpest, particularly by the accompanying trek oxen until, in the 19th century, trade cattle became the important spreaders of the disease. Gamgee (1866) asserted: "There are periods of recrudescence which are to a great extent accounted for by periodic and extraordinary movements of stock, as in times of war, or by an activated cattle trade; but neither cold nor heat, rain nor drought, storms nor severe frosts, affect the propagation of the malady, except in so far as they affect the movements of people and the traffic in cattle."

In October 1865, *The Times* carried a leading article "holding up the irrepressible Mr. Gamgee [as it called him] to the ridicule of all its readers in all quarters of the globe" (Constable, 1866a):

> The theory of importation from abroad is one which in former times was applied to all plagues whatever, but it has now in all other cases been deliberately exploded. What is said of Rinderpest was said in as many words of Egyptian plague. The plague used to visit us periodically, and was invariably ascribed to a shipload of infected substances. Yet we never get the plague imported now. That the Rinderpest of a century ago should be traced to foreign importation was perfectly natural, for nobody then believed in any other origin of plagues. This view is so conformable to our prejudices and old notions; so many matters are settled, so many obligations dispensed with by

> looking at this plague as a spark thrown among us to be stamped out with the foot, that it ought to be looked at with the greatest suspicion.

The Times leader vigorously attacked the importation theory, supporting the idea of a baneful atmospheric influence:

> ... we must say that every month's experience of this eventful year increases our incredulity respecting the theory of importation. That the conditions of the atmosphere are truly exceptional every man's senses will tell him, and we see the effect in all directions. Cholera is raging over a considerable portion of Europe, and the climate of England itself has been actually found compatible with the existence of yellow fever. Pigs have sickened. In France poultry have suffered from a strange disease, and it was feared that eggs, the last resource in cases of suspected food, would be no longer safe. Now, in the face of this universal testimony to the influence of some destructive element, is it consistent with philosophical principle to assume that the disease of horned beasts is utterly unconnected with the phenomena affecting all other animal life, and that though men, horses, pigs, sheep, and fowls would have suffered from the season, cows would have been absolutely exempt if a certain deck-load of cattle had not been landed at Hull from Revel?

Seventy years before, there had been more enlightened thinking in Germany. Stressing that it was "only and alone through infection [rinderpest] emerges and propagates," Faust (1797) stated that it was not known where it originated, but it was foreign to Germany and came each time from Hungary and Poland. Many scholars "were of the opinion" that it originated in Asia or Africa like smallpox and bubonic plague.

4. THE RUSSIAN FOCUS

Gmelin (1751–2) first described the "Russian disease," but this name was later applied to anthrax. Renault, Director of the Imperial Veterinary School at Alfort, argued that rinderpest originated in the steppes of eastern Europe which extend across the breadth of Asia from Manchuria west to the Hungarian plain, the Alfold, and it was from there that the plagues spread to western Europe. Despite what other French authors had written, it arose spontaneously only among the innumerable troups of cattle in the steppes of central Russia (Renault, 1856, 1860) (Figure 4).

In Russia, it was remembered that the disease was rare before the year 1842 when free trade in corn began with England. Odessa formerly exported much less grain to foreign countries than it was doing 20 years later. Transported in ox-drawn carts, every year more than a million wagons of grain, each drawn by two oxen, arrived at Odessa alone. This, it was alleged, regularly spread the plague, as was to be the case in southern Africa at the end of the century, when ox-wagon transport was again a major agent of dissemination.

Unterberger (1864) did not believe that Russia and its steppes were the source, asserting that it was a purely contagious disease in European Russia; a view also supported by Gerlach (1867):

> Beyond the Russian frontiers, and even in every part of that empire, the Steppes excepted, the Cattle-plague is evidently a purely contagious malady. It is never developed primarily with us, neither in our indigenous cattle nor in those originally from the Steppes, and it has not yet been demonstrated that it may be primarily developed in the Russian Steppes; the most recent observations even tend to prove that in the European portions of these regions, this affection is only present through the transmission of a contagium. Consequently the Cattle-plague is a malady which is perhaps primarily developed in the Russo-Asiatic Steppes—perhaps elsewhere—but is never seen in Europe except by the importation of its contagious principle.

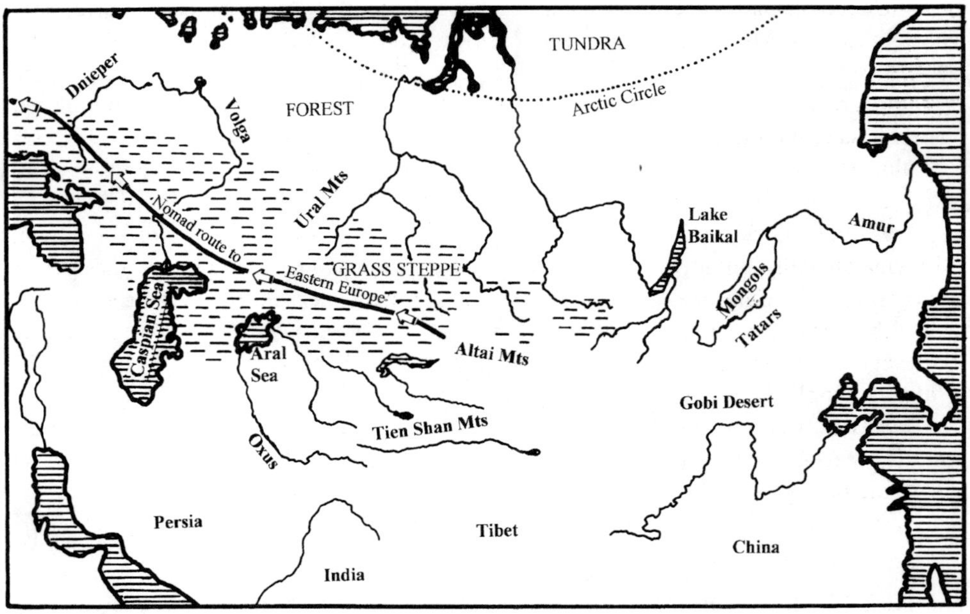

Figure 4. The Central Asian home of the grey steppe cattle.

As Budd (1865a) expressed it:

> Trade and war, those two inexorable tyrants, are the chief agencies by which it [rinderpest] is disseminated. In the course of ordinary traffic, vast droves of cattle annually leave the Steppe for the markets of the west. Too often, these droves carry with them the germs of the scourge ... from a single case of cattle plague, great outbreaks often spring, which, in a few months, spread over whole principalities, and count their victims by tens of thousands ...
>
> The Danubian Provinces and eastern parts of Austria are generally the first to suffer. To keep the great cattle plague at bay is at all times an object of national concern with the Austrian Government. Silesia, Podolia, Galicia, Moldavia, and Wallachia, are the seat of the periodical outbreaks of it, which sweep away myriads of oxen.
>
> In times of peace—by strict military cordon, and other measures directed solely against contagion—Prussia succeeds, to a great extent, in keeping the pest from her borders, and stands as a barrier between it and the North of Europe ... in time of war, when they can no longer be enforced, cattle plague has more than once invaded Prussia also, and has spread far and wide through the the north and west. In modern times, the results were first seen, on a large scale, in the wars of the first French empire.

Gamgee preferred to call rinderpest the Russian Cattle Plague because of its believed focus of origin in the Russian steppes. Although a Russian Commission in its report of 1865 concluded that its place of origin could not be ascertained, Gamgee thought differently: "The disease is always there, roving to and fro, winter and summer, amongst cattle of all kinds, sometimes killing many, and at other times attacking few ..." (Gamgee, 1866).

Simonds (1865) stated that there was evidence to the effect that the disease originated in the Asiatic part of the steppes and was spread to the European parts by infected animals. The marshy part of the steppes, the valleys of the Volga, and the Don were thought to be the centers. Asked whether it originally made its appearance there, Simonds replied he could not say whether it came from Asia, or whether it was a continuance of the plague that fell upon the Egyptians. He thought that no more was known of its origin than was known of the origin of smallpox or cholera.

Circumstantial evidence certainly pointed to the steppes of Russia as its center. The lower third part of the Dnieper, with its numerous affluents until it empties into the Black Sea, was surrounded by Russian provinces which bred about 8 million cattle constantly afflicted with rinderpest. Mainly Jewish merchants annually bought up large numbers and took them to fairs for sale, the most notable being Beltzy in Bessarabia, Elizabetgrad in Kerson, Balta in Podolia, and Berditchev in Volhynia. Balta had some 500,000 cattle at its fairs in a season. From these fairs, great herds were driven to Poland, Hungary, and Russia proper, but they rarely reached as far north as St Petersburg. The Russian provinces of Podolia, Ukraine, and Volhynia annually supplied Poland with about 30,000 head of cattle, the main avenues communicating the disease being through Poland and Hungary. Nearly 100,000 steppe cattle were believed to pass annually into Galicia and Hungary, and every 6 or 7 years, rinderpest appeared to ravage the latter country (Playfair, 1866), the only suggestion of a cyclical occurrence. Hungary supplied the London Metropolitan Market weekly with long-horned oxen.

Russia was probably the greatest cattle-rearing country in the world in the mid-19th century. In 1863, including Asiatic Russia, it was estimated to have 25.4 million head, nearly twice the number in the United States (17 million). This compared with France and Austria, each having 14 million, the United Kingdom 8.3 million, and the German States 8 million. All other European countries had far fewer (Lewis, 1866). In 1865, Ernes estimated that there were between 8 and 10 million cattle in the Russian steppes: "When they have devastated their grazing on one part they move further on; it has been stated many years ago that 100,000 are generally found on the roads coming to Perikop, Bessarabia, and Galicia, at one time, and the disease often breaks out among them while on the road."

Simonds reported in September 1865 that the cattle trade was believed to have extended into Russia, Hungary, Poland, and Galicia within "the last two or three months." This was because cattle could not be brought via Prussia on account of the controls there. Only after this route was circumvented by collecting the cattle at Revel and shipping them via the Baltic Sea to Copenhagen, and then to Hull or London, did the trade in Russian cattle begin.

5. THE RUSSIAN HERDS

It is convenient to explain away the origin of a disease as coming from the most remote and almost unknown places, and Siberia was still almost unknown to Europe in the early 19th century. This vast region was inhabited by pastoral nomads rearing large numbers of cattle, sheep, goats, and horses since at least the 3rd century BC. According to Barfield (1989), sheep were by far the most important subsistence animal raised and the mainstay of Inner Asia pastoralism; reproducing rapidly and using the most varied diet of plants on the steppe, they accounted for 50–60% of all animals raised on the Mongolian plateau. But although of most importance economically, it was the horse which held pride of place. It is not until 1598 that we have written records of cattle-keeping ("cattle" here possibly meaning sheep and goats as well as oxen) in the western Siberian lowlands in the region of the Irtish River, for, in that year, Prince Ivan Masalskoy attacked the chief, Kuchyum, and seized much cattle: "As for Kuchyum he kept in hiding and could not find a dwelling place as he was bereft by the settled Siberians of his whole family, his property and cattle, and people" (Armstrong, 1975). The Georgian Prince Yuri in 1605 informed ambassadors from Russia that his land (Georgia) had "innumerable" cattle (Allen, 1970). The Yesipov Chronicle (1636) tells us that in the steppes of central Siberia, near the Tunguska and other rivers, were most extensive grazing lands for cattle, and some of the people (the Ostyaks (= Ugrians) or the Samoyeds) drank blood "like water" from animals (Armstrong, 1975), which is done only with oxen. In 1599,

Sir Anthony Sherley reported Tatars with an abundance of cattle along the Volga in the region of Kazan (Ross, 1933).

One of the first modern travelers to provide us with an account of the region was the German scientist Adolph Erman, who traveled east as far as Okhotsk in 1827. He reported people driving huge herds of cattle in the region of Salekhard in northern Russia. In Kazakhstan near Omsk, on the River Ishim, the Kirghiz brought immense droves for bartering with Siberian traders, exchanging them for corn, leather, and iron goods. The people of Tashkent had never kept cattle but traded with the Kirghiz for both cattle and sheep. The Kirghiz, as with devout Russians, did not slaughter their calves and would not eat veal. At Irkutsk on the shores of Lake Baikal, cattle were obtained from the Buryats to the north, but principally from the Russians of Yeniseysk to the northwest on the eastern borders of the Siberian lowlands. The Yakuts on the Vilyuy River had such numerous herds of cattle that they were impossible to count, yet at latitude 62°N, they had far more difficulty in keeping cattle than any other people devoted to this husbandry, making long journeys for hay in winter and often feeding their oxen on willow and birch twigs from March to May. They had a tradition that the cattle were originally brought down the Lena in boats from near the source, west of Lake Baikal; but Erman considered that in view of the extreme climate under which they were kept, they must have stemmed from a wild breed which earlier occupied the region of the Lena River.

East of Yakutsk on the Amga River, people kept their cattle only for milk and butter, preferring to eat horseflesh. In the region east of St Petersburg, the 1864 Prussian report stated that cattle were kept only for the manure, the cattle themselves being considered unprofitable. Renault reported in 1860 that in certain provinces in mid- and southeast Russia, thousands of fattened sheep and cattle were brought to periodic markets and fairs to be slaughtered for their tallow, skins, and horns alone; although some meat was salted, many carcases were left to rot on the steppes. An official Russian report claimed that 6 million cattle were raised on the steppes and exported every year throughout Russia, Germany, Poland, Anatolia, and even to Syria.

The abbé Huc reported a trade eastwards to northern China in 1844, the Tatars bringing incessant herds of oxen, camels, and horses, while the Mongols used numerous two-wheel carts drawn by oxen for trade across the deserts of Tartary (Huc, 1850). However, the Ming in northern China, when they went to war against the nomad tribes on the northeast border in 1422, used 340,000 donkeys, 117,000 carts, and 235,000 cartpullers. There is no mention of ox-drawn carts (Barfield, 1989).

Atkinson (1860) was another early traveler who gave us some insight into this vast region in the late 1840s when traveling in present-day Sinkiang Province among the Kirghiz. The nomadic chiefs lived much as nomads do in Africa today, and a typical chief to the north of the Tibetan border at one camp had some 1,000 oxen and cows, together with 6,000 sheep and goats, more than 2,000 horses, and 106 camels, possessing two other such camps. In northeastern Sinkiang, another chief had an estimated 8,000 horses and 600 camels but had no idea how many oxen, sheep, and goats he possessed: "the plains were covered with camels, horses, oxen, sheep, and goats," recorded Atkinson, and the camels were sold to the Chinese. "A stream of clear water was winding in many a turn along the steppe, and thousands of cattle were scattered on its banks…" But Atkinson makes no mention of any trade in cattle westwards at this time.

In Russia, another school of thought believed trek-oxen to be the cause of the disease, supplying the Russian armies or carrying grain to the markets of Odessa. In the latter case, the oxen were forced to make long journeys in the heat of the summer across the arid steppes where there was no pasture or clean water. For 3 or 4 weeks in the hottest part of the year, the oxen making this journey had no other food than the withered herbs which they found by the roadside, and which became more and more scanty as they neared Odessa, while the ponds and pools by the roadside held only liquid mud. It was the effect of this on the oxen, it was alleged, which caused the disease, and those which returned spread it far and wide.

M. Dembinski, Professor of Analytical Chemistry and Natural Science, stated in a communication to the Lord Mayor of London that rinderpest originated in the steppes of Podolia from which considerable numbers of cattle were exported: "Deprived of congenial food and pure water on their transport through the steppes, and then arriving at marshy lands, the exhausted animals drank the stagnant water, which during hot weather, exhaled a pestiferous malaria, and infected them with a predisposition to the epidemic in question, which developed itself into a kind of fever on the voyage to England in a crowded condition" (Bourguignon, 1865). The same belief was quoted by Prince Manoakbeg, one of the principal landowners and cattle breeders in Bessarabia, that the sole cause of the disease was the long journeys during the great heat of summer across the arid steppes, where no pasture or wholesome water could be found. Returning to their several houses, the sickened animals thus spread the disease throughout the country (Murray, 1866).

Some peasants of the steppes of Kherson in southern Russia, where the disease was almost permanent, believed it to be the result of an inflammation caused by the cold; others that it originated in Taurida from the bite of a black spider. As herd after herd of cattle so bitten was driven to the northern markets, so the disease spread throughout Russia and was renewed in drought years.

In the 1870s, the disease was reported as enzootic in central China, and whereas the Caucasians stated that it came from Persia, the Persians laid the blame on China.

6. ANTHRAX IN RUSSIA

Wrongly often reported as rinderpest, anthrax, the bacterial disease believed by many likewise to have a spontaneous origin, was known as the "Siberian Pest" because of the frequent devastating outbreaks which killed numerous horses, cattle, and people there. The first reference is probably that of the Russian ambassador to Georgia who reported to the Tsar in 1603:

> ... we have questioned the couriers who have come to Vladimir from Astarakhan and the Terek and from Kazan and from Nizhni whether horses have been dying of any disease or whether there are epidemics among men. And the couriers have told us, your Majesty's servants, that it is healthy at Astarakhan and on the Terek and in Kazan and in Nizhni—thank God—there are no epidemics among horses or men, nor have men been stricken down; there had been an epidemic among horses in Nizhni before the day of St. Elija but it has passed since St. Elija's day and now, thank God, it is healthy. (Allen, 1970)

Gmelin first positively described it on the Irtish River in 1740, since when it has been known as a permanent scourge of western Siberia, breaking out annually in the summer. Many believed that a gnat was able to transfer the poisonous substance or "*Furia infernalis*." Erman (1848) described an enzootic focus in 1827:

> The district between Pokrovsk and Kainsk [east of Omsk in the Siberian lowland], is one of those low and broken parts of the steppe which are distinguished at once even in winter time, by wide tracts of reeds that force their way through the snow. These swamps ... render the vicinity exceedingly dangerous for summer residence, and cause much annoyance even to the traveller, at that season. The flies and gnats become then so tormenting, that it is impossible to venture abroad without a covering for the face and the cattle even are obliged to be smeared with tar. It is in the hot season too, that the terror of natives and visitors, the Siberian plague ... prevails. This malady is known to cut off frequently men and cattle in the course of a few days ...

In 1864, at least 72,309 horses, 60,000 cattle, and 302 people died from it. An outbreak in Tobolsk alone in 1867 had killed nine men, 356 cattle, and 4,270 horses when reported. In 1864, an official Prussian report on the cattle plague in Russia, compiled by two physicians sent to Russia by the Prussian government, completely confused it with anthrax (*yasva* in Russian). Although rinderpest was probably present as well, 1864 was a bad year for anthrax, described as particularly prevalent in the low-lying wet lands to the east of St Petersburg around the lakes, and along the Volga River in the southeast.

4

RINDERPEST AND THE PANSPERMATISTS

Theories of the Origin of Diseases

1. FROM THE BELIEFS OF THE ANCIENT GREEKS

Belief that organisms were generated spontaneously derived from the ancient Greeks, a writer in the Hippocratic corpus (*c.*300–200 BC) maintaining that all diseases were caused by the air (Jones, 1923), and Aristotle before this (384–322 BC) asserting that fleas and mosquitoes originated in putrefying matter. Allied to these beliefs was the doctrine of Anaxagoras (*c.*500 to *c.*428 BC) and Democritus (470 BC to 380 BC) that the elements were a mixture of the seeds of all things. This latter doctrine became the theory of panspermism, that the atmosphere was full of minute germs which developed upon finding a favorable environment. By the 18th century, these beliefs were still well founded among medical men and naturalists, to whom diseases arose spontaneously from *miasmata*, poisonous vapors which emanated from the earth and then became contagious, spreading in the air. But the discovery of the *Acarus* of scabies and other infestations in the second half of the 17th century was to provide the alternative doctrine of *contagium vivum*, that epizootic or epidemic diseases were caused by living organisms. This led to the "germ theory," described as the hypothesis that no life has ever been evolved (except in the remotest periods of the earth's history) otherwise than from a living parent or a living germ, as opposed to the spontaneous generation theory that life sprung *de novo* from molecular arrangements of the atoms of dead organic materials (quoted by Romano, 1997). The rinderpest outbreak in Britain in the 1860s focused attention on these theories which overlapped in the controversy which followed. Darwin's hypothesis of pangenesis, whereby every cell of an organism reproduced itself by contributing its share to the germ or bud of the future offspring (Darwin, 1871), led to a coining of the term pangermism, pangermists correctly maintaining that the germs of contagia resulted from preexisting germs. But this hypothesis, Brown asserted, was as unsustainable as it was undemonstrable (Brown, 1884). The term had but a short life because, by now, the understanding of contagious diseases was reaching fruition, yet long after the debate had ostensibly closed in the 1870s and bacteriology was a well-established field of research in Britain, in the early 1890s there was still no consensus of opinion that infectious diseases were caused simply by the action of parasitic microorganisms (Romano, 1997).

2. CONCEPTS OF CONTAGION

There had been some earlier dissenters from accepted theory, notable among them Girolamo Fracastoro (1478–1553), Italian physician and poet, considered to be the founder of the science of epidemiology from his creation of the doctrine of contagion by *seminaria*, "germs" or "seeds" of disease, which he regarded as inanimate substances able to multiply. Garrison (1910) accords Fracastoro's work the first scientific statement of the true nature of contagion, of infection, of disease germs, and the modes of transmission of infectious diseases. Fracastoro divided the last into disease infecting by immediate contact, diseases infecting through intermediate agents, and those infecting at a distance or through the air, in which last category he included pestilential fevers. He defined contagion as "a certain precisely similar corruption which develops in the substance of a combination, passes from one thing to another, and is originally caused by infection of the imperceptible particles" (Cave, 1930). The contagion which affects a distant object worked by the same means of the same common principle and by the same method as did other contagions. However, he could not decide on the origin of a contagion because they were so varied, and he asked whether it was due to vapors from the earth or engendered above.

But Fracastoro probably obtained his inspiration from Albertus Magnus of Cologne (*c.* 1200–1280), who had decided that plagues were not caused by divine will but by the contagious principle in the "corrupt blood" of sick animals, which could be passed from one to another by bites or injuries, close contact, and fomites, and in the air, the last being the most important.

A century after Fracastoro's death, the Dutch amateur microscopist, Anton Van Leeuwenhoek, began making his discoveries of microscopic living organisms which he termed "animalcules," while in 1686, another Italian, Francesco Redi, perhaps influenced by Fracastoro, elegantly refuted the theory of spontaneous generation, definitively formulating the concept of parasitism (Redi, 1687).

Fracastoro had some knowledge of a cattle epizootic, but the main challenge to panspermism was occasioned by the rinderpest epizootic which reached Venice in 1711, causing many distinguished physicians to devote their attention to it. Lancisi, the Pope's physician commissioned to investigate the disease, believed for example that the "seeds" of the disease were "tenuous corpuscles of the character of a particular poison" or "pestiferous ferment" which passed from host to host in the form of molecules, multiplying in the new host. His final opinion in his work on the disease, concerning the "manner and means by which the cattle plague insinuates itself into the bodies of the beasts by contagion or infection," followed Fracastoro's terminology, avoiding any suggestion that the agent of the disease might be living. But he recognized that it never occurred spontaneously, being always transmitted from beast to beast or by intermediaries such as hersdmen or dogs.

Wilkinson (1984) has pointed out that these ideas were so much more acceptable at the time that Lancisi's concept of causes was reflected, when not copied outright, in almost all subsequent literature until the rise of a proper science of microbiology based on sound experimental evidence in the 19th century. His description of the causative agent as "tenuous corpuscles with the innate quality of immediately creating and releasing in great quantity corpuscles like to itself in the body which it invades, and multiplying them by its contact" made his explanation closer to the truth in the case of viral diseases than the postulate of Cogrossi simply that the causal agent was a living entity.

Carlo Francesco Cogrossi (1682–1769), an Italian physician and university lecturer, linked the discovery of the scabies *Acarus* with a similar explanation for rinderpest, explaining his idea in his *Pensieri filosofici*, or *Nuova Idea Del Male Contagioso De' Buoi* (*New Theory of the Contagious Disease Among Oxen*) (Cogrossi, 1714). He explained how certain characteristics of contagious diseases could be explained by the hypothesis of *contagium vivum*, such as the receptivity of one species and the immunity of another to contagion; the receptivity of some individuals of the receptive species and the immunity of others. With people, he saw climatic conditions or the cleanliness of the people as controlling contagion, while isolation of infected persons could inhibit spread. An epidemic could be received by the arrival of an infected individual, more readily so when the meteorological reasons were favorable for the life and multiplication of the germ. Nothing prevented thinking that the cause of rinderpest could be identified in a microorganism, in the same manner that the cause of scabies was identified with the *Acarus*. The contagion could enter especially through the upper respiratory and digestive passages, and the multiplication of the organism within these parts, and in the saliva, made the saliva infective even to a healthy ox after it had been expelled on the ground by an infected animal. Hence, he differed significantly from Fracastoro in believing that the causal agents of infections were germs which were able to multiply and spread because they were "living" organisms. He was unable to examine any carcases in the 1711 outbreak because the "noble magistracy" deemed it necessary "and wisely so" to bury them hastily to avoid further contamination.

Antonio Vallisnieri, Chief Public Professor of Medicine in the University of Padua, to whom Cogrossi addressed his publication, approved Cogrossi's theory, stating that he had been meditating on the same idea, as both he and another (Bernadino Bono) had observed "with very sharp microscopes" the blood of oxen to be filled with tiny worms and wanted to conduct more observations. Bono had written to Vallisnieri that he had seen under his microscope the blood of oxen which died in the last epidemic and found "a great quantity of very tiny but lively little worms." Vallisnieri claimed to have made the same observation, but not in the blood of healthy oxen, and deduced that

> there can also be tiny worms, invisible if we do not assist our eyes with microscopes, and perhaps also some which cannot be distinguished even with microscopes, which likewise take delight in the same bodies or in their nourishing full pointfluids .... I shall always find it easier to understand that a living thing, rather than an inanimate thing, can pass from a man or animal to another man or animal, and that it can multiply with the immense fertility which is natural for insects, and more so for smaller ones than for larger ones ... time is needed so that these tiny worms may produce others, and these still others, and so on, until the host which has been nurtured in the vessels throws into confusion and disorder the whole composition of the fluids, and death ensues. (Vallisnieri, 1715)

Of endemicity, Cogrossi wrote:

> It may be also that when they [the worms] do not find circumstances suitable to their nature in those same countries (which can almost be called their native land), they stay on in so small and harmless a quantity that they do not make themselves felt unless because of poor food or air ... or other causes not of the body they are fomented and nourished and multiply their species ... perhaps their native land is beyond the mountains and the seas, and that even there they do not always display their ferocious nature, either because they are hidden away in some niche, or are not always so fierce, or are few in number, or for other reasons unknown to us which oppose their tyranny, or their portentous propagation, or their deadly communication.

This was Cogrossi's "new theory," promoted 150 years before his ideas were generally accepted, a refutation of the idea of spontaneous generation and the promotion that the contagion of rinderpest was spread by living organisms, even though he thought the symptoms of the disease derived "in the very manner in which the itch in man derives from the grubs which have been discovered on the skin." Likewise, the "human mind can fashion no more natural or more probable theory of the pest [i.e., bubonic plague]...why is it not permissible to suspect," he asked, "that in the epidemic among oxen the poisonous insects can pass from one animal to another of similar kind and through the fauces, the nose, and even the passages in the skin, creep into the blood and introduce there irreparable and fatal disorders?" Wilkinson (1984) points out that by 1713, in effect all the necessary ideas were therefore in place. The idea of living germs able to spread and multiply, transmitting contagious disease. Yet, with no further advances in microscopy, the "new theory" according to Wilkinson was to pass into oblivion for a century and a half.

An edict of the Dukedom of Trier in 1719 referred to rinderpest as a contagious disease, and Daniel Defoe strongly emphasized bubonic plague as being so, writing "... I look upon with Contempt ... likewise the Opinion of others, who talk of infection being carried on by the Air only, by carrying with it vast Numbers of Insects, and invisible Creatures, who enter into the Body with the Breath, or even at the Pores with the Air, and there generate, or emit most acute Poisons, or poisonous Ovae or Eggs, which mingle themselves with the Blood, and so infect the Body..." (Defoe, 1722a, 1722b).

In 1745, an English physician, Dr. Theophilus Lobb, wrote

> ... the dissolvent particles (however they came into the body of the first sick beast) like the variolus humour in persons who have the small-pox; and like the pestilential infection in people, assimilates, or transmutes part of the animal fluids into their own kind; which is a generation, and multiplication of infectious particles, some of which emitted from the body of the first sick beast is conveyed by contact, or somehow, thro' the air, into the bodies of the cattle infected by it, and so on. (Lobb, 1745)

In the same year, the Danish Academy of Sciences published three reports, the results of an official inquiry by the University of Copenhagen into the outbreak in Denmark.The first report, in December 1745, was by a layman, the Danish playwright Ludvig Holberg. Perhaps unaware of Cogrossi's work, he referred to as "Hartsocker's hypothesis" diseases such as bubonic plague being caused by little insects with a poisonous sting, found in the air particularly in Asian Tartary and northern China where the plague came from in the 14th century. He questioned this as a possible cause of rinderpest, believing the disease to be more likely due to bad feed (Holberg, 1746). Two of the reports were by physicians, and one, a pathologist J. B. Buchwald, supported the idea that the disease was caused by "insects." Hanno claimed to "demonstrate" that it was contagious but could not explain its true nature; "therefore we must be content rather with the knowledge of its effect and the help of experience" (Hannæo, 1746).

An anonymous writer to the *Gloucester Journal*, October 1747, was much closer to the truth when he wrote that he believed the disease was caused by "small live creatures, which the air, in infected places, abounds with, and are taken in, by men and beasts, by their breath and otherwise: the blood and humors of these last affording them a proper nest and nourishment, whereby they multiply prodigiously..." (Anon., 1747a).

Salchow (1755), opposing the Church teaching in Germany that rinderpest was God's will and could not be prevented, asserted that there were no miraculous causes and that it was the inevitable consequence of certain foregone natural events or changes in the humorous bodies. Often before an outbreak, there was a strong fog or haze, or an unfavorable stench. This he himself had witnessed at Ohldeslohe in 1745. The fog or haze was not a cause of the disease but held it. He was making the connection between damp, humid conditions, facilitating transmission, but claimed to demonstrate that it was an acidic poison of an arsenical nature, such as that produced by volcanic eruptions falling on vegetation. In this, he disagreed with Schreiber in the Ukraine, who claimed that the 1738–9 outbreak there was due to an alkaline poison. Bucard-Mauchart at Tübingen, from his examination of the bile, likewise favored the theory of an alkaline poison.

Plenciz (1762) also believed in the living nature of epizootic or epidemic disease, albeit referring to the disease agent as a "worm" but perhaps not in its literal sense, his tract on contagious diseases referred to by Wilkinson (1984) as offering a more closely reasoned discussion on the subject than perhaps any other contemporary treatise. Arguing against spontaneous generation as savouring of atheism, he wrote:

> Everyone knows that both plants and animals, and their seeds, consist of discrete parts and organs set skilfully and in due order in their proper places; so that it appears not only inconceivable but

> also impossible that such organs could emerge and develop from a disorderly and fortuitous movement of particles.... If therefore we posit as an infallible tenet, that all animals and vegetables take their origin from their own proper seeds, then it should be equally admitted that no seeds are produced *de novo*...

Cattle plague, he concluded, like smallpox and measles was infectious and communicable:

> If the matter of the little ulcers which attack the nose and throat, or of those abcesses, be examined with a microscope, always innumerable little worms are observed; which we can also see if the liquids which are contained in the vessels are subjected to such an examination...Whence it comes about that, if a portion of such material is applied in other healthy oxen as a grafting or inoculation of the same disease, that same disease is spontaneously generated; as customarily happens in the cases of smallpox and measles.
>
> From this again the material cause of cattle plague becomes clear. Because, if it can be communicated, spread and multiplied in the same way as smallpox, it necessarily follows that it likewise has the same elements, that is worm-bearing seeds. (Plenciz, 1762)

Denis Barberet, author of a prize essay of the French Royal Society of Agriculture published in 1766, compiled current views, making a distinction as was customary at the time between epidemic and contagious diseases, following the belief that epizootic livestock diseases were mostly due to contaminated food, especially rust-infected grasses and grain, while contagious disease originated in the air when it was charged with contagious miasmas (Barberet, 1766).

John Thomson (1859), Professor of Medicine and General Pathology at the University of Edinburgh, in his account of his predecessor Dr. William Cullen, the great Scottish physician and professor at Edinburgh, quotes from Cullen's late 18th-century lecture notes (Cullen died in 1790) that there were two kinds of contagions: specific and common. The former "it would appear" were "never generated anew," but the latter might "in any part of the earth, and under certain circumstances be generated." But in his *First Lines of the Practice of Physic* published in 1784 (4th edition), he is by no means so specific: "though we have endeavoured to show that fevers generally arise from marsh or human effluvia, we cannot, with any certainty, exclude some other remote causes, which are commonly supposed to have at least a share in producing those diseases" (Thomson, 1827). He then discusses the effects of cold, heat, and diet on producing fever in humans.

3. THE THEORIES FOCUS ON RINDERPEST

Camper did not agree with Cogrossi and Plenciz in the belief that infection was due to a species of insect or worm. In his *Leçons sur l'épizootie qui regna dans la province de Groningen en 1769*, he recognized the general contagiousness of the disease but dismissed Vallisnieri's theory of worms as the causative agent, stating that not only was he himself ignorant of the actual origin of the infection, but such matters were beyond his comprehension "and that of all mankind" (Camper, 1803). He argued later that diseases did not originate in the air or from earthquakes but could only be communicated or maintained by them (Camper, 1783).

The views of Cogrossi, Vallisnieri, and, later, Plenciz were in sharp contrast to some common beliefs. In Hungary in 1715–16, a "knowledgeable" naturalist had written that the cattle died from eating immense quantities of locusts and caterpillars, while in Poland and the Lowlands, some said that the deaths originated from human corpses. A general belief on the Continent in the 1745 rinderpest outbreak was that the deaths were attributable to the cattle of Bohemia being forced to eat rotten tree leaves during the siege of Prague, all the forage having been taken to feed the horses of the French army. Little different were Dr. John Barker's views in England. He dismissed the idea of infection and contagion, emphasizing the presence of a wet season before its arrival in each country. He thought that the severe frost of 1740 might have contributed to its intensity in England as the winter had adversely affected all vegetable life, but the rank coarse sour foliage suffered less than the sweet nourishing grass. Pastures had also been damaged by the 2-year drought, which admittedly prevented the coarse grass from growing too luxuriantly, but recent heavy rains had washed dressings into the soil, and the resultant rank growth was full of moisture and less nourishing, and the hay of poorer quality. The disease did not flourish on the poorer soils. Because it had increased with the frosts, the distemper clearly proceeded from crude aliment and not from infection in the air (Barker, 1746).

Barker was roundly criticized by an anonymous correspondent in the *Gentleman's Magazine* for claiming that the disease was not infectious. Most examiners were convinced of the distemper's infectious character. The legislature believed so, and the history of the disease bore out that belief. To propagate the contrary was a mistake and apt to prevent necessary measures. One might have expected that Barker would listen seriously to suffering farmers, Parliament, and his own medical brethren, but he had preferred to assail those who listened attentively to the "prejudices of farmers and cowkeepers." The writer himself believed that factors of air, soil, food, and management did not produce the murrain; nor did it generate spontaneously (J. S., 1746).

One 18th-century French opinion held that it was a virus of a caustic acrid and inflammatory nature which, received into the air and digestive passages, corrupted the juices and infected the chyle. People like Kunst (1763) in Germany believed that it was due to poisoning of the blood by bile. Another theory was based on Boerhaave's theory of inflammation, the cause of the disease being the arrest and stagnation of the blood in the extreme capillaries, followed by sudden inflammation and rapid gangrene. But Dr. Blondet, Doctor of Physic at the University of Montpellier and one of the French king's physicians in ordinary, clearly stated it was a contagious disease communicable by men or animals that had been among infected cattle, reiterating the various theories as to its cause, particularly that of Vallisnieri, ending:

> Here we must ingenuously acknowledge, that we are ignorant of the Cause of this terrible Distemper; and I believe there is nobody so bold as to affirm that he has a just idea of it. All that different Writers have advanced on the Subject, is nothing more ... than weak Conjecture. In this Point therefore let us confess our Ignorance, and by such an honest Confession we shall distinguish ourselves from those who are Fools indeed. (Blondet, 1748, 1751)

Others were not as pragmatic, and Dr. Richard Brocklesby, an English physician with medical degrees from Leyden, Dublin, and Cambridge, argued that the disorder did not come from the atmosphere. He concluded that contagions might be propagated by infected materials, of which the most likely were imported from Holland, and one beast infected its fellows (Brocklesby, 1746a). An anonymous correspondent in the *London Magazine* was of the opinion that all infectious diseases were caused by "oviparous animalcula." Being so, tar must be a poison to the "imperceptible Vermin," and so the administration of tar water, "by killing them, cure, or prevent, the Distempers they occasion" (Anon., 1748a). A correspondent to the *Gentleman's Magazine* referred to "some people ... teaching that the distemper proceeded from animalcula and not from infection." He asked therefore if it was not dangerous to leave the animalcula undisturbed, supporting the slaughter of infected animals (Anon., 1750a).

Sauvages (1746), professor at the University of Montpellier, came to the conclusion that no specific agent could be found, and there was no certain remedy. Nineteen of every 20 cattle infected would die.

Plenciz (1762) judged that "it would be a very good thing if certain doctors and surgeons experienced in their profession, should be appointed to undertake a comparative study of animals that have died of the contagion, and duly assess it; so that remedies may be applied more appropriate to the disease," but there was little further advance in the understanding of diseases for another century with belief in spontaneous generation still prevalent, for persons such as Lobb "Innocent of all knowledge of the nature of cells and micro-organisms, let alone viruses" as Wilkinson (1984) puts it, outstripped empiricism in his assumptions. The physicians with an intimate knowledge of contagious diseases could not reconcile the existing theory of contagion with the facts. A French physician, Michele Francesco Buniva, writing in 1793 and 1798 on rinderpest in Piedmont (Buniva, 1808), claimed proof of the contagious nature of the disease without committing himself concerning the nature of an agent. Faust (1797) stated that it originated only through infection, spread by carelessness in moving about to cattle markets, and this carelessness prevented its extermination: "This disease is not generated neither by cold nor heat, nor by dryness nor humidity, nor by rain, fogs, dews, nor pastures or feeds, nor by dirty grass or spoilt hay, nor by drinks, treatment, overwork, hunger, thirst, lack of care, and rest: it has no other source than the communication of the pestilential poison from a sick animal to a healthy one."

Johann Gottlieb Wolstein, a former German army field surgeon with battle experience in the Austrian army, who became a distinguished veterinarian setting up a veterinary school in Vienna, opposed the idea of contagion, believing that cattle plague was caused by environmental factors, especially conditions brought about by war. Arrested in 1794 for his protestant beliefs, he was imprisoned for 2 years, after his release moving to Altona near Hamburg and living in retirement. In 1799, he published a pamphlet on animal diseases with emphasis on rinderpest, stressing war as a cause: "The stench that the battle camp breathes out, makes the surrounding air unhealthy. The slaughter of the livestock, the spilt blood of men and animals remains lying unconsumed, and lack of any sort of burial poisons the air." The bad treatment of oxen and horses brought on disease, while incorrect bleeding and the use of so many wasteful and damaging medicines accompanied by long confinement also contributed. The so-called "whole stall feeding," in which the animals were confined to the stall and never let out, was also a cause. As a result of these factors, domestic animals were often infected, wild animals rarely so (Wolstein, 1799).

In 1795, an American, Dr. Mitchill, Professor of Chemistry in the University of New York, entered the debate by suggesting that contagion, and thereby fever, was caused by nitrous oxide; demonstrating its effects when generated in the stomach, inhaled, and applied to the skin. Frank (1802) accepted this as a likely explanation, theorizing that rinderpest was the result of a chemical reaction relating to the oxidation of nitrogen which created different degrees of acidity; this in turn was expressed as fever in the animal. Little notice was taken of the idea that contagion was a chemical process until Liebig resurrected it 50 years later, although its resurrection was short-lived.

4. REJECTING SPONTANEOUS GENERATION

Hurtrel D'Arboval, author of a dictionary of veterinary medicine and surgery, still held to the belief in 1816 that rinderpest was spontaneously generated among the Hungarian cattle breed, a theory upheld by Lorinser in 1831, although Nocard and Leclainche (1896) state that Sick demonstrated in 1807 that it arose solely from contagion even in Asia.

In 1840, Jacob Henle, considered the best microscopist in Germany, published a paper supporting ideas such as that of the London doctor, Henry Holland, a royal physician. Holland in 1839 had done nothing more than repeat ideas of a century before that epidemics were caused by generations of minute insects carried from one individual to another by the air.

Henle supported the idea that the material of contagions was living and comprised parasitic organisms or "infusoria," and each epidemic disease had its specific animal contagion or specific genera of infusoria, but he had not been able to demonstrate them under the microscope. Henle showed, according to his own words, that contagion was matter endowed with individual life which reproduced itself in the manner of animals and plants, multiplying by assimilating organic matter and existing parasitically on a sick body (Rosen, 1938).

When the plague spread to Bohemia in the autumn of 1844, although recognized as rinderpest by veterinarians, the medical authorities in Prague asserted that it had arisen spontaneously from a combination of causes analogous to those giving rise to typhoid and similar epidemics in man, and laxity of control allowed the disease to spread rapidly throughout the whole of Bohemia.

In 1771, Dossie, whom Smith (1931) placed "in the front rank of epizoologists," had written: "There is, however, a disease incidental to the beasts of this kind [cattle], which propagating itself by infection ...." But the traditional view, still current almost a century after Dossie wrote, was that "miasma" was noxious matter, originating externally and mixed with the air, which entered the body and produced disease even in infinite quantity, in the manner of poisons. Contagion, however, was a material formed by a disease and which occasioned the same disease in others. In 1847, William Sewell, Professor at the London Veterinary College, still lectured that the cause of epizootics was due to terrestrial influences, owing to the constant eruption of gases from the bowels of the earth. Ten years later, Simonds's definition of an epizootic was a disease which appeared suddenly caused by "something in the air" and was not contracted from another diseased animal. This was the manner in which he described it in his evidence before the British House of Commons Select Committee in that year in relation to the outbreak of FMD which had begun in 1839. Yet Simonds himself had conducted the first experiments in Britain in 1839, showing that it was transmissible from one animal to another by ingestion of milk from an infected cow, saliva, or matter from the pustules.

Little different to 18th-century opinion was Dr. L. S. Beale's explanation, although he made an important contribution to the advancement of germ theory in other proposals. Beale was Professor of Physiology and General and Morbid Anatomy at King's College, London and conducted the microscopical examinations for the Commission in Britain in 1865. Although correct in assuming the poison consisted of extremely minute particles of living matter which multiply in the blood, he postulated that they originated in the diseased tissues but did not state that they actually caused the disease; rather, they caused obstruction of the capillaries leading to complete stagnation of the blood.

In 1857, Farr, Registrar-General of Deaths, stated in his quarterly report, "The pathology of domestic animals is exceedingly imperfect, their diseases are badly characterized, and the effect of epizootic causes on the human race is little understood." Yet, William Dick, Professor of the Edinburgh Veterinary College, wrote the following year:

> When reports came that contagious epizootic had attacked cattle in Europe, I came to the conclusion that there was much unnecessary alarm. I inferred that the disease in that year (1842) was an epidemic produced by atmospheric causes operating on local influences, and the result proved me right. This can be cured by a little care, cleanliness, and attention and the results are seldom fatal ... (Dick, 1858)

Pleuropneumonia and potato blight, he asserted, were dependent upon the same cause, the "evidence" being that they both appeared at the same time of the year. Greenhow (1857) repeated "the best continental authors," who stated that the steppe murrain was spontaneously developed in Bessarabia, Podolia, and other provinces of Southern Russia.

In August 1862, Dr. William Budd of Bristol, an epidemiologist and probably a colleague of Farr, was appointed to a Committee of the British Medical Association set up to investigate epidemic and epizootic diseases and produced a number of works on epizootic diseases believing that important lessons could be learned from studying "epidemic disorders" in domestic animals. He was requested by the Council of the Association to prepare a plan for the investigation of epidemic and epizootic diseases, following a resolution passed by the Association that August binding it to undertake such a study.

Budd presented his plan in August 1864. In an address to the Association in July 1863, Budd had taken as his theme sheep-pox: "*Variola ovina*, sheep's small-pox; or the laws of contagious epidemics illustrated by an experimental type." Before the relaxation of import controls in 1842, the disease was unknown in Britain, but in 1862, an epizootic of major proportions erupted. Budd's analysis of the outbreak led him to the conclusion "Not less valuable is the whole history of this disease, for the signal record it bears against the doctrine of spontaneous origins applied to contagious disorders. Resting on purely negative data, this doctrine is a mere figment of the mind" (Budd, 1863). Diseases, "like animals and plants, however they may have once originated, are only propagated now by the law of continuous succession" (Budd, 1864). "Opinions like these," he continued, "are no doubt at present those of a small minority."

"A very large, and by far the most influential school in this country—a school which probably embraces the great majority of medical practitioners and the whole of the 'sanitary' public, holds the exact contrary, and teaches that sundry of these poisons are constantly being generated *de novo* by the material conditions which surround us" (Budd, 1863).

Pointing out that there was no proof of spontaneous generation, he urged that this question be investigated, proposing rinderpest, FMD, pleuropneumonia, and sheep pox as subjects. He chose rinderpest because he considered that it presented a striking analogy with human typhoid fever: "... typhoid fever of the ox [is] to human typhoid—not identical in either case; not intercommunicable; essentially different in species but so nearly related as to admit on a superficial view of being confounded each with its counterpart."

In August 1865, after rinderpest had erupted in Britain, he stated unequivocally: "It is in the body of the living ox, and by that most specific of processes which constitutes the disease itself, that the morbific agent is bred and multiplied, which sweeps through the herd with such fatal rapidity" (Budd, 1865a).

He acknowledged "unstinted help in theory and practice" from Simonds (particularly with regard to Simonds's studies on *V. ovina*).

In reply to a question posed by the Commission in 1865 whether rinderpest was spread by contagion, Simonds replied: "Decidedly so; that is to say, giving an extended meaning to the term 'contagion'; it is not necessary that there should be positive contact of animals." When it eventually appeared in Britain, many believed it to be due to spontaneous generation, but Simonds apparently never attempted to advance such an explanation, although two German veterinarians sent to England to report on the outbreak claimed that Simonds stated that although imported in this case, it might have arisen spontaneously from London dairies (Reed, 1866). In a talk given to the RASE on February 21, 1866, Simonds stated in regard to the question whether rinderpest might arise spontaneously from the atmosphere that he "could not for a moment suppose that to be the case. The origin of the disease was distinctly traced to animals that were brought to the Metropolitan Market from a foreign country" (Simonds, 1866). In the case of FMD, he had agreed earlier that it could develop in apparently healthy beasts solely from an atmospherical influence.

Perhaps he had never believed it and simply did not want to offend the leading figures in the profession who countenanced such views. His report on his continental visit in 1857 clearly showed that he had doubts about rinderpest being spontaneously generated when he wrote:

> Doubtless every disease has had its place of origin, and in it there may be persistent causes which keep alive, so to speak, the curse of sin. Such causes may possibly be found in these vast plains of

> Russia; and if so, here would be the natural habitation of the pest. Be this as it may, it is certain that in those countries which are contiguous to the steppes the malady has no such origin, and its appearance in them is invariably associated with the *recent* introduction of steppe cattle, and generally in the ordinary course of traffic. No disease which we have ever studied appears to be governed by such precise laws as this, with regard to the means of its extension; and it is difficult to believe that it should be spread by infection alone throughout Europe, and not be subject to the same law in the steppes.

But he concluded:

> When we observe a malady to be capable of being communicated from animal to animal by innumerable means of conveying the *materies morbi*, and when we take into account the varying susceptibility of animals to the immediate action of this matter, and also the further circumstance of its remaining dormant in the system for a fortnight, or possibly a longer time, we see many reasons for withholding our assent, without greater experience in the disease, to the opinion that the pest has a spontaneous origin in the ox of the steppes.

In an address to the Norfolk farmers in September 1865, he made his position quite clear on the subject: "Sometimes it went into Prussia, sometimes into Bohemia, and sometimes into Bavaria; but it never went there except upon the introduction of cattle in whose systems the disease was either incubated at the time, or the cattle themselves were positively in a state of disease." He went on to say that there were such men to be met with in the medical profession, and also in his own profession, who were called contagionists; and there were men not at all favorable to this view of disease being spread by contagion, or not exclusively so. It was difficult to draw a line between the two opinions, and perhaps the truth might lie in the middle: "but let them not lose sight of the fact that the disease is contagious, and it is by infection that it extends itself." In every case he had been able to trace in the country, the disease had come from the Metropolitan Market. When asked by a Mr. Wells whether he really thought the disease was spread by contagion or infection and in no other way, Simonds replied that when they looked at all the facts in connection with the disease, they were justified in saying that it spread only by contagion or infection.

In October 1865, *The Veterinarian* (of which Simonds was joint editor) carried an editorial clearly against spontaneous generation as a cause: "The proofs of the contagious character of a disease being complete, there exists no necessity for speculation upon other possible modes of extension; and the defenders of the theory of spontaneous origin should be prepared at least to show that the disease has been most rife in those situations where hygienic laws have been least obeyed."

In evidence to the Commission October 9, 1865, in answer to the question "Is it your opinion that the disease does not break out spontaneously?", Simonds replied, "It is decidedly my opinion, and I believe that that is confirmed by a great many authorities." Asked "You believe that the whole of southern Russia is the seat of the spontaneous generation of this disease?", he replied, "I do not believe that it has a spontaneous origin anywhere. I believe that the whole of the southern part of Russia is the home of the pest, and that it always abides there." Pressed again, "Do you believe that there is no instance of the spontaneous generation of the disease?", Simonds replied, "I believe so." Later, in answer to the question "Would not that [contaminated manure] give rise to spontaneous generation?", he replied, "The so-called spontaneous generation; it is merely transporting the seeds of the disease in another way."

Clearly, Simonds fully understood the nature of epizootic disease and had an understanding of it above that of the majority of the medical profession of the day. His statements convinced the Commission such that in their *Second Report*, the Commissioners spelt out that the evidence pointed to contagion as being responsible for the disease and not a spontaneous generation in the London cowsheds:

> The theory that the disease originated in the London cow-sheds is inconsistent, not only with the evidence appended to our First Report, but with what we have since received. The careful

> observations made by Dr. Ballard. Medical Officer of Health for St. Mary's, Islington, confirmed by those of the Medical Officers for Marylebone and St. Pancras, and the experience of other metropolitan districts, are in direct conflict with this theory, and point distinctly to contagion as the means by which the Plague was originated and propagated in London. (Bernard, 1866a)

At the Annual Meeting of the RASE in December 1865, Simonds reiterated that he still held that the plague was not indigenous to Britain, that it had no spontaneous origin, that it had been introduced from abroad, and that it spread, "as all contagious diseases did," by the conveyance, directly or indirectly, of morbific matter.

Gamgee's views were also directly opposed to theories of the spontaneous origin of diseases. He stated that 10 years' study of epizootic diseases had taught him that they did not arise spontaneously but spread only by contagion, following the lines of communication between different countries: "We may as soon expect the spontaneous outbreak of epizootic diseases as the spontaneous generation of wolves in Epping Forest" (Gamgee, 1863a). But only 2 years earlier, he had written: "Its spontaneous origin, regarding which as little is known as with other epizootic maladies, is evidently due to conditions of soil or atmosphere, though some have thought the cattle of the Russian Steppes could alone be primarily effected [*sic*]. Doubtless this is an error..." (Gamgee, 1861). After the outbreak, he stated: "This cattle-plague could no more have spontaneously generated itself in England than the mud in London streets can turn into living creatures."

But Priestman, the veterinarian who had been called to Mrs. Nicholls's dairy where the disease first broke out in 1865, and had then called in Simonds, when questioned at the first sitting of the Commission in October, stated that he believed the disease to have been spontaneously generated in London:

> QUESTION NO. 803: Do you consider this an imported disease, or that it has been generated in this country?
>
> PRIESTMAN: My opinion has always been that it has been generated in this country.
>
> QUESTION NO. 826: I asked you where you thought the disease was generated, you say that you think it is of spontaneous origin, may I again ask where you think that it originated?
>
> PRIESTMAN: In London.

However, Simonds, in reply to the question: "Some cases have probably come to your knowledge in which the disease has broken out spontaneously?" replied: "There have been many cases which have come to my knowledge in which I have not been able to trace the introduction of the germ what would give rise to the disease, but nevertheless I cannot come to the conclusion that these have had a spontaneous origin."

Gamgee (1865), asked why it appeared to originate in the Russian steppes, answered that he did not think that it developed there spontaneously but it was kept up in the world "like smallpox" by repeated recrudescences. "... I believe we may class it with hydrophobia and smallpox; not as originating spontaneously in any part of the world, but as being a transmitted disease."

Hugh Ferguson, Her Majesty's Veterinary Surgeon in Ireland, was to report to the Irish Government in September 1865: "Although by the majority of veterinarians and agriculturists malignant bovine typhus is considered to be invariably the result of contagion or infection, or both, such is not the case. Malignant typhus can be generated in an animal without being the result of either infection or contagion; but once generated here in a single animal it becomes rapidly extended and multiplied by contagion and infection." He recommended total prohibition of cattle imports from Britain to prevent its spread into Ireland, but cautioned "And even should importation be totally prohibited, it is by no means certain that the disease may not become generated from mere atmospheric influence..." (Ferguson, 1866).

5. CONVINCING THE CONTINGENT-CONTAGIONISTS

Until the 1865 rinderpest outbreak in Britain, debate surrounding the origin of diseases had been largely medical, although some hold that most English medical teachers and researchers were at least "contingent-contagionists" as to the causes of specific diseases (Pelling, 1978), and the general debate on causes was very much in terms of contagion versus miasmas. According to Fisher (1993), the spectacular virulence of rinderpest came as a shock to both professional and laypersons alike. Virtually all the great names in medicine in Britain gained first-hand knowledge of the disease, often causing them to revise their opinions.

A report from Hungary in October 1865 was explicit on the matter of propagation, stating that it took place in Hungary only by the conveyance of infectious matter or contagium. This contagium was developed in every infected animal at the beginning of the sickness and continued during the course of it. Its nature was partly volatile and partly fixed, and it adhered to the expired air, to the vapor of the blood, to the transpiration of the skin, and to all the fluids and remains of infected animals as well as all articles that had been in contact with it. It was operative to a considerable distance and when largely diluted, thus being easily spread by air and media of every kind. Once in a district, it was extended further and further by the general intercourse of persons and by butchers, tanners, quack doctors, and others; by giving food or drink touched by infected cattle; and by the free wandering of cattle "or by numerous other circumstances which are often difficult to be traced" (Egan, 1865).

In his report to the Commission, Beale noted that although undoubtedly many facts might be urged in favor of the view that contagium, the matter which was supposed to pass from one organism to another and which possessed the contagious property, really consisted of germs of low animal or vegetable organisms capable of living parasitically for a time in man and the higher animals, there were many broad facts opposed to this doctrine. The question to him was not whether the changes exhibited *may* be brought about by animal or vegetable germs, but whether all fevers, including rinderpest, could be proved to be due to this cause (Beale, 1866). He concluded that the contagium consisted of very minute particles of matter in a living state specific to the disease which could multiply rapidly and be transmitted through the air. In this, Beale was the first to claim that the contagium consisted of recognizable particles, not the postulated zymosis-producing particles of albumin developed by others, such as Farr, from Liebig's ideas. He claimed that the rinderpest contagium consisted of "minute particles of matter in a living state, each capable of growing and multiplying rapidly when placed under favorable conditions." He considered the "subtle poison" derived by uninterrupted descent from once healthy living matter within the body. Beale (1872) claimed that his conclusions were misinterpreted in the Commission's *Report* and criticized an assertion that the poison was always to remain undiscoverable by the microscope: "At the very time these sentences were written, the poisonous matter *had been made out* by the microscope ...." The contributor to the *Report* was of course correct, and in his assertion Beale had found no definitely formed substance that could certainly be said to be the cause of cattle plague.

Justus Liebig, a German chemist, postulated that the cause of disease was an organic substance in the process of decomposition conveying that decomposition to another body. He maintained that fermentation and putrefaction were chemical processes, in contrast to the "Panspermatists," who believed atmospheric germs to be responsible. But in France, another chemist, Louis Pasteur, upheld the explanation that an organized germ caused disease, although it was not first expounded by him. He conducted a number of experiments which established that fermentable liquors only changed and generated organic beings as a result of contamination with solid particles from the air, and not from spontaneous generation caused by gaseous elements, work which gained him the annual Grand Prize for Physiology of the French Academy of Sciences in 1859.

Beale's view was supported by William Crookes, FRS, who used the term "germ" freely in his contribution to the Commission report, but allowed the possibility that the infectious matter might be a living cell, thus supporting another line of thought on the nature of germs. Crookes supported Pasteur's theory, without acknowledgment, by stating that he preferred the word "infectious" to "contagious," because the latter was limited to actual contact, and popular opinion considered some diseases infectious and not contagious, while others were contagious but not infectious, ideas which implied a far more profound knowledge of the way in which diseases were transmitted than were yet possessed. Likewise, the words virus and poison were generally regarded as synonymous but should be considered distinct, for a poison could not multiply in a body as a virus did, rendering the liquids of an infected animal as virulent as the original germ. A poison might be organic, but it was not organized. Many considerations showed that the virus of rinderpest was a body similar to vaccine lymph and consisted of living cells possessing physiological individuality, which, if not exposed to extremes of heat, cold, or dryness, were capable of preserving their activity for a certain time outside of the living organism. When introduced into the blood, they required a certain time during which the germs multiplied until the ordinary symptoms of disease became manifest. The chemical theory was insufficient to explain the prodigious procreative power of the original particle, a power only belonging to the nature of an organized germ, capable of producing multiples of itself by a process of nutrition and subdivision (Crookes, 1866).

Sanderson, one of the leading research workers of the day, also supported Beale's ideas with the difference that he believed the particles to be independent of the tissues where they were found (Sanderson, 1866). Beale's work on rinderpest thus played a great part in stimulating an advance in germ theory (Crellin, 1968).

Despite such knowledge, and despite Henle's work, spontaneous generation was still widely believed in by many eminent members of the medical profession in Britain, as well as by veterinarians. John Parkin, MD, bluntly stated in an essay: "We may conclude that the bovine plague now prevailing in England has not been imported from Russia or any other country, and that it is of spontaneous origin" (Parkin, 1865). After drawing comparisons with the sudden outbreaks of other diseases such as cholera, pleuropneumonia, and potato blight, he continued:

> Although unacquainted with the remote cause, there cannot be much doubt with respect to the immediate cause. This is the presence of a deleterious agent in the atmosphere ... we can only infer that the agent is derived from some other source [than the atmosphere]—one with which only a portion of the atmosphere comes in contact. Finding, also, that the poison is more concentrated at lower than at higher altitudes—epidemic diseases prevailing to a greater extent in the former than in the latter situation—we may also conclude that the poison is extricated from the surface of the earth, along particular lines, and in certain spots of greater or less extent ...

It was hardly surprising, then, that a layman such as the London cattle-dealer, John Giblett, giving evidence before the Commission in October 1865, stated that he had published a letter in *The Times* on the cause of the then present outbreak, in answer to which he had received a hundred communications approving what he had written. This was that there had been a "baneful atmospheric influence," and the disease was not imported. He referred to FMD which, he asserted, was "no complaint in itself; it is a result of a complaint in the animal," a result of fever in the animal. Being hide-bound, the fever could not escape the animal except through the hooves, or the mouth and tongue; it was a "struggle of nature" to get rid of it in the extremities. It was much worse when the east wind blew, which to him was proof that it was atmospheric.

Some Austrian "medical men of position" connected "the seat of the disease with the peculiar electric disturbances of the atmosphere in the early part of summer, when it breaks out usually [in the steppe country of Wallachia, Moldavia and the Great Plains of Hungary]." This created "hunger typhus," a consequence of the nervous system being functionally disarranged.

Besides Gamgee, Budd was one of the few other persons in the mid-19th century known to be opposed to the theory of the spontaneous generation of diseases; a division of opinion which existed in both veterinary and medical circles, although Dr. John Simon, Medical Officer of the Privy Council and surgeon of St. Thomas's Hospital asserted, in evidence to the Commission, that there was no evidence of the spontaneous generation of diseases. Still, in 1865, a Yorkshire gentleman of leisure who used the Athenaeum Club as an address, H. Strickland Constable, while claiming that "cow doctoring" was in the stage of human doctoring a century before, fervently believed in the spontaneous origin of diseases, quoting in support Florence Nightingale (who considered the doctrine of contagion an antiquated myth) who "saw clearly that impurity of air and of habits will produce any disease." Ridiculing those (and here his ridicule was probably directed at Gamgee) who denied the possibility of spontaneous origin, he stated: "And this is the latter half of the nineteenth century!—and we talk of our scientific habits of thought and freedom from old woman superstitions!" Florence Nightingale he had hoped "had laughed us out of the old idea, that smallpox, measles, typhus, etc., were like dogs and cats, as she says, and organized beings that can only be reproduced from themselves" (Constable, 1866b).

Constable's relative, Sir George Cholmley, wrote to him: "People believe in contagion as they used to believe in witchcraft, and as fools now believe in spirit rapping." In a letter to *The Times* of March 31, 1866, criticizing Gamgee's book and his recommendation of slaughter as the only means of controlling the disease (*vide infra*), Constable quoted a number of cases of fatal human fevers having apparently arisen spontaneously, the inference being the disease had not necessarily come over Europe from east to west, as Gamgee had stated, but could arise anywhere. He quoted a "knowing old farmer," who stated: "I turns my beasts out every day; they kick about, and keeps themselves healthy. Why, we heard nothing of this plague, except in the town cowhouses, till cold weather coomed, and all the cattle were taken up and crowded together in yards and sheds."

In October 1865, a cavalry veterinary surgeon, W. C. Lord, had come very close to defining the true nature of transmission of such diseases. Having found protozoa in the intestinal tract of infected animals, he attributed these to being not a disease but "an exciting cause." Although these were the normal entozoa of a healthy animal, he concluded:

> I think the parasitic theory alone [i.e., that the parasites are an inciting cause] will explain how animals brought from Russia may remain well for an indefinite time, if not exposed to any influence of a depressing nature; and also how the same animals (although apparently well), and even sheep, may give the disease to others, which could not be the case if it were like any other disorder. It also explains why animals at a considerable distance from those infected may have the protozoa conveyed to them in the atmosphere, and thus become diseased.

Replace the word protozoa with virus, and you have an explanation of disease transmission as we know it today.

In the same month, the editors (Morton and Simonds) of *The Veterinarian* wrote: "The idea of the spontaneous origin of the 'plague' in our country is ridiculed by all who understand the question at all. The inquiry put to a foreigner excites a derisive laugh, as though the argument deserved no serious consideration."

In August 1865, Budd spelt it out as follows:

> 1. Cattle plague is a contagious (by the word "contagious"... I simply mean "communicable," whatever the medium, direct or indirect...) and eruptive fever peculiar to the cattle tribe, in

which the eruption, instead of occurring on the skin, as in smallpox, occurs on the internal surfaces of the alimentary canal and air-passages, nostrils included.

2. The poisonous germs thrown off by the eruption are cast out of the body in the discharges from these surfaces, which thus become the chief means of propagating the disorder...
3. The specific poison which causes cattle plague is, like the poison smallpox, the product of the disease itself; and the living body of the animal affected with it is the only known medium in which it is bred.
4. ... Cattle plague... like the living organisms which people the earth, having come into being under conditions of which we know nothing, is never known to spring up spontaneously now, but is self-propagated only. (Budd, 1865a)

British scientists had already observed that there was a relationship between minute particles and epidemic disease, but the Commissioners warily noted only that the growth of the poison seemed to indicate an independent living existence, "and on this point our knowledge is not yet sufficiently definite." Only by its effects on the living body could it be identified, but it required a peculiar condition before it could react. Hence, it could only multiply in the body of a bovine animal, or of a sheep, goat, deer, or gazelle. There was no satisfactory proof that it could be communicated to non-ruminants (Bernard, 1865).

Dr. Angus Smith likewise was still asking whether the cause of disease was an organic substance in the process of decomposition conveying that decomposition to another body (Liebig's theory), or whether it was an organized germ (Pasteur's theory) (Smith, 1866).

The Commission was to come to the conclusion in its *Third Report* (1866) that the manner in which the disease had broken out in the Paris zoo (the *Jardin d'Acclimatization*), introduced with infected meat from a railway van, and was eradicated; its repeated appearances in Aberdeenshire, and its absence from Ireland; and the manner of its spread in England and Scotland during all seasons formed conclusive evidence against an occult atmospheric condition (Bernard, 1866b).

The Commissioners stated "We have been able to find no evidence of spontaneous origin in England."

Yet, there were still those who remained unconvinced. In India in 1869, Dr. Mcleod, an Assistant Surgeon in the Bengal Medical Service, wrote: "My own impression, derived from a close study of all that has hitherto come to light on the subject, is, that these diseases are *enzootic* in the country [India], spontaneously generated by conditions inherent in the soil and climate, and not owing to importation from without; that in fact we are in the position of the Steppes, and have these diseases as permanent residents in the country" (McLeod, 1869). An army veterinarian, Hiram Farrell, in his report to the Indian Cattle Plague Commissioners in 1869, concluded that rinderpest was of spontaneous origin, "... once developed, it is propagated by immutable laws that govern all contagious diseases; the animals do not inherit the disease, nor do they carry it about in their flesh and blood, but, as I have stated, certain conditions of the soil and atmosphere are favorable to its development, especially where cattle are congregated in any large numbers" (Hallen *et al.*, 1871).

In June and July 1870, the question of spontaneous generation was reopened in Britain by Dr. H. C. Bastian, Professor of Pathological Anatomy at University College London, in a number of contributions to *Nature*, such that at the annual meeting of the British Association held in September 1870, the president, Professor Huxley, was still able to choose for his inaugural address "The question of spontaneous generation and the propagation of infectious diseases," a survey which he stated "has not taken us into very attractive regions; it has lain chiefly in a land flowing with the abominable and peopled with mere grubs and mouldiness." Huxley did not support the concept of spontaneous generation: "There can be no reason, then, for doubting that, among insects, contagious and infectious diseases, of great malignity, are caused by minute organisms which are produced from preexisting germs, or by homogenesis; and there is no reason, that I know of, for believing that what happens in insects may not take place in the highest animals …." Simonds, apparently having second thoughts about his previously more positive position, was now asking if pleuropneumonia was contagious.

In 1872, referring to the doctrine of spontaneous generation as "the belief of the vulgar," Beale wrote that it was "not a little remarkable that such a view, driven as it has been from one position to another during the last 200 years, should yet find advocates among the most active intellects of these days." In April of that year, a distinguished French veterinarian, Henri Bouley, stated unequivocally:

> "… there now exists no divergence of opinion as to the foreign nature of the cattle-plague in reference to Central and Western Europe, nor as to the mode in which it is propagated. We are now thoroughly convinced that outside the Russian empire it is never developed spontaneously … whenever cattle-plague appears elsewhere, it is because it has been imported by some means or another… it is kept up solely by means of contagion, … it cannot be perpetuated otherwise…it becomes extinct when it can no longer be communicated to other animals … contrary to the opinion of some physicians … the cattle-plague cannot become an indigenous malady in our country [France], under the influence of what is somewhat obscurely termed an epidemic nature … Contagion is thus the exclusive cause of the importation, the propagation, and the greater or less permanence of the cattle-plague in Western and Central Europe.

Despite such positive statements, the belief in spontaneous generation was still far from dead. At the House of Commons Select Committee hearing in 1873, the question was put to the Chief Constable of Cheshire, Captain Johnes Smith, by the Chairman William Forster in respect of FMD and pleuropneumonia: "Do you believe in the spontaneous origin of those diseases?" Answer: "I do; I believe they are very much like cholera and other diseases which visit us, and we know very little about those diseases; they travel with the atmosphere. I believe that there are atmospheric causes which we do not seem to understand. I believe that is one cause of those diseases coming and going" (Anon., 1873).

In 1877, Mr. Anderson, a member of the House of Commons Select Committee which sat from 11 May to 19 July investigating the 1877 outbreak, put the question to Professor Brown: "Does it not occur to you that the evidence in this case points rather to a spontaneous outbreak than to contagion?" Brown: "No, I am perfectly satisfied that whatever may have been the source of the disease it had no connection with a spontaneous origin." Mr. Anderson persisted: "But you have no evidence of any source from which it could have arisen?" Five questions later along similar lines in which Brown persisted it could not have arisen spontaneously: "It comes back to this, that you are quite certain that it could not have arisen spontaneously?" Brown: "Yes, certainly, for the reason that all the evidence which we have in reference to cattle plague clearly points to the fact that it never arises in a country unless it is introduced from some country where it has always existed."

"You think that no sort of bad drainage, or foul air, or over-crowding, or anything of that sort would produce it?" "Certainly not; it never arises in any part of Germany unless diseased animals are brought across their eastern frontier …."

The Committee asked whether the disease (among others) ever originated spontaneously in Western Europe, and Professor Charles Muller, Professor of Anatomy in the Berlin Veterinary College was to reply, "No, never. We are of the opinion that all contagious diseases of cattle are only spread by contagion, and that they do not make their appearance by spontaneous generation" (Muller, 1877).

But, John Giblett, the London cattle salesman who had given evidence to the Commission in 1865, still held to the belief that an unhealthy atmosphere engendered FMD, and it was also "easily carried on the wings of the atmosphere."

W. E. Forster, a Committee member, appeared also to believe in spontaneous generation when he put to William Stratton, a member of the Executive Committee for dealing with contagious diseases in Wiltshire, the question: "You do not think that because there may be a great deal of contagion there may be certain conditions of traffic of animals, or of the atmosphere that might bring about the disease (FMD) spontaneously?" "I am satisfied of the contrary." Forster persisted: "Are you aware that there are a great number of persons of great practical experience who do not agree with you upon that point?" Stratton replied: "All scientific authorities, I think, concur in my view of the case, and I believe that their authority must be very much greater than that of what you call a practical man, which perhaps is another word for an ignorant man."

Brown reported to the RASE (April 11, 1877): "A very wide-spread belief appears to exist in that most unfortunate of all delusions, the atmospheric theory of contagion; and so long as stock-owners entertain the impression that the disease is 'brought by the air,' it is quite hopeless to expect that they will avail themselvs of the means which they have at command for the protection of their animals."

In July 1885, a Danish veterinarian Engelsen, based in Russia, was still to write to his former superior Dr. Krabbe of the Danish Veterinary College in Copenghagen that rinderpest did not arise spontaneously from mouldy soil "as some [Russian] professors assume" (Berk, 1986).

6. SPONTANEOUS GENERATION LAID TO REST

It was 1888 when Wallace wrote, "There is no such thing as *spontaneous generation* or *special creation* of the virus of a common contagious disease" (Wallace, 1888). Both Budd and Gamgee had been ridiculed, and both lived long enough to know that they were right. Gamgee had written in 1866: "It is entirely *sui generis*, and never originates spontaneously beyond the Russian frontiers; within which it is probably not generated at any time *de novo*, but is kept up by constant reproduction, after the manner of other contagious maladies." Yet, in 1894, Pease still had to state "Rinderpest is a malady which always develops by contagion; this is pretty well established at the present day; it is absolutely false to say that it may develop spontaneously."

With heavy sarcasm, Constable had written in 1866: "I hear that some one has sent to Russia for a cattle doctor of that country… I should have thought this an unnecessary measure to take … All barbarous and semi-barbarous countries believe that diseases come from the Devil, and that the way to cure a cow is to burn a witch … is it quite certain that he will be allowed to adopt this mode of treatment in England? But, if not, where is the good of sending for him?"

Although the first veterinary schools had opened in France and Britain in 1762 and 1791, respectively, in general both Britain and America were far behind the Russians and Germans in their understanding and study of both human and animal diseases. By the late 1880s, there were four veterinary schools in Russia, the earliest dating from 1848, each offering 4-year courses. Three had full-time professors of pathology following the assertion of Rudolph Virchow, who taught pathology at the University of Berlin, noted for a classic study on thrombosis and embolism and for taking pathology along the new path of cellular pathology, that pathology must be an independent branch of science. Great importance had been attached since 1860 to the ability to distinguish between different diseases from postmortem examination, and pathology was considered essential to this understanding. Most of the teaching was in German, and there was a free exchange of ideas and staff with Germany, Austria, and Switzerland.

The slow march of ignorance from Fracastoro's foundations in the 16th century may be partly attributed to medical practitioners in general regarding the study of animal diseases beneath them; but both Ramazzini and Lancisi had considered that a comparative approach was essential to the understanding of all disease, criticizing those who opposed such views. In this, they opened the way to comparative study as well as underwriting a new approach to animal welfare. Camper was to follow, instructing his students in 1769 to watch over not only the health of their fellow citizens but equally that of all animals "useful to the community." Until the advent of veterinary schools, the study of rinderpest was undertaken by physicians and surgeons, a number of whom produced treatises on the subject. Many of these appeared anonymously because it was considered derogatory for medical men to write on animals, perhaps seeing it as contrary to the maxim of the poet Alexander Pope (1688–1744), "The proper study of mankind is man." But Brocklesby had the courage to argue that a physician needed more than a knowledge of the human body, as the health of mankind depended largely upon the materials used as food (Brocklesby, 1746a). An anonymous writer in the *Gentleman's Magazine* regretted that fashionable doctors feared to visit a cow lest they incur "the filthy appellation of cow-leech," but fortunately the profession had many "public-spirited members" (Philo-Boas, 1750). One exception also writing in the *Gentleman's Magazine* was Dr. J. Wall of Worcester, who requested readers to write to him with experiences of his remedy for cattle plague (Wall, 1751).

Although the first veterinary schools were established when rinderpest was a serious problem, they were conceived more out of a concern for hippiatry, linked to the use of horses in war, than concern for bovine diseases. Plenciz (1762) had written with respect to the rinderpest plague: "… such things … are widely committed for direction to stupid and untaught veterinarians who know nothing of anatomy nor of animal economy, much less of illnesses, as though it were unbecoming for a doctor or surgeon to handle things of a kind which the great Hippocrates himself was not ashamed to deal with …."

The value of studying animal diseases to understand their human counterparts was to be the subject of a medical thesis by Kolb in 1828, who bemoaned the lack of comparative studies of epidemic disease, referring particularly to rinderpest in Germany (Kolb, 1828).

From 1884 to 1886, a Russian worker at Karkov, Saveljef claimed to have cultured the bacterium of rinderpest, Engelsen writing to Krabbe in July 1885: "With Klebs 'sectioned culture,' he got a pure culture of the rinderpest microbe in gelatin and agar-gar. The microbe is half as big as the anthrax bacillus …." Fatally injecting dogs and mice with the culture, he did not inject any ruminants (Berk, 1986). In 1897, Professor Heinrich Hermann Robert Koch, a renowned German bacteriologist who had discovered the agar-gel method of cultivating bacteria and the aniline-dye method of studying them, invented the microscope illuminating apparatus, discovered the tubercular bacillus and the cholera microbe, and was to attain fame in South Africa for promoting inoculation against rinderpest, wrote to the *British Medical Journal*:

> The last great European epidemics occurred at a time when microscopy had not reached the stage when it could be of use in the investigation of the causes of disease … Of late years, however, the special methods of bacteriology have been applied, mainly by Russian observers, to the investigation of rinderpest, but have so far led only to contradictory results. No fewer than six distinct microorganisms have been described, each of which has been held to be the specific cause of rinderpest. To these have recently been added two further kinds of bacteria, the one discovered by Simpson in India, the other by Edington in South Africa. On the other hand, a research has just been published by Tartakowsky which gives the impression of having been conducted with perfect mastery of bacteriological technique, and at the same time with all the necessary self-criticism. Tartakowsky concludes as a result of many years of most thorough research, that with the cultural and optical methods at present available, specific organisms in rinderpest cannot be demonstrated. (Koch, 1897a)

Allegations that Koch was of a despotic disposition and spoiled by his early successes show through in the tenor of his writing. Whereas, on the basis of the above, Tartakovskii would appear to have been ahead of Koch in his thinking, being absolutely correct in his deduction, Koch seems to be implying that Tartakovskii was suggesting that rinderpest was not caused by a micro-organism and therefore must have been a believer in spontaneous generation of the disease.

Part II
THE HISTORY OF RINDERPEST IN EUROPE

5

RINDERPEST IN HISTORY FROM ANCIENT EGYPT TO THE 17TH CENTURY

1. INTRODUCTION

Records of cattle mortality in ancient and mediaeval times frequently refer to deaths of other domestic animals and of people, leading some authorities to suggest these early plagues were attributable to anthrax. Stock was often kept in wet and marshy conditions favorable to this disease. While anthrax may have accounted for some of the more localized plagues, there is no reason to suppose that rinderpest was not equally present. It is unfounded to assume that early plague records in Britain, for example, did not refer to rinderpest because the plagues died out on each occasion and apparently did not result in the establishment of enzootic foci. The cattle population was relatively small and dispersed, and experience from the Continent suggests that the disease does not become enzootic among deer, for example, although they succumb to it in bad outbreaks. Possibly, they react to it in the same manner as wild herbivores in Africa, the disease dying out when the cattle–game link is broken (*vide infra*). Thus, a failure for the disease to persist in a population of cattle does not negate the possibility of the disease having been rinderpest. Although care must be exercised in the interpretation of early anecdotal records, we can assume tentatively that many of the more devastating and widespread plagues *were* due to rinderpest, and this becomes true especially in the 5th century onwards after the irruption of the Huns into western Europe.

Particularly in mediaeval times, stock deaths, referred to as plagues, are often associated with famine years. Stock died regularly, it seems, from exposure and starvation during bad winters, although deaths were put down to disease or poisoning (or, more likely, witchcraft).[1] There was probably only limited storage of fodder for the winter, if any, and in a prolonged winter, the animals would suffer. Likewise, without water channeled from place to place, animals would suffer in drought years. Rinderpest and other contact diseases facilitated by conditions which weaken the hosts, by close herding and by crowding at waterholes in times of drought, could explain the frequent references to simultaneous plague and famine. Frequently also, cattle plagues are mentioned after very wet seasons. This could explain the often concurrent human mortality, for the very wet seasons would have caused epidemics of malaria, quite unrelated to the cattle disease. But in Britain, only between 1307 and 1327, when there was a period of excessively wet summers, is there any correlation with cattle murrains, four being reported in Britain and two in Ireland, and the death of cattle and sheep was

[1]It has been suggested that another cause of epidemics and epizootics and associated accusations of witchcraft was ergot *Claviceps purpurea* and other mycotoxins (Matossian, 1989), which develop in cold wet weather at temperatures between 17.4 and 18.9°C. A "nervous form" of cattle plague has never been proved to be due to rinderpest and may be ergot poisoning.

considered to have been a national calamity. Subsequent reports are not coincident with wet cycles, and for the "barren years" at the end of the 17th century, we have no reports of murrains in Britain. In the wet period of 1762–73, there was one outbreak of anthrax and an outbreak of introduced rinderpest in 1770. The introduction in 1877 was again during a wet cycle which had begun in 1875.

A number of references from Greek and Roman times involve mortality of both men and animals and have unwarrantably been considered by most authors to be anthrax because of the alleged associated human mortality.

We have it on good authority from the Russo-Turkish campaign of 1828–9 how one disease could lead to another (Chesney, 1854). The Turkish peasants, whose supplies were taken by the Russians, were reduced to such want that they died in great numbers, and at the same time, the cattle died of rinderpest probably contracted from cattle brought from Bessarabia for the Russian troops. The roads were covered with rotting carcasses, and the lack of hygiene among the Russian soldiers gave rise to typhoid, which created further great human mortality, reducing the Russian army in the Balkans to 12,000 men. Paulet (1775) notes that the years 1761–2 were fatal for herds in Europe, listing an epizootic among horses in Britain, two diseases among sheep in France, and anthrax among cattle in Lower Normandy; this is illustrative of the fact that there were many, often fatal, diseases acting concurrently and among which, rinderpest, from time to time, came to the fore. Between 400 BC and AD 1514, some 20 alleged epizootics have been recorded, at least four of which can be attributed with some certainty to rinderpest (Table 1).

Table 1. Cattle epizootics and significant events possibly related to rinderpest 431 BC to AD 1514.

Year	Place	Epizootic	Events Which Could Spread Infection
493 BC	Rome	Cattle distemper	
433	Rome	Plague struck men and cattle	
175	Rome	Epizootic among cattle	
90	Ireland	"Morreen of cows"	
88	Inner Asia	Epizootic destroys oxen	
47–45			Caesar's campaigns in Asia, Egypt, Africa
43	Italy	Plague killed "all the race of cattle"	
AD 325	Rome	Cattle epizootic	
370			Huns invade eastern Europe
376–86	Europe	Cattle plague all over Europe	
409	Rome		Sack of Rome by Huns
449			Invasion of Britain by Saxons
450	Northern China	Cattle epizootic	
451	Gaul		Huns invade Gaul
463	Eastern Europe		Uighur nomads move west to Pontic Steppe fleeing drought
546			Altai Turks conquer JuanJuan (Avars) who flee to Hungary
570	France/Italy	Nearly all the cattle destroyed	
591	Gaul	Fatal disease domestic/wild animals	
694	England	Great murrain of cattle	
695	Ireland	Murrain spreads from England	
696	Scotland	Murrain spreads to Scotland	
711	Spain		Muslim invasion
774	Italy		Lombards occupy northern Italy

Table 1. *Continued*

Year	Place	Epizootic	Events Which Could Spread Infection
751	Italy		Charlemagne conquers northern Italy
791	Hungary		Charlemagne destroys JuanJuan
793	Europe		Viking raids begin
800	Elbe & Weser	Pestilence of oxen accompanies Charlemagne	Charlemagne conquers Scandinavians
801	Italy/France	Pestilence of cattle	
802–3			Turks occupy Eurasian steppes
809–10	Europe	Plague spreads from east to west	
810	England/ France	Cattle plague	
820	Hungary/ Europe	Cattle plague re-erupts	
850	Western Europe	Cattle plague	
866	England		Danes invade East Anglia
869	England	A pest among cattle	
870	France	Serious cattle losses	
878	Germany	A cattle plague	
885	England		Danes from Belgium attack Rochester
886	France	A cattle plague	
892	England		Danes invade from Boulogne
	Germany		Magyar invasion
895–8	England	A cattle plague	
940–43	Europe	Widespread disease among cattle	
955	Germany		Magyars defeated at Lechfeld
960	Italy	Cattle disease enzootic	
980–82	England		Danish invasions
981	Ireland	Cattle mortality	
987	Ireland/ England	Mortality among all cattle	
1041	England	Cattle deaths	
1043	Saxony	Cattle plague	
1048	England	Cattle deaths	
1075	England		
1086	England	"A great animal plague"	
1096	Europe		First Crusade begins
1099			First Crusade takes Jerusalem
1103	England	Fatal disease of animals	
1111	England	Fatal disease of animals	
1131	England	Disease of swine and cattle	
1133	Ireland	Disease of swine and cattle	
1147			Second Crusade departs via Hungary
1148			Second Crusade returns in July
1149	Germany	Cattle plague	
1170–72	Germany	Cattle plague	
1187			Saladin occupies Latin states
1189			Third Crusade departs via Hungary
1191–2			Third Crusade returns
1202			Fourth Crusade departs
1204			Constantinople stormed
1219			Ghengis Khan moves west
1222	Germany	Cattle plague	Mongol invasion of Russia

Table 1. *Continued*

Year	Place	Epizootic	Events Which Could Spread Infection
1223	Europe	Cattle plague begins in east	
1230	Europe		Mongol invasion
1240	Europe	Cattle plague	Mongols attack Kiev
1241			Mongols cross Danube
1249	Europe	Cattle plague	
1258			Poland invaded
1278	Austria		Battle of Marchfeld
1299	Europe	Cattle plague	
1308	Ireland	Great murrain	
1315	Britain/Europe	Cattle murrain	Battle of Morgarten (Switzerland)
1319	Britain	Widespread murrain	
1321	Ireland/ Britain	Great losses everywhere	
1325–7	England	Devastating livestock epizootic	
1337	France/ England		Hundred Years War begins
1344			European forces attack Smyrna
1346	England	Cattle murrain	
	France		English defeat French at Crécy
1348–9	England	Cattle "perished in thousands"	
1349	Dalmatia	A "ferocious epizooty"	
1363	England	A great murrain	
1369	England	A "marvellous murrain upon cattle"	
1381			Mongols attack Russia
1382	England	Cattle murrain	
1385–7	Lombardy/ Dalmatia	Cattle plague	
1396	Europe		Invasion of Balkans
1407	Ireland	Great murrain	Slavs defeat Germans at Tanneberg
1415	France		English defeat French at Agincourt
1426	England		
1444	Europe		Invasion of Balkans
1473	Ireland	Great destruction of cows	
1480	Europe	Cattle deaths	Mongols approach Moscow
1481	Germany/ Switzerland	Lose a third of cattle	
1491	Ireland	Cattle epizootic	
	Poland	Great cattle losses	
1492	Spain		Spanish begin conquest of North African coast
1494	Spain	Mortality among wild animals	
1496	Ireland	Great destruction of cattle	
1499	Germany	A "dreadful murrain"	
1514	England	Cattle deaths	
	Italy	A cattle contagion	

2. PLAGUE IN ANCIENT EGYPT

Rinderpest is first suggested in the Ancient Egyptian Kahun Petrie papyrus dating from 3000 BC. One, Maspero, gives the symptoms in a bull of a disease which caused labored breathing, running eyes, inflamed gums, and swollen neck, likely to be fatal (Lepissier, 1971). The treatment appeared designed to lower the animal's body temperature, for if the animal went to water, it was to be immersed in it; if it avoided water, it was recommended that slices of cucumber be applied to its body. If this disease was rinderpest, it was enzootic in Egypt for almost 5,000 years to the turn of the 20th century.[2]

A number of indeterminate plagues affecting animals and men are reported in Europe from about 2000 BC, perhaps the earliest being from Ireland *c.*2048 BC. We next read of a plague *c.*1300 BC in Exodus (9: 3) the "Fifth Plague of Egypt" in the time of Pharoah IV:

> Behold, the hand of the Lord is upon thy cattle which is in the field, upon the horses, upon the asses, upon the camels, upon the oxen, and upon the sheep: there shall be a very grievous murrain.
>
> And the Lord shall sever between the cattle of Israel and the cattle of Egypt: and there shall nothing die of all that is the children's of Israel…
>
> …and all the cattle of Egypt died: but of the cattle of the children of Israel died not one.

It goes on to indicate that infection was spread throughout Egypt which "shall be a boil breaking forth with blains upon man, and upon beast, throughout all the land of Egypt."

3. RINDERPEST OR ANTHRAX?

The quotation from Exodus presents a recurrent theme concerning cattle plagues until rinderpest was positively identified in the 19th century, that many of the earlier plagues were anthrax, affecting man and other species as well as cattle. Here, the presence of boils is diagnostic of this disease. However, it seems more likely that there were both rinderpest and anthrax outbreaks (as well as of other diseases), and these were confused with one another. Although the presence of boils is characteristic of anthrax, anthrax is generally localized in occurrence and requires alkaline soil or water for survival. So, we have here elements which suggest rinderpest, and others suggestive of anthrax, both diseases no doubt occurring at different times. In the 19th century, Russia still witnessed devastating outbreaks of anthrax, affecting even fish, and even persons cutting hay contracted it. Paulet (1775) noted that of 92 continental epizootics between 810 and 1,316 of which there were records, 21 involved man as well as animals; and Buniva (1808) observed that of 20 epizootics reported in Italy and Sicily, eight involved man as well as animals. These figures at least show that many of the outbreaks were probably not anthrax. As stated above, the human deaths may have been quite unrelated. Humans could be affected by epidemics of malaria, cholera, typhoid fever, diptheria, and other fatal affections, regardless of what their cattle were suffering.

Paulet was well aware that there were many different diseases affecting animals and attempted a classification, identifying 43 different infections. Rinderpest he placed under "*Phlogoso-gangreneuses*" and anthrax in a separate group of "*Carbunculaires*." But for the disease described by Fracastoro, he classed this under "*Eruptives, exanthematiques*" as the "*Maladie eruptives des boeufs*," failing to appreciate the secondary effects that rinderpest may cause.

[2]Schwabe (1978) calls it malignant catarrhal fever, perhaps following an alternative suggestion of Smith (1931).

Youatt, in his seminal work on cattle which appeared in 1834, lumped several epizootic diseases together under the heading of "The Malignant Epidemic Murrain." He clearly confuses the rinderpest outbreak in Britain of 1745–58 with an outbreak of anthrax in France in 1757 for: "It spread from cattle to horses. The poor ass is said, for the first time, to have fallen a victim to it, and these animals perished in great numbers. The stags in the neighboring forests did not escape; and many flocks of sheep, over which these epidemics usually pass harmlessly, were swept away."

Vicq d'Azyr (1776) often confused anthrax with rinderpest (as well as FMD, but he claimed to be able to distinguish it from blackquarter). When an epizootic broke out in Normandy in the winter of 1775, appearing first in dogs which were believed to have transmitted it to cattle and killing pigs and many chickens, also affecting cats which did not die, he applied the same measures he had proposed for controlling rinderpest, ringing the affected area with troops and slaughtering all sick animals, paying owners one-third of the value of the slaughtered animal. Frequent outbreaks from the mid-18th century on of an epizootic disease in Sweden also appear to have been anthrax, as is clear from the story reported by Dr. Haartman, perhaps apocryphal, of the servant girl. Refusing to place some medicament into the mouth of a sick cow, her mistress did it for her and then playfully seized the breast of the servant girl with the hand that she had inserted into the animal's mouth. The girl was quickly overcome by a fever, her bosom swelled up becoming covered with boils, and she died.

Bourgignon (1865) also confused the two diseases when referring to mid-18th-century epizootics in Europe, considering in cases where man was fatally affected that the contagion "possessed venomous properties of an exceptional character."

Anthrax epizootics were characterized by long hot spells of weather accompanied by a "heavy atmosphere," that is, humid; and domestic animals which remained out at night were more likely to contract it than those confined in stables. Like rinderpest, some believed in its spontaneous generation, created by miasmata arising from bogs and marshes. It was less often found in sheep because they were pastured on drier ground, the disease generally making its appearance in wooded and boggy districts where numerous horses and cattle were kept.

4. FROM ROME TO THE HUN INVASIONS

After the Kahun Petrie papyrus, the first recognizable description of rinderpest is found in a Tamil treatise on the treatment of bovine diseases from the southern Deccan, written on palm leaves and estimated to date from the 1st millenium BC (Leclainche, 1955).

The next perhaps was about 493 BC, when Dionysius recorded in Rome: "A kind of pestilence also visited the population and destroyed great numbers of cattle; however not many persons died of it, the mischief going no further than sickness" (Cary, 1986). Livy, referring to a plague in Rome in 433 BC, mentions an epidemic which broke out that year struck men and cattle without distinction. Perhaps anthrax, or the result among men of eating rotting flesh, insanitary conditions created by the rotting bodies of cattle encouraging outbreaks of cholera, typhus, rats and bubonic plague; or malaria. During a drought *c.*429 BC, many cattle died of thirst, others of a mange, "and their diseases were by contact communicated to mankind." Perhaps sarcoptic mange or a reference to smallpox in man and the pustular eruptions in cattle sometimes accompanying rinderpest. In the summer of 399 BC, which followed an excessively cold winter, there was another great epizootic "noxious and baleful to all living creatures" suggestive of anthrax, which first descended on the fields and then spread to the City. In 175 BC, an epizootic began among cattle spreading to man the following year: "Those who were assailed by it did not easily survive the seventh day; those who had survived that length of time suffered from a lingering disease, usually quartan... The corpses, untouched by dogs and vultures, were consumed by decay... neither in this nor in the previous year, in spite of the great mortality of cattle and men, was a vulture seen anywhere" (Livy, *c.*27 BC). The disease attacking man appears to have been different to that affecting the cattle, more symptomatic of malaria. It was probably Mago who suggested keeping animals in small, separate groups, as a means of avoiding large-scale destruction in the event of major plagues. A Carthaginian author, his work on agriculture was written in Punic in 28 volumes. Captured by the Romans at the destruction of Carthage (146 BC), it was taken to Rome and translated by order of the Senate, being regarded as an authoritative work of reference for Roman agriculture. Drawn upon by subsequent Roman authors, Varro for example (54 BC), stated for many directions concerning the health of cattle, he made his herdsman copy out from Mago's books. Mago's text unfortunately has not survived.

Virgil, in the Georgics (31 BC), refers to plagues in Italy probably occurring in 43 BC, which "delivered over to death all the race of cattle." The lines:

> Flocks, herds, are gone: o'er yon deserted plains,
> On those lone steeps wide desolation reigns.
> The tainted air a dire disorder bred,
> Which raged, and with the heat of autumn spread.
> Wild beasts and cattle met an equal death
> Each pool, each pasture, felt the poisonous breath.

are certainly a good description of a rinderpest plague breaking out at the end of the dry season, although later allusions to men dying from handling poisonous fleeces suggest anthrax. But Virgil, like other authors up to the 18th century, is probably mixing different diseases together.

The anthrax bacterium is destroyed by putrefaction so that Virgil's reference:

> Aghast men earth on putrid bodies spread,
> And pile in pits the dead upon the dead.

would have been an effective means of limiting it.

In Ireland during Breasal's reign (*c.*90 BC) we read: "In his time there was such a morreen of cows in the land, as there were no more then left alive but one bull and one heifer in the whole kingdom," an event still commemorated in the name of Glensawisk, or Glen of the Heifer, in County Kerry (Mageoghagan, 1896).

What appears to have been rinderpest was reported throughout Europe from the beginning of the Christian era. Columella mentioned cattle plagues in his book "Husbandry" written about AD 55, recommending that diseased animals should be kept apart from healthy ones: "...when they are removed to a great distance, they must be brought into those places wherein no other cattle are fed, lest, by their coming, they bring the plague also among them" (Columella, 1745).

An apparently severe outbreak of rinderpest may have taken place in Rome in AD 325. By the time of Vegetius, in the latter half of the 5th century, epizootic diseases of animals fall under the general term of *malleus* or hammer, something which strikes, a word replaced in middle English by murrain, from the Latin *morior*, to die. Vegetius, last of the Roman encyclopaedists, who wrote a veterinary compilation *Artis Veterinariae Sive Mulomedicinae* (*On the Veterinary Art or Mule Doctoring*), which appears to have been a translation from Greek works of about AD 400 with much from the Roman Columella, refers to a disease of cattle which may have been rinderpest, or at least rinderpest was probably included amongst others such as FMD and anthrax. He wrote: "Because the animals infect the grass when grazing, the water supplies when drinking, the closed sheds, although healthy, the other cattle perish because of the air breathed out by the sick animals." Also:

> All these diseases are very contagious, and if one animal be seized by them they pass immediately to all; and so they bring destruction either upon whole herds or upon all those that are fully domesticated and trained to labour. Therefore it is that the animals which have been attacked must, with all diligence and care, be separated from the herd, put apart by themselves, and sent to those places where no animal is pastured, lest by their contagion they endanger all the rest; and the negligence of the owner be imputed (as is usually done by fools) to the divine displeasure. (Vegetius, 1528)

Nothing but deep burial of the carcass was permitted.[3]

We shall see that in spite of Vegetius's scepticism, the idea of divine displeasure was to persist until the very end of the 19th century. The officially backed offering of prayers against rinderpest in Britain in 1865 showed people's comprehension still wanting after 1,300 years.

5. AFTER THE HUN

Around the year 370, the Huns invaded southeastern Europe, driving out the Visigoths from present-day Romania, who fled across the lower Danube into Macedonia. But having settled in the eastern Roman Empire, the Visigoths then fell out with the Romans creating insurrections throughout the region. In the year 376, famine ensued, perhaps because no one could farm the land, but 374 was recorded as a drought year in England, and the drought may have been of widespread European occurrence. The famine was followed by plague "equally in ox and man, as well as in sheep" throughout the whole of Europe. Sanctus Severus, in his song *Carmen de Bovilla Peste*, asserted that in 376, the plague swept away all the cattle in Europe, "and all that work has produced, the length of the ages of life, was lost in two days, so quick is the course of the misfortune." His book contained an appendix on an epizootic of horses *De Equorum Epidemia*, thus presumably distinguishing between the disease affecting cattle and that of horses, the latter probably anthrax. Cardinal Caesar Baronius, an Italian ecclesiastical historian, writing in 1588–1607, probably quoting Sanctus Severus, stated that in the year 376, cattle died of a plague all over Europe: "... it certainly occupied the whole of Europe" (Baronius, 1739).

Coming from the east, for 10 years the plague ravaged Belgium, Flanders, Pannonia, Illyria, and the Roman countries, travelling from Hungary, through Austria, to Dalmatia, thence by Brabant to the Low Countries, Flanders, Picardy, and the other provinces of France. Baronius states that it was "a curse of animals," and although the death of men suggests anthrax, the bulk of the circumstantial evidence, its deadliness, and its progression, the movement of the Huns southward with their steppe cattle and an accompanying drought suggest rinderpest. People may have died from famine and from eating putrid meat. For the first time, mention is made of Hungary as the birthplace of plagues, a country which, writes Fleming (1871), was for centuries afterwards to bear this unenviable reputation.

In 570, another epizooty spread through France and Italy, destroying nearly all cattle. Saint Gregory (540–94), Gallo-Roman bishop of Tours, in his history of the Franks in

[3]Hussel (1960) states that Aëtius of Amida, the 6th-century Byzantine physician, in his compilation of earlier works, describes both rabies and rinderpest. But all he seems to be saying (Cornarius, 1549) quoting Rufus of Ephesus (1st–2nd century AD) is that human plague is caused by putrid miasmas as in wars, emanating from many things such as swamps, pools, abysses, affecting one if one does not go about in the right sort of air. Given the right conditions, it affects animals also. He raises the point that if the cause is in the air, the birds would be the first to die, if due to exhalations from the ground, then living quadrupeds would be the first affected; a point enlarged upon by Joubert (1567).

6th-century France, gives a contemporary description of the plague in Gaul in 591: "There was a great drought which kept back the pasture; as a result a serious disease prevailed among the flocks and herds which left few to transmit their kinds ... This pest raged not only among domestic animals but even among the wild beasts. For throughout the forest glades a multitude of stags and other beasts were found dead in pathless places" (Dalton, 1927). He goes on to say that the hay was ruined by rain and flood, implying that by the autumn, the drought had broken. Whereas bubonic plague raged in southern France in 590–91, Gregory does not imply that this was the same disease which attacked animals as some authors suggest.

A century later, in 694, the Irish *Annals of Clonmacnoise* record that there was "A great morren of cattle throughout all England," spreading to Ireland the following year and Scotland the next, where it raged for 3 years, causing people in Ireland and Scotland to commit cannibalism through want of food. This plague seemed to continue until about 707. Another cattle plague affected Ireland from around 770 until 778, but this does not appear to have been rinderpest (Mageoghagan, 1896).

The quantity of stock in Britain had apparently always been considerable, for it appeared striking to the Romans in 55 BC. Rearing cattle and hunting, to provide skins for clothing and milk and flesh for food, were the main occupations of the peoples in the interior. Later, cattle and hides were exported to Rome. One writer at the end of the 3rd century referred to the innumerable multitudes of domestic animals, "some with their udders distended with milk," although cattle were used mostly for traction. Beef as a food was a rarity, and until the Norman conquest, most people's meat probably came from pigs, sheep, and wild animals.

In Europe, as Charlemagne conquered the Scandinavians on the Elbe and Weser *c.*800, "So great was the pestilence of oxen in this expedition that scarcely in the whole army did one remain, but all perished; and not only there, but a plague among animals, causing a dreadful mortality, broke out in all the provinces conquered by the Emperor" (Fleming, 1871), following in Italy immediately after his crowning in Rome as Emperor. This clearly seems to have been rinderpest despite the fact that it was accompanied by many human deaths as well; but as Paulet (1775) remarked, "diseases seem inseparable from large mobs and from emigrations." Spreading to France in 801, it was possibly introduced with plundered cattle from Benevento in Italy. In Leyden the plague was ascribed to Grimald, duke of Benevento, allegedly sending out people with three wagon loads of magic powders to spread in the plains, mountains, meadows, and springs. Reported by Archbishop Agobardo of Lugdun (814–40), many people were seized and slain in retaliation, but most were tied to planks and thrown into rivers to drown (Kurze, 1895). In 810, it is recorded that in one night, 50 pairs of oxen belonging to a single abbot died of a sudden attack of plague, causing Charlemagne to prudently abandon his plan to attack the Gauls and Franks (Thorpe, 1969).

We first hear of cattle disease on the continent of Europe with clearer indications of rinderpest in 809–10, when a great plague spread from east to west from the Asian shores of the Black Sea, appearing in Hungary and Illyria and spreading rapidly throughout Germany, Austria, and Flanders; it was probably spread by Charlemagne with the movements of his army for, in 809, it left scarcely an animal alive in his Commissariat. Having crossed the Rhine to confront the Huns, he was forced to withdraw for lack of supplies. With the disease ravaging the whole of Europe, women and children were obliged to haul carts alongside horses and donkeys, and cultivation of fields was abandoned owing to a lack of work oxen. Before subsiding on the Continent, it reached Britain, probably with cattle sent to London from the Netherlands, where it created the "greatest mortality among horned cattle in Britain that is on record" (Jones, 1955). One account states that there was a mortality of cattle throughout Wales (Jones, 1971), while another refers to "all the island of Britain" (Jones, 1952). The Benedictine monk and chronicler Ranulf Higden recorded in his *Polychronicon* for the year 810: "A very great mortality amongst oxen laid waste nearly the whole of Europe, and more especially Britain" (Fleming, 1871).

Reported erupting again in Hungary in 820 and spreading once more through Western Europe, a pestilence of man and beast was also recorded among almost the whole kingdom of the Franks following excessive rains, perhaps an indication of famine or anthrax. Thirty years later, another outbreak ravaged France, Germany, and Spain, many provinces in France being reported as almost entirely cleared of their cattle. Frequent reports of cattle mortality followed, often in Ireland. Many of these were probably due to starvation from lack of winter feed, as well as anthrax and other ills, possibly including rinderpest. According to Henry of Huntingdon, in the winter of 869, there was a great famine, a mortality among men "and a pest among cattle" (Henry, *c.*1154). In 878, there was a plague in Germany especially near the Rhine. In 885, the Viking army which had been in Belgium, attacked Rochester and was beaten, leaving behind its horses. The following year, there was a plague in France and, in 895, great mortality among both cattle and men in England for some 3 years, disease having possibly been brought into England with the Danes from Boulogne in 892. We know that they brought horses with them from the record of their invasion in 885, and they may have brought cattle also as provisions both then and on subsequent occasions. In 940, we hear of another widespread epizooty among cattle in France and Germany which was probably rinderpest, continuing until 943 and extending to Italy, "but it did not last long in the latter country." Then, in 960, there is a reference to an enzootic cattle disease in Italy, allegedly introduced possibly by Romanus I, Emperor of the East; although the cause is assigned to the destruction of a marble ox head: "From that time, and up to the present, the breeds of cattle have not ceased to be destroyed in all parts of the earth wheresoever the Empire of the Romans extends."

6. MURRAINS IN IRELAND AND THE BRITISH ISLES

In 981, a murrain of cattle began in Ireland, called *moyle garie*, perhaps a mistake in chronology, for, in 987, there was once more "mortality upon the cattle in all the island of Britain" (Jones, 1952). Another version gives this latter as "in all Wales," while yet another reads: "A sudden great mortality which caused a slaughter of people and cattle in Saxonland and Britain and Ireland," here suggesting anthrax because of its involvement of human mortality. The Anglo-Saxon Chronicle records 986 as the year when "the great pestilence among cattle first came to England" (Garmonsway, 1953), "In this year the great murrain first occurred in England" (Whitelock, 1955). The chronicler Florence of Worcester, writing of an event which took place some 70 years before he was born, gives the year as 987 and states: "In this year two diseases unknown to the English in former ages, to wit, a fever among men, and a murrain among beasts, called in the English tongue 'scitta,' which in Latin may be said to signify a flux of the bowels, grievously troubled all England, and raged in every part of it beyond expression, causing great mortality among the inhabitants and the wholesale destruction of cattle" (Stevenson, 1853). Brompton (1652), another chronicler, recorded: "The 9th year of King Ethelred [II], some time ago two unknown fevers fell upon the people of England, certainly a fever of men, and a plague of animals which the English call Shitta, in true Latin called a flux of the bowels." The "flux of the bowels" is more symptomatic of cholera than anthrax, but this does not explain the deaths among cattle in which it is indicative of rinderpest, and for which people apparently had a name. There had been an invasion of Southampton by Danish pirate ships in 980 or 981; and Devon, Cornwall, and Dorset suffered raids up to 982, perhaps a source of introduced infection if the raiders carried livestock with them for food.

A cattle plague raged in Saxony in 1043, and in 1048, Florence (Stevenson, 1853) records that there was a mortality among men and cattle in many parts of England, but this may have been due to drought as, in some areas, there was much destruction of corn fields by "wild-fire." Brompton (1652) gives the year as 1046 and states "This year deaths of men and animals occupied many provinces of England." In 1075, the Council of London forbade hanging up dead beasts' bones to ward off cattle plague, "*quasi pro vitanda animalium peste*" (Kittredge, 1929).

In 1086, Florence recorded a murrain among beasts, and a great tempest followed the next year by many human deaths, first from fever and then famine, as well as fires consuming nearly all the principal cities of England; again suggesting drought conditions. The chronicler Knyghton (1652) recorded: "This year in England was a great animal plague, and bad weather, such that many died of fever and famine." These references indicate mortality from either summer drought or winter cold, but from the Norman conquest in 1066 onwards, there was frequent coming and going of troops between Britain and the Continent, and no doubt movements of pillaged stock which could have brought disease in their train. The chronicler Matthew Paris reported a fatal disease of animals in England in 1103. The year 1110 witnessed a very severe winter causing famine and great human mortality but also "a plague among animals, both beasts of the field and domestic animals, and a great destruction of birds" (Stevenson, 1853), given by Knyghton as occurring in 1118. In 1131, there was great mortality among swine and cattle in England which continued for some years, spreading to Ireland in 1133. The Anglo-Saxon Chronicle records: "In this same year, over the whole of England, murrain among cattle and pigs was worse than any within living memory; so that in a village where ten or twelve ploughs were in use, not a single one was left working; and a man who had owned two or three hundred pigs found himself with none. After that the hens died..." (Garmonsway, 1953). The high mortality indicates rinderpest, but the fact that hens succumbed suggests some other cause, although the deaths of the hens, which are highly resistant to anthrax, may not have been related. The plague continued in Ireland in 1134. The Annals of Kilroonan record the destruction of cattle in the previous year was of a magnitude "the like of which was not known since the former destruction of cows in the time of Flaithbheartach, son of Loingsech, and there were 432 years between them," which dates it, give or take a year or two, to the plague which began about 694.

That cattle might carry disease was known to the Anglo-Saxons, and we might say that they fixed quarantine at a period of 30 days, for the laws of King Ine of the West Saxons, *c.*690, stated that if anyone bought any cattle and then found any unsoundness within 30 days, he was to hand them back to the seller, unless the latter could swear he did not know them to be unsound when he sold them. This clearly was to accommodate diseases other than anthrax, which would have manifested itself more quickly.

As to the numbers of stock which died, there are few references to help us, but the chief occupation of the Anglo-Saxons was the rearing of cattle (Lappenberg, 1845). In 926, a treaty between the Saxon king Ethelstan and the North Welsh required the latter to pay an annual tribute of from 5,000 to 25,000 head of cattle, accounts differing as to the number. In 1066, just before the Norman Conquest, the Danes ransacked the country around Northampton, driving off vast numbers of cattle. In 1121, King Henry I accepted a gift of 10,000 cattle from the Welsh as a peace offering, and in 1211, King John likewise accepted a promise of 20,000 head; but the figures may be exaggerated, although one account circumspectly gives the first tribute as "about ten thousand head."

Between 1066 and the end of the 13th century, nine murrains are mentioned in chronicles. Three (1086, 1103, 1111) were described simply as "mortality of animals"; one (1131) as "mortality of domestic animals"; and in 1201, 1225, 1258, 1277, and 1283, the mortality largely affected sheep (Kershaw, 1973). The years 1086, 1103, 1201, and 1258 accompanied crop failures, so only 1111 and 1131 might be attributed to rinderpest.

7. THE MIDDLE AGES

Paulet (1775) states that in the 12th to 15th centuries, the chronicles are full of references to animal plagues but without the least description. In 1149, an epizootic appears to have raged in Germany and from 1170 to 1172, there was a plague of men and livestock throughout Europe; but there is nothing to definitely attribute these to rinderpest until the year 1223. From the beginning of the 13th century until 1241, the Mongol invasions from Asia through Russia to Silesia took place and probably carried rinderpest with them, especially in the years 1223, 1233, and 1238. In 1223, a great plague of cattle appears to have begun in the east and to have spread through Hungary and Austria, into Italy, Germany, France, and England: "It lasted three whole years, and the greater portion of the cattle died" (Königshofen, 1698). We can only suppose that for this and other plagues to have reached England, there must have been continual replenishment of stock from the continent. We know that William Marshal, Earl of Pembroke, sailed to Ireland in 1222 to engage an army with which he attacked Cardigan the next year. He was then called to see the Archbishop of Canterbury and sailed from Cilgerran to Ludlow, perhaps carrying infected Irish cattle with him for food. But there are no large-scale troop movements or invasions of Britain from the continent at this period. Walter of Henley, in his early English treatise on husbandry written sometime prior to 1272 (Lamond, 1890), was apparently unaware of epizootic diseases. He refers only to "murrain which falls sometimes on beasts" applying to all animals. He lists various means by which animals, particularly horses, may come to grief. Sheep deaths were the most common, and he referred to all sheep deaths as "murrain."

Erupting in Hungary in 1240, a cattle plague spread almost throughout Europe, with further eruptions reported for 1249 and 1299. The murrain of cows in Ireland in the Spring of 1286 was probably not due to disease, for it followed a cold winter and a scarcity of corn. In 1308, there was again a "great murrain" of cattle in Ireland. Perhaps this was the same epizootic referred to in 1316 as resulting from a long wet season, but the occurrence of another plague in Ireland about this time could suggest a widespread enzootic disease. During the period of 1315–17, there was persistent wet weather accompanied by extremely cold winters, causing great famine in northern Europe, including Britain, and murrains of epizootic proportions among oxen, cows, and sheep. Some hold this to be anthrax (Lyons, 1989), whereas others (Jordan, 1996; Kershaw, 1973) think it is rinderpest, as pigs were not initially affected and, according to Jordan (1996) "all the evidence" indicates that ruminants made up the affected species. The chronicler Holinshed records "by reason of the Murrain that fell among cattle, beefs and muttons were unreasonably priced." The wet weather having begun in Britain in 1314, the following year Edward II was already trying to impose price controls on livestock, suggestive that numbers had drastically declined, although perhaps mostly sheep at this stage. The deaths of sheep certainly seem attributable to the wet conditions rather than rinderpest, the sheep probably succumbing to liver fluke. Sheep flocks in Britain fell by as much as 25–70%. In northern France, one later chronicler added that the great dying of beasts was another calamity in addition to the harvest failures. The chroniclers summed it up as a "great pestilence of oxen and cattle" in Germany, France, and England, one claiming that it had raged over the whole of Christendom. A great murrain in 1319 which allegedly spread over the whole of Britain from Essex at Easter (probably imported from France for it is said to have affected the whole of Gaul at the same time, despite the fact dogs and crows which fed on the carcasses allegedly died) appears to have been rinderpest, chiefly affecting cattle. It reached the north of England and Scotland later in 1319, in the north-east first being heard of at the siege of Berwick in the late summer when nearly all the oxen being led to the siege suddenly died. By 1321, it was ravaging both Scotland and Ireland. Manorial records from every region of the country point to widespread destruction of cattle on an unprecedented scale during the accounting years 1319–21, and the hospital of St. James at Westminster claimed in 1319 that the numbers in the hospital could not be maintained on account of the "mortality of animals and the poverty of their resources" (Kershaw, 1973). Kershaw refers to the words of a poem written probably in 1321 as alluding to a cattle murrain:

> *A thusent winter ther bifore com never non so strong.*
> *To bind alle the mene men in mourning and in care,*
> *The orfe deide al bidene* [all the cattle died straightaway],
> *and maden the lond al bare,*
> *so faste,*
> *Com nevere wrecche into Engelond that made men more agaste.* (Wright, 1839)

But "a thousand winters there before came never none so strong" implies that deaths were due to winter starvation, although the writer simply may have been piling on the agony.

On three Huntingdonshire manors of Ramsey Abbey, at Broughton, 48 of 54 (89%) cattle died; at Upwood 45 of 47 (96%); at Broughton 56 of 64 (88%). Percentages we associate with rinderpest. In a letter to the king in September 1319, the abbot complained that a "sudden pestilence" had wiped out so many animals and was still so severe in the area that the Abbey had no longer the means to till its lands. Several of the manors recorded between 1318 and 1326 their smallest numbers of cattle and oxen from the mid-13th to the 15th centuries. The Crowland manors of Wellingborough in Northampstonshire, Drayton, Oakington, and Cottenham in Cambridgeshire also recorded lower numbers of oxen in 1321–2 than in any year since the first accounts of 1257–8. At Crawley, only four oxen remained at the end of the 1320–21 account, by far the lowest figure recorded between 1208 and 1449. The royal manor of Clipstone in Nottinghamshire lost 20% of its cattle herd in 1318–19, and almost 40% the following year. Of 186 head in 1320, only 64 remained, of which 32 were described as diseased (*morbosi*). More than 605 of the cattle at Sheen died in 1319–20, including 23 of 25 cows, and some 77% of a small herd at Teddington. On a Merton College manor at Cheddington, Buckinghamshire, 10 of 14 oxen died, and 6 of 9 cows. A few other animals were sold for fear of the murrain. At Thorncroft, Surrey, 7 of 15 oxen were lost by "sudden death," together with 4 of 13 cows, 4 of 9 calves and both bulls. At Cuxham, Oxfordshire, losses were 4 of 13 oxen, 5 of 9 cows, a bull, and 3 young bovines. The manor of Langwm, Monmouthshire, lost 9 of 33 oxen. On the Winchester estates in south and south west England, by April 1321, the total of 1,088 oxen had fallen to 500 in one year despite purchases of 111 in 1319–20 and 170 in 1320–21. Between 1319 and 1322, the oxen of Bolton Priory in North Yorkshire declined from 139 to 53, other cattle from 186 to 64, and from 80 or so to 27. The drop from 186 to 64 was explicitly recorded as being due to disease, although here, Scottish raids aided the depletion of stock (Kershaw, 1973). At the manor of Meopham in Kent, the cowman was dismissed after his herd was annihilated. Cattle herds on some Breckland manors declined in similar proportions. Near Ely, animal mortality appears to have been high judging from the amount of arable land under cultivation on one manor falling by 12–17% presumably due to a lack of work oxen. The flocks and herds of Canterbury Cathedral Priory had done well through the wet years but, in 1320, succumbed to the epizootic. Evidence from Wales also suggests a devastating cattle epizootic. In 1321, there came another plague in Ireland "the like of which was never seen before." Named the *moyle dawine*, it was still continuing in 1324, so this could have been rinderpest.

Fear of contamination was allegedly great in Britain according to Jordan (1996), although in Würzburg in 1316, it was reported that the poor were eating the carcasses of cattle which had died from disease. There is a record of a case in January 1316 in the English courts of a defendant grounding his defence in the expectation that the judges and jurors would view his refusal to skin an infected animal as justifiable. A not uncommon practice was to throw infected carcasses into deep abandoned wells. Jordan (1996) suggests that a rise in the prices of several types of farm implements in 1310–20 and 1320–30, commensurate with high grain prices, could be related to implements being destroyed for fear of contagion, wooden milk pails being a case in point. This is difficult to reconcile with what we know of an almost complete ignorance of contagion from at least the 17th into the 19th century.

The shortage of oxen following the 1320–21 plague obliged men to plough with horses, but this is often recorded only for those years, being replaced with oxen again as soon as possible. Kershaw (1973) points out how insignificant recorded cattle losses usually were in the 13th and 14th centuries, especially when contrasted with losses of sheep. This underlines the significance of the 1319–21 epizootic, which the low numbers recorded for the Ramsey Abbey and Crawley manors suggest to have been unparalleled in extent in the 13th to 15th centuries, contributing to the worst agrarian crisis since the Norman invasion. On the Earl of Lancaster's manors of Tutbury in Derbyshire, over 1,200 ha of land went out of use, as tenants could not be found to take it up because of a severe shortage of stock, caused by murrain and through recent political disturbances. It was found hard to find tenants for the bishop of Winchester's estates because "almost all the *averia* [cattle] of the area [were] dead." In the summer of 1322, pasture rents in at least 13 villages in Devon and Cornwall were lower than usual through lack of animals. Many other areas recorded difficulty in finding tenants for lack of animals. Almondbury near Huddersfield was but one example of many where pasture brought no income "because of lack of animals in those parts both through the common murrain and through the great depredation in the time of disturbance." Shortage of livestock was still noted in the Skipton area of the West Riding in 1325–6 (Kershaw, 1973).

In southeast England, at least in 1325 and 1326, there was both a great drought and a devastating livestock epizootic. In 1327, Christ Church Canterbury claimed to have recently lost 257 oxen, 511 cows and their issue, and 4,585 sheep "dead in the murrain."

In the early 14th century, probably beginning with the 1319 epizootic, the alien priories in Britain, drawing on their French experience, when murrain appeared appointed special officials termed *cadaveratores*, whose duty was to account for all animal deaths. Apparently, the only surviving record relates to an outbreak of disease at Combe in 1334 when Henricus Sewall and Henricus le Smyth, *cadaveratores*, reported on the deaths of a number of pigs and sheep, some of which had been devoured by dogs "which died of a common death near the side of the court and not elsewhere" (Morgan, 1946).

Repeated deaths took place over a period of 50 years from 1346 to 1411 in western Norfolk, as evinced by the Court Rolls of the manor of Heacham, in which reference is made to *Murina* or *Morina* in the presentments, becoming known as "the common murrain." Detailed accounts were maintained and all losses recorded on oath at the Manor Courts presumably to absolve the hersdsmen of blame. In 1372, *cadaveratores* first appear in the presentments and then are not mentioned again until 1377, from then on being the only persons making presentments. In 1360, there is a single reference to the *Veterinarius* to examine eight sheep dead of the murrain. From these data, Harrod (1866) postulated that a murrain had existed for 63 years, but the accounts only refer to occasional deaths: in all, a total of 1 bull, 92 bullocks, 43 cows, 111 calves, and 3,625 sheep (average 58/year) during the 63 years. Total stock numbers are given only for 1359 when there were 2 bulls, 18 bullocks, 33 cows, 7 calves and 595 sheep, of which 1 cow, 1 calf and 93 sheep died. And in 1395, there were 46 cattle, 8 calves, and 444 sheep, of which 2 cattle and 36 sheep died. Prior to the Black Death breaking out in London in November 1348, an above-average number of cattle deaths were recorded for 1346 and 1347, namely 10 and 9 animals, beginning with 1 ox, 3 bullocks, and 6 sheep in August 1346. The highest number of deaths recorded in the period was in 1405, when 31 cows, calves, and bullocks, died. Assuming the same stock numbers as in 1359, 52% died, which is not evidence of a rinderpest outbreak unless rinderpest was indeed enzootic. From time to time, other deaths were recorded: pigs, peacocks, swans, chickens, geese, and hives of bees, all listed under *Murina*, which can only be interpreted as "sickness."

But in the two years 1348–9, in other parts of England, another great plague allegedly attacked cattle, which perished by thousands. Herdsmen, panic-stricken, fled from their herds, fearing the disease was the Black Death and would be transmitted to themselves. Abandoned herds thus roamed the countryside, spreading the plague still further. Attempts to confine the cattle met with little response, owing to the belief that the plague would be communicated to man. Notwithstanding an abundant harvest, a great rise in the price of food followed, caution was thrown to the wind, and the remaining diseased cattle were slaughtered. Infected herds were allegedly separated from sound herds, while herdsmen who attended the former were not allowed to come into contact with the latter (Bernard, 1865). Knyghton (1652), however, who wrote up to 1394, refers to a plague of sheep, in one place more than 5,000 dying in one field and that the cattle died "in numbers beyond reckoning" not because of disease but because there was no one to look after them, the herdsmen having died of the Black Death. A horse previously valued at 40 shillings, could be had for half, a fat ox for 4 shillings, a cow for 12 pence, a heifer for 6 pence, and a fat sheep for 4 pence.

In 1349, a "ferocious" epizooty had broken out in Austrian Dalmatia, destroying nearly all of the animals, and in the same year, there was great cattle mortality in Devon and Somerset. An anonymous English chronicler in the *Annals rerum Anglicarum* records a great murrain of animals for 1363 in Britain, and John Hardynge's *Chronicle* states in 1369 that a third of men died "and likewise [there was] a marvellous Murrain upon cattle, so that the like had not been seen in many years before" (Hardynge, 1543). In 1382, the Abbess of the convent of Shaftesbury petitioned the king for help because of "the pestilences which have caused great mortality among their tenantry, and by the murrain of their cattle, which has destroyed a great number, and of great value, simultaneously and at divers times in all places...," perhaps again anthrax, as it is not mentioned elsewhere in Britain. In Europe sometime between 1351 and 1370, a terrible disease known as the Black Death, which killed men, horses, cattle, bears, wolves, foxes, sheep, goats, hares, and other animals, and had the hallmark of anthrax, was allegedly prevalent throughout Europe. Perhaps this disease was transmitted to Britain in 1363, or the chronicler may have been referring to the Continent. In 1385–7, a cattle plague is reported in Lombardy. Knyghton (1652) referred to an epizootic among deer in Britain in 1387.

There are few 15th-century references of epizootics; the accounts of the Bishop of Durham suggest one about 1426, but most come from Ireland. Ireland again suffered "a great murrain" in 1407 but also reported "foul and bad weather" that year. In 1450, there was also a great loss of cattle and many storms, followed in 1464 by an "epidemic cholera" in man and animals which appears to have been anthrax. In 1473, there was great destruction of cows in Ireland. From 1480 to 1482, there was famine in France, and women and children were obliged to draw the plough for want of draught cattle; many cattle deaths were recorded for Europe in 1480, and in 1481, Germany and Switzerland are reported to have lost a third of their cattle. There may have been another outbreak of cattle disease in Britain in 1480. In 1491, there was yet another cattle epizootic in Ireland and also great losses of cattle in Poland. A drought in Ireland the following year coincided with further cattle mortality. In 1494, there was a severe mortality among wild animals in Spain, and in 1496 once again, Ireland reported a great destruction of cows and other beasts and harsh weather. Prosecutions for witchcraft involving causing illness or death to animals in Europe during the 15th century suggest possible cattle epizootics at the end of the 14th century into the first decade of the 15th; in the period 1450–69; and again in 1475–99 (Figure 1 in Chapter 16).

8. THE 16TH TO 17TH CENTURIES

Beginning in 1499 with a "dreadful" murrain, in the following year there was great mortality among cattle in Germany, leading to a supposition that the pastures had been poisoned, and in the neighborhood of Meissen, some suspected "wicked knaves" were executed. That the mortality continued into 1505 suggests that it was probably rinderpest. A mortality among animals is reported for Britain in 1500, and at Magdalen College, Oxford, 607 sheep died.

An outbreak of rinderpest, "a mortal distemper," may have occurred in Britain in 1514. This was followed by a malignant murrain among cattle in 1517, but ravens and dogs which fed on the carcasses allegedly swelled up and died, and so this may have been anthrax. In 1523, an ardent agriculturalist, John Fitzherbert, produced *The Boke of Husbandrie*, in which, under cattle diseases, he dealt first with murrain:

> And if it fortune to fall murren among thy beasts, as God forbid, there be men enough can help them. And it cometh as a rankness of blood, and appeareth most commonly first in the head, for his head will swell, and his eye wax great, and run of water and froth at the mouth, and then he is past remedy, and will die mostly, and will never eat after he is sick. Then flay him and make a deep pit fast by, there as he dyeth, and cast him in, and cover him with earth, that no dogs may come to the carrion. For as many beasts, as feeleth the smell of that carrion, are likely to be infect, and take the skin, and have it to the tanners to sell, and bring it not home, for a peril that may fall. And it is commonly used, and cometh of a great charity, to take the bare head of the same beast, and put it upon a long pole, and set it in a hedge, fast bounden to a stake, by the highway side, that every man, that rideth or goeth that way, may see and know by that sign, that there is sickness of cattel in the township. And the husbands hold an opinion. That it shall the rather cease.[4] And when the beast is slain, there as the murren both appear between the flesh and the skin, it will ride up like a jelly and froth an inch deep or more. And this is the remedy for the murren. Take a small curtain cord, and bind it hard about the beast's neck, and that will cause the blood to come into the neck, and on either side of the neck there is a vein that a man may feel with his finger: and then take a blood iron, and set it straight upon the vein, and smite him blood on both sides, and let him bleed the mountnance of a pint or nigh it, and then take away the cord, and it will staunch bleeding. And thus serve all thy cattle, that be in that close or pasture, and there shall no more be sick by God's leave.

While it is not clear from his description whether he is referring to rinderpest or anthrax, the former is possible, for he also describes a disease termed "Crisen upon," that "no man can tell how, or whereof it cometh," that is, it was sudden. The symptoms were swelling of the head and especially running from the eyes, which close up, followed by death within an hour or two. The cause was a blister under the tongue which should be lanced. Only six other diseases were referred to: Lung sought (pleuropneumonia), symptom coughing, lasts

[4]In 19th-century Tyrol, the head of the first calf to die in a plague was placed on a pole. It was believed that the plague would then cease and break out in the direction the head was pointing. In Schleswig, the custom was to cut off the head of a live 2-year-old animal and tie it up in a window of the house with the eyes looking east, and no more would then die. In the Black Forest, it was the custom to hang up a calf's head in the house. In Albringwerde, a farmer cut off the head of a cow and hung it with the skin under the ridge of the house, claiming that since that day, there had been no more plague, the skull and skin still being present in the 1880s. Another custom was to cut off the head of the last remaining cow and place it in the loft. Persisting into the 19th century, the custom of storing the skull of an ox (or horse or sheep, depending on the plague) under the ridge of the house was widespread in northern Europe. Originally placed there in times of cattle plague, eventually other virtues of protection were attributed to it. In Aargau where the custom was to cut off a bull's head and place it in a box, an old statute forbade taking down such skulls for fear of alarming people; in Dottikon, for example, people believed the plague to be held in the skull by a spell (Jahn, 1884). In Wales at one time, a horse's head was hung in the stall to keep off cattle plague, allegedly a relic of ancient horse worship.

a long time, but few recover. Many cure it by placing an amount the size of a hen's egg of feiter-grass (perhaps setter grass *Helleborus foetidus*) bruised in a mortar, in a hole in the dewlap; Dewbolne or bloat; the Turn caused by a Hydatid cyst on the brain; Warrybrede or bots, the Foul or footrot, and the Gout.

A later writer, Leonard Mascall, in his *The First Booke of Cattell* published in 1587, clearly copying from Fitzherbert but also adding many other diseases and complaints of cattle, refers to displaying the hide on a pole in a hedgerow by the highway as a warning of murrain, "a common use of great charity." Again, from his descriptions, it is not clear whether murrain refers to anthrax or rinderpest, or both; but his reports of some men dying after flaying a carcass, and also the man who carried the skin to the tanner, the horse and the tanner himself, clearly indicate at least some cases of anthrax. He also gave the symptoms as swelling of the head, eyes staring and running, followed by frothing at the mouth, indicating that the animal would shortly die, adding that under the skin were bubbles like froth, some blackish, some blue, and some yellow. Under the causes of "pestilence" in cattle, other than chasing them, this was attributed to the dung of pigs and chickens in their stalls being eaten, especially did the dung of pigs breed a murrain.

In 1530, Fracastoro wrote of a disease in Italy in 1514, which led to much discussion in the 18th century (Fleming, 1871), for the symptoms he described appeared to be that of FMD and rinderpest also:

> I will mention also an unusual contagion which, in the year 1514, attacked cattle only, first observed in the district of Friuli; then gradually it was carried to the Euganeans, and thence to my own countryside [shores of Lake Garda]. At first the cattle abstained from food, for no manifest reason. But when the herdsmen looked into their mouths they saw a sort of roughness and small pustules on the palate and the whole mouth, and the infected animal had to be removed at once from the herd, or the whole herd became infected. Gradually this distemper descended to the shoulders and thence to the feet, and in cases where this change took place almost all the animals were cured, but when it did not, they nearly always died. (Fracastoro, 1546)

Thus, he is saying that those animals of which the feet were not affected died, which seems to indicate rinderpest. Elsewhere, he writes that the plague swept away not only "the wretched cattle, but also nearly the whole of the unhappy flocks of sheep…" (Fracastoro, 1530). Another source states that at this time, people in Venice and Padua were afflicted by an epidemic dysentery after eating the flesh of diseased cattle imported from Hungary. This was presumably flesh from cadavers, and the Hungarian connection suggests that the beasts had died of rinderpest. In 1514, the Council of Venice forbade the distribution of beef and veal, and of butter, milk, and cheese of any animal, under penalty of death as a consequence of the disease. Only the eating of mutton was allowed until the outbreak had entirely ceased. The archives of the Imperial Agricultural Society of Southern Russia mention that "cattle plague" appeared in Spain at this period, suggesting a Europe-wide outbreak, although there was no particular war in Europe at this time to which such a spread could be attributed. It was to be another 12 years before Suleiman, ruler of the Ottoman Empire, attacked and conquered Hungary. Large numbers of cattle were exported from Hungary at this time to the Venetian province, Austria, Moravia, Bohemia, Swabia, and Germany. In 1536, one man is reported to have offered 10,000 head for sale. The trade continued under the Turks, and in 1560, 7,418 head were ferried westwards across the Danube in one day (Bökönyi, 1974).[5]

[5]A record in the Augsburg archives dated 1526 notes: "Die hungarisch als die pesten Ochsen, so jetzo dieser Zeit zu gemeiner Stadt ham gebracht…dass zwischen polnisch und hungarisch Ochsen-fleisch ein grosser Unterschied sei…," which Bökönyi translates as "The Hungarian that is to say the best oxen, recently introduced to the city… that there is a great difference between Polish and Hungarian beef…" (Bökönyi, 1974), but which perhaps should have been translated as "The Hungarian that is to say the pest oxen…," already accused of carrying the plague at this time.

Rinderpest is alleged to have entered Prussia from the east and raged with severity near Magdeburg and in other districts in 1559, while another cattle plague is reported for 1562. In 1566, Franc de Bruges issued an Ordinance forbidding the loading of dogs with cattle skins to avoid spreading the contagious cattle disease.

Antonio Faccio recorded a plague of great severity confined to oxen in Italy in 1590 when another edict was issued in Venice, forbidding under pain of death the sale of beef, fresh cheese, butter, or milk (Ramazzini, 1711). The years 1595–7 were years of starvation over much of northern Europe usually attributed to harvest failure, but in 1598, rinderpest erupted over the whole of Germany. At the request of the Archduke Mathias, the College of Surgeons in Vienna produced a paper translated from Latin into German by Diomedes Cornarius, which concluded that the plague was due to astral influences, but the bad odour of corpses corrupted the air and thus spread the disease. The burning and burying of carcases were thus ordered, and throwing bodies into rivers prohibited; also, sick animals had to be separated from healthy. The plague spread to Upper Italy in 1599, where more than 13,000 cattle died, the origin allegedly being cattle imported from Hungary. Another account states that it began among cattle and goats "and by them [was] communicated to other animals," recognizing its contagious nature. Venice and Padua prohibited the import of cattle from Hungary and Dalmatia, whence they had received their supplies from time immemorial and as a result repeatedly suffered outbreaks of disease. In the Venetian States, it was prohibited under penalty of death to sell products of infected animals and ordered their destruction. An outbreak of human dysentery was thought to be related to the consumption of infected products. In Lyon, an Ordinance was issued on April 14, 1604, the health department ordering the burial of carcases 2 m deep covered in quicklime and prohibiting the slaughter of cattle which were brought in, without inspection by a master butcher in the presence of a commissioner of health. It also ordered the slaughter of stray dogs.

Toward 1609, the disease again invaded eastern and central Europe, first reported in Bohemia and the Palatinate, and in the following year, it reached Alsace, Saxony, Bavaria, and Holland. In 1616, it devastated the provinces of Padua, Treviso, Vicenza, and Udine, to such an extent that the use of beef and veal was again prohibited. Pope Paul V issued an edict prohibiting the slaughter of draught oxen which were suitable for ploughing, and calves were not slaughtered for some considerable time in order that the country's stock might be replenished. In 1617, there were widespread floods in Europe, and cattle died apparently from anthrax, people in the countryside who ate the meat of dead animals also dying.

In 1618, a fatal disease of cattle broke out in Spain, spreading to Malta, Sicily, Sardinia, and Italy, becoming enzootic for 22 years. The manner of its spread among the Mediterranean islands suggests that it was rinderpest imported to one island from another with infected stock, perhaps occasioned by the Thirty Years War of 1616–48. Cattle plague allegedly broke out again in Hungary in 1625, passing to Italy, introduced by Dalmatian cattle merchants, and raging there for some years, causing Venice in 1630 to change its source of supply. In about 1623, there was a cattle murrain in Scotland, and in 1630, a drought in the south of England. Reference to the fact that "the sickness" was rife in Walney (Westmorland) in 1631, and that the Flemings of Rydal bought no cattle at Ravenglass that year (Elliott, 1973), suggests a possible rinderpest outbreak in Britain. Perhaps it was the Italian outbreak which spread to the Jura Alps in 1640 and Saxony in 1643. In France, there is a report of persons being appointed in 1641 to examine military cattle supplies. Continuous continental epizootics between 1625 and 1645 led to a lack of work oxen for transport and ploughing.

Cardinal Gastaldi, writing of bubonic plague in Italy in 1656, noted that the outbreak was preceded by cattle deaths, another author adding sheep deaths. Hodges, writing first in Latin in 1672 referring to the Plague of London of 1665, wrote: "... the Year before the late Pestilential sickness, there was a great Mortality amongst the Cattel, from a very wet Autumn, whereby their Carcasses were sold amongst the ordinary People at a very mean Price; and a great deal of putrid Humors in all likelihood produced from thence: And this, in the Opinion of many, was the Source of our last Calamities; and many knowing Persons ascribed the Pestilence to this Origin..." (Hodges, 1720). Bates (1718) reported eyewitnesses asserting that there had been a great mortality among cattle in Britain in 1665, followed by human plague which was imputed to the putrefying carcasses. But Hodges went on to say that such food, "although very much corrupt, and liable therefore to excite Symptoms like to those in a Pestilence; yet they were not in Plenty enough to supply the whole Market; and therefore a Cause so private and particular, could not be supposed to extend to so universal an Effect." More importantly, an anonymous contributor wrote: "Upon comparing these different observations; I can hardly doubt, but that the forerunners of this distemper [the human plague] were already come into the Island, when there was a very great mortality amongst the cattle; from a very wet autumn, before the beginning infection mentioned at Westminster" (Anon, 1720a). There is no suggestion, therefore, that this was anthrax except that the damp conditions would predispose toward it; but in 1663, it is recorded that there was a great mortality of cattle and sheep due to the liver being eaten by small worms and, in some cases, the lungs, an apparent reference to fascioliasis or liver fluke, characteristic of wet conditions (Webster, 1799). Hodges goes on to refer to "...the frequent Mortalities amongst Cattle, which forego an Infection amongst Mankind..." Although this last statement could indicate frequent outbreaks of anthrax, clearly, eating rotten meat would cause sickness. Hodges implies that cattle plagues were of frequent occurrence in his time, but we lack such records. Muratori argued in 1714 that he did not think it probable that human plague was contracted from the cattle plague, as there were so many other examples showing that animal diseases were specific to them (Muratori, 1743).

Cattle had been a major import from Ireland from 1603 until the Cattle Act of 1666 prohibited further imports, southern England then depending principally upon Scottish cattle, Scotland being described as little more than a grazing field for England. Most cattle were driven south through Cumberland and then along major drove roads leading to the eastern and southern counties. In Westmorland, a cattle fair was held each September at Brough, the purchases then driven to Yorkshire and on to Norfolk for fattening, before finally going to the London market. Thus, with the cattle movement mostly one way, there were unlikely to be major epizootics, for even if infected cattle were imported from the Continent, they would not have been driven great distances inland. Famines in Cumberland and Westmorland in 1594–7 and 1623 are not correlated with any known cattle epizootics but were due to harvest failure caused by heavy summer rain.

Disquiet at the large numbers of cattle imported from Ireland, which glutted the markets and lowered the prices of English stock, was first voiced in the Commons in 1621, leading to a ban on imports in 1663, and again in 1667. Somerset accepted large numbers for rearing and complained that all the money went to Ireland, but it was not until 1656 that action was taken and an import levy of 2s. that had stood since 1627 was raised to 2/6d. No charge was imposed on Scottish cattle, so the Irish cattle passed through Scotland. In 1663, the levy was raised to 40s. for imports between July 1 and December 20 each year, and on Scottish imports from August 24 and December 20. This was seen as helping livestock breeders mainly in the north and west of England, rather than the graziers who were mainly in the south and east and would have benefited from a summer ban. The levy failed to halt the Irish imports, merely changing the shipping season, imports only falling from 61,000 to 57,000 between 1663 and 1665. In 1667, a total ban was imposed, but there was much smuggling. Even if the Irish cattle did not carry rinderpest at this time, they still carried other diseases, but these imports may explain Hodge's remark about frequent cattle plagues. More importantly, the English demand may have encouraged imports from the Continent leading to an outbreak of an unspecified cattle disease in some parts of England in 1673–4, which threatened a livestock shortage (suggesting rinderpest) and raised hopes of lifting the Irish ban; but there is no reference to any legal trade in the 17th century (e.g., Thirsk, 1985). Lapsing in May 1679, the Act was resumed in February 1681, between which dates there was a surge in imports, suggesting that demand exceeded supply and was not met by Scottish and Continental imports.

In 1679, a cattle disease raged in Austria, which was later likened to the rinderpest outbreak of 1713. In Germany, between 1665 and 1682, the Elector of Brandenburg ordered that stock owners be required to surrender all dead cattle to the public executioner and skinners; the public executioner using some of the flesh to bait wolf traps, while the skinners made leather buckets from the hides. But many owners ignored the order, and carcasses were either buried, thrown into rivers and streams, or fed to dogs. In 1689, the order was extended to the Duchies of Magdeburg and Pomerania and to the Principality of Halberstädt, evidence of concern at the continuing spread of a disease which was probably rinderpest. In July 1669 an instruction was issued at Stuttgart concerning cattle disease there and on the Danube, but referred to horses dying as well.

In 1682, a "murren" which spread from Italy to Switzerland, France, and Germany, known as the "angina maligna," the cattle dying in 24 hr, was reported by Dr. Wincler, chief physician of the Prince Palatine, in a letter to Dr. Frederick Slare, who communicated the letter to the Royal Society. Wincler wrote: "Those persons that carelessly managed their Cattle without a due respect to their own health, were themselves infected and dyed away like their Beasts." The following March, two travelers reported to Dr. Slare that the disease had passed through Germany to the borders of Poland, spreading 2 German miles every 24 hr: "it continually without intermission made progressive voyages and suffered no neighboring Parish to escape; so that it did not at the same time infect places at great distances [that is, it was a progressive advance and did not suddenly appear at different localities]. They [the travellers] added that cattle secured at rack and manger were equally infected with those in the Field" (Slare, 1683). Slare postulated that perhaps the infection was transmitted by some insect, the idea of infectious insects persisting into the first quarter of the next century. Whereas the first report seems to refer to anthrax, the second is more indicative of rinderpest. But referring to the cattle plague "now reigning" at several places, a pamphlet issued at Hildesheim in Germany in 1682 (perhaps written by Outhof) clearly describes "glossanthrax" or "tongue-cancer"; Actinobacillosis caused by *Actinobacillus lignieresi*, or "wooden tongue," the tongue rotting and falling out in 24 hr after the signs, spots on the tongue, first appeared, the animals dying in a few hours. A silver instrument was illustrated for scraping the pustules on the tongue (Figure 1), clearly derived from Pliny's suggestion (AD 24–79) of scraping the tongue with a silver spoon. This important little document, which preceded Lancisi's great work by more than 30 years, ranks among the first modern accounts of veterinary hygiene with the earliest modern instructions for preventing infectious disease transmission: that persons handling infected stock should wash their hands well between handling each animal and burn all used materials. For prevention of infection to themselves, it was recommended that citron, rue, juniper, or scorpion oil be spread on the hands, but poor people could use vinegar. It stressed that the hands should be washed before handling anything. If an animal died, it should be buried deeply with the skin, and other animals kept away from the burial place. Pigs and dogs should not be allowed near the stalls. It also provided recipes comprising mixtures of herbal ingredients which were to be widely recommended for the prevention and treatment of rinderpest for almost the next two centuries (Anon., 1682). Camper (1783) stated that the disease was a plague of cattle and horses, occasionally affecting pigs, and that Outhof, Vitet, and Haller, all called it "tongue-cancer." Canevese (1770) referred to its occurrence in cattle in Milan in 1682.

Figure 1. A silver instrument for scraping actinobacillosis pustules on the tongue.

Following two very wet years, Ramazzini observed a disease at Padua which raged from 1690 to 1691 characterized by blinding cattle and pustules on the neck, head, and legs. Entire herds died, and draught oxen collapsed in harness. Pigs died in herds "as if suffocated." The symptoms are indicative of rinderpest, and the death of bees and silkworms recorded at the same time must certainly indicate some concurrent unrelated fungal disease.

From his studies of Japanese records, Kishi (1982) concluded that rinderpest was responsible for destroying nearly all the cattle in Asia in the 17th century, but whereas it was widespread in Europe during this period, there were no major invasions of Asia or major Asian wars, and there is also no reason to suppose that its effects should have been more devastating in Asia than were witnessed in Europe.

Although the two diseases, anthrax and rinderpest, were to continue to be confused into the first part of the next century, with some exceptions, Ribbe (1814), for example, describing the differences both in etiology and in postmortem appearances and later, Bojanus (1820), rinderpest eventually came to be recognized as the dominant infection.

6

SAINT HERBOT SLEEPS[1]

The 18th-Century Ravages in Europe

1. INTRODUCTION

The outbreak seems to have lingered on in Switzerland, or repeatedly erupted, for in 1702 at Neuchâtel, the government guaranteed compensation for sick animals which were slaughtered (Jones, 1978). In the Prussian Dukedom of Trier, a Carrion-pit Ordinance was enacted on September 5, 1705, indicating the prevalence of some widespread epizootic.

In 1709, we have the first definitive reference to rinderpest, when one of the greatest epizootics on record, having ravaged the whole of Tartary, spread westwards through Russia, Podolia, Bessarabia, Poland, Moravia, Dalmatia, and Croatia. The time heralded the beginning of a period of great military activity in Russia, Peter the Great forging the Russian Empire, acquiring Estonia and other Baltic possessions, while losing to Turkey regions bordering the Black Sea. In the 1720s, he expanded east to the Kamchatka Peninsula and northwest to Finland and in 1722, defeated Persia on the borders of the Caspian Sea. Thus, during the first years of the 18th century, Russian military movements in Europe extended to all points of the compass, perpetuating a process first set in train by the Hun and followed by the Mongols. It was thus considered a matter of history that rinderpest followed the course of armies in times of war.

The wars of Louis XIV until his death in 1715 aided its propagation, and the armies of the British allies under Marlborough and Prince Eugène frequently carried it in their train, or received it in the capture of commissariat cattle from the French. Whenever Russian and Austrian troops penetrated westward in the 18th century, rinderpest spread with them.

In this period up until the end of the 18th century, rinderpest was never absent from some part of Europe on account of the almost continuous wars that were waged:

> The war of the Austrian Succession, on the death of Charles VI, in 1740, aided largely in maintaining and disseminating it as the Hungarians—who lent their assistance to Maria Theresa, the daughter of Charles—and the Austrians drew their supplies of cattle from Hungary. The movements of the hostile armies, accompanied as they were, by infected droves, diffused the contagion far and wide, and rendered futile all attempts to suppress it at the seat of war. And at a later period, Frederick the Great, contending repeatedly with the armies of Austria and Russia, carried the pestilence about in his own dominions, or was the means of importing new sources from the adjacent countries. (Fleming, 1871)

In the 8 years after the death of Charles VI, Europe's west and center lost 3 million cattle to the plague, and then Frederick the Great took the plague back with him to Prussia. At the end of the 19th century, it was to be spread once more throughout Europe by the Prussian army.

[1]St. Herbot, patron saint of cattle.

In 1711, it was introduced into Padua by Dalmatian traders with cattle from Hungary and spread over Italy, Germany, France, and Switzerland. Already having suffered the ravages of human plague, dread of plague-spreaders extended to rinderpest. The public authorities in Milan, for example, issued an edict warning against "a certain mixture, yellow in colour, scattered in the country enclosed in artfully cut leaves, which, following a careful examination, appeared to be a poisonous substance" (Mantovani and Zanetti, 1993). Between 1711 and 1714, it is estimated that another 1.5 million cattle died in western Europe, including England. In Silesia, 100,000 cattle are estimated to have perished, and in the Kingdom of Naples and the Papal States, 70,000 in 1 month. The Netherlands lost 300,000, and from November 1713 until November 1721, the Dutch placed a ban on the importation of foreign stock, losing a total of 200,000 cattle between 1713 and 1723. Its most important livestock husbandry area, the Province of Friesland, lost 66,000 head between December 1713 and February 1715. In Bavaria and Swabia, in the years 1713–16, 1729–35, and 1742–5, it completely devastated so many areas that it was rare to see a head of horned cattle. Reaching England in 1714, vigorous methods of control, consisting of the slaughter of all infected stock and prohibition of cattle movements, suppressed it by the end of the year.

Stähle (1981) used an analysis of mortality rates (the number dying of the disease/the number at risk; Halpin, 1975) to demonstrate that outbreaks in Bremen and Verden occurring in 1715–18, 1720–21, 1745–6, 1748–50, 1752–3, 1754–6, 1756–7, 1761–2, 1767–8, 1770–72, 1775–7, and 1779–81 were, by their similarity, most probably rinderpest. The "tongue-cancer" which, according to Stähle, appeared in 1732, affecting both cattle and horses with a lower mortality rate,[2] sick animals recovering quickly, was not heard of again until 1787 when it appeared in a decree of Bremen of February 28. But instructions issued at Hildesheim in 1716 (Anon., 1716a) included "cures" for "tongue-cancer," and an instruction issued in 1754 describes an anthrax outbreak ("pestilence of horses, cows and sheep") which originated in France, swept through the Rhine and Moselle in several days, and was now 2 hr distant (Anon., 1754b). Another pamphlet, issued at Hanover in 1767 (Anon., 1767), describes the "so-called fire-illness," another name for anthrax, and how to distinguish it from rinderpest: the first symptoms differed in that there was a swelling, particularly under the throat, also of the tongue and at the fore and hind legs. This suggests outbreaks of actinobacillosis and anthrax in 1716, 1754, and 1767, perhaps concurrent with rinderpest.

Gamgee considered that mortality among cattle throughout Europe had "perhaps never been equalled" (Gamgee, 1866). Germany alone lost an estimated 28 million head in the 18th century, and the whole of Europe about 200 million, although some estimates put it at only half a million.

2. THE GREAT 1709 EUROPEAN PANZOOTIC

The disease appeared first in 1709 on the banks of the Don and Volga. Highly contagious, it exhibited the effects of a peracute infection affecting large numbers of cattle and killing them suddenly and simultaneously, spreading quickly and uniformly. Extending to Lower Tartary and to Moscow, by the autumn of 1710, it had reached Kiev and Tcherkasi in the Ukraine. It was particularly virulent in Russia, especially in the provinces of Novgorod, Pleskow, and St. Petersburg, and the next year spread from Russia through Poland to Prussia, Brandenburg, and Silesia as far as Breslau, and to Moldavia and Wallachia. From Hungary,

[2]Horses were probably infected with the bacterial disease glanders *Pseudomonas mallei* and not *A. ligniersi* or "tongue cancer."

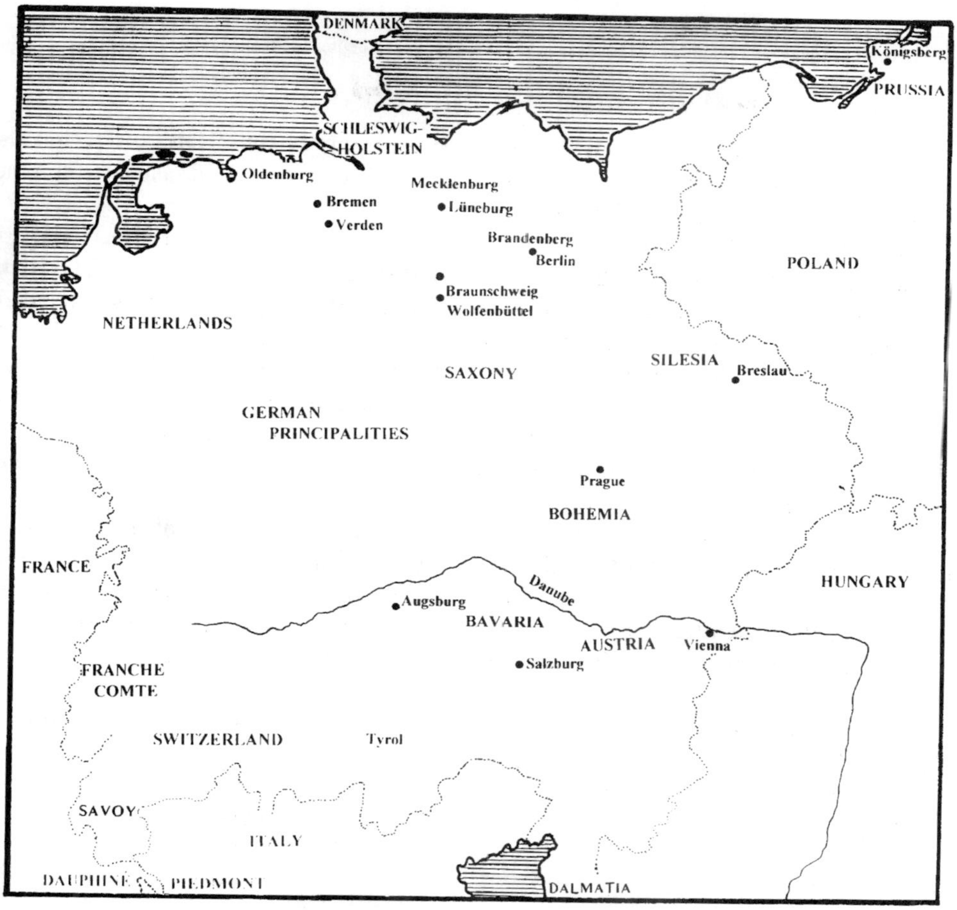

Figure 1. Mid-18th-century central Europe.

it spread into Austria, Bavaria, Suabia, and Switzerland, and from Dalmatia into Italy and the kingdom of Naples. In October, it reached Königsberg in Prussia, in 1711 Schleswig-Holstein, and at the end of the summer of 1711, Augsburg in Germany (Figure 1).

A Silesian, Kanold of Breslau, reported it prevalent in 1711–12 in Silesia, appearing in May with Polish oxen first at the village of Zundsfeld, spreading to Glokisch, Sackerau and others: "...they who had two or three herds of stock upon their lands could scarce keep a single animal, and large numbers of cattle were found lying dead upon the roads" (Kanold, 1721).

There were many suggestions as to its origin in the countries affected, but for centuries, the herds of Venice and Lombardy had suffered invasions of rinderpest through commerce in cattle across the Adriatic. Dalmatian cattle dealers who were engaged in carrying cattle to Italy from Hungary were often mentioned in connection with the disease (Gamgee, 1866). Gerbezii, writing from Bavaria, noted: "Every one agrees that the cause of this disease arises from infected cattle being driven from Hungary into Italy" (Gerbezii, 1713). Dr. Schroëckius was clear as to its origin when it manifested itself at Augsburg in 1711:

> ...it had proceeded by degrees from Hungary toward the Danube, attacked our territory [Germany] and produced great destruction to beasts.... And this was not caused by any foulness

> in the atmosphere, but by the contagion of oxen brought from the infected countries; and this was patent, because it first attacked those pastures adjoining the foreigners, and altogether spared those cattle to which no infected animals had approached, and which had been immediately separated from any in the same herd that were infected. (Schroëckius, 1713)

Kanold (1713) added that its spread was not through the movement of cattle alone, but also by people and dogs. Some villages fed their cattle in stalls, avoiding all contact with other communities, and thus escaped infection.

The outbreak was said to have reached northern Italy in 1711 with an infected herd in transit by sea from Hungary near the border with Turkey, through Dalmatia across the Adriatic Sea to Italy (Figure 2), allegedly spread by one straying ox, which was put into grounds in Agro Vincentino belonging to the Reverend Count Trajan Borromaeus, a canon of Padua. On August 17, 1711, a herdsman found a stray ox and brought it to the Count's estate where it was placed in a cowshed. It was soon reclaimed by a Dalmatian cattle drover, who stated that it had strayed from a herd belonging to the commisariat of the Austrian army. About a week later, the cattle in the Count's shed began to sicken and die, the disease spreading from thence rapidly through his herds into the suburbs of Padua, thence to Milan and Ferrara. In 1713, it appeared in the kingdom of Naples, spreading to Rome.[3] Wilkinson (1992) states, although probably apocryphal (or perhaps simply the defence of the drover) for an infected herd would be just as likely to spread infection as one animal straying from it, the story stressed the idea of contagion, the ability of the infection to spread rapidly from a single source. But Ramazzini stated in 1711 that it was well established and already published in printed form that one ox had

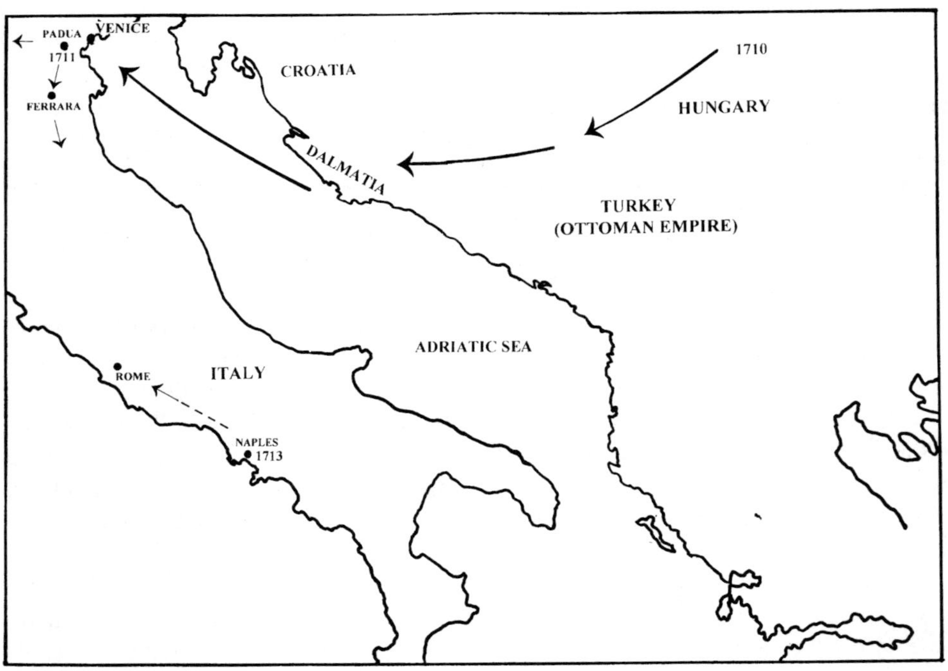

Figure 2. The alleged 1711 route of infection into Italy.

[3]An illustration of the plague appears in the *Insignia Degli Anziani de Commune Dal 1530 al 1796* in an allegorical painting depicting Felsina (Bologna) imploring saints to stop death striking the cattle at Bologna in 1713–14 and showing dead and dying grey oxen (Dunlop and Williams, 1996).

escaped from a herd of the traders, and Count Borromaeus had apparently tried to defend himself against the accusation that he had been instrumental in spreading the plague by harboring it, objecting that some herds were infected at a distance, while others closer at hand were unaffected. Lancisi replied that the sickness could be transmitted by cloths, clothes, other animals, or men; and Vallisnieri had reported that it could be carried by dogs. Instructions for treatment issued on September 2, 1713 gave remedies for pigs as well as cattle. On September, 24th Pope Clement XI sent a a personal letter to the President of the Highways, ordering him, as he was responsible for keeping roads clean, to ensure prompt and proper burial of carcasses, and giving him authority to publish proclamations and edicts, and award punishments, "granting you with a wide hand the whole jurisdiction, faculty, and necessary authority," broadly derogating powers against all persons.

Between 1713 and 1714, Lanzoni tells us, it raged to such an extent in the Duchy of Ferrara as to threaten the utter destruction of all cattle, and people feared that infection would be transmitted to them. Lanzoni considered it to be a malignant and contagious fever and quotes the "illustrious" Nigrisolus as saying that it was not only a malignant but also a burning fever (quoted by Anon., 1866d).

Passing to Sardinia and Piedmont, the latter region allegedly losing 70,000 head, the disease spread through Dauphiné into France where, in Lyon, the Ordinance of 1604 was renewed. Ordinances passed in April and September 1714 in France ordering the burial of infected carcasses in their skins had serious effects in some regions, causing fraud and the raising of prices of skins leading to public disorder. The disease then spread into Switzerland, the Tyrol, Germany, and Holland, from Holland, it was imported apparently into England, by then having invaded all principal countries of Europe except the State of Carniola which escaped through prohibiting the admission of cattle and other animals. It remained for 7 years in Italy, and Pope Clement XI, who between August 1713 and May 1714 lost 26,252 cattle, was so affected by the loss that he published stringent regulations for its suppression. However, first, on October 6, 1713, he led a big procession to the Basilica to ask God to allay the plague. His papal physician, Lancisi, stating that only careful prevention of infection would allay it, was charged with coordination and control of the disease, instituting measures which later commentators have praised as highly enlightened, not relying exclusively on prayers and appeals for divine mercy. Lancisi pointed out that only two herds had escaped at Rome "while in every other field or cowhouse the whole cattle were destroyed." One herd was that of Prince Pamphili at his villa at St. Pancras, the other that of Prince Borghese on his farm at St. Nicholas. Both were preserved by a most strict watch, refusing access to every person or animal and every kind of goods "with regard to which the least suspicion of contagion might be entertained" (Lancisi, 1715). Before 1711, the disease had been little understood, but now, with the Pope's intervention, all attention was focused upon it by the highest scientific and medical circles in Italy.

Kanold asserted that the outbreak of this plague in 1709 caused even greater terror in man than did the Black Death from which the populations of Europe were only just recovering after the great 17th-century pandemic, with people now starting to build up their flocks and herds:

> The first beginning or appearance of the disease in Europe, arose in that part of Tartary which lies on the borders of Asia; but whether it was originally generated in this corner of Europe, or whether it was brought thence from Asia, or yet whether perhaps it was an endemic disease…I am unable to decide with so much certainty…it seems that it appeared first in Astrakan, on the banks of the Don, or Tanais, and the Volga; from thence it spread to Kazan, in Lesser Tartary, and even to Moscow, when it caused great ravages in these governments among the horned and horse stock, but principally the former. From the symptoms of the disease itself, one learns at the time nothing,

> except that from the very first attack it was highly contagious; that it affected large numbers of cattle, and that very suddenly, killing them almost at once; and that it spread itself quickly in many neighbourhoods. So well marked were its characteristics, that its progress was easily noted. It extended itself steadily, but uniformly, over many kingdoms and lands during many years… (Kanold, 1721)

The reference to horses dying is confusing, and is repeated by Kanold with regard to 1710. But it may be simply that they contracted bacterial infections from the prevalence of rotting cattle carcasses. Certainly the geographic progress of the disease is indicative of rinderpest and not anthrax; but anthrax was also prevalent in the marshy areas of Russia and may have been spread contemporaneously with, and by the same means as, rinderpest.

People now began to look more closely at the symptoms, and Schroëckius noted that in affected animals, "the tongue was inflamed and covered with red vesicles" (Schroëckius, 1713). But the closest attention was given to it by Lancisi and Ramazzini. Ramazzini, for whom two eminent pathologists, Professors Molinetti and Viscardi, dissected a number of carcasses, provided a classical description of the disease, being the first to publish on it followed by Anton (1712) in Venice; Kanold (1713), Schroëckius (1713), and Cammerarius (1713) in Germany; and Lancisi following in 1715 in Italy. Despite the prevalence of human plague in Europe at the time, Ramazzini believed it to be an eruptive fever, similar to smallpox in man, calling it *variola bovina* or "cow-pox":

> It is evident that this disease, which has created such dreadul havoc among the whole bovine race, from its cold shiverings, followed by excessive heat throughout the whole body, is a malignant and deadly fever, as its accompanying symptoms testify. In the first place, there is intense anguish, heavy breathing and continual snorting (or groaning) accompanied by fever, stupor, and slothfulness or weariness; a continued running of ill-odored matter from the mouth and nostrils is observed, and most foetid excrement, sometimes mixed with blood, is passed; there is loss of appetite, and chewing the cud (rumination) ceases. On the fifth and sixth days, pustules break out over the whole body, and tubercles of a variolous character. On the fifth or seventh day, death ensues, which very few escape, and those more by chance than from the effects of remedies. We may reasonably suppose that the miserable oxen suffer much internal pain, when they lie groaning, or while they stand motionless with heads cast down toward the ground … In the carcases of as many oxen as were dissected by the eminent professors Molinetto and Viscardo, it was noticed in all, strange to say, that the omasum was hard and compact, and the leaves closely adherent; while all were of great size and emitted a horrid stench. In other parts, such as the brain and lungs, hydatids were found…. The tongue was covered with ulcers toward the root, and full of little vesicles on its sides. I know for certain, that that portion of the body which was observed in the abdomen (the stomach) to be hard and compact like stone, was primarily produced by a contagious miasma, which, while pursuing its own course of destruction, weakened and corrupted the gastric juice…. It might reasonably be expected that mankind would be left uninjured; for if, in the space of three months, the plague had attacked no other ruminating and horned animal, or in any way injured horses, pigs, or wild creatures, there is no reason why it should affect men, who are so different from these creatures. (Ramazzini, 1711)

Lancisi, however, referred to imprudent people contracting the disease by eating contaminated meat, quoting Deuteronomy: "Ye shall not eat of any thing that dieth of itself." However, Deuteronomy goes on to command: "thou shalt give it unto the stranger that is in thy gates, that he may eat it; or thou mayest sell it unto an alien…" (Deut. 14: 21). Far from observing the biblical command to this latter extent, Lancisi recommended prohibition of trade in meat, hides, and fat of infected animals.

Ramazzini's belief that it was similar to smallpox was supported by the Medical College of Geneva but disputed by the Philosophical College of Padua. The professor of practical medicine at Patavia, Marco Novara, issued a report with a further description of the morbid symptoms but expressing the opinion that the presence of pustules on the skin was a rare occurrence and that smallpox was not as fatal or as quick in its spread as this disease. But it was Ramazzini's likening it to smallpox[4] and recommendation that inoculation should be tried ("I therefore think that we should proceed in the cure of this disease as we do in the cure of the variolous fever of boys") which later led to attempts to control rinderpest by inoculation, attempts which ultimately proved the most successful manner of combating it. He predicted that the intensity of the epizootic would diminish with the onset of cold weather and, despite the public fear, considered that it would not affect humans. Workmen had closed their shops and streamed to the city cathedral in such numbers that never before had there been so great an assemblage of the people in one place, "nor so much reverence and religious feeling toward the Supreme Being" (Ramazzini, 1711).

Dr. Peter Anton also studied the outbreak in the Venetian territories in 1711. He compared the carcasses of healthy slaughtered animals with those dead of the disease and also bled 18 sick animals, all of which except one died a few days later. Quite in contrast to Ramazzini, he deduced the cause of the sickness to be a "preternatural" thickness of the blood, occasioned by the cold and wetness of the seasons which reigned the previous 12 months from October 1710 to November 1711 (Anton, 1712).

Attacking places missed the previous year, in 1712 the disease still raged in Russia, Poland, Siberia, and Turkey. Continuing its havoc in Italy, it spread into Moravia, Austria, Tyrol, and Switzerland, and in Germany spread as far as Fulda. The following year, it enlarged its boundaries in Germany, Russia, and Switzerland, while in many countries, although observers did not link the mortality to the same disease, pigs, sheep, and goats now perished. The Kingdom of Naples alone lost 50,000 sheep and pigs in a brief period.

The disease continued to spread in many areas because owners unloaded any sick or suspect stock. Although, as in the Roman States, markets and stock movements were forbidden, Lancisi discovered that cattle were traded secretly, and "... the butchers of the city, who had not yet discovered the deceit practised upon them, were led to buy these animals at a reasonable rate, and in this way were immediately disseminated the germs of the hidden contagion...." He also gave a good account of the symptoms, together with the first accurate description of the postmortem appearances (Lancisi, 1715). A number of other Italians gave exact and clear descriptions of the disease from other regions of Italy, and one was produced by the Medical Society of Geneva.

Lancisi believed that the disease originated with infected animals from Hungary and recommended that as soon as its presence was discovered, all traffic in cattle should be prohibited. Presenting his report to the Pope and an assembly of cardinals, Lancisi further advised that every diseased animal should be killed to prevent spread of the contagion, but the cardinals considered this too severe, and other remedies should be tried: "... and, in truth," wrote Lancisi, "they were greatly influenced in this decision by the number of people who pretended they had infallible cures for the affection. But the fact is, that in the cattle, as in the human plague, not every one who takes the disease dies of it. Some recover, thanks to Nature, rather than to the remedies which are resorted to."

At first, merchants contravening the bans on transport and sale of sick animals were to be imprisoned, some in chains to await later punishment; but when attempts to cure the disease had no effect, edicts were issued, forbidding bringing cattle into the city district of Rome under penalty of death for a layman, and the galleys for life for an ecclesiastic. The sale of hides was forbidden, and carcasses were to be buried in their skins covered with quicklime in deep pits. The Pope went further, ordaining that dead bodies be cut into quarters and buried with lime in holes 3 m deep, or 3.6 m if no lime was used, the holes to be dug at a distance from public roads. Carcasses could be burnt, provided it was done 16 km from a city or 5 km

[4] "... ac tubercula variolarum speciem referentia..." and "... uti fit in variolis..." "... and tubercules resembling a kind of variola..." and "... as in the smallpox..." (Ramazzini, 1711).

from a village. Laymen and ecclesiastics were liable to death or the galleys, respectively, if they allowed bodies to lie unburied or be thrown into rivers or marshes, and everyone had to notify a death immediately it happened. A reward of 20 Roman crowns was announced for informers and a pardon if they were accomplices, by which means apparently several transgressors were apprehended.

Inspectors were appointed to visit markets, and only meat branded by the inspector could be sold. Salting diseased meat was forbidden. In addition to these stringent methods, religious ceremonies were prescribed and prayers offered, the texts of prayers being distributed.

The objective methods as a result of Lancisi's wise counsels, relying on practical measures of isolation and slaughter, rid the area of rinderpest in 9 months. Lancisi summed up his steps for prevention in the event of an outbreak as follows:

> All roads and by-paths should be carefully guarded, so that no ox or dog be allowed to enter the country. Any animal so entering should be forthwith destroyed and buried. Should the pestilence, however, gain admission, the separation of the sick from the healthy must be enforced by decree. Indeed, in my opinion, by far the safest course is instantly to destroy the animal, and with the poleaxe, so that no infected blood may escape on to the ground; for, in attempting to cure the diseased animal, the veterinary surgeon may convey the plague to healthy oxen. The healthy cattle must be removed from their former pastures, which must now be regarded as contaminated. The diseased oxen should be kept in stables, to which no one is admitted except the veterinary surgeon or the herdsman. The fountains and vessels used by the animals should be frequently cleaned with quick lime. The clothes of the shepherds also should be fumigated. The dead carcases, from which not one hair is to be removed, must be buried in deep pits [here Lancisi probably took his inspiration from Virgil's *Georgics*]; and any saliva or secretions which may drop from them on the way to the pit are to be carefully removed. If any cows are infected, their milk is instantly to be thrown into a hole in the ground; and the severest punishment should be inflicted on those who disobey this order. The passage of all rustics and dogs should be forbidden.

Persons treating sick animals should wear waxed cloth overalls, and after finishing with the animal should take them off and leave them there, then wash the face and hands in vinegar, thus being able to treat other animals without spreading the infection. But it would be better to kill the animals than treat them.

The Papal Government also undertook measures to relieve the distress of those affected by losses, a number of edicts being issued by the Sacred College which Lancisi reproduced because he thought "they will be of great service to posterity, if a similar misfortune should ever again happen...." But they were, of course, either totally forgotten or simply ignored. Cognizant of the need to avert famine caused by lack of draught oxen for ploughing, the Pope reissued the edict of Pope Paul V in the 1609 outbreak forbidding slaughter of draught oxen suitable for ploughing, and a law was approved, requiring that every year in every 100 animals, 25 calves should be assigned to ploughing. The sale or export of suckling calves, young buffaloes, and heifers was controlled, and only old and sterile cattle could be slaughtered. The cardinals proposed that owners of uncultivated lands should be forced to cultivate them, and the area of cultivatable land in disease-free areas should be increased.

Lambs and wethers were unusually abundant, and in 1715, "multitudes" of cattle fed on the fields and meadows "as of yore." Provisions were cheaper in the Ager Romanus, where the outbreak had been, than in Picenus, where there had been no outbreak.

In other parts of Italy, the disease continued to rage for several years to come, killing 70,000 head at Turin. Spreading over Switzerland, it entered France on several fronts. Silesia lost 100,000 head, the Low Countries 300,000, and Piedmont 70,000. It spread to Livonia from Russia, and Liège from Holland. Mecklenburg was infected in 1711, and Prussia cordoned off its border with royal troops. It then experienced outbreaks in 1713–14, 1716–17, and 1722.

With the Dauphiné having been infected in 1714 by animals from Piedmont, Dr. Drouin, first surgeon of the bodyguards, and Dr. Hecquet, Dean of the medical faculty of Paris, were sent to study it; and in 1714, the medical faculty published a judgment "on the current publications relating to rinderpest." In 1715, the College of Physicians at Geneva discussed the nature of the disease and concluded, like Ramazzini, that it was due to a little smallpox.

3. THE PLAGUE REACHES BRITAIN

In July 1714, the disease was imported into England from Holland, although those favoring spontaneous generation were unconvinced of its importation. Writing in 1722 of the bubonic plague, Dr. J. Pringle, answering the proposition "That neither Goods, nor Persons, are capable of producing, or communicating the plague," stated: "Our last mortality among cattle, may properly be called a plague in their species, as taking its origin from Hungary, and thence making a gradual progress to Holland, with the usual velocity and symptoms observed in the Polish plagues afflicting Mankind. But whatever may be allowed in favour of a Communicative Theory on the Continent, there can be no Salvo for its crossing the Channel to Islington."

Breaking out at Islington, centring on three London dairies, it was investigated by a commission nominated by the Lords Justices, consisting of George I's recently appointed surgeon who had been a naval surgeon in Sicily at the time of the Italian outbreak, Dr. Thomas Bates; and four Justices of the Peace. Bates received his authority from Lord Harcourt, the Lord Chancellor. Bates's initial report, in a handwritten letter to the King, differs from his more detailed report published 4 years later in the Philosopical Transactions. Originally, he dissected only six cows and simply recommended separation of animals when any sickness was detected:

> Their Excellencys the late Lords Justices of Great Brittain &c were pleased to command that I should Examine into the Distemper amongst the Cows at Islington, by dissecting such as would be proper for that purpose pursuant to which, I had six of them opened, in their different degrees of Infection, and found the putrefaction of their viscera to increase in proportion to the time of their sickness.
>
> The two first that I opened had herded with those that was ill, and the symptoms of this Distemper was just become visible: in these the gall bladder was larger then usuall, and fill'd with Cholor of a naturall taste and smell, but of a greener colour. Their Pancreas was shrivell'd, their glandules tumified and obstructed. The glandules in their mesenterys were twice or thrice their naturall bigness. All other parts appeared as is usuall in a healthfull state.
>
> The third and fourth that I opened had been ill but three days; in them the Livers was Blacker than usuall, and had several Cystises formed, in which was a petrified substance like Chalk, about the bigness of Peas. Their Gall Bladders was twice their usuall bigness, and filled with Cholor of a naturall taste and smell, but of a deep green colour. Their Pancreas was shrivell'd and contracted, their Glandules very hard and Obstructed, and a blackish colour. The Glandular of their Mesenterys were many of them five times their naturall bigness, and of a black colour. Their Lungs was a little shrivell'd. Their Intestines was full of black spots; but their flesh was not alter'd in the colour.
>
> The two Calf that I opened were very near dying; in them I found the Liver to be blackish, much shrivell'd and contracted, and to have severall of their glandules distended, and formed into Cystises as big as nuts and Nutmegs, and filled with a petrified substance like Chalk. Their Gall bladders above three times their usuall bigness, and fill'd with Cholor of a naturall taste and smell, but of a very deep green colour. Their Pancreas was shrivell'd and contracted, their glandules very hard and Obstructed, of a black colour. The Glandules in their Mesenterys, were many of them distended to Ten times their usuall bigness, were as black as Ink, and in their pelvis a Yallow petrifaction the consistance of a sandy stone. Their intestines were the colour of a snake, their Inner Coat excoriated by their purging. Their Lungs was shrivell'd, with severall collections of a yallow purrulent matter in Cystises formed for that purpose many of them as big as a Nutmeg. Their flesh was very little altered in colour.

> In giving your Majesty my Opinion of this distemper I humbly beg leave first to premise that all Cows have Naturally a purgation of fifteen or twenty days in the spring, after which their Milk becomes sweet, thin, and of a blewish colour; which is allways allowed to be the most nourishing; but this spring being dry, and the Grass not having a due proportion of that purgative quallity which caused that evacuation; has made their Milk both thicker and less in quantity this summer then was usuall in other years: from which I humbly conceive this Distemper Originally to have proceeded; because the want of so naturall and large a purgation would certainly create a fullness of body, and an ill disposition in the blood, which obstructing so many of the Glandules in the Mesentery, would consequently deprive the blood of its supply; which in some ill habits of body, may very reasonably be allowed a sufficient cause to produce that putrefaction which made this distemper contagious—and I think there is but too much reason to fear that it will appear in most parts of Great Brittain. I further beg leave to observe, that in all contagious distempers, there must be putrefaction in the blood before the contagion can be communicated; and having not been able to discover in those cows that I dissected, any such putrefaction, till they had been ill two or three days, I most humbly recommend to your Majesty that an advertisment be published requiring all persons that so soon as they perceive any of their cows to refuse their meat, or to have any other symptoms of sickness, that they do immediately separate them from their others, as the most effectual method to prevent the ill effects of this Distemper.
>
> All which is most humbly submitted to your Majestys consideration by
>
> May it please your Majesty
> Your most faithfull and Obedient
> Subject and Servant
> Thomas Bates
>
> Sept: ye: 24th: 1714. (Bates, 1714)

Interviewing the three owners affected, the Commission found that one man had lost 120 of 200 cattle, another 62 of 72, and a third 38 of 87. The owners had been unwilling to admit their losses for fear that people would not buy their milk. Asked by the Lords Justices to provide recommendations for its control, Bates proposed that all cattle of the three affected owners be destroyed and burnt. Failing this, at least the sick ones and those unaffected be kept on the farms. The sheds where the infected animals were housed should be cleaned and not used again for 3 months; and the fields likewise not pastured for two. Those looking after sick animals should not have contact with other stock, which should be divided into separate herds numbering not more than 10–12 head.

These measures were confirmed by the Lords Justices, together with payment of half the value up to a maximum of £2 for every sick cow destroyed, and calves proportionately less. But because other stockholders had come to see the new disease, the contagion had already spread and began to appear elsewhere, until strict enforcement stamped it out such that its reach did not exceed 30 km westward of London. The measures proposed by Bates were then extended county-wide.

Already in Middlesex, Essex, and Surrey, by October it had spread to Norfolk, Suffolk, and Hertfordshire. But the measures, particularly those of isolating sick cows, had effect, and within 6 months, by December 25, the plague terminated. This was because no one would buy a cow in the country for fear of the disease, and the Justices prevented slaughter in the towns by having the markets inspected daily and meat which appeared suspicious condemned. The stockholders dividing their cows into small parcels did not lose the whole herd if one contracted the disease (herds of 100–300 had contracted the disease). Those who had no sick cows avoided all contact with those who had; and owners of sick cows, instead of keeping them as long as possible as they had done so before, brought them to be slaughtered at the first signs of disease. Bates thought that the purging in the "latter Spring" at the end of September might have caused a reduction in the disease, but the incidence increased at this time, and too many infected cattle were produced to be burnt, so he recommended burial in a hole 4.5–6 m deep and covered with quick lime. This is a considerable depth, and it is doubtful if any were buried deeper than about 2 m. It was further recommended that calves of sick cows also be slaughtered and buried, for which payments of 5–10 shillings were made. Bates drew up a recommendation for the Committee of Council to make it obligatory that carcasses should be buried 2.7 or 3 m deep, fearing that the numbers of rotting cadavers would produce some contagious disease among humans, as some people affirmed had occurred in 1665. Their Lordships deferred the matter, and 3–4 weeks later, the outbreak had virtually ceased.

Some owners, believing that the disease would become general, attempted to sell off their stock by taking it to distant markets, while others, like the original cases, did not immediately report their losses for fear of not being able to sell their milk. Owners were therefore informed that they would be paid the compensation only if they complied with the recommendations, although there was no legal requirement. Reliance was placed upon persuasion, the JPs calling the cowkeepers together once or twice a week and urging them to follow the recommendations. The JPs nevertheless applied the measures ruthlessly in Middlesex, appointing several butchers to watch near the cowkeepers' grounds and count their numbers of cows every morning, with orders to follow any sent to market and prevent them being sold by telling people what they were (Bates, 1718).

Bates, who, despite his former presence in Sicily, made no reference to the work of Lancisi or the Edicts of the College of Cardinals, reported that he could gain little assistance from questioning cow leeches or reading books, for the disease was unknown in the country; but the spring had been the driest in living memory throughout Europe, and Bates thought that the want of the habitual spring purging of cattle due to the fresh flush of grass might be the cause of the disease. Dr. Mortimer, Secretary of the Royal Society, recorded seeing many cows collapsing and dying in convulsions after drinking at ponds near Bethnal Green (Mortimer, 1745a), indicative of drinking after dehydration rather than rinderpest mortality. Some herdsmen denied that it was the murrain or cattle plague, declaring that it came from insects brought by the dry summer and easterly winds.

> The most general opinion concerning the cause of the distemper, was, that the cattle were first infected by drinking some unwholesome standing water, where 'tis probable some poisonous insects were lodged and bred; the summer having been extremely dry, attended almost constantly with easterly winds, the grass almost burnt up, and the herbs of the gardens destroyed by insects; but such as they were, (unfit for table use) were given to the cattle. There was likewise so great want of water, that many were forced to drive their cows five or six miles to it. (Bradley, 1721)

The success of the measures was apparently a close-run thing. For a time, the outbreak was contained within Islington and Haggerston, but then the Justices' powers lapsed for 5 weeks, and this, together with the popularity of a Dutch remedy, allowed the disease to spread. Only when the Justices' powers were reintroduced was the disease quelled.

Having lasted 6 months in all, a total of 5,857 head had been lost in Middlesex, Essex, and Surrey, but the losses were not so great in England as they had been overseas, and the disease did not last in its virulent form longer than 3 months.

The Treasury took the money for compensation out of the King's Civil List, possibly to counter the animosity toward the accession of the German dynasty in England, George I having arrived in England in mid-September 1714 speaking no English, an animosity which was probably also the reason why no laws were passed to control the disease. However, when the owners of affected cattle petitioned for full compensation, the State stopped the £2 payment on the grounds that it had not had the desired effect, but owners were granted leave to appeal for public funds. In total, government compensation in 1714 had amounted to £6,674. 1s. 1d., but the £2 for each cow was far short of the average value of £6, plus an estimated further loss of £4 on wages, rents and feed contracts already entered into.

John Morphew, a noted satirical poet, asserted it was the German accession which brought the plague into England, dubbing it the German Cattle Plague. He claimed the compensation destroyed more cattle than did the plague itself because owners knocked all their old cattle on the head in order to claim the £2, expressing it in a poem in 1717:

As soon as Britain had sustain'd
That fatal loss which Heaven had gain'd,
And parties squabbled to a madness
About their sorrows and their gladness,
A plague unprophesy'd succeeded,
That only reached the Horniheaded,
And like a fatal Rot or Murrain,
Turn'd all our bulls and cows to carrion.

The farriers now their skill employ'd,
But still the cows in numbers dy'd,
And with their horns and hides together
Were burnt, without reserve of leather.

Some cunning huxters, who had cows
Old, dry, and lean, not worth a souse,
Tho' sound in health, but scarce deserving
Of pasture, to prevent their starving,
These wisely knock'd 'em on the head
By night, when neighbours were in bed.
Next day assign'd their expiration
To this new fatal visitation:
So bore 'em to some distant pit
Or ditch, for such a purpose fit;
There, to the terror of our isle,
Consum'd 'em in their funeral pile.
Then like true hypocrites, put on
A mournful look, as if undone,
And claim'd the sum of forty shilling
For every cow of Heaven's killing-
A gen'rous bounty, that destroy'd
More cattle than the plague annoy'd.

Bates stands out ahead of his time for realizing the contagious nature of the disease and the most effective means of preventing its spread. It is not known whether he had access to Lancisi's work or whether he reached his conclusions independently of work on the Continent. But prompt action in slaughtering all suspect cattle was held up as a model to be followed elsewhere, particularly by the French. Referring to cures by drugs, Bates stated, "I think there is no one method in practice, but what was tried on this occasion, though I cannot say that any of them was attended with an appearance of success" (Bates, 1718).

4. CONTINUATION ON THE CONTINENT

In the autumn of 1714, France sent a representative to London to report on the control measures being taken, but had difficulty in adopting a slaughter policy in France because of cost and fear of angering the peasants. Orders not to skin dead beasts were rescinded because of apprehension of a leather shortage (Jones, 1978).

The following year, the plague appeared in Germany's dukedom of Bremen and Verden, continuing until 1718 with a loss of about 5,000 head; occurring again in 1720–21 with the loss of a further 8,000. It continued to rage in Holland until 1720, the Dutch stubbornly trying to find cures for it. Ravages also continued in northeast Germany, parts of Switzerland, Italy, and some districts of France. Quarantine regulations were instituted in France in 1714–15, and fairs and markets in affected areas were closed. In the Dauphiné area, the intendant, on his own initiative, began slaughter with a full compensation policy, but it was denounced by the government as too costly and likely to offend the peasantry.

The winter of 1716 was exceptionally cold, the River Thames freezing over in Britain, and these low temperatures may have assisted in reducing the incidence, for although pigs as well as cows died in Ferrara in the extreme north of Italy, it disappeared in Italy that winter and was dying out in France but continued in Holland and parts of Germany. The following year, it began to die out in Hungary, Prussia, Silesia, and Bavaria. In Brandenburg, tight quarantine regulations were instituted, administered by the public executioner. In 1716, Frederick William I issued a bulletin: *Gründliche Anweisung, woran man die leyder gar zu sehr eingerissene Krankheit des Viehes erkennen; das Gesunde davor praeserviren, und das Krancke curiren könne, A Basic Instruction on How to Recognize the Cattle Disease, How to Protect the Healthy Cattle from it, and How to Cure the Sick*, based upon the results of a questionnaire he had ordered printed and circulated in 1711. At Wolfenbüttel in the same year, a "short lesson" was issued on how to combat the plague now raging (Anon., 1716b), while at Hildesheim, a "thorough instruction" was issued, with the fear that the disease might be brought into Hildesheim as cattle were dying in the Dukedom of Bremen due to infected animals being brought among the healthy, noting that the disease was no different to an infectious inflammatory fever. The methods of prevention and cure, it noted, were in all existing management books and known to all householders and innkeepers, but they did not bother to take any notice of them (Anon., 1716a).

In the summer of 1718, it still raged in Puntden and Saxon (Switzerland), and the following year in Servia, Wallachia, and Bulgaria. In the Prussian Dukedom of Trier, it was reported as raging and worsening in September 1719. Provence in the south of France was ravaged in 1720 and 1721, and from the latter year to 1722, it was allegedly in the south of Sweden, continuing there until 1740; but a statement of the Swedish Government to the House of Commons Select Committee in Britain in 1877 stated that, although the disease was called cattle plague at the time, from its description, it was splenic apoplexy (anthrax), and cattle plague had never appeared in Sweden. If it was rinderpest, there were further outbreaks in 1745, 1750, and 1762–3, the latter with more than 100,000 deaths. In 1729, the King had solicited the assistance of the Swedish medical college, which made only vague recommendations. The Norwegian Government claimed it had never been detected in Norway.

At the end of 1722, there was an outbreak in Breslau spread far and wide by the sale of infected meat. In 1723, the provincial government of the Principality of Halberstadt, adjacent to Brandenburg, requested two physicians, Drs. J. W. Röseler and A. Ph. Bornemann, to undertake a study of the disease. They were instructed to obtain carcasses from the public executioner and conduct autopsies on them to determine the nature of the disease and whether or not it was contagious. Describing the visceral changes and the appearance of the disease, they concluded that fat cattle died sooner than lean, and bleeding only hastened death. The disease was easily communicable by the breath, and healthy cattle might become infected from people, dogs, and cats that had been in contact with infected animals. They also concluded that it stemmed in part from diseased or dirty pastures. Elsewhere, it was reported that people ate meat of slaughtered diseased animals without ill effect.

From 1726 to 1727, it still prevailed in Russia, Livonia, and Courland, and in 1728 was introduced into Brandenburg in Germany, thence spreading to Austria. In 1729 or 1730, it erupted again in Italy, Poland, Pannonia, Wallachia, Podolia, and Volhynia, apparently beginning among the wild boar and then destroying immense numbers of cattle. In the preamble to an edict of December 1729 in Brandenburg, the King observed that there was much complaint that people in town and country, "out of pride," frequently failed to report a death among their cattle, arguing that it was only an isolated occurrence (they were probably fearful of the restrictions which would be imposed if a death was diagnosed as rinderpest). The result was the disease spread out of control. It raged in Austria in 1729 and 1730, a lengthy livestock order being issued in the latter year. This was not of a penal but of a general and discursive nature, beginning by describing supposed causes of cattle disease: season, weather, lack of clean water, and bad diet. It noted that sick animals could infect healthy animals by exuding sour perspiration, which was taken in through the breath and saliva. Preventive methods such as not turning the cattle out when fogs and dew were present, burying the dead with the hides slashed, cleaning of stalls, and so on were listed. These were followed by preventive medicines consisting of the usual herbal remedies followed by curative methods, and finally measures to be taken in slaughtering and import of animals with no suspicious animals to be admitted.

In 1730, it appeared in Saxony, and in February 1732, according to a report of the Prussian King's Resident in Basel, it was present in the French district of Cannat, where some ascribed it to a poisonous mildew, others to a large mass of caterpillars appearing that year, the characteristic vesicles on the tongue presumably being thought to result from ingestion of the caterpillars.[5] In Bremen, Carl Koch, a cow doctor and an accepted authority on cure, introduced obligatory examination of the tongue. He instituted an oath on April 18, 1732 for all cow doctors to solemnly swear to examine the tongues of cattle daily and to notify the results daily. If there were signs of infection, the animal had to be examined for at least 3 hr. This disease, however, appears to have been the "tongue-cancer." Two decrees issued by the duke of Braunschweig and Lüneburg in April appear to refer to this disease (Figure 3). But whereas an outbreak of actinobacillosis was widespread in Europe in 1732 (Fleming, 1871), a report of a cattle epizootic in Languedoc states that above all, it attacked the lungs, yet also refers to tumors appearing, possibly indicating rinderpest but confusing it with preceding examinations of pleuropneumonia deaths (Anon., 1732).

While we hear no more of rinderpest from other countries over the next few years, it was still present, perhaps enzootic, in Poland and Germany. But in 1734, it was re-introduced into Italy by the war of Charles Emmanuel of Savoy against the Milanese territory, aided by France and Spain. Austrian armies invaded Italy, bringing with them Hungarian cattle. The disease soon spread over the whole of Italy. In 1735, it broke out on the banks of the Danube in France but was speedily stamped out by separating sick stock; but as France was at war with Austria, it soon spread to the Austrian armies and thence again to Italy. In Piedmont, it raged from 1734 until 1739.

In 1739, it was apparently present around Karlsrühe, although this may have been anthrax. Textors described symptoms and postmortem appearances, recommending that stalls be kept clean with fresh bedding and fumigated. Sometimes, the animals could be given a laxative of cooking salt or saltpetre, and should be well bled. Sick animals should be isolated, and dead carcasses buried. At the same time in 1739, a bad harvest in Germany was followed by a cold winter causing many stock deaths, the people resorting to feeding cattle with straw from their roofs and beds. So cold was it at the beginning of 1740 that birds were reported as falling out of the air. By the end of May 1740, in Verden 151 horses, 624 cows, 160 calves, 353 pigs, 5,061 sheep, and 7,297 lambs died.

[5]Some caterpillar species can cause blisters in the mouth. In Ireland, a species was regarded as injurious to cattle, and images of it were made to protect against it and cure the murrain. A large silver one used for that purpose was found in County Cork in 1845 (Burdick, 1905).

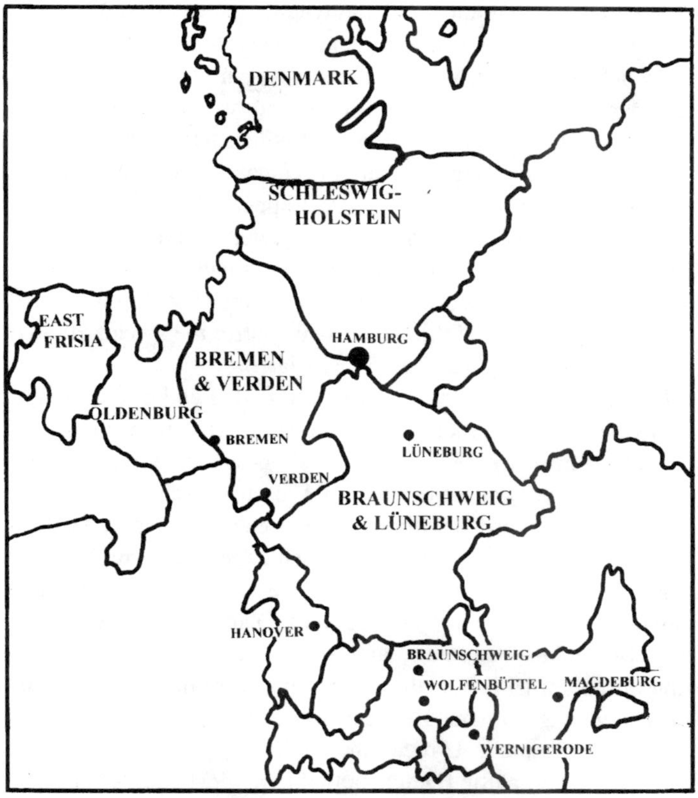

Figure 3. Some of the affected German principalities.

In 1740, the disease broke out again in Hungary, consequent upon the commencement of the War of the Austrian Succession led by Frederick the Great of Prussia, and was carried into Bohemia at the siege of Prague, and then into Bavaria. Its resurgence prompted the reissue of a booklet in Germany formerly published by the Medical Office of Halle, on how to recognize it and proposals for its prevention and cure: *Salutary Proposal on How to Prevent the Raging Pestilence and Which Preventive Methods are Useful.*

It then spread south into Switzerland, Piedmont, and Dauphiné, and northwards into Denmark, Sweden, and Holland.

In 1742, it was imported into Lorraine and the Vosges, and the following year accompanied the army from Bavaria into Alsace, also penetrating France through Switzerland. In France, the government emphasized the segregation of sick animals and disinfection of premises, using troops to maintain a *cordon sanitaire* around infected areas. In 1743, an Ordinance issued at Metz ordered that dogs on infected farms be tied up, otherwise they would be shot by the local authority. In the same year, some children in Augsburg drank the milk of a sick cow without effect, so the farmer slaughtered the cow and brought the meat into the village. All those who ate the meat, which was everyone in the village with the exception of one household, died, suggesting anthrax. In 1744, following the Prussian invasion of Bohemia and the siege of Prague, the disease again entered Piedmont and northern Italy, causing great losses, as well Germany, Franche-Comté, and Holland, destroying some 200,000 head in Holland in 2 years.

In Denmark, which it entered in 1745 and where by the end of the year, 50,000–60,000 head of cattle had been lost, it was believed to have been imported from Flanders on the raw hides of animals which had died from it, the first cows in Denmark sickening within a few days of the hides being unpacked. Holberg, who owned two large estates, commented on the almost 100% mortality rate and its rapid progress with 65 landholders affected in Zeeland within an area of 36 km^2. Some suggested that the calamity was due to a comet, but Holberg pointed out that it occurred long before the comet appeared (Holberg, 1746). More significantly, on many estates, the number of cattle had doubled in the past half century, with consequent destruction of the forests in which the cattle roamed. To this extent, the mortality of the cattle allowed the forests to recover somewhat (quoted by Wilkinson, 1992). By 1749, Denmark had lost 285,162 head, and the inability to control the spread of the disease resulted in its becoming enzootic, still widespread 35 years later.

It entered Holstein in early February 1745 and by the end of the month was present in Oldenburg, and widespread in Verden at the end of August. In March the following year, it was reported as almost everywhere in Bremen and Verden, some 31,617 head being lost but by August 1747 had apparently ceased. Yet from 1745 to 1772, barely a year was to pass in the dukedom of Bremen and Verden without an outbreak, the marshy lands and their neighboring districts along the dukedom's borders witnessing the highest mortalities of up to 100%. In the central area, outbreaks were rare, and the highest losses estimated at only 5% maximum. According to the *Gentleman's Magazine* of October 1744, in France's districts of Burgundy, Forest, Beaujolais, Lyonnais, and Dauphiné, it attacked cattle, calves, sheep, and hogs, creating havoc until 1747, with the French government using the army to maintain *cordons sanitaires* along its borders. Between November 1744 and the following August, Friesland lost 135,000 head, and in Utrecht, the loss of adult beasts was 34,000. It continued in Holland until 1765, the red cattle (Dutch cattle being mostly red or red and white belonging to the red type then common throughout central Europe) being so reduced in numbers that new stock had to be imported from southern Denmark and western Germany. The black-and-white pattern then came to the fore, owing to importations from Jutland (Figure 4).

Figure 4. An early depiction of rinderpest. Cattle dying in Holland in 1745, by Jan Smit (courtesy Atlas Van Stolk, Rotterdam).

The *London Gazette* of May 25, 1745 reported that on May 19th the disease had now reached within a German mile and a half of the Balliage of Pinnenberg contiguous to the territory of Hamburg, adjoining Danish Holstein: "The appearance ... at Hambourg gives an alarm lest it should gain further on that side of the country. Proper precautions are used to prevent its spreading and to get the better of it in its beginning, which is attributed to some infected cattle having passed through there. This evil has communicated itself likewise to the Danish islands of Zeeland and Fuhnen." It reached Constantinople the same month. Writing of bubonic plague there, Dr. Mackenzie stated: "The only antecedents, that I could observe to this malady, was a great murrain among the black cattle in May 1745 ...." (Mackenzie, 1765).

"A potent means of spreading any pestilence is war, which breaks through every precaution or measure suggested by human prudence," wrote the great Italian historian, Ludovic Muratori, tracing its passage through Italy with the war (Muratori, 1714). But Stähle (1981) considered that outbreaks in Bremen and Verden between 1715 and 1781 were due to the cattle industry, the disease first appearing in the marshlands bordering Oldenburg and Holstein which had a major livestock trade with Bremen and Verden. Large assemblages of Danish oxen took place on the drove roads in Holstein and Bremen from spring to autumn. Driven out to pasture in the spring, the fattened livestock were then taken to markets in the autumn and thence on to Bremen and Hamburg, the main consumer centers. Those not sold were brought back. Not until the start of the 19th century, in Stähle's opinion, did war carry the disease into Germany.

The disease was brought to Paris by a trader who bought 19 animals which had come from Picardy, where the plague was present. Soon, 26 of his cows died, and the disease quickly spread into other parts. The Magistrates set up a team of 10 doctors who examined all sick animals every day, trying numerous remedies, but the Magistrates rejected their proposal that the animals should be bathed, on the grounds that the water might then carry the contagion. Despite their efforts, only a few animals survived.

Between 1715 and 1770, Stähle identifed seven minor outbreaks and three major with durations of 5–15 months: 1715–18; 1720–21; *1745–6; 1748–50*; 1752–3; 1754–6; 1756–7; 1760–62; 1767–8, and *1770–72* (italic type indicates the major outbreaks). By calculating case mortality and mortality rates, he was able to discount outbreaks of FMD with rates of only 2–5%, and pleuropneumonia with rates of 30–50%. Stähle challenged the view that persistence of the disease was due to insufficient methods of control, superstition, and persistent attempts to cure it, arguing that it was predominantly due to the manner in which cattle were managed. Faust (1797) considered health passes to be useless, for, apart from being open to falsification, the testifier could not know if areas through which the cattle had passed were free of disease or if the cattle themselves were clean. He had argued, like Stähle (1981), that it was the cattle trade and cattle markets which caused it to spread.

A pamphlet issued in France in 1744, of which a translation was published in the *Gentleman's Magazine* in the same year, reiterated the measures given by Ramazzini and Lancisi for control (Anon., 1744). In the Dukedom of Trier, the Carrion-pit Ordinance of 1705 was re-enacted in October 1746, stressing that carcasses must be buried more deeply. In the Low Countries, the faculty of medicine of Leiden, and then that of Hardervijk, produced studies, as did Professors Brugmans of Leiden and Kraus of Utrecht, and in 1745 a commission of physicians. In 1749, the States General formed a new commission.

Paulet (1775), commenting on the outbreak in France, noted that there were two opinions with regard to its origin. Some believed it had originated in Tartary, "where it committed twice the damage of an ordinary plague," and penetrated Russia. From thence on one side, it passed into Poland, Livonia, the Duchy of Courland, Prussia, Pomerania, the Duchy of Mecklenberg, Alsace, and afterwards Holland and England, while on the other side, toward the east, it passed into Turkey, Bohemia, Hungary, Dalmatia, Austria, Moravia, Styria, and by the Gulf of Venice to Italy, Spain, France, then to Germany and finally Denmark, where an enormous mortality was reported.

A more popular opinion was that it had broken out in Bohemia during the siege of Prague, thence extending into Hungary, Bavaria, Styria, Carinthia, the Tyrol, Italy, and by the Alps into Provence. On the other side, it penetrated Alsace, Luxembourg, Franche-Comte, Lorraine, Flanders, the Low Countries, Picardy, and then France, the king's army returning from Baviere in Alsace, carrying infection across the Rhine in June 1743; although, as Paulet points out, it was already in the Vosges and other parts of Lorraine in 1742. It appeared in French Flanders at the beginning of 1744, dying out, or extinguished, in 1747.

7

RINDERPEST REACHES BRITAIN AGAIN

1. THE OUTBREAK

In 1745 rinderpest was carried into England, Ireland, and Scotland, according to one source, by English troops. But the Whig government blamed the Scots for introducing the disease when the Jacobites invaded England, reaching Derby in December, before retreating to be defeated at Culloden in April 1746. By May of that year, some 6,000 cattle were reported to have died in Argyllshire alone, and it is likely that infected English cattle looted by the Scots in their retreat were responsible. Contrary to conditions surrounding the outbreak 30 years before, the spring and summer were very wet and the ground very damp, followed by a dry and cold autumn. There had been severe frosts in 1740 with great drought in the seasons immediately following, until in 1745–6 there was much rain. This was claimed by some to have produced too rank a grass, causing the disease through a redundance of the "crude aliment" (Anon., 1746a).

The outbreak began almost certainly in the coastal marshes of Rochfort Hundred in Essex, and was probably brought to London in the autumn by the well-established Essex trade in veal calves (Broad, 1983). It is believed to have been introduced from Holland, one explanation, given at a meeting of the Royal Society by a Mr. Theobalds, being that it was due to two white calves imported from Holland by a farmer living at Poplar near London. An unlikely alternative explanation offered by Dr. Hoffman, "a learned Danish gentleman," was that it may have been by an infected parcel of hides. Hoffman informed the meeting that this was how it had been introduced into Denmark. A tanner allegedly bought a parcel of diseased hides in Zealand very cheaply because they were forbidden to be sold and should have been buried (Layard, 1757).

Appearing in the immediate neighbourhood of London in April 1745, the disease spread gradually throughout Essex and Hertfordshire. The government was informed of the outbreak in eastern Essex at its outset by John Milner, who sent one of his tenants to inform the Duke of Newcastle on April 9. On April 15, the Essex justices met at Rayleigh and took evidence from seven farmers who had lost a total of 138 animals. The justices reported that since farms were 6.4 km apart, and there had been no communication between them, the disease could not be contagious. Only one calf had been sent to market since the disease had arrived. Following this report, the Duke took no further action.

In November hides of infected animals were still being flayed and tanned, and it was not until November 25 in Middlesex that a Commission was set up which appointed various cowkeepers and butchers as inspectors. At least their mandate was more positive than was to be that of the Commission set up 152 years later, but it extended only to Middlesex. The mandate was:

1. To inspect cowhouses and separate sick from sound cows.
2. To see that all cowhouses and yards were kept thoroughly clean.

3. *To kill all sick cows and calves.* To slash their hides with several cuts from head to tail and round the body so as to render them useless, and then to bury them in graves 3 m deep, with 2 bushels of unslaked lime to each cow.
4. To certify the destruction of cows, for each of which the Treasury paid £2.
5. To see that proper returns were made by cowkeepers of their losses.

In Yorkshire, to encourage cooperation with government policy, 32 landlords promised their tenants an additional 10s. for killing their cows for every 20s. from the King, as well as assistance in making claims. In Durham, a number of gentry also supplemented the government compensation.

In November, a group of physicians planned to meet together every Monday until further notice, informing the public that factors other than the richness of the grass caused the disease, which might be communicated by animals or men. They recommended that every farmer should exclude strange cattle from his barns and pastures and, if his cattle became diseased, should bury those that died, as well as their litter and dung. Stalls should be cleansed by washing and steaming with vinegar poured on hot iron and then left empty as long as possible. No attendant on diseased animals should be allowed among healthy animals. They recommended that cattle should not be allowed to overfeed, "After the Cattle were once seized, Evacuations were found to be injurious: Few were observ'd to recover, which had either been plentifully blooded or purged: Pegging, as it's call'd, (or Rowelling) in the manner it was practised, had not been much Use ..." Nevertheless, they recommended rowelling and also a medicine comprising a pint of honey and vinegar simmered in a gallon of water. Quantities of saltpetre, camphor, and liquorice powder were then added to sufficient honey to make a ball, which was divided into four, one part being dissolved in a quart of the liquor and given every 6 hr. After 2 days, a mixture of Peruvian Bark and Virginia Snake-root, in $1\frac{1}{2}$ pints of warm beer was to be given instead. Later, they recommended doing away with the preliminary medicine, substituting rattle snake root for Virginia snake root, and administering this medicine after removing a quart of blood at the first seizure, "the loss whereof will not here be prejudicial, the Remedy supplying the strength which would be diminish'd by it ...."

> We urg'd the Necessity of keeping the diseased Cattle within Doors, and even in as warm Houses as possible, frequently steaming them with Vinegar, gently pour'd on hot Irons; that they should have warm Liquids, such as Water-Gruel, Barley-Meal boil'd in Water, and very thin Mashes, given them in Plenty, at least every two hours: As the Disorder abates, their Gruel, Mashes, &c. may be made thicker and more nourishing (Anon., 1745a).

Concluding that they had given sufficient attention to the matter, the physicians announced the discontinuance of their meetings in order not to clash with the Privy Council's investigation.

"Iater," an anonymous country physician (J. Davis) writing 2 years later, referred to the weekly meetings as still continuing, but the public was kept in ignorance of the fruits of their inquiry. He considered this most likely due to want of official recognition and encouragement of their researches.

A doctor signing himself "Rusticus" was taken to task by the editor of the *Gentleman's Magazine* for a letter stating: "The vigilance of the government, therefore, in taking this matter into consideration, and providing a remedy for the unhappy sufferers is highly commendable ...." The Editor noted:

> The letter-writer here commends, but with very little reason, the vigilance of the government. For this distemper of the cattle was known to be in Essex in the beginning of the year, and humble advice was offered to proper persons several months since. But, as our government has been for many years conducted, it is scarce possible, that it should exert itself in a vigilant or expeditious

> manner, because no board is to act until it receives orders from a superior, the direction of which, and consequently of all weighty affairs whatever, falls to the lot perhaps of one man, or, which is worse, a cabinet of two or three men. And from hence instances may be traced of divers fatal delays at the most critical conjunctures—As to this distemper, if magistrates of health had been established, as in other countries, due care would have been much sooner taken, before so many hundreds had been lost, which might have been prevented in some degree ...." (Urban, 1745)

2. THE FIRST ACCOUNT

Dr. Cromwell Mortimer, Secretary of the Royal Society, was the first to provide an account of the outbreak in a paper read before the Royal Society in November, but his remedies would have been scorned by Bates 30 years before. After recommending such cures as pouring a pint of warm vinegar containing an ounce of salt into the sick cow's nostrils, and feeding them balls of *Crocus metallorum*, he continues:

> As for the cause of this distemper, I am still at a loss. I think it cannot be owing to the food, because the cows which had it first in Essex eat only grass, turnips, and hay or straw... I am in doubt as to the air; the spring and summer were very wet and the ground very damp, the autumn was very dry and cold, the beginning of winter very damp and cold. The cows in Essex had the distemper in summer; it first began about London in autumn; it has spread itself equally among cows which have lain in the fields a-nights, and those which stood in stables or sheds; it spread itself in Essex, at first into such farms where they bought in strange calves, or lean cows, at market, which they did not know where they came from, but most probably from the hundreds where the disease first broke out; but how it got thither, whether by importing any cattle from Flanders, I know not, for surely there is too wide a tract of sea for any infectious miasmata to be wafted over to that part of the country by the winds!... but that it is contagious, and propagated by infected cows being mixed with well cows, therefore the not buying-in calves or strange beasts, but every farmer keeping his herd by itself, must be a great means of preventing the propagation of it. And housing the cows a-nights may be a proper preservative against it. (Mortimer, 1745a)

Mortimer observed that those cattle kept out in the cold air seldom lived beyond 3 days if they contracted the infection, whereas those which were kept warm in sheds and covered up lived for 5–7 days.

On January 9, 1746, Mortimer read a further account to the Royal Society with details of how he had drunk milk in his tea, after setting aside some for chocolate, from a cow which died 48 hr later. After describing some post-mortem appearances of the cow, he states:

> How it [the disease] got into the Park [St. James's] is very strange, there having been no fresh cows brought in there since the Welsh fair in August. And this is further very observable, that though this distemper seems so very infectious among the cow kind, yet I do not hear of any of the deer having fallen ill... I humbly therefore suggest whether it would not be the most likely means to put a stop to the spreading of the distemper, to forbid any cows or calves being brought to market, to be sold alive, or that any farmers should buy in any fresh cattle for six months, or till it is found that the distemper is entirely ceased; and that all fat cattle should be kept carefully separated from the cows and calves, and that under severe penalties. (Mortimer, 1746)

Not without good reason did Mortimer make his remarks, for a contemporary observer, Elizabeth Purefoy of Shalstone Manor in the west of Buckinghamshire, wrote on January 9, 1746 to her London agent, an innkeeper in Islington:

> I am sorry to hear there is such a bad Distemper amongst your Cows, I thank God wee have not one cow amiss at Shalstone, nor within twenty miles of us only at Shenley & Newnton Longueville; about 10 miles from hence one Mr. George bought a distempered cow of a London dealer & unfortunately putting her amongst his other Cows hee lost seventeen which were all hee had. There was the same case at Newnton Longueville, but I don't hear it has spread. Our butcher,

> Mr. Ben: King told mee hee had 40 fat cows, the worst of them worth twelve pounds & if ye Distemper shall fall upon his Cattle it would undo him.…Our parishioners have sent a certificate to London that our cattle of all sorts are in good health, so our butter and every thing wee send to Town from hence is highly accepted.

In March she was to write she had lost a cow on calving and "durst not buy any more these perillous times." By October 29, she regretted that she would be unable to supply the agent with butter any longer, for her two cows were inadequate and she "… durst not buy any cows, the Distemper being within 5 or 6 miles of us." A certificate dated November 28, 1746 addressed to one of her tenants, presumably to allow him to sell his cattle, was signed by Elizabeth Purefoy, her son, and the Rector, to the effect that all their cattle at Shalstone were in good health[1] (Eland, 1931).

3. THE PLAGUE PROLONGED

Dr. Theophilus Lobb, another contemporary observer, repeating the views of Bates, proposed isolation at the first signs of sickness and the immediate burial of any animals which died, with or without quicklime, 2 or 2.4 m deep. However, he considered epizootics among cattle to be a Divine judgment and a token of anger, and in order to propitiate the Almighty, people had to repent of their sins (Lobb, 1745).

One anonymous writer traced the disease from Africa through Europe into Britain, believing it was carried in the clothing of all who came into contact with cattle in any way. Consequently, no one should go near cattle "but those who fodder and milk them, and let these People never go to them in the same Cloaths they go to Fairs and Markets in," but each should have a special frock, hat, and even wig "else tuck all their Hair up under a Linen Cap made for that Purpose."

The situation was not helped by such persons as Dr. Barker, who published a tract under the anonymity of a "Member of the College of Physicians" in December 1745. Barker stated that the disease was not infectious but was due to the "moisture of the seasons of some time past and the great plenty of grass which these seasons have produced." Moist seasons stopped up the pores and prevented perspiration. As a result, the disease sprang up spontaneously in different places, and the advice given in the *Daily Advertiser*, to bury cattle which had died of the disease together with the skin and fat, was unnecessary. "Rusticus," writing in the *Gentleman's Magazine* to oppose the intructions which had been issued to kill all cows and calves as soon as they fell sick, was "fully convinced" by Barker that he, "Rusticus," had been wrong in thinking the disease to be contagious and "[I] am much obliged to the author for relieving me from the uneasy apprehension which I had entertained of eating beef, veal, milk, &c. all of which he has convinced me may be eaten as safely now as at any other time." Supposing, however, that the disease was infectious, the instructions for killing the cattle as soon as they fell sick, he argued, were "liable to censure, as being too general and unlimited." He could not suppose this advice was given by a physician as no physician ever proposed curing a disease by knocking his patient on the head.

The editor, of a more sage disposition, noted that several graziers, particularly those of two or three parishes in Bucks, slaughtered and buried in their skins their sick cows, being recompensed by the vestry, and kept the rest at a distance, thus stopping spread of the disease

[1]Until January 1752, the year ran from April to March; hence Elizabeth Purefoy's letter of January 9, 1746 is actually dated 1745, before the cattle plague entered Britain. It is usually impossible to tell in the literature whether quoted dates before 1752 have been corrected or not, although they have been so in the reprinted *Gentleman's Magazine*.

in the area. In response to Rusticus's claim that Barker had proved by "some very solid arguments" that the distemper was not infectious, the editor stated what Barker said did not amount to a proof, "... there have been much rank grass with wet seasons since 1714, and yet the cows were not affected. As to the recovery of some sick that were not separated from the rest, the like has been in all plagues.... As to his internal medicines they do not promise much effect" (Urban, 1745).

But in 1746, Brocklesby, whose conjectures concerning the cause and nature of the distemper were so "extremely jejune as scarce to deserve consideration" ("Several Hands", 1747), argued that it was best to bury carcasses 2.4–3 m deep, and they should not be covered with quicklime as a "fiery, incoercible spirit" instantly arises, which fills the air and might spread the disorder. Although burning was to be recommended, even this might cause people to panic by presenting "such frightful sights of woe." He wrote: "One beast goes—all sooner or later go. [The plague] will probably abate in a short time.... As an exotic plant flourish, it will dwindle, and die" (Brocklesby, 1746a). Of the same idea that burying in quicklime might be adverse, causing malignant particles to be sent up as it rendered the animal salts more volatile and pungent, was Dr. John Parsons (1746) who submitted an observation to The Royal Society by John Milner. The latter quoted Mr. Stallwood, a farmer at Hackney to whom the care of infected cattle was committed, who informed the Justices that he had buried 13 cows very deep with the legal quantity of lime of 2 bushels to each cow. The fermenting of the lime caused a strong scent of meat to arise, attracting dogs which tried to dig up the carcasses. Cattle which were buried with their hides slashed, but without lime, did not attract the dogs, which ran over the area (Milner, 1746).

One man allegedly saved the only cattle in his parish in 1747 by taking two or three heifers into a field remote from any road, and allowing no person to go into it other than himself. Whenever he went there, he changed every article of dress, or put on what he called his "Sunday clothes" (Thompson, 1866).

However, in 1748, a Derbyshire owner added to his uninfected cattle three from an infected herd before driving them to a Staffordshire fair. Here, they were sold and a certificate issued by a Land Tax commissioner on the oath of the farmer. The animals were then driven to Warwick with other beasts, where all fell ill also infecting many other animals in an area previously disease-free. In the same year, the Bedfordshire Justices complained that drovers and cattlemen organized their own markets in order to avoid checks and supervision on Wavendon Heath, just across the border in Buckinghamshire, even when the market in Leighton Buzzard was open. By the spring of 1748, there were numerous large Privy Council meetings to debate the cost and effectiveness of slaughter, and a number of physicians were called in to give advice.

On March 12, 1746, an Order in Council had stated: "If not timely prevented it will end in the destruction of all cattle," an admonition that the Council repeated the following year. Yet the plague raged until 1756 because attempts were made to try to cure it, rather than slaughtering all infected stock as had been so wisely done in 1714.

4. THE QUESTION OF COMPENSATION

Between 1745 and 1758, the government paid out £220,000 in compensation, of which £100,000 was paid by 1748. A correspondent to the *Gentleman's Magazine* signing himself "A Farmer," criticized the payment of £100,000 and a proposed reward for a remedy as in vain until better regulation of buying and selling stock was enforced. Since some still mixed sick and well cattle, he marvelled at how the mortality had not been greater ("Farmer," 1748).

One of the many anonymous pamphleteers pointed out in 1748 that most of the compensation under the first Order was paid out illegally because it was claimed by persons who did not slaughter the animal at the first signs of infection but only when it was ready to expire. After some time, a stop was put to these "sinister evasions," but then, although compensation of 10s. for burying the hide, slashed, was, according to the law, only given for animals which happened to die before the owners could slaughter them, it was being paid for all cattle which died of the disease and were buried with the hide slashed. Although this was amended to require a sworn certificate from a Justice of the Peace or Commissioner that the Order had been obeyed in all respects, the certificates were signed regardless. The writer estimated that for every animal slaughtered, perhaps 100 died, and compensation was paid. The last Order of March 22, 1747 (1748) he did not think had been observed anywhere. He recommended compensation be increased to half the value of the animal, but at the same time, the penalty for lack of compliance be increased from £10 to £50, or 12 months' imprisonment in place of 3 months (Anon., 1748b). At the beginning of 1750, he issued a second edition of his pamphlet, this time dedicated to the Archbishop of Canterbury and to the Lord High Chancellor. Almost identical to the first, it omitted a discussion on allowing the continuance of the jobbing of cattle, and also all criticism of the prayer which had been ordained and the proposal for a Fast Day. The writer noted that the Order of December 14, 1749 did not appear to be any better observed than the previous as far as slaughter was concerned. He ended with a sound attack on the JPs and other officials for failing in their duty, alleging that several magistrates had declared openly against the rule requiring slaughter, "They most certainly and plainly want a proper Resolution and Zeal, and a due Deference to the Law itself so far as relates to the killing of the Distempered" (Anon., n.d.[1750a]).

The compensation issue was raised again in 1750 when another correspondent to the *Gentleman's Magazine* pointed out that a farmer with 39 cows valued at £5 each, a total of £195, would receive on killing them £78 (at £2 each), which represented a loss of £117 plus the charge of £7.16s. for burying them. If the rinderpest was allowed to take its course, he would probably have salvaged 13, worth £65. He would therefore prefer to kill none, hoping to save more, and unlike new stock, the survivors would be immune for the future. He might therefore find it more profitable to disregard the order, and even if he killed them, he would not find men enough to bury them. But if the compensation was raised, fraud would flourish as already men had killed worthless cattle and claimed the £2. All traffic should therefore be stopped, but animals from Scotland, Ireland, and Wales were entering freely (Anon., 1750a). This was immediately taken up by one signing himself "Agricola," who feared lest misinformation might inspire mistaken action, prejudicing the farmers against regulation. He argued that cows were worth about £3.10s., and if a farmer had 40 worth say £4 each, and had to kill 10 to stop infection, he would lose about £22 compared with a loss of £136.16s. if he saved eight by letting the disease run its course. Amongst other things, he asked the correspondent how, since it did not happen elsewhere, the distemper would wear out in England without regulations? ("Agricola", 1750). Another argued that with a slack administration, an infected animal was always worth almost £2 alive and 10s. dead (for its hide): "Distempered herds are thus distributed over a whole County by persons buying cattle out of such herds for less value than the premium, who afterwards when they fall sick get the full premium" (Anon., 1750b). In Cheshire, few animals were valued at more than £4.10s., and in Lancashire none at over £4, whereas in Buckinghamshire and Hertfordshire, values ranged from £5 to £8.15s. and to £8, respectively. Thus, northern stock owners received a relatively higher compensation than those in the south. Whereas some critics of government policy advocated an end to all regulations, others in 1753 favored full compensation, as one anonymous pamphleteer put it: "... be the cattle of more or less value a sum that would ruin a great many farmers would not be felt by the Nation." A small rise in Land Tax or a Poll Tax would pay for it (Anon., 1753a).

A bill which would have paid full compensation for the first three animals passed the Commons in 1753 but did not reach the Statute Book.

Once a farmer had obtained a compensation certificate, he still had to turn it into cash. Farmers living near London were expected to collect the money from a London office, but in 1747, the office ran short of cash, resulting in numerous letters of complaint. In Buckinghamshire, Richard Grenville used his willingness to pay cash compensation in the county as an election policy. In some parts of the country, JPs tried to charge for issuing certificates until this was expressly forbidden in October 1747. In Shropshire and Cheshire, the authorities agreed to pay farmers from local funds and reimburse the county stock from London, and from 1749, the Shropshire Quarter Sessions ordered that half of the amount paid by the King should be added from the county funds. There is little evidence of cattle being overvalued, the Cheshire certificates giving age and type of beast, as well as often colour, distinctive markings, and names (Broad, 1983).

In some Cheshire parishes, informal mutual insurance schemes were set up in 1749 in the form of village cooperatives. In Great Budworth, the local gentry organized a list of all cattle in the parish with subscriptions to be paid by owners according to the number held. Any farmer receiving government compensation for losses would receive an additional 50% from the fund, while the gentry promised their own tenants an amount equal to the compensation, thus ensuring they received the full value. Similar schemes were set up in Nottinghamshire. The Great Budworth proposal was offered to farmers individually as it was found that too many difficulties and objections were raised at public meetings to such schemes, "If some of ye better end of ye farmers come readily into it, I think the rest will follow by degrees." On one Cheshire estate, the landlord paid his tenants about a third of their net losses but discriminated against those who had sublet their land or rented part of their holdings from another landlord. Most Cheshire landlords gave tenants some help, and there is evidence of the same in Lincolnshire and Buckinghamshire and on the Chatsworth estate. On the Tyrwhitt-Drake estate in Cheshire, tenant farmers lost 286 cattle valued at £843.18s. Of these, 216 were compensated for by government totalling £118.2s.6d. (Broad, 1983). But whereas they were paid twice as much by Tyrwhitt-Drake as the government compensation, this was set against rent arrears that were bound to arise during restocking, and the Estate demanded a 6-month credit on items produced by the tenants, such as cheese.

5. PERSISTENCE OF THE OUTBREAK

In May 1748, the Cheshire authorities petitioned government for stricter movement orders and called for a ban on all cattle movements from infected counties. The JPs wrote "there is no hazzard the common people will not run for the lucre of present gain, even be the advantage ever so small." The county disbursed £711 on border watches before the disease arrived, both Cheshire and Lancashire raising additional county rates to pay for their activities. Most counties appointed large numbers of inspectors who checked farms and markets, being paid initially from 1s.6d. a week in Bedfordshire, to 7s. to 10s. a week. London paid 10s. a day, and Middlesex 5s. Leicestershire paid 2s.6d. for every beast correctly slaughtered and buried rather than a wage, perhaps leading to excessive zeal on the part of inspectors. In 1756, the widow of a Bedfordshire Clerk of the Peace claimed £511.6s. for rinderpest work performed by her late husband since 1745. The bench was sceptical but granted her £175 because her husband had been a vigilant officer (Broad, 1983).

County authorities spent considerable sums printing large numbers of orders and certificates, movement certificates being required by the thousand. Market closures were widely advertised in local newspapers. Shropshire went much further and cock fights, stage plays, puppet shows, public dances, and meetings, were banned over wide areas: elsewhere both horse racing and foot races were banned.

In a pamphlet *The Case of Killing Infected Beasts Briefly Considered*, the author reported that where mortality had been heavy, grass rotted for want of mouths and hay could not be sold, while the noxious effluvia remained in the stall if not in the pasture. Men put too much faith in medicines and did not immediately kill their infected beasts. He quoted one parish where government measures were applied: four or five herds were affected, but the disease passed in 6 weeks and only nine or 10 animals died. The author then laid down several points:

- The distemper was infectious.
- Cattle should be slaughtered as soon as infected.
- No cure had been found although many had been tried.
- Cows did not double in value after recovery.
- The law requiring killing and burial within 24 hr was agreeable to the Magna Carta and the British Constitution.
- All laws should be obeyed, although some were unpopular.
- While cattle belong to individuals, the government may equitably order their killing for the public good.
- The distemper was not a divine judgement and should be combated in every way possible. Virgil and Columella had recommended the slaughter of infected cattle, and such a policy had proved effective in 1714 (Whitworth, 1750).

It reached Northumberland in 1750; and in November of that year, JPs reported great difficulty in enforcing the Orders, one having discovered some drovers who either had no certificates at all or only forged ones. In the spring of 1751, it broke out in Wiltshire and Oxfordshire. Raging in Yorkshire, Lancashire, and Westmoreland, it led JPs in the as-yet unaffected counties of Cumberland, Monmouth, Hereford, Gloucester, and Somerset, to forbid movement of cattle into those counties. Cumberland had guards placed on its roads to prevent the introduction of cattle, hides, carcasses, and tallow. The JPs of Somerset forbade the introduction of cattle from Wales, Wiltshire, and Gloucestershire.

By early 1751, Cheshire had lost 30,000 cows since the previous October. In October 1751, it still raged in Yorkshire, Westmoreland, Lancaster, Wiltshire, Monmouthshire, Gloucestershire, Dorsetshire, and Bristol; and in that month, it broke out again near London. At St. John's Gate, the issue of free powders was announced, of which the maker had "a strong opinion" would cure it (Anon., 1751b). In February 1752, it was still raging in Gloucestershire, Buckinghamshire, and adjacent counties and in parts of Lincolnshire, and York, Lancaster, and Derby. Its ultimate occurrence in Somerset caused JPs to go beyond the government legislation and order all infected animals in West Chinnock to be shot and buried 2.1–2.4 m deep. They also bought all cattle that had been herded with the diseased animals and had them shot and buried.

Affected cattle were now being shot by order of Quarter Sessions and paid for out of the county rate, and by March 1752, the disease had much abated, ceasing entirely in Wiltshire; but there were still some deaths in Somerset. Then on May 31, it appeared about 10 km east of Reading in Berkshire, where it had not been before. Elsewhere, a farmer was convicted of selling infected cattle and resisting officials. In November, Parliament, in considering means to prevent the spread, continued the existing legislation virtually without change.

Yet in April 1753, it erupted again in Wiltshire where, in 1745, it had little effect. Fifteen animals belonging to one man were ordered to be shot, and fairs and markets were prohibited.

Only fat cattle for immediate slaughter could be moved or sold. In September, it was carried to Suffolk by a farmer's drover driving four infected cows in that county, the disease spreading rapidly to 60 parishes within 10 days. In November, 97 cattle, probably imported, died on the road between Chatham and Canterbury. Robinet (1789) reports that men were infected, especially in 1752–3 in England, but this appears to have been diptheria. In February 1754, it was still raging in the North Riding of Yorkshire and in parts of Durham; and in December, several cows died in St. James's Park.

In 1748, one writer had urged that slaughter of suspect animals should be immediate and, to be effective, had to be applied nationwide (Anon., 1748b); but 5 years later, another anonymous writer, in an open letter to a Member of Parliament, argued that the law was being disregarded because the compensation payable for slaughter of £2 a beast was not worthwhile compared with a value of £6 to £8 for lean stock, and £10 to £12 for fat. Rather than slaughter at the first signs, the farmers waited to see which animals would die before killing them. He proposed that the full value should be paid, and, as it was a public concern, the cost should be borne by the public for a trial period of 1 year. The writer further proposed that sworn inspectors be appointed by the quarter sessions, and they should be responsible for the slaughter; also, owners should be obliged to report any suspicion of disease to them immediately. Regular accounts of all infected places should be forwarded to a collator in London, who would publish the lists in the Gazette, all drovers being obliged to observe these areas and avoid them (Anon., 1753b).

One writer asked whether foxes and other animals rambling at night might not convey the infection (Anon., 1753a), and in 1755, Malcolm Flemming reported that a case of infection he had seen between farmyards at a considerable distance from one another was effected by means of servants or of dogs moving between them. By this time, the disease had largely subsided but continued to break out intermittently, appearing in April 1757 in Kent and Shropshire. In June, an Act was passed "to more effectually prevent" the disease, and in Essex on October 4, all fairs and markets were prohibited, but cattle could be brought into the county, provided they had the necessary certificates of health. After this, the disease seems to have died out, no cases being reported in 1758, and on February 8, 1759, the King ordered a day of general thanksgiving to be observed, with a form of prayer distributed throughout the kingdom to be used in all churches and chapels.

Although not comparable with that of the mid-19th century, the cattle trade was large and complex, and the sudden ban on lean cattle movements in September 1747 left Welsh and Scots drovers stranded *en route* to markets with over 8,000 cattle scattered in droves mainly in the southeast and Midlands. The following year, eight Welsh drovers were quarantined in Kent, with herds ranging in size from 24 to 162 head. The interests of breeders and dairymen conflicted sharply with those of fatteners; for the latter, the free movement of stock was essential, while in the northwest, the breeders and dairy farmers were more concerned with reducing livestock movements as much as possible, showing a real sense of commitment to border watches and inspections not always apparent elsewhere. But counties near to London all had large numbers of cattle passing through on their way to markets and fattening grounds.

Few people were prosecuted, and convictions were far fewer than indictments, while several convictions were reversed on appeal. Broad (1983) saw JPs as unwilling to indict owners who had already been punished by the loss of their herds and also because so many escaped prosecution. The gentry around Northwich in Cheshire consistently opposed government policy, and the JPs there ignored their duties, some claiming "... it is lawful and reasonable for every person to do what he will with his own." Perhaps they feared antagonizing the farmers, some of whom in November 1749 threatened riot and direct action against inspectors. In Hertfordshire the Privy Council issued a special order banning Barnet Fair in 1751, but the

Privy Council secretary discovered by chance just before the fair that the local JP had not been informed of the ban. Barnet Fair was formerly held at Islington until the rinderpest in 1746 obliged the Welsh drovers to remove to Barnet (Newby, 1828).

In 1747, "Iater" had urged that in order to put a "speedy end" to the disease, a Board of Health consisting of noblemen, gentlemen, divines, physicians, lawyers, and experienced army officers, should be appointed to control the outbreak and frame regulations. These regulations should be executed by the Lords Lieutenant of every county, who should form a "Board of Lieutenancy" consisting of the greatest landed gentry, JPs, surgeons, and so on. This Board should be empowered to order destructions, pay compensation, inflict penalties, and regularly transmit to the Board of Health accounts of their work and the state of infection in their counties, as well as bills of mortality. Smith (1933) claimed that had this proposal been adopted, the country would have been saved millions of pounds.

In January 1757, Layard reported that the slaughter policy had resulted in a visible decrease in cattle, some persons having killed cattle suffering from other diseases in order to obtain the payment. As a result, payment was stopped. He recommended that one of the first measures to be taken should be to forbid all imports of cattle and hides from places overseas where the disease was present, and those ships which brought cattle and hides should have a bill of health. If necessary, the cattle should be placed in quarantine on arrival. Such ships should be examined by customs officers, and if they found horned cattle or hides on board coming from infected regions, they should immediately have them thrown overboard into the sea, as well as any horses, pigs, sheep, or dogs. As to the sailors, he considered their dress to be less likely to retain the infection because of the tar and pitch "sticking about them" and because of their frequent smoking. Nevertheless, before coming on shore, their clothes should be washed with warm water and vinegar, and smoked with brimstone and tobacco (Layard, 1757).

But Layard also proposed, quoting Columella, that the instant the disease broke out in a "village, pasture, or neighbouring town," the stock should be removed as far distant from the infected place as possible. If put into effect, such a method would clearly spread the disease more quickly. Although a law of 1747 specified that carcasses of animals dying of the disease had to be buried within 8 hr of death, Layard quotes the case of a farmer leaving the carcass of a dead cow for 12 hr before burying it, the disease spreading to the neighbouring farm.

The virus had already become attenuated in 1749 when it was reported that in the past $1\frac{1}{2}$ years, many more cattle recovered without treatment than did from the first appearance of the disease up to that time. The total number of cattle dying during the outbreak was roughly estimated at half a million (Bourguignon, 1865 estimated 300,000–400,000). Although there is no record of total numbers of stock which died, in the third year of the plague when Government ordered that every animal exhibiting the slightest sign of infection must be destroyed with compensation being paid to the owner, 80,000 were slaughtered, and the following year a similar number. In addition, nearly double that number allegedly died from the disease. Extrapolating from the amount of compensation paid on the Tyrwhitt-Drake estate in Cheshire and the total for the country, Broad (1983) estimates a total loss of 507,000 for the period 1745–58, with about 321,000 in 1747–8, the latter representing an estimated 20% of normal beef consumption for 2 years. One source gives 100,000 as perishing in Lincolnshire in the second year; 40,000 in Nottinghamshire and 30,000 in Cheshire in the third; with many other counties suffering similar losses (Playfair, 1866). In the fourth year (1748), cattle were being slaughtered at the rate of 7,000 a month until the regulation was revoked due to the abuses being practised. At the end of August of that year, the JPs of several counties certified to the Privy Council that the plague had ceased. In the following month, it was declared that it had erupted afresh around Burton-on-Trent in Buckinghamshire, and also near Camberwell in Surrey. In the same month, it was reported in Lancashire, York, and Durham. Until June 1747, it appears to have been confined to south of the Humber–Trent line and east of the Severn and Dee, but by October had almost reached the Mersey. By 1749, it had spread widely. In September, it continued to rage in Yorkshire and some parts of Durham, appearing in Cheshire and Lancashire, and in November, raging in Suffolk and on the Isle of Ely, causing the announcement in December that it was raging with a "malignity and violence little short of the first outbreak." Only Devon and Cornwall escaped, along with much of central and west Wales and Scotland.

The disease was to continue for another 8 years before finally burning itself out, as evasion by farmers, dealers, and drovers continued to perpetuate it. But then dealers, to avoid being landed with cattle which would become sick or die, developed a system of conditional contracts, withholding part of the purchase price for an agreed period, thus serving to make wilful evasion unprofitable and probably contributing in no small measure to the eventual termination of the outbreak.

Only a few cases were recorded in 1757 and none in 1758. In February 1759, a general thanksgiving was held for the cessation of the plague. Persistence of the disease for 12 years compared with the 6 months of 1714–15 is attributed to there being no figure like Bates to come forward and take the lead, although he was still alive and protested both verbally and in writing at the failure to implement proven methods of eradication. But more importantly, perhaps, Britain was in political turmoil in the crucial early stages of the outbreak, with an unpopular king and political occupation with the threatened Jacobite uprising, as well as the Austrian War of Succession.

8

FROM SEVEN YEARS WAR TO CRIMEAN WAR

A Century of Devastation Continues

1. ITS CONTINUED RAVAGES ON THE CONTINENT

The panzootic which covered the Continent beginning in 1745 lasted for 10 years and "swept away nearly the whole race of horned cattle throughout Europe." Between 1740 and 1748, it was estimated that western central Europe had lost 3 million cattle. By the end of 1769, total losses since 1711 had been put at 100 million for the whole of Europe with at least 10 million in France and Belgium, 14 million in Germany, and 395,000–600,000 in Holland. In 1769 alone, 115,000 were lost in Belgium and Luxembourg, and 43,500 in Holland and West Friesland. Paulet (1775) was more conservative, suggesting that 1.5 million head were lost in Europe in 1711–14, and from 1745 to 1755 a further 3 Million. (Faust, 1797 gives 51,022 lost in Friesland in 1769, with 17,237 surviving, 74.8% mortality). In Holland mortality averaged 73% (Table 1, Figure 1). By the end of the century the total losses for Europe since 1711 were estimated at 200 million, with 28 million for Germany alone.

Mortality in France in 1745 was 90% (Sauvages, 1746), but from 1750 to 1769, it is alleged that it was neither so fatal nor so generally diffused throughout Europe, and countries began to recover from its effects. However, records from Bremen and Verden show that in 1745–6, a total of 31,617 head died, in 1748–50, 18,339, and in 1770–72, 22,199; suggesting no decrease in virulence. Stähle (1981), however, claimed that by calculating the mortality rate, there was a decrease in rate due to the increasing number of young animals which grew up during the outbreaks and had colostral immunity. Also, surviving immune cows contributed to a lower mortality rate, whereas there were few immune young cattle between these two extremes.

It continued with varying severity in Europe and was repeatedly imported into Germany without becoming general. In April 1748, it erupted again in Schleswig-Holstein and in December passed into Oldenburg, continuing to rage with even greater ferocity, although the war of succession came to an end in 1748. By 1749, Denmark had lost 280,000 animals, and as the disease advanced into Jutland, many people killed their cattle before they could become infected. It entered Sweden where, in one province, only 2% of cattle survived.

Outbreaks in Schleswig-Holstein continued through 1749–55, becoming almost universal in some areas in 1756. From Holland, it entered Austria and French Flanders, but an epizootic in which deer and domestic animals died in the forest of Crécy in 1757 appears to have been anthrax. In September 1750, it appeared at "a couple" of cattle entry points in the Dukedom of Braunschweig and Lüneburg but seems to have been contained. Litthauen in Prussian Lithuania was not so fortunate and lost 145,596 animals in the countryside, plus 674 in the town. Galleski (1765) referred to 1750 as a bad year for Germany, with the disease being

Table 1. The course of rinderpest in Holland, April 1769 to March 1770.

Year	Month	Died	Recovered	Percentage Dying
1769	Apr	5,266	1,134	83
	May	3,115	745	81
	Jun	3,815	1,087	78
	Jul	6,515	2,280	75
	Aug	14,195	5,329	73
	Sep	30,374	12,367	72
	Oct	28,678	11,614	72
	Nov	21,599	7,885	73
	Dec	28,295	11,765	71
1770	Jan	12,255	5,343	70
	Feb	3,224	1,507	69
	Mar	1,896	535	78
		159,227	61,591	73

From Faust (1797).

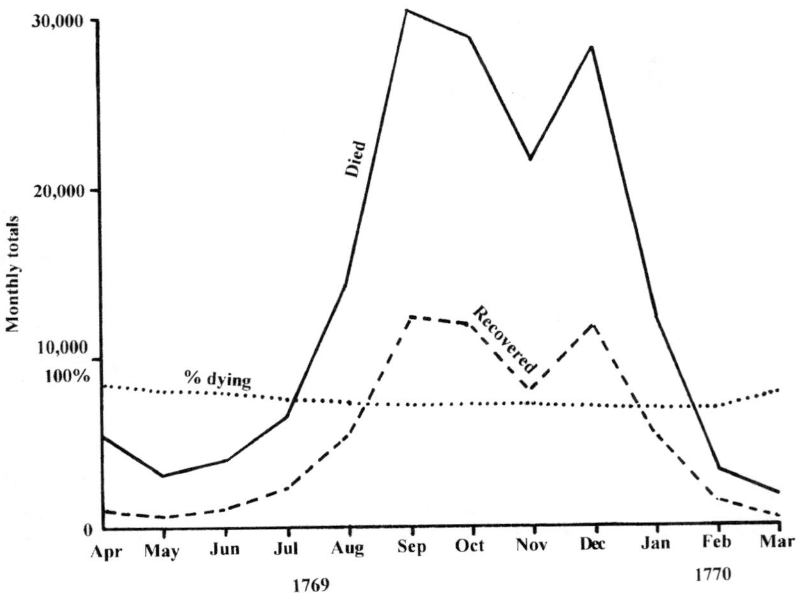

Figure 1. Mortality from rinderpest in Holland, April 1769 to March 1770.

widespread. He attributed this to the previous 4 years having been dry years, with moderate rains in 1750, and noted that the disease occurred only in certain years.

In 1756, it broke out in Eichsfeld with infected animals imported from Holland which the people were accustomed to buy in the autumn. To prevent its spread, all cattle suspected of infection were slaughtered and buried. The disease also reached the island of Minorca with cattle imported from Auvergne in France. With the commencement of the Seven Years War in 1756, the French army in Westphalia introduced it in the following year to Minden in Germany, and in Bremen and Verden, minor outbreaks occurred in 1757, 1758, 1759, and 1760, the whole country then becoming free until December 1760.

In 1750–51, it had reached Lapland, and in 1758, raging in Sweden and Norway (claimed later to have been anthrax in these countries), spread once more to southwest

Lapland, killing off vast numbers of reindeer *Rangifer tarandus* and reducing the wealthy owners of large herds to extreme poverty. In 1759, it reached Denmark again, where it was reported as far more lethal than it had ever been known in England or Holland or than it had been when infecting the same places before. Stricken animals died in 2–3 days, some even sooner.A report from Finland in 1761 stated that the disease had been present there for some years, where, it was claimed, men who attended the sick cattle and who flayed the dead were now being affected. A "stout young fellow" who, after flaying a cow which had died of the distemper, laid down at night on the skin wrapping it about him, the next morning was found quite stiff and dead. Whatever killed him, it could not have been rinderpest or anthrax. The "most ingenious Mr. Hartman" of Stockholm, considered this distemper in the human species to have perfectly corresponded with those cases of the "English sweating" (Anon., 1761).[1]

Although there were no wars at this time, at the beginning of 1761, it erupted again in Germany and at the end of the year appeared in Saxony and Austria, the next year raging in Germany, Austria, and Holland. In September 1762, it reappeared at cattle entry points in the dukedom of Braunschweig and Lüneburg but again was contained. Settegast (1774) refers to an outbreak on the Rhine in 1762–3, attributing it to bad seasons and lack of food. A local Swiss word, *louvet* or *lovat*, for a disease allegedly affecting both cattle and horses, probably referred to anthrax rather than rinderpest, but the "terrible outbreak" in 1761, the year rinderpest spread throughout the region, probably was rinderpest. Doctor Reynier, in a pamphlet *Le Louvet, Maladie du Bétail; Ses Causes, Ses Remèdes: & les moyens de le prévenir (Louvet, Illness of Cattle, its Causes, Remedies and means of preventing it)* (Reynier, 1762), attributed the epizootic to alkali salts in the animals' diets, but also criticized the way in which animals were treated, being overworked, their drinking water polluted, and feeding and stabling bad. He stated that peasant cattle were the most often affected, the disease being very frequent and deadly, killing many animals in the summer of 1761 and in 1762 twice as many horses as cattle. Whereas his analysis that it was more common in low swampy ground than on the high ground, and the winter of 1760–61 was very wet, describes conditions favoring anthrax, he notes a characteristic of the disease to be a swollen gall bladder, "*la vesicule du fiel engorgée*," a characteristic symptom of rinderpest. It seems likely, therefore, that both diseases were present.

Two years later, Britain imposed a quarantine against Germany, Denmark, and Sweden. An outbreak in Moravia in 1764 was followed by one in Schleswig, and in 1765, it was imported from Turkey into Hungary, spreading to Saxony and Austria again, probably by the same route it had done so 4 years before. France was infected also. In Italy, two broad sheets were issued, one for the recognition of anthrax (*milza* = spleen or "splenic apoplexy") with its methods of prevention and cure, the other for treatment of pleuropneumonia (*polmonea*). (An instruction for distinguishing anthrax was also issued at Hanover in 1767.) These are of interest in that the diseases were distinguished from each other, and presumably they were issued at this time to enable persons to identify rinderpest if it should occur in Italy (Figures 2 and 3). The distinction between rinderpest and pleuropneumonia had in fact been made for at least 150 years. Guiseppe Falcone, a Carmelite monk, in an agricultural treatise first published prior to 1597, wrote "Polmonera is a very bad disease: it soon separates the healthy from the sick, it is a contagious disease like plague among the bovine animals" (Falcone, 1612).

In 1766, the whole of Germany was declared "pure and healthy" until May 1767, although Nugent (1768) witnessed it at Strelitz in November 1766, the disease having come again from Turkey in 1766. In 1768, there was a fresh outbreak in the Netherlands, and it raged with renewed violence in Belgium, Holland, and the north of France, spreading also to

[1]The exact nature of the English sweating disease has never been determined but is thought possibly to have been relapsing fever.

GRATIS.

ISTRUZIONE

Per conoſcere certo male, che attacca la Milza nelle Beſtie bovine, preſervarne le ſane, e curarne le inferme.

QUeſto male, che da principio credevaſi d'Anticore, fu poi co' lumi Zootomici ſcoperto, e giudicato altro non eſſere, che una diſpoſizione infiammatoria del ſangue, che s'apprende principalmente alla Milza, ritrovata perciò in tutte quaſi le beſtie inciſe gangrenata, e fracida.

I ſegnali, che sì fatto male ordinariamente precedono, ſono un rallentamento dal cibo, ed un freddo ſenſibile all'eſtremità delle orecchie, che vieppiù creſcendo, e dilatandoſi và a terminare in un tremore univerſale, col quale cadono, e quaſi improvviſamente ſe ne muojono. Alcune poi prima d'eſſer ſorpreſe dal tremore ſi gonfiano, e talvolta dopo il gonfiamento migliorano, e ſi riſanano.

CURA PRESERVATIVA.

Primo. Si farà alla beſtia una cavata di ſangue in doſe moderata.

Secondo. Si farà una Decozione con Gremigna, radice d'Altea tagliata in pezzetti, e Malva in acqua comune, e di queſta ſe ne darà da bere alla beſtia due volte al giorno, aggiungendovi una metà di farina d'orzo, o ſegala, continovando per ſei, o otto giorni, con farne quella quantità, che la ſteſſa beſtia potrà giornalmente bere.

Terzo. Si prenderà un'oncia di Zolfo comune polverizzato, un'oncia d'Antimonio crudo parimente polverizzato, ed once ſei di Mele comune: ſi unirà tutto ad un quartajo di Cruſca di formento bagnata colla ſuddetta Decozione; e ben miſchiata ſi darà alla beſtia per tre, o quattro mattine.

Quarto. La ſperienza ha dimoſtrato utile di terminare la cura preſervativa con applicare alla gola poco diſtante dalla ſpalla ſiniſtra un cauterio, introducendovi dalla radice di elleboro, e mantenendolo aperto per lungo tempo.

Queſti rimedj diretti alla preſervazione delle beſtie, ſarà bene il praticarli in tutti que' ſiti ancorchè ſani, ove altre volte ſiaſi veduto lo ſteſſo male.

METODO CURATIVO.

Primo. Si uſerà tutta l'attenzione per iſcoprire, ed oſſervare i primi indizj del male, e immediatamente dopo ſcoperta la beſtia ammalata ſi ſeparerà dalle ſane, ponendola in ſito d'un ambiente temperato.

Secondo. Si farà ſubito ſalaſſare al primo indizio di male, replicandoci anche altre diſcrete miſſioni di ſangue, ſe il biſogno lo richiedeſſe.

Terzo. Se le farà ingojare una libbra d'oglio d'oliva, o di ſeme di lino, o di ravettone.

Quarto. Si prenderanno due metà d'orzo, o ſegala ammaccati, e bolliti in dodici boccali d'acqua comune ſino alla conſumazione della metà, e ſi darà da bere tepida alla beſtia mattina, e ſera, durante il biſogno, con unire alla bibita della mattina un'oncia di Nitro comune, e tre once di Mele.

Quinto. Il fieno, che ſi darà alla beſtia inferma, dovrà eſſer di quello raccolto nel meſe di Maggio, vecchio, e di buona qualità.

Seſto. Comparendo nella beſtia qualche gonfiamento, ſe le faranno aprire cauterj con irradicamenti, o ſetacci, invitandoli con eccitamenti, e frequenti fregagioni allo ſpurgo.

In Milano, nella Regia Ducal Corte, per Giuſeppe Richino Malateſta Stampatore Regio Camerale, e della Sanità. X 1765.

Figure 2. Italian broadsheet of 1765 on how to recognize and treat anthrax.

RIMEDIO

Per il male di Polmonea, o ſia Corata delle Beſtie Bovine.

SI piglia Once 4. Radice di Brionia.
Once 4. Radice di Genziana.
Once 4. Dittamo bianco.
Once 4. Radice di Elleboro nero.
Once 4. Radice d'Aſtrologia rotonda.
Once 3. Mirra fina.
Once 3. Aloè epatico.
Once 3. Marrubbio bianco.
Once 3. Radice di Scialappa.
Once 3. Finocchio ſeme.
Once 3. Dittamo Gottico.
Once 3. Tamariſco.
Once 3. Polmonaria.
Once 3. Agarico.
Once 3 Radice d'Imperatoria.

Si avverte, la ſuddetta doſe eſſere per dodici Beſtie.

Tutte le ſopraſcritte coſe vanno bollite in Acqua d'Orzo cotta, e fattane la decozione, ſe ne darà, eſſendo eſſa ancor tepida, alle Beſtie digiune in quantità di due pinte per ciaſcuna Beſtia, aggiungendovi tre once di Mele vergine per Beſtia.

Se poi la Polmonea ſarà morbida, o ſia umorale, alla ſuddetta Ricetta ſi aggiungerà Marrubbio, Aſſenzio, e Calino, ed il tutto ſi farà bollire in Vino roſſo, e puro in vece dell' Acqua d'Orzo, e nella medeſima quantità ſi darà da bere alla Beſtia.

La mattina ſeguente ſi porrà in infuſione della ſcorza di Fico, di Fraſſino, e di Liguſto, laſciandovela per ore dodici, e poi fatta riſcaldare, tepida ſi darà da bere alla Beſtia.

Tale rimedio ſi uſa non ſolo per curativo, ma per preſervativo ancora nelle dette Beſtie; e ſe poi ſi ammalano, ſi replica lo ſteſſo rimedio.

Figure 3. Italian broadsheet, probably of 1765, on how to recognize and treat pleuropneumonia.

Brandenburg and neighbouring countries where it was considered to have ceased. Some believed it had re-entered Holland with fresh hides imported from Hungary and Dalmatia, whereas others thought it had spread from Belgium.

It was much more virulent than before in Holland, carrying off in one year 115,665 cattle in the south and 162,276 in the north. In the first 5 months, 32,907 cattle died, and 10,405 recovered; in one instance, five in seven died. In Friesland from May to December 1769, 98,000 animals died, probably almost half the stock in the Province. In the Province of Holland from April 1769 to March 1770, 160,000 animals of a total of 225,000, excluding calves, died (71%). Between April 1769 and June 1770, 278,000 died of a total of 318,000 animals (70%). This compared with 50% mortality in Britain. Some 60% of cases occurred in winter, but this may have been because of animals being kept together in close contact in sheds during bad weather. Towards the end of the period, it had resumed its former virulence and spread rapidly, infecting areas it had not done so previously. The Dutch government offered a reward of 10,000 guilders to the discoverer of an effective remedy, a measure likely to have exacerbated the disease rather than control it. It invaded Luxembourg in June 1769 with apparently healthy animals imported from Liège spreading infection along the route.

From the start of the creation of veterinary schools in France, students were sent out to the provinces to control disease outbreaks, especially rinderpest. One lucky student who allegedly effectively controlled the outbreak in 1770 in the Duchy of Luxembourg, was presented by Prince Charles with a golden box containing a letter of exchange for 100 livres. Perhaps this was Gerardy who, on July 12, 1769, was ordered by the Provincial Council to inquire into the disease between Luxembourg and Bastogne and form a *cordon sanitaire* in each village both sides of the route. Guards were to be installed and all routes closed throughout areas where the disease was present. No beast was allowed to pass the cordon and none, sick or healthy, allowed to be brought in. All which were passed through, either secretly or by force, would be slaughtered. He went first to Steinfort, where he found nine cows belonging to two peasants had died. Forbidding any animals to leave the parishes the next day he arranged examinations of cowsheds in Sterpenich, Steinfort, Randelingen and Hagen, by three experts: a master butcher, a miller and one other. At Steinfort, five animals were slaughtered and buried at the charge of the inhabitants of Autel. An area marked by straw beacons was established around the village for the grazing of healthy animals. At Martelange an innkeeper had refused to lodge some cattle suspected of having brought the disease into the province. Near Bastogne the outbreak had ceased, but a cordon was established to prevent any communication with the few animals remaining (Theves, 1994).

An outbreak in November to December 1769 in Völckersen, Germany, was claimed by the people to have been spread by deer wandering the fields as no cattle had passed through the area. Junker wrote in Hungary in 1769 the disease had raged for 2–3 years among cattle and sheep there, and had been exceptionally virulent in the past year throughout the country. He considered Hungary was blamed for its prevalence because of its extensive marshes where the livestock ate poisoned grass. Koczians (1769) examined all possible causes of origin, such as insects in which he only considered ingestion, but finally concluded the outbreaks in Austria were due to proliferation of water mills. These caused flooding of the land resulting in obnoxious conditions for animals. Cothenius, in a memoire to the Berlin Academy of Sciences in 1768, believed the 1711 visitation had never been extinguished, persisting in Galicia, Hungary and Poland.

It then appeared again in Europe in 1770 continuing until May 1772, reappearing in French Flanders in March 1770 imported with an infected animal from Austrian Flanders where it was still present. Overall, 3 million cattle are estimated to have died in Europe in this outbreak.

2. THE 1769 INTRODUCTION INTO BRITAIN

Britain witnessed an outbreak of cattle disease in 1765, and the editor of *Notes and Queries* entitled an item by "W. D." in 1865, "The Cattle Disease, 1765, 1865," although it refers to the 1745 epizootic. Rowlin (1792) distinguished the 1765 outbreak from the 1745 cattle plague, apparently describing symptoms of anthrax. Following the new outbreak of rinderpest in the Netherlands in May 1769, an Order was passed in Britain forbidding import into Britain and Ireland of any cattle or cattle products from Dunkirk or any other part of Flanders. In December, import of hay and straw was also prohibited. But rinderpest was again introduced into England and Scotland towards the end of this year with Dutch cows imported into London. Spreading to Hampshire, Essex, and Suffolk, it was extinguished by January in the following year, 1771, the Government issuing orders to kill all affected cattle forthwith and those which had been in contact with them. In Hampshire, this led to a chase for seven Alderney cows across several counties, their secondary contacts also being slaughtered.

On January 9, 1770, the King opened Parliament with the words: "It is with much concern that I find myself obliged to open this Session of Parliament with acquainting you that the distemper among horned cattle has lately broken out in this kingdom notwithstanding every precaution that could be used for preventing the infection from foreign parts."

Orders based upon those of 1747 for Middlesex were issued, applying to neighboring counties also. This time, in order to prevent any escape of contaminated blood, Layard, who claimed it was "by mere accident" that the infection was discovered to have been imported from Holland to London, advised that cattle be slaughtered by strangling, and hides not slashed before burial. Also, all contaminated fodder and dung should be buried instead of burned.

In Scotland, making its first appearance in March 1770 at Portsoy, a seaport town in the Moray Firth west of Banff, it was allegedly introduced in hay or straw imported from Holland as packing, although an investigation by JPs in Banffshire was unable to discover "the remotest evidence of the least hay or straw having been imported about any packages" (Cottrell, 1770). The JPs ordered immediate slaughter of all cattle in the neighborhood, whether infected or not, fully recompensing the owners. Carcasses were buried 1.2 m deep and infected cattle stalls thoroughly cleaned. This was effected on March 21 and 22, 142 animals being killed. At the same time, all cattle within 3 km of Portsoy were inspected daily by four farmers appointed to the task. Instructions were also issued to fishermen not to smuggle goods from any ship which might arrive on the coast. All these measures were put into effect some 10 days before the official prohibition was enacted. For 8 days, the inspectors reported that all was well and then, on April 4, "to the unutterable astonishment of everybody," one cow died, and four more were reported sick 5 km from Portsoy. The next day another report was received. The JPs again ordered immediate slaughter of all cattle, both sick and healthy. One of the tenant farmers was forbidden to carry out his intended discharge of a cowhand for 1 month and ordered to keep employees on the farm as much as possible. An Order of the Privy Council issued for Southampton in December was extended to Scotland on April 5, empowering officials to assess the amount of indemnities to be paid and to order such steps as they deemed necessary to prevent the spread of the plague. The law officers also applied to Cullen to suggest what he thought might be useful. Cullen replied with a short *Memorial Concerning the Contagious Disease Affecting the Horned Cattle* (Cullen, 1770), which, together with an order of the Privy Council and other documents, was transmitted to the several sheriffs of the district. Following Lancisi, Cullen assumed that the disease could be spread readily not only by direct communication, but indirectly by substances which had been in contact with diseased animals and especially the clothes of persons which could spread it from one person to another and then to cattle. Cullen aimed to point out the less obvious modes by which infection might be communicated, as by fomites of various sorts, and to suggest means of obviating these (Thomson, 1859). Cullen observed, "All these facts are very well known to physicians, from the history of the plague, and other contagious diseases among men; and there can be no doubt of the same taking place also in the contagions among brutes," stressing complete isolation where the disease occurred. Stating that it had been found that dogs may carry the infection, he did not propose killing them but stressed that they should be kept away. The measures, whether those already in place or suggested by Cullen, were effective, and the disease ceased.

In May 1770, the Government was obliged to pass an Indemnity Act, as there appeared to be no legal power to issue the Orders it had given in February providing for compensation for slaughtered cattle. In September of that year, a continuation Order was issued, consequent upon another outbreak in Flanders. A total of £796 was paid out in compensation during the year.

In 1770, unidentified disease outbreaks occurred, without spreading, at Welford in Northamptonshire and near Southampton. A grazier recalled occurrences about 40 years before at Dalby and other parts of Leicestershire where it was called "Gangrenous Gulpy," perhaps "tongue-cancer."

3. EUROPE FROM 1770 TO 1800

In 1771, rinderpest was brought into Calais with a cow purchased in Bergues, and it raged in Brabant spreading to Flanders, Picardy, and other French provinces. In Austrian Flanders, the physician to the Empress, Dr. Le Cat, was ordered to investigate and concluded that its spread was due solely to the communication of the virus from one animal to another, recommending that all infected cattle be killed. Four hundred head were slaughtered, and the disease disappeared. In 1773, it was still reported as active in Holland and erupted again in Flanders, spreading to Picardy, which was just beginning to repair its losses. Allegedly, it was introduced by a sick cow brought from the Low Countries to Hainaut, appearing in the village of Groiser. It ravaged part of Flanders, especially Lille, and soon Soissons and Amiens, above all spreading along the Oise. The same year, the great Swiss physiologist and anatomist, Albrecht von Haller, published an account of what he considered to be rinderpest in Switzerland (Haller, 1773) but some now believe to have been pleuropneumonia. Gamgee believed it to be rinderpest and stated that Haller proved that it resulted from contagion (Gamgee, 1877).

The occurrence of an epizootic disease in Denmark in 1772 was not confirmed as rinderpest. During these years, outbreaks of pleuropneumonia, anthrax, actinobacillosis, FMD, and blackquarter were not infrequent, and sometimes their ravages may have been attributed to rinderpest. But in 1877, in evidence to a Select Committee, Gamgee stated that Britain had imported rinderpest from Denmark in about the 1770s, not once, but repeatedly. In 1865 in Edinburgh, two thirds of all animals examined for rinderpest were found also to have had pleuropneumonia, but in London, only one or two cases were found, although other animals had died of pleuropneumonia, the deaths of which had been attributed to rinderpest (Murchison, 1866).

In Bremen and Verden, Haltermann not only collected reports of outbreaks of rinderpest in the State from 1715 until the time of his writing but was the first to attempt to quantify progress of the epizootics both in time and space (Haltermann, 1773).

Rinderpest continued to be active in Holland in 1774. On February 5 an amendment to an Ordinance of January 7, concerning the presence of rinderpest in Luxembourg, issued at Namur in the Spanish Netherlands, allowed imported cattle for the provisioning of Namur brought by the "new route," or from Brabant, or by the Sambre and Meuse, and the length of the Sambre, to pass freely, even though they had to cross tongues of land or enclaves of Liège. As long as villages depended on these regions, if not infected or not suspected to be, raw skins and hides from the town of Mons destined for Namur were allowed, if accompanied by a certificate from the slaughterer at Mons, authorized by a magistrate, that the products came from healthy animals. In 1774, it reentered parts of France, Flanders, and Picardy. In August, it appeared suddenly in the south of France close to Bayonne, believed to have been introduced with fresh hides imported from Holland or from Artois and disembarked at Bayonne. One popular belief was that reins used on cattle made from the imported skins carried the infection; another was that it was introduced with milk cows sent by Hugenot refugees in Holland to others in the Midi after the Edict of Nantes (Reynal, 1873). Since 1740, the southern provinces, which had numerous herds of cattle, had been free of the disease, but the disease now spread to join other fronts of infection until it covered the whole of France, except for some provinces in the west, reaching the borders of Bordeaux. It was estimated that 150,000 cattle died in the southern provinces. In 1775, it was present in Savoy, prompting Voltaire to write from Fernay to Claude Bourgelat of the veterinary school at Alfort, on March 18:

> I wish that one would search for preventatives against the contagious diseases of our cattle when they are in good health, instead of when they are sick. One could then, in 100 cattle attacked, try a dozen different remedies and one could reasonably hope that of these remedies there would be some which would succeed.
>
> There is, at the moment, a contagious illness in Savoy, in a place of mine. My preventive is to have no communication with the sick, to keep my cattle in the greatest cleanliness, in large stalls well aired, and to give them healthy food ...
>
> The great misfortune of the peasants is their stupidity, and another misfortune is to be too neglected: one thinks of them when the plague ravages them, they and their herds; but provided there are pretty girls at the Paris opera, all is well. I would be very obliged, Sir, to well continue your kindness to me in communicating to the public knowledge of which they could benefit. (Courbet, 1927).

From 1775, the outbreaks in France no longer came from Italy to the south but from Holland in the north because of the immense stock trade there. An anonymous German writer (Anon., 1776) claimed that an outbreak began in Holland in 1775, spreading to East Friesland and from there to Oldenburg and Bremen, and reaching Hannover at the beginning of summer. A dealer in Stobel district had received about 40 animals which were mostly infected, obtaining a health pass from the clerk for 50 Reichsthaler. He distributed the animals in three parts: one to Hannover and one to Zelliche, and the strongest group, comprising 16 animals, was driven secretly at night to Hildesheim. Some of the animals were rounded up, and the drovers were still in prison awaiting judgment. Rinderpest then broke out in several different places at once, but some villages through which the cattle had passed were unaffected. Other animals went to Brunswick where they were slaughtered as healthy, and no disease broke out. In the summer of 1776 at Röhrfelde, 55 cows were grazed on common land. Although the herd in the town was infected, a milkmaid went illegally to the common and milked a cow there morning and midday. After 8 days, the cow sickened, but the herder, noticing it, isolated the animal, which then died, the remainder escaping infection.

Whereas an edict of 1756 ordered slaughter of infected animals, it was applied too sparingly, and beginning in 1776, Hildesheim lost 14,000 head, Hannover and Zelliche 8,000, and Bremen 12,000, to the disease. Hannover was always infected in the same way, by illicit movement of infected animals. A local judge in Barkhausen, Lower Bremen, wrote in 1775 that with 30 years' experience of rinderpest, he could state that it originated not in local and neighboring areas but with foreign fatstock from East Friesland traded at the October Bremen cattle fair.

It reached the district of Altmark in April 1775. One cow was reported infected on April 15 at Mahlpfuhl, dying on April 30. The remainder of the herd of 43 were then slaughtered, an edict of 1769 requiring that if it broke out in a herd of more than 12, all animals in the herd should be slaughtered. By about the middle of August, it was in the center of the Altmark district. The summer was extraordinarily hot and dry with drought conditions prevailing, which probably aided its spread. It was not known how it arrived in Mahlpfuhl, which was in the middle of the district, but from there, it spread to Mahlwinkel, Vethen (where the cattle were healthy up to June 27), Birkholz, Briest, Sethen, Rurförde, Zienau, and Letzlingen. The following summer, it had reached the district of Tangermünd and, lastly, Arneburg. In Drömling, it was reported that hardly had the disease broken out when a wolf was seen, the last having been taken in the area a century ago (Anon., 1777).

Faust (1797) observed that the Hamburg district escaped infection in 1745 and 1776–7 because of wise legislation, conscientious officials, and a cooperative public. Prevailing throughout Germany, the disease was described at Lindau by Kersting, at Pattensen by Dr. Lentin, and by another observer at Walsrode, finding it comparable with the Dutch, Danish, Swiss, and other outbreaks. In 1775, Joseph Fehr, a military surgeon, was sent to contain it at Münster.

Toward the end of 1774, a young doctor, Felix Vicq d'Azyr, was sent as commissioner by the French Academy of Sciences at the request of the French government, to travel to affected areas and inquire into the plague. He was witness to the fact that an infected calf was cured by careful nursing but, having recovered, communicated the disease to other cows. Only then was enforced killing of infected animals imposed based on Vicq d'Azyr's recommendation and that of Claude Bourgelat, who had established the first veterinary schools at Lyons and Alfort. Reimbursement was set at one-third the value of the beasts, but there is evidence of local and Church agencies supplementing the amount (Broad, 1983). When it broke out in the winter causing great mortality in Normandy, infected districts were encircled by troops and all infected and suspect animals killed and buried, Paris wits jesting that the army was being sent to fight sick cows. Stalls were disinfected and the disease controlled. Nevertheless, it continued to ravage other parts of France in 1775 and 1776. In 1775, the army blockaded Landes, Gascoigne, and Béarn, where the disease was raging. Any animal leaving the areas was shot and the owner severely punished. The population, who were obliged to feed the soldiers, rebelled, and many troops were killed.

Vicq d'Azyr was only 25 when, in August 1776, he published his book *Exposé des Moyens curatifs et préservatifs qui peuvent être employés contre les Maladies des Bêtes à Cornes (Account of Curative and Preventive Methods Which Can Be Used Against the Cattle Diseases)*. A short section had been issued in Lille in January the year before with instructions on the slaughter of all infected animals and their healthy contacts. This was to be carried out humanely by veterinary surgeons, farriers, or surgeons, under the direction of the military, and the carcasses buried. Farmers should be heavily penalized for not declaring sick animals. Cowsheds were to be disinfected, and skins disinfected with limewater could be saved (Vicq d'Azyr, 1775). Doctors ridiculed Vicq d'Azyr for his proposal of slaughtering infected animals, which they sarcastically referred to as an "ingenious" method of cure: "... one can only stop the contagion and destroy the disease, by sacrificing all of the sick ...," he had written. And: "None had ever cured the animal plague, and if one wanted to be truthful, one would agree that the few sick which escaped [death], owed it to nature ...." But elsewhere: "It is again well proven that the best of methods, if one considers them altogether, never cured more than a third of the animals attacked ...."

Slaughter was claimed by Vicq d'Azyr (1776) to have been an Austrian Netherlands initiative copied by the British. However, he stated, "Everywhere that one follows the most widespread system of slaughter as in England ... the disease has been completely destroyed." Opposing his recommendations for a "general massacre," which included killing all loose dogs and cats as well, his opponents probably also did not like his support for the idea that human and animal medicine were complementary.

His other recommendations were that chickens be shut up, and horses and sheep kept isolated. Also, where trade in cattle was permitted, this should be only with a certificate to the effect that each animal had been in good health for at least 6 weeks since arrival, and in the parish whence it came. He claimed that this would prevent smuggling animals out of infected areas.

In Austrian Flanders, his recommendations were taken much further. If the disease broke out on a farm, not only were all beasts to be killed which had been in contact with the infected beasts, but all house dogs and cats had to be killed, "even [dogs] in the court yards of owners," and thrown into the burial pit. In addition, from the appearance of the disease in a parish, all rabbits had to be killed.

Spain usually escaped invasions of the disease, but in July 1774, it arrived in the adjoining province of Labourd in southern France, and it was urged upon the Spanish government to prohibit import of cattle and sheep. Instead of acting as advised, the government sent two veterinary surgeons to investigate, who deduced that the disease consisted of a softening of the brain and recommended slaughter of all affected and suspected animals and their burial in deep pits, with compensation paid to the owners. This recommendation also was not acted upon, and the disease passed to St. Sebastien in the Pyrenees on the Spanish frontier. From thence, it spread quickly through Spain with, in some cases, total mortality of cattle in many villages and towns, as at Andoin in the province of Alva. The enzootic spread through Navarre, Guipuzcoa, Aragon, and the mountains of Santander and Pas (Villalba, 1803).

In the same year, an outbreak was said to have occurred at Laxfield in Suffolk, England, confined to one herd of cattle and suppressed by the enforcement procedures defined in the Orders of Council. There is no confirmation that this was rinderpest, but allegedly £1,684 was paid out in compensation by government in 1774–5, which, if correct, suggests a much wider outbreak.

As it spread throughout southern France in 1774 and 1775, Turgot, the comptroller-general who had been intendant at Limoges, introduced a military operation of slaughter preventing its extension to the rest of France (Montigny, 1775a); although in 1776, it was introduced to the island of Corsica. Haller wrote on September 7, 1777 to Vicq d'Azyr that in Switzerland, they prevented outbreaks by always shooting all infected animals.

In Germany, a broadsheet was issued on April 25, 1776 under the authority of His Serene Highness the Prince of Trier, instructing his faithful subjects to take notice of the various preventive remedies which had been used successfully for both healthy and affected cattle in neighboring states, and that everyone "for his own good be warned that all ridicule and prejudice be put aside and the prescribed preservative cure or remedy be followed precisely" (Figure 4). The Bügermeisters of Langwedel reported (April 27, 1776) cattle being moved from the pastures into "hospital-cottages." Because of the cold and rain, the cattle ran home easily and were then kept boarded up in the cottages. In June, the Rothenburg office wrote to Verden requesting that residents of an infected village keep their stock 200 paces distant from the Eversen, evidence of continued presence of the disease in Bremen and Verden.

The following year, in one of the more curious contributions to the literature of rinderpest, Dr. Johann Caspar of Ruess published a treatise on forensic medicine covering various forms of violent death with a view to providing a death certificate for the courts. To this, he added an appendix on cattle plague, apparently being opposed to slaughter and noting the renowned Swiss, Dr. Haller, who stated that there was no cure for infected animals and they should be slaughtered and buried intact, should have looked at methods of cure before recommending slaughter. Caspar described the symptoms and action to take if the plague struck but noted that it was very damaging to throw unslaked lime onto carcases in the pits as the fumes produced could carry the infection with them, and the lime should be sprinkled on. Authors did not describe other diseases to distinguish them from the cattle plague, so an illness of the ox described by Dr. Baron von Swieten was given, together with the remedies for it.

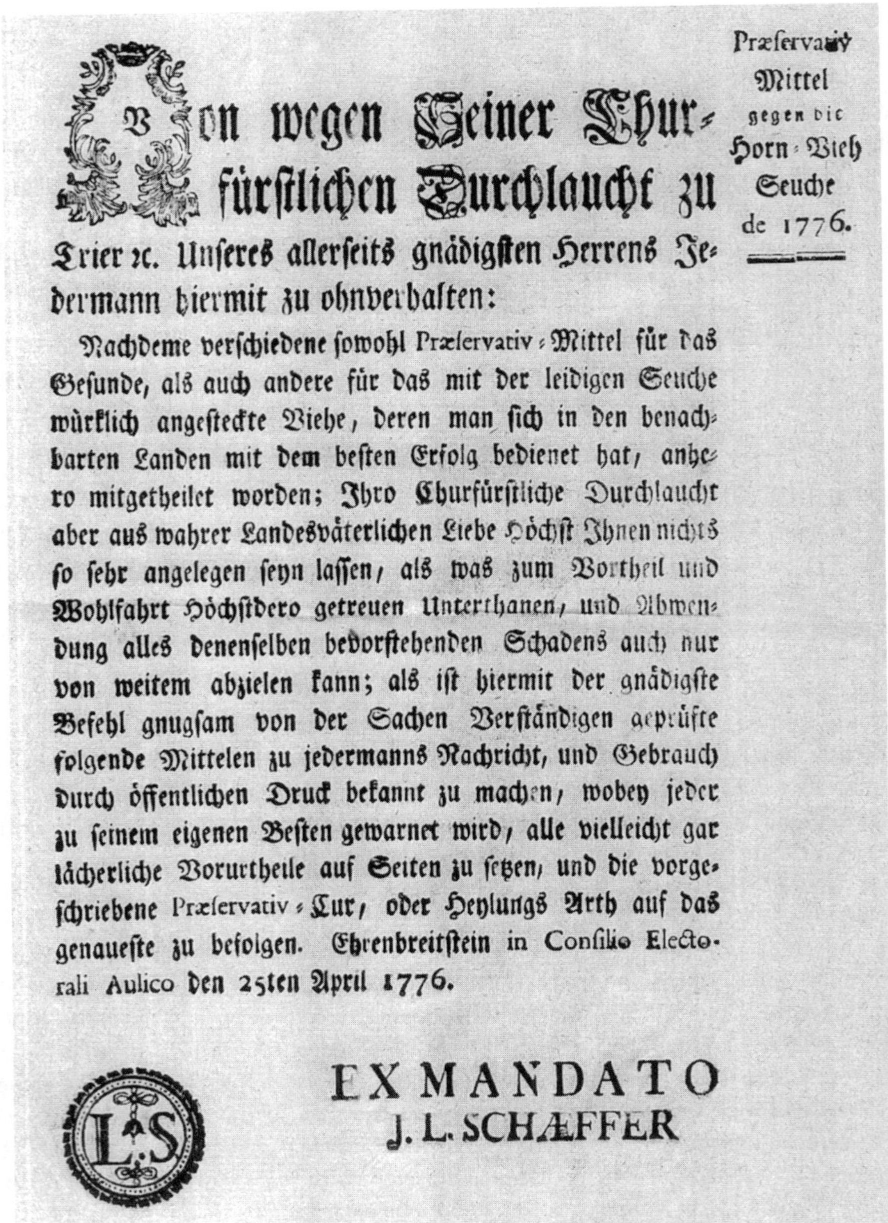

Præservativ Mittel gegen die Horn-Vieh Seuche de 1776.

Von wegen Seiner Churfürstlichen Durchlaucht zu Trier rc. Unseres allerseits gnädigsten Herrens Jedermann hiermit zu ohnverhalten:

Nachdeme verschiedene sowohl Præservativ-Mittel für das Gesunde, als auch andere für das mit der leidigen Seuche würklich angesteckte Viehe, deren man sich in den benachbarten Landen mit dem besten Erfolg bedienet hat, anhero mitgetheilet worden; Ihro Churfürstliche Durchlaucht aber aus wahrer Landesväterlichen Liebe Höchst Ihnen nichts so sehr angelegen seyn lassen, als was zum Vortheil und Wohlfahrt Höchstdero getreuen Unterthanen, und Abwendung alles denenselben bevorstehenden Schadens auch nur von weitem abzielen kann; als ist hiermit der gnädigste Befehl gnugsam von der Sachen Verständigen geprüfte folgende Mittelen zu jedermanns Nachricht, und Gebrauch durch öffentlichen Druck bekannt zu machen, wobey jeder zu seinem eigenen Besten gewarnet wird, alle vielleicht gar lächerliche Vorurtheile auf Seiten zu setzen, und die vorgeschriebene Præservativ-Cur, oder Heylungs Arth auf das genaueste zu befolgen. Ehrenbreitstein in Consilio Electorali Aulico den 25ten April 1776.

EX MANDATO
J. L. SCHÆFFER

L.S.

Figure 4. Broadsheet of 1776 issued by the Prince of Trier advising citizens to follow antirinderpest treatments.

In 1776–7, southern Germany experienced a very virulent outbreak. At Bückeburg from August to the beginning of December 1776, of 108 cows, 105 died, and three recovered, followed by more deaths to a total of 200. In the surrounding villages 1,003 died. In the town of Stadthagen, 311 died; but in the district, of 3,723 animals, only 213 died between October 1776 and February 1777 owing to legal action. Likewise, in Hagenburg district, of 520 animals, only 10 died. Table 2 shows some mortality rates in 1775 and 1776, which tend to dispute Faust's assertion that slaughter reduced the rate.

Table 2. Some mortality rates in Germany 1775–6.

Locality	Stock	Deaths	Slaughtered	Recovered	Uninfected	Percentage Dying	Percentage Slaughtered	Percentage Uninfected
Minden town	1,336	848	109	48	331	63	8	25
Dukedom	2,834	1,847	0			65	0	
District	9,952	1,738	252	249	6,717	17	2.5	67
Bielfeld town	576	21	6			4	1	

Faust (1797) records how a poor widow's only cow, the milk of which she depended upon for her children, died. Yards and villages were closed, children could not go to school, and people could not go to church. But after 1777, southern Germany became free of it, and the rest of Germany by 1782.

In Holland, slaughter was not countenanced, as there was no compensation, neither was inoculation, so the disease continued, with no strong central authority in the United Dutch Provinces to exercise control. Camper had attempted to introduce inoculation, but the objections of the peasants obliged him to discontinue. Then, in the 1770s, a Gröningen farmer, Geert Reinders, successfully inoculated the calves of mothers which had recovered from the disease as a result of which, inoculation was undertaken on a wide scale in the northern provinces. In Denmark, where the disease had apparently become enzootic, Layard, invited by the Danish government to advise on it, recommended inoculation, and the campaign was said to be successful. However, royal decrees against rinderpest were issued in 1775, 1777, and 1778. In October 1785, they were suspended, not to be reintroduced unless there was a reoccurrence.

In 1779, at Steyermark in Austria, 10,000 head of cattle had died, in Styria 90% of stock, while Belgium was reported as "nearly robbed of the whole of her herds." In the following year, France issued a royal decree controlling imports into France.

4. THE PLAGUE IN ENGLAND AGAIN

The eruption of the plague on the Continent and the appearance of d'Azyr's monograph in 1776, in which he claimed that the British had adopted slaughter with "the most complete success" by scrupulously translating and putting into execution edicts issued from the Juntos of Ghent and Brussels, prompted Layard to write to Joseph Banks, President of the Royal Society, to provide what information he had on the disease and ways of preventing it. In April 1770, Layard had been appointed foreign correspondent by the King and communicated to the Dutch, Flemish, French, and Swedish, documents concerning control of the disease in Britain; and in France, Flanders, and Picardy, the method of slaughter was successfully adopted. He recalled that in 1769, positive orders were issued that all cattle should be killed and buried properly "by which vigorous and salutary directions the distemper ceased intirely in a short time." This procedure, far from being copied from the Belgians, was, on the contrary, communicated to them, as well as to the Dutch and French, by Layard himself, who claimed that the "method" had been first adopted in England in 1747 (Layard, 1780). It had, of course, been implemented by Bates in 1714.

The year following publication of Layard's letter (1781), the disease was imported once more into England, apparently from Ostend to Margate, allegedly by the unlikely medium of two infected sheepskins abandoned on the beach sniffed and rolled on by a cow, but more probably by illegally landed animals. Not more than a couple of dozen cows died or were slaughtered altogether, an Order of Council being put into effect prohibiting movement and requiring the killing and burial of all infected and suspect animals.

Two years later (1783), a "distemper" broke out in Derbyshire, and an Order of Council was implemented, but the account by Dr. Kirkland (1783) of an outbreak in Caulk suggests that it was anthrax. The disease caused rapid death, the animals usually dying with swollen throats, and, as Kirkland observed, the disease was confined nearly to the spots where it began.

According to Broad (1983), later 18th-century outbreaks in Britain were kept secret, the Privy Council dealing with the county Lord Lieutenenat and one or two JPs only, often sending Orders in Council specifically to them rather than to the bench as a whole. But there is no evidence for such a "conspiracy of silence." It is hardly likely that Kirkland would have been able to publish his booklet if there was, and it is difficult to see how outbreaks would have been kept from the press of the day.

5. RESURGENCE OF WAR AND RINDERPEST

Greece, Smyrna, and the Crimea, suffered the ravages of rinderpest in 1783. Imported from Hungary, it also affected Verona and the Venetian Provinces, lasting until 1784, when it spread once more to Germany. A broadsheet issued in Germany about this time, probably at Augsburg, *Bei der vor etlichen Wochen unter dem Hornvieh sich geauserter Krankheit, ist zur praservation des gesunden Viehes nothwendig (Necessary Preservation of Healthy Cattle During the Several Weeks Before the Festering Cattle Sickness Ceases)*, stated that the plague had been spreading for some weeks and gave instructions for preserving the health of cattle, including treatment with various powders (epsom salts, saltpetre), purging, and advice concerning bleeding and the cleaning of cowsheds. A pamphlet was issued *Mittel zur Praservation fur der Hornvieh-Seuche (Means of Preservation from the Cattle Plague)*, probably of a similar date and place. By 1783, one Lutz had collected some 1,000 papers on rinderpest.

In 1781, more than 300,000 head of cattle were again lost in Holland. In the Province of Utrecht from 1769 to 1784, more than 400,000 animals died, and in 1786 it entered the Rhine and Bayern and Swabia. In 1787, Frederick Wilhelm II of Prussia commissioned a national school of veterinary medicine in Berlin owing to the failure of the German medical authorities to stem the serious losses of livestock and military animals from rinderpest and other epizootics (Schwabe, 1978). In 1789, what was probably rinderpest was reported in Croatia, Krain, and the Tyrol. Then, following the beginning of the French Revolutionary Wars, in 1793 it was carried into Lombardy by Russian and Austrian troops fighting the French and by the end of the year was once more in Piedmont, brought in by Hungarian cattle for the Austrian army, where it was described in detail by Buniva in 1793 and 1798, whose main interest was in emphasizing methods of control (Buniva, 1808). In 1796, the French army, retreating from the Austrians on the Rhine and Sambre-et-Meuse, brought it into France once again, and with its appearance in the Upper Rhine in 1797, more than 12,000 cattle died in a few weeks. With no control measures being applied, ration cattle became mixed up in the retreat. It then spread rapidly to Alsace, Lorraine, Champagne, Franche-Comté, and Picardy with a loss of 130,000 animals, eventually half of France being affected. When in the Lower Rhine 195 people died in the summer, the peasants believed the disease to be contagious. Two commissars, Huzard and Desplas, were appointed by the revolutionary government to combat the disease but without success. In 1797, an appeal was made to Faust in Germany who sent a copy of extracts issued in poster form from his book, *Über die Rindvieh-Pest*, titled *Noth-und Hülfs-Tafel zur Verhütung der Rindviehpest, oder Viehseuche*, which was translated and issued as *Avis au public pour prévénir et détruire l'épizootie ou la peste des bêtes à cornes (Advice to the Public to Prevent and Destroy the Epizootic or Cattle Plague)*, but the measures were not followed. In Haut Saône, the government returned to sealing off areas using the national guard, a measure which the Minister of the Interior had not been in favour of, regarding it as extreme. In order to maintain stock, the slaughter of calves and heifers was forbidden, and in the region of Paris, fairs and markets were suspended and health passes required for milk cows to enter the capital. The outbreak lasted in France until 1802.

While campaigning in Italy in 1795, Napoleon received an instruction from the Minister of the Interior to be on guard against infection in the herds brought for provisioning the army. He replied he could not take notice of the instruction because the symptoms of the epizootic had nothing different in common with that which reigned there and had deprived Italy of 200,000 animals. Spreading over the whole of Italy, in 1796 it killed 19,000 head at Bologna, continuing until 1801 and allegedly destroying 3–4 million cattle in the first 3 years alone. A broadsheet issued at Verona in June 1796 required that all human and animal carcasses, or parts thereof, be buried at least 2 m below ground. The penalty for contravention was a fine of 30 ducats, the money to be distributed among the poor (Figure 5).

In 1795, it erupted in southern Germany, brought in by the continual passage of herds of Hungarian cattle for the Austrian army on the Rhine, spreading to all the Lower Rhine provinces "with fearful rapidity." In Würtemburg, of 72,676 head, 46,381 (64%) died in 1796–7, and in Dillingen of 1,400 head, 1,360 (97%).

In 1796, a booklet on the current epizootic *Anleitung uber die Gegenwartige Hornviehseuche (Guidance for the Present Cattle Plague)* was issued by Dr. von Koferle, Oettingen-Spielberg council member in south-west Germany, which he began by stating that it was undoubtedly caused by the present war. He identified the symptoms and described post-mortem appearances from animals which he had dissected. He listed ways in which it was spread, on clothes and in the air, and gave methods of prevention of spread: keeping animals stalled, not allowing any dogs to roam, keeping stalls clean and fumigating them. He did not recommend harsh methods, such as treating with powders of juniper berries, sulfur, bay berries, gentian, linseed, and staghorn. An Italian, Atleta Filopatride, wrote a booklet in the same year, but it was not published until 1800. The author discussed the current outbreak (1796) in the Roman States and surrounding areas, and described the precautions to be taken to control spread, calling for areas with infected cattle to be fenced off. The clothes of people entering such areas should be fumigated afterwards, especially woollen clothing. Filopatride pointed out that Denmark and Holland had been trying to find a cure for the disease since 1781, but none had yet been found. Cases he quoted where people had become ill and died following contact with affected animals or their skins undoubtedly referred to anthrax.

In 1796, it entered France from the east and appeared again in Holland. Up until then, Westphalen and the Netherlands were free of it, but then the disease was reintroduced by the Seven Years War with a big influx of transport oxen from Hungary and Poland. Again, ravaging southern Germany, by 1797 it prevailed in the Danube, Bayern, Swabia, Franconia, and the Rhine. Appearing in Bergen, Westphalia, Cologne, and Clevischen, it was only 40 hr from the boundary with northern Germany. Faust pointed out that in Flanders, the slaughter of 424 animals saved a further 111,536; and in Castellani Courtray, 128 slaughtered saved 25,693. Of 3,399 animals in 10 villages scattered in Germany and Sweden, 48 animals died, 48 were slaughtered, and 3,303 were saved.

It did not reach Switzerland until 1798, then raging until 1801. In a little book on common diseases of stock published there in 1794, Muller, a veterinary doctor, distinguished *louvet* from "pestilential fever," classifying the latter under "contagious illnesses." Referring to its other names as catarrh and *peripneumonie* suggests that he was referring to pleuropneumonia (Muller, 1794). In France's Upper Rhine region, 12,000 animals and 195 people

NOI NICCOLO' FOSCARINI K.^r

Per la Sereniſſima Repubblica di Venezia ecc.
Provveditor General in Terra Ferma.

PEr ovviare al pericolo, che naſcer potrebbe dai traſcurati, o mal' eſeguiti ſeppellimenti dei Cadaveri; Commettiamo a tutti li Reggenti delle Comunità di queſto Territorio, che trovando nei reſpettivi Diſtretti il Cadavere, o le Viſcere, o qualche parte d' Uomo, od' Animale di qualſiſia ſpezie, debbano con la maggior prontezza farne il ſeppellimento a tale profondità, che vi ſiano ſovrappoſti almeno piedi ſei di Terra; indi portarne ogni volta la relazione a queſta Carica Generalizia.

Sarà levata agl' inobbedienti la pena di Ducati 30. per eſſer diſtribuiti fra i più indigenti del Comune in cui ſarà occorſa l' inobbedienza, e ſarà tenuto aperto Proceſſo d' Inquiſizione per diſcoprire, e punire i Contraffacenti; Et ſic ecc.

Verona 18. Giugno 1796.

{ K.^r FOSCARINI Provveditor Generale in T. F.

Giacomo Sanſoni Segretario.

IN VERONA, Dalla Stamperia Ducale di Domenico Carattoni.

Figure 5. Broadsheet of 1769 issued at Verona requiring proper burial of human and animal corpses.

died, the latter probably from typhus, although the deaths were attributed to the cattle disease. Altogether, in France, 130,000 animals died, and great destruction of cattle was caused in Bohemia, Westphalia, Poland, and adjoining regions in 1799, "... wherever during war Russian and Austrian parks of cattle followed the movement of armies the cattle plague appeared" wrote Gamgee (1866). Frank (1802) noted that in Poland, outbreaks first occurred in July and August. September experienced dull autumn weather, and the plague intensified, with animals dying so quickly that premonitory symptoms were not seen. Many animals driven out in the morning were dead by midday.

Faust (1797) summarized the ways in which infection was spread. The three most important in his view were:

1. Most frequent and most dangerous was the mixing of infected animals with healthy on roads, in stalls, or when grazing. One sick animal could infect many hundreds.
2. A sick cow left infection behind it.
3. People, animals, and things, were contaminated by contact with sick animals.

Other methods were:

1. Livestock owners gathered around the stalls of sick animals to gratify their curiosity and to watch methods of treatment.
2. Cattle dealers, butchers, and Jews (who played a prominent part in trade), coming from infected places into healthy stalls.
3. Quacks.
4. Farm hands and maids changing places of employment. (Presumably if an employer's stock died there was no longer work for them and they sought employment with those who still possessed animals.)
5. Ox carts and travellers.
6. Herders and others who had dealings with both sick and healthy animals.
7. Knackers who drove off diseased animals to skin and bury them.
8. Beggars and tramps sleeping in infected stalls and barns.
9. Horses, sheep, goats, pigs, dogs, and cats, coming from infected places.
10. Raw skins and other products from infected animals.
11. Hay, straw, wool, clothes, rags, tools, etc., from infected places.

Faust, who argued for a slaughter policy, proposed a countrywide system of zoning to control the disease, dividing it into five zones from the border, labelling them A–E, with various actions to be taken in each to prevent spread from the first. There should be a 1-km-wide livestock-free boundary between A and B marked by straw wisps on poles, and total prohibition of all movement from A into B and B into A. In B, stock should be confined and stall-fed. No stock should be moved from B to C and from C to B except for limited markets with restrictions in C for stock from B, but not more than 10 animals to be driven in one herd. No stock should move from C to D, but movement from D to C could take place. E was the free zone (Figure 6).[2]

In Britain, it was believed to have broken out at Frostenden in Suffolk in 1799, an Order of Council being issued for its control applying provisions of the Order of 1747. All fairs and markets within 16 km were prohibited, and compensation of half the value was paid for all animals slaughtered. In the following January, it appears to have broken out in the county of Orkney and was dealt with in a similar manner. However, the outbreaks were suppressed too quickly for their true nature to be determined, and Fleming (1871) points out that they may not have been rinderpest, people now attributing other diseases to it.

Despite its spread by war on the Continent in 1793, Frank (1802) noted that since it had appeared in the Ukraine in 1709 and spread to Poland, no ten years had passed without its presence in Hungary, Poland, Germany, and Holland. He attributed its persistence in South Prussia (former Poland) to a lack of observance of the Prussian laws, which were resisted by the Polish aristocracy. Dead animals were not properly buried and were dug up by dogs; skins were not treated but hung up to dry in the stalls or spread on the ground. The people ate the infected meat and secretly sold sick animals. The clergy taught that the plague was a punishment, and

[2]A three-zone system separated by stock-proof fences is required by present-day EC regulations for meat exported from FMD enzootic areas in Africa, such as Botswana. Zone A is the potentially infected zone, in B all stock are inoculated, and C is the free zone. No movement is allowed between the zones.

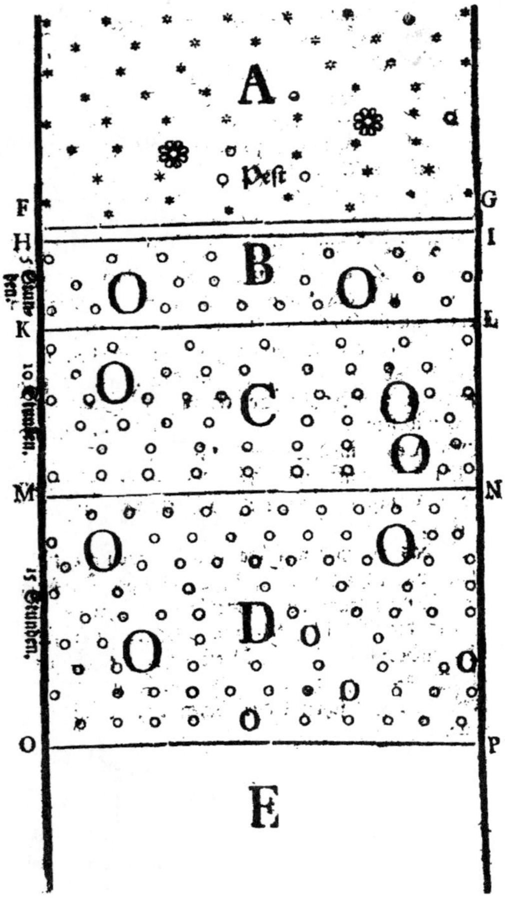

Figure 6. Faust's proposed zoning method for the control of rinderpest, 1797. For explanation, see text.

people must pray for deliverance. Stockowners drove their herds to the church to be sprinkled with holy water to drive out the demon whether rinderpest was present or not. The peasants were distrustful of State medicine but willing to waste their money on the advice of charlatans, shepherds, and old women. Isolation of infected communities did not work as little boys or old women were sent out to bring in supplies. If a nobleman arrived at the gate and was refused entrance, he would send a gift of some animals and was let in. Straw, hay, wool, and poultry were all sent to market from infected areas. After the cattle had died and the plague ended, the elbow-deep dung in the stalls was spread out in the yard, and when new stock was brought in, the disease broke out again. It was the Polish custom that ears, hoofs, horns, hair, or a piece of meat of dead cattle was left in the stall or thrown on the pasture. Frank had seen noblemen leave dead animals lying on their neighbor's border. He believed it could be controlled by inoculation but, having provided detailed instructions for cleansing stalls, then stated that the dung should be spread on outlying fields so that the air could carry away the poison.

6. CONSEQUENCES OF THE RUSSIAN WARS

Camper, who gave lectures on rinderpest at Gröningen University to audiences of regents and other prominent persons, as well as to students, believed it had become enzootic in Holland (Camper, 1803).

Having raged in parts of Poland from July 1800 to February 1801, the remaining European foyers of the disease in Westphalia, Bohemia, and Poland became extinguished in 1802, but while war was largely confined to the west, the plague returned in the Russian steppes, and in 1806 spread thence into the districts of Lithuania, Prussia, Silesia, and Courland. Napoleon's retreat after the battle of Eylau favored its spread, and it continued for 2 years, almost exterminating the entire cattle population. In 1811, following in the track of the Austrian wing of the allied armies, it was spread through Germany and the North of Europe. From 1800, it had been absent from France, although the country was still engaged in warfare, but was introduced again in 1813–14 when the French armies came into collision with the Austrian and Russian troops at the end of the Napoleonic War. The plague raged both there and in the Rhine provinces and Switzerland, encouraged by the movements of troops and their provisioning cattle, the German and Austro-Hungarian armies choosing steppe cattle for supplies in preference to others because of their strength and ability to walk long distances (Figure 7). On December 24, 1813, the disease was reported at 20 different places in Germany. It had been absent in Belgium since 1763, but from 1813 to 1815, several cases occurred between Namur and Luxembourg, said to have arisen from the passage of the Austrian army into France. The army's route was south of the Belgian border near to the above places, and the disease is reported to have shown itself along either side of the whole course of the military road. It was said that no disease followed in the wake of the Prussian army as its cattle were healthy, and thus a great part of Belgium escaped infection. This was not the case for France, and in 1814, it reached the walls of Paris. Stringent slaughtering of diseased animals brought it to a halt in most regions by 1815. The entire population of Paris and its suburbs, including the troops that surrounded and occupied it, fed on the meat for 2 months without harm. A thousand large oxen suffering from rinderpest were consumed by the allied and French armies and by the inhabitants near Strasburg, in 1815. It disappeared in France in 1816 after the withdrawal of the allied forces and was not heard of again in western Europe until 1827.

In 1825, there was an epizootic in western Siberia in the governments of Pskow and Novgorod, which reached Estonia and Livonia the following year, and in 1827, Courland. Then, with infected cattle sent to supply the army, the Russo-Turkish campaign of 1828–9 spread it into Bessarabia, Wallachia, and Moldavia. Again, it found its way into Podolia,

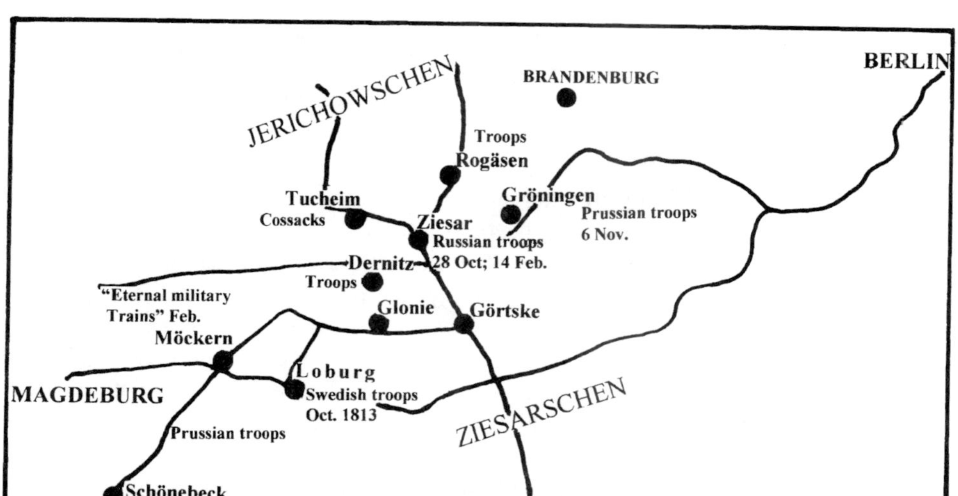

Figure 7. Spread of rinderpest in Germany by troop movements 1813–14.

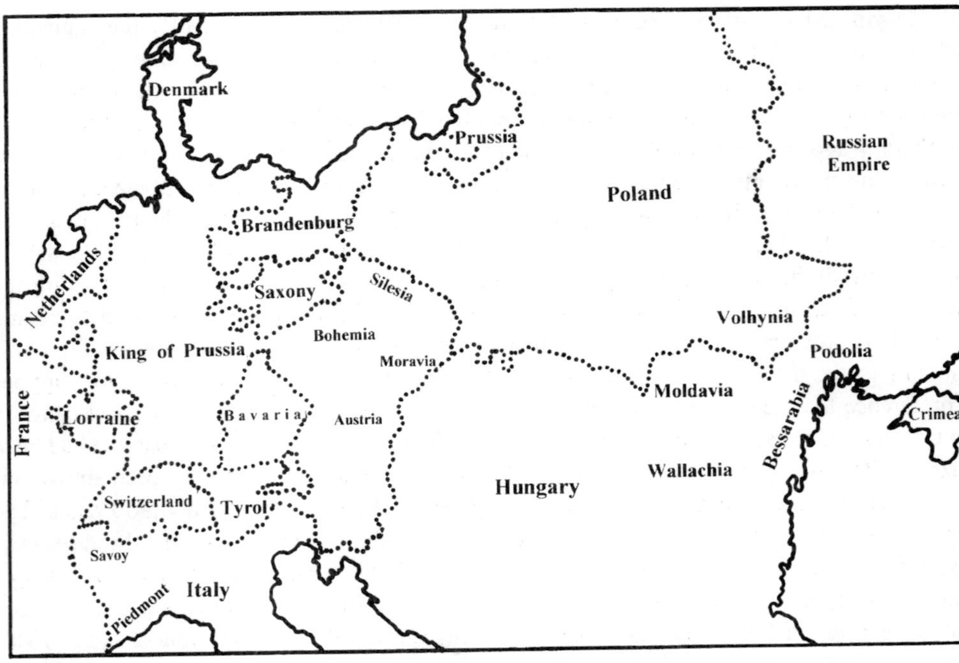

Figure 8. Countries affected by rinderpest in mid-18th-century Europe.

Volhynia, and then Prussia, Saxony, Hungary, and, in 1830, Austria. It was alleged to have entered Hungary with skins from Wallachia and Moldavia (Figure 8). Hungary lost 30,000 animals, Galicia, a province of Austria, 12,000, and Moravia, 9,000. It entered the principalities of Wallachia with the thousands of carts drawn by oxen which accompanied the Russian army. The winter of 1829 was very bad, and half the number of animals died from rinderpest and starvation. It became necessary to transport provisions from Bessarabia for the Russian army by means of forced labor, but the peasants, exhausting their own supplies, died in great numbers on the road, along with their cattle which died from rinderpest. This covered the road with carcasses, which Chesney (1854) attributes to causing a typhus plague among the Russian soldiers. The Turkish heavy guns, some 100, were drawn by buffaloes and oxen; but Chesney does not refer to any losses incapacitating the firepower of the Turks. From 1828 to 1830, southern Russia suffered much from the ravages of rinderpest, but so long as only indigenous cattle were affected, deaths were rare, compared with imported cattle from Austria of which less than one tenth survived. In Prussia, it reached its height in 1832, crossing the frontier in 1831 in the Polish Insurrection and continuing until 1833 with the loss of 1,000 head. In eastern Russia, it created great havoc from 1833 to 1835 and was very virulent in Hungary in the latter year. In 1836, it erupted in Lower Austria and passed from Bosnia into Dalmatia and was severe in Moravia and Hungary again in 1837, while cattle shipped from Dalmatia to the Neapolitan States spread the disease once more to southern Italy. Seventy of the Dalmatian cattle died soon after landing, infecting local stock. People in Italy still had not learnt from their experiences, inconceivably attributing the deaths to a change of climate and pasture and the salt water the animals had been compelled to drink on the voyage. So it was decided to move the remaining cattle to another pasture, thus infecting other cattle on the way. It also attacked some buffaloes but not as virulently as cattle.

Between 1844 and 1845, Russia lost one million cattle from an outbreak originating in Bessarabia. Tartary was subject to a serious drought in 1844, and perhaps related to this, in the autumn, the disease threatened many of the countries adjoining Russia. Spreading to Galicia, it passed through Moravia to appear in the kingdom of Bohemia in September, affecting oxen in Germany in December and spreading into a few districts in Hungary. In Moravia, only 163 animals recovered of 1,065 which contracted the disease, with 129 slaughtered.

Some young physicians who made a number of postmortem examinations in Bohemia considered the disease akin to abdominal typhus in man, believed to be noncontagious. Communicating their opinion to some members of the influential faculty of medicine in Prague, the latter conducted a number of further autopsies, coming to the same conclusion. The faculty therefore reported to the government that the disease was not contagious but could arise spontaneously, and the means the government had adopted, slaughtering all sick animals and throwing a cordon around infected areas, were useless to prevent it. The government acting on this advice, while not lifting restrictions entirely, relaxed them, with the result that the disease quickly spread. The Austrian government then sent Dr. Eckel, Director of the Imperial Veterinary Institute, Vienna, to Bohemia. Assisted by Dr. Langenbacher and Professors Röll, Hayne, Muller, Pillwax, Bruckmuller, and Bruhl, of Vienna, the outbreak was to become one of the best studied to that date, details of the origin, progress, and extension of the disease being recorded by this body of professors. Confirming the disease as rinderpest, Dr. Eckel recommended rigorous enforcement of former measures, which enabled the authorities to confine it to three districts. By the end of January 1845, after having appeared in the previous October, the plague was extinguished in Bohemia. Altogether 2,197 animals had contracted the infection, 1,122 had died, 945 were slaughtered, and 115 recovered.

In 1846, it had become widespread in Wallachia, and in the district of Ilfov of which Bucharest was the capital, 60–80 animals were dying each day.

In the summer of 1848, the plague spread extensively over Russia and that winter passed into Podolia, Volhynia, and Poland, then to the Austrian Galician frontier where the quarantine period for cattle at the border was increased from 10 to 15 days. In 1849, 25 communities were infected in Galicia, and the length of the quarantine period was extended to 21 days. Movements of Russian troops from Wallachia carried the disease into Transylvania at the beginning of August 1849, and from 1849 to 1851, it prevailed in Hungary again resulting from the passage of the Russian troops in 1849, allegedly destroying over 350,000 head of cattle. By the end of December 1849, it had infected almost the whole of the province. In March 1850, a further outbreak was reported in Galicia, and it appeared again in Wallachia, Bessarabia, Podolia, Volhynia, and Poland. The outbreak in Galicia apparently originated from a herd of 101 oxen from Bessarabia, one animal being taken ill 8 days after the purchase of the herd in a market on the Russian frontier. Of the imported herd, only one animal suffered, but infection spread to 157 others, of which 93 died. By March, it was reported checked in several districts in Transylvania. In April, there were fresh outbreaks in Galicia, but by June, it had almost subsided, and Transylvania was free of it. Then, it broke out in a district on the Moldavian frontier. In August, it was reported near Vienna, cattle coming from Hungary having been in contact with local cattle, of which nine died. It was not until September 1850 that an order was issued in Galicia requiring slaughter of cattle to prevent spread of the disease.

In 1850, it occurred in part of Hungary, but through 1852 to 1853, although present there, it was almost confined to Russia, especially bad in the south in the summer of 1853 during a period of intensely hot weather when there was apparently also an outbreak of anthrax, killing sheep, pigs, and birds.

At the end of 1853, it returned to the northern provinces of Austria, and the Austrian government commissioned Dr. Ritter von Koch to inquire into it. From October 1853 to April 1854, it can be traced into Galicia through Moravia, Bukowina, Silesia, Bohemia, Lower Austria, and Hungary, brought in by cattle from Poland and spread by the transfer of hides directly or indirectly to 14 villages. In the Balkan Peninsula, great losses were experienced in the region of Tschataldja. In early 1855, its effects were felt in Russia, whence it passed to Poland again. By the end of August, it was destroying immense numbers of cattle in Turkey.

7. THE CRIMEA

The declaration of war by France and Britain on Russia in 1854, and development of a battleground in the Crimea, meant that cattle from Turkey were exported to feed British troops there. With rinderpest destroying vast numbers of the animals, in December 1854 at the request of the French Government to assist in an inquiry into the causes of the disease among cattle in Turkey, Simonds sent out a questionnaire from the Royal Veterinary College to the military veterinary surgeons in the Crimea, requesting information on the disease. The purpose was to examine the possibility of controlling it "to save our brave army this essential portion of their food." There were 26 questions:

1. Does "Contagious Typhus," or any epizootic disease otherwise designated, exist among the cattle in the Crimea?
2. If any epizootic is known, can you say if it be peculiar to the country or not?
3. Is contagious typhus commonly prevalent in Turkey, or the surrounding countries?
4. Can the recent appearance of this or any other disease in the Crimea be traced to any special or direct cause, such as the importation of diseased animals, and so on?
5. Is it your opinion that the malady is contagious, and, if so, what proofs have you?
6. Has it attacked any other animals besides the Ox tribe?
7. How long a time has usually elapsed between exposure to infection and the appearance of the disease?
8. Has age any influence over the disease, and if so, what?
9. Does the congregating of cattle together, and exposing them to privations such as would necessarily be associated with a ship voyage favor the attack or otherwise?
10. What kind of food is given to the cattle, and what is the usual system of the management of the animals for the use of the army?
11. Which are the most susceptible to the disease, fat or lean animals?
12. Is the malady on the increase or otherwise?
13. What are the symptoms marking the commencement of the attack, and are they easily recognized or not?
14. What are the symptoms which accompany the progress of the disease, and particularly those that indicate the greatest danger?
15. Have any animals recovered in which diarrhea has shown itself in an advanced stage of the malady, and what is the character of the alvine evacuations?
16. Do you know of any instance in which an animal has become a second time attacked with a mitigated form or otherwise of the disease?
17. What is the usual duration of the disease when allowed to run its own course?
18. What are the causes which you gave found to be the chief obstacles to the eradication of the malady?
19. In what state or condition has the disease left those animals that have recovered from mild or severe attacks?
20. Has the malady steadily progressed, or have there been repeated outbreaks after intervals of comparative freedom from disease?
21. Has prevention been attempted by change of diet, situation, or management—by medical treatment, or any other means?
22. Can you form any idea of the number of cattle which have died, or have been killed or otherwise disposed of, on account of this disease?
23. What are the usual postmortem appearances?
24. Have you any means to suggest to arrest the progress of the pest?
25. Is the flesh of the animals which are slaughtered when first noticed to be ill, in a fit state to be used for food; and how soon does it undergo decomposition?

26. Can you trace any connexion between the existence of this disease and the outbreak of any epidemic among the Soldiers of the Allied Army?

It is doubtful whether many replies were received to this lengthy questionnaire from harassed veterinarians in the midst of a particularly grim battle. One succinctly explained conditions:

Camp 1st Division,
Feb 12th 1855

My Dear Sir,

I am sorry to say I can give you no information respecting the contagious disease of cattle referred to in your circular. We have quite as much indeed (speaking of Veterinary Surgeons in my own Corps) more than we can properly do, to attend to the horses under our charge. I myself have the artillery of three Divisions and the ammunition ponies besides being expected to look at all the government Battery horses, staff horses of each, and officers' horses.

We have never been called upon to attend any cattle, indeed they are slaughtered too soon after their arrival here to render such a thing necessary. In the French Service every Battery and Troop of Artillery has a Veterinary Supervisor attached to it, and doubtless they have a sufficient corps for every other purpose including commissariat, but with us these things are managed very differently.

I am sorry I can give you no information but bad answers would be worse than useless and therefore I abstain from giving any.

With kind regards to yourself and Messrs. Morton, Spooner and Varnel.

Believe me etc.
[signed] Hicks, William Veterinary Surgeon
Royal Artillery.

But the disease had already broken out, occurring in the French camp introduced with cattle brought 960 km from the interior. A few weeks later, it was in the British camp. The veterinarian in charge stated that he was very suspicious of a lot they had purchased "but we were obliged to have them," and in a few weeks, the disease manifested itself. In a letter to the *Cologne Gazette* reported in *The Times* of January 8, 1855, a correspondent wrote: "… on all sides between the torn tents, dead cattle, horses and oxen are rotting, and no one thinks of removing the pest-bringing carcases."

With the war at its height, the French and British Governments were recommended to stop the supply of fresh meat to their armies. Pickled meat was sent instead, but much of that supplied from Aberdeen was allegedly of animals which were by then dying of rinderpest in Britain.

George Fleming, veterinary surgeon with the King's Own Hussars, was to report:

> … I was encamped near a large division of the English Commissariat, and so great were the depredations caused by typhus and dysentery among the cattle, that I was frequently asked to give my opinion as to remedial measures. From what I could learn these animals were shipped in good health, and their appearanace certainly went far to prove that they had not been suffering from any chronic disease, but that their ill health had commenced with the hardships and privations of transport on ship board, and the inclement weather and insufficient food after landing. The French losses in the winter and spring of 1856 were much heavier than ours—apparently from the fact that their ships were more crowded, and cleanliness was less attended to.
>
> I took every opportunity of satisfying myself as to the contagious or noncontagious nature of the epizootic, and am decidedly of the opinion that there was nothing to induce anyone to consider the disease as contagious or infectious.

> No malady of a like character made its appearance among the oxen belonging to the Tartars, or the other landed proprietors, either in the valley of Baidar, or towards Simpheropol, so far as I could learn.
>
> The British troops provisioned on these cattle—I will not assert that the animals which had died were issued as food—were never in better health. (Fleming, 1864).

It was said in the space of 9 months, the French army in Samsoun lost 8,000 animals out of 17,500 and the British army 4,000 out of 10,000 (Simonds, 1865).

In Turkey, the disease was rife from June 1855 to January 1856, when it ceased altogether.

8. CONTINUATION IN EUROPE

Shortly after the Russian army was sent to occupy the Principalities, rumours of rinderpest became current, and by 1854, it had already made considerable progress in Volhynia and Podolia. It occurred in Hungary again in 1855, becoming widespread the following year. In France, the government had planned to hold an international agricultural show in Paris in July 1856. In view of the reports of the disease, it sent Renault on a tour of inspection of Austria, Hungary, Poland, Galicia, Moravia, Prussia, and the Baltic Sea area, to assess the risk if cattle were brought from these countries for exhibition. Renault concluded that he did not think there would be any danger, and the exhibition took place without any restrictions being imposed, some cattle even coming from Hungary. No outbreak was reported as a result.

Concerned at the incidence in Austrian and Russian Poland, the Prussian Government closed its border in March 1855 with detachments of troops but, hearing of further progress of the disease, sent a veterinary surgeon to investigate the Polish district for up to 5 km from the Prussian border, discovering that the disease was only 4 km distant. Despite the immediate implementation of rigorous precautions, in late November the disease crossed the border, appearing in the districts of Inowraclaw and then Gnesen, near the town of Posen. Measures were taken for closing the boundaries of the affected places, and special ones implemented for infected farmyards, by means of sentries posted under the supervision of gendarmes. Quarantine sheds were established, and superintendants and cattle inspectors appointed, provided with written instructions and bound by oath. All trade in cattle was forbidden within 5 km, all dogs were chained up, and every cattle owner within 3 km of an infected place was bound by law to give immediate notice to the mayor of the least symptom of disease among his cattle, who informed the Counsellor of the Administration of the district by express messenger. At first, preventive measures were entrusted to civilians, but they were soon replaced by military. Despite the measures, occasional cases occurred up to the beginning of 1856. With its decline in Poland, restrictions on trade across the border were gradually modified. Passes were required to cross the frontier, and all individuals had to be disinfected under the superintendance of a gendarme.

In April 1856, it broke out in Breslau, which had been free of the disease for 40 years. Investigation failed to trace its origin as the inhabitants were uncooperative, but it was believed to have come from Schrimm in Posen district:

> ... three different circles [administrative districts] were more or less affected, which circumstance arose from the want of experience of the district veterinary surgeons, none having had an opportunity of previously seeing the disease. Subsequently also about a month elapsed before correct reports were obtained from the Commissioners who were specially appointed for the investigation, arising from the great distances they had to travel and the difficulties which were in the way of their making post mortem examinations. (Lord Loftus quoted in Simonds, 1857)

The outbreak lasted 7 months, but of a total of 5,528 cattle, only 253 died, and 813 were slaughtered.

In September 1856, two men from Hamburg purchased 180 oxen in Hungary and sent them to graze on the islands and marshlands of the Elbe, where they transmitted the disease to cows sent out daily from Hamburg to graze; and thence it spread to Holstein. It was the occurrence of a murrain in the Duchies of Mecklenburg-Schwerin and Mecklenburg-Strelitz which provided the first accounts to reach England of a new outbreak. As a result, the Commissioners of Customs published an Order requiring a more rigid examination by the Inspectors of Foreign Cattle, "with special reference to a contagious disease called 'murrain,' which has recently broken out amongst the horned cattle of Mecklenburg." This elicited a complaint from Count Bülow, Minister to the Grand Duke of Mecklenburg-Schwerin, to the British Ambassador to Prussia, Lord Blomfield. Bülow stated that the "cattle murrain" alluded to had only manifested itself twice, for short periods, in 1855 and 1856. In each case, measures of such a stringent nature were taken as having effectually arrested the spreading of the disease, and since October 1 all restrictions had been lifted. The "cattle murrain," Simonds (1857) asserted, was only pleuropneumonia.

Erupting again in Russia in 1856, when the Russian army was sent to occupy the Danubian Principalities, the disease extended northwards to Lithuania and Courland. From 1855 to 1856 in the Austrian Empire, of a total of 270,548 cattle, of 56,784 infected (21%), 25,678 died, and 764 were slaughtered, a total loss of 9.8%. Continuing in Courland until January 1857, that province lost 2,000, and between May 5, 1856 and March 29, 1857, Poland lost 20,000. From April 1–15, 1857, of 14,574 cattle in Krakow, 53 died, and 84 were slaughtered. From the commencement of the outbreak, of 576 infected (4%), 361 had died, 169 were slaughtered, and 23 recovered, a total loss of 3.6%. Throughout the whole of 1856, as a consequence of the steady progress of the disease in Russian Poland, import into Prussia of cattle and animal products was strictly prohibited all along the Russian frontier of Eastern Prussia. On the occurrence of outbreaks near to the frontier in Tilset district, more severe measures were adopted. Herding cattle along the right bank of the Niemen was forbidden and all traffic between the countries suspended. No persons connected with cattle could cross or, if so, had to undergo quarantine, while mail passengers were fumigated at the borders. All cattle and horse fairs usually held in September and October in Gumbinen district were forbidden. Despite these measures, the disease crossed the frontier in March 1857, breaking out near Berlin in April in the villages of Bassnitzkehmen and Meldiglaucken, traced to some cattle from Galicia, all of which, and all cattle known to have been in contact with them, were destroyed (Simonds, 1857).

In February 1857, *The Times* reported "advices from Königsberg give unfavorable accounts regarding the cattle disease, which during the past two years has manifested itself in Poland, and other parts of Russia, and which now threatens to spread to Germany ...." A report in the *Morning Chronicle* of April 6, 1857 stated that all the herds were dead at Warsaw, whereas in reality, out of a total of 1,350 head, in September 1856 only nine had died of rinderpest, and 12 were slaughtered. On February 27, the disease attracted the attention of the British Government which sought information from its foreign embassies. From Warsaw, it was reported that it was "still prevalent," and "The infection is so virulent that it is said that a farm servant, who has been in attendance on sick cattle, may carry it in his clothes and impregnate other beasts with the fatal distemper." Measures to arrest it were very summary, and slaughter was indiscriminate, but in Prussian Poland, more stringent measures were enforced resulting in near elimination of the outbreak. The infection was stated to have come originally from the Steppes of the Ukraine, "whence vast numbers of cattle are yearly driven westward" (Mansfield, 1857). According to Crowther (1997), in order to raise funds to replenish his treasury, the Tzar of Russia began selling cattle in enormous numbers to Western Europe in the 19th century, recipient countries soon discovering that they were infected with rinderpest and banning their import.

A letter from the consul in Königsberg stated that the disease had been advancing steadily in a northerly and westerly direction from Volhynia and Podolia since 1854. The Prussian Government had now enforced stringent measures all along its borders and "no hides, calf-skins, wool, rags, or any article which is likely to have been in any way connected with cattle, and all Jews and other persons who are in any way suspected of having transactions with cattle ..." were permitted to cross the border, measures which, up to that time, had kept the disease out. The dealers had therefore routed their merchandise to Russian ports, and the hide and calf-skin export now went to Lübeck and thence eventually to Berlin, so that infection could be carried to the heart of Germany regardless of the protection of the eastern frontiers.

As a result of these reports, the RASE decided to send Simonds to the Continent to inquire into the disease and the measures of combating it. At the same time, on April 22, 1857, the Government took the precaution of issuing an Order in Council prohibiting importation of cattle, horns, hoofs, raw or wet hides, or skins, having an origin or having been in, or on, any ship having come from or been at, Russia and other countries bordering the Gulf of Finland (Prussia and Mecklimbourg-Schwerin), or any other part of the Baltic Sea between the Gulf of Finland and the territories of the Free City of Lübeck. Any such item arriving in Britain was to be destroyed, together with any hay or straw, litter, manure, and so on, that had been on the ship. At the time, Britain received annually mostly from northern Russia and Turkey a large number of dried skins and some salted. Salted skins usually came from southern Russia or the Danubian provinces. Large quantities of tallow, used in industry and for greasing ships' hulls, were also imported from central Russia, even, according to Renault (1860), during the Crimean war. Nine-tenths of imports were unloaded at London; the rest between Hull, Bristol, and Liverpool.

This action was followed by Hanover, Holland, and Belgium implementing like controls. France, having taken no measures against the disease and alarmed at the damage which these actions would do to trade, set up a commission which appointed Renault to visit the Dutch, Belgian, and French ports to look into the free import of skins and tallow from Russia via the German Baltic ports. He concluded that he could find nothing to justify Britain's action and those who had copied it, and nothing to interfere with France continuing its trade with Russia (Renault, 1860).

In March 1857, the disease was reported to have been introduced into Silesia with cattle from Galicia. It was traced to two herds numbering 44 and 37, bought by different landowners and distributed to their several estates. In the district of Breslau, eight fell ill and were immediately slaughtered but had already infected others, and a further 21 were slaughtered. Other cattle were not affected and the military cordon thrown around the place lifted 3 weeks after the last case. Twenty were brought to the district of Oppeln and distributed. When some fell sick, they were moved to a place where only sheep were kept and slaughtered there. On one estate, Ponischowitz, 20 oxen were brought and within a few days showed signs of infection. Having been put with 49 other cattle, the whole lot was slaughtered so that not a single head of cattle was left on the estate. The place was surrounded by a military cordon, but a carpenter's apprentice slipped past the guards to visit his father's house about 600 m distant, where he repaired a manger in the cowshed and changed his clothes. Subsequently, rinderpest broke out among the cattle there, which were all slaughtered and a military cordon placed around the farm. In all, it broke out in seven places, but stringent measures were adopted to wipe it out wherever it appeared, each place being surrounded by a military cordon until it was clear and it had not spread to any other Prussian province. In the same month, it was reported to be making great ravages in the Carpathian Mountains, and troops had been sent to halt it. In May, it was reported as almost extinguished in Prussian Silesia but still active in

Austrian Silesia and was reported at Olschau, where there was one of the biggest cattle fairs in Austria, but only seven animals on nearby farms are reported as dying or being slaughtered.

In May, it was reported that Simonds had forwarded his first communication at the end of April, asserting there was little doubt that pleuropneumonia had been confused with rinderpest, and he had to travel to Poland before he could find a case of the latter disease. His report was published at the beginning of June, and on November 4, the Order in Council was revoked.

In 1858, the disease was severe in Russia occurring in 47 jurisdictions in which 178,690 animals were affected and 118,315 (66%) died. Many different reasons were given as to its cause in a particular locality: the trafficking of herds, import of hides, the intense heat of the summer, insufficiency of food, bad water, and, in one case, giving cattle water used in washing salted beef. In Wallachia, however, the disease was very mild that year. In October, there was a fresh outbreak in Poland due to quarantine on cattle from Russia being reduced from 21 days to 48 hr. The Jews, who controlled the Warsaw meat market, had threatened to leave the city without supplies, and the Government was about to establish an independent company when the Jews persuaded the Governor of Lublin to permit the importation of 3,000 head, with the result that the disease broke out again and quarantine had to be reestablished. There was, however, provision for 2 days only of quarantine for cattle for the meat market, limited to specified towns and for the garrison of certain fortresses.

In December 1859, it reappeared in Galicia and from there passed to Silesia and appeared in cattle from the Ukraine in October 1860. It was still rife in Bohemia from 1859 to 1861 and appeared in Lithuania in January 1862, and in April 1863, it broke out in the south-west corner of Carniola, introduced by two oxen driven through from adjoining Croatia. In March 1862, it was reported as absent in Poland but in August the following year as having appeared in a number of localities there. It again caused great losses in the region of Tschataldja in the Balkan Peninsula.

In 1860, Russia lost 183,678 head of cattle to rinderpest. In July of that year, it broke out in Bukowina and Galicia, imported from Podolia, having reached Bukowina from Moldavia on one side and Bessarabia on the other, appearing in 28 places affecting 1,803 head of cattle of which 707 died and 52 were slaughtered. On March 7, 1861, both places were declared free of the disease, but 7 days later, it appeared again in Galicia imported with foreign cattle and continued until January 1862. A total of 1,089 cattle died, and 55 were slaughtered. Altogether, nearly 18% of the cattle exposed to the disease became infected, and about 68% died. In November, Polish oxen carried it into Bohemia, where, of 1,593 animals exposed to the disease, 39 died and 83 were slaughtered. In this outbreak, sheep were also affected, 10 out of 60 dying.

Polish cattle carried the disease to Moravia, where 15 animals died and 25 were slaughtered. It entered Lower Austria from Hungary, again only a small number of cattle dying. It appeared briefly on the Hungarian frontier but then created great ravages in Hungary itself, in 2 districts killing 2,500 of 4,000 head of cattle, in others killing over 6,000. This outbreak also killed 25–30% of infected sheep flocks.

Several outbreaks occurred in the winter of 1862 within easy reach of the Prussian frontier, but careful measures restricted it to Galicia and Poland. In October, it showed itself at Berguto in Dalmatia, having come from Herzegovina. An Austrian medical commission sent from Vienna to Sarajevo reported that it was not the Russian Cattle Plague, but it soon proved to be so. Between 1861 and 1862, 296,000 head were affected in Austria, of which 143,000 recovered, 152,000 died, and 1,537 were slaughtered. In Herzegovina, the outbreak was attributed to an excessive drought that year, some cattle watering only once every 3 days, travelling 24–29 km to do so. Many were alleged to have eaten poisonous plants because of a lack of forage.

In the same month (October), some infected Dalmatian cattle were shipped at Trieste for destinations in Italy, communicating the plague wherever they were landed. In August 1863, it was reported in the Campagna, more than 10,000 cattle being estimated to have died and no measures of an effective nature being taken by the Papal government to check it. "There have been religious processions, *neuvaines*, and the like, in the hope of averting the evil, but as yet they have not produced the desired effect," reported the *British Medical Journal*. By December, the city of Rome was without milk, but then due to a stringent slaughter policy, it was stamped out by the end of the year after some 22,000 cattle had perished. Renault, sent there to study it, contracted a fever and died on his return to France.

In Hungary, 25–30% of the sheep are alleged to have been killed by it in 1862.

In 1863, the Polish Revolution aided dissemination of the disease. At no time since the Crimean War was it more rife than in southern Russia, the Danube, and Poland, being more virulent in southern Russia than in the north, generally breaking out in the autumn. Again, it was kept out of Prussia, but not Austria, where it spread far and wide, destroying some 14% of the cattle. More malignant than usual, it attacked 26% of Hungarian cattle in infected districts with a minimum 65% mortality. Elsewhere, the number of animals affected ranged from 1 to 20%, and mortality from 77 to 100%. It continued in Austria killing 110,336 cattle in 1863, 44,447 in 1864, and 3,237 in 1865, making a total of 258,107 since 1849 out of half a million infected.

It was alleged, in 14 years beginning in 1849 when the quarantine period was lengthened to 21 days, that only once had the disease been introduced into Austria by cattle which had observed the quarantine regulations. All other cases were caused by smuggled cattle. Only completely healthy cattle of the major traders went into quarantine, sick or suspect beasts always being smuggled through. Austria received a total of 90,000 cattle yearly from Bessarabia and the Ukraine, of which it was estimated half were smuggled to avoid the long period of quarantine.

From January 15 to March 27, 1865, in the districts of Minsk, Vitepsk, and Mohilew; 47,000 cattle, 2,543 horses, and 57,844 other domesticated animals were slaughtered because of an epizootic disease, probably anthrax; but rinderpest assumed "formidable proportions" in Wallachia in 1865 although not spreading over the entire country, a great proportion of which was forest. Despite regulations and the fact that Ilfov was an officially declared infected district, several thousand wagon oxen entered and left Bucharest daily, all the requirements of the city being transported in wagons each drawn by 4–10 bullocks. The disease spread to Turkey, Austria, Bavaria, and then the Tyrol; with mortalities ranging from 65 to 100%. When it appeared in Belgium, it was quickly brought under control, at Hasselt 1,395 animals being slaughtered, of which only a quarter were infected. It appeared also in northern France but was soon controlled there by slaughter of sick animals and prohibition of all imports of cattle and cattle products. In January 1866, Prince Manoakbeg in Hungary replied in answer to questions by the Consul-General that rinderpest had been permanent in New Russia during the last few years, being sometimes in one district and sometimes in another.

According to Gamgee, "the ravages it created, and the misery it induced in various regions of Austria, Bohemia, Saxony, Prussia, Poland, Hungary, Silesia, and France, are beyond description." Only Spain (there is doubt about Sweden) from the beginning of the 18th century to the beginning of the 19th century, was unaffected, as it (like Sweden) did not import foreign cattle.

9

THE SECOND GREAT PLAGUE IN BRITAIN

1. CHANGES IN TRANSMISSION

Rinderpest appeared in Russia in the mid-19th century after every campaign in the wars with Turkey, with Russian armies being supplied by numerous wagons drawn by steppe oxen. But the role of war in facilitation of the spread of the disease became superceded by the growth of commerce. The more communications were multiplied and market stock became more abundant, so did cases occur more frequently. "The habitations and the markets the diseased cattle have frequented, the railways, vessels and roads they have traveled by, the pastures they have grazed upon—all become so many foci; from each of which the pestilence may radiate in every direction" (Fleming, 1875). The advent of the steamship was blamed because it could transport cattle cheaply, bypassing overland routes. Before the railway, one region of Hungary had immense holding meadows where sometimes 15,000–20,000 head of cattle were gathered for onward transport. By the 1860s, these meadows had been transformed into arable land, and the cattle dispersed directly to their destinations in small numbers by rail. Four distinct lines ran through Russia, one through Galicia into Bukowina through the center of the rinderpest enzootic region. Another ran through Austria and Hungary into the valley of the Danube where rinderpest frequently existed. And in the north, one line ran from East Prussia into Russia and another by way of Warsaw through Poland.

During the 18th century, rinderpest appeared in France about every 20 years until control measures were introduced in Germany on the land routes by which cattle were driven from the east but, "Now that the means of communication between different countries have become so rapid and so easy, the chances of typhus [rinderpest] overleaping or evading the barriers Germany has hitherto been able to present to its invasion have greatly increased" (Béhic, 1865).

In September 1865, the French Minister of Agriculture, Commerce, and Public Works, addressed a circular to all Prefects in which he stated: "Its [rinderpest's] most ordinary course, in past times, has been along the routes which have been pursued by the armies of Austria and Russia, whose victualling herds are principally composed of steppe cattle. Less frequently it has been introduced by the roads of commerce by land and sea." But by the 1870s, traffic in cattle was assuming greater proportions every year, and England, France, and Germany were compelled to seek supplies in regions "more remote and perilous" than formerly. The dangers were now appreciated, and even if the exact nature of the disease was not yet understood, the manner of preventing it was. In Germany, Adam had produced his *Veterinär-Polizei* in 1862, followed in 1869 by Haubner's *Handbuch der Veterinär-Polizei*. In France, Reynal produced his *Traité de la Police Sanitaire* in 1873. This was followed in Britain by Fleming's *Manual of Veterinary Science and Police* in 1875, putting forward measures of prevention and control as already recommended on the Continent.

Publications were not wanting. Apart from presentations such as that of Graf in 1796, *Abhandlung über die gegenwartige Hornviehseuche, ihren Ursprung, Natur, Vorbeugungs und Heilungsmittel (Treatise on the Present Cattle Plague, its Origin, Nature, Methods of Prevention and Cure)*, a pioneer contribution in German on the pathology of rinderpest was produced in Russia by Friedrich Brauell, a German and one-time holder of the chair of veterinary science at Kazan University, from 1848 pathologist at the Russian Dorpat Veterinary Institute. His book, published in 1862, was referred to by Sanderson as a "comprehensive and exact treatise on pathological anatomy" (Bernard, 1866b). Apart from a description of effects upon the visible mucous membranes by Jessen, who was described by Sanderson as having had larger opportunities of observing rinderpest than any other man in Europe, Sanderson considered all other descriptions until that of Brauell so indefinite that anyone who had not seen the effects could not form an accurate conception of their nature. Brauell had spent many weeks in the Russian steppes in 1861 making anatomical and microscopical observations, the latter the first thorough histological research into the disease. Likewise, the book published in German in 1864 from a paper in Russian published the year before by Joseph Ravitsch, a member of the Russian Imperial Commission on inoculation, *Latest Investigations of the Pathological Anatomy of Cattle Plague*, is regarded as one of the classical studies of rinderpest. The British Commission frequently cited it, stating in its *Third Report*: "The Russian pathologist, Ravitsch, has already described with great accuracy some of these congestive phenomena" (Bernard, 1866b). Despite this, the pioneering Russian work on rinderpest generally passed unnoticed outside of the Commission's 1866 reports.

Regardless of the advanced nature of veterinary research in Russia in the middle of the 19th century, Russian bureaucracy did not keep pace. *The Times* of September 10, 1865 reported of Russian Poland:

> Towards the end of June, the cattle plague, never entirely extinct in those tainted regions, visibly increased in all Lithuania, especially in the district of Kowno. After unconcernedly looking on at the destruction for several weeks, the authorities at length distributed among the country people a paper teaching them how to guard against the disease, and how to deal with the infected and dead animals. But the useful information being, unfortunately, conveyed in Russian to the benighted peasantry, remained a book with seven seals. The pestilence grew, the cattle were dying off fast in every village, yet the Russian police would not abate a jot of their dignity by circulating translations of their 'book.' In many cases the poor farmers were reduced to such straits that village elders were not ashamed of calling in the assistance of children, and applying to the rising generation, which has been furnished with Russian spelling-books, for an explanation of the mysterious yet valuable paper. But, of course, the handling of a Russian primer had not enabled the boys to interpret the contents of a veterinary treatise; so the cattle kept on dying, and continue to do so to this day.

Owing to the widespread outbreak on the Continent, in 1863 a Bill was introduced into the British Parliament as a precautionary measure relating to animals destined for Britain dying on board ship. The carcasses of such were to remain on board and not be discharged without written permission of Customs, which had the power to order them to be thrown overboard out at sea. Any failure to comply would result in forfeiture of the whole cargo of animals. As bills of lading were not legal documents, the measure would have been impossible to enforce, and the Bill did not pass into law.

2. THE 1865 OUTBREAK IN BRITAIN

On August 1, 1865, *The Veterinarian* cautiously reported that in the past month, an epizootic of a fatal character had "quite unexpectedly" shown itself in Britain:

> ...we have, as yet, to learn something more of its pathology. The general symptoms resemble a malady in the ox not altogether unknown to us, but they differ therefrom in many of its leading features...The bovine disease assumed at its outset such a fatal character that the attention of the

> Government was at once directed to it, and the Order in Council ... was at once issued, and one of us [Simonds] commissioned "to institute an inquiry into the nature, causes, laws of propagation, and best methods of treatment and prevention of the disease.

Inquiries would reveal that it was probably imported into Britain with Russian cattle brought from the Baltic port of Revel (now Talinn), and first noticed on June 27 in two English cows at Islington bought on June 19 in the Metropolitan Cattle Market. The following day, similar symptoms were observed in a cow at Hackney, purchased in the same place on the same day. It transpired that two Dutch cows at Lambeth, also purchased in the same place on June 19, had been affected on June 24. It then spread with great rapidity, breaking out in many other London areas and destroying large numbers of cattle. An estimate in one district was that more than 80% died or were slaughtered. There was at least one case of deliberate spreading of the disease, introduced into the parish of Paddington by a man who surreptitiously introduced a sick cow onto the farm of a man whom he had a quarrel with (Figure 1).

It then appeared in and near the Essex Marshes, and in early July Norfolk. A little later, it occurred in Suffolk and Shropshire. By the 1st of July, it had invaded Scotland with a calf brought from London, reported at Bogenjohn in Aberdeenshire. Under the Aberdeenshire Rinderpest Association, formed in August, stamping out by slaughter was practised with vigor, all animals which had not died on a farm being immediately slaughtered once the disease was known to exist there; the county was reported as the exception to the strong feeling initially throughout Britain against slaughtering (Anon., 1882).[1] Despite this, up to December 30, 265 animals were attacked. In Hampshire, where it had been present for the same length of time and no "stamping out" had been practised, 276 were attacked.

In August, Budd reported that on the 5th of that month a load of manure, which had not undergone any disinfecting process, was sent from one of the infected London dairies to one of the canals for despatch to the country: "Did it perchance meet any droves of cattle in its way through London?" asked Budd. "Is it at this moment traveling in some slowly moving barge, with its trail of infection behind it, to the meadows of Berkshire or Bucks? To what manure-yard next were the horse and cart sent which first bore it away?" (Budd, 1865a). It was to be April 16, 1866 before an order came into force prohibiting removal of dung from infected premises except under written permission for the purpose of destruction, although inspectors had been empowered under an Order of September 22, 1865 to disinfect or destroy within a district manure which might be deemed likely to propagate the disorder.

Giving evidence to a Commission in India in 1869, a butcher now employed by the Indian military stated he had seen two cases of rinderpest in England in 1865. "I was one of the men selected to go and assist a farmer making up his hay at Dover, and I saw two cows

[1]When Gladstone as Liberal Prime Minister received the freedom of the City of Aberdeen on September 26, 1871, he praised the county for the manner in which it had handled the outbreak:

> ... it was an admirable spectacle, when all over the country we were wandering and groping about, some proposing the most absurd measures by way of remedy and precaution, and others feeling themselves to be totally in the dark- ... when the gentry and farmers of the county of Aberdeen, associating themselves together, with nothing to rely upon except their own energy, except their own prudence and intelligence, devised for the ready, rapid, and complete extinction of that plague, the very remedy which, at a later period, after much ineffectual discussion, the Legislature found itself counselled by prudence to adopt ....

The Association formed a Central Committee controlling Parish Committees which met as soon as a case was reported, valued the stock and advised the owner to slaughter, then overseeing the burial, disinfection of byres and other precautions. A voluntary subscription of 1 penny per £1 from proprietors and farmers compensated owners' losses. When the disease broke out in Westerton, before the Association was formed, local farmers bore the loss incurred by slaughtering the affected owner's stock and were later reimbursed by the Association. In recommending slaughter of contact animals as well as those showing infection, the Association went in advance of the Orders in Council (Anon., 1882).

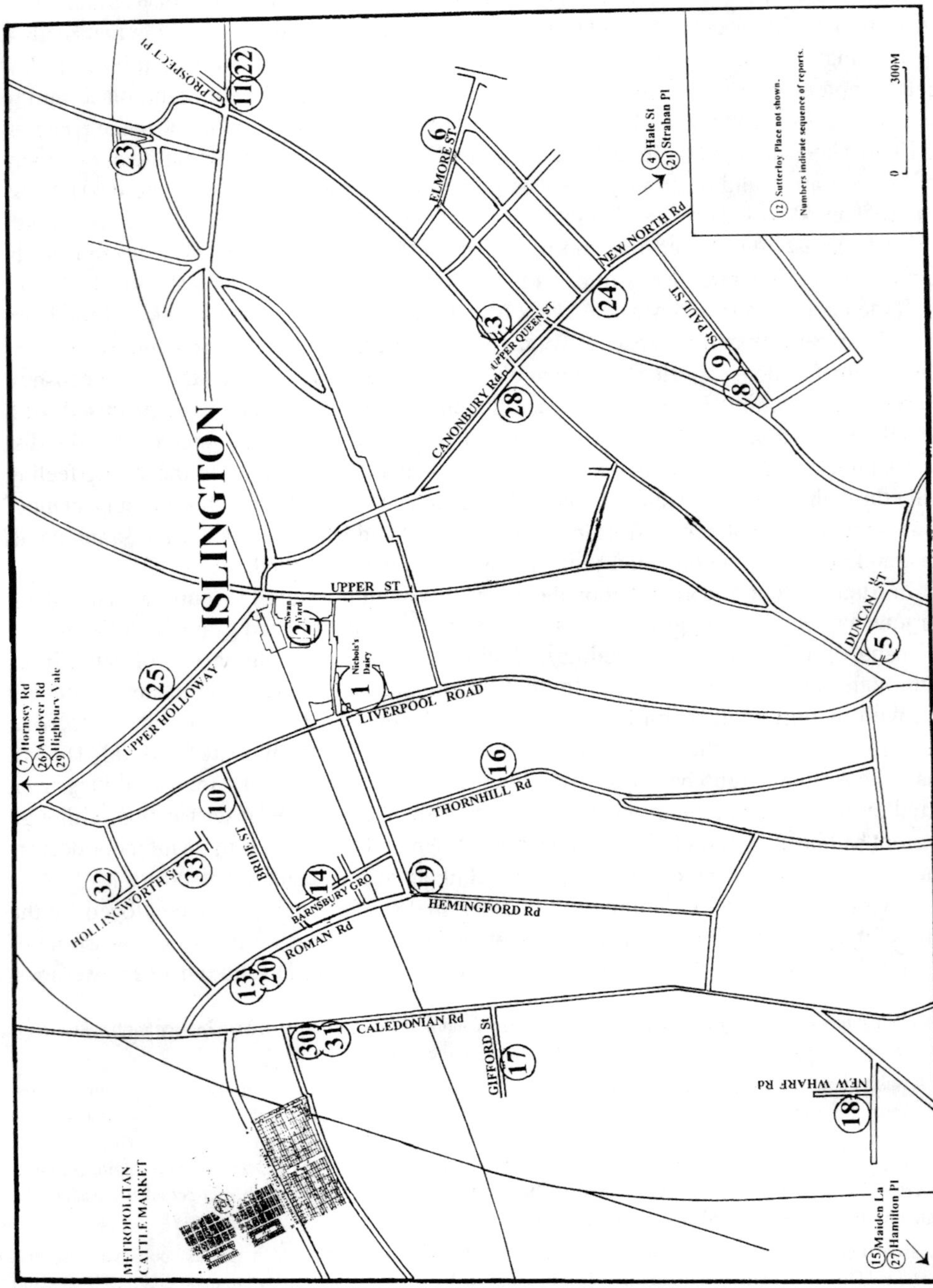

Figure 1. The spread in Islington (after Bernard, 1866a).

sick. I asked what was the matter with them, and they said it was cattle plague. I do not know what became of them" (Perks, 1869).

But such connections were yet to receive attention, and in September, Simonds stated that new foci of infection were usually traced to the purchase of sheep which had been in contact with infected cattle. Although as a result of his recommendations to the Privy Council, amendments to the Orders were immediately introduced to include other livestock, by October 14 it had appeared in 29 counties in England, two in Wales, and 16 in Scotland, and was still spreading. By October 28, it was recorded that 7,912 animals had died, 6,866 had been slaughtered, and 848 had recovered. But in one case, it was revealed that although 41 cows had died or were slaughtered, only 11 were reported, so the real numbers could have been much higher than official records showed.

By November, only four counties in England remained unaffected, and the disease was present in 19 of 33 in Scotland. Of cases reported up to November 4th, 44% of cattle on farms where the disease occurred were affected by it, and of this proportion 43% died. By the end of January 1866, these proportions had risen to 54 and 64%, respectively (Table 1). In February, it was reported that recoveries were 4% in London, 7% in Cheshire, 12% in Wales, 17.5% in Yorkshire, and 20% in Scotland, percentages which probably reflected methods of husbandry rather than stock differences in resistance.

A typical example of an outbreak was on Lord Granville's farm at Hendon, where 123 of 130 cows died. The nearest adjacent case was about 360 m distant, but Lord Granville's bailiff had purchased five Dutch cows in the Metropolitan Market which were placed in quarantine about a kilometre away. Within 10 days of purchase, one became sick and died 5 days later. Within another 5 days, all had died or were slaughtered. Every precaution was taken from the outset of the purchase to prevent contact with healthy animals on the farm, but 9 days later, five animals fell sick. The bailiff asserted that he had taken the greatest possible pains to avoid spreading the infection and did not go near the cows on the same day, or wear the same clothes, as when visiting the Dutch cows.

One London dairy affected was that in Highgate of the well-known philanthropist, Miss. Angela Burdett Coutts, who expressed more intelligently than most her concerns about the disease, stating her considered opinion that it was rinderpest. She posed a number of questions to Government which needed to be answered, such as whether the milk and meat of diseased animals were fit to be consumed: "At present no one knows where to turn for any advice," she wrote to *The Farmer's Magazine* on September 8, 1865, "... our Government should take the lead in guiding us how to act, and in assisting us to meet an emergency which is full of difficulty and danger to all classes." Her philanthropy was exemplified by her remark that she would have sent in the report on the outbreak by her bailiff earlier but had been extremely occupied with the various orders necessary to be given to avoid perpetuating the disease because her bailiff and herdsman "went to the seaside for few days." She did not let the outbreak interfere with her employees' summer holiday. As to the current theories that the disease might be due to the state of the cowsheds of London, or the state in which imported cattle were allowed to come into the market, she considered these a disgrace to the legislation and "wholly at variance with our professed civilization, not to say religion, that it should be possible even to advert to these as the most probable sources of this disease."

Some tried to lay the cause on the dirtiness of many of the London cowsheds, but its outbreak at Miss Coutts's model dairy dispelled the notion that it was accumulated filth which encouraged the disease. There was ample evidence to the effect that cleanliness was no guarantee against infection, and at the end of November, it was found that a cowshed of the "inferior sort" in Drury Lane was not affected, although the disease raged around it.

Table 1. Cases of rinderpest reported each week, England only, 1865.

Date	Cases	Slaughtered	Died	Recovered	Percentage Died	Percentage Recovered	Percentage Killed
June							
24	30	0	28	0	93	0	0
July							
1	147	103	21	0	14	0	70
8	575	317	148	0	26	0	55
15	321	168	117	8	36	2	52
22	475	114	346	7	73	1	24
29	712	387	254	6	36	1	54
August							
5	1,092	271	732	26	67	2	25
12	1,195	664	423	20	35	2	56
10	1,153	391	601	15	52	1	34
26	1,263	438	652	35	52	3	35
September							
2	858	290	438	18	51	2	34
9	736	293	329	9	45	1	40
16	961	382	377	35	39	4	40
23	1,040	364	468	72	45	7	35
30	1,262	425	728	103	58	8	34
October							
7	931	329	473	128	51	14	35
14	1,185	423	623	103	53	9	36
21	1,057	434	485	90	46	9	41
28	1,010	315	517	96	51	10	31
November							
4	1,588	504	707	126	45	8	32
11	2,173	570	1,068	117	49	5	26
18	2,611	640	1,460	149	56	6	25
25	2,840	661	1,465	223	52	8	23
December							
2	3,456	495	1,876	270	54	8	14
9	4,528	671	2,698	369	60	8	15
16	5,180	540	3,325	464	64	9	10
23	5,683	597	3,637	563	64	10	11
30	6,609	576	4,518	694	68	11	9
Totals	50,671	11,362	28,514	3,746	51	5	32

Source: Williams (1866).

Meanwhile, the Privy Council had approved a lengthy document by Dr. Thudichum, a medical officer employed by the Medical Department of the Privy Council, *The Principles and Practice of Disinfection*, "which elicited the ridicule of almost every farmer in England," according to one commentator; but to another, "should be read by all who are anxious to acquire information upon this subject, and who amongst us is not?" Whereas one of its recommendations was washing healthy animals with chloride of lime to prevent infection, Dr. Thudichum stated that all parts and derivatives from animals that had died or been killed on account of advanced disease were infectious. He believed that the intestinal discharges were the principal agents upon which mainly depended spread of the disease.

At the end of July, alarmed by its spread, Arthur Helps, Clerk to the Privy Council, had asked Simonds to prepare a memorandum on control measures which he, Helps, would forward to the RASE, the only body with sufficient influence to persuade the Government to take more effective action. Helps, Simonds, and Dr. John Simon, the Privy Council's medical adviser, agreed that the European experience demonstrated that the futility of attempting cure and restriction on animal movement was the logical basis of control. Simon was positive in supporting a total ban on the movement of cattle and the slaughter of infected animals, "no moderate or popular means will stop this disease," he declared later in evidence to the Commission.

The memorandum, dated August 30, contained the following recommendations:

1. Persons should abstain from buying "Store Stock" at fairs and markets, and should not buy from cattle dealers without a warranty against the disease.
2. All newly purchased cattle should be isolated for not less than 12–14 days.
3. Affected cattle should be isolated.
4. Severely affected cattle should be slaughtered and buried, and the skin placed in a disinfecting fluid.
5. No other cattle should be allowed near burial places for several weeks.
6. No person in charge of sick animals should be allowed to go near healthy animals, and all indirect communication by dogs or other animals between infected and healthy cattle strictly prevented.
7. All healthy cattle to be well washed and cleansed after removal from infected cattle.
8. Fodder and straw used by affected animals should be treated with lime or burned.
9. All manure in sheds or stables occupied by diseased cattle to be daily sprinkled with a disinfecting powder.
10. Buildings used by affected animals should be disinfected and whitewashed with quicklime before any other cattle are placed in them.
11. Cattle trucks, ships' holds, wharves, and so on, should be washed and disinfected after use.
12. No animals exposed to the disease should be allowed at fairs or markets within a month of exposure.
13. Apparently healthy animals in contact with affected animals which are to be slaughtered should be sent with all due care and caution to the nearest slaughterhouse if not killed on the premises, and the skins placed in a disinfecting fluid.
14. Animals should be looked after by protecting them from bad weather, feeding nutritious foods and given pure water.

A simple description of the clinical signs was given for wide circulation to ensure that the disease would be recognized, together with those of pleuropneumonia and FMD so that rinderperst could be distinguished apart.

The recommendations were published by the RASE but, as Simonds recorded, were most strenuously opposed by agriculturalists and the public at large, led by the Press: "and not only so, but by a general body of the Medical profession, by whom we were twitted as having no other remedy except the poleaxe for arresting the progress of the Plague." So great was the opposition that Helps informed Simonds on one of the latter's now daily morning visits that if the opposition of the Press and the public were not soon abated they would all have to resign. To Disraeli, he wrote that Government could not take strong measures which "are not to some degree warranted by public opinion."

In the Metropolitan District, considerable difficulty was encountered with regard to the interment of carcasses. At first, they were buried in the yard or near the premises where the disease occurred. This frequently interfered with the sewers, so it became almost impossible to find suitable burial places. For some time, large numbers of carcasses, which it was thought would do for manure, were sent to the Isle of Dogs; but farmers refused to take the manure, and so the Isle of Dogs residents refused any more carcasses. Daniel Stone, Professor of Chemistry at the Royal Manchester School of Medicine, saw the burial of cattle as a sad loss of manurial matter and an adverse reflection on science. He was to recommend to the JPs of Salford in February 1866 that instead of strewing lime on buried carcasses, the arteries should be injected with milk of lime, but how ordinary herdsmen were to do this was not explained.

3. THE CASE OF THE REVEL CARGO

Prior to 1865, there had been only one known import of stock direct from Russia, in July 1860. The infected shipment to which the origin was allegedly traced had been landed from the "Tonning" at Hull from Revel about June 1, 1865, and a number of cattle sent to various places in the north of England before the remainder was despatched to London. So it probably did not spread north from London as quickly as was supposed but radiated from other foci of introduced infection. It was traced because Simonds had been casually informed by an importer of cattle about the beginning of May that a number of animals was likely to arrive from Russia. In Simonds's words to the Commission: "The conviction at once flashed across my mind that it was not unlikely if we went on importing animals from Russia direct that we should not be very long without introducing the Rinderpest. I felt therefore a deep interest in the matter, and I determined to keep my eye upon this lot of cattle, and to make all the enquiries that I could."

Later, in evidence to the Commission, an importer, Mr. John Honck, explained how the shipment had come about. The summer before (1864) he had become acquainted with James Burchell, who stated he believed there was some good to be done from Russia with cattle imports, and he was going out there to see. He asked Mr. Honck if he would back him in importing some cattle if he found that the prospects were good. Honck did not commit himself but stated: "If you go and see I will hear what your report is first." Burchell went to Russia and returned about October, "... his representations were of such an extraordinary character ... and he made out the business in prospect to be so profitable, that ... Mr. Baker and myself were induced to come forward to carry out the contract which he represented he had made with the landowners in Esthland."

In his evidence, Burchell threw the blame on everyone else. Beginning with his problems at Revel, out of a total of 640 oxen collected together there, 46 came from St. Petersburg. From this mixed group, Burchell selected 321, of which 13 were of the St. Petersburg lot. The Estonian farmers' society which he was buying from forced him to take "all sorts of rubbish." Four he sold in the yard, two for having FMD, and two for being sick. He went to the burgomaster, who provided him with three arbitrators, but the society stopped his passports and brought him before the justices, doing everything they could to intimidate him to make him take the full number of cattle so he would not fall back on them for short delivery. The four cattle he sold they insisted upon him taking, "They stopped my passport and thought to frighten me very much. It is an awkward thing to be in a country like that, and to have your passport stopped, and you doing no harm." He had complained to the Government, but they replied that they could not interfere. Asked if he was aware one of the animals he sold had rinderpest, he replied he was not but did not think it would stand the voyage and would not have bought it if they had not forced him to do so. In answer to other questioning, he said he did not make any inquiry as to whether there was any disease among cattle in Estonia, but he knew of a case where one animal among 30 which were being grazed for him had caught the disease and died, and had been replaced by another. But he added that the disease did not affect the cattle in that country as it did in Britain.

With reference to an ox which became sick on board on May 28, and was sold in London on May 30, asked if he thought it had the disease, he replied he did and he had said so at the time. But it transpired that he did not know the disease other than by description, and he stated that at the time, he knew nothing about rinderpest, only learning and reading about it subsequently.

As to the ship's diversion, asked why the ship had touched at Copenhagen on its way to Britain, he replied that it had done so "On account of some scheme, I could not understand what it was, for the oxen were entered in the manifest as for Lowestoft, in order to cheat the Government ... but they were never intended to go to Lowestoft ...." Replying to further questioning, he stated, "Mr. Honck wrote to me, stating, 'I have ordered the things to come to Lowestoft, where we shall run no risk of the doctor.'" The ship was then diverted to Hull because the captain had orders to do so, according to Burchell "so as to avoid the doctors." Honck stated that the diversion was made because of the mixed reports from Burchell about the condition of the cattle, "partly because I thought that in the northern districts there would be a better market for lean cattle, and also as ... they ... might ... get the foot and mouth disease, for in Hull they have never taken any notice of that, but in London and Lowestoft they are very particular ...."

The cargo was unloaded on May 29. Of the total, 106 went to Derby or Leeds, 40 to Manchester, and the remaining 175 to London, where they were sold 4 days later. Twenty were sent on to Gosport, and the remaining 155 went to London butchers. A flock of 330 sheep shipped with the cattle went to Hull for slaughter.

Burchell swore that the cattle were not inspected at Hull. Whether they were or not, a certificate dated May 29, signed by two veterinary surgeons to the effect that the shipment was clean, was subsequently produced. But one of the veterinary surgeons admitted that the examination was not thorough enough to detect any serious illness. Of the 331 animals, none showed any signs except one which was ill on the voyage and apparently recovered. Burchell was adamant that nobody had inspected the shipment, "... because if they had carefully examined them they would have stopped this beast ... it was a most important part of my business to see that they did not cast any without due cause ...." He added, "I may mention that there was one sheep sick on board, and that sheep had to be carried on shore, and if the inspectors had examined it they should have mentioned it."

The story of the inspectors, Mr. Lockwood and Mr. Freeman, was quite different. Lockwood stated that when the Customs found there was a great number of cattle coming from Revel, the Customs ordered both to be there to examine them, "... and we did examine them thoroughly, and I think that I never saw in my life beasts come up in a fresher state than they did." Asked then if he had a special order from the Customs intimating that a certain cargo was coming from Revel and directing them to pay special attention to it, he answered in the affirmative. He was then asked if he had any objection to submitting the order and replied, "Not the slightest." Subsequently, he stated that there had not been any special order, the two inspectors had been there attending other cargoes both before and after the arrival of the "Tonning." Asked if any of the animals were sick, he replied: "They were not ... They were not all fat beasts, but they were in very fine condition; that is to say, fresh and healthy." He stood at the top of the ramp, examining the cattle as they passed while his companion did the same at the bottom. As to Mr. Burchell, he stated he saw "a party who ordered different people about; and at last he was ordered off the ship altogether, by Mr. Twidell, the shipping clerk of the firm of Brownlow and Lumsden [the firm which received the cattle]." Allegedly, the owner of the cattle came on board to tell the shipping clerk that he had discharged Mr. Burchell from his service, and he had nothing to do with the cattle. Burchell then hung about the pier and allegedly must have seen the inspectors examining the cattle.

This was contradicted by a partner, who claimed that the whole shipment was from Estonia. The point was immaterial, for, as Gamgee pointed out, the Russian cattle infect others when they do not appear to be suffering themselves. It was estimated that the overland journey from the point of origin took only 2 days by rail, and the sea journey another 8 days. A total of 10 days from the time the transit began. Another account states that the cattle were transported in four horse wagons, which must have taken a week, the distance being 320 km.

The plague had been in the neighborhood of St. Petersburg in November 1864, but not for several months prior to the transaction (Playfair, 1866).

Probably influenced by Simonds, who perhaps did not want to admit the truth of what Gamgee had forewarned, the Commission at its first sitting cast some doubt upon the Russian import as the origin of the outbreak. This was on the grounds that the bulk of the shipment came from Estonia, which was certified free of the disease. Simonds, who always chose his words with great precision, could only say that he thought it highly probable that it was introduced by the Revel cargo:

> ...we have no direct evidence that it was. We can only show that the animals came from Russia, and that they were the first lot that ever did come from Russia by way of the Baltic...We know, further, that these animals came from a district, or rather from a country, that is known to be the seat of the plague, and that within 19 days of their arrival we had the disease thoroughly established in the Metropolitan Market.

Read stated this was merely presumptive, especially as there was no connecting link between the sale of these cattle on June 1 and the existence of the disease in the Metropolitan Market on June 19.

No animal was shown to have contracted the disease at the Metropolitan Market from June 1 to June 19, the date of its apparent transmission to the Islington, Lambeth, and Hackney cattle, although a leader in *The Times* of August 15 stated that animals affected by rinderpest had been seen in the Metropolitan Market as early as June 12. As they were not Russian cattle, they must have been animals which had contracted the disease.

Playfair argued that it was 18 days after the sale of the Revel cattle in the London market that the disease appeared, and the only support for the Revel cattle being the source was that the purchasing agent fancied the sick oxen at Revel, and the animal on board apparently cured of its indisposition with brandy *might* have had the plague, his statement being an afterthought made following his quarrel with his employer. Playfair suggested that the disease was more probably introduced with imports from Galicia and Hungary. Hungarian cattle introduced the plague to Utrecht in May 1865, and it may have passed to Britain from Holland. The first beasts found affected with it in London were newly bought Dutch cows (Playfair, 1866).

4. Mrs. NICHOLS'S DAIRY

On June 19, Mr. Carvell purchased two cows, one of which showed signs of sickness 5 days later, spreading it to 12 other animals with which the two had been placed. Also on June 19, two cows were bought by Mrs. Nichols of Laycock's dairy, and one by a Mr. Baldwin of Hackney; all of which infected the dairies to which they were taken. Mrs. Nichols thought that her cows had been poisoned. A veterinary surgeon, Henry Priestman, was called to examine both her's and Mr. Baldwin's sick cattle, and had one or two animals killed to make autopsies. In 3 days, several died, and he examined 12 or 14 in this manner, but apart from observing that the internal appearance of all animals was the same, he did not reach any other conclusion.

On June 28, he was called to a case in Hackney which exactly resembled those at Islington; in this case, the animal died overnight. His autopsy revealed an internal appearance similar to the other cases. By July 3, he had examined 20 animals, which had died or he had slaughtered, and a further 20 were being kept in a quarantine shed.

At this point, Priestman approached Simonds, stating that he had a disease in some dairies which he could not understand: "it was quite foreign to me." The next day, he took

Figure 2. A 19th-century London dairy at Hungerford Market.

Simonds to see the viscera of a cow he killed purposely. He then showed him 20 cows under treatment, 10 of which he had condemned the previous day as hopeless cases.

According to Priestman, Simonds did not identify the disease, and Mrs. Nichols's cows were sent to the market or to slaughterhouses. She lost her whole herd of 93, bought some others, and lost those also. Priestman stated to the Commission that it was impossible to learn whether or not there were any cases prior to June 27, but a day or two before, he had passed over several cases in Mrs. Nichols's dairy as pleuropneumonia. Later, when Spooner saw it, he recognized it immediately for what it was without having had the benefit of a trip to Europe to study it as had Simonds.[2]

Baldwin's sick cow also came from the Metropolitan Market on June 19. It died on June 29. The next case occurred at Baldwin's on July 5 with another cow bought on June 19. Two days later, three cows and a bull fell sick. All of these were slaughtered. By July 14, the disease had appeared in nine other dairies throughout London, and soon nearly the whole city was involved (Figure 2). Maidenhead in Berkshire was affected immediately after the London

[2]Greenhow (1857) considered that the majority of symptoms reported in the 1745 outbreak in Britain indicated pleuropneumonia but that rinderpest was present as well.

outbreak due to cattle from the Metropolitan Market. Cattle bought on July 1 in the Market spread the disease into Norfolk.

5. THE OUTBREAK WIDENS

As Reed (1866) expressed it: "Gathering fresh force from every fresh center, the plague attacked the herds of the West Riding of Yorkshire, cleared 30 of the 200 cowhouses of Edinburgh, increased its hold of Devon, decimated the cattle grazing in the Isle of Ely, leaped upon Derby and Dorset, ... Worcestershire ... Cumberland, Northumberland and South Wales ...."

On July 22, the disease had appeared in Aberdeen, conveyed there from London by four infected calves. Three of the four having died, the fourth was taken to another farm and put to a cow but died a week later. Then, beginning on August 2, the disease appeared among other stock and was also introduced to another farm directly from London. On several occasions, it was stamped out by slaughter and disinfection of premises, only to appear again elsewhere. In Edinburgh, it was introduced by some low-priced foreign cows bought in London in July, the first symptoms appearing on August 8. In a very short time, a number of dairies in three localities became infected: "The diseased beasts were conveyed along the public streets in great numbers to be killed in the slaughterhouses, the carcasses carried in open carts to the tanners, and at a later date the affected animals were driven or conveyed in an open wagon to the Sanatorium .... Manure was freely taken to the country, and without any disinfection, in farmers' carts, the market was kept open without disinfection, though diseased beasts had been repeatedly exposed in it, and thus, not only did the disease spread rapidly in the town, but from this as a centre, it quickly spread in all directions into the surrounding country" (Gamgee, 1866).

The first sick animal seen by the Inspector of Markets in the Metropolitan Market was not until August 6, but at the beginning of August, a great number of sick animals were brought there, of which some were destroyed. The cattle inspector became alarmed at the number condemned, and so as a drove of 120–150 animals showed only the first symptoms of the disease and were considered fit for human food, they were allowed to leave the market as being too large a number to be destroyed. Later, Simonds advised that the Order in Council should be strictly adhered to.

On August 21, infection was introduced into Newcastle with a cargo of animals from Rotterdam, where the disease was now raging. The same ship, the "Lord Raglan," brought more cargoes of diseased animals from the same port on September 17 and October 8; but on these occasions, the inspector detected the disease and refused admittance of the animals. Another ship, the "Maas," a steamer specially constructed for the cattle trade, brought 280 oxen and several hundred sheep and pigs to Blackwall Docks on September 6. The shipment was examined by the veterinary surgeon who found one sick cow among the oxen and ordered its immediate slaughter. The postmortem examination suggested that the animal was in the early stages of rinderpest, and the oxen were placed in quarantine. The sheep and pigs were sent on to market.

In July 1865, 7,665 cattle and 2,773 calves had been imported into London, the total number rising to over 16,000 in October and November. Weekly returns of reported cases of rinderpest from June 24 showed an exponential rise with an almost constant ratio of deaths averaging 50%, not counting animals slaughtered, which averaged 32% overall, although the latter percentage declined as the number of cases increased (Table 1 and Figure 3a,b,c).

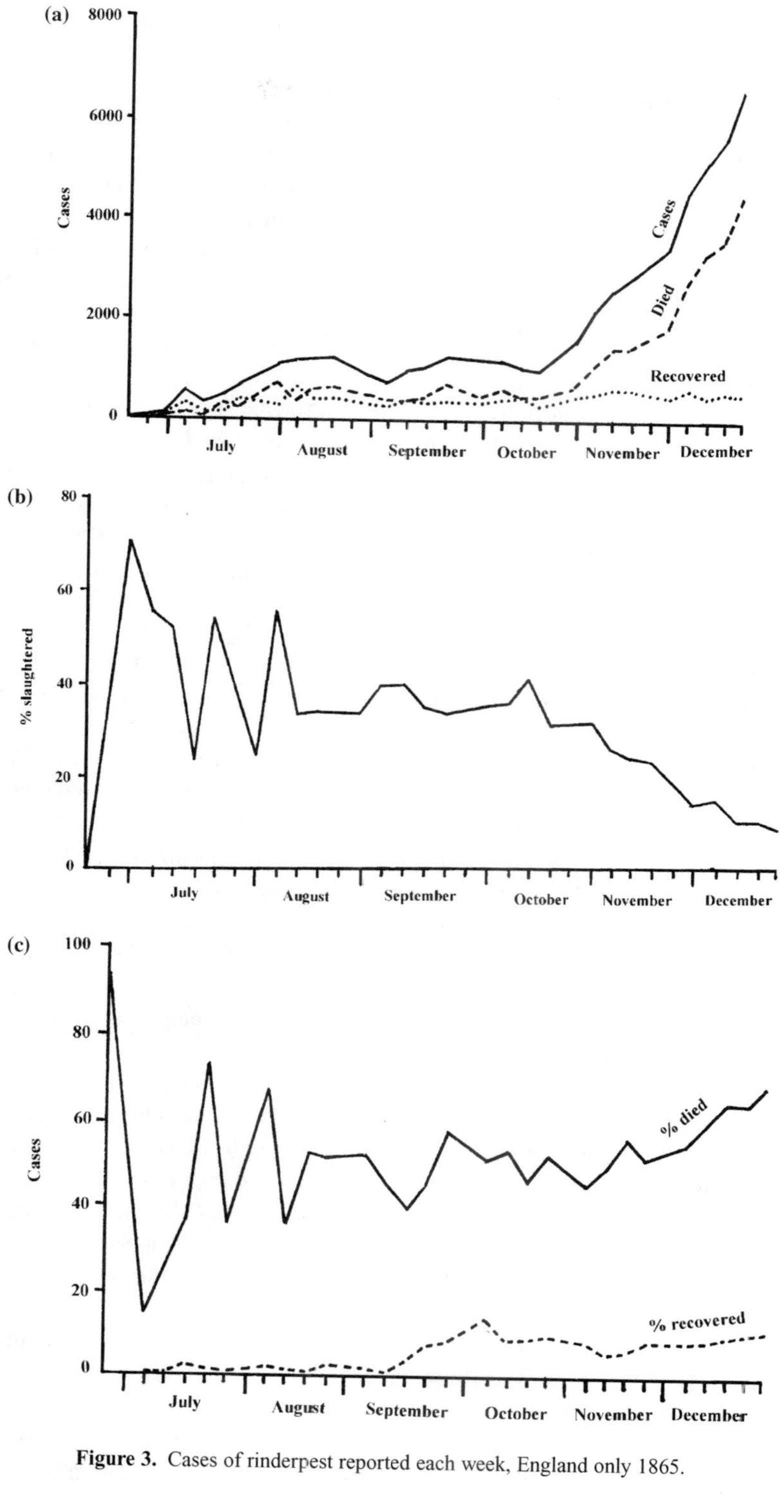

Figure 3. Cases of rinderpest reported each week, England only 1865.

Nearly every county in England and two thirds of Scotland were infected in the first 4 months.

The Farmer's Magazine of September 1865 had subjoined "some very sensible remarks from *The Times*, as to the commonly accepted theory of the plague having been brought from abroad, when the Veterinarians cannot as yet offer an atom of proof to that effect." These remarks were from an editorial written by the Very Reverend Dr. H. Wace, future Dean of Canterbury, who, together with the Reverends Dr. Woodham and J. Mosley, wrote most of *The Times* editorials in 1865. In a rebuke probably directed at John Gamgee, it stated "We should look for the causes of mischief at home rather than abroad, and should think more of our own cowsheds than the conditions of pasturage in Russia."

"Somewhere or other this plague must needs have been 'spontaneously developed,' so why not here?"

A parallel was drawn with bubonic plague believed also to have been periodically imported in the past: "That it should be traced to foreign importation then was perfectly natural, for nobody believed in any other origin of plagues...," although conceding that the disease was identical with the Russian disease of rinderpest, "It is so conformable with our old notions of pestilence to believe that the disease is none of our own, that we should be suspicious of so plausable [sic] a theory." Finally, "It is nothing whatever, notwithstanding its outlandish name, but typhus fever—the name indeed by which the French actually call it."

Following the article in *The Times*, Professor Murray of the Royal Agricultural College, Cirencester, was prompted to issue a pamphlet to set the record straight, the article having described the "Siberian boil plague" under the title of "The Kasva," which attacks all domestic animals and even man: "... so what '*The Times*' ostentatiously termed 'The Prussian Official Report of the Cattle Plague' was not an account of the 'Cattle Plague' at all, but by a gross blunder was published as such" (Murray, 1865). Murray stated cattle plague was propagated by contagion alone and could not have been introduced into Britain in any other way, "It is very clear that this disease is not of spontaneous origin." But it was not until Robert Lowe, Viscount Sherbrooke, an editorial writer for *The Times* for the past 10 years, produced an editorial on October 4, his first on the subject, that an end was put to the paper's crude anti-contagionist stance. Dwelling on the inadequacies of the medical and veterinary professions, Lowe censured them for their ignorance, not for their failure to find an immediate cure.

Several medical committees were appointed in the country, all of which recommended treatment; but Simon, questioned by the Commission as to how much infection was due to poor hygiene, pointed out that in human epidemic diseases, the best hygiene did not prevent susceptibility to a disease: "Typhus, you know, attacks the physician of the Fever Hospital as much as it attacks the poor people who are brought to the hospital," he stated. Good ventilation and drainage did not render people less susceptible to the cause, but might certainly render the cause less powerful. Yet in Cheshire, a Manchester physician lectured on cure of the disease.

Previous outbreaks seemed already forgotten, as too the lessons of Lancisi and of Bates in 1714. In the first month of its occurrence 2,000 cattle were reported to have died, and by August, it had spread from Middlesex to Surrey, Essex Marshes, Sussex, Portsmouth, Plymouth, Aylesbury, Norfolk, Suffolk, Warickshire, Kent, North Lancashire, Leicester, Peterborough, Nottingham, and Edinburgh (Figure 4). In September, it redoubled its violence.

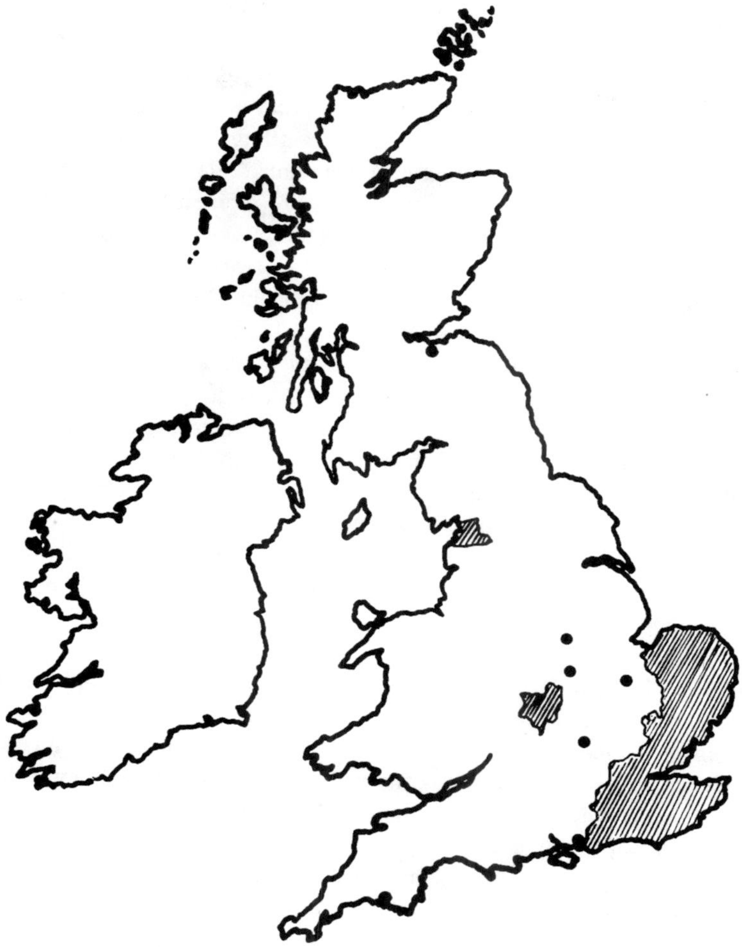

Figure 4. Extent of rinderpest infection by August 1865.

6. PROBLEMS OF CONTROL

One problem of control was that there were four types of inspectors. Those appointed under an Order in Council by the Clerk of the Council; those appointed by magistrates under the Order in Council; inspectors of fairs and markets appointed by the mayor and corporation; and inspectors of ports appointed by the Board of Customs. Such inspectors could be veterinary surgeons, farriers, cow-doctors, and others. But in Essex, qualified inspectors had been offered payment by the local authority of only half-a-crown per week (12.5 new pence); and magistrates had appointed butchers, tradesmen, publicans, and shoemakers, although within the Metropolitan District, they had no authority to do so. One shoemaker had been appointed inspector because he was a collector of poor rates, and a large cattle salesman declared to the Commission that he had ordered his men to use force if necessary to keep the man from examining his cows. A reason for its continued spread was considered to be inspectors executing slaughter orders going from one area to the next looking for diseased cattle. But other reasons for its rapid spread were thought to be the number of sales and transport by rail, rather than natural contagion; and because no compensation was offered by Government for stock liable to confiscation. As soon as an owner became aware of the fact that his stock might be infected, he sold off the animals in the markets, disregarding the consequences to others. *The Times* of September 8, 1865 reported that in the previous 2 days, 19 sick animals had been sent to the Metropolitan Market, and summonses were granted to the clerk to the market against persons sending in diseased animals. On October 11, the same newspaper complained of the veterinary profession relying on continental doctors who had dogmatized on the disease, although such continental authorities were far ahead in their knowledge compared with their British counterparts.

Another reason for its persistence was that farmers ignored advice given to them. The Government, for example, issued orders that the annual fair at Croydon should not be held in 1865 but the farmers determined to hold it and hired a private field for the purpose, leading Helps to ask: "What can we do if the farmers will not help themselves?" But by mid-June 1866, when the number of reported cases was in decline, we read that slaughtering throughout the country for the purpose of stamping out the disease had been conducted "with tolerable efficiency." According to Armatage (1894), "Officious inspectors armed by a magistrate's warrant, galloped furiously on horseback from place to place, carrying a gun over the shoulder and powder in the flask, ready to practise with an instrument probably never before used by them."

On October 17, 1866, Dr. Edwards Crisp stated that the best means of eradicating the disease was to prevent all movement of cattle and sheep for a month or 6 weeks, except under special circumstances and to establish dead meat markets in all cities and towns (Crisp, 1866a). On November 21, he noted with concern that Government had stopped movement of cattle but not of sheep, which showed rinderpest-like symptoms (Crisp, 1866b).

Toward the end of the first year, public opinion was already changing, following initial rejection of the RASE recommendations in August. In November, a deputation of the Farmer's Club, headed by Read, met with Sir George Grey, Secretary of State for the Home Office, representing the Privy Council. The deputation presented a memorial recommending eight points of action:

1. That for 2 months all traffic in cattle be prohibited, with the exception of fat stock sent to market for immediate slaughter.
2. That such animals be not then permitted to leave the city or town where the market is held, and be slaughtered within 8 days.
3. All imported foreign beasts be killed at the point of disembarkation.
4. All fat cattle be marked or branded before leaving any market, and those bought privately for slaughter in the country.
5. That infected districts be cordoned off and no animals allowed to leave alive until 1 month after the last animal affected by the plague has died, been killed, or recovered.
6. Importation from foreign countries affected by the plague be prohibited.
7. Hides of diseased animals to be disinfected before removal from slaughter houses and no manure to be removed from infected farms or cowsheds.
8. No cattle during the prohibited time be turned out upon any road, common, or unenclosed land.

In reply, Grey stated that much of what had been said had been anticipated in the new Order in Council, which had been drawn up after receiving many memorials from different parts of the country. The difficulty was to frame a general order applicable to the whole country, as some parts were presently free from the disease and would be greatly inconvenienced by such interference. Thus, the Lords of the Council felt it right to leave a great deal to the discretion of the local authorities. He thought it would be best to see first how this order worked.

At the beginning of December, the Bath and West Agricultural Society sent a deputation to the Privy Council under Mr. T. D. Acland, MP, to call on the Government to issue a general order to the effect "as very clearly put by *The Times* newspaper," that

> cattle, sheep, or pigs, exhibited for sale in the market or fair in or near any town, should be forthwith branded, and not allowed to leave the town alive; that cattle offered for sale in a strange place should be killed within 48 hours; that foreign cattle should be killed at the port of disembarcation, or only allowed to proceed to particular destinations under strict conditions; and that no cattle, sheep, or pigs should be allowed to travel upon any public road from any farm or place where there has been within two months any case of cattle plague.

In reply, the ministers pointed to the inexpediency of interfering with the provision trade of the country by any uniform or unbending rule; the desirability of obtaining the active concurrence of the local authorities in support of any measure, and that clause 18 of the last Order in Council gave absolute power to local authorities to prevent the entrance of cattle from any other part of the country into the district within their jurisdiction.

On December 14, a meeting chaired by Viscount Combermere was held at St. James Hall, Wakefield, attended by members of various local agricultural societies throughout the kingdom. The meeting supported a central authority with one law for the whole country: "... this meeting, seeing the daily-increasing ravages of the cattle-plague, is strongly of opinion that the present orders in council are not sufficiently stringent to meet the calamitous effects of that frightful disease."

By the end of the following January (1866), the Commissioners' *Second Report* observed that the disease had continued to spread and the "great breeding districts of the United Kingdom, into which, in ordinary times, cattle are but rarely and casually imported—Ireland, the North and West Highlands of Scotland, including the counties of Argyll, Ross, and Sutherland, and the whole of Wales, except Denbighshire and Flintshire" had hitherto enjoyed an entire immunity. They noted that it had moved irregularly through the rest of the country, seeming generally to have followed the highways and ordinary lines of cattle traffic. Whereas it often appeared to jump some areas, it was simply that the cattle responded more slowly in such places, but in contagiousness and virulence, there was no abatement.

In Argyllshire, a leaflet, dated February 20, 1866, to try and exclude the disease was issued by The Central Committee of the Commissioners of Supply in Edinburgh. It recommended that shepherds and others visiting surrounding districts should, if possible, leave their dogs at home, and change and disinfect their clothes upon return. No dog should be allowed to enter the County without being thoroughly disinfected. Supplies from Glasgow or elsewhere should not be packed in hay or straw, and persons receiving packages from steamers or otherwise from infected districts should have items and packing materials disinfected before stowing in farm sheds, and so on. It pointed out that there was reason to believe that infection had sometimes been communicated by tramps sleeping in byres or cattle sheds, and persons were requested not to allow strange men or dogs to approach their buildings or cattle, nor to encourage visits of hawkers, tinkers, or vagrants. Also, some towns and villages on the coast imported dead sheep from Glasgow and other places with the head and horns intact, and as the wool on the head could carry infection, the carcass should be imported without the head, or the horns taken off and the wool singed.

The disease was allegedly brought into Canonbie in Dumfriesshire from Cumberland by a worker who returned to visit friends in January, three affected cows being destroyed at the request of the neighboring farmers who apparently would compensate the owner themselves.

It was thought to have been carried to Kingussie in Inverness-shire by straw and hay gathered from north-going railway trucks, one cow belonging to a railway carter being affected. The cow and two pigs were immediately slaughtered and the cowshed, together with 12 tons of dung and a quantity of corn, burnt, being pronounced infectious. Aberdeenshire, having been "free" for a fortnight, reported a fresh case on Deeside, and 16 head were slaughtered. It was presumed to have been carried to the farm by dung carted from Aberdeen. Another outbreak occurred at Old Deer, infection believed to have been carried on a mat sent from Aberdeen in which hedge plants were rolled, the mat having come from England some time ago.

Sir J. P. Kay-Shuttleworth, an Argyllshire landowner, issued instructions to all his tenants recommending cleanliness, nutritious food, pure water, and so on, and allowing no vagrants to

Table 2. Cumulative totals of rinderpest cases for England and Scotland 1865–6.

Date	Cases	Died	Recovered
July			
1	177	49	0
8	752	197	0
15	1,073	314	8
22	1,563	671	15
29	2,320	957	21
August			
5	3,465	1,713	47
12	4,696	2,166	70
19	5,871	2,787	85
26	7,141	3,441	120
September			
2	8,063	3,906	138
9	8,910	4,280	147
16	10,264	4,869	206
23	11,597	5,465	289
30	13,263	6,314	422
October			
7	14,699	7,032	580
14	17,063	8,029	738
21	19,018	8,883	885
28	21,228	9,986	1,073
November			
4	24,386	11,625	1,418
11	28,147	13,757	1,744
18	32,463	16,390	2,132
25	37,474	19,296	2,703
December			
2	43,153	22,654	3,240
9	50,141	26,945	4,031
16	58,149	32,057	4,986
23	66,500	37,484	6,076
30	76,006	43,812	7,354
January			
6			
13			
20			
27	120,740		

Source: Williams (1866) (in part).

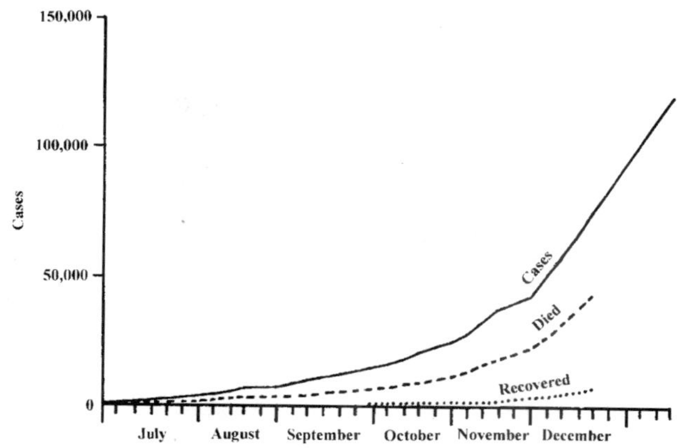

Figure 5. Cumulative totals of rinderpest cases, England, Scotland, and Wales.

Table 3. Counties with cattle populations 5 head per hectare and more, March 1866.

County	Cattle Density per Hectare
Herefordshire	5
Nottinghamshire	5
Shropshire	5
Cheshire	5
Staffordshire	6
Cornwall	6
Somersetshire	7
Lancashire	7
Derbyshire	7
Leicestershire	7

approach their farms, keeping away all packmen, pedlars, dogs, and other animals; to purchase no fresh stock; not to go near any infected farm or beast, nor allow anyone near who had been in contact with infection; and to be very careful about butchers and drovers. They should wash their shoes when they came home and scald in hot water any portion of clothing likely to have been exposed to contagion. Any animals showing any unusual signs were to be separated immediately and reported. If the disease appeared, they should try no remedy but kill at once any infected animal and bury it. He would supply chloride of lime and disinfecting powder: "For the rest, having done what we can, we may humbly look to Almighty God to bless our endeavours."

In March, the *Carlisle Patriot* reported rinderpest had appeared in Cadzon Forest among the famous breed of wild cattle belonging to the Duke of Hamilton. To preserve his valuable breed of Ayrshires, the Duke put them down his coal pits "where they are enjoying complete immunity from the Cattle Plague which is raging above."

The number of reported cases in England, Scotland, and Wales doubled roughly every 4 weeks, rising from 14,699 on the previous year's October 7, to a cumulative total of 120,740 by January 27, 1866 (Table 2, Figure 5). Yorkshire was the worst hit with a total of 19,331 cases reported, followed by Cheshire with 17,971. Of the total 73,750 that died of the disease, 16,742 were slaughtered, 14,162 recovered, and the rest were unaccounted for. This partly reflected the numbers of stock, for Yorkshire had almost 400,000 head, while in March 1866, the figure for Cheshire was 93,044 (Table 3).

A return of stock made on March 5, 1866 gave 3.3 million cattle for England, 541,000 for Wales, and 937,401 for Scotland, a combined total of almost 4.8 million. Sheep totalled 22 million and pigs 2.5 million. By 1877, the total was still to stand at only an estimated 7 million.

The average percentage reported attacked by the disease over the period was 48.6%, in the range 43–54% after deducting those cattle slaughtered as healthy. But as the number slaughtered decreased, so the percentage dying from the disease increased. Of those attacked, 11.7% recovered.

7. GOVERNMENT VACILLATION

On February 6, 1866, Queen Victoria informed the opening of Parliament that she had observed with great concern the extensive prevalence during the last few months of a virulent distemper among cattle in Great Britain, but it was very satisfactory to know that there were parts of the Kingdom not yet attacked. She trusted "that by the precautions suggested by experience, and by the Divine blessing on the means which are now being employed, its further extension may be arrested." The Divine blessing was not forthcoming, and the pusillanimity of the Government meant that the disease continued to spread.

In the House of Commons, Lord Cavendish in proposing the Address, stated: "The sympathy of Her Majesty, which had been so well expressed in the Address, must prove a strong comfort to all who suffered severely from the disease"; while in the House of Lords, the Marquis of Normanby contributed the following, as one reporter put it, "towards a solution of the perplexities of our case": "By the blessing of Providence, I am happy to say that it has not affected my own immediate neighbourhood, and therefore I personally know nothing about the subject."

The Earl of Derby summed up the situation rather more positively by declaring that there was

> a universal chorus of voices from both sides of the House, complaining that the course pursued by the Government with regard to the cattle plague was vacillating and unsatisfactory; that they were from first to last insensible of the magnitude of the evil; that they lagged behind the exigencies of the case; that they were endeavouring to see, before acting, how far public opinion would go along with them; and that they were throughout shirking the responsibility which belongs to the executive Government, and seeking to throw it on the shoulders of other persons, thus producing no uniformity of proceeding, but on the contrary, every variety of system.

There was no Ministry of Agriculture and no Minister in Britain in the 1860s that could be held responsible, affairs being directed by the Privy Council through the Clerk.

On February 12, a Standing Committee of the RASE had an interview with Earl Russell, submitting to him the following Resolutions:

> Resolved—That the Cattle Plague has increased and is increasing to an alarming extent. That the measures hitherto adopted have been wholly ineffectual to prevent its progress.
>
> That no method of dealing with the Cattle Plague, at the present time, will be of any avail unless it provides for:-
>
> 1st. The immediate slaughter and burial, at least 6 feet deep, of all cattle suffering from the disease; making compensation to the owners in such mode and to such extent as shall be considered advisable.
>
> 2nd. The rigid surveillance of all infected farms, and the immediate slaughter of all animals which, from time to time, shall show the slightest symptoms of disease.
>
> 3rd. The thorough disinfection of all infected premises, and a prohibition to remove therefrom all manure, litter, hay, or straw, for a period to be fixed; and then only subject to certain specific regulations.
>
> 4th. That the Government be requested to bring before Parliament a Bill to direct and

> empower the Justices in Quarter or Special Sessions to assemble immediately to carry out the above Resolutions; and in such Bill to make provision for charging the necessary expenses on the County Rate, and also for assimilating the action of Counties and Boroughs.
>
> 5th. That simultaneously with the destruction of all animals, the transit of all animals, whether by road or rail, be entirely prohibited, with such exceptions only as may be absolutely necessary.
>
> 6th. That during the existence of the Cattle Plague, all imported Cattle, Sheep, or Swine, shall be slaughtered forthwith at the port where they are landed; and their hides, skins, and offal disinfected there.

The Council also pointed out "the extreme importance of establishing permanent means for, and compelling the slaughter of, all foreign fat stock at the place of disembarkation; an efficient system of quarantine for store animals; and stringent restrictions upon the purchasers of stock after they are liberated from such quarantine, and that no exception be made as regards the Metropolitan Market." The Council had also since recommended the provisions of the Act concerning slaughter and compensation be extended for 6 months, and Government pay attention to complaints that Orders in Council were in some districts very inefficiently carried out.

The eventual reply was to draw the Council's attention to an Order in Council of May 9, 1866, which extended the provisions of the Act concerning slaughter.

On February 21, 1866, four members of the House of Commons asserted that sheep could not be affected by rinderpest, although the contrary was by then a well-established fact, and from mid-June until July 1866, cases in sheep were far more common than before. Yet many still did not believe that sheep could contract the disease, while others believed that it could be cured in them.

8. PERSISTENCE OF THE PLAGUE

An example of the ability of rinderpest to spread was provided by Market Drayton, where it suddenly broke out on a farm 12 km distant from the nearest infected spot. It transpired that a cow-doctor had been called in to a case of difficult calving, and although every precaution had been taken, he had carried the infection with him, and the entire herd was destroyed. However, whereas it raged with exceptional virulence in this district, some farms escaped altogether. In the parish of Hodnet, when the plague approached, some farmers took the precaution to lock all their gates and block up all approaches to their pastures, also displaying notices prohibiting trespassers and requiring all persons with business at the house to follow a prescribed roadway. Some young men were employed to keep away stray cattle and dogs and to ensure that the rules were observed. These precautions excited much ridicule, but allegedly there was not a single case of infection on the farms (Prosser, 1872). In the village of Wollerton, in the same parish, when the disease broke out in two small herds, although slaughter was not then obligatory, leading farmers of the village subscribed among themselves what they estimated to be the value of the stock and had them all slaughtered, the disease spreading no further.

By the end of December 1865, 73,549 animals were reported as affected and by the end of January 1866, 120,740; of which latter 73,750 died, 16,742 were slaughtered, and 14,162 recovered. On January 20, the *Saturday Review* noted: "The movement of cattle from place to place is the grand source by which contagion is propagated. If the plague is to be arrested, that movement must be stopped." But in February 1866, there were 13,113 cases in 1 week. In March, the total had reached 187,000, but with the introduction of statutory slaughter, the number of new cases in 1866 declined rapidly as follows:

- week ending February 23: 17,875
- week ending March 23: 9,388
- week ending April 20: 4,963
- week ending June 22: 679
- week ending November 23: 8.

The metropolis was twice apparently free of the disease. Detected there in June 1865, it prevailed until December 1866, during which time some 8,000 cows were lost, out of a total dairy population of nearly 16,000. Then, in January 1867, another outbreak occurred in Islington on Mrs. Nichols's premises where it had first appeared. The origin of its resurgence there was never traced. It next appeared in April in Limehouse, followed by occasional outbreaks up to the end of August. The cordon around the metropolis was kept in place for 2 years longer.

Despite its persistence (in the second week of May, it was reported in Cornwall in a cow belonging to Mr. Joll of Calstock, the cow being slaughtered), in May the ban on country cattle shows was lifted:

> exhibitions of animals, whether intended for immediate slaughter or otherwise, may be held in any county in England and Wales, with the licence of the Lords of the said [Privy] Council … provided that no case of cattle plague has occurred in the county where such exhibition is to be held within four calendar months prior … nor unless the exhibition is situated at a distance exceeding thirty miles from the outward boundary of any place declared to be infected …

No cattle could be brought from any place within 32 km of which a case of plague had occurred within the previous 4 months, or from outside of the county within which the exhibition was being held. Cattle had to be examined on entering and leaving the exhibition, and a licence was required for any animal to leave it.

In an open letter to the Lords of the Privy Council in May 1867, Crisp, who was given to letter-writing, this being numbered as his 16th on the issue of rinderpest, gave some of his views on the persistence of the plague:

> I need not again dwell long on the dilatory nature of all the proceedings for the suppression of this murrain. The passiveness of the Royal Agricultural Society; the non-assemblage of Parliament for eight months; the "patient" trusting to local authorities and their divided counsels; the suggestion of the Royal Commissioners that inspectors should not have the right to slaughter; homeopathic delusions and garlic disseminations … the order to tie up single dogs in Ireland and to allow their hunting in packs in England …

Although no cases could ever be traced to fox-hunting, at the Annual Meeting of the RASE the previous December Crisp pointed out that there were 200 or 300 packs of hounds "traversing the country" and that every one of those dogs should be tied up, policemen empowered to shoot stray dogs, and no hunting allowed for the present. Hunting had not been restricted, although it was believed that dogs could spread the disease, but when Crisp raised the subject again at a meeting in February 1866, at which Simonds was present, Simonds could not have endeared himself to Crisp when he replied that he thought "many eminent medical men would have stood in a better position had they not been so ready to put forward their opinions." He had no facts to show that cattle-plague was spread by hunting (Simonds, 1866). Crisp had also stated that if all "practical men" were shut up in glass cases for 6 months, and the editor of *The Times* with them, a great benefit would be conferred upon the community, "and we should be much more likely to get rid of the cattle plague" (Crisp, 1866c).

In the Consolidated Cattle Plague Order of August 1867, it was specified that a local authority could make such orders as thought fit to prevent the spread of the disease by dogs, and could order stray dogs to be destroyed or otherwise disposed of. Presumably, no local authority would have dared interfere with hunting!

Perhaps sensitive to Lowe's strictures in his editorial of October 4, 1865, the *British Medical Journal* was scathing of *The Times*, reporting on October 7, 1865:

> *The Times*, in its leaders, has written on this subject of the Cattle-fever, the most astounding nonsense—we might truly say, has done an infinite amount of mischief—ridiculing the proposals of real scientific experimenters; and accepting the phantastic, semi-scientifically worded, crudities of "S.G.O." as the exponents of its ideal of science. It must be a deep regret to all of us, that such should be the estate of men of science in this country; that the authority of *The Times* should be given to writers whose eloquent pens are guided by presumption and profound ignorance of the subject, on which they dissert with such pretension of knowledge (Anon., 1865e)

This was followed up on April 28, 1866 with:

> Science, we some months ago prophesied, would one day be justified in her dealings with the Cattle-Plague, despite of the abusive ignorance of *The Times*. The justification has come sooner than we expected. Up to February last the disease constantly progressed. But in February the legislature accepted in part the recommendations of science; and from that moment the disease has steadily declined… It is amusing to note how *The Times* now quietly sums up the mischief for which it is mainly responsible.
>
> These results confirm the view taken by the Royal Commissioners in autumn last. We could do nothing with the disease in the way of treatment; but the isolation of infected animals and the suspension of cattle traffic produced an instantaneous effect… We might have saved ourselves much loss if we had adopted the recommendations of the Commission at an earlier period; but the delay had this advantage—that it enabled us to bring many systems of treatment to the test of experiment. We have ascertained that this murrain is not the small-pox, and that it cannot be prevented by vaccination or controlled by inoculation. Homœopathy was of no avail against it; and, though one or two remedies were applied with some apparent success, it must be concluded, on the whole, that we know of no cure for the Cattle-Plague. Henceforth we know how only to deal with it. We must do as they do in Prussia and Hungary—prevent the dissemination of the poison to which we know no antidote. Had the murrain been "stamped out" last summer, we should have saved many thousand head of cattle; but we should not have been convinced of the incurability of the disease.
>
> We need not stop to point out the series of ignorant fallacies involved in this attempted exculpation of itself by *The Times*. It had already been proved to demonstration, over and over again, that there was no remedy of any avail in the treatment of the disease. …Our readers will no doubt do justice to the capacity of the writer who tells us that "homœopathy was of no avail against it." And we trust also that the public will do justice to the journal which, after ridiculing and falsifying the report of the French minister who adopted it, after persuading the public that the Cattle-Plague was born in the filth of London cowsheds, after abusing and vilifying medicine and science, now tacitly admits that medicine and science are quite in the right… If we have lost some 100,000 head of cattle, "we are now convinced of the incurability of the disease." Printing House Square has had its Hecatombs, and is at last convinced! It was our French friends who suggested that the British Lion is sometimes led by asses. What will they say when they come to tell the moral of this tale? (Anon., 1866e)

9. ANTAGONISTS OF POLICY

A gentleman of leisure, H. Strickland Constable, antagonist of Gamgee's slaughter policy, active in writing to *The Times* from the Athenaeum Club and producing scathing, ironical pamphlets, claimed in November 1865 that the disease was due to nothing more than cruelty and neglect "That at ordinary times would attract little notice." He quoted an example of 21 beasts driven 50 km in 13 hr, neglected for days together, and left almost wholly without food and water, when symptoms of disease appeared. The inspector appointed by the magistrates gave the word of command for slaughter to commence, and prussic acid was administered to the beasts. With the animals failing to die, the dose was doubled, which seemed to refresh them and make them more lively than before, so recourse was had to knives which "was attended with great success." The inspector "then demonstrated clearly and indisputably from the state of the intestines, the undoubted existence of Rinderpest…all the doctors still differed as to the nature of the disease."

He claimed little sympathy for the cow-keepers:

> If people shut up cattle in summer and give them rich food when nature intends them to live on grass in the open fields, they *must* suffer for it. If people crowd cattle in pens and ships for days together without water, and instead of feeding them, torture them, beat them, twist their tails, hoist them by the horns, suffocate them, and over-drive them, they *must* suffer for it.
>
> One account mentions three beasts dying … and finishes by saying, that they were taken with the disease immediately after being driven a long distance. I don't doubt it. For myself, I know if I were fed up till very fat in one place, and then all at once driven thirty miles in a hot sun, and frightened and beaten, and shouted at, and barked at, and kicked, and shoo-oo-oo-ooed, and sworn at, for three consecutive days by drunken drovers, I should be uncommonly astonished if I escaped the "Rinderpest." (Constable, 1865)

Attacking Gamgee's slaughter policy, Constable wrote that Gamgee said "That the disease is absolutely incurable, and that a beast must be slaughtered at once as soon as it shows symptoms of having caught it; that is, as soon as it shows symptoms of having caught a common cold, for the first symptoms of both are identical."

Constable listed 32 conflicting opinions and suggestions concerning its origin which had appeared in *The Times* in the few weeks prior to November 1865 and asked the question "Does the disease come from abroad or does it originate in England?" In a third edition of his pamphlet, issued in January 1866, he added a further four opinions. Earlier (1865), he listed 10 points:

1. When your cow catches a cold … don't shut her up in a box.
2. Don't make her drink spirituous liquors.
3. Don't inoculate her in the tail.
4. Don't prevent her walking about if she wishes to do so.
5. Don't deprive her of any nourishing food or drink ….
6. Don't bleed her.
7. Don't pour strong unpleasant drinks down her throat, out of a cow's horn.
8. Don't put her in a place where she can get no shade from the sun.
9. Don't send for Professor Gamgee.
10. Above all, don't cut her throat.

All except points 9 and 10 were good advice respecting treatment. While Constable may have been correct in claiming that animal doctoring was at about the stage that human doctoring had been a century before, this applied to the profession at large and not to persons like Gamgee, who were trying to place it on a proper scientific footing.

In October 1865, Lt. Col. B. G. Layard, Chief Constable of the East Riding of Yorkshire and a descendant of Daniel Layard, reissued a part of Layard's original essay of 1757, quoting many of the quaint remedies, such as "…take of Aloes, aniseeds bruised, rhaponicum or monk's rhubarb…or…the best Turkey rhubarb…senna and liquorice-root…," or "Take of sage, rue and rosemary leaves…four cloves of garlic…," but also stressing the value of inoculation when the disease had become enzootic, and the surest method of extirpating the disease was slaughter of infected cattle without effusion of blood, that is, by strangling.

That a Chief Constable should attempt to promote cures was perhaps surprising when it is realized that this was contrary to the law, for the law required that affected animals be slaughtered.

10. THE END IN SIGHT

The Commission was satisfied that rinderpest had entered Britain with imported cattle, for the first known cases were all in animals collected from different parts of England and Holland and brought to the Metropolitan Market on June 19, 1865. These were purchased by different dairymen and taken to five sheds in different parts of London, namely Islington, Hackney, Lambeth, and Paddington. There was no rinderpest in the parts of England whence these cattle came and none in the sheds to which they were taken. In addition, the length of the incubation period as well as the absence of any probable cause, negatived the idea of a spontaneous origin simultaneously in the five sheds. The cattle must have caught the disease while in the Metropolitan Market, the most likely place in England for rinderpest to be brought to from abroad and also an unlikely place for it to spring up in.

In July 1867, stock owners had still been lobbying for slaughter of imported cattle at the docks, or at least for fat cattle with quarantine for store cattle. One importer wrote to *The Times* in May 1867:

> ... we formerly got our supplies from Holland, then Holstein, Oldenburg, and Portugal; as our wants increased, and the price of meat advanced, we got further supplies from France, Spain, Prussia, and Denmark; and within the last two years, our wants have been such that we have drawn our supplies from Sweden, Poland, and Austria, as far as Galicia and Styria; and yet, with all this extension of area, we do not get as much as we require, and, I think, before long we shall be forced to go into Russia, as far as those districts where cattle are produced for their fat alone, and where cattle plague is supposed to have originated.

But there was no evidence of an expanding meat market being responsible for the introduction of the disease into Britain. The number of imported cattle at the Metropolitan Market had been declining since 1855 but then rose sharply in 1863, almost doubling the annual import, tripling the following year. It then fell sharply following the introduction of rinderpest. But the number of cattle on offer at the market had remained fairly constant to 1862, rising in 1863 almost 12% on the previous year, to peak in 1865. So perhaps declining imports had put pressure on British producers to maintain the market supply, which suddenly resulted in an upsurge of imports as the new, cheaper, Russian avenues were exploited (Table 4, Figure 6).

Conditions of hygiene in Britain were appalling. In July 1865, Gamgee had told the First International Veterinary Congress that Britain had a more extensive traffic in diseased animals than any other part of Europe he had visited:

> Foreign traders think that anything will sell on the London market, and British farmers and butchers believe that the stomachs and constitutions of the inhabitants of that metropolis are proof against the most active animal poisons ... an animal dying from anthrax, whose entrails readily kill pigs, dogs, or other creatures partaking of them, is sent to our city markets as readily as a healthy one, I can furnish no more striking illustration of the condition of the British meat trade.

Britain was highly regarded, however, for the purity of its stock, supplying the Continent with breeding animals, but not without disease also.

The plague continued to rage on the Continent, but the last case in Britain occurred in September 1867 after losses totalling 278,943 head of cattle (some estimates putting it as high as 420,000). Following its cessation and the prayers of thanksgiving which followed, a number of memorial stones and plaques were erected around the country. One such exists at Castle Eaton in Wiltshire on a pair of cottages known as "Plague Cottages," recording that the plaque was erected in gratitude for the preservation "of this and the adjoining properties from the fatal Cattle Plague AD 1866" (Jones, 1997). (Frontispiece.)

The reasons for the failure to contain it have been given as the ineffectiveness of the Orders in Council issued from the outset. Commercial interests were powerful and strongly

Table 4. Metropolitan market stock on offer in November.

Year	Cattle	Imported	Sheep	Imported
1854	23,422	7,120	12,1031	16,604
1855	27,411	7,367	97,460	17,094
1856	25,444	6,102	105,750	16,380
1857	25,383	4,409	103,120	17,830
1858	24,856	4,787	114,643	18,258
1859	26,492	5,927	120,840	21,907
1860	25,400	6,961	103,600	22,723
1861	26,590	5,295	109,370	27,833
1862	30,139	6,839	110,020	28,577
1863	27,704	11,020	99,130	30,347
1864	32,600	17,137	114,300	34,792
1865	36,820	16,254	167,230	52,517
1866	24,660	13,278	95,800	38,389
1867	24,080	10,761	109,960	33,202

opposed to import controls and slaughter on a massive scale. The cattle dealers gained the support of the farmers, who had not been promised any compensation for cattle ordered to be slaughtered by the inspectors (Wilkinson, 1992), although this was eventually introduced. Fleming (1871) calculated that cattle losses due to imported pleuropneumonia and FMD in the past 30 years "since our ports were thrown open to foreign cattle [in 1842]," had amounted to an average of £2.8 million per year. In 2 years, rinderpest had cost the country £5–8 million. In 1865, Britain had been importing 5,000–10,000 animals from overseas each week for several months of the year, Dutch ports alone exporting 150,000 cattle and 250,000 sheep every year to Britain. The Commission had recommended that importation be restricted to a certain number of ports where all fat cattle could be slaughtered on arrival, and store cattle placed into quarantine, but this recommendation was never put into practice, and by 1877, the occasional limited outbreaks of rinderpest were costing the country far less than the accumulated financial losses from pleuropneumonia and FMD.

Cattle stocks had been so destroyed by the 1745 rinderpest outbreak that it took 30 years to recover the national herd. Because of the cessation of imports, the stock became progressively more healthy toward the end of the 18th century, such that by 1801, the population was estimated at about 11 million, climbing to some 19 million in 1840; while the price of beef declined from 5s. 8d per 14 lbs to 3s. 8d, a fall of 35%. But by 1854, following the opening of free trade in 1842, the total number of cattle for England and Wales had declined to an estimated 3.4 million, with about 4.5 million in Scotland and Ireland; and a combined total of 7 million in 1865.[3]

In about 1839, some Dutch cattle were smuggled into Cork, bringing pleuropneumonia to Ireland, and after the embargo on the import of live cattle into Britain was lifted in 1842 by Sir Robert Peel, pleuropneumonia and FMD were introduced, followed by sheep-pox in 1847. Pleuropneumonia caused substantial losses totaling 160,000 cattle between 1854 and 1860. The Member of Parliament for East Norfolk had been justified in referring to the policy of free trade in stock as a "free trade in disease," which served to raise the price of meat due to losses from disease, rather than lower it. Following the outbreak of rinderpest in 1865, once restrictive measures were placed upon the movement of cattle, and fairs and markets prohibited, this led to a drop in mortality from pleuropneumonia of 63.5% of all mortality causes in

[3]In 1987 the total number of cattle for the whole of the British Isles was just over 12 million, and in 2000, it was estimated at 11 million.

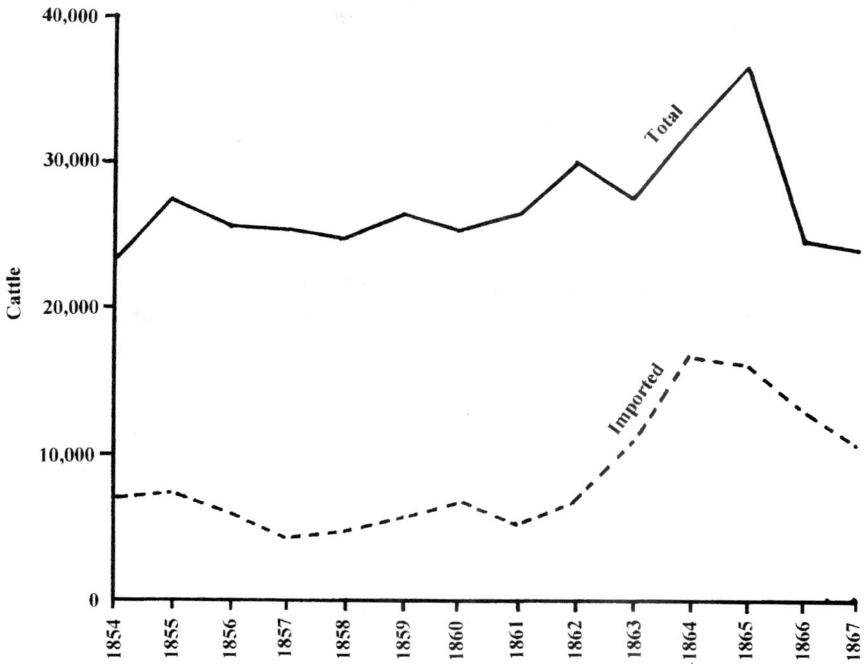

Figure 6. Metropolitan Market, stock on offer in November.

1859, to 1%, its lowest point, in 1866. Once the restrictive measures were lifted, mortality from pleuropneumonia began to rise again.

11. ALARM IN AMERICA

Alarmed at the outbreak of rinderpest in Britain in 1865, the New York Agricultural Society appointed its own Special Committee to investigate and report upon rinderpest, and propose a method of treatment in the event of it appearing in America. The Committee produced its *First Report* on March 29, 1866. This summarized the effects of the epizootic in Britain and, as the State of New York lacked any legislation for the suppression of diseases, proposed *An Act to Prevent the Introduction and Spread of the Disease Known as the Rinderpest, and for the Protection of the Flocks and Herds of Sheep and Cattle in this State, from Destruction by this and other Infectious Diseases*. Introduced into the Senate on March 31, it became law on April 20, but applied to New York State only.

A *Second Report* was issued on June 15, 1866 covering the history, pathology, and treatment of the disease. Homœopathic and other treatments were dealt with in detail, for slaughter, it was considered, could be effective only if it covered all infected cases. But if a quarantine did not have any effect, or if the disease was caused by atmospheric agencies, slaughter would recoil upon its abettors, championed by antagonists to scientific methods. Human epidemics marked

> the bold uprising and clamor of empiricism, and in its successive overthrows by the strides of pestilence they point to the modest but masterly persuasions and trials of science for true and enduring relief... so it always is in face of mortal pestilences that the howlings of terror, the onslaught of savage phrenzy and the fierce desolations of misguided zeal, precede the calm and

> benign intuitions of mercy and judgment, which make up what we call science, and give to it the radiance of a divine vision.

In other words, slaughter was an overreaction, and one should pursue scientific methods to combat the disease. But guided by religious belief rather than scientific principle, the author refers to "the still small voice" which followed the tempest, the earthquake, and the fire. It is perhaps fortunate that rinderpest did not cross the Atlantic, for the main treatment proposed was a course of sulfite of soda or bisulfite of soda (Conger, 1867).

A Washington Department of Agriculture report of 1880, briefly reviewing rinderpest, makes no mention of a Federal law, simply recommending the German legislation as a model of comprehensiveness (Appendix I) (Anon., 1880), although there was a *Contagious Diseases (Animals) Act* of 1878.

10

THE FINAL EUROPEAN OUTBREAKS

1. THE DUTCH OUTBREAK OF 1865

Rinderpest was reintroduced into Holland from England in August 1865, with 60 cases noted in 12 localities by the end of the month. Toward the end of November, it was recorded in 83 places and 1,000 cattle were affected. In the following 3 months, it covered the whole country, infecting as many as 7,700 animals in 1 week, killing 115,000 in total. Military and naval cordons were thrown around the Southern Province. In the province of Utrecht, the action of the authorities being resisted by force necessitated the use of military detachments to effect orders. In some cases, troops, beaten off by large bands of peasants, had to take the cowsheds by regular siege in order to slaughter infected cattle.

The disease passed from Britain to Holland with 23 Dutch oxen belonging to a Mr. Defries. These had been exhibited at the Metropolitan Market on 3 days between June 22 and 29 without effecting a sale and, in between times, were kept in Mrs. Nichols's fields, where the disease existed but at this time was unrecognized. William Simmonds, the inspector, not to be confused with Professor Simonds, considered the cattle to be in poor health but was powerless to prevent their re-export. Returned to Holland on July 2, after being placed in a field at Kethel near Rotterdam, within a few days 21 had died. At first, it was claimed that the outbreak had been virtually contained by throwing a cordon around the area, 5,000 cases being reported within the cordon, compared with only three outside, but it quickly overran the whole of South Holland. Soon afterwards, infected cattle were imported back into Britain.

An official Dutch report issued at the beginning of September was equivocal: "How the disease reached England, whether the said oxen had been in contact with diseased cattle there, and had in that way brought it over to this country, and whether they actually suffered from the same disease as that now raging in England, are points not yet satisfactorily determined." It further stated that slaughtering would not always be sufficient to eradicate the disease: "The newly purchased cattle might again take the disease." Furthermore, slaughtering meant that every opportunity for scientific investigation was taken away, and also no immunized animals would be preserved, animals which had the greatest value. If 20–25% recovered, they would have a greater value than those which died in those districts where cattle were bred.

Dutch Commissioners sitting in reference to the plague stated that it was increasing in the isolated districts, with many more cases in December–January but less severe. They complained of the carelessness of the owners who put fresh cattle into infected sheds without carrying out any disinfection, and spread infected manure on the fields.

In January 1866, it was reported to be increasing and the country suffering severely, the increase attributed to the removal of cattle from the pastures to their stalls causing close contact. As in Britain, the number of cases doubled each month, but the rate of mortality in Holland was lower. The Dutch Government placed its chief reliance on the maintenance of a cordon around the most infected Province, but the railway was allowed to carry cattle through the infected area for export to Britain. It was not until March 24, 1866 that an Order in Britain prohibited import of cattle from the Netherlands, and in June it was extended to "horn, hoof, raw or wet hide or skin."

Schrader, a veterinary surgeon employed by the Hamburg Government, suggested that the outbreak in Britain probably originated from Holland in the first place. In the spring, a considerable number of Hungarian cattle passed from Vienna into Holland through Germany by land and river, at a time when rinderpest had broken out in the region of Vienna. In May, numbers of cattle in Holland were affected near Utrecht. It was therefore feasible that it passed from the Dutch ports to Britain in May. Schrader did not believe that it could have been carried from southern Russia to Revel and then to Britain because of the distance, although the opening-up of railway lines traversing southern and central Germany now enabled cattle to be railed from Lemburg and Vienna to Hamburg and Rotterdam.

Subsequently, a law was passed restricting importation and transit of cattle and placing a strict cordon around the infected district. By October 28, 6,093 cattle had been affected; of which 2,121 died, 1,361 were slaughtered, and 1,752 recovered.

It was absent from Holland for a year before breaking out again at the end of August 1868 in southern Holland.

2. THE OUTBREAK IN BELGIUM

Both France and Belgium had prohibited imports from Britain, but the disease soon spread to Belgium from Holland and then to the north of France, which had been free of it for 50 years. The first cases were noted in Belgium about the middle of August, and an Order was issued on August 28 ordering destruction of all diseased animals. Trade between Belgium and Holland had greatly increased, imports rising by 25% each month as owners got rid of stock, but once it was known to be rinderpest, the frontier with Holland was immediately closed. On August 30, 1865, an Order was passed prohibiting imports from Britain, later extended to cattle products and sheep. On September 22, all fairs and markets were stopped except where the sale was for slaughter and the animal accompanied by a certificate of fitness.

But on October 6, the Minister of the Interior complained to the governors of the provinces that a Proclamation of September 22 was not being enforced; store stock markets had been kept open under the pretext that they were for butchers' stock, animals had been admitted without proper certificates, and certificates had been given without proper control. The Minister directed that gendarmes should exercise "severe control," and 4 days later, it was reported to the King that the measures had been successful and suppression quickly followed. An Order of October 12 prohibited importation of sheep from Holland.

Of 309 cases, 17 died, and the remainder were slaughtered. On one farm, the whole herd of 40 cattle was slaughtered in a period of 2 weeks. Three weeks afterwards, the disease broke out among sheep on the same farm, and the flock of 112 was sacrificed, 11 as diseased and the rest as suspected. "The carcases of the latter, being perfectly sound, were sent for sale to England" (Barron, 1866). In the early part of 1866, it broke out again, and 98 animals were slaughtered.

3. THE OUTBREAK IN FRANCE

Alarmed by the outbreak in Britain, the French Minister for Agriculture, Commerce, and Public Works had circulated a warning on September 11, 1865 to all Prefects with strict instructions for combating the disease should it appear. The leading French exponent of comparative medicine, Jean-Baptiste Chauveau, professor at Lyons, visited England in the company of a veterinarian, Henri Bouley, professor at Alfort, to assess the ravages of the British outbreak and plan preventing it reaching French ports by instituting quarantine measures.[1]

When it appeared in France in September 1865 via Belgium, steps were immediately taken to arrest its progress based upon a system of compensation whenever animals were destroyed. The Minister of Agriculture addressed a detailed circular to all Prefects, instructing them to inform all mayors of the actions necessary to be taken, listing the laws empowering the actions. These actions were that all cases of disease, or suspected cases, had to be reported by the owners or custodians of the affected animals. The mayor was then to have the animal or animals visited by a veterinary surgeon. It was to be stressed, that there were severe penalties for contravention of the instructions. If the veterinary surgeon established that the beasts were diseased, the mayor had to ensure their isolation. The mayor also had to send a daily report on cases to the Prefect, and it was also his responsibility to publicize an outbreak by the posting of placards to that effect. Once identified in a district, no animals could be taken out to fairs or markets or removed for any other cause. Unlike an unfortunate drover at Carlisle, who was imprisoned, in France the owner was responsible for the acts of the drover. Dogs had to be tied up and free-roaming dogs destroyed. On the first appearance of the disease in a district, the municipal authority could order immediate slaughter of the affected cattle and those stalled with them. Reports had to be sent to the Minister who would have compensation paid. Slaughtered animals were to be buried under at least 2 m of earth with the skins slashed. Throwing quicklime on the skin was recommended. However, owners could take the carcasses to knackers' yards or boilers if they were in the vicinity. Forage from the vicinity of sick animals, and manure, had to be buried. Cowsheds had to be disinfected afterwards. Articles used for diseased animals, such as halters, were to be burnt or disinfected.

As a result of these instructions, only 43 animals were lost. Nevertheless, it was unexpectedly again introduced in November 1865 into the Jardin d'Acclimatisation in Paris with two gazelles sent from London, but prompt action prevented spread. Placed among 135 different species of animals, 34 caught the infection and all except an aurochs which recovered, were killed immediately, plus two dogs and two peccaries which were not ill. Bouley visited London and, with Gamgee, endeavored to trace the origin of the disease. The gazelles had been purchased from an animal-dealer, Mr. Jamrack, and apparently contracted infection in a railway cattletruck in passing from London Bridge to Newhaven.

In September 1866, the disease was introduced into Switzerland and Bavaria from the Tyrol. The French Government immediately reinstated its prohibitory measures enacted the previous September, December, and the following April, against importation of cattle.

4. THE OUTBREAK IN IRELAND

In September 1865, a false alarm had been raised in Ireland by the Reverend M'Groarty, a priest in Donegal. He wrote several letters to Belfast newspapers, describing an alleged outbreak of rinderpest in his district. Hugh Ferguson, Professor of Veterinary Surgery

[1]Bouley commented:

> If England had not been disarmed by her laws it had been possible to have arrested its [the plague's] march by energetic measures like those adopted on the Continent, where the Austrian and Prussian sentinels, always vigilant, guard Western Europe, and protect it against the invasion of this Russian scourge. But in England, as it appears, the bovine race must perish rather than a principle. The maxim is sublime, no doubt, but it will be productive of great disasters (Anon., 1882).

to the Royal Agricultural Society of Ireland, and the best-known veterinarian in the country, eventually confirmed that it was not rinderpest but "only the ordinary murrain," presumably meaning FMD. The export of cattle from England to Ireland had been prohibited by an Order of August 25, 1865. On September 7, the export of skins, hides, horns, hoofs, "or any other part," had been prohibited, but this was amended on August 13 to exclude skins and hides which originated from India, South America, and Australia, on the grounds that these countries were free of the disease, a statement which falsely applied to India where rinderpest was enzootic. On January 20, 1866, an Order was published in Ireland providing for inspectors to examine areas, for areas to be declared infected, and empowering inspectors to slaughter cattle declared by them to be infected. It also provided for dogs within an infected district to be kept shut up or chained and prohibited the import of dogs from Britain accompanying drovers or other persons in charge of stock. On May 22, 1866, a Consolidated Order for Ireland was issued, which provided for all cattle in a district declared infected to be branded, and branded again when the area was announced clean. By the end of May, a great number of alleged cases of rinderpest had been reported from different parts of the country, but some were found to be other diseases. However, as a result of a visit of inspection to the north of Ireland by the Government inspector, Professor M'Caul, it was decreed that no cattle could be imported from Ireland to Scotland by the port of Glasgow unless accompanied by a certificate signed by a veterinary inspector.

Owing to doubt expressed by dealers and cattle owners, Professor George Brown, Chief of the Veterinary Department and responsible for animal diseases, was called in and confirmed a considerable number of cases near Lisburn in the district of Drumra, the disease having been imported despite prohibition of imports from Britain. Its origin was not traced but believed to have been carried by persons from diseased animals in Scotland and not by animal contact. Some cattle dealers were in the habit of bringing over animals from Belfast to Scotland and had been mixing with diseased animals in Scotland. One, who owned a number of animals near Glasgow, had gone on his return to Ireland to the byre with a view to purchasing the cow, which was the first animal reported affected. Appearing in the north, the disease was carried to the Midland Counties in Enfield district after having ceased in the north. Ferguson reported that the man on whose farm it had broken out had resolved to take all of his cattle to the fair at Kells but, having been away, when he came back, found them all dead. How they had contracted the disease was never established. Brown's version was somewhat different, stating that it was stamped out on the farm by slaughtering the animals, and then upon its appearance on the next farm, all animals, including sheep and goats, were slaughtered and buried without delay. Altogether, some 50 animals were lost (Anon., 1873). The stringent measures, described as "continental in their severity," stamped it out within 3 months. Later, Ferguson claimed that he had immediately slaughtered the affected animals and then called in Brown because of the opposition that the slaughter had aroused and the abuse of Government at public meetings, people disputing that it was rinderpest. Brown, however, stated that some of them were slaughtered under his direction. In 1875, Ferguson, speaking of FMD, told Irish farmers when the disease broke out on their farms that the best thing to do was to keep the diseased animals at home and send the rest to market (Read, 1877).

5. THE 1870S

From 1861 to 1867 in Hungary, 680 communities having together 908,209 cattle, were affected. Twenty-five per cent of the cattle were attacked, and 145,474 (63.9%), died or were slaughtered. From the middle of 1867 to the end of 1873, it appeared in 430 communities possessing 425,922 cattle, infecting 7.8% of which 20,258 (60%) died or were slaughtered. At the beginning of 1867, it broke out in Bohemia, became more widespread in Austria, Galicia, and Moravia, was widespread in Turkey, and broke out in Wallachia, Dobrudsha, and Roumalia, mostly in regions bordering the Black Sea. Egypt prohibited the import of cattle from Turkey, and Bavaria closed its borders to all traffic, using military cordons to enforce the closure. But 36 outbreaks were reported in Steinach on May 13, and by May 29, several places were reported infected in the Bavarian kingdom, but most restrictions were lifted by July. On the Prussian border in February 1867, travelers on the railway were obliged to disembark and subject to such strong fumigation that they were then refused readmission to the carriages. At the beginning of April, there were outbreaks of rinderpest in the Thuringia and Franconia provinces of Germany. By the end of May, 135 farms in 40 places were affected, and 791 cattle, 11 sheep, and 29 goats had been lost. By August 1869, the disease prevailed over a wide area in Eastern Europe, especially active in Hungary, Transylvania, and Galicia. It had also broken out in a part of Poland through which ran the railway line from Warsaw to Russia. It was still present in Romania and had recently broken out near Constantinople. Toward the end of 1869, it appeared in a very severe form in many places in the western and southwestern districts of Poland near the Prussian and Austrian frontiers, and had also entered Silesia. All animals affected were reported to have died, 600 in one small district alone. On November 27, the Polish frontier was closed and a military guard appointed by the Government of Silesia to prevent further cattle entrance, the disease having been traced to Pilica, despite the precautions it extended to Sohrau in Silesia. Cattle fairs and markets were prohibited and quarantine regulations strictly enforced. The plague also assumed serious proportions in Bukowina and Galicia, as well as still being present in Pesth (Hungary).

In November 1869, it was reported to have broken out in St. Petersburg and eastern Russia but had entirely disappeared in the southern provinces allegedly due to inoculation. Slaughter was neither recommended nor adopted, the separation of sound cattle and isolation of the sick being the preventive measures of choice. Medical treatment was used and, from time to time, inoculation. Much stress was laid on disinfection, and the food and water of cattle were made subjects of care and attention. But the disease continued to take its toll, and in early 1872, a Commission was appointed to inquire into the causes which produced and propagated rinderpest and suggest a remedy. Introduced into Silesia at the end of 1869, in Upper Silesia at the beginning of the 1870s, 127 farms in 36 villages were affected in 2 months, and 1,077 cattle had died. In February 1870, it was reported to be still very rife in Poland and in many parts of Galicia, Bukowina, Transylvania, and Hungary. Romania and parts of Turkey were suffering severely. There had been no diminution in infected districts in Asia Minor. In 1870, Germany lost well over 7,500 head. By February, the Prussian Government had found it necessary to impose controls in Prussian Silesia, guarding the frontier with troops due to spread from the western and southwestern parts of Poland. Prussia was threatened not only in the southeast but also in the northeast, the plague extending in Russia as far north as Knovno, and troops were sent to guard the frontier in East Prussia. It was still raging in Poland, Galicia, Transylvania, and Bukowina. By March, it was declining in parts of Asiatic Turkey due to the diminution in numbers of cattle.

In April 1870, it was reported in the English press to have broken out in France near the Swiss border, but this was considered to be FMD. In June, it reappeared in parts of Poland and still persisted elsewhere, except in Hungary, which was thought to be free of it. In August, it was introduced into central and western Europe again in the Franco-German war, following the Prussian army commissariat into France as far as Bar-le-Duc, almost to the walls of Paris. It was also raging between Berlin and Dresden, and had broken out in various parts of Saxony. It had spread from Prussia to Pomerania and Mecklenburg, and, following in the wake of the army, a third of the cattle had been destroyed at Ingelheim in Hesse-Darmstadt. It was also present in Bavaria, Rhenish Prussia, Baden and Wurtemberg, Alsace and Lorraine, and the Franco-Swiss border. It continued in Galicia and Transylvania, Romania, and some of the Turkish provinces, and was extending along the Asiatic shore of the Black Sea and around Trebizond.

All import of Russian cattle into Austria was prohibited and, from Austria into Germany, all cattle of the grey or steppe race. The Russian frontier ran from Memel on the Baltic to Cracow, and no cattle along this frontier could be introduced into Germany. From Cracow, the Prussian frontier bordered Austria, Silesia, Saxony, and Bavaria, where an empire regulation prohibited import of grey oxen.

Appearing in France in August 1870, 8 days after the battle of Woerth-Froeschwiller, it was introduced with steppe cattle purchased by the Prussian Administration from Russia, which allegedly did nothing to check the spread of the disease and may even have encouraged it. The ration herd destined for the population of Paris was destroyed and that for the army reduced from 7,000 to 2,000 head. The cattle of the armies of the east and of the Loire, and those of the Jewish traders which followed them, swarmed with disease, infecting 43 administrative departments. Although hostilities lasted only 6 months, the epizootic lasted 2 years and probably played a significant role in Germany's decisive defeat of France.

Cattle from infected provinces would have to pass through Germany to reach a shipping port, and western Europe relied chiefly on the system of cordons which Prussia established as head of the North German Confederation, when rinderpest approached the frontiers. As long as these States remained free of the disease, countries to the west of them were comparatively safe. But if the disease entered Germany, it became a source of danger to the rest of Europe. By September 1870, it was present in Berlin, allegedly introduced by Podolian or Steppe cattle, but they could equally have been Polish or Galician. It quickly spread to Saxony, which lost 2,000 head, then the Bavarian Palatinate, and then to France. Its significance was overshadowed by the Franco-German War holding the lives of thousands of human beings and the fate of nations in question; and the devastations of the plague were alluded to by many as simply one of the contingencies of war.

In 1871, the disease was present throughout those parts of France occupied, or crossed, by the German armies, and it was possible to trace the outbreaks with the army movements. The district of Ain, north of Lyon, was infected by the supply herd of a cavalry corps retreating from Swiss territory, the herd of 100 apparently healthy animals having been bought around Clerval (Doubs), close to an infected area. The animals were all slaughtered by mid-February at Polliat.

Laws in Germany were very stringent, but war prevented the adoption of control measures, and by the end of the year, the disease was widespread in Europe, a situation which had not occurred for 50 years. Passing onward with the German army, it entered the valleys of the Marne and the Seine and, according to some reports, had entered Paris, somehow passing the Prussian lines which beleaguered the city. During the siege of Paris, the considerable herds of cattle kept there were free of infection, but as soon as the siege was lifted at the end of January 1871 and fresh animals brought in, some of which came from the enemy army, the disease entered with them and spread rapidly among the 6,000–7,000 cattle there. As a result, the animals were slaughtered for food, as it was considered that this would cause no harm to humans (Figure 1) (Bouley, 1871). Around Metz, it also broke out among the sheep, and cattle were dying so fast that they could not be buried. To save as much food as possible, the Germans slaughtered the uninfected animals and salted the carcasses. In Prussia, it reappeared in many places where it was thought to have been wiped out, and it had also broken out in Rhenish Prussia and increased its spread in Galicia and Transylvania. It was present in Austria,

Figure 1. Rinderpest stricken cattle being collected in the streets of Paris, 1871.

Belgium, France, Germany, Italy, Russia, Switzerland, Turkey, and Persia at Caratépé on the shores of the Caspian Sea, where it appeared in July or August. About the beginning of March 1871, it was introduced into Switzerland, appearing in Neufchatel, believed with contraband cattle or hides from France. It was immediately stamped out in 16 places with the loss of 203 head of cattle but, shortly afterwards, appeared near Berne, supposedly introduced by persons who had come from Neufchatel. It was again stamped out.

Holland and Belgium had closed their frontiers to stock and product imports from Germany and France and kept the disease out. Britain established restrictions which were more stringent than those enforced in 1865–6. Orders of Council were passed on September 9 and 20, 1870, which provided for the slaughter at the place of landing of all cattle, sheep, and goats brought in from any port of France or the States of the North German Confederation. In the Port of London, stock could not be unloaded unless there was a special reception area where they might be confined and slaughtered. In March 1871, Britain prohibited the import from France and Belgium of all cattle, meat, and hides. But in India, where rinderpest had been rife in 1868–9 and where there was little control, the trade in hides in 1870 was reported in the Bengal Presidency as having increased because of increased demand in Europe, the trade encouraged by the opening of the Suez Canal at the end of 1869. About two thirds were exported to London direct but, in many cases, trans-shipped via Bremen and Hamburg. The remaining third went to America which took mostly buffalo hides (Hallen *et al.*, 1871). Whereas brokers claimed an ability to easily identify skins from animals which had died from rinderpest, infected skins would have been mixed together with others before being rejected as of inferior quality, although hides are unlikely to have been a source of infection. One hide broker claimed very few such skins were brought down to Calcutta, but went on to state that hides from Burdwan had been very scarce in Calcutta this year (1870), "notwithstanding the cattle disease which has been prevailing there. This is an exception to the general rule, for when disease is raging in a district we look for an increase in the number of 'dead' hides from that district" (Hannah, 1871).

By December 1871, several areas of Germany had been cleared of the disease, but it persisted elsewhere, reappearing in some. Prussia, Mecklenberg, and Saxony were declared free, but East Prussia was still threatened by reinvasion from the Russian states and Poland, which were still overrun. It had been reintroduced into Hungary, and it was firmly established in the northeast of France, but Belgium and Holland had still managed to keep it out. Apart from the military causes, its persistence and further spread in France were occasioned by persons who made use of it as a source of profit, buying contaminated animals in the infected districts at a low price and selling at profit in uninfected districts.

In 1870–71, when German troops invaded Alsace-Lorraine, there were 6,104 cattle and 160 sheep. When the troops returned again, there were 582 cattle and 944 sheep. Losses elsewhere were also very great, following the invasion by German troops, Lorraine and Haut Rhin losing 6,400 cattle and more than 3,000 sheep.

From 1870 to 1872, Belgium suffered on many occasions from the introduction of rinderpest, but in every instance, it was promptly stamped out. These outbreaks appeared to result from the inefficient manner in which regulations in the French Department adjoining Belgium were enforced, and the disease was more frequently traced to men than animals. In the 3 years, 478 cattle were slaughtered, and 51 died of the disease.

During 1872, two outbreaks were reported in Germany, one at Lockstedt near Hamburg and an isolated case at Koblentz. In the Hamburg outbreak, it had existed for some time before measures were taken to stamp it out with the loss of 105 cattle and 15 goats. The strictness of the methods was described by Professor Pritchard of the Royal Veterinary College:

> The thoroughfares surrounding the buildings were closed by barriers guarded by police; other policemen surrounded the houses contiguous to the cowshed, and prevented the inhabitants entering or leaving unless in dire necessity, such as for instance a person being ill and requiring medical aid, and then the parties were only permitted to leave after being disinfected thoroughly. The cowman was confined to the shed day and night, and disinfectants strewed on the ground around the shed ... the four cows were destroyed in the shed by 'pithing', so that as little blood as possible might escape. They were then conveyed, covered over in a cart, to a field where cattle had been buried at the previous outbreak. The conveyance was preceded and followed by policemen to the field, where, having arrived, the carcases were examined by the veterinary surgeon, and then buried in quicklime in a trench seven feet from the surface, and to mark the grave it was surrounded by posts and rails. Everyone on leaving the field was disinfected ... [with] chlorine gas, and their boots rubbed over with chloride of lime. The cart was left in the field until such time it could be well cleansed, and the horse not allowed to return ... the labourers ... were required to change their clothes and boots, ... [which] together with the tools used, were suspended for the night in chlorine gas. The shed ... was disinfected ... [with] chlorine gas, and great care was taken either to destroy or disinfect all matter that had been within a dangerous distance of these animals. (Pritchard, 1872)

Although always present in Russia, it seldom reached the district near St. Petersburg but did so in February 1872, lasting until the following April, and then apparently appearing again in July. In the first outbreak, the British Consul in St. Petersburg was not informed until the middle of May, after the disease had disappeared. The next report was received on July 31, after a shipment from Kronstädt had been landed in Britain.

In 1872, it was present in the following districts of Turkey: Albania, Bosnia and Herzegovina, Montenegro, Romania, and Roumelia. In 1873, it spread through Austria and Hungary and reached Bukowina, Dalmatia, Carinthia, Salzburg, Croatia, and Slavonia. There were two outbreaks in Germany in Prussian Silesia and Bavaria which were contained. In Austria, a military cordon was thrown around the city of Fiume to stop importation of cattle and products but later removed. A number of stringent regulations imposed in Austria did little to prevent import of the disease and check its spread. It was not known whether this was due to relaxation of the law and it not being effectually carried out, or that imports by sea remained open. In December 1872 to February 1873, an estimated 1,000 cattle were slaughtered around Vienna to suppress the outbreak. Hungary stamped it out by July 1873, but it reappeared in October, ceasing in November. In August, it reappeared in Warsaw at 10 localities. Sporadic occurrences in Russia affected 18,000 cattle in 27 governments, of which 14,000 died or were slaughtered. It also occurred in Syria, Macedonia, and Thessaly, and in March on the Island of Corfu, followed by other Ionian Islands. The next year minor outbreaks occurred in eastern Prussia (Lyck), Austria, and Dalmatia.

In March 1873, a proposal was submitted to the French Government to indemnify cattle owners against their rinderpest losses in conformity with the law of June 11, 1866, although formalities prescribed by the law had not been fulfilled on account of the war, explaining:

> ... this scourge, having come in the train of the German armies, penetrating everywhere with them, broke out in many places at the same time. It took by surprise the husbandmen in the midst of dangers of all kinds, of threats, of requisitions in kind and in money, bearing upon persons as well as things; it manifested itself at the time when the public authorities were dispersed, the administrative services almost if not altogether disorganized, and, therefore, the formalities to be fulfilled, even had they been known to the mayors and their subordinates, presented too many difficulties to their being strictly observed.
>
> It smote the rich as well as the most humble husbandman; in certain villages entire stables were laid waste, and the losses sustained were to many greater than those resulting directly from the requisitions of the enemy.

The French Ministry of Agriculture reported that 93,836 head of cattle had disappeared between 1870 and 1871, of which 55,401 were slaughtered as diseased or suspected, and 38,435 died. The worst affected departments were those most overrun by the German army, the Meuse and Ardennes, the former losing 13,507 head and the latter 8,899, but Government figures showed losses of only 9,688 and 6,637, respectively. Although a total of over 10.2 million francs had been paid to those able to complete claims, a further 3 million francs was required. From the middle of 1861 to the end of 1873, Hungary lost 165,732 head, valued at over 16 million schillings, the policing measures costing in excess of 1 million schillings.

In 1874, there was an outbreak in Italy, and the disease continued to exist in Greece and both Turkey in Europe and Turkey in Asia. In the following year, there were two outbreaks in Germany's East Prussia and Saxon Silesia. The former was bordered by Poland on three sides, occurring in January on an isolated estate where it was quickly suppressed, three animals dying and 16 being slaughtered. Germany was officially declared free of rinderpest on February 12, but then the second outbreak was notified on April 16, confined to one farm where it was extinguished, Germany declaring itself again free of rinderpest on May 7. Losses in Germany were estimated at 7,473, not including those of the army commissariat which were considered to be "enormous."

Spain and Portugal alone escaped infection in these 19th-century outbreaks. Spain rarely imported cattle and then only in very small numbers as comparatively little beef was eaten. However, there were considerable exports from the north and Galicia, the Galicia and Asturias cattle being exported to France and England.

6. OUTBREAK IN BRITAIN AGAIN

Cattle had been landed in 1869 and 1872 from infected districts in Germany before information was received from Germany of the presence of the disease. In the first case, the cattle were not infected, but between July 17 and 29, 1872, there were eight instances of imports to Britain of rinderpest-infected cattle. Rapid action was to contain the disease, and only 72 cases of infection were reported. Shipped from Kronstädt and Hamburg, all of the cargoes included a number of cattle of Russian origin mixed with German cattle. Rinderpest was prevalent in some of the Russian Baltic provinces and surrounding districts, especially Petersburg and Novgorod, which exported cattle to Kronstädt, where these infected cattle originated. The first instance was detected by a vigilant veterinary inspector at Deptford on July 19 in one animal in a cargo of 25 Russian cattle from Kronstädt. Some animals appeared more fatigued than would be expected, and one died during the night. The inspector suspected rinderpest, a fact confirmed by the postmortem. Gamgee's thermometer test was used on the remainder and suggested that three more had premonitory symptoms. The whole of the cargo was therefore slaughtered and the carcasses of the diseased animals subjected to a special steam apparatus preventing the disease spreading from the landing places.

An Order in Council (No. 351) was issued, prohibiting imports originating from Russia and ordering slaughter at the ports of all sheep and goats coming from there. During the week, other infected cargoes arrived at Newcastle (two), Hartlepool (two), Leith, and Hull, as well as again at Deptford. It seems on this occasion that when the animals had been landed at Deptford, they were not killed for a day or two, the last on the Thursday, and the bodies not destroyed or boiled down until the following Wednesday. There being insufficient space for burial, they were left lying at Deptford with lime sprinkled over them while awaiting boiling. The vigilance of the inspectors at Hartlepool, and Newcastle-upon-Tyne allegedly prevented any spread of the disease, but Simonds reported that on arrival at Hartlepool, he found it necessary to have all the animals buried as quickly as possible. He sent for the representative of the local authority, who was not at home, and then sent for the resident magistrate, who declined to act without first consulting his colleagues. Simonds therefore took it upon himself to dispose off the carcasses (Anon., 1873).

On July 23, the SS *Benachie* docked at Leith with 50 oxen, of which the veterinary inspector found three suffering from rinderpest and prevented the vessel unloading. The following day, he was instructed by the Veterinary Department of the Privy Council to slaughter all animals on board, disinfect the healthy carcasses, bring them ashore for sale, and throw the rest overboard at sea. The vessel steamed out, and the work was accomplished, but with the apparently healthy carcasses being brought back, a further instruction was received to throw those overboard also, and the vessel took them out to sea again. The carcasses were washed ashore at different points for many miles along the coasts of the Firth. During the 2 days that the *Benachie* lay at Leith, constant communication was kept up by drovers and others between the vessel and cattle grazing in the vicinities of Leith and Edinburgh, but infection did not spread (Walley, 1879).

On the morning of July 25, 1872, a ship from Kronstädt, the *Joseph Soames*, came into dock at Hull with 56 Russian bullocks on board. Two excited the suspicions of the Customs Officer, who boarded the ship and immediately informed the acting inspector for the Veterinary Department, who, according to Williams, Secretary of the Veterinary Department of the Privy Council, went on board under special instructions. Under the *Foreign Animals Order* of 1871, it was the role of the Customs Collector to board ship and prevent landing of stock and products in the event of disease, with power to seize and detain at his discretion. The veterinary inspector's duties began only when the animals or products were landed, and thus he did not normally board ship.

The Privy Council being informed, Simonds was requested by the Vice President to assist and arrived at Hull a few hours before Brown, who was in Lincolnshire when he received a telegram concerning the cargo on the same day and immediately went to Hull.

The following morning, Simonds, Brown, the Customs Officer, and Freeman, the inspector for the Local Authority, went on board. According to Simonds, he inquired of the veterinary inspector the night before whether Freeman knew of the affair, there subsequently being some dispute over this point with the Local Authority, and was informed this was so. By this time, 18 animals were affected. The port was crowded with vessels, and there was no place where the animals could be slaughtered without taking them through the streets of the town. Neither was it possible to burn them on the dock wall. Brown suggested they be weighted with pig iron and thrown overboard at sea, but the Customs Collector refused to allow the vessel to be moved as it was in a defined area, and the owners of the iron were not willing to part with it. It was therefore decided to slaughter the animals on board, put them in old lighters, take them more than 4.8 km out to sea, and sink the lighters at the mouth of the Humber. It transpired that the lighters could not be sunk, and they were set adrift, one with nearly all of its carcasses eventually being cast ashore on parts of the Lincolnshire and Norfolk coasts. The other was sighted offshore at Dimlington on July 29 and washed ashore at Huttoft on the Lincolnshire coast on August 2 with 40 carcasses on board. A total of 55 carcasses washed ashore, leaving only one unaccounted for, assuming all were from the *Joseph Soames*, and were buried by the Receiver of Wreck subsequently leading to a dispute with the Local Authority concerning who should pay, the latter refusing to do so, blaming Customs for not having carried out the letter of the Act and the Town Clerk claiming he was not kept properly informed by Brown. No rinderpest occurred in any part of the country where the carcasses were stranded (Anon., 1873).

But seemingly as soon as Simonds left Hull (he went to Leith and then Newcastle), on July 27 despite the prohibition on imports, about a score more of infected Russian cattle were docked, a few hundred paces from the cattle market. On the following Monday, cattle bought at the market were moved to Patrington and Bridlington. Rinderpest then appeared in these places and subsequently spread. Cattle sold at Hull market the following Monday, 2 days after the dumping at sea, took the disease into two districts in the East Riding of Yorkshire. Blame was placed on the washing ashore of the carcasses from the *Joseph Soames* but Simonds emphatically denied that these could have given rise to the outbreak. It was therefore concluded that infection must have been spread by persons or things which had been in contact with the diseased animals on board ship.

In a number of cases, there was evidence that sick cattle were knowingly loaded at Hamburg, and some died on the voyage and were thrown overboard. According to Simonds, it was clear that the disease had gained a foothold in the region of Hamburg with the importation of cattle from Kronstädt. As a result, a further Order in Council on July 31 revoked a previous Order allowing Schleswig-Holstein cattle to pass inland after 12 hours' detention at the place of landing if healthy. This Order had been passed after constant representations that cattle from Schleswig-Holstein were safe and, upon the German Government undertaking to provide a certificate that the animals really were Schleswig or Holstein animals, that the certificate was endorsed by a representative of the British Government, and that a bond was entered into that these conditions were fulfilled. Another Order required slaughter at the port of entry of all sheep and goats brought from any place in the Empire of Germany.

Mr. Nissler, a cattle dealer of Kronstädt, had purchased a large number of cattle at St. Petersburg market in the summer, whence they were sent to Kronstädt. About 50 were shipped to Lübeck and from there sent by rail to Hamburg, then driven to a field at Lockstedt on the Holstein frontier. Forty were then shipped to Hartlepool, Newcastle, and Hull. Of the remainder, one died, and the rest were butchered. At Lockstedt infection was passed to a large herd of German cattle in an adjoining field, one of which died on July 26. Finally, all were slaughtered on August 5 as well as seven in an adjoining field. Subsequently, 38 of the Russian cattle were sent to Berlin, where they were seized and destroyed after rinderpest was detected among them. The rest were sent to Britain, except for about 50, which were returned to Kronstädt and sold. Then, between July 17 and 29, eight more ships carrying infected cattle from Kronstädt or Hamburg came to Britain.

Infection first appeared in the East Riding of Yorkshire on Mr. Berryman's farm at Yapham, near Pocklington. An animal in a herd of 22 became ill and died. After another fell sick, the veterinary surgeon, who was also cattle inspector for the district, was called in. He declared the case to be rinderpest and ordered slaughter of all infected animals. A postmortem examination confirming the diagnosis the Privy Council was informed on September 4, Brown being called in, and all remaining cattle immediately shot and buried in quicklime. A police notice was then issued, declaring a circle of 1.6 km radius from the infected field in Yapham, a proscribed area, into or out of which all cattle movement was prohibited. People were also prohibited to enter within the radius, and all dogs therein were ordered to be destroyed. This was followed by an Order in Council on September 7 forbidding all fairs and markets or sales of cattle within the city of York and the East Riding of Yorkshire, except under licence from the Privy Council. All movement of cattle within the district was prohibited except by licence. However, before the disease was diagnosed, some cattle in adjacent fields had been removed by the owners, and as they had not been in contact with the diseased animals, it did not appear that slaughter could be ordered, but they were all kept within the boundary of the infected district. It is believed that some of these cattle were infected, extending the outbreak 8 km from Yapham before the outbreak was finally extinguished at the end of October. Later, Brown stated that if the three herds in fields near Mr. Berryman's cattle had been slaughtered at once, the disease would have been stopped the day after he arrived at Pocklington, but the local authority was not empowered to order this (Anon., 1873).

On August 19, Berryman had bought at Humanby market 22 of 26 cattle offered for sale by Mr. Woodcock of Bridlington. These were driven the 58 km to Yapham in three stages. It transpired that Woodcock had a disease on his farm for some time and, suspecting pleuropneumonia, disposed off his stock. His farm had in fact been infected with two heifers bought in Hull market on July 29 by a butcher at Patrington. A valuable bull he possessed, 2 days before it was found to be sick, serviced two neighbors' cows on August 31, both of which became infected. Animals herded with one were unaffected, the sick cow being slaughtered. Ten animals herded with the other became sick and were slaughtered. Five other farms were affected, which had stock that had been pastured at Bridlington. Infection had already been communicated to a number of other farms by Berryman's herd before rinderpest was diagnosed, in several instances appearing to pass over animals in adjacent fields to others in the next meadow, one case being nearly 1.6 km distant. But a number of people had visited Berryman's sick animals out of curiosity and were probably the means of spreading the infection.

Despite experience gained in the 1865–7 outbreak, there was considerable opposition from owners to the stamping-out system. It was commonly pleaded that compensation was insufficient to replace the stock, and pastures would be wasted, or, if a dairyman, his trade would be interrupted. The answer that this would happen anyway if the animals were infected failed to carry conviction, owners noting that some animals escaped it and wishing to try remedies. The authorities were strengthened in their conviction that slaughter was the only option, and the *Contagious Diseases (Animals) Act, 1869* should be strengthened to allow slaughter in the immediate neighborhood of an outbreak, referred to as the "Prussian system," or at least isolation.

Altogether, there had been 23 outbreaks involving 289 cattle, of which 72 were infected, 21 animals died, and 51 slaughtered. A further 171 were slaughtered to prevent spread of the disease, and 46 animals in contact with it escaped infection. It was commonly remarked that in Yorkshire, the infection had spread more slowly than in 1865–7, and the period of incubation was often longer. But the animals were not crowded to the same extent as in the previous outbreak, being in open fields and, as soon as rinderpest was suspected, had been isolated. When the disease appeared in a herd, all were generally slaughtered and buried, whereas the apparent extension of the period of incubation had been due to previous cases not being reported until infection was advanced.

Lionel Scott Pilkington, alias Jack Hawley, of Hatfield, Doncaster, lost his herd of 22 apparently in this outbreak. In 1875, he was, according to his wishes, buried in the grave of his cattle. He had interred them in rows leaving a space in the middle for himself and had an apple orchard planted over them (Anon., n.d. [1875]).

Following the outbreak in England, import of cattle from Russia by sea was prohibited into Prussia or Germany. On the Austrian frontier, a regulation in place from 1871 also prohibited introduction of any cattle of the grey steppe race.

The laxity of German inspection was shown by the fact that, in 1876, no less than 1,656 diseased cattle (pleuropneumonia and FMD) were received in Britain in that year alone. In February of that year, Denmark, which had almost doubled its exports of cattle to Britain in 1875 to over 50,000 head, passed an Order announcing specific measures with regard to the export of cattle to Britain. Namely, the cattle had to be examined by a veterinary surgeon before being loaded and issued with a certificate to the effect that they were free of symptoms of any infectious disease. Article III stated: "No cattle intended for shipment to Great Britain must be shipped until the veterinary police has given permission for cattle to be brought on board, which permission can be granted only when the vessel has been properly disinfected...." Each animal also had to be marked.

The total recorded cattle numbers for Britain, excluding Ireland, show that the outbreak had no appreciable influence on the total (Table 1).

Table 1. Total cattle numbers for Britain, excluding Ireland.

Year	Total
1871	5,337,759
1872	5,624,994
1873	5,964,549
1874	6,125,491
1875	6,012,824
1876	5,846,302

Questioned by the Select Committee in 1873, the Secretary to the RASE, H. Jenkins, considered that the live cattle trade from Russia would not develop much more because it was cheaper to send the cattle to Vienna and Berlin, where the price of meat was now as high as in London.

7. THE 1877 GERMAN OUTBREAK

In February 1877, the Danish Government issued a circular to all Prefects that, in view of the plague existing in Germany and England, lest it make its way into Denmark in spite of the precautionary measures adopted, any stock of cattle in which it was detected was to be slaughtered immediately in its entirety. In 1872, an outbreak at Hamburg, introduced by cattle from Russia landed at Lübeck and railed to Hamburg was speedily contained. In 1875, there had been two outbreaks in Germany, the following year none, but the 1877 outbreak, which could have originated only from cattle unlawfully introduced into Prussia, was considered the worst Germany had experienced since 1870–71 during the Franco-German War, the biggest European outbreak of the 19th century. Other cases were not confirmed as rinderpest: in 1874 in Beuthen, Upper Silesia; and in 1875 cases in East Prussia and Saxony.

It had broken out in the Oppeln District of Silesia near the Polish frontier toward the end of 1876 among some oxen smuggled from Poland, first confirmed there on January 5, 1877. Immediate action was taken, 69 cattle being slaughtered in two villages, but as five animals had already died, it was evident that the outbreak had occurred some time before it was discovered, and indeed infected cattle had stood in the market at Breslau on January 8. It was considered very common for Germany to have an outbreak on its eastern frontier and to stamp it out at once without interfering with trade, but in this instance, it spread to Berlin and Hamburg. On January 10, it appeared in the market at Altona, west of Hamburg, with cattle bought in the Berlin market, spreading to Hamburg on January 13 and continuing to infect some 50 localities affecting a total of about 1,000 cattle and several hundred sheep. On January 16, it was reported in the market in Breslau, and by the January 22 was still spreading. By February 5, it had reached Dresden and Cologne. Two days later, it was reported from Potsdam and Emden, close to the Dutch frontier. Lasting 3 months, altogether in this outbreak, some 100 sheep were attacked. Total losses were 1,238 cattle, 337 sheep and nine goats, and 94 cattle. Under strict legislation which had been in force for many years in Prussia before Germany became united, all imports and exports of cattle to and from Germany were prohibited, and all movement throughout the country stopped. Wherever infection was located, a cordon was thrown around the place for 20 km, within which trade in cattle and the transport of dung, fodder, straw, and other litter materials were allowed only by special permit. Watchmen, who had to keep themselves isolated from the inhabitants, were placed on infected farms to prevent any persons, other than those legally authorized, from entering or leaving. Only those engaged in destruction of the disease, law officers, doctors, midwives, and clergymen, could be authorized and had to be thoroughly disinfected upon leaving. The Hamburg cattle market was allegedly burnt down in its entirety, and a large military force of cavalry and infantry sent to the frontier to back up the officials.

The outbreak in Germany was not considered to be due to movement of troops in Russia in their war with Turkey, although Russian troops had mobilized on the Pruth and contiguous districts, and may have been supplied with extra quantities of meat which included infected meat, and it was believed to have followed the Russian army into Romania. But trade was now the primary cause of outbreaks as the facilities for transit increased every year from the extension of railway lines into areas of enzooticity, while Britain drew supplies from countries more distant and over which she had less control. When Simonds had visited Galicia in 1857 to observe the outbreak for the RASE, there were no railways beyond Lemburg and Prussia as then existing, so one knew when rinderpest was approaching the frontier. By 1877, there were three or four railways from Germany into Russia, and in consequence, Germany was not informed of the disease early enough before it was brought through by rail.

But on Prussia's western frontier with Poland, smuggling was rife due to a large mining population creating a high demand for meat. The system was well organized, seldom leading to confiscations. Smugglers in Poland bought cattle which had come from further inland at markets near the frontier and placed them in the care of peasants in frontier villages, who then ran them across when opportunity presented, usually employing children. In this instance, infected animals had been smuggled over the border by one man who was arrested and imprisoned. He had driven them across the country, stopping at several markets and taking the disease to Breslau, Berlin, and then Hamburg and other places. Import of cattle from Russia or Poland was entirely forbidden under German law, and the white or grey breed formerly imported to Moravia to be fed in the distilleries forbidden from any part. Three or four other persons were shot breaking through the cordon, the orders of the soldiers being to shoot any man who refused to obey after being challenged three times to halt. In one case, near Breslau, one of the persons shot was a deaf boy.

The Prussian inspector at Altona, who did not act promptly enough, was punished and dismissed, and the Hamburg veterinary inspector who allowed the cargo to be loaded was subjected to a disciplinary hearing.

The final case was on March 12, the German Empire being declared free of infection on April 10, 1877; but the following June, it was still prevalent in Russian Poland, where it spread more widely than on the last occasion.

Following information on the continued spread of the disease in Germany, importation of cattle, meat, and hides, from Germany and Belgium into Britain, had been prohibited on January 27, Belgium being included as a country of transit. Although in evidence before the Select Committee in June 1877, Septimus Lambert, a cattle salesman from Manchester, stated that German and Dutch dead meat was imported during the plague and landed principally at Sunderland, West Hartlepool, and Grimsby.

In 1877, Virchow sponsored in the Prussian Chamber of Deputies a resolution urging the government to enforce strict international undertakings to control spread of the disease.

8. THE LAST OUTBREAK IN BRITAIN, 1877

Walley (1879) reports how an infected cargo of animals originating from Sweden via Malmo and Copenhagen, was landed at Leith in 1876, a sheep and an ox being found ill, and the total number was less than that shown on the manifest. Having slaughtered the two animals, and a cow found to have swallowed a nail, and finding that no fresh symptoms developed in the remainder after 2 days, he released the cargo, and there are no reports of the disease having spread. But at the beginning of 1877, the disease broke out in England for the last time to date, imported in a shipment of cattle from Hamburg. In a letter to *The Times* dated Hamburg, January 15, it was stated:

> ... on January 13th a train of about 100 oxen arrived from Goslar, in the Hartz mountains, for shipment to London. The beasts were found ill by the town veterinary inspector, and on the 14th rinderpest of an aggravated form broke out among them, so it is said. They have all been killed and the carcasses destroyed. The street in which the stable they were kept in was situated, has been cut off from all communication, and a cordon of soldiers placed to guard it.

It was also on January 15 that the steamship *Castor* landed 39 infected cattle at Deptford, some of which were dying from rinderpest, one actually succumbing on the landing stage after it was driven off the ship; but it was not until the evening of that day that rinderpest was confirmed. Two hours after the landing, the veterinary authorities received notification of the outbreak in Hamburg from the German authorities, and after Brown had read of the German outbreak in that morning's newspaper. At the same time, another infected shipment was landed at Hull.

Investigation revealed that rinderpest had been declared to exist in certain stables in Hamburg on January 13, the day after cattle had been shipped to Deptford and to Hull. It transpired that a veterinary inspector at Altona saw the first cases of the disease on the evening of Saturday January 13 and suspected rinderpest but did not confirm his diagnosis, simply telegraphing his opinion to the government in Schleswig. On the Sunday morning, rinderpest was confirmed, and the police inspector telegraphed the government. Being

Sunday, a telegram was not sent to Deptford until Monday April 15, arriving at 1 p.m., 3 hr after the cattle were landed at Deptford and 20 hr after the landing at Hull.

Those going to Hull were in the steamship *Hansa* (elsewhere Brown stated it was the *Leopard*). The *Castor* had started with 40 animals, which had been officially declared by the inspector at Hamburg on January 12, as all having been examined and found healthy, one of which died at sea and was thrown overboard. Of 18 animals which went to Hull, two were slaughtered as infected with FMD but which was, from the symptoms described to Brown, rinderpest.

A good deal of time was lost allowing the cattle to be landed at Deptford and the vessel to moor at Horselydown at the owner's wharf, there being no provision under the law to allow seizure and quarantining of the boat. The majority of the 39 animals landed at Deptford were kept alive, some for 8 days, but 27 died before the rest could be slaughtered. While the boat was moored, people went on and off without being disinfected, the boat itself not being disinfected for 4 days. But according to Brown's account, as soon as rinderpest was detected, the animals were isolated and locked in one detached block of buildings where no one could enter or leave without the sanction of the market superintendant.

> A wide roadway separated this from any other portion of the market, in addition to which quantities of quicklime were placed at the different doors, so that anyone entering or leaving the premises must walk through it. On the following day the work of slaughter and the destruction of the carcasses was commenced. Special men were told off for the purpose and not allowed to leave the building without changing their clothes; and a fumigator was at work disinfecting the clothes they had worn while at work. The mode of destroying the carcasses adopted was to quarter them and subject them to the prolonged action of steam in a large iron cylinder, until very little of the carcass remained. (Anon., 1877)

A Reuter's telegram dated Berlin the evening of January 17 stated: "The *Official Gazette* announces an outbreak of cattle-plague on the property belonging to the Concordia sugar refinery, near Brieg, in the province of Silesia."

On January 17, local authorities were circulated with a letter from the Veterinary Department, simply stating that a cargo of animals infected with rinderpest had been landed at Deptford:

> I am to suggest that the inspectors of your local authority should be instructed to make strict enquiries as to the existence of any unusual form of disease among animals in their districts.
>
> In the event of any doubtful case occurring, it is desirable that the advice of a qualified veterinary surgeon should be obtained without delay.

No systematic inquiry was ordered, and no outbreaks were reported by the Local Authorities, but because of further spread in Germany, an Order was passed on January 19 coming into effect on January 27, scheduling Belgium and France and placing them under the same regulations as for Germany, prohibiting landing of cattle, meat, hides, and other products. Denmark and the Netherlands were informed that they would also be scheduled unless they prohibited import and transit of animals from Germany, and subsequently this was done.

On January 30, the disease was reported in London's Limehouse, although the symptoms were unusual, with absence of the indications in the mouth regarded (wrongly) as characteristic of the disease. The temperature, instead of being elevated, in some animals was below normal, and in others only slightly above. These uncharacteristic symptoms had also been found in the outbreak in the east of London in 1867. Upon discovery of the infection in Limehouse, a cordon was thrown around the Metropolis. On February 2, markets and sales in the Metropolis were forbidden, except by licence. It appeared that a drover who had been at Deptford when the *Castor's* infected cargo was unloaded had transmitted the disease, which fuelled the argument by some that slaughter at the point of disembarkation was insufficient. During February, there were 11 reported outbreaks in Poplar, Bow, Milwall, Bromley, Stepney, and Hackney, and on March 8, more in the same or neighboring districts. On March 26, it suddenly appeared in the western district of Shepherd's Bush without any apparent connection with other centers of infection, believed to have been carried by a man inoculating cows against pleuropneumonia.

On February 3, a further letter from the Veterinary Department was circulated to Local Authorities:

Sir, I am directed by the Lords of the Council to inform you that cattle-plague is extending in the metropolis, and it is feared that infected animals have been moved into the country before the passing of the Metropolitan Cattle-Plague Order of 1877, which prohibits the movement of cattle, sheep, or goats out of the metropolis.

Under these circumstances, I am to request that you will urge upon your local authority the importance of directing their inspectors to institute strict inquiry as to the sanitary condition of the stock in their districts, with a view to the immediate detection of any outbreak of cattle-plague which may occur.

I am further directed to advise that competent veterinary surgeons may be deputed by your local authority to act in this emergency with full power to order the immediate slaughter of any animal affected with cattle plague.

The first outbreak outside of the metropolitan boundary was at Stratford in Essex, extending to several dairies. Although first recorded on February 7, it was not reported until February 9. A meeting of the Executive Committee was not held until the February 15, at which resolutions were passed prohibiting movement of cattle, except by licence, and the holding of fairs. A letter was read from one of the inspectors for the Ilford division whom the Clerk of the Peace had instructed to obtain the advice of a qualified veterinary surgeon in any doubtful case. The inspector explained that in his area of 53-km circuit, there was not one qualified veterinary surgeon. On February 13, he reported on its steady progress and in another letter alluded to the fact his pay of 17 s. per week was insufficient to pay his trap hire alone. The Committee thereupon authorized payment of fees to veterinary surgeons and further allowances to the inspector. In the event, only six outbreaks were reported among 23 cattle, of which eight were attacked, one was dying, and the rest were slaughtered.

On February 17, it was found in the dairies at Hull, believed to be originating from the shipment landed from the *Hansa* in the middle of January, the infection conveyed from slaughtered animals by one of the handlers. The Privy Council took action, stopping fairs and markets in the East Riding of Yorkshire, especially the great York Market, due to be held on February 22. All fairs and markets were also stopped in Lincolnshire, following an outbreak on a farm near Great Grimsby on March 8, more than 48 km from the nearest center of infection. The origin could not be traced but was believed in some way to have been carried from Hull. A statement that it originated in bones from Germany was discounted as the only bones on the premises had come from Amsterdam and had been lying on the wharf at Grimsby for several months. Infecting six dairies, the Hull outbreak continued until March 22 when it was believed to have ceased, but then was to erupt again on May 4 at Sculcoates, some distance from the previous outbreaks, finally declared free on June 22. When it broke out in Hull, the Justices for the North Riding at great expense to the ratepayers, immediately took steps to prevent its introduction by placing a double cordon of police around every possible outlet throughout the whole Riding. In all, there was a total of seven outbreaks among 65 animals, of which 19 were attacked, two died, and the rest were slaughtered. The origin of the Sculcoates outbreak was never traced.

In April, a Veterinary Department *Memorandum* noted that the disease had not been stamped out chiefly because efforts had been mainly confined to dealing with cattle.

The danger of communication by persons, other animals such as dogs, and substances and utensils which had been on infected premises, had not been fully appreciated.

On April 7, shortly after publication of the *Memorandum* the disease was reported at Willesden, outside the Metropolitan boundary, and for nearly a week, no action was taken to contain it. It was believed that infection was carried by a butcher who had been engaged in dealing with the diseased animals at Shepherd's Bush and had come to examine the animals having previously been provided for slaughter, a number suffering from pleuropneumonia. The owner reported the first sick cow by letter on April 2 claiming it to be affected by rinderpest. The inspector visiting the farm the following day found the animal dead and conducted an autopsy, claiming that it was not rinderpest but diarrhoea induced by a change of food. It was not until 4 days later that he decided the remaining cows were suffering from rinderpest and reported accordingly. In all, 90 cattle were attacked, 15 died, 75 killed, and the remaining 68 slaughtered as healthy cattle. What was surprising was that in the course of his duties, the inspector had not spread the infection elsewhere. Parliament (April 10) reported that the Middlesex Local Authority had taken all possible precautions and, among other things, stopped all fairs and markets and the transit of cattle. There was no truth in the report of an outbreak in Oxfordshire. The next day, claiming it was concerned that the disease had been lurking about the Metropolis for 3 months and apparently had a hold in the dairies in the East End and in one of the leading dairies in the West End, an Order of Council was passed whereby the Privy Council took over the functions of the Local Authority within the Metropolitan Police District. In fact, this appears to have been in response to a large deputation to the Lord President from the RASE, and another from the Shorthorn Society, urging the necessity for immediate action on the part of the Privy Council. Under the Order, the Veterinary Department was instructed to stamp out the disease without delay. Accordingly, 12 Veterinary Inspectors were appointed with protective clothing, who, in the event of rinderpest being found on any premises, were to remain in charge until slaughter of the animals and disinfection of the premises had been completed. They were also charged with preventing movement of animals, including dogs, and verifying the disinfection of clothes of persons, milk cans, pails, and all other items used on the premises; the disinfection and destruction of fodder and manure; and affixing a notice at the entrance of infected premises forbidding entry without permission.

The liberal application of a strong solution of chloride of lime by means of a large brush to a person's clothes soon deterred sightseers. Brown stated that, having tried this plan in an infected district in northern Ireland in 1866, he was "fully impressed with its deterrent influence," the injury done to the clothes not being compensated for by the information gained (Brown, 1878). Until now, the Chief Inspector had been constantly compelled to witness the entrance and exit of sightseers to infected premises, on one occasion counting 10 persons entering a shed where several infected animals were standing. After handling the diseased animals, the men had left without any notice being taken of their movements. On April 11, another outbreak occurred in a dairy in Hull. On April 12, another outbreak at Willesden was reported, followed by others on April 19, 23, and 24; the last affecting one animal in a herd of 124 of which all were immediately slaughtered.

Altogether, there were five outbreaks in Middlesex affecting 425 cattle, of which 103 were attacked, 17 died, 86 were killed, and 322 healthy animals slaughtered. At this point, movement of cattle within the county except for slaughter was prohibited for 1 month, the same applying to the Metropolis where another outbreak occurred on April 17 in Ledbury Mews among a herd of 14 cows, but not confirmed as rinderpest until April 21. The infection had apparently been introduced into the herd on the owner's farm at Notting Hill when he held a sale on April 16 at the time infection was present at Willesden, before moving these 14 to Ledbury Mews. Another outbreak occurred on April 25 exactly opposite the infected permises.

The restrictions in the Metropolis stopping movement and compelling slaughter of animals in the market area excited more opposition than any measure which had been enacted since the *Contagious Diseases (Animals) Act* had been in operation beginning in 1869. "One deputation after another attended at the Council Office to represent in forcible terms the ruin which impended if the obnoxious Order were not immediately revoked; the entire stoppage of the Metropolitan Market, and the starvation of the population of London were among the least of the evils which were threatened" (Brown, 1878). But Government stood firm, and only two cases occurred in the district during the month, on May 1 and 15, while none of the predicted disasters came to pass.

Despite the cordon of inspectors established around London, men termed "blockade runners" smuggled bullocks though the cordon into the country. It was not until May 9 that an Order was issued by the Privy Council to the effect that the Veterinary Department of the Privy Council Office should slaughter every animal affected with rinderpest in the City of London and Metropolitan Police District, and had the discretion to slaughter those which had been in contact with affected animals. Compensation was to be paid by the Local Authority.

From July 1, 1877, the Cattle Plague Order, 1877, by which Local Authorities were empowered to make regulations for their respective districts, was revoked. The Privy Council Order for the East Riding of Yorkshire had expired, as had that of Middlesex, while that for Lincolnshire was revoked. The only remaining restrictions are the cordon around London, which was to remain in place until the effect of restocking the infected dairies had been seen, and the Order prohibiting fairs and markets in the Metropolis except by licence was also still in force. By July 6, 1877, although the country had not been declared free of the outbreak, there was no known case, but on July 14, another outbreak occurred at Bethnal Green. The previous nearest cases had been almost 1 km away on May 17 in White's-row, and before that on March 24 in the Whitechapel district; the two outbreaks being almost equidistant from the point where the carcasses of diseased animals were destroyed, Barber's the horse-slaughterer's.

On the evening of Friday July 13, an anonymous caller at Bethnal Green police station reported that milk sellers in the neighborhood had declined to sell any milk from premises in Norfolk Street, as the cows there were very ill, and the owner intended to have some of them killed that night. The police sent for an inspector, who telegraphed Brown after observing from the symptoms that it was the old type of cattle plague. Doubting the truth of it, Brown did not order immediate slaughter but first directed Professor Simonds to examine the animals. Subsequently, Priestman and Rayment also saw them, all three confirming rinderpest. It was considered of such importance that an extraordinary meeting of the Select Committee was held to take evidence, but at the time, no facts were known. Subsequently, it was determined that the first cow attacked was purchased by a drover at Romford market on June 27 and taken to a shed used as a slaughterhouse in Whitechapel, where rinderpest-infected animals had probably been killed. Finding that the cow gave promise of milking well, instead of slaughtering it, the drover sold it to a dairyman in Bethnal Green, where it was placed with nine others. A week later, it was observed to be ill, and in another week, the other nine were found to be suffering from rinderpest of an advanced stage.

Although this demonstrated that 2 months was insufficient time for infection to disappear, before the 2 months had elapsed, petitions were sent from various parts of the country to the Lord President begging revocation of the prevention of removal of cattle and sheep from the Metropolitan Market into the country, all similar to the following:

> To His Grace The Duke of Richmond and Gordon, K. G. &c., Lord President of Her Majesty's Most Honourable Privy Council.

> The Humble Petition of the undersigned Farmers, Graziers, Butchers, and others, Sheweth:
>
> That a period of nearly two months having now expired since the last case of Cattle-plague in this country, the necessity for maintaining the restrictions against the movement of Cattle from the metropolis no longer exists.
>
> That those onerous restrictions are a source of very heavy loss, and great inconvenience, to your Petitioners, many of whom experience the greatest difficulty in obtaining a sufficient supply of Cattle, &c., for their requirements.
>
> That in the opinion of your Petitioners, the Metropolitan Cattle Market should be a market solely for the sale of Cattle and Sheep intended for immediate slaughter, and that all Cattle leaving the market should be marked with a Broad Arrow, and should be killed within seven days, a Killing Licence being issued with each animal, upon the system adopted subsequent to the outbreak of Cattle Plague in 1865, but such Licence to be returnable at any Police Office within 200 miles of London.
>
> Your Petitioners therefore humbly pray that your Grace will take steps as speedily as possible to revoke the regulations which prohibit the removal of Cattle and Sheep from the Metropolitan Cattle Market into the country, subject, if need be, to such rules as your Most Honourable Council may deem fit to impose.
>
> And your Petitioners will ever pray, &c. (Brown, 1878)

By August 14, this last locality was declared free and all Orders for the Metropolis revoked, the Cattle Plague Order of 1872 on September 7 and its Amendment being revoked 10 days later.

Lasting from January 31 to July 15, altogether in this outbreak there were 47 separate cases all affecting dairy cows; 263 were infected, 35 died of the disease, 228 were killed as being sick, and 835 were slaughtered as a precaution, a total loss of 1,099, the disease being stamped out in every case by immediate slaughter at a total cost of some £13,000. In addition, 39 infected foreign cattle were landed at Deptford, of which 26 died and the remainder slaughtered. The measures taken were mild compared with the Prussian system, although far more stringent than those applied in the previous two outbreaks. But it took 3 months of deputations from Chambers of Agriculture, Farmers' Clubs, from Quarter Sessions, County Committees, and independent Members of Parliament before the Privy Council would exercise its powers under the Act of 1869. Only after 3 months did they do this successfully, reticence to act stemming from being loath to interfere with the workings of the Local Authorities. Whereas the outbreak was confirmed on January 31, an Order was not passed until April 12, coming into force on April 16 with the Privy Council taking the place of Local Authorities in the Metropolitan Police District and within a radius of 24 km outside.

Generally, Local Authorities did not take advantage of the powers conferred upon them, and infection was finally stamped out at comparatively trifling cost due to the administrative ability of Professor, now Sir George, Brown. Only one animal, a mere 0.09%, escaped the disease, compared with 46, or 15.9%, in the previous outbreak.

A majority of persons who gave evidence to the Select Committee were in favor of a dead meat trade, but objection was lodged by the Cattle Trade Committee of the Borough of Newcastle-upon-Tyne to any interference with their live trade, claiming that it had to accommodate a total population of 359,589 persons spread over a large area, and without importation of foreign cattle, it would not be possible to feed the population. A total of 104,876 foreign cattle, discounting sheep and swine, were imported in 1876, a number which was increasing annually. Even *The Times* had stated after the 1872 outbreak that all live imports should be stopped and nothing but dead meat imported, but Simonds was adamant that it was not possible.

On three occasions in the 12 years to 1877, it had been proved that rinderpest was not spread in a district by movement of infected animals as commonly as by persons and substances which had been in contact with infected animals: "hence, while power to slaughter and isolate animals in any direction round the infected premises is necessary, it is even more important to control the reckless actions of persons who ignorantly and for the mere gratification of their curiosity pass from diseased to healthy animals without taking the slightest care to avoid communicating the infection" (Brown, 1878). Another problem in this outbreak had been disposal of manure from infected premises which the law required. Eligible pieces of land were found where the manure could have been buried, but owners of the land declined in most cases to permit it, and it was found necessary to cart a large quantity of disinfected manure to a wharf and place it in barges to be towed out to sea and thrown overboard.

9. THE FINAL EUROPEAN YEARS

There were several outbreaks in Austria during 1877, but the numbers of cattle involved were small, and from 1880, Austria was free of the disease. In October, it was introduced into Hungary again with Romanian cattle. An outbreak in Germany in the same month was believed to have originated with cattle from Bohemia. This was of concern to Britain as cattle were sent via Hamburg or Geestemunde from the breweries of Silesia, Moravia, and Bohemia.

In the Bulgarian war of independence of 1877–8, rinderpest was introduced by Russian or Turkish transports into what were then the Eastern Provinces of Turkey in Europe, and after the war, the disease spread once more in eastern Europe to Romania, Serbia, and Albania, whence it passed to Anatolia and Russia, parts of Prussia, and the east and south of the Austro-Hungarian Empire, destroying almost two thirds of the country's cattle before it was brought under control. From December 31, 1878, Britain prohibited the import of cattle from Belgium and Germany and any sheep or goats which had been on board with cattle. It last appeared in Silesia in 1881, and then the German federal area became free of it. From 1881, Hungary also was free of it but in 1883, introduced into Varna on Bulgaria's Black Sea coast with cattle brought from Russia as breeding stock, it was soon eradicated by the slaughter of all affected and in-contact animals.

In January 1877, rinderpest was present in all provinces in Russia extending from the Baltic to the Black Sea, but there was no report of any action being taken against it. In the Black and Baltic Seas area and the provinces bordering Prussia and Austria, 63,983 animals were reported as affected in the first 11 months, of which 29,470 are said to have recovered (46%). At the end of 1879 to the beginning of 1880, there was an outbreak in Cyprus, probably due to cattle imports from Turkey. In 1880, vigorous methods to combat the disease in Europe were begun, pushing it to the southeast to the regions of the Caspian and the Caucus. In 5 years, the number of reported cases fell by 98% (Table 2, Figure 2). In European Russia, 919,550 deaths were recorded in 1884, but by 1896, Russia had eliminated it.

Table 2. The anti-rinderpest campaign in Russia.

Year	Cases
1884	919,550
1885	350,000
1886	280,000
1887	76,000
1888	87,000
1889	19,000

Source: Leclainche (1955).

Figure 2. The anti-rinderpest campaign in Russia 1884–9.

Apart from protecting its frontiers, Russia then maintained two serum production centers to serve the regions neighboring China and Mongolia, and for Transcaucasia, vaccinating animals imported principally from Mongolia and controlling any outbreaks which might occur. In Mongolia, there were 25,093 deaths reported in 1906, and in 1907 in Transcaucasia, 812 communities in six areas were infected.

In 1895, an Estonian, Constantin Blumberg, in a book on autopsy techniques (the Russian Ostapenko had produced an earlier one in 1892), pointed out that an accurate post-mortem diagnosis was the only basis on which Russia could hope to control its devastating epizootics. Blumberg was the first teacher of veterinary pathology at the Kazan Veterinary Institute, opened in 1874 in the city of Kazan on the Volga River. It had been sited there, Kazan province being the main cattle entry point, to provide a veterinary *cordon sanitaire*, and as a watch point on the route for the movement of commercial cattle herds from Asia into Europe.

From 1896 to 1914, the disease appeared only once in southern Russia, in 1908 three infected transports from Petrawlowsk in Siberia reaching St. Petersburg abattoir, but it was quickly controlled, although it was in the abattoirs at Petrograd brought in by Transcaucasian cattle on the railway. After the outbreak of war, the frontier controls held for several years, but following the revolution, they disappeared, and the disease crossed the Caucuses in 1917. By 1918, it had reached Bulgaria and Romania, where several foci were rapidly extinguished. In 1918, it invaded the Don valley and in 1919 reached the Ukraine, the whole country being infected by the herds accompanying the bolshevik armies. Serum production centers were set up, but the veterinarians were badly equipped and not well received in many places. Gradually, their numbers were built up, and 12 were operative by 1922. Control strengthened, eventually eliminating the disease throughout Russia by 1928. In 1939, it was reintroduced, apparently from Mongolia into Primorsky Kray, bordering the Sea of Japan, but quickly controlled. There were further outbreaks in the region in 1944–6 when movements of infected cattle from Japan are thought to have introduced it into Manchuria, infecting Primorsky Kray in 1944, and the Amur region in 1945 and 1946, where it was quickly suppressed in 1945 but then became widespread before being controlled in 1946. Until 1960, there remained free movement across the border with China. Further west, it entered Tadjikistan reportedly from Afghanistan in 1944, Mongolia in 1945, and Turkmenistan in 1950.

Remaining in Turkey, it was repeatedly introduced with imports from Asia, producing very virulent epizootics each year. Beginning in 1897, systematic attempts were begun to eradicate it by serum injection. In the district of Tchataldja (Roumelia), epizootics had occurred in 1853, 1863, 1877, and 1888, and then none until 1897, despite the presence of the disease near Constantinople. But in November of that year, a peasant brought a cart drawn by two oxen to Constantinople, which he stalled in a stable among sick animals, the origin of which was unknown. On his return, his oxen died, and the disease exploded in two villages, one where his first oxen had died, the other his home village which he reached before his remaining ox died. It spread widely before appearing to cease in February 1898, but in fact infection remained in several places. In one village, for example, it was concealed by the inhabitants for fear of isolation measures, but in some villages, it was the custom of the inhabitants themselves to maintain cordons during outbreaks to protect their stock. Mortality ranged from 81.5% of pure bred cattle, to 65% among grey cattle and buffaloes. In 1894, it broke out in the province of Aïdan with the loss of 50,000 animals; the following year, 5,000 died, and in 1896 and 1897, 3,000 each year until, in 1898, it was reduced to 500, representing 70% mortality, nearly all the mortalities being black cattle. In the area of Yozgat in 1898, over 30,000 head were lost.

The last years up to 1899 witnessed numerous epizootics in various provinces, thoroughbred cattle generally experiencing 80–95% mortality and requiring more serum than black cattle, which averaged 44.4% mortality in the range 30.3–67.8% in those villages where outbreaks were studied. Grey cattle were usually either resistant or contracted a curable infection. Buffaloes seemed intermediate in susceptibility between black and grey cattle (Réfik-Bey and Réfik-Bey, 1899).

In 1912–13, the Second Balkan War spread the disease once again in eastern Europe. The Turks reoccupying part of Thrace, the 4th Erzerum Division in July 1913 brought with it infected cattle from Asia. In a short time, it spread over the whole territory reoccupied by the Turks, and then fugitives from Thrace brought it with cattle into the southeastern districts of Old Bulgaria and over the Maritza river up to Grimildjina on the Aegean Coast. A total of 2,886 cattle died, and 1,339 were slaughtered before it was extinguished the following spring, a serum-producing station having been set up at Burgas on the Black Sea coast to combat it and a military cordon controlling borders. Several instances occurred in Constantinople following the importation of Asiatic cattle.

With the outbreak of World War I, a possible spread from Russia into Poland, Galicia and Bukowina, was seen as a threat by Hungary, and a publication describing the disease and its symptoms was produced by Hutyra and Marek (1916), professors at the Royal Hungarian Veterinary College in Budapest.

In 1917, following relaxation of control after the Russian Revolution, the disease crossed the Caucasus and entered the Ukraine, following the bolshevik armies into central Europe. The bolshevik advance in 1919 infected Latvia and Lithuania, but control was gradually effected. Prussia was seriously threatened but saved by energetic measures at its frontier. In 1920, the bolsheviks carried it into eastern Poland. An international effort, comprising 15 veterinarians from Denmark, as well as others from Czechoslovakia, Hungary, Romania, Yugoslavia, and France, together with equipment and serum supplies, and followed eventually by Russian refugee veterinarians, who were able to replace the temporary assistance, finally eradicated the disease from Poland in 1923. Altogether, 5–6 million cattle were lost in western Europe in this outbreak.

In 1918, a small outbreak is said to have occurred among civilians' cattle just behind the World War I fighting lines in Italy but was eradicated by prompt action of the authorities. Greece was infected in 1920, 1923, and then December 1926, the last apparently from Syria due to quarantine regulations not being rigidly enforced.

At the end of June 1920, Belgium was infected from an unexpected source. A shipment of zebu cattle from India, destined for South America, was held in transit at Antwerp. While in quarantine, several died, but rinderpest was not suspected, and the sheds were not disinfected. A few days later, some slaughter cattle from North America were stabled there before being distributed, some to the abattoir in Ghent where they came into contact with cattle imported from Germany, of which some were dispersed to 80 farms and others sold to dealers, resulting in rapid spread of the disease throughout Belgium. On July 10, several animals were slaughtered urgently at Gand, Namur, and Anderlecht. Rinderpest was not suspected at Gand, and no disease was seen at Antwerp. Not until July 20, was it recognized as rinderpest, and an association made with the imported zebus. By mid-August, it had reached its maximum with 222 farms infected in 85 districts, declining to four in October close to the French border. In all, 279 animals died, 490 were affected, and 1,859 slaughtered. Informed officially only on August 5, France closed its border to all animal and animal product imports, and implemented strict controls on movements, as well as censusing all cattle, sheep, and goats, in a protected zone near the border. Poland witnessed an epizootic in 1921, which led to an international conference at Warsaw to collaborate on control endeavors.

By 1930, the only region in Europe infected was Turkey, especially eastern. In 1932, Curasson referred to an "alert" regarding its occurrence in Africa, noting that the two regions that the disease held sway were in Africa and Asia. Then in December 1946 and January 1947, two cargoes of slaughter cattle from the Sudan introduced the disease into Malta. Ten per cent of the cargo was actively ill on arrival, and the disease quickly spread to cattle on the island. Vaccination of the island cattle gave disappointing results, probably because the live vaccine used was inactivated by long delays at ambient temperature at Cairo Airport. The disease continued to spread, a major factor being the continual supply of slaughter cattle from the rinderpest-free countries of Tripolitania and Cyrenaica. Imports were suspended in July 1947, and all infected and in-contact cattle on the island slaughtered. Malta was declared free of rinderpest at the end of August 1947. Then until 1969, when there was an outbreak in Turkey again, the last recorded outbreak in Europe was in the Rome zoo in 1949, imported with wild ungulates from East Africa.

In October–November 1989, it appeared on three farms in two districts of Georgia close to the border with Turkey, whence it is believed to have originated (or from Iran). It was eradicated by early 1990. Indications are that the border vaccination programme had lapsed. Molecular evidence indicates that the Georgian virus belonged to the Asian lineage. Turkey still witnessed sporadic outbreaks of the disease in the 1990s, attributed to reinvasions through cattle trade. After a 22-year absence, it broke out in October 1991 in the Van and Hakkari provinces in the east of the country, spreading throughout. The last recorded incidences were in eastern Turkey in 1994 and 1996, but it was unclear whether these were new introductions or a persistent enzootic infection. In March 1999, Turkey was declared provisionally free of rinderpest.

Seemingly eradicated in Asiatic Russia in 1967, after 24 years of apparent freedom from the disease, an outbreak was reported in July 1991 among cattle in Mongolia, introduced with young Russian cattle from the Chita region which moved south in mid-June to their regular summer grazing in Mongolia. There was no sign of the disease in the Russian herd of origin, but among the migrants, 174 of 902 head developed clinical signs of rinderpest, and 110 died (64%). In October–December, the disease was observed in predominantly unvaccinated domestic yaks in the nearby Tuva Autonomous region of the Russian Federation (50°N 97°E) on the northern border of Mongolia in the alpine conditions of the Tuva and Chita Republics Region, which had been grazing earlier in Mongolia where they had been in contact with cattle vaccinated following the outbreak there. Spreading to cattle in the second half of 1992, the outbreak was eliminated by 1993, total mortality in Tuva and Chita amounting to 2,800 yaks with mortalities of 17.4% in older animals and 63.6% in animals of less than 1 year; and 10,000 cattle representing 5–30% mortality. The mortality in Mongolia is not known. Bovine viral diarrhoea (BVD) is thought to occur from time to time, as among yaks in the Moron Region in 1990, conceivably misdiagnosed in reality perhaps being rinderpest.

One Siberian roe deer *Capreolus capreolus bedfordi* died after drinking milk from an infected cow. The outbreaks did not persist, and eradication was principally achieved by slaughter, vaccination, and quarantine.

In August 1998, rinderpest was reported in a herd of cattle in the village of Simonovo in Amur Region close to the border with China (55°N 127°E) (Figure 3), a region of about 300,000 cattle. There had been no livestock imports or exports in the region for 5 years, and there are no livestock markets. All cattle are slaughtered at appointed abattoirs and examined before and after. In April–May, 30,000 cattle were vaccinated in Amur Region, and toward the end of May, 58 of the 164 cattle in Simonovo. On June 17, an unvaccinated animal showed symptoms of rinderpest, dying 3 days later. A second case recovered, but then fresh cases occurred almost daily until August, rinderpest being confirmed in July. Molecular characterization identified the virus as an Asian Lineage virus virtually identical to the Kabul (Afghanistan 1961) virus and the K37/70 vaccine, which had been used since 1978 to maintain a 30 km-wide vaccine buffer zone along the borders with China and Mongolia, but which

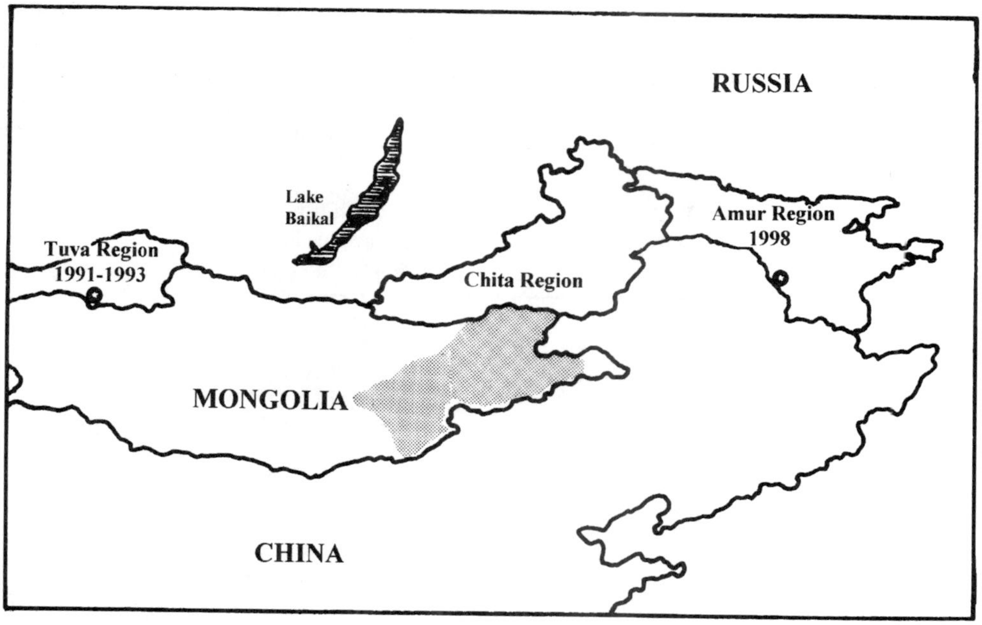

Figure 3. The 1990s outbreaks of rinderpest in Russia and distribution of the Mongolian gazelle.

recently had only achieved an average of 34% coverage because of poor compliance by farmers, although the vaccination is free. In some years, one million doses have been used in the border districts without any obvious signs of reversion. Routine tests are conducted on the vaccine, and experiments on three cows with the same batch after the Simonovo outbreak failed to induce an adverse reaction, although the animals seroconverted, showing that the vaccine took. It is safe to use on yaks, considered the most susceptible of all domestic species.

The virus is thought unlikely to have arisen from a focus persisting in a completely subclinical form in domestic or wild animals within the Russian Federation for so many years without having been noticed. The Veterinary Department is well represented and active down to the village level in the region. Livestock density is low in the area, there is low contact between villages, lack of cattle movements, an absence of markets, and a long winter housing period. Susceptible wild ungulates in the area are forest-dwelling, essentially non- or short-migratory species, roe deer, red deer, European elk, and wild boar, at densities unlikely to be capable of maintaining infection, and no mortalities are known other than due to classical swine fever in wild boar 10 years ago. However, in the wider region, two highly susceptible species occur, saiga antelope *Saiga tatarica* and Mongolian gazelle *Procapra gutturosa*, and it is alleged that between 1913 and 1945, rinderpest was frequently introduced by migrations of Mongolian gazelle (Roeder and Reichard, 1999), of which there was an estimated total of some 2 million. Lhagvasuren and Milner-Gulland (1997) report that in 1974, in eastern Mongolia, about 140,000 gazelle were killed by an unknown disease, and the same epizootic, but with fewer deaths, has occurred regularly since (Table 3, Figure 4). Partial investigation has shown that the gazelle died from hemorrhagic septicaemia (Dash and Sokolov, 1986), an infection with which rinderpest has been confused. In Kazakhstan, hemorraghic septicaemia allegedly caused the deaths of over 70,000 saiga antelope in May 1981, and 270,000 in May 1988 (Bekenov *et al.*, 2001), some 1,300 km west of Tuva. Mongolian gazelles also allegedly suffered an influenza epizootic about 1988 (Lushchekina *et al.*, 1988).

Extensive flooding occurred in northern China at the time of the Simonovo outbreak which could have caused migrations of what are termed locally "running species," such migrations being not uncommon in response to forest fires and floods. But there is no confirmation of such migrations having taken place. An unusually severe winter in 2000–01 killed many Mongolian gazelle in eastern Mongolia, and over 80,000 moved north into the Chita Region, an area where "few had ventured in past decades" (Kirilyuk and Laurie, 2001). Illegal hunting began immediately, with the poachers "loading full trucks." In February 2001, FMD

Table 3. Estimated numbers of death from epizootic disease of Mongolian gazelles in eastern Mongolia 1974–94.

Year	Deaths
1974	140,000
1980	15,000
1983	40,000
1984	2,000
1985	20,000
1986	15,000
1989	5,000
1991	3,000
1993	15,000–20,000
1994	3,000

Source: Lhagvasuren and Milner-Gulland (1997).

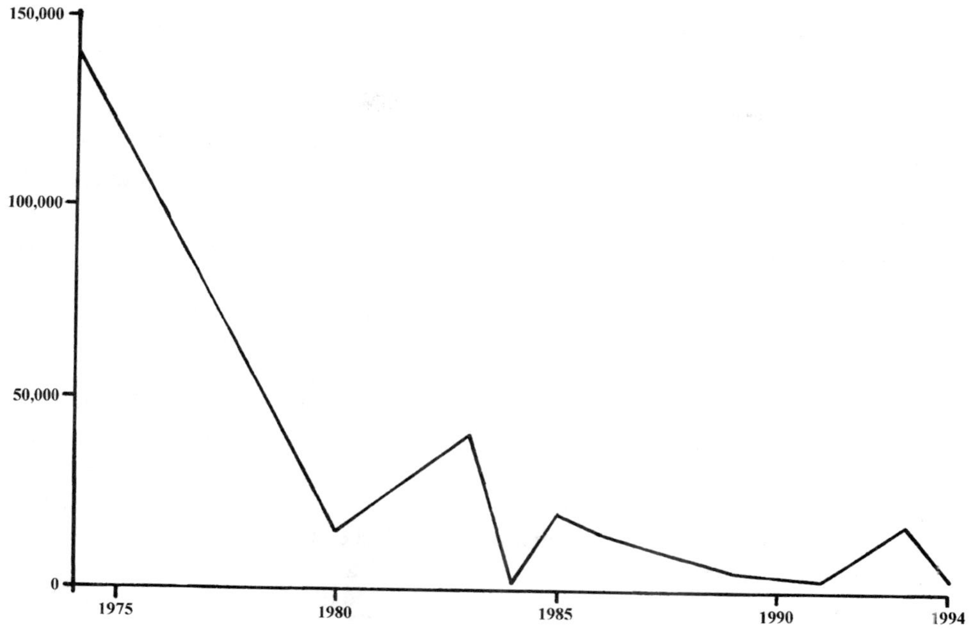

Figure 4. Estimated deaths of Mongolian gazelle from epizootic disease in eastern Mongolia 1974–94.

was reported in eastern Mongolia, and livestock slaughter and controls were imposed, but as of mid-March, it had not been reported in the gazelle, although allegedly many died of it in 1963–4 (Sokolov and Luschekina, 1997). Blood samples from 26 hunted and 56 neonatal gazelles taken in 1998–9, and from 25 sheep and goats, proved negative for rinderpest (Deem *et al.*, 2001). Surok or marmot *Marmota marmota* and suslik or Mongolian ground squirrel *C. c. ramosus*, have also been proposed as potential hosts, both rodents occurring in large numbers in Mongolia and the neighboring Russian Federation, Inoue and coworkers (1929) having experimentally infected the ground squirrel. But it is considered unlikely that these animals could have sustained the virus for 50 years without its presence being noticed in local cattle and wild ungulates.[2]

Likewise, transboundary spread from a neighboring country is considered unlikely. It is not believed to have been present in Mongolia since the 1930s, and there have been no reports in China since 1955, vaccination ceasing in 1956 except for limited vaccination in 1994; and the Russian Federation had no known connection to the nearest enzootic focus believed to be in Pakistan. Although some 1,500 km from the 1993 outbreak, the Amur outbreak was still closer to that than to the focus in Pakistan.

The available evidence, mostly circumstantial, suggests a common cause between the outbreaks, beginning with that of Georgia in 1989. Whereas Russian scientists seem to favor some possible cryptic wild animal source, and there is a long-held suspicion by the Central Asian countries and the Russian Federation that a reservoir of infection still persists in Mongolia and northern China, molecular characterization of the virus with its close resemblance to the Kabul 1961 virus would seem to discount this, and timing of the onset following

[2]A rodent reservoir would not be surprising for, in recent years, it has been revealed that cowpox virus reservoir hosts are wild rodents. In Britain, it is enzootic in the bank vole *Clethrionomys glareolus* and wood mouse *Apodemus sylvaticus*. Cattle are not the main hosts of the virus. Antibodies have also been found in 15% of susliks and 18% of gerbils examined in Turkmenia (Chantrey *et al.*, 1999).

vaccination is suggestive. At Simonovo, the virus was isolated from sick cattle some 3 months after the vaccine had been used in the village, but infection following vaccination is most unlikely to exceed 21 days, and therefore the vaccination alone, had the affected cattle been vaccinated, which is uncertain, could not explain the presence of the virus at that time. If the virus causing the outbreak originated from a source of virus which had been circulating in the field, it would have been expected to have evolved appreciably since 1961 when it was present in Afghanistan and 1998, showing a greater difference from the Kabul virus than it does. It is also considered unlikely that a virus would persist in the field unchanged over 10 years to affect cattle in sites separated by some 9,000 km. The most likely cause of the outbreaks therefore would seem to be a rare reversion to virulence of the attenuated vaccine virus. Possible laboratory contamination of the vaccine with virulent virus is ruled out, as one would have expected more adverse reactions (Roeder and Reichard, 1999). There is insufficient information on Mongolia to discount with certainty a persistence of the virus. Confirmation of hemorraghic septicemia in the Mongolian gazelle especially requires clarification, and clandestine movements of illegal hunters as agents of spread from some unlocated focus has yet to be considered. For the moment, information is wanting on which to base a definitive statement as to the origin of the virus causing the outbreaks, but clinical and serosurveillance in the 21st century suggests that there is no longer infection in Mongolia and Kazakhstan (Roeder, 2002a).

It was a cause of surprise, and concern, that in 1989 and again in 1998, the virus should still be appearing in its traditional region of enzooticity. Mechanized transport replacing oxen, and improved methods of feeding troops necessitated by their more rapid mobility transporting prepared supplies in trucks in the 20th century, meant that rinderpest would no longer be spread through Europe by armies with their trains of oxen, although it had remained enzootic throughout the continent of Asia from the Aegean Sea, the Caspian and the Urals, to the shores of the Pacfic Ocean. But in the 1990s, wars, in the form of civil strife, were to be a major factor preventing final eradication of the disease in Africa.

Part III

CONTROL MEASURES, LEGISLATION, AND EFFECTS

11

ARGUMENTS AND ENMITIES

Simonds and Gamgee

1. SIMONDS'S 1857 VISIT TO EUROPE

The three leading agricultural societies, representing England, Scotland, and Ireland, were the only bodies which viewed with alarm the spread of rinderpest into eastern Europe in the 1850s, sending Simonds, the leading veterinary expert of the day, to Europe in April 1857 to report on the state of affairs. Simonds could understand no language other than English but was accompanied by a British-based German veterinarian, William Ernes, to act as interpreter. The tour has been described as one of the most important epizootiological investigations in veterinary history (Pattison, 1990). At the same time (March 7), the General Board of Health commissioned Dr. E. Headlam Greenhow, lecturer on Public Health at St. Thomas's Hospital, to investigate "murrain" in horned cattle and the effects of the consumption of diseased flesh on human health (Greenhow, 1857).

Traveling through Belgium, Holland, Westphalia, Hanover, Hamburg, Lübeck, Mecklenburg, and Saxony, without encountering the disease but making observations on pleuropneumonia along the way, Simonds was informed in Prussia that he might encounter rinderpest near Breslau in Silesia and proceeded to there, where: "The Baron met us by somewhat facetiously remarking, that, fortunately for Prussia, but perhaps unfortunately for us, who had traveled so far to study the nature of rinderpest, it had no existence just now in Silesia."

Simonds then proceeded to Galicia, first encountering the disease at Lüdowica at the foot of the Carpathian Mountains. Here, one animal had just died and, together with nine others, had already been buried. Permission to disinter the carcasses was refused as the authority of the official accompanying the party did not extend to that region. Continuing through a mountain pass to another village, Zabrzez, here the party were 12 days too late and continued on to Kamienica, where the Austro-Polish authorities, who gave Simonds their full cooperation, had established a base to deal with the outbreak rife in the area. Hitherto free of rinderpest since 1846, the appearance of the disease on March 10 in both this village and Zabrzez followed the purchase of 10 steppe oxen at a fair at Jaczmirz, 32 km distant. These had originated with the large droves from Bessarabia at the fair totalling 3,000–4,000. Simonds bought two infected animals to better study the course of the disease, but subsequently, cases occurred daily so he was able readily to observe it in all stages, making observations of the course of the infection, studying in detail eight cases, and conducting autopsies. Owing to the prevalence of the disease in this area, Simonds did not consider it necessary to proceed to Russia (Figure 1).

When, shortly after his return, he was called upon to give evidence to a Select Committee of the House of Commons, Simonds stated that he had been able to observe the

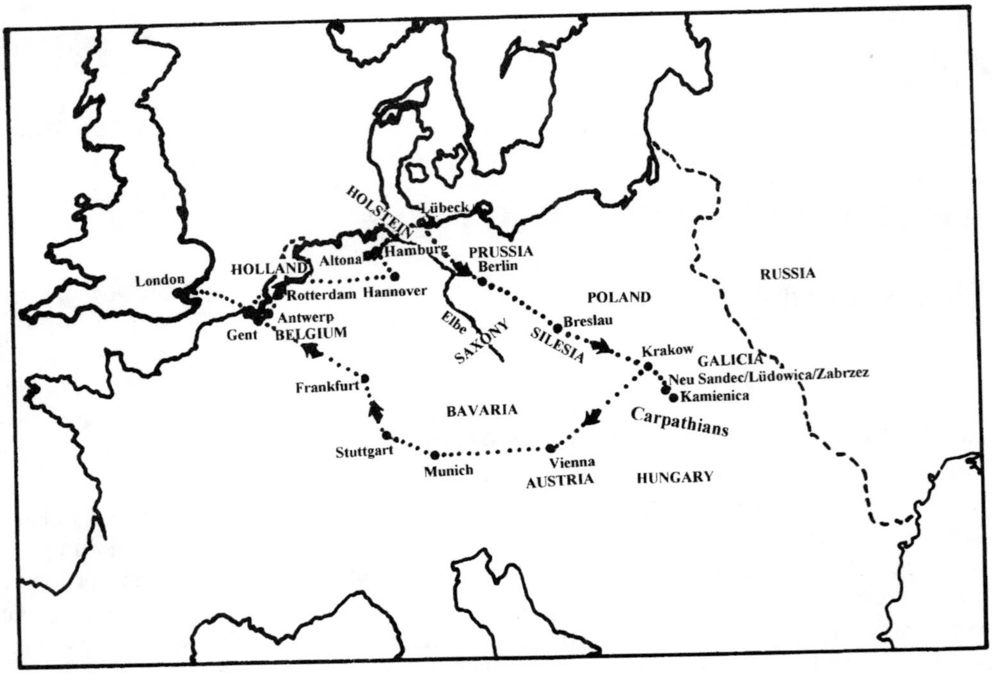

Figure 1. Simonds's European itinerary, 1857.

disease in animals from start to finish. He did this by buying two oxen, which showed early signs of infection, and then watching them, conducting postmortems after they died. "Subsequently," he stated, "I had many opportunities of observing other cases both in the latter stages of the disease as also in the first." Thus, he and Ernes should have been well qualified to recognize the disease when it broke out in Islington 8 years later. Simonds's report, published in the *RASE Journal* in 1857 and in 10 separate parts in *The Veterinarian in* 1858–9, only considering that the plague could travel through Europe to reach Britain, expressed confidence in the protective arrangements made by those states contiguous with the infected zone and allayed all anxiety: "Rinderpest is a disease which specially belongs to the Steppes of Russia ... in general terms, Rinderpest has not existed in Central and Western Europe for a period of forty-two years." He ascertained the disease was not present in those localities from which cattle were imported into England: "no fear need be entertained that this destructive pest will reach our shores ...." He had not foreseen the development of railway links across Europe and continued:

> Its present great distance from us would, of itself, afford a fair amount of security; but when we add to this that no cattle find their way from thence, directly or indirectly to the English market and also that in the event of the disease spreading from Galicia, it would have to break through hundreds of military *cordons*, one after the other, before it could possibly reach the *western* side of the German states; and, moreover, that for years past commerce has been unrestricted with regard to the importation of skins, hides, bones &c., of cattle from Russia and elsewhere, all alarm, we believe, may cease with reference to its introduction into the British Isles. (Simonds, 1857)

But earlier in the report, he had written: "...our own investigations have proved that no fear need be entertained *at present* (my italics) of 'the great cattle-murrain' visiting our shores." Clinical signs were characteristic and usually obvious, he stated. Medical treatment was useless. But: "... even should a revival of the trade in cattle ... take place to England through Holland, by means of the Rhine, there would be no risk of introducing this disease thereby, unless circumstances should unfortunately arise, by which it became as rife as it was in the years 1813, 1814, and 1815." That cautionary qualification was omitted from his "Conclusions."

Earlier in the report, he blamed the British Vice-Consul at Lübeck, Mr. Blackwell, for being unnecessarily alarmist. In a despatch dated May 30, 1856, Blackwell had written: "...although the disease which has broken out among the horned cattle in Mecklenburg seems to be regarded as the pulmonary murrain [pleuropneumonia], it may perhaps be the real steppe murrain, which is now raging in Poland to a fearful extent, notwithstanding the stringent measures that have been adopted by the Russian Government for putting a stop to it." On September 17, he wrote: "...as this highly contagious murrain has spread from the Steppes of South Russia, through Poland, Prussia, and Mecklenburg to Holstein, to a district from which the English market is supplied with cattle, I must beg leave to call the attention of her Majesty's Government to my Report of its origin, progress, treatment, symptoms, &c., transmitted on the 30th of May last."

Simonds wrote:

> These statements could not fail to add to the alarm which was originally felt in this country; and when it is considered that for several months afterwards scarcely a week elapsed without intelligence reaching us that the "cattle murrain" was spreading, the surprise becomes the greater, rather than otherwise, that some measures of a preventive nature were not earlier adopted by Her Majesty's Government. It is true that the reports from other British consuls did not fully bear out Mr. Blackwell's statements ...."

The government had introduced an Order in Council on April 2, 1856, prohibiting cattle imports from Lübeck, but Simonds found that rinderpest had been absent there since 1815, and regulations which had recently been enacted were for pleuropneumonia. Neither were any cattle exported from Lübeck by sea, nor imported from the Baltic states or elsewhere. He concluded that all countries of northern and western Europe from which cattle were exported to England were perfectly free from rinderpest.

Greenhow's report, antedating Simonds's, makes no mention of the latter's investigation, considering the disease in Europe to be pleuropneumonia, and he believed, like Simonds, in view of the controls in place in Prussia, that the import of rinderpest into Britain was unlikely, but it would be advisable perhaps to have a permanent cattle import ban on countries where rinderpest occurred (Greenhow, 1857).

The Veterinarian, of which Professors Simonds and Morton were the senior editors, stated that Simonds's report would be referred to in years to come both with satisfaction and profit (Anon., 1859). Simonds hailed with satisfaction the removal of all restrictions on the importation of manure, hides, hoofs, and horns from the Baltic ports (Simonds, 1857), a region close to the zone of the disease in Russia. Everybody, including agricultural societies, pedigree stock-owners, and the Government, were satisfied (Smith, 1931). But when giving evidence shortly afterwards on June 16, 1857 to the House of Commons Select Committee examining *The Sheep &c. Contagious Diseases Prevention Bill*, that had been given a second rereading on June 5, Simonds replied in answer to questioning that cattle plague should certainly be named in the proposed Act. He was certain that it had occurred in Britain in the 18th century, and it might well reach Britain again in the event of it reaching Prussia "and subsequently Holstein, or any other place from which we receive our supplies." By 1861, completely at variance with his first, optimistic, report, Simonds was lecturing his students at the London Veterinary School emphatically declaring his conviction that nothing could save the country from a visitation of rinderpest unless the authorities realized the probability of such an event arising and adopted vigorous measures to prevent it. It had raged in Poland "to a fearful extent, bidding defiance to all preventive means adopted by the Russian Government to limit its spread." Hungary, the Danubian Provinces, Turkey, Southern Russia, and the Crimea had all suffered considerable losses. Referring to its previous appearance in Britain in the 18th century, Simonds noted:

> If in those times, when the importation of cattle into our sea-girt isle was jealously guarded against by high protective duties, the disease made its appearance here; can we reasonably expect it will be otherwise now that these duties are removed, and it has become a matter of necessity, as well as policy, to allow the free entrance, among others, of each kind of animal used as food for the people?

In 1863, Simonds stated that it was no part of the duty of the veterinary profession to interfere with importation of animals from abroad. By September 1865, he had changed his views on that also and urged Government to adopt control measures.

In an address to Norfolk farmers after its outbreak in Britain, he stated that he was perfectly satisfied that the disease would spread in the places where it now existed, and sooner or later, unless stringent measures were taken, the whole country would be affected. He did not come to this conclusion simply because he viewed the disease as one which spreads its contagion, or because, as some thought, it was atmospherical, but from positive experience gained in other countries and the records of the 1745 outbreak. He pointed out that when the circumstances which had then prevailed were considered, and the small amount of traffic in cattle compared with that in 1865, the fact that the disease lasted for 12 years did not offer much hope for its removal now.

Despite his changes of opinion, Simonds was a man who normally chose his words with great precision, as for example when under questioning by Lowe while giving evidence to the Commission. Lowe referred to the animal which had died before being shipped by stating, "That was probably not more than the average?" Simonds replied, "One animal can scarcely be called an average out of a cargo of 321."

In his autobiography published in 1894, Simonds wrote: "It was scarcely to be imagined in those days that competition in the cattle trade would lead to animals which had been exposed to infection being imported by sea from a Baltic Port, as was the case in 1865," although Gamgee had stressed otherwise. In a letter to *The Times*, Gamgee had stated that there was no need for the disease to break through the cordon in Europe referred to by Simonds. It could come to Britain by sea from the Baltic.

2. JOHN GAMGEE

The only member of the veterinary profession who felt unable to accept Simonds's original optimism, John Gamgee, 20 years Simonds's junior, who, frustrated with what he saw, opened his own veterinary school in Edinburgh in 1857, was castigated for his opposition and branded an alarmist. It was asserted that all his endeavours from the commencement of his career in veterinary science had been directed to the complete overthrow of old-established doctrines and practice, something Gamgee would not have denied. His campaign against the sale of diseased meat led in 1862 to his appointment by the Privy Council to conduct an inquiry into the subject of "Cattle Disease in Relation to the Supplies of Meat and Milk." But when rinderpest broke out not long afterwards, he was not invited to join the Commission and was not called upon to give evidence until the seventh day. However, he received acknowledgment in the Commissioners' *Third Report* as the first to begin inquiries in Britain as to the earliest signs indicative of the disease; namely that a rise in temperature preceded any other symptom, enabling the disease to be detected at least 2 days earlier than had hitherto been believed.

While Simonds had been visiting Europe, reassuring the RASE, and giving evidence to the House of Commons Select Committee on the *Contagious Diseases of Animals Bill*, Gamgee was doing his own lobbying. The first was a pamphlet as a letter to Sir George Grey, the Home Secretary, titled *Cattle Plague and Diseased Meat in Their Relations With the Public Health*. In it, Gamgee referred to his many warnings about the risk of the introduction of rinderpest into Britain and the dangers to human health of the consumption of diseased meat, complaining that still no action had been taken by the authorities. He drew attention to the defects in the institutions; the insufficiency of the legislation, and the lack of competent officers to control the introduction of rinderpest. Cow houses in London were not inspected, and diseased animals were mixed with healthy animals in the markets. Diseased carcasses were trimmed to be sold for human consumption. He suggested that imports of live cattle from Holland and Germany be prohibited and hides and carcasses of diseased animals destroyed. This was followed by a second "Letter" consequent upon an Order having been issued controlling the import of diseased cattle. But still nothing was being done to control the sale of diseased meat for human consumption (Gamgee, 1857a). A similar "Letter" was addressed to the Lord Provost of Edinburgh: *Diseased Meat Sold in Edinburgh and Meat Inspection in Connection with Public Health, and with the Interests of Agriculture* (Gamgee, 1857c).

His surgeon brother, Joseph Samson Gamgee, lobbied the Home Office, describing the conditions under which London obtained its meat. Eventually, a Parliamentary Committee was appointed to report on the issue. In its report, published in 1858, it concluded that Government always had it in its power to prevent extreme evil consequences from the consumption of diseased meat, and any further steps would be an undue interference with legitimate trade. At the end of 1862, John Gamgee lectured to the Royal Society of Dublin on disease and mortality among cattle at which a Mr. Ganley, a salesmaster, stated that unless some means were devised to give some compensation to the farmer for diseased cattle, it was impossible to prevent him selling them or the butcher from killing and selling them. Every one of the salesmen sold diseased meat, otherwise the farmer could not pay his rent. Disease was so prevalent that he could not live were he to submit his cattle to destruction. Gamgee replied that he would have gone 1,500 miles to hear this confession. Asked at the first International Veterinary Congress at Hamburg in July 1863 what Britain did with her diseased meat, he replied, "Eats it." In October that year, he organized a meeting of all the cow-keepers in London at which he put the facts to them so they could hear it from him direct. It was a tribute to Gamgee that they "…never denied, never winced, never swore, but, charmed by the charmer, behaved gentlemanly and patriotically" (Anon., 1863).

Nonetheless, his attitude on the meat question and his attempts to stop the importation of diseased cattle made him a large number of enemies where vested interests were concerned. Not least among these were members of his own profession who hated him for his ruthless efficiency (Smith, 1931). Many of his students had considered that the time and labor devoted to lectures on rinderpest were wasted effort, contending a disease of Russian origin, almost exclusively confined to the Continent, could only interest veterinarians in the countries where it actually occurred.

The Board of Health had already commissioned Greenhow's report a month prior to Gamgee's first letter to Grey. Greenhow makes no mention whatsoever of either letter and concludes that there was no proof that the consumption of diseased meat in Britain was injurious, although there were diseases on the Continent which could affect humans. Furthermore, any investigation into rinderpest was unnecessary, as full reports were available from German authors (Greenhow, 1857).

But due to Gamgee's persistence, two bills, the *Diseased Cattle Act* and the *Diseased Cattle and Unwholesome Meat Importation Bill*, were put before parliament in 1864, one of their most vigorous opponents being Hall Maxwell, secretary of the Highland Society. At a meeting in March, the expression of opinion was "Hang Gamgee and oppose the Bills" (Farmer's Friend, 1864).

The year before, in two letters to *The Times* of November 10 and 13, which have been described as "momentous," Gamgee warned both Simonds and the whole country that rinderpest would enter the country with a cargo from a Baltic port:

Sir,

The letter which appears this day in your columns, showing the great advantages of conveying foreign cattle to London via Harwich, induces me to ask what additional precautions are being taken to prevent the introduction of diseased animals from abroad seeing that there is a prospect of the "largest portion of the cattle trade from the North of Europe" never coming under the notice of the inspectors at the port of London. A full cargo of bullocks, calves and sheep, is said to have been landed at Harwich at 12 o'clock on Sunday night, and they were in the metropolitan market by 10 o'clock on Monday morning. Now, unless I am very much mistaken as to the time required to truck and convey the cattle from Harwich to London, the landing, watering, feeding and inspecting of the full cargo must have taken place before 6 o'clock on the Monday morning—that is to say, when dark; and if this be the usual practice I know not what is to prevent a destruction of British live stock to an extent even exceeding the frightful proportions already attained.

I do not seize my pen to write without weighing the effect of restrictions on trade; but, Sir, I am receiving letters by every post from all parts of the United Kingdom complaining of the condition of the lean stock, for which fabulous prices are being paid, and which is at the present time distributing diseases in all directions. As yet there is not a farm that I have visited in England or in Scotland on which cattle from the great Falkirk Trysts, and many country fairs, have been taken within the last two months without suffering from the foot-and-mouth disease. On not a few of the farms the lung disease has already appeared, and in many more it must shortly manifest itself with its usual virulence. I have seen not a few lots of cattle bought at £12., £13., and £14., a head, which, three weeks or a month after purchase, have been reduced in value by at least £1. per head. When we consider that those animals are to be fattened by next spring the loss of so much time is a very serious matter; and knowing how universally this loss is sustained, I venture to suggest that we should not overlook the fact but do all in our power to study how we can encourage trade and protect our cattle from deterioration and destruction. The price of meat is high, and still higher must it go. There is no alternative. Our foreign importations have led to a steady rise in price for years past and must still favour that rise if we do not check their plague-disseminating tendencies. Are our farmers likely to get the full benefit of high prices? When I look at the reports of markets in your columns I see that, after all, those engaged in the fattening of stock in this country cannot make much money after paying as they do for store cattle and incurring the losses which, at the present day, almost inevitably attend the purchase of stock. Many stockowners have written me of late on this subject, and there is no doubt that they are getting fully alive to the important question of preserving the health of their cattle, and they ask me, how is this to be accomplished?

I hold that the abundant supply of wholesome animal food at a moderate price is so important a question to the people of this country that there should be inquiries constantly going on as to the condition of our live stock traffic, and as to the influence exerted by that traffic on the production and distribution of disease. The extraordinary prevalence of the foot-and-mouth complaint at the present time, which exceeds the formerly unexampled frequency of the malady twelve months since, depends entirely on the condition of foreign stock. This is at the root of the evil. We could not have the disease without its introduction. We never had it before free trade in live stock, and we have it not at present where cattle are only sold and exported from breeding districts. It is perfectly true that many Scotch, Irish or English cattle leaving their original homes in perfect health prove diseased in the fairs and in the farms they are taken to; but why? The answer to this question has been frequently given this year by practical men who have spoken at various meetings when the subject of cattle disease has been discussed. They unanimously say, "Look at the cattle trucks." They are never cleaned out from year's end to year's end, and after conveying one lot of diseased cattle the mass of filth bespattered on the sides and lying thick at the bottom of the trucks is charged with poison which readily infects any animals coming into contact with it.

Foreign stock brought into this country diseased or simply infected would not injure British animals, if we learned how to keep it by itself. Our farmers, as a rule, refuse to buy the foreign animals for store purposes, and the great majority of them is sold at once to be slaughtered. This being the fact, and as inspectors never can altogether prevent the admission of infected herds into this country, why do we not dispose of our foreign stock in special markets or in special divisions of the present market, where practicable, and have proper accomodation for their slaughter, or proper means for conveyance beyond the markets, without coming in contact with British stock? Of course, those interested in the foreign trade will pooh-pooh all this; but we cannot allow these gentlemen to continue as they have done, defying interference and destroying the produce of these islands. Twenty years' experience should open our eyes; and what does it teach us? It clearly demonstrates that had we never imported foreign cattle we should now be paying less for meat than we are doing. Our breeders, graziers, cattle dealers and others would never have known what it was to lose as they do at present. The production of live stock would have kept better pace with demand and our people would be better fed. I quite admit that if our importations had been conducted on a system such as would have prevented the spread of contagious disorders they would have done us almost as much good as they have done harm. And now, with the wholesale destruction by disease, we could not suddenly check such importations. Let us, however, by all means see see how we can counteract the very serious evils attending them. They are not likely to be counteracted by the rapidity of transit achieved on the Harwich route, and there is too much cause to fear that the promptitude with which animals are landed and trucked to London is quite incompatible with their most superficial examination with a view to prevent disease.

I do not wish to prevent our reaping the advantage of the rapid transit of stock; but it appears to me that we altogether neglect precautionary measures which are essential for the preservation of the health of our own animals when the foreign trade is acquiring great proportions, and can be carried on with increased facilities. We must watch narrowly unless small-pox in sheep should again reach us. The disease is spreading daily in Russia and Mecklenburgh. I receive weekly reports which prove that outbreaks of this disease are more numerous this year than last; and it is not unlikely that if we continue to neglect ordinary precautions, as we do at present, we shall soon have even more serious occasion to repent it than for the past. Let our markets rise a little more, and what is to prevent Russian cattle being shipped for our ports? The few Portolian oxen I have seen from time to time in the London cattle market have been fattened in Russia; but the time seems to be fast approaching when enterprising traders will accelerate the travelling of stock from the markets in eastern Europe, where it is purchased at a very trifling cost, and whence we may expect the transmission of plagues which would nearly create a famine in these dominions. I am not dealing with imaginary evils when I have before my eyes, as I write this letter, a statement of the losses sustained by the Russian plague in the Austrian dominions. During the last fourteen years, 500,000 animals have been seized with the disease, and of these 270,000 died. Even that loss is comparatively small when contrasted with the loss we have sustained during the last eighteen or twenty years through the lung disease; but if Russian cattle manage to reach our markets-and why should they not?—we shall certainly have more than the Austrians, whose organization for the prevention of disease among cattle is perhaps the best in Europe.

Prudent men must agree with me, that under existing circumstances we need something more than the rapid transmission of stock from foreign to British markets. We have already too much occasion to lament the introduction of foreign diseases into these islands, and we must watch lest we henceforward find that, whereas wheat barely remunerates the tiller of the soil for his labour, disease among livestock may irretrievably ruin him.

I am, Sir, your obedient servant,

John Gamgee
Professor of Veterinary Medicine

New Veterinary College, Edinburgh, Nov. 7th 1863.

This was followed by his second letter published 3 days later:

Sir,

In my letter in *The Times* of today, "Russia" has been printed twice instead of "Prussia." It is in the last-named kingdom that the sheep-pox has been committing serious ravages for many years past, and pehaps never more so than during 1863. The Podolian—not "Portolian"—oxen have been fattened in Prussia and not in Russia.

Many of your readers may not be aware that every Monday morning the London prices are telegraphed to Hamburg, Rotterdam, and other ports, whence the information is transmitted to such distant markets as those of Vienna and Berlin. If prices suit, cattle and other animals are trucked from these cities for London. The Austrian and Prussian stock is fattened to a great extent on the refuse of distilleries, sugar and starch factories, &c. Among the animals thus fed are Polish and even Russian cattle. These animals, have, however, passed a rigid quarantine on crossing into Austria and Prussia, and we have been protected from the Russian plague by the precautions taken in central Europe to check its spread. A century since we suffered, and so did most countries in Europe. It was in consequence of the losses by this plague that veterinary colleges were first founded and liberally endowed abroad. In this country the most energetic measures were adopted to arrest the progress of the plague, and with the best effect. Our ports were closed to foreign cattle, and remained so till 1842, with some trifling exceptions in 1839–40. Over the whole country precautionary steps were taken to exterminate the disease, and in this our forefathers succeeded. How different now! We do not prevent disease, but eat all our diseased cattle. Instead of veterinary surgeons being extensively employed to check the progress of maladies, the butcher's services are in requisition when animals are seized with sickness. It is a common occurrence for a veterinarian to be asked, when visiting a case, if it be serious, and any decidedly unfavourable prognosis is sure to be followed by the animal being slaughtered, dressed, and despatched to London.

I could scarcely have hoped for so fortunate a coincidence as the publication of my letter in the same impression of *The Times* as the remarks of our Paris Correspondent on the importation of cattle and sheep from the Danubian provinces and from Roumelia into France. The project is abandoned, but why? Simply because there remain to be finished 750 miles of railway on one route, and 550 on another. When these railways are finished, or perhaps before then, there can be little to prevent the markets in the West of Europe from being placed in direct communication with the markets in the extreme East. Last century wars and revolutions led to the free transport of cattle in the rear of armies, and thus the Russian and other plagues broke out in all directions. Steam has proved a more certain means for encouraging the transmission of contagious disorders than war and as yet we have not sufficiently considered its influence for evil in this respect. Your correspondent's observations are quite sufficient to show that I was not dreaming of impossibilities when sounding a note of warning in my letter of the 7th; but if we are receiving Spanish cattle in very fine condition by sea, what is to prevent a fast and suitably built steamer landing at a British port cattle shipped at Memel or Libau? Cattle are very cheap in Russia, and the Russians are turning their attention to the management of stock. If we do not husband our resources in the way of producing and fattening animals in the British Isles, we must depend more and more on the produce of Eastern, plague-breeding plains, and we may find ourselves at no distant period much worse off for animal food than we are at present, notwithstanding every facility being afforded for free trade in stock.

Dearly-bought experience has demonstrated to the Austrians, Prussians, French and others, that the importation of cattle requires constant supervision. Your own correspondents have this year furnished valuable information concerning the cattle plague as it spread through Turkey across the Adriatic and into the Roman States. The area over which this fearful malady has extended of late has been far wider than usual, and it has been found essential on many occasions to enforce a strict quarantine, and otherwise to interfere with the trade in cattle.

For the safety of British stock, and for the instruction of British stockowners in general, information concerning the health of animals in different countries should be published periodically, and it would be easy to devise a system of showing the diseased and the healthy regions of Europe in maps, which any cattle dealer could consult. On such maps the course in which plagues spread, and perhaps even the extent and direction of the cattle traffic, could be indicated. In a short time such a mass of information would be collected as to show how we should act to encourage the trade in healthy animals, and effectually to guard against the traffic in diseased ones.

I am, Sir,

and so on. Edinburgh November 10.

3. A CONFLICT OF OPINIONS?

When, in June 1865, Gamgee's views were vindicated and rinderpest did appear in Britain by the very route he claimed it could, it was Simonds who was appointed chief inspector and professional adviser to the Veterinary Department of the Privy Council set up to monitor the disease. But neither Simonds, Gamgee, nor anyone else, could persuade those in authority to tighten cattle import regulations.

Pattison (1990) states that Gamgee only repeated what Simonds had been saying since his visit in 1857, that rinderpest could invade Britain; but neither Simonds's original report nor his later autobiography bears this out. Simonds, claims Pattison, was not in opposition to Gamgee, merely less strident in his warnings. It seems more likely that he had second thoughts about what he had written in his original report, perhaps after the development of rail links or after remonstrances from Gamgee who may have pointed out to him that he was putting the whole country at risk.

Where Simonds and Gamgee differed was that, whereas Gamgee considered that only total control would be effective, Simonds was against closing the Metropolitan Market because of its importance and thought that it could be limited to the sale of cattle for immediate slaughter. He was also emphatically against halting imports and the slaughter of all imported cattle. His recommendation was that cattle imported from countries where the disease was known to exist should be placed in quarantine. Animals from Hungary or Galicia, where the disease existed, had their own chracteristics and could be recognized easily as coming from those countries. Quarantining them would probably mean importers would cease to import from such countries.

Gamgee was not an outspoken critic of Simonds, merely stating in his opening address of the first session of the Albert Veterinary College on October 2, 1865,

> There is great reason for regret that Professor Simonds, who saw the disease early in June, and the Government, who issued an Order in Council on the 24th of July, did not adequately warn the farmers of the appearance of the disease. Acting under the advice of Professor Simonds, the Government said, full five weeks after the malady broke out in London, "that a contagious or infectious disorder, of which the nature is at present uncertain, has lately appeared ..."

In his evidence to the Commission, he declared that Simonds had stated to the Metropolitan Plague Committee that Government did not see how it could prevent the removal of healthy cattle, but diseased cattle might be kept on the premises where they were found. Gamgee had seen a number of outbreaks arise from infected cattle leaving sheds to which they had been confined and being taken to markets and other places. Since, if they followed the Orders in Council to the letter, dairymen must lose the whole of their stock once infected, this inclined them to evade the law. He criticized the lack of authority in controlling movements of stock and in securing at once the slaughter of animals. His answer for control of the disease was a voluntary national system of insurance, accompanied by strict seizure and slaughter.

Gamgee was much more outspoken of his criticism of Simonds when called before the Select Committee of 1873. Referring to FMD, "... Mr. Bruce actively supported me in 1863 and 1864 [that FMD was an imported disease], when Professor Simonds said, 'This disease is not of foreign origin; it may arise spontaneously here.' With his support ... we were defeated." And of pleuropneumonia:

> For some years Professor Simonds was experimenting in this country, and I followed his views, and he induced me to believe that inoculation for pleuro-pneumonia was an absurdity. I had no practical means of ascertaining whether inoculation was good or not. Professor Simonds performed some experiments, and he said that it was not useful, and I, very foolishly, for the only time almost in my life, followed his advice, and therefore remained in the dark for some years.

He also stated in regard to FMD, of which Gamgee claimed to have said: "We have now got rid of foot and mouth disease, and we shall not have it again until it is re-imported." Simonds responded: "This disease is not of foreign origin; it may arise spontaneously here." And with regard to port inspections: "I am not saying a word against Professors Brown and Simonds, for whom I have the greatest respect as men, however much I may differ from them as veterinarians." He was also outspoken against Dr. Hall Maxwell "and other witnesses," who were brought to give evidence against him at a Select Committee hearing, claiming that they "were responsible for the introduction of cattle plague in 1865, according to my view of the matter" (Anon., 1873).

Long before this, frustrated with the lack of response to his views in Britain, in 1863 Gamgee proposed an International Congress of Veterinary Surgeons to be held that July in Hamburg. The circular of invitation which he sent out from Edinburgh noted:

> There has been during the last twenty years, an increased prevalence of contagious disorders ... Europe is now suffering to a considerable extent from the system by which one people attempts to save itself from loss by disposing of diseased and infected stock to another ... with progress in the commercial relations established between different countries, with greater facilities for transit and sale of stock so as to meet the demands of Western Europe, there has been, during the last 20 years, an increased prevalence of contagious disorders ...."

The Congress was a resounding success, attended by 99 continental authorities, all leaders in their field. But only one English veterinary surgeon other than Gamgee attended.

The British Government took no notice of the meeting, and it was not reported on in *The Veterinarian*. It was doubtless for the latter reason that Gamgee felt compelled to publish the proceedings *in extenso* in his book *The Cattle Plague*. The depth of the discussions, especially with regard to rinderpest and its prevention in western Europe, suggests that Gamgee's opinions might have indicated the manner of preventing rinderpest reaching Britain in 1865 had the authorities been prepared to listen to him.

So successful had the Congress been that a second was held in Vienna in 1865, encouraging such distinguished continental patronage that on this occasion, Simonds attended, accompanied by Ernes as translator. A third was held in Zurich in 1867 when Simonds attended again as the British Government's representative but, being alone, was unable to understand a word of the proceedings. A colleague gave a synopsis of Simonds's address in German, which began with an apology that he, Simonds, understood neither French nor German. But Ravitsch, for example, the equal of anyone in European veterinary medicine at that time, spoke fluently in French, German, and Russian. As Saunders (1980) pointed out, the few veterinary pathologists working in Russia in the 19th century, such as Semmer and Ravitsch, were much better educated than their English-speaking counterparts. The exception was Gamgee who, born in Florence, was fluent in several languages. Following his mentor Brauell's pioneer work on the pathology of rinderpest published in 1862, Eugene Semmer, a Latvian, published a short book in 1875 (in German) on its pathologic anatomy; following this in 1881 with a long review article. Largely unknown in Britain at the time, today the contributions to the pathology of rinderpest of Brauell, Ravitsch, and Semmer, classic as they were, are still mostly forgotten because they were not accessible in English.

4. THE ATTACKS ON GAMGEE

Gamgee had spent 2 years on the Continent studying rinderpest before Simonds visited eastern Europe. He knew more about the epizootics of Europe than anyone else in Britain and considered Britain especially prone to invasion due to the absence of the most elementary protective measures and that stock were imported from countries daily exposed to infection. To promote his views, he began a journal, the *Edinburgh Veterinary Review* and also contributed extensively to *The Times* and the *Daily Telegraph*, as well as lecturing to the public in Glasgow, Edinburgh, and London. His urging the need for meat inspection and legislation controlling imports made a host of enemies among those interested in the cattle trade, from landowners to butchers, whose views weighed heavily with Parliament, and earned him, as he stated, "torrents of abuse," as well as imputations as to motives and gross misrepresentations.

One detractor, signing himself "Anti-Clap-Trap," issued a printed card bearing a poem titled "Gammongee Repentant," imputing that Gamgee was lying to make money (Figure 2). Another pamphlet by "Worrems M.R.C.V.S." appeared, in which Professor Jammygee is said to have made a great noise on purpose to get his name in print and to excite public attention to his new veterinary college. Charlie Slaughter, "Slop butcher," discusses with Mr. Trax, a farmer, whether cattle trucks should be improved, with hot bottles for cows' feet and respirators for their noses. Professor Slimons, Mr. Giblets, and Mr. Mayyou have equally derogatory things to say. Yet another pamphlet read "Murder of the Innocents so strenuously advocated by a learned professor is a very costly, cruel, and contradictory exercise of the healing art. But it is so much to be regretted that it has been so prominently set forth and so much extolled in a work entitled 'Report on the Cattle Plague or Rinderpest' by which much error has been disseminated and a great evil entailed by a certain 'no-cure' professor."

Gammongee Repentant.

"When the devil was sick, the devil a monk would be."

One night Gammongee took a fit of the gout,
And 'twas in that organ, of *pigs* called the snout;
He'd also a malady raging within
His lateral region—*The want of the Tin!*
In short, the archquack believed he was dying,
While conscience raked up his swindling and lying.
To his *uncle* above he'd oftentimes go,
But dreaded to meet with his *uncle* below,
And vow'd that, if spared, and the sins of his youth
Remitted, he henceforth would stick by the Truth.
When, lo! the old lady, who lives in the well,
Appeared, as if conjured by some potent spell;
Her beauty was somewhat by anger deformed,
And more like a Fury than Virtue she stormed
At Gammongee stamping, she cried "Fie! for shame!
How durst such a scamp as you mention my name,
You base, you contemptible, wretched high-flier,
You both were begotten and born a Liar."
She said, and evanished in a whirlwind anon,
And next in the drama appeared—Mrs D—n.

ANTI-CLAP-TRAP.

Figure 2. From the James Beart Simonds Collection. *Scrapbook Volume I.* The Royal Veterinary College.

Gamgee was also alleged to have misled parliament, but this accusation seems to stem from evidence to the Commission of a cattle salesman, Hermann Gebhardt, before Gamgee had been called to give evidence. According to Gebhardt:

> It was gravely stated by Professor Gamgee when the disease first broke out, that week after week Russian cattle found their way to this country by way of Hamburg; but I can state upon authority [from the largest salesman in Hamburg] that this is false, there has been only one lot arrived about 10 years ago, from Russia, excluding this last cargo, and they were so very bad they could hardly be sold, and they never sent any more.

When Gamgee opened his New Veterinary College in Edinburgh, he stated in his opening address that 10 years would decide whether his fears and anticipations were well grounded or not. Eight years later, they were proved right. On November 10, 1863, he published his letter in *The Times* predicting what Simonds had so confidently asserted would not come to pass. One month before the outbreak, in May 1865, he wrote in *Baily's Magazine* (although it was held over for publication and did not appear until after rinderpest had broken out): "Steamers may, however, some day convey Russian herds of cattle through the Baltic to Leith and Hull. Instead, then, of seeing the healthy animals which are now coming from Jutland, it is not unlikely that the Plague may visit us … Long may we be spared any such infliction …."

Ironically enough, although it has never been admitted and fortunately for him not realized by his adversaries, it was probably such advocacy, specifically in his letters to *The Times*, which put the idea into the heads of cattle dealers, and thus within 19 months of their being written introduced the plague into Britain. It was in November 1863 that Gamgee's letter appeared in *The Times* and the summer of 1864 when the speculator, Burchell, approached the importer, Honck, with his proposal for an exploratory trip to Russia. Unfortunately for Gamgee, his lobbying did not produce the legislation which could have prevented introduction of the disease, and it was he, rather than Simonds, who was probably thus indirectly to blame for its importation. In a report to the Emperor Napoleon III, the French Minister of Agriculture stated

> … its introduction into England has arisen from the fact of speculators in cattle having found it profitable to draw their supplies from the Russian provinces, and to transport them by steamboats to the English markets, which have offered them sufficiently remunerative prices. Germany having thus been turned, and the voyage from the Gulf of Finland to the docks of the Thames having required less time than is necessary for the period of incubation of typhus, it is in this manner that cattle carrying within them the germs of this ruinous malady have been able to be introduced into England ….

It was estimated that a profit of the order of 100% could be made by purchasing cattle in Russia at a low price and then shipping them by way of the Baltic when the ice broke up, and it was Gamgee who had publicly made known in *The Times* that it could be profitable.

When the plague did appear in England, Gamgee's references to it in his opening address at the new Albert Veterinary College produced a scathing leader in *The Times* of October 6:

> The address of Professor Gamgee … instead of exhibiting the modest diffidence which is the best spur to future excellence … displayed the most absolute confidence. There was not one word about the uncertainties, the peculiarities, and the diversities of disease, but a round, dogmatic declaration that the speaker and the veterinary surgeons knew all about the cattle plague, and that everybody else was ignorant of the subject. Instead, again, of referring to the labour and success of former inquiries, the address eulogised nothing in the past but Professor GAMGEE'S own efforts, investigations and successes; instead of holding out the prospect of meeting disease successfully, it insisted that nothing could be done but accelerate its ravages by slaughtering every animal that was attacked. The address, in short, was contemptuous toward the past, confident in the present, and despairing of the future…
>
> After such an exhibition of the wisdom of veterinary surgeons, one may surely be permitted to ask whether the Government intend to persist in the course to which these egregious authorities have committed them. One thing at least is certain—that the measures which have been adopted have completely failed to arrest the spread of contagion throughout England and Scotland. Nor, as our correspondent, "S.G.O." lately pointed out, could it have been expected that they would be successful.

Gamgee replied (October 12th):

> I trust to your sense of fairness for the publication of the following remarks … I may be possessed of a delusion, and it may be little satisfaction to the world at large that the highest authorities in Europe on veterinary matters are not more sane than I am as to the origin of the cattle plague in Great Britain. I must confess that this is some consolation to me, and when those who are better

able to speak and write than myself put the whole question on its proper basis, it is not improbable that you, Sir, may acknowledge that there has been some method in my madness ... My object is not to mislead, but ... to direct public opinion on this subject ... I have the advantage of having predicted that this disease would reach us through the Baltic...The first cargo which came through the Baltic reached England only a fortnight before the steppe murrain was first seen in our markets, and we can now witness in this country, renowned for its beauty and the health of its stock, ravages of a character and extent unknown since the "rinderpest" died out of Great Britain last century ..."

He ended:

... I beg to state that my only reason for opposing the attempts to cure the disease is to preserve the largest amount of stock from infection. If the public, if the stockowners wish their animals to be treated, and if our Government consider it right and proper not to make a vigorous attempt to exclude the plague from these islands, I, for one, am prepared to treat and record recoveries. I have no fear of a certain measure of success and as large a measure of success as other people, and perhaps while helping to save the lives of some scores or hundreds the country will learn that the horned stock of this land cannot on such a system increase and multiply.

The Thunderer had joined the attack on Gamgee the day before:

Professor Gamgee, like the cattle plague itself, is irrepressible. As is the case, according to his own account, with animals attacked with that disease, he is wholly insensible to curative measures. We have "exhibited" to use medical language, various remedies for the delusion with which this gentleman is possessed, but it has been to no purpose whatever. He originally broke out, soon after the appearance of the Cattle Plague, in the Marylebone Institution, and he reappeared with all the old symptoms a few days ago at the Albert Veterinary College. We then endeavoured to administer a strong dose of argument, and we were hoping that the symptoms were at least suppressed, when, partly to our disappointment and partly to our amusement, the Professor suddenly breaks out again at Sheffield in the congenial air of the Social Science Association, and the symptoms are more violent than ever. We should be very unwilling to contemplate his being subjected to the same treatment which he proposes for the cattle, and though the disorder in his mind must now be in a very advanced stage we shall yet attempt once more to administer an antidote. Though it may be ineffective in his case, it may check the spread of the disease amongst those to whom he may have communicated the contagion.

The Professor holds fast, in the first place, to the theory of importation, and, when we have observed that he holds fast to it, that is about all there is to be said. He betrays a sublime indifference to all the arguments which have been adduced on the other side; and having simply recapitulated to his own satisfaction all his old familiar assertions, and having omitted to take any account of the objections raised to them, he concludes, like the Chairman of the section, Dr. Lankester, that his arguments are "unimpeachable"...The public, who are engaged in no such crusade against the Russians as this indomitable Professor, are somewhat less satisfied, and Professor Gamgee's own statement of the case would be quite sufficient to make them sceptical. As a specimen of the reasoning of veterinary surgeons it is worth careful preservation.

But despite its attacks on Gamgee, on October 18, 1865, *The Times* carried an article recommending the killing of all stock at home if they were to be moved at all.

Government turned a deaf ear to Gamgee, and, as Smith (1931) stated, it was difficult for it to do otherwise when it had before it the considered opinion of Simonds, both in his *Report on the Cattle Plague* (1857) and in the evidence which he put before the House of Commons Committee that year. There was also the report of Greenhow which independently supported Simonds's original views. Appointed adviser to the Privy Council in July 1865, Simonds was the first witness to be called before the Commission and gave his evidence, together with Ernes, on October 9. But with no end to the plague in sight, some opinion was turning in favour of Gamgee. *The Scotsman* of January 6, 1866 reported:

There is in Scotland at least a conviction that had Professor Gamgee's energetic counsels been taken at first we should not have been in these alarming circumstances. It has been the fate of Mr. Gamgee to see a little too clearly, and to speak a little too soon, for the class in whose interests

> he labours. He declared the disease imported; after much controversy, that is admitted on all hands. He urged the suspension or rigid supervision of the import of foreign cattle; the commissioner and the farmers have practically recommended the stoppage of the entrance of live cattle from abroad. He strenuously prescribed isolation and slaughter, as the sole effectual cure; and for the "killing cure" many of the farmers are now only too willing. He has long, almost if not quite alone, insisted on the necessity and advantage of a National Cattle Insurance Association, as a safeguard against the ravages of this and other epidemics; and the most influential meeting of agriculturists that has ever been held on the subject of the plague has taken up the idea, and resolved in favour of just such a system of insurance. It is something for any man to have been right, and to have pointed the way, in a time of emergency and tumult of opinion-to have silenced and convinced opponents, and to have led those who have laughed at him. This Professor Gamgee has done, and it is just that the credit he deserves should be given to him.

The *British Medical Journal* pointed out that M. Bouley, who had visited Britain to study the measures being taken against the outbreak and then recommended how to control it in France, had been made an officer of the *Légion d'Honneur*: "It is only justice to our own scientific countrymen to say, that all that M. Bouley has said and done, and advised his Government to do, in averting the plague, has been long ago, over and over again, said and done and advised by Mr. Gamgee" (Anon., 1866e).

5. SIMONDS'S EXPLANATION

Simonds was to write later (1894) that on July 4, 1865, he was called to see some affected cows at an Islington Dairy and, to his "astonishment," recognized indications of cattle plague, although Priestman claimed that Simonds had not recognized the disease when he first showed him the affected cases in Mrs. Nichols's dairy. A day or two later, Simonds made a post-mortem examination which "fully confirmed" his opinion. But it was not until July 10, 6 days after his visit to the dairy, that he reported the matter officially to the Privy Council, although *The Royal Veterinary College Annual Report of the Governors to the Council of the RASE for 1865*, stated that it (the disease) was recognized by Professor Simonds who *immediately* brought the matter to the knowledge of the Government. Asked if he was sure it was the cattle plague, according to his own words he replied that if he had not been sure, he would not have been there (at the Privy Council). Simonds was requested by Helps to look into the matter further and he produced a report of his inquiries on July 14. The Lords of the Council then instructed Helps to arrange for an Order in Council to be drafted, embracing Simonds's views. After examination by the Lords of the Council, this order was issued on July 24.

At the time, he wrote in *The Veterinarian* that the bovine disease (not naming it) "assumed at its outset such a fatal character that the attention of the Government was at once directed to it"; and an Order in Council was issued on July 24, 1865, which referred to a "contagious or infectious disorder of which the nature is at present uncertain." But in September in the same journal, he wrote: "Toward the latter part of June the malady gained a footing in the London Dairies." In August, he had written: "Immediately on the appearance of the disease the Government took action, and issued an Order in Council," reprinting an Order dated August 11. Simonds wrote in the Editorial of *The Veterinarian*: "The general symptoms resemble a malady in the ox, not altogether unknown to us, but they differ therefrom in many of its leading features." In September, an editor, probably Simonds again, wrote: "Not wishing to be alarmist, we did little more in our last number than announce the fact that a fatal bovine disease had made its appearance, and that the attention of the Government had been directed to it." But the Order of August 11 described the disease as "A contagious or infectious disorder... of which disorder the nature was uncertain [i.e., on July 24], but which has since been ascertained to be of a typhoid nature and is generally designated as Cattle Plague."

In evidence to the Commission, a cattle salesman James Brewster alleged that he had spoken to Simonds in a market and, referring to the outbreak of rinderpest, had asked him, "Have you ever seen anything of the kind before." Simonds had answered, "No, I have never seen anything like it."

Smith (1931) points out that these confusions are due to the fact Simonds did not recognize the disease when he first saw it and so did not report to the Privy Council until a week later. When he did report, he stated that the nature of the disease was uncertain, and it was not until 11 August that Government was able to state what it was. This may have been attributable to Gamgee, who was unable to make firsthand inquiries in London until July 29, when he immediately diagnosed the disease as rinderpest and was thus the first to do so. Simonds, however, perhaps not wishing to admit to the true nature of the disease, especially in view of Gamgee's previous pronouncements, did not make up his mind about it until almost the middle of August, and it was on August 11 that the first effective Order was published. But the following day after he had examined Mrs. Nichols's cattle, he met George Brown, who had recently resigned the Chair of Veterinary Science at the Cirencester Agricultural College, and confided to him in confidence that he feared cattle plague had been imported into England and, if so, veterinary assistance would be required.

Simonds explained what happened in his evidence to the Commission. On July 4, Mr. Priestman, a veterinary surgeon, and Mr. Nichols, the son of Mrs. Nichols, brought him the stomach and intestines of a cow which they believed had died of poison and asked him to inspect the rest of the animals, which he did. He satisfied himself before he left the premises that poison was not the cause, but he took for analysis some water from the well and samples from a cow which he had killed. "In the course of a day or two I satisfied myself that it would be quite a useless expense to have the water or contents of the stomach analysed." He had come at once to the conclusion that the cattle were dying "of a special disease," having many of the characters of rinderpest. In response to further questioning Simonds stated that he satisfied himself "within a very few days," but there were some points of difference to that which he had seen in Galicia which needed investigation. He ascertained on the next day that it was not an isolated case, but there were others at Hackney. He summarized: "…my attention having been first called to it on the afterpart of the 4th of July. I occupied five days in investigating the matter thoroughly to my own satisfaction, and on the 9th of July my mind was fully made up." On the morning of July 10, he informed the Privy Council. If anything, Simonds could be accused of thinking before he acted.

Simonds again explained what happened at a meeting of the Central Farmers' Club on November 6 in London at the Salisbury Hotel:

> I received information as to the affection on the 4th of July. I went to the dairy of Mrs. Nichols under the impression that her animals had been poisoned, that being what I was told me by the veterinary surgeon in attendance, and by Mrs. Nichols herself. Investigating the disease on the spot, I saw enough to make me hesitate at once as to whether that was an ordinary case of poison; and before I left the premises my mind was made up that we had a new cattle disease in the country. I did not say to Mrs. Nichols, nor to the veterinary surgeon in her presence, that I had any doubt as to the nature of the affection; but no sooner was I alone with the veterinary surgeon than I said to him … I feared we had a new disease to deal with … The next day I ascertained that other cases of disease had begun, and within five days after that I made a communication to the Government. During the five days which I spent in investigating the disease I drew up a circular, which I sent to every veterinary surgeon, in order to ascertain whether he had had any instance of such an affection; and I got returns from all parts of the country, showing that there was nothing of the kind.

Simonds had to be absent in Sussex for 24 hr after his visit to Mrs. Nichols, and during that time, Priestman had located other instances of the disease in dairies in Hackney, Islington and Camden Town. After examining these and other cattle, and conducting a number of autopsies, by July 10, Simonds was certain that the diagnosis was correct and informed the Privy Council Office.

Simonds continued:

> You can readily understand that the Government were not very likely to give full credence at once to my statement. The thing came upon them by surprise, and at first they naturally hesitated as to whether they should or should not deal with this affection ... when I went to the Government at a former period with regard to the smallpox in sheep, a gentleman whom I saw said to me, "You will excuse my asking you the question, but are you positive in this matter?" and on my replying that I was positive, he shrugged his shoulders and exclaimed, "Would to God that we had some despotic power!" For want of such power the Government could not take the bull by the horns, and they cannot do so at the present time: their hands are tied....In the present case, they have proceeded cautiously, and perhaps they have done all that, as a constitutional Government, they ought to have done under the circumstances ... I say—say without fear of contradiction—that the Government has acted nobly considering how its hands have been tied up ...

The latter elicited cries of "No, no" from the audience.

Professor Varnell, in an introductory address to students at the Royal Veterinary College in November 1865 was to add that the veterinary surgeon who attended the original deaths brought some of the organs of one which had died to the College for the opinion of the professor of cattle pathology. Varnell was present at the examination and asked Simonds if he thought that there was any analogy with the disease he had seen on the continent: "He answered at once in the affirmative, but he said, it would be necessary to see more cases before we set forth that such was the case, and that no time must be lost in prosecuting further investigation."

Bouley, who was sent over to England in the middle of August by the French government to investigate the outbreak, and who doubtless consulted closely with Simonds on the issue, reported back that although the first Orders which emanated from the Privy Councilled to the supposition that they were ignorant of the nature of the disease,

> Professor Simonds...was not at all deceived on that point ... he recognised the certain signs of the contagious typhus of horned cattle ... which he had studied eight years before in Galicia. The opinion of the learned professor, however, did not prevail ...." Reed wrote in March 1866 that Simonds was "well aware of the impending danger" and accordingly sounded an alarm, "This timely proceeding gained for him the title of "Alarmist." (Reed, 1866)

Typical of attitudes of the day was the remark of Constable in his letter to *The Times*: "Professor Gamgee adheres to the opinion, long exploded amongst educated physicians with regard to similar diseases amongst men, that this malignant cattle fever is only taken by contagion..." (Constable, 1866a) He proceeded to issue three editions of a pamphlet on the cattle plague attacking Gamgee's ideas. On June 1, 1866, he brought out another scathing and satirical pamphlet attacking Gamgee's book and the deliberations of the first two international veterinary congresses reported in it. Mocking veterinary surgeons, politicians and the law, it was titled *Observations Suggested by the Cattle Plague, about Witchcraft, Credulity, Superstition, Parliamentary Reform and Other Matters*. This rambling essay purported to draw parallels between the credulity of those who believed in witchcraft in the past and the current opinions, namely Gamgee's, on the non-spontaneous origin of rinderpest. Contributing nothing of value, Constable's works nevertheless illustrate the kind of perverse opinion Gamgee was confronted with, opinion held by educated and often wealthy people.

Ironically, Constable was exhibiting the same simple credulity of the peasants whom he condemned for their beliefs about witchcraft.

Gamgee met the attacks with courtesy, so much so that Constable wrote:

> I feel much respect for him. It is only his logical faculty I do not think much of.... *The Times* had a leading article, holding up the irrepressible Mr. Gamgee (as it called him) to the ridicule of all its readers in all quarters of the globe. Two days afterwards appeared a letter in *The Times* from Mr. Gamgee, expressed with the utmost courtesy, and not showing the smallest particle of irritation. The qualities Mr. Gamgee thus shows himself to possess, are worth all the "logical faculty" in the world....

Simonds appears to have hesitated too long after the initial outbreak at Islington, the result of which was "disaster on an almost unprecedented scale," as Wilkinson (1992) has expressed it, although Fisher (1993) considers that the delay was probably irrelevant, considering the inadequacy of enforcement mechanisms current at the time, the Privy Council lacking both power and staff to enforce compliance with orders. The only criticism Simonds seems to have come in for related to his supporting the view in the early stages of the disease that it was solely confined to animals of the bovine tribe. According to Gregory, by propagating these views so decidedly, he brought upon himself much undeserved odium from both the public and some members of the veterinary profession: "The professor, with a candour and straightforwardness which characterises him, on his ascertaining these facts by his investigations in Norfolk, at once acknowledged his error. This, I believe, may be said to be almost the only mistake of any importance that Mr. Simonds has made since the first outbreak of the disease..." (Gregory, 1866).

Gregory held that it was Simonds who first called the attention of the Government and the veterinary profession to the "vast and vital importance of the subject...He urged upon the Government again and again the necessity of prompt and vigorous action, but they were as slow to act as the people were to believe...Every act of the Government has been most reluctantly wrung from them by the necessities of the case." Since this was published in *The Veterinarian*, of which Simonds was a senior editor, it can hardly have been uncomplimentary, otherwise it would not have been published. But supporters of Simonds consider that when the disease entered Britain in 1865, it was brought by Russian cattle imported directly, a circumstance which, although predicted by Gamgee, was nevertheless impossible in 1857 when Simonds made his original report.

The outbreak did nothing to diminish Simonds's reputation, and after the disease was eradicated, a Veterinary Department of the Privy Council was formed, of which Simonds was appointed chief officer. But *The Lancet* of May 25, 1867 urged that the veterinarians be removed and medical men of high scientific standing appointed in their place.

6. A DISILLUSIONED PROPHET?

After the rinderpest broke out, Gamgee delivered a series of public lectures in London at which he did not forget to remind the obstructionists, in the forms of the cattle dealers and dairymen, how much of the trouble was due to their selfishness and greed. The impression has been given that, like many other disillusioned prophets, after the rinderpest plague debacle, Gamgee emigrated, leaving for America in 1867 and eventually turning his genius to other subjects. These other subjects were physical researches into thermodynamics and the condensation of gases, among his interests being the production of ice sheets for skating rinks and meat refrigeration (Wilkinson, 1992). But while ice sheets for skating rinks may have been a sideline, in developing meat refrigeration methods Gamgee was continuing to seek practical solutions to the problem of animal disease control by dispensing with the export of live animals. As he stated: "Since I found the cattle dealers were all opposed to foreign stock markets and the dead meat trade from abroad, I felt it was essential, in order to save this country, that we should work out the subject of meat preservation, and by that means effect the transportation of the cattle in a dead state, which would stop the importation of disease at once…" (Anon., 1873).

This was probably a misstatement of his intentions, for he went to America at the invitation of the United States Government to study the Texas Fever outbreak, originally thought to be rinderpest.[1] This involved working also for the Pork Packers Association of Chicago and studying the transport of cattle on the railroad, and while in Chicago, he brought to a successful conclusion an experiment in meat refrigeration. However, he had not been ignored by the veterinary profession at home and was still on the council of the Royal College of Veterinary Surgeons, on which he sat in the company of Simonds.

Returning from America on a visit in August 1869, he reported to the council meeting that he had been engaged in investigating diseases for the American Government and had brought back a number of photographs and exhibits for the College. A vote of thanks was accorded to Professor Gamgee "for the interest he has taken in the affairs of the College during his absence from England." But this marked the end of his connection with veterinary science, and he returned to what had been his first love, engineering, becoming a prolific inventor (one of his inventions was a washing machine) and an expert on refrigeration (Thompson, 1974).

Since rinderpest had now been wiped out (it did return briefly on two more occasions), and considerable advances in meat hygiene and cattle management had taken place, there were perhaps no challenges left in Gamgee's eyes in the veterinary world. He returned to give evidence to the House of Commons Select Committee in 1873, and again in June 1877. At the latter hearing, he stated that he had devoted his whole life to the question of cattle diseases and had "never ceased to labour at the means whereby we might secure the extermination of cattle diseases."

In response to the question whether he was still of the opinion that slaughter at the ports should be obligatory unless the dead-meat trade was adopted, he replied:

> …I naturally desire the most absolute freedom of trade consistent with the prevention of contagious diseases. Taking all the causes of cattle mortality in this kingdom together, for one animal that dies from all those combined causes, at least three British animals are killed by imported international cattle diseases which did not exist prior to the importation of live cattle [in 1842]… It is perfectly clear… such a state of things ought not to be sanctioned.

Foretold by Gamgee, who first demonstrated the method of chilling meat with cold air in 1869, beginning in October 1875, Britain started to receive chilled beef from America, with a first shipment of just over 16 tonnes, rising to over 3,886 tonnes in April 1877. His interest in refrigeration had as much to do with his ideas on the prevention of the importation of animal diseases as anything else:

> I may state that in 1865, having had rather painful experience of the difficulties in this country of securing such measures as would prevent contagion, I formed the opinion that by the conversion of the live-meat trade into a dead-meat trade, which demanded the preservation of meat, we might ultimately extinguish cattle diseases. It was that conviction that led me to devote myself, as I have done ever since, to those chemical and physical, and engineering questions, which I may say have had some influence in bringing about the results that we have witnessed." (Gamgee, 1877)

However, whereas it was considered that the transport of meat from America placed beyond all doubt the possibility of substituting a dead meat for a live meat trade, it had not led by

[1]The causative piroplasm was isolated in 1888 by Smith and Kilborne.

1877 to the necessary developments for its reception in Britain. Meat which arrived in first-rate condition was treated as if it were an indestructible commodity, and as a result, tons were wasted, "a modicum of the care which was exercised in bringing the meat across the sea would have sufficed for its safe transit all over England" (Brown, 1878).

One of his main objects in visiting the Continent in the 1850s had been to become thoroughly acquainted with the subject of contagious animal diseases and methods of controlling them in each country, because there were no such controls of any kind in Britain, despite the presence of contagious diseases and the huge losses which resulted from the importation of pleuropneumonia. Gamgee visited the countries where free trade did not exist and governments protected stock owners from disease.

As a result of what he had learnt, Gamgee began his campaign in Britain against rinderpest, but since the leading veterinary figures were still holding backward views, he was ignored. Simonds, for example, did not admit to the contagiousness of pleuropneumonia for 11 years after its importation, while Professor William Dick, principal of the Veterinary College in Edinburgh which he had set up, and an opponent of Gamgee, refused to acknowledge the contagiousness of any disease. Dying in 1866, in his will, he had inserted a clause that no person who had attended Gamgee's rival veterinary school could ever hold a post in his college. But one member of the 1865 Commission, Read, realized Gamgee's outstanding capabilities, stating that Gamgee "was a living monument of departmental incapacity and national ingratitude."

In February 1866, Gamgee published his lengthy work *The Cattle Plague*. In it, he criticized the futility of the Government measures and the absurd recommendations as to treatment, especially by medical men. He also criticized the public opposition to slaughter and control measures and Government's opposition to destruction with compensation. In his preface, he writes that he undertook the task of producing the book "from a desire to prove of service—a desire which has haunted me since first indicating the existence of the Cattle Plague on reaching London last July." Followed clearly by a dig at Simonds:

> A month had then been lost in warning the country of the great calamity which had befallen it—a calamity at that time capable of speedy and certain remedy. Not only was it then impossible to entertain any doubts as to the consequences of past negligence, but no sooner had I commenced to draw as faithful a picture as I possibly could of the disease—of its nature and prevention—than I found myself attacked on all sides …

Gamgee stated that the points on which he was mainly attacked were, at first, the foreign origin of the outbreak; the impossibility of its spontaneous origin in the London dairies; the insufficiency of any cure; the folly of establishing cattle hospitals; the importance of regulating and controlling the traffic in stock; stopping all movement of stock where the disease appeared; prohibiting imports from Holland, and establishing foreign stock markets and slaughterhouses.

He further recommended a system of national insurance or indemnity. Although the former was not adopted, a number of insurance companies offered private schemes. He also recommended the slaughter of sick and infected animals and opposed the recommendation of the majority of the Cattle Plague Commissioners of stopping the traffic in stock in the country without eradicating the disease.

Whereas, at first, up to the time of the *Report* of the Commissioners appearing in print, any suggestion of coercive measures was met with "resolute opposition," by March 1866 people were beginning to change their minds, and some were ready for anything. In January, a number had been still speaking of "non-interference," and Gamgee stated that physicians "of great eminence and influence" had spoken to him of the folly of advocating measures incompatible with the existence of any but a purely paternal Government.

As Wilkinson (1992) explained, Smith (1931) blamed Simonds in no uncertain terms for his failure to anticipate the outbreak which occurred. But one defender of Simonds claims that he deserved major credit for handling officialdom and eliminating cattle plague, while Gamgee, who Smith praised for his part, "played the minor role of publicist for control measures already advocated and fully understood by Simonds and others" (Pattison, 1990). There is no doubt that Simonds understood the contagious nature of the disease and the value of quarantine, but he failed to take the immediate action necessary when, as the Government's chief veterinary inspector, he came upon the first cases in June 1865 in Mrs. Nichols's dairy in Islington. Perhaps, his convictions earlier expressed led him not to believe what he saw.

Gamgee, however, had warned precisely of such a sequence of events as then took place.

But if Gamgee did not speak out against Simonds, this was not so regarding Brown. In his evidence before the Select Committee in 1877 on the subject of FMD, Gamgee stated that if the import of foreign cattle was prohibited, the disease would die out without any restrictions being placed upon the home cattle. A Committee member, Mr. Jacob Bright, interjected: "You are aware, I daresay, that men who can speak with quite as much authority as yourself have expressed quite an opposite view before this Committee?" Gamgee replied: "I know that the gentlemen who you state have quite as much authority as myself in the matter have been the advocates from the first of everything that has been contrary to that which I advise; and, if the history of my counsel and the history of their counsel be taken, I am quite willing to abide by the result."

"I speak of our first witness, Professor Brown?"

Gamgee: "I do not consider Professor Brown an authority on contagious cattle diseases...I do wish to be respectful to everybody; but it is a matter of vast importance as to the views that ought to prevail in this land of ours, and I believe it will be found that during 25 years I have never erred, and I challenge anybody to state one instance where, if they had followed my advice, they would not have been, at all events, four or five years in advance of what they were."

Sir Arthur Peel: "... confining ourselves for the present to foot-and-mouth disease, you will possibly not dispute the historical statements, at any rate, of the report signed by Professor Brown?" Gamgee: "I have always disputed many of the historical statements of their reports" (Gamgee, 1877).

Assessing Gamgee's role, Francis (1963) deduced he was pre-eminent in his knowledge of infectious diseases, possessing a clear understanding of the subject more than a decade before Pasteur experimented in this field in France. Gamgee had an essentially modern grasp and understanding of the whole compass of veterinary education, veterinary medicine, and veterinary science, and it took almost a century for his concepts to be brought to fruition. Most of his ideas have now been put into practice. Barber-Lomax (1968) considered him the outstanding figure of 1860s medicine and science, there being few at the time, either in the medical or veterinary professions, who had a clear understanding of the cause of epidemic disease. In the long run, it was Gamgee's methods which were employed to stamp out the disease.

12

LEGAL MEASURES OF PREVENTION IN THE 18TH CENTURY

1. THE PAPAL STATES

When rinderpest appeared at Padua in 1711, Lancisi blamed its spread on the throwing of dead animals into the Tiber. To combat its spread, he proposed closure of all stock fairs and markets, separation of sick from healthy beasts, publication of orders against the smuggling of stock, prohibition of drovers leaving their grazing areas and entering healthy places, and killing without exception all unleashed dogs.

Lancisi's proposals were at first rejected by the cardinals as we have seen, but concerned at the destruction of his cattle, the Pope eventually acted on Lancisi's advice, promulgating stringent regulations for suppression of the plague in the Papal States. From the first edict on August 16, 1713 to the last on March 24, 1714, 16 were issued and three proclamations were given; the earliest edicts incorporating the most draconian penalties to be instituted for offences relating to rinderpest until those of Frederick William I in Germany in 1716. All specified that they applied to any person of whatsoever position, degree, or condition, however privileged or exempt, both ecclesiastical and secular, and punishments for contraventions of up to the galleys for life for clergymen and death for laymen. Four edicts were exhortations to prayer.

The first edict of August 16 prohibited movement of bovine animals, cattle, and calves, from the Maritime Province and the Campagna, including Rome, and any person or group of persons moving animals into Rome or from pastures in the Maritime Province and the Campagna without a licence of the judicial and administrative council or the Lord President of the Commissariat. Cattlemen, herdsmen, and keepers, looking after sick animals, were not to mix with healthy herds. All incidents were to be reported. In case of infringement or simple omission, offenders were liable to forfeiture of stock and a penalty depending upon the circumstances of up to the galleys for life for ecclesiastics, and death for laymen. No excuse or claim of ignorance of the edict as it was published in Rome would be accepted. All would obey as if each had been personally informed. The second edict, passed 1 week later, referred to unburied carcasses and consumption of meat of dead animals. It prohibited skinning or allowing to be skinned, of carcasses, preservation and sale of the skins, and consumption of the meat. Carcasses were to be buried entire immediately in a deep pit and covered with lime. If carcasses were found unburied, persons, including those nearby, would be proceeded against. The following week, on August 30, another edict prohibited bringing or introducing cattle to fairs and was followed 3 days later September 2 by specific reference to the imminent fairs of Farfa, Madonna of the Oak of Campitelli, and Saint Eustachio in the Terra di Poli; prohibiting introduction, or allowing introduction, to them or their proximity, of any cattle for sale or any other pretext. It reiterated the maximum penalties of the galleys for life for a cleric, or death for a layman, and confiscation of the stock.

On September 15 a lengthy edict was issued additional to those of the August 16 and 23. Its first provision restricted movement of cattle along roads without a written permit. Any found being moved into the Territory without a permit were to be shot immediately by the guards and buried, those responsible for moving them being liable to 7 years in the galleys. Those moving cattle under licence could not leave the specified road, graze the animals, or take short cuts, but had to follow the main road accompanied until out of the territory by a mounted guard, who would be permanently stationed on such roads. Second, any cattle found unaccompanied were to be shot immediately and buried. Third, any straying cattle were to be shot immediately and buried, and the owners or herdsmen informed. Fourth, cattle could not be sent to pasture without a permit. Any animal brought by a single person without a licence was to be shot and the person responsible to receive 7 years in the galleys. Healthy animals could not be grazed or watered where infected animals had been.

The edict also drew attention to a bulletin addressed to all owners, herdsmen, and others, which defined where each herd of livestock could water, according to a written instruction. In the case of infringement, the stock was to be shot and the herdsmen responsible subject to three lashes without pardon, or some greater punishment if considered necessary. If local beasts were brought outside the appointed time, they were to be sent back; if foreign beasts, they were to be shot. For not watering at the appointed time, the penalty was three lashes and any other punishment deemed necessary. Any livestock watering from another Territory were to be shot, and those who brought them sentenced to 7 years in the galleys and any other punishment considered necessary. Two hours were to be left between watering each herd, and watering had to take place between sunrise and 9 p.m., outside which times the cattle could be bitten by a fly which might also have bitten an infected animal and then taken shelter in the watering ditch. It was the duty of all to inform others if they were watering stock in the same ditch. Every evening, water points were to be cleaned, including around the edges and on the banks. Shot animals were to be transported immediately in closed carts drawn by a buffalo, never by an ox, and thrown into a dry ditch made for the purpose, covered in quicklime and then with beaten earth with a depth of 12 handsbreadths.

The edict ended with a warning that it would come into effect within 3 days of being posted on the two principal doors of the castle of St. Peter in Formis and local churches, and proceedings would be taken against all persons contravening the provisions, without pardon.

On September 25, another lengthy edict was issued concerning oxen used for ploughing in view of the serious damage to agriculture foreseen from spread of the disease. Reference was made to the punctual observance of dispositions of the Pope's predecessors with regard to concessions of privilege in favor of those who practised agriculture in the district of Rome and the provinces of Lazio, Campagna, and Maritime, especially that published by Pope Paul V on October 16, 1611. Knowing how essential plough oxen were for cultivation, sale, and purchase of plough oxen to butchers in Rome or any other place, and their slaughter, even those untrained or under 10 years of age, were prohibited, under penalty of confiscation and a fine of 500 crowns. Merchants and others possessing flocks of white cows were ordered to keep them and in March to May each year raise for use in agriculture 25 calves for every 100 cows. No livestock was to be moved from the territory or any district, or from a slaughterhouse, without a written permit from the Lord Prefect of Annona. Butchers had to obtain a licence to buy animals for slaughter. It was stressed that the prohibition did not apply to the buying and selling of Camarrone and other oxen not used for ploughing, and those not fit for ploughing could be bought for slaughter.

An edict on October 26 forbade butchers to sell the meat of oxen, cows and buffaloes, and was followed on October 28 by another referring to that of August 23 forbidding skinning of carcasses and allowing the meat to be consumed "with apparent danger to personal and public health." It had come to the notice of the Food Council that carcasses were being thrown into ditches and into the Tiber. It restated that everyone must bury carcasses in pits 10 hands-breadths deep, well covered with quicklime, or under the edict of September 30, carcasses could be incinerated. Anyone skinning or throwing carcasses into the Tiber, lagoons, or other rivers, ditches, marshy places, or water would be punished as previously decreed but also fined according to the status of the person and the circumstances, one third of the fine going to any informer in confidence. On October 31, an edict ordered immunity and a reward to informers against those not burying carcasses according to the law.

Beginning November 4, emphasis switched to the protection of agriculture, with an edict on the production of seed and checking the supply of white cows. This was followed on December 9 by a reminder of the need for cultivation of the land and the ploughing the next March. On January 23, 1714, a check was ordered on wheat and fodder sown in the district of Rome the previous year and of the ground to be cultivated for future sowing for 1714. On March 24, it was noted that a large number of ploughs remained idle, with serious damage to agriculture and fear of famine as time was approaching to plough for 1715. The edict recalled that of December 9 last to continue the necessary gound culture, distribution of wheat for sowing, and disbursement of money by the Lord Prefect of Annona to the neediest farmers and other debtors who had received a diminution of tax payment. A committee was set up to propose the most suitable means to assure the harvest for 1715. The fixing of the price of wheat at 7 crowns was lifted, and in the year 1715, there would be no obligation to sell one fifth of the wheat in Rome at that price. A sum of 100,000 crowns was to be available for distribution. A board of three prelates was set up to arbitrate on claims against each other by tenants and farmers. These provisions were not intended to be used as a pretext to delay any due payments, or dues for rents or subleases, before the effects of the plague had begun.

2. FRANCE

In France, legislation enacted was of several kinds: laws of the king's Council of State; the Court of Parliament; royal orders; general orders; and later, orders of the Directory (of the revolutionary government). Later still came decrees and ministerial orders. Penalties were much higher than in Britain, ranging from fines of 200 livres in 1746 to 500 in 1775.

Until 1714, dead beasts were skinned and the carcasses simply left lying in the fields. Based on recommendations made by M. de Montigny of the French Royal Academy of Sciences, the first law to be brought in to combat the epizootic, passed by the king's Council of State on April 10, 1714, simply ordered that carcasses be buried a meter deep with skins intact. The penalty for not doing so was 100 livres, of which half went to the informer and half to the nearest hospital. For a second or subsequent offence, the fine was doubled. On September 16, the selling of cattle at markets in areas where the disease was present was prohibited, also movement of infected or suspect cattle to areas where the disease did not exist. For these offences, the animals were confiscated and the fine raised to 1,000 livres. In order to conserve stocks, in 1720 the sale to butchers of calves under 8–10 weeks old and cows of less than 10 years was forbidden.

On January 6, 1739, a royal order was passed, prohibiting the import of cattle and merchandise from Hungary and parts of neighboring countries, imports from Germany and dependent countries with permits of health being allowed. Such permits also applied to any traveller from Germany and other frontiers. A decree of March 13, 1745 provided a remedy which included the actinobacillosis cure of scraping the tongue with a silver spoon until it bled. Because the number of beasts in the kingdom had considerably diminished from

March 24, an order of the Court of Parliament now forbade selling to butchers calves under 6 weeks of age, or cows under 10, and buying such animals for slaughter. On July 19, 1746, a lengthy decree of the Royal Council was published, stating that the disease was being maintained by those who, when they suspected their animals were sick, in order to avoid a low price for them took them to fairs and markets where the disease was not present. Other animals were sold at a low price because they were sick to buyers from uninfected places, who bought them because they were cheap. The precautions to be taken were reviewed and, placing responsibilty mainly upon the authorities to ensure provisions of the law were carried out, the law decreed sick or suspect beasts must be branded "**M**" (*malade* = sick) and kept separate from healthy beasts. Branded animals found outside the area set aside for them were to be taken before the nearest Royal or Seignorial Judge and slaughtered in his presence. Certificates were required for the movement of animals from uninfected areas, and the highest penalty, 1,000 livres, was payable by authorities who issued certificates with false information, or even higher punishment could be awarded. In all cases, persons were to be held in prison until their fines had been paid in full.

When the disease threatened France again in 1771, the legislators were considerably more active than they had been 25 years before. Following its reported presence in neighboring countries, regulations were brought into effect on January 31 ordering the separation of suspect from healthy cattle. The first animal reported to die by the owner had its value paid in full, but if the death was reported by another person, the owner was fined 100 livres, of which half went to the informer. All animals were then to be confined and branded either "**M**" or "**S**" (*sain* = healthy). If cured, they were to be branded "**G**" (*guérit* = cured) and could be released. Animals were slaughtered only if the law was broken. Those which died were to be buried in their skins, 2.5 m deep, either burned or covered with lime. Infected parishes were to be isolated with signs at entry points proclaiming the infection. The sale of animals to other parishes or to strangers, or the purchase of animals, was forbidden. It was followed on March 13 by a decree of the Council of State, modeled on the law of 1745 and following a plan of precautions proposed by the veterinary school.

That these precautions were ineffective is shown by the law of December 18, 1774 which stated that the disease had now been widespread for 8 months in three provinces and was beginning to spread into others. It decreed that infected areas were to be inspected and the first 10 sick animals found slaughtered and buried with their skins. Owners were to be reimbursed one third of their value. In the Midi, a law was passed, implementing orders of the Council of State and those of the provinces making it an offence for persons to claim falsely to be able to cure the disease. In French Flanders, regulations ordering slashing the hide and burying carcasses in their skins were ignored by magistrates, and traffic in infected skins was openly conducted. On January 8, 1775, in order to help prevent spread of the disease by the movement of draught animals, the import of donkeys and horses for draught work was encouraged by payment of a subsidy of 24–30 livres for each animal sold in certain markets before February 20. After this date, the subsidy was reduced until April 20. By January 30, it had been deemed necessary to extend the order of December 18, 1774 to the slaughtering of all sick animals, the owner being paid one third of their value.

On November 1, 1775, much more stringent enforcement was ordered, and all "officers, soldiers, Cavaliers and Dragoons" were empowered to use force and to imprison those who resisted orders. They were also empowered to move sick or healthy beasts from one area to another, and Commandants in Chief had the power to assemble communities and arm them to assist the troops to execute their orders by force. The law forbade persons without a written permit to drive animals from one place to another, or transport skins or other items capable of carrying infection, in default of which the fine was 500 livres.

On January 10, 1776, all previous laws and regulations were combined in a new order. Bringing animals into healthy districts without a permit could now be punishable by imprisonment and banishment as well as confiscation of the animals. This was followed on February 14 by an order forbidding the treatment of sick animals under a penalty of 300 livres for the owner and 200 livres for the person conducting the treatment. There followed an Order of the Council of July 16, 1784, a Decree of the Constituent Assembly of October 6, 1791, and a Decision of the Executive Directory of the 27th Messidor in the Year V (June 19, 1796).

While accepting that the laws were perfectly homogeneous with a solidity of principles and in harmony with modern knowledge, Chauveau (1871) referred to their "dark chaos." The laws of 1774 and 1775, for example, ordered slaughter of sick animals with compensation being payable; but not until 1866, in referring to indemnity, did a law imply that slaughter extended to suspect animals as well. In requiring that infected animals be confined, the law of 1771 did not make it clear whether healthy animals which had been in contact with diseased were included also. The law of June 1796 (27th Messidor V) required that persons selling or buying animals in infected regions, while being liable to a fine of 500 francs, at the same time had to bring the animals before a justice where they would be killed in the latter's presence; an action which could spread the disease further.

3. GERMANY

In Brandenburg, the Elector Frederick William issued several edicts between 1665 and 1682 concerning disposal of dead cattle. A Patent of July 24, 1684 specified that the public executioner was to remove reported carcasses and bury them within 24 hr, or else the owners could do so. Such orders were extended to the Duchies of Magdeburg and Pomerania and the Principality of Halberstadt by the Elector's edict of April 22, 1689. These edicts had no penalty attached for noncompliance, but the threat of the spreading epizootic caused King Frederick I to issue stringent protective measures on December 7, 1711:

- No cattle could be imported into Brandenburg from Poland, Silesia or East Prussia, unless quarantined for 8 days at the border and no animals having died during that period.
- In East Prussia, any seller had to produce an affidavit certifying the place of origin of the cattle had been free of disease for 3 months.
- No sick or suspect cattle could be placed on common pasturage.
- If any cattle died in stalls, the others were not to be put out to graze without an affidavit from the local authorities certifying that there had been no reoccurrence of the disease in such stalls for 3 months. Without such affidavit, neighbors should keep such cattle from the fields. If the owner persisted in pasturing them, the neighbors were permitted to shoot them and bury them at once.
- All cattle which died were to be buried 5 ells ($\approx$3 m) deep with the hides and fat intact and the carcasses covered with quicklime.
- Places where cattle had died, whether in field, or stall, were to be disinfected. In fields, an area of 1–2 square rods (25–100 m^2) ploughed under. Buildings and utensils were to be washed with hot lye.
- Infected animals were to be isolated immediately from others.

The edict also directed that printed questionnaires designed to throw light on the nature and symptoms of the disease be circulated in infected areas among farmers and skinners, and the answers printed in a bulletin. At least one dukedom, that of Braunschweig and Lüneburg, followed up with its own edict the following year as a preventive measure. This edict appeared to refer to anthrax, specifically mentioning past disease among horses and pigs, and it specified rules for horses rather than cattle. But this only suggests that the most recent experiences in the area had referred to anthrax. Faust (1797) claimed, however, that these first instructions were useless as most people could not read. In 1714, the new king, Frederick William I, reaffirmed the instructions in order to "prevent the ruin of the land." In 1716, he issued a further law requiring that cattle imported into Prussia be marked on the right horn with the letters "**FW**," and accompanied by a certificate attesting that they were healthy and from a healthy region. If the disease appeared, it had to be reported to the authorities and sick animals separated from healthy. Burial areas were forbidden and taking animals to market prohibited. Cattle slaughtered by order had to be branded on the horn, watched for 3 days, and their carcasses inspected. But above all, punishments for infringement of the law were branding and forced labor for life, or death by hanging after flogging. In extenuating circumstances, offenders could be branded and subject to hard labor for life. A further edict ordered the cordoning off of infected areas with royal troops.

By September 1717, it was reported that the disease had ceased in East Prussia and neighboring countries, and restrictions on movement and sale of cattle were lifted. But a month later, the King regretfully noted that the disease still occurred in Holstein and Bremen to the west, and restrictions were reimposed limiting importation from these areas under a royal edict of October 11. In September 1719, the Dukedom of Trier issued an edict noting that the disease was raging and daily worsening. Referring to the need first to ascertain if it was contagious, methods were then announced for the control of smuggling and disposal of infected meat. All livestock had to be kept separate and dead animals buried unskinned in remote places at least 1.5 fathoms deep. Half the value of slaughtered animals would be paid, and the poor were required to pay only half of the knacker's fee.

Further edicts were issued by William in 1720, 1721, 1722, and 1724, followed, on December 24, 1729, by a major summary of all procedures on prevention, recognition, and remedies. It directed that whenever the disease was reported in the neighoring territories of Poland, Frankfurt-on-Main, and the Palatinate, provisions of existing decrees were to be activated immediately. All cattle deaths were to be reported to the village magistrate, and by him to the county commissioner. If failure to do so resulted in spread of the disease, the penalty was corporal punishment with hard labor on fortifications. An important addition was the requirement of a postmortem examination of the viscera by a physician as soon as the authorities had been informed of the death of an animal, to determine if death was from contagious disease. If contagious, healthy cattle were to be separated and removed from infected barns, as were all feed, litter, and manure which had been in contact with infected animals. Herders and caretakers were not to wear wool or fur, but clothes of linen or leather. After deaths had ceased, these men were not to care for healthy cattle until they had decontaminated the stalls, soil, feeding, and drinking equipment, as well as themselves and their clothes.

County commissioners and local magistrates were made responsible for preventing persons of affected villages traveling to other villages. Those whose cattle were first affected were likewise to be kept isolated from their neighbors. In infected and surrounding villages, dogs were to be chained, and those found loose shot. The county commissioners were required to post peasants as guards around infected villages to keep them isolated, and all egress was prohibited. Food and fuel for the villages, and cattle fodder, were to be furnished by the county, if necessary. Guard duty was rotated among the surrounding villages. Daily reports were to be made through provincial governments to the College of Health in Berlin, itemizing the number of cattle before the disease, number sick, number dead, number recovering, and number uninfected. After the disease had finished, buildings had to be disinfected. Any feed remaining, such as oats and barley, could be fed to horses or sheep, as it was not observed that the disease was infectious to other animals, but not transported or used near cattle. Before the guard could be removed and a quarantine lifted, the local residents had to certify in writing that all disinfecting procedures had been carried out and recovered cattle quarantined for 14 days.

Another edict was issued in 1730, followed by one on March 26, 1732 stating that previously published remedies had been ineffective and publishing those being tried elsewhere on the continent. It was noted that those cattle which recovered did not become infected a second time.

With the coming of spring and the market season, the government of Pfalz (middle Rhine) issued an edict on March 6, 1744 urging that all cattle transports be inspected and cattle brought in without a pass stopped and suspicious animals seized. On April 25, another edict reopened trade for cattle coming from areas free of the disease for 3 months if accompanied with a pass signed by two community superintendants, noting that the disease was still reappearing in places where it had formerly ceased. Sanitary measures were stressed, cleaning stalls and all items, changing clothes after contact, keeping healthy animals away from burial sites, and filling hollows in the ground where water collects and infected animals had passed over. Finally, it called on border officials to be vigilant and the notice to be displayed in public places.

The German princes produced numerous other edicts of their own. Between 1745 and 1746, the Duke of Braunschweig and Lüneberg issued 22 edicts and official instructions to combat the disease, and it does not seem to have appeared in the principality until September 1750. Whereas instructions referring to a decree of 1716 related to the obligatory use of setons, and the use of various quack remedies, no doubt the restrictions on import of livestock, meat, and skins, and the control of stock movement, contributed in no small measure to preventing the disease appearing sooner. Edicts and other instructions summarized in Appendix II, the first issued on March 15, 1745, show a considerable effort on the part of the Duke to protect his principality. Many methods he insisted upon were to be repeated into modern times: good feeding, cleanliness, warmth, isolation of sick animals, adequate burial of carcasses, personal hygiene, and so on. Whereas the suggested potions were known not to be effective, they were ordered in the absence of any known alternatives (Figure 1).

On March 12, 1745, the Count of Oldenburg, King of Denmark and Norway, issued an Ordinance ordering the shooting of all cats and dogs on infected farms and confinement of pigs, sheep, and geese, which, along with all other animals and people in contact with them, had to be washed all over with vinegar water. A place could be declared free of infection after 14 days. Two intelligent men were to be appointed to supervise slaughter of cattle and calves to ensure that they were not infected. An owner was obliged to show that animals for slaughter had not been sick within 48 hr, and a dealer within the past 3 months. They were not to be from suspected areas and had to be kept for 48 hr prior to slaughter in the presence of a supervisor.

On October 4, 1746, the Dukedom of Trier re-enacted a Carrion-pit Ordinance of September 1705. Carrion-pit supervisors, and particularly the public, had objected to burying carcasses, and it was ordered that contraventions should be punished but payment made for burying cattle and horses. Each place had to have its own carrion pit, and supervisors should not throw carcasses into water but bury them deeply. The Ordinance also drew attention to the fact tethered animals could be attacked by wolves and other carnivores, causing them to break free and roam uncontrolled.

In Hannover, an edict was issued on February 14, 1756, requiring slaughter and burial of all infected cattle and those which had been in contact with infected animals. An ordinance

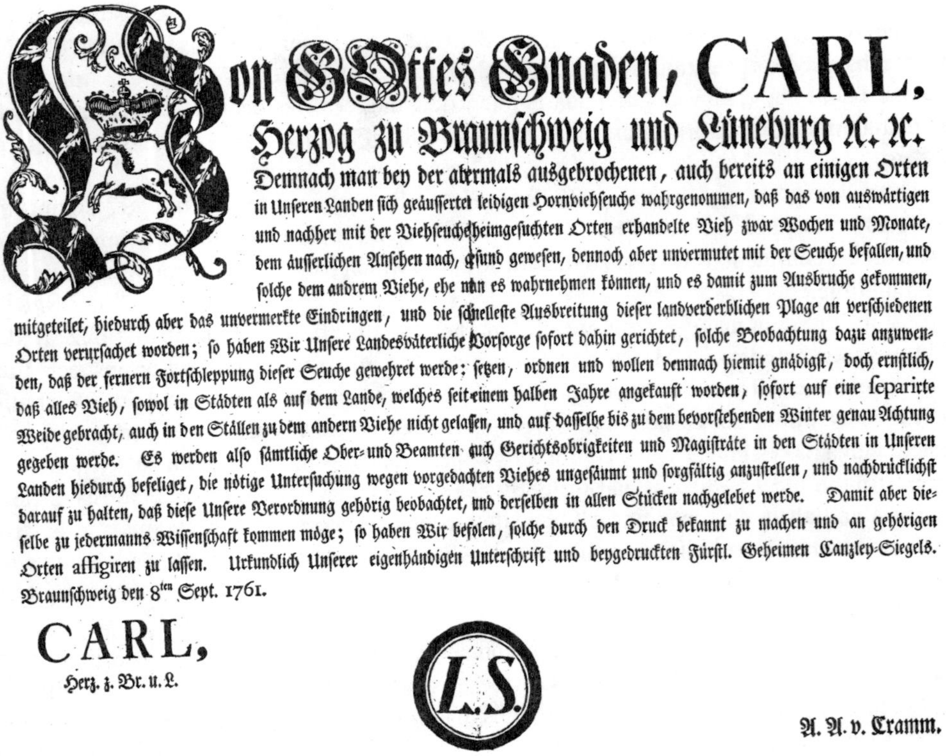

Von GOttes Gnaden, CARL,
Herzog zu Braunschweig und Lüneburg rc. rc.
Demnach man bey der abermals ausgebrochenen, auch bereits an einigen Orten
in Unseren Landen sich geäussertet leidigen Hornviehseuche wahrgenommen, daß das von auswärtigen
und nachher mit der Viehseuche heimgesuchten Orten erhandelte Vieh zwar Wochen und Monate,
dem äusserlichen Ansehen nach, gesund gewesen, dennoch aber unvermutet mit der Seuche befallen, und
solche dem andrem Viehe, ehe man es wahrnehmen können, und es damit zum Ausbruche gekommen,
mitgeteilet, hiedurch aber das unvermerkte Eindringen, und die schnelleste Ausbreitung dieser landverderblichen Plage an verschiedenen
Orten verursachet worden; so haben Wir Unsere Landesväterliche Vorsorge sofort dahin gerichtet, solche Beobachtung dazu anzuwen-
den, daß der fernern Fortschleppung dieser Seuche gewehret werde: setzen, ordnen und wollen demnach hiemit gnädigst, doch ernstlich,
daß alles Vieh, sowol in Städten als auf dem Lande, welches seit einem halben Jahre angekauft worden, sofort auf eine separirte
Weide gebracht, auch in den Ställen zu dem andern Viehe nicht gelassen, und auf dasselbe bis zu dem bevorstehenden Winter genau Achtung
gegeben werde. Es werden also sämtliche Ober- und Beamten auch Gerichtsobrigkeiten und Magisträte in den Städten in Unseren
Landen hiedurch befeliget, die nötige Untersuchung wegen vorgedachten Viehes ungesäumt und sorgfältig anzustellen, und nachdrücklichst
darauf zu halten, daß diese Unsere Verordnung gehörig beobachtet, und derselben in allen Stücken nachgelebet werde. Damit aber die-
selbe zu jedermanns Wissenschaft kommen möge; so haben Wir befolen, solche durch den Druck bekannt zu machen und an gehörigen
Orten affigiren zu lassen. Urkundlich Unserer eigenhändigen Unterschrift und beygedruckten Fürstl. Geheimen Canzley-Siegels.
Braunschweig den 8ten Sept. 1761.

CARL,
Herz. z. Br. u. L.

L.S.

A. A. v. Cramm.

Figure 1. An edict of the Duke of Braunschweig and Lüneburg of September 8, 1761, issued in poster form, stating that the cattle plague having broken out in some places, it is ordered that all cattle, both in cities and in the country, bought within the past six months, be immediately placed in separate pastures, and other livestock not allowed into the stall until winter.

of February 10, 1770 forbade until further notice the import of cattle from all suspect lands, particularly Oldenburg, Delmenhortischen, Jevern, Westphalia, and the Münster, Osnabrück, and Bremen city areas.

Following widespread resurgence of rinderpest in Europe in 1766, a renewed Prussian edict was issued in 1769 upon which the dukedoms again based their own controls. Summarized by Kilian (1934), the duchy of Mecklenburg-Strelitz issued a series of ordinances beginning March 12, 1770 and continuing until 1778. An exception to the Prussian law was that contravention did not carry the death penalty. To be announced from the pulpit as well as in a gazette, controls began with a requirement of health passes for all imports of cattle and untreated hides, followed by an 8-day quarantine on imports from clean areas. This was followed by prohibition of imports, cattle movements and markets as the situation worsened. On February 20, 1771, army cattle and calves were forbidden entry, and on July 13, pastors prohibited from vistiting communities where controls were in place, their duties to be undertaken by the schoolmaster. But it was not until early March 1776 that the disease entered the dukedom, and a detailed ordinance was issued on March 6, but by April 22, it was declared free of the disease, although further ordinances and instructions appeared until 1778. On May 7, 1778, an ordinance recommended inoculation followed by a confirmatory one on December 17.

At Hildesheim in November 1775, provisional regulations were issued on the actions to take if rinderpest appeared. Infected villages were to be isolated and necessities such as beer and bread collected at a boundary mark by horsemen without contact taking place. All cats in infected places were to be killed, and dogs shot, including hunting hounds, if necessary. This was followed by an edict the following April referring to edicts of the previous October 12, November 5, 14, and 18 exhorting compliance, and noting that in some places carcasses were not being buried deeply enough. A penalty of 50 Reichsthalers was instituted for officials who failed to ensure that such places were covered up and fenced off.

On May 13, 1780, Friedrich August, Duke of Saxony, Jülich, Cleve, Berg, Engern, and Westphalia, noting that cattle were dying in some places, issued a lengthy mandate beginning with a section on "cures," followed by an updated comprehensive compilation of the legislation referring to laws of November 21, 1712, December 23, 1726, January 16, 1724, April 8, 1732, October 16, 1745, November 29, 1746, November 29, 1749, January 20, 1753, November 29, 1753, and November 11, 1765. Among the many provisions, the import of garden plants and onions from neighboring infected areas was forbidden, and travelers required a pass stating that they were not from infected regions. The penalty for contraventions of any provision was 20 Thaler or 4 weeks' imprisonment, or both, or some more severe punishment.

In Bavaria regulations were passed on July 18, 1788, requiring inspection of all cattle and horses, and every person in all districts was to provide a list of his cattle. All cattle imported from other countries were to be inspected, and all cattle sent to fairs or markets inspected prior to entry into the fair or market. Grazing infected animals in open fields was forbidden, as was night pasturage. Cattle were never to be allowed to drink from dirty or standing pools. Diseased cattle had to be separated from healthy and their horns painted red, while all cattle in an infected district were required to be marked with the first two letters of the name of the place. No cattle were to be allowed into or out of an infected place. All carcasses of dead cattle had to be buried 2.5 m deep by specially appointed men who were not to approach sound cattle afterwards. The dung of infected cattle had to be burnt and no portion of a dead animal discarded. All dogs, cats, or pigs, which may have eaten anything from diseased cattle, were to be killed immediately and also buried 2.5 m deep. Only public slaughterhouses were allowed under supervision of properly appointed officers, and the secret sale of meat was prohibited. There were also strict regulations on thorough disinfection of premises once the disease had passed. Those cattle which recovered were only to be allowed on pasture after a period of 6 weeks. A further edict was issued in June 1796. Legislation to combat the disease was issued at Münster in October 1774, followed by a further edict in December 1779 stating that the measures of 1774 were being invalidated by persons not reporting infection of their stock. It imposed a prison sentence for failing to report the disease, with no compensation for loss, and specified prisoners were to be transported in linen frocks so that they did not spread the disease.

4. OTHER COUNTRIES

From 1714, an ordinance of the Governor-General of the Austrian Low Countries (Luxembourg) allowed importation of only those cattle accompanied by a certificate of health. If they were then seen to be sick, they were to be killed immediately and buried in their skins. When a contagious disease appeared, animals judged sick and incurable were to be slaughtered and not skinned. Those judged curable were to be kept confined until the healthy had been taken to another part of the farm. Importation of meat and skins, fresh or salted, was prohibited without a certificate.

A law, equally as drastic as some in Germany, was passed in Sweden on November 15, 1750, following its third epizootic since 1720 (later claimed to be anthrax), providing the death penalty for anyone guilty of malicious spreading of the disease, including the person's associates.

Equally strict laws were passed in Dutch Flanders and Brabant, where slaughter was mandatory from May 10, 1769. This was followed by a law of January 11, 1770 to the effect that if the disease was found on a farm, all animals, both sick and healthy, were to be slaughtered, as well as all dogs and cats found in the owner's house. Twenty-four hours after publication of the presence of the disease in a village, all domestic rabbits were to be killed, parents and masters being responsible for ensuring their children and servants, respectively, did the same.

In Denmark Royal decrees were issued from Christiansborg Castle in 1775, 1777, and November 30, 1778. The following May 5, a further decree was issued regarding shipping requiring all skippers arriving at any place in Denmark, Norway, or the Danish kingdom, to have a certificate stating that there was no infection in the provinces they had come from, or, where infection was present, not in the past 6 weeks within 8 km of the place they had come from, and the goods had not come from infected places. No goods, including fox and hare furs, were to be allowed in from an infected place, and no cattle, sheep, pigs, dogs, or other live animals except horses.

As elsewhere on the Continent, in Hungary many local decrees were issued by counties, towns and villages. The Royal Governing Committee issued instructions on November 8, 1735 concerning protection against the cattle plague, and eight decrees up to 1768. This was followed by a consolidating law and a law issued by Queen Maria Theresa. On October 20, 1789, the Royal Governing Committee issued instructions on the separation of animals, treatment, burning the skin of the abdomen with a red-hot iron, rowelling the dewlap with hellebore root, burying of carcasses, and prohibition of slaughtering.

5. LEGAL MEASURES IN BRITAIN

In Britain, a 1666 Act of Charles II solely for the encouragement of trade, prohibited import from Ireland "and other parts beyond the seas," "lean or fat cattle, dead or alive ... and fish taken by foreigners ...," and was not repealed until 1765 (5 Geo. III. cap.10). Although apparently forgotten by the beginning of the 18th century, it probably played a significant part in preventing the introduction of disease in the latter part of the 17th century and the early 18th century.

Policy in the 18th century was decided by the Privy Council and established by Orders in Council, the Privy Council calling upon advisers such as notable physicians when it felt that occasion demanded, with Parliament playing a minor and subsidiary role in decision-making. In 1746, the Commons spent a month in fruitless discussion on action to be taken regarding the rinderpest outbreak, a simple enabling bill finally passing responsibility back to the Privy Council. Later legislation rarely did more than provide statutory authority for existing Orders in Council. Although central government received reports from the counties and was receptive to local pleas concerning the practical implications of policy details, local initiative had only a small input in the development of policy. Government did not act speedily, the first law with regard to the outbreak of 1745 being on March 12, 1746 (19 Geo II c.5) a year later, based on recommendations of Bates in 1714. The seemingly strict provisions of Bates, especially that of slaughtering all affected cattle, would have appeared to be all that was necessary to combat the outbreak, but as people sought ways to circumvent the law, additions and amendments had to be added for 12 years before the disease burnt itself out.

Opening with the preamble "Whereas a contagious Distemper now rages amongst Oxen, Bulls, Cows, Calves, Steers, and Heifers, in several Parts of this Kingdom, which if it is not timely prevented, may end in the entire Destruction of such cattle," the Act continued with the observation that hitherto measures to cure animals had tended to propagate the disease, for, while the sick were being treated, it spread among the sound. Hence, as soon as cattle shewed any signs of the distemper, they were to be removed to some place distant from the rest and shot, or otherwise killed, "with as little effusion of blood as may be." The bodies were then to be immediately buried with the skin and horns, with at least 1.2 m of earth above them, the hides first having been slashed to render them useless. All hay, straw, and litter, which the infected animals had been in contact with, was to be burned, and persons who had been near infected cattle could not go near sound cattle in the same clothes.

Buildings where infected animals had been housed were to be cleaned, and wet gunpowder, pitch, tar, or brimstone burnt in the buildings to fumigate them, after which, they were to be washed with vinegar and warm water and kept empty for 2 months. Cattle which recovered could not be placed with sound cattle for 1 month, and then only after they had been well curried and washed with vinegar and warm water.

It was prohibited to buy, sell, or expose for sale, milk, meat, or entrails of any infected cattle, or to feed the same to pigs, calves, lambs, or any other animal. It was also forbidden to drive infected cattle to any fair or market, either in or out of the same county, or to any place away from their own ground where they were infected, and they could not be moved for 1 month if infected cattle had been there.

As soon as the disease appeared, persons concerned had to give notice immediately to the constable and the churchwardens and overseers of the town or parish, or to an inspector appointed by the Justices of the Peace. It was an offence to obstruct any officers in the execution of their duties. The officers were required to make an immediate inventory of infected and noninfected cattle when the disease was reported, and repeat it weekly, ascertaining that infected animals were killed and buried according to law. For every infected animal killed according to these rules, compensation was to be paid from the treasury of half the value of the beast, up to a limit of £2, or 10s. for calves.

The penalty for contravention of the Act was a fine of £10, of which half went to the informer and half to the poor of the parish, or 3 months' imprisonment. Rules and regulations were to be read publicly in all places of worship on the first Sunday in every month immediately after prayers, and every minister had to keep a copy available for consultation by the public.

It being questioned whether the King had the right to make such a law, a further Act in October (19 Geo II) empowered the King to make rules and regulations for, "... whereas Remedies for preventing the spreading of this dreadful Calamity cannot so effectually be provided ...."

This was an empowering Act, confirming many of the provisions already current: prohibition or preventing the buying or removing of cattle in specified places; burial of hides and skins with cutting and slashing of the hide; burial of animals which died or were killed; prohibition of import and export of cattle and products. The requirement for rules and regulations to be read publicly in church on the first Sunday in every month immediately after prayers was amended in 1746 to Sundays at the Minister's discretion.

In November 1746, a further Act clarified parts of this previous Order, doubts having arisen concerning one of the clauses. It was confirmed that the prohibition on driving, removing, sale, disposition, or other use of any cattle was "whether infected or not infected." It also empowered JPs to appoint salaried inspectors of houses, buildings, grounds and cattle, as assistants to the constables, churchwardens and Overseers of the Poor (20 Geo. II).

An Order of November 25, 1746 nominated certain JPs in the County of Middlesex to pursue the same methods used in 1714 and appoint inspectors. Nomination of JPs was to be followed for every Division without the appointment of inspectors due to cost. JPs were empowered to make any other rules or regulations as deemed expedient.

All expenses were to be paid out of the County rates or stock, and constables, churchwardens, Overseers of the Poor, and such others appointed by JPs were to make weekly returns of infected and noninfected animals in areas where the disease was reported, to oversee destruction and burial and submit claims for compensation. Payment was to be £2, or 5s. for a calf. They were also obliged to see that owners divided their herds into small groups at reasonable distances from each other as much as the land would permit; and dung of infected cattle was collected and deeply buried. All hay and straw used by, or near, or breathed upon, by infected animals, was to be burnt.

No person attending infected animals was to go near healthy animals in the same clothes. Buildings used by infected animals were to be cleaned and wet gunpowder, pitch, tar, or brimstone, burnt in several parts of the buildings, which were then to be frequently washed with vinegar and warm water, and cattle kept out for a minimum of 2 months.

The JPs were obliged to meet in groups in infected counties at least once a fortnight.

The disease having spread beyond Middlesex, royal assent was given to an Act on February 13, 1747 which came into effect on March 12, empowering the Crown to issue through the Privy Council rules and directions to prevent further spread:

- Cow-keepers must shoot infected animals and bury them 4 feet deep covered with lime with the hide slashed.
- All hay and litter used by diseased animals to be burned.
- Infected sheds to be washed and disinfected with burning sulfur, then washed with vinegar and water and not used for 2 months.
- Convalescent animals not to be mixed with sound animals for 1 month and then not until cleansed with vinegar and water.
- Flesh and entrails of diseased animals not to be given as food to other animals.
- No man whose herd is infected to drive any cattle beyond the boundaries of his farm.
- Local authorities such as churchwardens, overseers, constables, or cattle inspectors, who may be appointed, to be charged with the execution of the order and to report to each meeting of the JPs and furnish exact returns.
- Local authorities to persuade owners to divide up their herds into separate parts and to oversee the burying of dead animals and also all infected dung.
- Cattle traveling on roads to be stopped and examined.
- Houses, buildings, or yards used for cattle to be kept clean.
- Compensation for slaughtered cattle to be confirmed at the rate of £2 per beast, or 10s. for calves.

The requirement of the Order issued in March 1746 for reading Orders in Church was reissued on December 17. A second Order of September 8, 1747 came into effect from December 27, the disease having greatly increased "thro' neglect of officers appointed to enforce former orders." Through the neglect of officers and others, infected cattle had been driven to fairs and markets, and infected hides taken to them, by means of which the disease had been continued and spread. Very few or no inspectors had been appointed by the JPs. With the preamble "His Majesty… being desirous to comply with every Thing that may be for the Ease and Convenience of his Subjects, as far as the present circumstances of the said Distemper will permit…," it was therefore ordered buying, selling, and moving of lean horned cattle in England was prohibited for 3 months, later continued until April 8, 1748. The exceptions were fat cows and oxen for immediate slaughter, but they were required to be accompanied by a certificate whence the animal was brought, signed by a JP specifying the name and place of abode of the owner, and that on the oath of some credible person, it appeared that the beast and the herd out of which it had come was, and had been for the previous 6 weeks, entirely free from cattle plague. A similar certificate was required for raw skins. The oath could be administered by a rector, vicar, or curate, and attested by a churchwarden or overseer of the poor of the parish if no JP was resident within 4.8 km.

> And for the greater Certainty that the fatted Cattle are not infected, no Person is to buy, sell, or offer to Sale, any such fatted cattle, without delivering a Certificate, under the Hand and Seal of one or more Justices of the Peace for the County, Riding, Division, City, Liberty or Town Corporate, that the said Cattle and the Herd or Herds out of which the same were taken, were at that Time, and had been for the Space of two Months before the Date of such Certificate, entirely free from the said Distemper, and in the Possession of the Person selling the same. And for the better preventing the spreading of the said Distemper, by the Sale of lean Cattle, under the Pretence of their being fatted and ready for Slaughter, his Majesty orders that all Butchers, or others, who buy the said Cattle, shall cause them, to be kill'd within the Space of ten Days from their being so bought…

Churchwardens, overseers of the poor, constables and inspectors were required to stop any cattle or skins being moved without a certificate, and to take the offender before a JP.

"And whereas it has been found by experience, that nothing can contribute so effectually to prevent the spreading of the said distemper, as by killing the cattle when it first appears upon them, before it begins to be infectious, and communicates itself to the rest of the herd, so it is of absolute necessity that inspectors be appointed for that service…." The Order goes on to say that JPs are empowered to appoint one or more inspectors in their respective divisions where any infection is present, or shall appear, for the purpose contained in the Order of Council of March, who are to take care that such infected cattle are killed as soon as they appear to be infected, and to carry out any other instructions.

It was also prohibited to offer for sale milk, carcass, flesh, fat, or entrails, or any part of any infected beast, or to feed same to any other species of animal. Reiterating from the Order of March that carcasses must be buried complete with the hides slashed, it is promised that owners shall be paid a sum not exceeding 10 s. for the loss of each animal so buried which has died of the disease, and this money would be paid on demand to His Majesty's Treasury in Lincoln's Inn, without any fee or deduction by the solicitor, either by bringing the claims in person or by post.

On January 5, 1748, the Act of 1746 was continued and amended, empowering JPs to order reasonable salaries to be paid to inspectors. *A New Order of His Majesty in Council* was published on March 22, 1748, continuing to April 8 when they were to cease extant orders due to expire on March 24, after which new rules and regulations would replace them. Five Orders having been issued by April 16, the Lincolnshire JPs in the Division of Holland issued a broadsheet for public display summarizing the main points of each Order. One Order prohibited any ox, bull, cow, calf, steer, or heifer, whether fat or lean, to pass the Humber or Trent northwards nor the Rivers Severn and Dee westwards, JPs being ordered to cause a strict watch to be kept on all bridges, fords, etc. Where cattle were put out to pasture and the pasture was insufficient, they could be removed to alternative pasture in a different parish under a certificate stating they were in good health. It also removed prohibition on the removal and sale of fatted calves, provided they were free of disease. Where cattle were stopped, during examination of the offenders, churchwardens were empowered to place the cattle under the care of a proper person, and if found not to be infected but should not be removed, then they must be driven back the way they had come. Instead of requiring the oaths of two constables to the value of cattle shot and buried to entitle the owners to compensation, the oath of one would be sufficient if two were not available, provided the owner called a credible witness to make oath, whose credibility was certified by the churchwardens of his parish.

A further Order of October 27, 1747 permitted the removal of sound cattle from over-grazed fenlands to the owners' respective farms and enclosures, even though some animals in the fenlands might be infected, provided the owner entered into a cognizance of £10 that the animals so removed would be kept isolated for 2 months. It was also agreed that cows could be driven to a bull if both were certified healthy. It also stated carcasses must be buried within 8 hr after death instead of 3 hr. Yet another Order included a provision prohibiting the killing of cow calves by butchers in London, Westminster, and Southwark; but this allegedly proved so great a hardship to breeders that it was soon repealed. It also prohibited the selling of cattle before they had been in the seller's possession for 40 days. The latter clause was found to be ineffectual and amended in 1749 to include the provision of a certificate signed by a JP or Commissioner of Land Tax, issued gratis, specifying details and numbers of the cattle, place to be sold, origin, and other details: "and for the more effectual preventing the spreading of the said infectious distemper by persons, commonly called jobbers, who, occupying little or no grazing land, buy up cattle in infected places, and soon after dispose of the same …." It had to be sworn that the buyer rightfully occupied land sufficient to graze and keep the number of beasts on the certificate, over and above those already possessed. On arrival with the animals, the certificate had to be presented for inspection and marking. Butchers and others buying fat cattle for immediate slaughter were exempted, and the Justices were empowered to license any person without such certificate to buy and dispose of animals who entered into a recognizance with two sureties in the sum of £100 (23 Geo. II).

A lengthy Order issued on March 22, 1748 replaced that of 1746 with effect from April 8. Upon finding signs of the disease, all cattle owners or others were to separate immediately the infected animals and cause them to be shot or otherwise killed "with as little Effusion of Blood as may be," and the bodies buried entire within 24 hr of the disease being identified, 1.2 m deep with the hides slashed. Any animals which died were to be buried within 8 hr.

Overseers of the Poor in parishes and towns were responsible for reporting outbreaks to the Clerk of the Peace of the County, and every High Constable was obliged to signpost high roads in infected districts notifying the infected places. When the disease had ceased in an area for 40 days, the Minister, or Overseer of the Poor, had to sign a certificate to that effect and forward it to the Clerk of the Peace for deletion of the locality from the list.

No one could drive cattle within 1.6 km of an infected place, unless from one pasture to another, or to water or slaughter, but then not for more than 1.6 km. If outside 1.6 km of an infected place, a certificate was required to drive cattle for more than 8 km. No cattle could be moved between 8 p.m. and 4 a.m.

The sale of raw hides was forbidden without a certificate issued by a JP or a Commissioner of Land Tax stating that they were free from infection. JPs were empowered to prevent the buying and selling of cattle at fairs and markets where a danger of spreading infection was perceived. The buying or selling of milk or any part of an infected animal was prohibited, or feeding same to any "hog, calf, lamb or any other animal."

JPs were empowered to appoint salaried inspectors to examine houses, buildings, grounds, and cattle, where infection was likely to be.

Cattle taken to market required a pass issued by a JP or other competent authority for which the owner had to swear on oath the cattle were, and had been, free from the plague for a month. Passes were also required for the transport and sale of hides, except hides and horns of diseased animals. The latter had to be destroyed, for which compensation of 10 s. per hide was payable.

One anonymous pamphleteer claimed that the law was defunct in that persons were not slaughtering their cattle, but claiming and receiving compensation for those which had died or were killed when about to die (Anon., 1748c, n.d. [1750]). He alleged that one of the causes of the continuation of the disease was that not only had many clergy failed to carry out the instructions of the Order of Council, but even those who had did not act upon them. This, the writer considered, was in some cases due to the feeling among the clergy that it was a profanity. Animal plagues were regarded as evidence of Divine wrath incurred by the sins of the people and must therefore be submitted to with due humility. Another problem was that many magistrates were opposed to the destruction of affected cattle and so defied the law by pretending to a discretionary power. Furthermore, the writer claimed, false certificates alleging that destruction had taken place were frequently submitted by JPs and compensation paid. They supposedly even connived at the removal of hides from animals which died, leaving the carcasses unburied, or even having them thrown into rivers. But as a result of the measures in place, the plague was extinguished where they were applied with vigour, and at the end of August 1748, several counties were certified clear of the disease. Elsewhere, it lingered on to erupt again suddenly, spreading more rapidly than ever.

6. MEASURES IN YORKSHIRE

In the North Riding of Yorkshire, the General Quarter Sessions issued 15 local Orders up until February 1749 (Table 1).

As from April 19, 1748, Chief Constables in the North Riding were to be appointed Inspectors, making a report from time to time to their JPs of the decrease or increase of the disease when present. On April 29, they were ordered to set a guard at New Malton Bridge to prevent cattle crossing from the East Riding without a proper certificate issued if the disease was not within 8 km from whence the cattle were driven. This was extended to Kirkham Bridge on May 17, when all cattle fairs and markets were prohibited at Helmsley and Kirbymoorside.

Table 1. Orders issued in the North Riding of Yorkshire.

Date	Quarter Sessions	Places Affected by the Order
1748		
Apr 19	Thirsk	All
29	Brompton	New Malton Bridge
May 17	Hovingham	Rydale, Kirkham Bridge, Helmsley, Kirbymoorside
27	Brompton	Pickering-Lyth, Yeadington Bridge, Ayton Bridge
Jun 9	Hovingham	Pickering-Lyth
27	Hovingham	Lingham, Topcliff
Oct 4	Thirsk	North Riding
Nov 9	Sand-Hutton	Stamford, Buttercramb, Howsham, Kirkham, New Malton and Yeddingham Bridges; Bridges on River Derwent
21	Hovingham	All; New Malton
Dec 5	Whitwell	All
8	Thirsk	All
9	Bedale	All; Bridges on Yore and Swale
12	Guisbrough	All; Langbarugh, Whitby Strand
13	Brompton	Pickering Lyth, Harford Bridge
1749		
Feb 20	Guisbrough	Lanbarugh

From May 27, watches were placed on Yeadingham and Ayton Bridges. On June 27, the plague was reported raging at Lingham, leading to the closure of Topcliff Fair. In October, it was still raging in the East and West Ridings "as well as in other parts of this Kingdom," and all fairs and markets in the North Riding were prohibited. At the beginning of November, it was at Haybrigg Mill farm in the East Riding, within 1.6 km of the North Riding, leading to the sealing off of a number of bridges. Watchmen were to be paid each 10d. a day during daytime, and 1s. 8d. for two men at night. If cattle were brought to the bridges, and the certificate did not specify the colours of the animals, the name and place of abode of the owner, the name of the drover, the name of the person certifying the cattle free of infection, and the herds whence they came not having been within 1.6 km of any infected place, then the cattle were to be impounded and the drover taken before a JP or Commissioner of Land Tax for further directions. One JP reported having stopped five cattle driven within 1.6 km of an infected place in the East Riding without a legal certificate. The cattle were ordered to be detained until they had received a certificate of health. On November 21, the general Quarter Sessions at Hovingham announced that great abuses were committed by butchers and others driving cattle under 8 km without a certificate and refusing to state whence they came. It was therefore ordered that all persons driving such cattle without producing a certificate would be taken before a JP, examined under oath, and punished. The said court opined that certificates for driving cattle, or carrying raw hides, from the East Riding into the North Riding, although signed by a JP or Commissioner for the East Riding, were not legal unless backed by a JP or Commissioner in the North Riding, before the cattle or hides were brought into the North Riding. It was also ordered that raw hides could be sold only in the New Malton Market.

On December 5, it was ordered Chief Constables ensured that, beginning with the towns, every road leading from an infected place be marked one measured mile (1.6 km) from the place by a post to which was attached the names of the towns infected and to keep it maintained. No person in an infected place was to carry out of it any part of any carcass of beef or veal. Cattle were not allowed to stray and had to be herded to and from water. No infected animals were to be allowed to enter the street of a town or drink at any common watering place. Within 5 days of an outbreak being certified, Overseers of the Poor were ordered to make a census of all horned cattle within their Township, distinguishing which were infected and the number dying; the account to be sent weekly to the Chief Constable for forwarding to the Clerk of the Peace.

On December 8, at the Thirsk Quarter Sessions, it was ordered that no cattle or raw hides were to be taken from the south part of the North Riding, northwards, and, as at New-Malton, hides could be sold only in the usual markets. The following day, the Quarter Sessions at Bedale ordered that no hides could be sold in any part of the North Riding except at a market town on market day. Hides also had to have a certificate of health, and Chief Constables had the power to search any houses, shops, or other places where they suspected any hides to be illegally concealed. One Chief Constable in every Division of the North Riding was to constantly attend the markets and inspect all certificates of hides and skins brought for sale, and endorse the certificate when the hides or skins were sold. If unable to attend a market, the Chief Constable could appoint the Petty Constable of the town in his stead. As farmers and others had been exposing cattle for sale in villages or fields near market towns on fair days, it was ordered that this be prohibited.

Further orders were made concerning closing of bridges to cattle without certificates. On December 12 at Guisbrough, a number of farmers and others were appointed Inspectors at a payment of 21 s. a week. A certificate for moving cattle or hides could not be granted without the oath of a third person, not being the owner or buyer. Hides now had to be sold only between the hours of 10 a.m. and 2 p.m., at, or as near as possible to, the market cross in each market town. No cattle, or any part thereof, including tallow, could be brought, either with or without certificates, into Lanbarugh in the North Riding from anywhere east, west, or south. The Parish Officers of every township in Lanbarugh had to maintain a record of all cattle, with the owners' names and all alterations in numbers, so that no cattle could enter or leave a parish without the knowledge, or permission first granted, of a magistrate.

The Constable of Whorlton was ordered to shoot dead before noon the next day (December 13) two red oxen, driven from Cold-Kirby into Whorlton, the drover not having a certificate, and the oxen having been on December 8 in Yearsley where the plague then was. The oxen were to be buried as infected cattle. The Court would indemnify the constable against all Actions and Expenses which might happen to him. On December 13, two men were to be posted day and night at Seamer, on the road from the East Riding into the North Riding, to prevent cattle and hides being moved over Harford Bridge without a certificate.

The following February (1749), an Order was made forbidding the movement of pigs into Lanbarugh from any place east, west, or south. All hay and straw brought in by any "Fishermen, Badgers, Higlers or others" was to be seized and immediately buried by the Inspectors, or any Constable, Churchwarden or Overseer of the Poor.

7. COUNTRYWIDE MEASURES CONTINUE IN BRITAIN

Throughout the country, all fairs and markets, and all movement of cattle except for slaughter, were prohibited anew in 1749 for 3 months. Some relaxation was later permitted in that healthy cattle could be changed to other pastures and taken to bulls. Again, the disease was extinguished where the Order was properly applied but lingered elsewhere.

Because tanners were believed to have spread the disease by buying hides of infected animals, an Act was passed in 1749 requiring production of a certificate specifying color of the hide, name and abode of the owner, and health of the animal, before bringing the hides to the tanyard. Excise officers were empowered to search tanyards. The penalty for infringment was £10. The King could forbid the killing of any cow calves in order to encourage the breeding of cattle. No dealer could sell animals which had been in his possession for less than 40 days, and a certificate of date of purchase was required. Drovers whose cattle sickened on the road had to notify the authorities in order that the animals might be slaughtered and buried, with the hide slashed if diseased. Offending drovers were liable to a fine of £10 or 6 months' imprisonment. It was decreed that the King's Orders in Council had the same authority as parliamentary enactments. Infringement was liable to a penalty of £10, but obstruction was punishable with a fine of £50 if the action was brought within 6 months and in the county where the offence was committed (22 Geo. II c. 46).

At the end of 1749, due to the heavy losses being experienced, an Order was drawn up to come into force on January 14, 1750 whereby, for 3 months, movement of all stock was prohibited, whether fat or store, and slaughter permitted only within 3.2 km of where an animal might be. The Order did not apply to Somerset, Dorset, Devon, Cornwall, and Wales. It caused immediate uproar, and even before it came into effect, a Privy Council meeting, attended by the Middlesex JPs, repealed it on account of "the great inconvenience likely to happen from the said Prohibition, to the Cities of London and Westminster, and many other parts of the Kingdom." The Government then left it to the local authorities to work things out, and for some time afterwards, each county proscribed neighboring infected counties, refusing transference across county boundaries of cattle, hides, carcasses, and tallow, in effect implementing local campaigns against the disease with Government occasionally intervening to prevent fairs which were likely to spread it.

An Act of 1750 (24 Geo. II) empowered rectors or vicars who were rated for Land Tax at £50 per annum, to grant certificates of health for cattle as Justices were. Such certificates were not required if the buyer and seller lived within 8 km of each other, or within the same Parish or adjacent parishes, and the plague was not within 16 km of the seller's abode. Certificates of health were to be issued only where the cattle had been within the jurisdiction of the issuer for 40 days, and the buyer was obliged to keep the cattle in his possession for at least 3 months. The Act also empowered JPs to appoint salaried inspectors. Where JPs prohibited fairs and markets for a county at large, such prohibition extended to all Corporations within, or contiguous to, the County.

Sometimes, all fairs in the country were stopped for 2 or 3 months; at other times, stoppage was limited to country fairs, fatstock still being allowed to be exposed for immediate slaughter. The result was that some counties got rid of the plague, only to be infected again by others.

The Bill rejected in December 1749 was replaced by an Act in 1751 in which it was noted that people were driving infected cattle from place to place "ever more secretly" in order to dispose of them, and it had been questioned whether orders made by JPs prohibiting movement out of neighboring counties were lawful. It was established that any four or more JPs at their General Quarter Sessions, or any Adjournment thereof, were empowered to prohibit such movement if the disease was reported in the county. Magistrates could be fined £20 for issuing certificates in contravention of the requirements, and this penalty was extended to Commissioners of Land Tax, rectors and vicars (25 Geo. II). In 1751, a proclamation of August 9 forbade the holding of any fair likely to spread the disease, illustrating its continued persistence. An Order of Council forbade the sale of cattle at Barnes and Harlow Bush fairs.

The measures had partial effect only. In October 1753, a Suffolk farmer was convicted and fined £100 for buying and driving infected cattle, his servant being imprisoned for not being able to pay a £50 fine for intimidating and preventing an inspector and parish officers from killing four infected beasts. As a result of this farmer's action, infection spread to some 60 parishes.

In the 1753–4 Session of Parliament, the disease received much attention in the House of Commons, and a Select Committee reported on the causes of spreading the disease, producing 16 resolutions. These were agreed to by the House, with slight amendments to two (14 and 16), and a Bill was founded upon them. This passed the Commons but was rejected by the House of Lords, perhaps because of the approaching dissolution of Parliament, and a simple Continuance Bill was passed instead. The resolutions were:

1. The lodging of cattle driven for sale in places where infected cattle are or have lately been, may be one of the causes of spreading the distemper.
2. The mixing together of different herds of cattle not fit for immediate slaughter in fairs, markets, and other places for sale has been one other cause.
3. The mixing of infected cattle with sound on commons and common grounds has been one other cause.
4. Great difficulties and inconveniences have arisen, and may hereafter arise, from the want of proper powers to order the lodging of cattle stopped by virtue of His Majesty's Orders in Council of March 22, 1747–8.
5. The laws, rules, orders, and regulations for the killing of infected cattle upon their first being seized have not in general been punctually complied with.
6. Certificates for cattle to be driven beyond the distance of five miles, without a limitation of time beyond which such certificates shall cease to be of force, are subject to abuse, and may have been one other cause.
7. All contracts and bargains between the buyers and sellers of cattle, conditional upon such cattle, or any number of them, surviving a certain time, tend to the continuing and further spreading of the distemper.

8. The laws, rules, orders, and regulations now in being should be reduced into one Act of Parliament, subject to the alterations and regulations as His Majesty, by the advice of his Privy Council, shall from time to time think proper to order.
9. The notices ordered to be given of the breaking out and arising in particular places of the distemper by His Majesty's Order in Council of the 22nd day of March, 1747–8, should be published in the *London Gazette*.
10. No horned cattle should be driven or admitted into any pasture, close, or field where the distemper has been for two calendar months after such distemper has ceased, nor for six calendar months where such pastures, closes, or fields have not been cleared of all litter, dung, stones, or hay, and where such pastures, &c., have not been rolled, mowed, or fed bare with horses, hogs, geese, or sheep.
11. No certificates for cattle to be driven beyond the distance of five miles, conformable to His Majesty's Order in Council, should be of force for any longer term than ten days from the date of such certificate.
12. No two or more herds of cattle, consisting together of above 20 beasts, intended for immediate sale, and not fit for immediate slaughter, should be driven or admitted at the same time in the same pasture, close, field, fair, market-place, or other place of sale whatsoever, unless such pasture, &c. be so separated and divided as to preserve the cattle of such different herds from immediate contact.
13. No horned cattle should be driven to pasture in common or common grounds unless such cattle have been in the possession of their owners and free from the distemper 40 days at the least before such driving.
14. Proper powers should be granted for the lodging such cattle as shall be stopped by virtue of His Majesty's Order in Council, dated March 22, 1747–8.
15. All contracts and bargains between the buyers and sellers of cattle, conditional upon such cattle, or any number of them, surviving a certain time, should be void, and further penalties be inflicted upon all those who shall so offend.
16. Reward equal to the full value should be paid to the owner of every first, second, and third beast of the whole stock of horned cattle upon every farm or holding of the same person, that shall be seized with the distemper, and that shall be separated, killed, and buried conformable to His Majesty's Order in Council of March 22, 1747–8, provided such beast or beasts shall have been the real property and remained in the actual possession of such owner 40 days at the least immediately preceding such killing.

The amendments were the substitution of "disposition" for "lodging" in 14, and addition of "or upon commons thereto belonging" after the word "holding" in 16.

Following failure of the Bill, existing measures were simply continued, until in June 1757 a new Act was passed "more effectually to prevent the spreading of the distemper now raging amongst the horned cattle in the kingdom" (30 Geo. II c.20) in which previous enactments were reiterated, and new provisions were powers to prohibit the import and export of cattle, raw hides, and products. Tanners had to give notice to the Officer of the Excise District of bringing raw hides into their tan yards, and the said Officers were granted powers of search, which at night had to be in the presence of a constable or other Officer of the Peace. If any cattle sickened while being driven, the drover was to report immediately to the authority in the area, so that the animal or animals could be slain and buried. Failure to report a sick animal involved a fine of £10 or imprisonment for 6 months. Anyone obstructing, or colluding to obstruct, the King's orders, was liable to a fine of £50, but the old restrictions on tanners, drovers, and dealers still carried the same penalties as before. Whether the measures proposed by Parliament in 1754 would have been effective in bringing an immediate end to the outbreak can only be assumed, the disease continuing for another 3 years before finally burning itself out in 1757.

8. THE 1769 OUTBREAK IN BRITAIN

Following the ravages in Denmark and at various localities in Germany, an Order was passed in 1763 prohibiting import into Britain from infected countries of cattle skins coming via Holland and from Holland itself. Yet, in November 1769, the disease broke out again in Britain, several cattle dying in a number of parishes in Southampton. The parishes were extensive and the cattle widely dispersed, and people began to drive cattle out of infected areas toward London and other regions. To prevent spread of the disease in the county of Southampton, Orders in Council were passed on December 7, 14, and 22, 1769, and the provisions of the Act of March 22, 1747 (20 Geo. II) were confirmed in an Act of 1769 (10 Geo. III). The Act also confirmed that the Treasury should defray such costs as might be necessary, which action "could not be justified according to the strict forms of law, and yet were necessary, and so much to the service of the public, and their preservation against so great and dangerous a calamity...." It further made provision to indemnify all persons carrying out Orders in Council against all personal actions or prosecutions.

This was followed by another enabling Act (10 Geo. III) to prohibit the import and export of cattle, hides, products, hay, straw, litter, and so on, which had been in, or near, infected places. It confirmed all previous restrictions and covered also movement, burying, killing, and sale of cattle, and the destruction of fodder, hay, and so on. The penalty for infringement was £10, but for obstructing or aiding and betting the obstruction of an officer, £50. The penalty for issuing a false certificate remained at £20, but magistrates who knowingly permitted an animal to be brought into, or sold at, any fair or market within their jurisdiction in contravention of the Act were liable to a fine of £100. Rectors or vicars eligible to issue certificates of health now had to be rated for Land Tax at £100 per annum. This Act continued in force until the end of 1774 (12 Geo. III).

On April 7, 1770, an Order extended those made for Southampton on December 7 and 14, 1769 for preventing spread of the disease, to all counties. The Order of March 22, 1747 was revived with respect to slaughter and burial of cattle and preventing driving or removing cattle only within 1.6 km of any infected place. In May 1770, the sum of £796. 7s. 6d. was approved by Parliament "To enable his Majesty to make good the like sum, which has been paid to several persons in the county of Southampton, as a compensation, and in full satisfaction of their losses and expences, incurred pursuant to several orders of council, for preventing the spreading of the infectious distemper among the horned cattle."

In September, the disease having broken out near Ostend, an Order was issued prohibiting the importation of any cattle into Britain, or any "manner of" hides, skins, horns or hoofs, "or any other part of any cattle or beast, from Dunkirk or any other part of Flanders ...," referring also to places mentioned in an Order of May 1769.

Dossie (1771) argued that although the legislation was correct in ordering immediate slaughter of infected cattle if the disease should be introduced into Britain again, "Reason, and the legislature, both decree that measure ...," in practice the difficulty of execution was the general want of knowledge of the disease and the lack of ability to recognize it before it could gain a hold. "It is in vain that private persons should be enjoined by the Government to destroy their cattle and magistrates to compel them to it ... if they be wholly ignorant of any certain signs by which they can distinguish it. Yet this is almost universally the case at present...." The publication of instructions for the cure of cattle could have injurious consequences because it could cause people to neglect destroying afflicted animals in the hope of saving them by cure. But if, after destroying known infected cattle, the disease still spread over a large area, the best alternative was to attempt cure, as the destruction of cattle which a widespread enzootic would entail would cooperate with the disease itself in causing a scarcity of cattle.

Dossie pointed out that the Act of 1769 required many important measures to be executed by magistrates, inspectors, and others, should they believe the disease to be present in any particular place. But how could they judge? They needed to be informed of the symptoms of the disease. He proposed that simple accounts of the symptoms should be issued and widely publicized, together with the Orders; and every magistrate, inspector, or other official, should have a copy. As things stood, only a small number of people were even aware of the existing legislation. He recommended "A brief account of the most discernible and peculiar symptoms of the disease, and of the appearances of the inward parts of the beasts which die of it... should be drawn up... and expressed in the most plain and common language, divested of all terms of art...."

Altogether, five Orders were promulgated in 1770 dealing with quarantine and other preventive measures, and the Act of 1747, still in force under the 1769 Act, required that it, together with any implementing Orders of Council, be printed by church wardens and copies read in church by the minister the next Sunday after receiving them; and subsequently, on one Sunday in every month, the minister to keep the Orders available for reading by every person in the parish for as long as they were in force. The clergy were also required to offer up a daily prayer. All Orders of Council were first published in the *Gazette*, but few people in the country read the *Gazette*, and many parishes never received it. Not everyone went to church, and reading the Orders in church was often neglected. Dossie recommended that Orders should be displayed in a prominent place, such as the market place. He considered that in the last outbreak (prior to 1769), the continuance of the Orders for destroying cattle after the disease was widespread in the country had a bad effect, not only by contributing to a scarcity of cattle but also by the great room that the payments gave for fraud, with heavy cost to the public purse. Only when the disease first broke out in a limited area could slaughter be effective.

In 1774, following spread of the continental epizootic to parts of northwest and central France, a further Order was issued in Britain, which, referring to the outbreak of rinderpest in the region of Bordeaux, prohibited the import of cattle and their products; as well as hay and other foodstuffs used to feed infected animals, which could have come from Guyenne and Gascoigne. Further quarantine Orders were issued in 1775 and 1778 as the disease continued to rage on the Continent, especially active in France in 1777. In 1776, John Swaine published

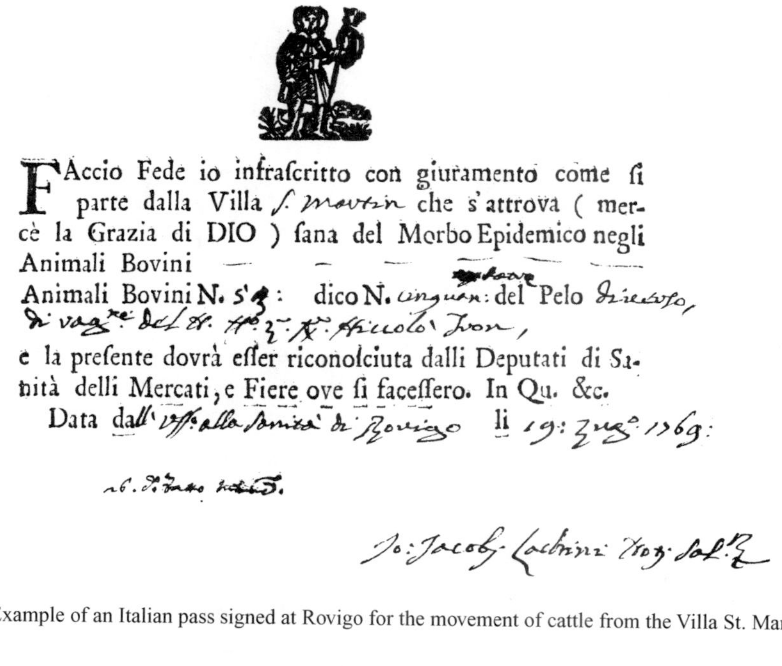

Faccio Fede io infrascritto con giuramento come si parte dalla Villa S. Martin che s'attrova (mercè la Grazia di DIO) sana del Morbo Epidemico negli Animali Bovini

Animali Bovini N. 5: dico N. [illegible] del Pelo [illegible]

e la presente dovrà esser riconosciuta dalli Deputati di Sanità delli Mercati, e Fiere ove si facessero. In Qu. &c.

Data dall' Uff. alla Sanità di Rovigo li 19: Lug.° 1769:

[illegible]

[illegible]

Figure 2. Example of an Italian pass signed at Rovigo for the movement of cattle from the Villa St. Martin, July 19, 1769.

Every Farmer his own Cattle Doctor. Advocating legislation for control of rinderpest through the exclusion of foreign cattle, he considered that Government would not do this, having opened-up the trade in 1765. It had temporarily restricted imports in 1774, but all quarantine orders were revoked in 1779 (Appendix III).

With the disease still prevalent on the Continent (Figure 2), an Order in Council was issued in 1783 implementing control measures, but the disease outbreaks in Britain this year were probably due to anthrax.

13

LEGAL MEASURES AFTER THE 18TH CENTURY

1. GERMANY

The 19th century saw more detailed legislation introduced in Germany, but penalties were less draconian. In Prussia, a law was passed on April 2, 1803, presenting a resumé of various antecedent measures to combat contagious diseases, with emphasis on rinderpest. Extremely strict regulations now came into operation, to be reissued in 1856.

Everyone was obliged to give notice of the smallest trace of sickness in stock, sick beasts were to be moved to quarantine stables along designated routes and slaughtered, or killed on the premises if necessary. Suspect beasts were to be slaughtered also. It was considered better that a beast sick of another disease be slaughtered than spend time on minute examinations. If the disease broke out in a solitary establishment where the stock was not above 10, all should be slaughtered regardless, likewise beasts which stood either side of a sick beast even though they appeared sound. It was stressed that animals be killed quickly without spending too much time on examination. Carcasses were to be buried immediately, and skinning was prohibited. As well as closing of infected and quarantine areas, hamlets also were to be isolated, the place "should be as if it were not in existence," no assemblage was allowed to take place, and even the church was closed. Infected premises had to be closed to outsiders except in harvest time, exceptions being persons involved in wiping out the disease and, if their presence was required, clergy, doctors, and midwives.

Simonds had been told in 1857 that if he went into an infected area, he must remain there until the plague had disappeared and 3 weeks after the death of the last animal. He would receive his food at the end of a pole, placing his money in a pan of water. The area of infection was defined as 3 German miles from its center.

Large cities or trading towns were subject to relative closure, meaning a system of guarding and control. Absolute closure meant that all movement into and out of a place was prevented. Trade might take place exceptionally within a closed area among noninfected farms under strict control of the police. Such controls were to remain in force for 2 months after cessation of the disease, and trade in cattle for a further 2 months only allowed with the permission of district authorities. All dogs were to be kept chained, and all cats and poultry shut up for 3 German miles around an infected place. Penalty for infringement of the regulations could amount to several years' imprisonment.

Disinfection rules were strict, persons being exposed to a "moderate fumigation" with chlorine gas for 10 min followed by changing their clothing and washing themselves thoroughly with soap. Clothes then had to be exposed to chlorine again and everything washable placed into lye and washed with it, then exposed to stronger chlorine fumigation followed by high temperature, before being hung to dry on separate poles and aired for at least 8 days. In stables, the floor, if wood, had to be taken up and burnt; if stones, washed and left in the open air for 4 weeks. The earth then had to be dug out for a depth of 0.6 m. All loose objects had to be burnt, stone walls washed, plaster scraped off, and walls re-plastered. Carts used to transport dead beasts, as well as the harness, had to be burnt, as well as dung carts. If not burnt, carts had to be washed with hot lye and dried in the open air for a long time. Cesspools on farms had to be filled in and covered with stones, and burying places overlaid with a pavement of stone to remain for 2 years.

The entry of cattle, raw skins, grease, and other products was now forbidden. Hard dry skins, but not frozen, were allowed. But if the disease was declared close to the frontier, dry skins and animal hair were also prohibited. These restrictions were not normally applicable to maritime imports, but if goods from neighbouring infected countries came by such routes, the authorities could use their discretion whether or not to refuse entry. This was reenforced in March 1836 and renewed in 1844. At the end of 1844, the disease reigning in Lower Austria and in Bohemia, the Bavarian government issued two orders prohibiting imports from the Austrian Empire.

In Saxony, a Royal Decree was issued in January 1860 as a precautionary measure against the possibility of the disease entering Saxony. When it appeared in Lower Austria in September 1865, this decree was invoked and a law passed providing for measures of control. Among other things, it provided for a pest commissioner with far-reaching powers. This commissioner, or a veterinary surgeon, could order the slaughter of cattle for purposes of examination. In conjunction with the police, he could close off infected areas and order slaughter of all infected or suspect cattle, carcasses of which had to be buried. Owners of cattle which died from rinderpest, or were killed on order, would be fully indemnified by the State, as they would be also for infectious articles which were destroyed. Owners of land set aside for communal burial of carcasses would also be fully indemnified, and all indemnities had to be paid without delay. An instruction issued with the Decree stated that the disease was not affected by temperature or state of the weather but was due to a contagion, imported with cattle from the Russian steppes. There was no cure and only two methods of arresting its progress. First, all cattle attacked by the disease or in contact with cattle suspected of having the disease had to be killed immediately. Second, the infected farm had to be shut off from all communication until it had been disinfected.

On December 6, 1866, the king of Bavaria issued an Ordinance in accordance with the governments of Würtemberg, Baden, and Hesse, for common action to prevent the introduction of rinderpest. During 1867, 45 proclamations and regulations imposing controls and then lifting them, especially with reference to neighboring Bohemia, followed. Whereas at first, the import of livestock and products, and their transit by rail, was prohibited, as well as the driving of army livestock, on May 8, it was announced that prolonged prohibition of animals and products from Austria and the closing of the border with a military cordon would have disadvantageous consequences, with lack of meat and indispensable raw materials, as well as price increases, and less damage would be caused by rapid suppression of outbreaks. The cost of maintaining the military cordon was also considerable. Transit of army stock and raw materials from disease-free areas was thus allowed on May 12 to be followed the next day by 36 cases of disease in Steinach, all sick animals being slaughtered. Controls were then imposed and lifted again whenever areas were reported free, import and transit of animals and products being permitted under permit stating that they were from disease-free areas. In all cases, import and transit of steppe cattle from Hungary, Poland, and Galicia were prohibited. The penalty for infringement of orders was up to one year's imprisonment or a fine of 400 guilders. The outbreak prompted a revision of the legislation with a consolidated ordinance on July 3 concerning control of imports of stock and products, including prohibition of trade in secondhand clothing and footwear. This was followed by another ordinance on July 10 covering measures to be taken if outbreaks occurred within the country. Karlsruhe followed with an ordinance of May 22, 1867 coordinating penalties with the other dukedoms for prevention if it should spread to there, in which event an ordinance of August 17, 1865 applied. An order followed on November 5 prohibiting incoming transport from foreign countries.

By 1870, the law appeared practically to be that of the North German Federation of April 17, 1869 (Appendix I). Each State had, to some extent, a separate jurisdiction, but the general regulations applied to the whole of them. These were most stringent and effectual in controlling outbreaks within the country itself, but, as in other continental countries time was required to put them into force, which led in two instances to the export of infected cattle to Britain. A Prussian law of June 25, 1875, *Law for the Suppression and Prevention of Certain Diseases Among Animals in Prussia*, specifically stated that it did not apply in any way to the procedure for the prevention and suppression of rinderpest.

Subject to special enactments in the event of an outbreak, it was not until June 1909 that there was a general law concerning epizootics, followed by another in December 1911.

2. AUSTRIA

Extensive regulations were produced in the Austrian Empire, dealing with methods of prevention by quarantine, controls over trafficking, and control of herd movements and of markets. In the event of an outbreak, special attention was to be paid to the frontiers, imports prevented and cordons placed along them. Any wandering, masterless dogs near the frontiers were to be shot. When infection was near frontiers, there were to be no markets within 3 German miles; no oxen used for draught, only horses; and no dogs allowed to accompany vehicles. If intercourse with an infected place was unavoidable, then on the return home, the shoes and clothes used during the journey had to be changed, hands and feet washed, and all approach to the children at home avoided. Any magistrate failing to report its appearance was to be immediately dismissed, declared incapable of filling the office, and punished under the Penal Code.

The frontier control regulations stated:

(a) Skins may only be imported when perfectly dry and hard. Fresh skins are allowed to pass the frontier only after having undergone the process of disinfection by the vapours of sulphurous acid.
(b) Horns and hoofs must be immersed for 12 hours in frequently stirred salt-water (5 kg of rock-salt to one pail of water), and then dried. The tips of the horns must be well washed with salt-water and well dried only.
(c) Melted fat or grease can only be admitted in casks; melted tallow in abundant wrappings can only be admitted after these coverings have been destroyed or cleansed on the frontier.
(d) Unmelted tallow and raw meat are to be refused entrance.

> If the malady prevails extensively in the neighboring foreign states [Russian Empire and Hungary], the authorities may for the time being prohibit the importation of cattle generally, and in fact all productions coming from infected places.... Persons coming from infected places, as also cattle-dealers, butchers, tanners, etc., are to be refused entrance, or must undergo the usual quarantine process, while all intercourse is strictly prohibited with the infected place itself.

The procedures to be taken in the event of rinderpest breaking out in a place were: report, followed by immediate action without waiting for the Plague Commission to arrive. There was then to be a medical examination of the affected stock, a census of all stock by the Commission involving house-to-house visits, and then procedures to prevent its further spread. The members of the Commission could not conduct a census of noninfected areas on the same day as infected areas but had to return home and carefully cleanse their clothes before proceeding the next day, if necessary sleeping in a special place. Where the disease was not widely spread, slaughter was recommended, the political Commissioner making the decision. The owner was then to be reimbursed by the State, but this was to be *minus the value of the diseased parts* and paid only where it could be satisfactorily established that no blame was attached to the owner for the animal catching the disease, the regulations having been correctly followed. Even then, if the cattle killed were in such an advanced state of the disease that they could not be expected to recover, compensation was to be one third of the value, and for cattle killed at the beginning where the outcome could not be known, two thirds. The products could be sold after due cleansing and their value deducted from the total sum. Where animals were slaughtered in cases of doubt, they were to be examined, and if the entrails showed no sign of the disease, the owner could be given the flesh for home consumption, the value of which would be deducted from the total estimate. Where several beasts were involved, the meat could be sold, smoked or pickled, and disposed of with due control, in this case the Commission selling the meat and paying the money earned to the Treasury. Where the plague was extensively spread so that slaughter would be ineffective, quarantines had to be established and the area isolated. Every house in which the plague had broken out was to be marked by some striking sign, and all intercourse of the inhabitants with those of healthy localities forbidden, enforced if necessary by the military and gendarmes. Visiting neighboring churches, schools, and pleasure resorts, and the carrying of grain to neighboring mills, was also forbidden. Doors of stables had to be closely secured that no dogs, cats, or fowl could enter. Dogs running about were to be caught and confined, or killed.

Despite the stringent law, the results were not satisfactory, rinderpest repeatedly occurring with disappearance for short periods of time only, and despite further stringent regulations in 1872, the disease increased. This appeared due to relaxation of the law and the ineffectual manner in which it was executed rather than any deficit in the law itself.

A law of February 1880 was similar to that in France, with the exception that the whole value of a slaughtered animal was paid. The law was renewed in August 1909, and in December 1909, the import or transit of ruminants and their products from Russia was prohibited, renewed in April 1930.

3. FRANCE

Following on from the laws passed at the end of the 18th century was an Ordinance of the King of January 15, 1815 under which the mayor, as chief of the municipal police, was the responsible executing authority, assisted as required by the police chiefs, rural police, gendarmes, national guard, and army. He was liable to severe penalties for negligence and issuing false certificates, the latter offence incurring a fine of 1,000 francs and even imprisonment (Chauveau, 1871), but in general, the French regulations were mild by comparison with those of Prussia or Austria. On September 5, 1865, the Emperor signed a decree forbidding importation into France of domestic animals which would present dangers of rinderpest unless subordinated "to such measures as may be necessary to prevent the invasion of the malady." The frontiers or parts of frontiers where this would apply were to be specified. This was interpreted in an order which forbade the introduction into France and transit of all bovines, their raw hides and other raw parts, from England, Holland, and Belgium. Animals from other sources were to be inspected and, if they passed inspection, kept in quarantine for 10 days. A number of the previous laws, beginning with that of April 1714, was still in force in 1865 and considered all that was necessary to prevent the spread of rinderpest. However, the laws did not cover imports by water, and this was provided for in a decree of September 5,

1865 which also specifically cited imports from Britain, Holland, and Belgium. Another law was passed on July 6, 1866 relating to the payment of indemnities for slaughtered animals. In addition, sections 459–462 of the French Penal Code applied.[1]

A law of July 21, 1881 and a decree of June 22, 1882 concerning the regulation of public administration replaced the existing legislation and consolidated the texts. Articles 68 and 69 of the latter stated: "When animals prohibited on account of rinderpest are presented for importation either by land or by sea, these animals are seized and slaughtered on the spot, ill or not. Equally slaughtered without compensation are ruminants which are part of a herd presented at the frontier before the prohibition and in which the presence of rinderpest is confirmed. In all these cases the bodies are buried with the hide slashed." On December 17, 1888, the importation or transit of grey or steppe cattle was forbidden, as well as ruminants and fresh skins coming from Serbia, Bulgaria, the Ottoman Empire, Greece, and Egypt, and live bovids and their products from the Austro-Hungarian Empire, Russia, Montenegro and Romania, an order not replaced until February 1921. Orders of January 12 and July 23, 1892 allowed the free circulation in France of sheep from Russia transported by sea on French ships carrying a French veterinarian, undergoing a quarantine of 3 days, having left the port of embarkation since at least 7 days; on August 24, the provisions being extended to Montenegro.

On June 21, 1898, sections 34, 35, and 46 of a "Rural Code" replaced provisions concerning slaughter, controls on stock movement, sale of diseased meat, burial of carcasses, use of products of slaughtered animals which had been in contact with sick animals, indemnities, and so on:

> When an order of the prefet states the existence of rinderpest in a community, the affected animals, and those bovine species which could be infected, even those which show no sign of sickness, are slaughtered by order of the mayor, conforming to the proposition of the veterinary health officer and after evaluation.
>
> In the case foreseen by the previous article, the sick animals are slaughtered on the spot or at the place of burial if transport of the carcase is declared by the veterinarian more dangerous than that of the live animal: the transport in view of slaughter could be authorized by the mayor, conforming to the advice of the delegated veterinarian, for those which have only been contaminated. Bovine and caprine animals which have been exposed to the contagion are isolated and submitted to sanitary measures determined by the public administration regulation rendered by the execution of the law.
>
> It is permitted for owners of animals slaughtered by reason of the rinderpest, compensation of three-quarters of their value before the sickness (maximum 600 francs).

Apart from general measures, specific measures were introduced on an *ad hoc* basis; for example, an Order of October 20, 1896 prohibited import of animals and products from all of Africa except for the French possessions, but later texts attempted to limit imports from French colonies also.

[1]Sec. 459. All holders or keepers of animals or beasts suspected to be infected with rinderpest [*maladie contagieuse*] who will not have notified on the spot the mayor of the commune where they are found, and who, even before the mayor has responded to the notification, will not have confined them, will be punished with imprisonment for 6 days to 2 months, and fined 16 to 200 francs.

Sec. 460. Equally will be punished with imprisonment of 2 to 6 months, and a fine of 100 to 500 francs, those who, in contempt of the defences of the administration, will have let their animals or infected beasts mix with others.

Sec. 461. If, from the communication mentioned in the preceding article, contagion results among the other animals, those who will have contravened the defences of the administrative authority, will be punished by imprisonment of 2 to 5 years, and a fine of 100 to 1,000 francs; all without prejudice to the execution of the laws and regulations relating to epizootic diseases, and the punishments which they carry.

Sec. 462. If the offences of police correction of which it is spoken in the present chapter have been committed by rural policemen or foresters, or officers of the police of whatsoever rank, the punishment of imprisonment will be one month at least and a third more above the highest penalty which would be applied to another culpable of the same offence (Code Pénal, 1810).

A law of October 6, 1904 covered contagious diseases in general replacing the Rural Code. In November 1917, import of ruminants and their products from French West Africa was prohibited, amended the following December to provide exemption for dried skins and wool. In December 1921, the ban on imports from Africa was lifted for South Africa.

When the disease appeared in Belgium in 1920, additional severe measures were introduced. An order of August 7, 1920 prohibited import and transit of ruminants and pigs from all countries except French colonies and protectorates, and included products of animal origin from Belgium. Movement of all animals between communs near the frontier was forbidden, as well as poultry, rabbits, milk, straw, dung, and so on. A further order of August 10 prohibited the importation of horses, donkeys, and mules. On May 10, 1921, importation of Belgian and Luxembourg cattle was reauthorized.

In May 1928, France signed an accord with Belgium replacing an agreement of 1895 (reconfirmed in 1900, 1906, and 1910), relating to the notification of contagious diseases in the frontier region. Another was signed with Germany in 1929 concerning sanitary policing, and with Italy in 1930, Italy having signed an agreement with Romania in February of that year.

4. HOLLAND

A slaughter policy had not been applied in the Dutch Republic during the devastating outbreaks of the 18th century perhaps because of lack of a strong central government. It was not until 1799, after the establishment of a unitary state, that a cattle fund was established, obliging all Dutch cattle owners to pay annual contributions to a state fund which would provide compensation for loss due to rinderpest. This opened the way for applying a compulsory slaughter policy in the 19th century.

When the disease broke out in the Southern Province of Holland in 1865, the Dutch government placed its main reliance on throwing a strong military and naval cordon around the infected province. No entry or exit of animals was permitted, the exception being through traffic on the Dutch–Rhenish railway which passed through the heart of the enclosed district. This railway carried a large number of cattle every week for export to England. In the northern province of Gelderland, regulations were issued on July 31, 1866, with amendments the following February, while in the province of North Brabant, the Duchy of Limburg issued its own order on November 8, 1866 on the disruption from the prohibition of manure, hay, and straw from North Brabant and Limburg.

Holland passed its own *Contagious Diseases (Animals) Act* in July 1870 under which regulations were issued in December 1870, detailing methods of disinfecting, slaughtering, and burning in the case of rinderpest: “Destruction of infected articles by means of burning or burying; infected hay, straw, reeds, dry manure, wooden flooring, if in bad condition, old woodwork, strongly infected clothing, and other objects not suitable for disinfection to be burned; or, if this should be dangerous, buried.” Even the upper layer of the road along which infected things had been brought (including animals) had to be collected, placed in a pit, and set on fire, the ash then to be buried to a depth of 1 m. Where cattle were buried without burning, the place had to be surrounded by a fence for one year. This was reinforced by a Decree of October 1872 ordering that cattle suffering from, or suspected to be suffering from, rinderpest, must be slaughtered and burnt or buried.

A decree of October 1873 enabling the authorities to make regulations for removal of cattle and control markets and sales was specially designed to check the spread of pleuropneumonia. A new instruction was issued in February 1877 forbidding removal without a permit of any animals in specified districts.

On May 20, 1890, legislation reinforced total slaughter of sick and suspect animals together with an indemnity of 100% for suspect animals and 50% for infected.

5. BELGIUM

In March 1866, a Royal decree stated that an indemnity for slaughter would henceforth only be paid where it was proved that the owner had kept the animal in a stable for at least 1 month. This was because the reappearance of the disease at Antwerp was occasioned by farmers who did not bother whence came animals they purchased, relying on the indemnity. The Government had previously issued a circular ordering destruction of diseased animals on August 28, followed by other orders on September 3 and 4 authorizing slaughter of suspect animals and laying down rules for slaughter and disposal, respectively. The legality of these measures was inferred from an Order of 1784 and the Rural Code of 1791 but contested under an article of the Constitution which stated that no one shall be deprived of their property unless in the forms established by law and for a just and previous indemnity, allowing for the destruction of animals deemed incurable but not of all attacked by or suspected of being attacked by a disease. These objections were countered by a very wide interpretation of some ancient laws, but nothing could be construed to prohibit imports and transit, and the Minister of the Interior took this responsibility upon himself.

On February 24, 1877, a revised law was promulgated which under Article 1 stated Government was empowered by Royal Decree to prescribe measures against rinderpest. The import and export of ruminants, their products, fodder, utensils and any items which might have been in contact with infection, would be prohibited. A census of cattle would be required in indicated areas and the counted animals branded. Any changes in stock had to be declared within 24 hr. No cattle or products were to be allowed in or out of infected places, the prohibition to remain for 30 days after the last case of infection and no animals to be introduced for 45 days afterwards. No animal could be introduced into places for fattening unless it had been subject to 15 days' quarantine in an isolated spot specially selected for the purpose, and not moved therefrom until a certificate of clearance was obtained from the Government veterinary surgeon. Fairs and markets were to be prohibited unless animals were brought for slaughter which had certificates of origin, and they could not be removed from the market afterwards. On report of the Government veterinary surgeon stating that they were infected, diseased animals were to be slaughtered by order of the burgomaster of the commune, members of the Agricultural Commission, the commissioner of the district, or the governor of the province. Suspected animals were slaughtered on the order of the governor of the province, and those which had been in the vicinity of infection, on the order of the Minister of the Interior. The rules applying to suspect animals also applied to cattle, sheep, and other ruminants suspected of having been fraudulently brought from an infected country. Slaughtered animals were to be buried according to rules prescribed by the Minister of the Interior, and meat from suspect animals could be consumed only if autopsy showed the animals to be uninfected. Two thirds of the value of the slaughtered animal was to be paid to the owner provided that the rules had been followed, and also for any items, fodder, and so on, destroyed by order.

Ministerial Decrees were passed to effect some of the provisions of the above law. On March 5, 1877, regulations for censuses in indicated communes were introduced, on March 6, measures to be adopted to prevent introduction of rinderpest were implemented, and on March 7, regulations controlling slaughter of diseased and suspect animals.

A law of December 1882 on sanitary policing of domestic animals was followed by an Order of September 1883, *Special Dispositions Regulating all the Measures Which Could be*

Rendered Necessary by Fear of Invasion or Existence of the Contagious Typhus, in which indemnity for slaughtered animals was set at one-third of the value.

In 1920, a Royal Decree was enacted concerning all necessary measures against fear of invasion by, or existence of, rinderpest. An indemnity was to be paid for slaughtered animals of 70% of the value, or 50% if the animal was ready for a butcher.

6. HUNGARY

In Hungary, a pamphlet was issued in 1816 concerning protection against rinderpest, followed in 1830 by a detailed description of the disease. From 1834 to 1850, a major epizootic spawned many decrees. Measures included throwing a cordon around a farm immediately the disease appeared. No man, woman, or child was allowed to enter or leave during the infection, and for 3 weeks after killing the last animal. Affected people could not attend church, and, where possible, roads were diverted from infected localities. In 1864, the laws were consolidated, and in July 1874, the Ministry of Agriculture produced a modern, comprehensive law, Act XX of the year 1874, *Being Statutes Concerning Measures Against Oriental Cattle Plague*. The regulation of isolation of endangered farms and communes was left to the Minister to prescribe in an Order issued in accordance with the principles of the Act. The Act was preceded by regulations issued by the Minister of Commerce on March 29 noting that with the advent of spring, the movement of cattle became more frequent and outlined precautions to be taken as rinderpest had now been stamped out in the country. But there were limited outbreaks in 1875, and a series of extensive detailed orders were issued, which came into effect implementing the Act if rinderpest was reported in an *adjacent* territory. The first thing was to stop all traffic from the infected area of ruminants and their raw products, and then to ascertain the extent of the disease. If it was limited to one village within 39 km of the frontier, then the Act came into force concerning supervision of frontiers: "Every arrangement concerning quarantine has one and the same object, namely, to examine the condition of health of the animals imported and the quality of the raw products, and consequently to admit only such animals as are healthy and such raw products as are beyond suspicion…." Quarantine was imposed on all animals coming from Romania and Bosnia, where there was no rinderpest, or it existed in a few small places at least 64 km from the frontier, and all had to be carefully examined. Drovers had to be isolated during the period of quarantine. Animals passing the quarantine had to be throughly washed down, branded at the base of the left horn, and a certificate of health issued with the route to be followed minutely described, any deviation being treated as smuggling and the cattle confiscated. Even so, cattle being driven to the next inspection or railway station were to be under the escort of a quarantine officer who would hold the certificates. The vacated stalls had to be throughly cleansed each time, washed with chloride of lime, and kept vacant for 15 days, and all dung and so on burnt. If the disease was epizootic in any adjoining country or broke out within 60 km of the frontier, a strict blockade would come into force. Cattle registers were to be maintained in frontier districts and annual returns made to the ministry. Standing regulations covered cattle passes, markets, slaughter, meat inspection, common pastures, inns, drove routes, inspecting committees, disinfection of railway trucks, and steamboats. Neglect of reporting any case of internal disease was stringently punishable because "There is no disease of animals in which the results of concealment are so serious… For a single case generates so much contagion… On the other hand, no other disease of animals is so easily extinguished, provided it be recognised at its first outbreak…." If an outbreak was confirmed, a Central Commission of Disease was to be set up to institute measures to combat the outbreak. The maximum penalty for infringements of the Act was a fine of 500 florins or imprisonment in default.

7. RUSSIA

In Russia, the measures adopted, directed against the Siberian plague (anthrax) as well as rinderpest, were:

- Early information through the police with a penalty for neglecting to report.
- Examination of reported cases by the civil authority assisted by a veterinary surgeon, who were to report on the nature and cause of the outbreak and the measures taken to arrest it.
- Cattle census.
- Separation of sick from healthy cattle.
- Special treatment centres for the sick.
- Designated places and precautions for sales.
- Disinfection of stalls, harness, and so on.
- Separate attendants for sick and healthy cattle.
- Smearing cattle with tar to prevent spread of the contagium by flies.
- Burial of dead animals with hides slashed.
- Prohibition of trade in products such as meat, hides, milk, tallow, etc.
- Burning of litter, dung, and so on.
- Confining of dogs, cats, and fowl.
- Designated byroads for the passage of cattle in transit.
- Cattle to be fed with hay damped with warm salted water.
- Rowelling to be used.
- Treatment of sick animals to be attempted.
- Precautions to be taken to prevent possible infection of the attendants.
- Prohibition of use of the meat and milk of sick animals.

Movement is not checked, and affected animals are not always slaughtered. If the disease breaks out on a journey, only the affected animals are detained for treatment, the remainder being allowed to proceed.

Although, in 1865, the law required that carcasses be buried in their skins, it was well known that it was never executed. Even if the police insisted upon it, the owner exhumed the animal at night and removed the skin, selling it as quickly as possible.

On June 3, 1879, a law ordered the slaughter of sick and contact animals, the rate of compensation to be fixed by local authorities.

8. GREECE AND TURKEY

At the beginning of 1870, the King of the Hellenes published a decree forbidding entrance into Greece of cattle and their products from Thessaly.

The Ottoman Empire, just emerging from its old system based on theocratic principles to that of a modern state, in the 1870s was in considerable turmoil. Lacking a constitutional government, there were no measures such as slaughter to control outbreaks. Instructions were issued by the superior Sanitary Council, of which a summary issued at Constantinople in April 1877 (in Brown, 1878) stated: "With regard to curative measures one cannot do better than consult the works on veterinary medicine, where will be found all that is interesting to know on the subject."

A circular of June 8, 1864 stated that raw hides from infected countries should be immersed in the sea for 48 hr and dried in an isolated spot, and on June 28, 1866, a circular required that epizootics be reported. A circular of December 31, 1867 stated: "Science being in possession of facts which prove that a healthy animal belonging to a kind other than bovine, and coming from a locality infected with an epizootic, say cattle-plague, can transmit the disease to other kinds of animals, when the disease exists somewhere, no distinction should be made, sanitarily speaking, between the different kinds of animals coming from the same place."

This was followed on January 14, 1868 by a circular stating "in order to restrict the propagation, while lessening as much as possible hindrances in the way of trade through the application of the restrictive measures found necessary…," which required that animals and their products from two villages, Cazas and Sandjacs, be accompanied by a certificate of origin as coming from an epizootic-free area and as such could be embarked at a port where the epizootic prevailed without

> incurring the application of the restrictive measures as to regulations at the port of arrival… the animals and their substances, such as hides, horns, hair, wool, bones, &c, notwithstanding the existence of the epizootics at the place of embarkation, are exempted from any purification or quarantine at the ports of arrival, and are immediately handed over for business purposes, provided they are furnished with the before-mentioned certificate determining their origin.

On April 28, 1873, a circular called attention to reporting on epizootics, requiring returns to state whether the authorities had or had not advised on the measures for repression, what these measures had been, and their result. Nowhere did the superior Sanitary Council specify any measures for repression.

9. BRITAIN

In 1842, seemingly indifferent to the question of disease, Government threw open the country to unrestricted importation of cattle from abroad. The shibboleth of the day was "free trade," although it led immediately to the importation of FMD, pleuropneumonia, and sheep pox *V. ovina*. An old Act relating to sheep pox was ineffective, and there were no other disease controls. The agricultural community were the first sufferers, but the general public shared in the misfortune by an increased cost of living and consumption of diseased meat. Those who raised their voices against the latter practice were either ignored or censured for interfering with legitimate trade (see Gamgee *supra*).

The importation of foreign stock was controlled by the "Board of Customs," which issued instructions to their officers to look out for disease. These lay inspectors had no knowledge of animal diseases, and so there was no effective control. A general Order in Council had been passed in the Isle of Wight in July 1856, empowering customs officers to seize any animals, not only cattle, which appeared infected with, or labouring under, any "contagious or infectious disorder." They then reported to the Customs Commissioners, who appointed a veterinary surgeon to examine the animals. His duty was to certify them if infected, whereupon the Commissioners would order their destruction. But the customs officers had to have the ability to recognize a disease in the first instance, and the Veterinary Department had no direct control over them as to what the examination of animals should be, and only indirectly by communications through the Customs and with their approval. In 1871, port inspection was transferred from Customs to the Privy Council, the Veterinary Department appointing their own inspectors, and in 1876, Acts of Parliament were in force when rinderpest appeared again.

Occasional outbreaks of sheep pox in Britain had led to the Act in 1848, renewed annually, to prevent the spread of infectious diseases among sheep, cattle, and other animals. Two Acts had been passed (1 & 2 Vict. c.CV, CVII), the first an enabling Act to prohibit importation of sheep, cattle or other animals for the purpose of preventing the introduction of contagious or infectious disorders, the second to prevent the spreading of sheep pox. However, the latter had a provision prohibiting any person bringing, or attempting to bring, any "Sheep, Lambs, Oxen, Bulls, Cows, Calves, or other horned Cattle ..." into any market, fair, or public place, knowing them to be infected with sheep pox or a disorder "of like nature." This Act was continued until the end of 1857 and not amended until 1866.

An 1848 Act made it illegal to expose for sale any meat unfit for human food, as did also the Public Health Act of 1848, but what was "unfit" was not defined.

In 1849, an Act was passed empowering the Privy Council to make orders and regulations from time to time "... to prevent ... the spreading of contagious or infectious disorders among sheep, cattle and other animals ...." These concerned:

- prohibiting or regulating the removal to or from such parts or places as designated of sheep, cattle, horses, swine, or other animals; and the meat, skins, hides, horns, hoofs, or any other part; or of hay, straw, fodder, or other articles likely to propagate infection;
- the purifying of any yard, stable, outhouse, or other place, or any wagons, carts, carriages, or other vehicles;
- directing how any animals, or parts of animals, dying in a diseased state; or any animals, parts of animals, or other things, seized under the Act, were to be disposed of;
- promulgating notices of the appearance of disease;
- making any other orders or regulations for the purposes of giving effect to the provisions of the Act.

When the general outbreak occurred in Galicia and spread through Lower Austria and Hungary in late 1853 and early 1854, it was brought to the attention of the British Government, which instigated inquiries into the measures taken to counteract the disease on the eastern frontier of Prussia. In July 1856, a Bill was proposed under the authority of the 1848 Act "to prohibit the importation of sheep, cattle, or other animals, for the purpose of preventing the introduction of contagious or infectious disorders," empowering Customs Commissioners to detain suspected animal cargoes and if found by a veterinary surgeon to harbor any diseased animals. The Commissioners could order the animals to be destroyed, and, if necessary, detain or destroy the rest of the animal cargo; and all straw, fodder "or other article," making owners liable for the cost of detention. The Bill failed.

In April 1857, an Order in Council was issued under the Act of 1849, noting a danger of infectious disease prevalent among cattle in countries bordering the Baltic Sea being introduced into Britain. It prohibited the importation of cattle, horns, hoofs, raw or wet hides, and skins, from Russia, Prussia, the Gulf of Finland, or any other part of the Baltic Sea between the Gulf of Finland and Lübeck, and the same from any ship that had been to these places. Any such animal or product arriving in Britain, together with all hay, straw, fodder, litter, or manure, was to be destroyed or disposed of as Customs might direct.

In 1857, another Bill dealing with the prevention of contagious diseases among cattle and sheep was proposed. Giving evidence to the House of Commons Select Committee discussing the *Sheep &c. Contagious Diseases Prevention Bill* in June 1857, Simonds proposed free admission of foreign stock without quarantine restrictions. His objection to quarantine was that a ship's cargo was usually made up of animals from different sources in the country of origin. If animals from one source were in the incubating, essentially noninfective, stage of a disease (he was specifically referring to pleuropneumonia), there would be only a slight risk that they would infect other animals during a voyage of a few hours. But if all the animals had to stay in quarantine after arrival, the incubating disease might become infectious and spread widely. Different diseases had different incubation periods. If animals were found ultimately to be infected, the farmer must, under a penalty, declare it and allow nothing to leave his farm. Asked whether he was in effect proposing that the farmer should place himself and his farm and cattle in quarantine, Simonds replied "I think so." He was then asked whether he was really acquainted with country affairs and the trade in cattle and replied: "Perfectly so."

On the question of the consumption of diseased meat, Simonds urged meat inspection to prevent it being sold as food. He was asked: "Have you been to Newgate Market on a Saturday?" and replied "Yes, frequently."

"Have you seen the description of meat that is sold there to the poorer classes?" was the next question.

"I have seen meat exposed for sale which was quite unfit for human food," he replied. Meat, he affirmed, in every state of disease.

Based on his evidence, the Committee came to the conclusion that it was impossible to prevent the spread of disease by legislative measures, and the Bill foundered. But all of Simonds's major recommendations to the Select Committee were eventually adopted and formed the blueprint for control in Britain of infectious animal diseases (Pattison, 1990).

In 1863, a Bill was proposed which obliged the Master of a vessel to report to Customs immediately upon arrival any animals in the cargo dying on board. No animals could be landed until permission was given in writing by Customs, which was empowered to order the Master to return to sea and throw overboard any carcases. Failure to comply with the regulations would result in forfeiture of the animal cargo. This Bill also failed. In the same year, yet another Bill, the *Cattle Importation Bill*, was placed before Parliament. This provided for an Inspector-General of Live Stock under the Privy Council, with veterinary surgeons as inspectors. It also failed, and was replaced with another which Gamgee had instigated, the *Cattle, Sheep, etc. Contagious Diseases Prevention Bill*, presented by Edward Holland. Gamgee argued that the domestic losses from imported diseases far outweighed the value of the cattle imported, but the consensus of opinion was that the proposed restrictions on trade would do more harm than the diseases themselves, and Holland was obliged to withdraw it, doing so on the understanding that Government would promote similar legislation in its next session. This was subsequently prepared by Sir George Grey and Mr. Bruce as the *Cattle Diseases Prevention Bill*. It assigned positions of importance to the veterinary profession throughout the country and would have provided an organized public veterinary service. But Simonds and the meat, dairy, and cattle trades opposed it and had it blocked, the Goverment stating in March 1865 that such difference and even conflict of opinion were exhibited before the Seelect Committee that "the Government did not intend to introduce any measure on the subject." *The Veterinarian*, probably under Simonds's pen, welcomed the withdrawal of these Bills, stating that it was not opposed to a really good measure, but the country was now left "in the same tranquil position it enjoyed before sensationalists appeared on the stage," a dig at the vigorous campaigning of Gamgee. "Greatly modified and improved as the Bill was by the select committee, it still contained some things highly objectionable…." The deficiencies were considered to be that while the Bill made provisions for recognition of certain diseases as being contagious or infectious, this did not include such diseases that might arise in the future. Also, while prohibiting exposure of diseased cattle in public places, it legalized sale of the same in markets or places which were set apart for the purpose, and it exempted FMD from the provisions. Objections were raised to the provision of cattle inspectors who would be lacking in the necessary qualifications, while at the same time, a meeting of farmers in Exeter Town Hall in February 1864 passed a motion praying for the appointment of an inspector to prevent the introduction of diseased cattle into the market. Other towns such as Worcester and Shrewsbury had already taken steps to have suitably qualified men appointed.

Also rejected was *A Bill for the Amendment of the Law Relating to the Importation of Diseased Cattle and Meat*. Importers, salesmen, and others, meeting at the Salutation Hotel in London on March 16, 1864, while not wholly opposed to the bills, considering any measure calculated to diminish fatal diseases of cattle would be a great boon, nevertheless considered that some clauses would result in losses to those connected with the trade in cattle. Like their Scottish counterparts, represented by the Highland Society, they were vigorously opposed to any control. An anonymous writer was concerned at their stance:

> ... the Bills may be thrown out; but we cannot think it. If they are, how bitterly will the northern farmers repent it! Time must prove to them what we have learned by dearly-bought experience—viz., that the direct importation of diseased animals is attended with the most serious consequences to the farmer. They only had direct importation last year, and there are prospects of its increase...But for these importations there would have been little or no need at present for the Cattle Diseases Prevention Bill. (A Farmer's Friend, 1864)

10. THE 1865 OUTBREAK IN BRITAIN

The proposed new Act was defeated by the cattle dealers and importers, as well as a number of veterinary professors. An inquiry before the Select Committee of the Commons in that year led not to strengthening of regulations but to relaxation. Previously, the presence of a sick animal in an imported cargo had been sufficient to have the whole herd condemned, but now it was ruled that inspectors should seize and slaughter only those animals which were actually sick, thus permitting infected animals to pass. The result was that when rinderpest was imported in 1865, there was no means by which it could be arrested. The case of sheep differed, and action was taken under the 1849 Act both in 1862 and in 1865.

Although the disease was raging with severity in several places by the end of June, the Privy Council issued its first Order in Council on July 24 using powers under the *Public Health Act* of 1848. This was nearly 6 weeks after the outbreak in Islington, 14 days after it was reported to Government, and 10 days after receiving Simonds's report. Declaring the disease to be of an "uncertain" nature, it simply required owners of cattle in the Metropolis to give notice to the Clerk of the Privy Council of the existence of any disease on their premises under a penalty of default of £20. The stated object behind simply making provision for reporting the disease was "for the purpose of proper regulations to be made for preventing or checking the further propagation and increase of such disorder" (Section 1). It also provided for inspectors appointed by the Privy Council with the power to enter premises and examine animals. It was not designed to check the spread of the disease but was chiefly for the purpose of ascertaining the extent of the disease in the London dairies. Although the two measures were allegedly based upon the recommendations of Simonds, when later asked by the Commission what was the effect of this first Order, Simonds replied that he thought it spread the disease rather more quickly from the Metropolitan Market than it otherwise would have been spread. Calling as it did upon cowkeepers to give notice of the existence of disease, after reporting any disease, as no compensation was payable, this caused them to send their cattle to the Metropolitan Market before they could be seized.

Following extension of the disease, on August 11 another order was issued, still applying only to the Metropolitan District although the disease had reached beyond that area. Now using the term "Cattle Plague," it made provision for the appointment of inspectors empowered to enter premises and prevent removal of any sick or suspect animals without a licence issued by the inspector. Animals suffering from the disease had to be separated as far as was practicable and all dead animals buried, their skins and the premises being disinfected. The penalty for non-compliance remained at £20, but owners sold on any affected stock as quickly as possible. A separate Order of the same date was issued for the rest of England and Wales. It defined a Local Authority, which comprised the JPs who were empowered to appoint inspectors with the same powers as in the first Order if the disease appeared in their districts. The word "animal" was defined as any cow, heifer bull, bullock, ox, or calf but did not include sheep. It provided for the reporting of any animal "labouring or suffering" under a "disorder." Inspectors had the power to enter and examine suspected premises. Persons within districts where inspectors had been appointed were obliged to isolate sick animals and required a licence from an inspector to send an animal to market or remove an animal from premises where sick animals had been reported. Every animal dying of "such disorder" or "slaughtered on account of" (no provision was made for ordering slaughter) should be buried on the premises or as near as possible; and if not with the skin, the skin had to be disinfected. Inspectors had the power to order the cleansing and disinfecting of affected premises.

On August 18, the provisions were extended to Scotland where the disease was now raging, and on August 25, exports of cattle to Ireland were prohibited despite great opposition from the Irish.

Five more orders were passed, including an important one on August 26, which extended the Orders for London and Scotland to the whole of Great Britain, and empowered inspectors to "seize and slaughter, or cause to be slaughtered, any animal labouring under such disease," extending their powers to fairs and markets. It also made it unlawful to send or bring *infected* animals ("animal labouring under such disease") to fairs or markets, and to move such animals by railway or sea, or along any highway, but not those which had been in contact with them.

Minor orders were then passed prohibiting the importation of skins to Ireland. On September 22, the former four orders were revoked, replaced by an amended and consolidated Order. There was an addition of two important provisions, one affecting the Metropolitan Market, the other giving local authorities the power to prevent animals or products entering a fair or market. The term "animal" was extended to include sheep, lamb, goat, and swine. Inspectors now had to report to their local authority and account for the number of days spent on duty and the number of miles travelled. More importantly, unless an inspector granted otherwise, dead or slaughtered animals had to be buried at least 1.5 m deep, in their skins and covered with quicklime or other disinfectant, but power to slaughter diseased animals was now taken from the inspectors except where owners refused to obey instructions as to isolation. Local authorities were now empowered to prevent animals or their products entering a market or fair within their jurisdiction, a prohibition which extended to the Metropolitan Cattle Market, and no animal could be taken into the Metropolitan District except to be sold there for immediate slaughtering. Every animal had to be marked for slaughter as soon as it was sold. Sick animals could not remain on unenclosed land, but oxen from infected herds would still be moved about under licence.

An Order followed on September 29, prohibiting export to Ireland of sheep or lambs and their skins from Britain, and an Order for the island and barony of Lewis prohibited import of all cattle.

On November 23, a further Consolidated Order was issued which closed some of the loopholes, but permitted animals to be sent by rail both into and out of infected districts. This was the eighth Order to be issued, but the increasing mortality occasioned by the disease and its widening spread showed that these Orders were powerless to arrest it. As the Orders of the Privy Council were operative only in so far as they did not conflict with the statutes, the Orders and statutes had to be consulted together.

Another Bill, *An Act to Check the Spread of the Cattle Plague in Great Britain*, was withdrawn in the Commons after having been returned with amendments by the Lords.

It was not until February 20 the following year (1866) that an Act received the Royal Assent. Although rushed through both houses of parliament in 1 week, and on the statute book before the end of February, it had taken a long time for Government to act positively. Entitled the *Cattle Diseases Prevention Act 1866* (29 Vict. *c*.2), it amended the 1848 Act "to prevent... the spreading of contagious or infectious Disorders among Sheep, Cattle, and other Animals," of which the Order had lapsed in 1857. This new Act included what had always been denied as impossible—provision for ordering slaughter and monetary compensation. An Order empowering inspectors to slaughter infected stock had come into force on August 26, 1865; this Act empowered local authorities to appoint inspectors to effect it. All animals affected by the disease were to be slaughtered and compensation paid not exceeding £20 pounds and not greater than half the value of the animal. In an address to Sir George Grey dated February 14, a "Land Agent and Valuer" quoting his own experiences, while supporting slaughter of sick animals, opposed slaughter of those apparently healthy animals that had been in contact with them, and also objected to the prohibition of all movement of store cattle for 2 months. He also wanted compensation to be set at two thirds or three quarters the value of the stock, paid from national funds (Sanderson, 1866).

In July 1866, a *General Cattle Plague Rate Bill* was proposed, whereby a proportion of rates obtained by the Poor Law Board should be assigned as a cattle rate to compensate persons whose cattle were slaughtered under the Act. Under the *Metropolitan Cattle Plague Order* of August 1867, expenses incurred in its execution by the Court of the Lord Mayor and Aldermen of the City of London (as the local authority for that area) were to be paid from the consolidated sewers rate. A receipt issued by the Parish of West Felton, Oswestry Incorporation, showed that on May 21, 1867, Mrs. Duckett paid 1s. $8\frac{1}{2}$d. in respect of the Local Cattle Plague Rate under authority of the Act at the rate of $\frac{3}{4}$d. in the pound. Carlisle Union in the township of St. Cuthbert's Within established a Cattle Plague Rate of $\frac{1}{2}$d. in the pound on the Poor Rate on July 27, 1867. On December 18, it was increased to 1 penny in the pound (the Poor Rate being 11d.) with half deducted from the landlord, and the owner was also granted a 15% allowance.

Government had wanted stock owners compensated through insurance schemes, the successful opposition that made compensation a charge on the rates being led by Lowe. The day the Act came into force, he wrote to his brother, a major landowner in Nottinghamshire, "I hope you will remember for future use that it is mainly to me that farmers have it that the rate is not put on *them* instead of on the County" (Fisher, 1993).

The Act also required that carcasses be buried at least 1.8 m in their skins and covered with quicklime or other disinfectant. The penalty for contravening any provision in the Act, or any order made by a Local Authority, was a fine not exceeding £20. If the offence related to more than four animals, the penalty was not to exceed £5 for each animal instead of £20. Under the previous Act, in cases of prosecution based upon informants, half of any penalty awarded went to the informer, but this provision was dropped making it less likely that people would inform. The Act applied only to Britain, a separate Act being passed for Ireland on March 6. An Order of Council was passed on October 4, 1866, continuing until February 1868, controlling import into England of stock from Ireland or Scotland.

On March 24, 1866, another Consolidated Order was issued. This Order executed the provisions of the Act and continued the prohibition of moving cattle (including sheep) by railway until April 16. Another Order, dated the same day, prohibited all imports from the Netherlands after March 28, and any cattle which had been on any vessel at the same time with any cattle which had come from the Netherlands. On May 9, an Order continued the provisions relating to slaughter until June 30.

The *Contagious Diseases (Animals) Act, 1867*, obliged Local Authorities to appoint at least one inspector, and it empowered him to order any animal to be slaughtered "for the Purpose of ascertaining the Nature of any Disease, suspected to be Cattle Plague, under which an Animal may be labouring…" (Section 15). It also empowered the Local Authority to declare an Infected Place, which was subject to the following restrictions:

- No animal could be moved out alive.
- No products could be moved out without a licence certifying that such had not been in contact with infection.
- No carcass or single piece of raw meat weighing more than 20 lbs could be moved out without a similar licence.
- No dung, straw, litter, and so on could be moved out without a licence.

Railway and other transport companies were obliged to cleanse and disinfect all carriages, boats, pens, and so on after use.

In August 1867, the *Consolidated Cattle Plague Order* and the *Metropolitan Cattle Plague Order* were passed, the latter prohibiting movement of cattle in and out of the Metropolis without a licence or market pass. No cattle could be moved out alive and movement of cattle through any part of the Metropolis or the City of London was forbidden, until in July 1868 the *Consolidated Cattle Plague Order* of August 1867 was amended, allowing cattle to be moved out alive in accordance with the Order's provisions.

Cattle from Spain, Portugal, Normandy, and Brittany could be landed at specified ports but moved from the place of landing, or placed in contact with other cattle, only after examination by an inspector approved by Customs.

The *Contagious Diseases (Animals) Act* of 1867 was replaced in August 1869 by the *Contagious Diseases (Animals) Act 1869*. This Act did not contain any detailed legislation for the prevention of the introduction of contagious or infectious disease of animals from abroad but conferred upon the Privy Council full powers to make any regulations which it might consider necessary for this purpose. The Act appeared to provide for an emergency by rendering it incumbent upon Local Authorities to appoint at least one inspector in every district, but inspectors were not required to have any qualifications, so detection of disease was in many cases left to persons totally incompetent for the work, and they were superseded by policemen who likewise knew nothing of animal diseases. Regulations under the Act provided for the whole cargo to be slaughtered at the place of landing if rinderpest was detected in one or more animals. All foreign animals landed in Britain had to be detained for 12 hr near the landing place and inspected by a Veterinary Inspector at the end of the period. The local authority was obliged to slaughter and bury every animal suffering from rinderpest and compensate the owner for the loss, but not exceeding £20 and not exceeding half the value of the animal immediately before being slaughtered. The Act provided that it *may* also slaughter animals that had been in the same shed or herd, or in contact with any infected animal, and *may* compensate up to £30 but not exceed three quarters the value of the animal. Whereas the Act provided that the Local Authority shall take a defined course of action and could be interpreted to mean the Privy Council may by Order assume the functions of the Local Authority, Brown (1878) considered that it was by no means clear that this was contemplated by the framers of the Act, and no such step had been taken or even suggested, the tendency in legislation being exactly the reverse. In 1865, the Metropolitan Police District was under the charge of the Privy Council, and all inspectors were officers of the Veterinary Department. Afterwards, the action of the Privy Council was confined to the Board of Works district, and in March 1866, the powers of the Privy Council were further curtailed, appointment of inspectors in the Metropolis and the general work of dealing with contagious diseases delegated to the Board of Works as the Local Authority.

Following the outbreak and spread of the disease on the Continent in 1870, Orders in Council were passed on September 9 and 20 under the 1869 Act. The first Order specified the regulations applied to cattle brought from any port in France and all had to be slaughtered within 10 days after landing. The second Order extended this to sheep and goats brought from any port of the States of the North German Confederation or from France and only allowed landing of cattle, sheep, and goats at places approved by HM Customs, from which no animal could be moved out alive.

In December 1871, the General Orders were consolidated in the *Animals Order* and the *Foreign Animals Order* of 1871. The latter applied regulations to cattle coming from the Empire of Germany but subsequently permitted cattle bred and fed exclusively in Denmark, Schleswig, or Holstein, to be landed at places other than those set apart for slaughter. The Order of Council of December 8, 1871, prohibiting landing of cattle from France or Belgium, and animals which had been on board at the same time with such cattle, was revoked with respect to animals from Belgium by Order of July 15, 1872. Following importation of rinderpest, the *Schleswig and Holstein Cattle Order* of 1872 was revoked on July 31, 1872, and after disappearance of rinderpest in France, the Order of Council of December 8, 1871 was revoked in respect of France. On August 20, 1872, the remainder of the *Foreign Animals Order* of 1871 was revoked. On July 11, 1872, an Order was issued, empowering local authorities to make regulations for burial of carcasses in accordance with the *Animals Order*. Provision was also made for disinfection of carcasses of animals which died or were slaughtered on board vessels within 4.8 km of shore (the limit of jurisdiction), and it was made an offence to throw the carcass of any animal affected with a contagious disease into any inland water, or into the sea within 4.8 km of the shore. This was followed by the *Markets Order* in August, which gave local authorities power to make regulations requiring owners and others to clean animal markets, fairs, and so on, at their own expense, and disinfect them, but only 23 of 404 local authorities had made regulations under this Order during the year.

The *Cattle Plague Order* of 1872 came into operation on September 9 following the appearance of rinderpest in the East Riding of Yorkshire, ceasing on December 1. There was very little practical difference in the restrictions under this Order from those in 1865–7. The last Order relating to rinderpest was a slight amendment to the former, allowing sales of cattle under certain circumstances (Order No. 359 of September 17).

The question of compensation was debated again by the Select Committee in 1873 examining the *The Contagious Diseases (Animals) Act, 1869*. Williams argued that the full value of an animal should not be paid, as this would encourage slaughter of any cattle whether affected by disease or not.

11. THE PLAGUE IN IRELAND

In Ireland, a committee was convened in December 1865 to draw up plans and legislation to prevent introduction of rinderpest. Ferguson dissented from the report of the committee and issued his own. One contested point was the proposal that, before embarking for Britain, drovers and others would be issued with a change of clothes, and on return, they would change back into their original clothes. Ferguson argued that the returned clothes could provide a hotbed of infection. Another proposal he disagreed with was that six medical men, two veterinarians, and two stockmasters should be sent to England to study the disease. Ferguson argued that only qualified veterinarians who had passed an examination should be sent. Medical men did not, as a rule, understand animal diseases. He proposed that slaughter was essential:

> However seemingly unjustifiable, on the score of its being inconsistent with modern medical science and the spirit of enlightenment, the immediate slaughtering of all animals on the discovery of their being affected with Cattle Plague may appear, the measure is to the fullest extent justified by an impartial examination of statistics ...The percentage of recoveries is too small, and the disease itself too terribly contagious, infectious, and accumulative, or spreading, to justify either any attempt at treatment, or leaving the animals to recover or die unassisted by medical interference.... (Ferguson, 1866)

The Chairman of the Committee, Lord Naas, was not pleased with Ferguson's dissension, and wrote in reply on February 7, 1866:

Sir,

I have received your note of the 3rd instant, requesting me to forward a report signed by you, on the subject of the Cattle Plague, to the Lord Lieutenant-I beg to decline to do so. Two months have already elapsed since the Committee presented their report to His Excellency. On referring to our minutes of proceedings, I find that during the sittings of the Committee, of which you were a member, you took no exception to some parts of the report which you now declare to be injudicious, and against which you "emphatically protest" [the appointment of persons as Cattle Plague Inspectors who had not taken an examination]. In the divisions that took place on two most important questions, that is, that of making all animals unremovable from an infected district [the Prussian system of total isolation proposed might be deemed as totally incompatible with the spirit of the laws in the country whose interference with might be objected to as establishing a dangerous precedent for the purpose of restricting personal liberty for "a far less worthy object"], and that of the compulsory slaughter of diseased beasts [Ferguson claimed compensation should be from a universal tax and not from the poor rates of the union in an infected district], you recorded your vote in favour of these propositions with the majority of your colleagues. I cannot, therefore, avoid coming to the conclusion, that the course you have taken in drawing up a report in opposition to that laid before the Lord Lieutenant so long ago, and on which the Government have already acted, is as inconvenient as it is unusual, and is calculated to do harm. I cannot think it to be part of my duty to enter upon a discussion of the important matters treated of in your paper. You decline to submit your opinions to the consideration of the Committee by whom they would have been fully attended to, fairly discussed, and decided upon.

You have, however, ample opportunities at your disposal for giving to your report the utmost publicity, and the public will then be able to judge for itself whether the report of the Committee was, as you allege, "premature," and whether, as you further say, the measures it proposed are "inadequate," and are "incompatible" with "established facts."

Ferguson's report was published separately (Ferguson, 1866), seven of the 10 committee members signifying their approval of it, Lord Naas and Lord Athlumney dissenting.

The Act of 1848 (11 & 12 Vict. *c.* 107), continued by 16 & 17 Vict. *c.* 62, was extended to Ireland by Act 19 Vict. *c.* 4 of March 6, 1866, The *Cattle Disease Act (Ireland), 1866*, as a precautionary measure, the plague not having reached there. The Act prohibited landing of cattle from any part of England, and provision was made for disinfecting drovers and persons engaged in the cattle trade. A veterinary department was established and inspectors appointed. Despite these precautions, the disease broke out shortly afterwards in May 1866 in a remote part of Northern Ireland at Drumra, probably introduced with smuggled cattle. By July, it had been stamped out with the loss of about 50 head.

A Consolidated Order was issued on May 22, which, among other things, required that all cattle in a district which had been declared infected be branded, and again when the district was declared clear. An Order of June 25 required that drovers, cattle purchasers, and other persons dealing with cattle, harvest men, or farm laborers, coming from Britain, be disinfected upon arrival.

12. THE 1877 OUTBREAK IN BRITAIN

In 1875, *The Animals Order* of 1875 revoked as from the June 23 all former Orders in force relative to contagious or infectious diseases among animals, except Orders relating to particular ports, replacing them with this new Order. Passed under *The Contagious Diseases (Animals) Act 1869*, it embraced other animal diseases as well as rinderpest, covering transit both by sea and rail, with provisions for disinfecting and cleansing as well as supply of water for animals in transit. Scheduled countries were: the Austro-Hungarian Empire, Belgium, France, the German Empire (with the exception of Schleswig and Holstein), the Hellenes, Italy, and the dominions of the Sultan (Turkey). Cattle from these countries could be landed only at specified ports and within specified areas of such ports. Cattle from the dominions of the Emperor of Russia, and cattle, sheep or goats having been on board a vessel at the same time as such Russian animals, were prohibited from being landed in Britain. Manure and hay from Russia were also prohibited; while meat, hides, fat, hoofs and horns, could be landed only at the specified ports and released after certification as unlikely to introduce contagious or infectious diseases.

The *Contagious Diseases (Animals) Act 1869* was in force in 1877, when there was a renewed outbreak of rinderpest in London on January 15. On January 16 and 18, Orders were introduced, controlling import of sheep and goats from Germany, Belgium, and France, followed on January 27 by prohibition of cattle, or sheep and goats which had been on the same boat with cattle, and cattle products, from Belgium and the German Empire. On February 2, the *Metropolitan Cattle Plague Order* of January 31, came into effect, prohibiting removal of cattle, sheep, and goats from the Metropolis, and markets and sales in the Metropolis except by licence; but at the same time, licences were granted for the holding of the Metropolitan and Deptford Markets. On February 8, the prohibition on imports was extended to the Netherlands. By February 9th part of the Foreign Cattle Market was declared free of infection, but the following day, the disease was found to have crossed the metropolitan boundary into Essex appearing at Stratford, and the *Cattle Plague Order* was passed on February 13, extending the provisions of the former Order more widely. As infection spread to other parts of the country, local authorities acted by prohibiting the movement of cattle and the holding of markets in certain districts. In Yorkshire, an Order passed by the Privy Council on February 20 prohibited fairs and markets in the whole of the East Riding upon representation from some local authorities, followed by Lincolnshire on March 6. An Order of March 8 required every Local Authority to maintain a *Cattle-Plague Record*, and on March 12, an Order restricted movement of cattle in Lincolnshire.

In consequence of the continued existence of the plague, an Order was passed on April 12, in which the Privy Council took the action of Local Authorities in controlling movement and fairs and markets in the metropolitan area, having power to deal with the disease in the whole of the City of London and the Metropolitan Police District. Coming into force on April 16, the Order specified that any person having an animal suspected to have rinderpest was to keep the animal separate "as far as practicable" from other animals and report the matter to a police constable, who was to inform the Clerk of the Council at the Veterinary Department. No animal, including a dog, could be moved alive out of a building or enclosed space in which rinderpest existed or had done so within 7 days. When an inspector affixed a notice to premises declaring them infected and forbidding entrance to unauthorized persons, it was an offence to enter in contravention and liable to a penalty of £5. Any person having been in such a place was obliged to have their clothes disinfected upon leaving. Any carcasses had to be transferred in a special van for the purpose which was to be disinfected immediately afterwards. All utensils used in or about an infected premises had to be disinfected, as well as the building or enclosed space. Any dung, hay, or straw could be moved out for the purpose of being buried or destroyed with a certificate signed by an inspector that it had been disinfected.

One man, apparently a curious onlooker at an infected establishment, refused to submit to the disinfection and was subsequently summoned and fined.

Three Orders were passed on April 27. The first extended the period of control in the Metropolis for Order 397 of April 12 from 7 to 28 days for a building or premises in which infection had been present. The second, to run for 1 month from May 1, prohibited movement of cattle, sheep, or goats out of Middlesex not within the Metropolis, except for slaughter under licence to the Metropolitan Cattle Market, or a slaughterhouse within the Metropolis, or within Middlesex under licence to a slaughterhouse. No licence was necessary to move cattle not more than 450 m from lands to other lands in the same occupation within Middlesex. The third Order, also to come into effect on May 1 for 1 month, extended the above provisions to the Metropolis, adding that cattle exposed for sale at the Metropolitan Cattle Market could not be moved out but must be slaughtered within its limits.

On May 9 an Order made slaughter compulsory within the Metropolis for any affected animal, and discretionary slaughter of those which had been in contact with infected animals. The Local Authority was to pay compensation.

An order of May 29 removed controls in Lincolnshire, but on May 31, it was necessary to prohibit movement of cattle, sheep and goats out of the county of Middlesex except into the Metropolis. Transit by rail was permitted. On June 14, the *Cattle-Plague Order* was revoked, and a restriction imposed on the import of fresh meat from Belgium or the German Empire, followed by a number of Orders deproclaiming addresses as infected areas, and a partial lifting of restrictions in the Metropolis on July 16 and 18. Then on July 18, part of the Metropolis was once more declared an infected area and movement of cattle prohibited, restrictions remaining in place until August 14 except as regards the Metropolis, where cattle, sheep, or goats could be moved in by rail, but limitations on moving stock out still remained. On November 21, all restrictions were finally lifted. In all, 46 Orders were passed between January 16 and December 4, a number defining infected premises and those after November 21 defining areas of ports where foreign cattle could be landed.

A Select Committee to inquire into the plague and the importation of live stock had been appointed on April 24 following a resolution forwarded to the Privy Council by the RASE. It was of the opinion that the Privy Council could not undertake the work of local inspection throughout the United Kingdom and agreed that the appointment of inspectors should rest with Local Authorities, subject to Privy Council supervision. More importantly, it recommended total prohibition of importation of live stock from European ports, and enforcement of uniform compulsory measures throughout the United Kingdom for the suppression of contagious diseases of cattle. One of its most important recommendations was to accept that rinderpest could be eliminated only by slaughtering *all* cattle on the infected, or suspected, premises, as well as those on adjoining premises within the outbreak zone, and compensation should be paid out of central funds rather than by Local Authorities. This should be the full value for those slaughtered suspected of the disease up to a limit of £40 per animal. It also proposed that importation of all animals from Russia be statutorily forbidden, and live cattle from Germany (except Schleswig-Holstein) and Belgium. Sheep should be allowed only if slaughtered at the port of unloading. But whereas some, such as Brown, lobbied for prohibition of all live cattle imports, the Committee accepted that cattle from unscheduled countries, such as Spain, Portugal, and Denmark, be freely admitted if the precautionary regulations in force in the countries were acceptable:

> Having, then, carefully considered the evidence, your Committee have arrived at the conclusion that compulsory slaughter, at the port of debarkation, is not likely either to discourage foreign importation, or to diminish the supply of our large towns, or generally raise the price of meat. The change, however, would be a considerable interference with the present system of trade carried on by butchers and salesmen. For this reason, and because proper arrangements would in many cases have to be made at the ports selected for debarkation, your Committee feel that it would be well to postpone the commencement of the change for a time sufficient to enable the necessary preparations to be carried out. (Recommendation 33, Anon., 1877)

Brown considered that one of the lessons learnt from the outbreak was the impossibility of effectively and economically dealing with rinderpest by "means of intermittent and irregular action" on the part of Local Authorities, which had to make onerous and useless regulations, unnecessary if there was central control; "in a state of general uncertainty prohibitive orders are scattered in all directions instead of being concentrated at the points where they would be most effective" (Brown, 1878). This was echoed by a Select Committee examinee who considered that regulations were better carried out by the Privy Council than by local authorities, citing an example that when the plague was in the east of London, there was only one officer to conduct examinations, and before a total prohibition was placed upon movement, many removals of infected animals probably took place. Another supporting central authority control gave the example of Northumberland and Durham. Northumberland county prohibited all live animals coming out of Newcastle market, while Durham county permitted it. So butchers purchased their cattle in Newcastle and took them across the River Tyne to Durham and then into Northumberland, bypassing Newcastle.

Suggested improvements were that the Privy Council replace Local Authorities in issuing regulations, and police should execute the Orders and the Act, replacing inspectors. Responsible to the local authority, that is, the courts of quarter sessions in the counties and town councils in boroughs, the police would have power to call in veterinary surgeons for advice. It was also considered necessary that Orders in Council should be imperative and apply to the whole of the United Kingdom. James Howard, representing the London Central Farmers' Club, stated that he would like to see the Act altered by changing the word "may" to "shall." The compulsory slaughter Order for the City of London of May 9 stated: "The Veterinary Department of the Privy Council Office shall slaughter every animal that is affected with cattle-plague, and may slaughter any animal that has been in contact with any animal so affected." This was interpreted as meaning that slaughter in the first instance was obligatory, but discretionary in the second.[2] In the *Contagious Diseases (Animals) Act* of 1878, the relevant slaughter provision was to state (Section 15): "The Privy Council shall cause to be slaughtered (i) All animals

[2]The definition of "may" and "shall" in law was decided before the House of Lords in 1880 in the landmark case of *Julius* v. *The Lord Bishop of Oxford*, enabling words being always compulsory where they are words to effectuate a legal right.

affected with cattle plague; (ii) All animals being or having been in the same shed or stable, herd or flock, or in contact, with an animal affected with cattle plague." But, "The Privy Council may, if they think fit, in any case cause to be slaughtered; (iii) All animals suspected of cattle plague, or being in a place infected with cattle plague; (iv) All animals being in such parts of an area infected with cattle plague as are not comprised in a place infected with cattle plague (but in this last-mentioned case subject to such regulations as the Treasury from time to time think fit to make)."

It was also contended that if full compensation was paid for slaughtered animals which had been in contact with diseased animals, it would remove the owner's excuse for concealing the fact. As to whether this would induce owners to claim for unhealthy animals which were not diseased, it was considered that if regulations were strict enough, there would be very little disease and therefore few claims.

Brown's suggestion was that areas in which disease occurred should be declared infected areas where the veterinary authority could make any regulations thought necessary, thus freeing the rest of the country from restrictions.

In August 1878, the *Contagious Diseases (Animals) Act* of 1869 was replaced by a greatly improved Act which revoked all former Acts and applied to the whole of the United Kingdom. Covering other diseases also, a special section was devoted to rinderpest. Implementing the *Contagious Diseases (Animals) Act* of 1878, under the *Animals Order* of 1878 passed in August of that year, and covering Great Britain only, a constable to whom notice was given of an animal infected with rinderpest, or supposed to be so, was obliged to inform his superior officer, who informed the Clerk of the Privy Council by any rapid means. No animal could then be moved alive from the place until it had been visited by an inspector or other officer of the Privy Council, being guarded by the police until such arrival. No horse, ass, mule, or dog could be moved out alive from an enclosed place in which rinderpest existed, or had existed, within 10 days; and likewise no dung, litter, or fodder, could be removed from such a place.

An endeavor to get rid of the unqualified inspector was defeated in Committee; instead, an inspector could declare only that rinderpest existed in a particular place, and this then had to be confirmed by the Privy Council. The Privy Council was then obliged to cause all animals involved to be slaughtered. It also had the option to order slaughter of all animals suspected of the disease, or being in a place infected with it, and all animals in an area infected with rinderpest in parts where the disease was not actually present. Compensation was paid from Government funds for all slaughtered animals. Where an animal was slaughtered on account of being infected, the compensation was one half of its value before it became infected, to a limit of £20. In all other cases, compensation was the value immediately before slaughter, with a limit of £40.

There was also provision for the burial of carcasses washed ashore. These had to be buried by the receiver of wreck, wherever possible not on the shore, the expenses chargeable to the Local Authority.

It also ordered that foreign animals could be landed only at specified ports and had to be slaughtered before removal from the wharf, apart from animals (unless diseased) intended for exhibition or other exceptional purposes, which had to be landed at a port providing quarantine.

Passed at the same time (August 1878) was the *Cleansing and Disinfection Order*, which required that all forms of transport of animals, whether vessel or railway truck, be cleaned and disinfected as soon as practicable after carrying live animals and before any other animals were placed therein.

In 1880, the *Animals Order* was passed, embodying in one document various regulations to be observed in the event of outbreak of a disease. In 1884, enforced slaughter of most animals at the port of entry was introduced, but the live trade continued from North America although becoming overtaken by that of dead meat. As the supply of foreign meat increased in Britain, and consumer prices fell, public concern lessened, but from 1886 to 1888, the *Cattle Diseases Prevention Act* was built up, until in 1896 the *Diseases of Animals Act* came into being, finally instituting a policy of complete exclusion of live cattle imports because of losses due to pleuropneumonia from America and FMD. No subsequent Act, fortunately, has been put to the test with respect to a recurrence of rinderpest.

Following the formation of a subcommittee of experts in veterinary policing of the Economic Committee of the League of Nations formed in 1919, and the creation of the OIE with headquarters in Paris on January 25, 1924, numerous international agreements were undertaken. Whereas the latter office was more for collection and dissemination of information, the former fixed the general rules for international commerce in animals and animal products, and the agreements between the sanitary organization of neighbouring countries. Placed by itself in the first group of contagious diseases, special consideration was given to rinderpest, allowing for prohibition of importation of all ruminants and pigs from countries not completely or certainly without the disease for at least 1 year. It was conceded that experimentation had shown, as regards animal products, that the virus did not survive well in the open. However, the conditions of conservation of products could subtract from the usual influences of desiccation, light, and so on, and the action of antiseptics was insufficiently known. Taking into account also the resistance of other viruses considered to be fragile, such as FMD, it advocated extreme prudence. Nevertheless, if they fulfilled the usual conditions of transport hygiene and would be accepted by the country of destination, it directed that no control should be imposed for the transit of animal products without serious motive.

14

THE ROYAL COMMISSION, LEGISLATIVE FAILURE, INSURANCE, AND GOVERNMENT ACTIONS IN 19TH-CENTURY BRITAIN

1. THE ROYAL COMMISSION

Simonds's proposals for suppressing the 1865 outbreak were not adopted by Government, which ignored the veterinary profession, handing over the campaign to the Medical Department of the Privy Council. The Government appointed a senior officer of the Royal Engineers as Director, and a doctor as secretary. Placing political expediency and its popularity first, the Government would neither prohibit further stock importation nor make effective provision for slaughter, but instead, 3 months after the outbreak began the Privy Council, having failed to prevent extension of the plague, as one observer put it, recommended to the Queen a Royal Commission, which was appointed on September 29 to investigate the origin and nature of the disease and frame regulations for its control, "ascertaining as far as possible the mode of treatment best adapted for the cure of the affected animals, and the regulations which may with the greatest advantage be made with a view to prevent the spreading of the said disorder, and to avert any future outbreak of it." The Commission consisted of 12 men who knew nothing of animal diseases in general, let alone rinderpest. The only veterinarian on the Commission was Professor Charles Spooner, professor since 1853 of the London Veterinary College, who had no knowledge of cattle epizootics (although later he identified rinderpest when Simonds did not). Except for Spooner, all members of the Commission were laymen representing landed interests and medical men. Among the latter were Dr. Henry Bence Jones, a physician also noted for his chemical research; Richard Quain; Thomas Wormald, a surgeon from St. Bartholomew's Hospital; Edmund Alexander Parkes, prominent in the field of public health; Robert Ceely an authority on vaccination; and Lyon Playfair CB, LL D, FRS, Professor of Chemistry in the University of Edinburgh, Vice-President of the Royal Society, Edinburgh and Liberal MP. Representing the landed interests were Earl Spencer; Viscount Cranbourne, later Prime Minister as the Marquis of Salisbury; Robert Lowe; Clare Sewell Read, Conservative MP for East Norfolk and well-known agriculturist; and John Robinson M'Clean. Budd, who, together with Gamgee, persistently opposed attempts at cure, and whose proposals put forward in early August had been ignored, refused to sit on it. As a result of the Commission's inappropriate advice, the plague spread in every direction, obliging the Commission to turn eventually to veterinary advice.

Gamgee, the most knowledgeable man on the subject and probably the most well known through his previous lobbying for controls, who had actually predicted the event, was totally ignored during all this time. That he was not included on the Commission has been described as a "national misfortune." Until the Commission took note of what was in effect Gamgee's veterinary advice, the disease continued to spread because its members urged treatment of affected animals and condemned the policy of slaughter.

Sitting initially from October 9 to 20, it examined 45 witnesses and received a deputation from the Lord Mayor of London, producing its first report at the end of October of that year. But as one commentator noted, throughout October "during the laborious daily sittings of the Commission, the plague continued to extend..."

The Commission's *First Report* (September 1865) has been described as "a sad failure." Its conclusion concerning the strict measures taken on the Continent to control the disease, particularly slaughter, was, "They are measures indeed which never could be enforced—they involve sacrifices to which no people could be reasonably asked to submit—unless in the presence of a dreaded enemy, and under a sense of overwhelming necessity" (Bernard, 1865). Constable satirically summed it up, "We have thoroughly investigated the matter in all its aspects, and we have come to the conclusion that we do not know how the disease originated, nor how it ought to be treated, nor can we agree upon the measures that ought to be taken to prevent its spreading" (Constable, 1865). Spooner signed without objection a report which enabled the disease to spread, leading Fleming to comment: "Veterinary Medicine in Britain must ever lament that the respected name of one of her best professors should be found appended to such a recommendation."

In its *Second Report*, issued in March 1866, the Commission passed a unanimous resolution: "To restrict importation absolutely to certain ports; to cause all fat cattle to be slaughtered at those ports, and all store cattle to undergo a period of quarantine; hides and skins, not salted, to be disinfected at the ports of importation." But this resolution was never implemented in its entirety, even after more outbreaks in 1873 and 1877. In evidence to the Select Committee in 1873, James Odams related how, in 1866 in the Metropolitan Market, he saw Simonds making an autopsy of an animal "in the presence of several scientific men." He was so struck with the malignant nature of the disease that he approached Simonds, stating that he had a number of water side sheds which were unoccupied as a consequence of the American War and would be happy to allow the Government to use them for isolation for 1 or 2 months while the outbreak existed. Simonds referred him to influential members of the Government, and Mr. Ward Hunt with Lord Burleigh and the Marquis of Exeter went and looked at them. They then brought the matter before Parliament, and during discussion, one Member facetiously stated that he had met with some old woman on the banks of the Thames who had some petroleum barrels to sell, and the matter was set aside.

If stock were examined at disembarkation, this could not necessarily prevent importation of the disease, for it was the practice to throw sick animals overboard and not declare them. On arrival, the remaining stock might appear healthy but may have been in contact with infected animals.

Lowe, as a politician with an interest in public health matters (who, the next year, was to cause the defeat of the Reform Bill and bring down the Liberal ministry by his shrewd and brilliant speeches), led the questioning of witnesses and later received most of the credit for the first report and the thrust of its eventual legislation, exerting a decisive influence on later debates upon rinderpest in the Commons. An important witness, who greatly influenced the outcome of the Commission, was Sir John Simon, closely connected with most of the eminent physicians and scientists sitting on it. While modestly disclaiming a professional knowledge of rinderpest and stating that the Commission should refer to the veterinary expertise for countering the disease, Simon stressed in his evidence that rinderpest was highly contagious, capable of being communicated from sick animals in various direct and indirect ways. This was significant in view of the large body of anti-contagionist medical opinion. He urged rigorous scientific inquiry into the nature of the disease, emphasizing the importance of learning from "more accomplished" Continental authorities.

Government policy favoured as much decentralized control as possible, and the Commission had to reflect this, although Lowe personally detested democracy, and the Commission has been seen by one modern commentator as a significant step forward in the emergence of the central administrative state (Burn, 1964). The slaughter of cattle was generally opposed on grounds of practicability rather than principle, the majority favoring stopping movement, with a minority, representing the landed interests, opposing even this measure on grounds of impossibility of successful implementation. Thus, the first and main recommendation was that the movement of cattle from one place to another within Britain be suspended for a limited time, with the option of renewing as necessary. This it believed to be the only certain means of eradicating the disease, but, in the event that Government would not follow this course due to the opposition it would arouse, a number of alternatives controlling movement and slaughter of cattle were proposed. However, if it appeared in Ireland, despite the ban on exports to that country, then immediate slaughter was recommended. In Britain, the inspectors' powers of seizure and slaughter which had come into effect in August 1865 were recommended withdrawn, except in cases where regulations were not being complied with. Although, in principle, a system of compulsory slaughter should be complemented by a system of compensation, the objections to this deriving from the public treasury on an extensive scale appeared to be insurmountable. The recommendations were expressed as follows:

1. The movement of cattle should be suspended. Failing that,
   - for an initially fixed period, no lean or store stock should be sold at any fair or market;
   - cattle should only be moved for immediate slaughter to a market or slaughter-house and only under licence, but the transit of cattle over any public road or in any coasting vessel should be absolutely prohibited;
   - every animal sold for butcher's meat to be slaughtered within a fixed period and cattle brought to a fair or market should not be allowed to leave the area alive;
   - infected areas should be "proclaimed" and only meat allowed out of the area under strict safeguards.
2. Inspectors should be empowered only to separate diseased from healthy stock or seize and slaughter stock if regulations were not being complied with.
3. If the first recommendation was adopted, then imported cattle should be slaughtered at the port of landing and landing allowed only at specified ports. Otherwise, if import is allowed and the cattle come from noninfected districts, then they may be sent by railway to any market in Great Britain.
4. No cattle should be turned out onto common or unenclosed land.
5. Periodical returns should be obtained of cattle and sheep in every parish.
6. Arrangements should be made for its control in Ireland in case it should appear there.

The difficulty of recommending any stern measures was underlined by the fact that, weak as these recommendations were, even so there were abstainers. Four Commission members, led by the two most influential, Earl Spencer and Viscount Cranborne, together with Read and Jones, dissented from the main recommendation, supporting only the alternative proposals. M'Clean, dissented from the entire report. Earl Spencer, Viscount Cranborne, Read, and Jones disassociated themselves from the recommendation of stopping all movement of cattle: "The sudden transformation of the enormous cattle trade by which the large towns are supplied into a dead meat trade, would involve difficulties and dangers of the most formidable kind." Interference with the foreign trade would cause the price of meat to jump. These factors would lead to evasion of the restrictions. They therefore agreed with the alternative proposals. Furthermore, Earl Spencer and Read considered that store animals should be allowed to be moved, provided they had a certificate from a JP certifying them free of disease and having been for a certain length of time on the farm of the vendor.

M'Clean considered that the magnitude of the calamity in no way justified the interference; economically the restrictions were unwarranted, losses in the 4 months amounting to only 0.1% of the national herd. Britain had a population of about 7 million cattle, and less than 1,000 had died of the disease in 4 months. During that time, healthy cattle from abroad had been imported at an average rate of 1,000 per day, so net gain had been 16.6 sound cattle for every one that had died of the disease. Comparisons could not be drawn with 1750 for, compared with other branches of national industry, the cattle trade was now of much less importance, while the numbers of people, their wealth and means of purchasing animal food had greatly increased. Consumption of animal food was not now confined to the wealthy, but had become a necessity for the working classes, and demand was greater than the home supply. Prohibition on import of foreign cattle would affect Britain's commercial relations with foreign countries and create distrust both at home and abroad as to the safety of investing capital in establishing cheap and regular routes to provide food for the community, as well as possibly causing foreign Governments to impose restrictions on export of food from their countries. Agricultural produce to the value of more than £40 million had to be imported yearly, and any interference with this could cause an incalculable amount of injury. It would occasion great and immediate suffering to the labouring classes, many of whom would be thrown out of work, while the price of food would be enhanced, and many able heretofore to eat meat would be deprived of it. This would interfere with the value of labour and affect competition with other countries by increasing manufacturing costs (M'Clean, 1865). In the event, the restrictions were eventually imposed and nothing so disastrous as M'Clean's predicted catalogue of economic disasters took place.

In the Press, only *The Daily Telegraph* critically examined the *Report*, commenting:

> We cannot understand how or why a body of commissioners, including a number of accomplished medical men, should have taken the opinion of non-professional persons as to the origin of the Rinderpest in this country, or to other points on which none but veterinarians of skill can speak with authority. Should we value what any number of non-medical men might say on the origin of diphtheria, typhus or cholera? We consider it waste of time to listen to such witnesses as Mr. John Giblett. He was asked what he knew of the first manifestations of the disease and he replied: 'I believe that there has been a baneful atmospheric influence ... In a hide-bound animal, having fever, it cannot escape except through the hoof, or through the mouth and tongue.'
>
> This gentleman may be able to talk about markets, and the purity of foreign cattle which he imports; but surely we might have been spared the infliction of lessons in pathology from him or from Messrs. Gibbins, Honck, Hicks, Gebhardt, and other cattle importers, who dogmatize so confidently on the origin of the plague in London, and contend that it was not imported. Against this evidence we have that of many competent professors and practitioners of the veterinary art; and it is to be regretted that much labour has been spent in vain by the Commissioners, and much valuable time lost, in prosecuting very useless inquiries amongst a class of people who knew nothing about the Rinderpest before it reached this Country, and have been very careful not to learn too much about it since.

The Times maintained its stubborn ignorance, noting that it was agreed that the disease had been imported, but "the famous importation theory, in the specific shape which it assumed, has not been sustained by the evidence produced." The Commissioners, it added, "do, in the main, adopt the views which have found most favour in our schools of veterinary science."

In their *Second Report* (January 1866), the Commissioners stated that there was but one class of precautions likely to be effective. That was to restrict importation to certain ports, except in case of "stress of weather," where proper accomodation could be provided and where all fat cattle would be slaughtered and all store cattle undergo quarantine. Undried as well as salted hides and skins from any part of the Continent should be disinfected at the port of importation.

Reviewing 18th-century measures, the Commission concluded that the plague was extinguished where local authorities acted upon the Orders with vigour, but continued where they were lax and restrictions with regard to the sale or removal of cattle were frequently evaded. It considered that almost every measure in 1745–57 was tried in vain, the disease then apparently dying out for want of susceptible animals, although communication in England was at that time more difficult. "England has now to contend with the Plague under disadvantages never experienced by any other country. The density of her population, the large quantity of her horned stock, and, above all, the enormous facility of communication by railroad ..." The Orders issued to date had not arrested the march of the plague, "nor can we persuade ourselves that they will materially serve to arrest it, now that it has spread so widely." The Commission came to the conclusion at its first sitting that it would be "a long time before the rules are understood," and the period in which they were violated by ignorance would be succeeded by a period in which they were evaded by design. But its members can have had little understanding of Russian and western European methods when they stated: "England is probably the worst country in the world for the working of a system of certificates, permits, licences, and passports ...."

It concluded: "The present calamity has shown how defective are our general precautions—if any precautions can be said to exist—for the detection and prevention of Contagious Cattle diseases" (Bernard, 1865).

The research commissioned at its first sitting appeared in print in May the following year in the Commission's *Third Report*. By this time, the epizootic was on the wane (although it was to linger for several more years), and essential policy decisions for its eradication and prevention had been embodied in the legislation. The results therefore received little attention and have been slightingly dismissed by some modern veterinary historians taking the narrow view in alleging that the scientific reports added little useful knowledge of the disease, but "provided a fine series of coloured plates illustrating the lesions." This view had been shared by Fleming, head of the Army Veterinary Service, who wrote: "Regular medical men and medical critics have lauded this third report as an immense achievement of high scientific attainments. In vain have I searched through it for a single practical fact worth the paper it is printed on and which was not known to veterinary medicine" (Fleming, 1866).

But those who contributed the research had scant time to conduct it and at least provided information which could be consulted in the future if need arose. More importantly, it contributed to the change in ideas of germ theory which the epizootic had stimulated. The importance of the research lay in the fact that it had helped change the focus in medical science from chemical to biological processes and thus brought British science into line with Continental directions, introducing British scientists to what for many was a new world of achievement (Fisher, 1993). Furthermore, people such as the microscopist Beale, continued their researches, Beale eventually producing *Disease Germs: Their Real Nature* in 1870. In contrast to the veterinarians, many medical men saw the *Third Report* as a model of advanced medical research, and the *British Medical Journal* commented: "... in truth, we know of no disease affecting man, whose characters have been better studied ..." (Anon., 1866g). Burn (1964) saw the gathering of research by the Commission as an important advance in the formulation of government policy based upon expert scientific opinion. However, the conclusion of the *Third Report* that the most important discovery made in Cattle Plague was Sanderson's ascertaining the fact that, after infection, the blood contained the poison of the disease, so that serum obtained from it would give the disease by inoculation, was knowledge which had been available for 120 years and, if now forgotten in Britain, was common knowledge to Continental scientists.

2. LEGISLATIVE FAILURE IN THE 1865 OUTBREAK

According to Burn (1964), the administrative history of the 1865 outbreak cast many of the actors in roles much different from those usually assigned to them in mid-Victorian Britain. Tory magnates and squires ridiculed the administrative capacity of Quarter Sessions and demanded drastic action on a national scale. The Government and the Home Office, far from relishing an excuse for more centralization, defended to the last the local authorities as the proper administrative units. Burn saw the success of the agricultural lobby in a country more than half industrialized as striking.

In September 1865, Read, together with Simonds and others, had visited the Home Office to suggest that it would be desirable to prevent removal of diseased cattle to different parts of the country, but there was a disinclination on the part of Government to make such an order. A suggestion for quarantining stock on arrival from overseas was also discounted on the grounds that there was no accommodation for quarantine, and existing inspection was considered sufficient, but it was agreed that there should be better inspection of cattle from Russia.

The isolation of sick animals was often poorly applied. A case was reported in February 1866 near Chester where a sick cow was shut up but her manure thrown out where healthy animals walked about.

By the Consolidated Order of September 22, local authorities and, where necessary, a Secretary of State were empowered to close fairs and markets, thus bypassing inspectors. Generally dealers were in favor of prohibiting fairs altogether, and two large dealers in Norfolk urged the Cattle Plague Association there to do so. The cattle broker Giblett was in favor of closing fairs but not markets, and especially not the London market, which he considered would be so serious that once the public could not get meat, they would soon have it reopened. Another large salesman, H. Gebhardt, believed that stopping both fairs and markets for 2 or 3 months would not be injurious because a man could go to a private place and buy cattle, but he drew the line at stopping all trade.

An Order of October 31, 1865 stipulated that all cattle had to be in the possession of the vendor for at least 14 days prior to sale. Dealers thus became agents as they could not hold the cattle, arranging sales between sellers and purchasers who often had to buy sight unseen. Frequently, this involved two agents and thus double the commission charged. There was evidence in some cases in the eastern counties that unscrupulous persons had deliberately spread the disease by transporting diseased animals to the neighbourhood of animals ready for butchering, so that by communicating the disease to them, they could then buy them at a diminished price.

Inspectors were deprived of the power to slaughter except where their orders to separate sick from healthy cattle had been disobeyed or were refused. This was followed immediately by another Order in November which virtually opened up closed districts by permitting cattle to be sent through, or out of them, by railway. Or imported cattle could be sent to the nearest convenient railway station. The role of local JPs was replaced by county JPs, so the whole county was controlled, and movement of cattle into or out of it could be restrained. JPs could prohibit animals being brought into their districts or impose their own conditions but could not stop mere transit or egress by railway, nor were they authorized to prevent removal of stock from place to place within their jurisdiction, except to a market, fair, or place of sale. The Secretary of State could close any fair or market which the local authority had refused to if a neighboring authority complained.

Limitations of the provisions were exemplified by the case of the Reverend William Cockett, vicar of Uppersby, charged at Carlisle on December 23, 1865 on two counts of offending against Orders in Council. The first count was buying a diseased cow and not giving an inspector notice of it, the second sending to Carlisle for sale without an inspector's certificate another cow which had been in contact with the former. The vicar pleaded ignorance of the Orders in Council. He was fined £5 for the first offence and 5s. for the second.

In the case of Mrs. Carmichael's farm near Thirsk in Yorkshire, the owner "carefully carried out" the instructions of William Crookes on disinfection, and although nearly every other farm in the neighbourhood suffered severely, this farm remained free of infection. Crookes notes: "This exemption is the more remarkable, as on an adjoining farm, held by a near relative, many cattle died; and during the whole of the time there was frequent communication between the two families" (Crookes, 1866).

In the Lords' debate on the address at the opening of parliament in January 1866, the Duke of Rutland asked: "Yet what had the government done to stop the spread of the pest? Absolutely nothing, or worse than nothing. They had issued vague and contradictory orders, shifting responsibility from themselves to the people. Even at that late hour the government should awake from their slumbers and prohibit the movement of cattle throughout the country and their importation from abroad." In the Commons, it was argued by Grey and others that more stringent regulations would not be tolerated; the principle of compensation had been abandoned in the last century, and it was impracticable to supply London from the dead-meat market alone. But the government was charged with vacillation and dilatoriness, and Sir John Trollope asserted that local authorities had been confounded by the numerous and discrepant Orders. Lowe repeated the argument for action by central government on a national scale, and Henley, concurring, said that movement of all cattle from the London area ought to have been prohibited at once.

Orders of Quarter Sessions were made in every county controlling cattle, but they varied much in detail. Some prohibited all movement, even from one part of the same farm to another if a public road intervened; others permitted traversing a public road provided that certain criteria were met. Most counties allowed movement for slaughter or breeding purposes under licence, but again, details of the licence varied considerably. In March 1866, it was noted that while farmers in Buckinghamshire and other counties were being fined for moving ewes across the road from one field to another for lambing, foreign sheep which had been in contact with foreign cattle were sent from the Metropolitan Market to Leicestershire. One farmer complained that the local authority orders meant that markets in one district were closed, while in the next they were open.

In February 1866, the Duke of Richmond referred in the Lords to a case in point in which the magistrates of a county assembled to put a stop to markets in their petty sessional division, but the mayors of the boroughs within the division declined to do the same. Appeal was made to the Secretary of State, who wrote to the mayors, who in turn replied refusing to comply. Some markets were closed for a few weeks, others for several months. The open markets were unusually full, but no right-minded farmer would buy stock for fear of introducing the disease onto his farm (Anon., 1866h).

The vagueness of the legislation did little if anything to arrest the spread of the disease. It was reported that in spite of every care, numbers of diseased or infected animals continued to be traveled about the country, and in March 1866, the *Leeds Mercury* was still able to carry an advertisement relating to Wakefield to the effect that every facility would be given for the removal of cattle and sheep. The general result of the Orders which had followed was to enlarge the powers of local authorities at the expense of the inspectors, replacing the former by new local authorities with a new sphere of jurisdiction. An Order in Council could only be applied to a town or place within certain defined limits. "… the Government was so vacillating and timid in its action, and the Orders in Council so perplexing, that neither magistrates nor inspectors knew from day to day what the law was on the matter, and frequent applications had to be made to Secretary Sir George Grey for explanations of them, and which were often as unsatisfactory as the orders they were intended to elucidate" (Smith, 1868).

Because the power to slaughter under an Order of August 26, 1865 was conferred upon local authorities and conducted without compensation, it was executed in an indecisive and capricious manner. Playfair (1866) pointed out that graduates of the veterinary schools had not yet rendered extinct the cow leech and horse doctor who were often appointed by local authorities as inspectors and who could not be relied upon in matters of judgment of disease. The Order was revoked on November 23 because of the opposition it provoked. During the passage of the Bill in the House of Lords, several Peers, especially those connected with Scotland, openly avowed to evade the law as far as possible. It was replaced with the provision that an inspector could direct that a diseased animal be kept separate from other animals, and the person having such animal was obliged to follow all directions given by the inspector, "… where the person having such animal in his possession or under his custody shall disobey such direction given by such inspector, then and in such case and such case only, shall such inspector have power to seize and slaughter…such animal." In that period 5,629 cattle were slaughtered in England by inspectors, 79 in Wales, and 1,704 in Scotland. The greatest number was 855 in Norfolk, followed by London with 850. Monetary compensation for slaughtered stock had been paid in the 1745 outbreak and it was the norm on the Continent. Simonds (1865) reported it was paid in Galicia in 1857 when he was there at a rate which he believed to be two-thirds of the value.

Thus, the Order of November 23 withdrew the power of indiscriminate slaughter but permitted slaughter if an owner failed to observe the statutory requirements. Inspectors were also empowered to see that premises were disinfected. Magistrates in Petty Sessions were given extensive powers to regulate or stop cattle trade in their districts, and Government was empowered to take control if magistrates were remiss in the discharge of their duties. But the order still failed to apply a general measure to the whole country. A later order extended the local authorities to Quarter sessional instead of Petty sessional divisions, thus enlarging the area of local action.

Under the *Cattle Diseases Prevention Act* which came into force on February 20, 1866 Local Authorities were to appoint inspectors and order slaughter of infected cattle on the disease making its appearance. The Quarter Sessions formed the JPs and others in each division into Committees to carry out the Act. Compulsory slaughter was not resumed until March 24 after a further 4,225 cases of the disease. In the counties of Salop and part of Stafford, the Local Authorities allegedly refused to implement the slaughter provisions due to a difference of opinion as to whether inspectors appointed by the committee should kill all animals infected with the disease or whether they should report to the committee, which would then make orders for slaughter as thought fit. At Longton, near Newcastle, an Order to slaughter the herd of 30 on a farm resulted in so many people gathering and becoming so threatening that the magistrates decided not to pursue the matter, and only eight animals were killed.

In Stafford, owners of slaughtered cattle laid a claim against the Local Authority for loss of cattle improperly dealt with by it. It was claimed that holders of stock who had sustained losses since the Act came into operation were morally, if not legally, entitled to compensation, as the Committee placed them in a worse position after the Act than previously inasmuch as they were deprived of the services of an inspector. They were left in the position that if they killed their animals without orders, they would not be compensated, but there were no inspectors to give orders for slaughter. Thus, some animals were left to die, spreading the disease, while others were killed by the owners themselves to prevent further losses. One stockholder went to a meeting of the local Cattle Plague Committee to complain of an inspector not carrying out the Act by slaughtering diseased cattle, as was done in other parts of the country, the inspector refusing to slaughter. The Committee informed the complainant that it would meet in a week when it had the inspector's reports, and there would be no compensation for cattle slaughtered before that time. "What is to become of all the cattle that are now suffering, they will all be dead in a week?" exclaimed the plaintif. "Well, all the better," a member of the Committee replied, "let them die, we shall not have them to pay for." The verdict of the Secretary of State was that, after making inquiry, the explanation of the chairman of the committee gave no reasonable ground of complaint.

On March 24, 1866, another Consolidated Order was issued of which *The Saturday Review* reported: "If the Lords of the Council were animated by a set purpose to drive the magistrates and farmers of England fairly out of their wits, they could not have hit upon a better means for the purpose than their order of March 24." It was "utterly vexatious and impracticable ..." defining "borough" differently to the definition in the Act and by imposing a form of licensing which could not be modified according to district, and stock could not be moved unless it had been free of any possibility of contact with the disease for 28 days. A declaration to this effect had to be signed by two persons each of whom had to occupy 100 acres. Whereas in some cases the number of acres held and the number of days of quarantine, could be reduced, the certificate from a JP had to be obtained to move an animal more than 500 yards along a highway.

Another Consolidated Order of the same date continued the prohibition of moving cattle (including sheep) by railway until April 16.

Under the Act, foreign animals were to be landed only at certain ports and not moved out alive, but exceptions were afterwards made for London and cattle allowed to move along defined routes to the City from the ports of Thames Haven, Harwich, and Southampton. However, the implementing Order (March 24, 1866) stated that imported cattle had to be marked by clipping the hair off the end of the tail (and could only be moved subject to licence); but this was the same mark as could be applied to cattle bought for slaughter in a licensed market if there was no other customary method of marking them for that particular market. This section was revoked in the *Contagious Diseases (Animals) Act 1867* and replaced with an empowering section to make rules and regulations as necessary for animals brought in by sea.

From March 1866 to the following July, resumption of slaughter, prohibition of fairs and markets, and other measures caused the number of cases to fall to 726. Although slaughter was now mandatory, within 2 months more than 12,000 animals appear to have escaped operation of the Act, for 6,039 were reported to have died and 6,035 to have recovered. All animals which should have been slaughtered. In the county of Cheshire, Orders of Council and the law were generally disregarded; in no week was the number slaughtered equal to the number infected, although in March, on average 4,000 animals a week were slaughtered. This led to greater than 52% loss of stock in the county, compared for example with less than 7% in Norfolk and 21% in Cambridgeshire. Cheshire was often cited as suffering more than any other county, but it was alleged that the dairy farmers displayed "more than the usual" obstinacy in resisting slaughter. A Cheshire farmer, Thomas Rigby, giving evidence before the

1873 Select Committee, claimed this was not so, there was opposition initially as the farmers could not believe the animals could not be cured, but the authorities were very strict in carrying out the Act. Questioned as to whether the Marquis of Westminster agreed to pay his tenants half the loss suffered from the plague on condition that they would not allow any animal to be slaughtered, he replied that he had never heard of it, although the late Marquis was very liberal to his tenantry in making up their losses. A member of the local authority did oppose the slaughter, but it was nonetheless carried out by the authority in the spirit of the Act (Anon., 1873).

However, a problem arose here for a cattle rate of 7d. in the pound was imposed by the Quarter Sessions on ratepayers to compensate owners whose cattle were slaughtered under the Act, but as the Act was not retroactive, those who had lost cattle before it was passed would receive no compensation for their losses. Many of those appointed to collect the rate therefore refused to do so for more than half of the ratepayers would be exempt as having lost cattle before the Act. It was therefore considered by the Chairman of the Quarter Sessions that it would be almost impossible to collect as there was a most determined opposition to it both in the towns and among the farmers, and the Boards of Guardians responsible for implementing the Poor Law rate resolved not to interfere with the collection. According to the Chief Constable, from February 23, 1866, 36,000 animals died for which the landlords raised a subscription to compensate the owners. The Secretary of State for the Home Department pointed out that there had been no legal difficulties with the Act in respect of collecting the rate generally, and until at least a portion was collected, the Government could not advance a loan to the County for compensation payments. Eventually, Government made available £266,000 to Cheshire for compensation, repayable over 30 years. But despite the laxity of the county in applying control measures, attempts being made to cure infected cattle, a meeting of the Cheshire Chamber of Agriculture in November 1868 considered this repayment an unjust imposition, and Government should be petitioned for a remission of the cattle plague rate. The motion for a remission of the debt being put to the Commons on July 13, 1869, it was roundly defeated by 126 to 85 votes. It was finally agreed that the County would pay back the loan by charging it to the county rate rather than imposing a separate levy.

In evidence before the Select Committee in 1873, the Chief Constable Captain Johnes Smith stated he did not think Cheshire farmers and landowners were more obstinate than the rest of England "led by the powerful "Times" newspaper were. That powerful paper wrote very strong articles against pole-axeing. The disease originated with us in 1865, and lasted up to February 1866, and up to that time the veterinary inspectors tried every possible nostrum to cure the beasts."

"Was it not a matter of common report ... that Cheshire was against putting the pole-axe remedy in force?" he was asked. "I do not think it was so more than other counties, or more than the general public, or more than the 'Times' newspaper. Professor Gamgee stood alone at that time," he replied.

At the same hearing, it was alleged that Cheshire cattle farmers were looked upon as very bad and slovenly: "Their farm buildings and the things relating to the management of stock are, as a rule, very bad, are they not?" "Yes, as a rule, they are very bad indeed," replied Septimus Lambert, a Manchester cattle salesman. "Would you apply that to the large farmers in Cheshire, as well as the small ones?" "Yes; upon the large farms, where they keep cattle; the buildings, as a rule, are very bad; but there are some very good ones, but as a rule, they are very bad" (Anon., 1873).

In June 1866, a cattle dealer sued another for damages, the defendant having sold the plaintiff a foreign cow that he had bought in London, stating that it was from his father's farm. The seller provided a warranty that it was free from rinderpest, but the cow subseqently died from the disease, as did five others which it infected. The plaintiff was obliged to sell 17 other cows at a loss. The jury awarded the plaintiff the price of the five cows as well as the one he had purchased, but the defendant appealed that the plaintiff was only entitled to the cost of the cow sold under warranty and not for those which caught the disease from it. The court, however, considered that the rule should be discharged, the seller being answerable for the consequences of his false statement.

At an early stage, Simonds had recommended to the Privy Council that cattle buying should be stopped for a certain length of time, as also import of foreign cattle. These recommendations were rejected on the grounds that Government might be called upon to frame some resolution which it could not practically carry out. The difficulties of implementing control were underlined by Brown in his evidence to the Commission. At the Barnet Fair, he had a staff of inspectors present for the purpose of inspecting the stock, together with the assistance of mounted police and a large body of men on foot. Slaughtermen were standing by at the different access roads ready to take charge of any animals which might be seized, and arrangements were made for burying any number of them. One result was that large numbers of cattle were not brought to the fair at all, the owners forming private fairs some miles way, and dead animals were found in the fields some 16 or 20 km from Barnet. The other result was that diseased animals which might have been detected were kept back while animals which may have been associated with them, or had passed through infected districts, were brought to the fair and appeared to have nothing wrong with them, subsequently transmitting the disease elsewhere. His opinion was that fairs and markets should be stopped altogether rather than instituting inspection. The penalty for contravention of the Act remained the same as 100 years before, at £20.

In his address to a meeting of Norfolk farmers in September 1865, Simonds had stated that the reason so little could be expected from Government was the lamentable condition of the legislation on these matters. The inadequacy of the legislation was again referred to by the Reverend Francis Lichfield in a letter to *The Times* of January 17, 1866. He had pointed out to the Home Secretary, Sir George Grey, that if a policeman found a drover driving a herd of cattle along the highway, even if rinderpest was visibly present, he could do nothing. The drover was not obliged to answer any question, and the policeman had no power to arrest him or seize and detain the cattle. Only an authorized inspector could interfere if there were symptoms of rinderpest. The cattle could therefore be driven through the country with virtual impunity.

It was not until the Order of September 22, 1865 that the 18th-century measure of burying the carcasses of dead or slaughtered cattle under 8 m of earth in their skins covered with quicklime or other disinfectant was reintroduced. On January 20, 1866, the depth was increased to 9.6 m, and skins had to be scored.

A visit by a fact-finding Irish mission in February 1866, accompanied by Professor Brown and the Cheshire County Inspector, found at Nantwich 22 carcasses had been buried in a small pit within 4.5 m of a stream. The stench was reported as intolerable so little lime if any could have been used. Elsewhere, close to a public road, 46 carcasses had been buried in a shallow pit, the covering clay raised into a high mound. Two miles further on, six cows had been buried within 3 yards of a house. All this demonstrated that there was no inspection of the burials.[1]

[1]In 1868, two persons were charged with digging up, or causing to be dug up, presumably for the bones, portions of animals buried under the *Cattle Plague Act*. The owner of the meadow who gave permission was fined £5, one of those who did the digging 30s., and two others 15s. each, including costs.

The disease was always 2 months or more in advance of the Orders in Council, and the Commissioners pointed out in their *Second Report* that such a mass of variations as existed was bound to cause considerable dissatisfaction and inconvenience. In March 1866, it was reported that one hardly ever opened *The Times* without finding some fresh instruction from Helps, Clerk to the Privy Council:

> Whose words of learned length and thundering sound
> Amazed the gazing rustics ranged around;
> And still they gazed, and still the wonder grew
> That one small head should carry all he knew.

In April 1866, the correspondent for *Blackwood's Edinburgh Magazine* reported:

> When we last met our readers, her Majesty's Ministers were engaged in attempts to cover their disgraceful shortcomings in the matter of the cattle plague during the recess. Their first attempt went to prove that all which could be done was done, and that anything like legislation on the subject would be at once absurd and mischievous. Failing in this, and discovering that Parliament was against them, they prepared a bill in a hurry, and in a hurry introduced it into the House of Commons. At once, and in the most unceremonious way possible, the measure was taken out of their own hands. It was cut, carved, altered, and improved to suit the views of a member of the Opposition. Another bill, intended to supply what was still manifestly wanting, ran its course, *pari passu*, with this, and both reached the Lords in such a state that not a single peer, either on the Ministerial or Opposition side of the House, would take charge of them. The result is, that the Government has been driven to do at last what it ought to have done at the beginning, and because of its neglect in not doing which, the legislature found it necessary to interfere. It has assumed, upon compulsion, the responsibility from which it shrank before the screw was applied. And now, at the eleventh hour, Orders in Council are issued, such as the Commissioners appointed to investigate the case long ago recommended, and which, had they appeared in time, might have stayed the plague, or, at all events, restricted its ravages within limits comparatively narrow.

At the beginning of November 1866, although rinderpest was still smouldering, restriction on markets was eased throughout the country, powers being provided for the granting of licences to hold markets in towns provided that rinderpest did not exist within 16 km. It was considered this would provide a measure of control because if rinderpest did then break out as a result of a market, it would be possible to trace whence infected animals had come. Under the policy of total prohibition, it was believed that many people were surreptitiously evading the law. The President of the Council at the Privy Council Office, the Duke of Buckingham, informed a deputation from the Central Chamber of Agriculture in December 1866 that he believed there to be a great deal of negligence on the part of the authorities. He did not think even in Holland that there had been any evasion of the law which had led to infected cattle being shipped to Britain, but there was reason to understand that the law was habitually evaded in Britain and in many districts the machinery for enforcing the restrictive system was no longer in force. In the Metropolis, notice had been given that the employment of inspectors would be discontinued after the end of December. He repeated that negligence on the part of local authorities created the most serious difficulty in dealing with the evil.

In September 1867, Government brought into effect *The Consolidated Cattle Plague Order* with a separate Order for the Metropolis, the *Metropolitan Cattle Plague Order*. The latter removed the restriction on moving cattle through the streets of London, and cattle could once more be landed and driven on the hoof through the streets to the Metropolitan Market. The stock owners were against these orders and wanted all foreign cattle slaughtered at the port of debarkation and no importation of foreign store stock without an effectual quarantine. It was pointed out that there was no point in separating foreign and local stock at the market and then allowing their dispersal afterwards throughout the country. The Metropolitan Market had instituted a separate area for imported stock but one had to pass through this area to the home stock section, so if there was infection in the former, it would be transmitted to the latter. At a meeting of the Central Chamber of Agriculture on November 5, the farmers resolved to send a deputation led by Read to the Home Office, to lobby for the slaughter of all imported fat stock at the waterside markets of the ports of landing. As store stock were subjected to 28 days' quarantine, it was felt that this should be allowed to continue but extended to sheep, as sheep were being imported at Great Yarmouth without any restriction whatsoever. They had to pass only an inspection by a veterinary surgeon and might then be driven over the whole country.

Another instance of the continuing shortcomings of the Orders in Council was reported in *The Daily Telegraph* in August 1868. A splendid heifer was sent as a gift from his Serene Highness the Duke of Saxe Coburg to Queen Victoria but was detained at Hull for slaughter under an Order in Council relating to cattle imports. However, another Order contained a clause permitting the outward shipping of cattle, so the heifer was shipped to Southampton where entry was permitted subject to 28 days' quarantine. After this period, it could be transported to the Royal Farm at Windsor under a certificate.

Such differences were still unresolved in the 1872 outbreak. In Forfarshire, the county local authority prohibited cattle from being brought into the county from any place outside its jurisdiction. After a large fair in Forfar, cattle which had been brought in, the market being within the limits of the burgh, could not be removed into the county again without infringing the Order, and a number who did so were prosecuted, one being fined a nominal 1s. But it was found that so many people would be involved that prosecutions were stopped and indictments against the others withdrawn (Anon., 1873).

In 1872, the Veterinary Department of the Privy Council collected together the legislation on animal diseases from 20 to 30 countries. The most efficient was mainly enacted after 1865–6. None appeared to be self-acting in respect to rinderpest as was the law in Britain, which otherwise frequently involved loss of valuable time in checking and afforded an opportunity for people to dispose of cattle before restrictions were imposed. This situation had resulted in infected cattle being imported into Britain from Germany before the latter had placed any prohibition on movement. The tendency of modern legislation was to increase the severity of the restrictions, both regarding introduction and spreading of the disease, exemplified by an 1870 Federal law of Switzerland which, apart from severe restrictions when rinderpest appeared, prohibited at all times movement of cattle and horses over 6 months old without a certificate of health. Switzerland, Belgium, and Germany all made use of the military to prevent introduction and spread of rinderpest.

On March 16, 1871, the Austrian Government had hosted the International Conference at Vienna for the attainment of a uniform proceeding against rinderpest, attended by delegates from Austria, Belgium, Britain, Germany, France, Hungary, Italy, Romania, Russia, Switzerland, Serbia, and Turkey. The Conference drew up six *Principles for an International Regulation for the Extinction of the Cattle Plague* (Williams, 1872), but it was considered hardly likely that countries such as Russia, Austria, Hungary, or Turkey would adopt all of them. The principles were:

1. Every country was to communicate direct by telegraph, information of every outbreak as quickly as possible and should publish a weekly bulletin on the state of the plague.
2. Every country should initiate a veterinary strategy.
3. Compensation to be paid for stock and goods destroyed by order equal to their value.
4. All objects used for transport of stock to be disinfected before being used again.

5. Conditions covering international trade when cattle plague has broken out.
6. Measures to extinguish the cattle plague.

The English delegate stated that what England desired before modifying her regulations were additional guarantees other than those that the conference had proposed, guarantees which would result in a greater and freer development of commerce and offer a more solid base to international relationships, whereas the measures proposed (which would free Russia, Austria, Hungary, Germany, Turkey, and Italy from the rigorous measures which the import of their cattle into England were subjected) would so increase the danger for England that still more stringent measures would be required (Anon., 1873). Britain particularly wanted to know if Germany would be prepared to carry out in conjunction with Britain any of the principles, especially the first. By the end of 1872, no definite answer had been received.

Of the suggestions which could contribute to improvement of British law, the Veterinary Department considered that the most worthy of attention were:

- Establishment of uniform international proceedings to prevent spread of disease from one country to another. By the end of 1873, all countries except Turkey had agreed to this resolution, so an arrangement for the immediate communication from one country to another of outbreaks of rinderpest was anticipated.
- Improvement to *The Contagious Diseases (Animals) Act 1869*. The regulations laid down by the Act and the Orders were rendered almost nugatory by the partial and inefficient manner in which they were executed as local authorities differed in their application. Other countries employed government officials either to see the law carried out in all its details or to see that local authorities really enforced it. The enumeration of cattle by the same officials was also considered desirable as a considerable amount was expended in obtaining a census, and this should be part of the duties of the inspectors. Isolation of fresh stock should be made compulsory as in Belgium, for although common sense dictated it, by no means had every farmer carried it out. Licences should be required for the movement of animals as in the time of the rinderpest outbreak of 1865. The power of slaughter should be extended by making compulsory the permissive enactment "may" concerning slaughter of animals which had been in contact with infected animals.

Suggestions that all imported cattle should be slaughtered on arrival were rejected on the grounds that this would encourage the import of diseased animals, as if they were going to be slaughtered it did not matter to the importer whether they were healthy or not. Thus, in 1872 at Deptford where all imported cattle were slaughtered, 2.7% were diseased. At London where slaughter was not required, only 0.008% were diseased (all diseases).

In 1873, a Commons Select Committeee, sitting from March 6 to July 24 examining 47 witnesses, was set up to consider the provisions of the Act regarding import of foreign animals and regulations issued by the Privy Council for carrying out the provisions. As rinderpest was no longer present in Britain, much of the evidence concerned FMD, pleuropneumonia and sheep-pox, especially with regard to the Irish trade and the presence in Ireland of FMD. But it made two particular recommendations. First, Section 66 of the *Contagious Diseases of Animals Act, 1869*, which gave the local authority discretionary power to order the slaughter of any animal which had been in contact with an animal affected by rinderpest, should be altered to make it the duty of the local authority to order such slaughter. Second, discretion should be given to the local authority to order the slaughter of any animals on any premises adjoining those in which the disease existed, upon a certificate of any inspector, being a qualified veterinary surgeon, that such slaughter was necessary to prevent spread of the disease, and compensation should be extended for any animal so slaughtered. Neither of these recommendations, requiring amendment of the Act, was put into practice.

3. INSURANCE

In China, a system of indemnity against loss of animals was introduced during the Zhou dynasty about 400 BC (Dunlop and Williams, 1996), but closer to home, mutual assurance societies for stock were in existence in Germany as early as the 12th century, becoming obligatory in Silesia for all cattle owners in 1765 under a law of Frederick the Great dated November 29. The Prussian government favored private associations, but the following century, a number of German principalities instituted communal offices of cattle insurance against epizootics. In Germany in 1796, D. D. Busch of Hesse proposed a livestock insurance fund, each owner contributing an annual sum to a general fund (Anon., 1797). In the Grand Duchy of Bade in 1834, Baron Elltrichshausen proposed that where cattle were most affected by epizootics, local loan banks should be instituted, but in the years of high cattle losses, the banks rarely obtained sufficient contributions to meet demand, obliging them to fix a rate of interest so high that for many farmers, it was prohibitive.

In France, the earliest known mutual aid association was that of the Brotherhood of the Nativity of Our Lady in the church of Saint Pierre-les-Dax, resurrected in 1553 in the brotherhood's statutes, requiring that if an ox of one of the brotherhood died, the abbott could instruct brothers to help bury the dead animal to avoid contagious diseases and infection "... *éviter à maladies contagieuses et infection*," with a penalty for refusal. A century later, on September 1, 1664, when cattle of the people of Izosse near Dax perished, families helped one another by contributing 30 animals 8 days after the loss. After the delay, it was permitted for the syndicate or the abbott to sell the pledges without any other form of authority. On September 1, 1747, cattle owners of Urgous (Landes) agreed that if an ox or cow died, the other owners promised to contribute a proportion of their stock or money to replace the loss up to 60 livres for each ox and 36 for each cow. But the following year, 12 such associations cancelled their agreements because of mounting fraud, many people buying oxen for 24 livres, causing them to die, and then demanding the 60 livres compensation (Courbet, 1927).

These *ad hoc* measures were replaced by the proposal of insurance against death in 1805 by Pierre-Bernard Barrau, an economist at Toulouse, who created the first mutual assurance company for this purpose (Barrau, 1805). This was to be the origin of all insurance companies in France.

In the Papal States outbreak of rinderpest in 1711, the Pope, to avert an increase in poverty, famine, and hunger in 1713–14, distributed abundant subsidies and those, who in spite of their misfortunes, plied their trades honestly, could apply to the Pope for them.

In Poland, compulsory insurance of all cattle against rinderpest was decreed on June 9, 1857, with the exception of steppe cattle imported for trade. The funds were collected by means of a levy on all cattle over 1 year old once every 5 years, or whenever it was necessary to raise a sum of 100,000 roubles to make payments in compensation for slaughter or death from the disease, and certain other expenses such as the cost of burial. Under this mandatory insurance act, any proprietor who opposed the extermination of his cattle was to have his protests ignored and be tried under the Penal and Criminal Codes for disobedience to the Government authority.

In Portugal the first *Society of Charity for the Loss of Cattle* was created in 1865 at Angra do Heroismo in the Azores, followed soon after by others in a number of districts in

Portugal. In Sweden, societies for cattle assurance began in the second half of the 19th century, and in Holland, assurance against cattle death formed the most imporant branch of the Agricultural Assurance, with a marked development from 1880. These were all private associations without government support (Courbet, 1927).

Informal mutual insurance schemes were first instigated in Britain in the 1745–58 rinderpest outbreak in the form of village cooperatives in Cheshire. At least 46 were known to have existed before 1860, among them the Mawdesley Cow Club in Lancashire, founded in 1807, but obviously not against rinderpest.

Private cattle insurance companies began to be set up in Britain, following the introduction of FMD and pleuropneumonia, the first in 1844, "The Farmers' and Graziers Mutual Cattle Insurance Company." All failed through the premiums not covering the claims (FMD and pleuropneumonia). Gamgee urged the creation of a national system of cattle insurance under Government control, but the objection was that losses could not be estimated, and so insurance could not be based on probabilities as in human life insurance. Whereas annual stock mortality prior to 1842 had been around 1.5–2.5%, including calves, by 1860 it had roughly doubled to 5%. Farr estimated it in June 1866 as 4–5%, but nevertheless considered that an insurance scheme was feasible. He had proposed a state scheme to Government but also a voluntary scheme to the RASE. Society members did not consider a voluntary scheme feasible, but nonetheless a number of voluntary associations sprang up.

On August 1, 1865, a private "National Association for the Prevention of Cattle Diseases" was formed at Gamgee's instigation, the role of which was to supply early information of outbreaks to members, distribute information concerning rinderpest and, by seeking assistance from the Corporations, provide relief for sufferers. In Norfolk, on August 12, the "Norfolk Cattle Plague Association" was formed, headed by Lord Leicester with a donation of £500. This was an apparently successful association with its own Medical Committee which did memorable service for the county, subscribers agreeing not to purchase fresh stock for at least 6 weeks, thereby helping stop spread of the disease. It appointed its own agents and inspectors to keep a check and provided instructions and advice for dealing with and preventing the disease. The outbreak lasted in Norfolk for $14\frac{1}{2}$ months with the loss of 6,380 animals before it was stamped out. The Association paid out £20,368. 6s. 10d. in compensation, roughly double that paid out by the Local Authority, namely £10,112. 2s. 2d. The expenses of the Local Authority amounted to £4,181. 7s. 6d., whereas those of the Association were £1,240. 11s. 9d., which latter included roughly a third of the amount for experiments. At the end of the day, the Local Authority had £2,275. 4s. 11d. on hand, compared with the Association's credit balance of £4,441. 4s. 9d.

The idea of a state scheme in Britain had already been rejected by Gladstone as chancellor on several points. He argued, among other things, that some private schemes had already been set up, and if they failed, the contributors would be obliged to meet both their own losses and, as taxpayers, the losses of others. A State guarantee would relax vigilance "and the ingenuity of self-interest." But his final point was that if the disease extended widely, the price of meat would rise, obliging the consumer to pay more, so those stockowners whose cattle did not suffer would profit hugely. How, therefore, could taxpayers be asked to pay a higher price for meat to compensate for cattle lost, while the landlords and agriculturists would profit?

The "North Staffordshire Mutual Cattle Insurance Company," established in June 1849, raised its premiums in September 1865 to include risk against the Murrain, charging higher rates if within 3.2 km of where "Murrain or other infectious disease" had recently appeared, and higher rates for cattle kept in towns or in the vicinity thereof. A leaflet was issued to subscribers in August to the effect that "those Landed Proprietors who may feel disposed to assist their Tenants, ... if they would pay the extra rate charged for Insurance ... the Insurance of Stock would become more general, and...in their opinion an adequate amount would be promptly realized to meet any emergency that might occur ...." The assets of the company to meet liabilities as they became due exceeded £3,000 and in the week of August 14, 35 new proposals had been received, amounting to £390.

A condition of the insurance was that no diseased animal should be knowingly introduced among cattle of the insured, and cattle should not be put into sheds or buildings which had been occupied by diseased cattle without first being thoroughly purified. An additional clause was that no cattle should be placed near, or with, insured stock which had been purchased in any fair or market during the epizootic, until 21 days after purchase and the cattle known to be free from disease. Any insured stock exposed in any fair or market or public sale also had to be segregated afterwards for 21 days.

In September 1865, a "County of Stafford Mutual Association for Protection Against Loss by the Cattle Plague" was also formed, subject to effecting insurance for at least 10,000 head paying a rate of 3d. in the pound with a maximum insurance of £18 per head. This was roughly a quarter the premium charged by the "North Staffordshire Mutual Cattle Insurance Company."

Another, smaller, Staffordshire association was also proposed in the same month, the "Leek and Moorlands Association for Protection against Loss by the Cattle Plague," covering an area within a radius of 16 km of Leek. Insurance had to be effected on a minimum of 2,000 head at a maximum rate of £15 for dairy cows, in-calf heifers, bulls, and feeding stock, £8 for store stock, and £4 for calves older than 6 months. In the event of loss, compensation would be paid at two thirds the average value of each animal certified to have died of rinderpest. Claims would be assessed at December 31, and if the total payout exceeded the funds in hand, the claims would be paid in a reduced proportion. Cattle among which the disease had appeared were not eligible, and the member must not have introduced any strange cattle onto his farm within 14 days of making application for membership, and must not so introduce cattle afterwards without the authority of the committee. Every member was obliged to observe any requirements of the committee for cleansing, limewashing, and ventilating his farm buildings and premises, and for separating and treating diseased stock, failing which all claims to compensation would be forfeited.

Another mutual association was formed in December for the County of Glamorgan, with a minimum insurance target of 4,000 head. The rate was 1% of the sum insured, with members' liability up to 20%. The maximum value of dairy cows and heifers in calf was £18, store stock £10, and calves less than 6 months £5. Cattle could be insured only if the disease had not appeared among them and had been on the proposer's farm for at least 21 days, unless the owner had taken satisfactory precautions. No strange cattle could be introduced without permission, but there were few other restrictions. The District Committee, formed of the members, had the power to inspect, or order to be inspected, the premises of members, and order them to be properly cleansed and whitewashed where necessary, and stagnant water removed. Any member purposely turning cattle out to graze on any public highway or turnpike road would cease to become a member.

In Aberdeen, the county agreed to provide a fund by a voluntary rate of 1 penny in the pound on agricultural rental and tenants, which raised £4,000. Claimants were indemnified two thirds of their losses when the animals died, or three quarters if they were slaughtered. A separate committee looked after each of 84 parishes and executed the orders of the central committee to slaughter the first animal detected as affected and to watch the rest, slaughtering all animals which had been in contact with the diseased if the disease spread. But in its first report issued in January 1866, it noted that of 226 animals slaughtered, the meat was sold after inspection, while another 143 were slaughtered and buried. By January 10, the outbreak had been wiped out at a cost to the fund of £3,875. But by February 1, it had broken out again almost simultaneously in four herds 11 km distant from each other. These outbreaks were quashed equally.

In the County of Dumfries, Fifth Division, a "Rinderpest Stamping Out Fund" was set up in February 1866 with a voluntary assessment resolved upon by the Joint Committee of Commissioners of Supply and Farmers in Dumfries, based on the landed rental of the County. This came to 2d. in the pound, of which half was payable by the proprietor and half by the tenant.

A "Norfolk Association" was formed in July 1865, which raised a fund of £9,000 by a rate on the Poor-law assessment supplemented by £5,000 from landowners. This fund paid two thirds the value of their cattle to sufferers. In December, another £8,000 had to be added to the fund. By the rules of the fund, the members were prohibited from buying any stock for 6 weeks, but only nine-tenths of the farmers were members. The remaining tenth (probably taking advantage of lowered prices) vitiated the rule causing the plague to spread throughout the county. The Mayor for 1865 finally closed the market, but it was reopened by the new Mayor the following year for fatstock for the Metropolitan Market. The result was that Yarmouth butchers received their cattle from Norwich via London. The Association finally gave up the idea of trying to stamp out the disease and turned to treatment instead.

In the end, Government paid compensation but did not introduce a national insurance scheme.

4. REACTIONS OF GOVERNMENT TO THE 1877 OUTBREAK IN BRITAIN

When rinderpest broke out again in Britain at the end of January 1877, as in 1865 Government hesitated to become involved. It possessed all the necessary powers under the *Contagious Diseases (Animals) Act* of 1869 but again preferred to leave action to local authorities, although when the disease was confirmed in London's Limehouse on January 31, it did respond by issuing an Order within 4 hr prohibiting movement of cattle and sheep out of the metropolis. In a statement to the RASE, Brown stated that it was clear that all of the Acts in the country were passed with the intention of their being carried out by local authorities, but if local authorities simply contented themselves with making regulations for use as soon as rinderpest was discovered, the probability was that the disease would not be detected in time to prevent its rapid extension. By February 7, Government had received the usual complaints from persons interested in trade objecting to the restrictions imposed, but it had firmly determined not to allow any relaxation of the regulations. However, it would be rash, Brown considered, to stop the cattle trade all over the country because the disease existed in two districts. Uniform measures could not be applied to the whole country without harshness and producing grave inconvenience, and he instanced the case of fairs and markets in Cornwall prohibited in the last outbreak on account of the disease in Yorkshire and Essex. It would be possible for the Privy Council to stop all movement of cattle in the East Riding of Yorkshire as it did in 1872, and if local authorities did not exercise the powers possessed by them for this purpose, it might be necessary for the Privy Council to step in.

A proposition by a member of the RASE that all cattle should be slaughtered at the port of disembarkation, in other words that there should be a dead meat trade only, was opposed on the grounds that in some areas, such as the east end of London, the employment of a large proportion of the population, and the supply of food to a still greater number of the working classes, was dependent on the foreign cattle trade. A government minister would wish to avoid an organized opposition which might take place in response to such an action.

The great fault of the *Contagious Disease (Animals) Act* pointed out by one RASE member was its permissive character, which caused varying action by different local authorities. Stock owners would be willing to submit to restrictions if foreign stock could be kept out of the country, but hitherto the Act had not been loyally carried out. The Privy Council already possessed enormous powers under the Act but would not exercise them, except when great alarm had been produced. As a rule, he did not believe in centralization, but no uniformity could be obtained without compulsion. Another member pointed out that whereas Government was empowered to prevent the disembarkation of animals from foreign countries, no power appeared to exist for preventing or controlling the disembarkation of passengers, drovers, crew, or goods, other than cattle, from vessels arriving with diseased animals on board, or for controlling the movements of such vessels.

Simonds considered that local authorities possessed sufficient power to control the disease, but it was not always sufficiently exercised. The first case he had visited was at Limehouse where he met an officer of the local authority, the Metropolitan Board of Works. Simonds informed him that in his opinion, it was cattle plague. The Privy Council Inspector was also there, but Simonds had been requested to confirm the Inspector's opinion. Three cows were lying dead, and one or two were in a dying condition. Nonetheless, he was asked by the local authority representative if the other cows could be moved. He feared that in some parts of the country, the inspectors themselves were not fully conversant with the causes of the spread of the disease, and local authorities were not sufficiently conversant with the precautions it was necessary to take. Simonds considered a "centralised power," the Government, to be a necessity (Anon., 1877). In the outbreak in Willesden, the owner suspected rinderpest on April 3 and called in the local inspector, who took 3 days to make up his mind, and the outbreak was not reported to the Privy Council until mid-day April 7.

In a letter to the *British Medical Journal* Sanderson noted, notwithstanding that the measures required for arresting the spread of the disease were well known, that the steps being taken were avowedly inadequate. The 1866 Commission had shown that the disease could be conveyed by many agencies, but the Veterinary Department, when the infected cargo was landed at Deptford, possessed no power of preventing the persons who had come into contact with the animals from carrying the infective material to their homes. The result was that the neighbouring district of the metropolis became dotted with foci of contagion, but there was no machinery for surrounding each focus with a cordon, and every diseased animal, and every animal which had been in contact, destroyed, and every article which had been in contact also destroyed or disinfected. Two things were wanted: money (for a staff of temporary inspectors) and power (Sanderson, 1877).

Lord Sandon had given notice to Parliament for a Committee to inquire into the outbreak, but Colonel Kingscote MP asked an RASE meeting what a Committee could tell them that they did not know already. He could not see that a Committee would do the slightest good in the world and what was the use of waiting until the Committee had reported, probably about the end of July? Mr. Ridley MP hoped Government did not intend to shelve immediate action by having a Committee; he believed it might be from a desire to find a loophole with regard to the import of livestock. But what they had to impress upon Government was to deal with a different subject altogether (i.e., control of the rinderpest outbreak), and the Government ought to take the matter into their own hands: "Let them [the RASE Council] submit to his Grace that while the Council welcomed the Committee in the hopes that it might do some good, they strongly urged the Government to supplement the action of the local authority" (Anon., 1877).

Following several weeks of discussion by the RASE, eventually a deputation was sent to the Privy Council on April 10, consequent upon which the Privy Council took control with an Order which came into force on April 16 within the City of London and the Metropolitan Police District. As soon as the Order came into effect, inspectors were appointed to implement it.

15

POLITICAL, ECONOMIC, AND SOCIAL EFFECTS

1. POLITICAL TURMOIL IN 18TH-CENTURY BRITAIN

When rinderpest entered Britain in February 1745, it was to be 13 years before it was extinguished, compared with 6 months for the 1714 outbreak. We have seen that failure to contain it is generally attributed to the Government's preoccupation with the Jacobite Rebellion (but this lasted only from December 1745 to June 1746) and the war with France and Spain. Yet the Privy Council issued one Order in Council for the suppression of rinderpest in March 1746 and further ones in December, and January, March, and April 1747. In addition, Parliament passed an Act on 13 February, 1746. Thus, attention was still given to the outbreak despite political upheaval.

The half million cattle lost, an estimated 321,000 between 1747 and 1748 alone, appears to have had little effect on the price of beef or the economy in general, otherwise perhaps there would not have been so much laxity in bringing the outbreak under control. There were no marked rises in beef prices during the worst years, prices remaining well below those caused by severe weather in 1740. Sales at Smithfield Market show a decline from a peak in 1735 to 1740 before rinderpest arrived, although they were lowest in 1748 and did not recover before 1755. Sheep sales, however, rose between 1745 and 1750 by 21% over the 1740–44 average, only falling significantly after 1756. It has been argued that the outbreak created untenanted lands in Cumberland and Westmorland which encouraged arable farming as grain prices rose locally in the early 1750s (Elliott, 1973), but elsewhere, there is no evidence of a switch into arable farming, perhaps because the 1740s were an era of general agricultural depression. However, one anonymous writer states that because of the danger of reinfection stock fatteners in 1748 bought very sparingly, and this rebounded on the breeders: "almost ALL the lands in England, have, for the two years last past especially, been worse to the occupier on this score by at least 20%, upon an average; over and above the still greater loss sustained by the mortality itself; which has been to particular persons, equal to 1, 2, or 3 years rent, in a very short space of time" (Anon., n.d. [1750]).

In northwest England, a great drove road ran through Hesketh and Plumpton in Cumberland, along which thousands of sheep and cattle passed annually driven south from Scotland. The outbreak of rinderpest and the consequent bans on cattle movement and fairs meant that the pasture lands which lined the route could not be let for grazing, which may have led to conversion to arable land, an upswing in grain prices coupled with a fall in stock prices encouraging the enclosure of arable lands which had formerly been common fields. The stoppage of Stephen Kirkby fair in 1746 and cessation of trade in cattle hit Westmorland as hard as it did Cumberland, the emphasis being on cattle rather than cultivation, with cattle reared in the fell townships which, as well as common cow pastures in the valleys, had access to vast areas of upland grazing. But factors such as a decline in stock prices from 1750 to 1790 slowed the rate of economic growth, and it was not until the middle of the 19th century that much enclosure of common fields took place. This was in contrast to the south of Westmorland, where considerable enclosure had already taken place by 1750 (Elliott, 1973). "Beverlac" (1865) reported that his grandmother and her friends were wont to give graphic descriptions of the plague, and the distressing consequences which ensued, but unfortunately such consequences of this outbreak are not on record.

Britain was in worse political turmoil in 1769, with talk of revolution and antipathy toward the king. Assailed by public hostility consequent upon dissatisfaction with the outcome of an election at Middlesex, this had resulted in a number of petitions to parliament to which the public expected a reply in the King's speech from the throne to the Houses of Parliament on January 9, 1770. The petitions were for the redress of grievances, removal of bad ministers, and banishment from the royal presence of certain counsellors. Petitions and remonstrances had been presented from the City, Westminster, Surrey, York, and other places. But instead, the King appeared to use the occurrence of the outbreak of rinderpest to avoid the issues, although the speech was probably composed by the Duke of Grafton. The King was perhaps more cognizant than most of the disastrous effects rinderpest could entail, worsening the present hostile atmosphere:

> My Lords and Gentlemen, It is with much concern that I find myself obliged to open this session of parliament with acquainting you, that the distemper among the horned cattle has lately broke out in this kingdom, notwithstanding every precaution that could be used for preventing the infection from foreign parts. Upon the first notice of its actual appearance, my next attention was to endeavour to stop, if possible, its farther progress; and, as the success of those endeavours must, in all probability, have been entirely defeated by any the least degree of delay in the application of them, I thought it absolutely necessary, with the advice of my privy council, to give immediate directions for every step to be taken that appeared most capable of checking the instant danger of the spreading of the infection, until I could have an opportunity of consulting my parliament upon some more permanent measures for securing us against so great a calamity: and to your immediate and serious consideration I earnestly recommend this very important subject.

A contemporary report states: "The public were much surprized at the silence concerning the petitions, and at the solemn mention of the horned cattle, which filled the place of that important business. It became even a subject of too general ridicule, especially as the existence of the distemper, or at least the extent or danger, did not dispose the people to more serious thoughts." The political commentator Junius, in a letter to the Duke of Grafton dated February 14, wrote: "Yet while the whole kingdom was agitated with anxious expectation upon one great point, you meanly evaded the question, and, instead of the explicit firmness and decision of a king, gave us nothing but the misery of a ruined grazier..." (Anon., n.d. [1874]). To make the matter more ridiculous, it was said that when mention was made in the speech of horned cattle, the Duke of Grafton and another peer, who had both recently experienced the infidelity of their wives, bowed to each other.

The Lords replied to the Address as follows:

> Most gracious Sovereign, We, your Majesty's most dutiful and loyal subjects, the Lords Spiritual and Temporal, in parliament assembled, return your Majesty our humble thanks for your most gracious speech from the throne.
>
> We beg leave to assure your Majesty, that it is with the greatest concern we have understood that the distemper among the horned cattle has lately broke out in this kingdom. We desire to express our gratitude for your Majesty's paternal care and attention to the welfare of your people, in the steps which it has pleased your Majesty to take, with the advice of your privy council, to check the instant danger of the spreading of the distemper, upon the first notice of its appearance; and to assure your Majesty, that we will immediately enter into the most serious consideration of this very important object, and will exert our utmost endeavours in taking such effectual measures as may secure us against so great a calamity.

The King replied:

> My Lords, I thank you for this affectionate and loyal address. Your resolution to enter immediately into the consideration of such measures as may best secure us against the spreading of the distemper among the horned cattle, affords me great satisfaction.

The Commons answered:

> Most gracious Sovereign ... We cannot but look upon it as a very serious misfortune, that, notwithstanding every precaution which could be used for preventing the communication of the infectious disorder among the horned cattle from foreign parts, that most alarming distemper appears to have again broke out in some parts of the kingdom: at the same time, we are truly sensible of your Majesty's paternal care and vigilance for the security of your people, in having given the earliest directions for every measure to be pursued that might be most likely to give an immediate check to the first spreading of the infection; and we will not fail to take this most important matter into our immediate consideration, and to make such provisions as shall appear best calculated to carry into effectual and complete execution your Majesty's salutary intentions; and thereby, as far as by human means can be accomplished, to guard against the danger of so great a calamity becoming general.

The whole social and economic dislocation which ensued in Britain consequent upon the 1745–57 outbreak remains to be assessed in detail, but as Mullett (1946) has pointed out, many of the solutions proposed show that medical knowledge and outlook were still steeped in ancient myth. Yet some represented a viewpoint indicative of a more progressive approach to veterinary science with the steady development of a more scientific viewpoint as the epizootic progressed and people realized that traditional cures were ineffective. But above all, Mullett considers that the Government deserved a tribute, despite Dossie's criticism, for its measures consistently reflected the most enlightened opinion. Farmers and laymen might protest or prefer traditional methods, and some physicians often supported or proposed questionable hypotheses, "but the government took a stand, irrespective of its popularity, in conformity with what now appears to have been the best advice, and except for a few modifications, which were apparently justified, courageously clung to that stand" (Mullett, 1946). Measures taken to combat the disease in the 1714 outbreak have been described by Thirsk (1985) as "a remarkable illustration of how effectively agricultural policy could be decided ... not by MPs ... in parliament, but by government officials, centrally and locally, using their influence without compunction in a spirit of benevolence and determination." In the 1740s and 1750s, the government took direct action, intervening in livestock markets, establishing quarantine arrangements, isolating infected places, and providing for slaughter of infected animals. Government action in the 1769 outbreak, despite the political turmoil, contained and extinguished the disease in 13 months, the outbreak lasting from December 1769 to January 1771.

2. THE 19TH CENTURY

The outbreak in 1865 came also at a time of considerable political disenchantment, on this occasion with the Whig government of the day in disarray following the death of Lord Palmerston and confronted with such issues as the reform of Parliament and the Irish question. Unlike the actions taken by the King in 1769, the political ineptness of the government in 1865 which had won election in July of that year on a manifesto of free trade, contributed in no small measure to the foothold of infection becoming an epizootic; showing that epizootics can have as much to do with political causes as natural causes.

After Queen Victoria's speech from the throne on February 3, 1866:

> The noble Lord who spoke first, attacked and exposed the disgraceful supineness of Ministers in their dealings with the cattle plague. There might have been some excuse for them while as yet the

evil was fresh. Wiser heads than theirs had been puzzled by it, and so might they. But that they should have shrunk from the responsibility of doing in England what public opinion compelled them to do in Ireland and in the west of Scotland, after their own Commission had reported, and the evidence of such men as Professor Simonds came into their hands—that was a subject not only of just indignation, but of astonishment, to all who reflected upon it. Therefore from both sides of the House—from their habitual supporters as well as from the Opposition—came complaints and remonstrances which threw poor Lord Granville quite off his balance, and caused him to do what men generally do under similar circumstances—a good deal to commit himself. He protested that till they had received the report of the Commission the Government were in advance of public opinion; and that if, after receiving it, they lagged behind public opinion, the circumstance was little to be wondered at, seeing that public opinion was so divided against itself. Hence they did one thing to-day and another thing to-morrow; issued Orders in Council which contradicted one another—which were either misunderstood and so disobeyed, or understood perfectly yet evaded with impunity. One thing, however, he made very clear—that, with a well-digested report before them, the Ministers, instead of acting, applied hither and thither for advice; and that up to the meeting of Parliament, or, at the most, a few days before it met, they had absolutely not made up their minds as to the course which they should pursue—whether by legislation or a persistence in measures which, according to their own acknowledgment, had led to no satisfactory results.

The reporter continued:

... Mr. Gladstone, the new leader of the House, crept into his place almost unnoticed ... The debate then began ... which, passing by every other topic in the Queen's Speech, fastened at once upon the cattle plague ... the utmost forbearance was manifested for a while. Scarcely an allusion was made to the supineness of Ministers in days gone by, while speaker after speaker expressed his opinion as the course which it behoved Parliament to adopt against the future. In an evil hour for himself and for his colleagues, Mr. Thomas Baring rose to defend his chief, and the Home Office over which he presides, from charges of neglect of duty which might be implied but were certainly not expressed. Immediately the tone of the debate became changed. Speaker after speaker censured and condemned, where previously they had been content to insinuate only distrust; and Sir George Grey himself made matters worse, by professing adherence to the general view of the case, on which he had acted from the outset. The results were, that though the bill which he promised was not rejected by an impatient House, he was given to understand, in rather unmistakeable terms, that the country would no longer be satisfied with half measures. Nor will it be, nor has the House been satisfied since Sir George brought in his bill—bearing upon the face of it marks of haste in the composition—which, leaving almost as wide a discretion as ever to "local authorities," fixed the measure of the compensation to sufferers from the disease at two-thirds of the value of the beast slain. So far as the matter of compensation went, the bill met with a favourable reception. But the proposal to act through local authorities was strongly objected to, as was the sanction given to a continuance of the railway traffic in live stock. The results were, that without absolutely rejecting the Ministerial measure, the House gave leave to a private member to introduce a bill of his own, which bill has gone forward *pari passu* with that of the Home Secretary, through all the preliminary readings, and into Committee. We venture to say that such a state of things was never heard of before in any country, the supreme authority over which is vested in a Parliament. If her Majesty's Ministers be incompetent to devise a scheme of legislation so eminently touching the very essence of government as this undertakes to do, they are unfit to preside over the destinies of the empire. If they be so conscious of their own deficiencies as to make way willingly for proposals emanating from a private source, the sooner the House of Commons gives them a hint to retire the better. But will a hint suffice? We suspect not. In the only division which took place in Committee, that of Thursday the 15th of February—when it was proposed to leave railways free to convey beasts to certain great borough markets—Ministers were beaten by a majority of eighty; while clause after clause in their bill has either been dropped or postponed, for the avowed purpose of replacing it by something better .... (Anon., 1866m)

Sir George Grey had proposed, and his bill affirmed, that compensation should be paid for every beast slaughtered by the authorities, at the rate of two thirds of the estimated value of the animal, limited to £20. "Mr. Gladstone was constrained, though much against his will, to pronounce in favour of the doctrine of compensation... He proposed that, instead of two thirds, one half only of the estimated value of the slaughtered animal should be awarded to the farmer; and amid the undisguised scorn of some, and the avowed regret of others on his own side of the House, he carried his amendment through the generous forbearance of the Opposition." There was then much debate as to where the compensation money should come from, a compromise finally being arrived at which left towns with a population of 10,000 and over to levy their own rates and provide for their own losses.

The Times (February 8) was moved to comment: "... if they [the Government] were once persuaded that the only way to really efficient measures lies through a change in Government, we cannot doubt that the excitement and energy which now flow in one stream would speedily be turned into another; and if the Administration will not do all in its power to destroy the Rinderpest, the Rinderpest will be found a very serious obstacle to the stability of the Administration."

After the Ministerial bill had passed the House of Commons but was still under consideration by the Lords, *The Star* newspaper commented:

> A Government in a panic is not an edifying spectacle or a beneficent agency. But a Government consenting to act as if in a panic, while all the time in possession of its right senses, is perhaps still less edifying, and is hardly more beneficent. Such is the position in which the Ministers seem at present constrained to place themselves in relation to the cattle plague. They do not appear to have really lost their senses. Mr. Gladstone and Sir George Grey talk coolly and soberly on the subject, and are evidently quite able to view all sides of it deliberately; but when they come to action, they allow themselves to show like the dumb driven cattle whom the poet contrasts with heroes in the strife. They have consented to hurry through the House of Commons a bill which involves principles hitherto unknown to English law and practice, without allowing the great bulk of the community time even to know what it is all about. Last night they allowed Mr. Lowe and his new allies to drive them to imposing the whole burden of the compensation to cattle-owners upon the public rates of the country. On Thursday night some of the Conservative members complained of being taken by surprise, when the Government was, for very shame's sake, compelled to moderate the scale of compensation in deference to the convincing illustrations of Mr. Bright, and the inexorable logic of Mr. Mill. But who, we ask, are the parties really taken by surprise in the whole matter? Surely the entire body of consumers throughout the country. We venture to say that if the consumers were at all homogeneous, to adopt a favourite word of Mr. Disraeli's, in their political composition, it would have been utterly impossible to pass so extraordinary and unjustifiable a compensation clause as that which has been adopted.

The Times continued its strictures:

> The Bill of the Government on cattle disease has had to struggle with more than ordinary difficulties. Among the principal of these may be counted the conduct of the Government itself... They had come to pretty clear views on the subject of slaughter and compensation, but they were still unwilling to admit the overwhelming importance of that prevention of traffic to which all other measures, how stringent soever, are only subordinate and auxiliary... Hence, while calling on the country for a heavy sacrifice, first in the destruction of large numbers of cattle not actually struck with sickness, and next in the levying of a rate to pay for their loss, the Government still adhered to the plan of leaving the most important matter of all, the prevention of movement, to the varying and often inconsistent views of quarter-sessions and municipal councils ...."

Suddenly, in the very climax of the discussion of the Cattle-Plague Bill, as the Blackwood's reporter put it, a declaration was made in both Houses that the Government's Irish policy had entirely broken down and the only hope of averting an armed outbreak was to suspend the Habeas Corpus Act, a motion passed virtually without dissent in both Houses. Thus since the Goverment had hesitated on this issue for some time, it appears that the Cattle-Plague

Bill was thrown up as a diversion in order that Government could pass through this motion without attention focussing upon it.[1]

3. DISAFFECTION AND THE PRICE OF BEEF

On Christmas Eve 1865, riots took place in Carlisle in protest at the slaughter of cattle, and perhaps also because a drover who took two cows found to be suffering from rinderpest to Carlisle market on September 5 was given 2 months' imprisonment.

Elsewhere, workers went on strike for higher wages because the plague allegedly put up the price of meat. Wholesale prices for beef at the major markets averaged $4\frac{1}{2}$d. to $7\frac{3}{4}$d./lb in November 1861, and $4\frac{1}{2}$d. to $8\frac{1}{4}$d./lb in November 1865, a rise of 6% in best-quality beef. But by the end of the month, due to increased supplies, the prices had fallen to $4\frac{1}{4}$d. to 7d., below those of 1861. Price increases were more marked in mutton which, from $4\frac{3}{4}$d. to $8\frac{1}{2}$d./lb in November 1861, increased to 6d. to 10d./lb in November 1865, rises of 21 and 15%. But this also suffered a drop by the end of the month to $5\frac{1}{4}$d. to 8d./lb. Averaged out over the period November 1861 to November 1866, whereas beef showed a continued steady rise in price, mutton jumped sharply in 1865 and then levelled. By November 1866, the increases were 3–14% for beef, good, and poor quality, respectively; and 10.5–13.6% for the same grades of mutton. Beef was $5\frac{1}{4}$d. to 8d./lb, and mutton $5\frac{1}{2}$d. to $9\frac{1}{2}$d/lb. In 1877, it was claimed that the plague had set a tendency for prices to increase which was still continuing (Rea, 1877), but this could have been due to increased demand and spending power (Table 1, Figures 1a–c). In the 10 years from 1851 to 1861, the population of London increased by almost 18.7% or 440,798 persons, but the number of cattle exhibited for sale at the London market had risen by an average of only 1,792 per year, or less than 0.8% and from 1861 to 1864, the supply of

Table 1. Average price of beef 1863–76.

Year	Price (d./lb)
1863	$5\frac{1}{4}$
1864	$5\frac{1}{4}$
1865	$5\frac{3}{4}$
1866	$5\frac{7}{8}$
1867	$5\frac{3}{4}$
1868	$5\frac{5}{8}$
1869	$6\frac{1}{8}$
1870	$5\frac{5}{8}$
1871	$6\frac{1}{2}$
1872	$6\frac{3}{4}$
1873	$7\frac{3}{8}$
1874	$6\frac{1}{2}$
1875	$6\frac{3}{4}$
1876	$6\frac{5}{8}$

Source: *The Farmer's Magazine*.

[1]The Reverend Grindle preached with great pessimism on the Day of Humiliation, March 7, 1866:

> This cattle plague—have we seen its worst effects when we have buried its victims?—not so, its most serious consequences are probably yet to come, in the scarcity of animal food it seems likely to produce—add to this a scarcity of other kinds of food caused by the great demand upon them; add that plague upon man, faint indications of which were visible last autumn; loss of employment for many; the ruin of many others; rebellion at home and war abroad.

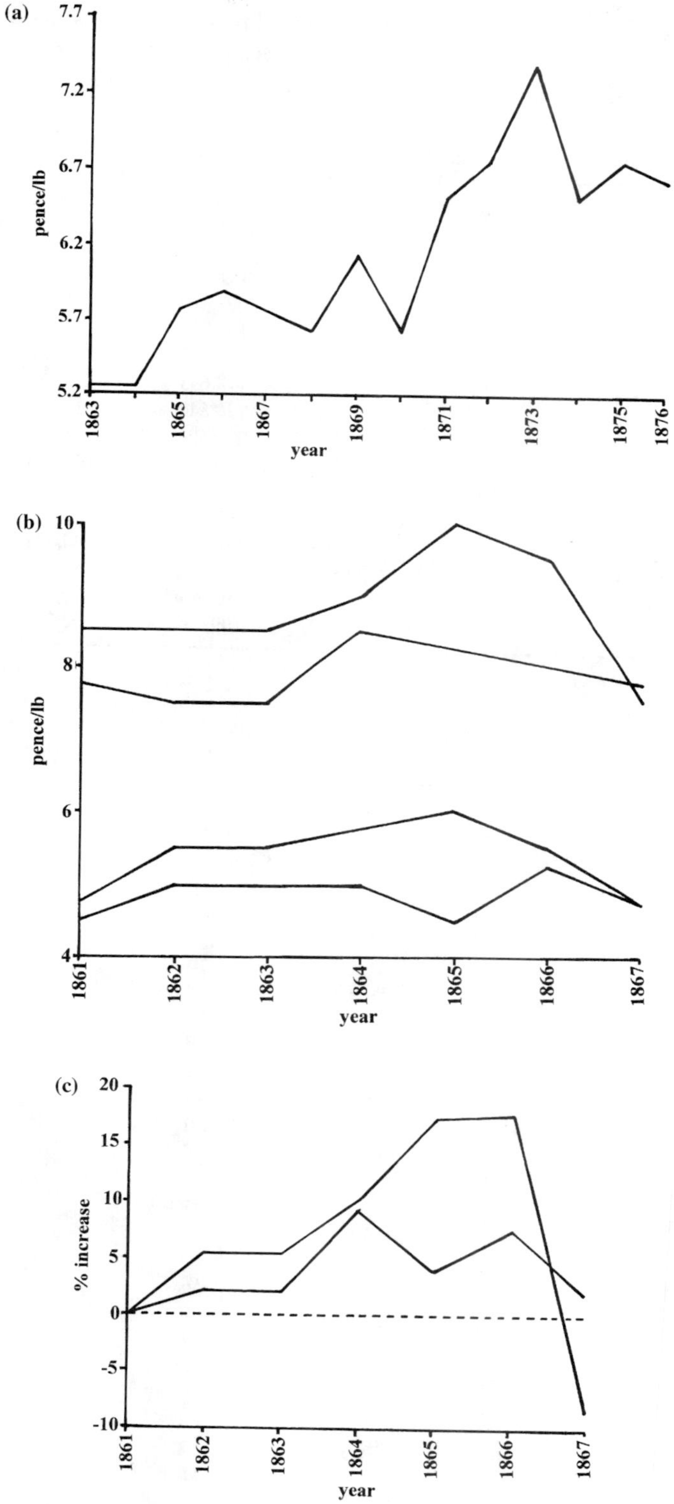

Figure 1. (a) Price of beef, (b) beef and mutton high and low prices, November, and (c) average beef and mutton prices, November.

home cattle decreased by 26,642. The important Christmas supplies show how this was replaced by imported cattle. The annual loss of stock from disease was 7 or 8 million pounds' worth (Tables 2 and 3, Figures 2a–c and 3).

Table 2. Metropolitan Market stock on offer per month 1854–67.

Year	Cattle	Total Imported	Home	Sheep	Total Imported	Home
November						
1854	23,422	7,120	16,302	121,031	16,604	104,427
1855	27,411	7,367	20,044	97,460	17,094	80,366
1856	25,444	6,102	19,342	105,750	16,380	89,370
1857	25,383	4,409	20,974	103,120	17,830	85,290
1858	24,856	4,787	20,069	114,643	18,258	96,385
1859	26,492	5,927	20,565	120,840	21,907	98,933
1860	25,400	6,961	18,439	103,600	22,723	80,877
1861	26,590	5,295	21,295	109,370	27,833	81,537
1862	30,139	6,839	23,300	110,020	28,577	81,443
1863	27,704	11,020	16,684	99,130	30,347	68,783
1864	32,600	17,137	15,463	114,300	34,792	79,508
1865	36,820	16,254	20,566	167,230	52,517	114,713
1866	24,660	13,278	11,382	95,800	38,389	57,411
1867	24,080	10,761	13,319	109,960	33,202	76,758

Source: *The Farmer's Magazine*.

Table 3. Christmas cattle supplies at the Metropolitan Market 1841–66.

Year	Cattle	Top Price (d.)
1841	4,500	60
1842	4,541	56
1843	4,510	52
1844	5,713	54
1845	5,326	56
1846	4,570	68
1847	4,282	56
1848	5,942	56
1849	5,765	54
1850	6,341	46
1851	6,103	50
1852	6,271	48
1853	7,037	58
1854	6,181	64
1855	7,000	50
1856	6,748	60
1857	6,856	56
1858	6,424	60
1859	7,560	64
1860	7,860	64
1861	8,840	60
1862	8,430	60
1863	10,370	62
1864	7,130	68
1865	7,530	64
1866	7,340	66

Top price = pence per stone (8 lbs).
Source: *The Farmer's Magazine*.

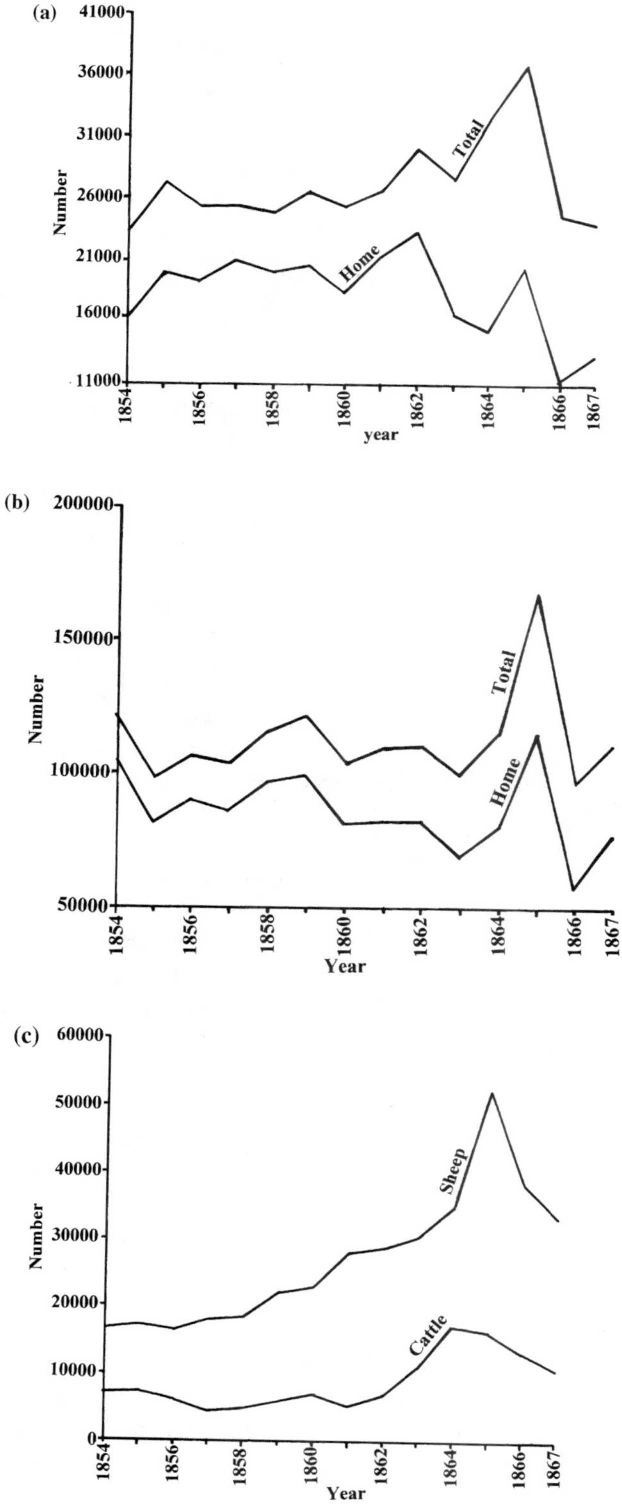

Figure 2. (a) Metropolitan Market cattle on offer, (b) Metropolitan Market sheep on offer, and (c) Metropolitan Market imported stock.

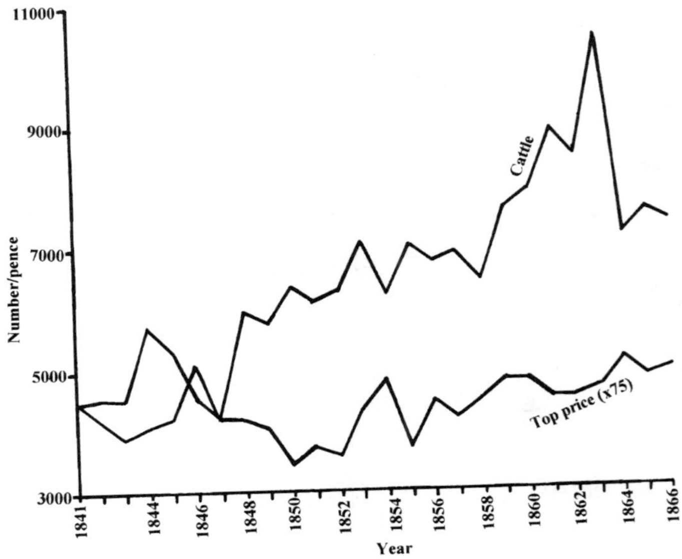

Figure 3. Christmas cattle supplies, Metropolitan Market.

Compared with a prior loss of less than 2.5%, in a report to the Privy Council in 1862, Gamgee claimed that 4–6% of stock died each year following the abolition of tariffs on live cattle and meat imports in 1842 and 1846 as part of the liberalization of trade undertaken by Peel's government. The average annual import of foreign cattle between 1853 and 1860 was 92,172, compared with a loss of cattle from imported diseases during the same period of 375,850. The total number imported he claimed was 553,043, and the loss from disease other than rinderpest was 2,255,100. It was believed that when the market was thrown open, a free trade in meat might exert the same influence on prices as free trade in grain, but for every imported animal, one or two home animals were to die from diseases imported with them. But Gamgee's figures were disputed by Brown (Anon., 1873), who claimed that the total loss would amount to only about 1% and that no records of losses from disease were kept until 1865, querying how Gamgee could have arrived at his figures. Gamgee explained that whereas before there were only limited cow clubs, after the importation of foreign cattle the agricultural insurance companies came into existence, and within a few years all were bankrupt. He obtained his figures from the insurance companies' payments for cattle losses, which amounted to not less than 5% as a rule, and sometimes over 7%. He stated that when he studied in London (1849–54), he saw 21 cows in a week carted away from one dairy of 200 head. This was from pleuropneumonia, a disease unknown at the time. It led him into the investigation of the insurance statistics. "... at the time that the Government endeavoured to bring in Bills to put a stop to these cattle diseases in 1864 ... they ... tried to make out that I had published statistics which were not actually warranted by facts ...." In reply to the question who "they" were, Gamgee answered Dr. Hall Maxwell and other witnesses (Anon., 1873).

Up to January 1, 1872, a cordon, very efficiently maintained by the metropolitan police, was maintained around London to prevent any cattle being taken out alive. This restriction was very much complained of and, with the opening of a Foreign Cattle Market at Deptford, was lifted. But during 1872, the increase in the number of cattle brought to London for sale was only slightly over 1%, or 3,025 above the previous average; which hardly entailed a loss of £2 per head on cattle previously sent to London, as was claimed in evidence before the Select Committee on the *Metropolitan Market Foreign Cattle Bill*, and disputed the claim that before the cordon was in place one third was reexported to the country for consumption.

In 1874–5, Germany repeatedly lobbied Britain to lift its ban on imports from Germany other than Schleswig and Holstein, twice declaring herself free of rinderpest after twice having outbreaks. Following a request in December 1874, the British Government declined to accede, replying after the first outbreak reported in 1875, which had gone undetected for 10 days:

> Their Lordships presume that the proposal of the German Government to remove the present restrictions enforced in Great Britain on this trade, is urged in the interests of German agriculture, upon the assumption that the cattle sent here will fetch a better price, or that a larger number of them will be sent. Upon the first point, it is difficult to form an opinion; but upon the second point ... Germany imports into her territiories as many cattle as she sends to Great Britain.

Under the restrictions 67,609 cattle were imported from Germany in 1875, many infected with pleuropneumonia and FMD.

In 1877, it was considered that the resultant prohibition of beef imports following the outbreak in that year had not raised the price. Mr. Rudkin, a member of the Common Council of the City of London and member of the Markets Committee for 20 years, considered that the country had never recovered from the effects of the 1865 outbreak (Rudkin, 1877). A flesher from Glasgow was also of the opinion that trade had undergone a great change since 1865, before then not slaughtering beef for the dead-meat market. Now there was a dead-meat market in Glasgow, and he bought and supplied the dead-meat market with cattle (Hamilton, 1877). There were also more cattle in Britain before 1865 than in 1877, declining from some 7 to 6 million and contributing to the general agricultural depression of the 1870s. In June 1877, the price of beef was about $1\frac{3}{8}$d./lb dearer than in 1862–3, the price in 1876 being $6\frac{3}{4}$d./lb.

Whereas figures show that there was some increase in the price of meat, Herrman Gebhardt, an importer of foreign cattle into Britain, claimed in evidence to the Select Committee that the price had not increased because it had been kept down by the import of American dead meat. Another argued that because of restrictions imposed when the cattle plague occurred, cattle losses were much less than before as the restrictions affected other diseases also. One disease so affected was FMD, which disappeared in 1866 apparently due to the restrictions in force.

In discussions at the RASE concerning allowing only a dead-meat trade, Earl Cathcart was opposed on the grounds that the poor were largely dependent upon offal for their meat. Another claimed that it would affect the fellmonger's trade, which was dependent upon importation of hides, horns, and hoofs, and it would be a serious thing to destroy such an enormous industry. An opposing point of view drew attention to the fact that many small farmers had never recovered from losses sustained in the 1865 outbreak, and as to fellmongers and other trades which might suffer, many trades had already been sacrificed for the good of the country. He especially instanced mail coaches at the time of the introduction of railways.

Britain had suffered considerable economic loss from opening up its ports in 1842 to free trade in foreign stock. The result had been the importation of diseases such as pleuropneumonia, and FMD caused great losses among British stock, sending up the price of meat and requiring that the market be continually replenished with low-grade foreign imports. Gamgee pointed out that if Britain had never imported foreign cattle, people would have been paying less for meat 20 years after the free trade was introduced than they were in 1863. However, had there been proper control of the imported stock, meat may have been cheaper still than that which a home market alone could produce. As it was, a vicious circle had been set in motion, whereby the cheapness of the foreign imports was counterbalanced by the increased costs of stock-keeping and the reduced head of stock due to losses from continually imported diseases. The rural community did not benefit from the imports, which were overwhelmingly fatstock or dairy cows for urban cowhouses, and it was grazing farmers from the East Midlands who had led the campaign in late 1865 for strict controls on internal trade.

If the Russian claim was correct that rinderpest was spread and maintained throughout Europe after 1842 when the free trade in corn with England began, none in Britain could have foreseen the enormous cost which Britain and the Continent would pay following Peel's measures. The Corn Laws had been introduced in 1815 during the Napoleonic wars, prohibiting import of foreign grain except when the price of wheat in the English market rose higher than 80s. a quarter (12.7 kg). In practice the price, and hence that of bread, fluctuated wildly due to speculators creating chaos among producers. Following pressure from the Anti-Corn-Law League, which was opposed to all duties on imported grain, Peel, as Conservative Prime Minister, in what was considered a bold measure, introduced a new sliding scale of duties for European grain, denounced in Parliament by Richard Cobden as an "insult to a suffering people." Although this opened the gate to European imports, the pressure continued, and Peel finally repealed the Corn Laws altogether in 1846.

4. FARMERS' WORRIES IN BRITAIN

In Norfolk, initially a 6-week prohibition on markets in 1865 was instituted but not renewed because dealers and some farmers were against it on account of the large root crop waiting to be eaten. But as one commentator remarked, it was better to lose your turnips than your cattle.

Prices of hay and straw fell very low because there were no cattle to consume it, so farmers who had lost their stock could not turn to selling their hay and straw. In Cheshire, many persons were reduced from comparative comfort to poverty or great distress from the loss of their cattle. Farmers who bought cattle at £12 to £14 per head, 3 weeks to a month after purchase found the value reduced to £1 due to disease.

One effect was to cause a glut of brewers' grains on the London market, formerly used extensively for feeding cattle in the cowsheds. Annual contracts were traditionally awarded at Michaelmas (September 29), but in 1865 the brewers were unable to dispose of the grains and were obliged to throw them away, some paying to have them removed. Feeding cattle on brewers' grains caused them to produce more milk but of lower quality. As a result of the glut, many country dairy farmers turned to using them, "and the anticipated great boon Londoners expected to derive by having 'pure country milk' imported into the metropolis will be reversed by the reception of 'London-made milk' from the country," as one correspondent expressed it.

The slaughter policy did not always lead to animosity. On June 8, 1869, at the Black Swan Hotel in Bedale, Yorkshire, about 150 farmers and landowners presented the local veterinary surgeon, Mr. J. Fryer, with a testimonal and presents as a token of appreciation of the services rendered by him during the the rinderpest. Mr. Fryer had always held that nothing could be done to save the cattle of the country except by stamping out the disease. The exertions of Mr. Fryer in this respect, convincing the cattle owners in urging them to raise a fund to purchase the diseased animals and destroy them, had prevented the disease spreading further westward.

5. TREATMENT OF CATTLE

Some good came to Britain from the plague, for it led to a more humane treatment of cattle. Already, in 1855, a new and cleaner Metropolitan Market had been opened, and a witness before the Commission in October 1865 stated that sanitary conditions had decidedly improved in late years. Now the market was washed down every week and disinfected. The earliest of 18th-century writers had stressed the necessity of keeping cattle healthy, and instructions issued by the Veterinary Committeee of the RASE in September 1865 urged that every cause which tended to weaken the animal's constitution should be avoided. Animals should be protected from bad weather and given nutritious foods such as cake or corn, and especially pure water. The cleanliness of cattle sheds was stressed, as was that of cattle trucks on the railways, and the accommodation of ships which carried cattle (Figures 4 and 5). Playfair (1866) considered that the mode in which cattle were transported on railways and steamers was a "disgrace to our civilized nation." Railway trucks were of the rudest description, the cattle driven into them by savage force, remaining there for 30–40 hr, sometimes 50 hr, without food or water. When driven into a siding in sight of water, the animals often became frantic in their efforts to try and reach it. Playfair noted, "A cabman in London is fined if he keeps his horse too long without water, but railway directors escape with impunity for their inhumane treatment of the cattle entrusted to their charge." Animals were sometimes so mangled during transit that the carcasses had to be condemned as unfit for food. As to steamers, two vessels reached Lowestoft in 1863 having embarked 608 cattle and 800 sheep. On arrival, 300 of the cattle and 230 sheep were dead, trampled to death when the cattle broke loose during the voyage.

Figure 4. Foreign cattle bound for London from Rotterdam, *Illustrated London News*, August 1865.

Figure 5. Landing foreign cattle at London, *Illustrated London News*, August 1865.

Surprisingly, there were still people such as Dr. Cullen, who the following year (1866) was to become the first Irish cardinal, stating that rinderpest was the judgment of the Almighty for the gross superstition in cattle worship, "or in other words, for the undue attention we bestow upon our animals." Another prelate, Dr. Cumming, expressed almost identical sentiments, "sentiments, however, which will find no echo in a farmers' club" (Howard, 1865).

Cattle were hardly treated with undue attention but on the contrary with considerable cruelty. Referring to an infected shipment which had to be slaughtered, a dealer informed the Commission that because the weather was so hot, and they had so little room, they could kill only 40 cattle a day over a period of 9 days, the cattle being kept penned closely together in the sun. Charles Stevenson, editor of the *North British Agriculturist* newspaper, expressing his opposition to private slaughterhouses, stated that he had been through a number in Smithfield "several years ago" and had never witnessed such scenes, "or could suppose to have been committed in any civilized nation; cattle were standing side by side, and men were busy in slaughtering them." There were, he added, perhaps 100 cattle in one such place. Another witness remarked on the distances cattle were made to travel. A lot in Lincolnshire was brought to a fair on a Monday having left the farm on the Sunday. During the night, they were moved on to Harborough. From Harborough, they were moved on to Northampton and during that night walked on to Wolverton. Then they went on to Banbury and, not being sold, were walked back to Wolverton, thus traveling well in excess of 130 km in 5 days.

In their replies to the Commission, the shipping companies appeared blameless. One representative stated that the vessels were specially constructed for the conveyance of cattle "and are most carefully fitted with stalls and pens for the proper separation and prevention of overcrowding of the cattle ...." And another: "The holds of the vessels appropriated for the conveyance of cattle are divided or stalled off by stanchions, rails and boards, and the compartments for sheep are formed of stanchions, rails and hurdles; by these means, and the ample space allowed for the animals, all overcrowding is entirely prevented." They claimed also that captains had vessels thoroughly cleansed after discharge of the cattle, and the holds from time to time whitewashed and disinfected. But the dealer Gebhardt reported to the Commission that the cheaper companies packed the cattle "as close as herrings." "Every steamer that carries cattle should have plenty of partitions, so that no more than three or four oxen stand in one place; and also they should have room to lie down. It would be much more profitable to the owners of cattle to pay higher freight and have their stock delivered in a good condition, than the present cheap freights, and delivered in a bad one." Ventilation was also very bad in some steamers. He quoted one instance of lots of oxen and sheep in fine weather being smothered from overcrowding and want of air. He was supported by another dealer who also claimed that the vessels were too overcrowded and urged that the practice of placing crates of small stock one on top of another should be prohibited.

But Gebhardt also blamed the regulations for causing suffering to cattle because of the time they had to wait for examination by the inspectors at the wharf without shelter: "... this [inspection] is only done in the daytime, cattle have sometimes to remain at the wharf from five in the afternoon till 9 or 10 o'clock next morning, some standing in the street, others in open places, without the slightest chance of receiving shelter, water or hay." They were not allowed to be driven on a Sunday, so if they arrived too late on a Saturday afternoon to be inspected, the time when most steamers arrived, they had to remain until Sunday night. If the cattle had a rough passage, when they were placed in railway trucks they were often too exhausted to stand and fell on one another, arriving in London in such bad condition that they had to be killed at the railway station.

In a letter to *The Times* in July 1869, a correspondent quoted a railway traffic manager as stating that there was "an increasing indifference—a callousness perfectly shocking—to the sufferings of cattle and poultry."

Like the shipping companies, the railway companies also appeared blameless in their replies. In answer to the question "are the cattle trucks on the Company's railway cleansed after every journey?" most replied "yes," except one which answered with an equivocal "As a rule the trucks, if fouled, are cleaned out." All claimed to clean their pens thoroughly "as often as required," "constantly," "after each lot of cattle has been removed from them," and so on.

If this was being done at the end of 1865, it had not always been so, as was shown by two circulars produced by the Great Western Railway from the General Manager's Office to all station masters. The first was dated October 1857 and noted that complaints were made that the cattle trucks were not regularly cleaned out, and "in some cases, in consequence of the accumulation of manure in the bottom of the trucks, the value of the cattle has seriously deteriorated." They were therefore ordered to clean all cattle trucks at their stations after cattle were discharged from them, and report any empty ones arriving which had not been cleaned. But in January 1861, the instruction had to be repeated, referring to numerous complaints of dirty cattle trucks. This time, any person sending away an uncleaned truck was liable to be brought before the directors.

But it was not until March 17, 1866 that an Order was passed obliging railway or other companies, or persons carrying animals for hire, within Great Britain, to "forthwith thoroughly cleanse and disinfect all pens, carriages, trucks, and boats" used by such Company or person, with lime water or other process approved by the Board of Trade. Such cleansing was to take place every time after use. The penalty for noncompliance was a fine not exceeding £5, but in evidence before the Select Committee in 1873, Brown considered the disinfection only imperfectly carried out. Williams, Secretary to the Veterinary Department, stated that there were "a good many" prosecutions for overcrowding on railways following an Order making it liable and that cruelty was decidedly diminished. In May 1870, the *Animals Order* listed stations at which water had to be supplied for animals being transported by rail, but Brown considered it was not having much effect because drovers considered it safer not to feed and water the animals in transit.

One of the worst carrying trades was that from Ireland, to which some improvement was made in 1874, with inspection taking place to check against overcrowding and proper fittings being installed in the vessels. Disinfection of railway transport was instituted in April 1873, and by the beginning of May, a number of successful prosecutions for noncompliance had already been made, although only a nominal fine of 1s. was awarded by the magistrates. Ferguson reported that it was the usual fine in that locality as there was a great number of magistrates sitting at petty sessions who were "very much interested" in the Steam Packet Companies, awarding such a fine for other breaches of Orders such as shipping cattle in contravention and not reporting outbreaks of disease (Anon., 1873).

In 1877, a Glasgow flesher considered that yet more improvement was needed in Ireland, and others were still complaining of cruelty. James Odams of the Central Chamber of Agriculture and a campaigner against cruelty to animals in transit stated before the Select Committee that the railroads offered "a premium for the inhuman treatment of animals" because they gave a rate per truck irrespective of the number of animals in it, and so the cattle were frequently jammed in the trucks "like herrings in a barrel … the manner in which they come is, I think, simply a disgrace to the times we live in, and an outrage upon humanity." In answer to further questioning, Mr. Odams admitted that things were very much better as regards the state of the trucks in that there was a remarkable improvement with regard to their cleansing, but the adaptation for the carrying of animals was not what it should be.

Considerable improvements came in 1878 with the *Transit of Animals Order* of the *Contagious Diseases (Animals) Act* of that year. Vessels had to have their animal carrying areas divided into pens by substantial divisions, each pen not to exceed 2.7 × 4.5 m, the floor fitted with battens or other footholds, and the place, if inclosed, to have adequate ventilation. Free water had to be supplied at cattle landing or embarking places. Railway trucks also had to have battens on the floors to provide footholds and be supplied with spring buffers. It was prohibited to overcrowd trucks so as to cause unnecessary suffering.

If these measures had any immediate effect, it did not last long, for in 1890 Samuel Plimsoll issued his publication *Cattle Ships*, "being the fifth chapter of Mr. Plimsoll's second appeal for our seamen (Published separately … on account of its pressing urgency …)." "More cattle can be got into a pen by putting them one heading this way, the next heading the other; but if a bullock falls asleep or is ill, and lies down, it gets up no more. It only lies down to die." But the problems now appeared to be with the Transatlantic trade rather than the European one, causing in 1891 a committee to be appointed to "give more complete effect to … The Contagious Diseases (Animals) Act 1878 … and to make regulations for protecting animals from unnecessary suffering during their passage by sea." And in 1898, it was necessary to appoint a committee to "inquire into and report upon the Inland Transit of Cattle," to consider "what further measures … can be taken … to protect (animals) from suffering and to ensure their delivery in the best possible condition."

6. EFFECTS ON LONDON

The 1865 plague was responsible for two major changes in London, perhaps the most significant resulting from a change in dairying practice. Previous to the outbreak, Londoners were supplied with their daily milk by dairy cattle kept scattered throughout the Metropolis on small dairy farms. Many could not be dignified with the name of "farm," but were simply sheds where milking cows were kept (Figure 2 in Chapter 9). The cows were held in the sheds for 4 or 5 months standing in their own dung without moving outside. One cattle salesman giving evidence to the Commission stated it was "a beastly concern," and a "filthy concern." Lord Granville's bailiff, C. Panter, estimated that 50–60% were grossly mismanaged. He claimed that a cow never spent more than 6 months in a London dairy because as soon as the milk yield dropped to about 5 quarts a day, the cows were sold to the butcher. Cows could be replaced two or three times in a year, a person keeping 50 buying 100 a year. But by the end of 1865, the state of the dairies was vastly better than it had been in 1862, the year the *Metropolis Local Management Amendment Act* came into force introducing a system of licensing for cowsheds by magistrates. The Local Authority became empowered to withhold a licence if it was not satisfied with the condition as to size, cleanliness, drainage, and ventilation. The regulations for St. Giles's District prohibited keeping of cowhouses under inhabited rooms, or in immediate proximity to bedrooms, the consideration in the licensing being human health rather than animal, but the conditions resulted in some improvement for the latter as it was considered that cows should be put in the best condition for providing wholesome milk. The most crowded locality for cowsheds in London at the time (1865) was in St. Giles's Parish and the upper part of Drury Lane.

The death of these dairy herds meant that milk had to be supplied from the country. By the middle of 1866, farmers, who 12 months before would have ridiculed the idea owing to the distance it had to travel, the practice often having been tried before and failed, were sending their milk to town daily. Once it was found to be a practicable proposition, the need for maintaining dairy fields in the Metropolis disappeared and the green spaces were given over to building, thus accelerating London's inner-city development.

In February 1879 the *Dairies, Cow-Sheds, and Milk Shops Order of 1879* came into force under the *Contagious Diseases (Animals) Act of 1878*. Cow sheds had to have lighting, ventilation, cleansing, drainage, and water supply, to the satisfaction of the Local Authority, and it was an offence if buildings used in connection with dairying were not so provided in a manner as was necessary or proper "for the health and good condition of the cattle therein." Public health was also improved by regulations governing cleanliness of milk, and that milk from diseased animals could not be used as human food and had to be kept apart from other milk. It could not be fed to other animals unless it had been boiled. Diseased or infectious persons were also prohibited from dealing with milk.

The other significant change was in meat marketing. Two dead meat markets existed, one at Newgate and one at Leadenhall. Previously, the majority of London's meat supply came from fatstock bought in the cattle market by London's 1,700 butchers, who, to keep the meat fresh, held the cattle live at their own slaughter sheds until they were ready to kill them. Not only did the butchers disseminate diseases by dispersing the stock throughout London, but many treated the animals with unmentionable cruelty in their unhygienic premises, and Government wanted to have public centralized slaughter houses. In their *Third Report*, the Commissioners recommended:

> ... it is now the time to carry out fully and permanently the changes required in the mode in which meat is supplied to our large towns ... We would gladly see an inquiry instituted without delay, to determine whether slaughter-houses might not be transferred from all our large towns to suburban points on the lines of railway. A few large slaughter-houses, properly arranged in the best principles, and provided with pastures and sheds where the cattle could rest, would take the place of the innumerable ill-kept and ill-tended places which exist in all large towns ... the saving of pain and torture to the animals themselves would be immeasurable. The Cattle Plague has been a great calamity, but it has put a stop, for a time at least, to much that was cruel and pernicious in operation: we trust that this wholesome effect will be lasting, and that it will pave the way for more extensive reforms.

However, improvement was only partial, and in 1921, Henry Salt was still campaigning against private slaughterhouses (Salt, 1921). The butchers strongly opposed the creation of public abattoirs and a dead meat trade replacing the live trade. They claimed that public slaughterhouses would destroy competition and lead to a monopoly, with an increase in prices to the consumer. The researches of Gamgee into the chilled dead-meat trade greatly changed the meat market, butchers giving up the practice of slaughtering at home. Whereas it was considered that there was enough meat without American imports, the trade nevertheless influenced a change in habits toward a dead-meat market, by the turn of the century imported dead meat supplying nearly half of the home market.

Reports of conditions in the Crimea in 1855, and Gamgee's lobbying, appear to have eventually effected a change in the feeding of the armed forces, for it was stated in June 1877 that the Army and Navy would not buy the chilled American meat, presumably because of deterioration after importation due to lack of chilling facilities. Allegedly, contractors were also not allowed to send it to poorhouses and prisons (Giblett, 1877).

7. RINDERPEST AND THE FRENCH REVOLUTION

Between 1711 and 1743, rinderpest epizootics were a significant factor in contributing to famine in Europe, rather than wars alone. Blondet had painted a miserable picture of the peasants' lot in France in the late 1740s resulting from rinderpest ravages: "Your pastures desolate, your lands uncultivated, your farms abandoned, sufficiently declare your distress" (Blondet, 1748, 1751). In 1770, there was famine, and allegedly 4,000 persons died in the area of Limosin and the Marche alone. A new outbreak of rinderpest, prevalent since 1769, must have contributed in no small measure to this and the riots which followed leading to the revolution, already perhaps heralded by the uprising in 1775 with the deaths of many soldiers in the districts of Landes, Gascoigne, and Béarn, in protest at the shooting of cattle.

In January 1776, along the River Garonne, it was decided to clear the area of cattle. Thus, when the disease broke out, both sick and healthy cattle which had been in contact with them, were slaughtered. The full value of the beasts was paid to the owners, half on the spot and the remainder at some later date. Other healthy animals were to be driven in separate herds inland in the infected region and sold to other people at the full value. Half of this was paid to the owners, the other half being paid if the animals died of the disease within 1 year. The animals were branded with an identification mark and a number. Destocked areas were to be disinfected, and the King encouraged the sale of horses and mules as draught animals in the destocked regions. Persons bringing cattle back into such an area faced a fine of 500 livres. Vicq d'Azyr, following the success of his survey and recommendations, together with the First Physician to the King, drew up a plan for a commission of 8, later 14, members to collect by correspondence countrywide data relating to weather, population, husbandry, and available medical care, and also to advise on matters of policing and welfare. All aspects that might bear on disease were to be recorded, and members could be sent to any part of France to report on outbreaks. Formally instituted in 1776, in 1778 the Commission became the French Royal Society of Medicine, dissolved by the revolutionary Legislative Assembly in 1793.

The outbreak, despite d'Azyr's recommendations, did not cease until about 1784. Only 3 years later, Arthur Young the British agriculturist, undertook his travels in 1787–9. He passed through the Garonne area but made no comments on cattle there and few elsewhere. But in the Bordeaux region, he observed, "These people, like other Frenchmen eat little meat…" In one town only five oxen were killed each year, whereas the same population in England would consume two or three each week. He did not seem to be aware that this was probably the result of years of rinderpest destroying cattle. Young mentioned an immense quantity of poultry in the Garonne region; a circumstance perhaps forced upon the inhabitants by the cattle removal policy. In his summing up of the causes of the revolution which broke out in 1789 during his travels, he laid the blame on the taxes of the peasantry, the harsh penal system and, among other things, the game laws. He made no mention of such events as the cattle destocking of the Garonne region, which must have been still remembered and resented by the inhabitants. He did refer to a local law passed in the district of Rouen which forbade the killing of calves "... it was a preposterous one and opposed by the administration; but it had its full force...." Such a law had been passed in 1745 to protect dwindling stocks due to rinderpest, and this law in Rouen probably had a similar basis of which Young was unaware. But other laws were much harsher than those imposed to control rinderpest. Thus, salt smugglers in districts other than Provence could be punished with death, elsewhere a fine up to 500 livres and up to 9 years in the galleys if armed. Unarmed smugglers carrying the salt on their backs could be flogged and branded, with 6 years in the galleys for a second offence. Even women and children were flogged for this offence. There was extraordinary leniency in the rinderpest control laws compared with many other offences, but nevertheless the peasants must have viewed the slaughter or seizure of their cattle as just another harsh measure imposed upon them, perhaps contributing in no small measure to the dissatisfaction leading to the revolution.

Vicq d'Azyr's instructions issued in 1775 on how to stamp out the disease provided for cordons of troops to surround infected areas with infantry soldiers in the interior of the cordon and cavalry on the outside, the task of the latter being to pursue drovers or hide merchants who had evaded the infantry guards. Yet in November of that year, it was stated that the method had not been successful, infection remaining in several parishes and also appearing in new places. This was attributed to persons not reporting the disease, suggesting a lack of cooperation because they did not want to be subject to control by the troops.

Young referred to the region around Poitou as a "poor, dull, and disagreeable country," "unimproved, poor, and ugly...." Ten years later, it was noted that hundreds of thousands of hectares of waste land had been put under cultivation, and La Vendée was a region of large farms and stock-raising. How much might the years of rinderpest ravages have contributed to the waste land which characterized so much of the French countryside in Young's time? The possible role of rinderpest and the imposition of control measures it occasioned have perhaps been neglected as a contributory factor leading to the revolution in 1789, totally ignored by its leading historians. Sée (1927) referred to crises in the French countryside in 1725, 1740, 1759, 1766–8, 1772–6, 1784, 1785, and 1789. We have seen that there were major outbreaks of rinderpest in 1745, 1765, and 1774–6; although the country then had an apparently clear period for 13 years up until the revolution. But the events prior to 1776 were probably the main cause of the impoverishment of the countryside and not, as Michelet (1847–53) claimed, the seizure of cattle in lieu of taxes until the peasant was left with nothing: "No more manure. The cultivation of corn... decreases in the eighteenth [century]... as the cattle may become extinct, so also the land now appears dead... men, women and children yoke themselves to the plough...." It was not taxes, but rinderpest, which had removed the cattle. Taine (1875) referred to many farms remaining uncultivated in 1750–60, many even deserted, quoting a source that one quarter of the soil was absolutely lying waste. But there had been an estimated 90% mortality of cattle in 1745. Lefebvre (1951) referred to drought killing off much of the livestock in 1785, Taine (1875) stating that it killed off all draught animals around Toulouse. Although this was not rinderpest, historians note other causes of cattle destruction such as drought, but not the equally or even more devastating rinderpest. Sée (1927) noted that the great drought of 1785 compelled farmers to sell a part of their cattle. If this was a contributory cause to the revolution, how much more so must 90% or more losses due to rinderpest have been? Labrousse (1943), in his analysis of the agricultural economy leading up to the events of 1789, failed to mention rinderpest at all; only Hibbert (1980), in his introduction to *The French Revolution*, briefly mentions that "in times of scant harvests or epidemics of murrain," many French peasants went hungry. In 1865, Budd surprisingly wrote:

> In 1811, this epizootic, following in the track of the Austrian wing of the Allied Armies, was sown broadcast through Germany and the North of Europe and brought to the walls of Paris.
>
> For the first time on record, the French, in addition to their other calamities, had the bitterness of seeing their herds destroyed by an epizootic which was as deadly, as it was entirely new to their experience.

8. THE NETHERLANDS

The great 18th-century losses to the important dairy industry of Holland affected all walks of life. Municipal and provincial authorities granted tax reductions to affected cattle owners as many were unable to pay. Receipts from the Frisian tax on fireplaces, heads, cows, horses, and arable land, fell by 53% in 1744–5, from 282,000 to 133,000 guilders. In the province of Gröningen, the loss in 1745–6 was 15%, revenue falling from 789,850 to 670,800 guilders (Faber, 1962). There is some evidence that the few who lost little or no cattle became very wealthy because of the high prices, but rising prices were an added burden to those who had lost all or most of their stock, reducing them to poverty. Many could not pay their land rents, and so rents fell and landowners suffered. As well as the loss of stock, there was a decreased output of milk, butter, and cheese, thus forcing up the price to the non-stock-owning consumer. The quantity of cheese brought to the weigh-houses in the northern part of the province of Holland decreased from about 8,400 tonnes in 1768 to 5,670 in 1770. Table 4 shows how butter prices fluctuated in the important dairy region of Friesland, and, later, the price at Leiden.

In the pasture districts, many cattle owners switched to sheep after the loss of their cattle, the quantity of wool brought to the weigh-houses in North Holland rising from 444.5 tonnes

Table 4. Average butter prices in the Netherlands 1741–70.

Year	Price (guilders)	Observations
Friesland		
1741	26.05	High following 1740 famine
1742	16.52	
1743	15.81	
1744	14.26	
1745	25.53	Rinderpest outbreak
1746	21.67	
1747	21	
1748	22.61	
1749	22.01	
1750	19.6	
Leiden		
1766		
1767	21.17	
1768	24.73	Rinderpest outbreak
1769	27.5	
1770	32.21	

Source: Faber (1962).

in 1768 to 558 tonnes in the peak year of 1775. Conversion to arable land in Friesland and elsewhere was mainly only temporary for cattle were needed to maintain its productivity with manure, although the amount of arable land remained higher after 1769, and stock levels stayed beneath former levels until 1805 (Faber, 1962). In the pasture regions, the nature of the soil did not allow the farmer to convert from livestock to arable farming, so his option was to turn to sheep, but these were not as profitable as cattle. Generally, the pasture areas seem to have been restocked with imports from Denmark and Germany, especially after 1748, but these sources would have depended upon how these countries had fared themselves. For the arable farmer, however, paying high prices for animals which were primarily for manure production was not a worthwhile option. The result was that in the pasture areas of Friesland, after the 61% loss of herds in 1769 of the total stock of some 160,000, the former level was regained in 6 years. In the arable region of the north of the province, this took 23 years.

9. OTHER COUNTRIES

Consequent upon Denmark's devastating outbreak in 1745, Holberg (1746) argued that it would not necessarily have as bad an economic effect as people proclaimed. Once the outbreak ceased, meat would become expensive so young animals should not be slaughtered, but farmers would change to rearing other animals rather than cease farming, and it could propel them into growing vegetables and fruit, while grazing meadows could be changed to grain instead of leaving them idle. It was better, he asserted, to lose cattle than horses which were much more needed. People could eat many other types of meat, and butter was not a necessity despite being almost the fourth most important item in a household. Other people lived without it, and one could use honey or syrup on bread instead of butter.

In the event, it did not change habits significantly in Denmark, but it did profoundly affect the lives of many peasants in Europe in a manner as yet little appreciated by social historians. In 1857, when Simonds observed the outbreak in the Austro-Polish dominions near the Carpathian Mountains, where peasant life had probably changed but little since the previous century, he noted that the disease prevailed chiefly among the cattle of the small proprietors and peasants: "... the daily lamentations of the poor women, to whom was confided the principal care, or rather, on whom was forced the labor of nursing and feeding the animals, at the losses they were sustaining, was most distressing to witness. Nearly the entire means of living of these small farmers depend on the well-being of their cattle ...."

In 1861, the plague in Egypt, by carrying off an immense number of camels, completely paralysed the trade of Upper Egypt through the high rise in transport charges which took place in 1864 and 1865 (Baker, 1867). In October 1863, Lady Duff Gordon wrote that everything at Alexandria was almost double the price owing to the cattle murrain and the high Nile flood, and the director of Briggs Bank estimated the loss in cattle already at £12 million. In February 1864, she stated that the loss of cattle had suspended irrigation to a "fearful extent," and the harvests of all kinds of Lower Egypt would be very scanty. In May 1864, she reported: "They are threshing the corn with donkeys, and men are turning the sakiahs (water-wheels) and drawing the ploughs, and dying by scores of overwork and want of food in many places. The whole agriculture depended on the oxen, and they are all dead" (Waterfield, 1969). But left without draught animals, this led to large orders in Britain for steam engines and steam ploughs, to the advantage of Britain's manufacturing industry. The Viceroy of Egypt considered that by compelling cultivators to adopt an improved system which would both civilize and enrich them, the plague had conferred a lasting benefit on the country. In October 1864, Lady Gordon reported that a Greek had bought up wheat at the beginning of the year at

60 piastres the *ardeb* and was now selling it at 170, and people had to borrow at 3.5% a month. Two men she knew were quite ruined and had sold all they had, the cattle disease forcing them to borrow at the high rates (Waterfield, 1969).

In Cambodia in the 1920s, the loss of a pair of bullocks or buffaloes represented 200 days' salary for a peasant. In order to restock peasants borrowed from Chinese moneylenders, and as successive epizootics affected their stock, they became completely entrapped.

10. THE SCIENTIFIC LEGACY OF RINDERPEST

Medical science benefited in 1865 with a grant of £5,000 by the Commission for research into rinderpest, the largest grant yet made available by the state for medical research. The disease had an influence on the leading figures in science and medicine of the day, some of the elite sitting on the Commission. But there was no immediate revolution in medical thought and science as a result of the outbreak, and it was the failure of medical science to produce alternatives that eventually led to the acceptance of a slaughter policy. An effect was to bring prestige to the veterinary profession in Britain, turning attention from horse doctoring to livestock and the nation's food supply. For the first time, there was Government acceptance of a veterinary dimension to agriculture. Rising prices and declining or stagnating numbers of stock due to disease were a major inhibition to agricultural adaptation in the late 19th century, but the shock of the plague brought considerable benefits in changing farmers' perceptions, inducing a new and positive attitude toward the general problem of cattle disease (Fisher, 1980). Although in 1873, John McBride, Professor of Veterinary Medicine at the Royal Agricultural College, Cirencester, complained to the Select Committee that the Government supported the medical profession but not the veterinary, when queried: "I understand you to say that since 1862 the veterinary professors have learnt nothing or scarcely anything with respect to the contagious diseases of animals?," replied: "Quite so" (Anon., 1873). Having answered in the affirmative the question: "Does it appear to you that veterinary surgeons are of much use anywhere?", he was then asked: "But when gentlemen come here and recommend that animals which are affected with various diseases should be slaughtered, and that animals in contact with those animals should be slaughtered, and that there should be an entire stoppage of trade in cattle between foreign countries and this country, does it not look like a declaration of complete impotency upon the part of the veterinary profession?" He replied: "Decidely not; it looks as if they had some knowledge of the value of pounds shillings and pence; our practice is wholly a matter of pounds shillings and pence" (Anon., 1873).

Public health benefited from the attention drawn to the filth of dairies and animal transportation, and from the decline in the London dairies in favor of rural practices. But although Britain was subject to a cholera epidemic in 1865–6 with the death of 15,000 persons, *The Times*, as Fisher (1993) has pointed out, devoted more than three times as many editorials to rinderpest in 1865 than it did to the cholera outbreak; and in 1866, more parliamentary time was spent discussing the question of compensation to stock-owners in Cheshire than on measures to check cholera. Cholera deaths were largely among the working classes, and when Simon attempted to secure a small sum for research on cholera, he was sharply rebuffed by Disraeli and the Treasury. The easy passage of the *Public Health Act of 1866*, which increased the powers of the Medical Department of the Privy Council in relation to local authorities, was probably more due to concerns over rinderpest than cholera. Bagehot considered the manner in which cattle disease legislation was hurried through Parliament savored of despotism: "The greater number of them [the landed interests] feared for their incomes" (Bagehot, 1872).

According to Fisher (1993), the 1865 outbreak helped undermine Lord John Russell's government, and it temporarily disrupted the pattern of agricultural change in Britain. It also provided the basic legal and administrative framework for future animal disease control. However, Worboys (1991) saw the outbreak as ultimately having a retarding effect upon veterinary science, for after the initial impetus given to it and the failure to find a cure resulting in restrictive legislation, all animal diseases tended to be viewed as imported. Thus, the accent was switched to controlling animal imports with further diseases being added to the legislation, rather than continuing research into origins and causes, viewed as only of academic rather than of practical interest. By 1891, research into animal pathogens had all but ceased and indeed was avoided for fear of undermining the legislation which was seen as the primary agent in preventing animal disease. A century later, one commentator was observing that the relative effectiveness of the vaccine in use for the control of rinderpest probably helped account for a lack of interest in the industrial countries in exploring the epizootiology and biology of transmission of the disease. Given the availability of the vaccine and the high standards of veterinary services and animal hygiene in the United States, Europe, and Japan, rinderpest was considered unlikely to be a problem there, and so research was not a high priority (Walsh, 1987).[2] In this, he was echoing Hutyra and Marek (1916) who considered that if areas in Europe were free of infection for a long time, experimental and practical vaccination would become of scientific interest only, because veterinary enforcers would cling to approved methods of containment, implementing border controls and using legal regulations and penalties.

Rossiter (1996) has pointed out that due to the widespread use of cell-culture vaccine in the 1960s, although suitable techniques were developed at the same time, enzootic rinderpest in unvaccinated cattle has never been studied, the best results coming from studies of wildebeest and buffalo in East Africa's Serengeti-Mara region. In fact, we were back to the situation when Gamgee insisted upon teaching control of the disease to reluctant pupils who argued that it was a waste of time because the disease did not occur in Britain.

But then during the 1990s, molecular biology focused on morbilliviruses created huge leaps forward. Second-generation recombinant vaccines became available, which can be used to protect animals from disease while enabling naturally infected animals to be distinguished from vaccinated animals. Molecular techniques improved the speed and accuracy of diagnosis, and it became possible to study in precise detail viral pathogens which were previously difficult to isolate in the laboratory, or from poorly preserved specimens collected in the field, leading to a new situation where the genomes of rinderpest and PPR, which are difficult to grow to high purity in culture, can be studied down to the nucleotide-sequence level, making it now possible to trace the movements of the viruses across continents. Barrett (1996) claims the "most dramatic achievement of molecular biology, at the ending of a century which has seen so many startling scientific advances, is the ability to rescue infectious rinderpest virus from DNA copies of its genes. This technique has given us the potential not only to produce a genetically defined vaccine but also to understand the molecular mechanisms which determine the pathogenic nature of the virus."

[2]The 2000–01 outbreak of FMD in Britain will probably demand a revision in thinking; the structure of the modern farming industry meant that it could not be controlled as effectively as in the 1967 outbreak.

Part IV
CURES AND REMEDIES

16

THE SEARCH FOR A CURE

Desperate Diseases Require Desperate Remedies

1. THE FIRST THREE THOUSAND YEARS

Five thousand years ago, the Ancient Egyptians recommended lowering the animal's body temperature for the treatment of rinderpest, but then resorting to magic. Making incisions in the nose and tail of the beast, one chanted to it, "he that has a cut either dies with it or lives with it." If, after this, the cow did not move and was inert to the touch with eyes closed, the eyes were to be covered with burnt linen to stop the discharge from them (Smith, 1933). Cutting the animal's tail was still practiced by some in Britain in the first 18th-century outbreak. By Roman times, concepts of veterinary medicine had developed to a state which was then to stagnate for 2,000 years.

Cato, in 200 BC, provided the following remedy against epizootic disease: "... give to the cattle a mixture of salt, laurel leaves, onions, cloves of garlic, incense, powdered rue, and burning charcoal, made up with a little wine." If a cow still became infected then: "... make it swallow an egg whole, and the next day give a clove of garlic beaten up in wine." In 1865, herbal mixtures, wine, and especially onions and garlic, were still being recommended as antidotes. Columella's remedy for "conquering and repelling" pesilential diseases in AD 55 was a drench composed of roots of "allheal" *Hypericum androsaemum* and eryngo *Eryngium maritimum* mixed with fennel seeds sprinkled with must boiled down, to which wheat flour and boiling water were added. If that did not cure the animal, it was recommended to pour into the nostrils a concoction of equal parts of cassia, myrrh, frankincense, and blood of a sea turtle, mixed with three *sextarii* of old wine. The blood of a tortoise mixed in old wine with fennel seeds, myrrh, and frankincense, was to be suggested again in the 18th century.

Allegedly, a remedy to protect cattle in times of epizootics in the Roman states in the 4th century AD was to brand sick animals on the forehead with the sign of the cross, using a red-hot iron. Although attributed to the 16th-century Baronius (Layard, 1757), all he wrote was that in the Europe-wide plague of AD 376, the plague was that easily cured by Christians by the sign of the Cross. Whether such sign (the sain) was made in the air or actually on animals is not said; nor indeed whether reference was being made to people or animals. But by such a great miracle, Gentiles were converted to Christianity (Baronius, 1739).[1]

The Anglo-Saxon Leech-Book, compiled about AD 1000, provided two brief herbal recipes. If cattle were dying, it recommended pouring into the mouth a concoction made from groundsel *Senecio vulgaris*, springwort (*Asarum europaeum*?), cockspur grass(?)

[1]Although later crossed out, the accounts about 1426 of Bishop Langley of Durham disclose that 9 pence was paid to Thomas Egliston for "signing" 16 head of cattle with St. Wilfred's signet to ward off the murrain, indicating belief in "signing" persisted long after.

Echinochloa crusgalli and cleavers *Galium aparine*. For "lung disorder," possibly an early indication of pleuropneumonia, then a concoction of holy water with a wort which bore black berries (vetch? *Vicia sativa*). Fumigate with a mixture of this wort, wild iris leaves *Iris pseudocorus*, fennel *Foeniculum vulgare*, hassuck (= hassock) grass *Elytrigia repens*, "cotton," and incense. "Make five crosses of hassuck grass, set them on four sides of the cattle and one in the middle; sing about the cattle Benedicam etc., and the Benedicite and some litanies and the Paternoster, sprinkle holy water upon them, burn about them incense and "cotton" and let some one set a value on the cattle, let the owner give the tenth penny to the church for God… do thus thrice" (Cockayne, 1864–1866).

In the Middle Ages, religious and superstitious practices largely replaced the medicinal methods of the Roman authors, and procedures followed into the 18th century such as collective prayers over animals herded together, or the gathering of animals at church doors, served to propagate infection. But realizing the danger, this was forbidden in 1745–6 by the Prince-Bishop of Liège, Jean Théodore, on pain of excommunication. As late as 1797, Faust wrote that no wise person now believes that three crosses on the stable door will prevent the disease. But by the 17th century, superstitions became much more vicious than simple offerings of prayers and marking of symbols.

2. WITCHCRAFT AND CATTLE PLAGUES

Of some 500 trials of both men and women for witchcraft in Europe between 1300 and 1500 held principally in Italy, France, Germany, Switzerland, and Britain, most relate to sorcery causing alleged death of persons, but a series in 1395–1405 refers to deaths of animals as well (Table 1). This is followed by some 25 other trials involving animal deaths during the century, 10 of which also included charges of "weather magic." So the deaths of people and animals, cattle being cited most frequently when specific reference is made although occasionally pigs and horses are mentioned (sometimes infants), were clearly linked to famine conditions engendered by climate and/or disease. Kieckhefer's (1976) graph of the number of trials shows several distinct peaks, all of which include prosecutions for animal illness and death. We can conclude tentatively that these peaks equate to years of epizootics. After the first indictments in 1395–1405, there is a gap of 50 years marked by a steady increase in the number of trials, but with no reference to animal sickness or death until 1450. From then on, there is a constant number, except for 1470–74, doubling in 1490–99, although other cases began declining after 1494 (Figure 1a,b). However, for the peak years covering 1450–69, there do not seem to be any continental European records of epizootics, the only records coming from Ireland for 1450, 1464, and 1473. In the next period 1475–99, 1480 witnessed a European epizootic with a third of the animals lost in Germany and Switzerland in 1481. Cattle deaths are reported from Poland in 1491 and severe mortality among wild animals in Spain in 1494. Then, coincident with the sharp rise in trials, there was a "dreadful" murrain in Germany in 1499 and 1500, and although records are lacking, it may be supposed the epizootic had persisted through the intervening years of this period.

Monter (1976) listed 833 trials taking place in Switzerland and the border area with France between 1525 and 1684. Unfortunately, the indictments are not given, but plotting the number of trials, we find that the peaks coincide with recorded periods of cattle murrain in Europe. The highest peak, at Geneva in 1565–9, comes within 10 years of the murrain beginning in Germany in 1559. Although the plague is not recorded for Switzerland, it conceivably spread to there, and as it persisted, more and more persons were accused of bewitching cattle. After a sharp drop, the number of trials begins to rise again in 1595–9, coincident with

Table 1. Witchcraft trials involving animal illness or death 1300–1499.

Year(s)	Place	Accused	Indictment	Sentence
1395–1405	—	Numerous persons	Inflicting illness on men and animals	Burned
1397	Appenzell	A woman	Inflicting disease on animals	Executed
1454	Lucerne	Dorothy Hindremstein	Inflicting illness on men and animals Inflicting death on animals	Burning (*in absentia*)
1455	Torcy	Two men, one woman	Causing illness and death to animals	—
1457–9	Faido	32 persons	*I.a.* affliction of men and animals	Most burned
1459	Andermatt	Catherine Simon	*I.a.* inflicting death on animals	Beheaded
1459–62	Arras	32 persons	Causing harm to men and animals (one case)	Burned?
1460	Lyonnais	Number of persons	*I.a.* inflicting illness and death on men and animals	—
1461	Lausanne	Man	*I.a.* inflicting death on animals	—
1466	Biel	Man	*I.a.* causing harm to men and animals, causing death to animals	—
1467	Biel	Man	*I.a.* infliction of harm on animals	—
1477	Lausanne	Man	*I.a.* killing people and animals	—
	Villars-Chabod	Antonia	Confessed to use of powders causing illness and death to men and animals	Tortured
1480	Lucerne	Woman	*I.a.* death to animals	—
1484	Ravensburg	Eight women	*I.a.* harm to men and animals	Two executed
1487	Zurich	Woman	*I.a.* illness to animals and men	—
1489	Lucerne	Woman	Illness to men and animals, death to animals	—
1490	Vivrais	Woman	*I.a.* killing animals	Burned
1491	Boucoiran	Martiale Espaze	*I.a.* death to infants and animals	—
	Zutphen	Three women	Illness to animals	—
1493	Fribourg	Woman	Illness to men and animals, death to animals	Burned
1498	Lausanne	Woman	Harm to men and animals	—
		Man	*I.a.* killing animals	—
		Woman	Killing children and animals	—
		Man	Killing men and animals	—
1499	Angermund	Two women	Harm to animals	—

I.a. = *inter alia*. Where the sentence is not given, it was usually execution. Most confessed to the alleged crimes after torture.

rinderpest appearing in Germany in 1598 and spreading to Italy in 1599. After a fall in 1600–04, there is a steady rise in trials covering the years 1605–19, peaking in 1615. The period 1609–18 was a time of widespread cattle plague in Europe, also extending throughout the Mediterranean. After this period, trials fall sharply in Geneva but continue to be high in Franche-Comté and Neuchâtel, suggesting that plague possibly persisted in these regions (Figure 2a–c).

The years 1480 and 1481 had been disastrous in Switzerland and Germany. Extensive flooding of major rivers took place on the Continent in 1480, followed by famine and epidemics. The harvest failed in 1481, and this was accompanied by a plague among animals destroying "one third" of them. All these calamities may have led to the preparation, at the Pope's command, of the infamous *Malleus Maleficarum*, which laid the blame for any unexplainable catastrophe on women practising witchcraft. But, for example, in Swabia at this time, on May 1 before dawn, women gathered leaves and willow branches from the woods to weave into wreaths, which were hung over stall doors to keep cattle safe from witchcraft for

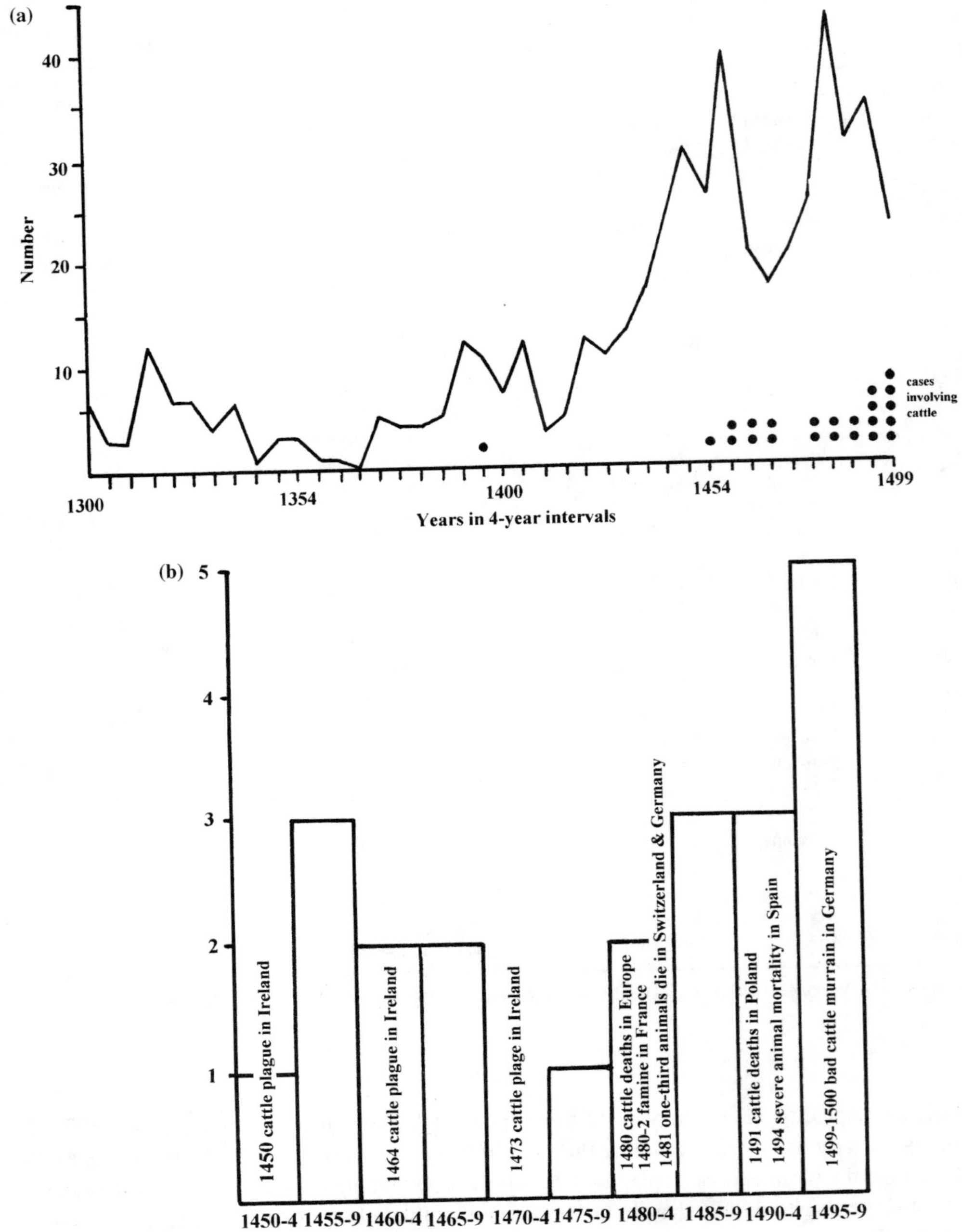

Figure 1. (a) Frequency of witchcraft trials and numbers relating to cattle deaths and (b) indictments for causing illness or death to animals by witchcraft.

the forthcoming year, this in itself being a form of witchcraft and thus identifying the woman's role in witchcraft.

The *Malleus Maleficarum* or *Hammer of Witchcraft*, published in 1486, states that in the diocese of Strasburg, a man affirmed that more than 40 oxen and cows belonging to him and

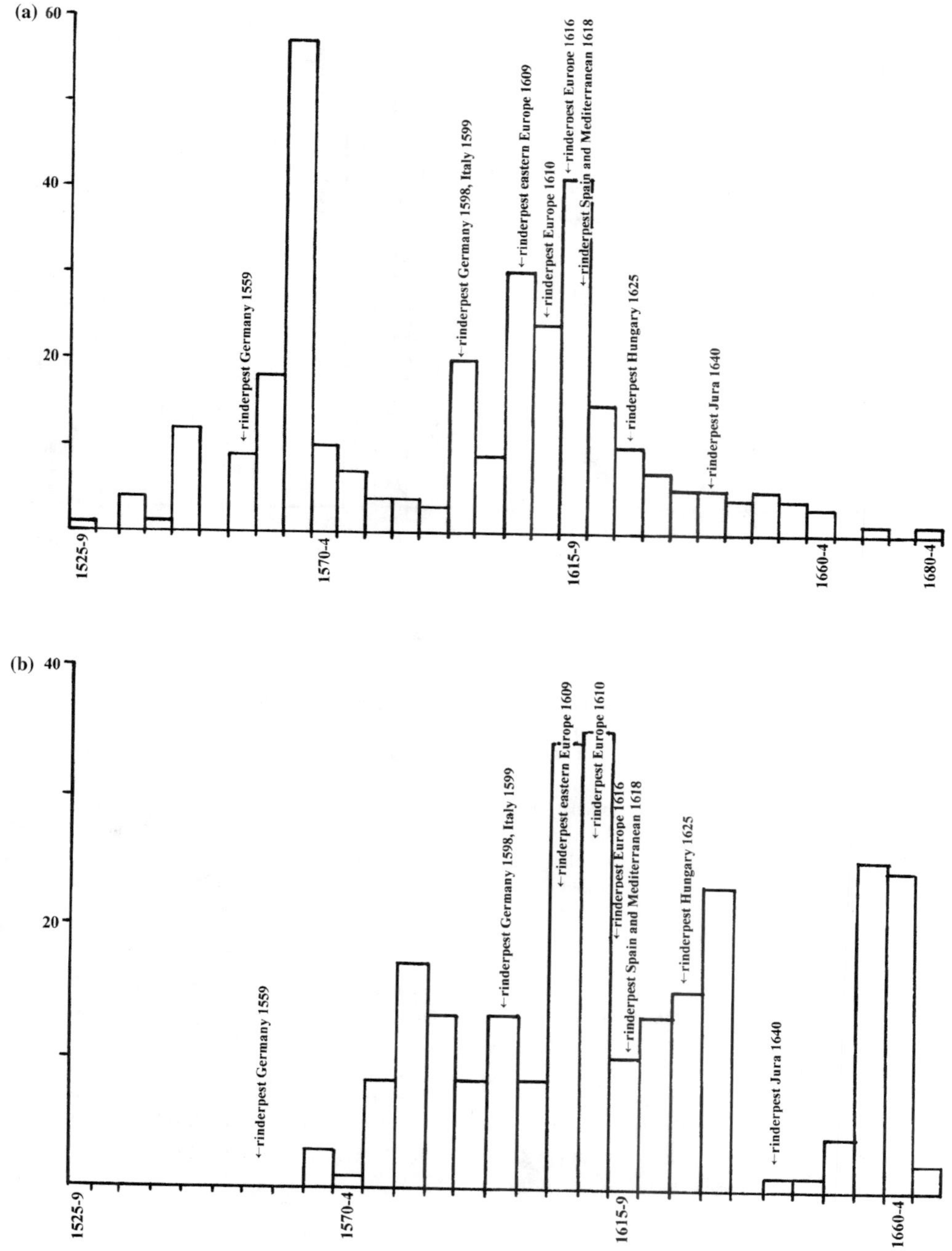

Figure 2. Witchcraft trials in (a) Geneva 1527–1681, (b) Franche-Comté 1599–1667, and (c) Neuchâtel 1568–1677.

others had been bewitched in the Alps within the space of 1 year:

> and that there had been no natural plague or sickness to cause it. To prove this, he said that when cattle die from some chance plague or disease, they do not do so all at once, but by degrees; but that this witchcraft had suddenly taken all the strength from them, and therefore everyone judged

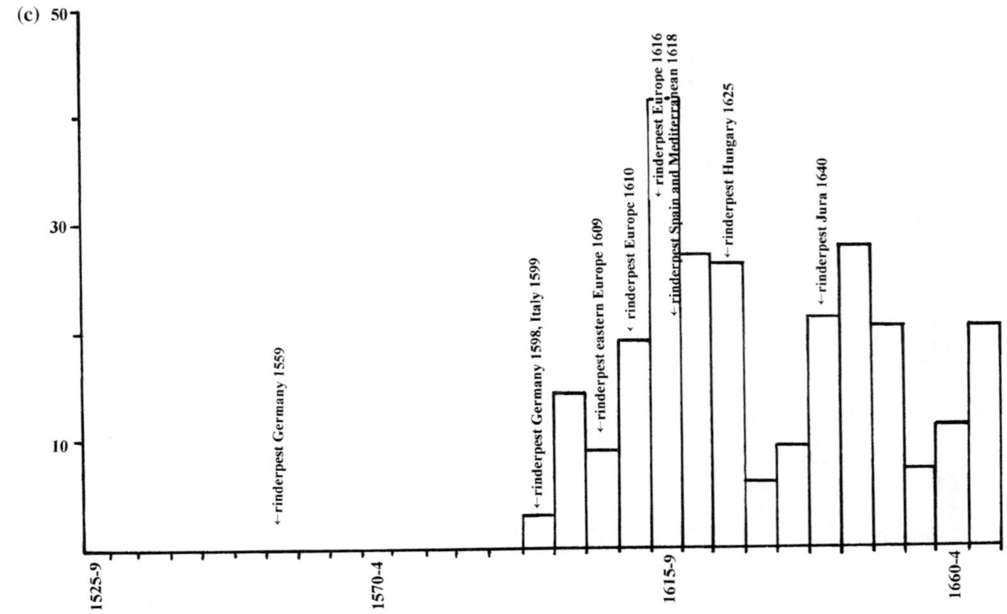

Figure 2. *Continued*

that they had been killed by witchcraft. I have said forty head of cattle, but I believe he put the number higher than that. However, it is very true that many cattle are said to have been bewitched in some districts, especially in the Alps, and it is known that this form of witchcraft is unhappily most widespread.

When an animal has been killed by witchcraft, and they wish to find out the witch, or to make certain whether its death was natural or due to witchcraft, they go to the place where dead animals are skinned, and drag its intestines along the ground up to their house; not into the house through the main door, but over the threshold of the back entrance into the kitchen; and then they make a fire and put the intestines over it on a hurdle. Then, according to what we have very often been told, just as the intestines get hot and burn, so are the intestines of the witch afflicted with burning pains...

... since often the whole of a person's cattle are destroyed by witchcraft, those who have suffered in this way ought to take care to remove the soil under the threshold of the stable or stall, and where the cattle go to water, and replace it with fresh soil sprinkled with Holy Water. (Summers, 1928)

In Britain in 1584, Reginald Scot in *The Discoverie of Witchcraft* exposed the fallacy of it and stated that to cure diseased cattle, one should look in B. Googe (translator of Conrad Heresbach's *The Foure Bookes of Husbandrie*, 1577):

or, if you list to see more ancient stuffe, read Vegetius ... or if you be unlearned, seek some cunning bullock leech. If all this will not serve, then set Jobs patience before your eies. And never think that a poore old woman can alter supernaturallie the notable course, which God hath appointed among his creatures. If it had been God's pleasure to have permitted such a course, he would no doubt have given notice in his word, that he had given such power unto them, and also would have taught remedies to have prevented them.

Few, it seemed, possessed Scot's wisdom, and outbreaks of "witch mania" in Britain occurred in 1582 in Essex, Scotland 1621–30, Lancashire 1633, Scotland 1643–50,

East Anglia 1645, Newcastle 1649, Kent 1652, and Scotland 1661–2. These cannot be correlated with cattle murrains; in Scotland, for example, cattle mortality followed peaks in witchcraft trials, but in 1593, George Gifford, an Essex clergyman, scoffed at the countryman's belief that "the country being rid of the witches and their spirits, mens bodies and their cattel should bee safe," suggestive of some correlation with cattle epizootics. Between 1560 and 1679, there were 487 witch trials in Essex and a total of 790 indictments between 1564 and 1664 on the Home Circuit (Essex, Hertfordshire, Sussex, and Kent), 63 of the latter referring to bewitching cattle. But the peaks in numbers of indictments are not matched by any known reports of cattle epizootics, and accusations of causing death of cattle mostly refer to only one or several beasts, with the exception of the indictment of Anne Waters, spinster of Godalming who, accused of bewitching 10 bullocks in 1581 in Kent, was pardoned. Another exception was the not-so-lucky case of the infamous trial of 13 women at St. Osyth in 1582, in which one of the charges was numerous villagers and their livestock had fallen ill under the witches' spell. These are the only cases which might be connected with a period of "murrain" in Kent in 1581, but the general lack of references to epizootics in Britain in the 16th century may be related to the dissolution of the monasteries and termination of the monks' keeping of chronicles.

In 1584, Elizabeth Brooke, spinster of Great Leighes, allegedly bewitched six cows belonging to one man, six of another, and two of another, together with horses and pigs, suggesting perhaps an outbreak of anthrax on adjoining holdings. She was hanged after confessing. In 1586, John Smyth, a laborer, allegedly bewitched eight cows, six calves, three hogs, and seven sheep, also suggesting an outbreak of anthrax. He was hung also (Ewen, 1929). In 1603, extensive disease among cattle was recorded but there are no indictments for witchcraft connected with it.

Richard Edwards of Manningtree, in evidence against Anne Leech of Misley in 1645, claimed the year before that he was driving his cows home past her house when a black cow fell down and within 2 days died. The next day, a white cow fell down within 5 m of the same place and died within a week. He caused both cows to be opened, and no trace of disease could be discovered which might have occasioned their death (in a separate trial, she was accused of bewitching Edwards's infant son, who died, and for this she was hanged). Margaret Moone was alleged to have confessed to a number of men giving oath before the Justices in 1645 that she had killed a cow of one of them and had two more "in handling." She had also killed a cow and a sow of another. Dorothy Ellis of Brentwood was accused in 1646 of causing the death of four cattle belonging to one Thomas Hitchal, apart from other crimes such as death of a child (Haining, 1974).

Helene Isbuster was tried in Scotland on August 13, 1635 for allegedly curing the sickness of Adam Lennard by passing it to another's cattle. As the cows and oxen of another died, so he recovered. John Brughe was also tried for witchcraft in Scotland but for opposite reasons. A warlock who learnt his trade from an "aged sorceress," Noance Nikclerich, he was repaid for his efforts on November 24, 1643 by being condemned to be strangled and burnt. Among his activities, about 20 years before, accompanied by Noance, during a murrain, he directed a tub to be filled with water, placing in it two "enchanted" stones, then had the cattle pass by as he sprinkled them with the water by means of a wisp. One of the oxen being unable to walk was dragged from the byre and, after they both sniffed its nose, announced it would not live and had it buried alive, the rest being made to pass over it, "and in that devilische maner, be charmeing" they were cured. But he also ordered the owner to clean his byres, stables, and house, and to "great chairges and expenss, caused him to build new houss in uther places for saiftie alsweill of him selff as his goods." He carried out the same cure for another charging him 12 shillings. For curing a number of sick oxen in 1634 that refused to eat by administering a pint of "new lettin goe wirt" to each ox with the words "God put theme in their awin place" repeated thrice, he charged 11 shillings plus two pecks of meal and three "tailyies" of beef (Reid, 1899).

In 1640, when a devastating epizootic broke out among cattle in the Jura Alps, Catherine Miget, "a poor woman," was tortured and publicly burnt as a witch in the belief that she had bewitched the water supply of Sancy in Franche-Comté and the herds of the district. When a murrain, which might have been anthrax, spread from Italy to Switzerland in 1682, some imputed it to the witchcraft of three capucins in Switzerland.

Michael Harward, who was practising as an "ox-leech" in Ireland in 1652 wrote:

> ... the natives, and many of our English-Irish luke-warm Protestants when any disease happens amongst their cattle do not look for help until some store of them be dead, and then to save those that are yet alive they make use of those that have charms, inchanted water, inchanted rings and bells.... Others drive their cattle some 10 some 12 or 16 miles to force them through a river which they call a murrain Ford, and some lose more by badly driving than by the disease itself. (Harward, 1673)

He added that for saving the life of a beast by such means they endangered the loss of their precious souls (through using witchcraft) "unless they truly repent, as I doubt few do.[2]

According to Burdick (1905), a rock crystal ball the size of an orange and encircled by a silver band, was preserved at Currahmore, seat of the Marquis of Waterford, a gift from Godfrey de Bouillon said to have been brought from the Holy Land and to have had great repute in healing cattle attacked by murrain. The cure was effected by placing the ball in a running stream through which the cattle were driven backwards and forwards. It was still sent for from distant parts of Ireland in the 19th century. A smaller one was in the possession of the family of Campbell of Glenlyon in Scotland still used at the beginning of the 19th century, while that of the family of Stewart of Ardvorlich was said to have been used in 1830–31. About 1629, a plague termed the "routing [lowing] evil" prevailed in Scotland, the symptom being that the cattle kept lowing until they died. In the trial of Issobell Young on February 4, 1629, it was revealed that some people travelled from East Lothian to the house of the laird of Leyis to borrow his curing stone for this disease, but the lady refused, giving them instead some flagons of water in which it had been dipped, which water was given to the sick animals to drink thus curing them. Such was affirmed to be the ordinary practice of "husbandmen of the best sort" in many parts of the kingdom, but now they were subjected to ecclesiastical censure, and these were appointed to undergo penance in the church of Dunbar. Known as the Lee Penny, this stone was "of the largeness and likeness of an English shilling," set in silver with some illegible characters. Placed in the end of a cloven stick and washed in a tub of water given to cattle to drink "infallibly cures almost all manner of diseases." Despite the censure of 1629, it was reported in 1702 that people came from "all airts of the kingdom with their diseased beasts" (quoted by Dalyell, 1834).

About 1757, in Yorkshire, a cow's heart was boiled and stuck with pins to check cattle plague the "hurt done" by a witch (Dawson, 1882). In the 19th century, it was reported in Britain that many cattle diseases were still regarded "by the weak and credulous" as being due to witchcraft (Clark, 1806), and in mid-19th-century Yorkshire, an ox heart was pierced with nine new pins, nine new needles, and nine new nails, then burnt at midnight accompanied by chanting of two verses of a psalm, a treatment also directed against the witch who caused the harm (Atkinson, 1891). In the 1865 rinderpest outbreak, Dr. Angus Smith reported that in one

[2]In the Parish of Tring, no farmer would employ John Osborne, considering him to be a wizard. In April 1751, he and his wife Ruth were ducked by a mob in the village pond, the wife drowning. Brown (1751) noted that it was an example of the infatuation of the people in those parts of Hertfordshire, who blamed the couple for the death of "cattle and such like."

cowhouse he visited, the treatment consisted of a cross marked on the wall in tar at the head of each cow (Smith, 1866). As late as 1895, rather than go to a veterinarian, some farmers still went secretly to a "white witch" (witch-finder) when disease affected their cattle to ascertain who had "overlooked" them, that is, had cast the evil eye on them (Elsworthy, 1895), one such farmer thus losing his stock to anthrax. Likewise, in Germany in the 1880s, when their cattle were sick, many still went to a person in the village "who understood" rather than to a veterinarian, believing witchcraft the cause (Jahn, 1884).

In a Russian method to prevent murrain entering a village, men and cattle were confined, while women in their shifts with their hair let down and carrying shovels and tongs yoked the oldest to a plough which was drawn around the village three times, followed by the rest singing; the malignant spirits being unable to cross the plough furrows; or a young girl might carry an image of St. Vlas, shepherd successor to the pre-Christian cattle-god Volos, while an old woman in her shift rode on a broomstick, the others shouting and singing; or other old women with lighted fir splinters circled a widow naked and with a horse collar around her neck, going to each farmyard and crying: "Ai! Ai! Cut, hew the Cow-Death! There she goes!" If a cat or dog ran out, it was taken for the cattle plague and killed. Another method was to light fires at midnight at the two ends of a street in a village, girls in their shifts dragging a plough to one of them carrying a holy picture. A blackcock was taken to the other fire by older women in black petticoats and dirty shifts, and carried three times around the fire. Then one of the women ran to the other end of the village with it while the rest cried out: "Cattle plague! Cattle plague! Spare our cattle! Behold we offer thee a cock!" The cock was thrown alive into the fire, the girls danced around it, and the plough drawn three times around the village. A similar ceremony was reported from Altmark in Germany, where they ploughed around the village then sat under the plough placed upright, enabling them to see the demons. In some villages, a piece of plough was hung over the doorway through which cattle passed, and no demon could then approach them (Burdick, 1905).

As late as 1894 in Hungary, Theresia Kleitsch was crucified at Rekesley, accused by her neighbors of bewitching the cattle in their stalls causing them to die of murrain, and at the end of the 19th century, in some districts of Dalmatia, in the case of an epizootic (or epidemic), it was still the custom to throw all women into the water on a specified day. Attached to ropes, those who sank thus proving their innocence of witchcraft were saved from drowning, and those who floated were hauled in and made to promise to renounce the devil or be stoned. Equally bizarre, but more innocent and led by the Catholic Church, almost 17,000 persons still took part in 1894 in the annual Springprocession at Echternach in Luxembourg, whereby everyone in the town joined in a peculiar energetic dance believed to prevent murrains and other ills, cattle as well as human (Evans, 1895).

3. NEED-FIRE AND FIRE FESTIVALS

Formerly an almost universal custom throughout Europe and Russia, annual fire festivals were held related to keeping witches and witchcraft at bay, and cattle were driven through or around fires in some countries as a protection against these supposed evils. When cattle plagues struck, they were put down to witchcraft, and the ceremonial ritual of driving cattle through the fires came to have a more specific objective. Old superstitions die hard, and at the beginning of the 20th century, in many parts of Prussia and Lithuania, it was still the custom on Midsummer Eve to light great bonfires on the high ground for protection against witchcraft, thunder, hail, and cattle disease, especially effective for the latter if, next morning, the cattle were driven over the places where the fires had burned. Estonians held this belief until recent times, the fire being kindled on St. John's Day, known in northern Europe as "John's Fire." It was often the custom until the 15th century to throw animal heads and bones into the fire.

Another fire, which may have been the origin of other fire festivals, was the "Need-fire." A custom denounced in the early Middle Ages by the Church as a heathen superstition (the Carolingian king Pepin III denounced it in 742), it was still practised occasionally down to the first half of the 19th century in various parts of Germany, England, Scotland, and Ireland (in Mecklenburg despite an ordinance being passed against it in 1682), and among Slavonic peoples to an even later date. In Britain the last reported case was in June 1826 in Perthshire, although the disease was not rinderpest:

> A wealthy old farmer, having lost several of his cattle by some disease very prevalent at present, and being able to account for it in no way so rationally as by witchcraft, had recourse to the following remedy, recommended to him by a weird sister in his neighbourhood, as an effectual protection from the attacks of the foul fiend. A few stones were piled together in the barnyard, and woodcoals having been laid thereon, the fuel was ignited by *will-fire*, that is fire obtained by friction; the neighbours having been called in to witness the solemnity, the cattle were made to pass through the flames, in the order of their dignity and age, commencing with the horses and ending with the swine. The ceremony having been duly and decorously gone through, a neighbouring farmer observed to the enlightened owner of the herd, that he, along with his family, ought to have followed the example of the cattle, and the sacrifice to Baal would have been complete. (*Mirror*, June 24, 1826)

Need-fire, as it was known among Teutonic peoples, or sometimes "wild-fire," and among Slavonic peoples as "living-fire," was usually performed when there was an outbreak of plague or epizootic cattle disease, for which it was considered an infallible remedy (Frazer, 1922; Jahn, 1884). Before it could be raised, all other fires and lights in the neighborhood had to be extinguished, not so much as a spark could remain. In some parts of the highlands of Scotland, this comprised all householders within the two nearest running streams. The need-fire was then usually kindled in the open air, sometimes at a certain spot; in Scotland, this was on knolls or small islands in rivers. The fire had to be produced by friction of two pieces of wood, sometimes of specified tree species. Various contraptions were used for the rubbing process, often a cartwheel or spinning wheel, called in Aberdeenshire the "muckle wheel." On the island of Mull, such a wheel was turned from east to west over nine spindles of oak wood. Certain persons (e.g., two brothers) had to kindle the fire, but perhaps the most extraordinary was on the Isle of Lewis, reported in 1716, where it was called Tin-egin, and had to be kindled by 81 (9 times 9) married men, who rubbed 2 great planks together. In North Uist, the 81 had to be first-born sons. On the Continent, it ranged from a young boy and girl to an old man and woman.

When the need-fire was kindled, using linen or rope (sometimes, it had to be used gallows rope), a bonfire was lit from it. As soon as the blaze had died down somewhat, by way of lustration and as a preventive, the sick animals were driven through it, sometimes two or three times through the smoke and flames so that some were scorched to death. Once the animals had been through, the young people would rush at the ashes and cinders and sprinkle themselves with them, those most blackened marching in triumph behind the cattle into the village. From the bonfire, people carried home live embers to rekindle the fires in their homes. These brands, after being extinguished in water, were sometimes placed for a while in the cattle mangers. In western Scotland and the islands, as soon as the fire on the hearth had been rekindled, a pot of water was boiled on it, and the water afterwards sprinkled upon the infected cattle. On Mull, when the bonfire was kindled, a heifer which had contracted the plague was cut up and burnt in pieces.

Once used for human sacrifice, in the Hebrides the very ancient druidic rite of Beltane fires kindled on May 1, mainly as a device against witches, was observed until the 18th century.

Lit on hill summits, among other ceremonies, cattle were driven round the fires sunwards to keep off plague for the coming year. Driving cattle through or between two fires was observed in Ireland until the 19th century, an Irish settler on the Isle of Man in the early part of the century introducing it to there, the Manx driving their cattle through the fire and singeing them a little. In mid-19th-century Ireland, bonfires were lit on St. John's Day and live hedgehogs, the traditional suckers of cow's teats, thrown into them, perhaps originally intended as a sacrifice. Cattle were also chased with wisps of burning straw. Kittredge (1929) considered that burning live creatures to stop cattle plague, present in 16th-century England, must have been settled practice long before. Toward the end of Elizabeth I's reign, a case was reported of an owner in Scotland, who, having lost many of his cattle, threw the carcass of the next one that died into a pit and burnt it; the rest then did well (Dalyell, 1834).

In 13th-century northern Europe, a variation on need-fire was to erect one or more phallic symbols at the fire. Dogs could be sacrificed, their testicles cut off, and the blood and contents spattered on the cattle herd (Jahn, 1884).

An interesting variation on need-fire in the mid-18th-century rinderpest outbreak in Britain was reported from Yorkshire in 1749, where, "on indisputable authority," an angel was seen to descend and set a large tree on fire. The neighboring cattle all collected around the burning tree, and those which were infected with the plague were immediately cured while immunity was bestowed on the others. The angel left a written instruction that the people should catch this supernatural fire and communicate it from one to another with all possible speed. If it should become extinguished, a new fire of equal virtue might be obtained by rubbing two pieces of wood together until they caught fire, "... so much is certain, that it [the story] has gained sufficient credit to have put the farmers actually into a hurry of communicating flame and smoke from one house to another through the whole country" (Anon., 1749b).

In 5th-century Syria, at the feast of St. John, large fires were kindled annually in several towns through which men, women, and children jumped in what was considered an ancient Asiatic ceremony of purification. In Germany, the "ancient heathen usage" of the kindling of the "Nodfyr" was transferred to St. John's Day because it was forbidden in the 8th century by St. Boniface. St. John's fires were also observed in Constantinople, one 9th-century writer regarding it as an ancient Grecian custom. Kanold (1713) noted that the "so-called need-fire" was not considered as effective as it was with those cattle which had grass or lime-tree rot mixed in their drinks. The belief still existed in Germany in the 1820s that people and animals which had leapt through the flames or smoke, were protected for a whole year from fevers and other diseases (Hecker, 1844). A similar belief in the preventive effects of fire was applied to Easter-fires, held on Easter Eve in Catholic countries. In Germany, the ashes from such bonfires were mixed with the drinking water of cattle as a preventive against plague, and, as with need-fire, cattle were sometimes driven through the dying bonfire. In Poland in the 1920s, the peasants still kindled fires each year in the village streets on St. Rochus's day, and drove the cattle through them three times as a protection.

Slavonian and Bulgarian peasants believed that rinderpest was caused by a foul fiend or vampire which could be kept at bay by interposing a barrier of fire between it and the herds, perhaps giving rise to fire rites. Although 19th-century Russian research into rinderpest was probably the most advanced in the world, during outbreaks in the more remote parts of Russia, villagers employed "plague-guards" on village approaches, who fumigated travelers' horses by burning smoke-producing substances all around the horses and coach. An American traveler, George Kennan, related such an incident in Siberia near the Yenisei River: "The cattle-plague was then prevailing extensively in the valley of the upper Yenisei, and it appeared that round this village the peasants had established a sanitary cordon with the hope of protecting their own live stock from contagion. They had heard of the virtues of fumigation, and were subjecting to that process every vehicle that crossed the village limits. The 'plague-guard' burned straw, birch-bark, and other inflammable and smoke-producing substances around and under our *pavoska* until we were half strangled and our horses were frantic with fear, and then he told us gravely that we were 'purified' and might proceed" (Kennan, 1891) (Figure 3). We can see how superstitions connected with fire enabled the peasants to relate to such a method, but regulations in force in 1865, unlikely to be read by peasants in remote places, stated that burning juniper, brushwood, dung, and so on as a fumigant, were useless, and even dangerous in enclosed places.

Figure 3. "Plague guards" fumigating travellers with smoke in Siberia, 1886.

Figure 4. Saharan rock-painting of 3,000 BC showing a cattle-cleansing ceremony in which they are being driven through an arch of mimosa leaves (or fire?). (After Kuper, 1978.)

Cattle-purification rituals are of widespread and great antiquity. Fulani pastoralists of western Africa build an arch of creepers of the sensitive mimosa tree, the leaves of which react to touch, through which the cattle are driven to protect them from epizootics and other ills, the ceremony apparently being depicted in Saharan rock paintings dating from some 3,000 years BC (Smith, 1992), although these might be depicting fire arches (Figure 4).

4. FAITH IN PRAYER

Since at least the beginning of the 18th century in Russia, outbreaks of anthrax which were not distinguished from rinderpest occasioned religious processions through the streets and offering of prayers in churches. In Tobolsk, a statue was erected to St. Nicholas, which allegedly began to work great wonders and attained fame among the orthodox. Sometimes, the cow death was reputed to wander though a village in the form of a black dog or cat, a mottled calf, or a tall shaggy man with hooves.[3] To expel the plague, a picture of St. Vlas was carried around the village, the people singing:

> Death, oh thou Cow-Death,
> Depart from our village,
> From the stable, from the court!
> Through our village goes holy Vlasy,
> With incense, with taper,
> With burning embers …
> Come not to our village!
> Meddle not with our cows,
> Nut-brown, chestnut, star-browed,
> White-teated, white-uddered,
> Crumpled-horned, one-horned.

[3]In southern Germany, cattle plague was believed by some to be caused by a rogue bull only half flesh, its middle comprising hanging skin. When it appeared, rinderpest ensued (Jahn, 1884). This belief could derive from finding a carcass which had dried out. The centre would be flat and empty, while the haunches would still stick out.

Ramazzini (1711) concluded that it was "manifest that the cause is none other than a punishment from God himself, justly enraged at our iniquities." He continued, "It is equally clear that the specific remedy for our distress lies in prolonged supplications in order to appease the wrath of Heaven." Between September 4, 1713 and December 1, four papal edicts were issued in Rome, ordering religious observances. The first commanded a call for prayers "to free us from this new and very serious scourge." On October 4, it was pointed out that despite the actions taken to control the disease, any human diligence was vain and useless without Divine help. A solemn procession was thus announced for October 6 comprising all cardinals, bishops, prelates, and secular clergy, and remission of sins would be granted to all those who accompanied it, fasting and confessing their sins on the following day, when after Holy Communion alms would be distributed to those poor appearing to be the most pious. On December 1, the faithful were instructed to gather at appointed churches at the next Advent, and until Epiphany, five Pater Nosters and five Ave Marias were to be recited every evening, and at night all the church bells in Rome were to be rung.

When, early in 1745, it entered Denmark across the southern border from Germany, one of the first actions taken by the authorities was to issue an edict ordering prayers to be said in all churches. The following year January 14, was appointed a day of public fasting and prayer. But the plague having killed an estimated 60,000 cattle up to the middle of December 1745, it continued to kill another 220,000 in the next 4 years. Salchow (1755) reported that several years before, a chaplain had promoted atonement and belief as a certain method of preventing the plague. In Britain, by a misreading or a misrepresentation of Leviticus (26: 22), Lobb wrote in May 1745 that God threatened the Children of Israel if they did not follow His ways, first with destroying their cattle, then sending a sword upon them, and then the pestilence, all of which manifested that to obtain deliverance from Divine Anger, people should turn to prayer.

Hervey (1747), in his *Meditations* exemplified the all-prevailing concern. Referring first to the sins of the times, he went on:

> What mean those peals and sobs which burst from the expiring cattle? What mean those melancholy moans, where the lusty droves were wont to low? What mean those arrows of untimely death, discharged on our innocent and useful animals?
>
> No wantonness or sloth has vitiated the blood of these laborious, temperate creatures. They have contracted no disease from unseasonable indulgences and inordinate revellings. The pure stream is their drink, the simple herb their repast. Neither care disturbs their sleep nor passion inflames their breast. Whence, then, are they visited with such terrible disorders, as no prudence can prevent, nor any medicines heal? Surely these calamities are the weapons of Divine displeasure, and manifest chastisements of an evil generation. Surely God, the 'God to whom vengeance belongeth,' has still a controversy with our sinful land. And who can tell where the visitation will end? May we turn from all ungodliness, before wrath come upon us to the uttermost, before iniquity prove our ruin.

In Osmotherly Church, Yorkshire, in 1747 a psalm was sung:

> No Christian's bull or cow they say,
> But takes it out of hand;
> And we shall have no cows at all,
> I doubt within this land.
> The doctors, though they all have spoke
> Like learned gentlemen,
> And told us how the entrails look,
> Of cattle dead and green.
> Yet they do nothing do at all,
> With all their learning's store;
> So Heaven drive out this plague away,
> And vex us not no more.

> This piece was so well received that after the service it was desired again by all the congregation, except for five farmers who wept, declaring that the lines were too moving. The minister, on going out, said to the clerk, "Why John, what psalm was that we had today—it was not one of David's." "No, no," quoth John big with the honour he had acquired, "David never made such a psalm since he was born—this is one of my own." (Hodgson, n.d.)

In 1748, a prayer was appointed to be used "by His Majesty's Special Command in all Cathedral, Collegiate, and Parochial Churches," as follows:

> O Gracious God, who, in Thy great bounty to mankind, hast given them the beasts of the field for their provision and nourishment, continue to us, we humbly beseech Thee, this blessing, and suffer us not to be reduced to scarcity and distress by the contagious distemper, which has raged, and still rages, among the cattle in many parts of this kingdom. In this and all other Thy dispensations towards us, we see and adore the justice of Thy providence, and do with sorrowful and penitent hearts confess, that our manifold vices and impieties have deservedly provoked Thine anger and indignation against us. But we earnestly entreat Thee, Almighty Father, in this our calamitous state, to look down upon us with an eye of pity and compassion; and, if it be Thy blessed will, to forbid the spreading of this sore visitation, and, in Thy good time, to remove it from all the inhabitants of this land, for the sake of thy mercies in Christ Jesus our only Saviour and Redeemer. Amen.

One writer noted that it "had been foolishly imagined by several" that the prayer was intended to supersede the Order as far as that part of it which required the slaughter of infected cattle. Others called for a solemn Fast Day to "deprecate this heavy Judgement for our Sins," without so much as thinking of reformation; and if the calamity be imputed to one's transgressions, then equally should such transgressions include disputing the Ordinances of the State. Also "besides our sin of disobedience in this particular, the frauds, prevarications, and real perjuries committed on the occasion, have not a little increased the measure of our guilt" (Anon., 1748a).

At the request of "some of my reverend brethren in the ministry," Samuel King delivered a sermon at Northampton on December 7, 1749 *On Occasion of the Present Mortality Amongst the Cattle*, with the theme "Behold the Hand of the Lord is upon thy Cattle which is in the Field" (Exodus 9:3). While he asserted it was "plainly and immediately the finger of God," a visitation upon people for sin, the sermon was more of an explanation of the plague in the scheme of things, not an exhortation to pray for the cattle to be saved, nor in fact were the people condemned with vituperative evangelicalism. King saw plenty of examples in the bible for the plague to be no surprise, "How do the Beasts groan? The Herds of Cattle are perplexed." (Joel 1:18). He asked for the people's compassion on the animals which were were suffering on their behalf but stressed that much worse things could have happened and might happen. Many similar sermons were delivered by clerics at the time.

In 1759, William Osmer, in *A Treatise on the Diseases and Lameness of Horses*, compares distemper in horses with that of cattle, claiming that they are identical. He recommends rowels and internal application of nitre and salt to cure cattle plague, stating that remedies employed satisfactorily for the horse must be good for cattle. Should rinderpest appear again, he recommends that Government should allow this treatment to be tested and reported upon, which, if successful, would spare the College of Physicians the trouble of any further meetings and the Bishops the trouble of composing pious forms of prayer "as if the Almighty Being (like an earthly prince) was to be influenced by human entreaties." Layard (1757), while referring to the fact the prelates had "drawn up a form of prayer to be used all over these kingdoms … A method on all accounts expedient, toward the success of our own endeavours," prefaces his remark by reference to heathens who made sacrifices and oblations to their gods to appease their wrath and avert disease.

In the mid-18th-century outbreaks in the Netherlands, opinion favored the view that the plague was a divine judgement on the sins of mankind. Days of public prayer were appointed, and protestant ministers summoned people to repent their sins, while Roman Catholic priests organized novenas and said prayers and orations in the cowhouses. Fighting the plague was seen as a rebellion against the will of God leading to resistance by the people to government and scientific activities, such as Camper's attempts to introduce inoculation. This opposition, however, had largely abated by the 1770s (Faber, 1962).

It was believed in the mid-19th century that one of the common causes of spreading the disease in Russia was the practice of calling priests and people together to pray in the cattle sheds, the assembled people moving from farm to farm. During the 1711 outbreak in Italy, Vallisnieri warned in an Italian journal in 1712 that one method of propagating the disease had been bringing cattle together from different stalls to have them blessed, confirmed by other various learned men in the same journal. The Archbishop of Toulouse, in a pastoral letter issued on Christmas Day 1774, urged the people to attend prayers and present themselves before the Lord to "move to pity his anger." Prayers should be attended even on weekdays in the evenings after they had returned from work, except those persons who had been in contact with sick animals unless they had taken strict precautions, because people and their clothes could carry the disease.

5. "EYE OF NEWT AND TOE OF FROG"

Attempts at treatment could be dangerous for the person attempting the cure, thus in Switzerland in 1539, Jeanette Clerc was sentenced to death for feeding a magical herb to her neighbor's sick cow, which had then dropped dead. Not all in 17th-century France believed witchcraft to be the cause of cattle disease, for at Agen in 1682, Jean Gayau published a pamphlet of "remedies and blessings" to be used against fatal cattle diseases. The remedy consisted of a mixture of garlic, dried grass, and wild artichokes, hung on walls or on old roofs to neutralize the poisonous matter in the air, while plantain, "pourriere," good vinegar, salt, and pepper were applied to the sores of the sick animals. Another remedy was a large spoonful of a mixture of equal parts of soot, gunpowder, brimstone, and salt, washed down with water. As soon as there was any suspicion of contagion in a herd, it was recommended that the tongue be examined for white, yellow, or black blisters (indicative of actinobacillosis). If present, these were to be rubbed, scratched, or torn, with a silver instrument (Figure 1 in Chapter 5) to cause bleeding, the matter then wiped away with new, unwashed linen, and a lotion of salt and vinegar applied. In Germany, a recipe it was claimed would prevent an infected animal dying contained garlic, sage, rue, lavender, "horse-pickle," vermouth, salt, and vinegar. To prevent infection, it was recommended that a mixture of sulfur, gunpowder, dwarf elder, and devil's bit scabious be placed under the tongue morning and night. Gunpowder, sulfur, and chimney soot were frequently cited, while rotten lime-tree wood added to the drinking water was also considered efficacious (Anon., 1682). Theriac was also recommended.

Ramazzini (1711) recommended that bleeding, "while not entirely wholesome is urgent," to prevent clotting, especially indicated because only the fat oxen fell sick; burning on both sides of the neck with a broad red-hot iron, the theory being the two wounds produced would act as "vesicatories" or weeping blisters; making holes in the ears with a round hot iron and placing hellebore *Helleborus foetidus* root in the hole, "as is advocated by the writers of veterinary medicine"; and a rowel or seton in the dewlap under the chin kept open with a stiff cord, the hole to be bored with a red-hot iron. The tongue and palate were to be rubbed with vinegar and salt. Medical treatment should employ such herbs as gentian *Gentiana lutea* roots, dittany *Origanum dictamnus*, red poppy flowers *Papaver* spp., coltsfoot *Tussilago farfara*, juniper berries *Juniperus communis*, burdock roots *Arctium* spp., and so on. In spite of some misgivings, 6 oz. a day of quinine at the onset was recommended. Hartshorn and broth of vipers were recommended animal derivatives, while mineral drugs were antimony, mercury, and rock oil (petroleum). At the onset of the disease, Ramazzini recommended 2 drams of spermaceti dissolved in wine, and antimony or *Antimonium diaphoreticum*. To prevent worms breeding, an infusion of quicksilver or petrol and milk was recommended. He also provided a recipe sent over from Holland, claimed to provide a speedy cure. The use of *Crocus metallorum*, a purgative which also caused sweating obtained from nitre and antimony by calcination, later to be advocated by Mortimer, was objected to by both Ramazzini and Lancisi as too irritating, while Theriaca, Orvietan, garlic, and onions were too hot unless corrected with acids. Lancisi (1715) published a simple remedy in 1713 consisting of a mixture of sulfur, coarse sugar, oil, salt, vinegar, and crushed juniper berries, in a hollowed out onion, one to be given to each animal, followed by a drink of pure vinegar.

In June 1714, a Dutchman, E. E. Raht, issued a recipe which, publicized in the newspapers, was issued as a broadsheet in German on July 1 allegedly having met with good

NOTIFICATION,

WAnn E. E. Raht von vornehmer Hand ein *Recept*, als ein ſehr gutes *Præſervativ* gegen die an verſchiedenen Orten unter dem Rind-Vieh graſzirende leydige Seuche, *communici*ret worden; als hat er dem *Publico* zum Beſten, ſelbiges durch die offentlichen Zeitungen einem jeden kund zu machen nicht ermangeln wollen, ob etwa bey dem Gebrauch der verhoffte *Effect* ſich finden mochte.

Præſervativ,
Fur das Horn-Vieh bey der itzigen umhergehenden anſteckenden Seuche.

Rec. Herb. ariſtolachiæ rotundæ.
Veronicæ a. m. viij.
Pulmonariæ
Hyſſopi
Scordii a. m. iiij.
Rad. Gentianæ
Angelicæ
Petaſitidis
Tormetillæ
Cardopat. 2. Pf. ſz.
Bacc. Lauri
Juniperi a Unciæ xij.
Miſce, Fiat Pulvis.

Hievon gibt man 3. oder 4. Tage aneinander zwey Loth dem Vieh ein, und wann in der Nahe das Vieh-Sterben *continui*ret, muſz man es uber 8. Tage wiederholen. Es kan dieſes Pulver mit ein wenig Waſſer angefeuchtet und ſo dann dem krancken Vieh eingegeben werden.

Item Theriaca Diateſſaron.

Hievon gibt man ebenfalls zwey Loth in wenig Waſſer zergangen dem Vieh taglich mit einem Horn ein, 3. oder 4. Tage aneinander. Das obige Pulver iſt doch dieſem vorzuziehẽ.

Figure 5. The "Dutch Recipe" or "Physician's Powder" broadsheet, 1714.

success, and was allegedly preferable to Theriac (Figure 5). Bates was ordered to conduct a trial of the Dutch recipe: "But the effect was answerable to my expectation, for in many instances, I was not sensible of the least benefit" (Bates, 1718). Like Lancisi, he was not a believer in cures. The "Physicians' powder" was given as: Herb. Aristoloch. Rotundae, Veronicae, aa m. viij., Pulmonariae, Hyssopi, Scordij. aa m. 4., Rad. Gentianae, Angelicae, Petasitidis, Tormentillae, Carlinae, aa lb. ss., Bacc. Lauri, Juniperi, aa zxij. Misce, fiat Pulv.

This was a mixture of dried and powdered leaves, roots and fruits, of birthwort *Aristolochia clematitis*, speedwell *Veronica* spp., lungwort *Pulmonaria* spp., hyssop *Hyssopus officinalis*, water germander or garlic-herb *Teucrium scordium*, gentian *Gentiana* spp., angelica *Angelica archangelica*, butterbur *Petasites* spp., tormentil *Potentilla* spp., carline thistle *Carlina vulgaris*, bay *Laurus nobilis*, and juniper *J. communis*, of which we can at least determine that "Misce, fiat Pulv." means mix and make into a powder. After first bleeding the cow, the concoction was to be given in warm beer or water 1 oz. at a time, three or four mornings successively, followed by rest for 4 days and then repeating with warm water if the disease continued. Bradley (1721) states that this recipe comprised most, if not all, the drugs used in the most serviceable medicines made use of in human plagues, the majority of which ingredients were known to be fatal to insects: "as strong scented roots and herbs; but above all, aromatic gums and saps of plants; as rue, garlic, pitch, tar, frankincense and olibanum. These ingredients are much used in France and Italy, by burning them and smoking such bodies, letters, or any other things as are brought from infected places …."

Theves (1994) claims that the recommended herbal ingredients were the best that people could do at the time to calm the digestive and breathing symptoms. The known effects of the most favored are given in Table 2.

Table 2. Known effects of some of the principal early herbal remedies.

Herb	Effect
Angelica	Soothes exaggerated intestinal peristalsis
Aristolochia	The acid aids phagocytosic action of white corpuscles
Laurel/juniper	Oils are diuretic, digestive and perspiration inducing
Carline	Depurative and diuretic
Gentian	Acts on the central nervous system
Hyssop	Oils act on the mucosa of the digestive and respiratory tracts
Petasitidis	Depurative and lightly antispasmodic
Tormentilla	Anti-diarrheaic and anti-inflammatory of mucosa
Veronica	Depurative and expectorant

A woman in Camberwell claimed to have cured six of seven of her cows in the 1714 outbreak by giving them once a week an infusion of rue *Ruta graveolens* and ale-wort (the fermenting infusion of malt).

In September 1714, the Duke of Bavière, Governor General of Luxembourg, issued an edict containing a recipe "to protect and cure cattle of the contagious sickness" which had previously appeared that month in *La Clef du Cabinet des Princes*. The recipe was an infusion of crushed garlic in good vinegar with some pinches of ground pepper with which the tongue and mouth were bathed, after having scraped with a spoon the buccal mucosa until it bled. Those who had used it had preserved their animals from the disease. As soon as the disease appeared, one performed the foregoing and then made the animal drink $\frac{1}{4}$ oz. of theriac in $\frac{1}{2}$ litre of good wine. Following this, the application of a seton of hellebore root would create an abcess the size of a hat which should be cut with a razor, the matter collected in a vessel and buried. The edict also included a remedy published in the August *Brussels Gazette* which was the "Physician's powder."

In Brandenburg, a 1716 edict contained recommended remedies among which were feeding the cattle early in the morning with fennel and dill seed spread on slices of buttered bread. Two or three times each week, a clove of garlic was to be thrust down the throat. Horses should be fed $\frac{1}{2}$ spoonful of powdered mustard seed mixed with a handful of moistened oats early each morning, presumably as a precautionary measure. From Wolfenbüttel, hanging a bag under the neck containing asafoetida and camphor was recommended as a preventive, as well as smearing the back with stinking stag-horn oil or oil of tartar (Anon., 1716a), which Dr. G. Hanno later considered as having no benefit, as was also the coat under the neck being spread with hart's horn oil (Hannæo, 1746).These remedies were also proposed by Friedrich Hoffmann in his *Friederich Hoffmanns medicinisches Gutachten über die in denen Jahren 1715 und 1716 grassierende Hornviehseuche, wier dieselbe sowohl vorzubeugen als zu curiren (Friederich Hoffmann's Expert Medical Opinion to Prevent as Well as Cure the Cattle Plague Raging in 1715 and 1716).*

A German broadsheet issued in 1719 recommended treating the ulcers by binding on a fried onion twice a day (Anon., 1719).

By 1729, a further Brandenburg decree admitted that all conceivable remedies had been applied without any apparent effect, and prevention rather than cure was recommended, so in a further edict of 1732, the remedies, being tried in France, Alsace, the Netherlands, and Baireuth in Germany, were suggested. For constipation, a purgative of Venetian soap, hazelroot, and savin, boiled in a half measure of beer or wine, administered warm, was recommended. For diarrhea, water in which linseed cake, wheaten flour, or white clay had been boiled, was claimed useful in driving out sweat and preserving the inner organs. A powder of eggshells, camomile, and yarrow flowers, lovage root, juniper berry, fennel, aniseed, scordium, saltpetre, and antimony, mixed with bran water and poured down the throat of a sick animal, would temper the acidity of the blood, improve sweating, and hinder sepsis. Boils or nodules on the skin should be treated with boiled onions, honey, and meal or brought to a head with pitch plaster, lanced, and healed with turpentine or pitch.

A lengthy Austrian livestock order of 1730 reiterated the usual herbal remedies under "preventive" and "curative" medicines but at the same time noted their limited success. It claimed that the best preventive was toasted bread soaked in a vinegar mixture of sheep's sorrel *Rumex acetosella* (ssp. *pyrenaicus*?), marigold flowers, Cardobenedictene leaves, masterwort *Peucedanum ostruthium*, juniper berries, and cattle mithridate, steeped in wine vinegar. Instructions for preparing cattle mithridate were given. Half a herring before the feed morning and evening was also recommended. Bleeding and the fumigation of stalls were also included.

In 1740, the booklet on cure by the Halle Medical Office was reissued, which reiterated most of the known remedies and potions with some variations in the herb mixtures. Fennel and dill seeds on bread and butter were recommended among the milder treatments.

In 1744, the Duchy of Luxembourg was again threatened with the disease, and the Austrian authorities published posters of a preservative against the disease which was a $\frac{1}{2}$ litre of cider vinegar containing a clove of garlic and as much bruised angelica, to be given each morning until the animal was better. A broadsheet published at Amsterdam in 1745 reported a remedy from Middelburg in Zeeland. This consisted of administering a mixture of white pepper and salt in two lots half the size of a large egg, followed by wine vinegar with more white pepper and salt. The animal was then to be driven about with increasing frequency with a stick between its teeth so that it could not close its mouth, until all the mixture had been consumed, the mucous generated by the irritant flowing out of the mouth. This was followed by a diet of turnip and other root juices. The instruction ended with an exhortation to apply the methods diligently and ensure that the disease was not brought in by negligence, and trust to God that He would bless the aforesaid method and diligence (Anon., 1745c).

A lengthy instruction produced at Wolfenbüttel was issued by the Duke of Braunschweig and Lüneburg in 1745 reiterating many of the proposed cures. Describing the history of the 1711 enzootic, with quotes from Ramazzini and Lancisi, it referred to the latter's recommendation of slaughtering infected animals as there was no known cure, and only a few survived more through luck than anything else. Kanold had approved strongly of this. It also noted that methods of cure in the press were so many and almost contrary to one another, and an efficient treatment was yet to be found. A reference to a renowned foreign doctor warning that no methods were of any use if the cattle were already sick probably referred to Dr. Mortimer. The duke therefore wanted to institute the best preventive methods, noting that the Italian outbreak of 1711 was allegedly due to a single ox. The manner in which infection could be spread was then listed: (1) through contact where infected animals have been; (2) infected blood from slaughtered animals; (3) not burying carcasses deep enough; (4) contaminated imported raw skins; (5) dogs; and (6) handling/nursing sick animals and picking up the infection on clothes.

The fumigation of sheds and stalls was recommended first, but also tying under each cow's neck a linen bag containing a mixture of pounded bay berries, juniper berries, fenugreek seeds, angelica, asafetida, and camphor, mixed with stag's horn oil. Other recommendations followed, including good feeding and good management, together with useful precautionary measures, such as blood-letting. Blood-letting at the tail had been recommended in a Hamburg newspaper (no. 49 of 1745), but additional blood-letting from veins under the tongue was recommended. The use of setons could not be recommended enough and had already been proposed in an ordinance of 1716. A Harlemmer newspaper (no. 11 of 1745) had recommended a piece of hellebore root as a seton, but this was difficult to retain in the wound, and so hair setons were recommended. Laxatives were only useful for animals already sick, but Lancisi recommended fig broth or cooked plums with salt and sennapods. Instead of sennapods, one could use tobacco. One should also provide rock salt for the animals to lick. Salted herrings dipped in tar, thrust down the throat, had been recommended by some, but it was probably the salt alone which had an effect. Tar-water was also recommended, and instructions given for mixing it. From time, to time a mixture of brimstone and antimony powder on a piece of buttered bread was healthful.

First appearing in Hamburg newspapers in March 1714 and after, in the cattle plague decree of 1716, and still recommended, was the following mixture: leaves of birthwort, or hollow-root *Corydalis cava*, speedwell, lungwort, hyssop, water germander; roots of gentian, angelica, butterbur, tormentil, berries of bay and juniper, which was the Dutch or "Physician's powder," minus the carline thistle root, rejected by Bates in 1714.

When tongue blisters appeared, they should be ripped open with a silver spoon or other utensil, and the wounds scraped until they bled (the treatment for actinobacillosis). They were then to be kept moist with well-water, or wine vinegar to which had been added pepper, salt, garlic, and herbs, then rubbed with blue vitriol and salt spread on them. Ulcers which had already formed were to be treated in the same way. In Alsace, a special instrument made of fine silver was recommended which had one edge serrated for scraping and the other ground for cutting. This had a wooden handle with a sharp pick at the end for ripping open hard and deep blisters and ulcers (as in Figure 1 in Chapter 5). The user was advised that none of the released matter should be allowed to pass down the gullet, otherwise the animal would die from absorbing it.

Another document issued by the same source in 1745 recommended fumigating stalls by burning sheep dung, as this had a very penetrating effect, causing cattle to sweat. Bleeding, rowelling, and the administration of theriac were all recommended. Then, one should give something warm to drink with three or four handfuls of stone flour, bruised grain, or wheat bran. If the animal was weak, one could add a pound of white clay or dried forecastle-stove glue. In the outbreak of 1682, a sure preventive was claimed to have been lime-tree resin in the drink, but this had not been proven effective. Saltpetre, antimony, Mary thistle or Cardobenedictine seeds, and unheated stag's-horn oil, all in warm beer, was recommended. After the first 24 hr and then every 6 hr, a handful of Wulfferlen or Great Lucian's herb should be given. If the animal became worse after 24 hr, two or three good handfuls of soot from the stove with two good spoonfuls of elderberry jam, stirred in warm water and vinegar, was recommended. If there was still no improvement, the saltpetre (etc.) mixture should be given with cascarilla bark.

Hanno, who was placed in charge of operations when the disease broke out in May 1746 in the State of Hamburg, produced a report based on his own observations in which he considered that cattle needed to take in certain herbs in the wild to maintain the body's balance. He drew parallels with the ergot outbreaks among people in 1596, 1648–9, 1675, and 1717, due, as he saw it, to an essential food being affected. Bleeding on one or both sides of the neck, removing 2–4 lbs of blood he saw as effective, but he had tried many herbs without the least benefit, although some showed a little. To deaden the inner fever, camphor, saltpetre, and occasionally saltpetre with cream of tartar, had good effect. Very good in some cases was flax oil, while useful against wind and diarrhea were chervil spp., "common pill," bistort root, larkspur spp., metallic sulfide(?), thousand-flower-water, camomile spp., eggshell, castoreum, cascarilla bark, asafetida, gentian root, birthwort and clover spp. For when the horns were affected, he recommended marsh-mallow root, salmon berry(?), and Eastern larkspur with St. John's Wort oil and Lily of the Valley, also stinking hart's horn oil.

In 1753, a publication issued at Munich announced some of the medically approved, and partly proven, recipes for prevention and cure. These were summarized as:

- bleeding;
- administering a purge of rhubarb, antimony, and salt;
- giving water containing wine vinegar and rue;
- as a preventive, a mixture of chimney soot and gunpowder;
- as a preventive, a mixture of antimony, saltpetre and Cologne chalk;
- as a preventive and a cure, a mixture of beer vinegar, wine vinegar, swollen crab's eyes, saltpetre, virgin honey, and hell-broth water;
- a preventive and cure used for cattle and horses "with proven effect" in 1689 (i.e., for anthrax): sickle-wort (scorpion vetch *Coronilla* spp.) beer, vermouth, chimney soot, garlic, rue, and salt; can be adminstered with flax oil on bread;
- as a cure, flowers of sulfur, triturated saltpetre, camphor, and diaphoretic fire powder, in water;
- as a cure, oil of turpentine, dragon's blood (*Dracaena draco*?), fresh egg, and mithridate, mixed together and placed in the ears and on a chunk of bread (Anon., 1753c).

"Cures" or "preventives" for anthrax and actinobacillosis were often mixed with those for rinderpest, similar concoctions being recommended, the distinguishing feature of actinobacillosis being the recommendation to scratch open the tongue blisters with a silver instrument.

Arguing that the disease was caused by an alkaline poison which had to be neutralized, Salchow (1755) reported on a number of allegedly successful antidotes. The first was used in 1745 at Holstein. This included administering a mixture of powdered rhubarb and saltpetre each morning, and twice daily a mixture of vitriol of tartar, saltpetre, citric acid, isinglass, and cascarilla, the animal to be kept warm in the stall, covered with blankets and sacks. In outbreaks at Aschersleben in 1751 and 1754, Salchow tried zinc sulfate with honey in warm water each morning, the cow being better after 2 days, but it had no effect if not administered at the outset. In the same locality, an old woman used the carpal and tarsal bones from a cow which had died of the infection, burned with "powders," administering two spoonfuls daily with warm beer. By this means, she claimed to have cured a cow in a few days, but only one of several. Schreber, in the foreword to a collection of Prussian laws on the disease, provided a "cure" which Salchow was unable to try "because of the obstinacy and prejudice" of the people of Ascherleben. The method was to burn to a powder lung, liver, and milk of a dead sucking calf, together with salt, and make tar pills of it. These to be given at the outset before inflammation set in. With all of the four "cures," only one or a few cows were cured. One man gave his healthy livestock sulphuric acid in the drinking water during outbreaks, allegedly with good effect, while some gave human urine in an old man's shoe. Salchow deduced that the only successful medicine was one comprising of distilled wine vinegar, melted beeswax, and human urine, boiled to a honey thickness, then adding French brandy wine and finally heating it to a dry salt, but his best advice was that of keeping the stalls clean and all other animals out.

Probably issued in Münster, a broadsheet of 1760 recommended a fresh preparation of roots of marsh mallow and tormentil boiled with senna leaves. When the scouring had passed, one should give cooked semis(?) leaves for 3 days. In Thuringia in 1764, it was reported that most lean stock were unaffected, so stock owners starved their animals, thus allegedly having fewer infections.

Caspar (1777) reiterated most of the standard methods of treatment: rowelling, herbal potions, and so on. An added "method" was to place horses in the stalls with sick or healthy cattle, as the steam from the horses hindered spread of the plague, horses being unaffected by it. He quoted the "renowned" Friedrich Hoffman's *Suis Consilius de Lue Boum grassante* (Hoffman was at the Medical University of Halle) as having a good recipe for purging (Hoffman, 1721). But he recommended that one should banish throughout the gross method of administering garlic, spirits, sulfur, theriac, asafetida, petrol, amber, and juniper oil, which help kill the animals.

6. MID-18TH-CENTURY BRITAIN

In Britain, the preamble to the first Act of March 1746, noted His Majesty, with the advice of his privy council "who have consulted physicians and surgeons thereupon, and they have given it as their opinion, that all the methods of cure which have been put in practice, both at home and abroad, have proved so unsuccessful, that they have rather contributed to propagate than stop the infection; for while means are using to save the sick, the disease is spreading amongst the sound, and is increased more and more, in proportion to the number seized with it...."

Typical of alleged remedies being publicized was a Swedish concoction claimed very successful consisting of roots of snakeweed, valerian, lovage, elecampane, masterwort, angelica, and carline thistle; and camphor, laurel berries, and agaric, pulverized and mixed with an equal weight of salt. After the cattle had been housed overnight, the next morning must be given to each, "on toasted Bread, when fasting, as much of the above Composition as can be taken with your Thumb and Fingers. The Beast must neither eat nor drink before Mid-day," and after that, the owner need not worry. Another medicine was composed of Venice treacle, lapis contrayerva *Dorstenia* spp. root, powdered rhubarb, saffron powder, camphor dissolved in wine, and oil of caraway, mixed and given in a drench, bloodwarm, with mint, wormwood, and rue water. Best when made into a ball, the medicine might be "faithfully prepared for one Shilling and Six-pence: for which Reason, the Author [of the article] has, for the Benefit of the Publick, ordered them to be sold for that Price ... three Doses will be sufficient for a Cure; and by Way of Prevention to the sound Cattle, one Ball may be divided into two Doses." Meanwhile the beast must be kept warm, given a mash of ground malt or boiled oats, and supplied with warm water (Anon., 1745d).

In the 1745 outbreak near London, Mortimer recorded that cow-keepers, at the first sign of the disease, shown by a sudden fall in milk yield and a refusal to eat or drink, gave a warm mash of malt, or a drench consisting of the juice of 2 oz. of caraway seeds boiled in a quart of water with a gill of white wine and a quarter pound of honey. The incisor teeth gums were lanced and rubbed with salt and vinegar. Bleeding and rubbing the nose with tar was also used. If they hung the head, running at the nose, then a bag of scalding hot malt was placed on the head, tied between the horns, which sometimes made them lift it up.

Apart from recommending balls of *c. metallorum*, and the pouring of warm vinegar with salt into the nostrils; for shortness of breath, Dr. Mortimer advised giving a pint each of whale oil and treacle of sugar with 4 oz. of flower of brimstone, all given in a mash of malt, bran and water, two or three times a day. For the scouring, he recommended first the *c. metallorum* purge, then every 6 or 8 hr a draught of the following: "Take whiting one pound, bruise it; pour boiling water upon it, a quart or more; let it stand to settle; pour off the clear water and fling it away; then put a quart of warm water to the wet whiting, and add bole armeniac, in powder, two ounces, Venice treacle one ounce, English malt spirits half a pint."

The few cows which had recovered from infection Mortimer noted were kept inside, blooded, fed warm mashes of malt and bran, and administered warm drenches of herb mixtures containing rosemary, wormwood, and ground ivy with honey or treacle (Mortimer, 1745b).

Dr. Barker's "rules" for the cure of the disease, which he published in his anonymous tract in 1745, mostly involved bleeding but also, at the onset, boring a hole in the animal's dewlap and inserting a piece of hellebore "in order to divert the humours from falling upon the lungs, which they would otherwise do in the course of the disease." Internal medicines he recommended should be of the same kind as that given for pneumonia, namely oxymel (a syrup made of vinegar and honey), or various herbal decoctions or infusions. One "successful" treatment was a powder composed of 1 oz. of saltpetre, $\frac{1}{2}$ oz. of spermaceti, and $\frac{1}{2}$ oz. of contrayerva, administered in a ball of honey. Dr. Barker's text was so highly approved of by four physicians of Worcester that they agreed to distribute it throughout the county. In a letter to the *Gentleman's Magazine* (Barker, 1746), he stated he could not say much about the success of the method as very few had made a proper trial of it, but the best success had been obtained by a woman at Kentish Town who followed all the directions thereby saving all her cows. Neglecting to take the same precautions the following autumn, she then lost them all. Barker claimed that this was convincing proof of the usefulness of his method, but "Several Hands" (1747) criticized the assumption, "For what Man, or Woman, in his or her Senses, who had experienced such singular Benefits ... would have failed to apply the same under like Circumstances the then present Year?"

The effects of some of these measures are illustrated by Layard, who states "Mr. Montgomery furnishes six instances out of seven, which were cured by Venice treacle; and tells us plainly, that the seventh was killed by giving a lump of tar with oil of turpentine, which checked the operation of the Venice treacle, by clogging up the stomach."

Some treatments were relatively humane and the accompanying cosseting probably assisted the animals in recovering naturally, the medicines then being claimed to have effected the cure as illustrated in an anonymous farmer's journal published in the *London Magazine* November 1745:

> Oct. 20, 1745. Late in the Evening four of my cows were taken with the Distemper that has been so fatal ... few of the farmers have sav'd one in five of the Cattle that have been taken; but most have lost their whole Stock.

21. I sent for a Cowleech ... who gave them Drink which he pretended was a great Secret; he bled them plentifully in the Tail, rowell'd or coaz'd them in the Dewlap, by my Order.

Half an Hour after the Drink I gave one Ounce of Salt-Petre, dissolv'd in Spa Water, to each Cow; which Dose I repeated twice more this Day, observing to be about five Hours between each Dose.

The Cows refus'd their Food, but, to prevent them from starving, I boil'd about three Quarters of a Peck of Oats in as much Water as a Cow may be thought to drink in one Day; which Quantity, at sundry Times in the twenty-four Hours, I gave to each Cow, in a Horn, Water and all, and a Malt-Mash of a Quarter of a Peck of Malt. I also gave each Cow two Quarts of Sugar-Sops, wherein much Cinnamon was boil'd, *viz.* one Quart about Noon, and the other in the Evening.

I kept them in the House warm litter'd, and kept milking what little they had, and threw it on the Dunghill. A Man and a Boy set up all Night with them.

22. They were very bad, cough'd much, run at the Eyes, and breath'd very quick. I repeated the same Medicines, and took the same Care; but they fell away surprisingly. A Man and a Boy still set up to nurse them.

23. They were somewhat better; I repeated the same in every Particular.

24. They were much mended, and began to eat a little Hay: I now left off the Salt-Petre and the Drink, but continued the Oats and Sugar-Sops.

25. They were so much better that ventured to drive them round some good Grass, about a Stone's-Cast from the Cow-house, where they continued picking about an Hour; during which Time, a Boy was constantly with them, to prevent them from laying down on the wet Grass, which I apprehended would kill them. This Day the Boy drove one of them into the Pond where it drank eight Go-downs: At Night I thought it would have died, but is since likely to recover.

26 and 27. These Days I continued to take Care of them without giving any Medicines, but gave them boiled Oats, good Hay, and Sugar-Sops, which they would eat very heartily. I have all Reason to believe they will do very well.

I had three more taken on the 21st, which were indeed very bad; but I think we discover'd their Illness a Day, at least, sooner than we did the first Cows; so that by early Application of the Remedies they have recovered their Strength and Appetites better than those first taken.

I know the Cowleech us'd one Quart of Wood Soot to each Drink, with sundry bitter Herbs, *viz.* Red Sage, Wormwood, Rue, and Smallage [wild celery *Apium graveolens*], which were boil'd in Ale, with half a Pound of Hogs-Lard.

N.B. I believe *Sperma Ceti* much better. (Anon., 1745c)

Eight of 13 cattle were allegedly cured at Bury St. Edmunds in December 1746 by dosing with the following: 1 gallon of new verjuice (sour fruit juice), 3 quarts of "foot" (ale dregs?), 1 oz. of saltpetre, 1 oz. of stone brimstone, 1 oz. of flour of mustard, 1 pint of tar, 1 pint of linseed oil, $\frac{1}{2}$ pint of honey, $\frac{1}{2}$ pint of goose grease, all mixed together, brought to the boil and stood until blood warm. A pint was recommended with $\frac{1}{2}$ pint more 6 hr later. The beasts were then to be walked about for a quarter of an hour, and those most infected kept inside at night and fasted. In the morning, they should be given another $\frac{1}{2}$ pint and walked about well. For a calf, a $\frac{1}{2}$ pint was enough (Anon., 1746b).

Apart from recommending branches of Scotch fir trees to feed upon, another remedy suggested by a writer in Shropshire was to make a small round pie with a stiff paste of flour and butter, as usually made for mince pies, and fill it with 6 oz. of tar. It is not stated how it should be administered, except one to be given to each animal. More unpleasant was to take a gallon of human urine, if a fortnight or 3 weeks old the better, and add 10 handfuls of hen's dung. Steep for 10 hr then strain and give each animal $1\frac{1}{2}$ pints, to which has been added a good handful of rue, bruised or ground. No doubt, the cattle would have found the recommendation of $\frac{1}{2}$ to $1\frac{1}{2}$ pints of good, home-made rum punch, more to their liking. This was recommended for all cattle as soon as disease appeared in any one of them and was considered to put an immediate stop to it. If not, it could be repeated. When the disease appeared at Sinceny in France in 1773, the Marshall claimed to have saved a significant number of animals, among his remedies being the use of chicken or pigeon dung soaked in human urine. At Rouy, it was claimed that of 50 infected animals, about half were cured by administering pigeon dung in wine. The use of bird dung probably stems from Pliny, who claimed that chicken dung could cure accidents caused by eating poisonous fungi.

A JP for Nottingham, Langford Collin, who employed a number of "sensible" men as inspectors, added his recipe which, if administered before the animal was infected, was "almost infallible" (it would have been illegal to recommend it *after* the animals were infected). He had tried it on 2 animals, of which 1 died, the survivor having been put among 20 which died. He had since tried it upon "many hundreds" with the desired success. This was simply a mixture of angelica roots, rae (rue?), vinegar, and aniseed, accompanied by bleeding on the second morning (Collin, 1747).

An anonymous contributor repeated in the December 1747 *Gentleman's Magazine* his recipe which had previously appeared in October:

> Take of the powder of elecampane, four ounces; Peruvian bark, Turmeric, æthiops mineral, Crude antimony and saltpetre, of each two ounces; camphire and assa-foetida, of each half an ounce; of the gall of any creature (especially beef's) one good spoonful; vinegar with a little flour, enough to make the whole of a proper consistence to be made into 14, 16, or 18 balls; which are to be given one every night and morning, washing it down with some warm water, wherein quicksilver has been boil'd, or infus'd; or with tar-water.

adding a note that in these contagious times, "it would not be a bad hint given to the ladies, to boil quicksilver in the water they make their tea with" (Anon., 1747b)!

A pamphlet dedicated to the President and members of the Royal College of Physicians at London was issued in 1747 "by Several Hands," a group of physicians who stated that they had never seen a case of rinderpest themselves but as there were such full and accurate accounts of it, they imagined themselves as capable, and altogether as well qualified to judge of it, as if they had performed or assisted at the dissections. They mainly repeated, or criticized, the various proposed remedies which had appeared in the *Gentleman's Magazine* up to December 1746, as well as those of Ramazzini, Lancisi, Mortimer, and Barker. The juice of elder and its bark, mixed with sugar, had been listed by Ramazzini and was much recommended by John Bouin, "that celebrated botanist" because Gesner (1516–65) had reported that, mixed with treacle, it cured numbers of people of the plague. They did not mention that Gesner, the great German–Swiss writer-naturalist, died of the plague himself. Although recommending that at first sign of infection, an animal should be kept warm and clean, their main recommendation for cure was 2 or 3 oz. of mercury every morning and evening. Alternatively, one could use purified nitre and soluble tartar, an ounce of each every 6 hr. It could be rendered better by adding an ounce of tar, the yolks of two fresh eggs, and a pint of bran water. Another alternative was 2 oz. of purified nitre, $\frac{1}{2}$ oz. of tartar vitriolate, and 2 oz. of crushed blanched almonds. They also could not "omit to advise the Use of Tar-Water, which we recommend to be made strong, and to be given in great Plenty, i.e. To twelve or sixteen Pints in twenty-four Hours Time."

Another contributor to the pamphlet scene was Dr. Wilkes, a physician of "great Practice, Fame and Repute" in Staffordshire, with *A Letter to the Gentlemen, Farmers, and Graziers in the County of Stafford* (Wilkes, 1747). While stressing, contrary to Barker's assertion, that the plague was contagious, and if there was doubt it was better to err on the side of caution, his suggested remedy for the "second stage" of the illness was one concocted of spirits with pepper, gentian, flour, mace, and cinnamon, which appeared in the newspapers in Birmingham and London in January 1747 in a letter of the Countess of Stamford, who had received it from Mr. Walker of Surrey and was quoting it "for the Good of Mankind." "What a monstrous Jumble or Hodge-Podge is this? Without the Shadow even of Rhime or Reason to authorize and support it," asked the "Several Hands," roundly criticizing Dr. Wilkes for quoting any recipe he could apparently lay his hands on. "Others (says he) advise you to put

Rue, Assa Foetida and Garlic into the ears, and then sew them up: Others to blow up the Nostrils with a Quill, as much of the leaves of Asarum powder'd, as will lie on a Shilling. This (says he, by Way of Note) purges the Head, and makes the Nose run greatly" ("Several Hands," 1747).

Many contributed their recipes to the *Gentleman's Magazine*, which simply printed whatever was submitted, such as a Norwich correspondent referred to by "Several Hands" who reported hundreds of surprising cures by blowing into the cow's nostrils "as much snuff (made of asarum leaves dried and beat to a powder) as would lie on a shilling." In a few hours, this would cause matter to run in large quantities "and thereby prevent its settling on the lungs" (Anon., 1747c). One farmer found a mixture of Venice treacle, brandy, and beer to be effective. When he administered tar with turpentine to a cow, it died; when he omitted it, his cows lived. Some of his neighbors, including the vicar, had the same experience (Montgomery, 1748). Another farmer, having lost several cows, gave them Jesuit bark in warm water and, finding they improved, he alternated between the bark liquor and red wine, and they survived. Another found the wine to be ineffective, but butter dock was effective. When fumigation with soot failed, one farmer tried bark and cinnamon water effectively; another used bleeding first. Yet another used bleeding, then administered a hot liquor of oak and burdock roots in ale, following this with milk pottage and the best hay in small quantities.

An anonymous pamphleteer wrote scathingly of remedies such as those appearing in the *Gentleman's Magazine* of October 1748, noting that although a particular animal sometimes recovered, a far greater number did certainly die of the disease. He criticized particularly the suggestion of Jesuit's Bark as a cure. The facts as stated in the *Gentleman's Magazine* were not sufficient in themselves to induce a belief that the bark would be found to be an effectual remedy, by way of cure or prevention: "For, if on account of imaginary preventions, and chance recoveries, the Bark is to be respected as an antidote, and a remedy; the same may be said for Tar-water, and many other medicines, which have only served, as it was not impossible this may do, to delude the people into a perseverance in wrong measures." Even if it could be found to be effective, he argued, it could not replace slaughter and other measures, because the disease was contagious. He also pointed out that advertising medicines for cure, as opposed to prevention, was "preposterous and audacious," as not to kill infected animals immediately was illegal and penal (Anon., 1748c).

A Derby farmer lost 22 of his 33 cows before trying a cure. He shut them up when first seized and gave them a little rye paste, which caused them to chew their cuds, which otherwise they lost. He then gave nothing but water to drink until he perceived that they grew worse, when he boiled colocynth *Citrullus colocynthus* and herb grace or rue, in water, adding sweet oil. Giving this to each animal, after 12 hr he gave only oil and then water gruel until the animal had nearly recovered, thus saving the remaining 11. Another writer emphasized the smallpox analogy, and since Americans had used tar water to cure consumptions and prevent smallpox, as the animal would not drink a sufficient quantity, he advised giving the tar itself made into a stiff ball with barley flour, oatmeal, or bran, administering one or two night and morning for a week (Anon., 1748d). "Mulo-Medicus" (a reference to Flavius Vegetius Renatus's *Mulomedicina*) suggested injecting a mixture of saltpetre and tobacco boiled in water into the neck vein after letting a quart or two of blood, "nothing can be done with greater ease, or at smaller expence."

"I cannot foresee any ill consequences attending this experiment: It may probably occasion some slight convulsions, and afterwards a profuse sweat, or some other evacuations; by which the beast will be freed from its disorder. It will also be one way of proving, whether the doctrine of animalcula be true or false; for if true, this bids fair for their destruction" (Mulo-Medicus, 1748).

A fly-sheet titled "For the Infection now Raging amongst the Horned Cattle, 1748–9," probably distributed in Yorkshire, began with stating that in the neighborhood of the distemper, it was not advisable to give cattle any hay, as any taking the infection when full of hay would certainly die. Thus, chaff and cut straw were recommended. "If a beast is attacked ... which may be easily discovered by their Eyes being Dim, and Cloudy, and soon after inflam'd, and streak'd with Blood," it should be bled plentifully and setoned in the dewlap. It recommended taking 5 pints of blood on the first day, 3 pints on the second, and 2 pints on the third, at the same time keeping the animal warm and giving plenty of warm water-gruel, or bran and water. Every 6 hr, it should be given a draught of saltpetre, honey, and cochineal, and if it did not dung, Glauber or Epsom salts were recommended. "Many Receipts have been tried to no Purpose, so that all that can be of Service to Cattle, is to keep them very open in their Bodies and thin their Blood." (Figure 6).

Dr. John Wall, a distinguished Worcester physician of wide interests who turned to a study of the disease to refute the charges of physicians' indifference to it, believed that continuance of the disease was due not to its incurability but to wrong treatment. The parts first affected were the glands about the throat, then the lungs, and finally the bowels. This closely resembled sore throat in humans, in whom strength must be kept up and perspiration induced. The same treatment should be used for cattle: warmth, cleanliness, liquid diet, and a drench of myrrh, vinegar, and Jesuit bark. If the victim was far gone, add wine or brandy and cascarilla (Wall, 1751). Jesuit's bark (quinine) was originally recommended by Ramazzini.

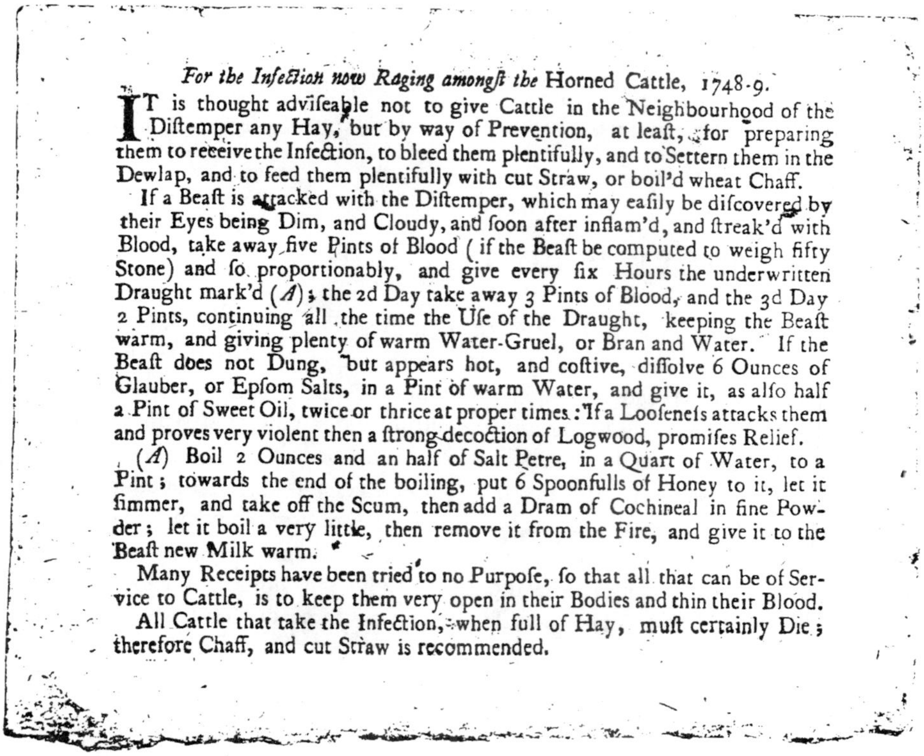

For the Infection now Raging amongſt the Horned Cattle, 1748-9.

IT is thought adviſeable not to give Cattle in the Neighbourhood of the Diſtemper any Hay, but by way of Prevention, at leaſt, for preparing them to receive the Infection, to bleed them plentifully, and to Settern them in the Dewlap, and to feed them plentifully with cut Straw, or boil'd wheat Chaff.

If a Beaſt is attacked with the Diſtemper, which may eaſily be diſcovered by their Eyes being Dim, and Cloudy, and ſoon after inflam'd, and ſtreak'd with Blood, take away five Pints of Blood (if the Beaſt be computed to weigh fifty Stone) and ſo proportionably, and give every ſix Hours the underwritten Draught mark'd (*A*); the 2d Day take away 3 Pints of Blood, and the 3d Day 2 Pints, continuing all the time the Uſe of the Draught, keeping the Beaſt warm, and giving plenty of warm Water-Gruel, or Bran and Water. If the Beaſt does not Dung, but appears hot, and coſtive, diſſolve 6 Ounces of Glauber, or Epſom Salts, in a Pint of warm Water, and give it, as alſo half a Pint of Sweet Oil, twice or thrice at proper times: If a Looſeneſs attacks them and proves very violent then a ſtrong decoction of Logwood, promiſes Relief.

(*A*) Boil 2 Ounces and an half of Salt Petre, in a Quart of Water, to a Pint; towards the end of the boiling, put 6 Spoonfulls of Honey to it, let it ſimmer, and take off the Scum, then add a Dram of Cochineal in fine Powder; let it boil a very little, then remove it from the Fire, and give it to the Beaſt new Milk warm.

Many Receipts have been tried to no Purpoſe, ſo that all that can be of Service to Cattle, is to keep them very open in their Bodies and thin their Blood.

All Cattle that take the Infection, when full of Hay, muſt certainly Die; therefore Chaff, and cut Straw is recommended.

Figure 6. A flysheet of 1749.

Despite the fact that 30,000 cows were lost in Cheshire in less than 6 months, a "never failing" remedy was announced in the *Gentleman's Magazine* by a local optimist, G. Venables Vernon (Anon., 1751c). The magazine continued to provide its readers with remedies, however bizarre, even though it was aware that they did not work. In February 1755, it published a remedy "from an honest farmer, who has tried it with success, and we shall therefore give it in his own words." Puncturing with a hot iron on the nape of the neck and in the dewlap, and rowelling with half a yard of flax dipped in tar, the farmer advised, "keep it in a fortnight or three weeks till the beast is got over it [the disease or the operation?]. I take it that the burning brings an humor to the part directly, and it runs before any digestion can be procured by buttoning, for they generally dye before they discharge anything." This was accompanied by the usual herbal concoction, in this case a mixture of rosemary, rue, lavender, madder, Venice treacle, diascordium,[4] and verjuice or cider, or red wine (Anon., 1755a). Later the same year, it published "An almost infallible Method of curing the contagious Disease of horned Cattle, by Dr. Fischer, Physician to the late Empress Anne of Russia, now in general Use all over Livonia":

> As soon as a beast is perceived to be infected, give him nothing to eat for three days together but wormwood, galengale, or some other bitter stomachic herb, and mix an handful of leven or yeast in his drink. Besides this, he must every evening be made to swallow a ball made up of powdered tobacco, meal and malt spirits, and every morning half an ounce of black soap, mixed with a bruised head of garlic; but if the scowering be violent, the soap must be omitted. During the cure the beast must be washed frequently with lye of wood ashes, his mouth cleansed with vinegar, and his head and neck covered up warm. This practice is said very rarely to fail.
>
> N.B. This method has the sanction of the highest authority from Russia, and may so far be depended upon as genuine. (Anon., 1755b)

In 1752, John Bowen, an Inspector in Salop, published his remedies for prevention, and cure. For prevention, he recommended a mixture of butter, rue, wormwood (*Artemisia absinthium*), garlic, and tar, and to cure scouring while stating "I conceive all heating Things applied inwardly to be prejudicial," recommended a mixture of Jamaican pepper and other spices.

7. CONTINENTAL REMEDIES CONTINUE

Sometime after 1750, a leaflet was issued at Lyon *Preservatifs Contre la Maladie des Bestiaux* (*Preventives Against the Cattle Sickness*) with the instruction that curates and officials were requested to read and explain it to their people. It began by stating that the first and most sure preventive was to carefully avoid all communication with infected places, observing exactly the laws and ordinances in force. Second, it was absolutely necessary to separate healthy animals from sick. The latter should be shut up in isolated sheds for treatment, the sheds kept clean and fumigated, all dung being removed when the animals were out grazing. When the animals returned to the sheds, two or three times a week their bodies should be rubbed with a bunch of straw dipped in the hot washings of ashes in which various herbs had been boiled. The animals should not be let out to pasture until dry and the mist had

[4]*Diascordium* was a famous remedy concocted by Fracastoro (1546) for bubonic plague comprising cinnamon and cassia bark, scordium, Cretan dictamon (dittany), tormentilla, bistorta (snakeweed or knotweed *Polygonum* spp.), galbanum (an unknown gum resin) and gum, opium, styrax (another unknown gum), acetosa, gentian, Armenian bole (an absorbent red earth), terra sigillata (sphragis or Lemnian earth), long pepper and ginger, white honey, aromatic wine, and sugar of roses.

lifted.[5] Ordinary fumigants were incense, juniper, sulfur, gunpowder, and pitch, mixed with strong herbs; to avert fire, they should be burnt without flame.

It alleged in 1714 that many people stopped the disease by thus cleansing and fumigating their sheds. In England "where this disease is frequent," the windows and doors of the sheds were closed and the walls smeared with liquid tar, except where the animals fed, while each morning, a little tar was placed above each animal's nose where it could not reach it with the tongue, avoiding the mouth and nostrils. Tar, suitably prepared by being mixed with fish oil, aspic, pitch, tallow, and other ingredients, could be bought in Lyon at the shop of M. Chassaignon, Spice Merchant, at the extraordinary address of "the Street of the Child who Pisses."

Before an animal fell sick, another preventive was to give it a drink twice a week of wine water containing an amount the size of an egg of flowers of sulfur mixed with cumin powder. Yet another was vinegar boiled with angelica root, garlic, sage, rosemary, lavender, thyme, rue, laurel, and other herbs. The muzzle should be sponged with the solution each morning and a little placed in the ear.

Doctors at the Hospital of Lyon allegedly cured four animals at the Golden Head in the following manner: they gave each animal an ounce of viper's powder and covered them with two or three hot blankets. The next day, the animals were washed with wine water or lye boiled with sage, rosemary, laurel, and thyme. A hot warming pan was passed under the blankets with leaves of sage and some sugar in the ashes. The animals were fed broth containing the same herbs and the addition of marsh mallow. Men should not use the blankets afterwards without first washing them. Any pustules on the tongue should be lanced with a spoon or piece of silver and wiped with a red cloth. The tongue should then be washed with vinegar, salt and pepper. If an animal had diarrhoea or passed blood, then for 3 or 4 days it should be fed only raisin pips and given oats steeped in wine to drink.

Another remedy was a seton in the dewlap of three finger-length pieces of dried wild hellebore root, Virgin's herb, or nettle, greased with white grease, sprinkling the animal morning and evening for 2 days with vinegar. At the same time, two people should rub the animal with very hot tiles on the loins and flanks, and then cover it well. After 48 hr, a swelling appears, when the root is removed and replaced with three greased stems of straw to prevent the wound closing, dressed twice a day until the animal is cured, which sometimes takes 2 months. All rottenness should be cleaned out of the wound with a wooden spatula and every morning the animal given an amount of Venice treacle the size of a nut in a glass of wine, injected into the nostrils, especially the left one. This treatment should be used as soon as the animal stops eating or refuses to work.

It stressed that it was very important to avoid all contagion not to expose sick animals, nor bury carcasses near roads, but far away from habitation; and to place spikes in the ground to prevent carcasses being disinterred by dogs and wolves (Anon., n.d.).

The anonymous instruction issued at Nuremburg in 1754 gave the following methods of prevention:

- Do not release cattle into the fields too early in the morning or collect them too late in the evening because of the presence of poisonous fog and dew.
- Do not bring wet grass into the stall.
- Do not drive through cold water in the winter on their way to the stall as this could cause fever.

[5]Among present-day Fulani of Nigeria, cattle are not let out to graze until the dew has dried off, neglect of this precaution leading to losses in cattle at the time of harvest (St. Croix, 1945).

- During prolonged cold rain and cold weather, especially fogs, give them herbs and herbal remedies.
- Provide clean water for drinking in the stall.
- Do not let healthy cattle near infected ones or persons who have been in contact with the latter.
- Do not allow foreign cattle to be driven through the area.
- Do not allow Jews [dealers and itinerant traders were usually Jews] and others near healthy cattle when there is the least suspicion they may have come from an infected area.
- Do not allow dogs, cats, or poultry near; and do not bring feed from infected areas.
- Do not drive out stock in fog or cold or prolonged rain.
- Fumigate the stalls and hang garlic and "devil's dung" (asafoetida) in them.
- Hang bags of herbs under the cows' necks.
- In prolonged rainy weather, apply a seton on both sides of the neck using hellebore root, or better a strip of leather or length of horsehair rope, with tar, Egyptian salve, or axle grease on the apertures.

If there is an infected one in the stall:

- Isolate it immediately and wash all utensils in soapy water.
- Bleed the healthy and bury the blood deeply.
- Apply setons not only on both sides of the neck but also at the hind end near the anus.
- Give purges each evening of a glass of wine vinegar with a spoonful of tree or flax oil in it.
- Give purgatives.
- When the loose faeces are evacuated, place into the anus two or three times a day a warmed mixture of bacon pieces cooked with mullein and flax in good milk or water and tree or flax oil. One could also throw colophene on cabbage leaves and saturate a cloth with the vapor, holding the cloth over the anus as long as it remained warm.
- For constipation, a ball of soap the size of a small apple could be placed in the anus. If this did not work, then an enema of mullein mixed with clean water or milk with wood oil and salt, was recommended.

But from general experience, rarely was an infected animal cured, and if lean or old, it was best to slaughter it. Carcasses of dead animals should not be dragged away but buried on the spot with the hide slashed. There should be a penalty for leaving them or throwing them in a pond or river, and they should not be burnt (Anon., 1754b).

Plenciz (1762) reported an Austrian plague remedy consisting of vinegar, spirits of wine, and boy's urine, "equal for man or beast." In the Duchy of Magedeburg in 1764, it was recommended that as soon as a cow went off its feed, a string of onions which would attract the poison be hung around the cow's neck and renewed each morning, the discarded onions being buried so that they could not be eaten. Bundles of onions could also be hung in the stalls. In 1766, a pamphlet by Clerc, a former French doctor of the royal armies in Germany, was distributed in France and Luxembourg giving the same remedies as were in use throughout Europe. In Italy, the experimental method in 1770 for curing the epizootic evil of cattle, of Canevese master of the stable of Count Arconati of Milan, was nothing more than a herbal remedy for actinobacillosis, accompanying the scraping of the tongue pustules with a silver instrument and fumigating the stalls. Setons were also suggested. There is undoubtedly confusion with rinderpest in some cases, and symptoms such as lachrymal discharge, drivelling from the mouth and hanging the head, are mentioned. For preventing and curing "the plague," he recommends giving green fennel in the mornings and taste of grains of "medemo" in the pastures (Canevese, 1770). The *Gazette de France* of December 1772 reported that surprising results had been obtained in curing and preventing infection by use of the roots of petasite or *herbe aux teigneux* (butterbur), the virtues of which had been known by doctors for a long time, especially in Germany.

Haller noted in 1773 that many new doctors had written on the cattle plague, but almost none had noticed the seat of the evil was in a myth. Some doctors recommended bloodletting, others acidity because sour apples in the Brandenburg District had allegedly kept animals healthy.[6] Others favored quinine or mercury, and yet others mixtures and old recipes. The elderly stood by a seton of hellebore. But both Holland and England had recognized the inefficiency of medicines and that there was little hope of a cure. Long experience showed that medicines were not a security. One could not be sure of their effects in animals as people had observed the effects for over 2,000 years in humans and knew the powers of each. This they did not know for animals, and cattle doctors followed the ancient Greeks almost unchanged, using certain set recipes (Haller, 1773).

Such pronouncements had little effect, and an edict issued at Hildesheim in 1775 added a recipe for good measure which comprised lungwort, mullein *Verbascum* spp., caraway, juniper berries, salt, honey, goose fat and gunpowder (Anon., 1775). Vicq d'Azyr (1776) listed every alleged cure he could find, while not accepting that any could be effective. When he arrived in the district of Condomois, he found people administering to sick cows a "*trés incendiaire*" concoction of gunpowder, pepper, sulfur, and herbs. As well as being "*trés incendiaire*," it was also "very murderous," and he had its usage stopped. He observed of various medicines: "Beasts which have taken a too great amount, have died in the middle of the most frightening convulsions." One treatment when an animal appeared weak was to administer an anti-poison and stimulant potion consisting of $\frac{1}{2}$ oz. of theriaque and 1 dram of viper powder mixed in four glasses of good wine. Spanish fly (ground dried *Cantharides* beetles used as an aphrodisiac) was also used. One peasant approach was to pour vinegar into the ears of a sick animal. "Honey, oil, vinegar, nitre, bran or flour, asafetida, water of life, ordinary water, more than that, exactitude and vigilance, such are the arms with which one can be sure to preserve the infection." He noted that veterinary medicine was "still in its infancy." It was to remain so for another century.

In Paris, treatments such as brandy mixed with gunpowder, and cordials, were given. Some dosed 5 or 6 cows with 350 bottles of wine in a short time, or over 40–50 L per cow. Another treatment was placing in the mouth twice a day a mixture of salt, peppercorns, garlic, and honey, while ears and nose were rubbed frequently each day with aromatic vinegar. One purgative consisted of a decoction of senna mixed with prune juice, and rhubarb or fresh linseed oil, infused in the drinking water, or applied as an enema in slightly salted warm water. At night, a mixture of nitre, camphor, red-lead, and rhubarb, in half a pailful of warm water, was recommended. The stalls were fumigated with burning junipers, sage leaves, rosemary, and absinthe.

An anonymous German writer (1776), in his *Beitrag zur Geschichte der Rindvieheseuche im Hannöverischen, Nebst zween Briefen über die Lentinischen Pulver (Contribution to the History of the Cattle Plague in the Hannöver District, Together With Two Letters on the Lentin*

[6] The full story was that a maid dropped half an apple by a prostrate sick bull which had not eaten for 8 days. Seeing the piece of apple, it tried to reach it, and the maid placed the piece in its mouth, the bull eating it. The bull was then given several apples, eating three on the first day and then more and more until it was eating cabbage and hay, recovering fully. It is clear that the bull had survived the infection and simply became hungry.

Powder), claimed that it was still worthwhile giving iron sulfate, second to arsenic, which had some good effects reported in Sweden, but the animals died when it was tried elsewhere. A powder known as "Lentin powder," produced for the government of Hannover by Dr. Lentin, municipal doctor at Clausthal, when analysed appeared to consist of nothing more than sugar, camphor, and a grain of lapis lazuli (blue copper carbonate). This was followed by Konrad Friedrich Uden, writing anonymously in his *Beiträge zur Gescichte der Hornviehseuche in einigen Kreisen der Altmark, und deren Erkenntniß und Heilung (Contributions to the History of the Cattle Plague in Some Districts of the Altmark, and its Knowledge and Cure)* (Anon., 1777). He considered that it was an eruptive fever and the same as chickenpox. Supporting the use of Lentin powder, he argued against fictitious remedies, noting that each doctor had their own opinion as to the nature of the disease. Lentin argued that the disease originated from an error of the blood which affected a particular type and mixed up the secretions, causing high acidity and putrid dissolution. Therefore, it was necessary to dilute the blood and moderate it. Good digestion was necessary, so roughage had to be cut by a third, and spirits, particularly malt, were especially recommended to dilute the blood. This should be followed by a medicament. Mercury had been recommended for smallpox by Huxham and others, but it did not enter the blood. Camphor was also recommended by Huxham, Haller, Hoffman, and others. He therefore produced his "Lentin powder." Uden wrote to Lentin, inquiring what the ingredients were, and Lentin replied: 4 grains of mercuric chloride, 12 plummet of sugar, some camphor, coloured with red mercuric sulfide, and some drops of stag's horn oil. It was essential to administer it according to the stated timescale. On the basis of the report of a correspondent who gave it to one cow in December 1776, which survived in the midst of an outbreak and was still healthy in June 1777, Uden thoroughly recommended the potion.

Uden noted that a surgeon came to Mahlpfuhl after rinderpest broke out in April 1775 and instructed that sections 24–29 of the Royal Instruction of 1769 should be followed, keeping a diary of events with number of cattle dying, number recovering and other details. A preventive was a mixture of jalap root (*Ipomœa purga*), saltpetre, cream of tartar, and fennel seed, while for a cure, a mixture of barley, cream of tartar, and juniper berries, was recommended. But all of these remedies had no effect. People heard that kitchen salt was a preventive and, after one cow died, gave salt to their cattle in stalls, but all died. From Prussia came a "wonderful" mixture of 18 ingredients, but Uden left the reader to judge its usefulness for them. In October, there was a rumour throughout the province of a stranger in Magdeburg who could cure the disease through secret magic arts. Unable to verify the truth of the claim, Uden invited him to Altmark but received no reply.

It was Galleski, in his *Angestellte Untersuchung über die Rindviehseuche Worin deren Wahre und Eigentliche Ursachen Bemerken und die Mittel sie Abzuwenden für Geschlagen Werden (An Employee's Investigation of the Cattle Plague Concerning True and Actual Causes and the Methods to Avert Defeat)* (1765) who stated that giving common cooking salt as practised in the Ukraine was the best preventive. This was argued at length by Baron Hüpsch (1776) *Patriotische Vorschläge die Ausbrietung der Jetzt Herschenden Landesverderblichen Hornvieseuche auf eine Leichte und Wohlfeite Art zu Verhindern, Nebst Einigen Nenen Undeigenen Beobachtungen über die Besten Bewahrungs und Heilungs = Mittel (Patriotic Proposal Concerning the Diffusion of the Reigning National Ruinous Cattle Plague of an Easy and Cheap Method to Prevent, Together With Some New and Personal Observations on the Best Preservation and Cure Methods)* (1776), claiming that natural substances were the best means of combating the disease. Needham (1772) in France argued for the beneficial properties of rock salt, or better, sea salt, which he considered not only a preventive but also a specific against rinderpest. In his *Gedanken über die Unzulänglichkeit aller Vorbauungs und Vorkehrungs Anstalten Gegen die Hornviehseuche (Thoughts on the Deficiency of all Precautions and Preventive Institutes Against the Cattle Plague)*, an anonymous German writer in 1782 stated that he had once thought rock salt, recommended by many, to be helpful, but from experience had now come to the conclusion that it had no effect. He also noted that he did not know on whose authority pieces of burnt infected stock were recommended as a medicine. Reuss, writing anonymously in 1784, noted that setons were not approved everywhere, and it was possibly better to make the poisonous matter flow out from the nostrils by inserting snuff. Another possibility was to burn the belly with a red-hot iron so that the poison in the abdomen could flow out.

8. PROTESTS AND PARODIES

Dossie wrote:

> The remedies, which have been tried for the cure of the murrain, have been in general as inefficacious, and absurd, as those employed for its prevention ... The far greater part of these medicines, like those used for the prevention of the disease, have been taken up on the authority of the antients, and the earlier writers; and consist of a jumbled variety of those medicaments, which are deemed antidotes of poisons, or alexipharmics ... Under this class have been given theriaca, mithridate, diascordium, opium, camphor, balsams, frankincense, myrrh, juniper berries, camomile flowers, marigold flowers, feverfew, rue, sage, fenugreek, madder roots, grass roots, horse-radish, bay leaves, mustard seed, snake root, contrayerva root, turmeric, savin, moth maslein, spearmint, calamus, aromaticus, garlic, onions, leeks, testaceous powders, sulphur, vinegar stalks, honey, raisins, figs, blood of a tortoise, and eggs. (Dossie, 1771)

The "antiseptics" which some physicians recommended based on more modern notions, that is modern for the mid-18th century, sound no better: sulfuric acid, vinegar, verjuice, sour dough, buttermilk, common salt, and ammonia. Others recommended nitre, cream of tartar, acids, mucilages and oils; yet others mercury and antimony.

Theriac or Venice treacle, popular in Roman times, was prepared with honey, and usually opium, with cardamom as a particular component and 45 to over 100 other ingredients. Used originally for wild animal bites, presumably rabies, it was an application of the "hit or miss" principle. If one ingredient did not work, another might. For centuries, it was prepared in public with elaborate ceremonies. In Germany in 1745, two types were available: common theriac, *Theriaca diatessaron*, and cattle theriac, *Theriaca propecoribus*. The price of the former with camphor was 2 groschen 2 pfennig, the latter 2 pfennigs dearer.

The many remedies employed were ridiculed by an anonymous contributor to the *Gentleman's Magazine* parodying Virgil's *Georgics*. In *A Pastoral*, Diggon Davy is burying his favorite cow when neighbor Colin Clout (perhaps referring to the JP for Nottingham, Langford Collin) comes by:

> Diggon:
> How could I hope, where such contagion reigns,
> Where one wide ruin sweeps the desart plains,
> Where ev'ry gale contains the seeds of death,
> That Diggon's kine should draw untainted breath?
> Colin:
> But have you, Diggon, all those methods try'd,
> By book-learn'd doctors taught, when cattle dy'd?
> Or, tho' no doctor's remedies prevail,
> Does the good bishop's fam'd Tar-water fail?
> Diggon:
> Each art I try'd, did all that man could do;
> Med'cines I gave, like poison, med'cines slew:
> The bishop' drink, which snatch'd me from the grave,
> Giv'n to my cow, forgot its pow'r to save.

Dejected first she hung her drooping head,
Refus'd her meat, and from her pasture fled;
Then dead and languid seem'd her plaintive eye,
Her breath grew noisome, and her udder dry.

Scorch'd with perpetual thirst, short sighs she drew,
Furr'd was her tongue, and to her mouth it grew:
Her burning nostrils putrid rheums distill'd,
And death's strong agonies her bowels fill'd;
Each limb contracted, and a groan each breath,
Lost ease I wish'd her, and it came in death. (Anon., 1747d)

A Scots physician, John Brown (1735–88), proposed that there existed only two forms of disease, sthenic, denoted by excitability, and asthenic, denoted by debility. Thus, there should be only two types of treatment, sedative in the case of the former and stimulant for the latter, the two chief remedies being opium and alcohol. This system, opposed by followers of Cullen, was known as the Brown or Brunonian system. Although Frank (1802) in Germany seems to have been the sole writer to introduce it into the context of rinderpest, claiming it justified methods of building up an animal's strength, while not totally embracing it, welcomed the return of practitioners to stimulant treatments. Dossie was apparently following it when he believed that the only way to treat an animal was by means which kept up its strength, Fleming agreeing also that this was the proper treatment. Believing apparently in the chemical theory advanced by Faust (1797) that nitrous oxide was the infectious substance, Frank recommended asafetida as the main method of control because of its large quantity of hydrogen which bound with the contagious nitrous oxide. However its high price prevented its use, as large amounts were necessary. Dossie recommended the astringent, febrifuge, gummous parts of vegetables, and vinous liquors. Peruvian bark (cinchona bark of which quinine is an important constituent) was considered the best so far employed, but its cost was uneconomical. Substitutes were white willow bark (which contains salycilate the effective ingredient of aspirin) and tormentil root, the latter important for checking diarrhea, often considered a cause of death. Despite recommending methods of treatment, Dossie asserted that the foremost preventive method, should the disease break out, was immediate slaughter of infected animals.

By 1780, slaughter was an accepted practice for stamping out the disease in Britain, yet when there was a brief outbreak near Margate in 1781 a correspondent in the *Annual Register* recommended that cattle be treated by bleeding and giving drinks, the use of medicines having frequently been more "prejudicial than salutary," causing greater loss to the cattle owner through the expense of the medicine and the loss of the cattle.

In 1789, Robinet in France produced his *Manuel du Bouvier, ou Traité de la Médecine Pratique des Bêtes à Cornes* or *Herdsman's Handbook*, which was more like a veterinary textbook. While admitting that there was no specific cure, he favored setons as a treatment "in place of uselessly tormenting it [the sick animal] by internal medicines and excitants," and a gag, these to be accompanied by abundant drinks of whey or acidified water, or a "refreshing and anti-septic" beverage of a decoction of fresh roots of sorrel and stawberry, leaves of sorrel and lettuce, with mineral water or saltpetre solution. Alternatively, one could use fresh marshmallow roots with sorrel leaves and wine vinegar.

Erasmus Darwin, scientist, poet, grandfather of Charles Darwin, considered to be one of the foremost physicians of his day, added his remedy in 1796: "I suppose about sixty grains of opium, with two ounces of extract of oak-bark, every six hours, would supply them with an efficacious medicine; to which might be added thirty grains of vitriol of iron, if any tendency to bloody urine should appear…." He did also recommend slaughtering all cattle within an 8-km radius if an outbreak should occur.

An Italian writer observed in 1796 that Denmark and Holland had been trying to find a specific cure since 1781 without success. In the mean time, he recommended that affected cattle be given liberal quantities of wine every morning, laced with garlic and pepper (Filopatride, 1800). In Germany, Faust (1797) noted that smoking stalls with devil-muck (*Succisa pratensis*), old shoe leather, horns, and hoofs was superstition. Stalls required complete cleansing.

9. BLOOD-LETTING

A typical treatment was:

- blood-letting at the neck;
- blood-letting at the ears;
- give a drink of six egg yolks, 1 oz. of flowers of sulfur and muscade in a mixture of wine and linseed oil;
- alternate with stimulating drinks.

If any animals survived, the treatment was claimed to be effective.

Farriers and cow-leeches had informed Dr. Mortimer that a cow could lose nearly 2 gallons (9 L) of blood without fainting. He recorded the case of a cow which showed symptoms of running at the nose and shortness of breath which was immediately bled by the owner taking 5 quarts (5.7 L) of blood from the neck, and given a warm mash of malt every 6–8 hr. The next day, the owner cut her tail and let it bleed for 2 hr. The following day, he removed 2 quarts (2.3 L) of blood from under the tongue and continued bleeding it every 14 or 15 hr seven times. "She soon recovered, now eats heartily, and has not slunk her calf" (Mortimer, 1745b). Bleeding in small quantities, Mortimer noted, was not successful.

Bleeding had been strongly recommended in the anonymous paper published in 1745 known to have been written by Barker, who later admitted to it in the *Gentleman's Magazine*. He advised early bleeding and even daily bleeding for three successive days, but bleeding had no effect in the last stages of the disease.

"J. S.," who praised Barker's paper in several letters to the *Gentleman's Magazine*, condemned the "FOLLY and OBSTINACY" of those who did not attempt to prevent infection by following his, J.S.'s, instructions, to keep the animal clean and dry if there was any suspicion of it falling sick; then bleeding, 3 quarts (3.4 L) for the first 2 days, then 2 quarts (2.3 L) every third day four times; drenching the animal every day with a solution of saltpetre with sulphuric acid (or vinegar or verjuice); and feeding hot mashes two or three times a day, and offering as drink water-gruel with verjuice or vinegar. Setons would be an additional security. But these measures were only effective if used at the first signs of the disease, and "J. S." agreed with slaughter if the disease had reached its height. Covering himself, he ended, "I must once more remind the publick, that if my directions be not punctually obeyed, or if any other supposed remedy be added to what I have advised, the trial is not fairly made, and I cannot be at all answerable for the consequences" (J. S., 1747a). Following up with yet another recipe which had "often been successful," J. S. recommended the addition of cutting the inward side of the animal's tail and applying a poultice of garlic and rue (J. S., 1747b).

Doctor Noah, in an anonymous tract addressed to JPs and Commissioners of Land-Tax in Suffolk, *c.*1747, "Printed and Sold by J. Rose, over-against the Angel above Hill, and at his Shop opposite the Peacock in the old Butter Market on Market Days," extolled bleeding as a cure. On the very first appearance of the disease, Dr. Noah suggested taking 1 gallon of blood twice within 24 hr, morning and evening for 3 days. He quoted a "Gentleman of Figure" as having also used a "perspiring" medicine with success, consisting of angelica root, vinegar, and aniseed, boiled together and a pint administered morning, noon, and night for 3 days. Whereas the Gentleman of Figure recommended bleeding on the second morning, Dr. Noah advised doing this first. If angelica root was not available, burdock roots could be used in winter, or leaves in summer. The nostrils of animals should be smeared with tar heated up in vinegar. Many hundreds of beasts were claimed to have been preserved from infection, although grazing common pastures where other beasts, not given the treatment, had died of the disease or were slaughtered for having contracted it ("Physician", n.d.).

In Germany in 1745, bleeding was recommended on both sides of the neck, removing half an *Oessel*, or 4 quarts (4.6 L); with another $\frac{1}{2}$ quart from the tail vein, and three or four spoonfuls from the veins under the tongue. A German mandate of May 13, 1780 of the Duke of Saxony recommended bleeding at the first sign of illness taking 2–4 lbs (1–2 L) three times at 12-hr intervals from the neck and under the tongue, "Bleeding and the use of hair-ropes constitute the most essential part of the cure and cannot begin too early …."

Dossie considered that where any attempted remedies had any effect upon the animal, either by stimulation or by producing or reducing inflammation, they only served to weaken its resistance to the disease, accounting for more deaths than if no treatment had been given at all. Foremost perhaps amongst such debilitating "treatments" was bleeding: "Whoever considers the effects of bleeding on the habit and nature of the distemper, as deduced from the symptoms, cannot doubt but that this evacuation must have largely contributed to augment the extraordinary number of those beasts which have died when subjected to medicinal treatment, compared to those which have been left to the favour of nature." Two quarts (2.3 L) of blood, morning and evening, were not thought too much.

In the terminal stages of the disease, the blood is thick and slow to clot with poor serum separation, indicative of severe dehydration. Bloodletting could only serve to worsen the animal's condition, but it was still being practised in the Phillipine Islands (in conjunction with inoculation) in 1909 (Ruediger, 1909).

10. OTHER METHODS

Hanno (1746) recommended slitting along the underside of the tail in the middle from its distal end to its first vertebra, which did not injure the cow but drew out the blood and humours. In 1754, clearly following Ramazzini's recommendation, a very beneficial method in Germany was claimed to be to burn the neck of a cow with a wide red-hot iron (Anon., 1754b). Cauterization on both sides of the neck was more specifically recommended, or piercing the nostrils and ears and pouring in two or three spoonfuls of vinegar. These methods were confirmed in the Halle booklet of 1740. Paulet (1775) cautioned that when cauterizing the neck by the side of the vertebrae, one should take care that the muscles were not severed, or the animal would be crippled.

Having their origin in antiquity, more conventional for the time were rowels, otherwise known as setons or pegging, in Germany also known as a hair rope, bommel or triel. Setons were pieces of cloth or threads inserted under the skin to keep open a channel to the outside so that the supposed noxious matter could drain out. Often, the wound was kept open with a mixture of turpentine and egg yolk. The recommended method was to use twisted or woven horse hair the thickness of a good piece of twine smeared with olive oil and inserted through the dewlap by means of a "pack-needle," the ends then tied together. If the seton moved, one could sprinkle Christian or black sneezing root (hellebore) powder on a portion of the hair rope and pull it through. Some recommended a piece of hellebore root instead of a hair rope, but this was difficult to keep in place, and so hair rope was preferable. The function of the powder, or hellebore root, was to provide an irritant, causing the wound to weep more and thus appearing to discharge the "poison." Hanno considered a hellebore seton to be without benefit (Hannæo, 1746). The duke of Saxony's mandate of 1780 recommended inserting a seton at the first sign of illness and after 12 hr either applying another or burning a hole in both ears or into the dewlap, hump, or neck with a red-hot iron. Turpentine and egg yolk, mixed with Spanish fly, should be rubbed into the wounds. Hellebore root, red gentian *Gentiana* ssp. or Florentine iris root (orrisroot) *Iris florentina* can be placed in the holes in the ears.

A French variant on rowelling used around Paris in 1745 was to bore a hole in the dewlap with a red-hot iron and loosen the skin around it to form a pocket. Into this was placed a root of black hellebore rolled in a digestive compound, one third of cantharides powder and the rest some suppurative ingredient. This was claimed to be the "only successful remedy" as it caused a "plentiful suppuration," continued for at least 15 days by means of hog's lard or any common suppurative (Anon., 1745d). Another French method, proposed by Dr. de Sauvages, was to make incisions in the skin and insert pieces of the inner bark of wild currant *Ribes nigrum*, then with the finger squeeze out the matter which formed. Reynier (1762) recommended setons, noting that their use was general practice in England and in the English colonies in North America under the belly of horses and cattle. He disapproved of hellebore, gentian, or pyrethrum roots, as they were too strong. A hair rope greased with Egyptian unguent or a mixture of black hellebore root with cantharides powder and honey was recommended.

The rationale behind burning on the neck and rowelling stemming from Hippocrates's assertion that diseases needed an outlet for the poison to drain away had been explained by Ramazzini but was explained also in 1746 by a contributor to the *Gentleman's Magazine* signing himself "Bucolicus": "Burning with a hot iron on each side of the neck, and rowelling the dewlap, will answer the purposes of blistering; for nature requires an outlet, whereby the peccant matter ought to be discharged ...." The rationale behind the following, which appeared in the same journal in 1748, is obscure: "As soon as you perceive the distemper on any of your cattle, first cut off the tip of both ears, burn behind the ears also with a red hot iron, and every morning, for three days following, bleed under the tongue, plentifully, giving the following cheap remedy, morning and evening, for nine or ten days ...." Then followed a mixture of a large spoonful of tar, an egg with shell, honey, linseed oil, sharpest vinegar, small beer, grains of paradise *Amomum meleguetta*. The treatment continued: "Tar the noses, breasts, and sides of the cattle once a week, and also wash their necks and legs every second or third day with stale urine ..." (Anon., 1748e).

A firm believer in the efficacy of setons, Dr. Hird (1753), in a rambling essay mainly drawing parallels with human epidemic diseases, recommended inserting them in the dewlap for at least 2 months as a preventive, even if the disease was not present. It was the practice to pull the cord backwards and forwards several times a day to keep up a discharge of liquid matter. A decree issued as a precautionary measure by the Duke of Braunschweig and Lüneburg dated August 28, 1745 commanded that all cattle owners should employ setons as specified in an earlier (1716) decree, noting that not everyone did so because of a conceited belief that they caused abortion in pregnant cows. This was contrary to medical reports. Although a seton might have side effects, these were of minor importance compared with the cattle plague.

The "Several Hands" (1747) "with all the Penetration that we are Masters of," could not see that setons would be of use in a sick animal which rarely survived the seventh day: "Nevertheless those, who have more Faith in them than we have, may freely make use of them; but then certainly the properest and best Places to set them in, are on each Side the lower Part of the Bosom, and in the Inside of each Thigh."

While not appearing aware of rinderpest, Downing (1797) recommended insertion of a small slice of bearsfoot (hellebore) with a piece of red dock *Rumex sanguineus* var. *sanguineus* root either side into the hole: "if it takes the desired effect, it will swell very much and discharge plentifully."

Belief in the efficacy of setons probably originated with the alleged survival of the single cow of Count Borromaeus which had been the only one in his herd with a seton. Their use is first mentioned by Columella as *auriculæ rescindendæ*, herdsmen boring a hole in the ear within an incised circle and inserting a piece of lungwort gathered with the left hand before sunrise.[7] When the circle of flesh rots and falls away, it has taken the poison with it. As a result of this primitive belief, cattle were condemned to suffer agonies for centuries.

Less painful for the animal was a French method of boring a hole in the cow's manger and placing some mercury in it, covering the mercury with asafetida, and then stopping up the hole with a peg of white thorn, not touching the ingredients. An English schoolmaster possessing three cows placed mercury girdles around the neck of each, and all allegedly survived (Anon., 1748e).

Recommended by a peer of the realm as a successful treatment, a grazier claimed that balls the size of a walnut each made of 2 oz. of asafoetida and of garlic, and a handful of rue, beat together into a paste, placed one in each ear which was then stitched up and left until dissolved, were effective (A. B., 1750). One man had holes bored near the base of the horns of two sick cattle, which produced a discharge of matter and the cows recovered: "As I was walking near this city [Norwich], I saw some persons burying a cow that died of the distemper … happening to break off the horns, they observed them to be quite hollow, and the pithy matter decayed. This gave me a hint to try the … experiment" (Anon., 1750c). Another tried this on sheep with allegedly beneficial effect, while another found it completely ineffective on large herds of cows (Anon., 1751d). In France, it was considered that those animals which secreted most copiously from the mouth seemed less seriously affected by the disease, so much so that in March 1745, the parliament at Rouen passed a law that infected animals should have a gag placed in the mouth for 3 hr every day to encourage salivation.

Methods of combating human plague in 17th-century Italy were not very different from 18th-century methods of combating rinderpest. Disinfection of the pest houses set up for isolating plague victims involved washing the walls and floors with vinegar, applying lime, and fumigating by burning sulfur. In France in the 1760s–1770s, Professors of the Faculty of Montpellier laid stress on clean and fumigated sheds, fumigating with the smoke of juniper, fir, and other aromatic woods. No fumigants were known which specifically affected the "rotten air" of epidemic pestilences, but fumigants corrected the stench of infected air and "modified its effects on the nerves." They also recommended changing the straw each day.

In Denmark, in the 1759 outbreak, many peasant farmers attempted to prevent infection by fumigating their cattle with tobacco smoke, continually smoking in the cow houses and even sitting up smoking all night amidst the cattle. It was noted that scarcely any of the cattle escaped infection and death. In Finland in 1761, it was reported that the liver, spleen, and lungs of dead animals were burnt, and the ashes administered to the sick cattle on account of the "volatile salts" in the ashes. Now, it was stated, they were beginning to burn whole carcasses for the same use.

Blondet considered that the disease was caused by "morbific matter," which was prevented from escaping through the skin due to the weight of the atmosphere pressing upon it and the skin's toughness in cattle. Having calculated the pressure on an ox with a skin surface area of 3 m^2 to be 38,790 kg, to assist the escape of the humours, he proposed placing the

[7]Paulet (1775) seems to have mistranslated lungwort as hellebore.

animal in a hot vapour bath, or bagnio, for 2 hr or more, with only the head exposed. By this means, the skin would be softened, the pores opened up, and the poison could escape. Cupping glasses were also recommended, lighting tow inside them before placing on the skin to create a negative pressure. He also calculated how much lightening of pressure could be obtained by this means (Blondet, 1748, 1751).

Yet another remedy suggested in 18th-century France was acupuncture. M. de la Coste, a well-known doctor at Montignac, proposed its use at the onset of the disease. His method was to make a considerable number of acupunctures over the body not more than 5 cm apart with a sacking needle reddened in the fire, puncturing as far as the underlying fatty tissue. At the same time, the unfortunate beasts so treated were to drink every morning a mixture of vinegar with camphor and other substances dissolved in it, and, three times a day, a concoction of herbs.

In Flanders, a recommended "cure" was to bury sick animals in a mudhole for 9 hr with only the head exposed. After this, they were allegedly taken out quite well, the cause of the disease having been that the spleen and lungs had dried up. "Several Hands" (1747) suggested that swimming the cattle to exhaustion would be better, a method practised by farmers in Northamptonshire.

In 1745, Lobb proposed that sick animals be collected into "pest houses" for treatment, an idea taken up by a correspondent in the *Gentleman's Magazine* in 1750, who stated that opportunities for experimentation should be provided by having in each county a place set aside for sick cattle where they were placed on payment of £2 by the government. For every animal which then recovered, the owner should pay £1 or £2, while dead animals could be examined by doctors (Philo-Boas, 1750). Dr. Wall proposed that where the disease was violent, all sick beasts should be placed in a pest house under a doctor's care, and the symptoms described so that treatments might be improved (Wall, 1751). This differed from the German method, whereby the sick cattle were isolated in a sickbay in each village or farm, not collected together at some centralized institution.

In 1747, an anonymous contributor suggested using electricity passed through a tube inserted into the cow's anus. This would permit the propagation of electric vapor through the whole intestinal tract, and conversely, projection into the mouth and thorax would affect the entire digestive and respiratory systems, with the most beneficial consequences (Anon., 1747e). E. Litton stated in 1750 that the disease was a punishment intended for the extirpation of moral evil, but he had studied a communication on the electrical treatment of the disease by a "famous Venetian lawyer," Pivati. This was to place drugs in a glass cylinder "agitated" by an electric machine,[8] and the "effluvia of medicinal substances are with irresistible force transmitted into every part of the animal machine." The cure would be effected in only a few hours or even minutes. Litton, although not having used the method himself, recommended that it be used with camphor, balsam of Peru, or turpentine. He was "ravished with contemplation" of the possible vast advantages the method would bring.

In Swabia as late as 1796, references, in poetical form, were made to "many a man" inserting a candle into the anus as a preventive, and to a village burying a bull calf alive (Froehner, 1930). Burying an animal alive to check cattle plague was an ancient Germanic rite. Near Stuttgart during an outbreak, an old woman advised that the parish bull should be buried alive, and wreathed in flowers, it was led in ceremony into a pit. Three times it broke out, but the third time, it choked. It was the custom in Karuthen to bury a sick animal alive if plague broke out, and in Scandinavia, a live cow was buried to prevent plague. In Albringwerde, they buried a slaughtered cow in a medicinal spring. In Aargau during an outbreak, a farmer,

[8] A Frenchman, C. F. de Cisternay du Fay, had discovered in 1733 that there were two types of electric charges.

alleging that the cattle were dying of witchcraft, claimed that the people had to lose one more but could select the worst, which, when it died, was buried under a stable threshold (Jahn, 1884). In Elizabeth I's reign, a man was brought before the Commissions for Ecclesiastical Causes for burying a live dog and a live cow. A dead creature might be buried feet up under the threshold of a stable or cow shed to protect the herd (Kittredge, 1929). In Northamptonshire and Cornwall, it was believed as late as the 18th century that killing a calf would drive away the murrain. In Germany, sometimes in time of plague, the first animal to die was buried with a willow shoot in its mouth. When the tree grew up, it was never cut in any way, believed to guard the farm from pestilence. In Ruthenia, Russia, the first victim might be burnt with a live dog, cat, and a cock, tied to its tail, while in Nijegorod, the plague was kept away by erecting an ash stake at a crossways and scattering about the village the remains of a dog that had been calcined for the purpose (Burdick, 1905).

17

REMEDIES IN THE 19TH CENTURY

1. RECIPES OLD AND NEW

Berkeley, Lord Bishop of Cloyne, "active in many fields but remembered chiefly as a philosopher," who, in 1744 started the European vogue of tar-water as a general medicine claimed to be successful in the treatment of over 150 different ailments, wrote to the *London Magazine* in 1747: "If I can but induce the general use of tar-water for this murrain, which is in truth a fever, I flatter myself this may pave the way for its general use in all fevers whatever." He thought that although people might be loath to experiment on themselves, they would willingly attempt so inexpensive a remedy on their cattle (Berkeley, 1747). Many had turned to tar-water in 1746 when T. Prior of Dublin, an admirer of the Bishop's method, quoted a source:

> Let the sick beast have poured down its throat a quart of warm Tar-water, made stronger than usual by stirring each gallon eight or ten minutes, and this to be repeated every hour or two for the *first* day, while the beast is awake. On the *second*, let one half of the former quantity be given; and on the *third* day, half of that which was given on the second: which last quantity is to be continued 'till the cure is perfected; during which time, the beast should be housed and lie warm.

He added, "I have no experience of the success of this method, as there is no infection of that kind in this kingdom," having prefaced his remarks with the statement that the disease now prevailed in some parts of England. He also suggested that it might be advisable to daub the nostrils, ears, etc. of all the cattle, whether infected or not, to prevent catching or communicating the infection by air, "And also to make the beast swallow one egg-shell full, or two of *crude Tar*" (Prior, 1746).

In a letter to Prior, commenting on his book, Berkeley stated that he knew that tar was used to preserve cattle from contagion, "and this may be supposed to have given rise to that practice of drinking Tar-water for a preservative against the small-pox." Others resorted to a $\frac{1}{4}$ lb. of tar every day by way of prevention, another "perfectly curing" a milch cow by giving it over 3 gallons (13.6 L) (Anon., 1746d).

Over a century later, when rinderpest broke out again in Britain, there was little change in the majority of attitudes and beliefs. Tar-water soon cropped up under the title "An efficacious remedy for the cattle disease" in the September 1865 issue of *The Farmer's Magazine*:

> An efficacious remedy for the cattle disease is tar-water, made of the best Barbadoes tar, and capsicum. Boil one ounce of capsicum in four gallons of water. When the decoction has cooled down to 100 degrees of heat, pour the same on one gallon of tar ... Then pour off the liquor, and administer to each beast from one to one-and-a-half pints, three times a day, until the whole four gallons are taken, which seldom fails in performing a cure. Smear a little tar on each nostril, and upon every foot, and between each hoof ...

The same journal reprinted an extract from Prior's 1746 account which recommended tar-water without any additions. But J. Weatherfield, writing in *Lloyd's Newspaper*, seemed unaware of this knowledge when he wrote: "Every person knows the conservative properties of tar, how it preserves cordage and timber from decay… I am strongly of the opinion, that if the cattle were well dosed with this simple medicine, they would be able to resist this infectious malady. Since I last wrote to you, I have added to my list thirty more cows, and they are all in good health, not one having been attacked." He added a PS: "A person keeping cows at Bermondsey saw the report in your paper. He used the tar and charcoal, and though the disease appeared among them, they are all well now."

Lund (n.d. *c.*1868) noted that many "old-fashioned" Yorkshire farmers gave their horses and cattle tar mixed with meal in a ball when anything was wrong with them: "Mr. Weatherfield's remedy, it appears, is an ounce of tar mixed with powdered charcoal into a ball." As a drink, he provided a recipe of tar mixed with three yolks of egg, treacle, salt, and gruel.

On the first night a meeting was called to discuss the plague in Britain, belladonna and arsenic were suggested as cures. In the 1860s, although people like Gamgee stood out as having a foot in the modern world, the majority of veterinary surgeons still practiced ideas current a century before. Hanging bags of camphor or strings of onions around the necks of unaffected animals was still recommended. Lord Faversham stated that the former was tried in the north of England, and in all cases with which he was acquainted, the animals escaped the disease. He had tried it himself on two farms, and so far none of the animals had contracted the disease which was within half a mile of one of the farms (Faversham, 1866). In 1865, a cattle doctor in Britain was reported as treating a cow for "inflammation in the lungs" by pouring down the animal's throat half a stone (3.2 kg) of shot, "to keep the lights down."

In Scarborough, another was reported as treating a cow for rinderpest by forcing down it large quantities of sand mixed in a thick gruel. The rationale for this treatment was that the internal passages were in too great a state of lubrication, so sand was administered to dry them. The cow died, but whether of rinderpest or from ingestion of sand is not recorded. Lund in his *England and America's Cattle Doctor or a Reformed Treatise* and so on, *c.* 1868, quotes Layard (1757) for a description of the symptoms and a Mr. Rowlin. To Lund, "This disease was formerly deemed incurable; but of late years it has yielded to proper medicine and judicious treatment." Quoting from the many bizarre recipes promoted in 1866, he began apparently with the method for actinobacillosis: "After cutting away and scarifying… wash the wound with luke-warm water and salt…." He then recommended a mixture of powdered gentian root, powdered bayberries, turmeric, Venice treacle, cream of Tartar and soft sugar, administered in "good household beer." To prevent infection, a decoction of angelica, rue, vinegar, and aniseed should be given. His sensible recommendations were the necessity of cleanliness in cowsheds, keeping them well drained, airy, light, and roomy. He was against bleeding, recommending instead gentle purges, keeping the animal warm in a clean, fumigated shed. Under the heading of Rinderpest, he writes "see Murrain," but does not refer to murrain, a disease "more or less destructive from the time of the Pharaoh," as rinderpest, recommending different potions for treatment: "In all cases of putrid or malignant fever, efforts should be made to supply the system with caloric (by the aid of stimulants,) promote the secretions, and rid the system of morbific materials."

In evidence before the Commission in October 1865, Dick stated that he believed that rinderpest could be treated, and it would disappear "in a little time" when the cold weather set in. It was pointed out to him that this statement was contrary to the evidence given by others and was asked if it did not disappear, what else could he rely upon? His "plan of treatment," he revealed, was to begin with the administration of mild purgatives, followed by stimulants, and after that, tonics. Where the disease was a little more advanced, he began with a dose of oil, followed by lime water and tincture of opium. But asked whether he had discovered any method of treatment satisfactory to himself, he replied, "Not thoroughly so, but I think it approaches to it." The percentage he claimed to have cured was "very small; not above 10 or 15%." Little wonder John Swan, a livestock agent, believed that farmers should be left alone to look after it themselves as the veterinary surgeons did not understand it:

"Do you believe that the farmers are more skilled and able to manage these matters than veterinary surgeons are?" he was asked by the Commission. "I do in this particular disease," he replied.

Giblett, when asked how he supposed the disease could be got rid of, replied: "Believing it to arise from a baneful atmospheric influence, I have the full hope of the cold weather causing it to disappear." Question: "Suppose that it does not?" Reply: "It is a dispensation of Providence." Asked if Providence must be allowed to have its way, he replied: "We have not had much assistance in the matter from the medical profession, certainly...."

Colonel Talbot, who kept a dairy herd of 100 cows within 10 km of London, wrote to *The Times* on August 29, 1865, stating he always kept his cows inside. Hearing the plague was raging in the district he gave each cow daily half a pint of prepared charcoal mixed with her grains and flour; and 1 oz. of niter dissolved in cold water, in half a pint of water every other morning. He painted the cows' noses with Stockholm tar each morning and used Burnett's disinfecting fluid liberally in the sheds, which were freely ventilated, and he proposed fumigating by burning tar on hot bricks in them every day.

"I do not mean to say that these means are a preventive, and am very far from being presumptuous enough to say that my turn is not at hand."

Wise Colonel Talbot, for shortly afterwards, his cows suffered severely.

A surgeon working in India claimed he had met with considerable success in treating the disease in Scotland. On first appearance, the sick were segregated in a hammel (a small shed with an open yard in front) of which several isolated ones were set aside for the purpose. A saline purge of Epsom salts with tartar emetic in barley water was then given and the animal clystered with a mixture of castor oil and turpentine in barley water. The purges were repeated in half quantities with calomel added, and a large ammonia blister applied on the underside and on the right side of the chest and hypochondrium. Fomentations to the throat were also useful, changed very frequently. Utmost cleanliness was observed and the hammels frequently fumigated. As soon as the animal was able to eat, warm mashes were given, and to the convalescents cordial drenches (Hallen *et al.*, 1871).

Setons were still much in evidence. In 1866, a farmer near Chester was reported as having setoned a cow with some irritating substance "in a most barbarous fashion" such that the shoulder was swollen on one side as if a natural collar was present. A veterinary surgeon working in India in 1865 stated that he was convinced that setons would serve in "counteracting the tendency toward congestion of the lungs...."

"The operation of 'pegging,' as practiced in the agricultural districts of England, would be attended with beneficial results in this disease...'Pegging' acts as a counter-irritant by the action of hellebore-root introduced into a wound made in the dewlap; and, therefore, my plan would be to operate on *all* the cattle immediately that murrain appears among them. I do not imagine that this would prove a *specific*, but I believe, from the great irritation caused by the hellebore, it would certainly render the animals less liable to the complaint, especially if the cattle attacked showed symptoms (as they often do) of inflamed lungs. In England "pegging" is largely practiced by cattle proprietors, particularly on their young stock, to prevent pleuro-pneumonia.... The irritation caused by the hellebore is very great, and the dewlap swells extensively..." (Dawson, 1865a).

Forming investigative committees on treatments in London, Norwich, and Edinburgh, the medical profession performed no better than the veterinary. The *Lancet*'s first mention of the rinderpest outbreak (August 19) was to confirm that it was definitely the "steppe murrain," asserting "lightly to set aside the unanimous opinion of continental observers as to the mode by which rinderpest spreads beyond the Russian frontier... would be at least highly imprudent" (Anon., 1865b). Later, it was to chide *The Times* for its obsession with spontaneous generation and dismissed the analogy with typhoid fever as specious. At the end of September, it still considered experience thus far merely confirmed the conclusions of continental observers but began to waver on the doctrine of contagion in the face of public opinion and hostility by publishing details of claimed cures. By December, it asserted that Murchison's analogy with smallpox deserved "serious consideration." As Fisher (1993) records, for a month it became almost enthusiastic over the possibility of a remedy in vaccination.

A French doctor who had taken up residence in Britain, H. Bourguignon, recommended such measures as washing a sick animal's head several times a day with vinegar and water. A drink to be given between the fourth and seventh day of the disease if its temperature dropped greatly was 2 pints of red wine containing $\frac{1}{2}$ oz. of cinnamon and 1 oz. of extract of gentian, to be drunk in 2 draughts. "Should the animal fall into a state of lethargy, you must have recourse to strokes of fire, according to surgical usage." At the "period of exhaustion," beef tea, fat soups, milk and farinaceous drinks should be given. His description of medication ends with the advice to "farmers and graziers," "... and, above all, arm yourselves against the insidious pretensions of quacks and charlatans, whatever mantle they may put on to hide their ignorance" (Bourguignon, 1865).

Dr. J. Collis Browne, whose well-known anodyne for human use was to sell (in modified form) to the present day, was another to appear on the scene with a remedy. A retired army surgeon who claimed great experience of fevers and cholera in India, he was known for his patent anodyne "Chlorodyne," claimed effective against consumption, coughs, colds, asthma, bronchitis, neuralgia, rheumatism, spasms, and so on, and to produce a quiet refreshing sleep. A concoction of chloroform, morphia, Indian hemp, prussic acid, and other ingredients, it was doubtless the same he now claimed to be efficacious in the treatment of rinderpest.

Following the loss of her herd, Miss Burdett Coutts obtained permission from the Privy Council to conduct experiments to effect a cure at her personal cost. "Her expressed motive was to avert in some measure the impending ruin of small cow-keepers, whose livelihood might depend altogether on the health of a few cows." She placed the experiments in the hands of her physician, Dr. Forshall, who had already witnessed another Dr. Browne's apparent success with a number of cows. Dr. Browne claimed that he could cure 90% of cases if the first symptoms could be detected. Of 101 treated between July and October 1865, he claimed that only 24 had died. However, only 46 remained, because others had been disposed off. Dr. Browne did not reveal his recipe, but it seems it comprised liberal quantities of calomel, yeast, castor oil, porter, port, brandy, and whisky (Playfair, 1866). Owing to the liberal quantities of whisky, brandy and port wine, the animals became "all excited" and died.

Mr. C. Hunting, in an address delivered to the North of England Veterinary Medical Association at the Crown and Thistle Hotel, Durham, on October 20, 1865, stated: "And where the life of the animal is of more importance than the expense of treatment, good sherry, in half-pint doses, is invaluable. But I protest against what is called "bold practice," *viz.*, giving a pint and a half of brandy or whiskey at a dose, for in ninety cases out of a hundred it would do harm instead of good." One member present remarked he had visited several byres in which a string of onions was suspended around the neck of each cow as a preventive.

In a letter to *The Times* September 1865, "M. D." of Sussex stated that as a typhus disease, the blood was not "decarbonized" and hence circulated as "black blood". His cure was a mixture of equal quantities of potassium chlorate, sodium chloride, and potassium nitrate, dissolved in hot water to which dilute nitro-hydrochloric acid had been added.

The Chelsea Vestry Clerk reported to the Commission on December 2, 1865 that they had found great benefit from treating cows in Chelsea with a quart each of a mixture of half a pint of best whisky and 1 oz. of sweet spirits of niter, mixed with arrow-root gruel.

On December 26, 1865, *The Times* reported with all seriousness a "Strange cure of the Cattle Plague." A cow was in an advanced stage of the disease and ordered to be destroyed. A neighbor, Mr. John Parker, begged to be permitted to try a few experiments and ultimately was allowed to do so. He took a knife and made an incision across the loins from one side to the other, then along each side of the ribs to the shoulders. This was followed by a deep cut in the neck penetrating close to the windpipe and gullet. The wounds were then filled with salt and saltpeter. Next morning, they were washed with salt water and rubbed with vinegar, then filled with "green salve," a preparation apparently well known to farmers of the day. Surprisingly, the poor cow immediately after showed signs of improvement, the flesh healed, and the animal apparently recovered, being reported as "now healthy and eats well."

Some promotion of alleged preventives and cures was subtly conducted. Thomas Greig, who claimed he had never met Mr. Joseph Thorley and had no connection with him, gave an "Address on the Cattle Plague" in Perth in May 1866 and wrote letters to the local press praising the qualities of "Thorley's Food." This was a cattle "condiment" supposedly marketed for over 14 years. Greig produced several alleged testimonials to the effect that those who used it regularly on their cattle had not suffered any losses from rinderpest. Greig also issued a brief pamphlet on his lecture: "Many of you must have seen a letter that appeared in the Perthshire Advertiser and the Perthshire Journal newspapers on the 22nd. February, calling attention to the announcement as *a great fact* put forth by Mr. Thorley—'Out of the hundreds of thousands that use my Tonic Stimulating Condiment, not one has lost an animal for the last six months.' "

This was followed soon after by a leaflet *A Tract on the Cattle Plague. Prevention Made Easy*, which claimed "No quackery—of that the reader is fully assured; no barbarous veterinary jugglery—this shall be made apparent." It stated a grateful farmer who had just preserved his cattle by using "Thorley's Condiment" had newly named it "Thorley's Essence of Animal Life."

"But a still more important claim is now advanced for Thorley's Condiment, it being the only real remedy yet discovered for the Cattle Plague."

Joseph Thorley was still promoting his condiment as "A Sure Preventive Against Cattle Plague. Thousands of Testimonials to corroborate this great fact" in 1878.

Dr. Parkin, who argued for the spontaneous origin of epidemic disease from cracks in the earth, proposed that the best preventive was different forms of carbon because of the relation of the disease to the earth: "Hence I have laid it down as a law, regulating the march of epidemic diseases, that they prevail to the greatest extent in alluvial and tertiary formations, less so on secondary formations, and least of all on primary rocks." So he suggested that the affected animal be made to drink water containing bicarbonate of soda and tartaric acid. If the cow would not drink it, then it should be poured down its throat or applied as an enema, care being taken to stop the gas escaping. Brewer's yeast could be substituted if necessary. In addition, one of the other forms of carbon should be resorted to, and since in its pure form it only existed in the diamond, hydrocarbons should be employed, of which the most suitable was naphtha, administered as 2 teaspoonfuls every 4 hr in mint or camomile tea. Failing this, a tablespoonful of cod liver, or olive, oil could be substituted. Another person claimed the infallibility of mixing sulphuric acid in the drinking water, while yet another preventive was once more claimed to be salt.

Dr. James Tucker, Medical Officer for Sligo in Ireland, in a paper dated 1866 considered that the value of garlic, onions and asafetida, was in the fact they contained sulfur, which was anti-parasitic. He also considered salt most important in combating disease for its oxygenating influence on the blood. Iron solutions were also important to health as the minute sparks of electro-magnetic iron were the centers for the construction of the red globules of the blood.

But no doubt the hapless cattle preferred Dr. Druitt's suggestion (in a letter to Sir Fitzroy Kelly): "Whisky, sweet spirits of nitre, beef-tea, and other nourishment, should be given freely." Or the remedies recommended by the Edinburgh Committee, such as hot coffee: "infusing two ounces of ground roasted coffee for a quarter of an hour in a chopin-bottleful of boiling water. It must, of course, be allowed to cool somewhat before being administered, and may be given in the above quantity every six hours." Better still was "two chopin-bottlefuls of good Scotch sweet ale three or four times a day."[1]

As in the 18th century, numerous "cures" and "preventives" were published in letters to the press. Mr. Wilson of Croydon recommended 1 pint of fresh yeast in 1 pint of beer in 2 quarts of gruel three times a day. The cure of a Mr. Graham of Capellie, near Glasgow, consisted of: 3 tablespoonfuls of sulfur, 3 of niter, half a tablespoonful of ginger, and 1 lb of treacle, in warm water. Some other suggestions were bruised oak bark placed in the drinking water, powdered ammonium carbonate, sulphuric acid, sodium hyposulphite (the "fixer" of photography), vinegar, and quinine.

Mr. Heatley of Norfolk recommended the following:

> Take of spirit of turpentine, 1 ounce; castor oil, 7 ounces. To be given first in one dose, and not repeated.
>
> Take of sesquicarbonate of ammonia, 1 drachm; powdered Peruvian bark, 1 drachm; powdered ginger, 30 grains; peppermint water, 6 ounces. To be given in one dose three times a day.
>
> Take of mutton suet, 1 ounce, boiled in half a pint of new milk, strained and given in one dose every night.
>
> Take of brandy 4 ounces, given in half a pint of linseed tea every morning.
>
> The doses may be increased or decreased according to circumstances.

But most famous for his "cure" was Mr. M. Worms of Ceylon, whose name a reporter of *The Farmer's Magazine* stated would go down in posterity as one of the country's greatest benefactors, while Simonds (1866) noted that the attention being paid to it made it not unlikely that it would influence the decision of Parliament itself. Worms proposed an "Indian" cure which he had seen used in Ceylon comprising onions, garlic, ginger and asafetida; producing a shortage of onions and asafetida in the country. Noted to be successful in a few instances if it was used in the early stages of the disease, the recipe was: 2 lb of small red pickling onions (not the green variety), 2 lb of ground ginger, 2 lb of garlic and 2 lb of asafetida, ground up and boiled into a concoction. One pint of this was to be administered to a full-grown animal with some cayenne pepper.

Worms was a wealthy gentleman of some distinction, a former gentleman coffee planter in Ceylon and connected by marriage to Baron Rothschild. His remedy was first tried at Datchet where a herd of 19 supposedly infected cattle were put under the treatment by a veterinary surgeon who was inspector of the Windsor district. Seemingly improved by the treatment, a paragraph appeared in *The Times* and other newspapers to the effect that all 19 animals had been cured. Shortly afterwards, they were all dead, Mr. Worms attributing this to some kind of mismanagement or the fact that the remedy had not been administered in large enough doses.

[1] A chopin-bottle was an English quart (1.1 L).

Lord Leigh, having the disease on his estate at Stoneleigh and reading of the "cure" in *The Times*, sent for Worms, who went down to Stoneleigh and applied the treatment to a cow said to be in such an advanced state of the disease a grave had already been prepared. The cow was said to have recovered after his treatment. Twenty-four others were then treated and also said to have recovered. However, the first cow was already beginning to recover when Mr. Worms saw it, but afterwards had a fatal relapse. Of the 24, not one apparently had the disease in the first place, removal from the rest of the herd protecting them.

Mr. S. Brodhurst Hill of Bache Hall, Chester, claimed to have followed Mr. Worms's treatment with great success. His herd of 28 cattle was attacked at the end of January 1866. By February 12th he had lost or had killed 12 animals, when he began to use the remedy. Of the 16 remaining, all but one were attacked, and only two died. The remainder were now taking their food well and milk from the cow was being used.

At the end of February, Worms's treatment was applied to Baron Rothschild's herd at Mentmore with apparent success, a veterinary surgeon, George Lepper, claiming to have saved 12 of 13 cattle and having been equally successful applying the remedy on other farms. Mr. Lepper's father was an examiner at the Veterinary College, and both he and his son were Inspectors for the Aylesbury District. With such credentials behind the claims, Simonds hastened to Mentmore to check for himself and concluded that none of the 10 cattle he examined showed signs of having suffered any serious disease and no symptoms of rinderpest. Of eight other cows subsequently picked from the herd as having the disease, Simonds found four were perfectly free of it although regarded as convalescent cases. Altogether Simonds found five animals which were undoubtedly infected. Within 2 days, one was dead and another dying, while four of the animals said to have been cured now had the disease. Eventually, only 66 animals survived out of a herd of 119.

> In the remainder of the herd a fearful havoc had likewise taken place. In one extemporised yard, out of 24 cows 10 were dying ... In the 8 other yards were 35 heifers, all of which were affected, several were dying, and 3 were dead. In sheds were found 5 heifers in a sinking state. In courts lay the bodies of 13 cows and heifers which had been shot, and two more animals were then under sentence of death. None of these animals were destroyed by the orders of the local authority ... All treatment had been abandoned for some time, and the animals left to their fate. The disease had unhappily spread to other homesteads on the estate. (P. H. F., 1866)

At a meeting of the RASE, Lord Cathcart stated that a neighbor of his who had lived in Ceylon said that the treatment advocated by Mr. Worms was by no means considered a specific. Rinderpest was raging in Ceylon at the present time and cattle actually dying on Mr. Worms's own farm (Cathcart, 1866a). Worms's remedy was met with much scepticism by the *British Medical Journal* which noted, "The Times, nevertheless, still tries to force the garlic down our throats ...."

But Mr. Worms was unwilling to accept the criticism, the *British Medical Journal* noting that like most other inventors of cures he retired from the scene only still more than ever convinced by his failures of the infallibility of his remedy, stating:

> I do not profess to have found the *elixir vitæ*; but, notwithstanding the "sincere regret" of Professor Simonds "that another supposed means of arresting cattle-plague by medical treatment should have proved abortive," and despite the learned dissertation in Professor Gamgee's interesting book, in which he proves the Rinderpest to be incurable, I still venture to hold my opinion, based upon twentyfive years' practical experience, that, if my system be fairly tried, it will prove as efficacious here as it did in Ceylon.

Backing Mr. Worms's remedy in the House of Lords, Lord Ellenborough referred to a letter from Warsaw "which distinctly declared that the disease had been cured ...," ending, "We should be guided by experience, and not by learned disquisitions, which required that a man must have read extensively before he could understand them. We never should get rid of the plague of cattle-plague, till we had got rid of the plague of trusting to professors rather than to common sense," to which the *British Medical Journal* retorted, "The English of this is, that there is no hope for the cattle of the country until common sense, as represented by Lord Ellenborough, takes the medical direction of cattle-diseases…" (Anon., 1866i).

An "Oxonian," reporting on its use in a letter to *The Times* (February 22, 1866), administered a pint to a sick heifer followed by a tablespoonful of cayenne pepper, whereupon the heifer "leant against the wall and seemed to breathe with greater difficulty"! Receiving a similar dose daily, 4 days later, the heifer was surprisingly reported as greatly improved and eating hay.

A correspondent to the *Manchester Examiner and Times* of February 1866 wrote:

Sir, A few days since, on my return from Manchester, my man was in great trouble, one of the cows having taken the cattle plague. She had been removed into an empty stable, and the symptoms seemed to confirm his opinion, viz., a running of saliva from the mouth and nose, painful breathing in gaps, and wildness about her eyes as if in pain. I sent my servant to Mr. Greaves, the veterinary surgeon, of Altrincham. But both he and his assistant were out. I then thought of a prescription given to the public by Dr. Caleb Williams, of the York Friends' Retreat, which had effected "miraculous cures":- tincture of Catechu 1 oz., prepared chalk 1 oz., sulphuric ether 1 oz., tincture of opium $\frac{1}{2}$ oz. Give in a gill of starch every three hours, and if needful, blister the throat with turpentine.

I tried this plan, and in 20 minutes the breathing of the animal became natural, the pain left her, and the running of the mouth and nose ceased. I kept her away from the other cows for 24 hours, but she is now with them, and as well as ever.

T. S. J.

Bowdon, Feb. 20, 1866,

In their *Third Report*, the Commissioners stated that no drugs could be shown to have any effect on the disease, and there did not appear to be any advantage obtained by giving large quantities of stimulants. However, in Scotland, feeding soft mashes of digestible food to small groups of less than 30 affected cows did apparently show a recovery rate of 62% (of 200), or 38% overall (of 503) when groups of 30–80 were included. Cottagers' cattle (95) were reported as showing a 73.7% rate of recovery. Any success of Mr. Worms's cure, which was at one time widely used, therefore undoubtedly lay in the liquid food recommended, rather than the onions, garlic, ginger, and asafetida.

Another suggested possible cure was lime water. A Cheshire farmer, Mr. John Maddocks of Whitby, who had lost 17 cows, drove 5 young infected cows into an orchard to kill and bury them there if they showed no signs of recovery. Refusing water, they made their way to an old pit formerly used for slaking lime from which they drank eagerly, and then began to rapidly recover. Mr. Horne, veterinary surgeon of Little Sutton, subsequently pronounced them out of all danger.

A tenant farmer at Crown Point lost some 1,500 sheep from a disease alleged to have been rinderpest, while another who lost 2 from a flock of 287 had them treated by a Mr. Woods, who claimed to have saved all but 5 by an undisclosed method. Dr. Crisp stated that he was proud to say that in his own profession, if a man had a secret remedy, and kept it, he would be at once scouted by almost every member of that profession. He believed that it was all twaddle to talk of a remedy; but if there were one, it should be divulged at once so that the public might have the benefit of it (Crisp, 1866a). It was considered by some that the disease was not rinderpest, and there had been none in the district at the time. But Simonds stated that he would stake his professional reputation upon the fact the 1,500 sheep *had* died of rinderpest, for although the sheep were mismanaged, no mismanagement would kill a sheep in the 4 days many of them had died.

In India in 1868, while admitting that medicinal agents were inefficacious, the army veterinary surgeon Farrell nevertheless claimed to have cured a number of animals (which were probably those already challenged by the disease) using the following methods: in the first stages, for constipation, a mixture of common salt, ginger, and rice water was administered. This was followed by stimulants, firstly rum or native spirits, then a mixture of antimony, niter, and black salt, or camphor, cannabis, and gentian. To this was added sometimes catechu or opium. If distension of the abdomen was present, asafetida, cannabis, and ginger were given. When recovering, iron sulphate and cannabis or gentian, the latter two items made into a decoction with water. The animals were also made to inhale fumes of sulfur twice a day, done by throwing sulfur on a pan of live coals and holding it under the animal's nose for 10 min at a time. Fumigation of cattle sheds with doses now and then of 2 drachms of black salt or 4 drachms of common soda was recommended as a prophylactic.

A year later, following operations in the Jessore District, Farrell reported that he was conducting experiments with carbolic acid following instructions in a circular letter from the Bombay Government: "The chief effect it has is to disinfect the alimentary canal and the excretions, thus rendering the virus of the disease innocuous to the animal by not poisoning its blood, and to neighbouring animals by disinfecting and deodorizing the secretions ... but it will never prove a specific for the distemper" (Hallen *et al.*, 1871).

In a circular issued widely in August and September 1869, Farrell instructed giving as a purgative common salt or Epsom Salts in rice-water. When the purgative had acted, a mixture of 10 drops of carbolic acid (Calvert's No. 5) with 1 oz. of gentian solution should be administered. In the absence of carbolic acid, one could add asafetida, and in the absence of gentian, black salt. When dysentery set in, the animals should be given a mixture of catechu and opium.

2. REMEDIES ON THE CONTINENT AND ELSEWHERE

Beliefs on the Continent showed no change from a century before, and despite the relatively advanced state of Russian knowledge of the disease, among "simple remedies" listed in Russian regulations were: giving an animal that had ceased to chew the cud a salt herring; giving an animal that was very hot, very hot *kvas* grounds or beer yeast, with saltpeter or common salt. For the same purpose, one could use brine in which cucumbers had been salted, or cucumbers themselves mashed to the consistency of porridge. If the feet were benumbed or swollen, incisions should be made in them, and when blood flowed, the incisions washed with vinegar and corn brandy. When swellings were present on the neck or between the legs, roasted onions should be applied at once, mixed with equal parts of garlic and mustard or horseradish to cause suppuration.

No doubt, if they could choose, cows would have preferred the treatment of the Kherson peasants in southern Russia. Their most common approach was to make the animal sweat, taking care to prevent it taking chill for 2 or 3 days afterwards, and then changing its pasturage or putting it in a warm building in winter. Sweating was usually produced by covering the animal with heavy cloths and making it inhale fumes of hot bran, salt, and steamed hay. Another method was to apply friction with cloths or brushes, covering up the animal afterwards. But perhaps the cows would have most preferred the third method, which was to rub them well with vodka and salt, especially along the spine and on the belly. One farmer tried hot vapour baths, and some peasants recommended injections of warm tobacco water but admitted that it was unsuccessful in bad cases. Another claimed to have had success by giving drenches of warm milk in which garlic had been boiled. Sweating was a method promoted by a Mr. Graham in Britain in the 1745 outbreak, although it did not produce favourable results.

A superstition among some Russian peasants held that the disease was stopped by being transferred to a neighbour's cattle, a belief which probably encouraged its spread.

In the district of Jassy in Moldavia, one peasant treatment was to make a hole in the point of the affected cow's horn a little larger than a goosequill, pour in a little mercury, and then seal the hole with wax.

At Odessa, recommended treatment was to apply a seton and a mustard plaster to the breast of the animal, the seton being steeped every day in fresh turpentine, while all swellings should be burnt with a red-hot iron immediately they appeared. Four or 5 lb of blood should be let. Internally, a pint of infusion of linseed mixed with olive oil was recommended three times a day, but when the disease increased, sulfuric acid and strychnine should be added to the drinking water sufficient to give it an acid bitter taste. In addition half a drachm of spirits of ammonia in half a pint of water should be given three times a day. Three ounces of glue dissolved in a pint of tepid water would stop diarrhoea!

Dr. G. H. Masius, professor of Medical Science at Rostock, published his recommendations for prevention in 1807. Against the disease he prescribed:

- The application of Glauber's salt as a laxative. If this did not work within 5–6 hr, then an enema of water and salt, or a suppository of honey and salt, or soap and salt covered with oil, was recommended.
- A seton at the sternum.
- Administering dilute sulfuric acid mixed with bran was very good.
- Fumigating the stall by "Humboldt's method," which was sulfuric acid poured on salt. Also, one could use a butt of oak bark chippings in water, stir daily, and spatter the walls, bedding, and livestock: "The smell of this tanning-liquor is stinging, refreshing and destroys the infectious substance."
- Grooming by rubbing the animal with straw. "Swimming the livestock is excellent as a precaution, as it has a beneficient effect upon the skin."

But the most successful precautions were:

- Cleansing the mouth and tongue of sick animals three or four times a day with a mixture of salt and honey in vinegar.
- Pouring into the nostrils vinegar and honey.
- At the first signs of sickness, not giving any dry food but rye bran in water and to drink much water boiled with linseed oil.
- At the commencement of illness, give a twice-daily enema of linseed oil and saltpeter, "Experience shows the enema is very beneficial in this illness."
- Stalls must be very airy, and if the animals are in the field, keep them out of the heat of the sun.
- Place blankets over them to cause perspiration. Some have found lubrication of the back and loins with oil very useful.

In Poland in 1857, it was claimed that placing pieces of old iron in drinking water was a successful preventive on account of cattle which watered at a chalybeate or iron-impregnated spring having escaped infection. Dr. Joseph Pessina, professor at the Veterinary Medicine Institute of Vienna, had published in the *Vienna Almanac* of 1802 various dosages for acidified ironwater using hydrochloric acid and cooking salt, in 1808 issued as a tract *Anleitung zur Heilung der Viehpest mit der eisenhaltigen Salzsäure (A Guide to the Cure of the Cattle Plague with Chalybeate Acid-salt)*. Referred to in the *Hannover Magazine* in 1803 as the Pessinasche method, Pessina states that acidified ironwater had been proposed by Professor Reich in 1797 at Nuremburg, although Reich had not tried it, basing his suggestion on experience of mineral acids in human medicine. Pessina recommended the usual additional treatments such as enemas, herbal concoctions, and stall fumigation, and stated that relapses were best avoided by inoculation, suggesting that the not inconsiderable advantages of inoculation could be happily combined with the described method. He finished his publication by offering to supply by mail ready-made acidified ironwater with salt in any quantity for 45 kr the pound. A publication by the Director of the Hannover Royal School of Medicine referred to the method but noted that "a visitor" claimed it had not worked in Hungary in the time of Prince Esterhazi (Havemann, 1814). However, Ferdinand Airer of the University of Göttingen recommended it, based on his visits to Austria and Hungary, where, he claimed, it had saved four fifths of the animals (Anon., 1813). The latter publication, issued by the temporary Government Commission at Munchausen, recommended treating infected animals with a 30-cm-long seton of hair rope soaked in turpentine threaded under the skin of the right flank, and administering a mixture of boiled willow bark, wild chestnuts, and sour vitriol. In response to queries, in the *Schwerin Advertiser* of December 24, 1814, G. Rüst, horse doctor and veterinarian, provided instructions on how to apply oxygenized hydrochloric acid, which was simply to add the correctly stated amount as given, to an infusion, and to give the animals flour-water with kitchen salt. In the *Mecklenburg Advertisements* of February 1814, Justice Jargow cautioned against placing too much faith in acidified iron-water. He had consulted a local apothecary of standing, Mister Bahlmann, who pointed out that they did not know much about the substance. Dr. G. R. Frank had several experiences of its use and recommended a different method of preparation to that of Dr. Pessina. It could have no effect because the animal might not receive the right quantity, or if there was no difference between the two methods of preparation, one must get false results. Frank prepared his substance by mixing black magnesium oxide, sodium hydrochloride, and sulfuric acid in $1\frac{1}{2}$ times its volume of water. For use, it was diluted with $\frac{1}{2}$ to 3 oz. of water: "[It] surpassed all my expectations, I and the bystanders were astonished. It was as if the breath of a benevolent fairy had been poured out over the suffering animals" (Frank, 1802). Unimpressed, Zimmerman (1814) argued that these methods had not been tested for long enough to have confidence in them. Pessina had made few attempts and, in the first, 2 of 21 animals had died, in the second 13 of 72. Frank had used only 20 animals, but hardly one had died.

Tolberg in Germany, while advocating inoculation as the only cure (Tolberg, 1808), arguing that he had said before that it was the use of drugs which had brought the disease to its present position, nevertheless described a number of the other usual treatments which should accompany it both before and after. A piece of bread toasted in spirits placed in the mouth once or twice a day was still one of the best. If the animal did not eat but had a dry cough and was not salivating, then place in the mouth a mixture of garlic, water pepper, and sea radish, wrapped in a piece of cloth and dipped in vinegar, or "devil's dung" with honey. So that the animal cannot swallow the bundle, place in the mouth a piece of wood like a bit to which the bag is tied, fastening the ends to the horns or neck and leaving the animal to chew on it for quarter of an hour each morning. If pocks on the skin become large, cut them open and burn with a red hot iron, "Those animals of which this operation becomes necessary suffer much, and recover more slowly. The wounds fester for a long time afterwards ...."

The *Schwerin Advertiser* of December 17, 1813 carried an advertisement from Fr. Windel, veterinary surgeon, offering his cure for 8 marks a piece. Of 20 sick animals, he promised, hardly one has succumbed and not half die, if the disease is not already too far advanced. In the same year, Dr. G. Mähl, Professor of Chemistry and Pharmacy at the university in Rostock, produced a pamphlet extolling the virtues of fumigation with hydrogen chloride gas created by pouring hydrochloric acid on kitchen salt (Mähle, 1813).

Ribbe (1814) stressed that medicines were only weakly effective and therefore wasteful. He quoted his former university tutor Professor Sick, an expert on rinderpest, as stating that the highly vaunted strengthening medicines were useless. A bare, good, and efficient diet with beverages was the best. Use of medicines could spread the disease more rapidly through being carried about by the persons attending the animals. Ramazzini, Vicq d'Azyr, Sauvages, Camerarius, Haller, and Adami were all correct when they stated that all medicines were ineffective. Fumigation with hydrochloric acid fumes showed little benefit. The Prussian and Austrian states quarantined incoming animals, but frequently thousands of foreign trade oxen arrived together, and it was impossible to pick out a single sick animal. It was possible to hinder its spread through proper attention and by slaughtering sick animals at the first sign, burning carcases on the spot. But Ribbe considered that slaughter ordered under the law, as by Frederick II, was a most damaging punishment as other animals which would not die were unnecessarily slaughtered. In Prussian Silesia, insurance had existed for many years, but owners were not obliged to join the scheme, and if they did, only received one third the value of an animal slaughtered.

In September 1865, an official Dutch report recommended a treatment of hydrochloric acid mixed with treacle, and a thick concoction of linseed. If the animal improved, decoctions of bark, sulfuric ether, and other stimulants could be employed.

In Bavaria in 1865, information on treatment issued in 1796 was reissued. This noted among other things that when an animal was sick, a leather soaked in vinegar and Spanish fly, or a cord of the same, also the winter aconite *Eranthis hyemalis*, should be placed on the protuberant part of the chest between the forelimbs for 3 weeks, and the blister caused washed with tepid water every morning. "Still more efficacious is the burning with a hot iron the spot where the leather is set." Seven "recipes" were also added, which were designed to strengthen or invigorate the animal. These included concoctions of the powdered roots of angelica, enzian, rhubarb, and pure tartar; asafetida mixed with gum arabic or camphor dissolved in spirits of wine added to wine vinegar; camomile "or any other bitter smelling herb and blossoms; for instance, wormwood, centaury, sage, etc.," infused with boiling water and the infusion added to wine vinegar.

In the district of Alexandria, a decoction of hashish (hash-ha) and chicory (hinbia) was favoured, using $\frac{1}{2}$ pound of hashish and 2 lb of chicory boiled for half an hour in 2 gallons of water. A large wine-glassful was then administered every half hour until the animal improved.

3. ALLOPATHY AND HOMEOPATHY

The chief remedies employed in homeopathic treatments were preparations of arsenic, *Rhus toxicodendron*, phosphorus, phosphoric acid, and sulfur. Other preparations employed to meet different indications were aconite, belladonna, creosote, nitric acid, sulfuric acid, strychnine, digitalis, *Thuya occidentalis*, veratrum album, and similar drugs. Interest was sparked off in Britain by a contributor to *The Times*, Mr. Tattersall, who provided a "positive cure" sent to him by a dragoon colonel in Hungary, comprising a dose of arsenicum (third potency or strength), consisting of eight or 10 drops given every 36 hr to prevent the disease breaking out. Any animal with the disease should have the same dose every 10 or 15 min. "And there" (stated *The Times* later), "if the pretensions of the system can only be established, will be the value of homeopathy, which professes not only to cure, but to prevent. If the prophylactic dose of 'arsenicum' should be found actually efficacious, our cows may be sent to market with perfect safety. The antidote will operate like vaccination in the case of small-pox, and in the end the disease itself ought to disappear."

In Holland, experiments were conducted by M. Gandy of the veterinary college at Brussels, and M. Sentin, a homeopathic chemist, in homeopathic treatments of 80 cattle certified by a veterinary surgeon as having rinderpest. They claimed that 60 recovered and 20 died. The treatment consisted of administering a mixture of arsenicum, phosphorous, phosphoric acid, *rhus tox.* (a drug obtained from the poisonous sumach herb), and sulfur. Allopathic remedies were claimed to have saved 45% of infected animals. The principal of these was dilute muriatic acid combined with linseed tea, sometimes gentian, tormentilla, and ginger. Another was sulfuric acid mixed with an equal part of sulfate of quinine. For external use, washing three or four times a day the eyes, ears, and nose with carbolic acid was recommended.

Simonds claimed that a great many of the so-called cures effected by such methods were not cases of rinderpest at all, but all disorders were now being called rinderpest. With genuine rinderpest, not more than 7 or 8% would recover, and they would do so anyway without treatment.

A report from Hungary in October 1865 noted that "not one of the remedies recommended sometimes as specifics by the allopathists has ever done anything of importance in the majority of cases, nor has homeopathy led to the discovery of any specific" (Egan, 1865). However, the same report alleged that in the early stages of infection, four to eight drops of essence of bryonia repeated at 2–4 hr produced a striking improvement in the majority of cases after 24–48 hr. When "especial excitement and restlessness" were present, this was changed to doses of belladonna, with similar result. Phosphorus to the sixth degree of dilution was also found to be useful in some stages, and when the foetid diarrhoea stage was reached, winter aconite or sometimes sublimate of mercury was useful, as in an outbreak in 1850 was solanin. Muriatic acid and arsenic, as recommended by Gunther, were found to be ineffective.

Considerable success was claimed in South Holland by two Belgian gentlemen, MM Seutin, a highly respectable chemist of Brussels, and Gaudy, a veterinarian, for a "not rigidly so" homeopathic treatment. In the judgement of the Dutch Government, however, the difference in the rate of recovery was not conclusive, and the gentlemens' contract was not renewed. Of 125 animals in a highly infected district, four took the disease in the first 15 days. Later, the others did also, but M. Seutin ascribed this to neglect of the farmers in carrying out his instructions as the system was "troublesome, especially with cows at grass." Subsequently, every animal in the district which contracted the disease was slaughtered immediately.

But the unfavorable reaction by the Dutch Government did not deter an English homeopathic physician, Dr. E. Hamilton, from visiting MM Seutin and Gaudy to examine their method. An association, the *Association for the Trial of Preventive and Curative Treatment in the Cattle Plague by the Homeopathic Method*, was formed in Norfolk under the presidency of the Duke of Marlborough for trying it out on an extended scale in Norfolk, Cheshire, and Yorkshire. In the Norfolk trial, out of 45 experimental cattle, one did not appear to contract the disease, four did and recovered, and 40 died. Better success was obtained at York by Dr. Pope, a surgeon to the York Homeopathic Hospital. In this trial, 117 animals were treated at various localities with 50% recoveries making allowance for "a few cases of mismanagement." He claimed that certain features of rinderpest were very like those of scarlet fever and because of this seemed to rely upon belladonna as of prime value. Apart from looking after the animals, the treatment comprised administering "belladonna, to subdue the general congestion; rhus toxicodendron, to allay the muscular twitchings; caustic ammonia, to abate the abdominal distension; turpentine, to check hematuria; phosphoric acid, mercuris solubilis, and arsenic, to control diarrhoea; mercuris corrosivus, to check dysentery; and arsenic to rally and buoy the system ..." (Reed, 1866).

Mr. George Hope of York claimed to have used Dr. Pope's remedy with great effect. An animal was so far gone with the disease that the inspector had been sent for to order her shooting and burial, but as it was late at night, the order could not be carried out until the following morning. So Mr. Hope persuaded the owner to allow medicine and gruel to be administered during the night, giving it belladonna. By morning, the animal had so far rallied that "all thoughts of destroying her were abandoned, and she made a complete recovery."

The *British Medical Journal* was scathing in its condemnation of homeopathy and *The Times*'s support of it:

> It is pitiable to turn from these, the plain teachings of scientific sense, to the miserable tale of homœopathic twaddle which is being told out at Norwich, under the direction of noble lords, London infinitesimalists, and a local reverend and "skilful homœopathist, the Rector of Lyng" who, it appears, instead of tending his own special human flock, is busy administering "arsenicum" to his neighgbours' cows.... This homœopathic experiment at Norwich has, of course, turned out a miserable failure.... Lord Bury admitted that homeopathy had failed, but then said he "It has not been properly tried." Lord Bury attacked *The Lancet*, stating that he had never heard such a tissue of nonsense in the whole of his life: "We know not if his lordship be deaf; but if he is not, and could hear his own observations, we venture to say that he would have heard a very fair share of nonsense from his own lips" (Anon., 1865f).

Thomas Baldwin, Superintendent of the Agricultural Department of National Education in Ireland, noted that homeopathy had not cured a single case of pleuropneumonia and there was no reason to suppose it could do better with rinderpest. Homeopathy, he asserted, could not have the least positive effect in curing rinderpest, "It is to my mind inexplicable, that intelligent men who have had any knowledge or experience of contagious diseases, could expect that the *infinitesimal* doses given in homeopathy, would counteract the virulent poison on which the very existence of Cattle Plague depends" (Baldwin, 1866). But homeopathy did not necessarily do any harm, and the animal was usually well nursed.

In Norfolk, "the first and last costly and extended series of experiments in homeopathy that were ever made in this disease" were instituted in conjunction with the London Homeopathic Society. The Lord Lieutenant of Norfolk offered a prize of £100 to the person who should be the most successful in curing the disease. The result was reported as a "complete and total failure," not a single applicant applying for the prize (Smith, 1868). Dr. Hamilton, the champion of homeopathy, had declared that 45% of diseased animals could be cured, but, as one commentator put it, he "signally failed to convince us of the truth of his system when fairly put to the test" (Gregory, 1866).

4. "HYDROPATHY", DISINFECTANTS, AND FUMIGATION

Dr. J. J. Reuss, in a work on human eruptive fevers, claimed that many a curable animal had been sacrificed unavailingly, quoting an instance of a cow taken to be slaughtered, which, when it lay down, had cold water poured over it, and it recovered. Contrary to the anonymous instruction published at Nuremburg in 1754 that one should not swim cattle through cold water, he noted that another Nuremburg author in 1797 observed that cattle herds which had to swim through water several times daily in the 1796 outbreak did not become infected. He therefore doused a sick cow with water which "seemed to recover" but eventually died. His prognosis was that the treatment had not been given early enough. He recommended dousing with water acidified with hydrochloric acid and a decoction of flax oil, while giving plenty of fresh water to drink; between dousings, rubbing the animal with straw; and on the head between the horns placing a cloth soaked in hydrochloric acid-water, keeping it wet and dousing the animal every 2–3 hr. The disease was very similar to petechial fever, and the only treatment beside cooling was hydrochloric acid. When there was severe dysentry, a measure of Theriac was recommended, but severe dysentery was very rare if the different prescriptions were followed exactly (Reuss, 1814–1818).

Professors Varnell and Pritchard of the Royal Veterinary College, London, tried what they termed "hydropathy." Two infected cows were well douched with cold water and then wrapped in wet sheets over which dry wraps were placed and closely bound. Of one of them, they recorded, "Even before this operation was completed she became partially unconscious, and lay down, in which position she remained for half an hour." She then warmed up and stood. She was then well rubbed and covered with dry rugs, fed some gruel and then given a draught containing 1 oz. of opium and an ounce of camphor in water. On the following day, she was much worse and again given the cold water treatment, which made her just as bad as before, but this time, she continued to deteriorate, dying on the third day. The second cow, it seems, behaved similarly. The professors concluded, "We are inclined to think that a modified form of this treatment might be beneficial."

A variant on this was published by Constable in the form of a letter from the Reverend J. P. Marriott near Lutterworth:

> There has been such a curious illustration of the wisdom of keeping beasts cool in the plague, that I have been wishing, day after day, to tell you about it.
>
> John Farrow, of Alkbro', in Lincolnshire, after having lost every one of a lot he had shut up and kept warm, found another lot beginning to fall ill of the complaint. In moving them from one field to another, one fell into a drain and got considerably chilled, but was observed to go on better from that time. Observing this, and being forbidden by the Inspector to bring the animals along the road up to his buildings, J. Farrow determined to try cool treatment, and with that view, made a sort of slight cover … so that the wind blew through in every direction. This was one of the bleakest fields on the whole estate … at the end of six weeks, out of twenty that had fallen ill, only three had died … after the bitter cold of the 11th and 12th of this month [January], when the frost, and snow, and wind that came suddenly was enough to try anything that was easily damaged by a low temperature. I asked him particularly, how had they borne it? and he said, they were not a bit the worse, and were getting well without any check.

In support of the "cold treatment" Constable adds a tenant of Sir. R. Sheffield's, not having room for a lot of diseased beasts at home, tied them to posts in the open field and every one of them recovered (Constable, 1866a).

Dr. Kidd, in a letter to *The Times*, wrote:

> The experience of a doctor in the treatment of the cattle plague in one of his own cows may interest some of your readers, especially those who have lost faith in the poleaxe practice of the veterinary surgeon … In three hours, by the labour of four men, the cowshed was converted into a vapour bath by removing from the scullery the ordinary copper and cementing it on a furnace, which was extemporized with a few bricks, while three 4-inch stoneware drainpipes made a flue. Day and night the copper was kept boiling. In a few hours the atmosphere of the shed became warm and moist. Two men remained up all night giving phosphorous every two hours alternately with the arsenic. Towards morning the breathing became relieved. From 56 in the minute it fell in 12 hours to 40 (the next day to 32). On Friday we were in great hope, but on Saturday morning the cow calved, and the men thought her dying, if not dead. Then I got them to pour down her throat four bottles of Barclay's stout in the course of eight hours. This the cow drank with infinite relish. To our surprise the calf was born alive. The poor cow seemed nearly dead, but, determined not to give her up, I ordered the gruel to be made with old ale, the bottled stout being also continued. Little by little she revived, and has gradually recovered.

About January 1866, the Medical Officer for Health for Edinburgh issued a brief *Suggestions for the Prevention and Treatment of the Cattle Plague*, proposing a vapor bath as treatment, an idea first mooted by Blondet in 1748 (Littlejohn, n.d.).

Alexander McDougall issued a pamphlet *The Cattle Plague: Its Prevention and Cure*, which was a plug for the use of his disinfecting soap and powder, arguing against the usefulness of chloride of lime in the guise of a paper on cattle plague. All mention is avoided of the McDougall's product until the rear cover, where an advertisement refers to the Commission Report which recommended use of a disinfecting powder, which was McDougall's Patent Disinfecting Powder. Details and prices of powder, soap and fluid, are then given (McDougall, n.d.). The president of the Lancashire Veterinary Medical Association succinctly observed in an address in April 1866:

> Mr. McDougall believed that all that the cattle required was to thoroughly disinfect them, so he washed their skins with his powder, drenched and glystered them; but they have continued to die notwithstanding. Mr. Grace Calvert, the eminent chemist, suddenly professed to discover a new power in carbolic acid, and the cow-houses from John O'Groat's to the Land's End stink with its fumes. The malady refuses to be charmed away by chloroform or chlorodyne …

Chlorodyne was Dr. J. Collis Brown's panacea.

Crookes (1866) was a great believer in the healing powers of carbolic acid. He claimed that Professor A. H. Church, not Grace Calvert, was the first to recommend it. In one of his experiments, Crookes injected an infected cow with "as large a dose as it would bear," stopping the operation at the first sign of trembling. "I had just finished injecting the second dose when trembling commenced. It was rather violent for a short time, but soon went off …." The cow bore "without inconvenience," 6 oz. of a 4% solution containing 105 grains of carbolic acid and actually recovered from the rinderpest. Other animals were not so fortunate. Crookes noted: "In dealing with the Cattle Plague it is possible to try testing experiments of a nature wholly inadmissible where human beings are concerned …."

In November 1865, Dr. Dewar, MD of Kircaldy, published a letter in the *Scotsman* advocating use of sulfurous acid gas as a fumigant obtained by burning sulphur, accompanied by "strict sanitary principles." In a further letter the following January, this to *The Times*, which did not publish it, he reported a farmer who had lost 30 head to the disease but none at his homestead where he practiced sulfurous fumigation twice a day, although he had mingled with, and touched, his diseased stock three or four times a day. Dr. Dewar obtained agreement from the Edinburgh Cattle Plague Committee to place two yearling steers he had purchased from an infected herd and supplied with food from the proximity of cattle dying of the plague, and subjected to regular fumigation, in another infected herd or in a contaminated byre or shed. But he seems never to have received permission to move the steers which he reported as still alive after 40 days, although the rest of the original owner's stock had all died. Not only was it claimed that sulfurous acid gas prevented cattle plague, but it also appeared to be effective against almost every human complaint: gangrene, tuberculosis, diarrhoea, catarrhs and many others, and, in cattle, pleuropneumonia. Dr. Dewar's method was recommended in the *Daily Review*, and the papers on it were sent in July 1866 to the American House of Representatives which recommended distribution throughout the country, but after this, we hear no more of it.

Mr. Kirwan J. Fernie of Norway wrote to the Earl of Clarendon in January 1866, also pointing out the benefits of sulfurous acid gas, believing that cattle plague and some other diseases were due to a "sudden decomposition of the blood" and noting how the people of Birmingham were less affected by cholera because of the sulfurous acid gas in the atmosphere. Flour of sulfur rubbed on a sick animal's tongue once or twice a day he believed would be an effective antidote.

The Russians placed great faith in their drink *kvas* as a disinfectant! This was the usual beverage of Russian peasants, made of stale rye bread mixed with malt on which boiling water is poured and the mixture left until fermented. Regulations concerning the disease stated that persons who had been dealing with infected cattle should not be allowed near healthy beasts until they had changed their clothes and washed their hands first in a solution of potash, then in vinegar followed by *kvas*. The stalls of both sick and healthy cattle should also be fumigated with vinegar or *kvas*, poured on the heated bricks or stones of the floors.

5. SANATORIUMS

In Britain in 1865, there was considerable support for "sanatoriums," animal hospitals where the animals could be detained and a cure attempted, like the idea of "pest houses" mooted by Lobb in 1745. At a special meeting of the Yorkshire Veterinary Medical Society held in the Queen's Hotel, Leeds, on August 29, 1865, the conclusion was reached that all animals suffering from the disease which members might be called upon to inspect, should not be condemned, but placed in a sanatorium and treated medicinally, as wholesale condemnation would be a blot upon the profession. Although strongly supported by the Metropolitan Cattle Plague Committee, the creation of sanatoriums was opposed by Simonds, who regarded the idea as impracticable: "... if founded they would be one of the greatest evils possible to conceive, by tending to perpetuate and keep alive the contagion on which cattle plague depends," he wrote to the Privy Council in September.

"Day and night plague carts would be traversing the streets, and day and night the seeds of the disease would be disseminated far and wide. No precaution which could be taken in the continuous removal of diseased animals could guard against this."

One of the arguments used for the sanatoriums was based on the supposition that rinderpest was curable. Nothing could be more fallacious in Simonds's opinion. The founding of such sanatoriums throughout Britain would convert the country into one great pest-house, prolonging and worsening the problem.

On October 18, the Commission had received a deputation headed by the Lord Mayor of London, accompanied, among others, by the district medical officer Dr. J. W. Tripe. The mayor complained that nothing had been done in the way of a curative process, and cattle were being slaughtered in large numbers. He considered that as it was possible that a remedy might be found, "it would be a sad reflection on the medical men of this day if they should not find something as a remedy for this disease," and supported the idea of sanatoriums for diseased animals. The case for the latter was put by Tripe, who began by stating that he believed it "the almost unanimous opinion of the medical profession, that treatment in this disease would avail." And, "I believe treatment is most likely to be successful in the commencement of the attack." The Commission, however, was cognizant of the impossibility of a cure. A railway traffic manager stated, "Surely science is in a much more advanced state now than it was in 1744, and I cannot think that a disease like the present, any more than typhus fever, can run its course unchecked by science," To which Playfair rejoined, "Are you aware that the highest scientific authorities in Russia, all throughout Germany, in Holland, in Denmark, and in France, have devoted themselves to this subject for upwards of a century and have yet found no cure for the disease, or means of stopping it, except extreme measures?"

6. RESORT TO PRAYER

In September 1865, the Archbishop of Canterbury authorized, and Queen Victoria ordered, a *Form of Prayer* to be used in every church in the land, beseeching God to have mercy and stop the plague. Introduced by an Order in Council, the prayer, to be used on and after October 8, had already been circulated under the heading: "A prayer suggested for use in schools, households and families within his diocese, under our present trials, by Charles Thomas Lord Archbishop of Canterbury."

The prayer read:

> O Lord God Almighty, whose are the cattle on a thousand hills, and in whose hand is the breath of every living thing, look down, we pray Thee, in compassion upon us, They servants, whom Thou hast visited with a grievous murrain among our herds and flocks. We acknowledge our transgressions, which worthily deserve Thy chastisement, and our sin is ever before us; and in humble penitence we come to seek Thy aid. In the midst of judgement, do Thou, O Lord, remember mercy ... stay, we pray Thee, this plague by Thy word of power, and save that provision which Thou hast in Thy goodness granted for our sustenance. Defend us, also, gracious Lord, from the pestilence with which many foreign lands have been smitten: keep it far from our borders,[2] and shield our homes from its ravages; so shall we offer unto Thee the sacrifice of praise and thanksgiving, for these Thy acts of providence over us, through Jesus Christ our Lord. Amen.

With its content of sin, providence and judgment, the prayer was seen by some as a retrogression in rational church policy, fuelling an ongoing debate started in 1853 by the Home Secretary, Lord Palmerston, concerning the efficacy of prayer. Palmerston asserted that the cholera outbreak of 1849 was due to a lack of observance of public health practices and nothing to do with divine will or judgement; "grime and filth," he stated, bred pestilence.

This attempt to pursue the policy of an earlier church-directed culture roused much secular criticism (Turner, 1974); *The Times* was to devote space in seven issues, and the *Pall Mall Gazette* six, to the debate which followed. On October 9, the *Pall Mall Gazette* noted that many laymen did not favor the prayer for removal of the cattle plague, not through godlessness but as an acceptance of God's will, and argued that this point of view was not incompatible with the use of prayer. But in its editorial the following day, it was harshly critical of the content of the prayer, for the Archbishop had confused the ownership of the stricken cattle which actually belonged to the farmers rather than to God. The prayer approached "sublime selfishness" in requesting the disease to be kept only from Britain. It accepted that the Archbishop had not meant what he said in his prayer, but "we have some right to expect that our common sense will not be outraged, and that we shall not have words offered us which, if we interpret them, carry blasphemous and unworthy meanings" (Anon., 1865c). The *Spectator* joined the criticism, stating "No one would dream of asking the Archbishop to write a prayer in a commercial crisis, attributing a crash in the City to God's judgement for our sins" (Anon., 1865d). *The Times* editors, while not criticizing the Archbishop's prayer, used the occasion of a pastoral letter by the Irish Cardinal Cullen, in which he had blamed the cattle plague on the sceptical spririt of the age, for a biting anti-sacerdotal and anti-Catholic attack (Anon.,1865e). In its final article on the subject on October 17, *The Pall Mall Gazette* remarked, "... these panic-stricken supplications have a tendency to foster that epicureanism in theology to which we are all of us sufficiently disposed." The contents of the prayer, as Turner puts it, thus stood in stark contrast to the sense and morality of secular life so congenial to both scientists and broad churchmen (Turner, 1974).

Professor Tyndall, physicist at the Royal Institution, who argued the impossibility of nature's laws being influenced by any prayers of men, was provided with a new opportunity

[2]A prayer against the human plague epidemic, to be read on a day of "General FAST and Humiliation" in December 1720, read "Preserve us from the evil we are afraid of, and suffer not the contagion of it to come near our Borders" (Anon., 1720b).

to air his views in letters to the *Pall Mall Gazette* of October 12 and 19, commenting on the original editorial. The special prayer served to consolidate ignorance and clerical dominance, and to further practical superstition when the methods of practical science should be pursued (Turner, 1974).

By November 1865, the prayer was being said every Sunday, as Constable put it, "not that the murrain may cease, but that it may be kept far from our shores" (Constable, 1865). In December, the controversy moved into the monthly journals, Playfair commenting in the *North British Review* that if the plague was a chastisement for violation of the laws of God, then it was our duty "to discover, understand and obey them" (Playfair, 1865). In the same month, an article appeared in *Macmillan's Magazine* by a liberal clergyman, J. Llewelyn Davies, who argued that prayer for the alleviation of financial loss to cattle farmers made prayer a matter of calculation instead of spiritual petition. But such criticisms had little effect, and the bishops continued with their prayers.

On January 12, 1866, Samuel Waldegrave, Bishop of Carlisle, preached on *The Cattle Plague: A Warning Voice to Britain from the King of Nations*. The Archbishop of Canterbury then petitioned Cabinet to appoint a day of national humiliation, which was refused, Cabinet referring to the use of the prayer already ordered by her Majesty; and so the Bishops of Oxford, York, Ely, Bristol, and London appointed their own days of humiliation, as did the Free Church of Scotland and the Chief Rabbi of Britain. At the village school of West Hendred in Berkshire, it was held on March 2. The Master, William Ingram, wrote in the school log: "A day of Humiliation—Children attended divine service: morning at 11, and in the afternoon at 4 o'clock. We only had singing practice from 3 tile [*sic*] 4 P.M. The Cattle plague was the chief cause of the day being kept." In Warwick, a special *Hymn on the Cattle Plague* was published, presumably for use in the church of Warwick St. Mary, to be sung to the tune of "Batty."

The *British Medical Journal* was scathing in its criticism of prayer:

> There is one side of the National Cattle-Plague Humiliation Fast which has been overlooked by our preachers; but it is one which the occasion was especially fitted to exemplify. We mean the humiliation which ought to fall upon those of our rulers who had not enough sense to seek and follow the advice of men of science in a matter where scientific knowledge was especially oncerned, and would have been especially effective in preventing the spread of the plague. It seems to us that the degradation and the humiliation of our country in this crisis are, that it loved and preferred money-making interests and quackery, rather than science and truth. Science saw clearly enough how to arrest and stop the plague, and would have done so in its beginning, as she did in France, if her voice had been listened to. Those who ought to wear sackcloth and throw ashes on their heads are those who listened to and followed the falsehoods and calumnious insults on medicine day after day uttered by the *Times*. They who said these things, and they who allowed them to be practised, should stand degraded and humiliated before the country. (Anon., 1866j)

The approach posed a problem for Arthur Stanley, Dean of Westminster Abbey, the leading spokesman for the broad church, as to him the national church must encompass the leading intellects of the nation and not employ its dignity and power to block or discourage intellectual discourse and discovery. He therefore consulted his friend, Professor Tyndall, who replied by letter: "With regard to humbling ourselves in the expectation that a single beast the less will die I would say in all frankness that I consider it a mere pagonism [*sic*]." Referring the Dean to Davies's article, he ended his letter: "I hope and think ... that you will not pray as others will for the staying of this plague, but will ask on the contrary for strength of heart and clearness of mind to meet it manfully and fight against it intelligently" (Tyndall, 1866). The Dean thus preached on the text "Thy will be done," telling the Abbey worshippers not to regard the plague as a judgment, but rather as God's stimulus to the activity of those scientific researchers "by which his supreme will on the works and laws of nature was made known to us." Those who were not scientists should encourage the latter and not scoff at them, for they deserved "not our scorn but our highest honour." But at the same time, Archibald Tait, Bishop of London and later Archbishop of Canterbury, was preaching the opposite, attacking the "shallow speculations" of those who questioned the physical efficacy of prayer (Anon., 1866k).

A number of tracts appeared in support of prayer, such as *Prayer and the Cattle Plague* (Anon., 1866l), *Prayer and the Cattle Plague: Two Sermons* (Smith, 1866), *The Secret of the Cattle Plague* (Swayne, 1866), *Prayer and Providence: Two Sermons Designated to Vindicate the Use of Prayer in Relation to Plague and Pestilence* (Moore, 1866), and *Darkness that May be Felt: A Warning to England at the Present Crisis* (McCaul, 1866).

A Jewish form of prayer against cholera and cattle plague was also to be used in all synagogues in Britain.

Early in the New Year of 1866, special sermons were delivered in all churches. In Chester Cathedral, the sermon was titled *Divine Providence in its Relation to the Cattle Plague and Prayer*; in Carlisle Cathedral, the Dean, unoriginally, preached on "God's visitation is upon us." But the rector of Old Swinford in Worcestershire preached:

> Humble myself before God I will and more especially at this season of Lent... But I will not humble myself, I will *not* bow down my head, nor shall the expression of my countenance be sad, because the cows have got the rinderpest, and it seems likely that beef-steaks will be scarce... I will pray, as my Lord has taught me to pray, to my Father which is in Heaven to give us this day our daily bread, but except in obedience to the lawful commands of my Sovereign *I will not pray for beef.*

A correspondent in the *Cornhill Magazine* for March 1866 wrote with great sarcasm:

> If we had in the first instance besought God for what we should stand most grievously in need of ultimately, we should have prayed Him—to infuse a spirit of courage into the Queen's Ministers, so that they should not continue to behave as if the penalty for failure would be the loss of their heads instead of their places—a spirit of unanimity into the Commissioners, so that they might neither confound the dull nor anger the wise, by reason of the opposite nature of their suggestions—a spirit of decency into the Butchers, so that they might resist the temptation for turning a national calamity into an occasion for wholsesale robbery—a spirit of moderation into the Inspectors, so that they might neither infect nor slay more than should be necessary to earn their salaries and extras—and lastly to send a spirit of patience into All Men who should be required to have dealings with the above-mentioned persons in whatever capacity.

The Governors of the Royal Veterinary College ended their Annual Report for 1865 (May 1866) with the statement:

> In conclusion, the governors would express their hope that it may please the great disposer of events to mitigate the infliction of this calamity, and to permit the exertions which have been and are being made by the Royal Veterinary College, by the Royal Agricultural Society, by the Legislature, and by the Government, such a measure of success as may at no distant period relieve the agriculturists of this country and the nation from the loss and danger to which during the last eight months they have been exposed.

This reads more like a prayer than the conclusion of a report and suggested that the Governors had little confidence in objective method. Playfair ended his pamphlet on the cattle plague with the words: "By the public prayers which we now make that this plague may be removed from us, we hope to have our minds enlarged, so as in some measure to comprehend the wisdom of the Creator, and to follow His rules with simple obedience. By this means we may again place ourselves in harmony with the laws which govern the animal economy (Playfair, 1866).

Most churchmen appeared to side with Tait, and on August 9, a new prayer for relief both from cattle plague and impending cholera was issued:

> A Special Form of Prayer to Almighty God; to be read … in the Morning and Evening service, In all Churches and Chapels in England and Wales, and in the Town of Berwick-upon-Tweed, On Sunday the Twelfth Day of August 1866, and whenever Divine Service is celebrated, During the Prevalence of the Cholera, and of Cattle Plague, in this Country, instead of the Prayer now used for Relief from the Plague existing amongst Cattle, and for Protection against the Cholera.

Whereas this new prayer did not ask that the plague be restricted to others, many phrases of the former prayer were retained. It remained a prayer of penitence, seeking forgiveness for transgression. The peoples' iniquities had "justly provoked Thy wrath," and their transgressions "worthily deserve Thy Chastisement."

On November 14, 1866, *The Times* reported that as of November 10, the prayer against the cattle plague had been suspended, substituted by "a form of prayer and thanksgiving for relief from the plague among cattle and for protection against cholera."[3]

In evidence to the Select Committee in the 1877 outbreak in Britain, Brown stated that priests in Germany, one of the classes of people permitted to visit infected premises, were considered the readiest means of spreading the infection as they were the most consulted by the poor. If a man had a sick cow, he would be sure to seek the priest's advice about it. This was emphatically denied by Professor Muller of the Berlin Veterinary College.

When rinderpest, famine, and human disease devastated Ethiopia in 1889, the emperor ordered public prayers for 2 months, the whole urban population coming out of their houses at 6 p.m. and praying toward the sky. On March 24, 1896 the government of the South African Republic declared an official "Day of Prayer" to be held on April 5, not only because of the rinderpest, but also on account of the drought and locusts. Over a 3-year period, six "Days of Prayer" were ordered by Proclamation to avert the "Wrath of God." When the epizootic was raging at its height in British Bechuanaland, all Cape colonial officials halted work on October 15, 1897 in Vryburg, the day solemnly recognized as set aside for Humiliation and Prayer. Roscoe (1922) found in East Africa in 1920 that many pagans thought cattle disease to be a direct result of the constant disregard of their old customs, a belief akin to that of retribution for offending God. The prayer of a Nuer priest in the Sudan on the occasion of a sacrifice during a cattle epizootic bore elements of almost Christian-like prayer: "What, god? Take this ox, it is thine. Thou art master of the world, the world is thine and it is thine affair alone. What, god? Do not kill all the cattle. What will our children eat? No one will live. Oh god, take thy cow and spare a portion of the herds. God, I am a man of the cattle and I have come to sacrifice this least; take it, it is thine" (Evans-Pritchard, 1956). In Lower Egypt in 1864, a cow was reported cured by giving it water weighed gainst a copy of the Koran, but one man was crying over a heifer decked with many amulets which was dying of the disease (Waterfield, 1969).

[3]In March 2001, a correspondent wrote to the *The Oxford Times*:

> Sir—A notice in our church porch informs us that the Archbishop of Canterbury and the Bishop of Oxford wish a bell to be tolled at 12 noon every Sunday during the present foot-and-mouth crisis in the farming community. These two headline-grabbing, urban clerics say that then the farmers will hear what the Church of England is thinking about them.… It would be much better for our "top" urban, unthinking clerics to stick to prayer—gimmicky gestures are no use. Prayer can achieve a great deal and we are praying for the farming community already in this parish. (Serjeant, 2001)

In a Lords debate on April 4, 2001, the Bishop of Oxford queried the FMD slaughter policy, stating: "The current crisis suggests that something has gone fundamentally wrong in our whole understanding of the relationship between human beings and our fellow creatures on this planet." The controversy over slaughter as a means of control of epizootic disease still continues after almost 300 years.

7. A DISEASE WITHOUT CURE

In 1715, Lancisi had written:

> My opinion is this (setting aside the prevention of its contagion, which *mehercules*! I would say, is the most excellent and only mode of averting the disease), that we must endeavour to preserve the oxen from being infected, by giving them proper diet; and that when they are infected, the only thing which can save them from death is still a proper diet… as to venesection and violent remedies, they are always hurtful in contagious diseases; and the sentence of Hippocrates may here be well called to mind: "So act that if you do no good, you at least may do no harm."

Lancisi recorded that they had found no remedy which would work, and many were useless: "The truth is, that in the cattle pest, as in the human plague, not every one that takes the disease dies of it. Some recover, thanks to Nature rather than to the remedies given to them… In our experience at Rome (1714) many remedies we found useless, many hurtful, and some few seemed useful" (Lancisi, 1715)

When Simonds visited the Carpathian Mountains in 1857, his hosts had repeatedly emphasized that rinderpest was impossible to cure, and Simonds dismissed treatment as irrelevant in his report, but few took much notice. In his address to the Central Farmers' Club, he summed up the situation:

> I am one of those who from the beginning have never felt any great confidence that this disease would yield to medical treatment. There are certain diseases-special diseases belonging not merely to cattle but to sheep, and even the human subject-which for all practicable purposes have ever been looked upon as incurable. Thus cholera is incurable, yellow fever is incurable, small-pox is incurable. [Cries of "No, no" from the audience.] Men who have had a medical education say yes; it is only those who possess a partial knowledge that say "no" (laughter). It is a most amusing thing that farmers—I say this with all due deference; for I have a great respect for farmers as a body—it is, I say, a most extraordinary thing that farmers and shepherds should set up their opinions on this question against those of the veterinary profession. [A voice: "And parsons do so."]….

Gamgee (1866) stated that it was difficult to classify "the many extraordinary methods of treatment" which had been suggested since August 1, 1865. One was the vapor bath, said to have originated with Thaer, the father of scientific agriculture in Germany, who claimed to have used it successfully in 1828, and recommended by a Russian authority: all right if you only had one cow, but impracticable if you had a herd to deal with. It was thoroughly tried by a Mr. Sweetland of Hendon on eight cows without success. A chamber was prepared, with the floor rising at one end, the animal standing with its head at the upper end in the fresh air, while boiling water kept the temperature of the chamber at 120°C, the animals being covered with blankets making them sweat profusely. The process, lasting a half hour, was repeated from time to time. The Edinburgh Committee claimed success by also administering doses of turpentine, coffee, and ammonia carbonate. Mr. David Christy of Colchester, while applying arsenicum, belladonna, bryonia and phosphorous, to infected animals, covered them in wet sheets with several dry wool sheets over them, twice a day for $2\frac{1}{2}$ hr at a time, allegedly saving 15 of 22."

Gamgee noted that electricity, the electrovoltaic pile, cautery, cod liver oil balls, bleeding to fainting and then giving brandy to revive, calomel and fumigation with mercurial vapor, and prescriptions based on cayenne pepper, were all advocated. Earlier, in 1862, he had stated in a lecture to the Metropolitan Association of the Medical Officers of Health that he and others believed that dosing cattle with salts of arsenic and antimony, jalap, cantharides, calomel, and aloes gave rise to ill effects in humans consuming the flesh of such animals. To a certain degree, this danger was ameliorated by the trade in the meat of diseased animals, encouraging the slaughter of sick beasts instead of attempted treatment.

The Edinburgh Committee's answer to many of the supposed remedies was that the results of the experiments "whether positive or negative, will be important."

In their *Second Report*, the Commissioners concluded: "The only means of combating the disease which remain consist in the stoppage of movement of stock and of things likely to be vehicles of infection, in the isolation and slaughter of infected cattle, and in the use of disinfectants."

In Prussia, the regulations of 1803, reissued in 1856, strictly forbade treatment of diseased cattle and the recommendation and public advertising of remedies.

In the 1870–71 outbreak in France, phenolic acid was popular as a supposed cure, injected at the rate of 10–15 g per day. Sodium arsenate at 2–3 g per day was also used. Experimentation was to continue into the next century, and in 1912, Walker, in India, claimed that injections of potassium permanganate lowered the mortality rate from 54 to 10%. The following year, Holmes claimed that potassium permanganate was ineffective, but a success rate of 6 in 14 with an intravenous injection of iodine. In 1918, Boynton in the Phillipines claimed to have experimented for $6\frac{1}{2}$ years with 20 different chemicals, finding none effective. The list reads like one of the 18th century rather than the 20th: eosin, medicinal methylene blue, cacodylate of soda, atoxzyl, quinine sulphate, camphor oil, creolin, potassium permanganate, ergot, iodine, potassium iodide, gentian, arecolin hydrochloride, nuclein, formalin, chloarzene, castor oil, alcohol, extract of nux vomica, and extract of gentian. In 1932, Curasson reported that no medical treatment had been found to be truly effective.

18

THE SEARCH FOR A CURE CONTINUED

Inoculation and Vaccination

1. THE 18TH-CENTURY INOCULATION EXPERIMENTS

Dr. Edward Jenner, who wrote on such subjects as geology and the migration of birds, is generally credited with the discovery of vaccination for smallpox in 1796. Apparently struck by the fact that milkers in Gloucestershire were unaffected by smallpox, his discovery was to use the analogue, cowpox, rather than the virus of the disease itself. Thus, Jenner introduced vaccination in place of inoculation or variolation, the injection of the causative virus itself which resulted in a milder form of the disease, but had the disadvantage of spreading the severe form to those not inoculated. Smallpox inoculation had been practised since antiquity in China, and its use was well known in the Near East. An account was published in England in 1714 (Timoni, 1714), followed in 1717 by the letters from Constantinople of Lady Mary Montague. Its use was popularized in England in 1721–2 and before the middle of the 18th century was reported as well established.

Doubt had been cast upon the efficacy of cattle inoculation as a preventive for rinderpest by the French physician, the Marquis of Courtivron, in two memoirs read before the French Royal Academy of Sciences in 1748. Together with a former army surgeon, Pelversier de Gombeau, the Marquis had conducted experiments in infection by application, digestion, and inoculation. He noted that in some areas of France, even after recovering twice from the disease, many animals died on the third occasion they were exposed. But in Brunswick in 1746, of 19 animals inoculated, nine allegedly survived, mentioned only by Johann Erxleben (1771), professor at Göttingen University, who claimed that as only four in nine died after inoculation, compared with seven in nine uninoculated animals, inoculation was justified. The inoculation was conducted by threading under the skin from top to bottom of each hindquarter a length of cotton soaked in infective discharge, knotting the ends on the outside. On the fourth day, the animal was bled and on the fifth given a purgative. From the sixth day, the animal fell sick and was then treated like naturally infected animals.

In 1754, as well as Sir William St. Quintin inoculating eight calves, at Malton another Yorkshire farmer claimed successful inoculation of an old ox which then resisted infection placed successively with two herds, which died (Anon., 1754a). These cases were related to a Mr. Dobson of York by an "eminent physician," who also described the method to him. Dobson passing this on in conversation to a Mr. Tyson of Tottenham, a farmer and higgler, the facts were reported by one signing himself "T. S." in the *Gentleman's Magazine* of November 1754 as Mr. Dobson's method, stating "Mr. Dobson declared that he made the experiment upon his own cattle when the distemper was very fatal in all the adjacent parts, and that, by this method, he preserved nine in ten of his whole stock" (T. S., 1754). The method related by

"T. S." was to bleed the cattle and administer two or three doses of cooling physic. An incision was then made in the dewlap and a piece of tow dipped in matter discharged from the eyes or nostrils of an affected animal inserted into the wound and left there for 2 or 3 days. The distemper would then appear, and the beasts should be turned out into dry pasture until the crisis was passed, giving no dry food but now and then a warm mash.

Dobson repudiated this claim to fame in the December issue of the magazine, noting that he was surprised to to find his name in the magazine and sorry that it had been used to introduce a method of inoculating cattle so different from that he had given in conversation, which was:

> First bleed the beast, and from that time keep him from hay, and all dry meat, till he is quite recovered; and in order to keep his body open, give him scalded bran, or chaff. At the end of three days make an incision in the dewlap, into this wound put a piece of tow, dipped in the morbid matter discharged from the nostrils, or eyes, of an infected beast, then stitch up the wound, and let the tow remain till the symptoms of the distemper appear, when the tow must be taken out; after turn the beast out to grass.

In spite of Dobson's correction the error has been perpetuated in the literature ever since, inoculation being referred to as the "Dobson method" (Reich (1797) refers to him as "Dodson," repeated by Fleischman (1800), Gamgee (1866), Friedberger and Fröhner (1886–1887) (who also wrongly give the date as 1744), Nocard and Leclainche (1896), Hutyra and Marek (1916), Curasson (1932), Scott (1970), Theves (1994)], only one author pointing out the error (Faber, 1962)).[1]

The following year, on April 1, a Dutchman signing himself "O.O," wrote from Amsterdam to the *Gentleman's Magazine*, claiming the method unsuccessful. A young ox had successfully been inoculated following the instructions given, but this was then extended to 17 cows at Beverswyk near Haarlem, bought by a group of Dutch merchants in Amsterdam for the purpose of experimentation, only three of which survived (O. O., 1755). They then purchased some more cattle to be inoculated under the supervision of the principal surgeon in Amsterdam, Dr. M. W. Schwencke, Professor of Anatomy and Surgery at the Hague. Having experimented with bad effects, Schwenke then varied the procedure with six young animals by inoculating in the buttocks infectious matter from the eyes, nostrils, and mouth of

[1]Theves (1994) repeats Koch (1891) and Leclainche (1955), claiming that the first attempt at inoculation as a cure or preventive was undertaken by Ramazzini in 1711, but Ramazzini did not undertake any experiments (he was almost 80, nearly blind and in ill health). Although he wrote: "*Crediderim itaque in curatione hujus malignae febris, ea Methodo procedendum, quæ a bonis Medicis servatur in curanda variolosa puerorum febre* ...," loosely translated as "I therefore think that we should proceed in the cure of this disease as we do in the cure of the variolous fever of boys," he makes no reference to inoculation but goes on to refer to warm remedies *remediis calidioribus*, and recommends bleeding and setons, used since Greek and Roman times: "... *aures quoque rotundo ferro perforandæ, indita postea radice hellebori, ut Veterinarii Scriptores suadent; similiter sub mento perforandia palearia, injecto postea funiculo, quem setaceum vocant; indiget enim Natura aliquo emissario, per quod venenum istud exantlet* ..." Which translates as "...also one could drill through the ears with a round iron and afterwards insert hellebore root as the Veterinary Writers recommend. Furthermore one can drill through the dewlap under the chin and afterwards pull a rope through it which one calls a stiff hair, for Nature needs some outlet through which it may excrete the poison." And: "*Non inutile igitur foret sanis Bobus, palearia sub mento candenti ferro perforare, atque funiculo indito, apertum, illud diù servare, ut per illud paulatim egerantur impuritates illæ;, quæ hujus contagiosi fomitis sunt soboles, & in capite potissimùm acervari solent*" (Ramazzini, 1718). "It would be beneficial therefore to perforate the dewlaps of the healthy oxen under the chin with a red-hot iron and insert a rope in the perforations to keep them open a long time. Thus may be drained off the unclean matters which are the issue of this infectious disease and which usually accumulate in the head." referring to Hippocrates. "All suppuration makes relapse unlikely, for it is at the same time ripening, crisis, and apostasis." (Epidemics 6.3.4). And: "Developing sores and swellings bring fevers to their crises. Those in which they do not occur are without crisis. Those in which they persist have the surest and quickest relapses." (Epidemics 6.3.21) (Smith, 1994).

a sick beast, leaving them free to graze in the open. As soon as they began to void hard excrements, considered a forerunner of the disease which appeared about the third day after inoculation, they were given Epsom Salts to keep them loose. They fell ill on the sixth day and appeared quite recovered by the 20th (Schwencke, 1755). He conducted further trials 2 years later (Schwenke, 1757).

The same year (1755), Dr. Malcolm Flemynge, a Scottish physician of good reputation, without conducting any experiments himself, proposed inoculation of calves for: "I apprehend that inoculation will stand the better chance of bringing on the distemper if the subject it is performed on is as young as safety will permit, the vessels being then most absorbent, and the animal economy most easily put into disorder." He concluded: "I heartily wish success to it, and that it may do the highest honor to the worthy, public-spirited man who first proposed it, and even put it in execution with some success" (Flemynge, 1755). This was a reference to Dobson, but it was another who was to obtain credit for introducing its use, Dr. Daniel Layard, a country general medical practitioner at Huntingdon.

Whereas Flemynge can be said to have laid down in theory the basis of a preventive treatment, Layard applied this theory to practice. Layard also saw a close analogy between rinderpest and smallpox, proposing inoculation by placing in the cut in the dewlap a piece of tow dipped in the matter of a ripe boil from a recovering calf as a means of prevention where there was acute risk of infection. But he did not recommend introducing the disease by inoculation into clean areas. He had observed the cattle plague at close quarters in 1756, although only 15 cases on a neighbor's farm, and the following year produced a booklet recommending inoculation (Layard, 1757), becoming the accepted authority on rinderpest for over 20 years.

In a letter to the Royal Society, Layard considered that the Marquis of Courtivron must have been misled by people who had been deceived into buying cattle which they were told had obtained immunity, as had sometimes happened in England. But furthermore, he did not believe that the French experiments in inoculation had been properly conducted, namely using a slip of raw hide of a freshly dead animal, or putting a pustule in the neck. The correct "English manner" was to have inserted cotton thread or silk soaked with matter into incisions on the shoulders or haunches. Nevertheless, despite defending inoculation, "nor do I find any reason to alter my opinion," Layard ended his letter by hoping that the legislature would prevent the importation of diseased cattle and hides, "the only means of naturalizing and perpetuating a dreadful distemper..." (Layard, 1758).

Layard stated that the disease bore all the characteristics of smallpox and was communicated by contact "by the air conveying the effluvia, which also lodge in many substances, and are thereby carried to very distant places." He noted that those averse to inoculation refused to acknowledge that it was similar to human smallpox "and have very idly asserted that the only intention of declaring the contagion to be a sort of small-pox was purposely and with no other view than to promote inoculation for the small-pox."

By no means did everyone agree with Layard's views on inoculation. The President of the Royal Society, the Earl of Macclesfield, queried its value. Before inoculation could be practised, "... it is necessary to bring proofs, that this disease is not susceptible more than once; and also assurances, that a recovery from the distemper by inoculation guards the beast from a second infection." Layard replied, quoting examples of immunity. At Godmanchester in 1746, eight cows survived on a farm, which was visited by the disease again in 1749, 1755, and 1756. The original eight cows survived each visitation, although they mixed freely with sick cattle, which died. Various observers, ranging from farmers to the Dean of York, Doctor Fountayne, attested that cows which recovered from the disease were not attacked a second time. He had five beasts inoculated by means of a skein of cotton dipped in infectious matter and passed through the dewlap like a seton. One cow, near term, died. The rest, after passing through the disease, recovered and did not experience a second infection despite being herded with infected animals. Bewley, a surgeon in Lincolnshire, inoculated with infective nostril discharge three 2-year-olds for a farmer, which recovered after experiencing mild symptoms of the disease and did not take the disease again, despite being herded with infected animals 12 months later. Layard inoculated eight, of which five died, and one he slaughtered for examination. Many other cases, taken together, show that more cattle were killed by the inoculation itself than if they had contracted the disease naturally (Layard, 1758).

About early 1755, Binkhorst in Leiden inoculated three 1-year-old calves with secretions from a recovering animal, and all had died. Funds were then raised for the purchase of cattle for inoculation experiments to be conducted in Beverwijk by Corneel Nozeman, a teacher in Haarlem, as well as Agge Roskam Kool and Jan Tak, Leiden physicians. Aware of Binkhorst's work, but claiming later that they had no prior knowledge of the report in the *Gentleman's Magazine*, they inoculated 17 animals of which only three survived, one having had the disease but nevertheless strongly affected by the inoculation (Anon., 1755c). Inoculation was seemingly a variation of rowelling, a thread soaked in the nasal or eye discharges of sick animals being threaded into the buttocks. In some cases, omasum contents were used as an inoculum. Nozeman and his coworkers concluded from this that inoculation could not be recommended but wondered whether better success would be obtained using a weaker inoculum, and also by adequate treatment of the inoculated animals (Nozeman *et al.*, 1755).

In October 1755, Wahnschaff, bailiff at Weferlingen in Germany, carried out a number of apparently successful inoculations for Lieutentant Colonel von Arnstedt of Brumby, but the number of animals treated is not given (Tolberg, 1808). The method was reported to the Halberstädt Office of War and Crown Lands which circulated it to all Magistrates, where it was put on one side.

Also in 1755, Eelko Alta, a clergyman in Boosum, Friesia, conducted a trial on his own cattle, but all the inoculated animals died. He considered that this could have been due to the animals being housed in stalls with naturally infected animals and resumed his trials in 1759. After bloodletting and thorough purging, he inoculated two subjects with a very small amount of nasal discharge from infected animals using a needle and thread passed under the skin. Both animals became ill but survived. However, when he exposed them to sick animals, they became ill again but in a milder form. He thought that this was due to not inoculating sufficient substance in the first place and believed that one could inoculate too lightly or too heavily. He thought that with adequate procedures, it would be possible to create gradually a more resistant breed of cattle, having noticed the higher resistance of calves from cows which had recovered from the disease. In the published report of his trials, he refuted the ethical and theological objections to inoculation which had been engendered by debate over smallpox (Alta, 1765).

Salchow (1755), without doubting the verity of the English and Dutch experiments, raised objections, asking if any animals had died afterwards and gone unreported, and was it really the same disease? He drew attention also to the experiments with vaccination against smallpox in New England ordered by the King of Great Britain, which had disastrous results. Much later, however, he came to support inoculation (Salchow, 1780).

In 1758, Johannes Grashuis, a doctor in Hoorn, also attempted inoculation, experimenting with six cows. A mild infection was induced, but when the animals were then again exposed to infection, four died (Huygelen, 1997).

In 1763, another Dutchman, Jan Engelman, a Haarlem doctor, appears to have been the first to note the close similarity between rinderpest and measles, comparisons hitherto having been made with smallpox. Engelman described 15 points of analogy but thought there were no solid arguments in favor of inoculation for rinderpest because it was not caused by a "pox poison." He added that if it was caused by a measles poison, and if one could safely inoculate against rinderpest, could one then not inoculate humans against measles? Although he wrote that no experiments in humans had been attempted, Francis Home had conducted measles trials in England in 1759 (Home, 1759).

In an anonymous publication issued in Braunschweig in 1763, inoculation was advocated because, even if the losses were high, the quarantine period could be much shortened since all animals would go through the disease at the same time, whereas in naturally occurring outbreaks, measures had to remain in place for several months (Anon., 1763). Having attended an autopsy in 1759, the author noted that there was no similarity to chickenpox, tending to the belief that the disease was due to a poisonous miasma spontaneously generated when various conditions came together. Nevertheless, he believed that it could be disrupted by inoculation, there having been so many attempts that its efficacy was no longer in doubt, and the 1755 experiments in England had been lauded in the *Hamburg Correspondent* for their effectiveness. His recommended method was to starve the animal for 24 hr, giving only warm flour-water to drink, and then bleed. This was followed by administering a concoction of juniper berries and saltpeter. After one more day, further bleeding is carried out from the neck vein and then the vein pinched together and plugged with infected lint. The method of tying a seton in the dewlap he considered was not so difficult but was not as efficient. In his tests on six animals, three died. Another trial was conducted with six animals as in England, of which three also died. The Duke of Durchlauchten then requested in August 1760 that eight of his animals be inoculated, which the author did with Herr Böhl, the prince's apothecary. Of these, four died, and a sick one was slaughtered for autopsy. He considered that more might have survived had they been younger animals, not starved so much beforehand, and useless medical purges not given to them. This writer's recommendations for inoculation do not seem to have been accepted, as inoculation is not mentioned as a possible method in any of the instructions issued by the Duke of Braunschweig.

After a lull due to the severe losses being experienced, interest in Holland in inoculation revived toward the end of the 1760s. Experiments conducted at Utrecht in 1769 caused deaths of inoculated animals within 14 days, whereas in naturally acquired infection, the longest period before death was 17 days. In England, symptoms generally appeared on or before the fifth day, but some took 10 days, and death from natural infection took place in 12–13 days.

In February 1769, Camper gave a lecture on rinderpest at Gröningen, stressing the need for inoculation trials. He and his colleague, Professor Wouter van Doeveren, had come to the conclusion that rinderpest would remain enzootic, and so quarantine, isolation, and slaughter policies would be ineffective. Camper considered that one should not be too discouraged by earlier results because when smallpox inoculation had started in England, morbidity had also been high, mainly because incisions had been made too deep. With experience, results had improved (Camper, 1803).

That April, Camper and van Doeveren began a trial with 75 cows; 39 died, and the remainder recovered (van Doeveren, 1769). Camper, aided by Wijnold Munniks, then turned to an experiment in Friesland where a violent epizootic was raging. A total of 112 yearlings as inoculated, of which 45 survived and proved immune to later challenging. This they considered encouraging as half of this number would have survived a natural infection. Their method of inoculation was a thread soaked in nasal or eye secretions, passed through the skin of the buttocks or elsewhere. On the fifth or sixth day, the animals were given a purgative. Layard attributed the deaths to "the obstinacy and interruption of the peasants, the badness of the situation, and inclemency of the weather."

Several others pursued trials in Holland in the same period. Spruit, a surgeon, had variable results (Spruit, 1769), but two doctors, Schutt and Römer, had disastrous results in five experiments in Utrecht (Schutt *et al.*, 1770). Johannes de Vries, a doctor at Leeuwarden, began trials in Langweer at the request of the authorities, but 39 of 53 animals died. This was attributed to working in a contaminated environment.

Because of the losses, public opinion turned against inoculation once more. Henricus Vink, Professor of Anatomy and Surgery at the Rotterdam Medical School, gave a series of lectures at Rotterdam in October 1769 opposing its use and strongly disagreeing with Camper and Munniks. Munniks had claimed that even if post-inoculation deaths were high, farmers still gained from the survivals because they sold them at a much higher price. But this led to fraud, for one could not tell whether an animal was immune or not unless it died.

In various experiments in Friesland between November 1769 and January 1770, a total of 330 animals were inoculated, of which only 156 recovered (de Vries, 1994). The Friesian authorities concluded from this that in November 1769, the cause of the cattle plague was God's displeasure with the sinful behavior of the people, and proclaimed November 15 a day of prayer and fasting (Huygelen, 1997).

In spite of the losses, Geert Reinders, a farmer in the province of Gröningen, decided to continue experiments. He contacted Camper and, with the help of Munniks, conducted several trials in 1769–70 with variable results. Of 150 inoculated cattle, 64 died. They concluded that infectious material taken from a sick animal on the second, third, or fourth day always produced rinderpest; inoculated animals which had shown typical symptoms were always immune to reinfection when the symptoms had been mild. The manner of inoculation and the various attendant treatments had no effect upon the results. Table 1 lists the earliest recorded inoculations against rinderpest.

In 1771, Dossie wrote: "We may, on the whole, conclude that no general effectual aid is to be obtained against the murrain, in a preventive intention, by medicinal means." He was against the practice of inoculation as a remedy:

Table 1. The earliest recorded inoculations against rinderpest.

Year	Name	Place
1746	?	Brunswick, Germany
1754	Sir William St. Quintin	Yorkshire, England
	Malton farmer	Yorkshire, England
1755	Alta	Friesia, Holland
	Bewley	Lincolnshire, England
	Binkhorst	Leiden, Holland
	Dean of York	Yorkshire, England
	Kool & Tak	Leiden, Holland
	Nozeman	Beverwijk, Holland
	"O. O."	Amsterdam, Holland
	Schwenke	Beverswyk, Holland
	Wahnschaff	Weferlingen, Germany
1758	Grashuis	Hoorn, Holland
1759	Anon.	Braunschweig, Germany
1760	Anon.	Braunschweig, Germany
1769	Camper & van Doeveren	Holland
	Camper & Munniks	Friesland, Holland
	Spruit	Holland
	Schutt & Römer	Utrecht, Holland
	de Vries	Leeuwarden, Holland
	Reinders	Gröningen, Holland
1770	Tode	Avnöe, Denmark

> The fact is well known that beasts which have had the murrain from accidental infection, like mankind with regard to the small-pox, do rarely take it again; and it being presumed on this ground that the analogy betwixt these two diseases still holds good in other particulars, and, consequently, that the communicating the murrain by inoculation would have the same consequences on the cattle in preventing their receiving the infection again, and in rendering the symptoms of the disease proportionably milder and less fatal, it was imagined that the use of this method might be, in a considerable degree, a substitute for a preservative from infection itself.

He went on to state that experience had shown this to be erroneous, and there was one essential difference between the murrain and smallpox inoculations. With the murrain, cattle could still become infected again, experiencing just as violent a reaction as non-inoculated animals in spite of the fact that this was rarely the case if they had recovered once from the disease. He pointed out that more cattle died from rinderpest after inoculation than would have done so from infection taken in the natural way.

Dossie also pointed out that inoculation would introduce the disease into whatever areas of the country it was practised, and if not kept up, when outbreaks occurred, there would be great mortality among susceptible animals which had not been inoculated. This was occurring with respect to smallpox due to its extensive dissemination by inoculation: "… a far greater number of persons have died of the small-pox in England within the four or five years last past [1762–7] than have ever been known within a like period, for a considerable length of time." He claimed that Camper's statement that the first attempts at inoculation against smallpox in England were attended with the same lack of success was wrong, inoculation against smallpox had been as successful at the start as subsequently:

> Some few, out of great numbers have, at all times, died of inoculation for the small-pox: but never, in any proportion to the beasts … which … have died from inoculation for the murrain. It was the striking examples of success, which could alone have introduced and established the use of inoculation for the small-pox here: and it would have been rejected with horror, and prohibited by authority, had a similar failure to that, which has been experienced in the murrain, been found in the result of the first trials.

Dr. Falke also raised his voice against the system, basing his opposition on observations made in England, Denmark, and Holland. Many cases were known of cattle becoming infected after having been inoculated, there being a number in Holland at the time of the first experiments conducted by Dr. Nozeman in 1755, and those of Professors Grashuis and Camper.

In Copenhagen, the disease having become enzootic in Denmark with widespread ravages, inoculation was conducted from 1770 to 1772 by the Royal College of Economics and Commerce at royal expense on the island of Avnöe (Figure 1) following instructions provided in a letter to the royal physician from Camper dated May 11, 1770. Twenty stalls were constructed with 2 animals in each, 80 animals being inoculated each month. Thread of two or three thicknesses infected with nasal discharge was passed through the skin in the inguinal region and the ends knotted together. In the first year, 42 of 61 animals died, in the second 91 of 160 became infected, 1 died, and 68 did not show any symptoms. In the third year, 123 of 169 animals recovered, 2 died, and 44 showed no signs of disease (Tode, 1775).

In Sweden, inoculation had been strongly opposed by Peter Jonas Bergius, Doctor of Medicine and Professor of Natural History at Stockholm, who claimed that rinderpest was similar to human plague, and since plague could infect one several times over, cattle could not be protected by inoculation (Bergius, 1772). Haller (1773), who confused pleuropneumonia and rinderpest, stated that people were content to mitigate the disease by inoculation as a remedy for cure could not be found.

Reinders resumed his experiments in 1774, inoculating calves of cows which had recovered from rinderpest, having observed, and informed Camper and Munniks several times, that

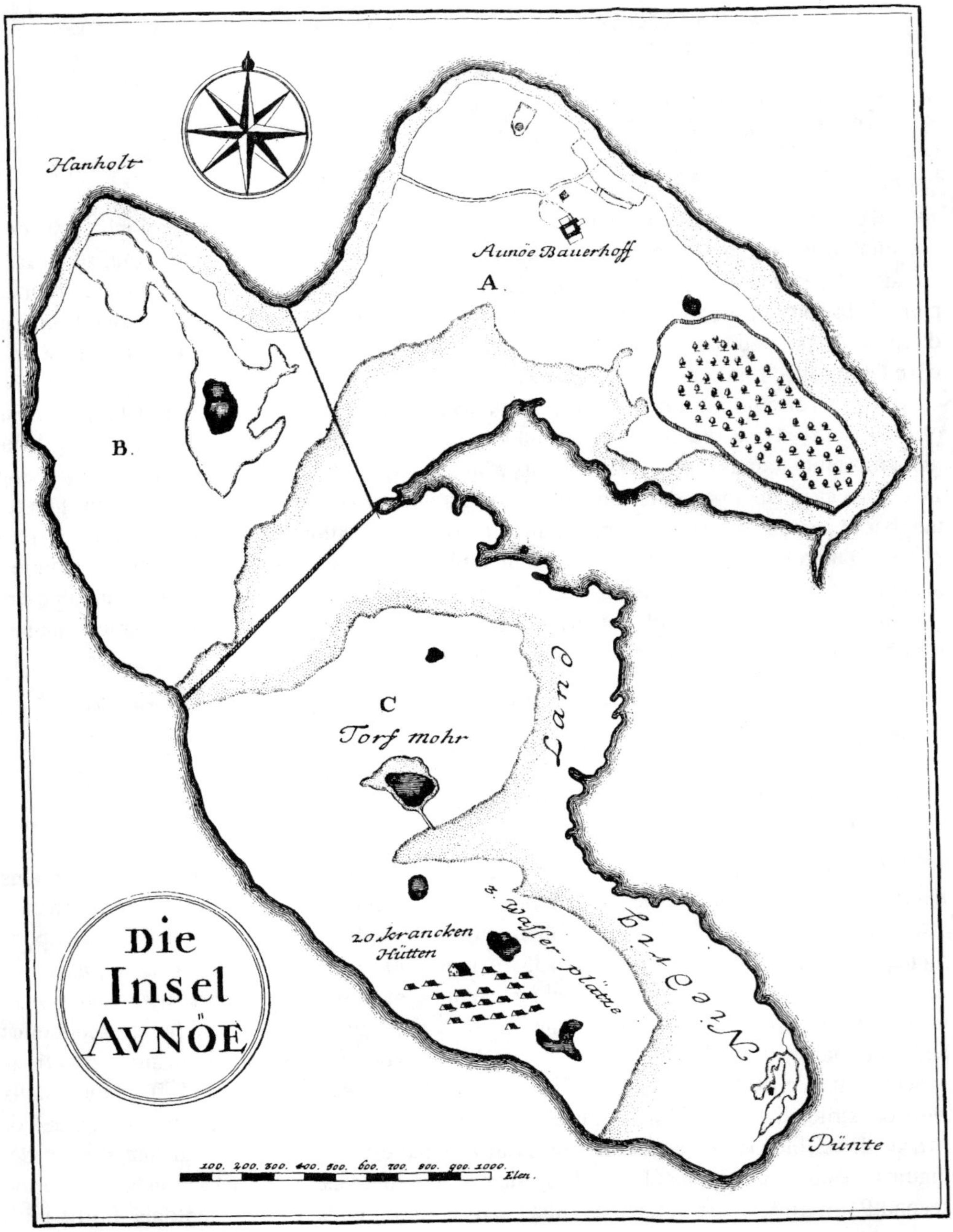

Figure 1. Isle of Avnöe, site of the Danish inoculation experiments 1770–72. The inoculation huts are shown in the extreme southwest.

calves from immune dams did not become susceptible to the disease before they were put out in the open in the spring, being probably the first to make use of maternally derived immunity (Huygelen, 1997). Initially, Reinders thought that the decline of resistance in the calves was due to exposure in the open air after being let out of the stalls in the spring but soon realized that this factor was unrelated. He then inoculated calves of 6–8 weeks of age from cows which had the disease 5 years previously. The calves showed only mild reactions. Some were then reinoculated after 4–5 weeks when out in the pastures and showed mild symptoms on the sixth day after inoculation. These were later shown to be resistant to infection, but he found that in order to increase the chances of success, he had to inoculate the calves two or three times at different ages. Three inoculations were recommended: the first in the stall at about 6 weeks of age, the second when the calves had been outside for 3–4 weeks, and the third in August or September (Reinders, 1774, 1776). Reinders invited Camper and Munniks to repeat his experiments, but there is no record of them having done so, possibly due to hostile public reaction. Reinders had been attacked by zealots, who accused him of blasphemy because he acted against God's will.

Reinders's example was, however, followed by a local doctor in Zwölle, Jan Hendrick Stolte, using 120 calves from immune dams. Twenty died, but four of these deaths were believed to be due to other causes. Of the survivors, 48 were inoculated a second and third time without showing any effects (Stolte, 1777).

These successes resulted in people taking up inoculation on a wide scale in the northern provinces (Bruins, 1951). Camper summarized Reinders's results in a paper he read before the Royal Society of Medicine in Paris, but without reference to Reinders himself. Camper was now able to press ahead, and a band of skilled inoculators was created which branded cattle to indicate their immunity. In 1778, he suggested the establishment of inoculation institutes to supply fresh lymph. An Inoculation Society was formed at Garnwert, which proposed to supply calves recovered from the disease, based upon conditions drawn up by Reinders (Reinders, 1778).

In 1777, trials were resumed in Denmark by Peter Abildgaard, later Dean of the Copenhagen Veterinary School. Of 1,199 inoculated animals, 38 remained healthy, 580 showed symptoms and recovered, and 581 died (Abildgaard, 1795).

After an outbreak in 1776, trials began in 1777 on a large estate owned by Herr von Bülow of the Royal Danish household. On some farms, the disease was associated with a low mortality, and neighbors had brought their cattle to these farms to be infected with the mild form. Trials began using this inoculum, and at first, mortality was high, but in a subsequent trial, only three in 25 died. Continuing trials in 1778, von Bülow observed that calves from immune dams reacted only slightly, if at all, even after three inoculates. Inoculated animals became sick on the ninth or tenth day, instead of the fifth when more virulent material was used. A total of 42 of 177 (22.6%) inoculated cattle died. Claus Dethloff of Oerzen, Master in the dukedom of Mecklenberg-Schwerin, then began his own experiments with a relatively avirulent strain from cattle in which >50% survived infection. In 1778, of 3,806 animals inoculated, 344 died (9.0%). A farmer on royal estatates in Glambeck, Jacob Brakenwagen, then set up a vaccination center with an insurance scheme. From September 5, 1778 to November 8, 1779, 511 animals were inoculated, of which 53 died (10.4%). In 1778 and the early part of 1779, 4,075 head were inoculated in Mecklenberg, of which 438 died, 290 others remained sick, and 160 had not contracted infection; but in some parts, scarcely 1% died. Yet at Oerzen from the beginning of 1778 to the end of 1779, of 269 inoculated, 94 died, and 41 were still sick, causing the Privy Councillor of Oerzen to recommended against the method. But at Glambeck from December 23, 1780 to October 19, 1781, of 600 inoculated, only 62 died (10.3%). Some batch results were as in Table 2.

The Prussian Royal College of Health, in an opinion on inoculation dated February 12, 1778, stated that it was still doubtful whether inoculation against smallpox was beneficial, inoculation importing the disease into places where it had not been present, and so inoculation as a method seemed partly very dubious and partly entirely inapplicable. As a result, the College recommended against its use, but some some rinderpest experiments were authorized by Prussia in 1779–81, although inoculation did not become widespread (Dieckerhoff, 1890),

Table 2. Inoculation trials in Mecklenberg.

	Inoculated	Died	Percentage Dying
First trial	18	6	33.3
	23	8	34.8
	16	5	31.3
	20	9	45.0
Total/Mean %	77	28	36.1
Second trial	57	0	0
	43	0	0
	100	3	3
	78	0	0
	20	1	5
	12	1	8.3
	35	0	0
	16	0	0
Total/Mean %	361	5	2.0
Grand total/Mean %	438	33	7.5

despite a series of articles appearing in a local magazine in Hannover promoting it with a call for a state-owned inoculation institute (Wens, 1987). But it then became widely practised in Mecklenberg, and with official encouragement, private insurance companies were created which provided inoculation in special institutes. Owners bringing cattle to the institutes were invited to estimate their value; the animals were then placed in stalls and inoculated. If an animal died, the owner received the estimated value as compensation; if it survived, he had to pay half the value before receiving the recovered animal. By 1779, the price had been reduced to one third of the value and overall the losses did not exceed 10%. Following approval with an ordinance of May 7, 1778, inoculation became widespread in Mecklenberg-Strelitz. Thomas Nugent, an English traveller, wrote approvingly noting inoculation gave great benefits but was underestimated, for whereas the disease often killed 80–90 in 100 animals, after inoculation, hardly 10 in 100 died, and it was the surest method to banish rinderpest (Nugent, 1781).

Von Oertzen, and W. Schumacher, officer in the dukedom and member of the Danish Royal British Agricultural Association in Celle, reviewed the experiments in 1777 and 1779 (Schumacher, 1779; von Oertzen, 1779). The method was to bring 15–40 healthy cattle together in a stall, excluding pregnant cows and calves under 6 months. The animals were then inoculated by an incision in the upper part of the flank, threads infected with nasal discharge being laid in the wound, which was then covered with a plaster left in place for 6 days, care being taken to prevent the animals licking it (Figure 2). Storage of the inoculum was a problem, keeping only 5–6 days in summer and 2 weeks in winter. Von Oertzen wrapped infected threads in wax paper and put them in an apothecary's box. Schumacher recommended burying them in a dry place in the ground.

In 1777, a prize essay competition was launched by the Berlin society *Naturforschender Freunde* with a prize of 20 ducats provided by Mr. Dohmherren, on the subject "What on the whole is the epidemic in the so-called epidemics?" It was won by Camper with a plea for generalized inoculation, and recommending that governments create and maintain non-immune herds to provide a source of inoculum. He declined the prize, stating that the honor was sufficient, and a second prize essay was launched to answer the questions:

- How long can the virus remain virulent and provide danger of infection?
- Roughly how long is it in the animal before it becomes evident?
- What precautions can be implemented between infection and outbreak?

Figure 2. Schumacher's method of inoculation on the flank, 1793.

This was won by Dr. Weiss of Leer in East Friesland, also advocating inoculation. He tried various medical treatments, but none was as effective as inoculation, claiming to have saved 49 in 100 animals in 1774, but argued, like Camper, the necessity of slaughter and deep burial of sick animals as the most effective method of containing the disease. In 1779, he claimed to have successfully transmitted the disease saving 38 of 44 animals by inoculating with threads impregnated with infected matter dried and kept in a glass tube for 6 years, then moistened with steam. It seems more likely that these animals were infected from another source (Weiß, 1783).

Also in 1779, Salchow, who earlier had not supported inoculation, suggested that rinderpest might be completely wiped out by it. In the same year, J. Kersting, a veterinarian at the Hannover Veterinary School created in 1778, was sent to Gartau to study the methods used in Mecklenberg, his report being published after his death (Günther, 1858). During his stay, he witnessed the inoculation of 135 animals, seven of which died. He considered that nasal secretions should be taken separately and the donor animal identified, only taking material from those which had shown mild symptoms. He also stressed the need to use animals in good health for propagation of the inoculum.

By April 1780, Layard himself had reached the conclusion that where the disease had become enzootic, as in Denmark and Holland, inoculation of herds at risk held out some hope of control. But where there were only sporadic outbreaks, as in England, in view of the opposition to inoculation the only solution was to slaughter affected animals and limit the importation of cattle to those which had recovered from the disease and were therefore immune (Layard, 1780).

From April 14, 1781, King Friedrich II of Prussia allowed inoculation under certain restrictions, following its successful use in Brandenburg in 1779, but in the German-speaking countries, many remained opposed, such as Johann Wolstein, Professor and Director of the Animal Hospital in Vienna and founder of the Vienna Veterinary school. He also opposed slaughter, but did not suggest viable alternatives (Wolstein, 1782). During the Franco-German wars at the end of the 18th century, inoculation could be used in Germany only with special permission (Kausch, 1793). An anonymous German writer in 1782 claimed that rinderpest was to livestock what smallpox was to people and believed that it could be cured, inoculation being an effective method but needing to be applied to all, and the present system needing to be changed.

In 1780, the Amsterdam Society for the Advancement of Agriculture published the results of a survey of inoculation trials. From 1777 to 1781, of 3,796 inoculates, 3,373 (89%) had survived (Poel, 1949). This compared with survival after natural infection from 1769 to 1776 of 126,307 out of 438,045 (29%). In 1784, more than half of the calves born in some parts of Friesland were inoculated.

In 1784, Christian Friedrick Reuss, a professor of medicine at Tübingen who wrote on many subjects, published anonymously his *Rindvieh Arzneybuch Sowohl vor die Gewöhnlichste Viehkrankheiten, als auch vor die Sonstige Viehseuchen* or *Cattle Medicine-Book for Common Livestock Illnesses as well as other Livestock Plagues*, quoting recipes mostly from the Braunschweig Beiträge of the 1760s. Under the heading of healing and preventive methods for mildew diseases, he stated that inoculation was useless. As Mr. Sagar found that an animal of 2–3 years had the disease more than once, "Inoculation is therefore cast aside." The method he described as passing a needle with thread sullied with matter from the nose of sick cattle through the skin two fingers apart, the thread inoculating the matter into the wound, and the animal becoming infected at most after 6–7 days. Under "Different approved methods in livestock epidemics," he stated that inoculation had already many a good effect as a preventive. He did not agree with those who claimed the disease to be a smallpox and outlined the differing symptoms. If it was a smallpox, one could propose inoculation, but it had been attempted by inoculating healthy cattle with cut-out pocks and never reproduced the infection. That one could with matter from the nose and mouth did not prove that it was a smallpox.

When the disease reappeared in Gelderland in 1797, Rudolph Forsten, Professor of Medicine at Harderwijk, recommended it under certain conditions, but stated that slaughter could be very efficient, as shown by results in other countries (Forsten, 1797). Faust (1797) was unconvinced of the benefits. He pointed out that Prussian law carried a penalty of 1 year's imprisonment for anyone who caused an outbreak of the disease through using inoculation. He made four points against its use:

1. Only an inconsiderable number of town and countrymen could use it.
2. Many places unaffected by the disease could become so by introduction of the disease with inoculation.
3. Inoculation decreased the cautiousness of prevention.
4. Extermination of the disease in Germany would be impeded by inoculation.

But in the same year, Gottfried Reich, Professor of Pharmacy at Erlangen, conducted an inoculation experiment in which almost all the animals survived and argued strongly for its use (Reich, 1797). Also in 1797, August Christian Reuss, a Wirtemberg doctor, published a small anonymous pamphlet suggesting that animals should be inoculated under the skin using in-oculate from those which had suffered a mild attack. The recipient first had to be prepared with herbal remedies and setoned, then inoculated copying the method for inoculating against smallpox, making little shallow cuts on the neck near the seton. The inoculum should be collected on the point of a sharp knife and then placed on the wound, moistening it with water if it had become dry. When the animal reacted, it was recommended that it be treated with various herbal remedies to ensure its good recovery (Anon., 1797).

Gottlieb Frank (1802) considered that rinderpest could be exterminated in Prussia with inoculation. Although many objections had been raised, and it was not known if an animal could be infected twice or not, hardly one in 100 inoculated calves died. He proposed a 3-year plan, with calves inoculated every 2 years. The best time to do this would be during March to May, when owners could do without their cattle for the 15 days necessary, and draught oxen were resting. Special inoculation sheds should be built where the animals would be kept, and a cattle plague fund set up to pay for their construction and operation. In 1802 the Leipzig Economics Society had as its annual prize essay subject "Rinderpest and its Cure". All the prize winners were cautious or dismissive concerning inoculation. Gottlieb Walz (1803) recommended inoculation when the disease was already present, but policing methods at the start of outbreaks. In 1808, Dr. Tolberg of Schönebeck issued a pamphlet *Ueber das Einzige Mittel bei der sich jetzt Nahenden Viehseuche das Rindviehe zu Retten (On the only Method to Save the Cattle Nearly Exterminated by the Present Cattle Plague)*, arguing for inoculation which, although not saving all, greatly decreased mortality. Everyone had their own cure, but, he asserted, "There are so far no drugs which prevent the disease and which can heal infected livestock." His method was a piece of coarse woollen thread moistened with infective discharge from the nose, threaded through the skin of the dewlap or inguinal region. He believed that one could also cause infection by cutting the skin and inserting pus into the cut, or even pricking with an infected inoculation needle. If all healthy cattle were inoculated everywhere the disease even started to break out, then, it was believed, the disease would be gone in 28 days.

A communication by C. L. Karsten of Rostock, Germany, dated December 23, 1813 and published the following year, *Wie Sichern wir uns Gegen die Wiederkehrende Rindvieh-Pest? (How do we Protect Ourselves Against the Recurrent Cattle Plague?)*, reviewed the Danish inoculation trials of 1778–9, concluding that inoculation was the only sure method. In a publication issued in 1813 by the temporary Government Commission at Munchausen, it was stated that with inoculation, the "long sought remedy was found" and could provide a big benefit, but was not effective with already infected animals (Anon., 1813). Sick (1812) who had used inoculation in 1801, considered all treatment useless and inoculation little better than slaughter. He recommended quarantine where possible. Zimmerman (1814) argued that inoculation was not a sure method and that policing provided "a wholly reliable and unerring method to fully exterminate the evil in a short time", but Bojanus (1820) argued when outbreaks were of isolated occurrence it was better to inoculate infected herds than slaughter them.

According to Koch (1891), only 13,000 animals in total were inoculated in Europe in the 18th century, with an average loss of 25%.

2. INOCULATION IN 19TH-CENTURY EUROPE

In Hungary, it was considered better to adopt inoculation as a preventive rather than rely upon cordoning off areas, and it was thought by a correspondent in October 1865 that it would soon be rendered obligatory by law.

In the 1830s, Ignatz Havas, chief medical officer of the county of Tolna, had repeatedly called attention to the inefficacy of medical treatments and launched an inoculation trial, obtaining the inspiration for his trials from pox vaccination. He reported his results to the county assembly on February 12, 1839. The previous December 12, using a sharp knife, Havas scraped off the epidermis on the internal surface of one ear and the inner side of the left leg of eight animals, taking care to ensure that bleeding did not ensue. He then rubbed into the prepared skin surface a mixture of mucous, saliva, and tears of animals which had died of rinderpest. Kept stalled, warm, and well fed, eventually all contracted rinderpest of a somewhat benign form, recovering within 12 days. Havas then infected another 30 animals with discharges from the previously infected 8. Of these, 14 contracted the disease, and 1 died. Finally, using matter from the infected animals of this second group, he infected two more groups each of four animals, poorly fed and kept in cold surroundings. Six contracted the disease with more adverse reaction than in the previous trials, the course of the disease lasting substantially longer. Thus, of a total of 46 animals, 29 took infection, and 1 died. Of those apparently immune, it transpired that the older animals had recovered from a previous epizootic, and the younger animals had survived *in utero*. Since the young animals were least affected, he proposed that in the event of a rinderpest outbreak, young stock should be inoculated first, and discharges taken from them on the eighth or ninth day used to inoculate other animals (Havas, 1839). The Szekszard County Council published Havas's report as a booklet, *Hiv Atalos Jelentése A' Marhapestist Enyhitõ Bánásmódrol (Official Communiqué on the Healing Treatment of Cattle Plague)*, sending a copy to all counties and to the Hungarian Agricultural Association, calling on them to conduct similar investigations, which none took up, but in 1847, Dr. Barrasch inoculated 2,500 head in Hungary, of which only 75 died.

While no reference was made to the earlier work of Havas, in 1861 and 1862 it was alleged that inoculation produced several successful results in the neighborhood of Pesth. Effected by scraping with a sharp knife the inner soft surface of the skin of the left forefoot and that of the ear, without drawing blood, the surfaces were then rubbed with a rag dipped in lymph from convalescent beasts in which appetite and rumination had returned. Discharges from the mouth, nostrils, and eyes were collected in phials, which were then sealed and kept in a cool place. So kept, the lymph was found to remain viable for 40–60 days (Egan, 1865).

In Austria, regulations concerning cattle disease in 1865 permitted inoculation for rinderpest with the consent of provincial authorities, but only in localities where the disease had spread extensively and a great number of cattle would die. Such inoculation was only for the purpose of shortening the duration of the outbreak and possibly the period of cordoning off an area, but it was prohibited where the disease had just broken out, when the official measures for combating it would be implemented. All other recommended measures still had to be used when inoculation was allowed.

Much earlier, D'Arboval (1826) in France was unequivocal in his condemnation. He asserted that when inoculation was successful, it was because it was effected at a time when the disease had practically ceased, while if practised at the beginning of an attack, the results were serious and fearful, and often worse than the disease itself: "The practice of inoculation as referable to this disease ought, therefore, decidedly to cease, for it tends to favour the propagation of this epizootic, to perpetuate its existence, to augment instead of diminish the mortality, and, at all events, to introduce another disease as murderous as that whose ravages it professes to restrain." But during the losses in France in the 1870–71 outbreak, Jacques Valserres and others attempted to resuscitate the practice of inoculation by re-publishing ideas promoted in 1746.

3. THE RUSSIAN EXPERIMENTS

At the beginning of the 19th century, a series of persons recommended inoculation to eradicate rinderpest in the Russian Steppes; Frank (1802), Walz (1803), Viborg (1809), Jessen (1834), and Spinola (1846). Spinola stated that the main advantage of inoculation lay in shortening the duration of an epizootic but could only advantageously be resorted to when whole countries were so overrun with the disease that it was no longer controlled by slaughtering stricken animals. It was, moreover, particularly advisable in countries in which circumstances rendered it impossible to exercise complete control over movement of cattle.

In 1847, the Russian Government sought the advice of various European veterinarians, who recommended improvement of veterinary police methods and the destruction of infected animals. The serious losses led to the appointment of a Russian Rinderpest Commission in 1852 "to inquire into the best means of extinguishing the Cattle Plague," which, after considering reports from veterinarians in Russia and other countries, reported "a system of quarantine could only be adopted in Russia when the boundaries of the native haunts of the disease were known." The greatest difficulties presented themselves against the extension of veterinary police measures, especially in the steppes, and existing regulations could only be carried out very imperfectly. The Commission drew attention to the favorable results of inoculation in the past and recommended prosecution of further experiments. In the same year, Jessen published a pamphlet renewing his 1834 recommendations for the establishment of inoculation institutes in the steppes, suggesting that rinderpest could be completely wiped out. Once the required number of inoculated cattle existed, no dealer would buy susceptible animals, and thus exportation of the disease from the steppes would be prevented (Jessen, 1852).

The Czar granted permission for a series of experiments, which began in 1853 at Kharkov and Koursk, followed by Cherson and Viatka. Previously inoculated animals were inoculated again to see if they were immune. Of 4 provinces, in the first 2 attempted, 999 of 1,059 inoculated cattle recovered, but results were unfavorable in the others. Further experiments were conducted the following year in Cherson, Kasan, and Dorpat.

Although against inoculation, Professor F. Unterberger undertook trials in Cherson, as he wanted to study the disease in the south of Russia and try inoculation with lymph from animals exhibiting a marked cutaneous eruption. Unterberger wished to try with rinderpest the idea suggested by Pessina in 1802, who had recommended cultivation of sheep-pox.

Using virus in lymph kept for 24 hr and some almost 11 months only from animals which showed an eruptive stage, of 64 inoculated, not one died of rinderpest, a single animal apparently contracting the disease but recovering. Twenty animals inoculated in 1853 and 11 in 1854 were sent to an area in Germany where rinderpest prevailed, and not one contracted the disease. Further experiments with exposure to infection of inoculated animals were tried, and all met with success, yet subsequently many took the disease by natural infection, and some died. But before he was aware of this, Unterberger considered his results of further experiments to be so favorable that they would not be received without misgiving and therefore used 32 animals not known to have had the disease as controls, injecting them without result apart from 2 calves becoming emaciated. This result cast doubt upon the previous results, for there was no difference in animals which had been inoculated before and those which had not been. Other experiments continued to give favorable results, but eventually Unterberger came to the conclusions that: (1) inoculated cattle in which the operation did not give rise to symptoms of the disease could not be regarded as protected; (2) the loss occasioned by successful inoculation was relatively great; and (3) mitigation of the virus had not proved to be of value. In other words, the virus did not lose its virulence to any marked degree by being stored. It had been hoped that by weakening it first through storage, only a mild infection would be produced. He wrote: "In my opinion, therefore, protective inoculation ought not to be employed, even in the Steppes, without the special permission of private stock owners, mainly for the reason that the inoculated disease is quite as contagious as that which is produced by natural infection." But it could be used advantageously in infected places where it was impossible to institute other measures of prevention, and if stock owners were compensated with the full value of any animal dying from the inoculation (Unterberger, 1864).

In 1853, the Kharkov Veterinary College conducted a number of experiments with "removes," passing the virus through a series of cattle to discover if cultivation of the virus by transfer from inoculated to non-inoculated animals was possible. These were generally considered to have some promise, it being claimed in a series of 10 removes that the operation of the lymph in the last 7 was much milder and weaker. At the same time, Professors Prosovew and Solotowsky of the Imperial Medical Academy in Petersburg conducted some inoculation experiments with calves in the district of Wjatka with apparent success, using 4-day-old lymph from a calf which had died (Solotowsky, 1856). A further series of inoculations by removes, beginning with matter 3 days old also promised success. Gamgee considered that these experiments belonged to the most rational ever undertaken in Russia, but the results were published only in Russian. They demonstrated that if rinderpest appeared after inoculation, its incubation never exceeded 10 days.

In 1855, both Unterberger and Jessen, who had conducted inoculation experiments near Odessa in 1853 by order of the Government, carried out experiments with success at Baraboi involving 98 animals. Various treatments involved inoculation of the virus using blood from Bessarabian cattle showing skin eruptions and mouth lesions. Some previously inoculated cattle were injected with 2-day-old and 2–3-day-old virus. Three bullocks had not been infected previously. None showed any signs of the disease following inoculation. Two calves were then inoculated using blood from an inoculated animal, and both died. Virus of the third remove was used on 51 animals, of which 11 took the disease, and two died.

In 1856, a number of experiments were conducted by a veterinarian Kobuschew on the Khirgiz Horde with some successes before bad weather interrupted the work. Many other inoculation experiments were carried out with varying results in the Kharkov and Kurskis districts but ceased in 1859. Rinderpest, however, "which never ceases in those regions," continued. Unterberger questioned the use of inoculation, where even in the steppe regions of southern Russia animals were still very susceptible in herds which had been infected for 10–20, and in one case, 40 years.

In 1865, Ravitsch, following the results obtained at the Inoculation Institutes in the provinces of Cherson and Orenberg in 1863, came to the conclusions that: (1) inoculation with old lymph was inefficacious; (2) inoculation with fresh lymph occasioned very serious losses; and (3) cattle which recovered after inoculation could not be infected again.

The Commission agreed with the conclusion that experiments had shown the virus not to be susceptible to mitigation by passaging. But Jessen, a member of the Commission, disagreed with all of Unterberger's conclusions. He believed that inoculated animals, even if they did not exhibit marked signs of infection afterwards, might be protected. The value of inoculation was proved, he claimed, by the fact that between November 1857 and December 1861, 719 animals were inoculated, of which 696 were thereby infected, and only 39 died, or 5.5%. In the Steppes, public opinion was favorable to inoculation (Jessen, 1865).

Jessen disagreed also with Ravitsch's findings. In a village near the Inoculation Institute, which had the disease for some time, 80 animals were inoculated, of which only 16 died. Jessen protested also against the conclusion in a paper appended to the report, of Professor Pelikan, another member of the Commission, that the cases of mitigation of the disease alleged by the advocates of inoculation were due to the mildness of the prevailing epizootic, or the fact that the animals had already been rendered immune by previous infection, and the question was definitely settled on this score.

At the Second International Veterinary Congress in 1865, Jessen stated that "the great task" of inoculation had not been rightly understood everywhere. The purpose was to be able to export Russian cattle which could neither take the infection nor spread it, a purpose which could be achieved only by inoculation. As Steppe cattle were less susceptible to the disease than other cattle, if they were the source from which all epizootics sprang, then if those exported to Prussia and Austria were made insusceptible to the disease by inoculation, they would be unable to communicate it and it would eventually cease to exist in Europe.

Dr. Gerlach proposed a vote of thanks to Jessen for his "zealous endeavours on the question of inoculation," which Jessen declined as being inadvisable to express thanks to individual members of the Congress. But he supported the vote of thanks which was carried for Her Imperial Highness the Grand Duchess Helena Paulowna for assistance rendered in carrying out inoculations in Russia at Karlowka, the Grand Duchess's large estate in the province of Poltawa where an Inoculation Institute had been set up in 1857, "thus contributing to the furtherance of veterinary science in general, and to the knowledge of the pathology of rinderpest in particular."

At Karlowka, there were three droves of cattle:

1. a drove of 14,000–15,000 head of steppe cattle belonging to the peasants;
2. 8,000–9,000 head of improved pure steppe cattle of "enormous size" used for draught; and
3. a smaller drove of acclimatized English Devonshire dairy cattle.

Rinderpest frequently infected the peasant herds at Karlowka and was communicated to the draught cattle causing almost 32% mortality in both, and, in turn, infecting the dairy herds which suffered 60% mortality. In 1857, the draught cattle were inoculated with a loss of just over 5%, but subsequent infection among the peasant herds was not transmitted through them to the dairy herds, which were not inoculated because of the high losses which would have taken place. The peasant herds continued to be ravaged by rinderpest until the owners were convinced of the benefits of inoculation and voluntarily had their herds inoculated.

Experiments were also conducted by the veterinary surgeons Raupach and Raussels. Those conducted by Max Raupach in Karlowka from May to September 1864 were considered to have provided the most favorable results yet obtained in Russia. Of 349 animals inoculated, all fell sick, only 10 were not severely affected, and only 20 died, or just over 5%. In all, 1,189 animals had been inoculated at the Institute since its inception, each animal being kept there for 8 weeks. Not a single one had died from rinderpest after release, although many were kept with herds which had been ravaged by the disease, there having been three outbreaks since 1857. Allegedly, virus of $1\frac{1}{2}$ years old had been used in some of the experiments and was protective. The Russian Government had "spared neither labour nor expense in sending learned, competent, celebrated men into the steppes, in order to study the Rinderpest." In the past 12 years up to 1865, attempts were made "in the far East [of Russia], in the steppes of the half-savage Khirgisians, as well as in the highly civilised provinces of the Baltic; in the South, on the shores of the Black Sea, as well as in the North and in the centre of the empire." Many thousand head of cattle had been inoculated. Prince Manoakbeg reported in 1865 that the Imperial Government had undertaken to introduce vaccination throughout the infected districts by competent veterinary surgeons.

But based on this Russian work, Gamgee (1866) considered that inoculation could never be resorted to with success, even in Russia, as a means of wiping out rinderpest. At the Third International Veterinary Congress in Zurich in 1867, delegates expecting to receive results of what they termed was the largest series of experiments in the world, were presented with a biased presentation of the results together with the wrong conclusions drawn from them (Saunders, 1980).

Bouley (1872) posed the question of whether the disease could be neutralized in Russia by a general compulsory inoculation through the whole region of the Steppes. Despite the fact that the Steppes were of immense extent and that the herds might well be termed innumerable, the majority of male beasts were oxen, and if castration was possible, inoculation was equally so. But after long and costly experiments, the Russian Government had renounced it, and veterinary opinion in Russia was divided as to whether it yielded results of real economic advantage. The results of experiments gave a mean mortality of 13%, even among Steppe herds, rising sometimes to 50%. Inoculation in Central and Western Europe over a period of 120 years gave a mean mortality of 18–19%. In 1866, France was estimated to possess 12.7 million head of cattle, so inoculation could produce an immediate loss of some 2.4 million (Bouley estimated 10 million head with a loss of 1.9 million), but the disease itself in a recent outbreak produced only 35,000 deaths, including slaughtered animals.

In 1877, Simonds was to state to the Select Committee that in Britain inoculation had been found to be very destructive, and the many experiments in different parts of Russia had been abandoned because of the fatalities it caused. As is generally the case, the inoculated disease was not found to be milder than the natural infection. Ravitsch and others at St. Petersburg opposed the Dorpat workers' findings, claiming that inoculation was dangerous, killing more animals than would die if a herd were left uninoculated. Furthermore, inoculated animals would still often succumb to the disease when later challenged with infective material from sick animals. Eradication by slaughter was claimed to be the only effective method. The St. Petersburg "faction," as Saunders (1980) termed it, had its way, convincing the government of the uselessness of the research stations, which were then closed down. In 1875, Semmer criticized Ravitsch for having drawn the wrong conclusions with regard to the effectiveness of inoculation. Ravitsch had conducted experimental work at a research station, which was 28 days' travel from St. Petersburg over rough tracks on a sledge or in an unsprung carriage. His conclusion that immunization was not effective resulted in the closure of the station. Perhaps, he was secretly glad of that! Jessen also criticized Ravitsch's obdurateness, and Semmer criticized Jessen's successor, Unterberger, for contributing to the decision to close the rindepest vaccine experiment stations. But both Jessen and Unterberger apparently considered it essential to inoculate every head of horned stock in Russia, an impossible task given conditions existing at the time.

Semmer, regarded as an outstanding scientist by the whole of the international veterinary world, assisted by Mikhail Tartakovskii was employed in the official service of the Russian Empire to develop a prophylactic vaccine. Applying J-J-H. Toussaint's method of inoculation with heat attenuated virus,[2] he had established in 1881 with the aid of Professor Ranpalle that infective blood or nasal mucus heated for 10 minutes at 55°C completely destroyed the virus. But obstructed by the intrigues of officials, he records that his efforts to undertake field trials in the Ukraine in 1892–5 were subverted by "sinister forces," which he does not name. A strong believer in vaccination, his minority view was opposed by a powerful group of bureaucrats in St. Petersburg who recommended eradication by slaughter, a method that Semmer pointed out would not work in the steppes of Russia. He considered that the problem of vaccination could be solved in 2–3 years at a cost of 2,000 roubles, but limitations of space and finance enabled him to keep only half a dozen experimental cattle. He pointed out that certain government officials did not want to bring the disease under control because it would have put them out of a job, and an incredible series of obstacles were put in the way of his field experiments in the south of Russia. Even though Count Orlov-Davidoff provided a grant of 60,000 roubles to support research on vaccination enabling Semmer to

[2]Toussaint claimed in 1880 that he had succeeded in making sheep resistant to anthrax by inoculation with defibrinated anthrax-infected blood heated to 55°C for 10 min.

open up another privately funded station in the Ukraine, a group within government called the Veterinary Committee stirred up so much local opposition that Semmer was forced to close it, but not before he had discovered, in 1893, the protective powers of serum and of milk taken from a recovered animal. He found also that serum from a horse injected under the skin with rinderpest diminished receptivity to the disease for a short period in experimental cattle (Semmer, 1893). On January 31, 1894, Carl Mæhl, a Danish veterinarian working in Russia, wrote to Krabbe in Copenhagen:

> In the South [of Russia] there are difficulties in overcoming rinderpest, and those involved are still optimistically hoping to find an efficaceous vaccine to replace the rated and devastating slaughter programme. Professor Semmer, who was previously in Dorpat but is now professor at the Institute for Experimental Medicine in St. Petersburg, has published a statement to the effect that he is nearing that goal.... However the practical application of such a method will certainly be very difficult, for example annual inoculation of all cattle under the auspices of the many republics. The reigning government here has already created two experimental vaccination stations, one in Poland and the other in Caucasia. In the meantime, these optimisms concerning the advent of a vaccine are tending to cause the unpopular slaughter progamme for control of rinderpest to go into relative disuse. (Berk, 1986)

The export of cattle to continental Europe, and later directly to Britain, was an important source of income to the Russian Empire, and despite the behavior of traders who tried to offload their sick animals on the export market using intimidation with the cooperation of officials, if necessary, the Russian government was taking constructive steps toward elucidating the cause of rinderpest and controlling the epizootics, setting up experiments using several thousand head of livestock. From the results obtained, Semmer, Jessen, and Tartakovskii came to the conclusion that inoculation was the way to control the disease and recommended that government undertake a campaign. The evidence for the conclusion was, however, inadequate, and a more effective and less dangerous vaccine was needed which was not developed until several years later, but it was nevertheless a method which could achieve partial success (Saunders, 1980).

During these years of Russian experimentation, 1860–80, some leading British scientists still labored with archaic theories of "contagion"; indeed, writing of human disease, Brown (1884) stated that inoculation and prophylactic measures based on the theory of pangermism were in theory "defective or illusory, in practice dangerous or pernicious."

At the end of the century in Africa, it was realized that vaccination was the only possible means in such a country (although the disease was not known to have established itself enzootically at that time), just as Semmer had realized was the case for Russia 40 years before. Thus, based on Semmer's discovery, Kolle and Turner developed the serum-virus simultaneous method of immunization in South Africa (Kolle and Turner, 1897), a method which was to be used for nearly 35 years, but the opposite view was still defended in Russia in 1959 (Chernyak, 1959). Although Semmer had been right in principle that vaccination would control the disease, it is improbable that the method he had developed would have done so. However, had he been able to conduct the experiments he wished, then he may have achieved his desired end and pre-empted the later work of Koch and Theiler in South Africa.

In 1897, when Koch in South Africa hit on the bile method of inoculating against the disease, it was heralded as a breakthrough in the scientific world, considered as the beginning of modern work on rinderpest research. But almost 40 years of work on inoculation in Russia had been totally ignored. Even in 1991, one medical historian referring to research on contagious animal diseases ordered by the Commission in 1866 stated that pathological research on animal diseases in Britain at the time was in advance of continental efforts (Worboys, 1991). In 1898, at the 50-year anniversary celebrations of the Russian Dorpat Veterinary

Institute, Professor R. Schmaltz of the Berlin Veterinary College was moved to comment: "Something that comes from Africa and is amazing the world public as supposedly new—rinderpest vaccination—has been taught and practised here for a long time" (Saunders, 1980). Although it probably furnished the seeds to his discoveries, Koch had been dismissive of the Russian work, writing to the *British Medical Journal*: "Russian doctors and veterinarians alone have from time to time published researches on rinderpest occurring in South Russia; but these works are mainly in the Russian tongue, and, furthermore, as far as I can tell from those which have been accessible to me, contribute very little to our knowledge of the disease" (Koch, 1897a).

Hutcheon, South Africa's Cape Colony veterinary surgeon, although referring to disinfection, showed a knowledge of the Russian work and appreciated it, noting in 1896 "All scientific men, and especially the Russian scientists, to whom we are most indebted, for they have made a study of the subject for the last fifty years, as they have the disease always with them, hold that the disease is not difficult to destroy, as most of the disinfectants even in weak solution destroy it (Hutcheon, 1896a).

Other workers had now come on the scene in Russia: Nencki, Sieber, and Wyznikiewicz, who experimented with the South African developments. In 1896, Nencki and Sieber obtained a serum from sheep which immunized sheep against rinderpest. These workers considered that the bile method did not give constant results, inoculated animals sometimes contracting the disease in a fatal form and possibly spreading it. At other times, only a fleeting immunity, or no immunity at all, was produced (Nencki *et al.*, 1898). Serum alone, however, was reported on much more favorably, 50–100 cc of fortified serum depending on the weight of the animal, protecting an animal for 4–6 months. In doses of 100–200 cc given on the first or second day of the rise in temperature, it gave surprisingly good results as a curative agent (Nencki *et al.*, 1899).

4. THE INDIAN EXPERIMENTS

At the beginning of the 19th century, in India they were attempting the opposite to experiments in Europe, that is, to find a means of vaccinating humans against smallpox with vaccine from cows, and looked to rinderpest-infected cows to find it. Cow-pock was often regarded as a malignant disease in India and called by the same names as human smallpox (*bussunt, mhata* or *gotee*). It was often noted that rinderpest and human smallpox outbreaks occurred at the same time, although Indians did not consider rinderpest infective to man. In fact, some 50 years before, in the early 1780s, an Italian physician, Eusebio Valli, had tried inoculating against plague with rinderpest, "Who first before me has inoculated cattle plague against plague? Who has ever inoculated the plague in those people who had the vaccine immediately before? … And long before the discovery of the bovine one had proposed smallpox as a preventive and as a cure against the plague …," he wrote to a colleague Gioab in 1785. Later, he wrote to another colleague, Mazarowich: "… the plague (of the animals) or rather the bovine plague sores … is not contagious to man. Therefore inoculating the pus cannot be, as I was flattered to think, a preventive for human plague" (Giarola *et al.*, 1967).

In 1833, Mr. Furnell, Civil Surgeon at Silhet, inoculated some Indian children with matter from a cow affected with *mata* (rinderpest) without apparent effect. During his absence due to ill health, his assistant, D. Brown, decided to repeat the experiment using scabs from a recovering animal (the outbreak being on the wane) pulped with water. This he inserted into the arms of four Indian children who subsequently showed signs of "genuine vaccinia" (cowpox) with a fever on the eighth day more severe than he had usually seen in vaccinia. From the vesicles of these children, he inoculated several others, all of whom exhibited a similar reaction. This was then propagated through successive numbers of children, by Brown or Furnell, or native vaccinators under their control, for the next 2 months (October and November); but there was no follow-up of these subjects after the eighth day. Believing they were producing a genuine vaccinia, at the end of November Brown vaccinated Major O's child with the virus, at the same time sending a supply to an Indian doctor with instructions to vaccinate Captain F's infant and as many Indian children as he could. On December 1, Furnell, being "so satisfied with the appearance" of the vesicle on Major O's child, vaccinated his own baby from a native child vaccinated at the same time and with the same virus as Major O's. Both children showed fever on the eighth day and reacted with an eruptive disease. On December 20, Furnell's baby died, the ninth day after the appearance of the eruption, from some "painful affection of the throat," having been unable to swallow anything for 2 days. Of Major O's child, Furnell reported on the third day from the commencement of the fever "an eruption similar to that preceding small-pox appeared, and in four days from that she was completely covered with an eruption resembling the small-pox in its confluent form, which ran through the same course as natural small-pox." Brown believed that Furnell stated that the disease went through a regular course *like* smallpox, and thus continued experimentation for several weeks after the death of the child. Despite its results in the European children, the reaction was much milder in the Indian children, and he continued passaging the virus until the beginning of February, when one of the Indian vaccinators reported that one of the children he had inoculated with the virus had contracted smallpox. The appearance of the reaction was as seen in the European children, and the child died shortly after Brown saw it. There was no case of smallpox in the vicinity, and Brown therefore terminated the experiments, reverting to the common vaccine. He concluded that inoculation with matter taken from a cow affected with *mata* produced, in the first instance, a vesicular disease exhibiting all the characters of genuine vaccinia, but its protective power against smallpox had yet to be ascertained. Brown speculated that the virus of *mata* was of a compound nature which produced two diseases in the experiments: first, a disease of a decidedly vesicular character resembling genuine vaccinia; second, one of a varioloid nature resembling modified smallpox, which in turn caused a similar eruption in a vaccinated individual (Brown, 1842).

The question of whether the children died of rinderpest, smallpox, cowpox, or septicaemia from the vaccination remains unanswered.

Macpherson, Civil Surgeon at Moorshedabad, reported to the Medical Board in 1835 rinderpest had not occurred for the past 2 years and few cases of smallpox: "From these circumstances we may infer ... that the unknown causes which favor the development of the disease in the human subject, have the same tendency in cattle: in fact, that variola and *mata* or *goti* owe their origin to the same cause" (Ludlow, 1842). Lamb, Civil Surgeon at Dacca, was sceptical. Some time prior to 1836, he had attempted to repeat experiments by Macpherson, in which the latter claimed to have apparently communicated smallpox to cattle. "I attempted sometime ago, when small-pox was prevalent and fatal, to repeat Mr. Macpherson's experiments, but altogether without success." Two young cows were wrapped up for weeks in blankets or other clothes in which smallpox patients had died. They were then inoculated with smallpox virus and on the udder of one about a dozen pustules appeared, but he was unable to produce smallpox or cowpox from them "by repeated trials on the human subject," and in a few days, the pustules disappeared without leaving any traces. He was unable to discover instances of the frequent coexistence of rinderpest with smallpox among the natives (Ludlow, 1842).

In Burdwan Division in 1836, the Civil Assistant Surgeon of Hooghly, Dr. T. Wise, submitted a report to the Medical Board, Calcutta, following circulars that the board had issued in 1834 and 1836 with the aim of finding an indigenous source of vaccine virus for the treatment of smallpox, the first requesting information as to the existence of natural vaccinia in India, the second regarding the "smallpox" among cattle.

> With reference to these documents, the Board have now to request that you will call on all medical officers within your circle, but more especiallly those in charge of vaccine depots, to make the fullest inquiry into the existence of variolous disease among the cattle in this country, requesting them at the same time, if they have reason to believe that it does exist, to describe the appearances and habits of the disease at the greatest length. (Hutchinson, 1834)

In 1836, Hutchinson was to report, "It seems to be no longer doubtful that a disease called by the Natives *matah* or *gootee* does prevail among the cattle in this country, but it is characterized by symptoms of such severity, and is otherwise so unlike the vaccine disease described by Jenner and others as prevailing in some of the counties of England that the Board have very grave doubts of its identity.

"Other circumstances have led the Board to doubt whether a variolous disease as well as a vaccine disease may not prevail among the cattle of this country, and accordingly any information on this point... would be highly appreciated" (Hutchinson, 1836).

Wise, in his reply, noted that the variolous disease appeared at regular periods, but most commonly in the months of March and April, smallpox prevailed, and the natives inoculated their children. He described the appearance of the infected cow, stating that the disease spread by infection and was propagated through the air and by fomites. In some cases, 50% and often 70% died in the epizootics. It only occurred once in the same individual.

He continued:

> 3. In this disease we have the same epidemic nature as in small-pox, the same precursory symptoms and pustular eruption of the skin and mucous membranes, the same infectious nature and contagious properties of the morbid secretion, the same severe inflammation of the mucous membrane of the mouth from the contact of atmospherical air charged with viurus, and afterwards the effects of its extension along the intestines, the same pits in the skin in cases of recovery, and the same fatal nature of the disease.
>
> 4. As to whether or not this disease of the cow be the same as that from which Jenner obtained his lymph is a point of great difficulty, and requires a proportionate degree of caution in drawing conclusions. Before deciding on this important question, it is necessary to recollect that epidemics differ in violence and fatality in different periods of their course, and in considering therefore the identity of particular diseases, these peculiarities should not be forgotten; certainly regenerated lymph from the cow affected with the above disease, as successfully practised by Dr. MacPherson, is a convincing proof of both being varieties of the same disease. A boy was sent up to this station from Calcutta vaccinated with this regenerated lymph, which passed through many individuals, presenting the characteristic appearance of vaccination without any unusual symptoms beyond that of, in some cases a severer accession of fever than usual. In one case the sore did not heal up for a month after vaccination, and the operation required to be repeated to secure the infant from the natural small-pox.
>
> 5. From my individual experience, I am led to believe that the variola and vaccinia in this country are modifications of the same disease produced from unknown causes, in the same manner as inoculation by variolous matter produces in some cases varicella, and cases of this disease appear in certain individuals in variolous epidemics.
>
> The milder disease, in this view of the subject, is produced by its passing through the human system, and still more so when obtained in favourable opportunities from the cow. (Wise, 1836).

Murchison believed the disease to be smallpox (Murchison, 1866) because human beings had been frequently inoculated with the virus and suffered either cowpock or often fatal smallpox. A severe form of smallpox in cattle was allegedly not unknown in England, an apparent case occurring in a dairy in Gloucestershire in 1825 with fatal results.

The Committee of the Provincial Medical and Surgical Association, in a report on vaccination (Anon., 1838), had maintained that the malignant distemper among cattle was the same disease as that described by Ramazzini and other Italian writers in the early part of the 18th century, that is, rinderpest. The symptoms of *mhata* described in Moorshedabad in 1832 by Macpherson, Superintendent of Vaccination at Moorshedabad, and at Dacca in 1836, left "little doubt of their similarity to those of rinderpest." The report continued: "Whether the pustules are numerous or rare, the disease is almost equally fatal, and unless measures are taken to separate the diseased from the healthy, it speedily runs through the whole herd, sparing few.... Many die before the eruption makes its appearance, but the fever is known by the discharge of the saliva, &c."

In July 1863, Captain T. Nelson wrote to the Collector of Kurnool in the Madras Presidency, proposing inoculation as a protective measure, the Collector in turn recommending it to the Board of Inland Revenue with the remark: "... the disease is the cause of enormous losses to the agriculturists throughout the Presidency, and recommends the institution of experiments to test the value of inoculation." But the Board's reply was that it felt that the matter was not "ripe for action." The Government, however, moved that the Medical Department be consulted, and other Governments should do likewise. But in November, Assistant Surgeon Dr. G. Bidie informed the Deputy Commissary General that he regarded inoculation as dangerous, tending to increase, rather than diminish, the ravages of the disease. In March 1866, the veterinary surgeon appointed to investigate the disease in the Madras Presidency stated that he could not advise vaccination for want of experimental proof of its value. He had inoculated three animals, two of which died, and while desiring further trials, he adhered to his previously recorded opinion as to the danger of disseminating the disease by this proceeding (Hallen *et al.*, 1871).

In 1864, the Civil Surgeon of Canara in the Bombay Presidency stated that he feared inoculation would only spread the disease although at the same time believed it might possibly be practised from animals in which the disease may have occurred in a modified and mild form.

In 1869, the veterinary surgeon Farrell then experimented in Assam using one heifer and two bullocks. "I inoculated the bullock with the saliva of a diseased animal, mixed with the matter obtained from a pustule in the mouth of the same diseased animal. I introduced the saliva and pus mixed together into the inner part of the left ear by means of a lancet; 81 hours after inoculation the animal began to look dull and refused his food...." On the fourth day, the animal died with all the symptoms of rinderpest. He inoculated the heifer with some of the fecal matter of a diseased animal in the dysenteric stage into the point of its tail. Eight days later, it showed signs of the disease, developing fully 4 or 5 days later. It ran through all the stages and eventually recovered. More than 12 months later, it was inoculated again with matter from the pustule on the tongue of a diseased animal, this time into the ear, but showed no signs of having taken the disease. However, the inoculate had been kept for some months between two plates of glass. The third animal was inoculated in the ear with water in which in about 1 oz. a piece of skin from the inside thigh of a diseased animal with pustules had been soaked for 12 hr. On the fourth day, the animal looked dull and refused its food but then recovered and showed no further signs of sickness (Farrell, 1870).

Dr. McLeod expressed the opinion that the time had not come for the introduction in India of any system of inoculation. The slender prospects of much benefit being obtained from it were contraindicated by the large scale and great expense of the experiments which would be required to establish its possibility. Unless an efficient cordon was maintained around an inoculation area, the measure would only serve to spread the disease (McLeod, 1869). In 1894, Pease was to state that protective inoculation still could not be seriously considered in India (Pease, 1894).

5. THE SMALLPOX ANALOGY

The superficial resemblance of rinderpest to human smallpox led to the supposition in Europe that it could be inoculated or vaccinated against. In a letter to *The Lancet* of August 26, 1865 Murchison corrected "a grave pathological error" in rejecting the assertion that rinderpest was the precise pathological equivalent of human typhoid fever, his post-mortem dissections revealing morbid conditions more characteristic of smallpox. After noting 18th-century allusions to its similarity to smallpox, he stressed that there was a very close analogy between the two diseases. In further communications to *The Lancet*, he stressed a very strong analogy, if not absolute identity: "This resemblance holds good, not merely, to a great extent, in the cutaneous eruption of the two diseases, but in their symptoms and anatomical lesions, and in their extreme contagiousness and capability of propagation by inoculation" (Murchison, 1865b). Murchison also suggested that cowpox was a less malignant form of the disease which conferred immunity, but cowpox had been in decline, and the cattle were now "less protected than formerly against the variola in a severe form" (Murchison, 1865c). He proposed an experiment in which some cattle be inoculated with cowpox vaccine and others with human smallpox to see whether any immunity against rinderpest infection was conferred. This apparently led to a rush for smallpox vaccine, out of which it is alleged that the medical profession profited handsomely. A guinea was charged for vaccination of a cow with ordinary human serum, and £2 if direct from an infant. The idea of vaccination as opposed to inoculation was to use natural cowpox lymph, human smallpox virus, or human smallpox vaccine, instead of injecting rinderpest virus. A sudden and great demand for vaccine led to the marketing of a spurious material as used in the sheep-pox epizootic made from collodion and croton oil or tartar emetic, which caused a blister to develop.

As a leading figure in medicine in Britain, Murchison was appointed to research the pathology of rinderpest for the Commission and attracted support from two other commissioners for his ideas on vaccination: Ceely, an acknowledged authority on vaccination, and another physician, Quain. Thus, with the approval of the Commission, from October 1865 to early January 1866, physicians vaccinated cattle countrywide. Stock owners, especially in Cheshire, were only too willing to provide their animals for trials, and by the end of January 1866, it was estimated that 28,000 cattle had been vaccinated in Cheshire alone. Lymph from a child's arm produced the best characteristic vesicles, but any immunity that might have been conferred was only transient. C. Deighton, MD, of Settle, Yorkshire, wrote to the *Leeds Mercury*:

> Gentlemen,—Allow me to communicate through your columns the result of an experiment undertaken by me, at the instance of the West Staincliffe and Ewcross Cattle Plague Committee, with a view to test the efficacy of Vaccination as a protection against Rinderpest.
>
> A healthy yearling heifer, purchased for the purpose, was vaccinated on the 3rd of January [1866] with lymph from the arm of a child, the matter being introduced in two places at the juncture of the teat with the udder. On the 10th, the vesicles were so fully developed that other cows were vaccinated with the lymph taken from them. On the morning of the 15th, the heifer was tied up in a stall with a cow in almost the last stage of Rinderpest, the dead carcase of another cow which had just died of the disease being removed to make room for the new arrival. The remaining cow died on the following day.
>
> The heifer has now been ten days in the infected stall, and under the same roof with other cattle suffering from the disease, and remains perfectly healthy.

A Mr. Fletcher of Worcestershire, wrote that he had been in the habit of vaccinating cows from time to time during the last 30 years and had obtained "satisfactory results" in the ratio of 75%. Vaccine vesicles not unlike those in a child were produced, which reached maturity on the eighth to tenth day. Mr. Matthews, a surgeon of Nantwich, had vaccinated about

600 cattle and distributed about 2,000 quantities of lymph. He first used lymph from children but then lymph from cows successfully vaccinated. He made two punctures about 10–12 cm from the root of the tail. Of 162 cases in 7 herds the operation was successful in 133. Dr. Dobie of Chester, and Mr. Oldingham of Huntingdon, vaccinated in the ear and had succeeded in the majority of cases (Lund, n.d.).

Under the heading "What is the Cattle Plague?" S. Parsons, a London medical doctor, wrote to *The Farmer's Magazine* (December 16, 1865):

> After devoting much time to the study of the cause of the cattle disease, I have come to the conclusion that it is nothing more nor less than small-pox in a suppressed form, so far as the symptoms are capable of developing themselves, owing to the closeness of the skin of the animal attacked. The small-pox pustules are not visible to the eye, but they can be distinctly seen, after death, between the soft cellular substance and the internal skin, showing clearly the character of the disease.
>
> Medicines have proved of little or no benefit, and the remedy which suggests itself is vaccination—in my opinion the only means of arresting the disease.
>
> Upon cattle that are already attacked it would possibly have but little influence except in the early stage of the disease.
>
> Every sound animal should be at once vaccinated, to ensure its safety ... No time should be lost in vaccinating every sound animal, as the only chance of arresting the disease. (Parsons, 1866)

But the *British Medical Journal* (January 20, 1866) questioned whether vaccination was of value and hoped it would not divert attention from the main point of extinguishing the cattle plague by stopping the movement of cattle throughout the country.

Gamgee (1866) referred to the Russian work of Dr. Brauell of Dorpat, which demonstrated that rinderpest was not smallpox. Murchison had enumerated 12 lengthy points of alleged similarity, and Gamgee replied with 12 points of rebuttal, regretting that Dr. Murchison may have alarmed many. Gamgee stressed that vaccination could not prevent rinderpest: "He [Murchison] has only paved the way ... for another great disappointment."

In a letter to *The Times* of January 22, 1866, Gamgee regretted

> that hasty conclusions as to points of analogy between rinderpest and variola have led to many fruitless and expensive experiments.... I learn that Mr. Tollemache's heifer is dead. This may serve to convince people that I have not spoken against the practice of vaccination without good cause ... Some men have been making ten and twelve guineas daily out of the poor farmers....

Mr. J. Tollemache of Peckforton Castle, MP and High Sheriff of Cheshire and an active supporter of vaccination who had 10 of his own cows vaccinated, had written 2 letters to *The Times* on the subject. In his second letter, January 1866, he naively stated: "Many vaccinated cows have died, but not one, as far as I can hear, where the vaccine has taken."

On January 30, 1866, in a letter to *The Times*, Murchison conceded that rinderpest was uninfluenced by human smallpox, and in all probability, the two were distinct species:

Sir,—The points of resemblance between cattle-plague and small-pox are so striking that certain observers were led to hope that vaccination might protect cattle from the prevailing disease. The experiment, I believe, has now been fully and fairly tried, and, although the first accounts appear favourable, there is sufficient evidence that vaccination confers no permanent protection from the plague. It is well that this fact should be generally known by publication in "The Times." Rigid isolation and the suspension of all movement of living cattle must still be the preventive measures on which we mainly rely. I may further state that neither Mr. Ceely nor I ever maintained that cattle-plague and human small-pox were identical. We merely pointed out that the analogy between the two diseases was sufficiently close to call for the experimental inquiry referred to.

Yours, &c., Charles Murchison, MD

Experiments conducted in January 1866 showed that rinderpest could be communicated with perfect facility to vaccinated animals, either by inoculation or natural infection, and when so communicated, its character was not modified. Thus, it was considered that the question was conclusively settled, and only 3 of 26 experimental animals were used. The only member of the medical elite to publicly express scepticism on vaccination, Sanderson concluded from his experiments on 11 animals that it would be incorrect to say that in any cases, the disease was modified in the same sense in which smallpox was modified in vaccinated persons. Even in the mildest cases, the development and progress of the disease were essentially the same as in those which terminated fatally (Sanderson, 1866). In a report to the Commission, Sanderson stated that no mitigation of violence of the disease was observed, whatever method of inoculation was practised. The results of inoculations did not show any appreciable difference either as regards symptoms, progress or duration, between the disease produced by inoculation and that resulting from natural infection if two cases of equal severity were compared, yet the proportion of fatal and severe cases was much smaller in inoculated animals and often led to spontaneous recovery. But results were inconsistent. In some epizootics, inoculation on a large scale gave favorable results, while in others, mortality among inoculated animals was almost as great as among those infected naturally. Thus, introduction of inoculation in the 1865–6 outbreak as a general prophylactic measure was condemned by all authorities on the grounds that not only was the result extremely uncertain, but it could not be conducted without risk of originating outbreaks, the extent and destructiveness of which could neither be controlled nor foreseen (Sanderson, 1866b).

Bristowe, in his report to the Commission *On the Morbid Anatomy of the Cattle Plague*, while admitting a resemblance between the eruption of smallpox and that of rinderpest, stated: "I have failed hitherto to detect in the eruption of cattle plague either a distinct vesicle or a distinct pustule, and certainly nothing like the umbilicated pustule of well-marked cow-pox or variola" (Bristowe, 1866).

In his report to the Commission in 1866, Varnell stated that Simonds and Gamgee had always said that the disease was not smallpox, and vaccination would be futile. He added, "I feel bound to record the correctness of their views…" (Varnell, 1866). Although the experiments that he had conducted on vaccination were few, he felt that they conclusively proved the inefficacy of vaccination as a preventive against rinderpest.

These observations were either unknown to, or ignored by, Lund when he wrote in his book (*c.*1868): "Vaccination has been adopted and found to be a cure."

6. TO VACCINATE OR NOT TO VACCINATE

In Norfolk, vaccination signally failed. Several cows were vaccinated and vesicles formed, but every one of the cows took the disease in a severe form and was slaughtered. In their *Second Report*, the Commissioners noted:

> Vaccination is not a protection against Cattle Plague. Careful experiments… have shown that an animal successfully vaccinated, and in which cow-pox has been fully developed, may, within a few days after exposure to the infection of Cattle Plague, contract that disease and die of it. We have at present no evidence that vaccination fortifies the system against this new assailant in any degree. The experiment, however, is now being made on so great a scale (upwards of 27,000 having been vaccinated in Cheshire alone), and under such a variety of circumstances, that, should this statement require any material qualification, the public will soon have full information on the subject.
>
> The possibility of mitigating the virulence of the disease by inoculation with the matter of the disease is a subject of much interest, and has engaged our attention. Such inoculation has been tried repeatedly, and on a considerable scale, by the Russian Government, and the effect of a long series of transmissions has been carefully noted and recorded; but the experiment has not been

> very successful, and it has for the present been abandoned. The Dutch Government, though much pressed to introduce it in South Holland, has refrained from doing so for fear of the consequences. (Bernard, 1866a).

When an animal was inoculated against rinderpest, the period of incubation was generally shorter by about 2 days. Although symptoms were just as severe and mortality not lessened, an advantage was seen in that an infected herd could be made to pass through the disease in 8 days, instead of lingering for a month or two with increased chances of spreading the infection. But Simonds alleged that in the 1865 outbreak, inoculation tended to kill more animals than otherwise would have died from the disease (Simonds, 1877). In an address to the RASE in February 1866, he stated that it was almost useless to go into the question whether security could be given by vaccination; it had been tested and found to fail. Although vaccination for smallpox was well established in early 19th-century Britain, Farr considered that its use in 1838–9 could have led to an extension of the epidemic by spreading infection artificially.

The Commissioners, in their *Second Report*, concluded that all endeavors to discover a method of treatment had failed: "The only means of combating the disease which remain consist in the stoppage of movement of stock and of things likely to be vehicles of infection, in the isolation and slaughter of infected cattle, and in the use of disinfectants."

In echoes of the present, in 1872 Beale wrote:

> One cow lived amongst the cattle plague poison for weeks, and was inoculated with it, but never took the disease. Whether this animal had had a mild attack before, or whether her escape was due to some other circumstance, was not determined. But ought we not to work to find out how to induce such a disease mildly in other animals? Will the public neither supply us with the means to discover, on condition that if the discovery be made it be given freely to the world, nor offer a reward sufficiently large to encourage an observer to risk all he has, and spend many years in the prosecution of what may turn out to be an absolutely barren investigation? Many a scientific man would work for the best years of his life for the value of one monster gun, and many valuable investigations might be conducted for a sum equal to that spent upon a single discharge. (Beale, 1872)

But in 1879 Thomas Walley, Principal and Professor of Veterinary Medicine and Surgery in the Edinburgh Royal Veterinary College wrote: "I imagine the *fiasco* of vaccination will never again be revived in Great Britain, and the adoption of curative measures would be as insane a proceeding as it is to attempt to treat a rabid dog or a glandered horse…" (Walley, 1879).

7. INOCULATION FOR PLEUROPNEUMONIA

Early investigators such as Vicq d'Azyr in France, Abildgaard, Oeder, and Viborg in Denmark, Adami in Austria, Barrasch in Hungary, Camper and Grashuis in Holland, Bülow and Oerzen in Mecklenberg, Kersting in Hannover, Lorinser in Galicia, Namsler in Silesia, Nebel in Hesse, Pessina in Italy, Reich in Franconia, Sick in Prussia, Walz in Würtemberg and Kausch had all added to the knowledge of rinderpest inoculation before Jessen described the Russian experiments. There had been countless inoculations of cattle in Britain since 1754. Some already practised inoculation for pleuropneumonia, inoculating near the base of the tail matter from the diseased pleura of a dead animal, and allegedly mortality from the disease was then much lower, some 25% dying each year among non-inoculated cattle. In a letter to *The Times* of August 7, 1865, Dr. Seaman reported having witnessed inoculation used successfully against "cattle sickness" in South Africa. This apparently was a reference to the Willems method of tail inoculation against pleuropneumonia, developed by Dr. Willems of

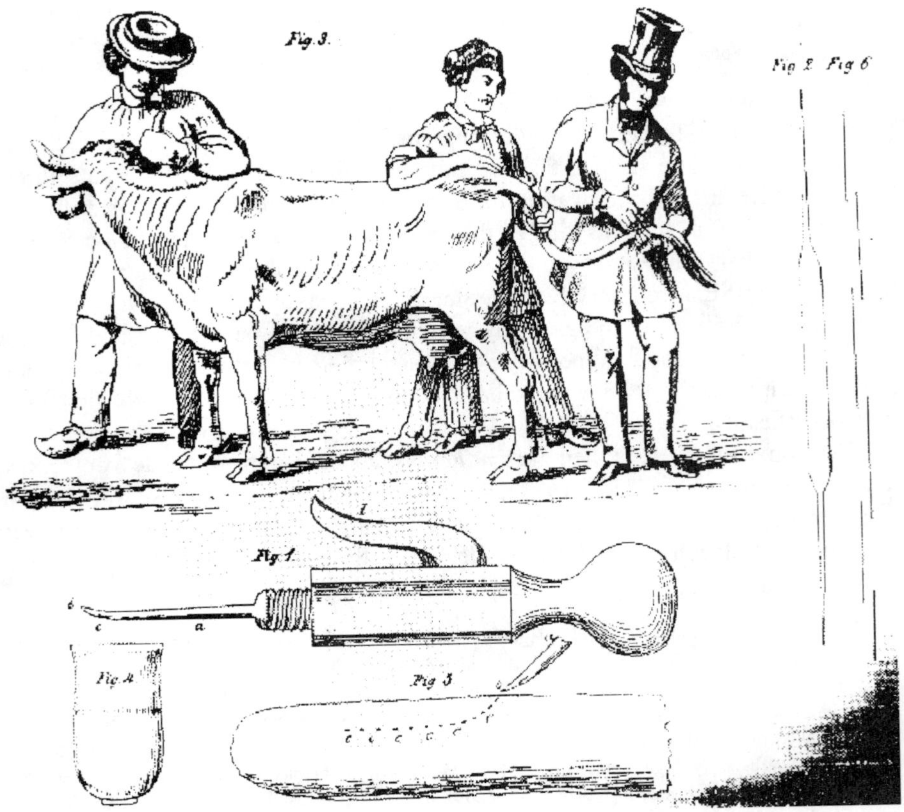

Figure 3. The tail method of inoculation for pleuropneumonia (Keulen, 1854).

Hasselt in Holland in 1852,[3] who claimed that it was unknown in Africa before it was introduced from Europe (Willems, 1900) (Figure 3). This was probably not true with respect to the Fulani of the western Africa Sahel who had developed their own version, soaking a piece of infected lung in milk for a couple of days and then rubbing the milk into an incision in the philtrum (St. Croix, 1945), a method also used by some of the Kel Tamasheq groups of the Sahara. Provost *et al.* (1959) state that the piece of lung itself was inserted. This could lead to such a severe reaction that it produced exostoses resembling a horn, and de Rochebrune suggested that these cattle belonged to a new race of bovines (de Rochebrune 1880, 1885, quoted by Huygelen, 1997). The Maasai of East Africa, after gashing the nose in a special pattern, then rubbed a piece of infected lung in the wound, attributing this discovery to Mbatian[4] (Schwabe, 1978), but who may well have received the knowledge via South Africa; however, they did not recognize any cure for rinderpest. The Fulani tried it with rinderpest without effect and abandoned the experiment. Croveri (1921) reported that some Somali herdsmen during rinderpest outbreaks gave their cattle milk to drink mixed with secretions or urine from sick animals, which showed a relatively attenuated form of infection. How old this practice was is not indicated. The Kavirondo of Kenya claimed sometimes to have had good results with a herbal concoction called *clunyasi luenombe*, maintaining that recovered animals were thenceforth immune, clearly a case of the animals already being immune.

[3]A Belgian Commission of Inquiry into the Willems method in 1852 concluded that it was "a cruel and a useless proceeding" (Greenhow, 1857).

[4]Mbatian was a seer from 1866 to about 1890.

19

THE BREAKTHROUGH IN AFRICA

1. KOCH AND THE BILE METHOD

When rinderpest swept Africa in the 1890s, it was by no means a disease unknown elsewhere, although not clinically diagnosed as rinderpest until it had reached Bulawayo in 1896 almost 8 years after the first reports of its presence (Mack, 1970). When it became clear in South Africa that the construction of a fence to limit animal movements would not check advance of the disease into the European cattle areas, experiments were begun in October 1896 at Marico in the Transvaal with immune serum inoculation by Arnold Theiler, a Swiss veterinarian, and Hubert Watkins-Pitchford, Principal Veterinary Surgeon for the Colony of Natal. Theiler had been operating privately until, in 1893, he was called upon to prepare smallpox vaccine for the Transvaal government. At a conference at Vryburg in August 1896 Watkins-Pitchford stated that the only hope of a remedy was in serum treatment, thus anticipating the breakthrough that was to come. Speaking for the first time at the conference, he exclaimed: "I think all the time spent in administering medicine internally for Rinderpest is not only wasted but very badly spent, the only possible hope of a remedy being in the serum treatment. I do not think any drug would check the disease when the organism is in the blood multiplying there while you are only acting on the alimentary canal." The serum-virus method of immunization, and the preparation of hyper-immune serum in goats, subsequently provided the control which had eluded workers in Europe for so long. In Bulawayo on March 26, Theiler had telegraphed his government that the Matabele uprising prevented him devising an inoculation against rinderpest, and on his return to Pretoria, he confirmed in a written report of April 10 his readiness to proceed with such work at a location suitably isolated to prevent any spread of infection. Finding Watkins-Pitchford in agreement at the Conference, on his return Theiler was able to persuade the Transvaal Government to take the proposal seriously, huge costs now being incurred by Government in disruption to transport from trek-oxen deaths. On September 13, 1896, Theiler wrote to his parents in Switzerland: "My next task is to devote myself to a series of experiments to determine whether it is possible to combat the disease with a serum injection. The Government has given me unlimited credit to achieve this object. The Natal Veterinary Surgeon will accompany me and be my coworker" (Gutsche, 1979). The site chosen was Marico in the Rustenberg district, to which Theiler, and Watkins-Pitchford loaned by the Natal Government, departed on October 1, 1896. Seven days later, Dr. Edington, the Colonial Bacteriologist, set forth from Grahamstown on a similar mission for the Cape Colony government to establish himself at Taungs in the northern Cape in a railway coach shunted into a siding. Beginning work on October 14, Edington was assisted by W. Roberstson, J. M'Namara, and J. F. Soga, the last a colored and the first South African to qualify as a veterinary surgeon (at the Royal (Dick) Veterinary College, Edinburgh).

But at the same time, the Cape Government appealed to Germany for help, requesting secondment of Koch at the instigation of Dr. J. Beck, an admirer of Robert Koch, the 53-year-old internationally acclaimed bacteriologist. While the Cape negotiated, the Free State evolved a similar plan to import a bacteriologist, and Natal attempted to import Dr. W. J. Simpson, who was dealing with rinderpest in Calcutta. But when all the states were requested to share the considerable cost of Koch, the individual endeavors were shelved (Gutsche, 1979).

Koch arrived at Capetown with an assistant, Paul Kohlstock, and both their wives, on December 1, 1896, demonstrating less than 2 months later in January 1897 that injections of bile from infected animals could allegedly be used with a high degree of success as a preventive, thus overshadowing the work of Theiler and Watkins-Pitchford, Koch arriving just before their experiments were completed. But prior to his arrival, on November 3, when his coming was published throughout the country, Edington telegraphed the Cape Agricultural Department that he had isolated the "Rinderpest microbe," believing it to be a bacterium, and Dr. Hutcheon, the Cape Government Veterinary Surgeon, pronounced his belief that Edington would devise a serum treatment. Edington stated publicly that although Russian and other scientists had similarly isolated the "fatal bacillus," they had been unable to reproduce infection with it, but he would succeed. In the event, he informed Koch by letter on the latter's arrival that he offered him all the assistance and resources at his command and assisted Koch who, with Edington's help, was unable to reproduce infection from Edington's cultures.

News of Koch's appointment was praised in *Nature*, leading to a critical letter from Sir John Kirk, former British Consul at Zanzibar:

> With reference to Prof. Koch's present mission, I would venture to observe that the German Government does not deserve the praise given in your last issue. The German Government, like our own, has been guilty of gross negligence in not studying the nature of the rinderpest in 1891 and in subsequent years, when it killed off the cattle and wild game in the British and German possessions in East and Central Africa.
>
> Up to the time that the epidemic reached the Zambesi, much might have been done, had the nature of the disease been then understood, to limit its progress ...
>
> The attention of the British Government was, as you say, called to the danger in ample time, but it is doubtful if in Germany the danger was even realised .... (Kirk, 1896)

A laboratory had been erected for Koch at Kimberley in some disused mining buildings belonging to the De Beers Company, where he was to be assisted by Dr. George Turner, Medical Officer for Health for Cape Colony, and Edington. On his arrival on December 5, Koch viewed the installations with Turner, Edington, and Hutcheon, and commented favorably on them. In the evening, he visited Edington's station at Taungs. The next morning, he attended the postmortem there of a cow which had died during the night, 23 days after infection with rinderpest. Koch considered that the animal had died of a secondary infection. Later, when another died, it showed characteristic symptoms of an early stage of rinderpest. The next day, he went to the Taungs Reserve, where it was said there was a large number of sick animals, but all had been driven into the mountains to save them from the outbreak. Soga reported 20,000 already dead. In his first report to the Secretary of State for Agriculture at the Cape, dated December 9, 1896, Koch confirmed that the disease was rinderpest but, although complimenting Edington on his help and cooperation, did not think Edington had cultured it (Koch, 1897a).

In his second report, dated January 3, 1897, he first confirmed transmission by the insertion of infective nasal swabs and subcutaneous injection of infective blood. Placing swabs in the mouth or nose, only one in eight and one in three, respectively, were infected, but all of five injected animals took the disease. He also tried injecting bile because it was alleged that farmers in the Orange Free State occasionally used a mixture of blood and bile, and he found in the bile a bacterium identical with one discovered in Calcutta by Dr. Simpson, but as the cause of rinderpest, it was found to be negative. All attempts to find a specific microorganism under the microscope or by culture had been unsuccessful. Koch stated that his main task was to find an attenuated virus for vaccination, and he was about to determine whether bile had any helping attribute or not. He stated that rinderpest could be combined with Texas fever, so it was necessary to check all blood used in the trials.

On December 14, he had inoculated Cape and Merino sheep and Angora and Cape goats with rinderpest blood. All exhibited temperature increases and again at a second injection on December 17. On December 24, he injected a second generation. He stated that the temperature increase showed that it was possible to attenuate rinderpest in sheep and goats, and he would attempt to transfer it to cattle. He also proposed trying passaging with antelopes, pigs, dogs, donkeys, and mules, and wanted to determine which different genera were susceptible, especially the camel. On December 22 and 29, he made visits to the Orange Free State, where he saw numbers of sick and dead animals, collecting material for investigation (Koch, 1897a).

In his report of January 31, he noted that he inoculated a heifer with blood of a second-generation Angora goat passage to which the animal reacted and recovered, and he was therefore hopeful that he could produce an attenuated goat-passaged vaccine. He thus injected four heifers with blood of an Angora goat and Merino and Cape sheep which had been passaged five times. The four animals reacted, and three died in 7–8 days, those inoculated with sheep blood reacting so quickly and the postmortems showing such a pronounced effect that he doubted sheep could be used. The heifer injected with the Angora blood recovered, the heifer on which Cape goat blood was used dying, but it was a weak animal, and the postmortem examination showed that the effects were not as profound as those produced by the sheep blood. He was therefore still hopeful of producing a goat vaccine.

He tried also attenuating the virus by mixing rinderpest blood with different concentrations of glycerol, likewise phenol, injecting subcutaneously. The animals did not contract rinderpest, indicating that these substances had a destructive effect on the virus. He also dried blood for 4 days at 31°C and then injected it, producing no effect. Injections of serum from rinderpest-immune animals produced protection, but it was not known for how long this would last. He had also inoculated pigeons, chickens, and rabbits, and fed a Secretary bird *Sagittarius serpentarius* for a week on rinderpest-infected offal without effect. Dogs, mice, guinea pigs, and rabbits were all immune, only pigs apparently being susceptible (Koch, 1897b).

On February 10, he reported that the serum of cows which had survived rinderpest when injected subcutaneously had a certain level of immunity, but 100 cc was necessary, and the immunity was passive and of short duration. A mixture of serum and virulent blood, however, gave protection against 10,000 times the minimum deadly infective dose, and he believed that an active immunity was conferred. This required only 20 cc. He also reported that 10 cc of rinderpest bile injected subcutaneously conferred an active immunity and suggested using serum inoculation in a belt where cattle were not yet infected, and bile in the infected areas. He was convinced that 1,000 cows a day could be saved with it and was ready to give a course of instruction in its use at Kimberley. He considered it no longer necessary to experiment with the camel, as his investigations with cattle had led to 50 favorable outcomes (Koch, 1897b).

Bile stations were set up and men sent to Kimberley for training in the method, but Koch was unable to complete his experiments, in March being ordered to India by the German Government to advise on human plague, and he nominated Wilhelm Kolle as his successor.

Theiler and Watkins-Pitchford had tried to make their anti-rinderpest serum active by subsequently injecting the blood of an infected animal, whereas Koch injected bile. Theiler complained privately in a letter home that Koch's discovery was the same as his, "with this difference that he at once announced it while I continued to await the result of further experiments.

I have committed the grossest stupidity in not publishing the matter while Koch has concerned himself with ensuring that within a few days, the whole world knows" (Gutsche, 1979). The basis of Koch's "discovery" remained the initial serum injection, and Theiler was sceptical of the bile aspect. "I therefore think," he wrote presciently, "that Koch has been rash in making common property of his first favourable results before he had proof that it could be repeated wholesale." Later, he explained how he and Watkins-Pitchford had set up an experiment to inoculate 10 cc of blood, serum, and gall. But no animals were available at the time, and when they were, "we completely forgot to inoculate with gall which we had found to be so swarming with bacteria that we simply did not believe in success—and now, there it is!"

Controversy was rife after Koch's announcement, the discovery alleged to be entirely accidental. It reached the point where Turner asked Koch point-blank whether it had been a fluke, as so many asserted. Koch answered:

> I was testing the virulence of the various fluids of the body and, amongst others, used gall from an animal dead of the Rinderpest, fully expecting that it would communicate the disease. I injected 10 cc into an animal which was none the worse for the experiment; on the contrary, when it was subsequently injected with rinderpest, it resisted its action. I immediately tested the bile on six animals and obtained a similar result. (Gutsche, 1979)

In a letter to Kolle, he confided that he had more luck than ever expected. The fortifying of an immunizing serum with a subsequent injection of virulent blood was orthodox serum therapy well known to those working in the field, "but quite new and having no connection with serum-therapy is the new method discovered by me." But Didier, Professor of Medicine at Marseilles, believed that the cause of bubonic plague was concentrated in the bile and in the outbreak at Marseilles in 1722, transmitted the plague to dogs by injecting them with bile of persons who had died of plague (Didier, 1722). The Marquis of Courtivron had demonstrated in 1748 that bile was infective by giving to a calf in some milk bile from an animal dead of the rinderpest. The calf died on the eighth day (Courtivron, 1748). These works were probably familiar to Koch, leading him to experiment with bile.

The same day, November 17, Koch set sail from Europe, the State Secretary at the Cape received a letter from the French consul in Pretoria concerning an individual in Johannesburg who was friendly with a French bacteriologist and pupil of Pasteur willing to come to devise a cure as he had done for other animals. It transpired that there were really two pupils, one a professor, and within a week, the Transvaal Government had applied to France asking for them to come as soon as possible, according to Gutsche (1979) playing power politics to assert its independence of colonial collaboration. At the end of January 1897, Drs. Jean Danysz and Jules Bordet, bacteriologists from the Pasteur Institute, arrived and were installed in a trial station at Waterfall, Pretoria. Theiler stated in his account that he was ordered to place himself at their disposal. Recalled to Pretoria at a few hours' notice, he left Watkins-Pitchford to continue the work.

Danysz and Bordet found that defibrinated blood from salted oxen could be used as a preventive, and in larger doses as a cure. As a preventive, two injections each of 100 cc were required, and as a cure 200–300 cc, sometimes followed by a second dose of 100–200 cc. With numerous experiments on local farms, they assured themselves of the lasting efficacy of this treatment with a mortality of only 15%. The team submitted its report on May 21 under the three signatures of Danysz, Bordet, and Theiler. A committee was formed to test the claims, and the team submitted a comprehensive report on June 18. This gave full credit to the Theiler/Watkins-Pitchford pioneering work on which the subsequent studies were based. Its immediate issue only in Dutch meant that people like Hutcheon were unable to appreciate its significance. An International Conference was therefore proposed for August, hosted by the Transvaal Government.

Koch's bile method was really simply an improvement of a method already developed by some farmers in the Transvaal known as the Waterberg or Grobler method, which Danysz and his colleagues stated was the only known method of inoculation against rinderpest at the end of January 1897. A mixture of blood, bile, and peritoneal fluid of sick cattle was used as an inoculate. The mixture was stood for 3 days, and then clean woollen wads soaked in it which were then placed under the skin of an animal.

> This process was subjected to endless improvements, each one improving it after his own fashion, and it was generally very difficult to know what were the ingredients of the mixture and in what condition it was made… the method was attended by varying results. In some instances all the inoculated animals succumbed after from 10 to 20 days after the inoculation, in others mortality was from 10 to 95 per cent, and in the remaining instances the inoculation seemed to have no result—the animals were not sick and thus were probably not salted. (Danysz *et al.*, 1898)

Koch studied this procedure and largely used it. Asked by Theiler whether he would recommend the widespread use of bile inoculation, Koch replied, "No, only where the disease had appeared" (Gutsche, 1979).

Whereas Hutcheon advocated stamping out as the best remedy, although fully concurring with immunization once a suitable inoculate had been found, Watkins-Pitchford considered that the only possible remedy *was* serum treatment, in which he was supported by Theiler. As a result, Watkins-Pitchford was permitted by the Transvaal Government, which rejected slaughter as an option, to work together with Theiler in the Transvaal on developing a remedy, work beginning under very difficult conditions on October 10. On December 12, in a letter to his wife in England, he wrote: "I am reporting by this post to the Government that our investigations are very promising and that we [i.e., he and Theiler] have produced a curative serum which will confer immunity." And on December 19, in a letter to Lloyd, Commissioner of Agriculture in Natal, he wrote that he and Theiler had "been able to discover a process by which undoubted immunity could be conferred." But on January 31, 1897, Koch reported on the use of serum, and on February 10, on serum/virus and bile. In a report to Lloyd dated February 15, the *Preliminary Report on the Rinderpest Investigation in the Marico District, S.A.R.* (South African Republic), Watkins-Pitchford referred to his letter of December 19, stating that he and Theiler were experimenting with serum/virus, and six animals had shown proof of the method's success: "I informed you that we had been able to discover a process by which undoubted immunity could be conferred." And, "It is a coincidence that the line of treatment adopted in November and December last by my colleague and myself is a similar process to that by which Dr. Koch produces his results." But he suggested that the results should not be published until furthur tests had been made. The announcement was published in a Natal Government Notice on April 5 and appeared in the *Veterinary Record* under Watkins-Pitchford's name in August. In September, *The Veterinarian* printed it with a tribute to Koch omitted in the previous version. But the work of Watkins-Pitchford and Theiler was ignored, probably because literally hundreds of so-called cures were being claimed by people all the time, and the international stature of Koch was such that the Cape Government, having engaged him on an annual salary of 10 times that which Watkins-Pitchford was drawing, had to listen to him.

The Marico camp was closed at the end of March 1897, and Watkins-Pitchford returned to Natal. Verney, who had come out from Britain to assist Watkins-Pitchford and Theiler in November 1896, stated that the closure was "owing to the behaviour of the Transvaal Government" (Verney, 1898). But Watkins-Pitchford, in his report of March 1897, wrote that he had intended moving camp to Rustenberg (British Bechuanaland) as continuance at Marico was out of the question: "Clean cattle could not be obtained for many miles, and the postal and commissariat difficulties were so great over the ninety-six miles of rough country which intervened between us and the nearest point where we could obtain supplies, &c., that it was imperative to move camp and come further down country" (Watkins-Pitchford, 1897). Theiler indicated his disapproval with the remark that he was "ordered" to assist the Frenchmen. Of the four assistants, Baxter, Stapley, Verney, and Webb, who had joined them shortly after the establishment of the camp, Verney and Webb returned to the Transvaal after a brief attempt to continue at Rustenberg; while Baxter and Stapley were sent to learn from Koch at Kimberley. Later (1936) Verney wrote:

> Conditions there were very trying and difficult, and except for one minute building of wood and iron the rest of the Camp was composed of tents. We worked there for about three months and it was under very difficult and trying conditions that Theiler so impressed one with his untiring energy, his intense keenness, optimism, and above all his great knowledge of the problem we were trying to solve. (quoted in Curson, 1936)

Watkins-Pitchford wanted to open a field station in Natal before rinderpest spread to there from the Transvaal, and introduce some infected animals to produce hyperimmune serum. This aroused opposition from the farmers in Natal and in parts of the adjacent Transvaal where the disease was absent. He was told that he could introduce infected animals to a station set up on the border when the disease was within 40 km, but the first outbreak in Natal was at Dundee on July 18, well south of the border.

On January 22, 1897 the *Eastern Province Herald* reported that Koch had found out enough about rinderpest to know that the discoveries of Dr. Simpson in India, and of Dr. Edington, were premature. He was abandoning hope of cultivation of a vaccine but would endeavor to discover a counter-poison, already having had some success in that direction.

In his report to the Cape Government, sent from Kimberley on February 10, 1897, Koch stated:

- 100 cc of serum from a salted animal was a preventive against a small quantity of virulent blood of an animal suffering rinderpest.
- Healthy animals injected with 10 cc of gall, taken from an animal that had died of rinderpest, would offer resistance to the disease 6 or 8 days after inoculation.

Koch wrote: "In my last report I was already able to inform you that blood serum from beasts that have recovered from rinderpest possesses a certain preventive property for healthy animals, when these are inoculated with it…." Danysz and colleagues pointed out that he thus confirmed the fact already discovered by Mr. Semmer (in Russia) "about a year ago."

Koch had found that 10 cc of bile taken from a diseased animal on the sixth to eighth day of infection injected into the dewlap of a healthy animal conferred immunity from the sixth to tenth day on for about 4 months, but it could be as short as 3 weeks or as long as 6 months. The disadvantages were that 3–7% of herds had to be sacrificed to provide the bile; the bile kept for a few days only, and the inoculation gave variable results, some animals failing to react. If bile was taken from animals which had died of the disease, only 17–26% was usable. Another disadvantage was that between bile collection and immunity appearing, there was a lapse of some 3 weeks. Koch's assumption that bile could not confer the disease was proved to be erroneous as, after inoculation had taken place, it broke out in many herds apparently free of infection. Nevertheless, using this method, several thousand head of cattle were saved in South Africa, in Basutoland 70,000 of 100,000, before better methods were developed. But initially, there were problems in finding suitable bile donors in some areas, and the injection of thousands of cattle would require many syringes, an item in short supply. The Cape, for example, gave only 35 to the Transvaal. An innovative Johannesburg watchmaker bought up all the bicycle pumps he could find and converted them to syringes, embellishing them with silver plating (Gutsche, 1979). Syringes were still in short supply in mid-1897, Scully (1913) recording that one of the greatest difficulties was their supply. Those sent by the Veterinary Department, the only kind available, were very brittle, breaking at the juncture of the needle with the nozzle when a beast struggled: "I sent despairing telegrams in every direction begging for fresh supplies... But none were to be obtained in South Africa." With only three syringes left, Scully solved the problem by apportioning to himself the supply of another magistrate sent to him by mistake, which were of an improved type.

Danysz and colleagues, in their report ending June 15, 1897 stated in their letter of March 17 to Government that they wrote that their main object was to procure a serum which could be applied to a herd in which the disease had already made its appearance and which would not aggravate the condition of the animals already affected.

> The result can now be looked upon as attained, and the use of blood from salted animals, as procured by us up to the present, can be applied in practice if a number of salted oxen were placed at our disposal sufficient to supply the quantity of blood which would be required from us. ... The results we had obtained ... were as follows: A. An injection of serum counteracts an injection made twenty-four hours later of a small quantity of virulent rinderpest blood. B. A previous injection of active serum allows the animals to offer resistance to the disease ... (Danysz *et al.*, 1898)

Koch discovered also that 100 cc of defibrinated blood from an animal recently recovered from rinderpest would temporarily protect another. This was subsequently confirmed by a number of other workers, but it was found that blood could transmit other infectious diseases. Attempts were made also to induce natural infection by smearing infective discharges on the noses of protected animals to induce a mild attack with a resulting lengthy immunity. But although a large number of animals were salted in this manner, losses amounted to 22–24%.

2. DOUBTS ON KOCH'S METHOD

In September 1897, Koch wrote to Kolle that the administration was sitting between two stools, as it had in his time: "because they listen to the hot air of Edington and his colleagues and are hoping to come across a magic substance that will banish rinderpest without too much expenditure of money or effort" (Mack, 1970). On his way to India in April, he wrote a discursive account to the *British Medical Journal*, stating: "Although I believe that I have solved the main problem in its essential points, there yet remain a number of secondary questions to answer, and above everything the results obtained at the experimental station have to be tested in actual practice." And: "I must regard it as a further fortunate circumstance that right at the outset I discovered an absolutely reliable method of inoculation" (Koch, 1897a).

Although it was alleged that more than 2 million bile inoculations were carried out with success between 1896 and 1898 (Hutyra and Marek, 1916), the method was to be discredited within months of Koch's departure from South Africa. It was not known how the bile could protect an animal, as it did not kill the virus. Filtered bile from sick animals was non-infective but also non-protective; yet matter remaining on the filter, even after washing twice with saline solution, although not causing infection, was protective.

Hutcheon (1902) was to write later that it was well known that

> Dr Koch entertained the opinion, and it was shared by all of us, that when cattle had been inoculated with good bile or serum, and were able, ten or more days afterwards, to resist an injection of 10 to 20 cc of virulent blood, that an *active* immunity had been conferred upon them, although they had not given any fever reaction to the blood.

> Practical experience in the field, however, very soon taught us that unless the injection with virulent blood were followed by fever reaction, it did not strengthen or prolong the immunity conferred by the bile. This misconception led to the heaviest losses after bile inoculation—both with pure and glycerinated bile–that we experienced in our operations.

3. EDINGTON AND GLYCERINATED BILE

Organized research into stock diseases began in South Africa with Edington, a medical graduate from Edinburgh, when in 1891 he set up the Colonial Bacteriological Institute at Grahamstown for that purpose. In his camp in the Taungs Native Reserve in the converted railway coach, aided by his assistants, he began experimenting under what he described as difficult conditions in the midst of rebellious natives. It was here that he devised inoculation with glycerinated blood followed by virulent blood, and glycerinated bile followed by virulent blood, there being a widespread belief that Koch's method introduced the disease into clean herds. Raupach had stated in 1864: "Glycerine modifies and then destroys the virus, as in the case of pleuro-pneumonia," but Koch dismissed the Russian work, claiming himself to show that addition of glycerine destroyed the virulence of infective blood. Edington developed the glycerinated bile method, by which infectivity of the bile was destroyed. If 0.1 cc of virulent blood was injected after an inoculation of glycerinated bile, followed 10 days later by a further 1.0 cc, a strong immunity was produced. But heavy losses sometimes occurred on account of there being no reaction following the injection of blood due to a varying degree of immunity produced by the bile, leaving animals susceptible to later infection.

In a letter to *The Lancet* of November 5, 1896, Maberly stated "the hopes entertained at one time of the method of glycerinising bile as advocated by Dr. Edington have proved fallacious," prompting a reply from Edington the following February 11 as these remarks had been taken up by Lord Lister. He pointed out that using his method, there had been only an 8.24% mortality in 393,777 cases, and immunity had persisted for at least 15 months. Edington claimed that when Koch discovered the immunizing properties of fresh bile, he, Edington, then determined to ascertain whether such bile could not be used in respect to the supposed presence of an immunizing substance. The results of experiments gave excellent results and were then proved on a herd of 126 animals at Belmont. Since then, the method was that almost solely used throughout the Eastern Province. He stated that the Cape Colony owed Koch a debt of gratitude for discovering that inoculation of fresh bile produced an immunity that might be maintained for several months, "which was the first real step toward protective inoculation…." But, he added, pure bile could transmit the disease, being dependent on the type of bile used contrary to the assertion of Turner and Kolle, the inhibiting substance to the immunizing property tending to disappear in the bile as the disease advanced. Edington claimed the discovery that the virus tended to attenuate by keeping, and this applied to virus in the bile. If the bile was used fresh, it could transmit the disease and possessed only a feeble immunizing power. Turner, in claiming that glycerinated bile had only a feeble immunizing power, "did not use glycerinated bile as I use it or as farmers used it" (Edington, 1899). He recommended in the case of a further outbreak inoculation with glycerinated bile and subsequently a further inoculation with Koch's fresh bile method.

Turner and Kolle (1897) also attacked Edington's findings, stating that Koch had first tried the method but abandoned it as useless. The only advantage of glycerinated bile was that it could be kept longer, Edington's other claims having no validity.

Nevertheless, Hutcheon was recommending the method in 1903, glycerinated bile now having been found effective after 3 years' storage, claiming that its chief advantages over serum were that it could be prepared immediately upon the outbreak of the disease at any center where serum was unobtainable and required very little skill to prepare it. Its advantages over pure bile were its keeping properties, and it could be used to more advantage in infected herds (Hutcheon, 1903). But the same year, Hutcheon gave up its use because it conferred only a passive immunity like serum but cost a great deal more, and there was no means of standardizing it or determining the length of immunity it conferred. The strength of the bile could be determined only by direct experiment. Once a serum-producing station was set up, serum became the preferred method. A dose of 20 cc of serum protected for 10 days, while 100 cc afforded immunity for 2–4 months. By allowing serum-inoculated animals to mix with infected herds in the hope that they would take the disease mildly, in several herds the disease was stopped temporarily. But many animals did not take the disease, and when it appeared again, there was heavy mortality. Serum and defibrinated blood exerted the same influence, but only 20 cc of the former compared with 30 cc of the latter was required, and blood could not be kept for long. Serum to which carbolic acid had been added was said to keep its immunizing properties for up to 9 months.

As a curative agent, serum was of little use once symptoms of the disease had appeared, but large doses of 100 cc and more could protect some animals if administered as soon as the temperature rose. Whereas serum alone became the preferred method once a system of production was instituted, it was foreseen that with eradication of the disease, the stored serum would eventually lose its efficacy, and there would then be no source of supply to deal immediately with any fresh outbreaks.

4. THE SERUM-SIMULTANEOUS METHOD

In December 1897, Turner and Kolle were given the Government cattle at the leper station on the rinderpest-free Robben Island to experiment with, and developed the serum-simultaneous method, inoculating a small dose of infective blood on one side after injecting serum on the other. Their findings were published in 1898 in a report in Afrikaans to the Transvaal Government, a German version coming out the same year (Kolle and Turner, 1898; Turner and Kolle, 1898). This resulted in 90% of animals having a more or less severe attack of the disease with a loss of only 1.5%, while those which did not react were immune for several months. During the experiments, Turner drank a pint of untreated milk each morning from the inoculated cattle to demonstrate that it was harmless. A second series of experiments was conducted at the Government farm near Capetown and also showed that the action of immunizing serum on the rinderpest virus was more effective in the body than it was *in vitro*. It therefore stimulated the body's immune reaction, not simply having a passive neutralizing effect.

The serum-simultaneous method was then widely adopted, live sheep being used as a means of conveying infected blood. In Matabeleland 3,141 head were treated with a loss of only 25 animals. Eventually, in Rhodesia alone, over 100,000 head were treated with a loss of less than 1% and the permanent protection of over 90% of them, thus eradicating the disease there. But because a number of animals experienced only very slight or no reaction, Hutcheon advised giving the infective blood 48 hr before the serum, administering the serum as soon as the temperature rose. But more animals were lost this way, and it required securing the animals twice.

There was no evidence of the disease being spread by the serum-simultaneous method. Individual herds were inoculated in many different places, and the disease had been practically eliminated in the Cape and Tulbagh districts by the beginning of 1899. But Theiler, after using the method for several years, abandoned it because of the mortalities caused due to the virulent blood being infected with other diseases such as piroplasmosis, trypanosomiasis, and babesiosis. In the 1990s, it was found that pathogenic trypanosomes can cause a delay in response to rinderpest vaccine, and in India, calves were predisposed to rinderpest due to immunosuppression by *Theileria* infection. The serum-simultaneous method was, however, the method that survived elsewhere in Equatorial Africa, Asia, and Europe, until it was replaced in the 1920s.

5. THE CREDIT GOES TO KOCH

Both Watkins-Pitchford and Theiler were given credit for serum by their contemporaries such as Hutcheon, Verney, Turner, and Kolle; but their failure to *publish* first meant that they could never receive the recognition which Koch did. An article in the *Lancet* gave credit only to the medical men who had been involved, attributing both the serum and serum/virus inoculation to Koch, "Who only abandoned them because of the great success of his bile method." Kolle and Turner "... brought the serum treatment to perfection ... the French savants ... first proposed how to produce active immunity" (Maberly, 1898). Not a word was mentioned of Theiler or Watkins-Pitchford who had passed on to them their results. But Danysz, Bordet, and Theiler, in their report up to June 15, 1897, stated: "Shortly before our arrival in South Africa Messrs. Theiler and Watkins-Pitchford discovered the same properties [i.e., serotherapeutic] in the blood. At Kimberley Professor Koch arrived at the same conclusion, but without ascribing to the serotherapeutic method that which we think is due. He declared it of less value than the process of salting with gall" (Danysz *et al.*,1898).

Danysz and Bordet, who continued working in South Africa for a period, obtained credit for what was known as "The French Method," receiving a £6,000 bonus. Theiler was given only £250 from Bordet, and Danysz, who promised an equal amount, never paid up. Neither shared any of the bonus with Watkins-Pitchford.

In 1901, Watkins-Pitchford decided to launch a claim of priority for the discovery of serum prophylaxis and therapy, but his documentation had been destroyed in an apparent arson attack on his office in the Natal Town Hall. He pressed the Natal Government, which set up a committee of investigation in June 1901; but beyond placing the findings of the Commission on the Table of the Natal House, no notice was taken of it except that its findings were published in the *Natal Agricultural Journal* (October 25, 1901). It concluded: "Mr. Watkins-Pitchford, in conjunction with Mr. Theiler, was the originator of the serum method," and this was "so important as to merit recognition." An anonymous contributor to *Black and White* of November 4, 1899 had stated "he discovered and was the first to use serum ... against the disease." In 1925, Watkins-Pitchford wrote to Sir Frederick Smith, the veterinary historian: "Theiler was associated with me in this work, although the serum system, especially the so-called simultaneous method ... rests with me ..." (Mares, 1990).

Watkins-Pitchford was allegedly bitter against Theiler, whom he claimed suppressed his work (Mack, 1970), although in his report, he stated, "I must take this opportunity of expressing my thanks and obligations to my fellow-worker, Dr. Theiler, whose steady and laborious application to the scientific treatment of the disease, under circumstances most adverse to scientific work, is worthy of much regard" (Watkins-Pitchford, 1897). Theiler, in his publications on the Marico work (Theiler, 1897a, 1898), mentions Watkins-Pitchford as sharing in it. Verney, Watkins-Pitchford's assistant at Marico, gave equal credit to him, Theiler, and Koch (Verney, 1898). Watkins-Pitchford also claimed to have put in a claim for the million-rouble prize offered by the Russian Count Orlov-Davidoff for the first person to find a cure, but the Royal College of Veterinary Surgeons would not support his claim, and the prize was never awarded. It was in fact for 10,000 roubles, and not one million as Watkins-Pitchford claimed, and the closing date was January 31, 1897, before he announced his results (Mares, 1995).

Verney wrote highly of Theiler, his untiring energy, and his knowledge of the subject, of his being entirely conversant with the Russian work of Semmer which formed the basis of the experiments, "There was no literature on rinderpest which he [Theiler] had not perused, and he was entirely *au fait* with the work the Russian worker (Semner) [*sic*] had done, and which formed the basis of the numerous experiments we carried out, and it was here the foundation work of serum and virulent blood was established" (Verney, 1936 quoted in Curson, 1936). But the method finally evolved in the Transvaal became known as "The method of the French doctors," referring to Bordet and Danysz (Mack, 1970).

Theiler's report to the Transvaal Government was dated June 30, but it appears that no notice was taken of it, perhaps because of that of Danysz, Bordet, and Theiler which appeared on July 21 and was put into effect.

At the International Conference convened at Pretoria from August 2 to 13, 1897 to promote 'the French method' of Danysz and Bordet, the Natal Commissioner of Agriculture C. B. Lloyd left a document to be read after he and Watkins-Pitchford were obliged to leave:

> The Congress will observe that Drs. Danysz and Bordet from the time of their arrival were placed in the favourable position of having at their disposal the result of four months' continuous and successful labour conducted by Messrs Theiler and Watkins-Pitchford. I have now read the Report which has been presented to the Conference by Drs. Danysz and Bordet and so far as I can see, there is but little difference between the results communicated in that report and those obtained by Messrs Theiler and Watkins-Pitchford in 1896. The latter gentlemen obtained certain results but could not take advantage of them, Mr. Theiler because he was recalled by his Government and Mr. Watkins-Pitchford because, up to the midde of June last, there was no Rinderpest in Natal. No doubt the Congress will appreciate the labour performed by each of the experts in the case of the sero-therapy of the Rinderpest in South Africa.

Verney wrote in 1898, "In attempting to find something valuable as a prophylactic or curative agent for rinderpest, Messrs Watkins-Pitchford and Theiler paid particular attention in the direction of serum-therapy, a treatment destined to save thousands of cattle in South Africa." He explained that the methods in use were "Dr. Koch's or the bile method, and the serum or Watkins-Pitchford–Theiler method." The wish of the Veterinary Department had been to use serum, but at first this was impossible as there were no "salted" cattle in Natal, and consequently bile had to be used. Stapley and Baxter having observed the method at Dr. Koch's camp at Kimberley, the method was applied as instructed, but "as time went on the work did not progress as satisfactorily as one could have desired, and experience has shown us that the bile method as taught by Dr. Koch is a failure." It had been the means of introducing rinderpest into areas where none had existed and had thus caused the loss of thousands of cattle, "... one cannot help being surprised at the great errors which Dr. Koch committed regarding its effects and actions." But he ends his paper with the conclusion "The results achieved are due to the valuable experiments carried out by Dr. Koch, and Messrs Watkins-Pitchford and Theiler" (Verney, 1898).

Watkins-Pitchford (1897), in his report, stated:

> Working upon the same lines of treatment as that adopted by my colleague and myself in October last, the success of Dr. Koch is remarkable ... However it is produced, the success of this famous *savant* must be a matter of immense importance and rejoicing to the whole of South Africa. Those alone who have seen the fearful ravages and devastation caused by the disease can fully appreciate the immense benefits which the country will reap from the successful labours of the great scientist.

The situation was clearly explained by Hutcheon in 1902, who wrote that in January 1897, in the early stages of his experiments at Kimberley, Koch "verified the fact previously discovered by Semmer, Nencki, Sieber, and Wyznikiewicz [at St. Petersburg in Russia], that serum obtained from cattle which had recovered from rinderpest—when injected into susceptible animals—gave them an immunity against the disease when given in large doses of 100 cc or more, but this immunity was only passive in its nature and temporary in its effects." To increase its immunizing powers and reduce the dosage required, Koch added 1% of virulent blood. This mixture acted much more satisfactorily, but just at that time, he discovered the efficacy of bile, which he considered was both safer and more effective than serum, and so abandoned furthur investigations into serum.

> Contemporaneously, experiments with serum were being conducted by Messrs Watkins-Pitchford and Theiler ... on similar lines. Just at this time Drs. Danysz and Bordet ... arrived in the Transvaal, and commenced inoculation experiments ... they devoted their attention principally to the perfection of the serum method of treatment, both as a preventive and curative agent; and the result of their experiments was very satisfactory ... The principal advance which [they] made in the application of the serum treatment to rinderpest, was the conception of the idea of communicating the infection to the serum-inoculated cattle, so that they contract a mild form of the disease, under the modifying effects of the serum, from which they recover and become salted. (Hutcheon, 1902)

Hutcheon (1902) also referred to Behring's "important discovery" being generally known at the date of the outbreak of rinderpest in South Africa (Emil Behring, professor at Marburg and the founder of serum therapy had been a student of Koch). This was that if toxins of a pathogenic organism were injected in proper quantities for a sufficient length of time into a higher animal, the blood serum of the animal acquired specific anti-toxic properties, but the toxins did not neutralize, or render harmless, the toxins of any other species of microorganism. It was also known that the blood serum of an animal which had been rendered immune against a specific toxin, when given in gradually increasing doses: (1) if injected into a susceptible animal rendered it resistant to that disease; (2) injected into an already affected animal, brought about a cure, and (3) when injected before, at the same time, or shortly after the toxin, tended to neutralize, or antagonize the latter.

In stating this, Hutcheon was probably quoting from the report of Danysz, Bordet, and Theiler, who wrote that they had found that the best method of grappling with rinderpest had as its basis the principle of serotherapeutics, "a principle, as is known, with which the world has been acquainted for many years." They then went on to explain the principle, ending with the words: "To everyone who is acquainted with this important principle, it must naturally occur to apply it to rinderpest, and try to protect the animals against the epidemic by means of injecting serum procured from cured and salted oxen. Mr. Semmer had made known in Russia that the serum of salted oxen protects the healthy oxen against rinderpest to a certain degree" (Danysz *et al.*, 1898).

Watkins-Pitchford had of course first promoted the idea of serum therapy, but Hutcheon did not apparently see that any one man could be singled out as a discoverer of the remedy; rather, it was a coming together of minds, with Koch developing, rather than discovering, the most useful method. In 1902, Hutcheon wrote that if rinderpest should become enzootic again in South Africa, prevailing in a virulent form over wide areas, an immunity could be conferred only by an injection of (1) pure bile or (2) glycerinated bile, followed by pure bile, or (3) simultaneous injection with serum and virulent blood. But there were strong objections against the use of virulent blood because it transmitted other diseases. But he then went on: "It is upon serum, however, that our principal reliance must be placed ..." (Hutcheon, 1902). This was because it could be produced in large quantities and kept for long periods; its strength could be ascertained before being issued for use; its action was immediate; it was also curative in early stages of the disease; and it did not communicate other diseases.

At the end of 1902, a serum-producing station was set up in Orange River Colony, and moved to Cape Colony at the end of the Boer war. It was not until October 1902, with the creation of this station, that Hutcheon was able to attack the disease in a systematic manner. It was then prevalent in 27,235 cattle spread among 22 districts. Using serum inoculation, the number of infective centers declined until the disease finally disappeared in April 1903. An outbreak involving several thousand native cattle belonging to the Barolong tribe was also dealt with in the same period. In some of the latter herds, where the disease had assumed a chronic form, a second dose of serum was given. The last dose of serum was issued by the station on April 2, altogether over 89,000 having been issued, the bulk to the Herschel and Kimberley Districts. In July 1903, the station was closed, although some serum was stored. With hindsight, Dr. P. J. duToit, Director of Veterinary Services in the 1930s and later, considered that the attempts to vaccinate had been useless, and disappearance of the disease was due largely to exhaustion of the supply of susceptible cattle (Scott, personal communication).

6. THE DEVELOPMENT OF INOCULATION IN INDIA

Inoculation produced as severe a reaction as natural infection and was not recommended by the Indian Plague Commission; nonetheless, it was introduced in 1899, following a visit to Muktesar in June 1897 by Koch in connection with the work of the German Plague Commission. While at Muktesar, Koch was asked to demonstrate his bile method and did so with the help of Dr. Lingard, the Government Bacteriologist. But because the virus was obtained in the hottest month of the year, it soon lost its virulence. Koch also inoculated some animals earlier than the method required because he had to leave. The results were thus considered by Lingard to be inconclusive. Improved laboratory facilities being completed by 1899 experiments were continued by Dr. Leonard Rogers in the absence of Lingard on sick leave in England, using bile, inactivated vaccines, serum, and virus serum alone.

Koch's bile method gave immunity to plains cattle for $3\frac{1}{2}$–5 months, results agreeing closely with those obtained in South Africa, Russia, and Turkey. But it gave little or no immunity to the Kumaon Hill breed of cattle. As immunity began only after 10 days, the rapid spread of the disease among hill cattle negated its use. Raymond (quoted by Rogers, 1900) alleged that bile from buffalo calves in Bengal gave immunity lasting up to 1 year, but this was not confirmed. Filtered bile had the same advantages and disadvantages but could be used 7 days earlier, 75% of all biles could be used, and double the number of animals could be inoculated. With glycerinated bile, about two thirds of plains cattle obtained immunity for 3–4 months, the remaining third obtaining a much longer protection after a somewhat severe reaction. But the lag period of almost 3 weeks from collecting the bile to the achievement of protection negated its effectiveness. With hill cattle, it gave only one satisfactory result in 12, and nine animals died following injection of virulent blood 10 days after the glycerinated bile injection, so it was almost as fatal as inoculating with the disease. Inoculations of serum alone had the same advantages as bile but gave full immunity immediately, thus gaining 12 days, although immunity lasted only a few months. In the serum-simultaneous method of Turner and Kolle, the hill cattle required many times the amount of serum in proportion of the weight of the animal than did plains cattle. Although the method probably conferred a lifelong immunity to those animals in which it was successful, it was not reliable, but where there was no reaction, this could be overcome by a second inoculation of blood 7–10 days later.

Practical objections to the bile method were that it required the sacrifice of some 7% of healthy herds to provide the bile which would be impossible due to religious objections. The use of bile from dead animals was not possible as one could not know exactly on which day they had died, and even if it was known, villagers might object to a postmortem.

7. SUCCESS IN THE 20TH CENTURY

Bile was taken on the the sixth, seventh, or eighth day of the disease, and 10 cc injected into susceptible animals. In a few cases, some suffered an attack of rinderpest, but the immunity, even without an attack, lasted from 10 days to 4 months, apparently according to whether a passive or active immunity was conferred. The bile was supposed to contain immune bodies of which the transference gave some temporary protection, and virus which produced the active immunity. Subsequent tests elsewhere 30 years later gave variable results (Daubney, 1928), seemingly depending upon the varying virus content of the bile during the course of the disease.

At the end of the 19th century, French workers in Turkey also worked on the South African developments, Nicolle and Adil-Bey (1899) of the Imperial Bacteriological Institute of Constantinople reporting that 25 cc of serum from a salted animal, fortified at once by a single injection of 400 cc of virulent blood, would protect against an injection of virulent blood in a few days and against natural infection. This method was used on healthy animals in several outbreaks with mortalities of 0–0.14% (0–1.7% in the incubation period), compared with mortality among herds previously of 44.4–81.5%. Although the immunity passed off in a few months, it was sufficient to protect herds throughout particular outbreaks (Réfik-Bey and Réfik-Bey, 1899). The bile method was distrusted and not used if serum was available.

In the third edition of his *Manual of the More Deadly Forms of Cattle Disease in India*, Hallen stated that a preventive inoculation was available which was extremely effective (Hallen, 1903). Immunity did not last for long but was immediate. Permanent protection was afforded by another method in which animals were made to suffer a slight attack of the disease for a few days, but great care and accuracy were required in determining the dose. Lingard conducted a number of experiments between 1900 and 1905, based on Kolle and Turner's work using Koch's fresh bile method, to confer immunity on plains bovines and hill buffaloes but not on the more susceptible hill breeds of cattle. He also tried Edington's glycerinated-bile method, the serum-simultaneous method, serum alone, and defibrinated blood; each of which had its disadvantages. Finally, he experimented with milk, egg yolk, fish sperm, spawn cell protoplasm, mucin, vegetable seeds, yeasts, and cattle testes. The best results were obtained from a solution derived from macerated cattle and buffalo testes. A preparation of this conferred a marked resistance on plains but not on hill cattle, but the serum of plains cattle which had been treated with the preparation conferred a resistance on hill cattle. Oil of pumpkin seeds produced similar results (Lingard, 1905a, 1905b).

Between 1900 and 1905, 468,853 doses of serum were prepared with mortality of inoculated cattle only 0.43%. Lingard proposed that a serum-producing laboratory be set up at Muktesar, and in 1914–15 its production was 1,388,560 doses, with an average issue of over 500,000 cc of serum per month.

Following a widespread eruption in Egypt in July 1903, serum was at first imported from Constantinople for use in the serum-simultaneous method using it at a ratio of 25:1 cc, but this resulted in losses of 80–90% due to a piroplasmosis. Serum totaling 25,000 doses was then imported from South Africa before a serum-producing institute was set up in September of that year at Abbassia, south of Cairo, operating until September 1905 when the outbreak terminated (but it later reopened and was active in 1925). During the course of its initial operation, 418,320 doses of serum of 50 cc each were produced. In 1904, a total of 125,000 cattle were lost, and deaths continued, ranging from 8,335 in 1908 to 50 in 1924. It was found that Egyptian cattle had a partial immunity to the disease not considered due to previous infection (Todd, 1912). From 1912 to 1926, a total of 700,000 vaccinations were effected, of which 90,000 were serum inoculations accompanied by smearing infective nasal discharge on the nasal area. Overall losses during this period from vaccination were less than 1%, but from 1926, losses were entirely due to vaccination. Imports of cattle from the Sudan, Syria, and Asia Minor meant a constant threat of infection, which could only be met by vaccination of the entire cattle population.

With all methods, there were problems in the amounts required, and there was difficulty in keeping serum, especially in a tropical climate, although in 1914, Holmes at Muktesar found that the serum did not become altered in potency by short exposures to high temperature, and it could also be sterilized without detriment (Holmes, 1914). Head (1906) reported that heat seemed to have no effect on the strength of the serum, as most of what he used to combat an outbreak in the Sudan had been carried for months on camels in the sun with the temperature reaching 46°C in the shade. Following Semmer establishing in 1881 that the virus was killed by heating at 55°C, Piot, a veterinary surgeon in the Egyptian Administration, announced in 1883 that he had succeeded in attenuating the virus by heat to the point where an inoculated animal received only a slight attack of the disease and thereafter became immune. Gordziaklowski in Poland then treated 28 military cattle in 1921 at Brzesc-Litowski where rinderpest had erupted, by inoculating with heat-treated blood. Three died, and the remainder allegedly became immune, for, inoculated 2 days later with virulent blood, all survived. Gordziaklowski then treated another 120, allegedly with the same success. But Piot-Bey (1922) disputed the results, claiming that heated virus was entirely innocuous. He explained the three deaths by the fact that the premises were infected. Curasson (1932) states that, following orders of the Governor General (apparently apprized of this work), it was tried by several veterinarians in French West Africa and shown to be totally ineffective.

From 1906, a group of Japanese workers working in Korea made a breakthrough successfully using inactivated virus obtained from infected pulped spleen, although the work was not published until 1918, and then in Japanese (Kakizaki, 1918). Using highly susceptible Korean cattle, Kakizaki's aim was to attenuate the virus for vaccination purposes. He claimed that Takazawa (apparently unpublished) was the first to diminish the virulence by passage through pigs. Kakizaki added glycerine to spleen and defibrinated blood, but the proportion of glycerine was not stated. When injected, no dangerous ill effects were produced. The avirulent virus in glycerinated infected spleen appeared to retain some antigenic properties even when kept in glycerine for over a year. The paper was summarized in English in the *Tropical Veterinary Bulletin* the same year (1918), but Curasson (1932) wrongly states that it was not known to western veterinarians until 1924 when a paper on the same subject appeared in France. In this year, Kakizaki published a review of his work, his experiments covering an immense amount of ground, but few experiments were carried out in each section, and in many instances, there were no controls. He found that an emulsion of spleen in glycerine may retain its antigenic properties after it has become non-infective, but its use appeared inferior to the serum-simultaneous method. In 1926, the work of Kakizaki and Oizumi was analyzed in the *Bulletin de l'Institut Pasteur*, showing that the method had been widely used in Japan since 1906, but the method of inactivation was not clear (Kakizaki *et al.*, 1926). At the same time (1926), two French workers successfully used formalised (i.e., treated with formalin) rinderpest virus obtained from spleen (Curasson and Delpy, 1926), these experiments having been inspired by the success of using formalised FMD virus by French workers. Daubney in Kenya also successfully used formalised spleen preparations, vaccine prepared from formalised spleen pulp conferring a high grade of immunity upon inoculated animals provided that sufficient was administered. Blood and liver pulp did not provide immunity, but kidney and lymphatic gland did to a lesser extent (Daubney, 1928). In Niger, a French veterinarian, Doutresoulle, inoculated calves of immune dams but the wide variation in the immune status of the dam resulted in variable results (Doutresoulle, 1924). Kakizaki and coworkers also claimed success with a vaccine developed from extracts of thymus and tonsil superior to that derived from spleen, which was similar to lung extract. They claimed to have used their vaccine in Manchuria in 1924 where, in one village with 7,098 animals, during a 1925 outbreak, only one died (Kakizaki *et al.*, 1928).

In the Phillipines, 43,000 cattle were vaccinated between 1914 and 1920, but there were large losses of suckling calves. American workers in 1928 announced the use of a vaccine in place of serum comprising a preparation made from infective lymph glands, spleen and liver, which gave promising results, a single dose of 20 cc giving immunity to carabao and 10 cc to ordinary cattle (Brown, 1929). But it was the development of caprinized rinderpest virus in 1927 by Dr. Edwards, then Director of the Imperial Institute of Veterinary Research, India, that was the first important advance in the development of a highly efficient rinderpest vaccine. Toward the end of 1926, he had begun experiments to "fix" the rinderpest virus in sheep and goats with the aim of producing an inoculate free of other infections, such as piroplasms. In the course of these, he discovered prolonged passage in goats attenuated the virus. There was now no need to inoculate the large amounts of serum, up to 300 cc, which had been required, although he advised continuing with a small amount when injection of the attenuated or "caprinized" virus was given (Edwards, 1930). This was used largely to control outbreaks from 1935 in a wet form and, together with the development of lyophilization or freeze-drying techniques in the late 1930s, revolutionized the control of rinderpest and led to a dramatic fall in global incidence of the disease, opening a way to its control in Africa (Daubney, 1953; Edwards, 1949).

But tested in Kenya by Daubney, the Indian Muktesar goat virus after 9 years' adaptation still required a simultaneous serum injection to be considered safe for East African cattle. Daubney therefore developed a Kenya goat adaptation in 1938 after 160 passages, rejecting the Indian virus after 149. Continuing passaging of the Kenya attenuated goat virus (KAG) was maintained and was to give protection for the whole of East Africa over the next 20 years. But as the quality of European stock rose more and more, the KAG vaccine was considered too severe, and the answer was found in lapinized (rabbit) vaccine brought from Nakamura in Japan in 1947, a freeze-dried product being developed at Kabete in Kenya by Brotherston. Daubney had suggested freeze-drying in 1937 as likely to be useful (Macaulay, 1973).

Nakamura had visited Muktesar from Korea in the 1920s and was impressed by Edwards's technique of serial passaging in goats. On his return to Korea, he initiated similar trials using a local bovine strain of the virus. As adequate numbers of goats were not available, he tried rabbits instead, with great success. Serial passage in rabbits also fixed the strain, and after 600 passages, it was even more attenuated than Edwards's attenuated goat strain and Daubney's KAG. In the 1930s, Nakamura further attenuated his lapinized vaccine by serial passages in embryonated hens' eggs. His live lapinized rinderpest vaccine was first used to control rinderpest in Mongolia in 1941 (Scott, personal communication).

But prior to Nakamura's development at the outset of World War II, considerable alarm was raised at the possibility of biological warfare, leading to many conferences in America which resulted in a study commission nominated by the American Secretary for War and the Canadian Minister for National Defence, specifically to research defence methods against rinderpest. Work began in early 1942 at a secure location in the mouth of the Saint Laurent in Canada, with the joint aims of (1) preparing large quantities of vaccine according to known practice and (2) researching a more economical vaccine which also involved less animals in its preparation. The team consisted of six U.S. Army Veterinary Corps officers, one U.S. Navy Medical Corps Officer, two Canadian academics, and a body of technicians from different countries. The research resulted in the development of an "avianized" vaccine cultured with chicken-egg embryos. Research continued in secret until 1946, when large quantities of vaccine were released to the United Nations Relief and Rehabilitation Administration (UNRRA) which were used in China and Africa (Stasiw, 1947). Despite its initial promise, avianized vaccine had certain drawbacks; it was necessary to demonstrate the presence of the virus during the course of production, and a level of virulence persisted which prevented use with very susceptible animals. Some workers also found difficulty in maintaining the source virus. As a result, lapinized vaccine became the vaccine of choice.

Attenuated viruses do not spread infection because they are not expired in the air or passed in the feces, and having lost their affinity with the epithelial surfaces, they induce the body's immunological reaction without destroying its tissues. Control was further aided in the 1950s by development of the Plowright tissue culture adapted virus (PTCV), a lyophilized modified live virus vaccine based on the Kabete "O" strain of virus (RBOK) at the 90th to 120th bovine kidney cell culture passage, which is the most attenuated and provided immunity for the practical life of the animal. But vaccination of each animal must be repeated annually for at least 2 years to ensure complete coverage of a herd, and calves are not vaccinated before the age of 6 months as the PTCV is neutralized by the colostral antibody. In Africa, more than 120 million cattle were vaccinated with PTCV, and global incidence of the disease reached its lowest known level in 1976 when its presence was reported from only three countries (Scott *et al.*, 1986). One advantage is that rinderpest exists as a single immunologic type, and vaccines, irrespective of their origin, protect against all known strains of the virus.

However, the vaccine required sophisticated facilities for its production and had to be kept refrigerated, drawbacks to operating in a country such as Africa. Transported long distances in tropical temperatures, there is a risk of injecting deactivated vaccine. Vaccinating half-wild cattle, which often start fighting when herded together, could also result in unsuccessful vaccinations, an animal perhaps escaping before a complete dose was administered. Because animals under 6 months of age are protected by maternal immunity, which can neutralize the vaccine, it is also necessary to repeat vaccination annually until the disease is eradicated. The disadvantages of production and the need to keep the vaccine refrigerated prompted a continuing search for an improved vaccine, and a new hybrid virus was created by genetic engineering of vaccinia virus, the same virus used as a smallpox vaccine which had attracted the attention of Jenner in 1794.

Microbiologists in the veterinary field were quick to use rinderpest virus in their early recombinant trials. Plowright's cell-culture-adapted rinderpest virus vaccine had all the desirable properties of being safe to handle because it was noncontagious and had a known high immunogenicity. Research groups in America, Europe, and Japan simultaneously chose vaccinia virus as the vector because it was large, and most of its molecular biology was known, but it was wrongly considered to be safe, although its use in America, Europe, and Japan was banned because of the risk of generalized vaccinia and death in unimmunized persons. All the groups successfully produced vaccinia-rinderpest recombinant vaccines expressing F- and H-genes of rinderpest virus that protected cattle against challenge with virulent virus, but multiple injections were necessary, and the duration of immunity conferred is not lifelong. At the United States Department of Agriculture Plum Island Centre off Long Island, the only facility in the United States permitted to work with exotic animal viruses, the hybrid virus had transplanted into it the genes which code for two proteins to coat the outside of the rinderpest virus, expressing the H protein gene of rinderpest virus, using the LC16mO attenuated smallpox vaccine strain as the vector. Administered by a simple skin-scratching method, cells at the inoculation site become infected with the engineered virus and produce rinderpest proteins, stimulating an immune response which gave total protection in experimental trials. All cattle vaccinated with either recombinant or with combined recombinants produced neutralizing antibodies against rinderpest and were protected against it when challenged with more than

1,000 times the lethal dose of the virus (Yilma *et al.*, 1988). One advantage was seen as the scab produced by vaccination of one animal could then serve to vaccinate 200,000 others, and the scabs could be stored dried at room temperature. But this stimulated a future President of the New York Academy of Sciences to comment in *Nature*: "... in plans to immunize hundreds of millions of cattle with recombinant vaccinia expressing rinderpest antigens with perhaps 100,000 vaccinators, there is bound to be adventitious infection of human beings with the recombinant vaccine. It will be essential that they are preimmunized against vaccinia virus." But when vaccinia virus had been administered to U.S. military personnel, a number of whom were later found to be HIV-positive, one of the latter died of generalized vaccinia, and he added "... one cannot help but wonder ... why anyone in the world is being vaccinated against smallpox" (Bloom, 1989). Thus, use of vaccinia-based recombinant vaccines (or any other virus-based recombinant vaccine) in Africa is opposed on grounds of human safety as any immunocompromised person (e.g., HIV-positive) might react to the vaccinia component. In the United Kingdom, meanwhile, a capripoxvirus recombinant vaccine expressing genes from the virus of PPR was produced and later a capripoxvirus recombinant using rinderpest genes, which if necessary could replace the vaccinia-rinderpest recombinant because the capripox virus does not infect man. Other recombinant vaccines which are undergoing trials have since been developed in Japan (Asano *et al.*, 1991) and the United States (Giavedoni *et al.*, 1991) using attenuated vaccine strains. One expresses both the H and F protein genes of rinderpest virus using the Wyeth smallpox strain as vector. The other is a mixture of two separate recombinants which expresses either the H or the F protein gene of rinderpest virus, using an attenuated vaccine strain of capripox virus as a vector. Not being a human pathogen, the same safety problems do not arise, but its drawback is that reduced protection is given when animals are first vaccinated with any modified live rinderpest vaccinia virus, indicating an interference due to pre-existing antibodies.

At the same time, in 1990, a thermostable Vero cell-adapted vaccine (i.e., using green monkey tissue cell culture) was developed from the RBOK strain (Mariner *et al.*, 1990). Stabilized with lactalbumin hydrolysate and sucrose, it was found to maintain potency for more than 20 weeks at 37°C, and an expiry period of 30 days at average ambient temperature was established tested successfully under field conditions in Niger. This overcame the logistical problems of maintaining a cold chain in countries such as Asia and Africa (Giavedoni *et al.*, 1991) with an estimated saving in the latter country of $3.6 million per year. It has thus become the vaccine of choice because it stimulates early interferon production within 2–3 days of being injected, long before antibodies appear, which recombinant vaccines fail to. While such a vaccine is available, there is a reluctance to commit resources to testing new vaccines in the field, but Barrett (1996) sees the role of recombinant vaccines as providing an easy stepping stone before vaccination is finally abandoned in a country because their seroconversion can be distinguished from that of natural infection, which is not possible with an ordinary vaccine.

Experience in Kenya and Tanzania in 1997–8 showed that immunosterilization could be readily achieved by two consecutive rounds of vaccination within 6 months of all cattle of all ages. Recent success in rescuing live negative-strand viruses from DNA copies of their genomes has made it possible to consider the use of these viruses as vectors for the delivery of foreign genes (Yamanouchi *et al.*, 1998). Unlike the genomes of positive-strand viruses, these genomes are noninfective and can create long-lasting immunity. Attempts are being made to develop a FMD recombinant vaccine using the rinderpest virus genome. The other advantage is that marker genes can be placed into the genome, making it possible to easily distinguish vaccination from natural infection, and thus using this type of vaccine would not interfere with serosurveillance, enabling detection of hidden disease without placing a large population of potentially susceptible animals at risk by ending vaccination. In 1998, the development of an enzyme-linked immunosorbent assay (ELISA) test which previously required chilled serum, using dried blood on a filter paper, which withstands heat and light and does not require cold storage, seemingly revolutionized detection of rinderpest antibodies in animal blood (Heller *et al*., 1998), but some results have indicated that the test may not measure antibodies against African Lineage 2 virus as efficiently as those produced against the vaccine strain, placing use of the test in doubt. Limited experimental data suggest that the ELISA test is adequate, but further tests were awaited (Anon., 2000a).

Scott (personal communication) considers that recombinant vaccines have arrived too late. Mass vaccination campaigns have been stopped worldwide, and strategic vaccination using live virus vaccine is permitted only to create cordons around the few pockets of known infection. The majority of countries previously infected with rinderpest have declared their intention to eradicate the disease, which commits them to stopping the use of rinderpest vaccines after 2 years if no infection is detected during that period. Cessation of mass vaccination recognizes the fact that vaccination programs which achieve immunity rates below 70% cannot be relied on to result in eradication of rinderpest from extensive pastoral areas. In free areas, such vaccination does not prevent the spread of rinderpest but serves to slow it, making the disease more difficult to detect. However, it was recommended in June 2002 (Anon., 2002a) that recombinant and other new generation vaccines which enable the differentiation of vaccinated animals from those infected naturally, be investigated with a possible use to eliminate remaining foci of infection identified in Somalia.

Part V

THE HISTORY OF RINDERPEST IN ASIA AND AFRICA

20

RINDERPEST IN INDIA

1. INTRODUCTION

Despite the prevalence in India of a number of different stock diseases, rinderpest caused more deaths than all others put together, and annual cattle losses in the 1860s were estimated to run into hundreds of thousands. A century later, this had been reduced to 200,000. The interest of India in the study of rinderpest is that it provided an example of an enzootic situation, with epizootic proportions almost concurrent with the 1865–7 outbreak in Britain. Beginning largely in 1869, in 1870, epizootics were reported in Mysore, Madras, Bengal, Oudh, the North-West Provinces to the foot of the Himalayas, Ceylon, and Burma. The maps shown in Figures 1 and 2 reflect the sphere of British reporting influence, rather than the subcontinental incidence of the disease.

Cattle murrain was attributed to any of three diseases, that most closely resembling rinderpest being an unidentified pox virus *puschima* with mortality rates of 50–95%. Often regarded as malignant, a disease called cowpox and known by the same names as human smallpox (*bussunto, mhata*, or *gootee*),[1] could not have been cowpox, now known to be restricted to Europe and adjacent regions of Russia. In Calcutta, *gootee* was distinguished from *puschima* (bloat), which occasionally attacked cattle but was more fatal than *bussunto*, and death was quicker than from *gootee*. FMD was always distinguished. Rinderpest was recognized by various names all over India, Hallen (1871) listing 121, far outnumbering names for other diseases. In the Bombay area, it went under the name of *hiney bialee*, the great, or worst, disease.

In 1838, the Committee of the Provincial Medical and Surgical Association maintained that the malignant distemper among cattle was the same disease as that described by Ramazzini and other Italian writers, that is, rinderpest. The symptoms of *mhata* described in Moorshedabad in 1832 and Dacca in 1836 left "little doubt of their similarity to those of rinderpest." The report continued: "Whether the pustules are numerous or rare, the disease is almost equally fatal, and unless measures are taken to separate the diseased from the healthy, it speedily runs through the whole herd, sparing few. ... Many die before the eruption makes its appearance, but the fever is known by the discharge of the saliva, &c." (Anon., 1838).[2] In 1848, Dr. William Gilchrist published *A Practical Treatise on the Diseases of the Elephant, Camel, and Horned Cattle, with Instructions for Preserving their Efficiency*, founded on his work at Hoonsoor Cattle Farm. In it, he pronounced the *burra azar* of Madras to be the same as the murrain of Europe, which, during its epidemic visitations, "occasions as extensive a

[1]The modern renderings are *gobussunta* and *mata*.

[2]Rinderpest attacks the immune defences, activating any latent pathogens such as pox and *Dermatophilus* spp. If these skin lesions are present, the prognosis regarding rinderpest is good.

Figure 1. Some first records of rinderpest in India and neighboring countries.

mortality amongst the herds of Asia as amongst those of the Western World." He believed that it might occur at any season, no one part of the year being more influential than another for its development (Gilchrist, 1848).

Twenty years after Gilchrist published his book, the outbreak in Britain became familiar to many. In October 1865, Dr. C. Palmer stated: "I am convinced that the disease now raging in England is the same we had here in 1864. I am persuaded that it is a disease which for years existed extensively in India, but which has up to this time not been recognized, and I am hopeful that it is amenable to treatment." (Palmer, 1865). Mr. Gudgin, a veterinary surgeon, after studying an epizootic in Burma in 1866, arrived at the same conclusion. Another, Mr. C. Barrow, veterinary surgeon to the 19th Hussars, who had observed rinderpest in Cambridgeshire as an inspector in 1865, concluded that an epizootic in Benares in 1869 was analogous and, based on his previous experience, made proposals for the military cantonment as recommended by the Royal Commission. But another veterinary surgeon, Mr. Stanley, allowed only that it somewhat resembled rinderpest, owing to its milder character and absence of some characteristic phenomena.

The periodical prevalence of rinderpest was known from the 18th century, and while there was an apparent growing incidence, this was due to greater knowledge and awareness from the 1830s rather than a real increase in frequency, although some observers suggested

Figure 2. India and neighboring countries *c.*1871.

that outbreaks had become more frequent due to facilities for locomotion, and an increasing great demand for hides thought to disseminate the disease. A change to arable agriculture had meant that by the 1860s, stock were severely mismanaged and weakened in their capacity to resist disease. One villager stated that there was almost always cattle disease in some village or other, and it came in a more severe form every 3 years. A Calcutta dairyman asserted that *gootee* came on every 2–3 years, and all agreed that it broke out in the dry months of January to March. Another informant, aged 70, stated:

> We have *gootee* every year and I lose about the same number every year; 10 or 15 in a herd of 60 to 70 cows generally recover. The disease comes one year in a mild form and in another in a severe form.... *Gootee* has been known in this place from time immemorial; but for the last 10 or 15 years it has been more severe and frequent. In this locality the number of cattle is being reduced. This is owing to the heavy taxes imposed upon us every day. (Hallen *et al.*, 1871)

Of one village in Furruckabad District, of 128 head of cattle in September 1868, 100 were attacked, 4 died, and 2 were sick. Villagers stated that they had never seen the disease in the village before, one man being 60 years old, but there had always been FMD. The cattle grazed with those of an adjoining village which had the disease first. When infection broke out, the villagers almost always knew whence it had come; usually, its prevalence in an adjoining village was communicated by the association of cattle on the pastures, but they invariably claimed that they had no idea of its cause, only one bizarre example being quoted by the Indian Cattle Plague Commission of it being produced by a tiger having urinated in the vicinity of a cattle herd. But many farmers considered that the poison could be carried by the air and the disease then spread by contagion. One belief was that it was spread by kites and vultures, another by rivers. Wallace (1888) dismissed both ideas, and neither was considered by the Commission. Pease (1894) stated that there were certain districts where the disease never ceased from one year's end to another, sometimes apparently dying out then bursting out again with renewed intensity. Furthermore, cattle owners throughout India and in the remotest parts of the Himalayas were aware that an animal, once attacked and recovered, did not take the disease again, and such animals were worth two or three times the value of unprotected animals.

The Assistant Superintendent of Police, Sasseram, Shahabad District, reported in 1870 in that district that it was largely confined to villages at no great distance from the hills. One village informant stated that during floods, the cattle and buffaloes were sent to the Chynepore (Rajmehal) Hills for pasture, where sometimes in the rains they got *gootee*, and many died, the disease being more general and more severe there. Other cattle brought the disease back to the villages. Sultanpore in Bhaugulpore Division, which grazed its cattle in the Nepal Terai in the first half of the year, was "famous" for *gootee*. People of Kishengunge believed that the hills and Terzai, in the neighborhood of Darjeeling and Julpigoree, were the foci of the disease which each year spread out regularly from northeast to southwest.

In Bancoorah District, when stock were attacked in a village, the people moved the apparently healthy animals to another, thus spreading the disease if any were already infected. In one case, spread was attributed to the cattle of pilgrims, in another to those of traveling merchants, and in yet other cases, climate was blamed. A case was reported in Assam Division of spread by a man who had lost two of his bullocks from the disease which was raging in his village. He went to purchase another bullock in a village 5 or 6 km distant where there had not been any plague. He looked at one and handled it all over but did not buy it. Three days later, the bullock was taken ill, and the disease spread through the whole village. Examination of the carcass of the bullock confirmed that it had died of rinderpest (Farrell, 1870).

A senior official replying to a questionnaire of the Indian Cattle Plague Commission in 1870 stated: "It seems that no epizootic disease has of late years attacked cattle in these parts, except the well-known diseases of small-pox and *khooree* [FMD]. The former disease is locally known as *matah* or *morai*" (Bernard, 1870). The inference from this statement is that the diseases were so common that no notice was taken of them, although enzootic effects ranged from benign to highly virulent, with mortalities from 5 to 100% averaging 50% of those attacked, and with a duration of 2–15 days.

Its mildness in most years meant that it did not attract attention. A frequent periodical visitor to all districts, in some alleged to occur annually, it was most common in the rainless and hot months, although one opinion was that it appeared at the end of the rains, another during the months of December to February. Delay in appearance in any one area meant that it was more virulent when it did strike, presumably a reflection of the increased number of susceptible young animals. In the Himalayan foothills, the natural topography presented a barrier to rapid spread, and indigenous breeds were intermediate in susceptibility between cattle of the plains and those of western Europe. Hill cattle, buffaloes, goats, and sheep, indigenous to the tracts lying between the snow mountains and the outer Himalayas, were more susceptible than the same species of plains animals, respectively. Pease (1894) stated that he had seen the disease clear out the cattle in a small state, leaving less than a dozen animals alive having spread through over 30 villages. It was not uncommon for whole villages to be left without cattle. The hill breed was reported the most susceptible of those cattle indigenous to India. Of imported animals, English, Aden, and Persian, found principally in the Bombay Presidency, those attacked all succumbed, while amongst Sind bovines introduced into India, mortality was 80%. Plains sheep rarely succumbed to an attack, while all

Tibetan sheep brought over the Himalayas to the Muktesar Laboratory by Bhutanese succumbed to inoculation. Hill buffaloes, much smaller than the plains buffaloes, were closer to hill cattle in susceptibility than to plains cattle (Lingard, 1905a).

The wide variation in mortality rates may have been influenced often by the condition and management of the cattle, as well as by seasons. In one Division in October 1870, the magistrate reported: "The late excessive rain has caused great mortality amongst the cattle all along the road, simply from exposure and want of shelter. They have not died from any of the diseases under report" (Campbell, 1870). A District Assistant Commissioner reported in April 1869:

> The cattle at this season of the year are most wretched objects from want of sustenance, as all the grass is parched up. They are barely able to crawl about from place to place, picking up the few *mahooah* blossoms that are left after the owners of the trees have picked up what they could—blades of grass here and there—and in their weak condition they are unable to withstand any sickness, and die off rapidly.

In August, he reported, "The severe drought this year greatly aggravated the distemper, and it commenced with alarming violence" (Cosserat, 1869). Ploughing by bullocks was conducted from May to August and working them when infected would have aggravated effects of the disease.

The Civil Surgeon, Canara, Bombay Presidency, described the manner in which cattle were kept in a native's cow house in North Canara:

> The animals crowded as closely as they can be packed with, beneath their feet, and forming their bed, the accumulated dung and manure of months, the mud floor saturated with the urine of years, perhaps of generations, with often no space by which the fresh air can find inlet, or the foul outlet,—often this mess in the crowded, low, unventilated, mud walled rooms, in which the inhabitants themselves live…, a cow-house in North Canara presents such a glomeration of filth and evil smells that one's wonder is raised that any animal can exist in it for a single day…When we add to this that the wretched animals when released from their nightly durance flock eagerly to the *place d'aisance* frequented by their masters, there to feed greedily on the human ordure, I need add nothing more to the causes which combine to render the murrain and other forms of loathsome disease among the cattle prevalent. (Anon. in Hallen *et al.*, 1871)

If London dairies were considered insanitary in the 19th century, those of Calcutta beggared belief. Thirteen dairymen were charged in August 1870 with keeping their lands in a filthy state:

> The defendants have kept up immense open cess-pools of cow dung, urine and other refuse in a state of putrescence, full of maggots and surcharged with noxious vapours. They have likewise kept up uncovered drains full of cow dung, urine and night-soil, which connect the several cess-pools with each other, and convey a seething and sluggish stream of putrefaction to the public drain, which is being rapidly choked up.

In many epizootics, more than one disease prevailed simultaneously: FMD, pleuropneumonia, anthrax, hemorraghic septicaemia, and blackquarter, as well as the condition of bloat. Sheep-pox was present among sheep. From the province of Oudh, it was reported that there was so much sickness and death among cattle throughout the province that the outbreak of rinderpest in 1870 did not cause any unusual mortality.

It was often noted that bovine and human outbreaks of disease occurred at the same time, and Dr. Brougham was reported as stating numbers of cattle died from smallpox which they contracted from people (Oliphant, 1870). In Nowgong (Assam Division) in 1869, there was terrible human mortality from cholera at the same time as the rinderpest epizootic, probably attributable to the Kallong River having been contaminated by throwing into it thousands of dead cattle. During an outbreak of cholera in 1818, the *Bombay Courier* reported: "We are informed that no rain had fallen at Seroor, when the epidemic [of cholera] made its appearance.

The cattle were dying for want of forage" (Anon., 1818), and the carcasses had probably, as in 1869, contaminated the water.

2. THE INDIAN CATTLE PLAGUE COMMISSION OF 1870

Because of numerous outbreaks of cattle disease in the years leading up to 1869, and influenced by events in Britain, the Indian Government appointed a Commission to inquire into the subject. The President was to be J. H. B. Hallen, Staff Veterinary Surgeon of the Bombay Army and newly appointed Principal of Dick's Veterinary College, who returned to India immediately. Assisting him were Assistant Surgeon Dr. Kenneth McLeod, Mr. A. C. Mangles, Mahomed Ali Jan, and Babu Hem Chunder Kerr, Deputy Magistrate and Collector. Following the resignation of Mangles in June 1870, his place was taken by Mr. J. G. Charles, Officiating Joint Magistrate, Patna. The Commission assembled in Calcutta on December 21, 1869 with 6–8 months to carry out its task but did not complete it until January 31, 1871. Its mandate was: "The contagious character of these murrains, and their connection with, and analogy to, rinderpest, should specially engage the attention of the Commission."

The examination was limited to the northern region of India: Lower Bengal, the North-West Provinces and Oudh; but in view of the presence of rinderpest elsewhere, cattle being reported as dying in their thousands in the Punjab and murrain reported as severe and wide-spread in Assam, information was also gathered from Madras, Bombay, the Central Provinces, Burma, and the Andaman Islands.

The Commission had the English documents at its disposal, including the *Third Report* of the Royal Commission (1866), but difficulties of operation were exemplified by the account of Dr. H. Farrell, a veterinary surgeon on secondment from the Bengal Army who had been a demonstrator at Gamgee's Edinburgh Veterinary College for a session before entering the service. While investigating in Assam Division traveling through the district by elephant, the beast bolted, obliging him to return part of the journey to obtain another. Suffering from repeated attacks of malaria, he referred to some of his problems in his report:

> Personally, I had a good many difficulties to contend with... from the perplexity and unusual expense attending the maintenance of one elephant and three ponies for baggage... travelling through many parts of the province, from almost impassable roads, is anything but pleasant, and at some places shelter is of the vilest description, so that I found it impossible to keep up the complete diary contemplated... consequently this report fails in many particulars which I would have brought forward under more favourable circumstances.... (Farrell, 1870)

The Commission concluded that topography had an important bearing on the prevalence of the disease, hills, forming an obstacle to cattle association. Where outbreaks occurred in the hills, these could be traced to imports into the hill station. During nonworking seasons when they were not wanted elsewhere, cattle were taken in droves to more fertile tracts. Such grazing grounds, where cattle were herded and grazed together and then afterwards dispersed to their villages, provided a nursery for contagious diseases. The whole tract of the Himalaya Terai, the Orissa Gurjats, and the Rajmehal Hills, illustrated this.

Watering places were also a means of spreading the disease. The investigators saw a small pond of foul water in which cattle were wallowing, while on its banks stood others suffering from rinderpest. Carcasses were more often found near watering places, animals suffering rinderpest having an abnormal thirst and making for water.

The influence of roads in spreading the disease was well known in Ceylon and Darjeeling. In Palamow, the trade of nearly the whole of the southwestern frontier was carried on by means of pack bullocks which traversed the country in all directions on tracks through areas where cattle were herded for grazing. Wherever they went, traders invariably camped on the same spot, so each successive party camping there ran the risk of infection if an infected train had passed through.

Cattle fairs were another important factor in spreading the disease. Animals suffering from rinderpest were seen side by side with healthy animals brought in from a wide circle and destined to travel out to other widely separated places.

Although no specific case of it causing rinderpest was demonstrated, the trade in hides from dead animals was extensive. Sun-dried or imperfectly cured hides from animals dead of rinderpest were traded all over India with the exceptions of Assam and the Sundarbans as the Assamese did not wear shoes, so there was no local hide trade or export of hides from Durrung. Thom related a common rumor undoubtedly invented to explain the spread of the disease that one of the richest Moslems in Calcutta made his fortune by purchasing for a few pice dozens of skins of animals which had died of rinderpest. These were imported into other countries and carried the infection with them (Pollok and Thom, 1900).

Its history in the Andaman Islands disproved allegations of spontaneous origin, proving that importation was "the main, if not the only, cause of outbreak of rinderpest and foot and mouth disease even in tropical malarious localities where endemic diseases are rife among men and even sheep" (Hallen *et al.*, 1871).

The Commission's recommendations were general in tone and mostly limited to "watching," such as: "Large fairs should be subject to skilled inspection and sanitary precaution, and smaller fairs carefully watched, especially in murrain times" (Recommendation VI, Hallen *et al.*, 1871). Other measures were, a census of stock before and after epizootics, herding and droving to be regulated in times of epizootics, hides destroyed or disinfected in times of epizootics, and a veterinary school established for training agents who would investigate and report on murrains and apply preventive and remedial measures, the last not being specified.

The most valuable contribution to the report was that of Dr. McLeod, despite a lapse into a belief in spontaneous generation. His exhaustive account gave a detailed analysis of preventive methods undertaken in Britain, Europe, and Russia. His recommendations were:

- Gathering, systematizing, and publishing information.
- Inspection when disease had been reported.
- Duty to give notice inculcated.
- Skilled examination and reports by competent persons of its nature and behavior.
- Proclamation of infected areas.
- Isolation of sick animals and separation of healthy.
- Prevention of markets and cattle movement in infected areas.
- Burial of carcasses after slashing the hides.
- Disinfection, including the destruction of dung, fodder, and anything by which the disease might be spread.
- Medical treatment combined with sanitary precautions.

When a better body of knowledge had been built up, then the course of preventive or restrictive action would become clearer.

The Commission concluded that the murrain was identical with that occurring in Britain in 1865–7 and was the Steppe plague or "murrain" of Europe and Russia, but rather than being similar to the disease occurring in Britain, it was exactly the same in every respect as the Russian disease in Russia. Mortality, for example, was low, overall probably not exceeding 5%. Higher in some places and in some outbreaks, it was considered doubtful that it could ever be reduced in them to less than 20%. Inoculation as a method of control was not recommended because of the difficulty of implementation, many places being almost inaccessible and even the total number of stock unknown.

The Commission's recommendations, aided by those of Professor Robert Wallace of Edinburgh University who conducted a survey of Indian agriculture in 1887 (Wallace, 1888), resulted in the creation in 1889 of the Imperial Bacteriological Research Laboratory at Muktesar, Kumaon, later the Imperial Institute of Veterinary Research and then the Indian Veterinary Research Institute.

Following the Commission's Report, Hallen produced his small manual (Hallen, 1871), containing practical management advice for officials rather than the peasantry.

3. HISTORY OF OCCURRENCE

People declared that a murrain carrying off hundreds of cattle was not a new event and was told them by their fathers and grandfathers, accepted as a feature of rural life. Cattle had been owned by Hindus in India since Vedic times, pre-500 BC, and prior to the advent of Buddhism, they were sacrificed and eaten at feasts and festivities. In the North-West Provinces, rinderpest had been known from an early period, said to have been in the Bhabur forests "from of old" where it was present in 1867–8 with about 50% mortality. It was even known in the Snowy Range, sometimes carrying off large numbers of *jooboos* (a cow–yak hybrid), but believed to be taken up by Bhutanese traders from the lower country.

The Madras Presidency was the first to have a skilled agency investigate the disease and devise methods of combating it, putting into operation a legislative enactment, although it lasted only for 3 years. The history of the disease in this area can be traced to 1791–2, when cattle of the army of Lord Cornwallis at Bunjator and Seringapatam were severely attacked by a murrain. Then, about 1795–6, there was a great epidemic of smallpox in and about Calcutta in the months of January and February, during which mortality was great among men and cattle of all sorts; and cattle disease occurred in the District of Seebsaugor in 1797. A 78-year-old Calcutta dairy owner stated in November 1870 that *gootee* had been prevalent since his boyhood, but not so frequent formerly as in the past 25 years. He had also heard of its past existence from his father and grandfather (Hallen *et al.*, 1871).

Table 1 summarizes reported outbreaks between 1800 and 1868, but in many regions the disease was enzootic with annual outbreaks, while in others, it persisted epizootically for several years, so the table simply illustrates its widespread nature.

In Gujranwalla, Lahore District, there was oral evidence of its occurrence in 1805, and it was known to the people's forefathers. In 1810, in the extreme south of India during an outbreak of cholera, there was great mortality among cattle, and in 1816: "It is a curious fact, that in the towns near the hills, where the epidemic [of cholera] was so fatal, a disease occurred among the cattle, which kept pace with, and often exceeded in mortality that of the human species" (Chalmers, 1823). Dr. Jameson, Secretary to the Medical Board of Bengal, reported that in 1815, half the cattle in the lower region of Tipperah were carried off by a cholera-like disease, suddenly expiring after being seized with vomiting and convulsions; and in 1816, there was an unusual mortality among black cattle, sheep, dogs, and other domestic animals in Bengal. Cows aborted (Jameson, 1820).

In Amritsar Division, before its presence in 1870, it was reported by a 60-year-old to have occurred in 1820, 1840, and 1860. One man had seen eight or nine outbreaks in a lifetime of 35 years, losing 11 cattle in 1861. As in August 1870, rinderpest and pleuropneumonia seemed to exist together in a number of outbreaks in this region. In Moorshedabad, after 1833, although a great number of animals were said to die annually from disease, there was no

Table 1. Reported occurrences of rinderpest in India 1800–68.

Year	Town/Village	District	Division	Region
1805	Gujranwalla	Lahore	Delhi	Punjab
1810	Hissar	Hissar	Hissar	Punjab
1811				Mysore
				Madras
	Hoonsoor			Madras
1815		Tipperah	Chittagong	Bengal
1817	Calcutta	24-Pergunnahs	Presidency	Bengal
1818			Assam	Bengal
1820			Amritsar	Punjab
1825		Durrung	Assam	Bengal
1828				Bengal
1830	Hissar	Hissar	Hissar	Punjab
		Jhung	Mooltan	Punjab
			Assam	Bengal
		Saugor	Jabalpur	Central Provinces
1833		Moorshedabad	Rajashaye	Lower Bengal
		Sylhet	Dacca	Bengal
1836				Lower Bengal
			Chota Nagpore	Bengal
		Lohardugga		Bengal
		Singhboom		Bengal
1840			Amritsar	Punjab
			Oudh	Oudh
1843				Bengal
	Calcutta	Calcutta		Bengal
1845	Bongong	Tirhoot		Bengal
1848		Durrung	Assam	Bengal
		Kamroop	Assam	Bengal
		Gowhatty		Bengal
1849			Benares	North-West Province
	Tezpore		Assam	Bengal
1850			Oudh	Oudh
1851			Umballa	Punjab
			Cooch Behar	Bengal
1852	Hoonsoor			Madras
		Kamroop	Assam	Bengal
1853		Nowgong		Bengal
1855			Benares	North-West Province
1856		Moorshedabad	Rajashaye	Lower Bengal
1857			Benares	North-West Province
	Calcutta			Bengal
			Bengal	Bengal
1858			Bihar	Bengal
		Hurdui	Oudh	Oudh
1859			Oudh	Oudh
	Calcutta			Bengal
1860			Amritsar	Punjab
		Sylhet	Dacca	Bengal
		Shahjahanpur		United Provinces
			Cooch Behar	Bengal
1861		Durrung	Assam	Bengal
		Maunbhoom	Chota Nagpore	Bengal
			Jhansie	North-West Province
		Darjeeling		Bengal
		Bengal		Bengal

Table 1. *Continued*

Year	Town/Village	District	Division	Region
1862		Jhung		Punjab
			Chota Nagpore	Bengal
		Bengal		Bengal
		Doon	Meerut	United Provinces
		Bowur		United Provinces
1863		Lullutpore		North-West Province
		Kurnool		Madras
	Coimbatore			Madras
		Orissa	Orissa	Bengal
1864	Hissar	Hissar	Hissar	Punjab
		Moorshedabad	Rajashaye	Lower Bengal
			Patna	Bengal
	Belvidere	Bengal		Bengal
1865			Benares	North-West Province
		Bengal		Bengal
		Backergunge		Madras
		Darjeeling		Bengal
		Duars		Bengal
		Rungpore	Rajashaye	Lower Bengal
1866			Cooch Behar	Bengal
	Ootacamund			Mysore
1867		Durrung	Assam	Bengal
		Hurdui	Oudh	Oudh
	Mynapuree		Agra	United Provinces
1868		Durrung	Assam	Bengal
		Bhabur	Kumaon	North-West Province
		Furruckabad	Agra	North-West Province
			Cooch Behar	Bengal
		Rungpore	Rajashaye	Lower Bengal
		Dinagepore	Rajashaye	Lower Bengal

report of rinderpest until in 1864, when it was stated that it had been present in 1856, suggesting that it occurred from time to time but was not reported. An inquiry in 1864 showed that cattle murrains throughout Lower Bengal were frequent and fatal, with an epizootic in 1836. One man of 90 in 1870 reported seeing it four times in his life, the first when he was 10 years old, that is, 1790.

More attention was directed toward rinderpest in Calcutta than elsewhere in Bengal. Farrell (1871) noted that the first reported severe outbreak was in 1817, and it had appeared at intervals ever since. In 1844, Dr. Duncan Stewart reported a murrain breaking out in the dairies of Calcutta every year or two and sweeping off numbers of dairy stock (Stewart, 1844).

In Assam Division, cattle disease was known to be present before the British occupation of 1824, the earliest record being 1797. In 1818, the Burmese invaded Assam with a large army, and soon after their arrival, a great plague broke out among the army cattle destroying them and spreading into the country. There is then brief reference to a plague in Durrung District in 1825, and again in 1850 when Bengalee buffaloes are said to have been first imported. Plagues were then reported in 1861, 1867, and 1868. Before British rule, there was no regular communication between Assam and Bengal, and the only cattle which came into the country were a few buffalo sent to the Rajah by the Rajahs of Gowalparah and Cooch Behar.

In 1828–36, a disease, termed *bosonto*, described as a species of dysentery, attacked the cattle of Bengal, one European owner reporting that his animals were affected three or four times in the course of 7 or 8 years (Piddington, 1836). His method of treatment was fumigation with muriatic acid fumes. In 1843–4, the outbreak was severe, beginning in September 1843 and lasting until the following June with a recorded mortality of 2,233.

In Saugor, Central Provinces, the Superintending Surgeon Mr. Campbell reported in 1830: "... during the prevalence of the epidemic (small-pox) in this district, great numbers of cattle died of a disease which the natives distinguish by the same name which they use of variola, namely *matah*."

In Dacca Division at Sylhet, rinderpest was reported in September 1833: "In the rainy season of 1833, the district of Silhet having been flooded four times successively, the cattle suffered severely from want of pasture. Hundreds died from starvation, and the survivors reduced to mere skeletons, were, in September of that year, attacked with *mata*. This epizootic was not confined to that district, but extended its ravages to the adjoining district of Kachar" (Brown, 1842). The symptoms were described by Lamb, Civil Surgeon for Dacca:

> The disease among cattle, known under the appellation of *bussonto* or *matah, sitlah* or *gootee*, chiefly prevails during the months of December, January and February, though not quite unknown at other seasons of the year... Cattle are said not to be subject to a second attack of *matah*; sometimes the disease is exceedingly fatal, at other times it is comparatively slight, without much reference to season or apparent cause; it is, however, at all times greatly dreaded, and with reason, as it is too often the cause of utter ruin to villagers whose whole wealth is in their cattle. (Lamb, 1836)

From 1836 to 1859, there was no record of any cattle murrain from this region until rinderpest occurred in 1860, but after 1860, there was again no reference.

In Chota Nagpore Division, *matah* was known in 1836 to rage every 2 or 3 years. Inquiry was started in 1864, when it was found that every year large numbers of cattle were swept off by it. One report was that *bussunto* was said to have always been in the district, and there were always many deaths from it, but for the last 2 or 3 years, it was more severe than formerly. In Lohardugga District in 1836, it was reported that sometimes whole herds were nearly wiped out in a very short space of time, and in Singhboom in 1836, immense herds were reported swept off.

In the North-West Provinces, Benares Division, there was a traditional and unrecorded history of cattle diseases, epizootics occurring in 1849, 1855, 1857, and 1865. A cowherd in Benares District reported having seen it four times in the 20 years prior to 1869. In 1849, he had 40 head of cattle, and 18 died in 8 days. There was another outbreak in 1855. The third was during the mutiny in 1856. A great number of cattle was collected for Government purposes, and many of them died. There was now the outbreak of 1869.

Stewart, whose interest was in finding a cowpox vaccine to vaccinate against smallpox, produced a report in 1844 to which he appended a section on the disease among cows. He had made inquiries over the past 15 years and noted that it was generally believed that *Matta* prevailed every spring in Bengal but not within 32 km of Calcutta. In Penang, "a few years ago," Dr. Oxley reported a fatal disease destroying a great number of cattle, poultry, sheep, and pigs. The same occurred in Ceylon and in many parts of India. Early in 1843, cattle in and about Calcutta were dying in large numbers from "small-pox."

> I visited all the (*Gowkhanahs*) cow-sheds I could hear of, and inspected very many sick cows; but not one had any pustular eruption which I could discover on any part of the body. I saw several young, fat, clean, and healthy looking cows die, after about three days' fever, with convulsive tremors, bloodshot eyes and parched mouths. These cases the natives persisted in still calling *matta*, but of the violent "inward" form, "not permitting the discharge of humours on the skin." I observed in others, and generally for a few days, signs of fever, such as panting, lowing, thirst,

> loss of appetite; then followed violent purging, profuse salivation and pituitary discharge from the nostrils; but on the fourth day ulceration of the tongue and pallet [*sic*], with bloody purging followed, and death occurred usually on the fourth or fifth day. (Stewart, 1844)

He was thus possibly witnessing two types of disease, or a particularly virulent strain of rinderpest as well as the more usual form.

Mortality among cattle began at much the same time as smallpox did among the inhabitants, probably reflecting a seasonal stimulus. In January 1844, it became very considerable in the town, affecting the supply of milk and butter, but was reported to be much greater in the suburbs where all the slaughterhouses were situated. "It was not generally known that the mortality was at the same time very great among fowls and pigeons ... we saw the dead fowls taken out by dozens of a morning, said to have died during the night of the universal small-pox" (Stewart, 1844). The Indians considered that *matta*, or smallpox, affected both man and cattle.

Mr. Greenfield, a Calcutta cattle dealer, stated that he had not any disease among his cattle since 1834 when he came to the area. "This year [1844] I have lost 60 cows with disease that has been all over our neighbourhood, something like strangles or glanders: ... they were all suddenly taken ill and seldom lived more than three days, many only one; no purging, no discolour of water appeared. I have seen the pox in England among cows, but this is nothing like it ...."

John Stalkartt first experienced *gootee* in 1845 at Bongong in Tirhoot, where he had about 400 bullocks, three quarters of which died.

In Gowhatty, rinderpest occurred in 1848–9 with only three in 80 animals surviving. Raging elsewhere, it was alleged that buffalo and tigers were killed by it. Since 1848, it had not been as bad as it was in 1870, but the country had never been entirely free of it.

In Chundpore village in Boolundshahar, it showed itself every 3–4 years, sometimes mild and sometimes very virulent, mortality ranging from 10 to 30%. Villagers were insistent that it could attack an animal more than once, sometimes after a lapse of several years. Some claimed that an animal could get it twice again. FMD, they claimed, could attack the same animal annually.

In Kumaon Division, the disease was well known, coming almost every year. When regular, it was less fatal, but if 2 or 3 years elapsed, it was very virulent. Animals were not separated on account of a belief that the sooner an animal contracted the disease and recovered, the better.

4. OCCURRENCES IN THE 1850s–1860s

In 1849–50, so many cattle died at Tezpore, Assam Division, that Colonel R. Campbell reported that it was impossible to remove them. The Secretary to the Board of Revenue, Mr. Grote, estimated total mortality at 120,000. Beginning in Kamroop in 1852, the disease gradually worked its way up the valley through Nowgong and Seebsaugor to Luckimpore. The Reverend Higgs of Debrooghur was informed in 1870 that the district had been visited by the murrain every 20 years, but since 1855, it was more or less annual. The number of deaths in Nowgong in 1853 was put at 21,625.

In Cooch Behar Division, there were plagues in 1851–2 and 1868–9, one belief being that the increase in communication brought and spread the plague. In 1860, there was a great loss of cattle in the hills, the disease appearing from the Terai below. During the time preparations were being made for the Sikkim campaign in 1861, a fatal disease broke out among the Commissariat cattle bringing stores up from the plains, spreading also to several Sikkim villages. Called *yor*, rinderpest was well known in Sikkim, appearing periodically.

One informant stated that it broke out among the British army cattle at Rungheet. His 30-odd cattle were 24 km from the camp, and there was no road to his farm, and no cattle were brought there. But the disease attacked his cattle, killing all but one animal. Villages closest to the camp were not attacked at first, the disease appearing in those some way off, and then others nearer the camp were affected. In the British camp, sheep and goats died also.

The Superintendent for 11 years prior to 1870 of the Calcutta municipal cattle stock, Mr. T. L.Taylor, stated that the first outbreak of plague he could remember was in February and March 1859, since when it had been of almost yearly occurrence with the exception of 1862–3, but no subsequent outbreak was as bad as that of 1859. His predecessor had told him of a bad outbreak in 1857 when the municipality lost 200 head of cattle. Each outbreak generally lasted 6–8 weeks, making its appearance in February or March, sometimes as early as January. At another municipal cattle shed, in 1870, five of 100 were affected, of which two died and three recovered. In 1867, the loss was considered to be very bad, 30–35 of 100 dying.

In 1861, 22,661 cows and 7,701 buffaloes are reported to have died of *bussunto* in Maunbhoom, Chota Nagpore; the following year, 26,032 cows and 6,293 buffaloes; and in 1863, 18,908 cows and 54,459 buffaloes. Many cattle died the next year, the epizootic continuing into 1866 when mortality was estimated at 22,000.

In 1861, *gootee* prevailed so violently among the Bengal municipal cattle that the services of a veterinary surgeon were considered necessary. It had occurred as early as 1857 and, since then, was annual. Between 1862 and 1863, many gentlemen's cattle were affected, but peasant losses were not reported. In 1864, it broke out at an exhibition at Belvidere much to the discomfort of the governor, who was presiding and therefore ordered an inquiry. With the disease being found to occur frequently throughout Lower Bengal, Dr. C. Palmer was appointed to report on the whole subject. He demonstrated that *gootee* was rinderpest but wrongly concluded that *gootee* and *matah* in the districts were FMD. It next appeared in the cold weather of 1865 in one of the districts and continued until 1866. It created great devastation in March and April in land adjoining the Sundarbans. In Jessore District, where it continued into 1867, the estimated loss was 1,200 head of cattle, plus great numbers of pigs, buffaloes, and deer. In the district of Nuddea, following an unusually high inundation of the river, another fatal disease broke out among the cattle, characterized by a swelling of the throat and neck. Hallen *et al.* (1871) noted that in the absence of a description by a skilled observer, it was difficult to pronounce decidedly on the type of this disease, but the symptoms are indicative of anthrax.

In Madras, rinderpest was present in 1863, allegedly killing half the cattle, and in 1865 was severe in Backergunge district where it had been imported from Jessore across the Balissur river. It raged until September, when it seems to have declined. Mortality in the whole district was estimated by the Magistrate, Mr. Sutherland, to be 30,000–40,000, bringing ruin to thousands of peasants.

A most virulent type broke out in the hills above Coimbatore in August 1863, only 2% of attacked animals recovering. Infection had been introduced by cattle sent down to Coimbatore to fetch machinery, one of which was attacked 8 days after the return, followed subsequently by 13 others. Of the first five attacked, four died. On the outbreak of the disease, the Indian cattle owners immediately drove their animals many kilometers away. Mr. MacIvor showed the superior efficacy of removing the healthy stock as opposed to the sick.

In November 1865, the manager of the Bullock Transit Train, Darjeeling, finding that plains animals did not stand up to the hill work, had some 40 hill breed cattle brought from the Morung. A few days after their arrival, two became ill, dying within 36 hr. By December 15, all had died. In April 1866, rinderpest broke out among some cattle the Commissariat had purchased from Bhutanese, and in a very short time, nearly all of these had died. Some of the sick animals had been continually wandering about on the cart road, causing the Bullock Transit Train manager to lose another 150 draught cattle, as well as some milk cows and calves. The Darjeeling Municipality then took over the Train, and at one depot, so many cattle died that the Train had to be closed for a while. One informant stated that the disease had occurred annually in the hills since this time.

Cattle disease was reported in 1866 as not uncommon in the Duars and Cooch Behar, proving very fatal in the Eastern Duars in 1865. Carcasses of dead cattle were to be seen in every village, and no effort was made to remove them. Local people in Darjeeling said that the disease was always brought by cattle from Nepal and its interior. It was not known to be brought from Bhutan. The reason why it was thought to come from Nepal was that newly imported cattle were always first to contract the disease, which broke out immediately after the arrival of new cattle, then spreading to the others which had been on the hills for some time (White, 1870).

In an outbreak in Rungpore in 1865–7, a belief prevailed that it was brought there by the Commissariat cattle of the Bhutan war, but the Commissioner noted that the disease was previously known in the district. Rinderpest was reported prevalent in the Terai in 1863, and the cattle of Rungpore and Dinagepore districts, where there was a wide prevalence of disease in 1868, like those of the district of Purneah, Bhaugulpore Division, were driven annually for grazing to the Morung or Nepal Terai. Coming from every part of the district in January and February during the cold weather, cattle from hundreds of different places concentrated there to graze, leaving at the beginning of the rains in the middle of June. In one area of 1,600 ha at Fazelnugger, there were 80,000 cattle. These were kept in herds of 200–300 each collected from distances up to 200 km away, sent up from Lucknow northwards to the Terai in November, and remaining until February (Figure 3).

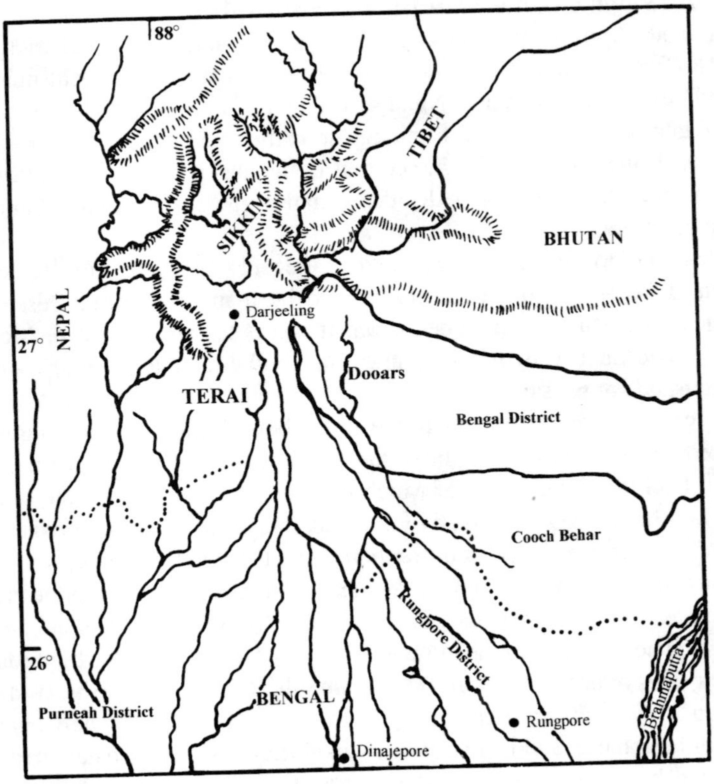

Figure 3. The Terai, an epizootic center, 1860s.

Both rinderpest and FMD were perennially present on these grazing grounds and carried back by the cattle to their various villages. Very few young cattle escaped exposure to the disease, and they thus acquired a protection, which accounted for only a proportion of stock being seized with the disease in any one season. It was believed that those animals which brought back the rinderpest usually recovered, but those which became infected later often died.

In Orissa District, no attention was paid to the disease until 1864, it having "raged fearfully" the year before in the Tributary Mehals running west of Cuttack for 240 km. More than half of the cattle in the forts south of the river Mahanadi are said to have been lost. In the district of Balasore, it was said to be well known, committing great ravages every 2 or 3 years. A possible outbreak of rinderpest occurred among artillery cattle at Cuttack in 1868, and following segregation, only one further animal died.

In Chittagong District, where cattle were few, rinderpest was unknown, but in 1864, it was reported as having been very destructive for many years in Noakhally District.

In March 1866, J. Thacker, a veterinary surgeon appointed in 1864 to investigate cattle disease in the Madras Presidency, saw the disease at Ootacamund and confirmed his opinion that it was the true rinderpest of Europe. Herds on the Nilgherry Hills were periodically attacked by it, but the hill tribes stated that the disease was not known on the hills until Ootacamund became a European settlement.

5. 1868–1870 EPIZOOTICS

In the Diamond Harbor subdivision of Bengal, the disease assumed such a violent form in June 1868 that Farrell was seconded to investigate. He recorded 4,927 deaths from August 1868 to April 1869, while the Deputy Magistrate, Babu Hum Chunder Kerr, reported 26,151 deaths for the year beginning January 1868.

From rough rounded estimates of the numbers of animals infected in a village and the number dying, it would appear that there was about 65,000 head of cattle in the subdivision. In villages of 200–350 animals, some 60–62% became infected, and 41–42% died. In a herd of 50–60, 51% were infected, and all died. Owners of small numbers of 4–16 head reported some 84% infected and 58% dying. The smaller numbers probably reflected a more accurate estimate of the proportion affected and dying, the apparent lower mortality among larger numbers reflecting greater generalization.

Farrell's numbers led to criticism by the Officiating Commissioner of the Presidency, Rivers Thompson, who questioned his results. As to a proposal by him that a veterinary school should be established in Calcutta, Thompson considered the proposal premature: "I do not suppose such a proposal as that referred to need be seriously entertained" (Thompson, 1871).

However, Farrell was reporting only the cases reported to, and seen by, himself. These he gave as 6,735, of which 4,927 died (73%), and 27% he claimed to have "cured," stating that in the course of 2 weeks, mortality declined from more than 800 a week to less than 300. Kerr's figures were almost all rounded to the nearest 5 or 10, and over the 289 villages reporting this could have resulted in a large discrepancy.

In 1867, there was unusually heavy rainfall and exceptionally high inundation, and in 1868, a serious rinderpest outbreak appeared in January in the village of Bansbeeria, Bengal Division, allegedly introduced with cattle bought at a fair at Oolooburia or Oolooberiah. Oolooberiah was a village situated on the right bank of the Hooghly river about 32 km from Calcutta, through which ran the high road to Midnapoor, Balasore, Cuttack, and other places. A cattle, goat, and poultry market took place there every Saturday, with cattle brought from the west and south and sold chiefly to the dairymen and butchers of Calcutta. Small purchases of draught bullocks also were made by peasants of the surrounding districts. The infection spread south and southwest, with one estimate putting the loss over the year at 26,151. A quarter of the paddy lands lay idle for want of draught buffaloes. In 1869, it was severe in Jessore and, at the end of the year, appeared in villages in Chooadangah, a subdivision of Nuddea district, spreading to reach a peak in March 1870. Most villages in the subdivision commonly experienced rinderpest of a comparatively mild type in March, which subsided in April.

In January 1868, it appeared in a village in Sultanpore Thannah, the inhabitants having bought a few bullocks from the Oolooburia Haut on the other side of the river where a very large number of cattle, mostly from Midnapore and Orissa, were sold each week. The epizootic reached a height in July and August, then declined until it had almost disappeared by the beginning of November. Although villagers believed the disease to be contagious, they had no means of separating their stock. Most animals that died were oxen, which formed the majority of their horned cattle.

Symptoms of the disease allegedly varied from place to place. In one region, an infected ox would pass watery stools, cough violently, salivate, and die in 6–8 days. In another, it would first cough, then start shivering, then discharge from the mouth, eyes, and nose, pustules then forming on the body. They would then pass watery stools, gradually sinking within 5–10 days from inception of the symptoms. The oldest man in the area could not recall a previous epizootic like it, but others reported that now and then, there were epizootics which had, to a certain extent, proved destructive.

Along Diamond Harbor Creek, where the disease ran along the bank, it was allegedly of a more virulent type, more fatal there than in those villages more distant from the water, affected animals dying in 48 hr. In the village of Raipore, all were exterminated in 1 week.

Carcasses of dead animals in all the villages were thrown into the nearest ditch or canal, and Farrell found bodies in every stage of decomposition not 45 m from the houses. Going under the names of *gootee* or *bussunto*, he considered it to be rinderpest. Death took place on the eighth or ninth day after infection. The deaths which took place in 48 hr he did not question.

Farrell asserted that there was no drug which was specific for the disease and no known antidote. Although he stated that good hygienic management was of much more value than resorting to medicinal agents, he nevertheless attempted treatment with the usual concoctions of antimony, and so on, and herbal remedies. He placed some villages in quarantine and prohibited all contact between them and healthy places, suggesting that this should always be done at the first outbreak.

A divisional magistrate, A. Smith, considered Kerr's report to be more reliable than Farrell's. Kerr had received information from the outset, whereas Farrell obtained his from natives, who probably knew little beyond what occurred in their immediate neighborhoods. But according to Farrell's informants, the disease was not new in that part of the country south of Calcutta, dating back far beyond the memory of the oldest people. Of periodical occurrence, it appeared here and there every year and, some 15–20 years ago, appeared to have caused great mortality. But generally, cases were so few that they did not attract attention.

Kerr, who was a member of the Commission, challenged Farrell's report, stating that the disease had been brought over the river in January 1868 with a few head of cattle purchased at the Oolooberia market. It was immediately after the arrival of these cattle that the disease broke out at Bansberia and spread rapidly south, becoming general over nearly the whole of the subdivision. Floods submerged a great part of the subdivision in June 1868, depriving cattle of food for 3 or 4 weeks, causing an accelerated mortality when they fed on rank fodder after the water had subsided. Smith, the Magistrate, stated that mortality at this time was attributed to starvation.

In Debrooghur, it was first observed about the middle of May 1869, a sepoy having brought some cows into the station from Golaghat, an infected district. At the beginning of July, some 70 head of cattle per day were dying, and of about 958 in the station, 614 had died by the middle of November, the disease also spreading to surrounding villages. The inhabitants then began driving their sick animals into the station to save themselves the trouble of burying the carcasses. In an attempt to preserve the cantonment from cholera, carcasses were buried in pits dug by the military authorities, and many thrown in empty wells, covered with lime and a layer of charcoal. Regimental elephants were at first employed to drag carcasses to the burial pits the moment a death was reported, regimental sweepers tying the carcasses to ropes suspended from the elephants. But, it being the height of the rains, water was found within 0.9–1.2 m of the surface, stopping all digging operations, and carcasses were ferried on rafts into the middle of the Brahmaputra river and dumped. Disease had been introduced into the cantonment of Debrooghur in 1854 by sepoys in the same manner as in 1869, but this seems to have been FMD for it differed from the symptoms exhibited in 1869 by great soreness of the mouth and feet, although death usually ensued in 2–4 days (Higgs, 1870).

In Seebsaugor, where the disease was very virulent, cattle owners were in the habit of clandestinely removing all the cattle they could to Debrooghur for sale, and from there, they were sent to the other side of the river to graze.

In 1867, it was present in two villages in Meerut Division. The people here segregated sick from healthy cattle, but in one village, 200 died in 2 months. In February 1870, it was introduced into a Meerut village by a cattle dealer, who brought five bullocks, selling one which the next day was sick, recovering after 7–8 days. But the bullock put with it died in 2–3 days, and the disease spread to other stock. Several villages claimed that they had not heard of the disease before and had no name for it, the disease being very benign. But both rinderpest and FMD prevailed simultaneously in the region, and reports confused the two.

In the Terai Pergunnahs, it was common after the rains to the end of June 1868. The disease was stated to be not new. In 1868, the jungles were full of cattle due to drought, and in Bijnour District, where severe diseases had broken out, the Commissioner sanctioned the employment of people to prevent infected herds being mixed with healthy.

An Englishman who had been at Locknathpore, Nuddea Division, for 9 years, stated that he had never seen such a long drought as that in 1870, nor *bussunto* so fatal. He had seen *puschima* twice after the inundations of 1861 and 1869.

Farrell believed that the mortality in 1869 in Jessore District within a circumference of 16 km was greater than 1,000. It had appeared as an epizootic in 1864 but remained limited to a small part of the district. It occurred annually in January to March, sometimes also in August and September, with sporadic cases here and there throughout the year with a mortality no greater than 5%. In 1869, it broke out with great virulence in mid-February in three or four areas simultaneously, spreading rapidly and causing a mortality of about 30%.

In Nowhatta, Jessore Division, the disease appeared in February 1869, extending to many villages inland from the river, raging with great virulence to the end of May, dead cattle numbering thousands, and some owners with only five or so animals left out of herds of 300 or more.

In May 1869, an English landowner in Jessore, H. J. Rainey, claimed that mortality was usually about 5% per annum, rising in epizootics to 10–20% or more. He reported on a disease that he supposed was prevailing in the Sundarbans, describing the characteristics of rinderpest (Rainey, 1869a), and in September remarked on Farrell's report of June 30 distinguishing two separate diseases (Rainey, 1869b), which Farrell nevertheless believed to be one and the same. *Puschima* or *puchima* meant "coming from the west" but was the same disease as *gootee*. In some places, it was called *dukhina*, meaning "coming from the south," and in others, *googlanee*, from the grunt of the sick animals.

In 1868, there was an epizootic in Kamroop and the Khasi Hills which gradually spread to Central and Upper Assam. Rinderpest was worst in 1869 in Assam, 69,967 cattle being reported to have died from when the disease first appeared in mid-April up to August 31. In Assam Division, it was called *mour*, distinguishing it from *gootee* or *matah*, and the people believed that early separation of healthy stock was essential. Within living memory, the country had never been free of the disease, epizootics allegedly coming about every 5 years, with major ones every 8–10 years spreading far and wide and attended with great mortality. But Farrell thought that the official figures for the 1868–9 outbreak of a total of 177,659 deaths were probably double, or even four times, the real number. In one instance, an "over zealous" policeman had simply collected the names of several villages which he had not even visited and added imaginary numbers to them, there being no rinderpest present at all. In another case, a group headman reported that a crow, a snake, and an eagle had died of the disease. Farrell remarked that he was either a wit or a cynic.

Present in the neighboring districts of Jounpore and Ghazeepore, the disease erupted in Benares in October 1868, and the following year it became a subject of attention when in March it broke out in the military cantonments among Commissariat slaughter cattle bought at Kat Berhampore Fair, Shahabad District, on February 10. The cattle arrived at Benares 13 days before disease broke out.

Four animals died on March 13 whereupon the Commissariat Officer immediately set up a committee to examine the cattle and take such steps as might be necessary to check the spread of the disease. Veterinary surgeon C. Barrow of the 19th Hussars confirmed rinderpest and recommended separation, slaughter of infected animals, burial, quarantine, and continued inspection. Accordingly, 17 animals were slaughtered, the skins destroyed, the carcasses buried, and the sheds disinfected. The disease ceased on March 27. In all, of 230 cattle, 93 were attacked, 35 died, and 18 slaughtered. A second batch of cattle was bought at the same fair on April 10, marched to Benares, and placed with the other cattle, appearing after examination to be healthy. But disease broke out 4 days later. Of 340 animals, 77 were attacked, and 37 died. Also, 18 sheep were attacked with similar symptoms, and nine died. Diseased animals were not destroyed on this occasion, with separation, cleanliness, and treatment by bleeding, purging, and administration of tonics, being resorted to. It was then ascertained that disease had broken out in several places where cattle had been purchased at the fair, where disease showed itself in April.

The matter was brought to the attention of the Viceroy, who ordered an inquiry, and a station Committee of five members was convened. They came to the conclusion that the disease was rinderpest and was enzootic, making its appearance at intervals in all, or nearly all, districts of southern India, including Benares. They could not recommend "stamping out" but urged segregation and legislation.

Toward the end of 1869, it broke out in the districts of Cuttack and Pooree (Orissa), continuing into 1870. In both these areas, rinderpest and FMD appear to have occurred simultaneously. In 1868, it was widespread in Tipperah, with deaths totaling 3,000–4,000. The following year, it widened its area, appearing also in Noakhally, and a partial outbreak in Chittagong being reported as rather bad in the hill tracts. In 1870, it was reported as minimal in Tipperah and absent in Chittagong.

In Seebsaugor, Upper Assam, from April 1869 to January 1870, 112,415 animals died. It was alleged to have originated in Assam from the importation of Bengalese buffaloes, normally there being no imports of cattle. Droves of as many as 900 buffaloes were brought in, taking up to 3 months on the journey, but at the time, none were reported as sick or dying. However, from Luckimpore, the Assistant Commissioner reported that at the beginning of March 1869, a batch of buffaloes passed through, one dying close to his camp on the banks of the Dikrang. In Durrung, where 200–500 were imported from Bhaugulpore, Purneah, and Mozufferpore in 1860, 1865, 1867, 1868, and 1869, large numbers died immediately after arrival in 1869, the disease having been severe along the route they had followed. In 1867–8, a large number died within a few kilometers of Tezpore. During March and April in 1868 and 1869, five herds mumbering 80–150 in each arrived in Nowgong from Rungpore, Dinagepore, and Purneah in Bengal, among which were several diseased animals. At the time of departure from these districts, rinderpest prevailed there. Many animals died on the 1–2-month journey to Assam. In March 1869, some 300 were brought to Golaghat, and a few days later, rinderpest broke out.

The earliest official record for the disease in Allahabad District was 1868, but residents stated that it had always existed within living memory. The year 1869 was one of great drought and scarcity of fodder in the North-Western Provinces, and a large number of cattle died in Allahabad Division from this alone; but in the districts of Humeerpore and Jounpore, disease also carried off the impoverished cattle in great numbers, deaths in Humeerpore totaling 15,598. One official stated that the disease was always present in the District but more prevalent some years than others. It also varied in virulence, sometimes sweeping away all of the cattle of a village and at others causing only slight mortality, only 10–15% of those attacked dying. Some villagers believed that it occurred every 3–4 years. In 1869, it began at the end of the cold weather and continued into the rains. In 1870, both rinderpest and FMD prevailed throughout Jounpore.

In Humeerpore, it was always present to a small extent in various parts of the District but appeared to have increased and become virulent in cold weather as fodder became scarce. An inhabitant aged about 60 stated that he had witnessed five epizootics in his lifetime, the previous occasion about 10 years ago. It varied in virulence in different years, and he believed that the disease was carried by the wind coming generally from the west. In the village of Moniejar, near Humeerpore, of 1,000 head of stock, it killed 400–500 in 1869. One villager said that he had seen it five times, the first 30 years ago. In Rampore village, east of Humeerpore, of 250 head, 100 died. It began in May and lasted about a month. A villager said he had seen it 10 times, and it occurred every 4 or 5 years.

In Lower Bengal in 1870, frequent outbreaks of *matah* and *bussunto* were reported as prevailing there. In the same year, it was found to be widely spread in several districts of Lucknow Division. In 1871, an outbreak was reported in Madras Presidency.

In 1890, pack animals carried it from the Kulu valley in the Punjab into Lahoul, and it affected yaks there in all the villages to Kyelang, leaving only one or two animals alive. In 1891, large numbers of muntjac perished from it in the hills beyond Simla (Pease, 1894).

At the end of May 1899, there was a widespread outbreak in Aligarh District with numerous villages affected, but sporadic in occurrence with a few cases not severe. It was also reported as very prevalent at this time in Bulandshaher district. One village in Aligarh lost 100 of 1,000 head in 3 months, another 10–20 out of 200 in $2\frac{1}{2}$ months. One village, which had not had the disease for 7 years, apparently not uncommon, lost 30–40 out of 800, while another witnessed mild outbreaks 3 years in succession. In at least three quarters of the villages in Aligarh, the disease had been spread by wandering cattle dealers, who passed from village to village in the hot weather with diseased herds numbering several hundred, mostly purchased in the Punjab fairs. Camping on the village outskirts, they watered their cattle in the same stagnant pools as those of the villages. One village had been unaffected for 10 years until cattle dealers brought the disease. Rogers, offering to inoculate the diseased animals of one trader, was refused, the man stating that he was going to sell them, so why should he have them inoculated? It appeared that this cause of spread was prevalent in a number of other districts also. Next to cattle dealers, Rogers considered butchers most frequently responsible for spread of the disease (Rogers, 1900).

6. POISONING

Epizootics were sometimes complicated by a widespread belief that *chamars* (leather workers) poisoned cattle in order to obtain hides. Farrell suspected poisoning in some cases, and in one village where he had suspicions, he fed a small portion of the contents of the fourth stomach of two deceased animals each to a dog and a goat. The dog died in 6 hr, the goat the next day, "both expiring in great agony." None could provide evidence of this in the Nuddea region, and as one farmer said, he would be able to recognize from which disease or otherwise an animal had died. But in some districts, hide traders provided an extensive system of advances which encouraged poisoning, arsenic often being used producing characteristic symptoms, the cattle dying within a few hours or sometimes 2–3 days.

The District Superintendent of Police in Sultanpore had heard that people sometimes gave healthy animals the scabs from infected beasts. A villager in Burdwan Division saw a shoemaker sprinkling gut contents from dead cattle on the grass of the path through which the village cattle passed out to the grazing grounds. At this same village, a shoemaker was caught in the act of cattle poisoning and sentenced. In Lohardugga, a new epizootic appeared among cattle in 1869 breaking out when rinderpest was at its height. The vomiting and death in 2–4 hr suggest perhaps poisoning. In Shahabad, an informant reported that a *chamar* was caught poisoning a buffalo in 1860, but the matter was "hushed up." It was also said that *chamars* got demons to bring on murrain. There were many cases of poisoning with arsenic in Benares Division, 27 persons being punished for the crime in 6 months. In Azimgurh, some 500 poisoners were arrested with 20 earthen mangers of arsenic, which in 1856 led to a law in Bengal against cattle poisoning, subsequently replaced by the institution of the Penal Code.

21

INDIA AND THE EAST

1. MORTALITY RATES

The Indian Cattle Plague Commission recorded mortalities of the total number of stock and, where figures were available, mortalities of those infected, the second category giving a much higher percentage. In other countries, usually only the first figure is available. In India, owing to the enzootic nature of rinderpest, the second description of mortality is more similar to the levels reported for Europe and Africa.

Table 1 summarizes some of the observations where the original number of stock was known. One must bear in mind that the Commission claimed that landowners and peasants exaggerated their losses, but the table reveals that on average in epizootics, 30% of stock were infected, and 17% died, representing 60% of those infected. Discounting the extreme case of the Calcutta dairies with their very high rates, the percentage infected ranged from 2 to 51%, the toll in the dairies reaching 90%. The percentage of stock dying ranged from 2 to 32%, with 59% in the dairies. The percentage of infected animals dying ranged from 32 to 84%, with 65% in the dairies, the very high figures probably reflecting other causes such as starvation.

In Seetapore District, of 47 villages at Baree between 1858 and 1862 of a stock of 68,544 cattle, 2,525 were attacked (3.7%) of which 1,309 (52%) died, or 1.9% of the stock. There was little difference in rates of infection and mortality a decade later in an outbreak of 1869–70. Of 304,907, some 6,542 were attacked (2.2%), and 4,878 died (74.6%), or 1.6% of the stock.

In Dinagepore District, *gootee* was said often to sweep off herds of cattle. Unless there were seasons of great and exceptional mortality, the people endured their losses uncomplainingly. In April 1870, of a stock of 252,219, 44,759 were attacked (23%), and 22,362 died (50%), or 9% of the total.

An informant from the district of Gowalparah, Cooch Behar Division, stated that in general 60% of stock was attacked, and 40% of those attacked died. Most susceptible were animals under 3 years of age.

Rinderpest erupted in Sylhet District in March 1869 according to the Superintendent of Police, whose figures the Commission did not consider to be absolutely correct, attacking some 42,000 head, or 42.2% of stock, of which 13,700 died (13.8% of total stock or 32.6% of those infected). Raging until the following May, it was estimated that final deaths would total 18,477, or 18.5% of total stock and 44% of those infected. The total stock in all villages was estimated at 99,500 head. Of 10,000 head attacked in three stations, 8,000 died.

In Dacca District, in three villages of 170 cattle, 50.6% were attacked, and 27% died, 53.5% of those infected. Commenting on this report from the Deputy Magistrate of Mymensingh in August 1869, the Commissioner of Dacca, F. B. Simson, stated that the whole country was under water, the cattle existing only on mounds where they were crowded together, and so no proper sanitary arrangements could be made. The carcasses were thrown into the water unskinned, and the heat and damp at that time of the year prevented the

Table 1. Mortality summary 1858–70.

Year	District	Stock	Percentage		
			Infected	Infected Dying	Stock Dying
1858–62	Seetapore	68,544	52	4	2
1867–8	Bijnour	46,880	73	26	19
1868	Calcutta dairies	2,068	60	85	52
	Diamond Harbour	13,402	74	43	32
1869	Calcutta dairies	1,984	65	90	59
	Mirzapore	347,754	32	10	3
	Benares	200	73	30	11
	Benares	607	84	27	23
	Dacca	11,780	75	24	18
	Dacca	170	54	51	27
1869–70	Seetapore	304,907	75	2	2
	Nischindpore	12,802	38	64	24
	Pooree	1,531	40	51	21
1870	Bograh	2,060	52	23	12
	Dinagepore	252,219	50	18	9
	Bancoorah	5,909	65	26	17
	Hooghly	4,745	65	22	14
	Banda	11,927	61	7	4
	Banda	2,700	57	13	7
	Allahabad	1,923	63	34	21
	Futtehpore	10,284	66	15	10
	Oonao	22,328	70	10	7
	Chooadanagah	12,159	47	15	7
	Total cattle	1,138,883			
	Mean %		60	30	17

successful preparation of hides. The inundation in 1870 was the highest in living memory of the oldest inhabitant, and many cattle probably died from starvation, others being very thin and weak. In November 1870, the District Superintendant of Police for Mymensingh noted that the disease generally prevailed about that season and was expected to abate. By December, reports alleged that it had ceased.

In Bancoorah, in 7 months ending June 1870, of a stock of 33,168, 14,209 (44%) were attacked by diseases, and 1,045 (34.3%) died, 77% of those attacked. Taking *bussunto* alone, of a stock of 5,909, 1,562 (26%) were attacked, and 1,008 (17%) died, or 64% of those attacked.

In the Hooghly magistracy between February and September 1870, of a total of 4,745 cattle, 1,030 were attacked (21.7%) and 674 died (65.4% or 14.2% of total stock). Figures for the Hooghly Division give 985 dead of a stock of 3,447, or 28%. Some peasants lost their entire stock, others almost so: "I had 7 cattle, *viz.*, 2 cows and 5 bullocks. I have lost all of them of *gootee*. I lost them in Falgoon (February). I lost them within one month. The disease attacked one after another…." And, "I had 20 head of cattle, i.e., 4 bullocks, and the rest cows, out of which I lost 14, leaving 2 bullocks and 4 cows; all these died of *bussunto* in Falgoon (February) and Cheyt (March). They died within a month. One bullock is still ill and looks as if he is dying. Every one of my cattle had *gootee*. Five have recovered and one, as already said, is still sick…." Another family, with 200 head, lost about 125 (Hallen *et al.*, 1871).

In Burdwan Division, 2,829 deaths were reported in 1870. In one village, Deegnuggur, about 1,000 families had 2,500 or more cattle, of which about 250 died of rinderpest in February to March. Toward the end of April, the disease appeared to be on the wane. They obtained their cattle from the Beerbhoom District and neighboring villages but did not frequent any fairs or markets. Plague appeared in January in the village of Bhagee, 6 km to the west. The people at the time of the disease went to cut wood in this village and apparently brought the disease back with them. In the village of Racona, about 175 cattle were lost of a total of some 300 owned by 150 families. They contracted the disease from the village of Konderpoor, 1 km to the north, as cattle from both villages grazed the same pasture.

In Luckimpore, by August 1869, where 11,162 cows of a total of 19,361 were reported to have died (58%), 2,427 of 3,192 buffaloes died (76%). It was reported that no disease of this description had appeared before in the region within living memory, but other sources claimed that there had been an epizootic prior to an outbreak of 1853, when two or three outbreaks had occurred at intervals of 7 years. Since 1861, it had been enzootic. The military Assistant Surgeon believed that the mortality among cattle in the district was 10–12% per year, rising to 40% in the epizootic. In the military station, about 60% were attacked, with deaths of 85–88% of those attacked. He observed that plagues appeared in May to August but after the first outbreak were unaffected by seasons.

Commenting on its contagiousness, he referred to a number of bullocks brought into Debrooghur from a place where the disease did not exist:

> Every care was taken to prevent their catching the disease ... merely stayed for a quarter of an hour in my compound, where there was a large number of diseased cattle, though special care was taken that none of them should come near these bullocks. However, the ground and paths were covered with dysenteric evacuations of diseased cattle. After their return to their factory, they were attacked by the disease in five days. (White, 1870)

The general conclusion was that in Assam, stock numbers diminished by over a half in the epizootic, and 66% of cattle attacked by the disease died.

In Jounpore, Benares Division, an estimated 12,000 head as lost, thought to be from poisoning as well as disease, but the cases of poisoning were probably a minor part of the total. In Benares District, up to July 1869, deaths totalled 4,123, or about 1% of cattle in the district.

In Banda District from April to December 1870, of 14,869 animals, 861 were attacked by FMD (5.8%), 167 died (19.4% or 1.1% of total stock), and 40 were sick. This compared with other villages with a stock of 11,927 of which 864 (7.2%) were attacked by rinderpest, 416 (48.2% or 3.5% of total stock) died and 110 were sick. In Kirwee Subdivision of 400 in villages attacked by FMD, 20 animals were attacked by rinderpest, and none had died. In other villages of 2,700, 350 were attacked by rinderpest (13%), 198 died (56.6% or 7.3% of total stock), and 15 were sick.

A total of 15,598 deaths from rinderpest (3.2% of total stock) was reported in May to September 1869 for Allahabad District (Humeerpore itself losing 1,188), which had already lost 18% of its stock, while 7,912 died of starvation, 785 from FMD, and 1,042 from other diseases and causes. "Usually, the great majority of animals recover [from rinderpest], but this year, being weakened by insufficient food, the majority attacked died." The number of deaths was considered to be some 3 times normal mortality. Two adjoining regions at the same latitude were distinguished by one being composed of black soil, and as a result, there was less famine, but deaths from rinderpest were similar in each (Table 2).

In a village in Futtehpore District, it was more virulent in 1865 than 1869. A 55-year-old informant claimed to have seen three epizootics in his lifetime. In March and April 1870, of a stock of 10,284, some 1,541 were attacked (15%), 819 died (53.2% or 8% of total stock), and 192 were sick.

In one area of Meerut in March 1870, from combined infections of rinderpest and FMD, of 10,746 animals, a total of 1,184 died (11%). In another area, out of a stock of 2,075, 1,535 were attacked (74%), 369 died (17% of stock or 24% of those attacked).

Table 2. Comparison of deaths from rinderpest and starvation in Allahabad Division May–September 1869.

		Deaths			
District	Total Stock	Starvation	Percentage	Rinderpest	Percentage
Julalpore	100,739	11,437	11.4	5,620	5.6
Momdhah	68,563	798	1.2	4,962	7.2

Table 3. Deaths at Hissar Cattle Farm 1860–69, as a percentage of total stock.

Year	Total Cattle	Rinderpest (%)	Other (%)
1860	11,362	4	26[a]
1861	8,674	1	7
1862	8,882	0	3
1863	9,961	1	5
1864	11,175	2	11[a]
1865	9,723	8	8
1866	9,959	0	9
1867	10,135	10	8
1868	9,468	3	4
1869	9,968	1	28[a]
Mean	9,931	3	11

[a]Drought years.

From 1860 to 1869 in Hissar, of a total stock averaging 9,930 head the average annual loss was 1,439 (14.9%), but this was from several diseases (Table 3). Rinderpest alone accounted for 21% of total deaths.

In Umballa Division from June to November 1870, of 30,867 head, 4,608 (13%) were attacked by both rinderpest and FMD, and 1,146 died, 3% of stock or 27% of the those attacked, while 1,216 were still sick.

In Sealkote, an epizootic began in March 1870, and up to July, a total of 3,516 head were attacked, 2,327 were dead (66.2%), and 255 still sick. In the Punjab, 46,525 cattle are reported to have died in Hafizabad in the year up to March 1870. In the district of Baraich, Nanpara area, in 219 villages, a total of 9,838 animals were attacked up to October 1870; 4,176 (42.5%) had died, and 1,997 were still sick.

Occasioning great losses, the disease was well known and widespread in the Bombay Presidency, going under many different names. In the Balaghat region, it was considered more prevalent and more fatal than in the coastal areas. In Sind 6,828 cattle died in 1862, and in an epizootic in 1866–7, 1,100. In Jerruck District in 1866–7, it was reported that 3,518 cattle, 2,829 buffaloes, 20 camels, 13 horses, 2 donkeys, and 446 sheep and goats died in one area, while in another, totals were 5,556 cattle, 403 buffaloes, 202 horses, 150 donkeys, and 4,696 sheep and goats. Whereas rinderpest seems to have been partly responsible, clearly another factor was also present. The mortality was reported as prevailing chiefly in flooded areas, flooding being unusually high that year in some regions.

In the Thurr and Parkur Superintendency, the last reported epizootic up to 1870 was in 1860 when 90% of cattle died, whereas in the Dharwar area, rinderpest was reported as enzootic, prevailing throughout the area and at all times of the year but greatest during the hot season. From March 1869 to July 1870, of 1,922 cattle attacked, 1,175 (61%) died.

At Port Canning, 34 km from Calcutta, there were over 7,000 head of cattle, herded in groups of 200–300. The number reared was insufficient to meet demand, and so in 1868 and 1869, 1,000–1,500 were imported from markets in the vicinity of Calcutta. Only 5% died from *gootee*, which always followed a fresh import. The English manager of an indigo concern claimed that the average mortality was 5–10% and never more than 25% at very worst.

In the Punjab in 1899, the average infection and mortality figures for three areas for a total of 1,012 head of cattle were an infection rate of 43 with 18% of total stock dying, or 37% of those infected. Although the percentage of total stock dying is the same as for the wider figures for 1858–70 (Table 1), the percentages of those infected dying is much lower and numbers of stock infected higher.

Overall losses for India in 1870 were put at about 1 million cows and buffaloes per year. By 1899, this figure was put at upwards of 100,000 (Rogers, 1900). This appears to have been an underestimate rather than a fall in mortality.

In the United Provinces, considerable abatement of the disease was reported for 1915–16, the number of deaths dropping from 19,839 in 1914–15, to 3,767, with few areas seriously affected; but the following year, recorded deaths rose again to 7,563. In 1917–18, there was a slight overall decline in deaths, but the following year, deaths rose to 11,281. In the Central Provinces, rinderpest was reported to be very prevalent in 1918–19 but mild in form in most districts with relatively effective inoculation. In the North-West Frontier Province, there were 928 recorded deaths in 1917–18. Deaths rose slightly the following year, but a great deal of progress was reported in inoculation, Indians now readily bringing forward their animals for treatment. In the Punjab in 1918–19, there were over 20,000 deaths. In Assam in 1916–17, there were 20,817 deaths in 348 outbreaks spread over 19 Divisions, and 29,986 cattle were inoculated. The following year witnessed a bad outbreak in Goalpara with 14,294 deaths. In 1917–18, deaths totalled 29,808, the following year only 7,239. In Bihar and Orissa, 19,144 deaths were reported in 1916–17, with 1,829 outbreaks touching most districts. In 1917–18, it was not so virulent, but there was considerable mortality. The efforts of the Civil Veterinary Department were hampered in Cuttack district due to local religious beliefs, but in 1918–19, there was a decrease in incidence.

In the Madras Presidency, there were 26,215 deaths in 1917–18, compared with 22,268 the year before, and the following year 30,733, the disease apparently increasing in virulence each year over the past 5 years. Infection was considered without doubt to come from Mysore, brought by cattle returning from grazing in the Anantpur district or brought for sale to one of the numerous cattle fairs. The serum-simultaneous method of inoculation was not practised to any extent owing to a shortage of staff, and the serum-alone method was found to be ineffective. Indians absolutely refused to allow inoculation by smearing the nostrils of animals with infective discharges.

In 1916–17, there was considerably less mortality in Bengal than in previous years. In 1917–18, it was 21,126, the following year 10,772. In Bombay, there was a marked decrease in incidence in 1918–19.

Altogether, in India, a total of 43,058 deaths were reported for 1915–16 among uninoculated cattle, and 396,293 serum-simultaneous inoculations carried out. In 1917–18, there were 67,491 recorded deaths of uninoculated cattle, and 495,117 inoculations were conducted. In 1918–19, there were 75,186 deaths. Overall, from 1912 to 1921, the number of deaths per year oscillated around 100,000, while some 800,000 animals had been inoculated.

In the 1950s, the loss was still calculated to be about 200,000 per year spread over more than 8,000 outbreaks. Following an eradication campaign beginning in 1954, in 1979–80 mortality had fallen to 2,295 deaths with 162 outbreaks (Adlakha, 1982).

2. OTHER ANIMALS

Reports of epizootics affecting animals other than cattle, while indicative of rinderpest in the cases of ruminants, when carnivores and birds were involved, suggest other infections, probably contracted from eating the carrion. Thom wrote of Burma: "These cattle, when they die, are dragged away about 200 yards from the village... tigers drag away some of their bodies and devour them at their leisure" (Pollok and Thom, 1900), but White (1869) did not believe that dogs or jackals contracted infection from eating diseased flesh.

In December 1870, a herd of buffaloes brought from Bengal passed through a tea estate in Assam, and about 2 days after a disease broke out among the pigs, every one of 40 dying. The disease then spread rapidly among the cattle, 137 having died by the end of the following January, after which the disease disappeared entirely. The total number of cattle exposed to risk was not given. White (1870) reported that pigs were attacked at Luckimpore, the disease running a more speedy and fatal course than in ruminants.

Goats and sheep were reported as suffering much from rinderpest in 1869 and 1870 in Furruckabad District, and a farmer at Chooadanagah saw sheep affected in an outbreak in 1870, four which contracted it from being housed with sick cattle aborted but recovered. If infected in the dry season, he estimated some 20–25% to die, but if during the rains, nearly all succumb. In Moorshedabad District, in one village, "five or six" sheep were lost, compared with 56 cattle in an outbreak of *bussunto*. In another region, a farmer claimed that sheep contracted *gootee*, and if 10 or 12 were infected, one or two would die; he had not heard of it in goats, but others reported goats and dogs affected. Two goats were seen to die allegedly of *bussunto* in a village in Burdwan Division.

The Civil Assistant Surgeon at Kamroop, H. R. Purves, stated that goats placed in sheds with affected cattle contracted infection, and at Sonaree in Assam, a tea-plantation manager stated that he had bought a herd of goats at Seebsaugor among which the disease appeared, killing them all, followed by all of his cattle. Rogers (1900), inoculating against rinderpest in the Kumaon Hills about 1899, met with an outbreak introduced into two villages by goats, and found that the disease was not infrequently spread by them in the Kumaon Hills; also possibly by Tibetan sheep which passed up and down the hill paths carrying goods from the foot of the hills "which is seldom free from the disease." During the last hot weather and rainy season, the disease was carried in this way up to the borders of Tibet, causing considerable loss of cattle and goats. Some hill sheep were allegedly much more susceptible than the plains breeds. In experiments at Muktesar, Rogers found goats to be much more susceptible than sheep, the disease often being more severe and fatal, and more easily induced by natural infection, bearing out the assumption that it could be spread by goats. Goats easily contracted the disease from cattle, and it could be conveyed from them back to cattle again.

An outbreak some 130 km from Muktesar in 1899 began in a small village 8 or 10 days after four goats belonging to some wandering Bhutanese died of rinderpest close by, the village cattle watering at the same place as the goats. In some 5 weeks, the disease had affected almost every animal in the village with 90% mortality of those attacked, only eight animals being left of 58. The survivors were said to have had the disease before. Spreading to another village 3 km distant, of the total of 365 animals all but 10 had been attacked, some of which had the disease as long as 10 years before, 20 which had been isolated from the village at the outset, and 28 others. Of this 307 attacked, 90% had died, and others were still ill. This high and rapid mortality contrasted with the plains, where only 10–20 animals out of several hundred were carried off over 2 or 3 months.

Buffaloes, deer, and pigs, were reported as dying in very large numbers in Assam in 1854–5, although in cattle, this disease appeared symptomatic of FMD; but in Seebsaugor district, Assam Division, in 1856, deer and other wild animals were found dead in the jungle with the same symptoms as cattle which had died apparently of rinderpest. In Kamroop in 1868, disease appeared with great severity in April, May, and June, affecting not only cattle but also sheep, goats, pigs, and wild animals. "Dr. Purves' description of the disease leaves no doubt that it was rinderpest" (Hallen *et al.*, 1871).

Pease (1894) wrote

> ... the malady is very commonly observed in the buffalo, amongst which it is very fatal. The yak also suffers from a very severe form of the disease, as do all his crosses with the cow, and it is very fatal in them ... Antelope also suffer from the disease, as do cervidae generally. The gooral and kakar, or barking deer, have been seen dying in large numbers from the disease in the Himalayas. The aurochs, peccary, gazelle, and stag also contract the disease, as also the bubalus armi, and the mithun and tsain.

Until the end of the 19th century, India had abundant wild bovid populations. Brander (1923) recorded approximately 1,500 head of game comprising 11 different species in one evening. Edwards (1930) referred to the wild herbivores as only "sometimes attacked" by rinderpest and that there were no reservoirs of the disease "as far as we know." Gupta and Verma (1949) considered that there was little information of its effects on wild animals, but in an outbreak in 1946 at the Lucknow Zoo contracted from nearby cattle, mortality was 100% among sambar, blackbuck, and four-horned antelope; spotted deer 88.4%, nilghai 75%, hog deer 66.6%, and barking deer 50%. In 1943, there had been an outbreak at the Alipore Zoo, Calcutta, in which deer, gayal and other wild animals were affected.

Buffaloes were reported seldom, if ever, to recover if attacked. In Durrung in 1867, buffaloes and deer were attacked and died in large numbers, especially the former. Other cloven-hoofed wild animals were attacked also. A tea planter stated that he had heard of deer, wild buffaloes, and tigers being scarcer than formerly, "this may be owing to disease; the natives say so" (Bruce, 1870). Indians stated that wild buffaloes, tigers, leopards, deer, "and all wild animals," were found dead at the time. Among bison in the Bengal Billiga-rungun hills, two thirds were alleged to have died, Sanderson finding many of their remains when hunting there in 1869 (Sanderson, 1878). In 1877, it broke out again among them, introduced by famine-stricken cattle which were driven into the jungles for pasture. Pollock (1894) quoted an Indian as saying that great numbers of bison died in 1867 at Annamullays, but many wild elephants suffered also. At Kamroop in 1867, it was reported that buffaloes, tigers, deer, and other animals had been found dead. The death of carnivores "and all wild animals" suggests that these deaths were caused by anthrax, but one Indian informant stated that a tiger which had allegedly died of cattle disease had ulcers on the mouth, and its thighs and tail were covered in excrement (Burooah, 1870).

At Burpettah in Assam Division, a tea planter reported seeing several deer dead in the jungle during the 1869 outbreak, and in some villages, all the pigs were carried off with symptoms almost the same as those in cattle. At Luckimpore, pigs, dogs, tigers, deer, wild buffaloes, and other wild animals were reported to be affected. The Assistant Commissioner reported that even fowls and ducks were stated to have died, but he had not seen any cases himself (Campbell, 1870). In Sylhet, Dacca Division, an epizootic in 1869 killed nearly the whole of the herds of wild buffaloes. The Commissioner when hunting noted a total absence on plains particularly suited for them, for the few survivors when seen were immediately killed by the natives. The herds of tame buffaloes were very thinned, and their numbers severely limited, but apparently not quite as badly affected as wild buffaloes.

Rainey stated once or twice, if not more often, that when hunting in the forest of the Sundarbans, he had come across carcasses of deer evidently destroyed by some plague, and that before, the disease had shown itself among cattle. "*Puschima* (choleraic diarrhea) is also designated by the natives *dukhina*, because it travels from the *dukhin* or south; hence may it not be possible that the plague originates with the (wild) deer, and from them communicates itself to the cattle in the Soonderbun grants, and gradually spreads northward and eastward?" (Rainey, 1870).

Farrell reported examining a tame deer which had died and presented the same appearance as animals affected with *gootee*. He saw it in a tame pig and heard, but had not seen, that pigs, wild boars, tigers, and jackals had died of it.

Sir Samuel Baker reported a few years prior to 1890 that there was an epizootic among bison in the Reipore district of India which spread into neighboring districts over a wide tract of country, causing great mortality, but no deer were affected. On occasions, he had found bodies of buffaloes in great numbers scattered throughout the country in all directions "causing a frightful stench" (Baker, 1890).

In Dehra, Meerut Division, in 1862 a fatal disease similar to rinderpest broke out among elephants at the same time there was an epizootic among cattle, but no details are given (Hallen *et al.*, 1871). A few years later, elephant herds in the Mysore Kakanote jungles were attacked by a disease, Sanderson recording that the mortality was not great and the disease soon abated (Sanderson, 1878). In the Punjab at Baharalla, Jhung District, camels, sheep, and goats were attacked, and 500 died of a disease called *thuddee*, considered to be rinderpest.

In Mysore in 1903, bison, sambar, and elephants were found dead in considerable numbers in the jungles (Stebbing, 1911), the elephant deaths suggesting anthrax as the cause.

Brander (1923) wrote that "until a few years ago," the State Forests of Khandwa District lying north of the Nerbudda River contained a small herd of gaur until they were wiped out by rinderpest: "The admission of village cattle to a restricted area which had been long closed to grazing was probably the determining factor in their disappearance." Known from actual records, and from experience of some places such as Yeotmal and parts of Mandla, buffalo had disappeared during his generation: "Inhabiting country as they do, they are constantly liable to infection from village cattle, and although this would tend to create a certain amount of immunity it has proved insufficient. Bison, on the other hand, owing to their habits often escape infection, but are less immune and show less resistance to the disease than buffalo." Rinderpest sometimes destroyed the stock of sambar deer in a particular tract, whereas barasingha were liable to suffer from epizootics of rinderpest but were far more resistant and suffered less mortality than either sambar or bison. In 1975, mass deaths of bison were reported in Periyar Wildlife Sanctuary believed to result from rinderpest transmitted by cattle (Anon., 1975).

What these limited observations show is that, despite its prevalence, rinderpest was apparently not enzootic in the wild ruminants.

There being no strong conservation-oriented body of opinion in conflict with veterinary opinion and a lack of a European-based stock industry demanding government action, it was always taken for granted that cattle infected the wild animal populations, rather than vice versa, as was the opinion originally held in Africa.

3. ATTEMPTS AT CURE

Several ancient treatises were compiled for cattle ailments, most important being the *Salihotra* about 600 BC. Although lost, an extract is quoted in the *Agni Purana*. Describing various herbal cures for different ailments, for diarrhea two kinds of turmeric or the Patha (a plant commonly called acindhi) should be given. In every disorder of the bowels, dried ginger and Bhargi, a shrub, should be given:

> I shall now speak of the Dhupa or incense made with Devadaru (*Pinus deodar*), Vacha (orris root), Mansi (spikenard), Guggula (agolochum), Hing (asa foetida) and mustard seeds, which does good

> to cows, and destroys the diseases occasioned by the unpropitiousness of the planets. The bell which is suspended from the neck of the cow should be exposed to the exhalation of this incense ... in the case of epidemic diseases, Vishnu, together with Lakshmi, should be worshipped; a level square piece of ground having been selected for sacrifice, a lotus consisting of eight petals should be drawn within it, over the filaments the names of the following gods should be written, namely Subhadra, Ravi, Akasa, Viswarupa, Siddhi, Riddhi, Santi, Rohini, and Indra, and they and others should be worshipped on the petals of the lotus, commencing from the east side. The Dikpalas should likewise be worshipped on the petals of the lotus, and oblations should be offered in the fire. The sacrificial fire should be kindled with the wood of the Kshira tree (*Mimosa kauki*) and mustard seed, rice, and a hundred Suvarnas (a certain weight of sixteen Mashas), bell metal, and other metals should be given away to the Brahmanas. On that day the cows should not be milked, but set at liberty.

At least this was harmless, unlike a Hindu cure for murrain, which was to hire a man of lower caste to take the evil upon himself and bear it way into the jungle. His face was turned away from the village and branded with a red-hot sickle, and as he walked away, the murrain went with him (Burdick, 1905).

Little wonder, in most places in the 19th century, Indian peasants admitted to being unable to cure rinderpest; "there is no medicine for *matah*" was a frequent village confession. Feeding with rice gruel or other soft food was sometimes undertaken, but usually the methods were fanciful. More often, cattle were left to recover on their own, although a Mr. Liebert in reply to the Commission stated that he would leave treatment to the natives: "They have so many good remedies of their own that I consider interference on this point useless" (Liebert, 1869).

Fumigation with resin was practised in some places. Worship of the goddess Shitolah and sacrifices at her shrine were much more approved of, and in some cases, the aid of the village *chamar* was invoked to repel some demoniacal influence. Considering the numerous and fanciful "cures" for elephant diseases, the lack of interest in cattle is surprising. Only in the Bombay area do we hear of 32 native modes of treatment, among them elephant's dung mixed with water and powdered tiger's tongue. In Meerut Division, villagers in one region caught and killed jackals and buried them on the pasture as a remedial measure. Burning tiger's flesh, making the smoke enter the nostrils of the sick animals, was another. Peasant owners in Diamond Harbour reported that no particular treatment was adopted, but the goddess Shitolah was worshipped.

In Raipur District in the Central Provinces, the cure for *chhye*, considered to be rinderpest, was to slit the ears or cut off one inch from the tips, then beat them with a switch until there was a copious flow of blood. A ball of salt, horseradish, root bark, chillies, butter, milk, ghee, eggs, and red ochre was then administered. No cure was attempted for *matah*, although *chhye* appeared to be just a more virulent variant.

Mr. Lamb reported in 1836, "The treatment is sometimes guided by this indication afforded by nature [i.e., diarrhea], and a variety of vegetable purgatives are employed, but more generally is limited to absurd mummery and fanciful magical incantations." In Futterpore District, Allahabad Division, villagers reported that they buried the carcass of the first animal which died of rinderpest in the belief that if the animal which dies first is buried, the disease would be checked. Sanderson (1878) reported that infected herds were frequently driven into the jungles where tigers were known to be, with the belief that if the tiger could be got to kill a beast, the disease would leave the rest. It was believed that tigers would not normally attack infected animals, and Sanderson deduced that when one recommenced attacking herds, the disease was probably on the wane, thus leading to the superstition. As to the reluctance of tigers to attack infected animals, this he attributed to the stench. He asserted that tigers would not touch the carcasses of animals which had died of rinderpest, but this was contrary to Thom's experience in Burma.

In 1811, and again in 1852, the treatment given at Hoonsoor military depot was frequent injections of a solution of sugar of lead into the nose and ears, applying alum water to ulcerated parts, and the use of camphorated oil. In an epizootic in Backergunge in 1816, the sick cattle were allegedly cured with opium and other human remedies (Jameson, 1820). In Maunbhoom in 1869, the Civil Medical Officer, Babu Woodoy Chund Dutt, set up a "cattle hospital" in a shed to which seven animals were admitted for treatment. Two died, and five were discharged "cured," after which the "hospital" was closed. His treatment at the commencement of the disease was leaves of the *muddar* plant (*Erythrina fulgens*) ground together with 21 black peppers, a medicine much used by the natives and which he considered a useful febrifuge and diaphoretic. In the second stage of diarrhea, he gave a mixture of catechu and opium in fermenting boiled rice water. Lieutenant Blathwayt, Assistant Commissioner of Golaghat, reported in 1869 that he had tried to begin treatment of sick animals according to Dr. McLeod's method, setting aside a building as a hospital, but the Assamese refused to send any cattle. Farrel reported that resistance was general throughout the province, with very few exceptions. At first, it was partial, but when it became known that a Government officer was deputed to examine into and treat the disease, the people positively refused any assistance, however small, stating that if English medicines were used, their god would destroy all the cattle as a punishment.

In the district of Hazareebagh, the Civil Surgeon S. Delpratt, reported that no remedy was known, and no medicine had been administered, "... sacrifices of goats to *Kali* and *poohjahs* have been made, but without any favourable response having been vouchsafed by the goddess."

Mr. Williamson of Behalee (Assam Division) in 1870 reported administering a tola (180 grains, the weight of a rupee) of opium in a $\frac{1}{2}$ quart of rum with good results, Mr. Lyell supporting this opinion. In Chittagong, after losing two cows, when the third was attacked, the owner administered aconite and belladonna alternatively for 5 days, the cow recovering perfectly, allegedly from the treatment. In a station in Chota Nagpore, following an outbreak among Commissariat cattle in 1858, they were treated with tartarized antimony without success.

When the disease broke out among Commissariat cattle at Gwalior in 1870, Major Wroughton administered a mixture of ghee and black pepper twice a day for 3 or 4 days. On the third or fourth day, this was changed to a mixture of carpopogon (*Mucuna puriens*) and henna (*Lawsonia inermis*) in water for 3 or 4 days. Of 26 cattle attacked, 7 died (27%), so Major Wroughton recommended his treatment as being efficacious. Dr. Corbyn claimed he had often treated the disease with success. His remedy was an initial purging with a mixture of Epsom salts, sulfur, niter and ginger, and strong sulfur ointment rubbed all over the body. If the mouth was sore, it was to be rubbed several times a day with a liniment of powdered ginger, powdered caraway, and gentian *Gentiana chiryata*. The cause, he considered, was exposure to sudden changes in the weather and feeding on wet land.

The veterinary surgeon at the Home Stud, Kurnaul, H. Dawson, recommended inserting a trochar to allow the escape of rumen gases if the belly was very distended. If no gas escaped, the contents of the rumen were to be removed by enlarging the incision through which the trochar was inserted and removing them by hand. A purgative of Epsom salts should be given in all cases. He was also a believer in rowelling, and considered that healthy stock should be rowelled as a precaution. His instructions were circulated to all Commissioners in the Punjab with a recommendation that they be translated.

Contrary to the customary recommendation of isolating sick cattle, in 1811 the Commissary General at Mysore, Colonel John Hill, stated that healthy cattle should, if possible, be moved from the neighborhood of infected localities; noting when casualties among private cattle which were not moved were counted by thousands, the Government herds very frequently escaped unscathed. But later in Lower Bengal, it was considered that removal of healthy cattle posed insuperable difficulties owing to the large numbers, and apparently healthy cattle could carry the infection with them, spreading it to a fresh locality. Stamping out was also impractical among the large herds kept in Lower Bengal.

The Collector of Kurnool, forwarding Thacker's report on cattle disease in the Madras Presidency to the Board of Inland Revenue, was hopeful as to the success of medical treatment, intending to have Thacker's directions for treatment translated and widely circulated. But he considered that nothing short of a special legal enactment was required to ensure the application of preventive and sanitary measures. Thacker's most radical measure was to adopt Colonel Hill's recommendation of removing the healthy cattle away from the sick, rather than removing the latter. This he considered easier to effect and more successful, but on one occasion when the disease had broken out in a herd and he advised the owner to remove the healthy animals, one of the latter was taken 24 km distant and placed among a herd of 120 healthy cattle. Within 2 weeks, 70 were affected by the disease. He claimed that he had treated 2,541 sick animals, of which 1,763 recovered, and 778 (30%) died; compared with 80–90% deaths if left untreated. But 30% was an average level of mortality for the more benign infection. Claiming that the main object was to support the strength and assist the powers of nature, he recommended slight laxatives and gruel in the early stage, then if there was no purging, a mixture of camphor, datura, niter, gentian, and arrack, in wheat gruel. If diarrhea persisted, after 24 hr, 2 drachms of gall-nut should be given.

A small farmer in 24-Pergunnahs, babu Sumbhoo Ghose, in evidence to Kerr stated that he had always owned 30–35 head of cattle. In the month of February or March 1866, people in a village about 3 km distant from his house purchased 12 or 13 bullocks at Oolooberiah market. Four or 5 days after the people purchased the bullocks, one showed signs of *bussunto*. The other cattle then became ill, and the disease gradually spread. A sick animal sank in 5–7 days, and if it recovered, it would take 15–17 days. The outbreak continued for upwards of 2 years.

> The disease is infectious and yields to no medicine known. I myself took much care of my cattle at this time, and out of 20 head of cattle attacked, I lost only three.... I used to clean the cowshed twice and I removed the healthy to another room, though in the same compound. I did not use any medicines, nor did I know of any. People in every part of the country knew the disease to be contagious, the disease was everywhere, and they could not think of separating the healthy from the sick. Besides, they had no means of doing so. The cattle, as a rule, graze during the day in the fields and are kept in the house at night; but during the epizootic, people did not follow this practice freely, and sick cattle were never taken out of the house. At the commencement of the disease, carcases were thrown into the village bhagar [carrion pit] as before, but latterly, on receipt of orders from the authorities, those who could afford, buried them, and those who could not, threw them into the river or canal.

In one village, both sick and healthy cattle were kept together in a shed with a dead one buried at the door to keep away the disease. The owner informed the Magistrate that he would separate them but did not, and all died.

In Furruckabad District in 1869, fumigation with the smoke of burning straw was practised by the inhabitants of two areas against FMD (without effect), but fumigation was never reported for rinderpest.

A Bengal Government circular letter issued in August 1869 directed a trial with carbolic acid (Anon., 1869). The Assistant Surgeon at Debrooghur complied with the instructions on 10 sick animals, of which 6 died and 4 recovered. He was not inclined to attach much importance to the trial, observing that from scientific journals reporting on its use in Europe, the results were not such as to justify its being regarded in any measure as a specific (White, 1869).

4. PREVENTIVE LEGISLATION

In 1866, the Madras Council passed Act II of 1866 for the prevention of the spread of cattle disease in the Madras Presidency, but after 3 years' operation, it was declared to be neither "expedient nor necessary," and the stamping out, limitation, and treatment of murrains was to be entrusted to the knowledge and personal efforts of individuals. The Act, which included camels, provided for the establishment of hospital pounds; obliged an owner under heavy penalty to report to the pound-keeper attack by contagious disease; directed the pound-keeper to examine the animal, order its removal to the pound, and disinfection of the shed in which it was kept; gave the pound-keeper power to examine suspect animals and contaminated premises, and act accordingly; provided that the expense of food and treatment may be defrayed by the Pound Fund or by the owner; gave the pound-keeper power to destroy diseased animals; directed burial and did not admit compensation; rendered the levy of fees for care and treatment lawful and sale of the animal to realize them; prohibited the sale or removal of diseased animals without a licence from the pound-keeper, or any animal which had been in contact with an infected animal; attached a penalty to disobedience of the pound-keeper's orders or opposing seizure of cattle under the Act; provided a remedy for improper seizure; placed fees and fines to the credit of the Cattle Pound Fund; and gave the Governor in Council power to enact by-laws as necessary.

The Act contained no definition of which diseases should be considered contagious and infectious, this being left to the hospital pound-keepers to decide. Being unskilled in veterinary art and having less experience than village cow doctors, it was considered by one Commissioner that they could be liable to corruption, extorting money by pronouncing an animal to be infected when it was not so. This could be overcome by defining the diseases and making the pound-keeper responsible for any illegal exercise of his power.

In January 1869, the Head Assistant Collector of Kistna District complained of the difficulties of carrying out segregation and recommended the Act be amended accordingly. As to the advisability of applying the Act, the Collector of North Arcot earnestly deprecated the measure. The Collector of Coimbatore did not consider the Act necessary, the people themselves being willing to carry out measures of segregation and treatment. But the Board of Revenue recommended that Government continued the measures "now in operation with so much success" in preference to appealing to the Legislature. On June 26, 1869 the Government decided that the Act, or any similar enactment, was not called for at present.

There was thus no specific legislative control in existence, except relating to public health. Slaughterhouse proprietors could be prosecuted under Act XXI of 1841 if they caused a nuisance. Some considered that certain actions could also be taken under the Penal Code if instructions were not complied with to prevent cattle in a village in which the disease appeared wandering among those of other villages. Requests were made countrywide by the Commission for comments from magistrates and officials on proposed methods of control. Replies were conflicting, some arguing that the Indian Penal Code was sufficient, others not. But as one official put it, if legislation had been found to be necessary in England, where enlightened self-interest might be supposed to have considerable influence, then it must be more essential in India with its ignorance or disregard of the most ordinary sanitary laws and utter recklessness as to how individual action might affect the general mass (Eden, 1869). Most considered enforced burial of carcasses impracticable, even though people would be prepared to do this, as well as segregate sick cattle. But Hindus would oppose slaughter and burning carcasses on religious grounds. In one Jessore District village in 1869, all the cows of a peasant had died except one, which was so ill that Farrell recommended its slaughter, the owner agreeing. Subsequently, criminal charges of slaughtering cattle for their hides were laid against him by people in the village. It transpired that this action had been instigated by the *chamars*, whose trade had stopped because of the burial of animals intact.

The station commander of the 42nd Regiment Assam Light Infantry at Debrooghur imposed orders in 1869 to the effect that any sick cattle found in cantonments would be confiscated and the owners fined under Act XXII of 1864. Owners allowing their cattle to die in cantonments, or in the civil station, and adopting no measures to ensure speedy disposal of the carcasses, would be prosecuted under Section 268 of the Penal Code (PC).[1] A JP at Debrooghur, owner of a tea estate, stated that he saw that measures were carried out which the Deputy Commissioner had adopted to prevent the spread of the disease, but, he claimed, these were not strictly legal. These measures were for police guards to be posted to prevent the passage of cattle at certain points; the burial of dead cattle; that all owners of cattle should report outbreak of disease and mortalities to the nearest police station; quarantining of affected areas and throwing a cordon of police around them to prevent the egress of cattle; and forbidding the disposal of carcasses into the river upstream of a certain point.

An opinion from the Bengal Judiciary Department was that Section 269 and following of the Penal Code referred only to acts likely to spread infection dangerous to human life, and the definition of a public nuisance under Section 268 was not sufficiently appropriate. A magistrate could not take order for the prevention of acts as referred to under Section 62 of the Criminal

[1]*Penal Code.*

Sec. 44. The word "injury" denotes any harm whatever illegally caused to any person, in body, mind, reputation or property.

Sec. 188. Whoever, knowing that, by an order promulgated by a public servant lawfully empowered to promulgate such order, he is directed to abstain from a certain act, or to take certain order with certain property in his possession or under his management, disobeys such direction, shall, if such disobedience causes or tends to cause obstruction, annoyance or injury, or risk of obstruction, annoyance or injury, to any person lawfully employed, be punished… And if such disobedience causes or tends to cause danger to human life, health or safety, or causes or tends to cause a riot or affray, shall be punished…

Explanation: It is not necessary that the offender should intend to produce harm, or contemplate his disobedience as likely to produce harm. It is sufficient that he knows of the order which he disobeys, and that his disobedience produces, or is likely to produce, harm.

Sec. 268. A person is guilty of a public nuisance who does any act or is guilty of an illegal omission which causes any common injury, danger or annoyance to the public or to the people in general who dwell or occupy property in the vicinity, or which must necessarily cause injury, obstruction, danger or annoyance to any persons who may have occasion to use any public right.

A common nuisance is not excused on the ground that it causes some convenience or advantage.

Sec. 269. Whoever unlawfully or negligently does any act which is, and which he knows or has reason to believe to be, likely to spread the infection of any disease dangerous to life, shall be punished…

Sec. 270. Whoever malignantly does any act which is, and which he knows or has reason to believe to be, likely to spread the infection of any disease dangerous to life, shall be punished…

Criminal Procedure Code.

Sec. 62. (1) In cases where, in the opinion of a District Magistrate,… or of any other Magistrate specially empowered… to act under this section, immediate prevention of speedy remedy is desirable, such Magistrate may, by a written order stating the material facts of the case and served in manner provided by section…, direct any person to abstain from a certain act or to take certain order with certain property in his possession or under his management, if such Magistrate considers that such direction is likely to prevent, or tends to prevent, obstruction, annoyance or injury, or risk of obstruction, annoyance or injury, to any person lawfully employed, or danger to human life, health or safety, or a disturbance of the public tranquillity, or a riot, or an affray.

(2) An order under this section may, in cases of emergency or in cases where the circumstances do not admit of the serving in due time of a notice upon the person against whom the order is directed, be passed *ex parte*.

(3) An order under this section may be directed to a particular individual, or to the public generally when frequenting or visiting a particular place.

Procedure Code (CPC), as this had to be directed at individuals and not to the community at large. Actions of the police not having the sanction of law could not be enforced by the executive authorities in the case of resistance. Others disagreed with these interpretations, claiming that Section 62 of the CPC *could* be used, and punishments awarded under Sections 188, 269, and 270 of the Penal Code.[1] Under Section 62 of the CPC, a Magistrate could issue an order directing a person to take certain order with any animal under his charge when such directions tended to prevent injury to any person lawfully employed, but injury was not defined. However, it was defined in Section 44 of the PC, which could provide a Magistrate with large powers to prevent the spread of infectious disease among cattle.[1] As to whether allowing diseased cattle to go abroad amounted to a public nuisance or not under Section 268, there were no Indian precedents on this point, although there were in English law. But the police had no power to interfere and could only report the offence or prosecute. Although the law of nuisance was wider in England than in India, it had nevertheless been necessary to pass special legislation in England to control contagious diseases of cattle. In Luckimpore, several men were punished for disobeying orders prohibiting transfer of cattle from infected and healthy localities. On appeal, it was ruled that the orders were illegal and the men released.

An Order could be issued under Section 62 of the CPC and then a summons under Section 188 of the PC, but all would plead ignorance of the Order. As it was quite possible that a person might have been ignorant, he would be warned and discharged. Before being caught again, any damage would have been done.

An Act for the Prevention of Cruelty to Animals could be applied to enforce cleanliness and proper treatment, but the Conservancy Inspector of the Health Officer's Department of Calcutta apparently had the power to seize diseased cattle, and if loose on the street or running about unclaimed, he could prosecute the owner if known. Sick animals which he seized he could have destroyed after obtaining a warrant. In the districts of Bengal (1869), an order of the Lieutentant Governor provided stringent rules for preventing the removal of hides from carcasses and to bury carcasses intact, requiring all district officers to enforce this. In Assam Division, the Assistant Commissioner reported in August 1869 that he had ordered separation of diseased cattle and burial of carcasses intact 1.2 m deep. At the end of January 1870, the Assistant Magistrate, Chooadangah Subdivision, issued orders for the burial of carcasses 0.9 m deep. The *moochees* (shoemakers) petitioned against it, and "respectable natives" objected on the grounds of inability to get shoes.

Among proposals for control was a rinderpest quarantine plot fenced and employed as a sick-bay assigned to every village or group of villages where sick animals could be promptly segregated. Slaughter was so antagonistic to public feeling that it was not recommended. In Bengal, people were against the flaying of carcasses of animals which had died of rinderpest for fear of releasing the demon which was supposed to have taken possession of the beast, so ordering the slashing of hides was considered unnecessary.

The Collector of Kurnool had proposed an Act providing for the erection of hospital sheds, cure of the sick, early report of the disease under penalty of neglect, and disinfection of infected premises. Breaches of the Act should be made punishable under Sections 269 and 270 of the PC and the Act made not compulsory, but capable of application on requisition of the local authorities.

A draft act was submitted to the Government of Bengal by the Secretary of the Agricultural and Horticultural Society, but was an empowerment act only based on the English legislation. In Orissa Division, it was felt that the Madras Act of 1866 was more suitable to the country. Charles considered that any legislation having for its object the repression and prevention of rinderpest should be introduced with great caution: "I believe it to be doubtful whether much good would result from a very strict law on the subject until the people have been trained to see the necessity for it" (Charles, 1871). Asked if it would be better for Government to interfere, a cultivator in Assam replied: "I should like the Government to interfere, so far as it did not concern me."

From another source, the objection raised to legislation was with police who did not know whether any disease which showed itself was infectious or not; the law, however mild, would be used "as an engine of grievous oppression" (Boddam, 1870). Others emphasized that trained staff would be required to implement any control.

In some areas, sick and healthy cattle were not separated because the owner had only one shed. But in other cases, it was claimed that cattle were not separated on religious grounds, owners fearing that by doing so, they would offend the goddess Sitla (Shitolah). As concerned slaughter offending the beliefs of the Hindus, Stalkartt considered this a very shallow objection as so many Hindus now ate beef and knowingly sold their cattle for slaughter; also, they should traditionally slaughter cows at the Dewalee Pooja, but as cows were dear, they sacrificed buffaloes instead. But Farrell advised against stamping out on the grounds that the disease was not imported but of spontaneous origin, and if enforced contrary to the people's beliefs, there would be unpleasant consequences: "Mild persuasion and tact with the natives will always have more effect than forcible measures in matters mixed up with their religion" (Farrell, 1871). The Magistrate of Tipperah, E. D. Lockwood, in his report to the Commission in August 1869, observed that there would be great difficulty in stamping out the disease in a Hindu village, for the pious Hindu would rather see his family perish before his eyes than have recourse to such an unholy expedient: "The Mussulman community also would resort to a thousand subterfuges rather than give up the faintest hope of saving their cattle."

Extraordinary ignorance was shown in reply to the Commission by Colonel Haughton, Commissioner of the Cooch Behar Division. It was his belief that infectious disease among cattle in India was comparatively rare, and murrain in some localities was due often to causes such as caterpillars or other insects in the grass, or poisoners who poisoned the food or inserted poison through wounds "and the like." "My opinion is that there is not at present sufficient information before the public on the subject of cattle disease to warrant any legislation at all" (Haughton, 1870).

Some years later, people were expressing the opinion that any interference with the peasant's management of cattle would result in another mutiny. Wallace (1888) wrote that this was absurd; his inquiries revealed that few would even grumble if regulations were passed by Government on enlightened lines which would prove of benefit and blamed desk-bound bureaucrats for opposing any measures of improvement. But he considered that the Madras Act of 1866 (unaware it had been repealed) went too minutely into detail for the present state of knowledge. It was sufficient, where disease had broken out, that Government should place areas under quarantine and suspend bullock traffic on highways where it was known that such traffic could spread the disease.

5. ECONOMICS

The value of cattle had risen in the 10 years to 1870 from 10 to 100%, but cattle disease was only a local and transient factor affecting prices, although a bullock or buffalo which had recovered was more valuable. In the Dum-Dum cattle market (Bengal Division), a butcher stated that the price of slaughter cattle had increased in the 10 years to 1868 from 8 to 10–12 rupees. At Fort William (Calcutta), a meat agent stated that cattle had become cheaper again because troops now landed at Bombay instead of Calcutta, and now not so many ships came, so there was less demand for meat.

Asked why he thought the price of cattle had greatly increased in Assam, the Assistant Surgeon replied that he did not think it had anything to do with the rinderpest plague but was to do with the large amount of money thrown into the province owing to the tea enterprise, although cattle were very little employed in the venture (White, 1870). He did not expound further, but doubtless increased employment opportunities among the peasantry created more and larger cattle owners. The Reverend Higgs, Deputy Commissioner Debrooghur, considered it wrong to measure the effects of cattle plague by the ruin that would result in England or some other European country. In Assam, the loss was confined to comparatively few individuals who only suffered from loss of their customary draught of sour milk. He thought that if all cattle in Assam were to die, scarcely any loss of capital would accrue, and he did not believe agricultural operations would be affected. The 1869 rice harvest was the most plentiful in 20 years, and yet the cattle plague was at its height at the busiest time of the year for cultivation, being the most severe that had occurred within the memory of most people (Higgs, 1869).

The traders in hides considered that hides from animals which had died of rinderpest were of no use to them and easily identified, although a hide merchant in Purneah reported that in the past 2 years to 1870 he had been buying 3,000 hides a year, as opposed to 1,000 or less per year before that. This was due to the increase in price on the Calcutta market. In Cawnpore, it was reported in 1869 that a great number of hides of animals which had died of rinderpest had been brought over the Ganges, but there were few the following year. A Calcutta hide broker stated that trade in hides in the Bengal Presidency had increased very much during the last 2 or 3 years to December 1870, and exports from Calcutta had increased from $2\frac{1}{2}$–3 to about 4 lakhs per month owing to increased demand in Europe, trade having been encouraged by the Suez Canal opening in November 1869.

6. THE LATE 20TH CENTURY

Effective control began in 1934 with the development of Edwards's caprinized vaccine at the Indian Veterinary Institute at Izatnagar, but a National Rinderpest Eradication Program did not begin until 1954. As a result of mass vaccination, followed by annual vaccination of calves, the establishment of controls on interstate borders, and vaccination stations on international borders, incidence of the disease was reduced considerably, but total eradication was not achieved. Figure 1 shows how mass vaccination has an initial startling effect upon incidence, but it then becomes increasingly difficult to eradicate remaining foci. Vaccination of cattle let out to graze in forest grasslands was practised with success in restricted areas, Coimbatore in the south and Mandla in Madhya Pradesh; but in 1969, Seshadri wrote that he doubted if such vaccination was then being widely employed anywhere.

During 1971–2, the number of vaccinations was increased from around 26 million in 1958–9, to 44 million, reaching 50–59 million in 1975–8. But the number of foci rose by 140 to 160 per year in the population of about 240 million cattle. In 1981–2, about 190 foci were declared, mainly in the southeast. At the same time, from 1970, there was an increasing number of outbreaks in sheep, goats, and even pigs, especially in the southeast where the disease had become enzootic, representing 28% of the 331 reported foci in 1981–2. In Tamil Nadu, it affected sheep and goats but not cattle, which seemed immune, suggesting that the virus was maintained exclusively in the former, a situation until then not reported elsewhere (Khera, 1979) (although long suspected). Raising the level of vaccinations to 20% of the cattle population did not eradicate the disease (Plowright, 1985).

In 1991, The South Asia Rinderpest Eradication Campaign was introduced by the Food and Agriculture Organization (FAO). In 1985, rinderpest still persisted in the tip of Peninsula

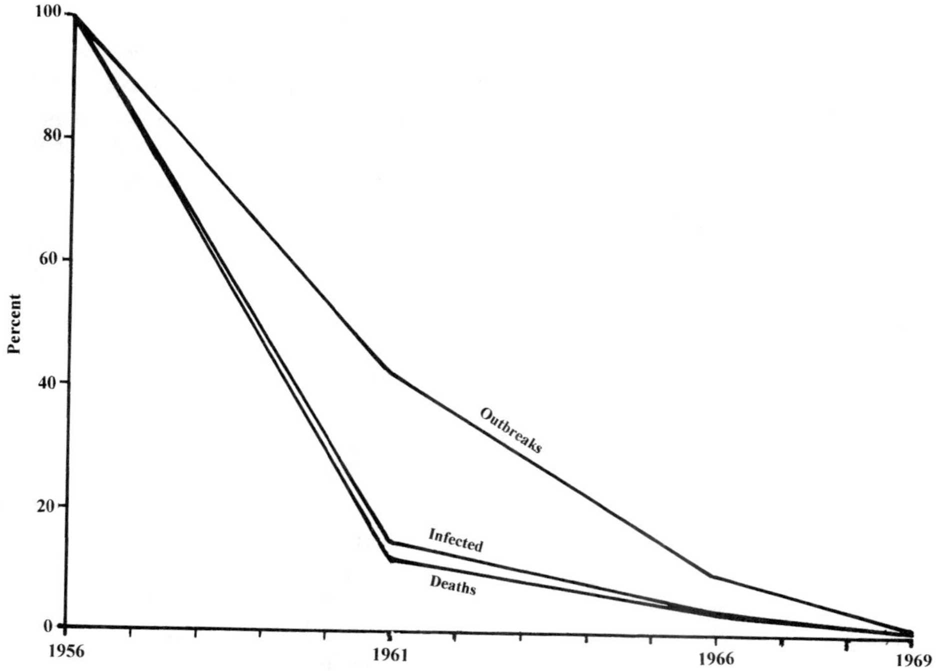

Figure 1. Initial progress of the rinderpest eradication programme in India. Smoothed data based on figures in Adlakha (1982).

India (formerly Madras Presidency), and in 1992, a National Project on Rinderpest Eradication drawn up in 1987 was launched with the aid of EEC funds. This resulted in a drop in reported outbreaks in cattle from 103 in 1993 to 29 in 1994, 10 in 1995, and zero in 1996. From March 1998, India declared itself provisionally free of rinderpest, but it is considered that there is no conclusive evidence that the steady decline was due either to vaccination or to herd immunity (Rossiter, 1996).

In October 1995, Afghanistan was infected via the Punjab through the North-West Frontier Province. Although a vaccination campaign was undertaken, this was relatively ineffective at first, as the veterinary personnel looked for classical rinderpest symptoms, whereas many animals exhibited only mild effects and were not vaccinated, and cattle owners did not appreciate the implications if their cattle were not vaccinated. But the outbreak was contained by February 1996 mainly because numbers of stock in the country had been limited by war. Restocking from Pakistan in several regions led to the 1995 outbreak. In 1994, Pakistan opened up a new trade route to Central Asia, livestock from Pakistan passing through Afghanistan to Tajikistan, Uzbekistan, and possibly beyond, instituting the possibility of a reintroduction of rinderpest into its historical region of enzooticity. A serological survey of 951 animals in 1998 detected 11.8% exposed to infection. This was followed by a more thorough survey in 2000–01 in which of almost 20,000 animals, 0.25% tested positive, and in 2001, a total of 3,564 samples from buffaloes and cattle in the suspect areas were tested by two different methods all proving negative.

In Pakistan, rinderpest is known to have occurred in cattle and buffaloes since the beginning of the 20th century, and two epizootics were recorded in the 1950s when "hundreds of thousands" of animals died. Since then, it continued to smoulder, although for some years, it was considered that rinderpest did not occur in Pakistan, but what was classified as a "rinderpest-like disease" is now known to have been classical rinderpest, diagnosed in Sindh Province each year since 1991. In 1993, an epizootic was reported in Punjab State. The following year, a separate epizootic appeared in the Northern Areas in Gilgit and Hunza Valleys. This outbreak killed an estimated 40,000–50,000 hill cattle within a year before being controlled by vaccination and has been described as perhaps the best example of cattle plague in the world for 20 years in demonstrating its devastating potential. Both epizootics had their origins in the south through trade in cattle and buffaloes. The disease had been maintained for many years by constant circulation within the Landhi Dairy Colony, and possibly other colonies, which supply milk to Karachi, but it is not clear why the disease was able to persist in Sindh State, and possibly other States. Buffaloes and cattle are intensively stall-fed throughout the period of lactation, and then most are slaughtered, but a proportion is returned to the villages for breeding. This provides a constantly changing population which continually receives rinderpest-susceptible animals into the dairies, but those being returned to the villages are probably immune. Although the epizootic died down, there were reports of rinderpest in sheep close to Jammu and Kashmir in 1995, and an outbreak close to Islamabad was reported in 1996 (Rweyamamu, 1996). The discovery of a goat with mild rinderpest might have had important epizootiological significance, but when other sheep and goats from the same village were tested for antibody a few months later, less than 5% incidence was found. This left unanswered the question why, if the virus could infect some goats, it did not spread among others which had been in contact. In June 1997, an outbreak occurred among buffaloes in Rawalpindi introduced with 1 cow buffalo brought from 400 km away. In the year 2000, 3 outbreaks were detected on dairy farms near Karachi and inquiry revealed a small number of unreported outbreaks had occurred in northern Sind during 1998 to 2000. It is still not clear how this infection persisted but by March 2002 it was declared there was growing confidence the virus was limited to the Indus buffalo tract of Sindh Province (Roeder, 2002a). A survey of 2,697 animals in 2000–1 in Baluchistan, the North-West Frontier Province and the northern areas of Jammu and Kashmir, revealed that only 0.3% were positive, but there was concern at the risk of spread from the Indus valley because in this final stage of the eradication process, all countries had ceased routine mass vaccination in favor of surveillance and elimination of residual foci of infection. The risk was from movement of livestock for slaughter to feed troops and civilians, and breeding stock required for rehabilitation of agriculture in Afghanistan as well as development projects in Pakistan. "Never have the countries of South Asia been so vulnerable to a resurgence of rinderpest," stated one source. But the last reported outbreak in Sindh Province was in October 2002, and by November, it was announced: "It is conceivable that Asia is now free from rinderpest for the first time in millennia" (Roeder, 2002b).

7. OTHER ASIAN COUNTRIES

7.1. Andaman Islands

The Andaman Islands were first stocked with cattle sometime between 1858 and 1864. In September 1868, rinderpest appeared among the Commissariat cattle introduced with cattle from Burma, but great vigilance prevented its complete spread. Nevertheless, of a total island stock of 555, some 325 animals were attacked, and 120 died. In October, a further batch of cattle arrived from Burma with one dead on board and another animal obviously suffering from rinderpest. The animals were kept isolated, and the disease did not spread. On December 25, an infected shipment again arrived, this time from Calcutta. One cow died the day after arrival, and seven more during the next 2 weeks. Once more, precautions prevented its spread, cattle being disembarked at an isolated spot and kept there until no more disease was apparent. Thus, between September and December 1868, rinderpest was imported three times but as of 1871, had not been recorded again.

7.2. Burma (Myanmar)

The first report of an epizootic disease in Burma (reports of deaths of cattle and other animals in the military expedition to Arracan in 1825 would seem to relate to starvation) is of an outbreak of fatal disease among pigs in Malacca in 1840, followed by a disease which then spread throughout the country "killing the buffaloes and other horned cattle by hundreds." But reporting upon it, Dr. Oxley stated:

> The epidemic having swept off nearly all the swine in the settlement, next fell upon the horned cattle; it is doubtful whether the disease which raged amongst these latter took its origin from the swine. I am rather disposed to believe that it was brought here by a gentleman who purchased, cheaply, a number of sick cattle at Singapore; the disease broke out immediately after their importation, and, when I heard of their introduction, I immediately prognosticated the result, from what I had witnessed a few years ago at Singapore. (Oxley, 1841)

He noted that carcasses were scavenged by tigers and wild pigs, "and not the least remarkable circumstance in the history of this epidemic is the effect of the above food upon these wild animals, the bodies of several having been found lying dead in the jungles. I have positively ascertained that the bodies of at least four tigers were found dead in the above manner, besides several hogs."

During an outbreak of rinderpest in 1864–6, veterinary surgeon T. P. Gudgin of the 2nd Dragoon Guards, established that it had been known for more than 60 years, in some places appearing annually, in others after lapses of 3–18 years, but apparently becoming more frequent probably due to increased trade and movement of cattle. It was severe in 1864, and still more so in 1865 with an estimated toll of 100,000 cattle. Gudgin considered it analogous to the disease in England. Mortality was estimated at about 85% of those attacked.

An army officer and an Assistant District Superintendent of Police wrote that the cattle of Burma were very small and died in hundreds each year from rinderpest "from which Assam is never free" (Pollok and Thom, 1900). They also wrote of seeing numbers of gayal "dead of the murrain of which Assam is never free," and came across many skulls of sambar killed either by tigers or by rinderpest. In 1916, a sporting barrister wrote that a large number of wild buffaloes had died in the Irrawaddy delta from cattle disease "which raged a few years ago, amongst domestic cattle" (Christopher, 1916). The banteng, Peacock noted, were very much subject to cattle diseases which annually ravaged the domestic herds which fed on the fringes of banteng habitat, "With the advance of civilization these herds of cattle are on the increase, and, as a result, many famous *saing* [banteng] grounds are now only poor travesties of what they used to be" (Peacock, 1933).

In 1912–13, some 20,000 cattle died of rinderpest but in 1916–17, the total recorded was only 2,058, although Rangoon in 1917–18 had its worst visitation for 25 years. In 1918–19, a severe outbreak caused an estimated 25,000 deaths, and in 1920, 25,000 were reported in Lower Burma alone. Serosurveillance in the 21st century has detected no sign of infection.

7.3. Thailand

From time immemorial, trade buffaloes and cattle moved south through and from Burma into Thailand, seeding rinderpest almost annually. A serious incursion occurred after World War II affecting almost the whole country except for the south. An emergency eradication program was drawn up with three stages. In the first stage, veterinary cover in the north of the country was withdrawn to work in the central area until the disease was suppressed. In the second stage, all staff were moved to the north and worked there until the disease was eliminated. Finally, in the third stage, staff were concentrated in the northeast until the disease was eradicated. The plan envisaged a period of 5 years but was accomplished in $3\frac{1}{2}$ years. Infected animals were slaughtered, the owners receiving an indemnity of up to three quarters of the value of the animal. In-contacts were vaccinated with goat-adapted virus which proved safe for cattle but was lethal for buffaloes, and the Nakamura III strain of lapinized rinderpest vaccine was used to protect buffaloes without mishap. Rabbits for its culture, however, were virtually unobtainable, and the problem was resolved by passaging the lapinized virus in indigenous pigs and harvesting the spleens 4 days later, the harvested spleen pulp being freeze-dried. This lapinized-pig vaccine protected buffaloes but not cattle.

Thailand remained free of rinderpest for 8 years until 1958, when the disease in neighboring countries broke through the border. Emergency measures were reinstated with support from FAO. Infected animals were slaughtered and those at risk vaccinated, and by 1959 the infection was eliminated. The danger of reintroduction was appreciated, and therefore 50-km-wide buffer zones were established along all borders (Scott, personal communication). As in Myanmar, serosurveillance in the 21st century has detected no sign of infection.

7.4. Indo-China

A decree of March 10, 1898 applied the French law on animal health of July 21, 1881 to Cochin China, and in January 1908, an Order of the Governor General stopped the slaughter of sick and contact animals, and payments for slaughter ceased, while an amendment of the French law dated January 8, 1909 made illegal the slaughter of rinderpest-infected animals to prevent people trading the meat and skins in Cochin China. But, as Jauffret (1928) pointed out, these laws drawn up for metropolitan areas in France were inapplicable under the conditions pertaining in Indo-China. A further regulation of October 2, 1916 also applied the 1881 French law.

Annam, Tonkin, Cambodia, and Cochin China, were all affected, the disease usually spreading from Laos in the north to Cambodia in the south, and from Cambodia eastwards toward Cochin China. An enormous mortality of pigs in Tonkin and North Annam was reported in 1897 by Camboulives (Carré and Fraimbault, 1898). Some rare foci were known in Cochin China in 1909, but investigation in 1922 revealed that the disease was almost unknown to the people and practically negligible in its effects. It was always the least affected country in Indo-China, outbreaks being short and benign as the Cambodian cattle usually imported were fairly resistant and destined for slaughter (Jauffret, 1928). Baradat (1929) agreed that it was partly true that Cambodian cattle were resistant, especially with regard to bullocks crossed with zebu strains, and in the northwest, where these crosses predominated, rinderpest was less fatal than in the rest of the country. Buffalo, however, showed no difference in resistance to those in other countries. Carré inoculated cattle with infective virus in 1898 obtaining 5% cattle and 10% buffalo mortality (Carré and Fraimbault, 1898). Serum inoculation was conducted at Phnom Penh from 1913 on cattle exported to the Phillipines.

Cambodia was particularly menaced by imports from Siam and Laos where control was insufficient, creating outbreaks in 1915–16, 1921–2, and 1927–8. At the beginning of 1922, Phnom-Penh was affected simultaneously by rinderpest and FMD, although the nearest known center of rinderpest contamination was 200 km distant. Whereas birds of prey were considered responsible, it seems more likely that an infected animal was clandestinely introduced. At the end of 1922, a serious outbreak was introduced with animals from South Annam, which were very susceptible, and in 1923, about 100,000 head were affected. If, among animals brought from Siam, one was found infected, the entire herd was sold immediately at a low price, spreading the disease, the peasants having no understanding of contagion, using only prayers and spells when animals were sick; and when rinderpest appeared in a village, the animals were left to wander in the forest. It was periodically more frequent than in the Phillippines and between epizootics every 5–6 years appeared almost annually in the frontier zones through the long migrations of cattle seasonally moving between the riverine plains and high ground. Pigs, also affected, were transported everywhere as they formed an important meat supply, rendering containment of outbreaks difficult. In the dry season, small foyers of disease burnt themselves out. The average mortality in epizootics was 20–60% of cattle and buffalo infected; in some villages, cattle mortality was 50%, while among calves, it reached 90%. Molinié (1930) reported that in lower Laos, until 1920, outbreaks were short and benign but since then had followed with strong regularity, the years 1926–9 being particularly disastrous, with rice cultivation in entire regions reduced to an absolute minimum because of a lack of draught buffaloes. In contrast to India, the plains animals were more affected than those in the highlands.

In 1928, there were cases of simultaneous rinderpest and hemorraghic septicaemia among buffaloes in Cambodia, where, in 1927–8, numerous skeletons and carcasses of wild animals such as gaur, banteng, and deer were found.

At the end of 1929, centers of infection were present in Tonkin, Cochin China, Cambodia, and Laos, infection being continually imported from China and Siam. In Cochin China, there were localities where the disease disappeared for long periods of time. In Phyen, which had a vast reservoir of cattle, in 1932 it had not been known for 20 years. In the south of Annam, it had not been known for 8 years, and in central Annam not within living memory (Curasson, 1932).

7.5. Dutch East Indies (Indonesia)

In the Dutch East Indies, rinderpest was known from 1879, and severe epizootics on Sumatra and Java in the 1880s stimulated the recruitment of nine veterinarians from Holland who conducted eradication of the disease on both islands by wholesale slaughtering of infected and suspect herds. An outbreak in Java in 1879 lasted until 1883 with the loss of 223,443 animals. In 1889, 22,000 animals were affected and 20,000 died (91%). In 1898–9, there were outbreaks in Borneo, Sumatra, and Madura. In 1902, a controversy developed over whether the disease was rinderpest or hemorraghic septicaemia. In 1907, it was eradicated in Java, but in 1909 and 1910, there were unconfirmed rumors of continued existence of the disease in Sumatra in young animals and imported stock. In 1929, it was reported as enzootic, and particularly prevalent after hot weather and the monsoon.

In 1964, a fatal epizootic attacked banteng, buffaloes, and cattle in the Jembrana District of the island of Bali, covering the whole island in 8 months. Between December 1964 and September 1967, 60,000 deaths were recorded. Rinderpest was suspected but not confirmed; nevertheless, 258,000 cattle were vaccinated, with vaccine imported from Japan. Indonesians investigating a renewed outbreak of disease in Jembrana in 1972 concluded that it was not rinderpest but possibly due to a rickettsia. A later study by an FAO team doubted it was rickettsial and favored an unidentified virus.

7.6. Ceylon (Sri Lanka)

Oral records of murrain in Ceylon extend back to the pre-14th century Sinhalese period, but the earliest established record of an epizootic was in 1842. In 1868, the Ceylon Cattle Plague Commission was set up to investigate epizootic diseases, finding rinderpest to be the most common and asserting it to be identical with that of Europe and India. Up to 1840, it appeared to be relatively benign but, since that year, made its appearance with greater frequency and with increased virulence. This could have been related to the rapid extension of coffee planting and associated activities. The average annual mortality over 3 years to 1867 was 70,000 head, representing an estimated 7% of the island's stock, and therefore probably not much greater than the expected annual mortality. The Commission made a number of proposals for a legislative enactment relating principally to quarantining and treatment of sick animals.

In 1916–17, 876 cases of rinderpest were recorded, rising to 2,076 the following year, then falling to 161 in 1918–19. Later in 1919, the disease was introduced from India to Colombo, and radiating outwards along the main lines of traffic, it caused 2,751 deaths.

The country was free of infection for decades until it was reintroduced with Indian live provisions in World War II, and again with live goats by troops of the Indian Peace Keeping Force in 1987 after 40 years of freedom, but was considered controlled by 1992. Vaccination ceased in 1998, and the country is now declared provisionally free of rinderpest.

7.7. China

Rinderpest was rampant in China in the 18th century and thereafter considered enzootic (but probably was so before), yet dissemination was not widespread for cattle were not used as draught animals. In 1860, there was an outbreak in Hong Kong among cattle collected for the provisioning of troops attacking Peking. In the spring of that year, when a large expeditionary force to north China was being prepared, many cattle were assembled at Hong Kong, brought down rivers in sampans from long distances in the interior of China. Many were attacked with acute dysentery upon arrival, often dying in 12 hr. The disease was allegedly not known in the districts from which the cattle had been brought, but one Chinese drover asserted that the disease, although not often seen, was known as *lo-peng* or "belly sickness," caused by fatigue and bad weather. The cattle continued to die on arrival at the depots, and 70–80 in every 200 died on the 10–12 days' voyage to North China. When the survivors were disembarked in Manchuria, the disease disappeared, and when they re-embarked, it did not reappear, many of the animals eventually reaching Peking. A large number of Japanese bulls, reputed as very hardy, also died when embarked for China. At one of the Hong Kong depots, while the imported animals were dying, a battery of Indian gun bullocks kept a few meters away did not contract the disease, either then or during the campaign. Neither did oxen employed in the surrounding rice fields, nor cattle kept by Europeans in Victoria, a kilometer or two distant. The Chinese ate all of the dead animals without any ill effects.

Many animals were reported dying within the municipal limits of Shanghai in the autumn of 1868. The cause was known to the Chinese as a common disease called *ma-ping* or "blood disease." (In Szechuan in the 1920s, it was known as *gan chwang bin* or "enlarged gall bladder disease".) As soon as owners were convinced that an animal had this disease, it was slaughtered for food, so it was difficult to determine the extent of infection. It seemed to have entirely disappeared from Shanghai and its neighborhood before the beginning of 1869, but was allegedly of annual occurrence thereafter although not reported until March 1872 when the municipal market inspector and dairyman reported it among his cattle, 25 of 38 animals dying. Dr. Thin, who was interested from a meat hygiene point of view, then uncovered many more cases. He believed, contrary to a belief that it was imported from France, that the disease was probably enzootic in the Great Plain of China (Henderson, 1873). Edington (1896) reports an outbreak in 1892 allegedly imported from Japan, with 9,716 animals dying or killed in 1 month. At the beginning of the following century, Keylock (1909) tried to protect the Shanghai cows and animals required for export by the serum-simultaneous method, but it was a financial failure. There were problems also with bovine piroplasms in the virulent blood injections. He switched to serum alone and claimed satisfactory results in 11,000 cattle.

In Shaowu, in the eastern province of Fujian, outbreaks were reported in 1912, 1914, 1915, 1918, 1920, and 1921. Bliss (1922) considered it the greatest economic drain on the agriculture of China affecting all parts of the country, droughts and floods notwithstanding. In Shaowu, outbreaks occurred every 2 years, sometimes in successive years, and lasted 6–9 months beginning in July or August, or in winter, when the cattle were in their poorest condition:

> An epidemic with a large number of cases and few recoveries begins during the extreme heat of July or August. The number of cases and the mortality decrease markedly during the mild October only to increase during the chilly rains of November. There is another decrease when the atmosphere clears with the frosts of December, followed by a large increase during the season of cold rains in February and March, and finally the epidemic dies out completely with the coming of mild weather in late April or early May.

In 1919, an outbreak was reported in Honan, and it was widespread and especially virulent in the northern provinces of Honan, Shantung, and North Anwhei in 1924. In Honan, farmers frequently noticed that their oxen were sick after returning from several days' trip to the market or after a family pilgrimage to distant temples during the Chinese New Year, the disease apparently being scattered by much visiting during the leisure months of winter. Gibbs (1925) explains that in summer, the farmers kept their animals at home, and the disease gradually decreased until by mid-summer it was difficult to find any cases. In most places, it started up again soon after the harvest was gathered and trips to market begun, culminating in an epizootic soon after the New Year holidays. The Chinese claimed that if an animal survived for 11 days, it would recover.

Brown (1929) referred to "terrible epidemics," which killed thousands of cattle each year and cost China millions of dollars, in Szechuan, mid-western China, the problem being particularly acute because of the dependence on water buffalo. The annual loss to the salt industry in the Tsiliutsing area alone was estimated at a quarter of a million dollars. In 1924–5, a severe epizootic led to the loss of 50% of most local stock and 90% of imported breeds. In 1926–7, the losses were "enormous." Reporting beginning January 1, 1929 showed that owners were losing 50–66% of their herds. In 1919, the means of control were not known, but although inoculation became developed after the experiments of Bliss in 1921–2, the cost of serum, manufactured by a Shanghai veterinary firm, Keylock and Pratt, was prohibitive for most peasants. It was reported again in Hong Kong in 1930 and after World War II was widespread in mainland China, but UNRRA mounted a control campaign with freshly harvested fluid Nakamura III lapinized vaccine, the disease last being reported in 1955.

7.8. Japan

At least two large outbreaks of a disease, which appears to have been rinderpest involving entire western Japan, occurred in the 17th century. The first appears to have been in Nagato Province lasting from 1638 to 1642 and followed an outbreak in Korea which ended in 1637. To assist rehabilitation, the Mori clan issued an edict on June 23, 1643 exempting farmers from paying tax on cows, and trade in cattle without permission of the lord was prohibited. In the second outbreak, also in Nagato Province, 3,906 of 4,311 (91%) of cattle died between 1672 and 1673; and in Yamaguchi prefecture between April and June 1672, 48,848 cattle were reported to have died, 405 survivors having been isolated in the mountains. Again, this followed the termination in 1672 of an outbreak in Korea. Documents of the time refer to the *annihilation* of cattle (Tables 4 and 5, Figure 2). Christianity was prohibited in Japan in 1633, and under the influence of Buddhism, farmers in the Yedo area abstained from slaughtering animals. The Tokugawa Shogunate decreed an order of pity for animals in 1670, forbidding abandonment of live cattle, a measure which would contribute to the high mortality rates. Dead cattle appear to have been buried together in certain areas (Kishi, 1976). Kishi (1976) considers it unlikely that infected cattle would have been smuggled in from Korea as those who communicated with foreigners were executed, and there is no record of agricultural trade at this time, only exceptionally trade in luxury items. But there was a treaty of exchange of drifted fishermen, and he suggests that drifted vessels may have been the route of invasion (perhaps indeed if they carried pigs on board as food, although Kishi does not mention this possibility. His suggestion that fishermen's clothes might have carried infection is improbable).

From 1700 to 1717, government officials planned the restoration of agriculture in the area impoverished by rinderpest. Each of 5,230 farmers was obliged to plant 40 paper mulberry trees and submit 6,000 sheets of paper as tax after the growth of the tree. A fund

Table 4. Rinderpest in 17th-century Japan.

Year	Month	Region	Mortality
1638	Apr–Jun	Western Yamaguchi to whole of prefecture	
	Sep	Oita and Kumamoto prefectures	>90%
1639	Sep	Hiroshima prefecture	>50%
	Oct	Kagawa prefecture	100%
1640	Aug	Shimane, Tottori, Hiroshima, Kyoto, and Wakayama prefectures	>90%
	Sep	Northern Mie prefecture	49%
	Oct	Saga and Kumamoto prefectures	100%
1641		Entire Kyusha region	100%
1672	Apr–Jun	Yamaguchi prefecture	>90%
	Jun–Sep	Oita prefecture	100%
	Jun–Oct	Hiroshima prefecture	>50%
1672–3	Jun–Sep	Ehime prefecture	(10,438)
	Aug–Jun	Kumamoto prefecture	100%
	Aug–Apr	Iki Island	>90%
	Sep	Shimana prefecture	78%
	Sep–Nov	NE Yamaguchi prefecture	90%
1673	Jul–Oct	Kochi and Tokushima prefectures	100%

From Kishi (1976).

Table 5. Some recorded mortality rates.

Date	Place	Number Dying (%)	Number Recovered (%)	Number Uninfected (%)	Total Cattle
1672					
Sep 30	Hane-cho, Shimane	97 (77.6)	11 (8.8)	14 (13.6)	125
Nov 30	Kuga-gun, Yamaguchi	3,008 (89)		390 (11)	3,398
1672–3	Yamaguchi	39,115 (91.5)			42,743

From Kishi (1976).

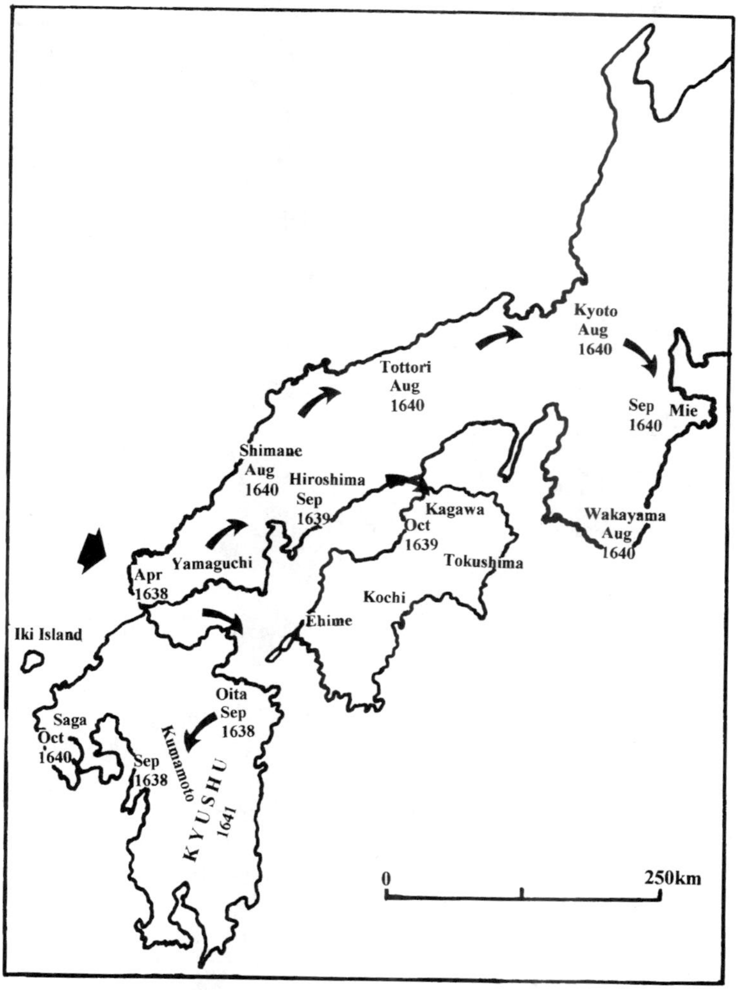

Figure 2. Rinderpest in western Japan in the Yedo era 1638–41. After Kishi (1976).

was also set up, granting loans without interest for buying cattle, with the loan repayable by annual instalments over 10 years, and a livestock market was established (Kishi, 1982). In 1718, an epizootic destroyed 8,800 cattle in Kagawa Prefecture. Published at Kyoto in 1720, a book *Gyuka Satsuyo* or *Compendium of Bovine Medicine*, described a disease *tachi*, which appears synonymous with rinderpest, as highly fatal and intensely infectious. The symptoms, including rough hair coat on the back, swollen eyelids, redness of the conjunctiva, purulent lachrymal discharge, and corneal opacity, were illustrated (Figure 3).

Probably imported from China or Korea, it was known in 1852, and warned of widespead outbreaks in China and Korea in 1869–71,the Japanese government prohibited import of livestock from Shanghai, Siberia, and the whole of Korea, this being the first administrative ordinance passed in Japan against livestock diseases. But withdrawn on October 5, 1871, rinderpest was imported in 1872. Between April 27 and September 11, in the Zabu and Komabano regions of Tokyo, of 116 cattle, 111 died (95%), 1 recovered (<1%), and 4 (3.5%) were uninfected. By 1873, a total of 42,297 cattle had been lost overall. From 1872 to 1911, there were 19 outbreaks, of which 13 originated from Korea, 4 from China and 2 from an

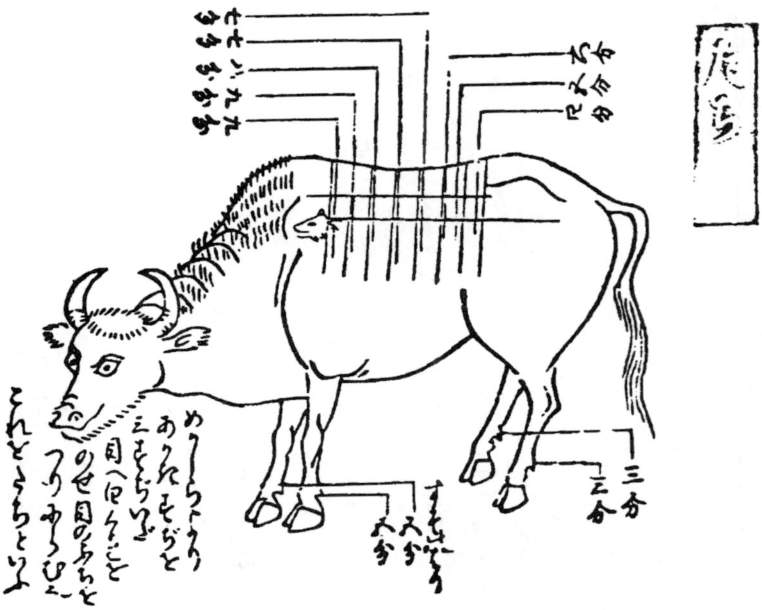

Figure 3. Illustration of a cow in the *Gyuka Satsuyo*, suffering from *tachi* or rinderpest, Tokyo 1720. From Kishi (1976).

unknown source. Appearing in 1892, it lasted until 1894 with the loss of 9,716 animals. In 1902, it was imported from Korea and eradicated by 1904 through widespread slaughter of infected animals but imported again in 1905 and then in 1908, with losses ranging from 177 to 3,331. From 1909, imports were forbidden and strict controls introduced, but it reappeared the following year, and again in 1920–22, in the latter 3 years, total losses amounting only to 39 animals (Curasson, 1932). It was last reported in 1924.

7.9. Korea

With symptoms indicative of rinderpest, epizootic disease causing 80% mortality among cattle was reported in Korea in 1541–2, 1577–8, 1636–7, 1644–5, 1668–72, and 1680–84 (Miki, 1962). It was then recorded in 1894. On the frontier with China, an outbreak occurred each year in the first quarter of the 20th century, caused by many large purchases of cattle from Manchuria where rinderpest was enzootic. The local Korean yellow cattle are hypersusceptible to rinderpest, and epizootics were devastating. In 1910, Japan took complete control of Korea, and Dr. Kakizaki developed the first chemically inactivated organ pulp vaccine, which cut the losses in the 1920s and 1930s. An outbreak beginning in December 1924 lasted 6 months with the death of 800 animals, inoculation programs protecting others. It then broke out in the province of Kankyo Hoduko in February 1925, but again, inoculation programs meant that there were few deaths. Nakamura's development of lapinized–avianized vaccine in the 1930s successfully protected the yellow cattle from further outbreaks.

7.10. Taiwan

Present from at least 1895, after a quarter of a century of effort, rinderpest was eradicated from Formosa in 1920. In October 1949, pigs imported from Hainan Island were placed in contact with cattle, of which 22 died before rinderpest was diagnosed. The surviving in-contact cattle were slaughtered, but the disease spread rapidly. Dissemination was curbed by purchasing and using lapinized rinderpest vaccine with which 87% of the animals at risk were immunized and the outbreak eradicated within 4 months. Strict quarantine measures are now employed when live animals and animal products are imported.

7.11. Malaya

In the first half of the 19th century, buffaloes and cattle were expensive in Malaysia because they were subject to frequent murrains. General opinion favored rinderpest, but many of the deaths were probably due to hemorrhagic septicaemia (Scott, personal communication). The ravages continued into the 20th century, and confusion as to the cause was rife. In 1903, the inoculation of a drove of cattle with serum from India and its subsequent exposure to the disease confirmed the presence of rinderpest. Serum supplies were erratic in World War I, and in May 1916, an outbreak among 54 animals caused 30 deaths. In 1933, a veterinarian with much experience in Africa was appointed, who eradicated the infection by 1940.

7.12. Nepal

Because of the annual movement of Indian cattle into the Terai, Nepal had been infected for as long as India. Rinderpest was reported as common in the hills, the Terai plain being regarded as a "home" of the disease. The Residency surgeon at Kathmandu reported to the Commission that the infectious diseases were generally brought into the valley by the herds of buffaloes and cattle that came from the Terai and Bhutan. It was reported to be very common in 1866 when it was termed *dosalin* or "plague." In the 20th century, enormous annual losses stimulated the formation of a Government Veterinary Service in 1939. At first, vaccine was purchased from India, but in the 1960s, a production plant was built in Kathmandu to back up the establishment of a 26-km-wide immune belt along the whole southern frontier. All bovids were vaccinated within this zone, and on entering or leaving. Incidence of rinderpest fell to seven outbreaks in 1969–70, and one in 1971–2. Illicit stock movements produced a few more outbreaks, the last reported in 1991. From Bhutan, the last report was in 1969.

7.13. Philippines

The first record in the Philippines was in 1882, which blamed the introduction on imports from mainland Asia. A serious epizootic followed the introduction of carabao from Indo-China in 1886. Spreading unchecked the outbreak caused enormous losses with 90% of animals dying. With time the disease became entrenched, epizootics occurring in cycles. In 1894 it was allegedly imported with cattle for American troops, but in fact was already enzootic. After the Philippines were ceded to America by Spain in December 1898, an eradication campaign was instituted, which dragged on for 30 years. In 1916, the disease had assumed serious proportions with annual losses of 18,000 deaths, reaching 26,951 deaths in 1917. Nearly every import of cattle arriving at two stations in 1917 was found to have infected animals, four lots from Hong Kong (a voyage of 60 hr) being badly so. In 1927, it caused 180,000 deaths.

Boynton (1916) noted that during outbreaks, pigs also became infected in certain localities. On the island of Romblon, where the disease had not existed before, and no cattle had been imported for some time, a serious outbreak among cattle followed the import of pigs from another island where rinderpest was present. First, the pigs died in considerable numbers, then cattle, and then carabao began to die. From 1917 to 1927, the annual loss of cattle was 18,000.

Standard control measures made little headway until Boynton developed an inactivated vaccine, refined in 1928, which led to eradication of the disease by 1936. In 1955, the President of India gave a group of Murrah milk buffaloes to the President of the Philippines which, quarantined on arrival, developed symptoms of rinderpest, and the group was slaughtered.

7.14. Tibet

A native of Lhasa in Tibet stated that sheep and goats in Tibet contracted rinderpest, as did "those wild animals which have feet like cattle." He had seen in the jungles of Tibet a kind of wild goat die of it, and 100 yaks on one occasion, as well as 10 wild sheep at a watering place in the jungle. Probably, the bharal or blue sheep was affected, as well as a species of wild sheep, perhaps the Marco Polo sheep, almost exterminated in the winter of 1897–8 in the Pamirs by rinderpest (Lydekker, 1924).

Although cattle were hardly ever seen in Tibet, the yak being the main domestic animal, a Tibetan cattle doctor testified to the prevalance of the disease, known as *chunneah*. Both rinderpest and FMD were well known, affecting wild as well as domestic ruminants. An informant in Lhasa stated that it appeared every 4, 5, or 6 years, and remained localized as long as no cattle or meat moved out from that place. The last outbreak he reported as in June 1870 and had known of its presence for the past 50 years or so. First appearing in about 1820, it was alleged to come from the west (Kashmir?), whence salt was brought on yaks and sheep to Lhasa from Sarka (85°N 30°E), about 240 km northwest of Mount Everest. Once a year, the inhabitants of Lhasa, Chornapary 290 km west of Sarka, and Janglachi 240 km west of Lhasa, met at the Sarka salt works. Some of the yaks and sheep returning to Lhasa laden with salt died on the road from rinderpest, although cattle travelling from Lhasa to Sarka were healthy.

8. THE NEAR AND MIDDLE EAST

Iran, repeatedly infected in its northern regions via the Russian Caucuses and Turkestan, experienced epizootics in 1910, 1918, and 1923, with mortalities reaching 80–85%, the latter two outbreaks originating from Azerbaijan. The 1923 outbreak was at first confused with anthrax; enzootic in Persia, it remained enzootic in 1935 in northern Iran and in Azerbaijan and Kurdistan (Rad-Pezechk, 1935). Greece witnessed an epizootic of rinderpest in 1924, believed to have originated from Syria, but the outbreak there did not occur until late the following year.

After the end of World War I and the separation of Syria from Turkey, there was no rinderpest in Syria until August 1925, when it was imported with three infected cattle from Turkey by French troops suppressing the Druse revolt. Not until December 29 was an order issued prohibiting cattle imports from Turkey. By the following January, it appeared to have been wiped out but in June 1926 appeared in the northwest border region and spread south, invading Palestine and Transjordan. On November 20, a new regulation allowed cattle imports under strict control. Owing to disruption caused by the revolt and corrupt officials, it was not finally controlled until October 1927, despite a proclamation of martial law. On September 7, 1926, for example, the loading at Damascus of 100 animals destined for French troops at Harwa, among which eight cases of rinderpest were discovered, was permitted, despite breaking the law. Arriving at their destinations, at one place, animals were allowed to drink at a pool used by those of the surrounding villages, and rinderpest soon invaded the whole of the most fertile region of Syria (Said, 1931). Ignorance of the most elementary precautions also aided its spread, the information service in one region giving an order on September 9, 1927 that people should throw the carcasses of animals which had died of rinderpest into the River Jordan, which carried the infection into Palestine and Transjordan. But owing to coordinated efforts of the French and British authorities, it was eliminated by October 1927 using serum vaccination with serum imported from Egypt. Because of its expense, this was used sparingly and was not all that effective owing to differences in the virus strains. Success was probably mainly due to the small numbers of cattle involved, the total number in Syria being only 20,600. Of these, 6,600 became infected, 3,293 died, and 1,100 were slaughtered. Despite these relatively small numbers, they represented a significant part of the Syrian peasant economy.

In 1969, a panzootic swept through Iran and toward the Mediterranean, affecting most countries, including Turkey. National and international action brought rapid control, but infection persisted in Lebanon, where it emerged again in 1982–3, infecting Syria and Israel. Iran was again invaded from the east in 1981. The next major introduction into the region appears to have been into Iraq in 1985 with 600 buffaloes from India, although it is considered that the buffaloes could have become infected en route. They were distributed widely in Iraq causing a country-wide epizootic, which, in Baghdad Alfedelia Dairy Village alone, killed half of the 30,000 buffaloes there. From then until 1994, an intensive campaign of vaccination appears to have eliminated it. An apparently separate persisting focus in the marshes of the Iraq/Iran border close to the Shatt el Arab was detected in 1994, with buffaloes entering Iran. An earlier epizootic in Iran in 1987 was also from this source. Iran was invaded in 1989 and again in 1994 from a source in the northern Governates of Iraq. In 1997, an outbreak of rinderpest in sheep and goats was reported in Iran, but by the end of 1999, the Near East was considered free of infection.

The countries of the Arabian Peninsula presented a unique situation. A large demand for fresh meat combined with the economic strength of the majority of countries generates a massive live cattle import trade. Although most animals are slaughtered, some are dispersed and repeated outbreaks of rinderpest occurred in Kuwait, Saudi Arabia, Yemen, Oman, and the United Arab Emirates, the last two regions experiencing outbreaks in 1995. The virus lineage appeared to come from South Asia, considered surprising, given the close proximity of the Ethiopian Afar enzootic focus (Rweyamamu, 1996). Controlled in most of the countries, Yemen constituted an area of enzootic persistence with a continued presence for more than 17 years, despite a cattle population of only 1 million, annual vaccination campaigns, and husbandry practices which restrict the cattle to small homesteads. Although quarantines exist at major ports, it is estimated that only about a third of imports pass through them, the rest being landed at many points along the coast. The continuous infection appeared to result from itinerant traders who trade stock from village to village and market to market, stock owners customarily changing their cows every 1 or 2 years. Animals taken to market and not sold are returned home after having been in the proximity of other stock, but the infection is now believed to have died out about 1997.

22

THE GREAT AFRICAN RINDERPEST PANZOOTIC

1. INTRODUCTION

From 1151 to 1169, in company with Ireland, Italy, Gaul, Sicily, Judaea, and Asia, there was "famine and pestilence" in Africa following severe winters and dry summers, where "Africa" probably meant Egypt, Libya, or Morocco. But we have little other information on pestilence in Africa until Bruce reported of Sennar in the Sudan in 1772:

> No horse, mule, or any beast of burden, will breed, or even live at Sennaar, or many miles about it. Poultry does not live there. Neither dog nor cat, sheep nor bullock, can be preserved a season there. They must go all, every half-year, to the sands. Though all possible care be taken of them, they die in every place where the fat earth is about the town during the first season of the rains.

We cannot determine what was causing such mortality but a few pages on he continues:

> Nothing is more pleasant than the country around Sennaar, in the end of August and beginning of September, I mean so far as the eye is concerned ... Every where on these banks are seen numerous herds of the most beautiful cattle of various kinds, the tribute recently extorted from the Arabs, who, freed from all their vexations, return home with the remainder of their flocks in peace .... (Bruce, 1790)

We know that rinderpest was introduced into Egypt from a ship at Alexandria in 1841 with infected Russian cattle offloaded for treatment after others had died. With no effective treatment known, the animals were marketed and dispersed. In a short time, the disease covered the whole of Egypt, allegedly killing 665,000 head of cattle but by 1843 had apparently burnt itself out. Toward the end of the year, large numbers of cattle passed from Nubia to Egypt to replenish the losses, and the disease erupted again, also killing buffaloes and large numbers of sheep and goats. In the next 3 years, a further 350,000 cattle are alleged to have died, representing 90% of the herds. The disease, which may well have been imported on previous occasions before 1841, then remained enzootic with an upsurge in 1863–4. On October 20 Lady Duff Gordon wrote from Alexandria that the cattle murrain was fearful and now beginning in Cairo and Upper Egypt, "The gazelles in the desert have it too, but not horses, asses or goats." In November, she wrote that, when asked for milk near Cairo, a woman exclaimed: "Milk! From where? Do you want it out of my breasts?" But she sought some from another family, no one refusing it if they had a drop left as they all believed that the disease was a punishment for churlishness to strangers, but by whom no one knew. At Abbassieh the same month, she felt there would be much illness from people burying dead cattle on their premises inside the town as it cost 100 *gersh* to bury an animal outside, "Hekekian's wife had seventy head of cattle on her farm-one wretched bullock is left, and, of seven to water the house in Cairo [by turning the water wheel], also one left, and that expected to die." On January 1864 she wrote of Benisuef "which used to be a great cattle place, not a buffalo was left, and we could not get a drop of milk. But since we left Minieh we see them again, and I hear the disease is not spreading up the river." Again, on January 29, she wrote that the disease had not extended above Minieh to any degree, and there had not been a case at Luxor. On February 19, she repeated that the disease had not spread above Minieh "or only very slightly," and again on March 16 repeated that it had not spread above Minieh, but lower down it still went on, and the destruction was fearful. Then on April 14 she reported that it had broken out "desperately" in Esneh, just above Luxor and more than 350 km upstream, "we see dead beasts float down all day. Of course we shall soon have it here." On April 18, "only a few have died as yet." Seventy-five head were reported dead at El-Moutaneh, and one Egyptian thought the outbreak less virulent than in Lower Egypt, "but dead beasts float down the river all day long." On April 23, it was reported to her that eight to 10 animals died each day, making 150 to date, compared with four each day at Luxor in a population of about 2,000. Of her own, only a few calves had died. The Pasha's agent had put himself in quarantine and allowed no one to approach his house. On April 30, she wrote that dead cattle floated down by thousands, and the Pasha's agent buried 1,000 at El-Moutaneh alone. On May 15:

> ... Yussuf, ... told me all the cattle were dead. Mustapha lost thirty-four, and has three left; and poor farmer Omar lost all-forty head. The distress in Upper Egypt will now be fearful. Within six weeks *all* our cattle are dead .... At El Moutaneh and the nine villages round Halim Pasha's estate 24,000 head have died; four beasts were left when we were there three days ago.

In June, someone had reported that in Alexandria, dead cattle were lying about the streets of Luxor, and the place was pestilential. Lady Gordon explained that the official in charge worked very hard, but disposing of 300–400 dead oxen a day with a limited number of laborers was no trifle: "if a travelling Englishman smells one a mile off he calls us 'lazy Arabs.' The beasts could not be buried deep enough, but all were carried a mile off from the village. I wish some of the dilettanti who stop their noses at us in our trouble had to see or to do what I have seen and done" (Waterfield, 1969).

Further upsurges occurred in 1881 and 1903, and it was reported as enzootic again from 1926 after eradication programmes.

In 1889, when many parts of Africa were in the grip of a serious drought which lasted several years, devastation struck its ungulate fauna. A great rinderpest panzootic swept the length and breadth of the continent in the space of 7 years, 1889–96, destroying populations of many species of artiodactyl both wild and domestic, and contributing to the worst famine Ethiopia had known within living memory. But not Ethiopia alone, for there was wholesale starvation and death among many livestock-dependent tribes such as the east african Wahima and Maasai. The Maasai tribe in Kenya was reduced by starvation, together with a smallpox epidemic which followed, to such an extent that when Europeans entered the coastal hinterland, they found great tracts of vacant land probably formerly occupied by the Maasai. Heaps of cattle and human bones were still to be seen in 1897 (Schillings, 1906). In southern Africa, the plague sparked off the Matabele rebellion.

Some effects of the panzootic have been summarized by Ford (1971) and Kjekshus (1977) from the writings of Baumann (1894), Lugard (1892, 1893), and other contemporary sources. Ford (1971) considered it the greatest natural calamity ever to befall the African continent, but we should guard against drawing an overly exaggerated conclusion as to its effects, such as that of Kjekshus (1977): "It broke the economic backbone of many of the most prosperous and advanced communities, undermined established authority and status structures, and altered the political contacts between the peoples. It initiated the breakdown of a long-established ecological balance and placed nature *again* at an advantage ..." (my italics). If it did indeed do all of these things, then it was probably not the first time that this disease, or some other, had done so, although Roscoe (1922) believed that this, and other outbreaks of disease, were due to the pacification of Africa by colonial powers: "Another disadvantage of this free tribal intercourse among pastoral tribes has been the introduction of cattle disease; tribal barriers used to prevent the passage of animals and goods from one tribe to another for trading purposes, and wild animals were never known in the past to have spread cattle disease over any wide area in Africa."

Chanler (1896) wrote that before the plague, the coast in the vicinity of Lamu was one of the richest on the east coast of Africa where many thousands of cattle, sheep, and goats were raised, "In the midst of this vast plain, now covered with tall grass, may yet be found stalks of wild millet and stunted Indian corn, which lend their testimony to the fact that what is now but a prairie had once been the scene of considerable agricultural industry." But this decline was due not to rinderpest but to prohibition of the slave trade, these lands formerly having been tended by slaves. It is erroneous to think that Africa had undergone long periods of stability, anymore so than any other country; and it is incorrect to state, as Davies (1979) has done, that the effects of the rinderpest panzootic had no parallel elsewhere. The European panzootic of the mid-18th century witnessed probably 40 times the number of cattle deaths in an area one-sixth the size of Africa, while in terms of human deaths, the 14th-century Black Death and early 20th-century influenza epidemics had no parallels, with an estimated 25 million and 20 million human deaths, respectively. Lugard (1893) was being emotive when he wrote:

> Never before in the memory of man, or by the voice of tradition, have the cattle died in such vast numbers, never before has the wild game suffered. Nearly all the buffalo and eland are gone. The giraffe has suffered, and many of the small antelopes, the bush-buck, and reedbuck—I believe—especially. The nsunu (Kobus kob) [Uganda kob] was affected only partially, and very large herds were left, both in Buddu, and near the Albert Lake; but Mr. Sharpe reports this antelope [puku or lechwe] as having been especially attacked in Nyasaland. The pig (wart-hog) seem to have nearly all died. The elephant, hippo, rhino, and—I think—all classes of hartebeest, wildebeest and waterbuck are exempt. It is noticeable that the animals nearest akin to the cattle have died—viz.: the buffalo, and the most bovine of antelopes, the eland. It is therefore extremely curious that the wildebeest has escaped. Goats and sheep have been exempt. The zebra ... has perished in Nyasaland, but is exempt in East Africa ...

But we know more about this panzootic than we do of possible earlier natural catastrophes in Africa; and the accounts of the abundance of game emanating from hunters at the beginning of the 20th century suggest that either certain populations had recovered their numbers very quickly, or many more animals escaped death than contemporary observers indicated. Unlike India, where its enzootic nature meant that there was no pattern to the 19th-century outbreaks, in Africa we are able to trace the passage of an outbreak from north to south and east to west, across the entire continent.

Enzootic in Europe and Asia at the time, it nevertheless ravaged Africa in an epizootic of apparently appalling magnitude, causing over 90% mortality in the flourishing cattle populations and creating unknown levels of destruction among the dense populations of wild herbivores. Its virulence has led to the assertion that it was characteristic of a new introduction, and it is indeed probable that the wild populations existing at the time had not previously been challenged; that is, they no longer possessed antibodies from possible previous infection, since these would have disappeared within some 20 years, except in the long-lived hippopotamus, where they were still found after more than 30 years (Plowright *et al.*, 1964). But the virulence of the disease was in fact no different to that experienced repeatedly in Europe throughout the plague's known history.

Eliminated south of the Limpopo by 1903, it remained enzootic in the tropical belt south of the Sahara. Widespread immunization, largely by goat-attenuated virus, was practised from about 1943 to 1963, so in many of the accessible areas, the only susceptible cattle were young, having lost the maternally derived antibody and had yet to be vaccinated. But wars and governmental mismanagement eventually led to a resurgence of epizootics after the 1970s which continued to provide foci of infection to the present.

In the oscillating pattern typical of the course of a virus infection, the disease may have repeatedly played a role in modifying the numbers and distribution of Africa's fauna, but the suggestion that the geographical distribution of most artiodactyl species in Africa today is largely set by the 1889 panzootic is open to question. Pearsall (1954), writing of western Uganda, stated: "Epidemics of these diseases [East Coast fever, rinderpest, etc.] have sufficed to determine the distribution of some of the larger mammals. Thus zebras and certain sorts of buck are absent from country which is now the Queen Elizabeth National Park in Uganda. They were exterminated by rinderpest." Zebra, however, are not susceptible to rinderpest, and the impala, which is the buck referred to, was not reported elsewhere as seriously affected. A casual reference to the former presence of giraffe in this region in 1889 by Parke (1891) suggests that this species may have been exterminated by the epizootic, but other species including hartebeest are more likely to be absent due to anomalies of topography and geological change. What is true is that subsequent to this panzootic, rapid changes in settlement and development patterns took place with the expansion of colonial occupation. The panzootic is of great interest to African ecology, not only because it apparently played a major part in limiting numbers of some species aided by the expansion in stock-keeping and settlement patterns, but more so because of the inference that was drawn—that the decline in herbivores led to the encroachment of bush and the spread of tsetse fly carrying trypanosomiases deadly to both man and cattle, thus triggering further malignant epidemics and epizootics.

2. THE ETHIOPIAN ORIGIN

The outbreak in 1889 has led to much speculation as to its origin. A theory it began in Eritrea was given weight by Sir John Kirk in a letter to *Nature*, June 10, 1896, based on Lugard (1893). Edington (1896) considered it to have been imported by an animal sent to the Italian troops at Massawa (Massowah), the disease then passing down the Nile with Arabian caravans. Lugard wrote:

> The Plague seems to have started in the East Coast, opposite Aden, and to have spread inland. It began at the end of 1889, and when I went into the interior in December of that year, it had not reached Masailand and Ukamba, nor yet in the spring of 1890. When I returned up country in the autumn of 1890, it had spread through these countries, and the cattle and buffalo were dead. It had preceded me through Kavirondo and Uganda, though—when Mr. Jackson passed down in the summer of 1890—it was only just beginning to show its effects. Beyond Uganda, I found it had just preceded me through Ankoli and Unyoro; and in the far heart of Africa at Kavalli, it had swept off every ox only a few weeks before I arrived (Sept. 1891). Passing southwards it reached the north of Nyasa [lake Malawi] about July 1892 .... (Lugard, 1893)

Mettam's observation (Mettam, 1937) that the Nile Valley as far as Khartoum had already been infected during the British campaign of 1884–5 introduced by animals purchased from Russia and Black sea ports seems to derive from Curson (1919), who suggested that the Expedition was accompanied by baggage animals purchased from Russia. But Colonel Butler, the Commissary officer, makes no mention of cattle, either as baggage animals or as food. There were 2,000 baggage camels (many of these carrying forage and which it seems unlikely would have come from Russia), 1,000 riding camels, and a number of horses. The main supplies were carried up the Nile in special boats stocked with rations for 101 days, which included preserved corned meat, preserved fresh meat, and bacon or boiled mutton. The Egyptian support group derived its food "from the grain and goats" of the districts through which it passed and was in the habit of "summarily impressing all the camels and donkeys of any village they came to for transport" (Butler, 1887). Thus, if the Expedition did carry rinderpest infection up the Nile, it would have had to have been transmitted by camels.

Lugard was considered the authoritative source until recent times, some of the most precise documentation charting its possible origins now having been given by Pankhurst (1966, 1985), describing its beginnings in Ethiopia. The Ethiopians traditionally blamed the Italian reinforcements landing at Massawa in November 1887, accompanied by cattle from India that could well have been infected (the Italians had been there since February 1885). The importation was allegedly conducted by an Italian called Andreoli. This was asserted at the time by the Ethiopian ruler's Swiss private adviser, Alfred Ilg, and a German traveller, Conrad Keller. Many Ethiopians believed that the Italians had spread it deliberately, a once widely held view being that they had inoculated three animals with the disease (Pankhurst, 1966). Zonchello (1917) reported that rinderpest appeared to have been first imported into Eritrea in 1887–8, its spread occasioned by the incompetence and dilatoriness of the Italian authorities in the administration of elementary prophylactic measures. Instead of slaughter, specific measures of treatment were attempted on a large scale with generally unsatisfactory results. Ferraro (1917) states that this was the "caudal inoculation" method practised as a method of vaccination (presumably copying the Willems method for pleuropneumonia), but the outbreak did not spread widely and was of short duration. During the Italian campaign of 1895–6 it remained circumscribed in the Agrodat District, reappearing along the sea coast in 1897, and, owing to a repetition of attempts at caudal inoculation, assumed a widespread character. In 1903, Zonchello states, it was decided to adopt the serum-simultaneous method of Kolle and Turner just at the time when Theiler and the Bloemfontein Conference condemned it, but Ferraro states that Conti had already tried this method at Asmara in 1900–01. It is probably from either the "caudal inoculation" method or the serum-simultaneous method that the rumor of the Italians deliberately spreading the disease arose.

The Italians seized the port of Massawa in February 1885 and in the summer of 1889 made their way up to the plateau, by the end of the year occupying an extensive stretch of territory, which included Asmara and the neighboring plateau. But the outbreak preceded their arrival. Bòttego (1895) states that it began in 1887, but the earliest references appear to be those of Roman Catholic missionaries in Hamasien Province as reported in *Les Missions Catholiques*, who referred to a serious epizootic after the rainy season of July to September 1888. From Massawa on November 16, 1888, one reported that the greater part of the cattle had perished in an epizootic (Bettembourg, 1888). On January 8, 1889 at Massawa, the newly arrived missionary Crouzet wrote: "For more than a year, all the scourges have swept down upon these unhappy christians. A terrible epizootic has carried off the cattle, now in many places…" (Crouzet, 1889a). On January 19, Mother Reeygasse wrote from Massawa, "… M. Picard wrote to me that the people, having nothing more to eat, to feed themselves boil the skins of cattle destroyed by the epidemic" (Reeygasse, 1889). On February 22, having visited Keren the week before, Crouzet wrote that the Fathers Joseph and Cazeaux informed him that they had possesed some cattle "in 15 days more than three-quarters died. If we wish to cultivate, it is necessary for us to replace these animals and pay 20 thalers each for them, because one can only find them with difficulty. Our flock of goats has disappeared, the epidemic has left nothing" (Crouzet, 1889b). But later, Ilg was to report to Wurtz, a doctor sent from Paris to study the epizootic, that only some species of gazelles were refractory, as well as sheep and goats (Wurtz, 1898). Travelling around Keren, Crouzet reported that when he recommended to people that they study religion, they replied by showing him their herds destroyed, the ground bare, their houses empty, and their stomachs hollow.

Apparently thus beginning in the north in Hamasen, a contemporary observer, Alaqa Lamma, recalled that "within three days," all cattle in the Province were "paralysed, refused to graze and died." It then spread southwards to Tigre and, during the rains of July to September 1888, swept across all the northern provinces, invading Begemder, Lasta, and Gojam, reaching the latter in October to December. From thence, it spread to Shoa (Figure 1).

On February 20, 1889 the Italian envoy Antonelli reported from Shoa that all Ethiopia was threatened with a crisis which might well be fatal to her economic life. Capucci, another eyewitness, estimated that 90% of cattle had perished (Scarfoglia, 1936), while Skinner (1906), the first United States envoy, put it at 92–93%, all of which estimates of mortality are usual for rinderpest. Afawarq Gabra Iyasus stated that many an owner of 1000 head was left with perhaps only 1 or 2, while Alaqa Lamma states that of his father's 300, only 1 survived. Travelling to Addis Ababa by way of Dabra Libanos, he passed through land from which the livestock had almost completely disappeared. At Salalé, he saw one calf, and in the mountains above Dabra Libanos, six old oxen. At Dabra Berhan, he found that a few cattle had survived here and there at the highest elevation. At Laja-gend, he saw five or six, but in the lowlands,

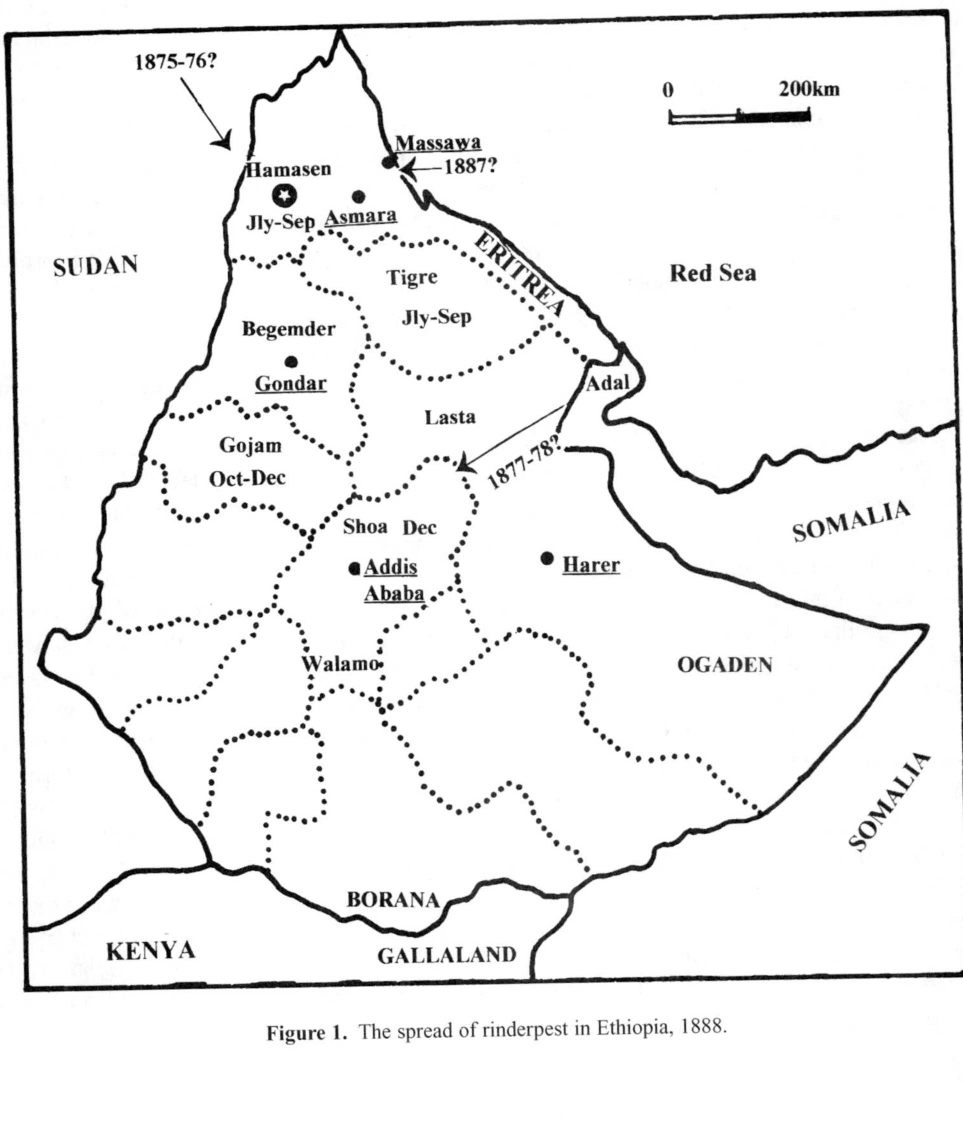

Figure 1. The spread of rinderpest in Ethiopia, 1888.

extermination was complete. Herds of 500–1,000 were reported dying in 1 or 2 days, and at Bulga, Wurtz reported that all cattle died within 8 days, while of the Emperor Menelik's several thousand, not one survived. Menelik is alleged to have lost about 250,000, while some of the richer Gallas lost 10,000–12,000 (Wurtz, 1898). The chief of Koa, in the province of Lasta, lost 56 of 57 oxen in less than 10 days, and only 2 or 3 of his heifers and some of his calves survived (Wylde, 1901). The province of Begemder was formerly renowned for cattle, but soon none were to be seen there. All were destroyed in Shawa, and both east and west of Shawa fared likewise. A vast number perished in the Danakil desert (Bòttego, 1895), and large numbers also in Kaffa in the west of the country. The final, overall mortality, was estimated at 90–93%, only herds in very high country, 3,000 m and above, being preserved and the epizootic at that altitude being less intense (Wurtz, 1898).

Pankhurst notes several earlier references to cattle plagues. Ethiopian legend refers to "beasts and cattle" perishing *c.* AD 831–849; but this may have referred to drought. At least 23 major famines were experienced between 1540 and 1800. Writing glowingly of the fertility of the country and its fine cattle in 1682, Ludolphus nevertheless remarks: "Africa was formerly Famous for the Heards and Droves of Cattle fed therin ..." (*sic*) suggesting that some change had taken place (Ludolphus, 1682). In 1829 in Shoa, there was famine due to failure of the grain crop, followed by the death of much livestock, but perhaps not due to disease, although it is recorded as an epizootic (Pankhurst, 1968). Although Johnston (1844) had recorded large numbers of cattle dying in Shoa about 1828–9, this was a drought year, and no disease is mentioned. In 1835, there was famine followed by cattle deaths.

During the Abyssinian Expedition, in January 1868 the British army imported 8,075 bullocks from Bombay for transport and ration purposes, leaving at the end of the campaign with 1,075. It is mentioned that there were 1,050 casualties among the transport bullocks between January and April, and the draught bullocks did not thrive well in the cold and wet climates which they encountered (Holland and Hozier, 1870). There is no mention of disease, although these animals had come from an enzootic rinderpest area experiencing an epizootic in 1867.

Drought was certainly not the cause of mortality in 1875 when the Earl of Mayo was prevented from entering Asmara due to a plague among the Shoho cattle herds, said to have broken out in Hamasen after invasion by Egyptian troops in the winter of 1875–6 (Bourke, 1876). The Earl was informed that an order had been issued throughout Ethiopia that no cattle were to travel, or be allowed to go to and from infected districts, and wrote: "This is worthy of the notice of our sanitary commissioners at home." When he reached the plain of Asmara (March 1875), he found notwithstanding the precautions that people had taken, that cattle disease was rife among their beasts, and he saw several lying down, stretched out, by the side of pools. At Massawa, he reported that hides had been accumulating there as the Egyptian government would not allow merchants to ship them to Suez because of the cattle disease in Abyssinia. The British *Annual Report of the Veterinary Department of the Privy Council Office for the Year 1874* quoted a report from Cairo that rinderpest occurred in Abyssinia during that year (Williams, 1875), thus predating the military invasion from Egypt of 1875, and the Italian invasion by 15 years. The report was not issued until June 29, 1875 and the information could have come from Bourke, although he did not report coming across cattle disease until March 1875. It was again reported to be present in Abyssinia in 1876 (Brown, 1877), and in 1877 as having ceased in Abyssinia and Massawa (Brown, 1878). Following extinction of rinderpest in Britain in 1872, the Veterinary Department's prime concern had been pleuropneumonia, and it seems unlikely that the Department would not have made every effort to confirm the true nature of the disease reported.

Some years later Simon (1885) wrote: "... Hamassen [in 1881] was almost abandoned ... the Egyptian army campaign in 1876 and the plague (*peste*) which followed it spread desolation and death in this country, where formerly assembled a dense and industrious population; the animals themselves were not spared; an epizootic came and cleared their ranks." It was in Egypt from 1872 to 1876 at Assiut, but rumor of an outbreak in Lower Egypt in late 1877 was denied (Brown, 1878).

Although Simon is possibly referring to bubonic plague as well, rinderpest was probably enzootic in Egypt at this time. It had apparently carried off an immense number of camels in 1861. Lugard (1893) wrote: "Not for thirty years has a plague like this been known in the country [East Africa], and even then it was not to be compared in virulence to the present one." His "thirty years" from 1890 to 1891 would coincide with the major Egyptian outbreak of rinderpest, although in 1892, he wrote: "the plague of some thirty-five years ago worked a great destruction."

Captain Martini (1880) reported that a cattle disease from the lowlands of Adal had spread to Shoa about 1877–8, and in 1879, a cattle epizootic is definitely referred to. Writing in December 1879 from Ankober, he commented on the high price of livestock since the epizootic communicated from the country of the Adals (i.e., the coastal region of Berbera) which had destroyed the livestock in some places. Simon's observation (Simon, 1885) strongly suggests that the Egyptian army spread rinderpest in the country in 1876, although its presence appears to have predated this by at least 2 years. That it did not develop beforehand into a panzootic may have related to climatic conditions. High levels of Lake Victoria indicate good rains from 1876 to 1880 and again in 1892–5, then a steady decline to 1902. Rainfall fell sharply in the subtropics and tropics about 1890–1910, while 1880–95 was very wet. Hence, wet conditions in 1876 may have meant a lack of concentration at watering points and widespread dispersal of animals during feeding, leading to a lack of infective contacts. Reintroduction of infection during drought conditions which were building up at the end of the 1880s found ideal conditions for spread, or the conditions may have encouraged radiation from already existing foci of infection.

The year 1888–9 was excessively hot and dry in Ethiopia, conditions considered to favor the spread of rinderpest through the closer association of animals at fewer watering points. Constable, in a letter to *The Times* of March 31, 1866 concerning the outbreak in England, pointed out that 1865 had been the "warmest and longest summer ever known in the memory of living man," and the summer preceding the 1745 outbreak in Britain had also been unprecedentedly hot. In Ethiopia, excessively dry conditions may well have sparked off the epizootic.

Vignéras, travelling in Abyssinia in 1897, reported that he came to a wide plain named Bourka, which, although then having few inhabitants, had a numerous population before the cattle plague which had come from Tigre. The plague spared some provinces, but for others, the consequences were terrible, people not being able to cultivate for 2 years for lack of draught animals, carnivores coming into the villages of Bourka devouring weakened people. Vignéras came to a pasture where some herds of cattle were grazing.

Menelik forbade the killing of horses and donkeys for food on pain of losing a hand, as these animals were essential for warfare. Subsequently, many Ethiopians would not eat zebra meat because the animal resembled a donkey.

"During 1889, all the horned cattle were killed. In a herd of the king in particular, of several thousand head, not a beast escaped" (Wurtz, 1898). Hyaenas and vultures refused to eat the carcasses, and all other ruminants, buffaloes, antelopes, "and even many lions" died, according to Wurtz. Indirectly, he considered that it killed about a third of the human population due to the famine it occasioned, followed by an outbreak of smallpox.

All the Galla lands in the south suffered, and there was great havoc among Borana herds (Cipolla, 1927). Wurtz (1898) stated that it spread to the Somali coast from Addis Ababa.

The Somali trading center of Lugh was affected, the disease extending as far south as Ganale (Bòttego, 1895). In some stockades of 300–400 animals, all died within 48 hr. But Swayne (1895) in 1892, a few kilometers west of Hargeisa, met hundreds of cows. He reports that cattle from the interior were largely exported to Aden to supply the garrison, in 1891 68,000 sheep alone, and vast quantities of hides were exported annually to America, and he noted that it was possible of late years that the Aden supply had been affected by the great drain caused by the Abyssinian foraging expeditions into Ogaden. These expeditions were no doubt to replenish Abyssinian cattle stocks. In Northern Somaliland, however, although outnumbered by sheep and goats, the principle domestic animal was the camel. This fact, and the arid nature of the country, was probably inhibitory to the spread of rinderpest. But rinderpest having destroyed the Galla herds, in October 1901 an epizootic was killing their goats [6°30′ N 39°E], and around Ido, 200 km to the north, rinderpest was killing the cattle.

Was there an initial virulence in sheep and goats as the missionary report seemed to suggest, species reported elsewhere as little affected and generally exhibiting benign infections? All that the missionaries actually stated was that their goats "had disappeared," which could have meant that they were stolen by someone who had lost his stock, but the context does not seem to imply that, probably meaning "gone." The Ethiopian chronicler, Afewerk Gabre Selassie, recorded the plague as affecting wild animals (Fusella, 1963), and when it reached the south of the country, this aspect was well noted.

In 1900, Baird wrote: "Buffalo, which used to roam in hundreds over this country, are now almost extinct, nor did I see any sign of hartebeest which are also said to have been common: both have been carried off by rinderpest. Indeed the only game seen up to this point was reedbuck." In April, 130 km north of Addis, he reported no sign of rinderpest, and cattle herds were numerous, but north of Dangalla, he recorded: "Round Libtan and during the whole of the latter part of the march we passed quantities of bones of cattle which had been killed by the rinderpest. Each village had its cattle-pen ... but these were nearly all empty. Those cattle which had escaped the plague—and there were a certain number of them—were fine beasts." Powell-Cotton (1902) reported that it had destroyed all the buffalo in the province of Gojam (although some existed to the west of Gondar), and Skinner (1906) recorded that "desert antelopes" had also perished. In southern Somalia, Bòttego (1895) reported that it created great mortality among buffalo and antelopes.

At the same time rinderpest struck, there was a harvest failure from drought, followed by outbreaks of locusts and caterpillars; the latter probably army worm, *Spodopterus*. The net result was allegedly the worst recorded famine in Ethiopian history, lasting from 1888 to 1892. Many human deaths due to smallpox followed the famine, and the incidence from typhus, cholera, and dysentery was greater than ever before. No oxen were left to plough in the earlier famine, and the people did not use hoes, so no crops were planted. Locust swarms were a common feature of northern Ethiopia, although they were believed to have been larger than usual at this time, but caterpillar invasions were rare. There appear to be no estimates of human mortality, but over 100,000 people are calculated to have perished in the 1972–5 famine (Rivers *et al.*, 1976). Prior to the former famine, in 1879, when King Tekla Haimanot rebelled against the Emperor John, the country of Shoa was laid waste, thousands of people and cattle were slaughtered, and the country had never properly recovered.

The combination desolated what was apparently once a rich and fertile country, so that Antonelli (1890) wrote of a journey between Harer and Addis Ababa in December 1890:

> Previously the country was inhabited; there were very beautiful fields of durra and barley, numerous herds of cattle, sheep and goats, and the whole area had an atmosphere of abundance and prosperity. At present it is one continuous desolation. Except between Harer and Burka where there are still a few inhabitants and occasional areas of cultivation, the remainder up to Menta and Cara in Minjar

> is absolutely a desert; no more inhabitants, no more cultivation, no more flocks, but low acacias and tall grass rendering the beautiful valleys of Chercher and Ittu unrecognisable.

This change, Antonelli noted, had taken place between June 1889 and December 1890. But even low acacias could not have grown up so quickly in a drought as to change the landscape in this manner. Wurtz (1898) noted that sites of former villages were now nothing but "thickets and grass," but this was almost 10 years after the event.

Remaining enzootic in many parts of the country in 1897, there was a renewed outbreak allegedly originating around Harer which spread first to Somalia, where both kudu and hartebeest suffered severely. It was probably this outbreak which led Ferrandi (1903) to report "enormous areas" of bleached cattle and buffalo bones in the Juba River region where there had been an "almost complete mortality" of cattle. By 1898, the cattle in Ethiopia still had not recovered their former numbers. In 1901, Wylde observed that it was "still very bad" in Shawa, and nearly all the cattle at Tanquille were dead. However, Begemder was restocked from Shawa at the beginning of the next century, and in 1900, infection was believed to be "less prevalent" in Shawa than elsewhere (Pankhurst, 1985).

3. THE SPREAD WEST

Mettam (1937) states that rinderpest was responsible for a West African epizootic in 1865, but this is generally attributed to pleuropneumonia. In July 1870, Nachtigal, journeying to Bornu in West Africa, first encountered evidence of what was apparently pleuropneumonia. He terms it variously *Lungenseuche, Rinderseuche*, or simply *Seuche*. The chief at Kuka regretted the fact that Nachtigal had visited him at a time when widespread cattle mortality had caused such damage to the prosperity of the community. The disease had come in a few years before from the *west*, and of his once 32,000 cattle, he had now scarcely one fifth. Three days before Nachtigal's arrival, anthrax had apparently broken out among the horses, killing many. In the low-lying country around the south-west of Lake Chad, Nachtigal found pleuropneumonia to be devastating the cattle. He conducted an "occasional" autopsy confirming the nature of the disease but was unable to identify the disease among the horses which was reported as usual in any year of abnormally heavy rains. At the end of 1870, he wrote that the cattle epizootic had persisted for 3 years (Fisher and Fisher, 1980). From Ethiopia, rinderpest may have spread rapidly west as well as south, but St. Croix (1945) believed it to have originated around Darfur at an earlier date, presumably travelling through the sahelian cattle belt to reach Chad by 1886 and Nigeria in 1887, where it exterminated the majority of cattle maintaining epizootic proportions until 1891, although some consider this also more likely to have been pleuropneumonia (Figure 2). St. Croix, however, was an Inspector of Livestock responsible for the anti-rinderpest inoculation campaign and fully conversant with both diseases. Stenning (1959) presumably follows St. Croix in dating its presence in Bornu from 1887 to 1893. He implied that it spread westwards from Kukawa on the west side of Lake Chad to Kano along the main trade route, which was straddled by nomadic Fulani pastoralists. Those further south did not suffer so much from the outbreak. Ford (1971) misinterpreted Thomé's remarks (Thomé, 1964) by writing that rinderpest appeared between 1884 and 1891 in the former French colonies. What Thomé actually wrote (apparently quoting Mettam, 1937) was that it entered Africa with the Nile expedition of 1884 and thus deduces that it must have entered Chad between that date and 1891. Nevertheless, a date of 1886 for Chad coincides with a possible extension of rinderpest from Egypt. Pécaud (1924) states that there is evidence to show that the outbreak was complicated in Chad by

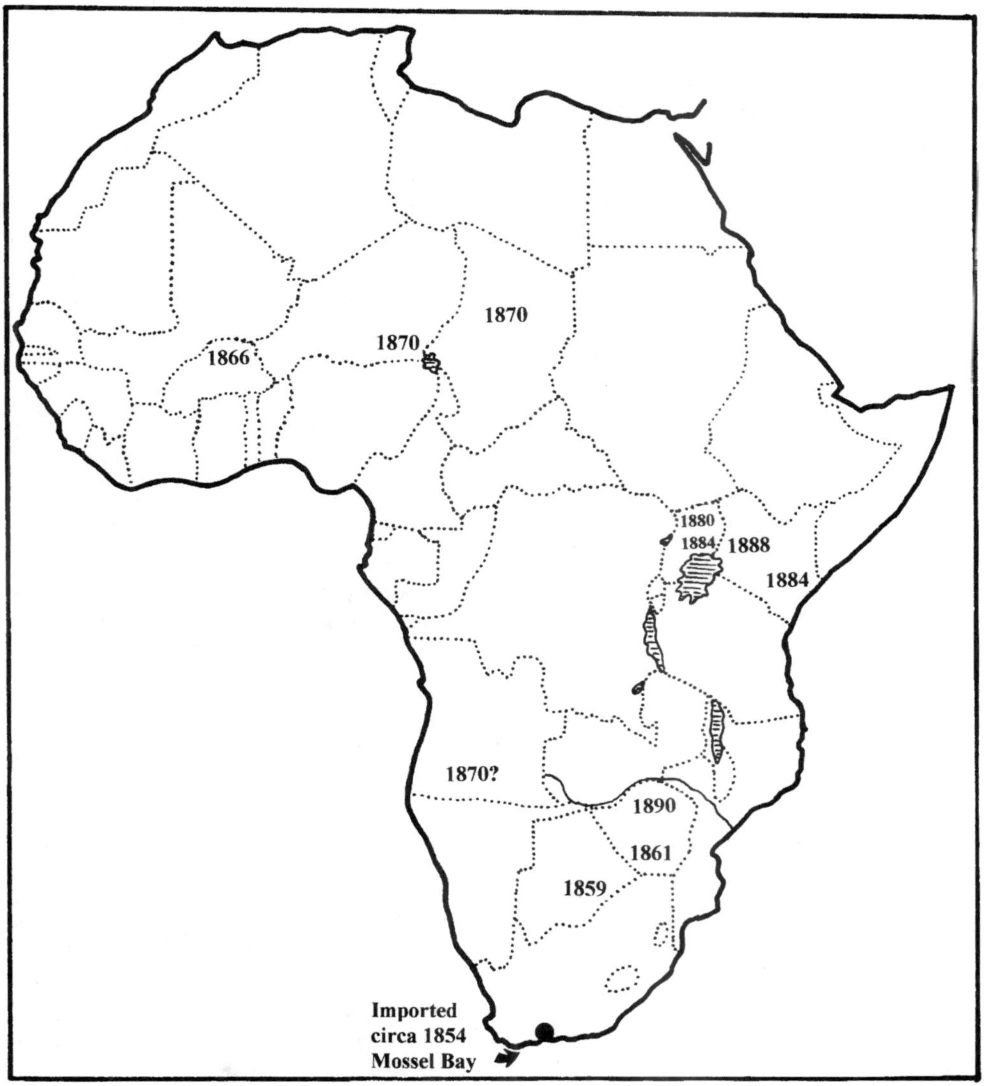

Figure 2. Reports of pleuropneumonia in 19th-century Africa.

coccidiosis. Native tradition dates the first appearance of rinderpest many years earlier, blaming Indian and Persian traders, but the 1887–91 outbreak was known in Hausa as *Sannu*, from the greeting used as an expression of sympathy.

What we do know is that Monteil (1895) first encountered rinderpest at Dori in Burkina Faso (formerly Upper Volta) at the end of May 1891, where he lost his baggage oxen to it. Monteil wrote that a *year* ago, every Peulh in "all the region to Sokkoto" had at least 100 head of cattle, but now herds of 500–600 were reduced to, at most, 10, 20, or 50 head. Although a pilgrim from Mecca stated that in January 1891 that he had not seen a single cow between Kano and Dori, Monteil recorded that the epizootic "exploded" in February of that year and was still continuing in May. It was stated categorically that it "came from the east."

Older Peulh stated that they had never seen such mortality of cattle, while Monteil noted that it was impossible to count numbers of deaths, but mortality was "immense," piles of rotting carcasses still being present. This was the first time that Monteil met the disease on his journey from the west, so the date of its arrival at Dori is well established, and an earlier date of 1884 for the region seems unlikely. The Peulh had experienced pleuropneumonia in 1866 and were able to distinguish it apart. Not only had Monteil described the symptoms of rinderpest, but he was well aware that it was this disease, well known in France. Binger, traveling from Bamako in Mali to Accra in Ghana between December 1887 and January 1889, came across several herds of fine cattle. The most he counted was 800 head at Fourou in the Ivory Coast in December 1887, all in good health. Ouandrama, in northern Ghana, was rich in cattle in January 1889 (Binger, 1892). He makes no mention of disease among them, but in 1892, the year following Monteil's report, General Achinard, commander of the west African military zone, was taken to task by the French Government for exceeding his military occupation budget, one of the causes being the difficulty of finding animals for troop rations.

In 1892, the slave trader Rabih invaded Bornu from the east, and in February 1893, Maistre (1895) reported numerous cases of an epizootic amongst cattle herds near the Taraba river in the Adamawa region of eastern Nigeria. A camp of Foulbé had their herds in a "sort of quarantine" outside a Hausa village. But he reported no disease among the herds of game which he met a few days later west on the Benue River. Apart from Monteil's statement that it killed roan antelope (*coba*) and buffalo, there are no records of its effects on the sahelian fauna other than later Peulh reports that it destroyed hartebeest (Curasson, 1932), but the consequences for the then numerous fauna of this zone must have been just as great as in parts of eastern and southern Africa.

Chevalier (1908) states that an epizootic occurred throughout Tchad in 1893 destroying a great part of the cattle. Wild buffalo and antelopes were equally affected, allowing them to be speared by the natives. In the following year, sheep and goats were ravaged by an epizootic. No disease was reported when he made his visit in 1903.

4. EASTERN AFRICA

Of the plague's progress south through the Sudan, we know next to nothing because the country was in the grip of the Mahdists, although Pankhurst and Johnson (1988) quote a report (Kolle, 1901) of an extensive outbreak of cattle disease between Wadi Halfa and Khartoum in 1888, and state rinderpest appeared in scattered but isolated areas in 1889–90, causing far less damage than in Ethiopia. We know from Slatin Pasha that famine was raging in Omdurman at the close of 1889. In March that year, when the epizootic was rampant in Ethiopia, the Mahdists invaded the western part of the Sudan, so if the disease was not already present, it could well have been carried there by the Mahdist forces who may also have initiated its westward passage. About this time, they also ravaged the Dinka tribe and brought back immense herds of cattle to Omdurman (Slatin, 1895). Wingate informs us that camels and oxen were almost exterminated during the long campaign against the Government, "... and in 1889 a cattle plague still further reduced their numbers" (Wingate, 1892).

Between June 1886 and March 1889 was a period of successive crises for the Mahdists on their eastern and western borders. In January 1887, the Abyssinians invaded the east with a great army advancing to west of the Atbara and then withdrawing after victory. In May 1889, a vast Mahdist army left Dongola for the north to invade Egypt but was defeated at Tushki. Three large armies were maintained, one each in the north, east, and west, which meant vast movements of men throughout Kordofan whenever they were mobilized. Coinciding with the beginning of a great famine throughout the Mahdist controlled area of the Sudan, a whole tribe migrated through Kordofan to Omdurman. The famine, due to a bad harvest, began at the end of 1889 in northern and eastern Sudan, and not until late 1890 was there sufficient grain for planting. Holt (1958) refers only to an absence of grain as a problem, making no mention of cattle scarcity as a contributing factor; but this famine followed the rinderpest outbreak, although, as Wingate reported, cattle stocks had already been largely destroyed by the Mahdists.

About 1890–91, the panzootic swept along the south-eastern border of southern Sudan apparently following the Pibor and Sobat river courses, affecting people south of the Sobat such as the Jikany and Murle around the Pibor, the eastern Lou and the Lak above the Sobat–White Nile confluence. It does not appear to have extended north of the Sobat nor to east of the Lou. Pankhurst and Johnson (1988) speculate that it could have been carried to southern Sudan by migrating game congregating on the tributaries of the Sobat and coming into contact with Nuer cattle. Their second possibility is that it was spread by the Ethiopian Tesemma's conquest of Buré, which is said to have spread disease into the Baro area, the Lou being hit by simultaneous plagues of rinderpest and smallpox, the two diseases associated with the Ethiopian famine. As this outbreak was contemporary with its appearance among the Maasai, these authors suggest that southern Sudan could not have been its main route to East Africa but is likely to have been a link in the chain of transmission to Equatoria and north-eastern Uganda. With the exception of the Lak Nuer country, it appears that rinderpest did not extend very far west. According to Littlewood (1905), the eastern Sudan was affected in 1898 by cattle from Eritrea passing down the valleys and the river Atbara.

About the turn of the century, a Nuer Lou prophet, Ngundeng Bong, foresaw a terrible cattle plague termed *diu*, which Seligman considers to be almost certainly the name of a sky spirit, rinderpest being called *jok* in other parts of the country. Had the prophet not gone forward to meet the disease and sacrificed oxen in front of it, the herds would have been wiped out. He took active measures to contain the epizootic through large-scale sacrifice of cattle and the effective isolation of people and herds from areas where the epizootic was raging. But as he had prophesied, the disease was severe. The Bari of southern Sudan claimed in 1929 that they once possessed many more cattle. If their cattle were infected with a serious disease, even if only one valuable animal was ill, a sacrifice was made, and a *gili* (a stake usually erected on a grave) erected as part of the cure. The owner might have to sacrifice an animal at the spirit tree and call the old men of his family to eat the meat, others fearing the disease would sacrifice at their spirit trees also. The owner might be told to break up his herd, distributing it in different parts of the country and abandoning his kraal for some specified period of time, usually until the rains broke (Seligman and Seligman, 1932).

In November 1889, when Jackson reached Mumia's on the Nzoia River in Kavirondo, western Kenya, he reported that within a few days, their splendid herd of cattle contracted a "lung sickness," which he termed pleuropneumonia, and began dying daily in twos and threes until none but a few emaciated old cows were left. By December the herd had been reduced to 77 "with no sign of the fell disease abating" (Jackson, 1930). This seems undoubtedly the first record of the rinderpest panzootic reaching Kenya (Figure 3).

In Uganda and Kavirondo (Kenya), it began to show its effects in the summer of 1890 and, by autumn of that year, had just reached Unyoro and Ankole, apparently spreading westwards. About July–August of the following year, it had swept off "every ox" at Kavalli, on the western side of Lake Albert. Emin received information of rinderpest in Ankole in January 1891, and in early April of that year, he found that it had reached the Kagera River via Mpororo. In June, it had reached Bukoba, where it was estimated that 400,000 head of cattle were reduced to some 20,000 in 1891–2 (Ford, 1971). Presumably, the plague had begun to affect the Nkore herds in 1890, and the chief then raided Mpororo and Rwanda to replenish his stocks.

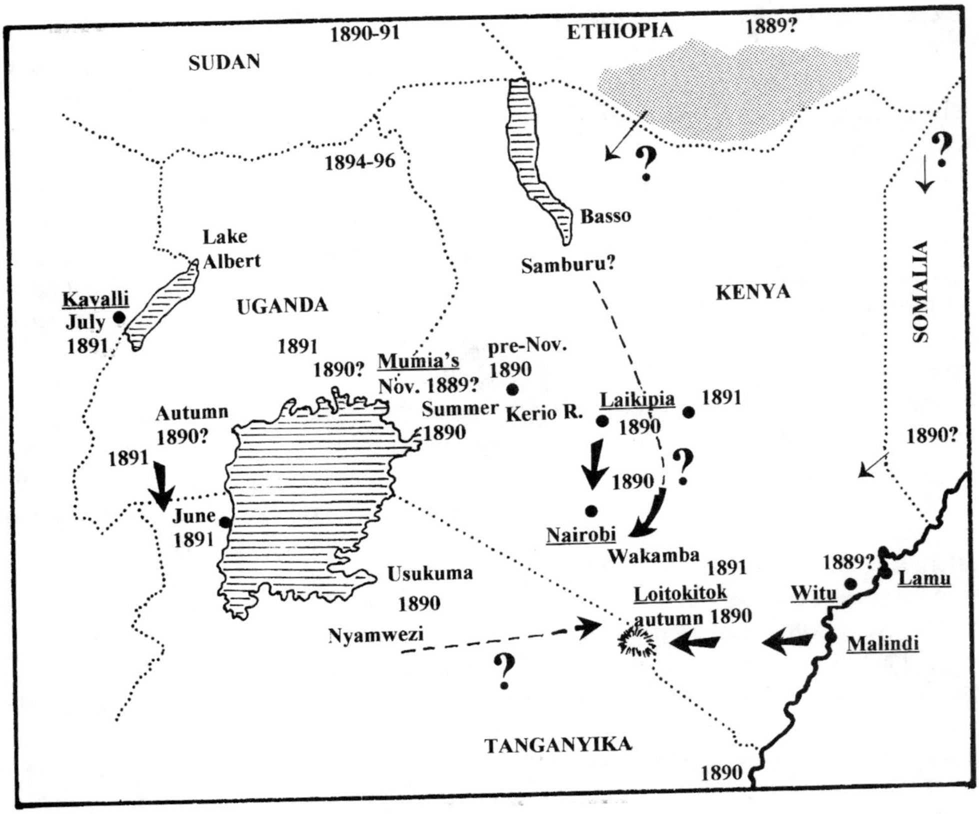

Figure 3. The spread of rinderpest in East Africa, 1890–91.

We have several first-hand accounts from East Africa, such as that of Lugard. When Lugard was on his way from the East African coast to Uganda in February 1890, he came across large herds of cattle among the Wakamba, one herd at Machakos estimated at over 1,000 head. It was not until he reached the Karien (Kerio) river in Kamasia on November 20 that he first encountered dead buffalo and noted: "The buffalo have got the cattle disease, and the guides say are dying in very large numbers. We came on several carcases, and the air is full of vultures" (Perham and Bull, 1959). On November 23, he recorded: "Constantly we pass dead buffaloes, carcases a month or so old, mostly uneaten by vultures or hyaenas ... This disease seems to have swept off the whole of the buffaloes. A porter in my caravan passed thro' here with Dr. Peters in his journey to Uganda. He says there were thousands and thousands of buffalo then, and that is only last March or April." What the porter most likely said was "*mingi mingi*," Kiswahili for "many, many," which might equally mean "hundreds."

Six kilometers further on, he camped at an "enchanting spot," but full of dead buffalo. Here, however, was an *enormous* (his italics) herd of hartebeest. On November 25, "game none, all gone to fresher grass, and buffalo all dead." Coming to a Kavirondo village on the November 27 he noted: "During this last year a terrible cattle disease seems to have raged, and all the cattle are dead and also the wild buffaloes." On December 4, he came to Mtindi on the Sio River north of Lake Victoria:

> ... the people, having no cattle, are dependent on grain food to live on. Everywhere in the paths are heaps of bones and horns of dead oxen. It ... has swept off every living ox, and the wild buffalo

> also, from Lake Victoria to Kamasia. Jackson's large herds died off by hundreds, but he says the disease did not affect the Kavirondo cattle. Since he left a few months ago, all have died. I wonder if [he] brought the disease from Lumbwa, it seems like it .... (Perham and Bull, 1959).

In his book he wrote: "When Mr. Jackson had passed here in June the large herds of cattle were still alive, though his own were dying of the plague. Now, in November, I did not see a single ox throughout Kavirondo, and their skulls and bones strewed the ground" (Lugard, 1893).

On December 6, he marched into Busoga: "Here we still found that all the cattle were dead, not one left!" On December 14, southwest of Bugungu he noted tracks of buffalo, "but they say ... are all dead. The cattle are all dead also." In Bunyoro, "all the countless herds of cattle have died ..." On the western side of Lake Victoria, east of Masaka, he found the plague to be less virulent, "for I saw a couple of small herds of cattle grazing—the first I have seen in Uganda, and around Sunga there were still some wild buffaloes left ... they say the wild pig have all died of the plague too ... I asked if any antelope had died and they said the Ngabi [bushbuck] had." On December 21, he wrote: "We go on to Ruanda ... There is said to be endless game there." The following day, a chief, Ntale, sent a message saying that his cattle were all dead, but later it was reported that a few fine cows (Ankole) were seen, and one of Lugard's entourage stated that, formerly having some 7,000 cattle, the chief now had only 300. On June 19, 1891, at Marongo on the borders of Buddu and Ankole, he wrote: "All the buffalo are dead, also all the eland and *all* the game they say (including Nsunu) which swarmed here, except these Roan, and Waterbuck, and a few pah [duiker]." On October 22, Lugard was at Kasenyi on the south-west shore of Lake Albert where antelope appeared plentiful but, "the buffalo are dead, all but a few (which have diseased feet) and have retired to the thick jungles. All that remain are Waterbuck and Mpallah, and very many of the latter have died also. I saw one pah also ... " (Perham and Bull, 1959). Previously, he wrote that he had been shooting at Kooroo (waterbuck) and Uganda kob. The reference to buffalo having diseased feet suggests an outbreak of FMD also. In his book, he wrote:

> The buffalo, however, has been almost exterminated by the plague. Recent writers have reported him as almost or quite extinct. This is not the case; for covering, as I did, great distances daily in pursuit of game on my downward march, I not unfrequently came on the freshest possible sign of buffalo, all the way from the Albert Lake to near the coast. The remnant that are left, however, have retired to the densest forests; the natives say they have sore feet.

He went on:

> It is strange that so little is known of the nature of this plague. It has been reported as a kind of anthrax by Dr. Charlesworth at Zanzibar, while others (and these apparently the majority) have described it as lung-disease. Dr. Mackinnon and Captain Rogers thus describe it from Witu. Mr. Sharpe from Nyasaland states the disease there to have been pleuro-pneumonia. (Lugard, 1893)

In 1892, Lugard wrote that the plague had carried off millions of cattle but greatly regretted that no case with pronounced and unmistakable symptoms came under his own observation for it had invariably preceded him. His own cattle had died "presumably of this disease, but in an attenuated form—debility and gradual wasting away and refusal of all food were the only symptoms" suggestive of trypanosomiasis and not rinderpest.

Of Ankole, he wrote that the plague seemed to have been very virulent, sweeping off almost all the game as well as the cattle. According to Johnson (1902), it carried off three quarters of the Ankole cattle, and the Bahima, who depended almost exclusively on their cattle for food, perished from starvation in great numbers. Tucker (1908) recorded in January 1891 at Kampala that a cattle disease had swept away almost every head of cattle in the country.

Roscoe (1922) wrote of 1890

> Within a short time there were thousands of dead cattle lying about in Ugogo and Usakama [Usukuma], far too numerous for any wild scavenger animals to devour, and no attempt was made by the people of the districts to bury or destroy the decaying carcases. The smell near any village wherever animals had crawled and died and had been left by the natives to decompose was beyond description. It was horrible to have to pass such places … for years afterwards bones lay about, bleaching and crumbling in the sun …

Entering Kenya from the north, the Turkana region appeared to escape its passage, while the Loyoro Dodoth drove their cattle into the Timau Forest and high up on Mount Morongole to avoid infection. The Rrapp Borana still had a great number of cattle in 1893, "despite the recent plague" (Chanler, 1896).

The disease probably moved south in 1890 through the coastal Galla herds into those of the Wakamba, whence it spread to other tribes, particularly the Maasai. It had not reached Maasailand in the spring of 1890 but had done so by autumn of that year, apparently first appearing among herds of the Loitokitok Maasai at the base of Mount Kilimanjaro, possibly carried there with infected cattle raided from Malindi. The Maasai termed it *olodua*, distinguishing it from *ol kipie*, pleuropneumonia, which first appeared in the second half of 1883. Chanler (1896) reported from hearsay that the disease had destroyed many thousands of cattle in the vicinity of Witu and Lamu in 1889 (but this may have been 1890), and in 1890, the Laikipia Maasai raided the Maasai near Nairobi, carrying off infected stock. As a result, most of the Laikipia cattle died. By March, it had reached the Kinangop on the slopes of the Aberdare Mountains 230 km north of Loitokitok, but it was reported that there was still "a vast quantity of cattle" in the Kedong Valley which lay south of the Kinangop (Waller, 1976). Nevertheless, tradition suggests that the herds were almost wiped out, and the Maasai were reduced to eating donkeys and even hides. Waller (1988) questions allegations that the disease was spread by Maasai cattle raids, claiming the raids can be shown to have taken place after rinderpest had already appeared, or were mounted in areas free of the disease. Hobley was told that stock raided from the Embu had introduced rinderpest into Laikipia, but the raid was in September 1891, by which time the disease had already appeared. One Maasai graphically described how the Engaruka Basin was littered with the remains of cattle and people: "so many and so close together that the vultures had forgotten how to fly" (Branagan and Hammond, 1965). The Keekonyukie herds disappeared almost overnight, the Il Dalalekutuk lost their cattle "within a week," and in Loita, it took 2 months. Whole camps were left with no more than half a dozen head and many families with nothing. One man lost all but three of a herd of several hundred near the Pangani River, and another at Elmenteita had only 10 survivors of a herd of 500. By the end of 1891, most of the Maasai cattle were dead or dying (Waller, 1988), and the disease had burnt itself out.

Waller considered it unlikely that Maasai would knowingly take cattle from a sick herd, and newly infected animals not yet showing symptoms would survive the return journey. Even if they did not, they would have transferred the disease to others, and the Maasai may well have raided those herds which still looked healthy but in fact were in stages of incipient infection. Waller considers that the fact that rinderpest appeared among Maasai herds in 1891 immediately after it had become widespread and virulent in game, especially buffalo and eland, is suggestive. The spread within Maasailand he considers likely to have been determined by herd movements and the pattern of settlement, herds being coralled together at night, while transmission between herds may have been facilitated by the sequence of seasonal concentration and dispersal, different groups sharing certain areas and water points. The disease spread at the onset of the long rains. When the Kisongo herds were infected in February, they were still concentrated close to permanent water along the Pangani River. A month later, when Smith and

Gedge each witnessed its onset in the Naivasha area, the herds were gathering there and beginning to move down from Laikipia into wet-season pastures in the Kedong Valley. Purko camps which had spent the dry season near permanent water north of the Uaso Nyiro may have contracted the disease from Samburu or Laikipiak herds which already had the disease in 1890 and brought it to the Central Rift when they moved southwards at the end of 1890 or early 1891. Given the greater dispersal of stock on the northern plains, the disease could have been spreading quite slowly in the latter part of 1890, but set against this is the high initial virulence making it unlikely that it could travel very far on the hoof. Most herds seem to have died where they stood, but not all animals were affected equally, and some did survive. Some areas, such as Mashuru in Kaputiei, apparently escaped infection entirely.

Hobley (1929), during his exploration up the Tana River, first came across evidence of the disease in August 1891 below Mount Meru on the Tana River:

> The buffalo were chiefly affected and they had come down to the river in thousands to die. Apparently when attacked by the disease they become consumed by thirst, and so congregated at the river which provided the only water supply for many miles; it was a tragic sight to see all these great creatures dead and dying. The stench was nauseating, and one could not have believed that there were so many carrion birds in Africa as were collected for the feast. The buffalo were also sometimes aggressive, and on several occasions a poor doomed creature would dash out of the bush and make a half-hearted charge at the caravan ... The giraffe, lesser kudu, and eland also suffered from this visitation ...
>
> Upon our arrival at Ngomeni we found the people in a terrible state of depression, for the rinderpest epidemic ... had recently attacked their cattle. This area had, up to then, been probably richer in cattle than most of the Kamba country ... the fell disease had spread with such rapidity that the disposal of the carcasses of the cattle was a task beyond their powers, and the desiccated remains of tens of thousands of beasts were piled up in the form of a wall, a few yards from the villages ... and the air night and day was pervaded by a sickly odour of putrefaction. The people appeared to be overcome with a hopeless apathy, and the elders sadly produced a small bunch of about twenty head, the sole survivors of the great herds.

Macdonald (1897) reported that in 1891, the Wakamba had considerable flocks of sheep and goats but relatively few cattle, being exposed to Maasai raids from both north and south; and at Kibwezi in 1892, drought was causing famine.

In the coastal region, north of Malindi, Fitzgerald wrote in 1892: "A Galla whom we met ... reported that in the country to the north there were numerous topi, and a few rhinoceroses, elephants, and ostriches, but that giraffes had suffered from the disease which had played such havoc among the wild buffaloes" (Fitzgerald, 1898). In December, F. G. Foaker, an officer of the British East Africa Company traveling down to the coast, reported that there were no buffaloes, the "cattle epidemic" having killed them all. But on the way to the coast, he met a large number of Wakamba from Machakos going to Mombasa to sell cattle, although he gives no indication of the number. The following February (1893) along the Tana River, Fitzgerald referred to a plain covered with good short grass, "... at one period the Wa-Galla used to feed their cattle here; but now they appear to possess none, a fact due in the first place to Masai raids, and second to the cattle disease ... ." And: "I was informed that in the year 1889 the Wa-Galla owned from 200 to 300 head of cattle on the swampy grass plains near Kau ... but all the cattle were destroyed by the disease."

Captain Smith, on a journey from Mombasa to Uganda noted on March 3 that his party had no trouble with the Maasai at Naivasha,

> What few cattle they had left were dying rapidly, the west and south-western shores of the lake being strewn with carcases ... the only obvious internal disorder being a considerable enlargement of the gall-bladder, which in some cases was as large as an ostrich egg. There are no cattle left in Kavirondo or among Kabra's people [Nandi], and disease is rapidly exterminating the buffalo.

From thence to Wakoli's (near present-day Jinja, Uganda), until his return to Naivasha, he does not mention the disease again until he notes that the Maasai at Gilgil and Naivasha had lost nearly all of their cattle. In a letter to Sir Francis de Winton, he noted that the cattle of the Maasai at Ngong (Kenya) seemed to be decimated by disease

> just as much here as on the over-fed and over-stocked pastures of Naivasha. These are not over-stocked now; the whole of the west shore of this lake is hardly approachable for the number of dead beasts. Neumann and I made many *post-mortem* examinations and there is no symptom of any disorganisation of the lungs, the only abnormal thing about the animals being a greatly enlarged gall-bladder. The beasts keep fat to the last, but for two or three days before they succumb are unable to do more than just keep on their legs. Kikuyu, Ukambani, and the whole coast-line, notably Lamu and the northern ports, have been visited with a cattle sickness, but in the case of the coast towns I believe this has been declared to be an affection of the lungs.

Of Mumia, a Kavirondo chief, he wrote:

> ... when Jackson was with him [Nov–Dec 1889], was a rich man; he is a poor one now, having lost every head of cattle he possessed from disease. This disease is destroying the buffalo; on Mau and along the Nakuru Naivasha Valley, where they used to be in great herds, not a buffalo now, only their skeletons, fifty and more passed every day. (Thomas, 1959)

Hinde and Hinde (1901) reported that after the famine in 1892, caused by the rinderpest, some Maasai sections moved near to the Nairobi river, where, reduced in numbers by smallpox, they had sufficient cattle to keep them from starvation. Almost all the cattle of the Loita Maasai were destroyed, so they attacked the Maasai near Nairobi, defeating them and carrying off the cattle. Later, some of the cattle were recaptured. The Wakamba stock was allegedly infected from cattle looted from near Lakes Stefanie or Rudolph. A story that Percival heard was that the disease came from Wanyamwezi with cattle looted from the south-east of Lake Victoria (Percival, 1918). Isolated regions such as Mounts Marsabit and Nyiro in the north of Kenya, and some parts along the coast, apparently escaped infection, later supplying cattle for restocking elsewhere.

Chanler (1896) writing in 1892 noted that the Wamsara near Embu a few years before possessed many hundred head of cattle, but they had been largely reduced by some sort of plague, he supposes pleuropneumonia, although he was well aware of the rinderpest epizootic.

Journeying westwards from Mamboya in eastern Tanganyika in August 1891, Ashe (1894) recorded:

> All along the line of march, but in Ugogo especially, we came on hundreds of dead cattle. The plague had swept the whole country from Uganda to the east coast, but strange to say, passing over the Sesse Islands in the Nyanza, as if the water were in some way a barrier to the infection. Buffalo, giraffe and other big game had also suffered severely from the disease.

The German expolorer, Oscar Baumann, reaching Tanganyika's Ngorongoro Crater in 1892, described the results of the cattle losses and famine upon the Maasai:

> There were women reduced to walking skeletons, out of whose sunken eyes looked the madness of hunger, children resembling deformed frogs rather than human beings, warriors who could hardly crawl on all fours, and moronic, emaciated greybeards. These people ate everything available; dead donkeys were a delicacy to them; but they also devoured the skins, bones and even horns of cattle ... Swarms of vultures followed them waiting for victims .... (Organ and Fosbrooke, 1963)

On his journey, he had seized in reprisal a "lot of cattle" from the Wambugwe some 60–80 km away at the south end of Lake Manyara, so destruction of cattle in the area was not universal. It was not until he met the Maasai in the Crater floor that he found any starving, but these were refugees from the Serengeti plains to the west. On his way to the rim, he recounts meeting some who looked fairly well fed as they still owned some small livestock and hunted wildebeest and rhino. "The abundance of game was magnificent. Large herds of antelope roamed around and long-maned gnus ...."

Stuhlmann recorded "abandoned villages were almost without exception the only traces of the Masai people ... All their cattle have been wiped out and the surviving people have taken refuge with the agricultural tribes at the edge of the steppe in Ndjoyi, north Ugogo, Nyangana, Hedya, etc." (Stuhlmann, 1892). In Usambara, Africans in the 20th century recalled that the rinderpest reached Mlalo in 1892, substantially reducing the herds of the Wambugu just before the Germans came. A German column on its way to Kilimanjaro in early February 1890 reported nothing amiss but on its return at the end of the month found the disease raging among the Kisongo herds on the upper Pangani River. The Tanganyikan chief Merere in Iringa allegedly had over 30,000 cattle, his herd being reduced to 3,000 (Schele, 1894). In Sukumaland, south of Lake Victoria, by 1893 only chiefs and wealthy people had any cattle left.

In November 1893, Scott Elliot (1896) noted that among the Wakamba, the government administrator, Ainsworth, had, by a system of isolation, prevented the cattle disease from entering the Wakamba country. In the Kedong valley, he found the land covered with Maasai cattle and goats, noting at the same time that the cattle disease had been deadly among the Maasai. In Busoga the following year, he noted that cattle were both plentiful and cheap, and in western Ankole around lake Albert Edward, cattle appeared to be easily kept over the whole country and were in the very best condition.

The German Vice-Consul at Zanzibar stated in 1892 that the disease had been prevalent in the district around 2 years. At one time, no cattle were to be had for several months, people being obliged to live on tinned meat (presumably meaning officials). On February 7, 1893 the German Consul-General wrote that the Governor of German East Africa reported on December 9 last great havoc had been done of late, not only on the shores of lakes Victoria and Nyassa, but also on the coast and through the intermediate parts of the country. A Veterinary Commissioner was to have been sent out by the German Imperial Government but never came. Writing at the end of 1893, the governor, Schele, noted that cattle were still dying at the coast, while the government was powerless, not knowing how to prevent or cure it. It was not possible to employ a veterinarian, and so the nature of the disease remained unknown (Schele, 1894). Smith (1941) was in error when he wrote "When Germany took over German East Africa, an attempt, which met with some success, was made to control rinderpest by means of immunising inoculations. During the Great War, however, that control was lost ...." The progam did not begin until 1913, and some of the first serum batch came from Kenya.

The Karimojong and Jie in 1876, and again about 1887, lost many stock, but this was probably pleuropneumonia. The Jie stated that such epizootics had not been known before (Lamphear, 1976). Barber (1968) places the first outbreak in Karamoja in 1894–6, 1896 being a drought year. Game was reported to be dying also.

In 1897, a renewed outbreak began at Nzawi in the southern part of Ulu, Ukamba Province, which, despite "all possible measures of isolation," (Ainsworth, 1905) killed some thousands of head of cattle in various parts. Buffalo, eland, wildebeest, and hartebeest also died but not in great numbers. It destroyed all the government transport oxen in Kikuyu, but in June a detachment of Royal Engineers from India headed by Major Herbert Austin, to survey the Uganda railway, imported from Bombay where rinderpest was enzootic, 100 bullocks and buffaloes as draught animals, five bullocks dying on the voyage and five at Aden, attributed in the latter case to the heat. He reported (1903) that his party left Kikuyu with 150 bullocks.

It is not clear if this number included the Indian animals which were drawing 25 Indian carts and 14 "Sclater" carts in the train. By the middle of the year, it had spread up the transport route from Tsavo with oxen to the Maasai herds at Naivasha where the "flocks and herds" were reported suffering from rinderpest near Lake Elmenteita, scores of animals dying daily. It was necessary for Austin to drive his bullocks through the area without allowing them to graze, making a forced march of 50 km from Naivasha. On his return journey in January 1899, Austin reported that the Maasai here had to a large extent escaped the cattle plague "which had lately devastated the oxen along the Uganda road employed in transport work." Ainsworth (1905) reports that he immediately warned the Maasai headman Olonana of the outbreak in 1897, who promptly moved all his herds to the south of Ngong, isolating them with a cordon of armed warriors and thus escaping infection, although Ainsworth suggests that the herds may have become immune from the last outbreak. But according to Waller (1976), it did strike them, Olonana allegedly losing 1,000 head. Abating at the beginning of 1898, it was immediately followed by a mild outbreak of pleuropneumonia. The administrator, Francis Hall, was allowed initially by the Maasai to inoculate some animals, again in 1899, presumably against pleuropneumonia.

Hall wrote in 1897 that the Maasai had vast herds of cattle, and at the beginning of January, Cook (1945) came across a number of Maasai driving large herds of cattle and goats which they had raided from the Wakamba. Crossing the Mau Plateau at the end of 1897, where bullock carts were used, Sykes (1903) reported the animals overtaken by "lung sickness," undoubtedly from his description rinderpest and not pleuropneumonia: "I cannot convey in words the ravages of this terrible disease—its suddenness, and rapidity of action. I need only mention that in a few days 294 out of about 300 [oxen] succumbed, and lay strewn about our path."

No lessons were learnt immediately, for in June 1897, Glossop (1906) reported a ship sailing from Berbera to Aden as crammed with sheep and cattle, "numbers of which died during the night...."

Although Ainsworth (1905) reported Ukamba Province clear of disease by the end of 1898, and only sporadic cases of rinderpest and pleuropneumonia up to 1905, it flared up again in August 1899 in Machakos and spread to Maasai herds at Ngong, attacking especially calves and cattle brought in after a raid, before being brought under control by isolation (Waller, 1988). Over the wider area, it persisted until mid-1900, when it burnt itself out. Ainsworth (1905) reports that the province was affected by famine due to irregular rains in 1898 until the latter part of 1899, which caused a great deal of disorganization amongst the tribes affected and a large amount of thieving of cattle and goats took place.

On November 1, 1897 from his headquarters at Machakos, the Commissioner of the East African Protectorate, Sir Arthur Hardinge, submitted to Prime Minister Lord Salisbury a set of proposed regulations which he had drafted on October 26 for Ukamba, to prevent the spread of rinderpest "now present among the Masai cattle" as follows:

> Whereas under the East African Order in Council 1897, Her Majesty's Commissioner has power ro make regulations for [the] peace [,order,] and good Government [of the Protectorate (?) and for the regulation of all matters relating to agriculture and public health within the Protectorate,] and whereas it is expedient to prevent the spread of cattle plague it is hereby notified that the Commissioner has in the Exercise of the powers aforesaid made the following regulations:
>
> 1. These regulations apply to the Province of Ukamba only.
> 2. From and after this date and until further notice all persons within the province either owning or having cattle in their charge, must at once report any case of sickness to the nearest Government Officer who will use his discretion as to whether or not the animal should be slaughtered.
> 3. All Transport riders will report their arrival in any district and intended departure from same to the District Officer and they must account for any cattle lost en-route.

4. Any Government Official may at his discretion order the examination and detention of any cattle where-so-ever found provided of course that he has reason to believe that such cattle have any sickness or have come from any of the proscribed Districts notwithstanding that the owners or drivers may have a pass for such cattle from the Officer in Charge of the District which they have left.
5. Any person convicted of having concealed sick cattle or of committing any other breach of these regulations shall be liable, to a penalty which may extend to one month's imprisonment of either kind or a fine of one hundred Rupees, or equivalent, or both.

These regulations may be—as the Cattle Plague Regulations (Ukamba) 1897.

Hardinge clearly seemed to have had some knowledge of rinderpest and measures which should be taken, but his regulations did not become law. Foreign Office comments, while noting that if the plague reappeared in a severe form, transport to Uganda would become almost impossible as local staff had no veterinary knowledge, a transport scheme just proposed provided for veterinary staff, and this need was emphasized by the fact that the railway mules were dying and sickening fast. What appears to have worried the Foreign Office was Hardinge's proposal to slaughter cattle, considered a very risky step to take, especially with the Maasai, noting that it was believed to have been to a great extent the origin of the Matabele rebellion, and asking whether Hardinge should be warned not to apply the provision. One annotator queried whether it was not a matter of agriculture or public health, in which case, the Commissioner had special powers to make regulations.

Ansorge (1899) noted that the caravan road across Mau in Kenya was "recently" littered with dead and dying bullocks belonging to the Government bullock transport, which consequently broke down. The Government veterinary surgeon whom he met at Machakos, presumably Veterinary Captain J. A. Haslam, was in no doubt that it was rinderpest.

Writing of the country near Bukoba on the western side of Lake Victoria about 1898, Kollman (1899) referred to the wealth of beautiful long-horned cattle, but it was a matter of daily occurrence that a few fell victim to a disease which had not yet been investigated. Sheep were the worst in resisting it, and goats and calves died more frequently than horned cattle. The natives considered it a type of FMD:

> The beasts must endure great suffering; they bellow with a rattling sound as they stretch their legs on the ground in a helpless manner ... A yellowish liquid flows from the mouth even after death. Many of the smaller beasts show spots on the hide when dead, as if they had rubbed the skin off, but I did not observe this in the case of the larger ones. For weeks together in Bukoba we daily lost four to eight of the smaller animals, and from time to time some of the full-grown cattle ....

And of the Wasukuma: "Herds of cattle are still very numerous, especially in Nera, notwithstanding the great destruction of cattle wrought by the rinderpest." It seems strange that Kollman did not refer to rinderpest, or discount it, as a possibility.

In his annual report for 1898–1900, the veterinary officer for British East Africa and the Uganda Protectorates, Robert Stordy, made brief reference that the disease had existed for the last "five or six years, during which much work has been done by way of prevention and suppression." So far, the disease had not been recorded beyond the Gilgil River, 16 km from Naivasha, and it was hoped that the regulations in force would keep the disease within its circumscribed area. Bullock transport was now operating from Nakuru Railway station (Stordy, 1901). If rinderpest appeared among their cattle, the Maasai were advised to construct special pens at a distance from any habitation and remove all sick cattle to them, which proved successful in limiting further outbreaks up until 1899, one sector placing a strong cordon of warriors around the grazing grounds to prevent communication. But it broke out again among their cattle in 1900, according to Hinde and Hinde (1901) "through no fault of their own."

Geil (1905) reported that across the Nyamgasani river in western Uganda, the Basongoro were formerly great cattle keepers, but during a widespread plague which "swept the Semliki

Valley a few years ago," most tribesmen lost their animals and up to this time had been unable to recover their herds. They were accustomed to speak of things as before or since "the cows died."

5. THE SPREAD SOUTH

From the Kilimanjaro region, the plague moved relentlessly south through Tanganyika. Major Von Wiessmann, who conducted an expedition to the interior in 1888–9 with cattle imported from either Aden or Bombay, having done his utmost to obtain suitable cattle locally (Littlewood, 1905), apparently lost the imported cattle to rinderpest, materially hampering his expedition. Le Roy (1914), arriving at Maboua on the eastern side of the Paré Mountains on September 14, 1890 was told that the cattle had been taken by a sickness, but traveling south about 5 days' later, he reached Same where he saw long moving black lines of cattle returning from the pastures and reported numerous troops of antelopes and the presence of buffalo. The cattle numbered about 2,000. In August 1891, Vrithoff (Alexis, 1894) reported that at Mpwapwa, the camp was surrounded by the carcasses of 2,000 cattle taken by a disease, "Here and there one met entire heads, feet, skins, etc. This place is not very healthy." On August 24, at Mouhallala, a day's march from Kilimatinde, he reported that there were few herds throughout Ugogo, a disease having killed 14,000 cattle in Ugogo alone. Lieutenant Prince, writing in 1897, estimated that 90% of cattle and half the game had died. Cattle were reported to have survived in only a few places such as on Mafia and other offshore islands, in east Sukumaland and further east, in the higher reaches of Mount Kilimanjaro, but not at Moshi, and possibly in parts of the mountain districts of Buha. Sharpe, crossing Tanganyika in 1892, noted that parts of the country "had been visited and portions had escaped" (Sharpe, 1893a). He reported that practically all cattle of the north Nyasa country were "cleared out," mortality being over 90%. Lugard (1893) noted that the Wankondé north of Nyasa possessed vast herds of cattle until rinderpest destroyed them at this time. Visiting the area between lakes Tanganyika and Mweru, Sharpe stated that he had no evidence of the disease affecting wild game until he arrived at the south end of Lake Mweru in October 1892:

> Here enormous quantities of game have died: for the first two days up the Luapula, where it passes through marshes, it is near the banks only of the river where good dry land is found. These banks are, or were, covered with game, mostly pookoo, buffalo, and lechwe. At the time of my passing up through these swamps the plague was at its height. Dead and dying beasts were all around. The first day I counted over forty dead pookoo within half a mile of my camp. Subsequently, on my return down the river, I saw scarcely any live game near the mouth, but the whole country was scattered with dead bodies ... Elephants, of which there are great numbers in these marshes, do not appear to have been attacked by the plague.

Sharpe further reported game in "unlimited quantities" in the area of north-eastern Zambia between Kazembe's village and Rhodesia. The plague is believed to have missed northeastern Rhodesia, and keeping west of the Luapula, it bypassed the Nyasaland Plateau passing from northwestern Rhodesia into Matabeleland. In Barotseland (western Northern Rhodesia), African cattle were apparently little affected, if at all, but in the Songwe valley in Northern Rhodesia in 1898, Grogan and Sharp (1900) noted that there were very few cattle:

> Before the rinderpest the whole of the Songwe valley was black with buffalo; now I do not believe there is a single beast except in some jungle two days' march to the north .... only a very few head survive of the countless herds of cattle that were characteristic of the Wankonde.

Among game, only eland and buffalo appeared to be attacked at this time (Pitman, 1931).

An outbreak followed in north-eastern Rhodesia in 1902, attacking buffalo and other species which had escaped in 1892. Sharpe (Pitman, 1931) records that before the rinderpest in 1902, buffalo existed in north-eastern Rhodesia in enormous herds on and around Mweru Marsh as he saw in 1901, "simply a black mass of buffalo as far as one could see on the plain, many thousands." And for both ordinary and black lechwe, great numbers existed on the flats east of Lake Mweru and the Marsh: "Before the rinderpest they were in herds of hundreds and hundreds." In 1902, it killed all of the buffalo on Chiluwi Island in the Bangweulu region.

In a letter to *The Field*, Sharpe (1893b) reported that zebra had died also:

> On the Luapula river, immediately south of Mwero, the havoc has been something terrible ... The disease has not touched the elephants, but among others it has spread like wildfire. As I ascended the river (by boat) dead and dying beasts were all around, and one was seldom away from the stench of dead meat. Buffalo have died there in great numbers. Between Mwero and Tanganyika I saw numbers of dead zebra. The disease has proved very fatal among the cattle at the north end of Lake Nyasa ... I do not suppose 5% of the cattle have been left. Throughout the plateau lying between Nyasa [Malawi] and Tanganyika I was informed that wild game were dying from the plague. At the African Lakes Company's Station, "Fife", halfway between the two lakes, meat from beasts which had been found dead in the bush was constantly being brought in.

Elsewhere, he was reported as stating that 90% of the natives' cattle had died, but it was curious to note that the disease had capriciously spared certain districts. The Lakes Company had lost nearly all their cattle at the north end of Lake Nyasa.

> Buffalo, zebra and antelopes have died in great numbers. In one small district Mr. Sharpe counted over forty dead cobus antelopes [puku or lechwe], and quantities of sick and dying who could not get out of his way. He also met clumps of dead buffaloes lying in groups under trees as though suddenly struck down by the disease: the stench of dead bodies rendered his camps most noisome in several parts of the Mweru country. (Anon., 1893)

The great mortality reported among zebra between Lakes Mweru and Tanganyika is also attributed to Crawshay, the official in charge of Rhodesia, the administrative station on Lake Mweru (Anon., 1893), while Decle (1896) asserted that his donkey died of it near Ujiji in Tanganyika (on the Malagarasi River). But the hard swellings on his donkey's head which he described are indicative of anthrax, and his report that north of Lake Nyasa, the rinderpest spread from the *west* killing all of the cattle, and also "the buffalo and destroying nearly all the game, zebras, wildebeest and even elephants; goats and fowls also died in great numbers ... all along the eastern shore of Lake Tanganyika the disease had passed," suggests an unusually widespread outbreak of anthrax apparently concurrent with the rinderpest.

In an earlier account, Decle (1893) did not know what the disease was and gave it its local name of *sotoka*. He reported that when he traversed the Tanganyika plateau east of Lake Tanganyika in May 1892, he did not see a single head or spoor of game, although the area was previously said to have been splendid game country. It was the same from Ujiji to Tabora and in the Wahha country. The disappearance was due to *sotoka*, which had raged for 2 years. He was told that it had begun in Ugogo and then traveled westwards to Lake Tanganyika, following the shores of the latter southwards, then taking an eastern course to Lake Nyasa:

> ... at the north end of which it seems to have stopped ... In Ugogo, in Unyamwezi, in Ujiji, hardly any cattle are left. On the other hand, the Wahha, close to the river Malaglarazi [Malagarasi], have still thousands of magnificent cattle. At the north end of Lake Nyasa the Wakondo were formerly celebrated by their numbers and beauty of their cattle. Now but a few head are left. I have not been able, so far, to find out what kind of disease the sotoka is ... When natives eat the meat of the animals that have died of it, they soon get covered with malignant carbuncles.

In this account, Decle describes his ill-fated donkey's first symptoms as swollen testicles, a characteristic of equine anthrax. Although he conducted a postmortem on the body, he himself did not contract an infection, a high possibility with anthrax. It is also surprising that the natives who developed carbuncles from eating infected meat did not die: "Whilst the sotoka was raging on the Nyasa-Tanganyika plateau, numbers of animals, from elephants down to the smallest antelopes, were daily found dead in the bush by the natives."

Jackson (1894) stated that "even elephants" were affected, "but as my informants could not speak from personal knowledge, but only from native reports, I am unable to vouch for their accuracy."

The Reverend A. Marensky wrote in September 1892 that in January it had carried off the cattle of chief Merere to the east of Lake Kilwa, reaching the shores of Lake Nyassa about Karonga in April, "When I came down the lake in July last, we had on board the steamer

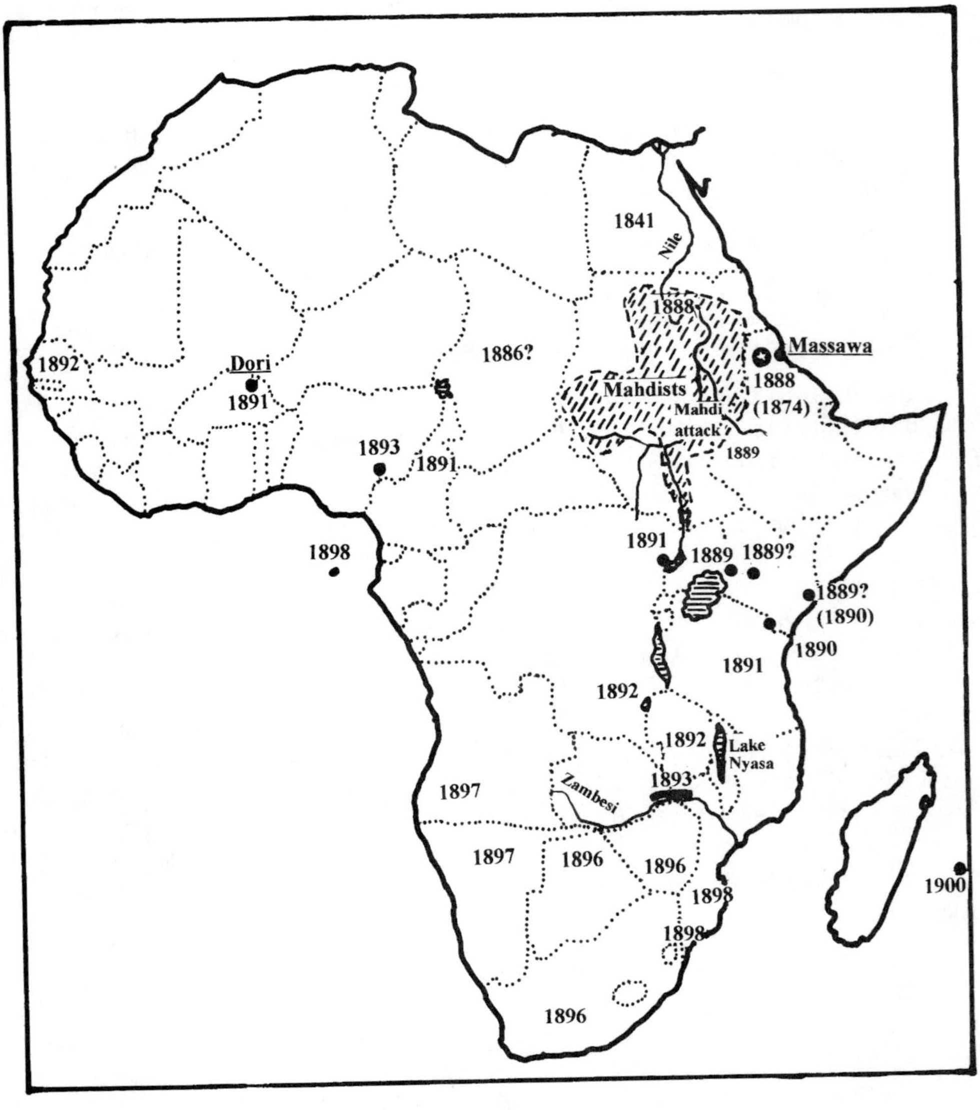

Figure 4. The spread of rinderpest in Africa, 1888–96.

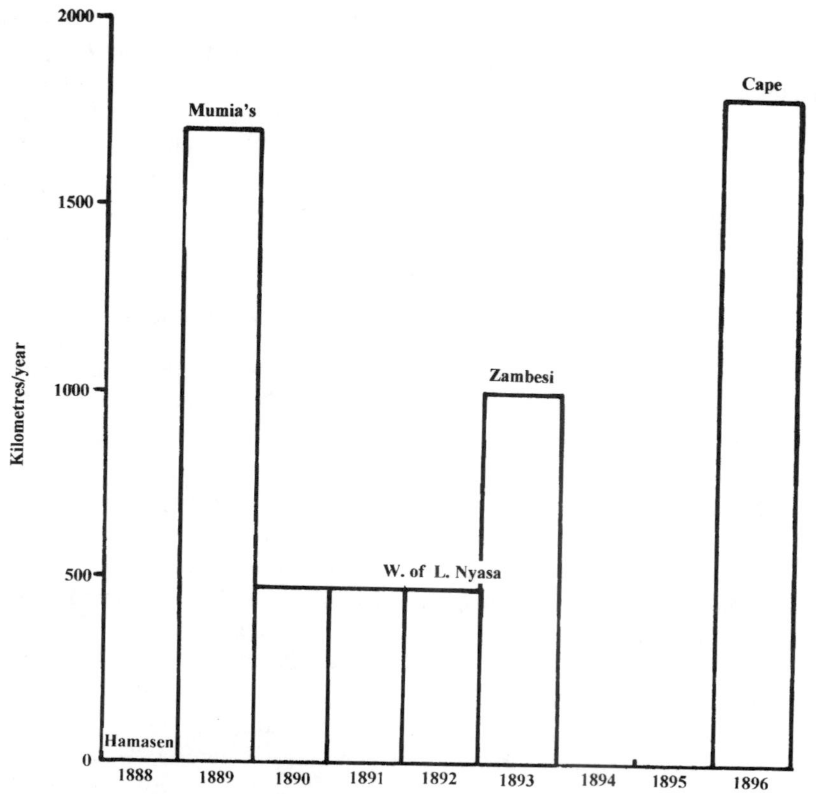

Figure 5. Schematic representation of the rate of spread of rinderpest south.

'Donura', belonging to the African Lakes Company, some twenty head of cattle, all of which died during the voyage" (Marensky, 1896). Contrary to Decle's assertion of it halting there, the rinderpest continued into the Ngoni country and then south to the Zambesi River, where it halted in 1893 (Figure 4).

Overall, it moved south from Hamasen to the Zambesi at an average rate of some 825 km per year, or almost 70 km per month. If it spread west to Dori from Hamasen, the rate westwards would have been over 1,330 km per year, almost double the southerly rate of spread (later, after it crossed the Zambesi, the inital rate of spread was almost some 14 times faster at 960 km per month or 32 km per day, roughly the equal of a daily ox trek). But if its origin was infection originating in Egypt passing through the western Sudan engendered by Mahdist activity, the rate would have been about 900 km per year, similar to the spread south, suggesting an Egyptian origin spreading to the Sudan as a more likely source of the original infection than from Massawa. Figure 5 shows a schematic representation of the rate of spread southwards.

6. CATTLE SURVIVAL

By 1893, Neumann, writing of the country near Lake Rudolph, considered rinderpest a thing of the past. Herds were already reported as building up again around Lake Naivasha, to the west of the Kinangop, but this was partly due to large-scale raids, and partly through regrouping of some Maasai who gathered their remaining stock together (Waller, 1976). Jackson (1930) reported that early in 1889, the western flank of Nzawi hill in Kamba and the valley below were thickly populated and the people rich in cattle, goats, and sheep. It was the same upon his return from Uganda in August 1890, but 4 years later, the population had vanished owing to rinderpest followed by a large-scale Maasai raid. Portal wrote that in 1893, the hills around Machakos were covered with great flocks of goats and sheep, intermingled with many cows. The epizootic:

> which two years ago destroyed nearly every head of cattle, all the wild buffaloes, and most of the wildebeest in East Africa, did not spare the Wakamba, who told us that they had then lost every bull and cow belonging to their tribe; but whether by honest purchase, or, as is more likely, by successful raids, they appear to have succeeded in again collecting a small stock for breeding purposes.

Of the Maasai, he wrote, "the terrible disease which slew every buffalo in their country did not spare the cattle … their cows and bulls died by tens of thousands, the whole race was reduced to the verge of starvation … a persevering system of raiding on their neighbours has enabled them to collect together a fair number of cattle …" (Portal, 1894). In September 1893, "many cattle" were observed on the Tana near the Grand Falls (Embu, east of Mount Kenya).

But either the devastation was not as complete as records indicate, or recovery was remarkably rapid, for despite the almost total mortality reported in the lowlands of Ethiopia for 1889, Wurtz stated that it was raging again with intensity in 1898, so herd numbers must have recovered somewhat. The Governor of Harer was allegedly able to seize 10,000 head of cattle in Ogaden in 1892, and 2 years later, the Emperor Menelik allegedly seized more than 36,000 in Walamo Province. Wurtz (1898) refers to the value of a "salted" animal being 10–20 times the normal price. Powell-Cotton (1902), travelling in Salali in 1899, saw numerous herds of cattle, and there was a good supply of cattle at the market of Debra Markos, capital of Gojam. The Derke islands in Lake Tana were celebrated for the "fine herds of cattle." Donaldson-Smith (1897) made no mention of the epizootic in his travels from Berbera in Somalia to Lake Rudolf in northern Kenya in 1894–5 and was astonished to find "great herds of fine cattle" at Sessabane (Sasa Baneh, Harer) on August 1, 1894, never having seen half so many cattle together before in Somaliland. He also saw herds of cattle near Ginir in Shoa. In January 1895, he passed through many droves of fine cattle along the Webbe Shebeli, and the Gere Gallas in the region of Lugh Ganana he reported as rich in cattle. At a Boran settlement called Lenja, east of Lake Stephanie or Chew Bahir, he saw larger herds than he had seen "anywhere outside of Texas. They must have numbered over ten thousand." At the northern end of Lake Stephanie among the Arbore, he saw "thousands of livestock of all kinds, except camels and mules and ponies." The posthumously published account of Count Bourg de Bozas (1906) relates that in 1901, south of Harer, the party traveled through a countryside of pastures with large humped-back cattle. East of Lake Zwai, some 200 km south of Addis Ababa, in December 1901, an enclave of big herds of cattle and "superb" goats was found. Writing in March 1902 of the Arussi country, traditionally a great province for cattle, de Bozas stated that the country was deserted by both man and beast, but the account is illustrated with a photograph of healthy-looking cattle in Guraghe. Some 250–300 km due south of Addis Ababa, the party commented on the abundance of cattle and the low price compared with Addis Ababa. In northern Karamoja, people had large herds of cattle and sheep, and in the Lado enclave, numerous herds of small cattle were raised.

In 1894, the rebel chief Kabarega at Mashudi in Uganda had over 300 cattle, but this was probably a small number compared with what he once possessed. In 1897, the Karimojong were short of stock, but the following year, it was reported that there were "immense quantities of livestock of all kinds" in Karamoja District; and the country to the west of Lake Rudolf was reported supporting "large herds of cattle."

In Tanganyika's Usambaras after the disease reduced the Wambugu herds about 1891, the Wambugu rebuilt their herds from stock which had survived in the Pare Mountains, but the Maasai then raided the Wambugu (Giblin, 1996). In northern Tanganyika in November 1896, a German punitive force against Meru and Arusha allegedly seized 10,000 head of cattle which the Chagga force took to Kilimanjaro. Despite the fact that the figure is probably at least a five-times exaggeration, clearly cattle were numerous.

Arkell-Hardwick (1903) reported that in 1899, the Boran possessed large numbers of cattle, but "many years before," the sheep and goats of the Rendille had been swept off by disease. Perhaps this was in the first wave of rinderpest. In 1899, Buxton (1902) found that among the Dinka on the White Nile, cattle and goats were very numerous, while Cook (1945) reported in 1906 that the cattle were few but of fine quality.

In February 1895, 400 head of cattle were captured from a Kavirondo village in a punitive expedition, and the Lango were reported as having many herds of cattle in April. The kings of Bunyoro were alleged to have possessed 30,000 head, and the total in the country was still regarded as large. In April, some 220 cows were captured near Lake Kioga. In May, a force captured 600 head in north Bunyoro. Near Masindi, 51 head were recovered which had been taken in a raid. Altogether, the punitive expedition rounded up some 2,000 head in Bunyoro, a number of which were of the Ankole breed (Vandeleur, 1898). In the Baringo area, expeditions captured a number of cattle, one chief having "various herds." Altogether, 300 cattle and 8,000 sheep and goats were seized from dissident Kamasia. Seizures were also made in Nandi in June, the Nandi possessing "a great many cattle." In July, 695 cattle were captured from rebels near the Ankole border (Ternan, 1930). British punitive levies supported by the Maasai seized some 2,150 cattle from the Wakamba in November 1895. Large numbers of sheep and goats were taken at the same time, and 29,306 from the Nandi in June 1897. In 1898, Ansorge noted that there were magnificent cattle at Singo but does not provide numbers. The cattle, of the Ankole breed, originally belonged to the Buganda king Mwanga but had been seized by the Government in a punitive measure in 1897 (Ansorge, 1899). Tucker (1908) reported that the Wasukuma near Lake Victoria in 1897 possessed large herds of cattle, while in 1899 in Ankole, he saw "numerous herds" of cattle on the skyline, and reported that the King of Ankole was said to have had 20,000 head. In May 1898, an expedition accompanied by 70 cattle with 26 calves and 138 bullocks traveled from Mumia's north to Save on the west side of Mount Elgon. In August 1898, an expeditionary camp north of the Suk hills had 115 cows and oxen "in grand condition," and an expedition to Lake Rudolf left this camp at Ngabato with 78 oxen and a few milk cows and calves. On the return journey in October it still had some 50 cattle including the milk cows and calves. The people near Moroto in Karamojo were reported to possess fair numbers of cattle (Austin, 1903). In January 1899 Austin (1903) reported the Maasai near Lake Elmenteita had recently returned from a raid in German territory where they had captured enormous herds of Maasai cattle and were now grazing magnificent herds of cattle and flocks of goats and sheep at the northern end of the lake. Approaching Tsavo in February he reported the land terribly parched and the country practically deserted "owing to the long-continued drought and famine."

Near Lake Kivu, Grogan and Sharp (1900) reported that in 1898, the Watutsi owned "enormous herds of cattle, goats, and sheep." At Lado near the Nile was a fine herd of cattle looted from the Dervishes, while east of the Gondokoro hills, according to reports there were many cattle. Among the Nuer, he saw large herds of cattle. In March 1901, Austin found that people in the Karuno country in southern Sudan possessed large herds of cattle. The Magois tribe which Donaldson Smith had found in the area had recently been driven west by the Turkana who raided their cattle. Later, bordering the Kerio River, Austin found that the Turkana possessed large numbers of cattle, sheep, and goats, as well as camels and donkeys

(Austin, 1902). Powell-Cotton (1904) found Suk grazing cattle near Lake Baringo but does not indicate numbers. In the Didinga country of northern Uganda, he seized 300 head of cattle in retribution for theft, and some of these he took with him as far as Nimule on the Nile with no report of disease. In further British punitive levies supported by the Maasai, 3,466 head of cattle were seized from the Nandi in July 1900, 1,300 from the Kikuyu in September 1902, and 1,087 in February 1904. Meinertzhagen (1957) reported that in July 1902, a Kikuyu chief had "many" cattle and in December seized 665 head in a punitive expedition. In March 1904 in the Fort Hall district, he seized 1,280 head. In July, he reported the country between Gilgil and Lake Elmenteita as covered with Maasai herds. In 1905, two lots of 5,000 were seized from Sotik and Kisii in June and September, respectively, and in October 16,213 head from Nandi. A furthur 36,205 sheep and goats were seized from the Nandi in October 1905 (Waller, 1976). These figures show that considerable numbers of cattle were to be found, but these large-scale movements of stock were undoubtedly a potent means of spreading the disease further.

In 1906, Wollaston (1908) found all the open spaces at Mbarara in Uganda crowded with herds of cattle which had been levied as a fine, while on the Kichwamba plain in western Uganda, he referred to the fatness of the great herds of cattle which fed there. South of the Rutshuru River, some of the people were very rich in cattle, while in Burundi in the Rusisi valley, great herds of fat cattle were grazing. The Kitosh in Kenya possessed herds in 1907, and the tribes to the southwest of Mount Elgon were rich in herds (Bary, 1910), but Stigand (1910) in April–May 1908 heard that there was a bad outbreak of rinderpest in the north of Borana. In 1910, Bland-Sutton (1911) reported that the Maasai in Naivasha Province had some 35,000 cattle.

In 1910–11, Kassner (1911) noted cattle to be numerous on the west side of the Kundelungu Mountains in the southern Congo. At Baudouinville, the mission kept hundreds of cattle in good condition. In the Rusizi Valley, there were "great herds of cattle," and in the Kivu district, they were plentiful. On the slopes of the Rutshuru valley, cattle were kept in great numbers, and on the eastern side, the Wahima and Watutsi possessed great herds. Near Fort Portal, he passed great herds, and on the eastern slopes of the Ruwenzori Mountains, cattle thrived in the "splendid climate," resembling a well-stocked farm in South Africa. East of Lake Albert, the people owned many herds and also west of the lake were rich in cattle. Among the Bari, south of Gondokoro, cattle were encountered in some places, but the people informed him that they formerly owned many which all died, leaving them poor. Kassner attributed this to the presence of tsetse fly. Dracopoli (1914) reported that the Somali at Goloshé Waré in Jubaland were rich in cattle and noted many between Gama Gar and Gulola. In a battle among the Ogaden Somali, 400 head were seized, while the Maghabul who wandered the districts of Joreh and Gulola possessed immense herds. At Tur Guda, the villagers had a large herd. Dracopoli also encountered a party driving cattle from Gulola intending to sell them in Nairobi. West of Lorian, he reported that the Boran owned enormous herds. The only suggestion of disease was that of trypanosomiasis suspected in several cases of cattle brought down from Lorian to Rumuruti and Naivasha. In Ruanda, it was estimated that there were some 2.5 million cattle in 1912, perhaps an overestimate as in 1965 the total only just exceeded 0.5 million. In western Africa in 1896, Dubois (1897) passed "wonderful herds of oxen" between Segu and the regions bordering Timbuctoo, and of the valley of Niger in this region, cattle rearing attained "an extreme degree of perfection." Foureau (1902) found in February 1900 in the Bornu region of Lake Chad thousands of cattle, and at Kousseri in the lower Shari region, the people had 15,000 head plus sheep and goats.

23

THE PANZOOTIC REACHES SOUTH AFRICA

March to December 1896

1. THE ZAMBESI CROSSED

> But a far more general plague than the small-pox, and a much more terrible scourge than the locusts, suddenly made its appearance, and dogged our steps. This was the *rinderpest*.
>
> No one who has not lived in Africa can form the least idea of this awful calamity. It mowed down the whole bovine race in its passage. Hundreds of carcasses lay here and there, on the roadside, or piled up in the fields. In vain did the natives gorge themselves, careless of the consequences. In vain did legions of vultures and beasts of prey gather to devour them. They could not overtake the quantity, and the carrion lay there, putrefying everywhere. More than nine hundred waggons, loaded with merchandise, without teams or drivers, stood abandoned along the Bulawayo road. In a few weeks—a few months, let us say—I am assured that eight hundred thousand head of cattle—some say nine hundred thousand—perished in Khama's tribe alone.
>
> Never within the memory of man had such a thing been seen. The Government grasped the situation from the beginning. But in spite of all the sanitary cordons, and the severest preventive measures, the scourge pursued its course relentlessly....

So wrote the French missionary François Coillard in 1897.

The Zambesi had proved an effective barrier to further natural transmission in initial expansion of the disease, but in February 1896, Rhodes's private secretary Robert Coryndon, who had been on a visit to the Luangwa River, reported to the authorities in Bulawayo that the disease had crossed the Zambesi. Mr. Isaacs of Mafikeng received a report on February 19, 1896 that cattle south of the river were dying rapidly from some mysterious disease. It is unlikely that the crossing was effected by eland and other large antelope, as Hole (1926) suggested. According to Brown (1899), little notice was taken of Coryndon's announcement as the general opinion seemed to be that the disease would be confined to the low, unhealthy belts of the country and would not affect the plateau.

Later in February, a herd of cattle was brought from the Zambesi region to Bulawayo where it was sold and dispersed in all directions (Theiler, 1897a). Gibbons (1898) dates its appearance in the Bulawayo region as "on or about" March 4, while Hutcheon (1897) gives it as March 3, and Sir Hercules Robinson, the High Commissioner, reported it as March 5 (Robinson, 1896), although the Bechuanaland Protectorate's Resident Commissioner at Mafikeng, Francis Newton, stated that it first appeared in the Bulawayo district on February 22 and was reported on February 29 (Newton, 1896). Thought to be similar to "Zambesi fever," within hours it was widespread, and the Chartered Company's Acting Administrator, A. F. Duncan, "sent telegram after telegram" to the High Commissioner at Capetown. Robinson referred him to the Cape Government which consulted Dr. Duncan Hutcheon the Colonial Veterinary Surgeon, who had taken over in 1880 from William Branford the first appointee in 1876, a former lecturer at the Dick Veterinary College. Hutcheon informed the

Company that they already had a qualified veterinary surgeon working for them as a telegraph operator, Charles Gray, who should be assigned to investigate it, but on March 4 urgently advised the Cape government to proclaim a total prohibition on movement of cattle from infected areas and to slaughter infected animals.

Gray had applied to the Chartered Company for employment as a veterinarian on January 1, 1896 but was offered the post of telegraph operator at Macloutsie, a position he held for 2 months. On March 18, he confirmed the disease to be rinderpest. By this time, it had reached Salisbury more than 400 km northeast of Bulawayo, destroying most of the cattle in the vicinity within 2 weeks.

Two weeks after its appearance at Salisbury, the commonage around the town which had been dotted with thriving livestock "presented the appearance of desolation and death. Fully ninety-seven per cent of the cattle seized with the disease perished" (Brown, 1899). Within days of the first announcement of the disease, the Bulawayo District was declared an infected area under the *Animal Diseases Act, 1881*; and on March 9, Robinson issued a Proclamation empowering the Administrator to appoint a Controller of Cattle to forbid movement of cattle, sheep, and goats, without permission, and order isolation or destruction of any infected, or likely to be infected, animals, to prevent spread of infection. He was also empowered to halt movement of wagons, and any cattle, sheep, or goats found trespassing could be destroyed by the owner or occupier of the land without compensation being payable. Any infringement of the proclamation, or wilful obstruction to its execution, was liable on conviction to a fine of up to £100 in respect of each offence, or imprisonment with or without hard labor for up to 12 months. The Controller of Cattle was H. M. Taberer. About the middle of March, Frederick Courtenay Selous, already a famous naturalist and hunter, was appointed cattle inspector for the district between the Umzingwani and Insiza rivers, doing a lot of riding about to try and stop the progress of the disease, an activity which probably assisted in its more rapid spread. The first Inspector of Cattle had been appointed in 1891 for the control of pleuropneumonia, followed the next year by another and the promulgation of the *Infectious Diseases (Cattle) Regulations*, which established a quarantine farm near Salisbury and ordered people to destroy or isolate cattle suffering from any supposedly infectious disease.

Toward the end of March, the *Bulawayo Chronicle* reported:

> Some time ago there were rumours of a peculiar disease which was killing game north of the Zambesi, and later on it was stated that herds of buffalo, roan, and sable antelopes were dying in a mysterious manner in the Inyorka country to the south of the Great River, but as it did not touch the domesticated animals no measures could be taken to destroy it. Within the last few days, however, information has come to hand that the native cattle are dying from the same disease, and from all appearances the disease was infectious. In Shiloh and around Inyati a large number of cattle have died, and the disease has swept off a herd at Thaba Induna …

Conveyed down the main routes by trek oxen, it left nothing but rotting carcasses behind, and although all movement of cattle from infected districts to clean areas was prohibited on March 21, the Government in Bulawayo decided to admit wagons into the country subject to strict supervision, the disease reported to be rapidly abating. On March 24, a general famine being feared, it declared the roads open for traffic so that as many wagonloads of provisions might reach Bulawayo while any oxen were still left alive to pull them. Visiting Mangwe at the end of March, Selous reported that all his Dutch friends had suffered terrible losses:

> some, who had been rich men a couple of months before, possessing several hundred head of stock, had now scarcely a beast left. All along the road, too, from Bulawayo to Mangwe … Hundreds of carcasses in every stage of putrefaction everywhere lined the track, whilst here and there were groups of empty waggons abandoned by their owners, who … had left the rest of their property standing uncared for in the wilderness, and walked away ruined men.

> At Wilson's farm, six miles from Bulawayo, where herds of infected cattle had been slaughtered in the hopeless endeavour to stamp out the disease, acres of carcasses were lying festering in the sun ... Strangely ... not a vulture was to be seen. I have heard it said that too hearty an indulgence in rinderpest meat in the early days of the plague killed all the vultures, and whether this is so or not, certain it is that these useful birds are now as scarce as cows in Matabeleland. (Selous, 1896)

On April 14, the Controller of Cattle at Bulawayo issued Government Notice No. 53, making it optional for owners to slaughter or attempt to cure infected cattle, but prohibited destruction of those apparently unaffected although in contact because "clean cattle can be protected by washing their mouths with a mixture of paraffin and salt and giving them an aperient." The notice also stated that "cures" of sick cattle had been effected by keeping them from water for 5 days while dosing them every other day with raw linseed oil, followed after another 5 days by small quantities of water.

On May 24, now campaigning with the Matabeleland Relief Force, Selous (1896) reported in the Insiza valley some 80 km east of Bulawayo "In many of the kraals were found large accumulations of dried meat, and many dried skins of bullocks, cows and calves, proving that the rinderpest had been brought into this district by the natives since the outbreak of the rebellion, and had been playing havoc amongst their cattle." But in the same area, one of the fighting columns captured about 700 head of cattle, the disease perhaps not yet being universal in Matabeleland, but it persisted in the north, and on September 20, Baden-Powell found several carcasses of kudu which had died from the disease. Near Enkeldoorn, Boer farmers made a laager and brought in over 2,000 oxen, of which by November 10, only 70 were still alive.

Wallace (1896) referred to the rinderpest having "done so much mischief among the cattle of Mashonaland and Matabeleland" noting that it "strongly resembles the cattle disease in Masailand mentioned but imperfectly described by Thomson, and it may yet be proved to be identical. If so, it seems to be spreading like a great and destructive prairie fire in a southern direction over the centre of the African continent, and Cape Colony ought to make preparations to resist it." Cape Colony was in fact not unaware of the danger, and Robinson passed the information received from Bulawayo to the Transvaal Government in Pretoria. President Kruger immediately issued a Proclamation on March 11, forbidding import of all cattle from Mashonaland, Matabeleland, and Khama's country (in the Bechuanaland Protectorate now Botswana) using a law for the control of pleuropneumonia:

> [I] ... hereby proclaim, declare, and make known that all traffic in cattle, including bulls, oxen, cows, calves, horses mules, donkeys, sheep, goats, game, and also dogs and other domestic animals, & c., coming from Matabeleland, Mashonaland, and Khama's country to this Republic is hereby strictly forbidden; that cattle which may come over the line will, if possible, have to be driven back or will otherwise be destroyed; that all transport or other traffic with or by means of cattle or transport cattle coming from Matabeleland, Mashonaland, and Khama's country, and vice versa, with or by means of cattle or transport cattle coming from Matabeleland, Mashonaland, or Khama's country over the boundary to the Republic, is hereby forbidden: special provision will be made as regards milk conveyance; that all cattle, as specified above, being and grazing in this Republic on the bank of the Limpopo, must be kept at a distance of at least three miles from all places where there are drifts through the river, and I proclaim, order, and make known further, that all landdrosts, assistant landdrosts, resident J. P.'s, mining commissioners, field cornets, assistant field cornets, commissioners of natives, collectors of taxes, police officials, and all other persons, will have to conduct themselves strictly in accordance with this proclamation and in accordance with further instructions and regulations to be issued under this proclamation, and I command poundmasters and other officials of the South African Republic to act as regards cattle suspected of suffering from infection in accordance with and in terms of Article 34 of Law 2, 1882 ....

A 5-km-wide strip along the borders was cleared of stock and an official "Day of Prayer" declared for Easter Day to pray against drought, locusts, and the cattle disease.

Unlike the Cape and Natal, the Transvaal did not have a veterinary service at this time, and Kruger called in Theiler, then a pig and poultry farmer near Pretoria, but it was not until April that he was to be made official Government Veterinarian of the South African Republic, although no staff were allocated to him. For 20 months, he conducted a campaign with which, according to Curson (1936), no parallel exists: “Not only had he to draft regulations, to amend some and repeal others … but he undertook research, attended committee meetings and conferences, answered correspondence, issued reports, advised and supervised such as he was never again called upon to do.”

But for now, Theiler travelled to Bulawayo, the journey taking 6 days. On the way, he first saw the disease near Fort Tuli on March 16. All oxtransport was forbidden to leave Tuli for 14 days in an attempt to stop the spread. At Bulawayo, he met Gray, and together they conducted a number of autopsies on March 21, 22, and 23. From textbook descriptions, together they confirmed that the disease in a herd at Thaba Induna was rinderpest. Theiler wished to return quickly to Pretoria to alert the Government, but because of the outbreak of the Matabele rebellion, his way was blocked, so he travelled by coach via Tati through the Bechuanaland Protectorate, witnessing for himself the devastation caused on the transport route. He reported seeing hundreds of carcasses, mostly in the dried-up river beds, while everywhere was the stench of rotting animals.

By April 17, more than 5,000 cattle had been shot in the Protectorate, 241 in Cape Colony, and 141 in British Bechuanaland. In Mafikeng district, of a total of 94,000 cattle, 80,000 died, 7,800 were slaughtered, and only 2,600, or less than 3% survived. In the Transvaal, no killing was authorized.

2. THE BECHUANALAND PROTECTORATE

During March, it continued its relentless journey southwards. This southerly spread appears attributable to transport oxen, the deaths of which created chaos to the 644-km transport system between Bulawayo and Gaborone. On March 10, Newton telegraphed Robinson at the Cape urging isolation of Matabeleland with strong border guards, asking that a veterinary surgeon be sent up at once: “I have no doubt that this has been done but I understand that Mr. Soga of the Cape service is now at Maritzani near here if the Cape Government could spare his services.” Soga, having completed a course in bacteriology under Edington at Graham’s Town, was at Maritzpi near Mafikeng to inoculate cattle belonging to Dr. Smart against pleuropneumonia. Mr. Wright of Maritzani informed Newton that cattle were dying everywhere, but he attributed it to the severe drought.

Although the plague was at first termed “Zambesi fever” (= East Coast fever?), Hutcheon was quickly assured that it was rinderpest and, in a letter of March 14, 1896 to the Under Secretary for Agriculture at Capetown, urged amendment of the provisions of Section 9 of Act 27 of 1893 to make compulsory the immediate destruction of all cattle which had mixed, or been in contact, with infected animals or with contaminated areas. Temporary measures such as quarantining had not only failed to halt the spread but propagated it:

> The only method which has ever proved effective in suppressing this disease in Europe has been the absolute stoppage of all cattle traffic within the infected areas, and the immediate destruction of whole herds of cattle in which the disease manifests itself. Any temporising measures, such as quarantining infected herds and regulating the movements of cattle within areas where the disease existed have not only failed to arrest the spread of the disease, but have proved the very means of propagating it …. (Hutcheon, 1896a)

A senior veterinary officer, Otto Henning, was sent by Hutcheon to Palapye where a quarantine line had been formed on March 18, by March 17, 100 infected animals having died or been shot there and 95 presumed cases shot. Arriving on March 22, he confirmed the disease as a virulent type of rinderpest, although on March 23, the Assistant Commissioner at Palapye reported that no native cattle had as yet been reported sick. Nevertheless, Henning immediately issued a number of regulations:

1. no animal liable to the disease to enter the Bechuanaland Protectorate until further notice;
2. all wagons traveling north or south to halt at the nearest outspan, north and south travelers separate from each other;
3. all spans of oxen traveling south to be destroyed immediately;
4. all owners to be compensated at rates of £6 for first-class animals; second-class £4.10s., third-class £3;
5. all cattle traveling northwards to be quarantined until (3) had been carried out, and then only allowed forward in groups of not more than 20 wagons, each escorted by two men;
6. quarantine stations were designated for (5) at seven points;
7. any ox traveling northwards which showed any sign of sickness to be destroyed with its companion immediately;
8. carriers whose oxen had been destroyed were to be enlisted as special police; and
9. contractors should be compelled to station mules along the line as soon as possible.

The recommendations could not be properly executed, however, because of the lack of sufficient police, and the Cape Government refused to pay compensation for slaughter of healthy cattle. Had Government agreed to pay compensation for slaughter of all cattle traveling southwards, the disease might have been arrested. Robinson argued in a telegram to Chamberlain (April 5):

> ... I am unable to support it [Newton's proposal for compensation], as it appears to me unjust to ask British tax-payers to compensate cattle owners in or passing through Bechuanaland Protectorate for destruction of cattle which, if not slaughtered, are bound to die; second, if compensation be once approved, the door will be open to wholesale fraud and abuse, and infected cattle will be sent into Bechuanaland Protectorate from neighbouring countries such as Matabeleland, Transvaal Republic, and Cape Colony, in which compensation is not given or proposed to be given; and thirdly, looking at the unenclosed state of the country through which hundreds of thousands of cattle roam at large, and to the fact that infection is carried by game, the disease, I believe, is bound to run its course (Robinson, 1896)

But on March 26, he approved limited proposals for compensation on condition compensation applied only to teams traveling south between Gaborone and Ramathlabama at that time, that it should be paid only if every ox in the span was free of disease at the time of slaughter, and the rates should not be higher than already proposed.

On March 27, Gray, now with the title Government Veterinary Surgeon, published a description of the symptoms of the disease together with directions for the disposal of carcasses, in the *Rhodesia Advertiser* at Umtali.

On April 3, Henning was met at Palapye by Theiler: "At the time of my arrival we visited an outspan, where the dead oxen lay in hundreds, namely along the shore of the river as well as in the river. The following days we saw thousands of sick animals and the misery was indescribable" (Theiler, 1897a). On April 6, Chief Khama selected some animals for them for experimental inoculation, but Theiler was unable to remain and continued his journey.

By the time of Theiler's arrival, the disease had already spread south of Palapye. Theiler reported that the coach in which he was traveling was drawn from Palapye by oxen until it reached the first outspan (20–30 km distant) where all the oxen developed rinderpest:

> From there to Palla there were dead oxen everywhere, and abandoned transport waggons. I met drivers who had coupled together two or three waggons and still had half a team of oxen, although these were of course already infected. They were hurrying in order to get as far south as possible, to reach the populated districts. I saw others who had lost their draught animals, and with women and children were in the wilds, exposed to hunger and thirst, several days away from human habitation. They all had to return on foot. Later the Transvaal Government sent mule carts to the Crocodile River, to bring home these people. In Palla I found 40 abandoned transport waggons, whose oxen were all dead, that is, 700 .... (Theiler, 1897a)

Puxley (1929) reported that at Palla Camp, all the transport oxen lay dead beside their wagons, on which were piled many hundreds of tons of urgently needed supplies for Rhode's great enterprise at Bulawayo cut off by the Matabele uprising. It was reported that eventually there were 64,000 dead trek oxen and 4,000 stranded wagons along the route.

Gibbons (1898) was a close observer of its passage at this time. At the beginning of 1896, he recorded that there was "very little game in the country now" in the region of what is today the Kafue National Park, Zambia. At Sesheke on the Zambesi, it had killed off "all but a dozen of the hundreds" of cattle there and crossed the Zambesi River. The 120 mission cattle were all dead. He heard that it had already crossed the Chobe River and was traveling south through the Kalahari. At Kazungula in the Protectorate, game was dying beyond the radius of a few kilometers, although as yet it had not appeared among the cattle. Traveling south to Palapye, it became a race against the rinderpest to get there before his trek oxen died:

> The rinderpest had been raging for about three weeks, having spread from Tati, so that when my start was made from the Zambezi this sweeping disease was north, south, east and west of me, the only untainted district being a narrow strip 400 miles long, through which the south route lay....This was the first occasion on which the oxen had been right into the infected area, where cattle were dead and dying everywhere. (Figure 1).

It is difficult to reconcile Gibbons's report that it had crossed the Chobe in a southerly direction with a westward spread from Bulawayo. That trek oxen regularly crossed the Chobe from 1893 suggests that an upsurge, perhaps resulting from drought, may have been the reason for its further spread.

> The day but one after leaving Tati my attention was drawn to a koodoo yearling standing behind scrubby bush on the very edge of the track. The oxen were pulled up alongside the brute, which was soon seen to be suffering from rinderpest, nor did the animal go away until startled by the crack of the whip as the oxen were once more started, but had allowed me in the meantime to examine him from a distance of six feet.
>
> Nothing of interest occurred during the last trek of 120 miles [to Palapye].... All along the road the putrefying carcases of oxen were strewn, and especially thickly round the few waters. (Gibbons, 1898)

His oxen were put into quarantine at Palapye on arrival, although he does not give the date, rinderpest reaching there on March 10. They are not reported as having succumbed to the disease in spite of the close contact they had, but taking such liberties with infected animals as Gibbons did, it is hardly surprising that the disease continued to spread. Although Gibbons claimed that Palapye was still free of disease on April 26, this conflicts with other reports of its presence there on March 10. Newton (1896) claimed that it was "fully developed" there by March 9 at the latest, and Willoughby (1896) reported that on April 21, Khama, chief of the Bamangwato, had cleared way all rubbish "and dead animals" because typhoid had been reported in Bulawayo. When rinderpest did reach Palapye, many sheep and goats died from it (Hutcheon, 1896b). On March 10, it was reported that a transport rider named Pretorius had just returned from Bulawayo, and one of his oxen died that morning,

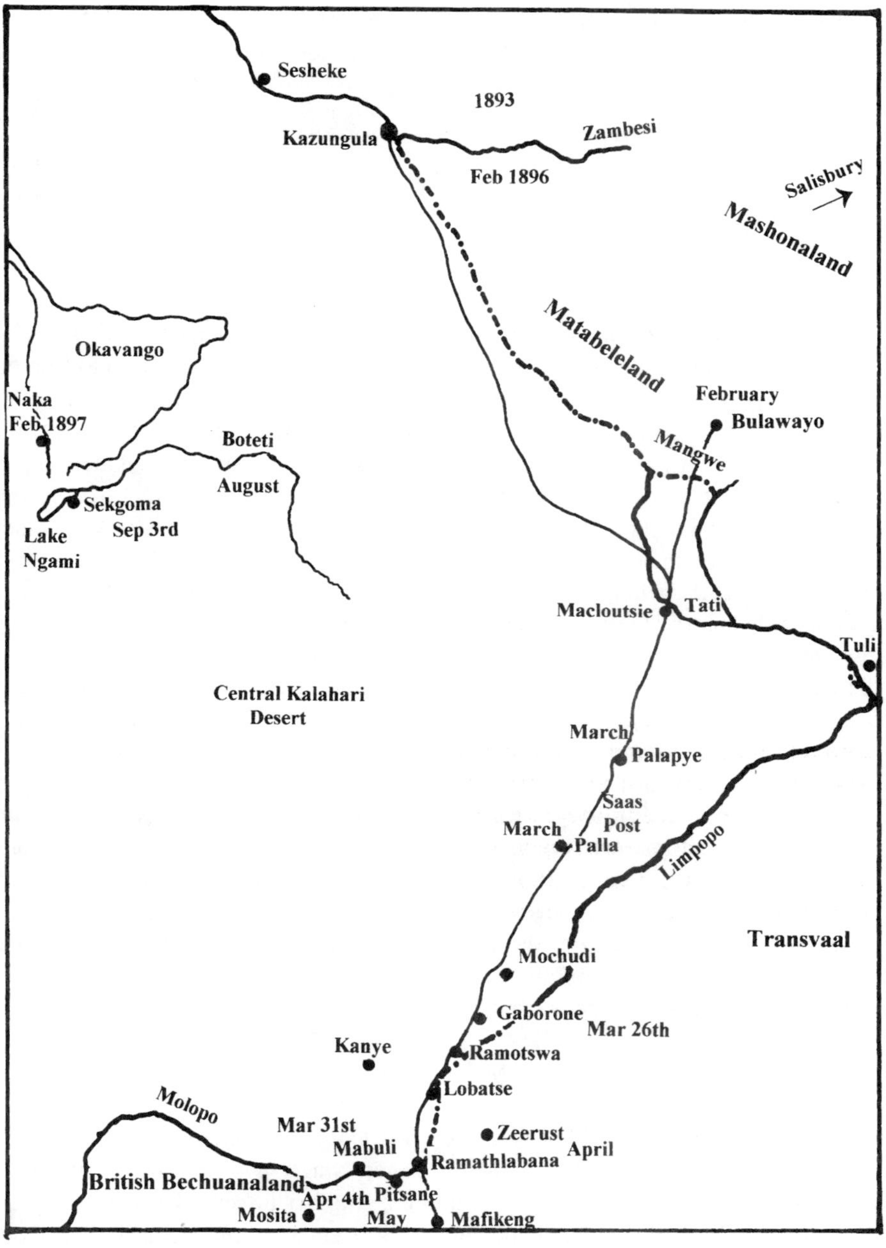

Figure 1. The spread of rinderpest through the Bechuanaland Protectorate, 1896.

while another was sick; also, a man named Van Zyl, whose wagons were 13 km out of Palapye, had lost 32 oxen in the last 2 days.

The *Bechuanaland News* of March 28 stated that it was "officially reported on Saturday" (March 21) that there were five cases at Palla Camp, and three spans of oxen had died 27 km north of Palla (at Saas Post on March 18). These were subsequently confirmed by Henning to be rinderpest. At this time, no cases were reported south of this point, and the disease was confined to transport riders' cattle, African cattle not having contracted it.

Selous later noted:

> ... in the case of the herd of buffaloes living between the Shashani and Tuli rivers, in western Matabeleland, up to 1896, which were said to number about forty head, it was reported by the bushmen that not one single animal survived the epidemic, whilst of all the cattle (approximately about 100,000) alive in Matabeleland in March 1896, not more than one per cent survived the disease. (Selous, 1908)

Two days after its appearance at Saas Post, March 19, the quarantine cordon at Palla was forced by about 100 wagons, African transport drivers driving their oxen southwards in all directions. This was not reported by Robinson to the Imperial Government.

On March 26, the disease appeared at Gaborone, 150 km from Mafikeng, and the following day Ramotswa. Newton now authorized destruction of all oxen traveling southwards in wagons between Gaborone and Ramathlabama, sending an urgent telegram to Robinson to approve compensation as absolutely essential. Robinson replied by return giving his approval provided that it was only for teams traveling south at the moment between Gaborone and Ramathlabama, and only for oxen free of the disease at the time of slaughter, appointing Captain Fuller as inspecting officer accompanied by Hutcheon. At the time, over 1,500 oxen were at Ramathlabama daily receiving fresh additions from the north, and it was decided that these should be placed in strict quarantine. However, on the same day (March 27) two spans arrived from Mochudi belonging to Chief Linchwe of the BaKgatla which had traveled through infected country. These were turned back about 3 km and shot the next morning after valuation, together with another span that had been traveling south.

Proceeding north, Fuller and Hutcheon met five spans traveling south which were destroyed, but as the Africans refused to bury the carcasses, an arrangement was made with a storekeeper for burial to include those shot near Ramathlabama. Further on, another span was encountered conveying a missionary and his family to Mafikeng. The span was destroyed and the missionary obliged to make arrangements to be collected from Mafikeng. On March 28 they met another wagon which had entered the Transvaal near the Limpopo and had been turned back into the Protectorate by the Transvaal police. This oxen span was also shot. Shortly after reaching Aasvogelkop, four spans were encountered and shot, one span belonging to Chief Ikaneng of the Ramotswa. On arrival at Ramotswa, they found that a span which had returned from Palla was infected, and the chief had ordered it shot, but in conversation with him, Hutcheon considered that he failed to appreciate the gravity of the situation.

On March 27, 52 oxen were destroyed near the border beyond Mafikeng, having arrived from the infected area.

On March 29, Fuller and Hutcheon reached Gaborone where they found rinderpest present among 117 animals which they then had shot. Ninety-two were oxen which had been driven through the cordon at Palla and recaptured. Another infected herd at Gaborone was also shot. On March 31, they came upon some sick oxen a few miles north of Aasvogelkop abandoned from a span going south to the Transvaal border near Lobatse. Seventeen oxen were discovered sick out of a span of 18 and destroyed.

It was then proposed to get the two chiefs, Bathoen of the BaNgwaketse and Montsiwa of the BaRolong, to move their cattle back north and south, respectively, in order to form a clear belt of 8 km with police stationed along its length from Lobatse to near Kanye. But the chiefs could not be encouraged to act with sufficient promptitude before the disease passed the line.

By March 31, the cordon breakers had carried the disease to the north side of the Molopo near Mabuli on the west and Kopani's Stad in the Transvaal on the east. On April 1, it was reported to be present at Skelvynstad and Linokwana in the Transvaal. One affected animal from the Molopo reached Mosita Native Location in the Mafikeng District on April 4, and a second outbreak was reported on April 14 at "Sunnyside," 6 km northeast of Mafikeng in Cape Colony. Here, 38 animals in a herd of 120 in an African kraal were found sick, and four had died. The Africans believed that the disease had been introduced by an ox from the Transvaal side, most probably from Kopani's. The healthy cattle were immediately valued and shot on April 15. Two adjoining kraals were placed in quarantine under strict guard and observation. The cattle soon showed signs of infection and were killed on April 18 and 22. A number of sheep and goats at these kraals were lying about not feeding, but it was not believed at the time that they could contract the disease and so were left. There were five other kraals in the immediate vicinity, but the disease did not spread to them, the reason believed to be that the cattle were held by a different tribe, and so there was no intermixing. In mid-March, two Africans who had breached the Palla cordon were committed for trial at Mafikeng for contravening the quarantine.

The breaking of the cordon at Palla was considered the main cause of the spread of the disease through the Southern Protectorate. The whole of the Protectorate now being regarded as an infected area, it was necessary to fall back on the Molopo, a line of 15 police posts each about 16 km apart being established from Ramathlabama to west of Mafikeng to guard the northern border of British Bechuanaland.

Species of game reported as most affected in the north of Botswana were eland, buffalo, gemsbok, and reedbuck, and in the south, steinbuck, duiker, springbuck, and kudu (Surmon quoted by Hutcheon, 1897); but Hutcheon makes no reference to the possibility of its spread by game and clearly did not think that game was a significant factor.

Newton had argued that it was better to spend £50,000 at once than suffer an annual loss for the whole of South Africa of double that sum. By February 1897, Hutcheon was to put the figure at more than 20 times that. Nevertheless, over 10,000 head of cattle were slaughtered in the first few weeks in the Protectorate, effected by a handful of police "almost without a murmur and without the slightest resistance, either from Europeans or natives" (Ashburnham, 1898). Compensation was set at an amount sufficient to make it in the interest of a farmer to report an outbreak at once, but not liberal enough to induce one to sacrifice his cattle deliberately or encourage carelessness in looking after them.

Hutcheon (1897) reported: "Within a period of 25 days (of the first reported occurrence in Rhodesia) therefore, rinderpest had traveled southwards at a rate of 20 miles a day, and had reached a point 16 miles north of the colonial border [i.e., Cape Colony] on the 31st of March." Owing to the limited number of police available for enforcement, measures were unsuccessful, and thousands of cattle were reported dying. No trekoxen were believed to be left in Khama's country. He himself was said to have had 300 or 400 trained trek oxen prior to the outbreak.

On April 23, the disease was reported 30 km west of Pitsane in the Protectorate. Hutcheon found it widespread and within 13 km of Pitsane, having evidently been in the area for some time. Although he was unaware, it had already crossed at the junction of the Molopo and Setlagoli rivers east of Mabuli. Hutcheon then warned chiefs and others in the region and, through Newton, had one of the chiefs supply 50 men, with another 50 promised, to watch and guard the river to see that sick cattle did not drink near healthy cattle. But it was found to be impossible to get the Africans to follow instructions, those living on the south side of the river continuing to visit their cattle on the north side, and not maintaining separate drinking places. As a result, the disease was reported at Pitsane on May 2. Many cattle were found to be infected there and between there and Mabuli.

The next plan was to move the police cordon back to a clear belt behind the Molopo Native Reserve, stretching from Mondosia Petsi to a line south of Mabuli. Hutcheon suggested three options to the Government:

1. slaughter infected herds as the disease appeared in them; or
2. slaughter the whole of the cattle within the Native Reserve; or
3. leave the cattle alone in the Native Reserve and form a clear belt behind them.

The last suggestion appeared the only practicable one at the time and was the course adopted. If they had tried to slaughter all cattle within the reserve, apart from confrontational problems, owners would disperse what they could outside the reserve, thus spreading the disease.

The belt chosen formed an effective cordon as it was waterless with very few stock, all those at the north drinking in the Molopo and those in the south at the Maritzani or Setlagoli rivers. Beginning at Rooi Dammatje on Madibi Spruit at the Transvaal border, the belt ran north and then southwest. Two cross-belts were also formed, and then a third in the Protectorate. Considerable delay was experienced establishing the line and getting it fully equipped, due to the scarcity of water and mule transport.

On May 19, the first outbreak occurred outside the Molopo infected area at a kraal in the Mafikeng Native Reserve some 19 km east of Pitsane, from where it was subsequently discovered the disease originated. The kraal owner had some infected cattle at Pitsane and was in constant contact with them. Although the outbreak was promptly dealt with, all the cattle being shot and owners compensated, several other outbreaks followed in quick succession. There being no pass law in Cape Colony, it was impossible to control African movements or to know where persons had come from. Nearly the whole of the cattle within the Mafikeng Reserve had to drink at the Molopo river, making it impossible to isolate suspect cattle, and the Africans opposed shooting of any herds before the disease appeared among them. The disease had also extended down the Molopo river westwards from Mabuli to the last European farm, after which there were comparatively few cattle, the majority belonging to Africans who took them to the north side of the river rather than have them shot.

On May 22, Baden-Powell (1897) reported that wagons and goods were stranded at Mafikeng waiting to go north because all the oxen on the road were dying fast, but every train was bringing up more mules and donkeys for use of the Matabeleland Relief Force. After Gaborone on May 25 "Every mile now began to show the grisly, stinking signs of rinderpest... Occasionally we passed a waggon abandoned owing to the loss of animals." At Palla were collected some 200 wagons which had lost their oxen, 3,200 animals being dead at that one place. In Khama's country, centring on Palapye, all along the road the carcasses had been burned by Khama's orders.

In the same month, R. S. Godley, a trooper in the Matabeleland Relief Force, rode in the relief column to Bulawayo, giving a vivid description of the continuing scene of devastation:

> Along the road over which we passed from Mafikeng to Khami hundreds of waggons were left abandoned, many with whole teams of sixteen bullocks lying dead and rotting in their yokes. Hundreds of others lay about the veldt. The stench from these putrid bodies was appalling, and poisoned the air for great distances... The vehicles were loaded with every imaginable kind of article, from ladies' stays and artificial limbs to grand pianos! It seemed a pity to allow all this to waste, so in spite of orders some looting of stores was indulged in by various members of the column. (Godley, 1935)

What was not generally reported was the death of people in the Protectorate, apparently due to an outbreak of typhoid. In June, an African from Basutoland living in the Protectorate wrote to his home newspaper: "... if you have 100 cattle in the kraal the night before, you wake up the next morning with nothing left. Then after the death of cattle, the disease enters among human beings. Oh! you have never seen anything like it before! In one day approximately 100 people die in every village" (Lefantiri, 1897).

In August 1896, the epizootic was reported to be spreading westwards along the Boteti river which was densely populated with game, especially buffalo. On September 3, Lugard, who had witnessed it in East Africa, in camp near Lake Ngami reported that it had reached Sekome's [Sekgoma] near the Lake, and would presently be in Damaraland and German South West Africa: "It ought to have been most easy to prevent its progress westwards, for there is little communication between Nyamiland [Ngamiland] and the rest of the Protectorate, since the Kalahari Desert, with its sand and thirst, intervenes. But nothing whatever has been done" (Kirk, 1896). Goold-Adams, traveling in the Okavango Delta area in early 1897, found that at one village, the men (Hambukushu) were away collecting meat from the carcasses of buffalo dead of the rinderpest, and learnt that the disease was approaching Nakaletswee on the west side of Lake Ngami, where he encountered it in February.

The route along the Boteti was through the Kalahari desert, used by only a few hunters, although we do not know how much tribal movement there was, so spread by ox wagons along the route from Bulawayo to Gaborone the disease traveled south much more rapidly than it was spreading westwards. In the last few months before August, 40,000–50,000 carcasses of oxen from the wagons transporting supplies to Bulawayo had been buried on the road between the southern border of the Protectorate and the northern border with Matabeleland. The majority died at watering points where they were outspanned, and hundreds (which were not buried) died some distance from the route. Bulawayo was in desperate need of supplies, requiring about 600 wagon loads. Mules and donkeys were found to be inadequate and anyway could not be used when the season of horsesickness began. It was proposed that fresh oxen be railed from the Cape to Gaborone, yoked and branded, and rushed to Bulawayo, not being allowed to return. The journey took about 1 month, and if they had rinderpest, those which were going to succumb would do so in the first 2 weeks. The risk was considered therefore to be with the carriers, rather than with spreading the disease. But the Transvaal objected that the oxen could revive the disease and cause it to spread across the Marico and Crocodile rivers: "The cattle would leave their droppings on the river banks. The wild bucks would nose these, and die in the river reeds, and the cattle from our side coming in contact with these carcases would be infected (Du Plessis, 1896)." By August, many cattle had been shot at Zeerust.

3. BRITISH BECHUANALAND AND CAPE COLONY

On March 21, a public meeting composed largely of farmers had been held in Vryburg, called by the Chamber of Commerce to discuss effects of the quarantine regulations on the Transvaal border which threatened the economy of British Bechuanaland. A resolution: "... seeing that Rhodesia must have supplies of the necessaries of life, [the meeting] resolves that the Colonial Government be requested to allow all cattle proceeding from Bechuanaland to Rhodesia to pass, but to impose the most stringent quarantine to prevent cattle coming from Rhodesia to the Colony." The disease, it was noted, was 800 km away.

The Prime Minister's reply was that the quarantine must be maintained, and transport for the moment confined to horses or mules. Efforts were being made to push on with the construction of the railway, expected to reach Ramathlabama in 3 weeks. A meeting of the Vryburg Chamber of Commerce was held on March 28 to discuss a resolution of the Mafikeng Chamber recommending that all cattle traveling southward from Bulawayo and within 8 km of the transport road be destroyed, owners being compensated. Vryburg's answer was that the latter was impracticable and reaffirmed its resolution that cattle should be allowed to enter Rhodesia from the south. It added that the High Commissioner urge the Government of the South African Republic to strengthen the cordon on the northern border of the Republic.

The reply was the High Commissioner had requested the South African Republic Government to strengthen the cordon on their own border. On March 31, regulation No. 1 concerning the cattle plague was partially suspended, allowing ox wagons to travel north.

Following the reports in March, a cordon of Cape police was established all along the northern border of Cape Colony, and if they were not sufficient, the Cape Mounted Rifles would be sent up, but shortly afterwards it was reported that the latter would not be required. Two hundred police were forming the cordon, and 200 more had been requested, 100 of whom would be on the Vaal River.

On April 4, it was reported by Captain Fuller that a sick ox had been discovered near Mosita in British Bechuanaland. Hutcheon ordered it to be shot and buried, and all its contacts, a herd totalling 400. As a result of prompt action on the part of the Field-Cornet in shooting the herd, there was no extension of the disease.

On April 17, an interstate conference was held at Mafikeng by the Secretary of Agriculture, Cape Government, attended by delegates of the different states and colonies of South Africa, to discuss the position and policy which should be adopted to prevent further spread of the disease. Edington contributed most of the discussion on policy. The delegates unanimously agreed upon slaughter provided that fair compensation was paid. The Cape representatives also proposed that a fence be erected along the whole of the northern boundaries of the Cape Colony and South African Republic, a distance of about 960 km, paid for by all parties, a point upon which agreement was not reached, although it was in the interests of all, including those unaffected countries, Natal and the Orange Free State. Guarding the line without a fence required 500 men costing £250 per day. With a fence, it was argued that one man every 10 miles would suffice instead of every mile. The cost of a double fence was estimated at £60,000.

The Orange Free State representative, M. J. Beuskes, pointed out that the gravity of the situation was not appreciated at Bloemfontein, and the Raad (parliament) would believe that their delegates had been talked over. But in the Transvaal, Theiler reported, strong quarantine measures had been adopted and a cordon established along the Bechuanaland and Protectorate borders, but there was no authority to slaughter cattle and compensate owners. The Chairman stated that the Colonial Government would go ahead with measures in the Colony whether they had the support of any other Government or not.

Measures taken by the Cape Government were:

- issue of a proclamation prohibiting entry of cattle into the Colony from any place north of the Molopo river;
- posting police along the Molopo to enforce the order;
- infected cattle in the Protectorate were shot.

On April 18, the districts of Mafikeng, Vryburg, Taungs, Kuruman, Gordonia, Hay, and the portions of Barkly West and Herbert north of the Vaal River, were declared infected areas, from which no cattle, horses, asses, mules, sheep, goats, dogs, and pigs could be moved to any other part of the Colony, and no cattle from farm to farm within the area, without a permit (Figure 2). This excluded Mafikeng whence all removal was prohibited absolutely. Infected animals had to be destroyed immediately, the carcasses buried 1.8 m deep and covered with quicklime or other disinfectant. No compensation was payable for infected animals destroyed, only for healthy animals certified by a Government veterinary surgeon or a Police officer destroyed to prevent the spread of rinderpest. By Act 27 of 1893, the penalty for contravention was a fine not exceeding £50, or imprisonment with or without hard labor for up to 3 months.

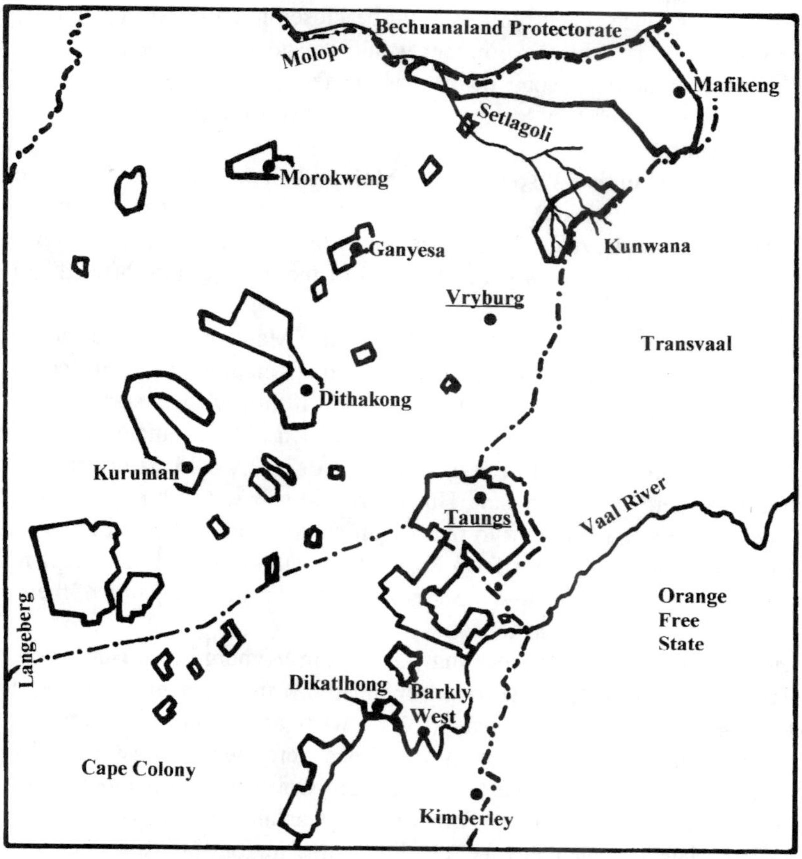

Figure 2. British Bechuanaland with native reserves and locations.

Then, the Minister for Agriculture, P. H. Faure, informed the Vryburg farmers on April 21 that it had broken out near Zeerust in a African kraal 6 km from Mafikeng. Mr. Barnard, an owner of 600 cattle, asked Faure about slaughter. In answer to the reply that if the cattle were not slaughtered they would all be dead anyway within 14 days, he stated that he would believe it when he saw it himself.

On May 9, a public meeting was held in the Vryburg Town Hall at which the ratepayers protested at the Council's decision to pay half the cost of special farmer constables to patrol each ward in the Division, a cost amounting to some £500 per month. The ratepayers had recently to pay a smallpox tax, trade was at a standstill, and many were without fuel because of the stoppage of wagon transport. The meeting supporting the objection a meeting was held with the Vryburg Divisional Council on May 11, the result of which was that the Council would do its best to consider the representations made. It pointed out that if the disease spread to there, Government would stop paying compensation.

In early May, it also broke out among trek oxen on the Harrismith townlands near the Natal-Orange Free State border, prompting a special meeting of the Natal Farmers' Conference. A resolution was passed, asking that all borders might be closed, to which Government replied with a "General Plan for Meeting the Dangers of Rinderpest." This plan was the construction of a fence from the extreme northern point of the Colony along the northeastern boundary toward the mouth of the Tugela river, which was already under way, followed by an inner fence. Each border would be closed upon the presence of the disease in a neighboring territory. The Government would consider loans for fencing, as the best protection was the immediate repair of all existing fences and fencing of those farms not already enclosed. Four Veterinary Surgeons were expected from Britain to assist the head of the Veterinary Department.

The next isolated outbreak south of the Molopo belt was on May 26 in the Setlagoli Valley due to eight animals which got past the guards when the belt was being formed and spread it to two other herds. After the destruction of the second of these herds on June 6 no further cases occurred here until a general inroad of the disease into the Setlagoli Reserve months later.

In May, Hutcheon discontinued shooting in the Mafikeng Reserve, not because of mounting opposition by Africans but because it was ineffectual in halting the spread, and the Africans could thwart control measures in many ways if they wished, such as by not reporting outbreaks soon enough. Hutcheon therefore persuaded Chief Montsiwa to accept a designated guarded infected area in the Reserve to which cattle would be driven after infected animals from their herds had been shot. Hutcheon paid tribute to Montsiwa and his people, who consented without much demur to remove all cattle 5 km back from the Madibi Spruit and to clear everything out of their farm Kgaba without compensation, except for the huts which were burned, even though these measures were not for the protection of their own cattle but those of European farmers further south.

On June 22, a further public meeting was held in Vryburg Town Hall mostly attended by Dutch-speaking farmers. The Chairman explained that the Government proposed erecting a fence along the north side of Griqualand West, whereas it ought to be sited north of the Vryburg district. Various proposals for siting a fence were then put forward and a proposal, which was rejected, that food aid be requested from Government for those in distress. The Rèverend D. J. Pienaar stated that as (British) Bechuanaland had been declared an infected area, there was no law to prevent anyone traveling with a wagon and oxen wherever they liked within it. This was considered regrettable by some members, and one thought that before anyone tried to move cattle on the strength of the Reverend Pienaar's statement, it would be advisable for them to obtain a written promise from him to pay the fine. A motion to have cattle movement made illegal was carried unanimously. At the end of the meeting, a collection to pay expenses was proposed, whereupon, according to the *Bechuanaland News* (June 27), "half the audience rushed to the door." The other half contributed about 50s.

On July 11, rinderpest was reported to have broken out on two farms 48 km from Vryburg at Boschpan near Genesa, 96 km from the nearest known infected area. Hutcheon went out and, confirming it, had 100 cattle shot immediately and 223 the following Monday (Figure 3). The outbreak was attributed to a trader, Thompson, driving a span of oxen from an infected farm to Vryburg and back, one of his oxen dying where cattle grazed near Boschpan. The news created consternation in Vryburg, many saying that Vryburg was doomed as cattle were the mainstay of three-quarters of the population. At a meeting of the Divisional Council attended by Hutcheon, he was very critical of the people's attitude, accusing them of carelessness. The sick cattle had been allowed to wander about without being reported, for the disease must have been there for more than a week.

On July 19, it was reported in three kraals in the Setlagoli Native Reserve in the Maritzani Valley, about 20 animals being found infected and shot. Investigating this, Hutcheon found the disease at another kraal also, where six animals were infected. As the first outbreaks in this reserve, he was anxious to stamp it out and had all the cattle from this other kraal, numbering about 400, rounded up and shot after valuation. But soon after, the disease appeared at several other points. Numbering about 20,000 altogether in the reserve, the cattle

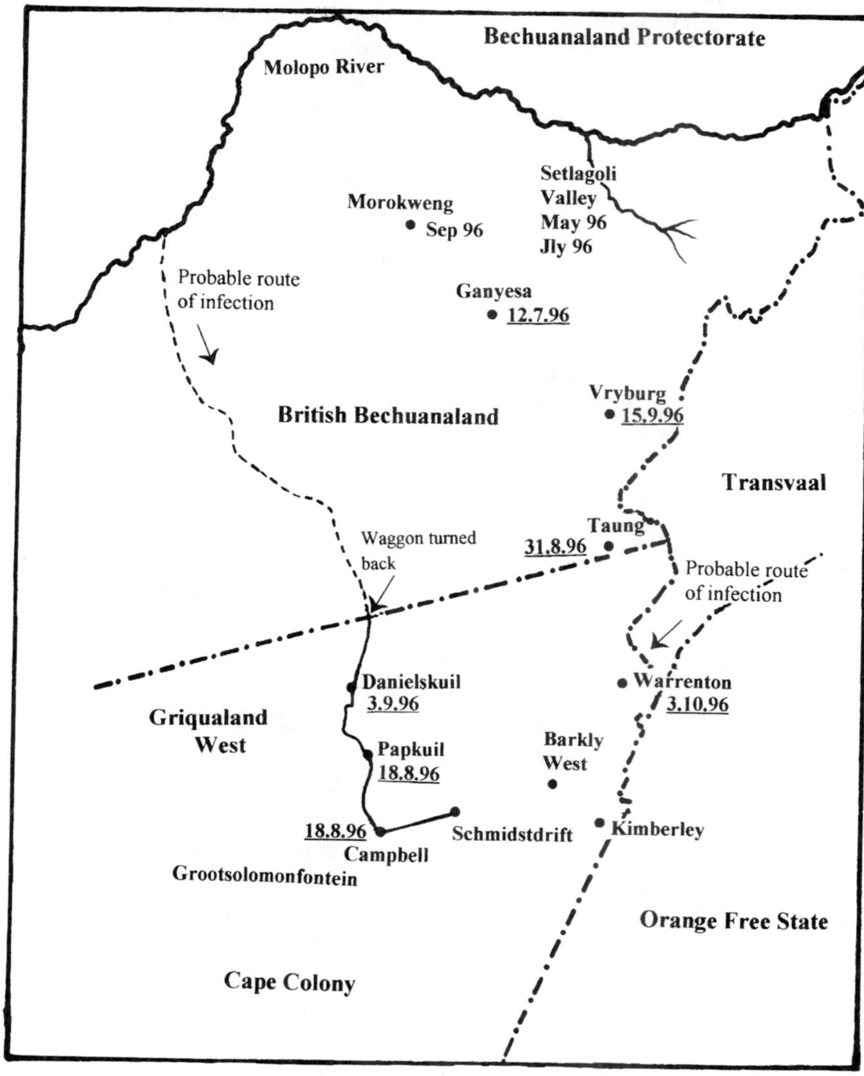

Figure 3. The spread of rinderpest in British Bechuanaland.

drank at only three main watering places, traveling long distances to and fro, making it impossible to isolate centers of disease.

After the first cattle were shot, the Chief Moshete strongly resisted any more killings. Hutcheon therefore visited him and Chief Phoi with a magistrate to explain the law and the reasons for the Government's actions. Eventually, both chiefs and headmen agreed to follow the law, and no further resistance to stamping out was encountered up to August 10, when slaughtering was given up as ineffective, the measure taken in the Mafikeng Reserve being adopted. It was also decided to place a fence around the Reserve, but before the fence construction could be started, the disease had already spread to the Setlagoli Crown Reserve, so it was decided to include this reserve within the fence as well. But still the disease spread too rapidly for this to be achieved, appearing here and there at isolated places as a result of Africans moving about to and from infected places. In one case, the kraal owner's sons had been tending sick cattle in the Protectorate which all died, and the boys returned home. Their father told them to wash themselves and clean their clothes before going among his cattle, but they ignored his request, and about 14 days later, the disease appeared.

The general opinion was that the disease was brought into the Setlagoli Reserve from the Transvaal and not via the Protectorate. Chief Moshete had fled into the Reserve with many of his people from the Transvaal, leaving a large number of his stock behind. Seeing the fence being erected on the boundary of the Transvaal, he slipped through a number of cattle about the time the disease appeared in the Khunwana Native Reserve directly opposite.

Commencing in July, Frances Macnab, a doctor, traveled from Kuruman to the northern boundary of British Bechuanaland with a span of oxen, seemingly oblivious to rinderpest, but made repeated remarks about the numbers of sick cattle and the high mortality, which she attributed to pleuropneumonia and an affection of the liver and spleen, explaining "The Boer looks upon disease as a decree of Divine providence, which it would be impious to arrest. Cattle diseases are consequently rampant throughout South Africa." But at one English farm, the owner had 800 or 900 good cattle, favoring Damaraland cows. No reference was made to rinderpest, but the owner stated that there had originally been no disease in Damaraland but it had spread to there from Bechuanaland [Protectorate].[1] Reaching the Molopo River, at New Barnet near Pitsane, Macnab came across a great many quarantined cattle said to have come from Lobengula in Rhodesia. Some were suffering "an awful disease." "It was said, I believe, by the Kaffirs that this disease was rinderpest. But the idea seems hardly probable" (Macnab, 1897). Later, she stated that there was a necessity for district veterinary surgeons: "This is just what is happening at last with regard to the rinderpest."

It was not until July 1896 that fencing was started in British Bechuanaland. Questions were raised in Parliament over the fact that the wire was bought by Government without calling for public tenders, and bought moreover from a member of Parliament. The first fences were built from brushwood as there was a shortage of barbed-wire which had to be ordered from Europe, and the first consignment was not expected until October. No tenders were invited for the fencing contracts, and Government was paying more than double the usual rate for the work, but the use of mules and horses for draught instead of oxen may have partly contributed to this. The Transvaal Government was not prepared to share the cost of a fence on the border, and the Cape Government intended to go ahead alone. The Member was Mr. "Matabili" Thompson, who was awarded a contract for 480 km. He had volunteered his services free for the overall supervision of the fencing projects and offered to remain in Bechuanaland until the work was completed, fencing going on simultaneously at over 12 different places, the contractors confident that they could finish in 3 months.

The inference of corruption appeared in the *Diamond Fields Advertiser* and was quoted in the *Eastern Province Herald*, the allegation being that when in Vryburg, Mr. Thompson had given contracts to a syndicate consisting of several well-known firms and individuals, including the firm of which Faure was a partner, he, and several other gentlemen, making quite handsome fortunes out of the contracts. Later, the Rinderpest Commission, of which Thompson was a member, called for tenders for several other fences which the firm tendered for along with others. Faure rebutted the allegations in a letter to the *Herald*, denying that he had given out any contracts, stating that only the Rinderpest Commission could do so. But mud sticks, and Molema wrote that it was a fact that "some European speculators in and around Mafikeng, became wealthy by this outbreak, as contractors in erecting cordons ..." (Molema, 1966).

The climax was to be a double barbed-wire fence along the Orange River for 1,568 km, completed in December 1896. Prior to this, police were stationed along the Orange River, and

[1]Damaraland does not appear to have become infected until early 1897 (Leutwein, 1906).

a wide zone created depopulated of all animals between Cape Colony and the Orange Free State. For a while, this was to prove an effective barrier.

On July 21, under Proclamation No. 243 of 1896, a Rinderpest Commission was established in the Cape with six members, chaired by Hutcheon. It met first at Mafikeng on July 28 and entered into contracts for the second fence, modifying the contract for the first. It decided on a double fence with a clear belt of 900 m between, from the Transvaal border at Marbogo Pan to Dikgalthon, with stamping out of all infection south of this line. The plan failed due to mounting opposition to the stamping out policy fuelled by a number of "cures" being published by men of authority, beginning with the Acting Administrator at Salisbury on April 16 who convinced Robinson. The Proclamation was followed by a new Act, Act No. 22 of 1896, which provided for the proclamation of infected areas, removal and slaughter of stock, with compensation payable only for the slaughter of uninfected animals, establishment of quarantines, and regulations controlling movement of people within infected areas. After stamping it out at Boschpan, northern Vryburg was confronted with a wave of infection from the Transvaal border, particularly in the region of Morokweng.

A second interstate conference to discuss measures to halt the spread was held in Vryburg on August 31, 1896 and was attended by delegates from Cape Colony, Natal, Basutoland, the Protectorate, South African Republic, Orange Free State, and German West Africa. The conference passed eight resolutions, foremost among them erection of a double wire fence 150 m apart along borders as the most effective method. The area between would be patrolled and all animals within it destroyed. Although the disease was already south of a line from Rooigrond in the east to Middleburg district in the west in the South African Republic, it was intended to stamp it out there once a fence had been erected. The conference also concluded that Africans carried the disease in their clothing, and a strict pass law was recommended, Africans to be disinfected if passing from one area to another and their passes stamped to this effect. It concluded that hides and horns could be safely disinfected with carbolic acid and export allowed. Sheep and goat skins should also be treated, and wool and mohair properly scoured. Hutcheon spoke at length on quack remedies, followed by Theiler and then Watkins-Pitchford, who advocated serum treatment. The Conference concluded that a series of experiments was required, especially with regard to determining the length of time a farm remained infected.

There was great concern to stop its advance south. On the day of the conference came the first report of its occurrence in the Taungs Native Reserve near Pudimoe, infected from the Transvaal by some Africans slipping in their cattle as the disease approached in the Transvaal. Hutcheon admitted to not pursuing the boundary fence with the Transvaal quickly enough because it was thought that the latter would continue the stamping out policy adopted with such vigour at the commencement, following Theiler's recommendation to President Kruger upon his return on April 9, which it did not. The chief Molala and his people at first strongly refused to allow any cattle to be shot. Arriving there on September 2, Hutcheon held a meeting with the Chief in the afternoon, the latter promising to let Hutcheon know his decision by the evening. No answer came so the Commissioner sent for 50 mounted men who rode all night and arrived early next morning. Accompanied by the Commissioner, and Thompson who had now arrived, a visit was paid to the Chief, who refused to see them. Returning to the hotel, Hutcheon sent a message to the Chief, stating that he was anxiously awaiting a favorable reply, but whether the Chief agreed or not, the diseased and infected herds of cattle would be shot. The mounted men were then paraded, and the Chief immediately sent a message stating that he was not going to resist the law but pleaded that his cattle not be shot. Hutcheon then went to the nearest infected kraals accompanied by the mounted escort and shot what affected herds he could find, taking all day to shoot just over 200 animals.

The next day, the entire Commission arrived and met the Chief, explaining the position to him. Eventually, Molala accepted in good grace and offered to send his son and some headmen to assist. Afterwards, Molala said to Hutcheon, "They tell me that you are a doctor, and that you are a great doctor, but can you do nothing but kill?" Hutcheon complimented the Chief on the incisiveness of his remark and invited him to some light refreshment, which the Chief accepted.

News of calling in of the mounted force no doubt led to the widespread rumour-mongering that the Chief was resisting, and the slaughter policy, had to be abandoned. Hutcheon pointed out that this was totally untrue. Leaving an assistant veterinary surgeon to continue the slaughter policy, he received frequent messages from Molala, asking if it was not time to stop the slaughter as it was having no effect, but he offered no resistance. When it was seen that the farmers in Vryburg would not support the policy much longer, Hutcheon called a halt to it in the Taungs Reserve on the condition that the Chief clear a belt of cattle 6.4 km wide round the east, north, and northwest boundaries, which he readily consented to and promptly carried out.

A problem in the Vryburg district was the drying up of water supplies, creating continual requests for permits to move stock elsewhere. At the same time, after the disease appeared in Griqualand West, the Commission became divided upon what course of action to take, and it was rumoured that Government intended withdrawing from British Bechuanaland. At a meeting of the Commission on September 9, Thompson proposed a resolution:

> That as rinderpest has made such rapid strides, and as it has already appeared in the division of Barkly West, is in my opinion fairly out of control, all forces that can be spared from Mafikeng, Vryburg, and Barkly West be concentrated on the south bank of the Orange River to be guarded from the eastward about Herschel onto the German territory, and that cattle be forced back from the Orange River, clearing a belt of about five miles, and that a fence be constructed at once, subject to the consent of Government.

As the resolution was not agreed upon, Thompson published it in the *Cape Times*, adding,

> I hold it as my opinion that the Rinderpest Commissioners are not acting in the interests of the country, and I refuse any responsibility in the result of their want of action. All cattle north of the Orange River are, in my opinion, now doomed, and steps should be taken to make a last stand on the Orange River. This does not meet with the views of the other members, and I have determined to make it known. About £300,000 has already been spent.

The publication of this led to many farmers believing that Government would abandon the Vryburg district, and Hutcheon considered that as a result, the disease was being spread deliberately with a view to obtaining compensation before payment was withdrawn, a number of suspicious outbreaks following immediately afterwards. It appeared on two farms on the northern border of Vryburg district and on three within 16 km of Vryburg toward the Transvaal. Only one animal on each farm was infected, and in one case, the farmer pointed out the infected animal before the veterinary officer could detect symptoms. Farmers and Africans who had previously strongly opposed slaughter were now quite willing to have their herds shot in order to obtain the compensation. Hutcheon therefore suspended slaughter and compensation, ordering strict quarantining in its place, and when it was reinstated, compensation was reduced by £1 per head.

Following its meeting, the Commission sent a letter to the *Diamond Fields Advertiser* signed by Hutcheon, recapitulating what had been done to stop the disease. Writing in no uncertain terms, Hutcheon, whom we may liken to South Africa's Gamgee, pointed out that the real obstacles to stamping out the disease were the apathy, laziness, obstinacy, carelessness, or pig-headed resistance of some of the farmers themselves. They not only failed to report cases but did not look after their stock properly.

On September 5, a Dutch farmer had written to the *Diamond Fields Advertiser*: "Indiscriminate shooting is monstrous and the Almighty will visit the land with an even worse visitation. They [the farmers] will be shot before they will allow their cattle to be shot."

"Rusticanus" followed up on September 12: "If all the cattle are shot, there will be an end to Rinderpest, of course, but an end to the cattle too! The disease can go no further …." He argued that 10 or even two cows surviving out of 100 was better than having the whole 100 slaughtered. In the same issue, "Farmer" urged Africanders to "lay another cornerstone to your nationality" and rise up and overthrow "the scourge [of the English] that is at your gates."

In the middle of September, large numbers of cattle began dying in the Morokweng reserve, and the Africans, as soon as they saw its effects, killed many of their sound cattle and dried the meat, also drying some of the meat of those which died. Captain Jenner was instructed to raise a burgher force to guard the Orange River from its junction with the Vaal eastwards to Herschel. In the northern districts of the Transvaal, the disease was reported to be spreading with fearful rapidity.

The disease breaking out simultaneously at two centres in Griqualand West at the same time it appeared in the Taungs Reserve led to many theories, such as infection being carried on the wind, and therefore fences and other methods were useless. But when the outbreaks were closely investigated, it was revealed that they were caused by a Mr. Maritz driving a span of oxen about, having picked up the disease from the Protectorate border. At Groot Solomonsfontein, the veterinary officer found that the majority of cattle had already died, and the carcasses were lying about with no attempt to bury them. The area was used as a public outspan frequented by many wagons passing to and fro, yet farmers had not even bothered to report the outbreak. As a result of the veterinary officer's efforts, the infection was stamped out by the middle of October. At another farm where the disease broke out, 20 cattle were found dead and 155 sick on October 17, the disease having been present about 3 weeks. Yet again, no report had been made. From here, it spread to African cattle in the Soedin location, then in all directions, but eventually was stamped out.

At another farmers' meeting in the Vryburg Town Hall, held on the morning of September 19, it was voted that Government be requested to maintain the present regulations in force and not retire from the country. There was no cure for rinderpest, and it was best to support the slaughter of infected cattle. Hutcheon, who was present, stressed the necessity of slaughter, and that if it was objected to, Government would have to cut off the Colony altogether, as it would be wasting money on other measures. To applause, a Mr. Jacobs asked whether Dr. Hutcheon's religion did not teach him that it was wrong to shoot cattle and that when the Almighty sent a plague as a punishment for sin, they must not anticipate it by shooting down healthy cattle. Hutcheon was well up to such arguments and, quoting from Genesis, stated that they were sent here to learn the whole of nature's laws and bring them under as servants to mankind. God placed man in the world as the supreme animal to rule everything else. It was man's function to overcome every difficulty on the earth, and if they neglected that, they were neglecting their duty, and instead of blaming God, God would blame them. This reply earned him a round of applause.

The meeting lasted 4 hr, at the end of which, a motion was put that Government be asked to continue the measures for stamping out, and the people would support the Government in every possible way. This was carried by a show of hands, but the minority anti-shooting party "who certainly made the most noise" according to the *Bechuanaland News* (September 26) claimed that they were in a majority and that the motion was not carried.

Another meeting was held in the evening at the same venue called by the South African League. The meeting supported the Government but considered no further measures to those already in place were necessary. It called upon Government for a reduction in railway rates to help people by making supplies cheaper, while fencing materials should be carried free.

Following criticism, the Commission was dissolved by the Prime Minister on October 1, leaving policy in the hands of local Rinderpest Committees, stamping out as a national policy being abandoned. The reaction to the cessation of shooting in the Dutch-speaking newspaper *De Express* was that: "Whatever may happen to the rest of South Africa, Bechuanaland is doomed." The greatest difficulty was now raised by the farmers, from whom Hutcheon and Fuller met active opposition at Dryhartson from October 3 to 5.

On October 3, it made its first appearance across the Vaal River in the Warrenton Ward of the Kimberley district, and the Commissioner of Police had the whole herd of 381 shot at once, leading to much press comment.

At a public meeting in Vryburg on October 12, the farmers were almost unanimous in appealing to the Secretary for Agriculture to stop the slaughter, and Hutcheon was instructed to do so. One farmer claimed that it was the "indiscriminate shooting of cattle" which spread rinderpest the most, as the carcasses could not all be buried and the microbes from them were thus spread far and wide by the wind. In the Transvaal, they shot only those animals which were sick and as many per day as could be buried. On October 14, a farmer found a sick beast on his farm used as an outspan 24 km north of Vryburg and immediately reported it. Finding it to be rinderpest, the whole herd of 585, considered one of the best in the district, was slaughtered. The infection was believed to have come from a neighboring area where it had existed for 2 weeks without being reported.

On October 24, the farmers held another public meeting at which resolutions condemning the stamping out policy were unanimously adopted; but 22 herds were saved plus the spread being contained. If it had been allowed to run its course, it would have spread throughout the district and it was estimated that over 70% of cattle in the area would have died (Figure 4).

Figure 4. Cattle shot in the stamping-out policy, probably in Vryburg District. © ARC-Onderstepoort Veterinary Institute.

Figure 5. A rinderpest guard. © ARC-Onderstepoort Veterinary Institute.

On October 27, another conference, chiefly of Cape delegates, was held at De Aar, and two days later, a regulation was proclaimed that carcasses must be buried or burned as near as possible to the spot where death occurred, under a penalty not exceeding £50 or 3 months' imprisonment with hard labor.

A farmer complained in the *Bechuanaland News* (October 31) that the native guards were largely responsible for spread of the disease:

> The very men who are supposed to guard and protect people do their best to spread the disease for the sake of the pay. They have nothing to lose and get £18 a month, so why should they bother about it? After my cattle were shot, a guard was stationed on the farm to disinfect the boots of anyone who passed. I suppose he did so sometimes, but as he lay in the cool all day, the chances were not in favour of his seeing many. When we were digging the trench to bury the oxen the 'guard' slept all day and went home at night. There is not the slightest supervision exercised over the gang of boys engaged in burying the cattle. When they had finished at my farm away they went without disinfection or anything else .... (Figure 5).

At the beginning of November, Thompson, who had resigned his post on the Commission in protest at the continued shooting of cattle, expressed his views in *The Times*:

> If the policy of killing cattle is pursued with the colonial natives...my long experience with South Africa and its natives leads me to the conviction that we shall have to face one of the biggest wars Africa has ever experienced. It will be found that the Zulus, Griquas and Basotho will join hands, and as it is a well known fact that we did not beat the Basotho, what could be done in the case of such a combination (Thompson, 1896)

On November 26, a public meeting was held in Kuruman District. Although holding by far the largest number of cattle in British Bechuanaland, it was still untouched by rinderpest despite the disease having broken out close to the northern border on the Mashowen River, and was in the neighboring districts of Mafikeng, Vryburg, and Taungs. The district was virtually sealed off and due to drought and locusts, the previous season's crops had been a failure. Some people lived up to 320 km from a source of supplies, and all carriage with oxen had been stopped as well as incoming supplies by rail bringing some people to the brink of starvation. It petitioned Government that in the event of rinderpest breaking out, the slaughter policy which it had abandoned in the north should be immediately adopted, with a three-quarters value compensation for animals killed; and all stringent measures should be continued. It also requested the appointment of a Commission in Kuruman and a qualified veterinary surgeon.

4. THE SOUTH AFRICAN REPUBLIC (TRANSVAAL)

Following receipt of the telegram from Capetown on March 6 that rinderpest, or some similar disease, had broken out in Bulawayo, the Government of the South African Republic issued a Proclamation on March 11, which: prohibited the introduction of all game and domesticated animals from Mashonaland, Matabeleland, and Khama's country; prohibited transport to and from such territories; ordered the withdrawal of all stock along the border to a distance 4.8 km south of the Limpopo River; and ordered that all stock suspected of suffering from a communicable disease were to be dealt with by the Pound Law (Law 2 of 1882, Article 34).

On April 1, the Transvaal authorities were informed that the disease had been seen in Marico district in Linokani's Location and also at Skelvynstad, but it was known to have entered the Transvaal before March 11 (Curson, 1936). According to Theiler (1897a), the disease must have occurred at Zeerust in the Transvaal on the border of the Protectorate at the beginning of April. A Proclamation of April 1 extended the prohibition relating to transport to Zeerust (Marico) and Lichtenburg districts. This was followed by another Proclamation on April 14 forbidding even movement within a farm of cloven-hoofed stock except under permit certified by a magistrate, JP, or Field Cornet, that the stock had not been in contact with diseased animals, and ordering notification to neighbors and isolation of suspected cases. Anyone contravening these provisions was liable for damages. Following a report from a customs official at Hendriksdal that rinderpest carcass meat was being imported, a Proclamation of April 15 forbade the importation of animal products and any materials which might convey infection, such as fodder, with a penalty for infraction of £7.10s. or up to 1 month's imprisonment. Confiscated material was to be buried at least 0.9 m deep or burnt.

On April 27, a further Proclamation extended that of March 11 to Bloemhof, and on April 28, an important Proclamation provided for slaughter and compensation. It also provided for a committee of three in every ward for the execution of duties such as: notification; quarantine of affected animals; valuation of herds from calves at 10s. to bulls at £5; disposal of carcasses; compulsory labor where necessary; placing of guards; disinfection; closure of roads; issue of permits; posting of notices; tracing of contacts; and prohibition of movement. It also allowed for the closure of a magisterial district to transport if more than three wards were affected. Contraventions were liable to a fine of up to £500 or imprisonment for up to 2 years. The Proclamation also made the first reference to a Government Veterinarian, although no such post existed, so Theiler (who had drafted the proclamation) immediately applied for the post and was confirmed on May 11.

In May, the disease broke out in the area between Pretoria and Johannesburg, stopped by the slaughter of 300 cattle, and on May 9 another Proclamation ordered complete cessation of all movement of ruminants for 1 month, those in transit remaining where they were unless a permit was obtained from the ward. Continuous supervision day and night was ordered for all herds and the appointment of vigilance committees in towns for the control of abattoirs, slaughter stock, and delivery of meat. A further Proclamation on May 16 tightened up measures generally, especially with regard to the responsibilities of the ward committees. Dogs, pigs, and game were to be destroyed and buried where the disease had appeared. The disposal of milk from suspected cases was prohibited and provision made for transport of meat to populous centers by train or horse and mule carts; but contradictorily, the government gazette, the *Staatscourant*, carried several notices of public auctions of cattle seized under the regulations. On May 27, a Government Notice prohibited export of meal and grain on account of spreading the disease to neighboring territories, and on June 4, the Game Law (Law 5 of 1894) was temporarily suspended with regard to ruminants. The Waterberg region where 15,000 cattle were said to have died allegedly became infected by game migrating away from infected areas, such that kudu were found in places where they had no longer been known to occur (Theiler, 1897a).

Two thousand carcasses were estimated to lie in the veld around Rustenberg. It was therefore decided to fence off the border, but because of a lack of wire, brush fences were made constructed of intertwined branches.

A further Proclamation appeared on June 8 consolidating previous orders and strengthening measures generally. An important added provision was the institution of cordons, the first immediately around a quarantined farm, the second 10 km back, with all movement of ruminants within the cordon prohibited for 1 month after the last case of disease. The duties of Rinderpest Ward Committees were clarified, and they were to make weekly reports to the Government Veterinarian.

The disease then appeared 10–15 horse riding hours beyond the fence line, although the first assumption was this was malicious spreading reports came from two different sides that it was probably caused by inoculates. In Lichtenberg, slaughter was at first successful in controlling the outbreak, but in the African reserve, thousands of animals coming to drink at the same watering places caused a new outbreak. Slaughter could not be used among the thousands of African cattle for fear of revolt.

In the second week of July, 299 oxen died in the Zeerust district, and 111 were shot. Nearly all oxen in this part of the district were dead, and the more wealthy cattle farmers in the Transvaal were selling off their herds to vanishing point, preferring to have the cash. In July, President Kruger held a conference at his farm, and this was followed up in Pretoria at the end of August by an International Rinderpest Congress to review the situation.

The Orange Free State was reported by the *Bechuanaland News* of July 18 as "not caring." After accepting the general principle of a rinderpest law, the Raad had then thrown it out altogether. Nevertheless, 27 oxen of a Free State Boer quarantining in the Transvaal were shot near Reitz after escaping and crossing the river. The Free State guards met with such opposition from the owner that one guard was seriously wounded.

Special regulations were passed for the supply of meat to Johannesburg (G. N. 228). Slaughter stock could be accepted after veterinary inspection at Vereeniging and Volksrust on the Orange Free State and Natal borders respectively, and consigned to Fordsburg where it was to be slaughtered within 48 hr. On August 3, a Proclamation authorized transport of goods under strict supervision, and a Government notice 2 days later authorized the erection of fencing, including over private property. A Proclamation of August 7 allowed farmers who had moved stock to the lowveld for winter grazing to return with it under certain restrictions.

Up until August, there were no reports of sheep or goats being affected, but the same could not be said of game. By then, there was very little left along the Crocodile River in Transvaal's Waterberg District. Kudu, warthog, and duiker lay dead, but there were no oxen carcasses. A South African Republic Government Notice (No. 377) of September 23 was to refer to springbok, steenbok, and reedbuck also succumbing. As a result of inoculation, by October the disease became widespread in the eastern part of the Transvaal, where Percival states that he saw the greater kudu "wiped out." Theiler (1897a) reported that species affected in the Transvaal were bontebok, blesbok, duiker, steenbok, waterbuck, reedbuck, springbuck, gemsbuck, bushbuck, and kudu. Unaffected were impala, blue wildebeest, black wildebeest, and hartebeest. Outside the region warthog and guinea fowl were reported to have died.

On November 3, owing to continued extension of the disease, a Proclamation discontinued compulsory slaughter of infected herds except in special cases. Furthermore, remedies might be tested in infected areas; ward committees and special guards were to be abolished from November 15; but there was to be compulsory herding of stock, and owners had to bury dead stock, new ward committees being empowered to commandeer laborers at 5s. per day for Europeans and 1s. 6d. for Africans. All horned transport was prohibited for 2 months except for despatch of slaughter stock from healthy areas to town under permit. Asian hawkers were forbidden to travel the countryside. No further fencing was ordered. Disease of stock was to be reported immediately to the field cornet, failure to do so entailed a penalty of up to £25. Finally, the Government intended conducting research.

Altogether, 39 proclamations were issued for the Transvaal alone up to 1899, although most were repealed on November 11, 1897 following the apparent success of inoculation campaigns. A Proclamation of November 1, 1897 imposed protection on rhinoceros and buffalo throughout the Republic, and kudu, blue wildebeest, waterbuck, and nyala were protected in the district of Vryheid for 5 years. On December 1, a Proclamation laid down conditions under which persons impoverished by the epizootic might receive financial assistance, the State having allocated £150,000 for the purpose.

5. CALLING IN THE WITCH DOCTORS

On November 5, 1896, the Cape Under Secretary for Native Affairs sent telegrams to the Civil Commissioners, Magistrates, and other officials in the eastern Cape and Transkei, stating that it was highly important that the natives should understand the ravages caused by rinderpest and the regulations for preventing its spread. Two representative Africans with a knowledge of cattle sickness and herbal remedies, that is, witch doctors, should be sent from each district to British Bechuanaland to see the disease for themselves and try their hands on sick cattle. The party of nine witch doctors and four headmen visited Taungs, arriving on December 3.

> That he [Molala] and his people looked upon the advent of our men with a certain measure of suspicion and distrust was evident enough to those who understand native character. None of the ordinary courtesies generally extended to strangers were shown ... The natives certainly were not friendly, and it was a relief to Dr. Soga and myself [A. K. Soga] when the work was done. (Soga, 1896)

The official accompanying the party noted that large tracts of country in the Taungs Reserve were now practically lying vacant except for a few herds of sheep and goats, the latter animals allegedly saved by Government forcing the people to bury the cattle carcasses. Almost denuded of cattle, only about 800 were left out of 20,000. From these remnants, 99 sick cattle were obtained with some difficulty for the witch doctors to practice their craft, some 20 others being taken away by the owners who feared that the flesh would be poisoned, and they would be unable to eat the carcasses. Every animal which died was autopsied for them to see the internal effects of the disease. Whereas it is not clear whether the Government had hoped that the Africans might have a cure, the exercise was of benefit in alerting the party to the disaster of rinderpest, and almost all agreed that should it appear in their regions, advocating and supporting the policy of stamping out, isolation of infected kraals, quarantining, and the burying or burning of carcasses. The representative from West Pondoland, however, did not think it neccessary to slaughter cattle to prevent it but thought that "some treatment" could be used.

All the men were fumigated and dipped on their way back, and their clothing burnt and substituted with khaki suits and other clothing. As to the results of the experiments, 27 animals died while the witch doctors were present, cure being claimed for the remaining 77. But within 7 days of their departure, all the remaining had died. Opinions among them as to the nature of the disease ranged from red-water, lamsickness (botulism), and a combination of all the diseases of cattle.

24

THE CONTINUANCE OF RINDERPEST IN SOUTH AFRICA

1897 to 1905

1. CAPE COLONY IS REACHED

It was believed that the last outbreak in the Kimberley District, which occurred close to the borders of the Orange Free State and the Transvaal at the end of January 1897, was a result of two cows being shot on different farms which all the surrounding farmers came to see and assist at the autopsies, declaring the disease to be gall sickness. They then left without disinfecting themselves, and the disease appeared on adjacent farms. It continued to spread until January 30, the last outbreak believed to be due to dead carcasses from the Transvaal side being washed down the river in flood onto the colonial side where cattle drank. An isolated case then occurred on February 20 caused by a farmer loaning a neighbor, whose stock had been shot over 2 months previously, two milking cows, which were returned daily. In all, the disease appeared on 16 farms where 29 animals died, and 2,951 were shot.

In the final case, some dairy cows contracted infection from grazing where diseased cattle had been. The owner surreptitiously handed them over to another, and after their arrival on the latter's farm, two calves became sick and died or were killed. The farmer then sent the rest back to the original owner, and a few days later, an outbreak occurred at both farms. The two herds were then shot, both men prosecuted at Kimberley and fined.

One farmer, D. J. Van Zyl, had 109 cattle shot by rinderpest guards on January 20 and 21 at the Orange River in the western Cape, the cattle having watered in the river at an unfenced point. As the cattle were healthy, the owner was not entitled to compensation. Left with only eight, the farmer petitioned the Colonial Government for compensation (Van Zyl, 1897). The disease was contained by slaughter in Griqualand West until the end of March, when slaughter was abandoned generally throughout the Colony and Koch's bile inoculation adopted. When it appeared in the Basuto African reserves of Herschel in that month, cattle owners refused to allow their cattle to be shot, demanding inoculation. This provided the first opportunity to try out Koch's method on a large scale, 20,000 cattle of 60,000 being inoculated, of which 14,000 survived. But outbreaks of rinderpest continued in the vicinity, tending to discourage others adopting inoculation. The logic of shooting sick animals eluded the Africans, and as inoculation sometimes spread the disease, they began to wonder why it always followed the government officials, asking how it was that "first comes the special with his bottles, and then—after, not before—comes the rinderpest?" (Anon., 1897a).

In May 1897, the Cape Government enacted legislation empowering compulsory inoculation (Act No. 2 of 1897). During a debate on June 3, a Member of Parliament urged removal of the cordon along the river, "that the frightful expenses be stopped, and the farmers be encouraged to attend to their cattle themselves. Let them be provided with the best bile for inoculation." As there was some distrust of inoculation, it was not enforced but was left to local Rinderpest Committees to decide whether to adopt it in their particular areas. But Hutcheon believed that the abandonment of slaughter in favor of inoculation was premature.

At Berlin, in the King William's Town district, a demonstration of bile inoculation was arranged for a number of Africans the recipients dying, and as a result, further inoculations were refused. Acceptance varied widely in Transkei and Pondoland; at Umtata the Chief inoculated his own cattle and did everything he could to persuade his people to follow his example, but most refused and lost their cattle; while at Elliotdale, two thirds were inoculated, and 75% survived the epizootic. At St. Marks, inoculation was opposed, and 85% of cattle died. In Basutoland in 1897, 34,000 head were inoculated with bile in one district alone saving 50% of the total. In a district where animals were not inoculated, 30,000 died.

Scully, Resident Magistrate at the time at Nqamakwe, Transkei, close to the east coast, provided a first-hand account of a Resident Magistrate's work in the outbreak. Referring to it as the most strenuous episode of a fairly strenuous life, he wrote: "We knew it was coming, and to a certain extent discounted its consequences." Early in June, he received a report that cattle were dying suddenly from an unknown disease near Xilinxa, 20 km distant, and at once requested the services of a veterinary surgeon, throwing a cordon around the locality. Soga arrived within a few days and pronounced rinderpest. The infection began to spread immediately along two lines almost at right angles to each other, the directions of an alternately prevailing wind. He records that excitement among the people became acute, fuelled by rumor that Government had introduced the disease to impoverish the people and force them to work at the mines. Learning "from secret sources of information" that the situation was growing dangerous, he sent his family into the Cape and ordered 100 rifles and a supply of ammunition:

> ... what was the best course to follow? Dr. Koch's great discovery as to the preventive effect of inoculation with the bile of infected animals had hardly as yet been tested. Experiments with bile had been made in the district of Herschel, but the results were indeterminate. Soga had gained experience of the inoculation process at Herschel, but had left there before any convincing results had eventuated.

But neglect spelt ruin, so he proposed to Government that inoculation be introduced, and it was sanctioned under the condition that the people were only to be "advised" to adopt it, and no pressure was to be exerted. Scully called a mass meeting and explained the situation. The result was that, after lengthy debate, the majority rejected inoculation, but one influential man, Henry Shosha, agreed, and a few others followed suit. The next morning, Scully purchased some sick animals and collected the bile, then riding to Shosha's village with Soga, he superintended the inoculation of "some hundreds" of animals. A number of other owners then agreed also to inoculation, Scully noting that it was then necessary to act quickly before they changed their minds. He records how he took two influential men to the kraal where he had bought the sick animals to show them the dead and dying cattle. One, who owned about 30 head, agreed to inoculation, the other owning 300–400 refused, and Scully predicted that his cattle would all die, little realizing his action in taking the men to the infected kraal would probably ensure this. Four weeks later, the latter's cattle were all dead, "My chance prophecy had come literally and appallingly true ..."

> The weather turned bitterly cold; snow fell in all the high-lying parts of the district. The cattle died like flies in early winter ... I spent all my time riding about the district ... I preached the gospel of the bile. At every principal kraal I emphasized my belief that inoculation or ruin were the alternatives ... I almost lived in the saddle, never heeding where I slept or what I ate.

As the disease spread, killing the uninoculated herds, opinion changed, and about two thirds of owners in the district visited the Residency, clamoring for inoculation of their herds.

To obtain sufficient bile, Scully made a levy of 10% on the herds. The third which had refused inoculation, having lost nearly all of their cattle, became extremely hostile, and some threw lumps of meat from their dead cattle into the kraals of inoculated animals. Scully assured the owners of the latter that they had nothing to fear. At length, he heard that one headman had killed a black bull, a sign of preparing for war. Scully detained him and awaited developments, but the people sank into sullen apathy, and the crisis passed. About 4% of uninoculated cattle survived, but Scully thought that had the plague occurred in summer when the grass was good, mortality might not have been so great.

It had been intended to hold a great feast on Queen Victoria's Jubilee Day (June 20, 1897), but this was cancelled. Scully records how orders such as the separation of cattle were obeyed after he punched a drunken headman on the nose. On another occasion, he was pursued by a lunatic with a club trying to kill him, believing he was responsible for the cattle deaths, but on his horse outdistanced the man. The outbreak lasted until the end of October: "Over two-thirds of the district the carcasses lay thick: it was difficult to get out of sight of them or to avoid the pervading stench."

Scully claimed that over 20,000 animals had been saved, but where magistrates in surrounding districts had obeyed Government's instructions not to pressurize people to adopt inoculation, all the cattle had died.

> I should say here that as soon as I had satisfied myself that inoculation was of at least some use, I decided to disobey the orders of Government ... I determined to use every available ounce of pressure toward making the people inoculate. I kept four horses under saddle, and rode continually from one part of the district to another, preaching, advising, threatening, and cajoling-doing everything in fact, that might be calculated to bring over the waverers ... if Government had possessed my knowledge no attempt would have been made to tie my hands. (Scully, 1913)

The Transvaal witnessed much activity following the report of Danysz, Bordet, and Theiler, published in English in the *Pretoria Press* and instructions for inoculation printed and circulated in Dutch. An information center was opened in Pretoria where the "French method" was expounded, and a conference of Veldcornets expected to carry out the procedure in their districts was convened from the whole Transvaal to learn about it, a demonstration being given with 70 cattle on the President's farm. Theiler described it to a secret session of the Second Volksraad. These activities were coupled with the promotion of another International Conference held from August 2 to 13.

Held in Pretoria, the conference has been described as "explosive." Factions developed, with Natal supporting Watkins-Pitchford (who was at first absent combating the outbreak in Natal), the Transvaal supporting Theiler and the French workers, and the Cape supporting Koch, although relying on Edington's glycerinated bile. The statement from Lloyd (probably drafted by Watkins-Pitchford) read out to the Conference probably had more to it than met the eye. The promotion of the Conference and the support of the French workers were largely nationalist politics being played by the South African Republic Government, but Danysz and Bordet demanded a reward from the Transvaal Government, and there was the possibility of other prizes. Defended by Theiler, the Conference accepted the role that he and Watkins-Pitchford had played, but all the Conference achieved was to establish the validity of both Koch's method and the "French method," and publicize the work of Danysz, Bordet, and Theiler. Koch, who had left the country, was attacked but vigorously defended by Turner, who claimed that Koch was attacked because he happened to be German. But it is doubtful whether Koch appreciated his qualification: "Let us hope that Koch is yet mentally unaware of all he has lost by being born under the German Eagle rather than the British Lion" (Gutsche, 1979).

Although by the beginning of 1897 the means of prevention were thus available and prophylactic measures saved 61.4% of the Cape's 1.6 million cattle, services just did not exist for such preventive measures to be practised in Africa on a wide scale, although work on inoculation soon extended to many other countries, as, for example, Eritrea (Memmo *et al.*, 1904).

The popular explanation for the entry of the disease into Cape Colony, breaching the Orange River line on March 24 in spite of the protective measures, removal of stock and hides from one place to another being prohibited, is given by Mettam (1937) as an African ox-wagon leader who found a sack on the roadside containing dried meat and a pair of blood-stained trousers: "Although he realised that he was breaking quarantine regulations he proceeded to don the trousers and consume the meat. To his surprise and to the horror of the authorities the trek oxen under his charge developed rinderpest and thus the invasion of the Cape Colony was accomplished despite all precautions." The two leading oxen developed rinderpest a few days after crossing the Orange River.

The story is probably apocryphal, apart from the fact that the virus is destroyed by drying, and so the blood-stained trousers are unlikely to have been infective. If indeed the items that this unfortunate man found had been infective, somebody else must have already introduced infection into the area. We are not told either where the trek oxen were coming from, except that they were traveling south across the Orange Free State toward Cape Colony. It is more than likely that the oxen had already contracted the disease.

On November 12, the disease was reported at Mowbray Park near King William's Town in the Eastern Cape, and also among Cecil Rhodes's captive elands at Groot Schuur, Rondebosch.

During 1896–8, 575,864 (35%) of 1,639,435 cattle died of rinderpest in Cape Colony proper.

In German South-West Africa, the Governor of Damaraland had prohibited introduction of cattle, sheep, goats, and animal products. Military posts along the borders had been strengthened and the frontier patrolled. A circular had been distributed in the native language giving details of the disease, and every civil commissioner had the right to destroy diseased animals immediately; as soon as cattle in Damaraland were infected, they would be killed. It was not until April 1897 that the first cases were reported, just south of Windhoek near the Schaf River, the disease then spreading throughout the country in a matter of weeks. While inoculation was used, reducing mortality to 20–50%, only the German settlers used it, losses among Herero cattle reaching 95%. Although Government gave priority to European herds, many Herero herds were in inaccessible areas.

2. THE PLAGUE ENTERS NATAL

By June 1896, some 856 km of fence had been erected along Natal's borders at an average cost of £80 per 1.6 km. In the early stages, the Government did not wish to alarm the African population for fear of an uprising, the Secretary for Native Affairs informing magistrates in April:

> I think it will be well if you will cause the Natives, generally, to be told through the Magistrates, that there is a bad cattle disease in the Transvaal, and the Government has stopped all cattle coming into Natal from the Transvaal. Natives near the Transvaal border should also be told that they are expected to give immediate notice to the Magistrates or Police if they hear of any cattle attempting to cross the Transvaal boundary. (Moor, 1896)

But it was September before any real attempt was made to inform African communities of the threat. Many Whites were of the opinion that if it broke out among the African cattle, a vigorous slaughter policy should be adopted, the Colonial Veterinary Surgeon advising that if compensation were paid, much of the African opposition would be overcome. The *Natal Witness* was critical of the soft-line approach and warned its readers "what there is reason to fear is a degree of tenderness for the native that will bring ruin to the white man" (Anon., 1896a). When it did break out, the first endeavor was to stamp it out, but owing to so many outbreaks occurring directly after and in so many different parts of the country, such a procedure was found impracticable (Verney, 1898).

The Department of Agriculture employed 883 men to serve in the campaign against the disease, the country being divided into 21 districts, each with an inspector backed by a total of 127 European guards and 679 Africans. The Europeans were armed, and both black and white employees were sworn in as special constables with authority to quarantine infected herds, shoot infected animals, and arrest people contravening regulations by bringing cattle across the border into Natal. Mountain passes between Natal and Basutoland were dynamited to make them impassable to cattle and ox wagons. Against his better judgment, Watkins-Pitchford was obliged to set up disinfecting stations (to which Hutcheon was opposed also) at main border crossings, fumigating the mail train and its passengers at the frontier, passengers being obliged to sit in a shed over a pit producing sulfur fumes. At Charlestown on the Transvaal–Natal border, passengers were "bundled off" the train to a center where they were "fumigated" with a light spray of bacterial disinfectant (Figure 1). Items left in the compartments were untouched, but unless they contained fresh meat, infection was unlikely to be carried by this means. Africans were stripped and queued for fumigation while their clothes were soaked with disinfectant, which they then had to don while still wet. The Secretary for Native Affairs pointed out to the Commissioner of Agriculture, who was responsible for the procedures, that this could be harmful in the coming cold season and the clothes should be dried first, a measure which was put into effect. Leading to much controversy, when loud objections were raised to fumigation, the Minister stated that it had been on Watkins-Pitchford's advice.

Figure 1. A disinfection tent. © ARC-Onderstepoort Veterinary Institute.

The Government was also attacked for its policy on fencing, there being insufficient materials to meet the demand despite quantities being imported from Britain and the United States in 1896, leading to exorbitant prices as Government bought up most of the stocks. Altogether, £76,715 was spent on fencing, disinfecting stations, and salaries (Ballard, 1986).

By May 1897, the Burgersdorp and Aliwal North areas of the northeastern Cape province were riddled with rinderpest, and the Transvaal was being steadily depleted of stock. Captain Carr, a resident of the Waterberg district, stated that he had lost nearly all his herd, one neighbor had lost 280 of 300 cattle, and another had only five of 100. Farmers were purchasing donkeys from the government on credit, and Kruger stated that in the Volksraad, they had been able to obtain only 1,300 of the 8,000 ordered from outside the Transvaal.

Bloemfontein in the Orange Free State had escaped infection until now, but on May 29, 1897, an ox, one of eight inspanned in a wagon, collapsed in the market square. A crowd soon assembled and members of the Raad joined the crowd. Dr. Krause, the District Surgeon, then appeared and pronounced the ox to be suffering from rinderpest, whereupon it was shot. The street was closed and everyone present disinfected by having Jeyes's Fluid poured over their boots. The town gates were closed and no animal traffic allowed in. Within 2 weeks, the disease was rife in the southern Orange Free State.

Almost a year after the first alarm, in late June or early July 1897, the disease broke out among trek oxen at Ladysmith but was not reported until, on July 16, it was confirmed at Dundee in Natal, 90 km south of the Transvaal border and 50 km east of the Orange River Colony. Legislation was already prepared and enacted immediately, declaring infected areas and placing restrictions upon cattle movement. A serum station was maintained at Marico, but badly needed further stations were not agreed for fear that they might spread infection. Simultaneous outbreaks were reported 80 km apart without any obvious connection. Verney (1898) later wrote: "For two years our borders had been protected by means of a double fence, regularly guarded, a procedure which cost the Natal Government nearly £100,000; and there is good reason to believe that had not rinderpest been sown broadcast throughout the colony by our malicious neighbours we might have been free of the disease today." As to obtaining "salted" cattle in Natal:

> one had to depend upon the honesty of the farmers in the adjoining colonies, but unfortunately in many cases it was evident that such a quality as honesty was not recognised. There is no doubt that many of the so-called salted cattle were simply immunised, and, in course of time, they developed rinderpest. The serum obtained from such cattle naturally was responsible for many deaths in the serum treatment.

Despite the amount of money spent and the effort put into the campaign, insufficient control was exercised over transport riders. It was pointed out that until the outbreak, hundreds of tons of forage and farm produce were brought to Ladysmith from Harrismith where the disease was spreading in every direction, and sent throughout the Colony. After it appeared, in 3 months it had spread over almost two thirds of the country, taking aback the authorities by its rapidity of spread (Figure 2). Of the 26 administrative divisions, 24 were infected between July 16 and October 30. Losses of African cattle, an estimated 90% of herds, were double that of Europeans estimated at 40–65%.

One observer reported near Greytown: "Cattle died in thousands, and as all sick animals made for water, streams were soon choked with rotting carcasses. The stench was terrific, and the bushes fringing the small streams flowing down the Greytown Hill were covered with a canopy of blue bottle flies in millions" (Lugg, 1970).

A stamping-out policy was pursued, slaughtering all infected and suspect animals, the majority being in the African locations. Because of the discontent this policy aroused, within

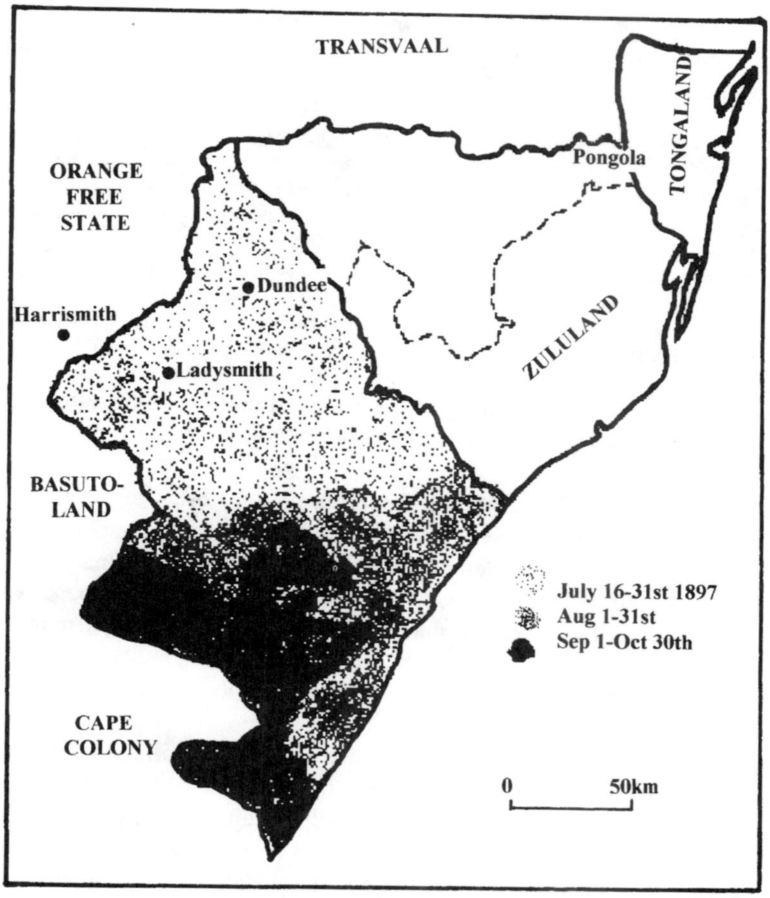

Figure 2. The spread of rinderpest in Natal.

10 days bile inoculation was introduced in the western African locations, Government recommending halting slaughter. Owing to its over-commitment, the Department of Agriculture had to call upon the Native Affairs Department for assistance with the inoculation programme and also advertised for missionaries to assist. The Africans were at first suspicious of inoculation, especially as a certain level of mortality was associated with it, and as it was not compulsory there was a fear that some farmers would refuse to inoculate. To counteract doubts, the *Eastern Province Herald* published two testimonials from farmers in the Orange Free State who had used the bile method. The first lost one animal of 327 from the process, the second 44 of 315: "It all depends on the bile. If the bile is clean and in order, you cannot make a mistake," wrote one. The latter's Africans inoculated 140 animals and lost five (*Eastern Province Herald*, July 26, 1897). In September, serum inoculation replaced bile inoculation as it had a better success rate, but by the time extensive programmes were under way, the majority of African stock had been lost, and there was a scarcity of "salted" cattle from which to obtain the serum. One magistrate blamed the high African cattle losses on the unwillingness of Africans to have their stock inoculated, and yet another reported that the serum was distributed to European farmers, African cattle being left to their fate. The Commissioner of Agriculture in his *Annual Report* for 1897 admitted that many Africans, especially those in the remote areas, never gained access to the serum. But whereas Africans grazed their cattle

communally in large herds, which was conducive to the spread of rinderpest, European farmers kept theirs in smaller herds on individual farms, separated and mostly fenced.

A correspondent to the *Cape Times* in the Transkei wrote, however:

> Inoculation from the outset was looked upon with suspicion by the natives even by those more advanced and most inclined to be loyal to the Government. A policy of procrastination was generally adopted and a waiting game played, each kraal hesitating until the actual disease appeared, and then, when too late, having the residue of their stock inoculated… As the good results of inoculation began to be realised natives appeared very anxious to adopt it, and for a time it was felt that the entire population would follow the lead; but at this critical moment blood inoculation proved so unsatisfactory that the native… immediately drew back, and the great majority of them now roundly assert that inoculation in any form is a failure, and further, that the whiteman has simply invented the process for the sole purpose of destroying their cattle. (*Eastern Province Herald*, September 24, 1897)

Since the outbreak in one area, the magistrate had been untiring in his efforts to check the disease; living almost entirely in the saddle he had done little or no bench work, a reference presumably to Scully.

As Chief Veterinary Officer, Watkins-Pitchford bore the brunt of the public's blame for failure to stop the spread of the disease, usually occasioned by farmers not observing controls and precautions, while serum inoculation was conducted without veterinary advice. As with Hutcheon earlier, Watkins-Pitchford was deluged with "cures" from farmers, which he was ordered to investigate. All were secret, many written in Dutch, and all demanded payment, 12 wanting it in advance (Mares, 1990). One "old man" claimed a bunch of onions tied on the head would prevent the disease; perhaps he was familiar with some of the early literature on rinderpest prevention.

The Natal press at first hailed Watkins-Pitchford as "our brilliant young scientist" (he was 30), who, arriving in Natal in July 1896, officially replaced the Principal Veterinary Surgeon Dr. Wiltshire. But the Government, providing him with only one damaged microscope, a few old instruments, and some dozen bottles containing drugs, in a place screened off at the end of a corridor in the Pietermaritzburg City Hall, was now unwilling to take his advice, and he was attacked by his own Minister abetted by factions of the press and farmers.

Further political interference came when a clerk was appointed to his office to keep cattle records on the direct instructions of the Minister. Suspecting him, Watkins-Pitchford obtained access to a locked desk and discovered two sets of different records being kept. The secret records included the sale of salted cattle to the old man claiming the "onion cure," and another secret file showed that the clerk had been reporting or spying on Watkins-Pitchford for the Minister. On July 12, 1898, Watkins-Pitchford called the clerk and the Minister's personal secretary into his office in the Town Hall and confronted them with the documents. Both denied the Minister's involvement. That evening, the wing in which Watkins-Pitchford's office was situated was burnt down, together with all the records. The clerk was retired without any official reason being given. Watkins-Pitchford alleged that arson was supected by the police, but there was apparently no investigation. Rinderpest now being in decline, the opposition forced a motion of appreciation through the Natal House for Watkins-Pitchford's efforts, but the Prime Minister only grudgingly conceded that Government would "recognize the services of this gentleman when the time came" (Mares, 1990).

In October, the Natal farmers called a conference which passed a resolution calling upon Government to appoint a Commission "to enquire how the rinderpest got into the Colony and how it spread, and, why the 'excellent stringent measures' adopted collapsed." In response, Government produced a report which placed the blame on the enormity of the epizootic and inability of any human agency to contain it before serum inoculation was introduced. But the *Natal Witness* considered that nearly every step the Government had taken had been forced upon it.

3. SWAZILAND AND PONGOLA

In July 1897, the *Times of Swaziland* wrote at length on the responsibilities of the Pretoria government in the event of rinderpest entering the country, appealing for the greatest vigilance in watching the course of the disease (Phoofolo, 1993), but nothing was done, and when it did break out in August, it spread rapidly over the country such that the *Times of Swaziland* reported: "In no part of South Africa has the rinderpest appeared with so much virulence as in Swaziland and in no part of the continent will more difficulty be experienced in fighting it" (Anon., 1897b). Before reaching Swaziland, it had been rife in Mozambique, killing unknown numbers of cattle and wild animals. It was first reported there on April 10, 1896 at Chimoio, introduced from Rhodesia by transport drivers, and noted as having attacked buffalo and other game. Grogan and Sharp (1900) reported in 1898 that at the Muda river, it had made a clean sweep "and the only thing we saw in the place of the thousands of buffalo of three years ago, was a young eland …."

In the August 1897, annual report of the Pongola Game reserve, which lay on the extreme eastern border of the Transvaal between the north bank of the Pongola River and the Swaziland border, the warden reported that rinderpest had not yet broken out despite being in almost every other surrounding district, and it had not yet reached the game reserves in British Zululand which had the most game to be found in southern Africa, with a lot of blue wildebeest (Van Oordt, 1897). His comments attracted the interest of Dr. J. W. B. Gunning, Director of the Zoological Gardens and Museum in Pretoria, who wanted him to census the animals before the disease might break out and, if it did, determine whether only cloven-hoofed animals were affected, as there was great uncertainty on this point.

In a report dated February 16, 1898, the warden stated that the disease coming from Swaziland had broken out since his last report on December 28, 1897 to the Ministry of Internal Affairs. The outbreak was short lived, which he attributed to immediately burning any carcasses, and he did not consider it to have been serious, but after the annual grass fires, he found more dead animals than he had thought existed. About a dozen mature bull kudus had died, but most affected were mountain reedbuck, with some common reedbuck. Many mountain reedbuck carcasses were found near the river where they had come to drink, and many were taken by crocodiles. Impala seemed unaffected, as apparently also were waterbuck, red hartebeest, tsessebe, blue wildebeest, bushbuck, duiker, steinbok, and klipspringer. Warthog and bushpig, he assumed, must have died, although no carcasses were found, but plenty of bushpig died on bordering farms south of the Pongola River. The reserve seemed the last place in the Transvaal to be affected, and although only brief, it had not returned, although apparently it had broken out again here and there along the boundaries (Van Oordt, 1898a).

On March 11, he answered Gunning's questions. It was difficult to state whether the animals which died in most numbers were more susceptible to rinderpest than others, as it was not certain that rinderpest was always the cause of death. It seemed clear both there and among surrounding animals that the kudu was the most susceptible, together with the nyala, and similarly, the mountain reedbuck, which was quickly infected. Three instances of dead common reedbuck were known, but mostly it seemed unaffected. A farmer reported them unaffected on his farm despite the presence of rinderpest. The impala seemed little or completely unsusceptible. Bushbuck had not been found dead, although Africans claimed to have found an occasional one, and the same with duiker and steinbok. Waterbuck, blue wildebeest, hartebeest, and tsessebe had not been found dead anywhere. In Natal, in the game reserve on the other side of the Lebombo river, kudu and other species died near blue wildebeest, but he had not heard that any of the latter had died (Van Oordt, 1898b).

4. REASONS FOR SPREAD OF THE DISEASE

Africans were often blamed for spreading the disease wilfully from motives of hostility or jealousy, for there were frequent inexplicable jumps, such as one of 11 hours' horseback ride in the Republic. An African had been arrested on suspicion, believed to be spreading it deliberately, as he was found carrying some substance which resembled liver compounded and pressed and had a piece of meat. It was intended to test the items for rinderpest. The African refused to say where he had come from, but it was believed that he was from an infected area or Bechuanaland. Of wilful intent, the explanation of the Under State Secretary for the Republic Mr. Krogh was: "Well you know the jealousies among Kaffirs. We believe that some big chief, seeing his own cattle dying off, while another chief has all his sound, cannot bear to see this, and so proceeds to infect the other's cattle ... he cannot bear to see another flourish while he himself goes to ruin" (Krogh, 1896).

But Hutcheon doubted whether Africans spread it wilfully, there were only one or two cases of spite to near neighbors, and he was satisfied that there was no evidence of passing the disease maliciously to European farms. The principal mode of spread was through ignorance and stupidity on the part of Africans. When an outbreak occurred and their cattle were threatened to be slaughtered, owners would try to move them away, thus spreading infection. In the case of the Mosita reserve outbreak, the area was almost fenced in when it was believed that someone, fearing he would be fenced outside away from his friends, rushed in with his cattle at the last moment, bringing the disease inside (Hutcheon, 1896a). One case of deliberate spreading in the Orange Free State was reported by the *Eastern Province Herald* of January 22, 1897, a rinderpest guard, Jacob Snaif, being sentenced to 5 years' hard labor and 100 lashes.

In 1899, there was an epidemic of stock theft in some areas of the Cape, and this could have spread disease. In British Bechuanaland, it was reported that most were thefts of one or two animals due to hunger. Two African women were arrested for stealing and eating a young donkey.

As Van Onselen (1972) has shown, despite a warning issued as early as 1893 by Hutcheon of the consequences if proper measures were not taken, the Cape Government took few, if any, precautions against the disease, and when action was taken, it was frequently jeopardized by lack of cooperation. Public hostility to slaughter policies and a lack of agreement on costs of compensation, as in Britain in 1865, were only two of the factors. Whites believed that Africans were spreading it deliberately, whereas the latter believed that it was Whites. Both groups did indeed aid its spread by movement of infected cattle, Whites especially continuing to maintain trade in cattle and often replenishing neighbors' losses. Hutcheon (1896a) stated that it was "most disheartening," and there was a continual cry to move cattle from here to there. Experience in Britain had shown that if every man guarded his own cattle and kept them from all contact with man or beast, the outbreak would be checked. Stating this at the second Vryburg conference and that the best thing would be to stop all cattle movement until the disease had been arrested, the Chairman simply moved on without reply to the subject of fencing details.

When the disease was in British Bechuanaland, the first aim was to stop it at the Molopo river, but this was found to be impossible because so many cattle drank there, coming from long distances (the Molopo is now a sand river). The disease prevailed for 3 months on the Mafikeng reach before working its way down on the Bechuanaland side, evidently now moving much more slowly. A belt of waterless land between the Molopo and Mafikeng native reserves, and the northern part of Vryburg district at the Transvaal border, was strictly patrolled. One outbreak affecting eight cattle in Setlagoli was quickly stamped out. One man moving down the Molopo gave rise to an outbreak at Mosita. Then, the old man's sons brought it into the Setlagoli reserve, and it soon spread. In almost every instance at the early stage of an outbreak, it was known exactly how the disease had been brought; for example, in the first outbreak at Sunnyside, among three kraals not 100 m apart, the kraals were destroyed. The neighboring kraals, 200 m away, were placed in quarantine for a couple of months and then restrictions lifted, the disease not appearing. Hutcheon did not believe that infection was carried by flies, despite a newspaper suggestion to that effect.

When the disease appeared in the northern Transvaal, the whole country was placed under quarantine for 1 month. Burgers were called up and paid to patrol a dividing line from south of Rooigrond in the west through Zoutspansberg. There were 300 such burgers in Rustenburg alone, each paid 10s. a day. All infected cattle were slaughtered and burnt, sometimes buried. A patrol also went about burying dead game. It appeared that the disease was being contained, and a brushwood fence was constructed, backed up by a wire one, guards patrolling between the two. Shortly after this, there was an outbreak in Rustenburg, the cattle being promptly destroyed and burnt. From the Crocodile river eastwards in Rustenberg district, cattle were removed 11 km from either side of the line and guards ordered to shoot every animal other than ostriches which came within 5 km of the outer line. Ostriches, of considerable commercial importance, were not believed to be capable of spreading the disease. Every person coming from the infected side was obliged to be cleansed. It was further proposed to move all farms adjoining the line further away. Once the disease had stopped, the infected areas would be burnt.

Despite these precautions, infection still broke out here and there beyond the line, so the belief was that it was caused by Africans wishing to destroy the Boer cattle.

A problem with fencing, for example along the Transvaal–Bechuanaland Protectorate border, was that often a European farm straddled it, and a farmer could be cut off from his stock. Hutcheon considered that, in conjunction with fencing, no centers of infection should be left. This was attempted at first, but in dealing with African cattle, the bill for compensation became more than the Government of the Cape was prepared to pay. "... if we had the money and the courage to stamp the disease out the moment it appeared in a herd of cattle by destroying it, there would be no infection to be carried" (Hutcheon, 1896a). In the Setlagoli reserve, it was pointed out that over 36,000 head of cattle drank at three points coming from up to 16 km away each day, and the Government hesitated to kill such a number.

In the Transvaal, a distinction was made between Africans coming from infected areas and those from uninfected areas. Subject to pass law control, an African coming from an infected area was dipped, and everything he was wearing washed with soap and hot water and Jeyes's fluid. Any milk, meat, or cattle products in his possession were burnt. After being washed, he had to remain until dry, and then his pass was marked and he could continue. Any African found in an uninfected area without a pass was taken before a Rinderpest Commission member or field-cornet, who disinfected him and then sent him to the public prosecutor where he could be fined £5, or given 3 months' imprisonment, and his pass endorsed. Where rinderpest had passed, there was generally famine, and 64 wagons were continually taking maize to such areas. This had to be paid for, so the Africans had to be allowed to leave the areas to go to the mines for work. Stopping all intercourse with infected areas would mean starvation, but the ox wagons were a weak link in the strategy.

On the borders of the Orange Free State, they took anything off Africans which might carry infection and burnt it. There was no pass law in Cape Colony, but it was considered that action could be taken under Act 22 of 1896, Section 7, which empowered the Governor to make regulations by proclamation.

Hutcheon considered the cause of its spread to be principally through African servants. It was next to impossible to prevent them visiting infected kraals where cattle were dying to obtain some of the flesh, and their soiled hands and dirty clothes became very suitable media for retaining and communicating the infection: "There is abundant evidence to shew that natives were the main cause of transporting the infection. In a large number of fresh outbreaks the first animals that became affected were suckling calves, the infection evidently being carried by the native milkers" (Hutcheon, 1902).

The Boers likewise were a problem in spreading the disease, for as soon as it broke out on a farm, those all around would jump on their horses and gather at the farm to inspect the disease: "They go into the kraal, handle the sick animal, and try their remedies on it, such as garlic and Cooper's Dip; and after assisting the owner in dosing the animal, go back home and carry the disease back to their own kraals" (Crosbie, 1896).

Hutcheon paid tribute to his staff, some of whom were from time to time in the saddle for 48 hr at a stretch subject to all sorts of discomforts including sleeping on the open veld with nothing but a blanket to cover them. He and Soga in April 1896 had spent much time along the borders of Cape Colony slaughtering cattle, themselves in the saddle for 48 hr at a time, and were frequently threatened with death by incensed Africans. Soga later, when assisting in inoculation, according to Scully (1913) suffered from stress. Soga

> worked in a most self-sacrificing manner. ... But after a time his mind seemed to give way under the strain. He took to drink and began obstructing the work. When I took him to task he threatened to shoot himself... Subsequently he took to wandering about among the kraals, raking up old feuds and talking to the people about war. I put a stop to this, and he sent a telegram to the Secretary for Native Affairs saying that I was grossly mismanaging things and that the district was being ruined... I sent him over the border into the Colony.

In 1907, Theiler, who was Swiss-born and had worked for the Boer Government during the Boer war, at one time providing information which led to the successful shelling of a British troop advance, was honored with a CMG by the King, whereas Hutcheon, who had done so much in the battle against rinderpest, died virtually penniless and unrecognized by any body other than the RCVS. In 1912, Watkins-Pitchford resigned his government post rather than work under Theiler at the now central laboratory of Onderstepoort, returning to veterinary service in the British Army.

5. STRANGE "CURES" AND STRANGER BELIEFS

The "cure" of the Acting Administrator of Salisbury, who convinced the High Commissioner of its worth, was a dose of raw linseed oil accompanied by keeping the animals away from water:

> I am pleased to inform your Excellency that we think a cure has been discovered for the cattle disease now prevalent here. The treatment is to keep affected cattle without water for five days at least, dosing them every other day during that period with one dose of three-quarters of a bottle of raw linseed oil, and at the expiration of the five days giving them as little water as possible for a few days. This treatment has resulted in one instance in the cure of eight oxen, and in another instance one ox has been cured. In the case of these eight oxen cured they were kept away from water for at least seven days, and the cattle inspector who effected these cures confidently states that if this treatment is properly carried out, it is bound to cure the oxen affected. It has further been discovered that cattle, which, although surrounded by infected cattle, on being treated as follows have not contracted the disease, *viz*:- their mouths are washed with a mixture of paraffin and salt, and an aperient administered, and Stockholm tar daily applied to the nostrils... (Anon., 1896a)

A rumor went around that paraffin was a safe cure, another potassium permanganate. Then a Mr. Prentice of Palapye claimed that potassium permanganate saved 98% of those treated. This was followed by a Mr. De Kock of Mafikeng, who claimed success with copper sulfate solution. Hutcheon's office was flooded with "secret" remedies which the authors wished tested, and a reward if they succeeded. Many suggested remedies were sent by persons who had never seen rinderpest and knew nothing about it. In the Transvaal, they tried the linseed oil, salt, and paraffin treatment, losing 97% of cattle. With potassium permanganate only 1% survived. They also tried an Indian remedy of boiling an ox which had died of rinderpest and feeding the soup to infected animals, a remedy allegedly supported by Joseph Chamberlain (Burman, 1971). The Reverend Zimmerman advocated skinning a sick ox, placing salt in the skin and surrounding it with a low fire for 24 hr. The idea was to drive the virus from the skin into the salt, three quarters of a cupful being given to infected animals with the result that not one survived, although some officials reported favorably upon it. Hutcheon stated that the most amusing thing about this remedy was that the salt was supposed to go straight to the stomach without the animal smelling it, so its nose was covered with a cloth moistened in a solution of carbolic acid while the salt was placed in the mouth. Hutcheon pointed out that the latter was the most effective way of making the animal breathe through its mouth, and the virus would be destroyed by the heating process.

At a mission station "somewhere near Victoria Nyanza," the missionaries dosed their cattle with quinine, and 9 in 10 survived, but Hutcheon thought that quinine would be no more effective than potassium permanganate. Another favorite preventive was to inoculate with a mixture of garlic and Stockholm tar; then it was rumored that animals salted for redwater fever were immune to rinderpest, and the Premier, Sir Gordon Sprigg, arranged for 20 such animals to be sent to the Taungs reserve. All eventually became infected, and all except two died. Then the Agent-General for Cape Colony in London sent out "Heath's remedy." Tried on some cattle in the Setlagoli reserve, the cattle suffered heavily. Mr. Heath later complained that his remedy had not been given a fair chance.

"William Cooper and Nephew" publicly proclaimed in the Transvaal that their powder containing sulfur and arsenic killed germs and would "absolutely prevent rinderpest." Theiler, like Hutcheon, had to waste time testing "cures." The Government Secretariat ordered Theiler to visit the tribal lands in the Waterberg district where the disease was raging more fiercely than elsewhere and put the "soups" and potions to test. Provided with a mule-drawn wagon, Theiler spent 6 weeks from June 20, 1896 traveling and doctoring African cattle with the various "remedies." The authorities wanted all "known and proven" farmers' remedies tested, and when Watkins-Pitchford joined Theiler in July 1896, they had to waste more valuable time before they could begin vaccination trials. At first regarded with suspicion and then understanding by the Africans, no remedy was effective, and ultimately they objected to treatment, hoping that at least some animals might survive by chance (Figure 3).

About the same time, exaggerated reports of the benefits of inoculation circulated. The most popular method was to take a tablespoonful of blood from the heart of an animal that had died of the disease, one of fluid from the rumen, a like quantity from the abomasum and of bile, and scrapings from the duodenum. Some also took peritoneal fluid. These samples were mixed together and filtered, and the inoculate was ready for use. Inoculation was as practised for pleuropneumonia, making an incision in the tail, sometimes the dewlap, and inserting a piece of wool soaked in the liquid.

This inoculation practice had already begun in the Vryburg District before slaughtering was abandoned. Hutcheon pointed out that if the inoculate was applied immediately, it generally communicated the virus, producing a heavy mortality, but if kept 2 or 3 days, as was usually the case, the practitioners of inoculation at first journeying from the Transvaal, the solution was putrid, killing the virus. When inoculated, it caused septicemia, from which most cattle recovered after a few days. It was then claimed that the symptoms had been a mild attack of rinderpest, and the survivors were now salted. It could take some time, if ever, for

Figure 3. The rinderpest in South Africa: physicking cattle. *Illustrated London News*, May 30, 1896.

such cattle to be exposed to infection, and while much was heard of the alleged benefits, nothing was heard of the reverses. If, however, cattle were already infected when the putrid virus was introduced, the inoculation appeared to be decidedly beneficial, the induced septicemia probably through the inflammation which it set up somehow affecting the rinderpest virus. Hutcheon therefore advised farmers who were set on inoculation to wait until the disease appeared in their herds.

The farmers were so intent upon inoculation that Hutcheon considered they would have done so against the law, spreading the disease, as indeed some already had. On September 19, the *Bechuanaland News* reported that a great deal of excitement had been shown in the western Transvaal by reports that inoculation was a remedy in the same way as it was against lungsickness. D. J. Oosthuizen, who demanded a good sum before revealing his method (it was apparently the tail-inoculation method with a mixture of infective matter *vide supra*), had tried it on two or three farms with great success and guaranteed any herd that he had treated. At one, he inoculated 235 animals of which only 13 died. The Transvaal Government were so impressed that they appointed a Commission to investigate and meet on the border. It was later reported by representatives who attended from Vryburg that the treatment simply made healthy cattle sick, and instead of 30% recovering, as had been claimed, all the animals died. At the end of October, it was reported that inoculated cattle were dying in hundreds at one location (Groot Chwaing). It had not produced any marked effects on Vryburg cattle, and opinions as to its efficiency or danger were much divided. Transvaal farmers were reported as now inoculating in the throat.

In the same month, a Transvaal farmer by the name of Uys claimed that 50–60 cattle of a herd of 320 survived because he had dosed them with tar and salt for some time before and then inoculated them. Visting Vryburg in mid-October, he claimed to be a member of the Schweizer Renecke Rinderpest Commission in the Transvaal and to have a telegram from

Dr. Leyds, State Secretary of the South African Republic, describing a successful inoculation, trial using his method. But following a request from the Bechuanaland Government to Pretoria for information on successful inoculation, the State Secretary replied on October 20 that, as yet, inoculation had no favorable result in the South African Republic.

Finally, the President of the South African Republic gave his approval to a mixture of Podophyllin,[1] resin, tobacco, paraffin oil, and a decoction of an incense plant. The result of publication of this remedy was that the chemists' supplies of Podophyllin were exhausted. Reporting this as "Oom Paul's 'Sure Cure'," the *Bechuanaland News* of October 31 asked, "What is 'the incense tree?' "

By February 1897, Hutcheon was to write that he was confident that Koch had found a reliable preventive remedy which was both effective and lasting (Hutcheon, 1897), but in November 1898, a Transvaal Government Notice (No. 543) published the use of snake poison as a cure, recommended by the JP for Belfast in the eastern Transvaal.

In Griqualand East, witchdoctors were reported busy spreading mischief. In one case, an African woman claimed to have died and risen again. During her sojourn in Heaven, it was revealed to her that the White man had in his possession two boxes designed to destroy the African. The first, which he had already opened, contained rinderpest. The second, shortly to be opened, contained locusts with horses' teeth. Another, a man, claimed a similar reincarnation, and his mission was to inform the people that when the rinderpest had passed, a disease was coming among the people. In order to arrest it, inoculation must be adopted for which purposes, to produce bile, an infant must be contributed from each village. The revelations of locusts and human disease were probably based on previous experience of such events accompanying cattle mortalities associated with droughts. Another superstition was that the White man had poisoned the salt and sugar, and many natives would neither purchase nor receive as gifts salt or sugar from the traders. In the Transkei and Pondoland before the epizootic took hold, a spate of wild rumors and prophesies was reported. Stories of persons, animals, birds, "and even stones" prophesied evils and wonderful things to come (Van Onselen, 1972).

6. THE FINAL OUTBREAKS

In the Transvaal, proclamations continued to be issued. On February 4, 1898, the importation of second-hand clothing, ox tails, and raw hides, unless disinfected, was prohibited, and on March 7, regulations were promulgated, restricting importation from adjoining states of cattle suffering from communicable diseases, which involved quarantine at the expense of the owner, but there was little or no compulsion due to a lack of supervision. With the decline of rinderpest, game was protected for long periods in several regions such as the Pretoria Townlands, Groenkloof, Rustenburg, Waterberg, Marico, Belfast Town, and Wakkerstroom Town.

On October 11, 1899, the second Boer war was declared, nullifying any regulations in force restricting movement of oxen or stock, and there were also unsubstantiated reports of deliberate spreading of rinderpest by the Boers, but the war itself overshadowed the disease. Isolated cases probably still existed in 1900, and in July of that year, rinderpest was apparently encountered on the Zeerust road among cattle brought from Bechuanaland, while several suspected cases were observed on postmortem examination at Pretoria in August. From October 1899 to the end of February 1900, Watkins-Pitchford was besieged in Ladysmith and

[1] Podophyllin—a bitter catharctic resin obtained from the dried rhizome of *Podophyllum peltatum*, a North American plant of the buttercup family.

unable to direct anti-rinderpest measures in Natal. Contemporary reports following the progress of the war make little mention of the disease, and although horses and mules were the main means of transport, ox wagons were still used. One private soldier wrote home on February 28, 1900 from Modder River Station: "The country is not worth twopence to any one ... the land stinking everywhere with dead cattle ... " (Friend, 1900).

The last case of rinderpest in Cape Colony had been in August 1899, but on March 3, 1901, known to be present as early as June 1900, it was reported to have become epizootic in German South West Africa, and British soldiers were sent to guard the drifts across the Orange River. In May 1901, it broke out at Mafikeng, and a disease also broke out among cattle near Maseru in Basutoland [Lesotho], General Kitchener sending for Turner to investigate it. Turner requested that he be accompanied by Theiler as he was not confident of diagnosis if it was other than rinderpest. Permission was granted by Kitchener and, traveling by train, coach and Cape cart, they reported at all outposts and were searched. They immediately diagnosed rinderpest, and a total of 40,000 animals was inoculated, with the bile method incurring only a small loss, no serum being available from the Cape as Hutcheon had been supplying Egypt. Turner was directed by Kitchener to start a serum station forthwith at Pretoria at Theiler's laboratory, but it was many months before an adequate supply was available. Immediately afterwards, the disease was discovered in military transport oxen at Ladybrand in the Orange River Colony. From these foci, it spread to numerous other centers in the Transvaal, Natal, Zululuand, the Protectorate, British Bechuanaland, and Cape Colony, mainly due to movements of military transport oxen and the capture and collection of cattle. By October, it had reached Pretoria. Borthwick, acting in the absence of Hutcheon, was preparing serum at new Cape stations at Aliwal North and Kimberley, while Turner, Theiler, and Phillips worked at Daspoort, without which the British army's drive to corner the Boer guerillas might have been halted, depending as it did on trek oxen. It seriously affected them first in August 1901, when 16,200 died of 95,700 (17%). In September about 18% died, in October 17%, and in November and December about 12%, a total of 67,500 deaths (Gutsche, 1979).

From June 1901 to February 1902, Hutcheon was absent on leave in England, returning to find many parts of the country once more overrun with rinderpest. The rest of the year was spent trying to eradicate it but was hampered by the military activities until peace was declared in May. By this time, the disease had extended over a considerable area of the frontier districts of the Colony, into Griqualand West, British Bechuanaland, Griqualand East, and the Transkei.

To assist with its suppression and several other diseases which were causing widespread problems, six additional veterinary staff had been sent out from England. One reported how, "due to the vigilance of the Boers, who were constantly moving about amongst the Natives in the [Taungs] Reserve, we were unable to carry out sytematic inoculation." In June, it was introduced among cattle on the Vryburg Commonage by Lord Methuen's transport oxen where the grazing was in poor condition as there were about 1,000 head of cattle driven there for protection by neighboring farmers and traders. Although inoculated, a number died, the veterinarian believing that this was due to debility, but probably the serum dosage he was using was too small. The same veterinarian noted that the disease did not spread to the African cattle; the reason was probably that they had gained immunity from the previous outbreak.

Hutcheon (1903) stated that the only known center of infection which had been left after 1899 was in Damaraland, German South West Africa, and it was difficult to explain how it had arrived in the Orange River Colony without leaving traces along the route, but he considered that an origin from Damaraland was the only rational explanation. Neither Hutcheon nor almost any others, both during the previous epizootic and now, ever attributed its spread to wild game, later to become the favored scapegoat to account for maintenance and spread of the disease. At a conference on animal disease held at Bloemfontein from December 3 to 5, 1903, when Gray of Rhodesia moved that "game carrying infection be destroyed to obviate this infection," the motion was not carried. Koch had raised the possibility of various species of antelope becoming infected and carrying the disease some distance leading to isolated outbreaks, although he also considered that it might be due to sheep and goats but had not been able to continue his experiments. It could have been maintained in a benign form in sheep and/or goats and may not have derived from the Damaraland focus. But that it first appeared among military transport oxen must lead to the strong suspicion (not voiced by Hutcheon) that they were deliberately infected by Boers in order to disrupt the British army's transport system. Hutcheon had been informed that it broke out in Basutoland on a farm, the owner of which had been to German South West Africa 3 or 4 weeks previously, and the local people attributed the outbreak to him. Herr Von Lindequist, representing German South West Africa at the conference, considered this improbable because at the time of the war, it was nearly impossible for anyone to come from their territory through Bechuanaland to the Orange River Colony. Had the farmer been an agent, however, it was not so impossible, and the Germans would naturally want to disarm all suggestion of complicity with the Boer effort.

The large majority of transport oxen would have had some immunity left from the previous epizootic, and the veterinary assistants who had been dealing with the outbreak in Hutcheon's absence found a serum inoculation of 20 cc dose sufficient to protect them. But when the same dose was used on farms with cattle born since the previous outbreak, such a small dose was ineffective, and it took a considerable amount of tact and persuasion after Hutcheon's return to induce White farmers and Africans to readopt serum inoculation, using a much larger dose of not less than 100 cc.

The last serious outbreak in southern Africa occurred in the Protectorate in 1901, introduced with cattle brought from Waterberg. The Acting Magistrate at Tsau in Ngamiland received a copy of the Proclamation in May and sent a copy on to Gobabis in Damaraland, pointing out that all stock movements into the Protectorate, including sheep, goats, and horses, were forbidden. The Anglo-German boundary survey operating along the Protectorate's western border at this time reported no signs of the disease.

Its spread in the Protectorate had been principally due to a scarcity of water, large herds coming from long distances to drink communally at the available supplies. That, as far as is known, it did not spread into the vast herds of wild ungulates in the Kalahari Desert, can be attributable to their relative isolation due to a scarcity of water in the desert, although many drank at the Molopo where cattle also watered.

Experience of localizing the disease in Britain may have influenced the decision to try to check its southward march by a double fence along the western and northern borders of Cape Colony and the Transvaal border, but the disease passed sections of the fence line before it was completed. Although fencing failed as a method of control, it initiated a culture of veterinary fencing in southern Africa actively pursued to this day, with disastrous results for wild animals in Botswana, causing huge losses by limiting migration and preventing access to water in times of drought.

The strain was so virulent that in the end, it more or less burnt itself out, despite vast populations of susceptible game animals in the Kalahari. Aided by vaccination and movement restrictions on stock, it was largely controlled in South Africa by 1898, and by 1903 it was finally eradicated in the Cape. The last South African case was at Gopani Location in the Marico District on August 30, 1903, 7 years 5 months after its first appearance there.

At the beginning of December 1903, Hutcheon convened a conference at Bloemfontein on disease among cattle, with primary emphasis on rinderpest and its final eradication, coordinating legislation and preventive methods throughout the southern African colonies and territories. It was attended by delegates from Cape Colony, Natal, Orange River Colony, Southern Rhodesia, Basutoland, the Protectorate, German South West Africa, Portuguese East Africa, and the Transvaal. Koch attended the early stages. By the beginning of December, the disease had been stamped out in the Transvaal, remaining only in Natal and German South West Africa, in Natal breaking out in isolation with no apparent explanation. Until the 1903 outbreaks, cattle had been exported from the southern part of German South West Africa to Cape Colony by agreement, and no case of rinderpest had been transmitted. The Transvaal, however, prohibited entry of cattle but had agreed to allow it consequent upon the posting of a veterinarian in the Protectorate to examine all cattle coming through. On the German side, all export cattle had to be quarantined first for 21 days and a veterinarian stationed near the border supervised this. Because of drought, cattle exports had been suspended and the Bechuanaland veterinarian withdrawn, but the Germans hoped that the import restrictions would be lifted after the last case of rinderpest had been extinguished.

In March 1904, the disease, believed to be still existing in Zululand, reappeared in two districts in Natal where it was quickly contained. It was reported at the same time at Windhoek in German South West Africa, the Herero rebellion in the latter area disrupting the authorities' final push there against the disease.

Contrary to the rest of the continent, where it remained enzootic, it was finally eradicated throughout South Africa in 1905, never to reappear, although there were vast quantities of susceptible wild ungulates in the region. Anticipating events in Tanganyika in World War I, on November 1, 1913 as a precautionary measure, the Protectorate passed regulations for the control of rinderpest (High Commissioner's Notice No. 58 of 1913). These regulations controlled movement of cattle and products in the event of an outbreak, disinfection, and inoculation under the control of the Resident Commissioner. Inoculation without the authority of the Resident Commissioner was forbidden. In the event, the southward spread was halted in Northern Rhodesia.

In 1900–02, the island of Réunion was affected, allegedly by forage imported from southern Africa, but it is difficult to reconcile this supposed source of infection with the fact that forage is unlikely to remain infective for longer than 36 hr, and a clandestine infected sheep or goat import seems a more likely explanation.

7. ANGOLA AND SAO TOMÉ

It has sometimes been stated (e.g., Ford, 1971) that rinderpest did not enter Angola, Fox and Varian (1911) being quoted as the authority. What these authors actually wrote was that it did not reach as far as the districts inhabited by the kudu near to the coast, "the rocky country near the port of Lobito," the range of the kudu extending to about 270 km on the railway line from Lobito. Thus, it seems that only a small part of Angola was actually unaffected, although it did not reach western Angola until late 1897 or early 1898. Mendes (1998) has shown that it was equally devastating over most of Angola as elsewhere, spreading also to the Island of Sao Tomé.

In September 1897, the Angola Official Bulletin (No. 36 of September 4) issued instructions on vaccination against the cattle plague by Artur of Paiva, Intendent at Humpata, recommending Koch's bile method as the only effective means. Most cattle were owned by Africans, but there were more than 500 colonists on the southern plateau of Moçamedes in 1884, mainly Boers who used ox carts, who implemented the recommendations. The unhygienic manner in which vaccination was applied caused many accidental infections, particularly by those carrying out the vaccinations. But the first reference to the presence of the disease was on May 30, 1898, when the provincial gazette reported that it was spreading with great intensity in the Benguella and Moçamedes districts, and to prevent it spreading to the area north of the Cuanza River, the authorities were to use all necessary means to prevent cattle crossing, impounding any found crossing illegally. In April, the health bulletin had reported a great death toll among cattle in the coastal town of Ambriz, but the disease could not be identified. On July 2, the same bulletin reported the cattle plague deaths in Benguella and Moçamedes. In August, it was reported to be almost extinct in Moçamedes, and the following month, restriction on the export of cattle, sheep, or goats was lifted, only to be hurriedly replaced after a reappearance of the epizootic was reported in November. This was due to a lack of effective control of the territory and consequent lack of information. In December, the disease killed almost all the cattle in Luanda and its suburbs, and continued outbreaks were reported the following January.

Evidence of continuing presence is provided by the provincial gazette of February 1899 in which export by the port of Luanda of domestic and wild animals, alive or dead, and of their products and items considered by the animal health committee to be capable of carrying infection, was prohibited. The following March, the official bulletin carried Royal Gazette No. 71, which ordered the execution in Angola and Sao Tomé of measures recommended by the Animal Health Advisory Committee to prevent spread of the disease to other countries, adjacent islands, and the Metropolis, repeating the prohibition of export of wild and domestic animals and their products. The baggage of any person embarking who had been in contact with animals in the infected areas was to be disinfected. The authorities were required to disinfect sheds or places where infected animals had been.These measures were to cease when the disease had not been reported for 40 days. The order had been taken directly from Portuguese national regulations, and the "competent authority" required to institute the measures could not be defined. Neither did a Committee of Animal Health exist in Angola. However, after March 1899, there were no further references to the disease, and there was no reappearance.

On the offshore island of Sao Tomé, a cattle industry was practically inexistent, but oxen were the main means of transporting goods in the city especially for carrying them to the port for export. The government and local health authorities apparently never even considered the possibility that the disease might spread from the mainland, but toward the end of 1898, a number of cattle imported by a butcher rapidly died. Other sick animals were isolated from areas where unaffected animals drank, and their litter burnt. Heavy penalties were instituted for anyone contravening the orders. A number of autopsies was conducted without revealing the cause of death, but the rumen and intestines were not examined. Registered mortality was between 96 and 97%. Initially, only one or two animals died, but then complete herds; of one herd of 50 in the city, only one animal survived. By January 1899, of an estimated 2,000 head in the city, just over 100 remained. Those who had possessed 30–40 animals were left without any. Regardless of the existence of the disease in Angola, the mortalities caused great speculation, people fearing that they would be next to be infected. The administrator convened a meeting at the Governor's Residence with the President of the City Hall and the Hospital director at which the cattle importer was asked to give an account. He claimed that he was convinced the animals he had imported were free of disease. He knew that during the voyage, some had died, but that always happened as a consequence of ill treatment during shipment. The animals were always examined by the municipal doctor or his delegate and had always been found to be in a fit state for consumption. The Health Committee admitted that it was unable to diagnose the disease and on December 30 resolved to ask for the services of a veterinarian.

Dr. Antonio Roque da Silveira, who had experience of the disease in the Transvaal, arrived on February 3, 1899, his arrival coinciding with a decline of the epizootic. Given private lodging, he received no help or collaboration from officials. Occasionally, he received news of sick animals but was unable to confirm any cases and saw only cured animals or those recovering. On February 17, the government asked him to return as he had not seen any cases of the disease, but Roque ignored the request, deciding to continue. Making private inquiries, these did not confirm the official explanation of the introduction of the disease, the first cases probably occurring in July or August 1898. All agreed that the disease came with cattle from the Moçamedes district of Angola, many authorities allowing import/export traffic in animals, the authorities' indifference to regulations being due to pressure from exporters and merchants.

Mendes (1998) points out that unlike in Angola, the whole economy of the island was affected, harming the interests of the rich and influential businesses with disruption of the export trade following the lack of transport animals. This led to pressure for a railway and proper roads for the island. The Isle of Principe remained free of the disease.

Before the disease broke out in Angola, the authorities there showed an extraordinary apathy toward it. Roque had been invited to represent Angola at the Pretoria conference in May 1896 but received the invitation on the opening day, while the journey to Pretoria took almost 3 days. At the next meeting, held in Vryburg, no one attended from the Portuguese colonies. In 1899, Roque wrote that although the method of vaccination was known at the time of the Angolan outbreak, saving 75% of animals thus treated, it was not properly adopted by the Portuguese. They should not repeat what happened "and is happening" in Angola where some hundreds of thousands of animals had been dying, the majority of which could have been saved at the time of the outbreak by vaccination (Mendes, 1998). Dr. Joaquim Bothelo, a health officer in Moçamedes, at first tried isolation without result and then successfully vaccinated 241 oxen with defibrinated blood, testing their immunity by placing them with sick animals. But he was an exception, and the authorities resorted almost solely to administrative measures of control, without effect.

The interest of the Angolan outbreak is that the Zambesi seemingly initially proved an effective barrier in its upper as well as lower reaches. The panzootic did not apparently reach western Angola until May or July 1898 from the south with cattle from Damaraland imported into Moçamedes. A failure to spread westwards north of the Zambesi could perhaps be explained by a lack of any significant east–west trade routes in this region, the trade limited to foot porterage carrying ivory and beeswax, and formerly commerce in slaves. Livingstone (1857) noted that the region west of the Zambesi up to Angola was devoid of cattle, but they did well in the country around Cassange. He also found cattle among the Bassongo, while the district of Ambaca was famed for raising them. But Pinto, crossing eastwards further south in 1878 noted only some small, poor, and weakly bred cattle among the Bihé. Thus, transfer of infection by cattle westwards was probably almost non-existent until the settled regions close to the coast were encountered. It seems unlikely that infection traveled westwards through game animals, for significantly, despite large numbers of susceptible game animals in Angola, as far as is known rinderpest never reappeared there apart from a localized outbreak in slaughter cattle in 1972.

25

RINDERPEST IN AFRICA IN THE 20TH CENTURY

1. PLAGUE IN EGYPT

Despite the precautionary measures in place at the Alexandrian Quarantine Station controlling imports from Asia Minor and Russia, in June 1903 the disease was again introduced into Egypt with cattle from Asia Minor originating in the Baghdad district, affected with a mild form of the disease which did not display typical symptoms, thus escaping detection on import. It carried off more than a third of Egyptian cattle in less than 1 year, the outbreak lasting until the end of 1905. One estimate gives a loss of 500,000 cattle (Arloing, 1905). According to Littlewood (1905), between August 1903 and August 1904, 364,647 cattle and 72,227 buffalo were lost (these figures including slaughtered animals), representing 38% and 10% of the total numbers, respectively, losses continuing in 1905. No sheep or camels were affected. Whenever the disease appeared in a district, there was much indiscriminate slaughtering, butchers and dealers visiting villages and urging people to dispose of their animals before they were affected by the plague. Healthy animals worth £10 were sold for £3 or £4. Also, large numbers of recovered or "salted" animals were sold to abattoirs as they were weak for some time afterwards and unable to work. The greatest mortality was in the winter months, but a high rate in the summer in one area was inspected by Littlewood in 1904:

> ... a more overworked and underfed lot of cattle it would be difficult to find. In one day's hard riding I only saw one bullock receiving an allowance of beans; all the other cattle were eating the worst and dirtiest tibn [chopped grain] obtainable. The healthy cattle were skeletons, and those affected with cattle plague never had a fair chance to recover ... many were worked until they were so ill that they could hardly move ...

Controlling movement of animals, necessary to prevent spread of the disease, was a new form of legislation, and of such magnitude that the authorities were unable to cope with it, an army of trained officials being required, while the people at first did not understand or believe in its importance. After initial closure of markets, in September 1903, two were reopened as a tentative measure, followed by others. While not immediately apparent, later the mistake became obvious as the disease spread. Permission was also given to villages to consume the meat of slaughtered animals; this led to butchers visiting infected villages to buy the meat at reduced prices, then re-selling it in other, until then unaffected, villages.

At the beginning of the outbreak, only small amounts of serum were available for inoculation, but imported animals did not seem to respond to it as effectively as the local breed. This was also found in cattle imported from the Sudan in 1904, and later in animals from Cyprus, which exhibited signs of rinderpest after inoculation. Later, due to heavy losses caused by blood infected with other parasites after the serum-simultaneous method was introduced, villagers refused to allow their animals to be inoculated.

In 1911, a total of 1,450 animals was reported infected. Up to 1912, the serum alone method of inoculation was used but found to be a failure and replaced by the serum-simultaneous method. A general program of eradication was then interrupted by the 1914 war, allowing the disease to reestablish itself. In 1914, there were 1,348 deaths, including 166 from serum-simultaneous inoculation, 290,726 cattle being inoculated. The following year, there were 4,833 inoculations and the lowest number of deaths since 1903, 358 plus 110 in the Alexandrian quarantine pens among cattle imported from the Sudan. In 1916, this fell to 18 deaths with 104 cases among imports from the Sudan. But in 1917, it spread over almost the whole of Lower Egypt, and the death toll rose to 4,794, plus 627 cases among Sudan imports.

After the war, recovery of the situation was effected and in 1920–1, no deaths were reported from rinderpest while 41,676 serum-simultaneous inoculations were carried out. But imports from enzootic areas of the Sudan were to mean that a low, sporadic incidence, was still recorded in the 1980s. In early 1982, there was an outbreak spread by cattle imported from the southern Sudan, 4,000 animals dying before it was brought under control.

2. WEST AFRICA

Allegedly free of rinderpest since 1893, in 1913, an epizootic of an attenuated form in Tchad, supposedly spread from Darfur, killed only young animals and adults weakened by diseases such as coccidiosis. The year 1913 had been one of drought and famine, which caused the inhabitants of north and northeast Tchad to migrate to less affected regions in the south such as Baguirmi and Salamat, carrying the disease with them. From the beginning of 1914, the whole of the country was affected due to migratory movements of cattle herds in the drought, the epizootic ceasing as soon as the rains came. Of the zebu cattle, only 10%–30% died, young bulls being most affected, but on the islands of Lake Tchad, mortality ranged from 50% to 80%. Buffaloes, large antelopes, and warthogs were also affected, but giraffes and small antelopes appeared not. In March 1917, another epizootic crossed the frontier of Massalit at the same time as entering Zaghaoua, becoming enzootic, and it is reported as having killed buffalo, hartebeest, kudu, roan antelope, and warthog. Giraffe, oribi, duiker, and gazelles were said to show no signs of infection (Pécaud, 1924). The Fulani stated that western hartebeest and topi were killed at the beginning of the 1915–17 epizootic, and Lelwel's hartebeest in Tchad in 1917 (Curasson, 1932). Many cattle had been immunized in 1914, and it spread only slowly westwards, killing an estimated 200,000 animals in French Equatorial Africa compared with some 500,000 in 1914. According to Delpy (1931), from 1915 the disease became enzootic in French West Africa.

In 1918, it again reappeared from the Sudan in a virulent form at first restricted to Ouaddai, but migratory movements then spread it further. Reaching the Chari River at the end of the year, it continued through 1919, mortality being estimated at 70%, or 200,000 head.[1] From 1920 to 1923, the disease became enzootic, with isolated outbreaks appearing here and there and further epizootics in French Equatorial Africa in 1921 and Mayo-Kebi in Tchad in 1928. Then from 1928 to 1964, outbreaks were reported almost annually, peaking in 1937 and 1951; but total deaths were known only from 1953 to 1964. Delpy (1931) reported that in French West Africa, more than 80,000 animals died each year of rinderpest when the animals concentrated near the Niger river during May to July. The Tuareg nomads of Mali managed to keep their herds free of infection for 3 years until drought in 1929 forced them to the Niger

[1]In 1918, the French law of July 21, 1881 was applied to the French colonies.

river where they came into contact with infected herds. The mortality of those animals older than about 3 years was higher than that of new calves, indicating that the older animals had lost their immunity during the 3 years.

A country of almost 1.3 million km^2 occupied by nomadic pastoralists with an estimated 1 million cattle in 1919, Tchad was served by only 2 qualified veterinarians up to 1933, rising to 8 by 1939, and 29 after 1945. Therefore, although some vaccination was conducted between 1913 and 1933 with glycerinated bile and intramuscular injections of phenolic acid, this could be on a token scale only. At the end of 1933, a vaccine production laboratory was set up at Fort Lamy to make formalised vaccine, as had been used in West African French colonies by Curasson and Delpy since 1926, when serum production centres were set up at Bamako, Saint Louis, and Niamey. Two other centers were opened in Tchad in 1936 and 1938. The establishment of these laboratories enabled vaccination campaigns to be undertaken. René Malbrant, who was in charge of the Fort Lamy laboratory, experimented with other vaccines, but until 1950, formalised vaccine was used almost entirely. From 1950 to 1953, production was switched to formalised aluminized vaccine, by 1953 half of it being caprinized. It was from 1950 that vaccination coverage of yearling cattle was considered to be satisfactory, but the success of caprinized vaccine leading to minimal losses resulted in herdsmen losing interest in having their stock vaccinated, bringing in only a few young animals each year. In 1962, a coordinated campaign was launched, with cattle in the western half of Tchad being vaccinated each year and almost 2.5 million vaccinations being conducted between 1962 and 1963. The following year, the figure rose to over 2.7 million, and in the final year of the campaign of 1964–5, it exceeded 1.9 million (Thomé, 1964). But from 1970 into the 1980s, Tchad was to be once more without proper veterinary care due to internal troubles.

Nigeria was engulfed with major epizootics in 1913–14, and again in 1919–20. In 1915, it was reported as the most important disease, but mortality was sometimes very low, ranging from 5% to 50%, the latter being exceptional. Africans stated that this was the first outbreak since about 25 years ago, and despite several reports in the first half of the year, there were none in the latter part. Owners did not pay much attention to a disease when it affected only a small percentage of their herds, and as a result, the disease was probably being kept alive without being reported. Among the Fulani, the 1919–20 outbreak was known as *Docchal*, a remnant, because it left only remnants of herds, there being so many deaths that even the hyaenas did not eat the carcasses.

In French West Africa, after the 1891–2 outbreak, there had been no rinderpest until 1915, when it spread from the east into Niger. A German prisoner of war veterinary officer diagnosed it as rinderpest at Gaya in southwest Niger, and it was established that it had come from Tchad, passing through upper Nigeria. It continued westwards toward the coast with local trade cattle, crossing the Niger river but was halted for some time on the west bank by *cordons sanitaires*. In Niger, certain areas were declared infected, and as a result of measures taken, Upper Senegal, Mali, and Upper Volta were preserved from it until 1916, when Upper Volta was infected via the regions of Gao and Dori in the northeast, and from the right bank of the Niger river. It then spread throughout the country by the trade routes, and infected animals sent by rail to Kayes in western Mali soon spread infection to Senegal.

It appeared to be assuming a more virulent type, in Senegal up to 1919 destroying about four fifths of the country's cattle. The epizootic moved backwards from Mali to Guinea and the Ivory Coast in 1917, finally reaching Togo.

Thus, it spread first through the areas of dense sahel cattle populations (there were an estimated 6 million cattle in French West Africa in 1918; at Timbuctoo, large herds were kept by Tuareg tribes and Fulani) and then returned eastwards through the more sparsely populated forest–savanna mosaic herds (Ford, 1971) (Figure 1).

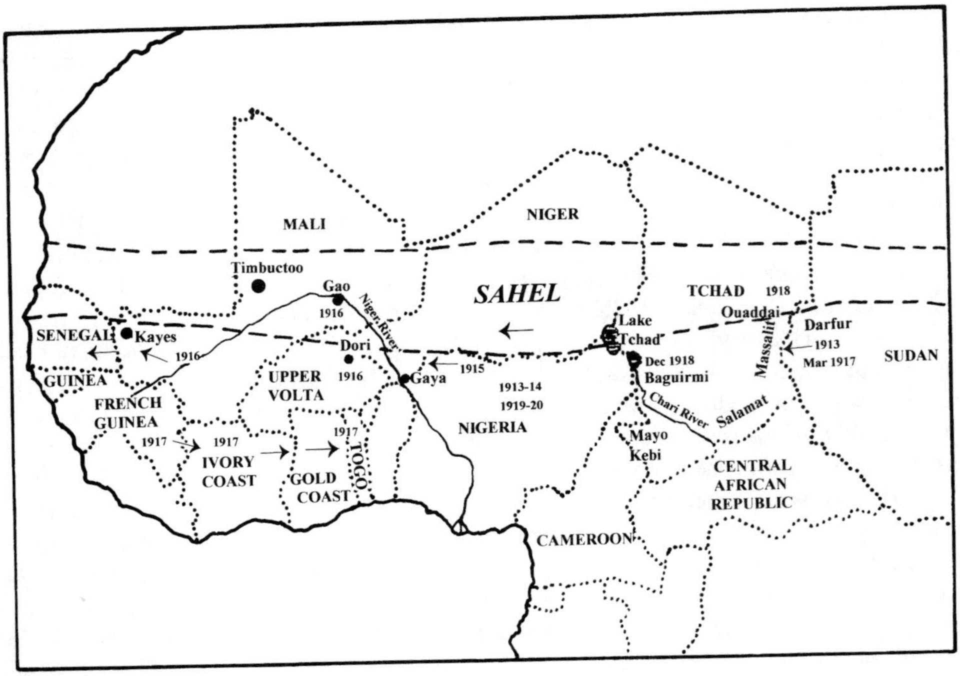

Figure 1. Rinderpest in western Africa, 1913–18.

In February 1925, rinderpest was reported from the Ouahigouya district almost in the center of Upper Volta killing 492 animals with 432 sick of a total of 4,521. In May, it continued to ravage the region, deaths totalling 732, and the Peulh herdsmen demanded vaccination for their cattle, but because of a lack of serum supplies, the administrator could only ask for patience. In June, he reported the disease as in regression, "as each year at the beginning of the winter." The number dying during June was 366, making a total of 3,093 declared victims to that date. A total of 3,337 animals had been vaccinated, but the following month, the rains having failed, the disease increased in intensity with the number of deaths rising to 688 for the month, declining in August with the onset of good rains, 150 animals dying. It continued to be reported from time to time until April 1928, when it was noted that 19 bottles of serum had been received, but 30 were required each month. In May 1927, the Administrator had noted that there was no serum available to vaccinate, but the Peulh practised vaccination according to their custom. Pleuropneumonia was present as well, and it is not specified which disease or diseases they vaccinated for. Rinderpest was not reported again until March 1932, when the region was declared totally infected, and people had difficulty paying their taxes (Marchal, 1980). St. Croix (1945) states that about 1935, an experiment was conducted in Yauri Divsion, Nigeria, a number of Fulani exposing their animals to cattle which had a mild infection causing no mortality, but some 4 years later, many of the cattle so exposed succumbed in another outbreak, and the experiment was not repeated.

According to Baier (1980), the major epizootics in Niger after 1913 (1914–15, 1916–17, 1919–20, 1921, and 1925) were a more serious check on the growth of cattle herds than the 1890s panzootic. Baier suggests that, according to oral sources in Damagaram (southern Niger), the former epizootic did not have the catastrophic effect upon the sahelian economy that Ford (1971), for example, suggests. Relative price data suggest that cattle were not particularly scarce by 1900. The epizootic was also possibly more severe in western northern Nigeria (Sokoto) than it was in Niger. Edwards (1930) noted that it was observed "some years ago" in West Africa that waves of rinderpest appeared with a somewhat regular periodicity in territories without artificial obstacles to its spread. According to Curasson (1932), Fulani cattle owners expected such waves every 25–30 years, always from the east, but there could not have been more than one before 1932, unless rinderpest was present in the continent prior to 1890, although the major epizootics are claimed as being in 1828, 1865, and 1892. As in the latter outbreak, the major part of the stock disappears, herds then being slowly rebuilt from the survivors.

In French Guinea, Claverie claimed in 1931 to have seen several cases of the spread of the disease by jealous or vengeful Foulahs, who threw a meaty bone from an infected carcass into the enclosures of healthy animals.

In Nigeria at Vom, there was an alleged outbreak in sheep in 1958 which spread to cattle (Johnson, 1958).[2] By 1983, the disease was again widespread in Nigeria, and outbreaks were reported in Niger, originating from the Sudan enzootic zone. It had also become enzootic in Upper Volta. It appeared also in Cameroon, some 1,500 cattle as well as antelope dying in the Waza National Park. An estimated 3,000 Buffon's kob, 80 topi, 40 reedbuck, and 10 warthog died, while roan, giraffe, and red-fronted gazelle were unaffected. The outbreak spread to the Adamoua region, which had not had vaccination since 1964, and by May, some 10,000 head of cattle had died. During 1983, buffalo, warthog, and giant eland were said to have died in the Benoué National Park following an incursion of infected cattle from the Central African Republic. Early the next year, buffalo, waterbuck, and warthog were affected in the Boubandjidah National Park, and buffalo died along the southern edge of the Faro Reserve near Dougoudougou.

3. SUDAN AND THE CENTRAL AFRICAN REPUBLIC

In 1903, rinderpest broke out in Eritrea and the following January passed to Kassala in the eastern Sudan, infected cattle from Eritrea apparently passing down the valleys and the river Atbara as in 1898. Four oxen were also reported to have died on the Nile at Mutmir about halfway between Berber and Khartoum, these having come from Kassala in December 1903 in a herd of 82. The herd had been driven over 400 km across the desert taking 30 days without any sign of sickness. An army veterinary surgeon, Lieutenant Head, was sent to investigate and, after isolating all those which had come over in this herd, slaughtered and autopsied 4 of 19 animals at Mutmir which were sick, but he was unable to detect any obvious signs of rinderpest. Four days later, an ox from the Nile, which had got mixed up with the sick herd, developed virulent rinderpest and was slaughtered together with all the contacts, totalling 17 animals. But in one village, the owners hid two oxen which were found by Head a week later, both suffering from rinderpest and both having spread infection to other cattle. At the same time, the disease broke out among local cattle in every village where any from the imported herd had gone, a distance stretching 160 km. Head therefore began a program of inoculation using the serum-simultaneous method after checking the blood for protozoal infections. Altogether, in this area, he inoculated 1,061 cattle with a loss of 7, or just over 2%. A total of 208 died from rinderpest, and 149 recovered. He then went on to Kassala to continue inoculation there, treating some 6,000 animals at Gedaref with the same level of loss.

[2]The confirmation of rinderpest in sheep was made by Dr. Nakamura using a dubious complement-fixation test. Hindsight favors the hypothesis that this was one of the earliest outbreaks of PPR in Nigeria. The cattle grazing the same paddock as the sheep were part of an assessment of the potency of a vaccine batch and had been challenged; the batch failed (Scott, personal communication).

He then proceeded to the Atbara where the plague was reported to be spreading. Some 25,000 cattle were examined as they came to water, and he inoculated about 5,000, the disease ceasing by mid-July. By mid-September, it ceased at Gedaref, and although a new infected herd managed to cross from Eritrea into Kassala, this was quickly controlled by inoculation. In total, 3,653 animals died of rinderpest, and 1,575 recovered. He reported that during the outbreak, one often found cattle owners sitting down watching their herds die and leaving the carcasses unburied saying that the plague was sent by Rubonah (God) as He was angry with them, so it was useless to do anything (Head, 1906).

The Dinka at Wau in the Sudan had thousands of cattle at the beginning of the century, but Landor reported that rinderpest had frequently played havoc with them, especially further east, although "the meat of these animals has on every occasion been greedily devoured." But Comyn (1911), writing about 1906, stated that at Dem Zubeir, it was said that 25 years ago, the lowing of kine was the predominant note along the watershed, "In our time not even sheep could live."

In February 1971, there was an outbreak among sheep and goats in the eastern Sudan near the Rahad and Dinder rivers causing 60–70% mortality among young animals and 20–25% among adult, but it did not spread to the wild animals in the Dinder National Park, 8 km to the south, and cattle in the vicinity were vaccinated. It is believed that this outbreak derived from goats brought in by Eritrean refugees. Cilli (1951) found that 16% of goats in Eritrea were completely resistant, supporting the hypothesis that the disease was enzootic in them. But whether the severity of the 1971 outbreak was caused by an increased intake of wild virus, or by a mutant strain developing in the refugees' sheep and goats, could not be determined (Ali, 1973).

In 1976, buffalo and eland were affected in the Southern National Park, and in the Kurun Hunting Block in 1979. In 1980–81, there was an outbreak among cattle between Nyany and Malakal, but no wild animals were seen affected. Although tiang, which were numerous in the area, seemed unaffected, 2 in 14 serum samples taken in 1982 showed neutralizing antibodies. Owing to an inability to maintain annual vaccination of young stock, the disease had spread countrywide, 20,000 head dying in the northern region up to June 1982. In 1983, there were 3,000 recorded deaths, and by August, the outbreak was apparently under control, but in 1983 and 1984, it was present in cattle at Wau, and there were unconfirmed reports of its presence in buffalo and giant eland north-west of Wau near to the border with the Central African Republic.

In the lesser-known Central African Republic, formerly French Equatorial Africa, although it must have mirrored the Chari region outbreaks in Tchad, rinderpest was reported first in 1939, then in 1946, 1956, and 1968, the latter two outbreaks apparently killing only game. The 1956 epizootic in the north of the country is alleged to have been very virulent, killing many adult buffalo. It probably had its origin in Tchad as the 1968 outbreak is known to have done, starting with a high mortality south of Haraze in December 1967. The first dead buffalo were seen on the Aouk-Aoukalé River in January 1968. Within the several weeks following, hundreds of animals died, especially buffalo and warthog, and possibly some giant eland and Buffon's kob. Local reports claimed that roan antelope, waterbuck, reedbuck, and duiker were also affected without always dying. In March 1969, seven giraffes were found dead where a number of buffalo had died, and on May 4 giant eland carcasses were found. In June 1970, a giant forest hog was found dead. The buffalo which succumbed were mostly aged up to 4 years, with few adults; but mortality was estimated at not more than 5–10%, or 10,000–20,000 head. Samples of 69.3% buffalo and 43.9% warthog were found to contain antibodies indicating exposure. Small samples of bushpig, giant forest hog, giant eland, hartebeest, topi, roan antelope, waterbuck, Buffon's kob, bushbuck, oribi, and duiker also showed positive reactions.

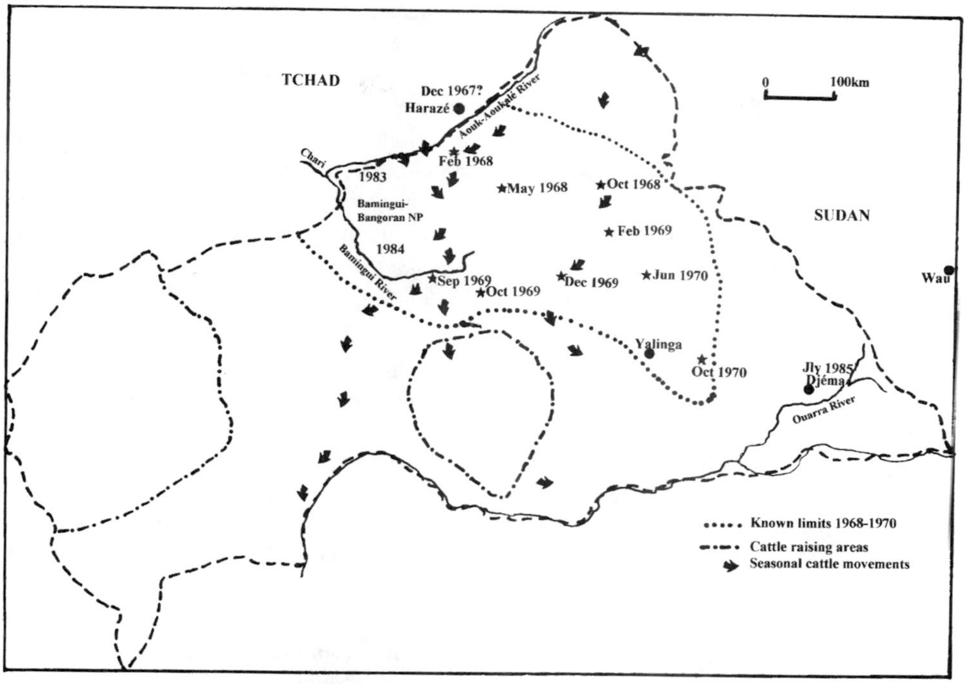

Figure 2. Rinderpest in the Central African Republic, 1968–85 (based in part on Thal, 1972).

In 2 years, this outbreak traveled a distance of 800 km from west to east and from north to south across the country, apparently following the rivers. But either it bypassed the cattle, which are mostly restricted to the north of the country because of the presence of tsetse fly, or it did not affect them (Thal, 1972).

Tchad was infected in 1983 from the Sudan, the disease spreading rapidly south, and nomad cattle from Tchad suffered severe mortality in the Central African Republic in 1983, some 2,000 dying. It was short-lived among the game in the north-west, mainly affecting female and young buffalo, losses being roughly estimated at 50%. In 1984, some giant eland, a few waterbuck, and warthog were seen to be affected. It then spread south along the stock route and eastwards into the Bamingui-Bangoran National Park, apparently increasing in virulence and badly affecting buffalo, eland, and warthog. In February 1984, it was killing young buffalo and warthog along the Bamingui River east of Bamingui, including some giant eland which seemed to be old males, and then continued rapidly east and south. In 1985, buffalo were reported dying on the Ouarra River in the south-east, the disease reaching Djéma in the east by July, buffalo and Lelwel's hartebeest being the species most affected. In 1984, it also broke out among cattle in the west but was brought under control by vaccination (Figure 2). In 2002, the western part of the country was declared provisionally free from rinderpest.

4. ETHIOPIA

In 1905–06, rinderpest was enzootic in most parts of Ethiopia, although Landor (1907) found in 1906 on the Baltchi escarpment east of Addis Ababa that the people had thousands of cattle, and around Buré in the west, while oxen were plentiful all over the country.

Zonchello (1917) reported that although the disease was prevalent in other parts of the Italian colony of Eritrea, a long strip of country bordering the Red Sea known as the Sahel district appeared to have been free of the disease until 1914, when it was introduced with cattle which had been first moved out and then back in again from the vicinity of an infected area. Following this, in the rainy season, there was a serious outbreak with over 95% mortality among the herds of the entirely pastoral population. Recourse was then made largely to raising sheep and goats, but in 1915 the disease broke out again and became enzootic. However, mortality progessively decreased until it rarely exceeded 50% and was sometimes nil. But local people reported eight different outbreaks between 1905 and 1914, with control attempted by serovaccination. In many instances, although results were satisfactory, treated animals created new centres of infection with a considerable spread. Failure to achieve better results appeared to be because the method was not rigorously or rationally effected, always carried out in a partial manner with the result that at least half the population of cattle remained susceptible. The British authorities in the Sudan protested against the use of the method on the Anglo-Italian frontier in the winter of 1914–15, accusing the Italians of maintaining and distributing the virus.

At the beginning of September 1918, there was a small outbreak among 135 Somali cattle at Bir in British Somaliland. By mid-October, 114 head had died. The source of infection was not discovered but was probably Eritrean cattle. The report stated that there was no authentic record of any previous outbreak in the country, but Africans spoke of isolated cases. Although other cattle were grazing the same area at the time of the outbreak, no others became infected. Movement of cattle was stopped and infected cattle isolated until the disease was brought under control (Adams, 1919). Maydon (1925), hunting in Ethiopia in 1924, reported in eastern Arusi, near the Webbe Shebeli river, one place where the ground was white with the bones of cattle due to a disease which practically wiped out the stock in the neighbourhood "some time ago" so that now villages appeared to barely exist.

5. UGANDA

Presumably, the Karimojong of northern Uganda replenished their herds after 1897 initially through raiding, although by 1910, they had recovered them by buying with ivory. In the south, Jack (1914) reported that sometime prior to 1907, thousands of cattle were paraded in Ankole. Melland and Cholmeley (1912) reported that in 1907, in the Lango country, large numbers of cattle were found. In 1908, the export of cattle from Uganda was prohibited, possibly to prevent the spread of rinderpest as much of the trade had been by Somali traders from Nairobi. A strong guard had been put in place to prevent trade animals crossing the Nile (Hattersley, 1908), but in 1911 an outbreak killed half of the stock of both Karimojong and Suk tribes.

In neighboring Belgian Congo, it was absent in 1911, when Roby (1911) reported that the government ranches, maintained for the purpose of feeding its soldiers, contained thousands of head of cattle.

The disease appeared in the Mengo district of Buganda Province in August 1913 and during the next 2 years extended into other districts. Oxen were used for transport in some areas, and Lorimer (1917) was shown a letter in 1913 from a transport clerk sending two bullock skins to the Director of the Uganda Transport in Kampala from animals which "were very sick," killed by lions near Fort Portal, Lorimer noting that the allusion was to the "great scourge of the country, rinderpest." On the road to Hoima, Lorimer remarked that rinderpest often killed off whole herds of the small humped cattle at a time, "so only the chiefs can afford to keep very many."

In 1915–16, an outbreak killed large numbers of Uganda kob. It appeared early in 1915 in the extreme east of Mengo District near the Nile and in May appeared in Bukoba, having spread from Bugerera. Between May and July, 5,007 cattle were inoculated in the Bukoba division, deaths among uninoculated cattle amounting to some 1,800. In July, it spread into Buruli, where buffalo and bushpig died in considerable numbers, and then continued rapidly into Bulemezi, inoculations being continued until October when outbreaks occurred simultaneously at other centers, probably due to illicit movement of cattle from infected areas in southern Bulemezi. In Eastern Province, the disease was prevalent since being introduced from Kenya in 1910, losses apparently confined mainly to young stock.

In 1915, 3,058 cows and 848 calves were inoculated by the serum-simultaneous method, with 196 and 210 deaths, respectively. In the early stages, potassium permanganate was used as a treatment in 41 animals, resulting in only 20% dying as against 70% of untreated animals. After March 1916, there were no outbreaks in cattle or game in Buganda, the disease eradicated by quarantine measures and inoculation. In 3 years, 12,000 uninoculated cattle are estimated to have died, and altogether, 23,000 were immunized, an estimated half to two thirds of the total number in the infected areas. In 1916–17, numerous outbreaks were reported in Eastern and Northern Provinces. In 1916, it struck the cattle in Karamoja again, but only 10% died, immunes presumably now being present in the population. The following years, 1917–18, it spread to Bunyoro, and all cattle within 32 km of the boundary were moved southwards. In the Eastern Province, numerous outbreaks continued, but bullock transport was maintained with immunized animals. It was reintroduced into Lango in June from the Northern Provinces, causing widespread losses but dying out by the end of December. This outbreak was allegedly traced to the importation by an Indian trader of some undried hides from the Gulu district. In Bunyoro, it was believed to have been introduced by game. In West Nile district, buffaloes were dying early in June near Pynodni on the west bank of the Nile, and it spread southwards causing the deaths of large numbers of game animals along the Kafu Valley, first reported among buffalo in the Chobe region whence it spread to the Kafu plains, killing warthog, bushbuck, duiker, kob, and reedbuck. Duke (1919) reported:

> The disease spread rapidly through the buffalo, which, as usual, migrated... Bushbuck and warthog were dying in numbers.
>
> As a direct result of the epidemic, the buffalo moved south along the Bugungu plain [Lake Albert]... They left a train of dead along their course, a striking feature of these bodies being that most of the skins remained intact and the bones were not generally pulled out by carnivores. Bodies of kob and warthog were also found.

Up to the end of March 1917, the epizootic in Bunyoro was confined to game in the tsetse fly areas but an outbreak passed through the Masindi Port country of the Northern Province. As well as cattle, buffalo, bushbuck, duiker, and warthog died in large numbers, the last three being the most noticeably affected as buffalo came and went according to local report. Waterbuck and hartebeest were little affected (Duke, 1919). An exceptional drought at the end of 1917 followed this outbreak, causing famine the following year.

It also spread west, appearing among game on the shores of Lake Albert near Butiaba in 1918, but in spite of the deaths of buffalo, in 1918 Duke found tracks of some 20–30 in the Lake Albert region: "Buffalo have survived in small numbers out of the numerous herds which existed throughout the district before the rinderpest. Kob and reedbuck also occur; bushbuck and pig [warthog] used to be common, but I saw none alive during my visit." Numbers of waterbuck had survived. The outbreak reached to within 45 km of Kampala, and Duke reported "... it is almost certain that, in their panic of the rinderpest, considerable numbers [of buffalo] must have trekked south into Buganda ... the main agent in this spread is beyond doubt the game, especially buffalo" (Duke, 1919). In Buganda Province, it appeared among game in Buruli and Singo early in April, having spread southwards through the game in Bunyoro since the previous July, crossing the Kafu River near to which buffalo, bushpig, bushbuck, Uganda kob, reedbuck, and oribi died in large numbers. Infected buffalo migrated westwards to Bugangadzi and southwards along the Myanja Valley. By the end of May, cattle in this area had become infected, in July spreading it eastwards to cattle in Bulemezi and in August to other areas. In September, it spread among buffalo along the Kitumbi River and later as far as those in Singo. It appeared to have spread south of a line between the Kafu and Katonga Rivers along the Kitumbi and Nabukazi rivers only with buffalo. In December, it crossed the Katonga River to Buddu and other areas, and from there was carried by buffalo to Toro in January 1919. In Singo, Bwekula, and Bugangadzi, it seems to have been spread mainly by game movements from infected to clean areas, as carcasses of game animals were usually found before the appearance of the disease in cattle. In Western Province, it appears to have been introduced in January by buffalo from Bwekula, large numbers of buffalo being found dead in the south-eastern part of Toro District in areas where they had not previously been known. In Buganda Province, an estimated 30,000 cattle died during the year due to the delay in reporting.

About August 1919, Roscoe encountered an epizootic in Ankole where it was "carrying off the cattle by the thousand." People were reduced to the last stages of famine through loss of milk, many dying from starvation, while others were wandering about looking for friends who might still have cattle, and yet others tried to live on a vegetable diet. But journeying toward Kigezi, sometimes he found large herds of cows and at one village recorded:

> These women were still fairly fat, though they had not nearly so many cows as formerly to supply them with milk, the plague of rinderpest having passed through the district and carried off a great number of their cattle. It was singular to find how some tracts of country had escaped that distressing plague, so that the cows were in good condition and the people comfortably well off, whereas in other places there was not a cow to be found.

At Kyagamba, he visited a Government inoculation camp. The herders were unwilling to have their cattle inoculated but obliged to comply. At Hoima, because the rinderpest had carried off many pigs (warthog?) and other wild animals, the lions were short of food and haunted the villages by night, even attacking stray people by day, so that the women were reluctant to go to the fields to work (Roscoe, 1922). In November, the border between the Belgian Congo and the Kigezi District of Uganda was closed on account of rinderpest and only the mails allowed across (Barnes, 1922) (Figure 3).

Ruanda-Urundi and the region of Lake Kivu were affected in 1920. In February, infection was introduced with two animals from a focus in Ankole by Bahima pastoralists north of the 10-m-wide Kagitumba River, to a village with 1,900 animals south of the river, where, by March 4, 20 had died and 288 were sick. Within 6 days, 2,800 animals on both sides of the river were infected or in contact. The disease was spread further south by porters who brought supplies and serum to the Belgian veterinary team combating the outbreak, probably by carrying back infected meat. Vaccination of all animals at risk, totalling 18,000 serum-simultaneous vaccinations, limited the spread, and by June 16, no further cases were reported. In the previously infected area, a total of 14,307 cattle remained of the former total of almost 16,000. A charge of 2 francs was levied on all animals brought in voluntarily but only requested if the animal was cured. In return, the authorities required animals were supplied free for serum production. In all, a total of 10,000 francs were collected. The game, numerous in this area, appeared unaffected (Carlier, 1920).

From the Lake Albert region, it probably passed through the area of today's Queen Elizabeth National Park and into the Belgian Congo's Albert National Park, where buffalo

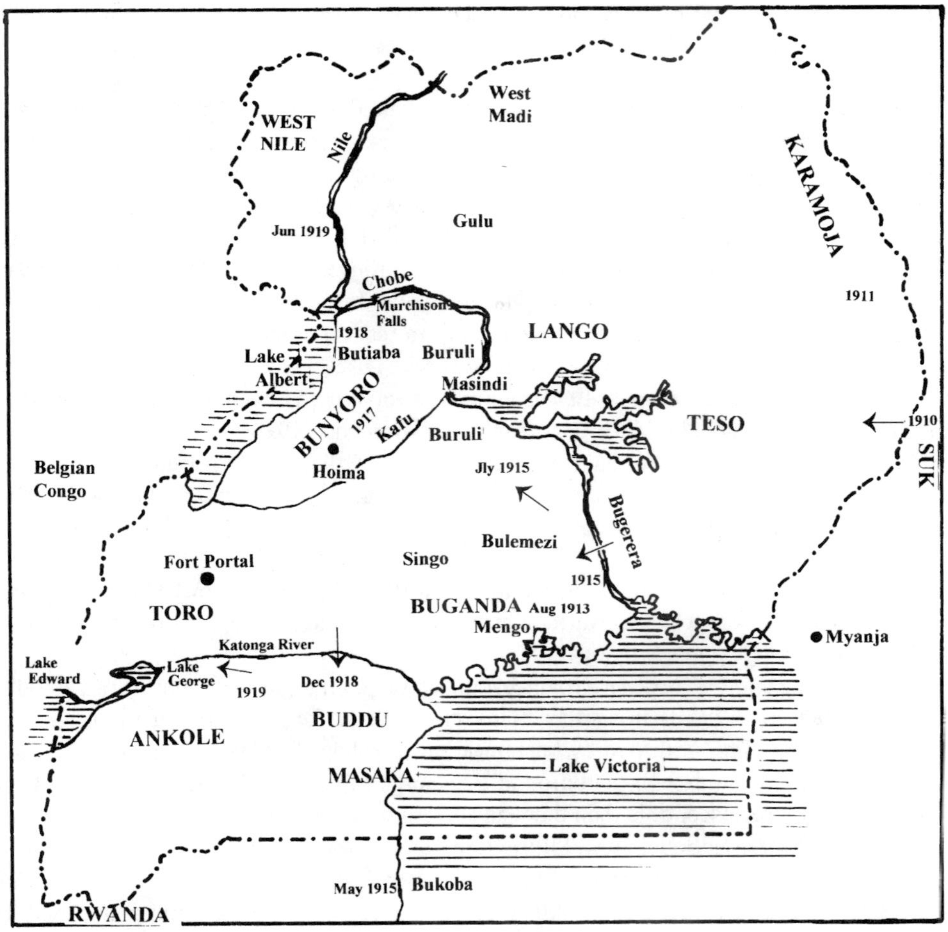

Figure 3. Rinderpest in Uganda, 1910–19.

were reported to be greatly affected in 1920–21. From 1918 to 1926, 250,000 cattle vaccinations were carried out in the Belgian Congo. In the latter year, there was an outbreak in Urundi.

Carlier (1920) laconically remarked that the disease remained enzootic in Uganda, and Simmons (1957) recounts how, beginning in 1919, the success of the serum-simultaneous method was due more to chance than skill and judgment in its use, for disaster might follow a break in the weather, bringing cold rains in the reaction period, or, conversely, no rain and a shortage of water and grazing could cause an unexpected movement of game carrying infection with it. As a result, after disastrous results in Western Province probably due to trypanosomiasis and the cattle being the highly susceptible Ankole stock, inoculation was halted, and all that the veterinary officers were allowed to do was confirm outbreaks of the disease and then quarantine the cattle in the area. This led to concealment of outbreaks as the Africans did not want their apparently healthy cattle moved to quarantine locations. After protest at the situation, Simmons was allowed to reintroduce trial serum-simultaneous inoculation in Busoga, but owners had to pay a fee of 4 s. for every animal released at the end of the 4–6-week detention. Following a mortality rate of 10–12%, confidence of the cattle owners was gained and inoculation extended to Teso and Lango Districts, where rinderpest was hampering the development of an increasing cotton industry as the crop was carried to the ginneries by ox wagon.

There was a further virulent epizootic in West Nile District in 1925 when kob were unaffected (Carmichael, 1938), and the beginning of the year witnessed a rapid extension in Teso district, but by the end of the year, all outbreaks in cattle had been eradicated. By 1926, the disease existed only in the Eastern Province on the Kenya border, the following year some 7,000 cattle dying compared with about 70,000 annually between 1912 and 1920 (Carmichael, 1973). But at the end of the year, a fresh wave came from the north, 150,000–180,000 cattle dying in the Eastern Province in 1928. The outbreak was checked by inoculation of some 160,000 cattle with serum produced in the field from recovered animals.

In 1926, there was another outbreak in West Nile District, estimated to have a population of 100,000 cattle of the very susceptible diminutive zebu type. The outbreak was controlled by serum-simultaneous inoculation (Richardson and Carmichael, 1926).

Simmons recounts, in the days prior to 1926, the difficulties experienced by staff who were subject to heavy casualties due to malaria, dysentery, and mental strain. One officer died of blackwater fever, another committed suicide, while two others were invalided, home, leaving only two survivors.

Kennedy (1929) reported another outbreak in West Nile District in 1929, buffalo being heavily affected. Bushbuck and warthog were also affected, while hartebeest, waterbuck, oribi, and duiker were apparently not so.

In Buganda and the Western Provinces, the cattle population increased from 220,000 in 1921 to 677,000 in 1929, when another major epizootic broke out among game in the west. Buffalo succumbed first, followed by bushbuck and warthog, and later eland and waterbuck. Many buffalo and Uganda kob died around the shores of Lakes Edward and George. Cattle were allegedly more susceptible than those in surrounding countries. Once again, it spread to the area of the Albert National Park, killing topi, giant forest hog, and warthog.

In 1930, it entered Ankole and Masaka from the north, where it is estimated to have killed 15,000 head of cattle, but 78,000 were inoculated in Ankole District and 225,000 in Toro, losses being contained. In 1931, there were only two outbreaks among cattle, 30,000 being inoculated, but in October and November 1931, there were numerous buffalo deaths to the south and west of Lake George, and the following year, buffalo were reported dying along the shores of Lakes Edward and George for a long period. It spread again into the Albert National Park, affecting particularly warthog, giant forest hog, and the normally refractory topi. In 1933, it was very virulent in South West Ankole. Here, 15,000 cattle died from inoculation, and it then appeared in the Akagera National Park in Rwanda, where eland and sitatunga were said to be greatly affected, numbers of eland dropping to less than 200 head. Based on later studies (1969), they probably never numbered originally more than about 600. It persisted among cattle and game in Ankole and Kigezi districts in 1934, and in 1936 was widespread throughout the year. Buffalo were infected in West Madi, and bushpig were reported as infected in North Teso. In February, it was introduced in the south into Ankole and Kigezi by illicit cattle movements northwards from Rwanda. Inoculation programmes eliminated it from Ankole in August, and from Busoga and Mengo Districts by November. In Buganda, game was affected on a large scale.

A further outbreak in Ankole is reported for 1937; and then in 1942, another wave entered Uganda from the Sudan, spreading once more to the Lakes Edward and George region which appears to have been free of outbreaks since 1934. Extending to the Albert National Park, 70% of the buffalo are reported to have died. It reached Ankole in 1943 having spread almost throughout Uganda, but was never out of control as far as the veterinary authorities were concerned, over 250,000 cattle being inoculated, now with the much more efficient KAG vaccine. By the end of the year, it existed only in game, according to Carmichael (1973) causing profound alteration in the symptoms seen in cattle, which later became infected with game-passaged virus, assuming a mild character with absence of characteristic mouth lesions, nasal discharge, and diarrhea, making diagnosis difficult and confirmed only by clinical tests.

Pitman (1942) wrote that about 1940, more than half a million cattle existed in the eastern province of Uganda between what he described as the "rinderpest-infected wastes" and Buganda. During droughts, there was a generally westerly movement of game animals from the arid eastern regions bringing them into contact with the cattle at watering points and periodically spreading rinderpest.

Another outbreak occurred in the Albert National Park in 1945 when buffalo, bushpig, giant forest hog, and bushbuck died, rinderpest being confirmed in buffalo. Limited incursion from the Sudan occurred in Uganda again in 1948.

In 1950, it was present in scattered cattle herds in Karamoja with a major outbreak in Teso early in the year. There were reports of buffalo dying in southern Buddu, and infection was transmitted to cattle by infected herds moving from Bukoba in Tanganyika. It was still present in unspecified game in November. In 1949, a programme of buffalo elimination, more likely to spread the disease by scattering them than contain it, was begun along the border with Sudan in the West Nile and West Madi regions in an attempt to control its spread from Sudan. A total of 300 buffalo was shot in 1949 and 599 in 1950, but the situation in game populations gave rise to anxiety in 1954, when the disease in buffalo spread from the Sudan border south as far as Murchison Falls National Park and Lake Albert and invaded Ituri Province in the Belgian Congo, which had been free of it for 10 years. Further minor outbreaks occurred in the Belgian Congo in 1956 and 1960, allegedly caused by infected buffalo and giraffe moving south from the Sudan.

Uganda then became free until 1966 when vaccination campaigns ceased and cattle movements became uncontrolled. In 1979, 27 outbreaks were recorded in which more than 5,000 cattle died, and by the end of the year, 1.5 million had been vaccinated, but vaccination was not maintained. In September 1988, there was an outbreak in the Murchison Falls National Park of northern Uganda in which several hundred buffalo died.

6. KENYA

There was doubt at the beginning of the century as to whether or not many of the infections attributed to rinderpest were indeed that, resulting in the "catarrhal fever-rinderpest dispute."

Stordy reported in 1898 a transport of oxen into Uganda had introduced rinderpest with a huge mortality among cattle and consequent famine among the natives, but this outbreak was now apparently extinguished. In 1903 the disease reemerged in Kenya and in 1904 there was an outbreak among cattle at Naivasha extinguished by veterinary teams. Percival (1928) reported game died in numbers on the Athi Plains in 1905 but this is not known to have been from rinderpest. Patterson (1909) noted in 1907 how greatly the herds of the Maasai in Kenya had increased in the past decade. Whereas before only a few cattle, sheep and goats were to be seen, he estimated he had seen at least 10,000 cattle and 30,000 sheep and goats in Laikipia. In 1907 a new epizootic threatened Kenya from the north but had not yet traversed the Northern Frontier District where infection was severe but little known of its extent, although in 1908 an outbreak occurred in Borana (southern Ethiopia) and thousands of cattle were reported dying daily (Pankhurst, 1968). Tests were conducted with bile and convalescent serum and attempts made to establish a serum production station in case it entered the European settled areas, but a source of infection could not be found to provide the serum and it was impossible to penetrate the vast extent of the Northern Frontier District. A test was also run on 20 cattle in the Lamu coastal area with serum from South Africa (Onderstepoort). At the end of 1908 there was an outbreak at Nakuru but with low mortality and attempts at Naivasha by Stordy to transmit infection with blood from these animals were negative. Dion Lardner (1912) reported the Maasai as possessing enormous herds of cattle at Lake Naivasha in 1909.

Theiler, who visited British East Africa and Uganda in 1909, considered that the infection he had seen could not be rinderpest but was a special type of gastroenteritis. Stordy obtained six sick animals from Nakuru and brought them to Nairobi. A very sick one slaughtered for autopsy showed throughout the appearance of rinderpest which Theiler had been familiar with in the Transvaal. From the slaughtered animal, 20 cc of blood was injected into the remaining five, and over 40 cc administered to two calves obtained from a European farmer plus two other calves. The animals showed a feverish increase in body temperature, but all survived. Fourteen months later, all were still alive at Kabete when they were injected with 2 cc of infective blood. All sickened, and 4 of the 5 cows died, and 3 of the 4 calves.

Montgomery examined sick cattle on the coast in northern Kenya and at Kisumu in 1909 identifying the disease as a coccidiosis. Passing rapidly through the Maasai herds on Laikipia in 1910 and causing heavy mortality in the Kavirondo district, in May 1910, Montgomery saw a new serious disease outbreak in Njoro and on the Nakuru-Naivasha caravan road confirming rinderpest by successfully transmitting it with 2 cc of blood. With Stordy and Theiler, he believed that there were two diseases in British East Africa, rinderpest and a coccidiosis, the latter identical with the German East Africa virulent catarrhal fever. In 1910–11, the position was considered serious, the epizootic having spread to Uganda and German East Africa.

At the Nakuru Show in December 1912, exhibits of animals were limited because all cattle were quarantined on account of rinderpest. Nevertheless, oxen were used for transport in some areas, Lorimer (1917) using them on safari from Londiani to Uasin Gishu in January 1913.

In that year, a second major epizootic began continuing until 1921, although, until 1918, Kenya itself had a notified outbreak only in the Nandi district.

A test of hyperimmune serum in Nyanza had proved successful, and Kabete began producing serum by 1912, nearly 24,000 doses being issued and outbreaks in Ukamba and the Rift Valley then being treated with success, although in some cases, a high mortality ensued. Cranworth (1919) attributed this mortality to depreciation of the serum, but it was probably due to complications caused by protozoal infections. Meikle and Meikle (n.d. [1914]) reported that the disease had been rife all over the Uasin Gishu plateau in 1913–14 with a death rate of only 5% due to treatment, whereas the Nandi in the same period lost 30,000–35,000 of their untreated cattle.

This second outbreak was said to have been fanned by World War I military campaigns with their transport oxen (although the disease had proved itself just as capable of spreading without such influence). Eland and giraffe were badly affected, followed by buffalo, bushbuck, and reedbuck.

In 1914–15, there were 12 outbreaks on European farms, the disease being enzootic throughout most of the African reserves, and more than 35,000 military cattle were inoculated by the serum-simultaneous method. In 1915–16, outbreaks occurred in several areas: Lumbwa, Molo, Londiani, Njoro, Nakuru, Naivasha, Kedong, Ngong, Nairobi, Thika, Machakos, and

Samburu in the Northern Frontier District. The serum-simultaneous inoculation method was used to control these outbreaks at Molo, Nairobi, Thika, and Machakos, the remainder by serum alone. At the Fort Ternan quarantine station, several animals developed anthrax after serum-simultaneous injection, in many cases infection apparently taking place at the site of inoculation.[3] It was also observed toward the end of the war outbreaks of fatal redwater fever appeared wherever the serum-simultaneous method was applied. The virulent blood which was injected was taken from native beasts with enzootic redwater, almost all native cattle having been exposed to the disease as subadults. The veterinary laboratory had thus been spreading redwater by the serum-simultaneous method throughout the high-grade European-owned herds. Once this was realized, the laboratory built up a herd of redwater-free cattle to provide a reservoir of rinderpest serum (Huxley, 1935).

The year 1917–18 witnessed 13 outbreaks and in one of these buffalo, eland, reedbuck and bushbuck were found dead leading to the assumption "there was abundant evidence to show that it [the outbreak] was due to game" (Stordy, 1918).

In 1920, a meeting was held in Nairobi of representatives from the three East African territories at which it was admitted that rinderpest was enzootic in the three countries.

In 1922, bushbuck suffered severely from an outbreak in an unspecified locality but probably the region of the Aberdares (Percival, 1928). Mallett (1923) reported an outbreak at Bardimat in the Narok district of Kenya. Four trek oxen had died, and many wildebeest carcasses lay along the road: "They lay there in hundreds." There was rinderpest in Sotik, and at a small bridge over a river were two manned sentry boxes to prevent any crossing of oxen.

Carnegie (1930) records how, in June 1924, cattle were inoculated on her Kenya farm, although inoculation was not compulsory. Bache (1934) reported that the Veterinary Department had "for many years" inoculated cattle all over the country as far as it was possible, and only immune (i.e., inoculated) oxen were allowed on the roads. She records that inoculations had to be paid for by native as well as European owners. The cattle were usually inoculated twice, first as calves at foot under 9 months old, second when young stock. After this, immunity was absolute, and they received the "coveted" A.M. brand. She states that it was a "double inoculation," that is, the serum-simultaneous method, carried out at the same time and the brand given 1 month after the second inoculation. During that month, the whole district was in quarantine as uninoculated animals were liable to contract infection from inoculates. Warning was given months before by the Veterinary Department that it proposed erecting a crush in an area, and farm owners then informed their labour to count the numbers of animals they wished to have inoculated and raise the necessary money. The Maasai, Wakamba, and Nandi, mindful of the catastrophe of the 1890s, submitted willingly.

Carnegie writes that grown cattle reacted worse than young, and apart from the risk of redwater, cows were liable to abort. Her herd had previously been inoculated after a lapse of nearly 5 years, the weather had been very hot and dry and the cattle reacted severely. Although the official death rate was under 4%, a considerable number of calves were aborted. Some 15 or 16 animals also needed hand-feeding after the inoculations, while most animals suffered from abcesses, "another unavoidable consequence of the inoculation." Most abscesses were behind the shoulder where the needle went in but might be anywhere. The suffering cattle were nursed for several weeks, and all but two recovered. Subsequent inoculations were nothing like as severe, and as a result, regular inoculations at 18-month intervals were carried out. In 1925, 180,000 vaccinations were carried out with 2% mortality.

[3]Some inoculators also contracted anthrax through the palm of the hand, made sore by repeatedly pressing the syringe plunger.

In 1934, widespread outbreaks occurred in game in Kenya, but only 64 were reported on European farms and 54 in African areas. These were brought under control with some 215,652 inoculations. Of these, 150,045 were double inoculations, the remainder spleen preparations (Macaulay, 1973).

Typical of contemporary beliefs, which were to persist even into the 1990s, was that of Bache (1934) when she wrote,

> From time to time in the past, native herds of cattle have been decimated by the scourge of rinderpest, which is carried by the game and by wild pigs.
>
> Rinderpest was formerly the chief agency for the keeping down of native herds to reasonable limits. Now that the Government has almost stamped it out... their numbers have so greatly increased that some of the Reserves are already overstocked.

7. TANGANYIKA

Two German veterinarians had been present in German East Africa since at least 1905 (one of whom was engaged in the inspection of imports) when they made the first reports of a rinderpest-like disease in Ugogo in 1905 known as *Sabatele or Sotoka pili* (Rinderpest again) famine, but it was not apparently until 1912 that the Germans established that certain districts were definitely infected with rinderpest. In 1907, the chief medical and veterinary officials submitted a memorandum to the German government arguing that the great potential of the cattle economy remained unfulfilled because of the effects of disease on the herds. As a result, an Animal Disease Institute was set up at Mpwapwa in 1912, by 1913 employing 15 veterinarians, five of them in a special task force to combat animal disease, and eight stationed across the country in the most important cattle regions. The following year, it employed 30 Europeans.

Melland and Cholmeley (1912) reported that in 1907, the Rukwa region was stocked with large numbers of sheep and goats but few cattle. From Rungwa to Tabora, there were practically none, but Tabora had a large supply of cattle in its vicinity; whereas from thence to the plains around Mwanza gulf, enormous herds of cattle were everywhere.

Tanganyika came to be regarded as a pivotal area for the maintenance of rinderpest because of the vast herds of susceptible Serengeti antelopes in the region adjacent to Mwanza, representing the largest concentration between the southern Sudan in the north and the Kalahari Desert in the south. While susceptible short-ranging species such as warthog and bushbuck were considered to present little danger, it came to be realized by the late 1920s that the disease could be spread quickly by migratory species such as buffalo and eland.

Its persistence in Tanganyika has been well documented by Branagan and Hammond (1965), who showed that successive outbreaks in the 20th century were apparently not the result of fluctuations in virulence or oscillations between immune and non-immune wild populations, but varying degrees of control maintained by the authorities through inoculation programs. Although it gained momentum throughout the country during World War I, more through lack of control than movement of transport oxen, it was brought under control by inoculation.

According to African reports, rinderpest in German East Africa had remained present since the 1890–92 outbreak (Ostertag, 1916). In 1905, a disease was observed among cattle near Lake Victoria which, in the next few years, spread to Mpwapwa, the Meru and Usambara regions, Kilimatinde and Tabora, Mangati, Irangi, Uhehe, Ugogo, and also Usuwi in Rwanda (Figure 4). In 1909, it was found on the island of Ukerewe in Lake Victoria. This disease was the object of much controversy, the "catarrhal fever-rinderpest dispute," still unresolved in 1915, for, although it was very similar in clinical and pathological appearances, there were

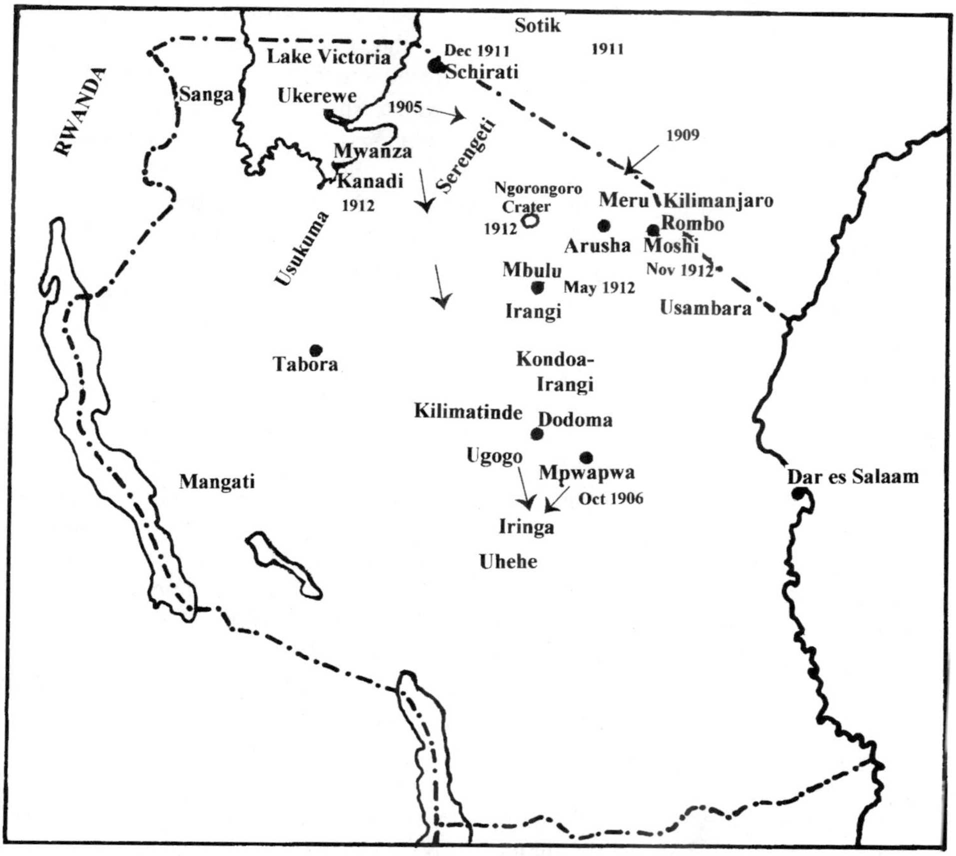

Figure 4. Rinderpest in German East Africa, 1905–13.

important differences. These were inflammation of the eyes, non-transference of the disease by vaccination and contact, low mortality, and an absence of mass infection in wild animals. Koch, when asked his advice in Berlin, was unable to give a positive answer.

In October 1906, the disease was identified at Mpwapwa in a herd brought from the far interior of the country. Dr. Lichtenheld, who identified it, considered that earlier localized outbreaks in Africa were not rinderpest but a special form akin to a virulent catarrhal fever which had already led to confusion in Germany. In 1907, Dr. Claus observed at Kilimatinde that 60% of affected animals suffered total cornea destruction with iris prolapse, attempts at transfer were negative, and mortality at Schirati was only 20%, while on Ukerewe, it was 15%. Various veterinarians conducted transmission experiments, including eight cows each infected with 100 cc of rinderpest serum from Cairo, without conferring protection. In 1911 and 1912, Dr. Manleitner claimed to have succeeded in transmitting rinderpest with blood, bile, lachrymal discharge, nasal secretion, and saliva, the infection having the appearance of virulent catarrhal fever. Further experiments in 1912 by Dr. Wölfel in the Kilimanjaro and Mbulu areas demonstrated that calves injected with 40 cc of Kabete rinderpest serum were protected against rinderpest, showing only a momentary temperature rise, while control cattle succumbed, thus establishing that rinderpest was present in German East Africa. The negative results of previous experiments could be attributed to the disease being enzootic and cattle having an immunity. Numerous new outbreaks at the end of 1912 and during 1913 affected both young and old animals.

Wölfel considered that rinderpest was introduced from British East Africa. In 1909, it had spread south from Kenya to German East Africa and west to Uganda the following year. In 1911, it was present in Sotik, and in June 1912, spread into the settled area near the Uganda road. In December 1911, a rinderpest-like disease broke out in cattle in Kissaka at the Mara in southern Sotik near Schirati on the German border but was not looked into as it was thought that pasture changes would check it. From this area in June 1912, two farmers transported livestock to the Kilimanjaro settlement where rinderpest was diagnosed, but by isolation and vaccination, its further spread was checked. When it reappeared in Uganda in 1910, the German governor Rechenberg decided to create a 50-km game-free zone along the German East Africa border, employing the army to shoot out the game, a policy apparently widely considered as folly (Koponen, 1994). It did indeed fail to prevent advance of the disease.

Dr. Manleitner's identification of rinderpest in May 1912 at Mbulu was later confirmed by Wölfel. Manleitner reported that in August 1911, an Indian in Mbulu bought a calf which African herdsmen had brought to Arusha, and placed it with an African herder. Two calves then died in the next few days, both showing affections of the eyes as well as other symptoms.

Occasionally, punitive expeditions against "insubordinate" natives in the Mbulu and Dongobeseh countryside seized herds of cattle, which were driven to an isolation camp for a short stay and then marched for 7 days to Arusha, where, on May 14, it was notified that two unhealthy animals had come in. In the first 5 days, 4 animals died, and up to June 19 , on a different pasture, 59 more died, with 8 more up to August 31, making a total loss of 73%. According to Africans, Indians, and Goanese, the disease was introduced to Mbulu by a Greek's ox transport. A third disease focus was determined in 1912 in Ugogo in the Dodoma region, undoubtedly introduced by nomadic Maasai herdsmen from the adjoining Maasai Reserve. The disease spread along the old caravan routes of Kilimatinde to Mpwapwa, Nyangallo to Bati, and later Nyangallo to Mpwapwa, and advanced along both of the southern main routes Kilimatinde-Iringa and Mpwapwa-Iringa. How long it had been present in Ugogo could not be determined, Africans stating that cattle had been dying for 5 years before 1911. But when the government veterinarian, Dr. Münchgesang, went to investigate in 1913, the outbreak was extinct, although numerous animals showed inflammation of the eyes and were very thin. At the end of 1912, rinderpest was reported at Kanadi in the Mwanza area, Africans claiming that it came from the English border near Mbalageti-Ikoma to the northeast, where hunters reported that many wildebeest and buffaloes had come into the area. It was also demonstrated that the disease had been introduced onto Ukerewe Island and into the Schirati region through livestock transportations from the Kilimanjaro settled area. In November 1912, it was reported at Moshi, introduced from Mwanza or Usukuma, and was also in African stock in the Rombo region. The latter was infected from the south, and from thence, infection spread north, showing itself first in a herd of the Maasai-Mangi Massaka whose pasture adjoined the southern Rombo area. In the herd of the Massaka, it appeared in nine cows which had just been bought from a Somali drover who had obtained them in Usukuma. It also broke out on the Siedentopf farm in Ngorongoro Crater (which, according to Kjekshus, 1977, had been abandoned by the Maasai after the 1890s rinderpest), but was controlled by an immediate vaccination campaign, only 23 animals of the total herd of 900 dying. This compared with 4,000 dying of the 10,000 wildebeest which grazed in the Crater.[4] It was also brought to Dar es Salaam but, through slaughter and vaccination, immediately controlled.

[4]The mean number of wildebeest using Ngorongoro Crater for 1970–79 was 14,497, and for 1980–89, it was 9,620. In 1987–8, there were 25,000 resident Maasai with 137,398 cattle (Perkin, 1995). According to Fuchs (1907), Siedentopf had more than 2,000 head of cattle in the Crater by 1905 and planned to expand the numbers to 5,000, but clearly something had affected them to reduce their numbers to 900 by 1912.

Prior to 1913, it was brought into Mbulu by cattle inoculated with rinderpest serum introduced onto the farm of a planter Heinrich Rhode at Temi near Arusha. At the same time, it broke out among African livestock in Ndoruma. It was then spreading in 1913 to Schirati, Mwanza, Mkalama, along the caravan route Mkalama-Singida-Saranda, and into the areas of Kondoa-Irangi and Dodoma. The mortality rate was dropping below 90%, in Mbulu ranging between 16 and 50% with an average of 30%, but the outbreak in Ugogo experienced 70%.

Drs. Manleitner and Trautmann observed that climate had an effect. In the dry season, less damage was caused, but heavy rain could cause up to 100% mortality, these differences probably being related to the animal's resistance becoming weakened in wet and damp conditions, but equally plausible was Ostertag's suggestion that the virus might become more mobile in damp conditions. Also, where redwater fever was present, rinderpest infection might allow latent redwater fever to develop so that the latter dominated the symptoms. The government veterinarians reported also that milk-calves did not succumb to rinderpest, which Ostertag regarded as "striking" and suggested that inoculation experiments should be conducted to confirm it. These were carried out by Trautmann in November 1912 at Msanga, injecting three calves with virulent blood and apparently achieving a reaction, but the results were inconclusive as he was obliged to interrupt the experiment.

Ostertag noted that a peculiarity of the current outbreak was that it proceeded without game dying, although there were incidents of game mortality. At Kanadi, the natives reported numerous antelope deaths in the adjoining savannah, and there was the huge mortality of wildebeest reported in Ngorongoro Crater.

The reports of 1912 and 1913 showed a very variable clinical picture, and no disease was apparent in sheep and goats. Immediately upon receipt of the reports in 1912, the infected districts were isolated, but there was great difficulty in carrying out the policy. Trading out of them of animal fat and hides was allowed only on condition that the fat was heated to at least 60°C, and the skins thoroughly sun-dried. Movement of animals was forbidden. Dung was removed from isolation kraals, but this was not possible everywhere as some tribes used the kraals for burying their dead. Flesh of diseased animals was allowed to be eaten, provided it was well cooked. Twelve thousand doses of serum were on hand from Egypt, and a further 15,000 doses were obtained from Kabete. Two centers were immediately erected for the production of serum. One was at Ngare Nanyuki on the east side of Mount Meru covering 1,000 ha (now part of the Arusha National Park), all game in the area being killed. Between January and August 1913, it produced 67,000 doses of serum. The other, at Mpwapwa, did not begin production until July, due to the difficulty in finding a suitable site. Serum-simultaneous inoculation was then instituted on a small scale in the Arusha, Iringa, Mwanza, Kondoa, and Dodoma Districts, and it was recommended that all cattle on the northern frontier be inoculated owing to danger from British East Africa due to the permanence of rinderpest in Abyssinia and Somaliland.

Ostertag (1916) claimed that the superiority of the serum-simultaneous method was demonstrated at the German East Africa Veterinary College as the introduction of piroplasmosis by this method was unimportant, nearly all cattle being infected. Also, rinderpest usually occurred in districts where tsetse fly was unknown or seldom found, and the Africans were very careful in guarding their cattle against it. As a rule, losses amounted to only 1% by this method.

8. TANGANYIKA IN WORLD WAR I

The inoculation program ceased soon after its introduction due to the outbreak of war, dashing Ostertag's hope that complete stamping out of the disease in German East Africa might be looked forward to. Instead, the disease spread almost throughout Tanganyika and southwards into Northern Rhodesia.

The first case south of the Central Railway occurred 50 km north of Iringa in November 1914 but was speedily eradicated by the Germans with the "serum alone" method and vigilant control. The nearest enzootic areas at this time were Tabora and Mpwapwa Provinces. As early as January 1915, the Colonial Office warned the Northern Rhodesia Government that there was a high incidence of rinderpest in German East Africa, and they were to stand by to prevent it spreading to there. It did not do so, but in May there was a bad outbreak of pleuropneumonia, and extensive cattle slaughtering took place. On July 25, 1916, a report was sent from New Langenberg to the Chief Veterinary Surgeon, Southern Rhodesia, to the effect that no further news of rinderpest in German East Africa had been received: "It is to be hoped that with the advance of our troops toward Iringa definite news of the progress of the disease will be obtained at an early date. Every precaution is being taken that cattle do not enter Northern Rhodesia from this territory" (Sharpe, 1916). It is possible that German columns retreating from Tabora brought infected cattle via Iringa to Mahenge Province in September 1916. Daye (1918) does not refer to rinderpest casualties but reports that the Germans slaughtered all the thin animals in a "scorched earth" policy taking only the best with them, an action which may have helped limit the spread of rinderpest.

It broke out again south of the railway at Iringa in January 1917 caused by military supply cattle collected from north of the railway, only a proportion having been given a passive immunity with serum. It was the rule for supply cattle to be given an injection of serum before being sent on a long trek, but this could not always be enforced.

From Iringa, infection appears to have been carried south to Njombe following the British lines of communication with a gun team of the ox battery of the South African Mounted Reserve, but infection may have been carried along the route previously by supply carriers fed on rinderpest infected flesh.

British forces began losing cattle in the Njombe area from January 10, 1917 onwards with what was thought to be rinderpest, and by March, it was well established. On February 22, German troops looted supply cattle from Tandala in the New Langenberg Province at the north-west end of Lake Nyasa, infected supply cattle having arrived there from Njombe. Looted cattle were then driven north and west to the south-eastern corner of Lake Rukwa and on toward Tabora, establishing the disease in Ussanga where only 600 of 3,000 cattle were left by the end of June. The outbreak among supply cattle at Tandala showed no tendency to spread. There was no evidence that game, even buffalo which were numerous in the New Langenberg area, contracted the disease, and there was no evidence in sheep and goats (Curson, 1919). Foci created at Usafua and Malema's in Usongwe district tended to die out and, in May 1917, were finally eradicated by the Rinderpest Commission.

The German commander, Wintgens, who had led the movement into Ussanga from Rwanda, was considered responsible for the danger threatening Nyasaland and Rhodesia for 18 months, which led to the despatch of a Rinderpest Commssion from South Africa headed by Gray in April 1917. Before the Commission arrived, the Nyasaland and Northern Rhodesia authorities created a 16–32-km-wide cattle-free zone south of the German boundary in the area between Lakes Nyasa and Tanganyika. A few outbreaks having been identified south of the Livingstone range, it was resolved to clear the disease from affected localities by the serum-simultaneous method and inoculate some 100,000 cattle in a 40–60-km-wide strip to the north of the German boundary. At the same time, Tanganyikan authorities carried out campaigns in areas south of the railway from Dar es Salaam to Lake Tanganyika, immunizing some further 100,000 cattle (Curson, 1919).

Inoculation camps were established to carry out the first plan, but the disease showed a tendency to die out, and the African cattle appeared to possess a high degree of natural immunity. Unanticipated difficulties led to the original scheme being abandoned, and instead of a belt of immune cattle along the southern border of the German territory, measures were taken to guard the New Langenberg district against introduction from the other side of the Livingstone range where it was more or less enzootic (Gray, 1917). The disease appeared to have been brought under control when it reappeared in the New Langenberg district south of the Livingstone range due to unauthorized movement of cattle undergoing reaction subsequent to inoculation, but again it was controlled. A further outbreak close to the Nyasaland border due to movement of two calves out of an infected area was also controlled. Other centers of infection being revealed, it was decided to revert to the original plan of creating an immune belt along the whole of the Northern Rhodesia–Nyasaland border. The appearance of the disease about 160 km west of New Langenberg, and mortality of eland, bushpig, and warthog in the Rukwa valley close to the Belgian sphere of operations, made this urgent. By May 6, a total of 27,152 cattle in the New Langenberg district had been inoculated. Various other temporary inoculation camps were set up until nearly 100,000 head of African cattle had been inoculated with a mortality not exceeding 5%, and at the end of July, it was decided that the programme could be terminated, a skeleton staff remaining.

Many military supply cattle which received serum inoculations during July–August 1917 and January–April 1918 before leaving stations in New Langenberg developed rinderpest within 7–10 days afterwards, attributed to them having contracted the disease prior to inoculation. In April 1918, some infected cattle were moved to Ukonde. In September, some movement of cattle took place in North Nyasa District across the Songwe River into the Bundali section of the Province of New Langenberg, resulting from calling up of Africans in the former area for military service. To avoid conscription, numbers migrated to the other side of the border, taking their cattle with them, moving stealthily at night and tying hides over the cattle's feet to conceal the tracks.

At the end of October 1918, General von Lettow-Vorbeck crossed the border with his German troops, traversing the immune and cattle-free belts near Fife into Rhodesia, pursued by British troops. About 800 head of cattle brought from Mereres country immediately north of the immune belt established by the Commission were taken across the border for rations, many coming from a district in New Langenberg where all had been immunized in May. But at least half of the herds came from rinderpest enzootic country in Wassanga and Wabena. These infected cattle had travelled through Rhodesia, by way of Fife, to Chambers river, South Kasama, and thence to Abercorn. Within 10 days of the German cattle being surrendered, rinderpest was diagnosed, four cases in cattle surrendered by the Germans, and three among cattle left by them at Kasama, South Abercorn. All non-immunized cattle and all inoculated with serum alone were then destroyed within a radius of 22 km around the Kasama quarantine camp. After the cessation of hostilities, several deaths were reported by British troops among animals driven after the column for food. These troops were met at Abercorn, and as soon as surrender had taken place, a quarantine station was formed. All susceptible in-contact cattle were destroyed and immunization of all other cattle within a 22 km radius carried out. Searches were made for all stray cattle along the routes taken. Unless game picked it up, as there were few cattle in the vicinity of the outbreak and none within 6 km of the road along which the herd had passed, it was considered that the outbreak might be arrested.

In November, an outbreak occurred at Isoko among some animals which had been brought from Nyasaland in June to test a reported outbreak among immunized animals. The outbreak was apparently traced to an animal which had not been inoculated, purchased in the Bundali region.

By December 1918, the southern areas were considered clean, and in March 1919, it was decided to close down everything at the end of the month if the situation continued to be

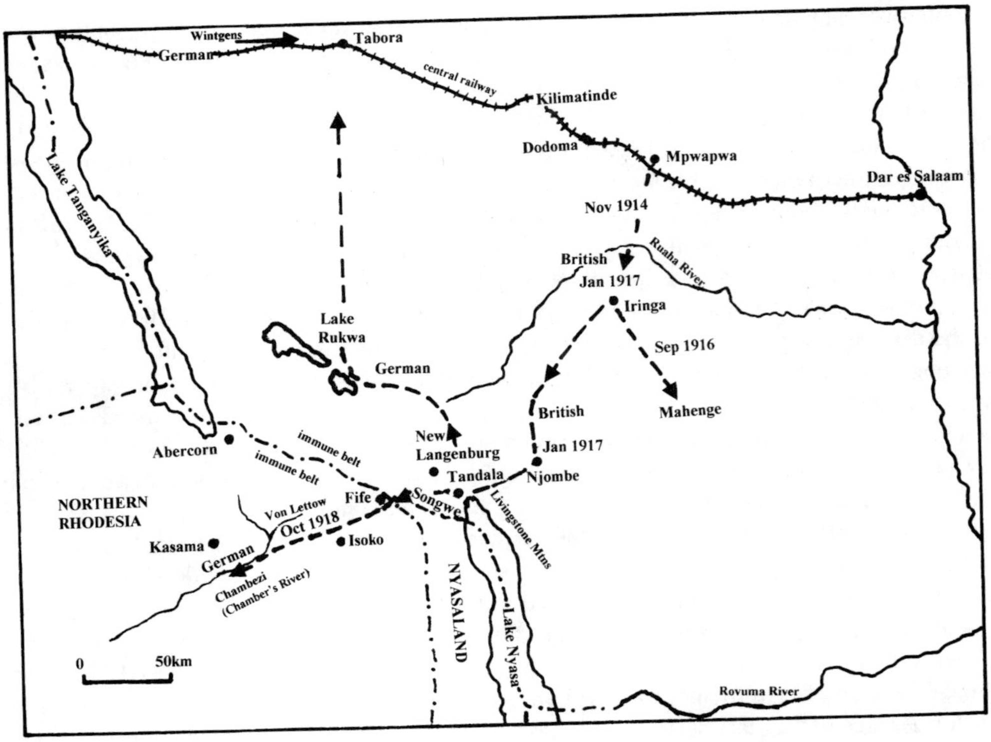

Figure 5. Troop movements and rinderpest spread in southern Tanganyika, 1914–18.

satisfactory; but pressure from Rhodesia and South Africa to prevent a southwards extension of rinderpest from Tanganyika was to continue for the next 50 years (Figure 5).

9. CONTINUED INFECTION IN TANGANYIKA

Curson (1919) reports that there were many difficulties maintaining the immune belt between lakes Nyasa and Tanganyika. The administration was unsympathetic toward the African, perhaps because British officials had only recently occupied the country and saw them as German collaborators. Malaria affected everyone, and it was a difficult country of mountains, valleys, and dense bush, peopled by tribes suspicious of the European, especially considering that both German and British had levied their cattle to feed the troops. There was also an African belief, probably deriving from the method of marking inoculated cattle by branding, that if cattle were branded with a spear, there was no need to take them to the European for treatment, and a shortage of food meant that deaths were concealed so that the flesh could be retained.

In 1921, there were 86 outbreaks among cattle, which Curasson (1932) attributed to troops from the Belgian Congo. The outbreak reached the Rukwa valley in southern Tanganyika, killing eland, bushpig, and warthog. Its further spread south was then halted by the establishment of a zone of immunized cattle and the elimination of game along the Rhodesia and Nyasaland borders, so that by 1922, the disease was restricted to the north of the Dar es Salaam to Tabora railway line. This occupied most of the staff in containing its spread during the following year, with a widespread outbreak in southern Dodoma district which was stamped out by the end of the year, its further southward extension prevented only by a swollen Great Ruaha River. In 1923, Tanganyika witnessed 70 fresh outbreaks in addition to 23 which had begun the year before, involving a total of 374,000 cattle of which 42,894 (11.4%) died. Generally mild in virulence with a declining mortality averaging only 4.3% in 1926, it remained widespread in the northern part of the Territory, although Iringa had remained free for 5 years, this year witnessing the first recorded occurrence of very mild rinderpest at Tabora. Its spread was attributable to illicit movements of cattle, infection of common watering places, and the infection of susceptible game. Movement of meat and skins of dead infected cattle was also partly responsible.

Then, in 1927, a number of Kenya Maasai with 45,000 head of cattle, some of which were infected with rinderpest and others with pleuropneumonia, moved to Loliondo in northern Tanganyika. From this focus, the buffalo of the Ngorongoro Highlands became infected, and these spread infection to Mbulu and Mbugwe to the south, and the southern end of Lake Manyara. At the same time, cattle carried it to Lolkisale in the southern Maasailand region. In 1929, its progress was aggravated by drought, spreading rapidly among the dense concentrations of cattle around watering points. It then spread north, affecting the whole of northern Maasailand, Arusha and Kilimanjaro regions, illicit cattle movements by the Maasai taking it into the dense Chagga settlements on the slopes of Mount Kilimanjaro. By 1930, the authorities had eliminated it east and west of the Rift Valley escarpment by means of quarantining and serum inoculation, reducing it to 11 circumscribed foci in the Northern and Lake Provinces, the number of cattle involved having declined from 240,000 in 1929 to 52,000 at the end of 1930. But it then reappeared in October of that year in wildebeest in the western Serengeti (Hornby, 1931), believed contracted from Maasai cattle originating from Kenya. Buffalo and giraffe were also affected. That it was rinderpest in the wildebeest was not clinically confirmed until 1933, when inoculation experiments established that wildebeest in the Ngorongoro Crater were infected (Cornell, 1934). But in 1931, the disease spread rapidly again in the Northern, Western, Lake, and part of Central, Provinces.

Owing to difficulties of storage and transport of serum (because of a wider prevalence of African horse sickness and cattle trypanosomiasis, unlike in South Africa, there were no horses and only limited ox transport), research was being undertaken into inactivated tissue vaccines, the first field trials of which were conducted in 1929 with some success.

In 1932, the highly susceptible Ankole cattle of Bukoba District suffered mortalities of 80%, sometimes 100%, but inactivated tissue vaccine reduced this to 6% in one herd, compared with 77% in a neighboring herd which the owner refused to have vaccinated. But the disease appeared to continue to smoulder in some vaccinated herds, so it was decided to restrict the use of vaccine to tradestock with serum-simultaneous inoculation, the preferred method for controlling outbreaks. Diagnosis was also now sometimes presenting difficulty, fresh outbreaks being initially very mild and classical symptoms only becoming apparent after 3–6 weeks.

Administrative difficulties caused by lack of funds allowed the disease to overwhelm control, a total of 21 veterinary officers in 1929 were reduced to 14 in 1934 and 12 in 1937. The years 1933–40 saw a slow loss of containment with the southward spread of the disease. At Loliondo, as professional staff were not available, nasal swabbing combined with serum inoculation was substituted for serum-virus inoculation, giving 70% success if virus donors were carefully selected. In 1933, there was an outbreak at the southern end of Lake Jipe in eastern Tanganyika brought in by Kenya Maasai, spreading to 12,000 head of cattle. In January 1934, it also became established among buffalo, eland, and lesser kudu. Moving south and west, buffalo and eland are said to have carried it to the area of Same, whence it spread to Naberera, and by November, 42,000 cattle were infected.

Meanwhile, in the triangle formed by Tabora, Singida, and Nzega, a "shuttle movement" of rinderpest was carried on by illicit livestock marketing and seasonal migrations of WaTaturu and WaSukuma herds, but largely controlled by the end of the year. It was reintroduced into Bukoba District by buffalo from Ruanda-Urundi, causing heavy mortality among Ankole cattle; and the following year, 1935, it spread extensively in Lake Province with many wild animal deaths in the western Serengeti. Near Ngorongoro Crater, this disease was "almost an annual occurrence, though in recent years it does not appear to do much harm"(Teare, 1935). In 1936, many deaths were reported in the western Serengeti near the Seronera River among 7–8-month-old wildebeest (Teare, 1936), indicating previous exposure and immunization of older age groups. Teare reports that five sick calves were shot and blood slides sent to the Director of Veterinary Services "who diagnosed that the mortality was probably due to rinderpest," something he could not do from blood slides. Teare stated that it was noticeable that only young animals showed signs of the disease. Reports were received also of eland dying in Arusha district, and dead eland, giraffe, and bush pig being found on the plains between the Naperera main road on the east and the Maasai district boundary on the west.

Policy was again revised, the decision being to leave the northern infected areas as an uncontrolled zone, south of which a belt of immunized cattle would be created, and south of this again, all outbreaks were to be eliminated as rapidly as possible. Control was effected by strict quarantine, serum-simultaneous inoculation of infected herds using goat blood virus wherever possible, vaccination of in-contact herds, and nasal swabbing with serum inoculation where there was no veterinary officer.

Wild animals were regarded as the decisive epizootiological factor which spread it southwards in 1936, and in the southern part of the country the greater kudu was often implicated in linkages. Most domestic cattle outbreaks were among zebu cattle which had not been exposed for 10 years and were highly susceptible, but mortality was generally about 20%. The following year, drought once more aided its further spread as both game and cattle assembled from long distances at watering points, but also its low mortality led to illicit stock movements as an important disseminating factor. By October of that year, it had crossed the Great Ruaha River, which was very shallow due to drought, but by the end of the year, the disease had been eliminated in the Southern Highlands Province.

26

CONTINUATION OF RINDERPEST IN AFRICA

1. THE CONTINUING FOCUS ON TANGANYIKA

In January 1938, an international conference was called at Nairobi which agreed to create a 64-km-wide belt of triple-immunized cattle to the south of the infected area in Tanganyika. Rhodesia, South Africa, and Kenya, all agreed to contribute an officer for 6 months. In May, low levels of the Ruaha River allowed it to be crossed again by infection as in 1936. The special campaign began in June 1938, and by December, a 128-km-wide belt of immune cattle had been created.

In February 1939, the conference convened again to review progress and recommended that a rinderpest "intelligence service" operate in southern Tanganyika which did not begin until December, when the disease was found in cattle and game in Mbeya District. Emergency meetings then called at Lusaka and in Tanganyika decided again, among other measures, to create a belt of triple-vaccinated cattle along Tanganyika's southern border between Lakes Nyasa and Tanganyika. Furthur outbreaks were also revealed, including buffalo in the northern Selous Game Reserve and in the Kilombero Valley. Infection continued to spread and, by the beginning of 1940, was only 60 km from the southern border with Northern Rhodesia.

Representatives of Tanganyika, Northern and Southern Rhodesia, Nyasaland, South Africa, and Kenya hurriedly met in January of that year in southern Tanganyika to decide upon an immediate action. With £55,000 guaranteed by South Africa and Southern and Northern Rhodesia, and the secondment of staff from these countries as well as the High Commission Territories and Nyasaland, approximately 1 million cattle south of the central railway were immunized by the end of the year, by which time there was only one confirmed case of the disease in the region, and that in a kudu. In April, the Adviser on Animal Health to the Secretary of State for the Colonies, Dr. J. Smith, was sent out to East Africa primarily to inquire into the situation in southern Tanganyika, arriving in May. A meeting of the participating countries was held with Dr. Smith at Dodoma in June, and a grant of £64,000 for 2 years' further intensive control work was approved from the British Colonial Development and Welfare Fund. In the same year, it was found that the danger of disseminating the disease by means of the new attenuated goat vaccine developed in Kenya (KAG) was negligible, and it was decided that all cattle behind the vaccinated border herds and as far north as the Central railway should be inoculated with the goat vaccine (Smith, 1941). Increasing efficiency in the use of this vaccine proved to be the decisive factor in the future control of rinderpest (Branagan and Hammond, 1965).

In June 1941, infected cattle and game were identified to the southwest of Lake Rukwa, and sick kudu had been reported north of the Tunduru-Maasai road in Southern Province, but the latter could not be confirmed. Another international conference was held in southern Tanganyika, which recommended construction of a 3.6-m-high elephant-proof wooden palisade across the 48 km of the Rukwa Valley, later extended to cover the 255 km from the northern end of Lake Nyasa to the southern end of Lake Tanganyika, with a 32-km-wide game-free strip on the southern side. Work was begun on this by the Northern Rhodesian authorities in August. As soon as the Lake Rukwa outbreak was discovered, another 42-km-long game-proof fence was constructed in the Saisi Valley where game was infected, and cattle both sides of the Northern Rhodesia and Nyasaland borders immunized. At the end of the year, the only known focus was in game at the northern end of Lake Rukwa. The following year, over 2.7 million inoculations were carried out and the disease brought under control. Annual inoculation of *all* susceptible cattle in important game areas was to represent the turning point, and by 1942, the disease was under control in all areas except north Mara, and it was thought that eradication would soon be complete. The 1943 Veterinary Department *Annual Report* stated that the day of final eradication of the disease from Tanganyika was "not far distant." Then, in 1944, owing to wartime staffing exigencies, control was lost once again, and infection appeared on the central railway at Tabora where it was of such mild character that it was feared it could spread southwards undetected. Further outbreaks occurred, 1945 witnessing an extremely virulent strain at Lake Jipe which spread among cattle and game in Tanga Province. Not only were KAG-derived immunities overcome in some adult cattle, but species of game usually regarded as fairly resistant exhibited mortality; such as waterbuck, oryx, Grant's gazelle, and impala. A year later, its effects in calves were so mild as to be almost undetectable. It was quickly controlled in cattle, but game apparently carried the infection across the Ruvu River, and in the northwest, large numbers of yearling wildebeest died in the western Serengeti, apparently spreading infection to cattle at Maswa; while buffalo became infected at Gelai and Naberera: "eventually game was largely responsible for infecting the whole of the Northern Province except Moshi." Legislation was passed prohibiting hunting in some localities because it was considered to have scattered infected game over wide areas. It was speculated that virulence might be enhanced by passage through certain species of game, but experiments were not possible due to a lack of resources.

At the beginning of the following year (1947), the disease was confined to Loliondo, Ngorongoro, and North Maswa, but an outbreak suppressed in cattle near the Kenya border was then spread rapidly southwards to near Lake Eyasi by game. Passing through Maasailand and Mbulu, it is alleged that migrating eland were primarily responsible for carrying infection to the large herds of game near Lake Eyasi, whence it continued to spread, reaching Shinyanga. By the end of 1948, it had spread over a wide area aggravated by severe drought conditions of 1948–9 and continued to extend slowly among cattle despite inoculation with KAG vaccine. At the end of the year, it was found that some batches of KAG vaccine had been contaminated with "wild" virus, which disseminated a virulent form, such that by April, every province north of the central railway was affected. It became widespread in the central region in 1949 because quarantine measures could not be enforced due to drought, cattle in many areas being widely dispersed in search of grazing and water.

It was confirmed in the Serengeti in November–December 1949, and suspected in 1951. In 1950, game carried the disease to isolated groups of cattle in the Malagarasy region and buffalo from Lake Manyara carried it to cattle in Mbulu. In 1952, buffalo and bushpig at Oldeani passed the disease to cattle, but infected Loliondo buffalo did not transmit it to the large herds of local Maasai cattle. It was confirmed in the Mara area in 1953, and suspected again in the Serengeti in 1954, confirmed in wildebeest, buffalo and impala, to the west of the Park.

In 1948, the Director of Veterinary Services concluded that the disease would probably die out where game concentrations were small but persist among large concentrations of protected susceptible species, as in the Serengeti, unless contact between game and cattle could be prevented. Echoing the Director, Branagan and Hammond (1965) claim that it was the game–cattle linkages which continued to nullify control measures. But the policy of *ad hoc* mass inoculation in outbreak areas as and when outbreaks occurred was clearly ineffective in eliminating the disease. Mild forms could pass unnoticed until it was well established in an area, permitting game to become infected and spread the disease further. Thus, beginning in 1950, *all* susceptible cattle in important game areas north of the central railway were to be inoculated annually, in place of simply containing outbreaks as had been previous policy. However, the Southern and Southern Highland Provinces were to be excluded so that any southward penetration of the disease would be immediately apparent.

Much of the failure to achieve the degree of expected herd immunity was due to the numbers of calves which escaped inoculation on account of the laxity of owners and local authorities, such that in 1954, it was found that as many as 25% were unvaccinated in some areas, and by 1960, there was active opposition to inoculation (Branagan and Hammond, 1965). In 1953, the situation deteriorated due to severe drought coinciding with a shortage of vaccine. In one area, where preventive measures were opposed by local people, some 6,500 Ankole cattle died, mortalities reaching 88%. In Bukoba District, where owners also opposed veterinary measures, a sixth of the cattle population was lost the following year. In 1955, increased supplies of vaccine resulted in almost 3 million cattle being inoculated, and the year 1956 became the first year since 1919 in which no outbreaks were confirmed. But the following year, game animals were reported dying in the Maji Moto area of Musoma District whence infection spread rapidly southwest into the western Serengeti and Maswa District. Mortality was heavy among buffalo and wildebeest, wildebeest dying in October in the Musoma and Maswa districts being reported as apparently all between 1 and 2 years of age. A few dead zebra foals were also found in the Maswa locality, but the causal factor was not confirmed (Swynnerton, 1958). By the end of the year, the disease had died out without spreading to cattle or other susceptible species.

In early 1958, it was confirmed in yearling wildebeest in Musoma and North Mara to the west of the Serengeti with no apparent spread to other species; at the same time in the Ngorongoro Crater Highlands in April, it affected wildebeest, buffalo, eland, and warthog, dying out by July. There were also rumors of disease in buffalo and eland in Ngorongoro Crater, rinderpest being confirmed there in a bushpig. Although mortality among buffalo was heavy, a good number survived, and by July, the disease had died out. Also in April, it was suspected at Lake Manyara and in July confirmed in buffalo at Karatu and Mbulu. In October, it was reported among buffalo, eland, and cattle at Lengai, this being the first report of cattle being affected. In January 1959, it appeared in young cattle at Engaruka in the immediate vicinity of a trail from the Ngorongoro Highlands followed by buffalo and eland some 2 weeks previously. In March, a large number of cattle deaths were reported in the eastern Serengeti. The buffalo carried it to Lake Manyara in August, where some 1,000 predominantly young buffalo allegedly died, although Game Department records give an estimated total loss of only 300, equivalent to about 10% of the population. Most affected was the Mto wa Mkindu herd, which lost much of its yearling stock, about 100 or 17% of the estimated herd of 600. Next most affected was the Mto wa Mbu herd of which only some 50 died, or 7% of an estimated herd of 700. The remaining 300 deaths were from the rest of the 1,700 buffalo spread throughout the reserve. It then moved eastwards among buffalo to and beyond the Tarangire Game Reserve where a few dead animals were recorded and infection estimated at 10%. But despite prolonged contact, there was no evidence of transmission to either cattle or other suceptible game: notably giraffe, wildebeest, warthog, and bushbuck. It did not spread to Southern Maasailand, having burnt itself out by March the following year with the last three buffalo deaths recorded at Endabash in Lake Manyara National Park (G. Child, personal communication). On the northern plains of the Ngorongoro Conservation Area in December 1959, there was heavy mortality among yearling migratory wildebeest with similar symptoms to rinderpest, but serological tests proved negative. Resident animals were apparently unaffected.

The extension southwards to Engaruka in 1959 necessitated the expansion of the annual immunization campaign to include Singida, Mpwapwa, and Kondoa Districts; where local opposition in some places was so fierce, the paramilitary branch of the police, the Police Field Force, had to be called in to assist.

It was thought to be present in wildebeest in the Mara, sick animals being seen in southeast Serengeti in March. There was no apparent disease in the first 8 months of 1960 but animals born early in 1960 and 1961 had acquired active immunity by 1965–6. This suggests that there was an epizootic in this group in the last weeks of 1961. Minor outbreaks occurred near Ngorongoro Crater in 1960 and in the west of the Serengeti (Figure 1).

In August 1960, small foci of infection appeared in calves in the Arusha and Moshi Districts considered to result from illicit stock movements originating northeast of Lake Manyara and in the Loliondo region. In the latter area, infection was shown to have been present in buffalo born before 1960. In October 1960, there was a severe outbreak among yearling wildebeest in the Serengeti, but this was the last.

In 1961, illicit stock movements during drought took the disease south to Monduli Juu, whence it extended further along illicit stock routes, necessitating a fresh immunization campaign. Mild strains were isolated from cattle which wildebeest had probably contacted in several parts of Maasailand in 1961, and yearling buffalo were affected around Ngorongoro Crater at the end of the year. By September, the disease was considered eradicated in cattle, but there was one confirmation in buffalo.

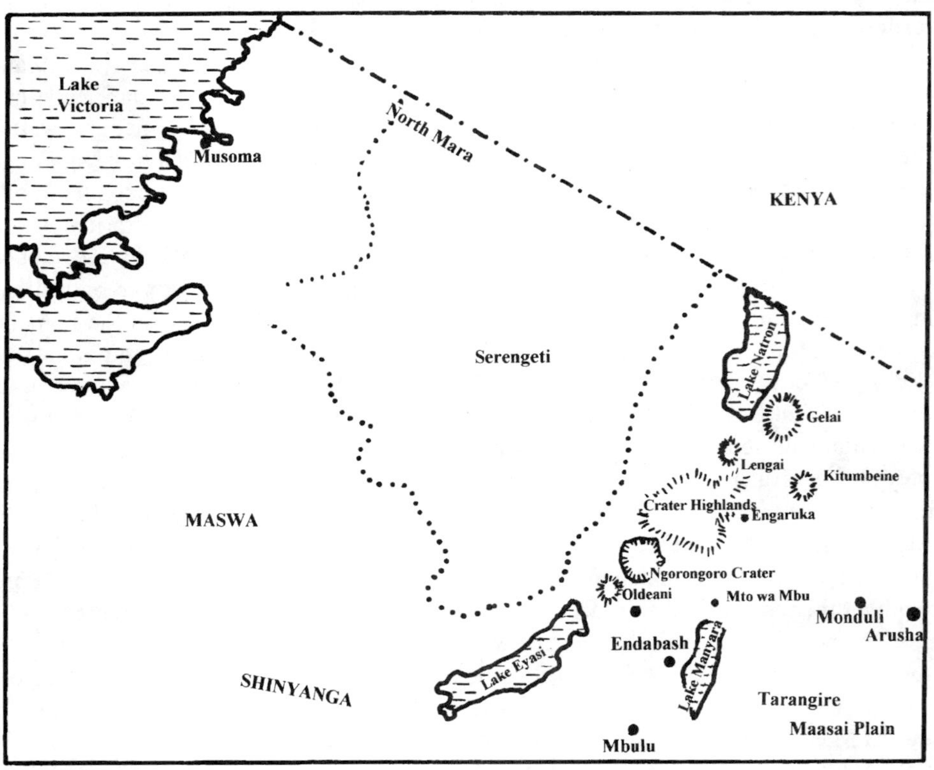

Figure 1. The Serengeti–Ngorongoro focus.

From September 1961, Tanazania was generally considered free of the disease, although Ngorongoro resident wildebeest were infected in the second half of 1962 but then remained free of infection. Sera from wildebeest born and collected in 1963 showed no rinderpest antibodies, and there were no reports in cattle or game in 1964. The number of cattle was estimated at 8.8 million, compared with 3.2 million in 1921. In December the following year, the disease was confirmed once more in cattle in Tanzania's Loliondo District but did not spread to contiguous susceptible game populations. The only other known focus at this time was of a different strain, 320 km to the north in Kenya's Isiolo District, suggesting that the Loliondo outbreak was evidence of an enzootic infection which had persisted in isolation since 1961 in an undetected form. In March 1966, a fresh focus was confirmed in the same district, where resident and migrant Serengeti wildebeest were concentrated, but there was no indication that these were infected. Antibodies were detected in 1–4-month-old calves born early in 1965 or 1966 but not in those born in January 1965. Antibodies persisted in 81% of animals born prior to 1962 (Table 1). No antibodies were detected in Ngorongoro calves born after January 1963, wildebeest populations of the Serengeti-Mara region of southern Kenya and northern Tanzania being free of infection for 3–4 years up to 1967, despite its presence in cattle in 1966. It was estimated that this represented some 200,000 animals at risk, or >60% of the total of migratory species, and probably a similar percentage of resident populations. It was therefore considered surprising that infection was not contracted from the cattle as migrant wildebeest grazed land from which infected stock had recently been withdrawn, and sick cattle, although quarantined, remained within territory occupied by wildebeest (Taylor and Watson, 1967).

Tanzania was almost free of rinderpest in 1965, but then stolen cattle from the Sudan spread the disease into Kenya and Uganda in 1982, and infected cattle looted in Uganda by Tanzanian troops carried the disease into Tanzania where it spread to wild animals in the Serengeti and Ngorongoro areas, and then to cattle further south. It resulted in a sudden and unexpected outbreak among buffaloes in the Serengeti National Park in May 1982, 60,000 buffaloes allegedly dying, together with warthog, eland, giraffe, and kudu. An attempt was made to localize the infection by ring vaccination of cattle, but it was not totally effective, and a number of infected cattle moved southwards in October. Peaking in 1983, it was checked by local vaccination campaigns, cattle in Malawi and Zambia along the borders with Tanzania being vaccinated in September 1983 to March 1984 to form an immune belt to prevent southwards spread. In Zambia, one cow tested positive before vaccination but could have originated from a vaccinated herd to the north. Five buffalo, three impala and one puku, sampled in September 1983 in the Luangwa National Park and Nsefu Game Reserve in Zambia, were all

Table 1. Rinderpest in Serengeti migrant wildebeest 1965–6.

Year of Birth	Percentage Positive
Up to 1959	83
1960	86.6
1961	63.6
1962	6.2
1963	0
1964	0
1965	0
1966	0

Source: Taylor and Watson (1967).

negative. No antibodies were detected in 111 adult wildebeest sampled in the Serengeti-Mara in 1984–5, and none in 14 from south of Arusha sampled in May 1984, but one eland showed a high level of antibody suggestive of recent infection, which could have been contracted elsewhere in 1982 or 1983.

With an estimated stock total now at 14.5 million in 1984, sampling in 1983–4 showed that five cattle tested positive from Mtwara district, four of which had been recently purchased from the north. Of 136 goats sampled in Mtwara, 14 in Mbeya and 15 in Iringa, only one from Mbeya was positive (Rossiter *et al.*, 1987).

2. CONTINUATION IN KENYA AND UGANDA

In 1955, several small outbreaks were reported in Kenya from African pastoral areas, some of which were places frequented by buffalo. Owing to the State of Emergency prevailing since 1953, FMD, and drought, the usual vaccination program was disrupted, and in some areas, absence of the disease for some time had resulted in a state of complacency in respect of vaccination.

The next confirmed case in a wild ungulate was in an eland near Kinna in Eastern Province in 1974; then, in 1978, of 26 buffalo sampled in the Narok district of southwestern Kenya, four were found positive, and in 425 sera samples from 18 different species collected between 1970 and 1981, 35 from 13 different species contained rinderpest neutralizing antibodies, indicating that these animals had been in contact with the virus.

In 1981, a high prevalence of antibodies was found in small ruminants sampled in western Kenya, but other studies reaching from the Uganda–Kenya border in the north to south western Kenya proved negative, and only one low positive titre was found among 528 sheep and goats sampled in north-east Maasailand in 1981–2 (Rossiter *et al.*, 1987).

No antibodies were detected in 42 giraffe sampled in northern Kenya in 1982, a species regarded as a sensitive indicator host. In 1983, antibodies were found in waterbuck and impala in Nakuru National Park, but none in 20 samples of each taken in March 1985. Seventy samples collected between 1983 and 1985 from six other species were all negative, but antibodies were found regularly in buffalo, the highest prevalence being in the Maasai-Mara Game Reserve where 31 of 35 buffalo sampled and born between 1978 and 1983 were positive. Infection was apparently highest in the east, appearing to die out in the west of the Reserve. Positive samples were detected in 12–15-month-old animals after each sampling in November 1983, January and December 1984, and February, May, and June 1985, indicating that seroconversions had occurred frequently and possibly continuously rather than in one outbreak. Many of the positive calves born in early 1984 were probably infected late that year. There were no reports of rinderpest in buffalo or other animals during the period, although many buffaloes died in the Maasai-Mara Game Reserve between October and December 1984, deaths being attributed to starvation caused by drought and army worm. Seven buffalo sampled in Tsavo West National Park were negative, but one of three animals sampled at Naivasha in 1984 was postive (Rossiter *et al.*, 1987).

A major wave of infection swept over Karamoja and Uganda's Northern Province in 1960–1, killing large numbers of eland, kudu, warthog, buffalo, bushbuck, giraffe, impala, oryx, bongo, and bushpig, and creating blindness in giraffe. But the infection did not move south. After the 1979 war in Uganda, it broke out again in Karamoja, and there were continued reports of small localized outbreaks in several parts in 1983–5, probably due in part to restocking with unvaccinated cattle to rebuild the herd which, because of drought and disease, had fallen from some 600,000 head in 1977–9 to 200,000 in 1983. In 1985, 27 of 53 unvaccinated cattle were positive, and 11 of 46 goats, suggesting that rinderpest was enzootic in the area; and there was an outbreak in West Lango following introduction of cattle from Karamoja by road transport, several hundred cattle dying. One of six goats tested in the area was positive. Although controlled by vaccination, in March it was reported that the disease had moved south to Luwero, north of Kampala, almost certainly through uncontrolled movement of market cattle. In early 1986, the disease was reported in western Uganda around Tororo and Mount Elgon. It spread to Jinja in 1987 and Arua in West Nile in 1989.

3. THE 1990s KENYA OUTBREAK

In 1986–9, an outbreak passed unnoticed, reported in the Nairobi area as malignant catarrhal fever and confused by the presence of theileriosis and other infections (Wamwayi *et al.*, 1992). A severe outbreak was contained in Tsavo National Park in 1993, but there was then heavy mortality in Tsavo and other parts of eastern Kenya in 1994–5, the outbreak killing several hundred lesser kudu, buffalo, impala, eland, and kongoni. Initially, the virus was thought to be of the Type 1 lineage originating from the southern Sudan and neighboring areas but was found to be completely different, confirmed as the Type 2 lineage found in West Africa up to 1983 and active in East Africa since the 1940s, related to previous viruses which affected game north and east of the Tana river in the 1960s. It was isolated from a reticulated giraffe in Kenya shot near Garissa in 1962, suspected to have originated in Somalia. Over the previous 10 years, only virus of the Type 1 African lineage had been detected in Kenya, Sudan, and Ethiopia, and it was thought that the giraffe lineage (RGK/1) had become extinct. Thus, a second main focus in East Africa was revealed after remaining undetected for 32 years, two distinct lineages co-existing for many years. The exact location of the focus was uncertain but believed to be north-eastern Kenya and southern Somalia (Barrett *et al.*, 1998). The genetic stability of the RGK/1 virus is not unique. If environmental conditions remain stable within an ecological niche, the virus could remain stable at a peak of genetic fitness, and as a result, a relative genetic stasis would be observed, despite the high polymerase error rate characteristic of such viruses. The virus was distinct from those of the Sudanese and Ethiopian Afar foci. Its outbreak resurrected the question of a possible game-adapted strain only eradicated by killing the game, but the pattern was classically epizootic and game sampled in adjacent areas appeared negative, an unlikely occurrence if rinderpest was enzootic in the region. The enzootic source is uncertain, as it is considered that such a virulent strain could not have persisted unnoticed in Kenya game for over 30 years, undetected in the JP15 campaign and 8 years of Pan-African Rinderpest Campaign (PARC), and serosurveillance of game in north-eastern Kenya has not implicated game in maintenance of the virus in a silent form. Reversion to a mild form could be a means whereby the virus escapes detection for many years, and cattle in eastern Kenya, Mandera, Wajir, Garissa, and Lamu districts as far west as Meru and adjoining Somalia were considered as likely enzootic foci. Here, often in combination with drought, it probably caused periodic reductions in game animals over the last 40–50 years. Each epizootic can last several years followed by gradual recovery of game populations given adequate rainfall. Serological surveys indicated that it had not been active in north-central Kenya from Samburu-Buffalo Springs northward through Losai to Marsabit. However, the mid-1990s epizootic spread westwards from Tsavo to Amboseli National Park and north to Kajiado and Nairobi National Park. It also spread south into northern Tanzania from Loliondo to West Kilimanjaro and to Lake Manyara, with Ngorongoro and Serengeti clearly under threat (East, 1997).

The source of the Tsavo outbreak was not determined but believed to be connected with illegal movements of large herds of cattle from Garissa along the eastern border of the Park.

One rumor blames Somali refugees moving south to safety with their families and livestock, the source presumed to have been infected cattle introduced west of the Tana river between Garsen and the Tiva river from Garissa, Wajir, or Somalia, during drought in 1993. That of the Meru outbreak was likely to be similar because there are no longer contiguous game populations between Tsavo and Meru. Each year, cattle are moved seasonally from Garissa District to near Tsavo National Park for grazing, and it is believed that in 1994 some carried the virus which was passed to wild animals already weakened by drought, occurring also in Meru National Park in the north. After apparently disappearing, it re-emerged in Nairobi National Park at the end of 1996, conservative estimates suggesting that of 300 resident buffalo, 200 died. It was effectively controlled through emergency ring vaccination of all herds in the southern districts, imposition of quarantine, and strict livestock movement control.

When first observed in April 1994 at Voi in Kenya's Tsavo East National Park, with buffalo exhibiting diarrhea and low mortality, rinderpest was not suspected especially as drought conditions prevailed, and Kenya was considered free of the disease, so serological tests were confined to FMD. A few weeks later in the north of the Park along the Tiva river and around Ithumba, numerous lesser kudu were reported blind and dying (one carcass per 16 km^2 was found by a helicopter survey of 400 km^2). The high mortality of lesser kudu in which the most commonly observed clinical sign was blindness caused by keratoconjunctivitis raised many questions. It was then discovered that many buffalo had died in late 1993 along the Tiva, a sand river. During the months following April 1994, all susceptible species in Tsavo declined by between 40 and 70%. Buffalo initially suffered a mortality of about 15% in October 1994 in Tsavo West National Park, falling in numbers from 35,000 in 1991 to 5,000 in 1997 from drought and rinderpest. In October, the disease was apparent among buffalo in Tsavo West National Park near the Taita Hills and within weeks had spread to the Chyulu Hills to the north, confirmed by serology in November. It appeared to have just reached buffalo in Amboseli National Park in January 1995. Sick buffalo were reported there in March, but mortality was low. Cases were then reported among lesser kudu and eland, the last case being a lesser kudu in May 1995. In that month, blindness in kudu, giraffe, and bushbuck was reported in Meru National Park over 300 km to the north where few buffalo were present due to seasonal displacement elsewhere. Rinderpest positive serum samples were obtained from kudu and buffalo in Mkomazi, Tanzania, in 1995, and 50% of goat samples from near Amboseli tested positive.

The next outbreak was at Nairobi National Park in October 1996; observed first in an eland, within a month it had spread to all the buffalo herds. Sick impala were also seen, characterized by malaise, weight loss, mild nasolachrimal discharge, and staring hair coat but no mortality. By contrast, in Tsavo, mortality among impala was estimated at 10%. There was no evidence of infection in warthog, a normally susceptible species. Hartebeest, gazelles (Thomson's and Grant's), reedbuck, and wildebeest also appeared unaffected. Persisting only in cattle, this was the first time the virus had become established in cattle in Maasailand since the late 1950s. Apparently mild in cattle, of which the majority was vaccinated, it was not so in wild ruminants. The estimated mortality for the largest species (those most readily counted from the air) from all causes (drought, poaching, rinderpest) between 1991 and 1997 in the 40,000 km^2 Tsavo ecosystem is shown in Table 2.

Buffalo and wildebeest numbers in the Serengeti-Mara region were relatively stable since the early 1980s, but in the 1994 drought, buffalo declined by 80% in the Maasai-Mara.

There was no evidence to suggest a major outbreak among cattle in Kenya before the Tsavo outbreak, but in 1996, the disease was confirmed in cattle in southern Kajiado, and by 1997 in northern Tanzania. Mild in cattle, it may have passed unrecognized elsewhere.

Table 2. Estimated mortality (all causes) of major wild ruminants in the Tsavo area of Kenya in the period covered by the rinderpest outbreak of 1991–7.

	1991		1997		
Species	Estimated Population	Standard Deviation	Estimated Population	Standard Deviation	Estimated Mean Percentage Mortality
Buffalo	34,583	±11,368	5,488	±1611	84
Eland	5,073	841	796	173	84
Giraffe	9,054	661	2,018	316	78

Source: Kock *et al.* (1999).

Kock (2000) considers that in vaccinated or partially vaccinated cattle, a strain of the lineage causes only mild disease which is often missed by herders and anyway is of no economic importance to them. There is some evidence of this strain causing epizootics among game in Kenya and Tanzania on a 10–15-year cycle. The last reported incidence in Kenya was in December 2001, two buffaloes in Meru National Park surviving infection with lineage 2 virus.

In February 1998, continuing presence of rinderpest was established in the southeast Sudan enzootic focus in an area of civil strife and uncontrolled livestock movements, Torit County, close to the Uganda border in an area between Torit and Laifon, involving the extensive pastoral herding systems which spread into the neighboring areas of Ethiopia, Kenya, and Uganda. The virus was shown to belong to African lineage 1 previously associated with outbreaks in southern Sudan and adjoining regions. Approximately 80% of animals of 6–12 months of age and 10% of adults showed symptoms with 50% mortality. Although the origin of the infection was not confirmed, it was thought possibly to be infected cattle brought in from a neighboring area, although there were no reports of outbreaks elsewhere; but for security reasons, the suspect area could not be surveyed until in 2001 it was established that the virus had probably been present since 1998 in animals of the Jie and Murle tribes in Eastern Equatoria and Jonglei. An outbreak of serious stomatitis-enteritis disease in the vicinity of Pibor in 2000–01 was identified by the pastoralists as rinderpest. Veterinary opinion was mixed, and investigations attributed the disease to schistosomosis, but the Operation Lifeline Sudan (OLS) Livestock Program, which FAO assumed responsibility for in late 2000, seized the opportunity to mount an intensive vaccination campaign among the almost 1 million head of Jie and Murle cattle between May 2001 and May 2002. This was the culmination of an effort which had begun in the early 1990s when the United Nations International Children's Emergency Fund (UNICEF) encountered resistance to its child vaccination program in southern Sudan: "Vaccinate our cattle and then you can vaccinate our children," they were told, "because if our cattle die, then our children will die anyway." Joined by the American Tufts University and nongovernmental organizations, UNICEF launched a livestock program within OLS, providing thermostable vaccine to respected herders who were taught how to use it in their communities.

Despite difficulties and operating in remote bush without roads or infrastructure, working on both sides of a conflict-racked area, the FAO campaign was completed successfully by the end of June 2002, and vaccination ceased throughout the whole of the Sudan. Checks up to November 2002 have shown no evidence of the disease, and all indications are that the African lineage 1 virus could now be extinct (Anon., 2002a).

In April 1998, a boat from Somalia carrying infected cattle was turned away from the port of Dubai and is believed to have unloaded the animals at Kismayu in southern Somalia.

In early 1999, there were reports of an outbreak among cattle in southern Somalia and high mortality among warthog. In the Lower Juba region, there was evidence of clinical symptoms related to rinderpest of a benign form which usually did not cause mortality in cattle. Livestock owners claimed to recognize both a mild and a severe form of rinderpest. For security reasons, control and investigative work in the area had to be suspended, and rinderpest was not definitely confirmed, but it is believed that rinderpest remained in the Somali cattle herds of southern Somalia and north-eastern Kenya where it has reappeared periodically in cycles of about 5 years. In November 2002, it was stated that an intense international effort was required to focus on this final area before the virus breaks out through the movement of nomadic herds and the export of cattle. The southern part of the Somali pastoral ecosystem could reinfect nearby areas, but trade could also carry it across the Red Sea to the Arabian Peninsula and even to Southeast Asia, to where traders were arranging to export cattle (Anon., 2002b).

4. CONFLICTING VIEWS ON CONTROL

There were two conflicting viewpoints concerning inoculation. One school of veterinary opinion was that the disease could be eradicated only by immunization programs if it were not for the large concentrations of highly susceptible game animals which coexisted with cattle and migrated freely in northern Tanganyika and into Kenya. An opposing viewpoint was that, if eradicated in cattle, it would disappear because it "burned itself out" rapidly in game (Plowright, 1963a). Reid in 1948 had pointed to a "smouldering reservoir" of rinderpest in game but considered that it died out in smaller game communities, as opposed to the large concentration of Serengeti wildebeest.

Whereas there is ample evidence of the occurrence of rinderpest in sheep, goats, and domestic swine, there was no evidence for these species contracting it in Tanganyika, and they were never included in any immunization campaign. Only in the 1980s were rinderpest antibodies found in sheep and goats in southern Tanzania. In Uganda, antibody studies suggested that the disease was enzootic in these species, the majority of infections in sheep and goats in Africa probably being subclinical or inapparent, confined to young animals in the enzootic areas (Scott and Brown, 1961).

Provost considered that there were only two possible explanations for the isolated instances of the disease in Kenya in 1974 and 1978. One was the natural occurrence of a hypovirulent strain in buffalo in Kenya, in which case there would always be a danger of it erupting and regaining its virulence in cattle. The other possibility was that buffalo were infected with PPR, which would be less alarming. The widespread prevalence of rinderpest antibodies in buffalo in the Maasai-Mara Game Reserve and surrounding area in 1983–4 was quite unexpected (Rossiter *et al.*, 1987), the seroconversions taking place 2–3 years after the last clinical outbreaks in the northern Serengeti, and 12–18 months after the last reported outbreak in Tanzanian cattle some 500 km to the southeast. The high titers to rinderpest virus found in these buffalo suggest that rinderpest, rather than some closely related virus such as PPR, was responsible, as antibodies to such viruses and to other morbilliviruses react much more strongly with homologous virus. In September 1998, 3,000 sheep and goat serum samples were tested in Tanzania for the presence of PPR, but all were negative, confirming the belief that the disease was absent from Tanzania.

Branagan and Hammond (1965) considered that, indirectly, drought was a most important factor in disseminating rinderpest due to the large-scale migrations in search of food and water it induced, heavy concentrations of cattle and game then occurring around water points; as well as natural river barriers being breached.

5. THE PAN-AFRICAN CAMPAIGN

Until 1950, when vaccination campaigns were increased, some 2 million head of cattle were dying annually of rinderpest in Africa, Asia, and parts of Europe; and in many parts of Africa, it still remained enzootic.

In 1962 an internationally funded and coordinated vaccination campaign, Joint Project No. 15 (JP 15), was initiated. It began operations in parts of Cameroon, Chad, Niger and Nigeria, in Phase II extending to Dahomey, Ghana, Ivory Coast, Mali, Upper Volta and other parts of Niger and Nigeria. In 1965, the project was extended to Gambia, Guinea, Liberia, Mauritania, Senegal, Sierra Leone, other parts of the Ivory Coast, and Mali. In 1968, it was extended to East Africa, to include Ethiopia, Kenya, Somalia, Sudan, Tanzania, and Uganda. In Ethiopia, 60 million vaccinations were carried out, but follow-up measures were inadequate, the disease remaining in highland areas. In Nigeria, the number of cases decreased from several hundred annually prior to 1962 to two outbreaks in 1963–4 and was allegedly eradicated entirely in 1972. In 1970, it broke out in 74 places in Upper Volta but was contained by 1974 with an overall total of 2,238 deaths.

At the termination of the campaign in 1976, over 70 million head of cattle in 22 countries had been vaccinated at a cost of US$51 million (Anon., 1983a), but internecine difficulties

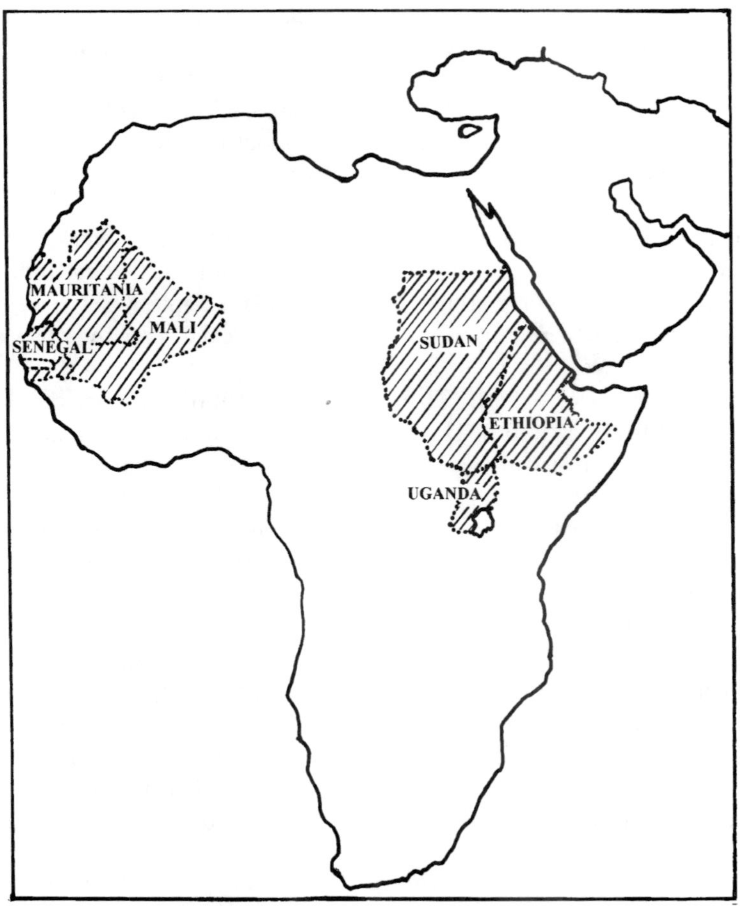

Figure 2. Known presence of rinderpest 1975–9.

in some countries, particularly Sudan and Ethiopia, meant that complete coverage could not be achieved, and the disease remained enzootic in some areas. Beginning in 1975, it spread progressively from a focus in southeast Ethiopia to cover the entire country once again and cross into the Sudan in 1977. In central Ethiopia in 1977, a rinderpest-like disease was reported among greater kudu, characterized by eye lesions and blindness, as well as nasal discharge and diarrhoea (Figure 2). From 1976 to 1979 in West Africa, rinderpest was reported only in Mali, Mauritania, and Senegal; but after 1979, the disease began to spread once more in Nigeria, Niger, Benin, and Upper Volta. This spread probably originated from Mali, developing among the huge mass of cattle which congregates in the dry season on its border with Mauritania near the Niger River. There was an upsurge in outbreaks in 1979 with 27 reported cases, compared with 7 the year before. In Upper Volta, no foci had been reported in 1979, but in 1980, there were eight outbreaks with further outbreaks in 1981–3 introduced by nomadic herders from Mali and Niger, causing a total of 671 deaths over the period, including animals slaughtered. In April 1981, national parks and hunting areas of Benin were declared infected and closed to the public. The virus was not virulent but spread rapidly eastwards in the Sahel. Then, provoked by drought conditions, the largely unvaccinated focus in southern Sudan erupted, a virulent strain moving rapidly west to Nigeria producing arguably, according to Scott (1993), the biggest cattle disease disaster of the 20th century. One third of the cattle belonging to the Fulani died, and many Fulani herdsmen committed suicide, while their relatives were obliged to seek famine-relief camps. In Nigeria alone, 500,000 cattle died, and 2 million were affected.

In the early 1980s, it was reported in cattle in eastern Ethiopia near the Somali border and also in the southwest. In 1982–3, outbreaks were reported in 50% of the administrative regions. With approximately 27 million cattle at this time, Ethiopia had the largest cattle population in Africa. Many countries failed to conduct compulsory annual vaccination of the calf crop, so that 10 years later, after the vaccination campaign in the mid-1980s, herds possessed very little immunity.

Re-emerging in East Africa in 1979, there were widespread outbreaks until mid-1983, a vaccination campaign eliminating it from Tanzania, but it remained enzootic in northeast Uganda. In Kenya, buffalo in and around the Maasai-Mara Game Reserve were exposed to infection in late 1984, but there was no known case of clinical disease in Kenya and Tanzania during 1983–5 (Figure 3).

Following resurgence, some US$2 million were spent on a further campaign in 1981 covering 10 West African countries. This was followed up by an international conference in Kenya in November 1981 at which a Pan-African Rinderpest Campaign (PARC) was proposed under the aegis of the Organization for African Unity (OAU) and the Inter-African Bureau for Animal Resources (IBAR), which would be conducted by FAO and the International Office of Epizootics (OIE) (Figure 4).[1] The aim (as proposed in 1948) was eradication of rinderpest in Africa with simultaneous vaccination campaigns in all countries with enzootic foci. Immunization against pleuropneumonia would be conducted at the same time. It was considered that ultimately some herds would have to be quarantined and animals slaughtered, and to ensure that suspected cases of infection were reported, it would be necessary to provide for compensation for the destruction of herds. A 4-year project, followed by a 6-year follow-up phase to eliminate residual foci, was proposed, commencing December 1984. But by June 1983, the finance required, calculated at US$83 million [but eventually rising to over US$90.1 million (Anon., 1983b)], had failed to be raised, reducing the campaign

[1]This was followed by the West Asian Rinderpest Eradication Campaign (WAREC), which ended in 1994, and the South Asia Rinderpest Eradication Campaign (SAREC).

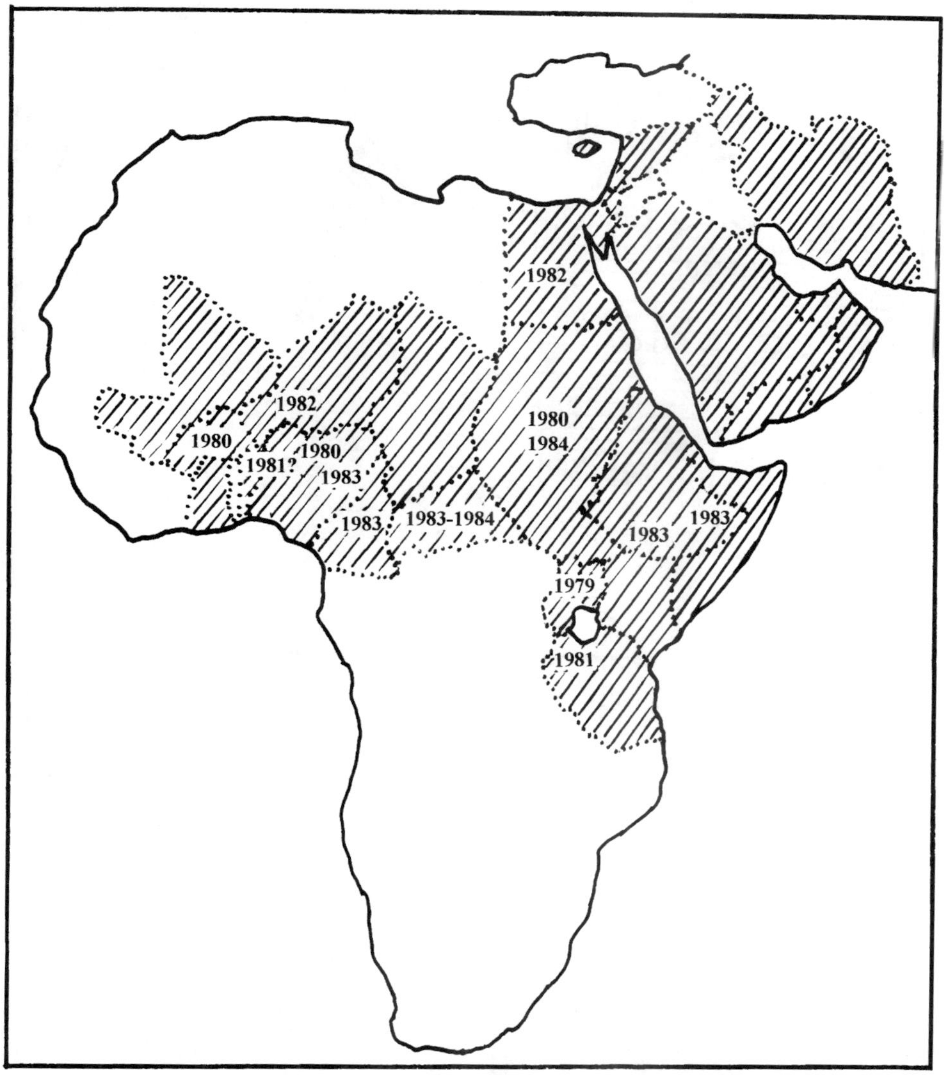

Figure 3. Known presence of rinderpest 1979–84, Africa and the Near East.

to emergency assistance. It was to begin field operations eventually in 1987, and by the year 2000, donor-financed operations had cost nearly US$300 million (Kock, 2000).

After the emergency campaign conducted in 1981, further outbreaks occurred in Benin, Ethiopia, Mali, Nigeria, Somalia, Sudan, Uganda, and Upper Volta. The following year, there were outbreaks in Chad, Ethiopia, Ghana, Mali, Niger, Sudan, Uganda, and Upper Volta. In 1983, the disease reinvaded Nigeria from both east and west, and it was demonstrated that two distinct viruses were involved. One source was probably the large nomad herds in Mauritania and Mali, the other Ethiopia.

In 1987, field operations for PARC began in 34 countries (Figure 5), largely funded by the EEC. America, through the US Agency for International Development, was unwilling to contribute, electing to concentrate efforts on developing an improved vaccine. The World

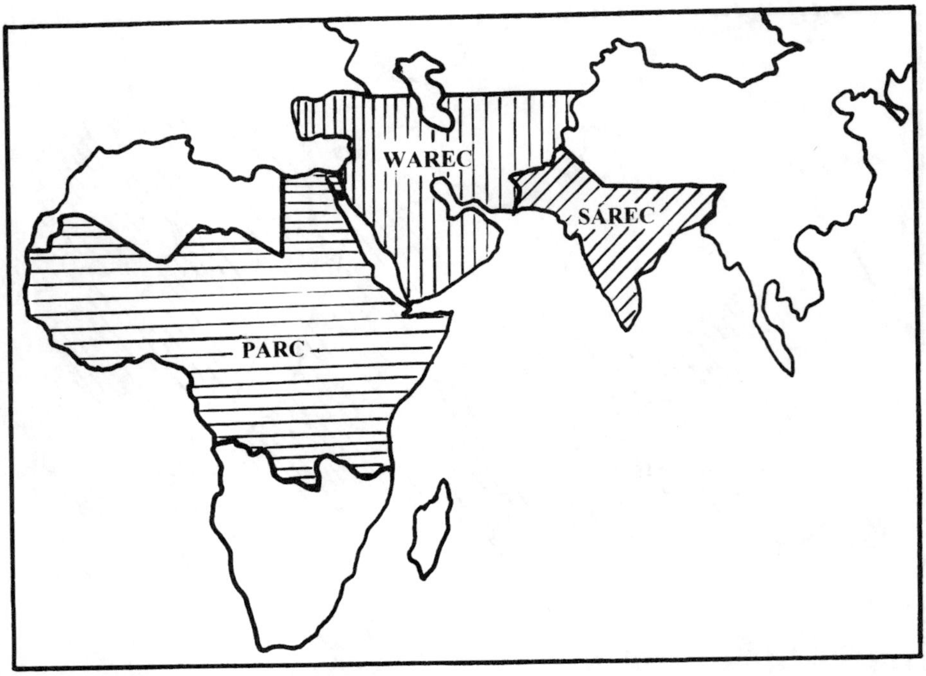

Figure 4. Zones of rinderpest vaccination campaigns, 1981.

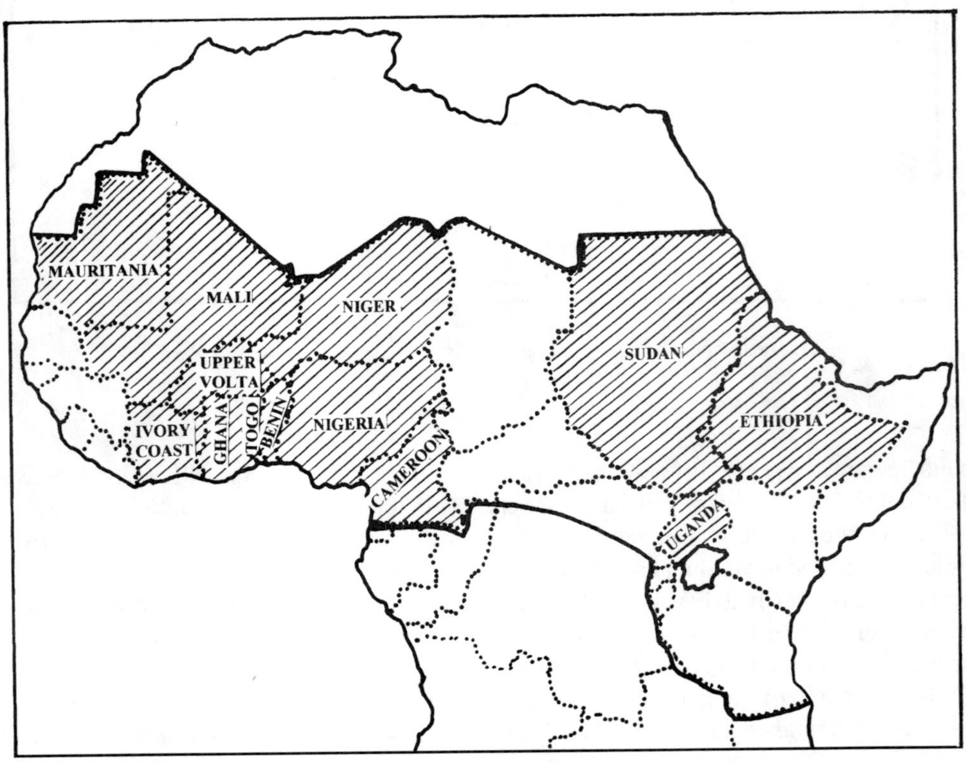

Figure 5. Countries reporting outbreaks of rinderpest in 1985–6 and area of the PARC vaccination project.

Bank also expressed reservations about the campaign, largely because the former campaign had been unsuccessful due to the weakness of the African veterinary services and the inaccessibility of areas where there was civil strife. The World Bank also saw question marks over whether game animals represented a reservoir of the disease and the role of PPR. But according to Scott (1993), initial results were spectacular, rinderpest not having been seen in West Africa since 1988. But it remained in the southern Sudan and neighboring areas of Ethiopia, Uganda, and Kenya. In August 1988, outbreaks were reported in Kenya in Nairobi, Kiambu, and Kajiado. Explaining the outbreaks, the Minister for Livestock Development stated that in recent years, inoculation had not been carried out properly due to a shortage of transport and insufficient equipment.

In 1993, Eritrea detected infection in the northern tip of the Afar focus, and in the last half of 1994, some 40 outbreaks occurred in the east of the Ethiopian highlands through trade with the Afar lowlands, but the outbreaks were quickly contained. Closure of markets resulted in a diversion of trade through markets in the west and the northern highlands, causing a small number of outbreaks there which were also quickly controlled. In early 1995, infection occurred in Asmara in trade cattle but was rapidly eliminated.

Outbreaks in Southern Sudan and Kenya in 1986 and 1993 were only mildly pathogenic to cattle, the strains being closely related to each other and to strains in Egypt and Nigeria, but distinct from the mild strain in East and West Africa in the 1960s which was closer to the Asian and Middle East lineages.

In 1994, FAO established a new priority program, *Emergency Prevention System for Transboundary Animal and Plant Pests and Diseases* (EMPRES), of which the initial thrust would be against rinderpest with the Global Rinderpest Eradication Programme (GREP) in conjunction with the OIE, and a target date for worldwide eradication by the year 2010. The United Kingdom's Pirbright Laboratory was designated the FAO World Reference Laboratory (WRL) for rinderpest. As a result of this program, the following countries declared themselves provisionally free of rinderpest: Gambia in 1990; Togo and Egypt in 1996; Senegal, Ivory Coast, Ghana, Niger, and Mali in 1997; Burkina Faso, Nigeria, and Tanzania in 1998. By the end of 1999, Eritrea, Mauritania, Niger, and Cameroon were added to the list. Ethiopia was also declared provisionally free but remained at risk of infection from neighboring southern Sudan and Somalia. Total eradication was deemed ultimately dependent upon eradication of the disease in East Africa where two, possibly three, discrete foci appeared to have persisted for many years.

The first focus covered a large area comprising adjoining areas in the south of Sudan, lowland areas in southwest Ethiopia, northwest and west Kenya, and northeast and east Uganda, where, in the last region, the large herds of cattle of the Karimojong, Turkana, and other Iteso tribes had carried infection for decades. The northern boundary was unclear. The disease spread repeatedly north along trade routes in the Sudan into the west of Ethiopia and into northwest Kenya. This was the source of outbreaks in Marsabit in 1987 and around Nairobi in 1988. This focus was related to, if not continuous with, the southern Karamoja focus in east Uganda, which extended regularly into western Kenya in West Pokot, and in 1994 probably into the Tororo District of southeast Uganda. The enzootic and epizootic limits of the lineage 2 virus in Kenya and Tanzania were established by the year 2000 (Kock, 2000). The second focus lay to the east of the northern Ethiopian highland massif among the pastoral Afar whose extensive herds range the lowland pastures as far as Eritrea. Seasonal migration and raiding bring their cattle into contact with those of pastoralists inhabiting the semi-highland areas and with a seasonal trade in draught oxen from the lowlands to the highlands. The Afar area had been free of rinderpest until 1976 when infection was introduced along the Rift Valley from southeast Ethiopia, resulting in an epizootic followed by a continuing enzootic presence.

The third possible focus had never been clearly defined, but in 1975, the disease re-emerged in southeast Ethiopia, first in lesser kudu, giraffe, and warthog, with mortality preceded by ocular discharge and blindness, in late 1975 appearing in cattle. It was believed to have originated on the Somali border. Subsequent research indicated that a focus was present south of the Bale mountains in southeast Ethiopia until 1993, but as of summer 2000, an absence of infective activity in both Ethiopia and Uganda was established, and active control measures demarcated and sealed off the pockets of probable infection in southern Sudan and Somalia. Stock within these areas and *cordon sanitaires* around their borders was vaccinated. Elsewhere, vaccination against rinderpest is forbidden by agreement, and countries intending to eradicate rinderpest under OIE procedures were advised to acquire legal powers to control imports and production of rinderpest vaccines.

For a country to become officially declared free of rinderpest, it must adopt the "OIE Pathway." That is, it first undergoes intensive clinical surveillance. If no rinderpest is detected for at least 2 years, it applies for "provisional freedom from disease" and must stop using vaccines. Clinical surveillance continues for a further 2 years, and if no infections are detected, intensive serological surveillance of unvaccinated animals is undertaken for at least another 2 years. If no seropositives are detected, the country is declared "Free from disease." Active serological surveillance continues, and if no evidence of rinderpest is found, the country will be declared "free of rinderpest" after 2 more years, that is, 7 years from the declared intention to follow the "OIE Pathway." Thus, to achieve the target date of global eradication by 2010, final foci must be made extinct by the end of 2003.

In the middle of the year 2000, PARC was replaced by PACE, the Pan-African Control of Epizootics (PACE) giving wider coverage of animal diseases. PACE is funded by the EEC and operates under the umbrella of the International Atomic Energy Agency.

Provost (1980) considered it puzzling that veterinary authorities had embarked upon the original JP15 campaign, for if, as was generally believed, wild ungulates were the reservoir of the virus, there was no hope of eradicating it by simply immunizing cattle. But the campaign was a success where applied, and eradication of the virus in cattle, sheep, and goats led to its spontaneous disappearance in wild ungulates. In Kenya, buffalo were tagged during the epizootic of 1993–7 and later re-examined and compared serologically with those born after the epizootic, which were negative, supporting the claim that the species does not act as a reservoir (Kock *et al.*, 1999). Earlier, Plowright (1963a) concluded that rinderpest of a mild type had been continuously enzootic in the cattle–game contact areas of Tanganyika and Kenya at least from 1930 to 1963, involving relatively stable strains of virus, and the Serengeti game population, even in the absence of cattle, was capable of continuously maintaining the virus, a conclusion which proved incorrect. It is now accepted that the absence of reservoirs in wild ungulates means that vaccination campaigns can eradicate the disease (Anon., 1997), but recently, Kock (2000) has suggested: "It may be the very effort at eradication through vaccination of cattle that is encouraging the evolution and persistence of this particular strain [i.e., the game-virulent lineage 2 strain]." Scott (personal communication), however, considers that GREP and the "OIE Pathway" are on course to the goal of total eradication. Should GREP succeed in eliminating rinderpest in domestic stock within its target date of 2010, the only possible continuing source of the virus might be in wild animals, and we can expect research to switch to a more detailed study of its possible presence in other species and the means of its maintenance and transmission, research essential to ensuring the permanent success of GREP (Figure 6). A start in this direction has already been made by the Zoological Society of London working in Kenya with the Kenya Wildlife Service. Among the project aims are: gathering information on the role of game in rinderpest; assistance to Governments in decisions relating to the OIE pathway for the declaration of freedom from

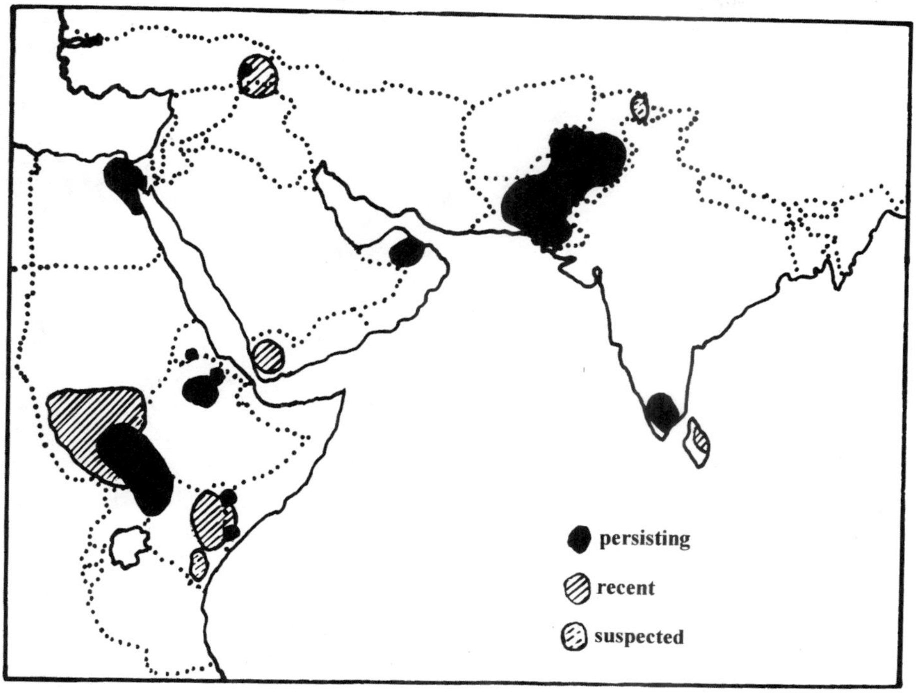

Figure 6. Global Rinderpest Eradication Programme target areas 1996 (after Rweyamamu, 1996).

rinderpest using wild animal sentinel populations; and improved knowledge of the significance of rinderpest to wild animal populations. This is achieved by two regional coordinating units, one in Kenya to cover eastern Africa, and one in the Central African Republic covering west and central Africa (Kock, 2000). In 1999–2000, a wildlife surveillance project was initiated, collecting sera of over 5,000 wild animals from Burkina Faso, Central African Republic, Chad, Ethiopia, Kenya, Uganda, and Tanzania. This again confirmed that wild animals do not maintain the virus indefinitely, and there was no spread by game outside the known remaining foci in eastern Africa.

27

ECONOMIC AND SOCIAL EFFECTS IN AFRICA

1. INTRODUCTION

Cattle are not the recent introduction to Africa that historians of a few decades ago supposed. We know now that in the African Iron Age, from about 200 BC to AD 400, sheep were kept in both Namibia and Zimbabwe, and cattle were present on the edge of Botswana's Okavango Delta slightly before AD 300. In the central Sudan, cattle were present by 3500 BC and in Kenya possibly by 4000, perhaps 5000 BC (Gowlett, 1988). Import of Asian humped-back zebu cattle is believed to have occurred 2,000–3,000 years ago, and this would have been via northeastern or eastern Africa. The impression is that pastoralism arrived in East Africa after the onset of drier climates to the north and spread to the south as drier climatic conditions permitted. Predating Bantu entry into the region, cattle are known to have been present between central Africa and the Cape from before AD 300, and may have been present at the Cape by the 4th or 5th centuries (Smith, 1992). Herskovits (1926) has evaluated in detail the former significance of cattle in African cultures in eastern and southern Africa, where, above all, they signified wealth, providing milk for consumption cattle were rarely used as a source of meat, and their loss following the rinderpest panzootic had its effect primarily upon the social order.

Although some authors (e.g., Kjekshus, 1977) have given the impression that before the advent of colonial rule in eastern Africa, the country was a land "flowing in milk and honey," cattle herds in the middle of the 19th century were not as large as at first appears. Speke (1863) and Grant (1864) may well have expressed astonishment at the thousands of cattle possessed by Rumanika in Tanganyika, one estimate being 10,000 on the banks of the Kitangule River, but they had traveled 1,100 km or more across Tanganyika without reporting seeing any at all. Stanley in 1876, when 16 km from Issosi traveling to Kasya in Karagwe, counted 32 separate herds totaling 900 head (Stanley, 1878); and between Mawangara and Ibanda in the Ruzizi basin, he passed over 4,000 head (Stanley, 1890). Parke (1891) wrote of Mbarara in Uganda: "There are thousands of cattle roaming over this region." Le Roy (1914) noted that in 1890, people in Tanganyika's Umba Valley kept very few cattle because they attracted the Maasai, while Stuhlmann (1894), describing Kafuro in 1891, states that Hima pastoralists drove their cattle out in herds of 100–1,000 head.

But Europeans were not used to seeing large herds of cattle and may have exaggerated their impressions. The total cattle population of England and Wales in 1864 before rinderpest broke out was estimated at 3.4 million, as in 1779, and that of Scotland just under 1 million (Gamgee, 1866). A herd of less than 30 per farmer was probably close to the norm, and the overall average was only 0.2 head per hectare, so herds of just a few hundred would have been viewed with some astonishment. It would appear that the numbers of cattle in Africa at the beginning of the century were relatively small compared with later years since, apart from the Sahara, there is no evidence of former habitat destruction in Africa on the scale which began to manifest itself in the 1930s from overstocking.

The number of cattle in Africa today, which in 1983 was estimated at almost 176 million head (FAO, 1984), is many times, probably 30, in excess of the number which existed before the panzootic. A cattle census undertaken by the Germans in Tanganyika in December 1901 suggested a total of 671,000 head, but a more accurate count in 1903 revealed 460,752 head, plus about 50,000 in Ruanda and Burundi, from which, Kjekshus (1977) estimates that Tanganyika must have had at least 4.5 million head in 1890. Even if this was a realistic estimate, compare it with that of 14.5 million in 1984, plus 1.2 million in Rwanda and Burundi (FAO, 1984). But a total of 4.5 million head is biologically improbable. If we assume a 95% mortality in 1890 and the cattle thereafter increased at only half of the potential maximum rate of increase, the total would have been about 3 million. A more probable figure would be 1.1–1.5 million, assuming a somewhat higher rate of increase and less than 95%, but greater than 90%, mortality. Koponen (1988) states that there is no reason to assume that the German figures were accurate, but in view of the fact that the figures were probably required for tax assessments, it seems more likely that they would be overestimated than underestimated. In 1904, Hollis estimated that the Maasai in Kenya had 50,000 head of cattle and half a million sheep and goats, in 1910 rising to 130,000 and 2.5 million, respectively. On Laikipia, it is alleged that it took eight seasons for the herds to recover (Waller, 1988). If Southern Rhodesia lost 100,000 head, and this represented 95% mortality, then the country could have had only a little over 105,000 head. But the Ba-Ila of Northern Rhodesia in the Kafue region are alleged to have possessed more than 70,000 after the panzootic, the biggest owners having up to 6,000 head each. Compare this with today's estimates of 5.5 million in Zimbabwe and 2.4 million in Zambia. The Bamangwato of Botswana allegedly lost upward of 800,000 (actually thought to be less than 9,000), but in 1984, there were an estimated 2.9 million head in Botswana. In Cape Colony, contemporary estimates gave the number of cattle, excluding those in the Transkei, as 1,639,435 before the rinderpest. By May 1898, 575,864 (35%) were lost to the disease. In the Transvaal, some 980,000 head were lost, claimed to represent 66%, thus giving an estimated total before the rinderpest of almost 1.5 million. Natal had a little under half this, where actual figures show a pre-rinderpest total of 738,450 head and a post-rinderpest total of 221,298 (Table 1), representing a 70% loss and not 46% as is quoted.

Copying Roscoe (1922), Ford (1971) advanced the theory, relative to the spread of the tsetse fly and trypanosomiasis, that prior to European intervention in Africa, tribes and communities tended to live in isolation from one another. If this was true, it is possible that if the

Table 1. African and European cattle units in Natal.

Year	African	European	Total
1895	508,938	229,512	748,450
1896	494,402	242,165	736,567
1897	114,829	125,992	240,821
1898	75,842	145,456	221,298
1899	122,007	174,024e	296,031e
1900	135,409	202,591	338,000
1901	140,577	212,817	353,394
1902	147,606	230,014e	377,620e
1903	229,554	247,211	476,765
1904	291,560	261,171	552,731
1905	329,323	265,838	595,151
1906	387,000	243,857	630,857

Source: Ballard (1986). e: estimated.

cattle population had been maintained at the level it was before 1890, the majority of populations would have been too small to maintain the virus in an enzootic form. This could account for an apparently less widespread nature of the disease if it did indeed exist in Africa south of the Sahara prior to the 1890s. St. Croix (1945) quotes a disgruntled old Fulani who once said to him: "I never saw all these different diseases until after the 'white men' came: my cattle were all right until everything was changed by them." But Ford's thesis of stability overlooks the turmoil in both western and eastern Africa probably occasioned by the slave trade. West Africa was consumed with the ravages of the slave trade in the 18th and 19th centuries, which spread to the east in the 19th century, while the Ngoni displaced from southern Africa allegedly fought their way north as far as Lake Victoria causing disruption on a grand scale. The Herero pastoralists of Namibia raided the Nama to the south for cattle, but with most of the cattle dying in 1896, they became hunter-gatherers. Long before this, raiding Arabs surged unchecked up the Nile and westwards into the central Sudan as far as Lake Chad after the downfall of Dongola in the 14th century and Alwa in 1504. In addition, cattle owning tribes across the Sahel and down the eastern side of Africa would have responded with cattle-raiding whenever drought or diseases other than rinderpest destroyed their herds. Oral histories of the Meru in northern Tanzania refer to continual wars about the middle of the 16th century forcing them into a mountain retreat, and further conflict in the late 19th century when they clashed with Maasai pastoralists.

It would be difficult to assess which regions in Africa suffered most from the panzootic as our knowledge depends upon reports made at the time, and these are sporadic, with the exception of the detailed studies in recent years of events in parts of southern Africa. But the greatest human suffering seems to have been in Ethiopia. However, we find that in many countries, rinderpest only compounded the problems of wars, drought, famine, locust plagues, and fatal human epidemics such as cholera and smallpox, all taking place at the end of the 19th century. Serious drought was experienced from north to south of the continent, and rinderpest was only one factor in the devastation wrought among African economies. Waller (1988) points out that there is little evidence from Maasailand to support the view that the panzootics precipitated a general ecological crisis, stock epizootics being less harmful in their long-term effects than either endemic disease or drought. In the absence of drought, recovery of herds is rapid because more food is available for the survivors. But along the Nile in the Sudan between Wadi Halfa and Khartoum, it was considered by Head in 1906 that only a third of the land was cultivated, compared with that before the Mahdists destroyed the people (Head, 1906). Baird (1900) referred to the southwest corner of Lake Tana as "a melancholy bleak country, harried successively by the Dervishes, the native Rases and the rinderpest." A village headman related how, 7 years before, the Dervishes had killed almost all the inhabitants, destroying a flourishing district.

There are conflicting accounts of the number of cattle which died, one being 2.5 million in South Africa and 5.3 million in Africa as a whole. We have seen how the numbers were grossly exaggerated in Botswana, but even so, these are comparatively low figures by today's standards, for although Mettam (1937) states that there were less than 2.5 million head in southern Africa in 1937, in 1984 there were almost 13 million in South Africa alone, and over 25 million in southern Africa as whole. Mack (1970) stated that 61.4% of 1.6 million in South Africa were saved. But Wallace (1896), who had just completed a survey of South Africa's agriculture leaving the country about March 1896 before the disease had reached Cape Colony, provided census estimates to show that the number of cattle in South Africa had been declining from 1,557,166 in 1891 to 1,290,461 in 1894–5, although rainfall had been normal over these years. But one of the worst droughts on record in 1896 would no doubt have drastically reduced numbers still further, even in the absence of rinderpest.

Lugard (1893) was mistaken when he wrote: "And the pity of it is, that in all probability this vast destruction of the sole wealth of these millions of human beings, and the terrible starvation and mortality among the pastoral tribes which have followed in its wake, might have been arrested by the ordinary precautions which civilised veterinary science would have prescribed..." for it was to be another 3 years after he wrote before a form of prevention was found. Even if it had already been known, difficulties of access and transport would have meant that inoculation of these vast herds was an impossibility at the time.

2. EFFECTS IN ETHIOPIA AND SOMALIA

In the 1880s, the Continent had suffered the pleuropneumonia panzootic, contributing with drought, famine, smallpox, and rinderpest (other epidemics in Ethiopia at the same time were typhus, cholera, dysentery, and influenza) in the view of Anderson and Johnson (1988), to making the 1880s to 1890s in northeastern Africa stand out as the most significant historical marker in the changing relationships between societies. But Pankhurst and Johnson (1988) argue that the great Ethiopian famine of 1888–92 was only one of a long series of natural calamities in the region reaching back in history. Ethiopia had already been ravaged in 1875–6 in a war with Egypt, followed by subsequent Egyptian harassment. A drought beginning in 1888 caused harvests to fail, followed by rinderpest, which killed off the plough oxen so that further crops could not be planted in sufficiency. Drought and rinderpest were attended by plagues of locusts and caterpillars (probably *Spodopterus* spp.), and in Amhara in 1892, thousands of rats were reported. With cultivation almost impossible, acute famine followed in which the starving peasants succumbed to cholera and smallpox, much of the country becoming depopulated. One Ethiopian missionary observed: "Thousands are dying. The famine is indescribable... Famine has spread over the whole of central and western Abyssinia..." (Aragawi in Pankhurst, 1968). In Shoa, roads were lined by dead and dying bodies. Thousands of famine victims abandoned homes and farms and made their way to the coast in the hope of obtaining imported grain. Taxes paid to the Emperor, particularly cattle tax which was not reimposed until 1903, were stopped, the peasants refusing to work. The Emperor himself set an example by hoeing and planting a field with his army and opened up his granaries. Reduced to eating roots of trees, people murdered for a scrap of bread. The potato, introduced 30 years before but ignored, became an item of food for the first time. The Emperor ordered caravans of rice from the coast, but little arrived at Shoa as they were pillaged by the Somalis and Danakils, who were also starving. With the exception of a few rare fields, the fertile plateau of Shoa was a wasteland.

Loss of oxen for ploughing was particularly serious for those Galla who had recently abandoned a pastoral life for agriculture, but knew of no implement other than the plough. There had been 3 years of famine in the Buré area, and in Harer, many people were dying. The pastoral people of Fogara had been mainly nomadic herdsmen obtaining grain as required from neighboring agriculturalists in exchange for livestock. Destitute of stock, they had to become agriculturalists themselves, a pattern repeated throughout pastoral Africa.

A missionary reported in March 1890: "Everywhere I meet walking skeletons and even horrible corpses, half eaten by hyenas, of starvelings who had collapsed from exhaustion" (Coulbeaux in Pankhurst, 1968). Adowa was reported as a deserted city like "a cemetery." Antonelli (1890), who had traveled between Harer and Addis Ababa a year or so earlier, was appalled by the change he found in 1890: "Previously the country was inhabited; there were very beautiful fields of durra and barley, numerous herds of cattle, sheep and goats, and the whole area had an atmosphere of abundance and prosperity." But now the country was "one of continuous desolation ... absolutely a desert; no more inhabitants, no more cultivation, no more flocks." The social disintegration led to the eating of traditionally forbidden food, abandonment or sale of children by their parents, self-enslavement, suicide, murder, and cannibalism, even reports of mothers eating their own children (Pankhurst, 1968). As was to happen elsewhere, the missionaries received a number of converts, largely in exchange for food.

Coming at a time when Menelik was rebuilding the Ethiopian Empire by expanding southwards, the plague facilitated the advance of his forces as the Gallas were too weakened to provide an effective resistance, but the Italians also saw it as their opportunity to advance, and they moved up to the plateau in the summer of 1889, occupying Asmara in August. The vacated lands were seen as "abandoned," causing the Italian Government to embark in 1890 on a policy of settlement which was not terminated until the final defeat by Menelik in 1896.

The plague and subsequent famine, lasting until 1892, led to a substantial increase in prices, and although these gradually fell, they never returned to their former level. In 1897, it was reported that cows which formerly cost $5–10, now cost $25. Cattle prices increased 30–40 times, and that of plough oxen 20–30. In the Galla country, the price of an ox rose five to seven times. Prices of all food increased, wheat some 150 times, affecting all classes of people. The rich resorted to selling their cows, goats, mules, and lands, and an export traffic in hides was considerably diminished. Baird (1900) reported that trade in hides had formerly been a flourishing industry but succumbed to rinderpest.

The Mursi, a pastoralist tribe inhabiting the Omo Valley in southwest Ethiopia, lost most of their cattle about 1891–2, becoming dependent upon cultivation, hunting, and fishing. Bòttego's expedition in 1896 reported very few goats and oxen (Vannutelli and Citerni, 1899).

Somaliland suffered likewise, and in the southern Somali region, many people died of apathy because they could not accustom themselves to a diet of food other than meat and milk. More than one rich Somali stabbed himself to death on finding that all his cattle had died overnight (Ferrandi, 1903).

3. SUDAN

The Sudan had been ravaged by the Dervishes from 1879, destroying its once extensive commerce, caravan roads becoming obliterated (Figure 1). Their final defeat only took place at the end of 1898 and, when rinderpest passed through, the Sudan was already a devastated and impoverished country. The cattle of the Shilluk and Dinka tribes which inhabited the east and west banks, respectively, of the Nile north of its junction with the Sobat, were driven off. The Bari once had numerous herds, but, pillaged by the Dervishes, they were forced to retreat inland about 16 km from the Nile, and only the chiefs were left with a few head. Grogan and Sharp (1900) in 1898 reported how their vast herds of cattle written about by Baker in 1861 had been looted and destroyed. But the Dervishes did not reach further south than Dufilé; beyond was untouched, as was the country east of the Nile, the eastern Dinkas still owning enormous herds, unlike the less fortunate western Dinkas. In March 1889, the Dervishes killed King John of the Abyssinians at Metemma and seized large numbers of cattle which had not yet succumbed to rinderpest, but which perhaps spread the disease westwards to the Nile. During the long war in the Sudan, camels and oxen were almost exterminated, and in 1889, rinderpest further reduced cattle numbers. Following a bad harvest in 1888, 1889–90 was a famine year, sweeping off hundreds of people at Omdurman where bullocks had already become scarce.

The Nuer who inhabit the central southern Sudan, claimed to have been much richer before the panzootic and their bride-wealth and blood-wealth were reckoned in 40, sometimes

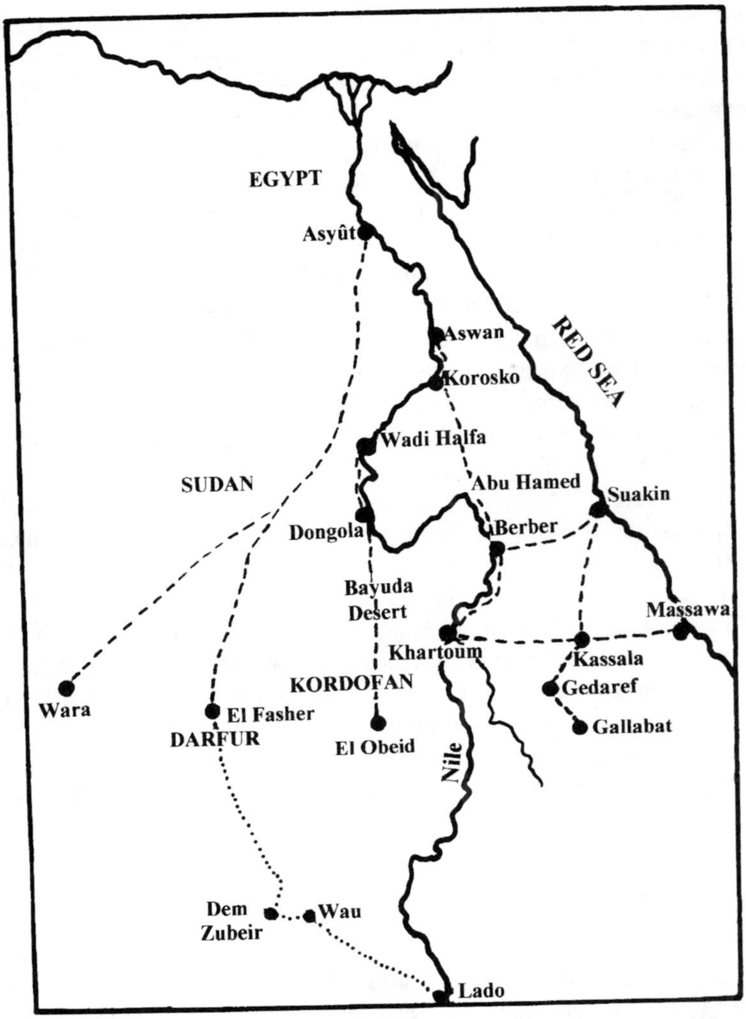

Figure 1. Principal camel routes in the Sudan before about 1888.

50–60, head of cattle. In the late 1930s, this had fallen to 20–30 for the bride-wealth, but the blood-wealth level was maintained. Rinderpest prevented a complete dependence on milk foods, obliging them to depend upon a mixed economy and was considered a worse calamity than crop failure, for if both occurred in the same year, people expected older age groups to be wiped out, although carcasses of rinderpest victims were eaten. Evans-Pritchard (1940) considered that it could not be exactly estimated what, and how extensive, had been the social changes resulting from disturbance of the ecological equilibrium caused by the panzootic. For example, raiding of their western Dinka neighbors to replenish herds probably intensified, thus affecting their relationship with them. But he takes no cognizance of the effect of the Dervish raids, which must have been directed more at them than at their Dinka neighbors. Among the plains Murle of the southern Sudan, a cattle-owning tribe, initiation of young warriors, an important rite in African societies, seems to have ceased around 1890, possibly due to the rinderpest, for considerable numbers of oxen were sacrificed in its attendant ceremonies. The reason given was that a monster termed *Mainlorien*, "the size of a small mountain," stalked the country one year, killing off the cattle. But Lewis considered that *Mainlorien* by itself could not be regarded as a reason for any basic change in their society, for there was subsequently a gradual growth in the number of marriage cattle, such that by the 1950s, the Murle were richer in cattle than most of their neighbors (Lewis, 1972).

4. UGANDA

Prior to 1890, Uganda had been ravaged by tribal warfare, and for nearly two seasons, little cultivation undertaken. South of Lake Albert, a country formerly rich in cattle had been devastated several times by raids from Bunyoro, while an area of 4,800 km^2 was practically uninhabited through fear of Mahommedan raids. The great plains to the north of Bulamwezi were "abandoned to game." Much of Singo was deserted, and the rapid deterioration of its plantations "accelerated by the depredations of vast herds of elephants." A third of the population had fled the country or had been killed. Those who remained were further destroyed by famine and disease (Macdonald, 1897). Thus, at Kampala in 1891, Tucker recorded that "the whole country was in a miserable condition. Many gardens had fallen out of cultivation owing to the wars, and food was consequently scarce. The plague had broken out, and whole districts were almost decimated. A cattle disease had swept away almost every head of cattle in the country." This made Captain Lugard's task of taking control of the country much more difficult for, "Distrust, uncertainty, perplexity and jealousy filled men's hearts and minds" (Tucker, 1908). In December 1891, Lugard reported of people in Ankole:

> These people are a solely cattle-keeping tribe and are themselves suffering greatly from hunger. Now their cattle are all dead they are forced to cultivate, but apparently they don't know how and produce nothing. Otherwise by this time their fields should be full of crops ... they are half-starved looking, most of them, and covered with Itch—a most filthy looking disease which is most contagious, and the body is covered with open sores like small-pox. (Perham and Bull, 1959)

Summing up, Lugard (1893) wrote:

> The enormous extent of the devastation it has caused in Africa can hardly be exaggerated. Most of the tribes possessed vast herds of thousands on thousands of cattle, and of these, in some localities, hardly one is left; in others, the deaths have been limited to perhaps 90%. In the case of the Bantu (or negroid) tribes, the loss, though a terrible one, did not, as a rule, involve starvation and death to the people, since, being agricultural, they possess large crops as a resource. But to the pastoral races the loss of their cattle meant death. The Wahuma, I was told, had perished in vast numbers with their animals. Everywhere the people I saw were gaunt and half-starved, and covered with skin diseases ... Not only had they no crops of any sort or kind, to replace the milk and meat which formed their natural diet, but many were unable to accomodate themselves to such a change, and all were completely ignorant of agriculture. The Masai are the same ...

Earlier (1892), he had written that the Wahuma were now largely dependent for sustenance on Bantu who had settled among them, although gradually learning to cultivate for themselves: "Very large numbers died with their cattle, and the residue, unused to a vegetable diet, which thoroughly disagrees with them, are a gaunt, ill-fed race, generally subject to skin diseases."

The effect upon the pastoral ruling Bahima of Uganda and German East Africa's Karagwe mirrored that upon the Maasai in Kenya, an epidemic of smallpox following the outbreak of deaths from starvation after they lost three quarters of the cattle upon which they depended almost exclusively for milk, the cattle rarely being eaten. Before the rinderpest, the Bahima only tended cattle, agricultural work being done by subordinate Wanyambo; but following the loss of their cattle, many were forced to undertake agricultural work also, destroying their old social order. Thirteen years later the human and cattle populations were considered to be not more than a third of what they had been before the rinderpest.

In northeast Uganda, the Karimojong suffered like other pastoralists, rinderpest devastating the cattle in about 1894, followed by a locust plague in 1895 and drought the following year, coupled with an epizootic which attacked sheep and goats. Thousands of Karimojong died of famine, while some emigrated to adjoining richer areas, or resorted to hunting and gathering in areas beyond their previous range. The Labwor also were dispersed by famine.

In northwest Uganda around Arua, the Lugbara were affected by rinderpest in 1912, which, together with human meningitis in the period up to 1918, caused them to revert to a cult known as the Yakan water cult, based upon drinking allegedly magic water before battle, which then made the drinker invincible. Armed with this belief, the people rebelled against the Government in 1919, causing the death of 11 policemen and a minor chief. Rinderpest occurred again in 1924–5, one formerly wealthy lineage having only three cows left of 120. As cattle represented wealth, such losses caused the redistribution of lineage power bases and necessitated acquiring cash for taxation and other needs by means other than the sale of hides. As the country was unsuitable for many cash crops, with the exception of tobacco, the people were obliged to sell their labor, so the loss of cattle led to a change in the economy of one of independence to a labor-based dependent economy (Middleton, 1971).

5. KENYA

The British East Africa Company, in its reports to Britain, appears to have misrepresented the situation, for *The Times* quotes

> An important evidence of the altered conditions of the country is supplied by the report that the Gallas, near Golbanti, on the river Tana, are re-stocking the country with cattle. These Gallas, through their fears of the Somalis and the Masai, have kept no cattle for a great many years; but the prevailing sense of security from the administration of the company is allaying their fears and giving them confidence. (Anon., 1893)

The Rendille of northern Kenya in 1893 had about 80,000 camels and countless sheep and goats, but their cattle, once numerous, had been reduced by plague to about 1,000. Both the Rendille and Samburu had grazed their stock to the northeast of Lake Turkana's northern shores about 1880–90, but because of the attacks of pleuropneumonia, rinderpest, and smallpox which struck the area between those years, they were forced to regroup farther south where they came into conflict with the Turkana from the west of the lake, who were expanding eastwards around the southern shore (Sobania, 1988).

Most of the country in East Africa suitable for stock was dominated by war-like pastoral tribes mainly of non-Bantu origin, while the majority of Bantu agriculturalists were largely confined to areas unsuitable for cattle. Within a few years, the entire situation changed, and long-term indirect effects compounded the change, such as bush encroachment in the absence of cattle, and disappearance of the tsetse fly, which, when it recovered, was able to colonize a far more extensive area, thus reducing possible cattle range.

Among the Maasai, when the cattle began to die, those villages first affected banded together and raided their, as yet, unaffected neighbors. Thus, as Chanler (1896) puts it, sprang up an internecine war, which destroyed nearly all the warriors inhabiting the country of the Maasai north of Kilimanjaro. Prior to this, there were two groupings, the larger occupying the Kinangop which was the most successful in the wars following the death of the cattle as it had sufficient sheep and goats to fall back on. The smaller group, numbering about 4,000, inhabited an area between the Njombeni range and Mount Kenya, and was already becoming merged as agriculturalists with the Kythere people inhabiting the area. Numbers of women and children during the famine joined the people in the Njombeni range, and some merged with Kikuyu and Kamba. Chanler considered that the total number of Maasai probably never exceeded 200,000–300,000, unlike the 2 million some claimed.

In 1891, the Purko clan defeated and practically annihilated the Laikipia clan, whose cattle had been wiped out. The Purko withdrew after the battle, leaving the whole of Laikipia unoccupied until it was taken over by European settlers.

Following the loss of some 90% of their cattle, those pastoral Maasai who were able turned to other pursuits. In the Loitokitok region, they engaged in hunting and agriculture. Some, including the Kaputie and Matapato in the area south of Ngong, were forced to beg from the Kikuyu agriculturists. Others, like the Purko, Loita, and Kisonko, relied on cattle-raiding to replenish their herds. It was not until after the rinderpest that some Maasai settled near the base and on the lower slopes of Mount Kenya, while others settled among Chagga, Meru, Taveta, and Arusha communities of Tanganyika as agriculturists. In attempting to restock, the Purko attacked many of their Maasai and non-Maasai neighbors, which contributed to a split within the Purko and a war that lasted until 1902. However, the panzootic ostensibly did not affect sheep and goats, thus saving many Maasai from starvation (Ostertag, 1916). But Waller (1988) reports the breaking down of a system of stock loans and transfers which normally was sufficient to tide families over short periods of local difficulty. Loan animals were repossessed, and the social relations which they underwrote collapsed or were repudiated when loans were refused, claims of kinship and clan being overridden to break out into open warfare. Waller considers that it was not the losses themselves which struck hardest but the civil war and turmoil which followed.

Maasai society rested on three complementary supports: family and kinship networks which upheld the houseold and its reproduction; the local structure of camps and neighborhoods which linked households together and promoted cooperative exploitation of pastoral resources; and the age system through which community affairs were managed. Each was broken down, and initiation, marriage, and the establishment of a family herd were all disrupted. Households lacked a head, sons lacked fathers, and late initiation or delayed marriage put young men at social and economic disadvantages, affecting their standing in the community.

According to the Maasai, the plague (as well as the smallpox and the coming of the White man) had been predicted by the laibon (soothsayer-chief) Mbatian at his deathbed about 1890 (probably having heard of its ravages in Ethiopia), "Now I see neither men nor cattle, the cattle have followed my children [who died of smallpox], and the land is empty and bare" (Hinde and Hinde, 1901). The death of the cattle led to an increase in hyaenas, which scavenged the carcasses. Once there were no more carcasses, they turned to people, perhaps having become accustomed to human flesh, and became a serious danger from time to time taking Maasai children. Percival (1928) reports that they acquired the habit of assembling in packs and running down game, considered now to be their normal behavior.

Tucker (1908) reported in November 1892 that on the outskirts of the Kikuyu forest, hundreds of skeletons of Maasai were lying about in all directions:

> Deserted kraals were dotted about here and there ... their homes had been broken up by small-pox and starvation. The cattle plague had carried off vast numbers of the flocks and herds of the Masai, and as they are not cultivators of the soil their only means of subsistence was gone—and starvation claimed its victims by the thousand.

It was little better at Lake Naivasha:

> To such sore straits were these poor people reduced, that they were willing to sell their donkeys for an amount of flour valued at one penny of English money each ... It was most pitiable to see these starving men and women, many of them little better than living skeletons ... It may be that

> the starvation which devastated the country at the time of which I write-through their entire dependence on their flocks and herds for the means of subsistence-will drive the lesson home in a way which nothing else could do [i.e., teach them the value of cultivation of the soil].

Yet in 1892, both Austin and Lugard remarked on the large herds which they encountered in the Rift Valley, in the latter case raided from the Kitosh and Kavirondo, which suggests that not all of the stock there and on Laikipia had been similarly affected (Waller, 1988).

The hunter-gatherer Wanderobo of Kenya, according to Chanler writing in 1893, in former times sold ivory to traders in exchange for beads, wire, and cloth, which they in turn exchanged with the Maasai for cattle, goats, and sheep. But since rinderpest had destroyed the Maasai stock and dispersed the people, the Wanderobo, in place of beads, wire, and cloth, demanded sheep, goats, flour, and beans (Chanler, 1896).

Roscoe (1922) recorded that there were many miles of grass plains where the Maasai formerly roamed "as lords" with their large herds of cattle, which were being turned to agricultural purposes by settlers. "There can be little doubt that they would have recovered their old spirit of independence, had not the British arrived on the scene at the time of their extreme need and given them new motives in life, while they were still in a pitiable condition, needing a helping hand to raise them from poverty."

Whatever the 1880s cattle epizootic reported by Thomson (1885) may have been, the effects were not as disruptive as the 1890s rinderpest panzootic; but the latter, as in Ethiopia, was accompanied by excessive drought, locust plagues, and a smallpox epidemic; compounding effects of the loss of cattle, and the weakened state of the Maasai is generally considered to have enabled British rule to take place without resistance. Lugard wrote: "In some respects it [the rinderpest] has favoured our enterprise ... The advent of the white man had else not been so peaceful. The Masai would undoubtedly have opposed us ..." (Lugard, 1893). As a result of the plague, between 1895 and 1904 the normally uncooperative Maasai entered into an alliance with the British authorities to further mutual interest. The British, hampered by lack of money and troops, could not afford to antagonize the Maasai who controlled their lines of communication and thus used them as irregular troops on punitive expeditions, allowing them to seize stock. By the time this alliance began to break down as interests diverged, the British had managed to secure their position in the country (Waller, 1976).

In 1898, famine struck Ukamba and the coastal districts resulting from irregular rains and compounded by stock losses from rinderpest the year before, followed by smallpox, Ainsworth noting that it was apparently a regular thing for smallpox to follow famine in Africa. He estimated that approximately 25% of inhabitants in Ulu and Kitui districts died, and some 15% in Kikuyu, despite the opening of relief camps; thousands dying before they either heard of them or could reach them. He observed that this famine did not affect the Maasai (Ainsworth, 1905).

Cranworth (1919) reports that after the panzootic, there was a period of exceptional fecundity, and by 1904, a careful estimate showed the Maasai to have 50,000 cattle and 600,000 sheep. By the beginning of 1910, these totals had reached 130,000 and 2,230,000, respectively.

Recovery among the white farming community was rapid, the main problem stemming from the tick-borne rickettsial disease East Coast Fever (ECF). At the beginning of 1904, there were four significant cattle farms, and by the end of 1908, probably 20, all more important than any in 1904 with improved numbers and quality of animal. The most marked growth was in Naivasha Province, which was practically free of disease, ECF affecting Mount Kenya and Ukamba Provinces. Naivasha developed with the sale of 751 cattle considered of exceptional quality seized on the Sotik Punitive Expedition. Following this, some 11,000–12,000 cattle were seized in Nandi, about 3,000 being sold at Naivasha in early 1906, attaining higher prices than the Sotik sales. In 1908, more than 1,000 cattle were seized from the Kisii and sold. These auctions led to regular annual sales at Naivasha sparking off a growing cattle industry. Prices rose steeply between 1904 and 1908, the price of cows with calves at foot almost doubling. In a review of the growing cattle industry over this period, no mention is made of rinderpest (Newland, 1908).

However, development of East Africa might have proceeded much more rapidly had it been possible to introduce ox-drawn transport in place of carrying everything on the heads of porters until the Uganda railway had been constructed. The British East Africa Company had attempted to introduce wheeled traffic using 30 or 40 Indian bullocks with 16 bullock carts, but all the bullocks died at the inception of the experiment allegedly due to rinderpest, although in 1913, an American hunting party went on safari from Nairobi with two carts, each drawn by 16 oxen. It returned with only four oxen, the rest having died allegedly from tsetse fly (Hepburn, 1913).

In November 1892, on a visit to England, after having witnessed the rinderpest in East Africa, Lugard made a plea before the Royal Geographical Society for a veterinary surgeon to be sent there to report on the plague. It was 5 years later Haslam, after attending the August 1897 Pretoria Conference with the Natal delegation, was transferred from India to supervise the large-scale animal transport used in the building of the Uganda railway. According to Macaulay (1973), he was murdered in July 1898 by Kikuyu near Kabete while investigating rinderpest among their animals, but the official report is very different. Toward the end of 1897, Haslam was seconded from the Railway Transport Department to institute measures to check the epizootic which had reached Kikuyland and Ulu (Ukamba), arriving at Machakos at the beginning of 1898. Having done all he could in Ulu, Haslam moved on to Kikuyuland to institute similar measures. He then went to Nairobi where he formed a camp and began a series of inoculation experiments. Before he had completed his observations, he decided to join a punitive expedition against the Wakikuyu of Muluka, who had murdered some natives sent out by the railway to buy foodstuff. The expedition started without him, and, while trying to catch up with it, he and some of his porters were attacked and killed by a band of the Wakikuyu (Ainsworth, 1905).

6. TANGANYIKA

In October 1894, Tucker witnessed the effects of the famine in Tanganyika near Mpwapwa, "... we found nothing but bare fields, deserted villages and ruined homes. Here or there might be seen a few starving women or children, but practically a once busy centre of life was uninhabited." The famine persisted into the following January, killing tens of thousands of men, women, and children, for, although the rains set in well, a remarkable plague of locusts consumed the new crops (Tucker, 1908). Vageler (1912) claimed that the Wagogo possessed as many as 10,000 head of cattle before an outbreak of rinderpest in 1905, while Fuchs thought that the entire cattle population in 1907 of Mpwapwa District was less than 10,000. In Karagwe, many individual Wahima possessed thousands of cattle when Stuhlmann first visited in July 1890, but the plague wiped out the herds shortly after, forcing the Wahima to become agriculturists. Kollmann in 1896 reported empty cattle kraals throughout the countryside (Kollmann, 1899). Traveling south from Bukoba, Stuhlmann reported that only a few head of cattle were left of the former large herds (Stuhlmann, 1894). In Kimwani district, the people apparently turned entirely to fishing as a livelihood, possibly due to invasion of tsetse fly, which would have prevented cattle rearing. A German official, Lieutenant Sigl, reported that in 1892 all trade in cattle in the Tabora market, where previously a daily sale of 10–20 was the normal occurrence, had come to a standstill after the rinderpest, and there was now hardly 100 cattle left in all of Unyanyembe of the former 30,000–40,000. The situation was typical of other districts.

Of the Maasai, Stuhlmann (1892) reported that all their cattle had been wiped out, and the surviving people had taken refuge with agricultural tribes at the edge of the plains in Ndjoyi, North Ugogo, Nyangana, Hedya, and so on. They were on the verge of starvation and quite indifferent, but some had made faint attempts to engage in farming. As Kjekshus (1977) speculates, the intense tribal raiding the Germans experienced in the first years of their colonization was the direct consequence of economic losses suffered initially from the rinderpest.

In Karagwe, the panzootic had an equalizing effect, forcing the ruling Wahima to depend upon the agricultural Wanyambo. Liebert (1898) claimed that far-reaching changes had taken place among the Wahehe where the remaining cattle became concentrated in the hands of chiefs and headmen, creating distinctions rather than having an equalizing effect.

7. WEST AFRICA

In West Africa, Fulani pastoralists were said to have become demented, many committing suicide, others roaming the country calling imaginary cattle, and assaults on persons for imagined provocation or suspected derisive remarks as to loss of cattle were common (St. Croix, 1945). Others left their families and wandered unclothed, heads unshaved, eating dust, and looking for their dead livestock (Stenning, 1959). Prior to the panzootic, there were many Fulbe in what is now the Sokoto Emirate and, to a lesser extent, in Gwandu Emirate, northern Nigeria. With the loss of their herds, many of these were obliged to turn to agriculture. Hopen (1958) suggests that before the outbreak, as a result of political insecurity in the region, cattle were kept closely confined for many years and were weak and emaciated, making them more susceptible to mortality from rinderpest. But whereas close confinement aids rapid spread of the disease amongst the confined, healthy cattle are affected just as much as debilitated cattle. Some herd owners lost all their animals, while others were left with one or two. Some bartered their slaves for a cow or calf for restocking.

West African pastoralists had already been considerably affected by the pleuropneumonia epizootic of 1867–72. In Kanem in November 1871, people reported to the traveler Nachtigal that their stock of cattle had been reduced to less than a quarter, but some places he found much less affected and still with considerable numbers. The epizootic raged much less in northwest Kanem than anywhere else in the western Sudan. Formerly nomadic, Shuwa, who had been pushed close to Lake Chad and onto its islands, found that their camels no longer flourished, and they became cattle herdsmen, until pleuropneumonia destroyed the cattle, and many then resorted to agriculture, becoming completely sedentary. Playing an important part in the economy of Bornu, they became responsible for supplying the markets with grain, butter, and wild rice. Writing about 1872, Nachtigal stated that the number of Kreda people in the Bahr el-Ghazal had greatly diminished since the cattle epizootic had reduced their stocks. He also reported an epizootic among hippopotamus in January 1872 on Lake Cha (Fisher and Fisher, 1987), but this was probably anthrax. Tribes on the Salamat river were obliged to pay an annual tribute to the King of Wadai of, among other things, 1,000 oxen, reduced after the epizootic to 500–600. The inhabitants of a village south of Abeshr had only become sedentary since 1870 following the loss of their cattle in the epizootic (Fisher *et al.*, 1971). The Wodaabe pastoralists of Bornu, after suffering heavy cattle losses, were then attacked by Rabih in 1892, compounding their economic difficulties, although they attributed their misfortunes mainly to Rabih, of whom it was said "Rabih, takes a hundred cattle to make one pot of soup" (Stenning, 1959).

8. SOUTH AFRICA

South Africa had already witnessed a series of disasters throughout the 1880s into the 1890s before rinderpest struck, beginning with late rains necessitating late ploughing, resulting in late crops, which were often destroyed by winter frosts. The worst drought in living memory before the ploughing in 1896 provided the right conditions for the hatching of locusts, which assumed plague proportions and devoured the young seedlings. When the rain came in the late spring, it came in torrents which destroyed the surviving ripening wheat crop as well as other young crops (Phoofolo, 1993).

According to Maberly (1898), South Africa was teeming with cattle in January 1896 (in fact, as we have seen, the numbers were already in decline). By early March 1897, total bovine mortality was estimated at 150,000 and 25,000 for the Transvaal and Orange Free State respectively, which included the majority of African-owned cattle. At Marapyane in the Transvaal in 1897, only 100 head remained from a former 2,000. Swaziland suffered worse than the Transvaal, as very little was done to arrest the spread of the disease.

With the loss of so much of the domestic stock of the Transvaal, the government suspended all hunting restrictions, allowing free public access to game. This action was strongly opposed by many Volksraad (Boer parliament) members, who were rich men and therefore not as adversely affected as the impoverished rural community. But the government felt justified in rescinding game protection because it enabled the poor to acquire food, while in addition, because game was implicated in the spreading of rinderpest, its destruction was condoned and even encouraged. But after extinction of the rinderpest and its accompanying reduction of game, the government proclaimed many tracts of state land closed to hunting for 5 years, although this protection collapsed in the South African War (Carruthers, 1995).

In the Transvaal, cattle which died, together with those which were slaughtered, are estimated to have totaled 980,000, and in 1897–8, 1.3 million in Cape Colony, not including the native reserves and Pondoland. In British Bechuanaland, cattle numbers were reduced by 35% from 1.64 to 1.06 million. In the Transkei, 90% of cattle succumbed.

For the African, cattle produced meat, milk, fuel (in the form of dried dung), and clothes; representing not only wealth but the "pivotal point of a complex and inter-woven social, political and economic system" (Van Onselen, 1972). But Africans were found to have surprisingly high cash reserves. Tax was fully paid in some Cape districts, and Africans also spent large sums on restocking despite prices being high. The Chief Magistrate of the Transkei reported: "It is surprising (notwithstanding the high price of cattle in the Colony) how many cattle, principally young animals, are being purchased from colonial farmers with money earned on the mines and public works, and brought into the territories" (Anon., 1899). But many were forced to become migratory workers or move permanently to the towns. The railway in the Protectorate was built partly by "rinderpest labour," but it was only with extreme reluctance, according to Van Onselen, that they would proceed to the Transvaal goldfields.

The ox cart with a team of 16 animals was the principal means of transport in southern Africa before the panzootic. By 1896, the railway from the Cape reached as far as Gaborone, and from there to Bulawayo, the "Road to the North", was 640 km. Goods for the developing colony of Rhodesia had to be carried along this stretch by ox cart, a journey taking 1 month covering 22–32 km a day, drawing a load of 2 tons. Maberly (1898) reported that hundreds of ox wagons could be seen along the road to Mafikeng and Bulawayo. The cost of transport along this route was 14 s. per hundredweight (51 kg) for 800 km. By July, the country did not have sufficient cattle for ploughing, and transport costs had risen to £6 per hundredweight, an almost ninefold increase, while the price of transport oxen had risen as much as 10 times for "salted" animals. After disruption of this traffic by rinderpest, the use of oxen for draught never resumed its previous scale for the railway was hurriedly pushed through to Bulawayo, reaching there in 1897 and Salisbury 3 years later. Its initial advance through the Protectorate would have been much faster, but many Africans refused to continue working on it because dishonest contractors failed to pay them. After the rinderpest losses, the cost of ox-drawn goods rose to £5 for 45.5 kg between Umtali and Salisbury, later declining to 10 s. At the height of the outbreak, when oxen were dying in their hundreds along the road, out of several lightly loaded wagons leaving the railhead, only one perhaps would reach Salisbury with the more valuable freight. But oxen were still required by farmers, and as late as 1904, the use of camels was suggested in order to overcome the shortage. The suggestion seems to have come to nothing, although they were later used by Protectorate police in the south of the country. Horses were used for riding and drawing light carts but were susceptible to African horse sickness, and replacement stock had to be imported. Mules likewise suffered from African horse sickness, and donkeys succumbed to foot rot. In August 1897, a donkey realized £30 in the weekly market, very inferior oxen guaranteed for 2 months against rinderpest reaching the same price. A lack of transport animals forced mechanization upon those who could afford it, extension of railways became vital, the bicycle business boomed, and electricity began to be applied to transport, particularly trams. The use of steam was extended to agriculture.

The urbanization of impoverished Boers is considered to have fuelled dissatisfaction leading to the Boer War, and direct financial loss to the South African Republic was estimated at £3 million in 1899 (Vogel and Heyne, 1996). Meat supplies for the Cape were imported from as far away as Madagascar, while white men in the north of the Transvaal asked to be gaoled for the sake of food. Natives ate disinterred carcasses buried many months before, and when a locust plague came, many were too weak to gather them. A relief fund was privately started, and Kruger received a deputation asking for action to relieve the famine in the north assuring it that everything possible was being done. Totally insufficient supplies of maize were sent and shamefully exploited, but the disastrous effects were generally overshadowed by the Boer War (Gutsche, 1979).

9. NATAL

In Southeastern Africa also, notably Natal, agriculture was already suffering a disaster before rinderpest invaded, occasioned by severe drought and locust plagues from 1894 to 1896. Apart from the destruction of food and cash crops, locusts removed animal grazing, so stock numbers were probably already in decline. Overall losses were 70% with some districts losing their entire herds. Mortality among African cattle in Umsinga District was estimated at 98%. Chief Bande, once wealthy with 760 head, was reduced to comparative poverty with only 29 animals surviving.

The mortality of transport oxen in the Transvaal and Orange Free State had immediate effects upon transport riders from Natal. Thousands of whites and even more Africans were employed in the haulage of goods between Natal and the interior, which was developing apace with the gold and diamond industries. By 1897, much of the traffic was consigned to the railways, but the countryside still depended largely upon ox transport to reach the railway, as well as for the movement of food and other goods among settlements. Havoc was wrought with this internal system, and because of scarcity of oxen, transport rates were doubled or trebled. Almost complete poverty was brought to transport riders stranded on the cold winter veldt in the interior, having lost a lifetime's investment in oxen (Ballard, 1986). In the Transvaal, the stamping-out policy meant that oxen were often destroyed, whether sick or not, and with the closing of the Natal border, no riders could return, even if they had healthy teams. One Natal African was reported in the *Natal Witness* in May as saying, "Up to the present we have spent most of our earnings in buying food at exorbitant prices, and we will not be able to hold on much longer, for we have now to send our crops by train to keep our boys alive at the wagons. This will be the last straw, for it will mean starvation in the future" (Gumede, 1896).

When inoculation was introduced in March 1897, although Africans had used a type of inoculation for 50 years against pleuropneumonia, some interpreted it as part of a campaign introduced by Government to spread the disease to force people into a state of poverty and compel them to work for Europeans at a very low wage. At Ngqeleni, Africans assumed a hostile attitude toward inoculation after they heard rumors from Fingoland that Government was poisoning the cattle to force the natives to go out and work. In Bizanza, it was rumored that it was an attempt on the part of Government to kill people by poisoning the meat and then inducing them to eat it.

As elsewhere in southern Africa, with the collapse of agriculture and the additional loss of income from transport riding, many Africans in Natal had to seek work in towns and mines. The disease also impoverished many hundreds living as tenants on European farms and either drove them off or made them more indebted to the owners. The magistrate of Ixopo Division noted in his *Annual Report* for 1897: "... I fancy the Europeans will find it necessary to reduce the high rents paid by many native tenants." Whereas the magistrate of New Hanover Division in his report hoped the losses sustained by the natives "will ultimately awaken the men to the necessity there exists for developing habits of industry ... their time seems to be spent chiefly in drinking beer and in breaking heads at faction fights ...."

The Europeans, suffering less stock loss, profited by selling "salted" animals to Africans, the price rising in Umvoti Division in 1898 by 100%, in some areas 400%. Consumers had to pay between 33 and 50% more for meat, and in 1897–8, the shortage meant increased imports of chilled meat and dairy products from Australia, North America, and Argentina (made possible by Gamgee's studies). Up until at least 1904, butchers had 5-year contracts to receive half of their meat supplies from overseas causing Natal's trade deficit to rise significantly.

African losses disrupted their entire social system for a fundamental institution was payment of the bride price with cattle, and marriages were postponed for lack of the bridewealth (Ballard, 1986). That many young men could not get married led to a growing discontent which some consider laid the seeds of the Bambatha Rebellion of 1906. Sanctioned by the colonial authority, the *Natal Code of Native Law* stipulated 10 head of cattle to be paid for the wives of commoners so the number could no longer be negotiated. The magistrate of Weenen Division commented in 1897: "... the disorganisation of the social system because of rinderpest ... since cattle is the basis of all their social contracts and ceremonies ... has been disastrous." Marriage formed productive units, cattle being acquired in exchange for the wife. If the wife bore daughters, this would bring cattle to the new marriage unit.

A serious effect upon the health of African babies and children ensued for those who depended in large part upon the system of *ukusisa*, a custom whereby wealthy cattle owners loaned animals to poorer people, who, in exchange for tending herds, were permitted to consume the milk and eat animals which died. The social and economic institutions of the African cattle system demanded a large degree of geographic mobility to function, and restrictions on moving stock had harmful repercussions on the socially extended and economically interdependent institutions which formed the core of the cattle-keeping system (Ballard, 1986).

Table 2. Prices of cattle in Natal 1895–1906 (£).

Year	African	European
1895	5	5
1896	6	6
1897	7	7
1898	13	13
1899	17	20
1900	20	16
1901	16	16
1902	15	16
1903	13	17
1904	11	16
1905	9	15
1906	10	16

Source: Ballard (1986).

The decline in peasant farmers after 1895 was also due in part to increasing demands for cheap African labor from the industrial and commercial farming sectors, and a high priority attached by European interests to dismantling the old systems of farming which were increasingly less capable of producing South Africa's food and other requirements on an internationally competitive basis. But it was rinderpest which played the major role in undermining the traditional African farming system of Natal.

In 1895, the African population had owned more than double the number of cattle Europeans possessed, and they fetched equivalent prices. With the epizootic, the Africans suffered much greater losses because prophylactic measures were initially made more available to Europeans, apart from a certain resistance to such measures by African stock owners. It was not until 1903 that African stock numbers once more exceeded European, but from then on, European prices were higher, the stock being of better quality (Tables 1 and 2).

Two years of locusts, severe drought, and rinderpest had destroyed much of the crops and nearly all of the livestock of the Africans, but with the death of their cattle, they turned to feasting and beer-drinking, which went on for days. The reported "sulky" and "restless" nature attributed to the African by the Europeans Ballard (1986) saw as an expression of resignation to calamity. But the legal requirement of dipping Africans in disinfectant when entering Natal led to intra-racial discontent, the more educated objecting to being treated like their less educated brethren (and probably like sheep also), and it was observed that the usual friendly politeness of the Africans toward Europeans was changed to an attitude of "semicontemptuous insolence." The Durban City Council was to continue dipping all newcomers until at least 1925.

10. THE BECHUANALAND PROTECTORATE

In the Protectorate, the panzootic dealt a final blow to a trade in wild animal products which had been in decline from 1855 due to overhunting of elephant and ostrich, and which had come to depend mainly on skins and heads of antelopes and buffalo, and skins of zebra. It particularly affected the San (Kalahari bushmen or hunter-gatherers) who brought in the products. Trade in cattle hides was also affected. By July 1896, large stocks had accumulated, and it was not possible to tell which hides were or were not from rinderpest infected animals. The Cape Colony therefore prohibited the import of all hides from the Protectorate lest the export trade to England be affected, and a proposal was made to raise the special police to the full strength of 50 to help prevent hides of any kind leaving the country. However, the Board of Agriculture in Britain considered that there was no risk from such hides in view of the time which must elapse before landing, and with dipping in carbolic acid solution, there should be no danger provided they had not been in contact with living animals. It was considered advisable that hides from the Protectorate be allowed through, but this was up to the Cape Government.

According to Percival (1918), "Khama and his people [the Bamangwato of Botswana] lost upward of 800,000." But the latter was a gross exaggeration of the Reverend William Willoughby of the London Missionary Society. Willoughby (1897) claimed that 60–70% of Bamangwato cattle died in April, 5–10% equalling 90,000 in May to early June, 70–80% at the end of June, 70% in July rising to 90% in September, 94–96% in October, and 96% or 800,000 by November (Figure 2). The true total was believed to be almost one hundredth of this figure, about 8,640–8,650 (Ashburnham, 1898). Likewise, probably exaggerating, at the beginning of June 1896, the London Missionary Society representative, H. Williams, estimated that of 10,000 herds of cattle, probably meaning heads, owned by Chief Sebele of the Bakwena, only 70 herds (heads) were spared. Another missionary estimated that by July, only two or three head of cattle remained alive for every hundred that had existed at the beginning of the year (Phoofolo, 1993). In his annual report for 1897, the Assistant Commissioner at Palapye stated that Chief Khama was said to have owned 300 or 400 trained trek oxen, afterwards left with insufficient to make up one span. Robinson reported on April 16, 1896 that Khama had already lost every one of his oxen. But Puxley (1929) reported that Khama had

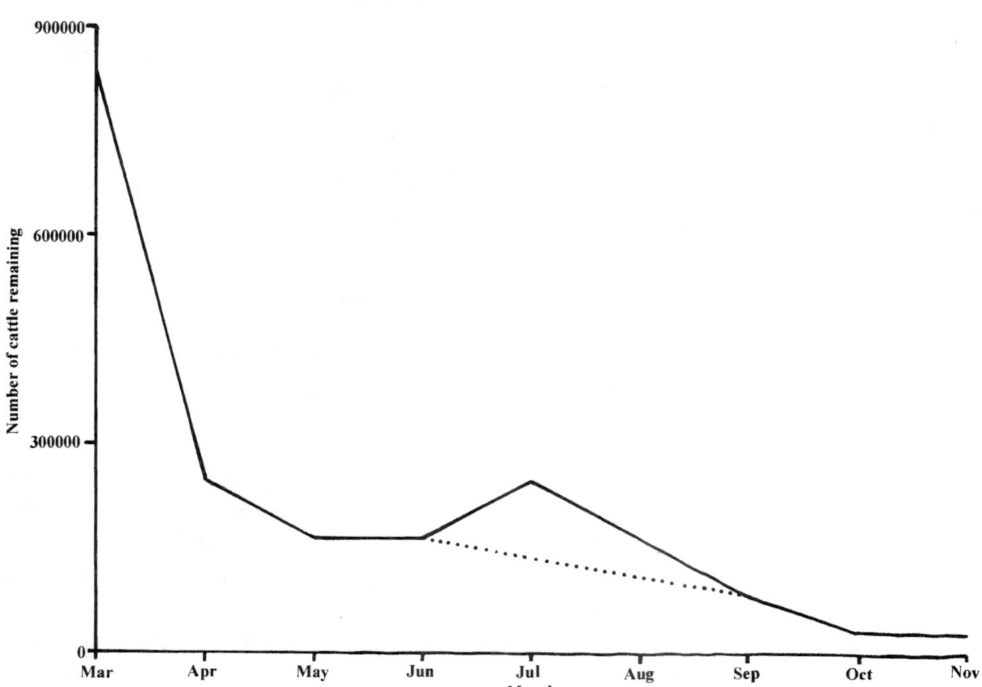

Figure 2. The course of rinderpest in Bamangwato according to Willoughby's highest estimate (Willoughby, 1897).

a few "salted" oxen left, which he exchanged for two salted heifers and shortly was the only man who possessed any serviceable beasts in that part of the country. Allegedly, his foresight resulted in his becoming a millionaire, probably the first very rich native African, but this is uncorroborated. The deposed chief Sekgoma, in a letter written in 1907 in which he attempted to recover some cattle allegedly stolen from him by San in 1903, stated: "I had bought 60 of these cattle after rinderpest and the other hundred were their progeny. I had bought them with money realised by elephants' tusks, skins, hides, horns, etc ...." It is not stated where he bought them from.

In the northern part of the Protectorate, the cost of living was enormously increased, and it was reported that Africans were spending money with reckless extravagance, buying meal, sugar, sweets, and jams at exorbitant rates. Local merchants reaped a rich harvest, but it was predicted that the lavish expenditure by the Africans of their cash would affect the trade of the district for some time to come. In some areas, traders increased prices of foodstuffs out of all proportion to the increase in transport rates. The Government dropped its tax demands during the crisis but considered the Northern Protectorate recovered well enough in 1899 to reimpose tax gathering.

As elsewhere, Africans depended upon their cattle for milk for their families. For the previous two seasons, crops had failed through drought and locusts, and after the rinderpest, many people died from scurvy believed to result from a meat diet, meat from dead or slaughtered cattle having been dried. But distress was not as great as had been expected partly because of the preparation of the dried meat which provided food for several months, but also because their reserve of purchasing power in cash, or in produce other than cattle, was greater than had been estimated. A small amount of food relief was given to destitute aged and sick persons, but there was no free grant of relief generally. Supplies of grain were provided by government at fixed points and sold at approximately cost price. Preparation against famine had begun as early as April 16, 1896 when Robinson proposed to Chamberlain that 30,000 bags of maize be imported from North America, beginning with an immediate order of 10,000 bags expected to arrive at Delagoa Bay on June 8 or 10. On June 29, Lieutenant-General Goodenough, Administrator and Acting High Commissioner at Cape Town, reported a conversation of Khama's:

> ... we think the best thing I can do is to trust to my own resources. I am very grateful for kind offer of help, but will do our best to conquer difficulties with which we are surrounded. The sickness which killed our cattle is still in our midst, and even goats were dying, but we are afraid to confide in horses and mules because they die here, and I do not know habits of mules. We see the English people express sympathy with us and we are very grateful, but we must do what we can with our own resources. We know there will be probably some hunger, but we know what hunger is, and the Bamangwato have some cattle left, and with help of these and by turning over the soil with their hands [i.e., using hoes] will doubtless be able to do a fair amount of ploughing. Perhaps Lord Loch's Committee might do something by sending out a few bags of superior quality of mealies or Kaffir corn for seed or some good pumpkin or melon seeds, but anything of this sort ought to be here at the end of October in time for ploughing season. (Goodenough, 1896)

The "London Relief Committee" offered free seed grain, but Khama said they would rather purchase it, and it was sold at the cost of transport. The London Missionary Society had written to the Colonial Office on June 2 concerning the possibility of impending famine in the Protectorate suggesting that the best relief would be a supply of remunerative work rather than food handouts, but stated that it had already launched an appeal. Government of course was well ahead of them. By July 8, 1896 it was already considered that the distress would not be as great as originally anticipated, and no further consignments of maize after the first would be necessary, one reason being that traders were also importing grain. At the end of March 1897, there was a considerable balance of unsold grain in stock, but it was pointed out that had Government not undertaken to provide grain, some traders would have made fortunes, and many natives would have starved.

Willoughby complained in June 1896: "God never made a lazier set than the Bechuanas, and if this hard year will compel them to work, it will be good." Later, he objected to the Society's distribution of food aid, as he felt it would fail to teach the people the timely lesson which rinderpest had been sent to administer: "The able-bodied should work or starve, they are the laziest set on the earth, especially the men," he wrote the following June (Phoofolo, 1993). Perhaps it was his admonishments which caused Khama to prefer to pay for the food aid. The secretary-general of the London Missionary Society allegedly welcomed the rinderpest among the Tswana, for it would drive them into the labor market. Other missionary groups were of like mind, and the *Christian Express* reported: "Undoubtedly the plague will act as a species of compulsion to the other acts and motives that are tending to drive native labour from the fields into the mines, the stores, the railways and the wharves" (Anon., 1897c).

Beginning in 1885, the Rolong had developed a long-distance cattle trade with Ovamboland, trekking cattle back some 500 km in herds of almost 4,000 across the Kalahari desert to south of Pitsane; as well as extensively engaging in transport riding on the pioneer's road to the north after 1889. In 1896, rinderpest brought an abrupt end to both enterprises.

For the period 1896–7, the Protectorate government spent £4,707 on maintaining an extra force of special police and in burning or burying thousands of carcasses in the vicinity of principal roads and rivers. Most of the chiefs and headmen eventually gave assistance in the destruction of carcasses and kraals. In August 1897, £11,181 was paid out for slaughtered cattle.

The Europeans, especially transport carriers, were reported as suffering the most: "Being less inured to hardships, and feeling the pinch of poverty more than the natives." Many had invested their last pennies in oxen and were left destitute. A few obtained temporary relief from government, but most preferred to look after themselves. Advanced money by traders to purchase salted oxen, with the now high price of transportation, the majority was soon able to repay the advances.

After the last outbreak of rinderpest in the Protectorate in 1901, it was estimated that about 175,000 cattle remained in the African territories.

11. SUSPICION AND REBELLION

Phoofolo (1993) saw the Cape Government's initial methods to control the spread of the disease by slaughter and cordons as "archaic," but whereas slaughter was recommended as early as 1711 in Britain and Italy, it is still the preferred method for disease control in the absence of a proven medical preventive, as it was with FMD in Britain in the year 2001. Provision was also made for a certain amount of slaughter in the 1987 PARC project. That the methods failed in South Africa does not mean they could not have worked effectively had there been greater cooperation and understanding as Hutcheon argued. When it came, even inoculation was to be treated with suspicion, some Africans believing that their cattle were being deliberately poisoned or that it was a device for more rapid propagation of the disease. In the Transkeian district of Nqamakwe, they suspected that their cattle were being inoculated with bad bile, while those of the Europeans were being treated with good bile, so they could sell the recovered animals to the Africans at enhanced prices once the latter's were dead. In September 1897, the resident magistrate of Kentani reported rumors current among the Mfengu that the government, being jealous of their rising strength, had introduced rinderpest, and then under the pretence of helping the people, introduced inoculation. The first inoculations were done well with good bile in order to encourage people, and then, when all had agreed to it, the cattle were to be poisoned. In this way, the tribes would be reduced to poverty, forcing the people to work for the white man at a wage of 6d. a day.

In Rhodesia, the Matabele had been becoming increasingly discontented with the white man. Having suffered a bad drought followed by a locust plague, when their cattle began to die off and then the British South Africa Company started to shoot infected cattle, this was the last straw. Senior witch doctors claimed that they had received a prophecy from the Supreme Being: "These white men are your enemies. They killed your fathers, sent the locusts, this disease among the cattle, and bewitched the clouds so that we have no rain. Now you go out and kill these white people and drive them out of your father's land and I will take away the cattle disease and the locusts and send you rain" (Ranger, 1967). When cattle first began dying in the north of Matabeleland, the *Umlimo* or god of the Makalakas who dwelt in the Matopos Hills forecast that the white man's blood was about to be spilled and claimed to have sent the rinderpest. But as Selous (1896) stated, he did not see how that could benefit the natives, although the white men were promised some equally terrible disease.

A few days after Theiler arrived in Bulawayo in mid-March 1896, on March 20, the Matabele attacked a small company of policemen, killing two. Within days, other war parties attacked isolated European farms and settlements, murdering a total of 244 white people, only those in Bulawayo and the mining town of Gwelo surviving. The uprising and disruption which it caused compounded difficulties caused by the rinderpest. Food supply was very low in Bulawayo due to a lack of transport and store cattle, creating a serious situation. Boyes (1911), on his way to Bulawayo to enlist, reported that food at Macloutsie was very dear. The price of a tin of corned beef had risen to 5s., and bread cost 1s. a loaf, but previous prices are not given. Baden-Powell (1897) noted the difficulties in June of obtaining sufficient supplies for the relief force, as it was necessary to lay down extra before the rains in November, when the roads would become impassable, and mules would die of horse sickness. The oxen were all dead, and those coming up from the Cape were not allowed to return, so it was very difficult to get contractors to tender for supplies, even if they could get them through. Selous (1896) considered that had the Matabele used a little more intelligence by conducting a coordinated campaign, they might have rendered the position of the Europeans absolutely disastrous.

Following suppression of the rebellion, the destruction of cattle by rinderpest meant that possession of cattle was no longer a question likely to excite resentment, but to Selous, the country would now remain at a standstill under a purely military occupation. Not only had the cost of living been rendered almost prohibitive through destruction of almost all cattle in Matabeleland and Bechuanaland and the substitution of mules and donkeys in place of oxen for draught purposes, but farming had also been rendered very difficult. Apart from a lack of store and dairy stock, no ploughing could be done without oxen, and no agricultural produce transported to market as salted and acclimatized horses and mules were too expensive for farm work: "The rinderpest therefore, has for the present put an end to all European enterprise in the way of mining and farming in Matabeleland." At the beginning of 1896, there were over 100,000 head of cattle in Matabeleland, but by the end of the year, Selous doubted if there would be more than 500 left in the whole country. He conceded that when the disease had died out and the railway had reached Bulawayo, the country would be gradually restocked, mining machinery imported, and the mines at last worked.

On April 13, 1896 the British Agent in Pretoria signalled the Cape that he had learnt from native sources of a probable uprising of the natives in the Protectorate. Mindful of the Matabele Rebellion, Chamberlain immediately telegraphed Robinson that he was anxious about the state of affairs, asking if troops should be sent up from Cape Town. Newton informed Robinson that he had no apprehension as to the Chiefs who were cooperating, although he thought it advisable not to press the shooting of oxen any furthur. Robinson replied to Chamberlain that it would be unjust to the Chiefs if he were to advise that additional troops were necessary, and the large force which was shortly to pass through to Matabeleland made an uprising there appear most improbable (Robinson, 1896).

Rinderpest is alleged to have been one of the causes of discontent leading to the Langeberg rebellion in British Bechuanaland (Saker and Aldridge, 1971), and it created a series of opportunities for political entrepreneurs to mobilize Africans elsewhere in southern Africa (Van Onselen, 1972). But the whole of British Bechuanaland was already humming with rumors of impending rebellion before rinderpest reached it in April 1896. In that month, isolated groups of white settlers went into *laager* (i.e., formed defensive camps together), and there was a run on ammunition in Vryburg. But contrary to white people's expectations, Chief Montshiwa's Rolong and Chief Molala's Thlaping initially cooperated with the government cattle-slaughter policy, perhaps through a lack of understanding of what was involved, described by Shillington (1985) as "government's ham-fisted attempts to halt the advance…." But Petlhu, a Thlaping leader, announced that for every ox the police shot, a policeman must be shot. Later, he was involved in the murder of the first whites.

Chief Galeshewe of the Phokwani Thlaping, who was hostile to the colonial authorities (and had spent 4 years in prison for an attack on Griqualand West in 1878), could not understand Molala's acceptance of the cattle-slaughter policy. He particularly resented the compensation which Molala received for the slaughter of 2,000 cattle. There was a widespread belief throughout South Africa that the government was deliberately introducing the disease to reduce Africans to positions of servitude. After abandonment of the shooting policy in the Taung Reserve at the beginning of October, in the middle of the month, the disease struck the Phokwani cattle. An Anglican missionary, Canon Bevan, recorded: "When the people saw that the loss of their cattle was imminent, they began to slaughter them freely; and for some six weeks or two months they lived upon fresh beef" (Bevan, 1897). He recorded that political disaffection rapidly increased as people's movements were restricted by the rinderpest regulations, and there was continuing drought. Because the cattle were all dead or dying, the land, which was hard and parched, had to be hoed by hand, a method not used for many years.

The rinderpest Regulations and the shooting of African cattle in the Taung Reserve were alleged to be the origin of rebellion by Galeshewe, although none of his cattle were specifically targeted as they did not have the disease until after shooting was halted on October 8, and the rebellion did not break out until the end of December. On November 27, the Border Siding police shot 17 Phokwani cattle which had strayed out of the Taung Reserve onto a border farm (Figure 3), 6 of which were infected with rinderpest. The chief, for a long time having adopted a negative attitude toward the police, may have used this shooting as an excuse. He demanded of the policeman in charge, Corporal Denyssen, by what right he had shot his, the Chief's, cattle, and the official policy was explained to him. The Chief then demanded compensation, which the Corporal, while sympathetic, refused to pay. According to Saker and Aldridge (1971), this incident precipitated the Langeberg Rebellion. Although the rebellion had originally been planned for the following May, Galeshewe felt obliged to act. In the previous months, he had declared loudly and openly that he would never permit his cattle to be shot and would have to be killed first himself. He was also incited by a Boer, a Transvaal Field Cornet named J. H. L. Bosman, who supplied arms and ammunition to the Tswana. When the government began slaughtering cattle, Bosman declared that this was nothing less than an attempt to wipe out the people, and in rebelling, Galeshewe thought he would have Boer support. After two attempts by the police to arrest him, police reinforcements

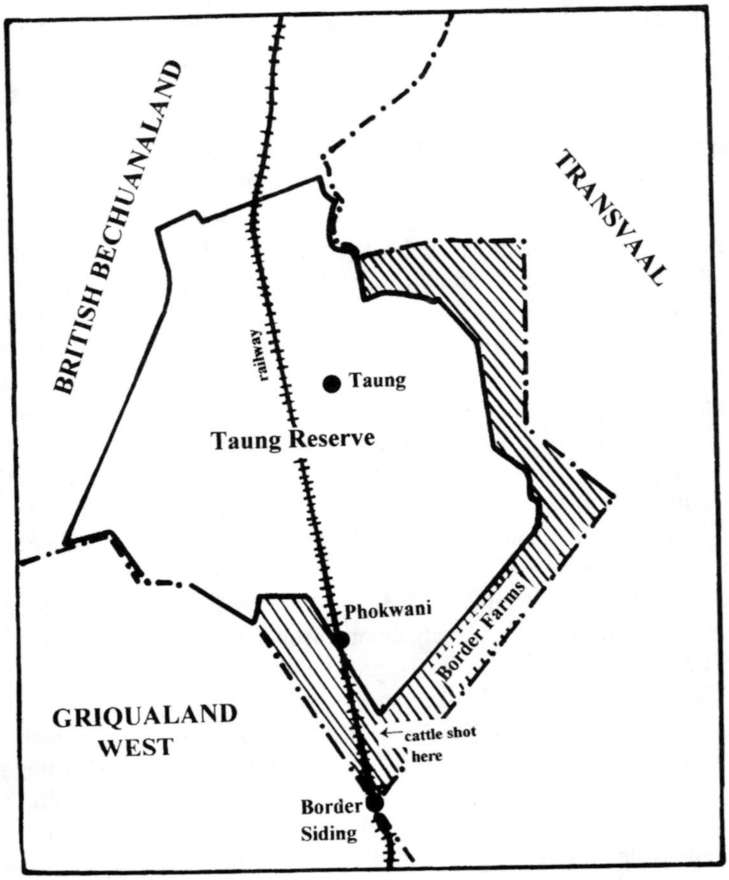

Figure 3. Area of origin of the Langeberg Rebellion dispute.

were sent to the area at the end of December to assist and also rescue a white trader and his family held hostage within Phokwani. Hostile bands of natives resisted the police, and volunteers were called in, swelling the numbers to over 400 and leading to a full-scale battle and the sacking of Phokwani with the death of 50–70 Africans. Galeshewe himself escaped to Langeberg leaving three white traders dead. The loot sold by Government on January 6 included 2,025 sheep and goats and 50 cattle, the last being all that had survived the rinderpest.

The day before, a trader was murdered by Kgalagadi, the son of a chief from the Mashowing River reserve, and two stores and a farm were looted. Government forces then cleared the Mashowing reserve, capturing the stock, but the numbers are not given.

In February, government forces were sent to Langeberg to arrest Galeshewe for the murders of the three traders. The refusal of the chiefs at Langeberg to hand him over led to a full-scale attack on the area with a scorched-earth policy, mainly conducted by a force of over 2,000 volunteers. Although the natives had some initial successes, holding out for 8 months and capturing 260 cattle from a farm near the Griqualand West border, the chiefs were killed or arrested and the inhabitants who survived the battle, numbering almost 5,000, removed. These uncontrolled movements of men and animals meant that any hope of implementing rinderpest control policies in the area was out of the question.

Shillington (1985) claims that the objective of the government and the volunteers was primarily to obtain land for white settlement. The "inalienable" Langeberg reserve was appropriated and sold off for farms and many of the captives made slaves on white farms. Although rinderpest may have exacerbated tensions, it does not seem to have been as important in creating discontent as purported, for we know that rinderpest officials were able to circulate in the reserves. The missionary J. T. Brown, in an interview with the *Cape Times* the following October, stated that in early January 1897, Chief Mongale escorted a group of white traders and rinderpest officials to safety at Kuruman. Because of overstocking and the drought, native cattle in Griqualand West had been dying of starvation for some time before the advent of rinderpest, and people had been selling off stock to buy food. Thus, when rinderpest spread through the area, the economic crisis in the native locations had already become so acute that Orpen, inspector of the Herbert locations, considered it was a toss-up between whether the remaining cattle would die of rinderpest or starvation. Between September and December 1896, 835 cattle were shot for rinderpest control, while 638 died allegedly of starvation. Although in Cape Colony, the disease reduced cattle herds by an estimated 35%, in Griqualand West because of the impoverished condition of the cattle, losses due to the disease were estimated at 90% or more. In the Herbert locations, the numbers fell from 4,500 in September 1896 to 462 by the end of the year. In the Barkly West African locations, some 14,000 cattle were reduced to 500 among a formerly pastoral people with a population numbering over 7,000.

Basutoland Africans believed rinderpest to be a product of the white man's activities. The precautionary measures to keep the disease out of Basutoland, such as restrictions on cattle movement, gave rise to a number of rumors: the government was importing loads of ammunition (perhaps this was so, for shooting cattle if necessary); men in Matatiele were being conscripted to attack Basutoland, this probably engendered by the enlistment of border guards for cattle control and the relatively large numbers eventually seen on the borders, of which Natal had 110 by the end of January 1897; and Basutoland was to be opened up to gold prospecting. But most of all, people were alarmed at the border fencing, for, prior to the annexation of Pondoland, the whites had fenced off the area.

In December 1896, one paramount chief's son, Makhaola, accompanied by a large band of armed Basuto, took up a position on the East Griqualand border where fencing was under way, halting the construction. When an assistant commissioner went to try and sort the situation out, he was forcibly detained. Alarmed by the situation, the Chief Magistrate for East Griqualand ordered the Cape border guards to disarm. Eventually, Makhaola was persuaded to leave the border but political tension remained high, and there were threats against European life and property at the village magistracies of Mefeteng and Mohale's Hoek. Makhaola now tried to elicit further support with the aim of overthrowing British rule, but was unable to get the backing of the older chiefs. The rebellion of Makhaola led to a commission to determine the exact position of the Cape–Basutoland border, and the veterinarian Soga made use of his report on the epizootic in British Bechuanaland to complain of slavery of the Vaalpense among the Barolong.

Tensions in Basutoland caused concern at the Colonial Office, the Earl of Selbourne, Permanent Under-Secretary for the Colonies, minuting: "If [Lagden, the High Commissioner] pulls us through this period of native unrest without a war, he will deserve very well of his promotion" (Selbourne, 1897). In June 1897, Lagden was knighted.

The discontent at the beginning of 1897 was also seized upon by a Griqua, Andries le Fleur, who agitated for "East Griqualand for the Griquas and the natives." Possible rebellion was nipped in the bud, although le Fleur was not apprehended until November by a newly formed Criminal Investigation Department. Released after trial, he continued his subversive activities and was arrested again after an attack on a farmstead and sentenced to imprisonment. But the problem was still present a year later. Van Onselen (1972) attributes these disturbances and the hastened establishment of a Criminal Investigation Department to discontent arising from the rinderpest.

The cattle-owning Ovaherero in Namibia, devastated by their losses, went to the mission stations where many were converted, and also began growing vegetables for the first time. But as with other pastoral peoples, it was not rinderpest alone which crushed them, for it was followed by a devastating malarial epidemic. But rinderpest ended the Ovaherero cattle monopoly, providing German settlers with an opportunity to establish a foothold in the cattle industry. Because many Ovaherero sought employment within the German economic system, from railroad construction to wood gathering, traditional Ovaherero society was destroyed (Ofcansky, 1981), but this led to a festering hatred of German rule which was to erupt in rebellion 7 years later when the Germans responded by almost totally annihilating the Ovaherero. Some in Germany thought that rinderpest had been very beneficial to the development of German South West Africa because it destroyed the cattle-owning economy of the Ovaherero, who opposed the German occupation.

In the Cape, officials had been threatened by armed men, and veterinarians engaged in slaughter were occasionally threatened by armed whites. Violence was often to attend inoculation campaigns, and in Uganda in mid-1929, trouble was anticipated in Buganda but averted by the Provincial Commissioner attending the operation. Then, during the latter's absence on leave, a religious sect known as the Malakites, led by a renegade protestant minister and a retired local chief, resurrected the protest causing the European inoculating officer to be speared in the arm and several Africans to be killed. A local Buganda chief apparently effected control of the situation and saved the European from possible death (Postlethwaite, 1947).

In Kenya in 1949, refusal to cooperate in rinderpest control measures by the Luo was allegedly due to a belief that the inoculation campaign was simply to obtain a count of cattle in connection with destocking, but early that year, a serious and embarrassing situation arose from the contamination of two batches of KAG vaccine with virulent rinderpest. The vaccine killed a high proportion of cattle herds on 20 European farms, and of native-owned stock in eastern Kavirondo, Kamasia, Maasailand, and Ukamba. Senior management of the production plant at the newly formed East African Veterinary Research Organization were cashiered. It was for this reason, that the vaccine killed a large proportion of cattle, that the Luo refused to cooperate. The Maasai elders simply said that we all make mistakes.

In July–September 1951, there was a rinderpest outbreak in areas adjacent to Kenya's Central Province, and Government ordered all African cattle in the Fort Hall District, a Kikuyu area, to be inoculated. Inoculation began, but many animals died, and James Beauttah, a central figure in the beginnings of the Mau Mau movement, believing that the inoculations were a ruse by Government to reduce the number of cattle, organized a campaign against them, delivering speeches and writing a pamphlet urging women in his district to march on the District Commissioner's office in protest. Shortly afterwards, several women in Kaharo village tried to prevent inoculations in their area and were arrested and imprisoned. Beauttah then wrote in a Kikuyu paper praising the women, which led the authorities to believe that he was behind the resistance and arrested him collecting money without a permit at a meeting, jailing him for 3 months. During this period, in early November in a series of demonstrations, large numbers of women stormed the inoculation areas and burned the cattle crushes, resulting in the conviction of 500 women (Corfield, 1960). Beauttah claims that a sample of the vaccine was sent to the British MP Fenner Brockway, who had it tested, confirming that it could kill cattle. In fact, the vaccine would have had to have been kept frozen, and it is unlikely that this was done. The Government accused Beauttah of instigating the resistance. In February 1952, he was sentenced to 2 years' imprisonment, it being believed that this was one of the earliest activities of the Mau Mau movement.

12. THE WIDER SOCIAL EFFECTS

Local colonial officials in southern Africa welcomed the destruction of the Africans' cattle "as a blessing in disguise" because it had the effect of driving more people onto the labor market. But William Hall, the Barkly West inspector, was incorrect in stating that they formerly lived from their cattle and had no call to work, "Rinderpest has changed all this, and they will have to work or starve." Rinderpest was not the sole cause of their distress and proletarianization. As Shillington has expressed it, rinderpest helped accelerate a process which overstocking and drought in 1893 and from 1895 to the end of the century, combined with coercive policies of colonization, had done so much to promote in the preceding decades. After the rinderpest, colonial officials reasserted their pressure, using the devastation of the cattle herds as both an argument and an opportunity to urge concentration of the natives into the north-eastern locations as "almost a necessity" (Shillington, 1985).

Mining companies like De Beers, while assisting in the fight against rinderpest, nonetheless profited from the labor made available to work in the mines, and as Phoofolo (1993) puts it, in the throes of the rinderpest crisis, the labor-starved Chamber of Mines, acting upon the general view that there would be no want of laborers, implemented a decision in November 1896 to drastically reduce wages in the mines by 30%, at the same time lengthening working hours.

Many other cattle-keepers had to turn to crops, and in the Transkei almost every piece of arable land was cultivated. Ploughing was done with whatever was available, in Basutoland using riding ponies. In some cases in the Cape, Africans paid Europeans to plough for them.

The panzootic probably eased land pressure for some African graziers who remained put. Van Onselen (1972) considers that it certainly contributed to the proletarianization of the African, for not only did they lose their accumulated wealth in the form of cattle, but their deserted lands were at risk. In Natal, the administration pointed out that it gave the colony a most favorable opportunity for delimiting African lands which had thus been vacated. It was not long before some officials saw the results as beneficial for, "... although reducing the native to poverty ... the native has now learnt humility to those to whom he is subordinate, and also the lesson that by work only can he live, and having learnt to work he is now a happy and contented man instead of the discontented, indolent, lazy and besotten being he was when the numerous cattle he possessed provided his every want" (Anon., 1899).

The panzootic did not produce any major structural changes in the emerging pattern of South African society but revealed, and often exaggerated, the political, social, and economic forces which were to play a central role in the industrialized South Africa of the 20th century (Van Onselen, 1972). In the opinion of Phoofolo (1993), the panzootic's greatest claim to historical significance in southern Africa lay not so much in its ability to lead directly to revolution, as in the unique manner in which it highlighted contemporary tensions and latent antagonisms which other intervening factors could have turned into explosive social and political upheavals. The striking nature of the calamity in southern Africa was that very little of the prevailing social and political tension did transfer into open rebellion, and those rebellions which occurred were fragmented and unsustained. There was not the cohesion between tribes which some feared, and the people, emotionally and physically drained by events, were resigned to their fate. But also many of the chiefs and elders were perhaps intelligent enough to appreciate that the Europeans were trying to prevent a national calamity in the shape of the plague. They were familiar enough with animal diseases and climatic disasters perhaps to know that they could not blame the government.

The long-term effect of the panzootic was more destructive to African society than the short-term effects of starvation and poverty, providing an opening for the European powers to advance and extend their rule throughout much of the continent, particularly in eastern and southern Africa. The alternative might have been a more bloody confrontation.

As with the creation of veterinary schools and services in Europe, rinderpest in Africa was responsible for the creation of significant institutions, perhaps the sole subject upon which countries united. In October 1948, another international African rinderpest conference was held in Nairobi with widespread representation by the African colonial powers. It recommended establishment of an African Rinderpest Organisation, emphasizing that complete eradication of rinderpest from the African continent could be achieved only by close cooperation between governments, game departments, and veterinary departments. The main recommendations were accepted at a further conference held in Johannesburg the following year at which the Commission for Technical Co-operation in Africa South of the Sahara (CCTA) was proposed, together with a Scientific Council for Africa South of the Sahara. The CCTA then set up the Interafrican Bureau for Epizootic Disease near Nairobi, which eventually became the Interafrican Bureau of Animal Health, serving as a coordinating and information clearing center. In 1964, the CCTA was absorbed by the OAU.

Unlike in Europe, where in Britain particularly it led to improved dairy hygiene, more humane treatment of cattle, and control of diseases through sanitary policing, nothing good can be said to have come out of the panzootic for Africa except discovery of the means to control the disease itself. In 1987, Walsh wrote: "The next decade will show whether FAO, the European Community, and their African partners can muster the will and resources to eradicate rinderpest in Africa." More than a decade later, it was still present, but recent control measures have demarcated and sealed off the pockets of probable infection in southern Sudan and Somalia.

No other single disease, human or animal, has had such an effect with such wide and diverse ramifications, from religious debate in 19th-century England, to rebellion and the destruction of power in African tribes. But because of its enzootic nature in Asia, such profound events are not apparent there; indeed, its presence may have been beneficial in preventing overstocking leading to land degradation and desertification, especially in India, centuries before European intervention.

28

THE EFFECT OF THE PANZOOTIC ON AFRICAN GAME

1. INTRODUCTION

About 1894, Heinrich Fonck, station commander at Moshi, wrote of Tanzania that one could "wander about the colony for the next ten years without seeing any more wild life than in two hours at the Berlin zoo" (Fonck, 1910). But in spite of the accounts of destruction of game in Africa, contemporary reports of sportsmen and travelers reveal that either the effects were not as catastrophic as painted or most species recovered their numbers remarkably quickly, even though before the disease was passing into South Africa in 1896 it had already begun to re-erupt in areas it had passed through as little as 4 years previously. It is conjectured that this recrudescence was caused by the virus becoming attenuated and smouldering on in an enzootic form in these localities. But when observers enthused over the quantities of game, this often included species unaffected by rinderpest, such as rhinoceros, elephant, and lion, references to which I have largely omitted from quotations of contemporary observers, with the exception of zebra, a species closely linked with the presence of wildebeest.

The effects on game are best illustrated from East Africa, most accounts stemming from Kenya because south of the Zambesi, although its effects upon domestic stock were better documented, game had already suffered half a century of hunting and displacement by European settlement, considerably reducing numbers. Vaughan-Kirby wrote of southeast Africa in 1896 just before rinderpest struck: "The disappearance of the game is not solely attributable to its extermination by the rifle; the advance of civilisation has driven it back, and caused it to retire from those parts where it was constantly hunted…." Its retreat to tsetse-infested bush where horsemen could not penetrate, and the occurrence of malaria in the summer months, were factors which in Vaughan-Kirby's view had preserved big game from utter annihilation. Findlay (1903) wrote: "There can be no doubt that Zululand was once over-run by an enormous number of noble animals… But the work of extermination went on there, as in other parts of Africa, until recent years, when measures were taken to preserve the little that was left."

In East Africa, hunting and displacement by European settlement had been relatively minor before the advent of rinderpest. By 1918, all but greater kudu and roan antelope were said to have made an almost complete recovery in numbers, although this was probably a misconception, enzootic infection depressing numbers below their potential.

2. SPECIES AFFECTED

Over 40 species of wild animal in Africa are potentially susceptible to rinderpest, but the response varies widely from subacute to peracute reactions terminating fatally, while blindness contracted by giraffe and both species of kudu is rare in cattle. The most peracute infections other than in cattle are found in buffalo, eland, kudu, and warthog. At first in the panzootic, only cattle are alleged to have suffered, although there is some doubt concerning sheep and goats in Ethiopia; but it was soon seen in the "then huge herds of buffalo" (Percival, 1918). Eland showed it early, and subsequently all ungulates in East Africa were affected; with the exception of gazelles, zebra, rhinoceros, elephant, and possibly hippopotamus.

A great falling off in numbers of oryx was noted in 1895–1900, but Percival (1928) had never heard of it suffering from rinderpest. According to an old report and African tradition, roan suffered severely in Kenya. Eland was thought to be almost exterminated, but by 1928, Percival could report that in Kenya it was as plentiful as ever, as in "the old days." For the first 15 years after the panzootic, it was a rare sight, in 1901 they were comparatively few, and the sight of a small band was an event to be noted. But its then rapid increase led to complaints of damage in the Rift Valley and on the Uasin Gishu plateau, necessitating withdrawal of special protection. By 1918, there were said to be hundreds on the Athi Plains, and by 1924, it was not unusual to see a herd of 100. Percival (1924) considered that during the outbreak, they suffered less in the bush country than on the plains. Despite subsequently many milder outbreaks of rinderpest, according to Percival (1928) one seldom now heard of any but young eland dying (indicating that the adults had been exposed to an earlier infection).

The greater kudu had been widely but thinly distributed in Kenya before 1896 but was then almost exterminated. Hinde and Hinde (1901), who knew East Africa from 1896, considered that the greater kudu might be extinct but at one time must have been fairly numerous "since decayed heads and horns are still to be found all over the plains." It was at least 10 years before it began to recover what numbers it ever had. A common species in South Africa, Stevenson-Hamilton (1929) considered that, although nearly exterminated, enough were left for the species to become, in less than 20 years under protection, the most numerous of all bush-loving antelopes next to impala. Warthog were very susceptible, according to Percival dying before there was any sign of the disease among antelopes. Giraffe sometimes died in large numbers, but in Tanganyika in 1922, it was reported as not detected in giraffe, although they were present in quantity in the Maasai cattle country. Being a relatively long-lived species, the lifespan sometimes exceeding 20 years, these giraffe could have acquired immunity from previous exposure. Wildebeest were said to be last to become infected in the panzootic according to the Maasai, only succumbing after almost all the cattle had died. Hartebeest were not reported as affected at that time, but were so in later outbreaks.

Of the earliest reports in Uganda, Lugard (1892) recorded:

> This plague has practically exterminated all the buffalo, which formerly were to be met with in great herds. It has also attacked much of the game, especially the eland, and, I believe, the pig [warthog]. Some of the small antelope (especially the bushbuck) have suffered; but the waterbuck, hartebeest and zebra, appear to have enjoyed complete immunity. The giraffe has died out; but the elephant, rhinoceros and hippo are untouched.

Other reports state giraffe being only blinded and recovering if not taken by predators. Blindness occurred again in giraffe in northern Kenya in an outbreak in 1960. Mettam (1937) reported that, although in Uganda waterbuck were seldom infected, they appeared to fear the disease, observed apparently trekking away from infected loci. One herd of over 280 was seen moving from north Singo where rinderpest existed, toward Lake Wamala where it did not.

Jackson (1894) wrote that buffalo had been almost destroyed by

> a kind of anthrax, the same disease which carried off nearly all the native cattle in 1891. The disease, I am told, was fatal to other species of game, including giraffe, eland, and lesser kudu, and even elephants; but as my informants could not speak from personal knowledge, but only from native reports, I am unable to vouch for their accuracy. However, game is still to be found in enormous quantities ....

In the 1929 outbreak in Uganda's West Nile District, buffalo suffered heavily, and bushbuck and warthog were also affected. Hartebeest, waterbuck, oribi, and duiker (also elephant, white rhinoceros, and hippopotamus) were reported unaffected. But even those animals affected survived in far greater numbers than they apparently did in the 1890s. Oribi, waterbuck, hartebeest, duiker, buffalo (not numerous), and kob (and white rhinoceros) were still found in the more open areas 3 years after the 1929 outbreak.

In South Africa, warthog, bushpig, buffalo, bushbuck, kudu, eland, duiker, reedbuck, waterbuck, gemsbok, blesbok, bontebok, springbok, and steenbok, were affected, but not wildebeest, hartebeest, or impala (Theiler, 1897a). Surmon reported that in the northern part of Bechuanaland Protectorate, buffalo, eland, gemsbok, and reedbuck were the most affected. In the southern part, it was kudu, rhebok, duiker, and steenbok (Hutcheon, 1902). The rhebok is not known with any certainty to have occurred in the country, and the reference may have been to the mountain reedbuck. That, with the exception of kudu, only small species are referred to as affected in the south, was probably because larger species had by then been shot out there. The last great springbok trek in South Africa took place in 1896, and it has been suggested that rinderpest may have played a part in this species never again recovering its former numbers. As rinderpest did not persist in southern Africa, the population was probably subsequently limited by an inability to attain its former levels in time before repeatedly being reduced by recurrent drought, habitat restriction and hunting, albeit by much smaller amounts each time (Spinage and Matlhare, 1992).

Table 1 shows species considered susceptible to rinderpest, although in some cases, this is not clinically proven. It is unclear whether it was rinderpest or anthrax which affected lechwe

Table 1. The known natural host range of rinderpest in Africa.

Order Artiodactyla	Reaction: Subacute	Reaction: Acute	Reaction: Peracute	Clinical Proof
Family: Hippopotamidae				
Hippopotamus	*	*		*
Family: Suidae				
Warthog		*		*
Bushpig	*	*		*
Forest hog	*	*		*
Family: Tragulidae				
Chevrotain	*			
Family: Giraffidae				
Giraffe	*			*
Family: Bovidae				
Buffalo	*	*		*
Bongo	*			
Sitatunga	*			*
Bushbuck	*	*		*
Greater kudu		*		
Lesser kudu		*		*
Cape eland		*	*	*
Giant eland		*	*	*
Duikers	*			
Grey duiker	*	*		*
Reedbucks	*			
Kobs	*	*		
Uganda kob	*	*		*
Defassa waterbuck	*	*		*
Puku				
Addax				

Table 1. *Continued*

Order Artiodactyla	Reaction: Subacute	Acute	Peracute	Clinical Proof
Roan antelope	*	*		*
Sable antelope	*			
Gemsbok				
Beisa oryx	*			
Black wildebeest				
Blue wildebeest	*	*		*
Bubal hartebeest	*	*		*
Bontebok				
Blesbok				
Topi	*	*		*
Impala	*	*		
Springbok				
Gerenuk				
Thomson's gazelle	*	*		
Grant's gazelle	*			
Klipspringer				
Dik-dik	*			
Oribi	*	*		*
Steenbok				
Rhebok				

In part after Scott (1970). Animals undergoing subacute reactions usually survive after a short convalescence; acute and peracute reactions are usually fatal, the latter within a week.

and puku to the north of Lake Malawi. There appears to have been an unusually widespread epizootic of anthrax concurrent with the rinderpest panzootic at the southern end of Lake Tanganyika, which has led to some confusion concerning species' susceptibility. Anthrax also manifests itself toward the end of the dry season and is not uncommon some years in Tanzania and Namibia, causing many elephant deaths in the latter country in recent years. Percival (1918) considered anthrax the commonest disease known among game, particularly on the Athi Plains and in the Southern Game Reserve in Kenya, but its characteristic is that it does not spread. In 1905, there was a very bad outbreak on the Athi Plains game dying "... till they lay so thick that hyaenas and vultures could not deal with them ... ." In the worst hit area, "thousands" died.

3. CONTEMPORARY OBSERVATIONS

3.1. Somalia

Accounts of the profusion of game in Northern Somaliland from that of Lord Wolverton who hunted there in 1892–3, Captain Swayne 1887 and 1892–3, and Prince Ghika 1895–6, would suggest that rinderpest was unknown there, perhaps promoting Adams's report (Adams, 1919) that there was no *authentic* account of an outbreak in British Somaliland before 1918. Wolverton moved with a large herd of cattle for supplies and stated that people of the district through which he was passing (the Ogaden about 200 km south of Berbera) depended upon their herds for food (Wolverton, 1894). However, Swayne (1895) stated that on the Webbe Shabeleh (Shibeli) river, about 06°N 43°E, the guide said (1893) that the hippopotamus was scarce and wary, "as the Adone negroes, during a recent famine, when nearly all of their cattle had died of disease" had killed them for food and greatly reduced their numbers: "The great epidemic of cattle disease which three or four years ago raged in Masailand and other parts of East Africa was also felt in Ogaden, the cattle and the koodoo antelopes dying of it in large numbers. It was felt as far north as the Marar Prairie [west of Hargeisa]." Referring to the greater kudu, he stated, "The large koodoos scarcely exist in the parts of Ogaden which I have visited. Either they never existed there, or, as my followers declared, they died of the great epidemic of cattle disease four or five years ago."

That the panzootic reached to the west of Hargeisa suggests that a great part of Somaliland was affected, yet this seems to have had little effect upon the game, and there is scant report of its effect upon stock. The principle game animals were Swayne's hartebeest, which occurred in vast herds on the open plains, Soemmering's gazelle, and oryx. The hartebeest was reported to have been affected in 1897, but not in the initial panzootic, neither were the oryx, Soemmering's, and other gazelles reported as being so. None of these species is water-dependent. Thus, the principle game species of Somaliland appeared to have been unaffected, and in 1887, Swayne saw vast herds of hartebeests, oryx, and Soemmering's gazelle, masses of game giving way before him as he advanced in an area where the plains were "covered with camels and ponies."

In 1892, the plains were still covered with the three principal species. Near Jig-Jiga, vast herds of hartebeest were seen "comprising several thousands," although Abyssinians came to Jig-Jiga every morning to water cattle (and horses) at the wells. In 1893, the plains near Jig-Jiga still bore masses of oryx and immense herds of hartebeest and Soemmering's gazelle.

Swayne also came across numbers of waterbuck, a species occurring only on the Webbe river, referring to them as very plentiful. He only ever heard of four buffalo in Somaliland, and these had strayed through 130 km of Galla country to the Webbe to an area where the species was otherwise unknown. The Gallas alleged that they were plentiful on a tributary of the Juba River. The warthog was common but the giraffe rare, greatly sought by native hunters, and only occurring on the other side of the Webbe. The numbers of lesser kudu, a species of localized distribution, he did not think had been "much diminished" of late years.

In 1895–6, Ghika (1898) reported that there was no big game between the coast and Hargeisa as it had been hunted out by officers from the Aden garrison, but elsewhere he met with immense herds of antelopes as Swayne had done, namely hartebeest, Soemmering's gazelle, and oryx, and there was no lack of other species of gazelle (Pelzeln's, Speke's, gerenuk). Dik-dik were numerous also. Greater kudu he reported as rare and localized, but Mounts Djigo, Dourdour, and Golis, had "many." Lesser kudu and waterbuck were also met with.

Donaldson-Smith (1897) did not mention hartebeest until he reached the Sidamo region in April 1895, noting many Coke's hartebeest and giraffe and that previously he had seen only Swayne's hartebeest, but he makes no mention of large herds. Near Lake Abaya, there were many hartebeest and gazelles, the former very numerous; and at the northern end of Lake Rudolf, vast herds of them came to drink every night. Elsewhere along the lake shore, he met large herds of hartebeest, Thomson's gazelle, and zebra. On the Gandabbela plain south of Harer, he saw large herds of oryx and zebra, while oryx, zebra, and Soemmering's gazelle abounded in the Ogaden. In the neighbourhood of Bari were many waterbuck, and they were numeous on the banks of the Web. In the region of Dawa, waterbuck and giraffe abounded, and giraffe were found in great numbers near El Modu. In the Sidamo region, he reported, "Never in my life have I seen more game than there was about the edge of these mountains. Hundreds of zebras, hartebeests, gazelles, and ostriches, as well as many giraffes and rhinoceroses, were to be seen on all sides." Then, in the Galana Amara valley (40°12′N 37°35′E),

"I never saw a greater variety of game at one time than I did in this march. Burchell's zebras, Coke's hartebeests, Waller's and Thompson's gazelles, water bucks (*Cobus defassa*), *Oryx beisa*, ostriches, warthogs, elephants and rhinoceroses abounded." At the northern end of Lake Rudolf (Ethiopia), there were many herds of topi and many giraffes and Grevy's zebra. The Rusia people here he noted were not in the same thriving condition as when found by Count Teleki in 1888 when they owned thousands of oxen, sheep and goats; all their cattle having succumbed to rinderpest. Donaldson Smith's only mention of the disease.

In 1897, Glossop (1906) found Soemmering's and Speke's gazelle, gerenuk, and warthog, plentiful west of the Goli mountains. Moving south, there were many Soemmering's gazelle and oryx; later, the country was "full" of them, with a few gerenuk. On the Toyo plains, he encountered a mass of buck "such as I have never seen before or since ... as far as the eye could reach were masses of fawn-coloured aoul [Soemmering's gazelle] interspersed with groups of hartebeest, sometimes forty to fifty in a group, but more often ten or twelve. ... One afternoon I got up to an enormous herd." Later, he counted 250 within easy range, "with, I suppose, four score or more no further than 100 yards from where I lay." Beyond the Toyo plains, he came upon a plain "which was a moving mass of Soemmering's gazelle, reminding one of the accounts of the sights that met the eyes of the early South African hunters."

The panzootic therefore appeared to have had little or no effect upon the majority of the vast numbers of susceptible game in Somaliland, which seems mainly attributable to their habits. For the most part inhabitants of arid or semi-arid country they seldom drink, and therefore probably had little contact with cattle. But in North Somaliland, in 1901, Dickinson (1908) was told that greater kudu, which were very local there, inhabiting the Golis Mountains, "were just thinking about recovering from rinderpest."

3.2. Ethiopia

In 1899, Powell-Cotton (1902) saw large herds of Soemmering's gazelle west of Harer in eastern Ethiopia, and near the Hawash valley, a herd of oryx was encountered. The plain "seemed full of game" with oryx, Soemmering's gazelle, hartebeest, zebra, and ostrich. Near the Hawash, numbers of greater kudu tracks were found, and some females seen. Buffalo tracks were also found, and oryx and warthog encountered. In January, Captain Wellby witnessed a hunt by Arusi Gallas driving an immense herd of hartebeest near Lake Zwai (Wellby, 1901).

De Bozas (1906), travelling south of Harer in 1901, noted that the fauna was abundant and varied (8°30′N 40°E), above all warthog, oryx, and kudu. Later, zebra, oryx, and kudu furnished an ample supply of fresh meat. Somewhere in the region of Bululta (6°12′N 39°03′E), tracks of a herd of buffalo were found in the forest. Lord Hindlip (1906), in 1902, encountered several lesser kudu near Lake Margherita (Abaya) and greater kudu at Lake Hora.

In 1904, Bulpett (1907) observed throughout the highlands of southern and central Abyssinia an almost entire absence of game, but it was numerous in the valleys of southern Abyssinia. In the region of the border between Abyssinia and the Sudan, south of the Adura, he reported tiang and defassa waterbuck as numerous. Nearing the Sobat were numerous herds of antelope, large herds of white-eared kob, and tiang, waterbuck, and giraffe in numbers. At the Gelo river, hartebeest were numerous, while south of the Gelo between it and the Akobo, game abounded in great quantity and variety. From the Gelo were numerous traces of giraffe and buffalo, a herd of the latter being encountered; while tiang, waterbuck, reedbuck, oribi, roan antelope, and hartebeest abounded. Across the Akobo river were found herds of giraffe and hartebeest in great numbers, and reedbuck were numerous. Numerous fresh tracks of buffalo were found on the Owag.

3.3. Sudan

Just above Dufilé on the Nile in 1897, giant eland was to be found but was a rarity. Few herds were met with, and they were extremely shy (Sykes, 1903), but this is normal giant eland behavior. On the White Nile in 1899, Buxton (1902) came across hartebeest, roan, waterbuck, and some sign of buffalo. In 1900, in Nuer country on the Bahr-el-Zaraf, Grogan and Sharp (1900) found numbers of hartebeest, Mrs Gray's kob, white-eared kob, reedbuck, waterbuck, and roan. North of Fashoda, they saw many herds of waterbuck, hartebeest, and roan. In May on the Akobo River in Southern Sudan, Austin (1903) saw "a fine herd" of buffalo, and some giraffes. In March 1901, he saw large numbers of hartebeest, gazelle and zebra in the Boma country, "the grassy plain was alive with game…." Southeast from the Karuno River, he came to a plain where he saw eland, gazelles, giraffe, elephant, and zebra. Making for Mount Nakua at the beginning of April, he crossed an open plain "swarming with hundreds of hartebeest…." At the beginning of May, approaching Mount Lubur, he saw a big herd of giraffe, in addition to zebra, oryx, hartebeest, gazelle, and ostriches (Austin, 1902). At Nimule, in 1901, the officer in charge claimed that the fauna was very rich, naming waterbuck, Thomas's kob, Jackson's hartebeest, and bushbuck (De Bozas, 1906).

Wellby (1901) reported that in the Boma country to the north-west of Lake Rudolf, immense herds of hartebeest swarmed. Traversing the country along the Sobat River the following July, he saw herds of hartebeest and waterbuck. The country near Lake Lamina was plentifully stocked with lesser kudu, hartebeest, and gazelle; and north of Lake Abaye, he reported kudu and buffalo. In northern Karamoja, he saw great herds of antelope and great numbers of giraffe. On the upper Akobo river, where large herds of giraffe wandered on the floodplains, he noted: "At present the country is given up to immense herds of antelope, elephants, rhino and giraffes."

Comyn (1911) about 1904 reported a huge herd of hartebeest near Turda in the Southern Sudan. At Kodok (Fashoda), where there was game in plenty, he saw buffalo. Near the Bahr el Arab, he saw many roan antelope and in one morning's march listed buffalo, roan, kob, dik-dik, hartebeest, waterbuck, and bushbuck. Buffalo were also met with elsewhere. On the Kuru in Dinka country, he came across a plain full of game of every sort: "To the left a small herd of buffalo … from the edge of the forest came a herd of waterbuck … Jackson hartebeest were in scores on my right, and a few single roan stood boldly out. Numbers of cob were scattered everywhere in herds and singly." At the Sopo river were waterbuck, hartebeest, and kob galore. Lloyd (1906) reported the "whole valley" of the Atepi in Bari as alive with game of all descriptions. He saw many tracks of buffalo, and close to camp were seen hartebeest, waterbuck, duiker, bushbuck, reedbuck, and warthog.

In the Zande country, Villelume (1909) in 1907 reported that the most common animals were antelopes from roan downwards in size. The bongo is mentioned as being present, but no indication is given of frequency. On the plains, kob were very numerous, as were elsewhere also wild pigs, giant forest hog, and warthog.

At Meshra-er-Zeraf, to Churchill (1908) the game reserve seemed full of game, and he shot a buffalo and two roan antelope.

In the region of the Lado enclave in 1909 were many waterbuck, kob, hartebeest, bushbuck, reedbuck, oribi, and duiker; and near Rejaf, the giant eland was encountered (Roosevelt, 1910).

3.4. Uganda

On the Semliki plain in 1900, Grogan and Sharp (1900) found "countless" reedbuck and herds of Uganda kob. During the rains, it was home to the "few survivors of the teeming herds of buffalo that formerly roamed all over this country." Buffalo were encountered near Mboga.

Wollaston (1908) in 1906 beyond Kikorongo in western Uganda saw swarms of small antelopes on all sides, the commonest being Uganda kob and reedbuck. The plain, now in the Queen Elizabeth National Park, was home to great numbers of antelopes, especially waterbuck. The Ishasha plains had antelopes almost as numerous as the Athi Plains, one of the commonest being topi, which occurred in immense numbers.[1] The reedbuck was very common and could be seen "in hundreds." Warthog, however, he recorded only at the south of Lake Edward.

Of the plain south of Lake Edward Jack (1914) wrote in 1907, "everywhere was animal life." Great herds of kob dashed past, and waterbuck, topi, and reedbuck were seen. But the first time he saw buffalo was while observing a herd on the eastern side of the lake, described as "a mass of black backs." At the northern end in the crater lake area (Kikorongo), he met enormous herds of kob, reedbuck, and waterbuck.

In 1909, Dion Lardner (1912) came across a large herd of hartebeest at Kisu, near Lake Albert, and reported them as very common at Ruengo in Buganda. Traveling from Mbarara to Kabula he recounted that all along the way, he came across countless game—herds of topi, and waterbuck, dikdik, and warthog in great numbers—while the whole district was marked by buffalo tracks. A plain in the region was the haunt "of myriads" of topi, exceedingly common in the district. Near Kirkoma in Masaka district, he came across five eland.

At Kasindi, Roby (1911) wrote in 1910, "As I lay there … I could look out across a great plain, with Lake George in the distance, and often did I see elephant and buffalo moving over the plain …."

Kob were only partly affected by rinderpest in Uganda, but hundreds were lying dead on one plain near the Mayanja river in Singo in a 1937 outbreak, while in another part of the country, "great herds" were so severely attacked that dozens of dying animals stood, heads lowered, too exhausted to move "even when touched by an observer" (Mettam, 1937).

3.5. Rwanda

In 1907, Mecklenberg (1910), between the Kagera and Kakitumba rivers, encountered a "strong" troop of eland. Toward the Kagera river, "numberless herds" of every kind of game were seen, mentioning zebra and eland, and later impala. Near the northern slope of Mount Ndama were herds of eland and topi, and close to the summit a herd of roan.

In northern Rwanda, Jack (1914) found topi the most common animal, in "immense herds." There were also large herds of impala, while eland and roan were found in small numbers. Oribi were seen everywhere, and frequently reedbuck. There were not many eland about, but he encountered a small herd.[2]

3.6. Belgian Congo

Moore (1901) in 1900 wrote that in the Rutshuru valley, he entered a land of game:

> Hitherto we had seen no more game than is usually met with in traversing any part of the great interior. There had been everywhere a certain amount, but, except on the plains north and east of

[1]In 1970, the total number of topi at Ishasha was estimated to be 3,796, with more than 2,000 in one aggregation (Jewell, 1972).

[2]In 1968–9, estimated numbers from ground counts were buffalo 6,100, impala 31,475, topi 8,425, zebra 5,525, eland 1,075, and roan 150 (Spinage *et al.*, 1972).

> Nyassa and south and east of Tanganyika, we had obtained nothing extraordinary in the way of sport ... plenty of eland, water-buck, reed-buck, zebra, roan antelope, and puku on the Tanganyika and Nyassa plateaux ... on these plains south of the Albert Edward Nyanza ... There were hundreds of head in sight, different kinds of antelope grouped together in various colored clumps ....

He then shot topi and waterbuck. Marching to Vitshumbi, he passed through herds of game "which surrounded us in all directions, and were so tame that we could walk right amongst them ...."

The following year, in the Rutshuru valley, Grogan and Sharp (1900) reported a considerable number of topi and Uganda kob. In 1907, Mecklenberg (1910) found the area covered with immense herds of antelopes: waterbuck, Uganda kob, reedbuck, oribi[?], bushbuck, and topi. Buffalo could be seen "daily in great herds in the bush." The Semliki plain was "alive with game," particularly Uganda kob and reedbuck.

In 1910, Roby (1911) travelled from Rutshuru to Vitshumbi: "Imagine an enormous plain, with thousands upon thousands of every kind of antelope—reed buck, bush buck, roans [topi?], and I know not what else. The whole place alive with them, flocks and herds of them in every direction." To Kassner (1911), in 1911, the Rutshuru valley was "a veritable zoological garden." Waterbuck, kob, and presumably topi (he calls them gemsbok) grazed in great herds everywhere. Duiker, reedbuck, warthog, and buffalo abounded. "It was a delightful sight to see thousands of these animals ... Up to the distant horizon great droves of them were grazing quietly all over the plains and slopes ...."

3.7. West Africa

Glossop (1906), in 1898 in Northern Nigeria, reported herds of roan. In Borgu, hunting kob, waterbuck, hartebeest, duiker, oribi, warthog, and roan: "There was so much game here ... that we shot pretty much what we could ...." At Bajibo, a number of buffalo were found to inhabit some islands.

Kumm (1910), in 1909, encountered buffalo on the Nigeria–Cameroon border, and one was "sure to find" them within a day's journey of Fort Archambault in Tchad. A large herd was encountered in the Shari valley, while near the Pata river, the main game appeared to be elephant and buffalo, antelopes being scarce. Game met in the Shari region were giraffe, hartebeest, warthog, waterbuck, and Buffon's kob. East of Ngore, the country was "teeming with game ... I saw five herds of cob, two herds of water-buck, one herd of Senegambian hartebeest, besides reed-buck and duiker."

3.8. Kenya

Lugard wrote of his travels in November 1890: "Through all this great plain [Kavirondo] we passed carcasses of buffalo; and the vast herds of which I had heard, and which I hoped would feed my hungry men, were gone! ... In Nandi, however, we came on a huge herd of hartebeest ...." Rinderpest had apparently not reached Lake Jipe by July 1890 as le Roy (1914) commented on the richness of game there, listing Grant's gazelle, waterbuck, red duiker, gray duiker, bushbuck, impala, dik-dik, klipspringer, sable antelope, kudu, hartebeest, and gnu, and encountering one buffalo. He noted that other travelers had not spoken highly of the area but had not camped there overnight.

In 1891, Hobley found a great number of carcasses of lesser kudu which had succumbed to rinderpest (Schillings, 1906).

Chanler (1896) reported in 1892 that the plague "which had lately ravaged the country from Lake Rudolph on the north to Kilimanjaro on the south, although it had exterminated vast herds of buffalo, and had even destroyed a large number of antelope, had apparently left the zebra untouched." Jackson (1894), writing in 1893, stated that "game is still to be found in enormous quantities," although "buffaloes too, have been almost destroyed… the same disease which carried off nearly all of the native cattle in 1891."

In the Tana region, Chanler in 1892–3, wrote of the river bordering what is now the east of Meru National Park, "our eyes were continually greeted with herds of game," but he lists only gerenuk, Grant's gazelle, waterbuck, zebra, and rhinoceros. East of Archer's Post roamed herds of oryx, Grant's gazelle, zebra, and ostrich, while he also saw giraffes and rhinoceros. Oryx, Grant's gazelle, and zebra were the species seen most often, species of low susceptibility or entirely immune to rinderpest, although somewhere in the Maralal region, he came across a herd of 12 buffalo.

In February 1893, Portal (1894) could report in the Lake Naivasha area: "on the road we had seen game of every sort and kind in vast numbers," and around Lake Naivasha itself "The prairies were literally covered with game, especially with thousands of… [Thomson's gazelle]. Zebras, too, were present in fair numbers… and a few hartebeest and Grant's antelope added to the variety…." Between Lakes Naivasha and Elmenteita, "we had seen game of every sort and in vast numbers…" (zebra, Thomson's and Grant's gazelles). Near Lake Nakuru, "the country was still full of game… the common hartebeest of the plains… was replaced by… Jackson's hartebeest…." In his diary for August 27–29 near the Ngong River, he recorded seeing "many hundreds of wildebeest, at one time some 300 in one herd, also many hartebeest and *Thomsoni*… Saw near camp waterbuck, mpallah [impala], and hartebeest… bush-buck. Saw Wanderobbo [*sic*] having drive of immense herd of wildebeest… Shot two waterbuck… saw hundreds of wildebeest… lots of zebra, hartebeest and wildebeest…." On the Athi River, he saw "large herds of mpallah and waterbuck, and one bush-buck…." On the "Mackenzie River" (the branch of the Tana forming the eastern boundary of present Meru National Park), he recorded dik-dik, warthog, impala, giraffe, waterbuck, and gerenuk.

In the same year, Colville (1895) described the region between Kikuyu and Kavirondo as an uninhabited tract, "Its only tenants at present are countless herds of game…." In western Kenya, he referred to "a plentiful supply of hartebeest" (for food). When Neumann (1898), describing the country south of the Athi River above its junction with the Tsavo in 1893, wrote:

> Here are great stretches of uninhabited bush country with a perennial river running through it, and hardly any animals… a very few waterbuck at intervals, an occasional lesser kudu or two, a little lot of impala, sometimes at one particular point, and rarely an odd bushbuck, are all that are seen, without counting the diminutive paa [dik-dik], and now and again a rhino, giraffe or hippo spoor

rather than describing a country depopulated by rinderpest, he described the arid Tsavo bush country as it was in the 1940s to 1960s, after recovery from rinderpest and before elephants opened it up. Visibility was impeded by the bush and the density of game was determined by availability of food in the dry season, with the addition of recurrent droughts in the region.

In 1895, Vandeleur (1898) considered rinderpest "seemed to have died out," and mentioned small herds of buffalo as seen "lately by government officials in various parts of the Protectorate." Neumann (1898), in his travels east of Lake Rudolf, wrote of it as a thing of the past which had almost exterminated the buffalo and greatly reduced the numbers of eland.

In 1897, on the Serengeti plains west of Voi, Tucker (1908) reported vast herds of game seen scattered about in almost every direction: hartebeest, zebra, giraffe, and various buck. In July, game was reported as plentiful at Makindu, consisting of hartebeest, wildebeest, impala, zebra, and ostrich, and at the Athi, game was abundant on both sides of the river.

On the southern spurs of Mount Elgon, game was very abundant in November, Austin (1903) seeing many hartebeest and some 20 giraffes.

In January 1898, Austin (1903) reported large numbers of game at Lake Nakuru, mentioning hartebeest and Thomson's and Grant's gazelles. At Lake Naivasha, hartebeest, Thomson's and Grant's gazelles, and impala, were numerous. In December, he reported seeing "for the first time" enormous herds of hartebeest on the western shores of Lake Rudolf (Austin, 1902). The following year, 1899, Arkell-Hardwick (1903) saw vast herds of game on the plains toward Ol Donyo Sabuk, including kongoni, Thomson's gazelle, impala, water-buck, and zebra. On the way to the upper Tana near Maranga, he saw great numbers of antelope and numerous herds of wildebeest, and on the way to the Uaso Nyiro, the quantity of game seen was "simply incredible. Vast herds of oryx, zebra and grantei … giraffes were also in sight …." A herd of about 30 buffalo was also encountered. On the plains eastward of a camp on the Uaso Nyiro roamed "vast herds of game—zebra, oryx, water-buck, and grantei." Opposite the camp, "a mighty herd" of buffalo marched along, but this, he recorded, was a sight that was now rare. In another area, immense herds of game were seen, including giraffe, buffalo, Grant's gazelle, and waterbuck. South from the Tana, at the river Thika-Thika, kongoni "once more appeared on the scene," and a roan antelope was shot, "the first we had seen during the trip." Besides kongoni, there were numbers of zebra, wildebeest, Grant's and Thomson's gazelles, and warthog. The next day, "Congoni were again numerous."

Percival (1918) recounts that in 1901, small, isolated herds of buffalo could still be found here and there in the bush, but the "whole of the animals inhabiting the open country vanished …." Hinde and Hinde (1901) reported that giraffe were not very uncommon in the Kitui district and in the neighborhood of Taveta. The wildebeest could be seen in thousands in a given neighborhood for a couple of months and then disappeared completely with the exception of a few solitary males, not being seen again for 6 or 8 months due to its change its pasturage. The oxpecker *Buphagus erythrorhynchus*, a bird which feeds on ticks, had become carnivorous since rinderpest destroyed the host animals, killing unattended domestic animals by eating the flesh to the bone and tearing holes in their backs and femoral region.

In 1902, Powell-Cotton (1904) reported as he neared Ol Donyo Sabuk that game became more plentiful, with many small herds of wildebeest and zebra, one herd of kongoni numbering over 200. At one camp, a herd of 40 eland filed past. On the sides of Ol Donyo Sabuk, he found "endless" fresh signs of buffalo. At the foot of Mount Kenya, he also found fresh sign of buffalo and encountered buffalo in the Obbo region. Near Lake Baringo roamed eland, Grant's gazelle, oryx, and giraffe. Eland, an absolutely protected animal, in western Laikipia and around Lake Baringo he reported far more plentiful than Jackson's hartebeest. On one day, he saw three herds, the largest numbering 148, the smallest 15, as well as several solitary bulls. Oryx were even more numerous, although herds rarely numbered more than 30, one double that. In the rains, the plain was reported by Africans to be covered with giraffe. Near the Uaso Nyiro, the country "simply swarmed with game," and Jackson's hartebeest and oryx were encountered as well as steinbuck, Thomson's gazelle, and waterbuck. On one occasion, he passed a large mixed herd of eland and zebra. At Mount Sergoit, troop after troop of game came in sight, on one occasion an estimated 200 hartebeest, 300–400, zebra and 20 eland passing his front. Other species encountered were reedbuck (once putting up 20), warthog, waterbuck, Thomas's kob, eland, and topi. East of Mount Elgon, all the country was "teeming with game." During a march, he encountered giraffe, reedbuck, oribi, waterbuck, hartebeest, and topi. On the Kedef river, he had good sport with hartebeest, waterbuck, bushbuck, Grant's gazelle, duiker, dik-dik, warthog, and oribi, besides which giraffe, topi, and reedbuck were numerous. In the Lake Baringo region, he came across five kudu in the district, where Teleki had found them in 1887: "I feared that disease might have decimated them, as had been the case in so many other parts of Africa, and even as far north as Abyssinia." A hill was pointed out as one where kudu were always to be found, and several were encountered there.

Between 1902 and 1906, Meinertzhagen (1957) conducted a number of game counts. Results of the most common rinderpest susceptible species plus zebra are shown in Table 2. On May 5, 1902, he counted what he could see from his carriage to the south side of the railway line crossing the Athi Plains but reported game as equally numerous on both sides. The Rift Valley in October 1902 he reported as "teeming" with game, mostly zebra, Grant's and Thomson's gazelle, and hartebeest. At Mbuyuni, west of Taveta, where he reported game as abundant, in April 1903, he saw a small herd of eland then another of about 100. Schillings (1906) reported that on the banks of the Njiri swamps (Amboseli) were clustered thousands of wildebeest and zebras, while there were also waterbuck, hartebeest, and gazelles. He was informed that giraffe had almost disappeared in 1896 due to rinderpest and could only rarely be found in out-of-the-way places, but he found no proof of this, the giraffe in his view being mostly affected by wanton hunting. He found it was not the case that eland had been practically exterminated, finding the biggest herd of all in Kikuyu country numbering 47 head.

In January 1903, Hindlip (1906) found several greater kudu near lake Baringo and buffalo on the Kerio River. Near Eldama Ravine, he encountered 14 roan antelope, while tracks of "pig," buffalo, and "probably bongo," were "visible everywhere."

Wilson (1913), in 1904, at the Mara River found in the open country, roan, topi, hartebeest, impala, waterbuck, wildebeest, and warthog, with buffalo in the hills:

> game was everywhere, literally in thousands ... the undulating prairie ... was literally peopled with game. An enormous herd of topi, some three hundred strong ... farther off a large herd of wildebeeste ... Thompson's gazelle ... roan moving slowly away ... Everywhere there was game-mile upon mile of smiling country covered with grazing herds ....

At Ndasegara, there were many eland about and between there and the Uaso Nyiro lots of buffalo spoor. The plains either side of the Uaso Nyiro "fairly swarmed" with game.

Table 2. Most common game counted by Meinertzhagen, Kenya 1902–06.

Date	Wildebeest	Hartebeest	Eland	Grant's	Thomson's	Impala	Giraffe	Zebra
1902								
May 18	760	845	0	324	142	46	18	4,006
June 8	967	846	46	932	546	146	19	2,430
1903								
Feb 28	1,247	946	134	832	621	887	32	1,465
Apr 12	0	264	146	94	59	0	34	426
May 12	894	276	46	326	426	184	18	684
Oct 23	0	88	46	186	109	88	34	198
1904								
Jul 20	1,267	1,654	42	432	234	66	18	568
1905								
Mar 22	876	543	7	169	174	98	11	974
1906								
Apr 7	186	702	0	734	167	16	0	1,644
Apr 23	0	68	18	36	0	17	18	0
May 27	456	745	14	224	88	87	5	486

The places were, respectively: Athi River Station to Nairobi, Athi Plains, Athi Plains, Maktau to Mbuyuni, Athi Plains, Pesi (where he also counted 436 oryx and 66 waterbuck among other species), Athi Plains, Stony Athi Station to Nairobi, Stony Athi Station to Nairobi, Maktau to Mbuyuni, and Nairobi to Stony Athi Station (Meinertzhagen, 1957).

In the Rift Valley near Eburru, Chapman (1908) reported the area as "literally teeming" with game, naming zebra, hartebeest, impala and Grant's gazelle: "... all these were literally in thousands ... the plains were alive with herds of the small gazelle (*Thomsoni*) ... Besides these were wart-hogs ...." At Elmenteita in September, apart from numerous zebras, he saw a herd of 32 eland mingled with hartebeests: "while the nearer foreground was alive with gazelles in scores and a few wart-hogs ...." He stressed that these concentrations were seasonal, and in February there was little to be seen. At Lake Baringo, while in March he saw very little, the government administrator there reported later in July seeing in one evening's ride 300 eland, 100 oryx, and 32 giraffe. Chapman's brother encountered a herd of about 50 eland in the area. In June 1905, south of Kitale, Meinertzhagen (1957) reported that he had never seen so much game, especially giraffe. Other game which abounded were topi, hartebeest, waterbuck, oribi, bushbuck "in incredible numbers," and a few zebra. In just over 6 km, he counted 124 giraffe, 232 topi, 167 hartebeest, 17 bushbuck, 85 waterbuck, 24 oribi, 7 warthog, and 62 zebra, among other species. Chapman encountered fresh spoor of buffalo near Menengai, but the Mau forest was the chief stronghold. In January 1906, on the Stony Athi, he saw a herd of 200 wildebeest, and hundreds of kongoni. Although he saw other wildebeest they did not seem numerous, and he had difficulty shooting one (Chapman, 1908). Kirkland (1908) also noted from the railway that wildebeest were numerous, although the hartebeest "was most in evidence."

At the end of 1907, Patterson (1909), on the Yatta plateau, counted roughly 300 hartebeest, while below the plateau, herds of zebra, hartebeest, Grant's and Thomson's gazelles, and a few eland were scattered about. A small herd of eland was also encountered on the Kauti plain, and there were many hartebeest. Near the Karusi river, in one walk, he counted 300 hartebeest, 50 impala, and 30 wildebeest, among other animals. In Meru country on the Jombeni plains, a herd of 200 eland was seen and herds of oryx, Grant's gazelle and some gerenuk. In Samburu, he saw "great herds" of oryx. Patterson noted that game in the Rumuruti district was getting rather scarce, partly owing to the shooting safaris, but mainly owing to the vast flocks and herds of the Maasai eating up the pasturage.

In 1907–08, Rainsford (1909) on the banks of the Uaso Narok which "teemed with game," saw oryx and eland, which were quite plentiful. Impala were plentiful, and there were flocks of Thomson's and Grant's gazelles. The greater kudu he reported as not common anywhere in British East Africa, while eland were much more plentiful than 3 years ago. Wildebeest were plentiful on the Athi plains and in all the country south of the railroad. The oryx was found in abundant numbers on the plains stretching to the base of Mount Kenya. At Sergoit lake in the Mau, troops of eland, herds of zebra, and Jackson's hartebeest trooped down in hundreds in the evening to drink. At Laikipia, a herd of 20 wild eland were grazing with the Commissioner's cows. In May 1907, Rainsford considered that 5,000 head of game could be seen at Sergoit at one time. Jackson's hartebeest dotted the whole land in vast numbers, eland trotted away, and waterbuck waded in the lake margin.

Madeira (1909), in 1908, noted that eland were said to have been almost exterminated in East Africa some 10 years ago by rinderpest, but protection had fostered recovery, and he found them in herds of considerable numbers all through the journey north of the Tana and in Laikipia, in bands of 10–25. At Ngare Rongai, oryx were somewhat plentiful, but north of Rumuruti, innumerable oryx were seen, as well as warthog, eland, impala, and Thomson's and Grant's gazelle. At Muhoroni, great numbers of hartebeest were seen, and numbers of topi, at one place, two large bands of waterbuck, some roan, and "literally hundreds of hartebeest."

Bary (1910) met Jackson's hartebeest around Mount Elgon. On the Uasin Gishu plateau, antelopes were abundant, and on the high plateau, an unbelievable number of zebra and hartebeest were encountered in thousands. Near Londiani, the only place in Kenya where they occurred, troops of topi were met with on all sides feeding on a plain.

Stigand (1910) reported game plentiful on the Baragoi plains, noting buffalo, oryx, eland, and giraffe. On the Laikipia plains, he had encountered eland and hartebeest. Tjader (1911) considered eland fairly common, easily found near the Sotik and Loita plains and still quite plentiful along the Athi and Tana rivers. On the Uaso Nyiro, Bronson (1910) found giraffe, Grant's gazelle, gerenuk, oryx, lesser kudu, and buffalo, although elsewhere, he stated that gerenuk and lesser kudu were now very rare in British East Africa. In the Rumuruti area, game was seen by the hundreds, and near Lake Natron, great herds of wildebeest. At Ngare Kiti, there were tracks of all sorts up to eland and some buffalo. On Mau summit, he passed through "almost solid herds of game" (particularly wildebeest). At Engabai camp near Isuria in February 1909, he records how, after firing a shot, there was a stampede of game "in thousands," wildebeest, topi, kongoni, zebra, Grant's and Thomson's gazelles, impala, warthog, giraffe, waterbuck, and oribi, estimated at 10,000–15,000 head. At Juja Farm on the Athi Plains, game abounded "in thousands." Buffalo sign was seen west of the Usao Nyiro, and fresh sign was thick on the Mara, but no buffalo were seen. At Kamiti Farm, the papyrus swamps were full of them. Several eland were encountered at Ngare Kiti. On Mau Plateau, the country was alive with game: wildebeest, topi, Grant's and Thomson's gazelles, buffalo, giraffe, and waterbuck. Eland were few and usually running with zebra, wildebeest, impala, or other buck. They were also met with near Ol Donyo Sabuk.

Roosevelt (1910) recorded that at Kilimakiu, a small herd of eland was encountered, and there were eland on the high downs not far from Meru, and on the Uaso Nyiro, a herd was found. East of Baringo, greater kudu was hunted, and on the Uasingishu plateau, herds of roan antelope were encountered. At Njoro, on the edge of the Mau escarpment, giant forest hog and bongo were found, the latter also near the top of the escarpment. The sable antelope, found near the Kenyan coast, were reported as "not many," but a small herd was encountered.[3] Up the Nzoia River in Uasin Gishu, game abounded on the plains, mainly zebra and hartebeest, but also oribi and reedbuck. In another area, it was mostly Thomas's kob and defassa waterbuck. From the Nzoia to Lake Sergoi, herds of game were encountered all along the route, hartebeest predominating, followed by zebra, topi, waterbuck, and oribi.

Barnes (1911) wrote of the Uaso Nyiro river's amazing wealth of game, in one morning seeing a herd of eland, some 20 giraffe, and gerenuk. Grant's and Thomson's gazelle, zebra, and oryx were "in great herds: there must have been hundreds, almost thousands, there together…." At the river's junction with the Isiolo, there was a small herd of about 15 buffalo. In the neighborhood of Embu, he found fresh buffalo tracks every day and encountered a herd of over 100.

The bongo of the dense bamboo forests are thought to have suffered heavily and according to the Wanderobo almost exterminated. This may have accounted for this species's comparatively late date of discovery in 1902. By 1923, Percival (1928) considered that it was as common as before its practical disappearance.

3.9. Kenya's Athi Plains

After the opening of the Uganda railway in August 1899, the Kenya visitor's first encounter with game was usually at the Athi Plains, more correctly the Kapiti Plains, of which the Athi Plains are a smaller northern extension. Beginning some 60 km before Nairobi and extending for 40 km, the area is an almost treeless level plain. Here, one first encountered the migratory herbivores.

[3]In the late 1970s, the estimated number of sable in the Shimba Hills was 200–250 (East, 1988).

Before the railway, the plains had to be crossed on foot, as Jackson (1930) did in July 1889 prior to the panzootic, when, because of the season, game was scarce: "One day, far away to our left, we saw a large and compact herd of wildebeest, and that was about all, except for a few Coke's hartebeest, Grant's and Thomson's gazelles dotted about…."

Macdonald (1897), crossing the Kapiti Plains in 1892 after the panzootic, reported that game began to swarm on every side. "Gazelle and hartebeest predominated," and on the Athi Plains, "game swarmed on every side." Hartebeest, gazelles, and a few waterbuck are mentioned. "Another long march through countless herds of game brought us to the edge of the Kikuyu forest." In August, marching to the Athi River, "we experienced the novel sensation of marching for miles through thousands of wildebeest." Along the Athi, the country was again reported as swarming with game, but only a solitary buffalo seen, while waterbuck and impala were shot. In November, Tucker (1908) found the plains alive with great herds of game, hartebeest and wildebeest being mentioned.

In January 1893, Portal travelled from the Kenya coast to Uganda to take over for the British Government from the East African Company. This, and disturbances in Uganda, led to a number of officers following the route between 1893 and 1894, several of whom recorded their impressions at a period following the rinderpest. Portal (1894), crossing the Athi Plains in 1893, reported the following animals seen, some singly, some in many hundreds: rhinoceros, buffalo, hippopotamus, lion, zebra, wildebeest, waterbuck, hartebeest, Grant's gazelle, Thomson's gazelle, dik-dik, duiker, and impala. "We could have got any number of hartebeest, zebra, wildebeest…." Thruston, following in September, reported the Athi Plains as

> probably unrivalled in the whole world… stretching away to the horizon we saw, scattered in vast profusion, many species of antelope, large and small. Hartebeests, various kinds of buck, herds of zebras and of gnus, frequent rhinoceroses and ostriches… lions… Not long since buffalo and eland abounded, but when we passed they were scarce. A terrible epidemic swept away most of them, as well as the cattle… One day, looking ahead, the country for miles was swarming with gnu, the grass was cropped short; there must literally have been tens of thousands of them… a portion of the herd stampeded… that portion of the herd took over half an hour to pass. (Thruston, 1900)

In November, Scott-Elliot (1896) noted that the abundance of game "was still extraordinary, in spite of the amount of extermination which has been practised by certain persons… hundreds of hartebeest, and a most beautiful little antelope (*Thomsonii*), besides wildebeest…."

At the beginning of October 1894, Vandeleur (1898) wrote that the land was: "… literally covered with game of all sorts; thousands of zebra were placidly feeding with innumerable herds of antelope of different species—wildebeest, hartebeest, a few mpala, and many gazelles (*G. thomsonii* and *G. grantii*), whilst away in the distance there were a few stately giraffe." Ternan (1930), crossing the plains in December, also reported great herds of game were to be seen:

> including zebra, Jackson's hartebeest [Coke's], Grant's gazelle, Thomson's gazelle, wildebeest, occasional ostriches and giraffe. On some days the whole plain as far as the eye could reach swarmed with herds of game… an epidemic of rinderpest had just previously caused great mortality among the herds of buffaloes which usually were to be seen in East Africa, and consequently buffalo had become very scarce… In addition to the more common sorts of antelope there were plenty of other varieties in different localities, such as eland, impala, kudu, oryx, and the little grysbok [grey duiker]….

Ansorge (1899) stated that on the old caravan route from Mombasa in the Tsavo area, one could see giraffe and eland, one of the rare mentions of eland at this time. But game had already become wary along the caravan route and kept its distance. The Athi Plains were to him the most magnificent game country in the whole world, "with its lions and ostriches, hartebeests and wildebeests." Wildebeest could be seen in herds of a few to many hundreds. Judging from the skulls which littered the plains, he thought a good many fell prey to lions; hartebeest skulls were also fairly common. But many of these skulls perhaps resulted from the panzootic. Other species he mentioned were impala and Thomson's and Grant's gazelle.

When Vandeleur (1898) returned in March or April 1895, the Athi Plains were so completely dried up that there was little game to be seen. In August, 3 years after his previous visit, Tucker (1908) found the same spectacle as he had before, the plains "studded with game in infinite variety."

At the end of 1897, Sykes (1903) reported that game swarmed, "... we saw tens of thousands of head of game daily. Antelope of all kinds ...."

In February 1899, Austin (1902) reported that game was extraordinarily plentiful there: hartebeest, wildebeest, and impala being hunted. Plenty of game, including wildebeest, was seen at Kiboko. But Buxton (1902) considered that wildebeest in Kenya had already been "badly thinned" (by hunting), although on the Athi Plains, he saw a trek which surpassed anything he had previously seen, the herd numbering at least 3,000. He considered that there must have been a great number of buffalo on the plains before the rinderpest because their skulls were to be found all over the place, but the buffalo themselves were now confined to one or two small herds. Eland had also suffered and were then rare, but a herd of 60 had been reported near Lukenya. He also came across hartebeest, roan, and waterbuck.

Wilson (1913) wrote of the plains in 1902:

> Great herds of Grant's and Thompson's gazelles, zebra, wildebeeste, and everywhere the dull red hartebeeste, commonest of African antelopes. But it is the quantity rather than the variety which holds us in astonishment. Mile after mile the country is thick with the grazing herds—there must be millions of animals; it is a thing of which it is impossible to have any conception ....

Meinertzhagen (1957) wrote that in July, the game there was astounding, with countless herds of zebra, wildebeest, hartebeest, and Grant's and Thomson's gazelle.

To Dickinson (1908), in 1904, the plains were thronged on every side with innumerable herds of antelope: "These herds of game may be counted, not by ones and twos, but literally in hundreds ... Zebra, hartebeest, and gazelles swarm around, with a few wildebeest ...."

In December 1905, Wollaston (1908) referred to swarms of game, huge herds of hartebeest and little groups of wildebeest, "two or three immense elands."

When Churchill (1908) arrived in 1907, the plains were crowded, "... one can see that it is the same everywhere, and can distinguish long files of black wildebeeste and herds of red kongoni ...." In December, Madeira (1909) wrote "... there were continuous herds on both sides of the railroad, covering the plains as far as the eye could reach. We saw a rhino, some giraffe, thousands of zebra, Coke's hartebeest, Grant's gazelle, Thomson's gazelle, wildebeest ...." In the same month, to Bary (1910), it was an extraordinary spectacle, with zebras, wildebeest, hartebeest, ostriches, and large and small gazelles. Fifty kilometres before Nairobi, he passed among great herds of zebras, ostriches, wildebeest, and hartebeest. But at the end of the following January (1908), the plains were dry, and the animals had migrated, only a few zebras being seen.

Roosevelt's account of his hunting trip in 1909 indicates a profusion of game. On the Athi Plains, "Hartebeests were everywhere." Giraffe, waterbuck, and impala were also seen from the train. "There were herds of hartebeest and wildebeest, and smaller parties of beautiful gazelles [Thomson's and Grant's] ...." With zebra, hartebeest were the most abundant animal throughout the whole Athi region. "There are places where all the game, perhaps notably the wildebeests, gather in herds of thousands at certain times ...." There were "immense quantities" of game on the plains, principally zebra, wildebeest, hartebeest, Thomson's and Grant's gazelles, but also impala, reedbuck, duiker, steinbuck, and dik-dik. "As we travelled and hunted, we were hardly ever out of sight of game; and on Pease's farm itself there were many thousand head, and so there were on Slatter's," although wildebeest were the least plentiful. On Slatter's farm, "All around us as we marched were the game herds, zebras and hartebeests, gazelles of the two kinds, and now and then wildebeests." Near Slatter's farm, warthog were reported as common throughout the country over which the party hunted. On another farm between the Ruwero and Kamiti rivers and on both sides of the Nairobi river, "game swarmed in almost incredible numbers ... a small herd of blue wildebeest ...," and there were "plenty" of warthog.

Dugmore (1910) in 1909 encountered a large herd of eland near Ol Donyo Sabuk mountain, and on the mountain itself, a herd of 28 buffalo. On the Athi Plains were immense herds chiefly of kongoni, zebra, Grants', and Thomsons' gazelles; "and in some parts the wildebeest."

Tjader's impression (Tjader, 1911) in 1910 was that hartebeest, wildebeest, and zebra had declined since 1906, finding only hundreds instead of "literally thousands," but this could have been due to hunting. In the southern part of Kenya, the wildebeest was one of the commonest animals, sometimes seen from the Uganda railroad in "hundreds at a time," but had either diminished between 1906 and 1910 or had moved elsewhere for the greater part of the year. His impression was that hartebeest had also greatly declined in numbers, "The enormous herds, years ago always found on the Athi Plains ... have certainly perceptibly decreased." The impala was very common in herds of 20–100. Bland-Sutton (1911) reported that wildebeest were often seen on the Athi plains but were not as common as hartebeest and suffered almost as badly as buffalo and kudu from the rinderpest in the last decade of the 19th century.

Treves (1910) was of the opinion that a greater variety of animals was to be seen from the railway line between Naivasha and Nakuru than elsewhere on the line, which in numbers might exceed the herds of the Athi plains. Zebra, hartebeest, and gazelles were in thousands, and many warthog were seen.

In 1913, to Hepburn, the game on the plains were in "literally thousands upon thousands." At Kapiti, "It is an understatement to say that at least one thousand zebras and one thousand hartebeests, together with much lesser number of wildebeests, Grants, Tommies, impala, etc., went by in the procession." Wildebeest were plentiful.

3.10. Tanganyika

Tucker (1908), traveling through Ugogo in Tanganyika in August 1892, reported seeing buffalo and antelopes of various kinds in considerable numbers. In 1899, on the Pangani River, Schillings (1906) encountered herds of waterbuck of 150–200 head, "Waterbuck in thousands sped hither and thither ...." A number of buffalo had survived the rinderpest in the Pangani swamps on an island, which had also preserved the waterbuck.

In the Rukwa area, Melland and Cholmeley (1912) encountered topi herds of 50 and 100 in 1907, and reported "huge herds," also encountering a good many waterbuck, hartebeest, roan, impala, reedbuck, and warthog, but no sign of buffalo.

3.11. Central and southeastern Africa

Moore (1901) stated that in 1896 in one morning, he saw several groups of zebra, some waterbuck, and eight kudu on the Shiré river plains between Lwandi northwards to Nyassa. In 1899, he wrote that on the Shiré river between Lwandi northwards to Nyassa, the plains were one of the richest game countries in the whole of British Central Africa.

On the Kafue floodplains, Harding (1905) reported that in June 1900, game abounded, and in a 3-hr morning trek, he saw eland, sable, hartebeest, wildebeest, and zebra, as well as reedbuck, steinbuck, oribi, and duiker. The Nanombi flats were "literally covered with game." In one herd, he counted 400 wildebeest and zebra and two other herds of about 50 each.

Kassner (1911) came across small herds of eland on the Lukanda River in northwestern Rhodesia in 1910. In the southern Congo in the region of the Lomembe River, hartebeest and sable antelope "abounded in plenty" at the beginning of 1911. On the western side of Lake Tanganyika, antelopes abounded, and he also saw buffalo.

In Mozambique's Kapoche district, in 1894, Foa encountered a herd of buffalo and later hunted a number in the Manyema district of the Congo (Foa, 1899).

In March 1898, on the Sabi river, Grogan and Sharp (1900) reported many impala and small buck but few large antelopes. On the Pungwe in May, a "nice herd" of buffalo was encountered, also 9 eland and 100 impala were found at the camping spot. On the Pungwe flats, he shot eland, hartebeest, tsessebe, waterbuck, reedbuck, oribi, and impala. Varian (1953) reported that the large herds of plains animals had suffered but by the early part of the century were numerous again on the Pungwe and Gorongoza flats. At the western extremity of the Gorongoza plain, Grogan and Sharp found enormous quantities of game: "thousands of wildebeest and zebra, and many impala, waterbuck and hartebeest…." Toward the Urema plain, they reported seeing 40,000–50,000 head of game, mostly wildebeest, "vast moving masses of life, as far as the eye could reach." Four buffalo were seen here. A "nice herd" was then encountered, and at the south end of the plain, two buffalo shot in front of hundreds of zebra and wildebeest. In the vicinity of Mount Chiperoni in Mozambique, spoor of buffalo, sable, and hartebeest was plentiful, and a herd of 20 eland found.

4. ASSESSMENT OF THE OBSERVATIONS

From 1892 in the wake of the panzootic until 1913, reports of the profusion of ungulates on the Athi Plains continue undiminished, even though most of the species reported are those susceptible to rinderpest, particularly the wildebeest. But in Meinertzhagen's counts between 1902 and 1906 (Table 2), the ratio of wildebeest to zebra, where both species were present, averaged 1 : 1.85, zebra being almost twice as numerous. In the Serengeti, in 1961, before the wildebeest population began to increase in the absence of rinderpest (see Chapter 29), the ratio was about 1 : 0.68, and by 1976 had become 1 : 0.0002 as wildebeest increased in numbers while the zebra population remained more or less static. In 1961, in the Nairobi National Park, which represents the northernmost limit of the Kapiti-Athi Plains migration, the ratio was 1 : 0.67, almost identical with that of the Serengeti in the same year. But after a serious drought in that year, the numbers of both species declined, the wildebeest by 87% and zebra by 63%, and so in 1966, the ratio was 1 : 1.9 (Foster and Kearney, 1967), similar to Meinerzthagen's 1902–06 ratio. It appears therefore, that in relation to zebra, the numbers of wildebeest Meinertzhagen observed between 1902 and 1906 were relatively low.

The alleged virtual disappearance of buffalo and eland apart, and the greater kudu and the roan probably never very numerous in Kenya in the first place due to lack of preferred habitat, the contemporary accounts offer no evidence of any great diminution in numbers of other susceptible species in the period immediately after the panzootic in 1892 as far as East Africa is concerned. But probably species such as wildebeest were reduced, for subsequent observations in the Serengeti have shown that enzootic rinderpest decreased the population by a factor of 5. In 1892, wildebeest were reported to occur in thousands, although the hartebeest is more often mentioned at first possibly because the males are sedentary, remaining after the seasonal departures of the females and wildebeest. Yet, Thruston (1900) reported the Athi Plains as "swarming with gnu" in 1893, but those contemporary observers who could make comparisons with an earlier visit considered that there was some decline, such as Tjader in 1906 and 1910, who believed that hartebeest had perceptibly decreased in numbers.

It is impossible to state with any degree of certainty how many wild herbivores succumbed in the initial panzootic, but there could have been some 500 million susceptible herbivores in 10 million km^2, with the possibility of 50% liable to an acute reaction. At a conservative estimate of 30, ranging to 90% mortality overall, this would indicate between 75 and 225 million deaths, compared with the estimated 200 million cattle deaths in 18-century Europe, an area of about half the size.

29

SPECIFIC EFFECTS ON AFRICAN GAME

1. THE ROLE OF AFRICAN GAME IN THE EPIZOOTIOLOGY OF RINDERPEST

In 1896, Nocard and Leclainche wrote without any foundation: “In Africa, the pest [rinderpest] has been carried from north to south of the continent by the wild animals, notably gazelles.” In truth, despite large populations of susceptible species, game was never widely entertained as being an important factor in the spread and maintenance of rinderpest in southern Africa, although in Uganda and in Tanganyika, it was considered to play an important role in rinderpest epizootiology. In April 1940, a conference was held at Blantyre, Northern Rhodesia, to discuss the relation of game to rinderpest. It was considered impossible to establish a game free area on the Tanganyika border, but it was recommended that a team of research workers be established to study the epizootiological and biological aspects. Smith (1941) stated that in the mean time, the periodic movement of game into and out of danger areas should be studied.

The importance of a particular species is a function of its innate resistance, population density and distribution, behavior, and degree of contact with other susceptible species (Branagan and Hammond, 1965). The disease burns itself out rapidly in those species with a low innate resistance and which, although of importance in dissemination, are relatively unimportant in maintaining it. It has been argued that the dangerous species (dangerous that is to cattle) are those with a high innate resistance which can perpetuate the disease unrecognized, but conversely such animals are unlikely to excrete much virus, and thus the chances of passing on infection are low. It seems certain that the disease dies out comparatively quickly in smaller game communities inhabiting bush and woodland. Hence, although bushbuck and bushpig are susceptible, these species do not travel far and, when infected, are unlikely to spread the disease. Eland, by contrast, travel long distances.

During 1969–71, 12 buffalo and 49 wildebeest from rangeland in Kenya were tested for neutralizing antibodies as well as a number of eland (26), Grant’s gazelle (58), Thomson’s gazelle (177), giraffe (44), hartebeest (104), impala (59), oryx (4), and waterbuck (24). Only one wildebeest showed a postive reaction, indicating that it had been exposed to rinderpest at some stage before 1964, the other species being negative (Fay, 1972). This seemed to confirm that the virus did not persist in game populations where the disease is controlled in cattle, but Rossiter *et al.* (1987) considered that an absence of observed clinical symptoms in buffalo showing a high level of antibodies in the early 1980s supported the contention that strains of mild rinderpest virus may persist unrecognized for months, or even years, in a cycle which includes game. Computer simulation indicated that it was improbable that the 6,000–10,000 buffalo in the Maasai-Mara Game Reserve at the time could maintain the virus by themselves. Either buffalo outside the Reserve or other neighboring domestic species were involved in maintenance of the virus, but buffalo densities outside this area were much lower, and there was no evidence supporting involvement of other wild species.

At an International Veterinary Conference in 1953, Wilde stated that where rinderpest had not existed for 2 or 3 years, the susceptibility of game animals (as in cattle) would be high, for the natural increase would have outweighed the residual immunity of those animals which had survived the most recent attacks. In enzootic areas where the disease was rarely absent from game and where repeated immunizations of cattle were conducted, the susceptibility level was dependent on natural increase, and no matter how frequently immunizations were carried out, it was almost impossible to create a state of zero susceptibility for more than a few weeks at a time. This difficulty was accentuated by the limitations of attenuated viruses which produced an immunity of only short duration in immature stock. Thus, it was possible in such areas for the disease to pass from cattle to game and back in a continual series of transferences.

Wilde concluded that if rinderpest was to be eradicated from areas where cattle–game contact was prevalent, there were only two alternatives. The first was complete eradication of the game:

> It is not, however, as unprecedented as one might at first think. Eradication of this kind has virtually taken place in the countries of Western Europe and the process is proceeding in many countries where the pressure of human populations on the land is driving the wild animals into extinction. The process can be seen in a fairly advanced stage in Nigeria where game is rapidly disappearing ....

His other alternative was some "policy of segregation," but before this could be envisaged, considerably more accurate information had to be obtained on the relationships between game animals and diseases of domestic stock. Whereas his examples of extinction relate to much lesser populations of game, in both variety and number, than exist in eastern and southern Africa, it is fortunate that work in the Serengeti region of Tanzania has shown contrary to the accent being placed upon game, it is upon cattle attention needed focusing.

Plowright (1982) classified susceptibility as shown in Table 1. Hartebeest, for example, are generally unaffected or at least lack visible clinical symptoms, but in Somalia in 1897 and West Africa in 1913–17, they were severely affected. Grant's gazelle is not usually observed to be sick but exhibited diarrhea and blindness in Tanganyika's Tanga Province in 1945.

Pitman (1942) wrote that the Uganda kob was very susceptible and about 1930 suffered disastrously around Lakes Edward and George. By the time he was writing, he considered that it had not recovered its former numbers, although by the 1960s, it was numerous. He claimed that in many parts of Uganda, it never recovered, illegal hunting pressure finishing the job of extermination. The giraffe, although susceptible, appeared "in recent years" not to have suffered undue mortality despite frequent epizootics.

The impala was reported as unaffected in the initial outbreak in southern Africa, while Poulton (1927) considered its reaction variable in Uganda. But clinical trials conducted on six animals in Kenya all produced a reaction. It was reported as affected in the outbreak in northern Kenya in 1960, and a sick impala tested positively (Scott *et al.*, 1960).

Earlier observers all reported hippopotamus as unaffected, but sera from 315 hippopotamuses collected in 1963 from the Queen Elizabeth National Park in western Uganda showed a number of positives. The animal being relatively long-lived, about 50 years, it was possible to show that animals had been exposed to infection in recorded epizootics in the area of 1920–21, 1931–3, and 1944–5. There were no positives among animals aged 11 years or under, but about 4% had been exposed in the age group 13–25 years. From 28 to 41 years, some 36% were positive, the earlier exposure being the greater. Antibody titers in animals infected about 30 years before were at least as high as those in yearling wildebeest which had been infected for only a few weeks (Figure 1) (Plowright *et al.*, 1964). If hippopotamus died in the water, a low rate of mortality might well pass unnoticed, while constant immersion of

Table 1. Susceptibility of African ungulates to rinderpest.

Very high	Buffalo Warthog Eland Kudu
High	Giraffe Bushbuck Bushpig Sitatunga Uganda kob Giant forest hog Bongo Wildebeest
Moderate	Reedbuck Topi Blesbok Bontebok Gemsbok Roan antelope Sable antelope Oribi (v) Impala (v) Springbok
Low	Waterbuck Duiker (v) Oryx Grant's gazelle (v) Dikdik Hartebeest
Very low	Thomson's gazelle Hippopotamus Gerenuk

(v): allegedly variable response.
Some species, such as black wildebeest, have not been included.

the head would remove any visible symptoms. Therefore, it may be more susceptible than had been assumed.

2. WILDEBEEST AND RINDERPEST IN EAST AFRICA

Anderson and May (1979) made the point that parasites were at least as important as the more usually studied predators in regulating natural populations. Rinderpest in wildebeest in the Serengeti was shown to be an outstanding example of this, having depressed the population to one fifth of its potential irrespective of the role of major predator populations.

The observations of yearling disease in the Serengeti wildebeest, which, post-1939, was the biggest single concentration of wild game in East Africa, focused attention on the role of wildebeest in the epizootiology of rinderpest. We have seen that wildebeest appeared to have

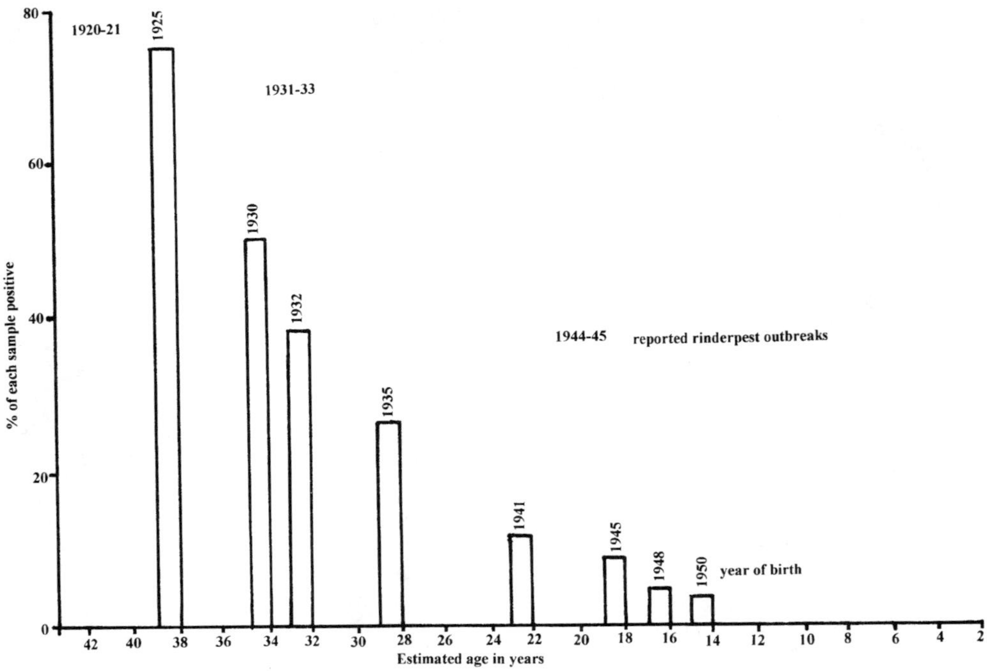

Figure 1. Occurrence of rinderpest-neutralizing antibodies in hippopotamus (after Plowright *et al.*, 1964).

been one of the last species to contract infection in the panzootic, the Maasai maintaining that it was only after their cattle had been almost wiped out that wildebeest became affected.

Although reports dating before 1957 are not consistent, mortality of yearlings probably took place every 1–2 years, earning it the name of "yearling disease." From 1957 to 1960, it was reported annually in the main wildebeest concentrations until the cattle–wildebeest link was to be broken. Plowright and McCulloch (1967) reported a severe outbreak with considerable mortality in October 1960 among yearlings in the Serengeti, an outbreak which had already declined considerably by November 6, but virtually all yearlings showed a harsh, discolored coat, with drying feces around the anal region and on the tail.

On the basis of the disappearance in the calf of maternally derived antibody, Plowright (1963a) and Plowright and McCulloch (1967) conducted studies on wildebeest in three areas: the Kapiti Plains–Nairobi National Park, the Loita–Mara–Serengeti area of Kenya and Tanzania, and Ngorongoro Crater. Sampling from 1959 to 1963 showed the presence of antibodies in the blood of wildebeest from the Kapiti Plains and Nairobi National Park in 21% of a sample of animals from birth to 4 months, and 20% of animals older than 3 years, with no immunes in between, suggesting that there had not been any rinderpest in the population in the past 3 years or more. In the Loita–Mara–Serengeti area, however, more than 65% of animals up to 4 months of age were immune and none in the 5–8 month age group, providing definite proof of infection in yearlings. In animals older than this, immunity ranged from more than 65 to almost 87% in animals of 3 years of age and more, some animals escaping infection in the first year apparently acquiring it during the second and subsequent years, suggesting the active presence of rinderpest during each of the preceding 3 years. In the Ngorongoro area, 95.6% of adults were immune but no 2-year-olds, while 33% of animals 5–8 months old were immune, but those born in late 1960 or early 1961 had apparently not

Table 2. Rinderpest neutralizing antibody in wildebeest, Mara–Serengeti area 1959–62.

Sample Size	Percentage Immune	Age (Months)
56	76.8	0–4
12	0	5–8
89	65.4	9–18
34	73.5	24
90	86.7	36+

Source: Plowright (1963a).

Table 3. Acquisition and decline of maternal antibody in wildebeest calves.

Age of Calf (Weeks)	Percentage with Antibody	Percentage with Levels > Dam
1	43	100
2–4	71	67
6	100	100
8	100	82
10	75	38
12–14	67	0
16	57	25
20	80	0
24	0	—
28	0	—

Source: Plowright (1963).

suffered exposure, suggesting geographical isolation of the Crater population. Both populations showed that calves of immune dams acquire colostral antibody immunity from the dam at birth, an immunity which persists for 24 weeks, in some animals only 10, but all lose it by 28. Passive immunity declines by 50% every 4.4 weeks compared with an actively acquired antibody (i.e., exposure to the disease itself), which may persist for many years. Thus, calves become susceptible in the first year of life. The disappearance by the fifth month differs from cattle in which calves retain antibodies for some 8 months, but the level of antibody in the serum of adult wildebeest, and therefore presumably in their colostrum, is lower than in cattle (Plowright, 1963a). In some calves examined by Plowright and McCulloch (1967), there was a failure of dams to pass on any antibody, although they had high levels. This failure was thought to be due to stress factors caused by the sampling disturbance but perhaps could be caused equally by large predators which are active among calving wildebeest.[1]

During the first month of life, the calf can have a concentration of antibody three times higher than its dam. By the sixth week, it can still be twice that of the dam, but in the eighth week, it has fallen below the maternal level and thereafter decreases rapidly. By the sixth and seventh months, all passively acquired antibody has disappeared, and by September of the year of their birth in the Serengeti, the calves are susceptible to infection (Tables 2 and 3), which, when present, caused the annual "yearling disease" (Figures 2 and 3).

[1]Release of adrenalin causes contraction of the intercellular spaces through which the large antibody proteins otherwise pass.

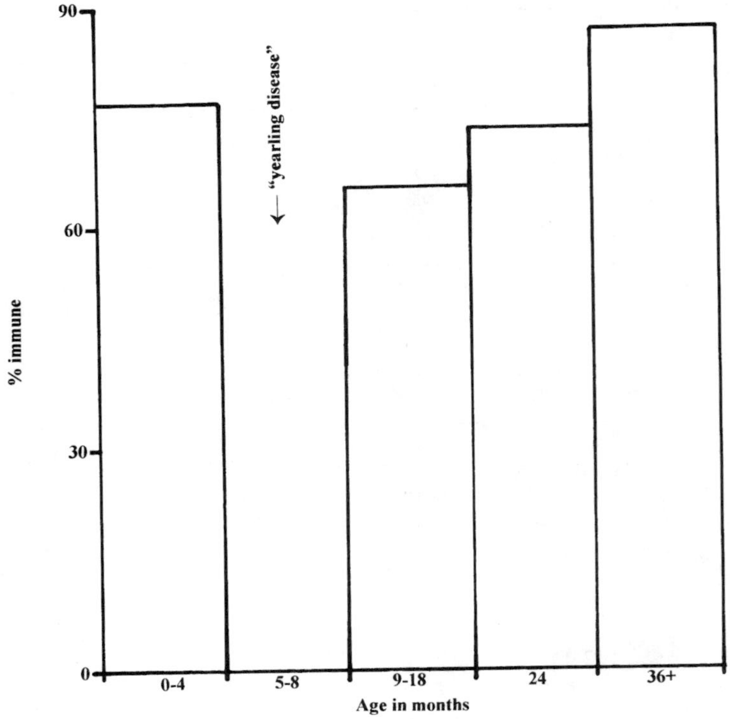

Figure 2. Rinderpest neutralizing antibodies in wildebeest (after Plowright, 1963a).

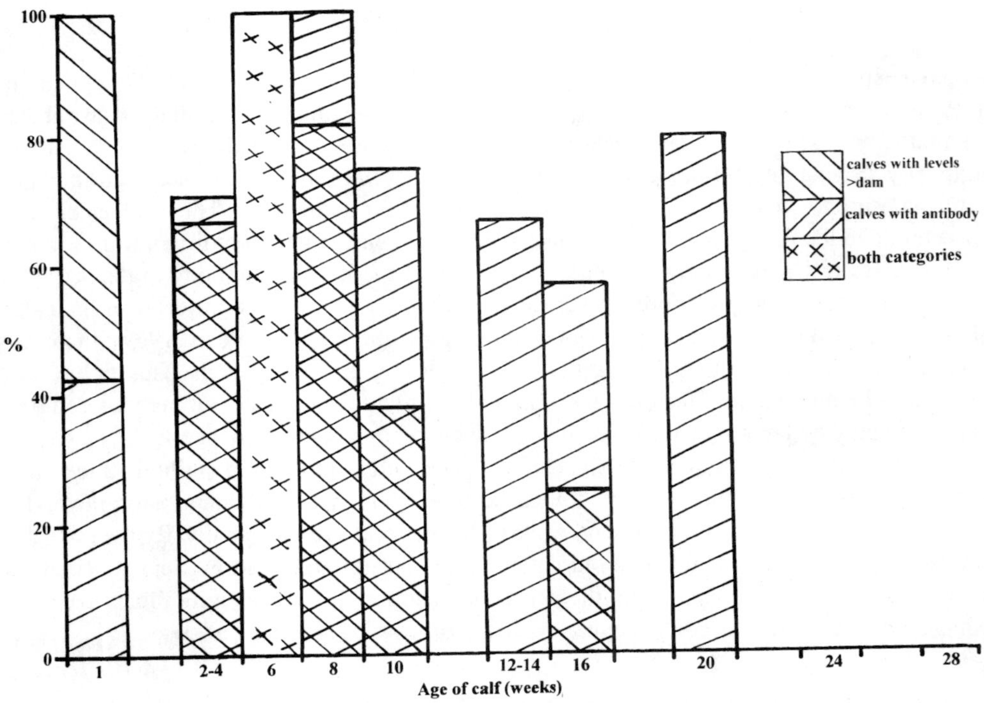

Figure 3. Acquisition and decline of maternal antibody in wildebeest calves (after Plowright, 1963a). Gaps are due to lumping of age groups.

In 1961 and 1962, infection occurred "silently" at Ngorongoro, no overt signs being detected, although some animals born in those years and sampled in 1962–3 showed serological conversion. Small numbers of apparently silent infections occurred in the Kajiado area of Kenya both before and after 1960, affecting no more than some 15% of the wildebeest population.

Thomson's gazelle, Grant's gazelle, and eland, in contact with wildebeest in the Serengeti, showed different levels of susceptibility. Grant's gazelle were apparently completely refractory, and a small proportion of Thomson's acquired antibody probably during the first year or so of life, but the majority escaped infection. However, the large numbers of Thomson's gazelle in the Serengeti meant that if the annual infection rate was about the same as that found in the 9–18 months age group, about 10.5%, this would mean that there could have been 50,000 infections a year, and the Thomson's gazelle could be an important contributor to the maintenance of the virus (Plowright and McCulloch, 1967). Eland were shown to be highly susceptible, the majority acquiring antibody during the first 2 years of life, while all adults were immune, although eland have often been observed to suffer heavy mortality from rinderpest, yet this had not been observed in the 2 years preceding sampling. Thus, the virus was of a low virulence not only for wildebeest but for eland as well. Eland suffered widespread infection in the Serengeti–Ngorongoro region before May 1962, and sick animals were seen on the Ol Balbal Plains in 1958. Eland calves born in 1960 showed serological conversion in approximately the same proportion as wildebeest at Ol Balbal or in the Ngorongoro Crater, but those born in 1961 showed a low incidence of past infection comparable with that of Ngorongoro wildebeest calves of the same year, and apparently much less than that of migrant Serengeti wildebeest born in early 1961. The mean level of antibody in the eland was somewhat higher than that observed in wildebeest (Plowright and McCulloch, 1967).

In 1965–6, Taylor and Watson (1967) examined sera of 149 wildebeest from the same population studied by Plowright. No evidence of antibody production was demonstrated in 60 samples of animals born since 1963, but it was present in 81% of animals born before 1962. It appeared that the last major infection among the Serengeti migrants was in 1961, as approximately 64% of animals born in January of that year experienced the disease, compared with only 6% born the following year. Similar results were obtained from the resident wildebeest groups at Kirawira, Mara, and Ngorongoro.

The last severe outbreak of clinical rinderpest had occurred in the main Serengeti wildebeest herds in October 1960, and although Lowe (1942) had reported that it soon died out in game if not supported by a cattle link and considered that this was probably true in areas where it had been more or less enzootic for a number of years, this interpretation was not generally accepted. Plowright (1963a) considered that it was impossible to say if rinderpest would disappear should the cattle–game contact be broken. He thought that the Serengeti populations were capable of maintaining the virus in the absence of cattle, but events proved otherwise. From 1961 to 1977, the wildebeest population increased fivefold, from an estimated 263,000 head (compared with over 800,000 head of cattle in the adjacent Maasailand) to 1,320,000. The population is believed to have stabilized at about 1.5 million in 1983.

Buffalo, last infected in 1963, increased from an estimated 29,500 in 1961 to some 61,600 in 1970, when their numbers began to stabilize. Both wildebeest and buffalo populations had therefore increased at a rate of about 10% a year. Rinderpest had probably been depressing the buffalo population since 1889. Other species are less well documented, but topi and Grant's gazelle both showed some increase; and the warthog, formerly regarded as common on the Serengeti plains, increased rapidly in numbers. But zebra, which are not affected by rinderpest, showed no change in numbers at all.

Analysing this, Plowright (1982) considered that 12 months after the last outbreak of the disease, there must have been a total susceptible population of at least 120,000–150,000 ungulates in the Serengeti region, and it was remarkable that the rinderpest which had infected the buffalo in 1963 did not spread to other susceptible species. In the closely allied virus, human measles, Plowright pointed out that breaks in endemicity occur when total susceptible populations fall below 500,000. The transmission chain is broken in smaller communities with fade-out taking place when there are fewer than 4,000–5,000 cases per year. This requires a total population of 300,000. He thus found it surprising that accumulations of more than 100,000 susceptible game animals in relatively small areas (with much more contact than human populations have) could not maintain the virus. The answer to this apparent anomaly might lie in the fact that human populations live in confined circumstances so that a fragile virus is not exposed to such rigours as one in the wild.

In the Serengeti region, Plowright postulated that at the periphery, where cattle–game contacts were frequent, a feedback mechanism of cattle–game–cattle may have operated in prior years, which was necessary to maintain the virus. When this link was broken and infection confined to small foci, the virus rapidly died out. What does not seem to have been taken into consideration is that a similar situation to that experienced in the Serengeti had occurred in southern Africa. After the disease was eliminated in cattle in 1903, although large populations of susceptible game animals, especially buffalo, kudu, eland, and blue wildebeest, still existed in the region, for example in Botswana's central Kalahari, the virus disappeared in the game, never to reappear. Yet in 1979, prior to a long and devastating drought, there were an estimated 262,000 wildebeest in Botswana's southern Kalahari desert (and a smaller population in the north), a population formerly undoubtedly much larger. One record claims that prior to the 1940s, there was an annual migration south across the Molopo river into British Bechuanaland, where the wildebeest would mix with cattle. In the recurrent droughts, there was also a migration into eastern Botswana's cattle areas causing cattle losses due to malignant catarrhal fever (MCF) (Spinage, 1992). The springbok apart, the other dominant herbivore in the Kalahari, zebra being absent in the south, was the hartebeest, which occurred in similar numbers to the wildebeest. In 1990, the population of buffalo in the north of Botswana was estimated at 32,000, a population which, at the turn of the century, extended south of the Boteti river, but in its present limits in the Okavango Delta region, the presence of tsetse fly excludes cattle, thus preventing a buffalo–cattle link.

A link with the disappearance of rinderpest in the Serengeti wildebeest may be the collapse between the 1960s and 1990s of the wild dog *Lycaon pictus* population. Whereas causes of the decline are in debate (Sinclair, 1995), it is believed that the species is susceptible to outbreaks of canine distemper contracted from domestic dogs in the surrounding areas. Scavenging rinderpest-infected carcasses of yearling wildebeest in the past may have provided them with enough temporary immunity to the allied canine distemper morbillivirus to withstand major mortality. Domestic dogs inoculated with rinderpest virus developed antibodies to it but not to canine distemper, but cattle can be immunized, at least temporarily, against rinderpest with canine distemper virus. The increase in wildebeest has also led to a more frequent transmission of MCF to cattle simply through increased competition for space.

3. BUFFALO

The buffalo is considered the most important wild species in the epizootiology of rinderpest in Africa (Scott, 1970), being highly susceptible to some strains, ubiquitous, and gregarious, and, although often sedentary in habits, able to exhibit considerable mobility when forced to do so by drought. Its requirements are similar to those of cattle, its range often overlapping, but although buffalo tend to inhabit those areas denied to cattle by trypanosomiasis, gaps in the distribution of cattle can be bridged by infected buffalo as well as *vice versa*. Eland, to a lesser extent by virtue of lesser numbers, play the same role. Both buffalo and eland herds are said to split up when infected and move more quickly and farther afield in search of water, prompted by diarrhea and the dry season. But Hornby (1939) considered that spread of game for more than 80 km had not been found. He thought wide dissemination was less likely in the dry seasons when game was restricted to the vicinity of water but was, on the contrary, encouraged by shooting, which caused dispersal.

An idea of buffalo former abundance was given by Gessi (1892) who, in the 1870s in the Sudan, saw herds numbering thousands which, when coming to a stream to drink, sometimes formed processions 3 km long. Contemporary accounts of the panzootic show how the species drastically declined in most areas. Although it apparently regained its numbers relatively rapidly, the experience of the Serengeti population shows that in fact it probably remained depressed in numbers throughout its range, which, by comparison with the reaction of wildebeest, was by about 52%. So its former abundance was probably only about double that reported after recovery.

Before the panzootic struck, Jackson (1930) in August 1889 near Kenya's Aberdare Mountains reported that there was "no doubt that there were plenty of buffaloes about...." The top of the Mau escarpment was then the "home of many large herds of buffalo, eland and zebra...." Near Mount Elgon at the beginning of 1890, Jackson recorded: "On the last day I saw giraffe, hartebeest, topi, zebra, warthog, eland, oribi, waterbuck, buffaloes, rhinoceros and elephants... During the rains and judging from old spoor, it was much favored by buffaloes in large herds...." One afternoon, he saw a herd of about 200 buffalo fleeing a fire.

MacDonald (1897), in April 1892, found fresh buffalo tracks in Mau forest, but Portal was to find "hundreds and hundreds of skulls, skeletons and scattered bones of the unfortunate buffaloes, which only two or three years ago used to range in vast herds over these mountains."

> Three years ago the magnificent African buffalo roamed in tens, and even hundreds of thousands over the Masai plains, over the Mau mountains, over, in fact the whole of what is called British and German East Africa; but now a traveller may wander for months in all the most likely or inaccessible places, and see nothing of the buffalo except his horns and whitened bones scattered over the plain, or lying literally in heaps near tempting springs and cool watering places... The stately eland, which was never so numerous as the buffalo, appears to have succumbed to the same plague, and the natives assert, though with what truth I know not, that there is not one left in East Africa. (Portal, 1894)

Later, Portal reported buffalo tracks as numerous around "Hargazo Falls" (near Bahadale) on the Tana, and buffalo and wildebeest were reported on the edge of the Kikuyu forest. Vandeleur (1898) reported that the Mau escarpment in 1894 was the home of a few buffalo "which have survived the rinderpest, and whose tracks can sometimes be seen."

Varian (1953) reported that a big herd of buffalo on the Bigameti River, a tributary of the Pungwe (Rhodesia), was reduced in 1896 from several thousand to about 100 and, instead of feeding in the open during the day, had changed its habits to emerging only at night.

Schillings (1906) wrote that in 1899, a few solitary old bull buffaloes survived in the Pangani marshes (Tanganyika), and some in the neighborhood of Lake Manyara. A small number could be found near Nguruman, a few were seen on the Mau escarpment in Kenya, and a few others were found in the Njiri marshes (Amboseli). Powell-Cotton (1902) reported that near the Awash in Ethiopia was a spot which the local headman said was the only place any buffalo were left in the area. Although heard in the reeds, none were spotted. West of Lake Tana, buffalo tracks were encountered on the Balarse River, as well as and several skulls which the natives said were from the rinderpest of a few years before. Tora hartebeest and roan were also encountered. He noted at Shimerler Jowee and toward Metemmeh tracks of a small herd at the former place, and he followed a herd of seven at the latter. "Very few now left, said to be plentiful in Walkait."

Grogan and Sharp (1900) recorded it still numerous in one portion of the Portuguese coast, but rinderpest had killed off nearly all the cows, which run in herds, the mostly solitary bulls escaping.

Hinde and Hinde (1901) reported that at one time among the commonest of animals in Maasailand, it had been reduced to two or three small and isolated herds. A considerable number lived in 1901 in the neighborhood of Sibaka, a small herd near Ol Donyo Sabuk, and a herd of perhaps 30 in the Kikuyu forest. A great number of giraffe were said to have been destroyed by rinderpest, but both giraffe and eland seemed to have recovered more rapidly than buffalo, although this might be an impression caused by drought, for animals which had scarcely been seen for some years had approached the trade route in search of food and water and thus appeared to be more numerous than formerly.

In 1928, Percival wrote that the buffalo formerly had died in numbers from rinderpest but now did not and, despite frequent outbreaks, was increasing and flourished, only a proportion of the young succumbing during outbreaks. In 1901, he counted not more than 40 on Ol Donyo Sabuk and in 1917, despite hunting, almost 400. In the Kapiti swamps in 1901, he saw about 30, whereas in 1917, there were not less than 200. At first, the impression during outbreaks was that not a single buffalo survived, but later Percival learnt from the Wandorobo that during outbreaks, survivors moved into the densest vegetation and thus were simply not seen. Near Fort Hall in 1924, they were numerous. On the coast south of the Tana river, they had always been numerous, and Percival found that they were really plentiful there in 1901. It appeared to him as if the 1891 plague had spared the coast districts (Percival, 1924).

In Abyssinia, Le Roux (n.d.) stated that the forest of Handeck, between the rivers Didessa and Angar, in 1901 was one of the last retreats of the buffalo, which was rarely met with. He claimed that, from its tracks, it was abundant there.

Meinertzhagen (1957), in July 1902, found buffalo tracks on the Ruiru River and in November recorded that buffalo were found in a swamp 3 km from Nairobi. In May 1905, he recorded that buffalo in the Nandi forest were reported by Africans as very scarce.

In 1903, Schillings (1906) encountered a herd of more than 60 in the open at Mombo, Tanganyika, a rare sight which he recalled was met with daily in those regions of East Africa before the rinderpest. Between Lakes Tanganyika and Nyasa, Torday (1913) recorded that herds of buffalo were met with every day. Eland also occurred in one area but were not easily found.

Dickinson (1908), in 1904, reported buffalo as being found all down the German (Tanganyika) boundary, particularly near Lake Natron, on the Ngong Hills, the Tana River, both sides of the North and South Uaso Nyiro rivers, and near lakes Solai and Baringo. In the Northern Uaso Nyiro and in Laikipia, they were rapidly increasing, more especially around Lake Natron. Eland he found on Laikipia but not in large numbers, and there were a few on the Athi Plains. Around Kilimanjaro, Arusha, and Erok, they were very plentiful, and in the Mara River district on the Tanganyika boundary, "they simply swarm—so it appears as if they are recovering from having been decimated by the scourge of rinderpest of ten years back." Lloyd (1906) reported the same year that for some time, there had been a large herd of buffalo in the Kafu river valley in Uganda, although he only saw their fresh tracks for a mile or more along the road. Wilson (1913) encountered buffalo tracks on the Kijabe river in Kenya in 1904, and the bush bordering the Uaso Nyiro river fairly swarmed with buffalo.

In 1905, in the Belgian Congo south of the Kasai river, Torday (1913) found considerable herds of buffalo between Luano and Gangan. Tjader (1911) reported that in 1906, once countless herds of buffalo in Portuguese, German, and British East Africa were substantially reduced some 18 years before by rinderpest, but in the last years, they had increased there again in great numbers. In Uganda, the Government had recently taken it off the list of protected animals, declaring it "vermin," but in Kenya, it was not quite so plentiful and was protected until 1909, after which a hunter was allowed two on a licence as it had greatly increased during the last few years.

Wollaston (1908), in 1906, reported that the forest buffalo "swarmed" in the forest north of Beni in the Congo and was especially numerous in the Semliki Valley, while the Cape buffalo was "not uncommon" about the eastern foothills of the Ruwenzori and in Ankole. In Bukoba in 1907, Mecklenberg (1910) saw some buffalo on the Kitangule, while others were sighted in Karagwe at Mtagata. A herd was also encountered near Mohasi in eastern Rwanda. Jack (1914) wrote in 1907 of encountering a small herd followed by a large herd in the region of Uganda's Lake George. At the southern end of Lake Albert, near Isura, another herd was encountered. In Kenya, by the Thika River, Patterson (1909) met a herd of some 200. Near Fort Hall, they abounded, and fresh spoor was found on the Uaso Nyiro river. Villelume (1909) reported forest buffalo in herds along the banks of the Congo river above the junction with the Kasai. They were met with a little everywhere.

Madeira (1909), in 1908, encountered buffalo at Punda Milia and near the Tana River; at one point on the latter, he estimated a herd to number 250, and a herd was known to exist at the junction of the Uaso Nyiro and Uaso Narok rivers. Rainsford (1909) recorded them as plentiful in the Kerio-Elgeyo country. More plentiful than a few years ago, the plague almost exterminating them in some districts, he considered they were becoming destructive. There were several herds near Nairobi, many on Ol Donyo Sabuk, and more in the Kedong. Near Fort Hall, there were several herds, and all down the Tana near Embu, they were numerous. They were common along the Uaso Nyiro in Laikipia, and the Kerio valley and slopes of the Cherangani were "full of them." Bary (1910) confirmed that they appeared to be numerous on Ol Donyo-Sabuk. Along the Thika River, numerous tracks were seen. They were also met with on Mount Elgon, and there were numbers on the Turkwel River.

Roosevelt (1910) refers to the panzootic as well-nigh exterminating the buffalo: "For years the East African buffalo ceased to exist as a beast of the chase. But all the time it was slowly regaining the lost ground, and during the last decade its increase has been rapid ... in many places the herds have become too numerous ...." But during his stay, he heard that there had been a fresh outbreak of rinderpest among them. On a farm by the Kamiti River, he hunted a herd of about 100, and they were plentiful on the west side of Mount Kenya. He also found them on the banks of the Uaso Nyiro feeding on the open plains and encountered some in the Lado Enclave.

Stigand (1910) reported their existence east of Mount Kulal, referred to by local people as "a few." In 1909, he wrote that the buffalo was sadly diminished in numbers since the rinderpest. Before that, it was said that countless herds could be seen on the Athi Plains, while below the Nandi Hills near Lake Victoria, numbers used to roam. Von Höhnel described great herds met with by Count Teleki near Baringo where now hardly a head was seen. Owing to strict preservation, it had recovered slightly of late and in some places was fairly numerous, but nowhere were such vast herds to be met with as in the days prior to the disease.

Near the Katonga river in Uganda, Dion Lardner (1912) came across a herd for the first time, and others were encountered between Kabula and Mbarara. They were reported by Africans to be exceptionally numerous at Gomba near Mbale, where he found a herd of over 50.

Cranworth (1919) recorded that after the epizootic among the buffalo in Kenya, 3 or 4 small herds survived which now existed in thousands, in 1910–11 more than 700 being killed. But they had never resumed their plain-loving habits, mainly restricting themselves to thick bush or forest. The eland, which had suffered almost as badly as the buffalo, had made almost as satisfactory a recovery. In the Tana valley, Hepburn (1913) found buffalo plentiful, and altogether 4 were shot, 2 from a herd of 25.

In 1911, Mecklenberg (1913) met them in Bagirmi (western Tchad). Forest buffalo were encountered on the upper Ubangi river, and a large herd of Cape buffalo near Zemio. Much spoor was seen near Tambura in the western Sudan.

4. BUFFALO RECOVERY

Percival (1918) recounted: "For many years there seemed to be little or no increase in the eland and buffalo; but suddenly the change came, and the increase was rapid." Within 10 years, buffalo recovered well, and at Ol Donyo Sabuk, an estimated 40 in 1901 increased to an estimated 400 in 1917, while at the Kapiti swamps, some 30 are thought to have increased to over 200. These estimates are fairly close, for, at an unimpeded maximum rate of increase of 0.151, 40 buffalo in 1901 would have reached a total of 379 in 1917, and 30 a total of 286. Elton (1927) commented that it took the buffalo about 35 years to recover its numbers. The recovery was in spite of a renewed outbreak of the disease in Kenya in 1897.

Pitman, in 1942, wrote that of the past 10 years or so, buffalo had recently shown a marked increase in numbers but had, in recent years, suffered severely from successive outbreaks of rinderpest. But no matter how severe its losses, it invariably speedily recovered its numbers. Repeated epizootics had affected buffaloes in Mabira forest over the past 20 years and were believed to have wiped out the marginal forest variation, an intermediate between the red forest buffalo and black Cape buffalo (common along Uganda's western border), now replaced by the latter.

Kennedy (1929) considered buffalo to be naturally a dweller in dense cover near water, but when numbers increased until an area was "saturated," considerable numbers were forced to live in the open "under relatively unsuitable conditions." Any decrease in buffalo would therefore be more marked in open bush country than in the dense coverts, as losses in the former would be accentuated by movement of the survivors to the riverside covert, to a certain extent replacing losses in the latter. Thus in the worst Nile outbreaks buffalo disappeared from the more open areas, but, with other animals, remained abundant in the coverts. Genetic studies show that the variation between populations at the regional level may be attributable to the fragmentation of a previous panmictic population (Simonsen *et al.*, 1998). Rinderpest, rather than human activity *per se*, may have been the proximate cause.

5. THE RELATION BETWEEN RINDERPEST AND TSETSE FLY

One of the most damaging indirect effects of the panzootic on game animals in Africa was an alleged relationship between the reduction in numbers of wild animals caused by the disease and the disappearance of the savanna tsetse flies *Glossina morsitans* and *Glossina pallidipes*, which carry fatal trypanosomiasis of man and cattle. This was to lead to half a century of game eradication programs in the belief that removal of game would mimic the effect of rinderpest by eliminating the tsetse's food supply, and thus control the spread of the fly which began at the end of the 19th century.

The first suggestion of a link was that of Fitzgerald (1898) writing of Kenya, when he stated "... the tsetse fly is said to be localized in a comparatively narrow belt, ... there is no doubt that the destruction of cattle and game during the epidemic considerably lessened the numbers of this insect pest." The other early reported cases of the coincident disappearance of tsetse fly from rinderpest affected areas were in the Transvaal and Rhodesia, but all accounts overlook the fact that there was a period of intense drought at the time which may have been the proximate factor influencing retraction of the tsetse fly's range. After the rinderpest panzootic, the fly was found to have disappeared from the whole of the Transvaal and the adjoining Portuguese Territory. Fuller (1923), while pointing out that there had a been a gradual shrinkage of range over a period of some years, stated that its disappearance from the eastern low country of the Transvaal could be "definitely attributed" to the rinderpest epizootic. Fuller also stated that experiments had shown that rinderpest infected blood did not affect the fly but he provided no details. Writing in 1910, Howard (1911) stated that in the Inhambane and Lourenço Marques districts of Portuguese Territory, the fly began to disappear after the epizootic.

Stevenson-Hamilton (in Swynnerton, 1936) described how part of the Kruger National Park occupied a portion of the Transvaal which was infested with *G. morsitans* before the rinderpest panzootic. Until 1858 to 1866, it was heavily settled, which may have partly accounted for the fragmented nature of the fly belts on both sides of the Portuguese border. Civil war took place during these years, and then, beginning in 1870 and continuing until 1897, there was heavy destruction of game. When the rinderpest came, impala survived in some numbers taking to the dense thickets, eland were already greatly reduced, and it was believed that there were about 200 buffalo in the Sabi area which number was reduced to 15. He deduced that only a certain limited type of vegetation which remained green all year near water holes would harbor tsetse year round, those in other areas migrating in the dry season. This limited vegetation type was also the home of the 200 buffalo until they were almost exterminated. This, it was thought, could have been the insect's undoing, although it did not apply to the great Rhodesian belts evacuated by the fly at the same time.

Similar occurences were recorded in Uganda with a great diminution in fly numbers in the Masindi fly-belt following the 1917–18 epizootic. In 1914, flies were fairly numerous throughout the area. In 1917, the wettest year for 4 years, game died throughout the long grass season, the epizootic still being in full swing in November when already an absence of flies on the road was noticed. Buffalo had fled to Buruli, south of the Kafu river, and Bugungu. An exceptional drought then began. In 1918, Duke found that *G. morsitans* was considerably reduced although not exterminated, but a reduction, if any, was less apparent in Bugungu. Waterbuck had survived the epizootic in considerable numbers. Buffalo had also survived in some numbers, although the number of carcasses on the Bugungu plains signified a heavy death toll. But in northern Buruli, the fly was apparently more numerous than it had been in 1914.

Duke concluded that the drought, beginning in November, led to an early drying of the grass and beginning of fires with an unusually complete burning of the bush. Thus, first, the fly's food supply was reduced at the most critical season of the year; and second, the adverse conditions of deprivation of shelter and destruction of breeding grounds by the fires were intensified by the prolonged drought. These factors could explain the diminution in numbers of the fly. The Bugungu country, being uninhabited, was less subject to burning. Surveys in 1927 failed to find a single fly in this area (Carmichael, 1938), perhaps suggesting the existence of a critical level for a population to survive in a habitat which Swynnerton considered was not the most favourable to *G. morsitans*.

During the 1919 rinderpest outbreak in South Ankole, a retraction of the fly toward the Akagera River was reported, and in 1925 in West Nile District, there was a marked diminution of the fly, game numbers still being low and fly scarce in 1929. But in other areas visited in 1926, there was no apparent decrease in the fly, although the areas had been subject to rinderpest.

There is no record of any disappearance of *G. palpalis*, the riverine or lacustrine fly, and, using this species for experimentation, Duke (1919) deduced that rinderpest-infected blood did not affect it. Carmichael (1938) attempted to verify whether or not *G. morsitans* was affected by virulent blood. In his first experiment, he fed flies on cattle infected with virulent blood from a sick eland. The flies fed to within an hour of death of the host despite the odor associated with rinderpest-affected cattle, without any apparent ill effect. He next fed pregnant flies on sick cattle, again without any demonstrable effect. His third experiment was to compare the success of pupal emergence from areas where wild animals had been dying of rinderpest with those in areas where rinderpest was absent. Again, there was no demonstrable difference. His final experiment was to determine what effect the virus had on the fertility of *G. morsitans*, but the results were inconclusive. He recovered only 7 pupae from 1,500 female flies feeding on rinderpest infected cattle in an insectary, and none from the control enclosure. All in all, he found it impossible to determine whether rinderpest infected blood had any effect upon *G. morsitans*, either directly or by way of reproduction.

Stevenson-Hamilton, warden of the Sabi Game Reserve in South Africa, in a letter to *The Field* in 1907 reported the disappearance of the fly after the passage of rinderpest from the low-lying bush country of the eastern Transvaal and in the adjoining Portuguese territory:

> The game in general would not have appeared to have suffered to the same extent as it did further north, and there still remained a few buffalo. Since the establishment of the game reserves, these animals, in common with other game, have increased very considerably and still inhabit their old haunts; but there has been no resurrection of the fly ... Cases are quoted of districts where the buffaloes were exterminated, but the fly remained. Here is an instance of the contrary. The theory that the fly disappeared because the game was killed out, which is sometimes advanced, does not appear to hold good, at least in this case.

Commenting upon Stevenson-Hamilton's report, Selous (1908) considered that the buffalo had not recovered its numbers in 11 years as it should have done in the Sabi, taking into account that it had been protected first by war and then by legislation. He believed that the reduction or disappearance of buffalo alone was responsible for the disappearance of tsetse fly, and not other game:

> ... in all the countries I have enumerated [valley of the Limpopo; between the Lebombo and Delagoa Bay; the Zambesi valley west of the Victoria Falls; both banks of the Chobe; many large areas on the Boteti, Tamalakan and Mababe rivers and many districts north of Lake Ngami (Botswana); areas south of the Zambesi] immediately upon the extinction of the buffaloes the tsetse flies, which in many places had previously existed in veritable swarms-as, for example, in the valley of the Chobi River, rapidly commenced to diminish in numbers, and sooner or later completely died out, notwithstanding the continued presence of all other kinds of game native to the country.

It was not known whether *G. morsitans* or *G. pallidipes* disappeared from the north-eastern Transvaal, although the country suggested *G. pallidipes* habitat. Between the Olifants and Crocodile rivers, there had been three distinct fly belts in an area of about 96 km by 190 km. By 1896, eland had been nearly exterminated in the area by hunting, and buffalo reduced to less than 100, but there had been no diminution in the fly, which existed in large numbers up to the outbreak of rinderpest:

> During the early stages of the rinderpest, that is during the cold weather of 1896 (the disease having been most likely introduced into the game country by the cattle of the farmers coming for the winter grazing), fly was observed to be present on the Sabi some miles below ... [Komatipoort]. The game was dying fast about October ... In the following December ... in the height of the hot weather ... Mr. Ingle visited the Sabi (where fly had been observed the previous June), and was surprised to find none present ... the fly had wholly or nearly disappeared from all the low

> country ... between October, 1896, and June, 1897. Since the latter year there has been no tsetse reported in the Transvaal ...
>
> After the rinderpest, it was found that the last of the eland in the North-East Transvaal had disappeared, and that the buffalo were reduced to a herd of about twenty, which remained in the densest part of the Sabi bush—the heart of the former fly country. A small number of kudu and bushbuck survived in the same locality. Impala, wildebeest and other species, native to the district, appeared much as formerly. (Stevenson-Hamilton, 1911)

In 1956, he reported the fly still absent.

Buffalo, kudu, and eland, previously abounded along the Chobe River, but early in 1899 during a 3-month survey of the area, Stevenson-Hamilton found that rinderpest had "apparently made a clean sweep of the buffalo and kudu." About 1895, this same area had been infested with fly, but now he found none. The local natives kept only a few goats and he saw no cattle until he left the Kwando River. Buffalo, kudu, and eland were entirely absent, but wildebeest, tsessebe, lechwe, and zebra were numerous all along the Chobe and Kwando rivers. In 1904, he was informed within the 2 years 1902–04, the fly had returned from a source in certain islands in the Chobe swamps, where lechwe and bushbuck had escaped the rinderpest (Stevenson-Hamilton, 1957).

To Fuller (1923), "the benign effect of rinderpest, where nagana [cattle trypanosomiasis] is concerned, goes unquestioned." But he noted that a point usually lost sight of was that tsetse had disappeared from a relatively greater area before the rinderpest, and there was sufficient evidence to show a contraction along the west and north of the Transvaal beginning long ago "was gradually consummated." There was much reason to believe that the fly was extinct in half to two thirds of its former range well before 1888. According to tradition, the fly finally disappeared in that year "when the buffalo went north" due to harrassment by farmer-settlers. Fuller considered that in the greater part of the affected territory in South Africa, the disappearance of the fly did not coincide with any abnormal climatic condition. His final conclusion was that complete eradication of wild animals did not appear essential for suppression of the fly, but a decided reduction thereof followed up by constant disturbance of the animals which escaped to such an extent as to bring about changes of behavior and shyness tended to lead to disappearance of the insect (Fuller, 1923).

From some areas, there was reliable evidence that the rinderpest panzootic had no effect upon the numbers of fly. Sharpe (1893b) reported that when it visited Nyasaland, there was a noticeable decrease in game, especially buffalo, in some of the fly belts, but he had never noticed that this had any effect upon the fly. Gielgud (1901) noted that in Northern Rhodesia, despite almost total destruction of the buffalo, "These flies have not disappeared, although, perhaps, not so numerous as formerly."

Swynnerton (1936) concluded that although flies flourished well on other game species, buffalo was the favoured food and the one to which they were particularly attracted. With disappearance of the buffalo coupled with warthog, its other preferred host, both animals which showed placid tolerance to the tsetse, the flies might not immediately adapt completely to other large mammals, especially if they were not abundant and in view of the fact that these other species, such as hartebeest, are much more intolerant of the fly's attentions and might move away from its range. If it was only the great numbers of buffalo previously present that had made the concentration sites primary foci of the fly, in the subsequent year also, any offspring that survived the rains would also be stranded in the dry season. In areas in which the focus sites generally offered no great advantage one against the other, a reduced game population incapable of occupying all, might be attracted to different sites each dry season due to early grass, burning, or showers, and thus might bring about extermination of the flies over a number of years. Another suggestion was that surviving game might have developed a shyness as when hunted, visiting the vleis at night when the fly is not active, and leaving at dawn, so the fly did not get the same opportunity of feeding. While some fly belts disappeared, others were only slightly affected or not at all, and probably rinderpest contributed to the reduction of belts that were reduced coincidentally with it due to some other factor. In the Transvaal generally, owing to the small size of the fly foci and their wide separation, any disaster or even an unfavorable weather cycle might have destroyed them. In Tanganyika, the areas denuded of flies were restocked as they were adjacent to the main fly belts. In the Transvaal, this was impossible.

In attempting to emulate the apparent effect of rinderpest on the fly, Jack (1923) wrote

> it was thought that a similar phenomenon might be brought about by a sudden and intense onslaught on the game within the limited area selected. This did not occur, however, and the difficulty of eliminating game within a limited well-watered area, adjacent to heavily stocked game country, is one of the important lessons learnt.

Nevertheless it was to be the preferred method of attempting control until recent years.

In 1983, Owaga (1985) discovered that African buffalo urine was an attractant to tsetse flies, furthur research establishing it as a potent olfactory attractant for *G. pallidipes*, enhancing trap catches of flies by a factor of 10, 7 times more effective than cattle urine. Thus, Selous's insistence on the link being with buffalo appears to have been vindicated after almost a century of scepticism, although other wild species had not been tested. Whereas stale buffalo urine appeared more attractive than fresh to *G. pallidipes*, subsequent observations on the behaviour of *G. morsitans* (raising of antennae in response to the olfactory stimulant) showed a clear preference for fresh urine (Saini, 1986). If urine is the attractant, it is clear why tsetse flies would vacate areas when buffalo disappeared from them.

An apparent contradiction was that while the fly disappeared from some areas, it began to expand rapidly in others. It is thought that destruction of cattle led to an increase in bush in former pastoral areas, providing a new habitat for the fly. Thus, as the fly continued to expand its range, so did demands for the destruction of game animals increase.

30

DID RINDERPEST EXIST IN AFRICA BEFORE THE PANZOOTIC?

1. INTRODUCTION

There is no justification for the oft-repeated remark that the virulence of the rinderpest panzootic in 1889 indicated that the disease was new to Africa (De Leo *et al.*, 2002; Gulland, 1995; Mack, 1970; Mettam, 1937; Sinclair, 1979). As the history of rinderpest in Europe shows, its virulence is no indication whether or not it was present before. A classic morbillivirus, it causes epizootic and panzootic havoc in unchallenged susceptible hosts. If sufficient hosts survive to produce offspring, the disease becomes enzootic, otherwise it disappears or "burns itself out." A mortality of 90–95% among cattle after repeated waves of infection was the norm; while in an outbreak in 1746, the province of Schönen in Sweden suffered a 98% level of mortality. The claim that it was new to Africa has embroiled the subject in the realm of politics; for example, Anderson and Johnson (1988) assuming that it was new to Africa brought in by colonial penetration, claim that its advent conformed to "ecological imperialism," creating biological changes which follow contact but precede political domination.

2. HISTORICAL PLAGUES IN AFRICA

Rinderpest is believed to have existed in Egypt since the third millenium BC, and in the 12th century AD, we learn that there was "famine and pestilence" in the Mediterranean and Asia as well as in Africa (where Africa probably meant Egypt), following severe winters and dry summers. One cannot consider Africa as isolated from what passed in the Middle East and Europe. We know that merchants from Cairo traded with Timbuctoo in the 16th century (Lange, 1972), and for centuries, wealthy Egyptian landowners purchased immature cattle for fattening from Anatolia and eastern Europe, on many occasions importing rinderpest. One outbreak at the beginning of the 19th century spread from the stockyards in Egypt, killing off oxen which worked the irrigation pumps and rice polishers. The result was the creation of the first modern veterinary school in Africa. Another outbreak introduced with animals from Turkey in 1841 killed some 800,000 cattle, including an alleged 665,000 head owned by the Pasha, forcing him to send 3 regiments of horse artillery to his farms to plough them (Morcos, 1953). In 1862, an outbreak in southern Italy spread to Egypt the following year, killing 80,000 cattle and thousands of sheep and goats, an outbreak which Curasson (1932) thinks could have spread to West Africa (one source gives 600,000 cattle, 100,000 buffaloes, 3,000 camels, and 1,000 horses dying from epizootic diseases in 1863–4 (Anon., 1865g)). In 1882, Egypt was again infected, this time with Russian cattle provisioning the English army.

Turkish rulers of Egypt sent regular military expeditions up the Nile to the Sudan seeking slaves and gold, and Scott (1992) concedes that they may have introduced rinderpest. There is definite proof of a cattle trade link between the Sudan and Egypt existing in 1843, and this could well have meant that traders carried the infection then prevalent in Egypt back with them, perhaps in sheep or goats used for food on the journey back to the Sudan. In 1798, there were regular trade caravans from Darfur (in the region of Jebel Marra) to Egypt, composed of up to 2,000 camels (Browne, 1799), although on the return journey, they were not usually composed of more than 200 camels, plus a number of Egyptian donkeys. These caravans were of ancient origin, offering a direct link with Egypt into the heart of the Sudan and the whole of western Africa. Camper (1783) was the first to suggest in his prize essay written in August 1778 that the disease might come from Africa and be carried by camels. He wanted to buy a dromedary to try inoculation experiments, but the price was too high: "It was however very disagreeable that I could not plan the inoculation of the dromedary, because without doubt the only likely manner this cattle plague is brought is by camel caravans from Asia or Africa." He could not say if it infected buffaloes, aurochs, bisons, and others, "especially Africa's own oxen," because he had not performed any tests; but its control very much depended upon finding out if it had its origin in Africa or Asia.

Boahen (1964) shows that the Sudan and western Africa were criss-crossed with caravan routes in the 19th century, traversing the Sahel cattle-raising zone (Figure 1). Given that the camel might carry infection, there was no obstacle to the spread of rinderpest to the west and south of the Sahara, although caravans could have carried it with infected live sheep and goats taken along as food. It seems unlikely that such a contagious disease would never have penetrated these routes.

In Mogodor (Barbary), it is recorded about January 1800: "after the plague subsided, a murrain attacked the cattle, and great numbers of all kinds died; so that they became reduced in the same proportion as the race of man had been reduced before" (Jackson, 1820). Rinderpest was present in North Africa in 1805, 1828, and 1865, and it may have invaded West Africa from Lower Egypt in 1842–3. Oral history relates that it ravaged Mali, coming from the east about 1805, and Senegal in 1828 (Olivier quoted by Curasson, 1932). In 1822, Hecker wrote of the 1349 cattle plague in England that the cattle fell by thousands "... and, as

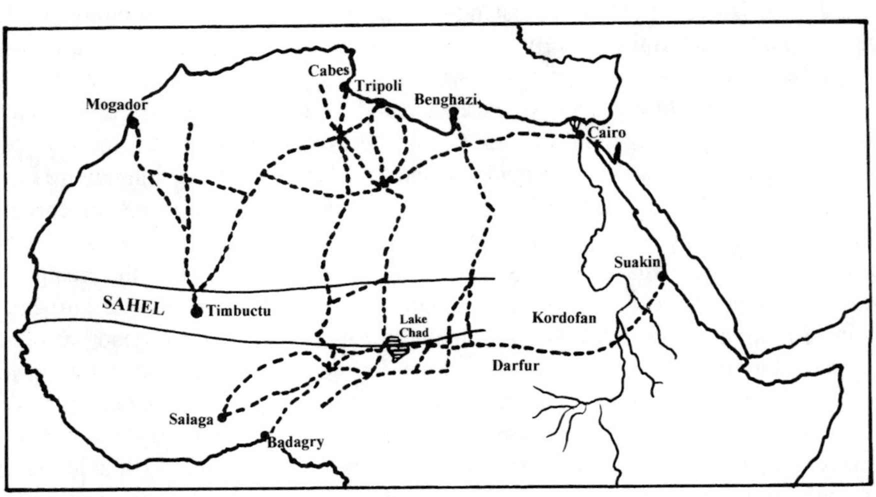

Figure 1. Main North African caravan routes in the 19th century (after Boahen, 1964).

has likewise been observed in Africa, the birds and beasts of prey are said not to have touched them" (Hecker, 1844).

Curasson (1932) believed that rinderpest existed in Senegal from the first quarter of the 14th century, which, if so, meant that the whole of the Sahel must have had contact with it. He also believed that there was an outbreak in Senegal in 1828. Suffice to say, there is little justification to consider all plagues prior to 1889 as other than rinderpest, especially when it was enzootic to the north in Egypt, to the east in India, and probably also Oman, which latter had regular intercourse with Zanzibar and other centers along the East African coast. The distribution of zebu cattle along the East African coast is considered primarily due to importation from Arabia and India after the 7th century AD. Fitzgerald, writing of events in 1891–3, stated that East African coastal arabs were deterred from possessing many animals "through dread of the cattle disease that swept the coast *periodically* [my italics] from Somaliland" (Fitzgerald, 1898).

Severe cattle epizootics are reported in Tanganyika in the 1870s and 1880s in an area reaching from Ubena in the south to Unyaturu, Usandawe, and Bukerebe; to the western and eastern shores of Lake Nyanza, where Fischer encountered an epizootic in 1886; and Maasailand in the north where a German veterinarian, Sander, writing in 1893 suggested that "sadoka" had first broken out in 1876 or 1877. In Ubena and Usandawe, the disease was known as *sotoka*, which appears to describe anthrax although used in the East African Kiswahili language for rinderpest, but Richter (1900) stated that *sotoka* referred to a "lung disease," and rinderpest was called *mubiamo*. In Usandawe, it was remembered as having been worse than the rinderpest panzootic of the 1890s, the epizootic about 1875 creating widespread famine. According to Hurel (1911), the disease known as *sotoka* was responsible for the depopulation of the interior of Bukerebe island. Fitzgerald (1898) noted that at the time the panzootic was raging on the mainland of East Africa, he believed that frequent cases of anthrax occurred among horses at Zanzibar, but there is no indication of anthrax being confused with rinderpest. In Tanzania's Lake Manyara National Park, three outbreaks of anthrax between 1961 and 1984 repeatedly killed large numbers of impala, but the dense herds of other susceptible species, buffalo, wildebeest, zebra, and elephant, were hardly affected. This may have been due to a difference in behavior, impala drinking from stagnant puddles, other species from the lake. Whatever the cause, this disease did not spread through the other populations.

In north-east Uganda, about 1876 and again around 1887, epizootics, the like of which informants stated had not been known before, killed much stock among the Karimojong and the Jie, while in the north of Kenya, the Samburu herds were attacked by an epizootic in the 1870s.

3. PLEUROPNEUMONIA OR RINDERPEST?

Joseph Thomson, traveling through Maasailand in October 1883, was "greatly astonished to observe the dried carcasses of numerous cattle which dotted the entire district" in Laikipia. Asking the Maasai for a guide, they replied "What! ... do you not know that our cattle are dying in hundreds on all hands?... Round about the kraals the scene was simply fearful—hundreds of animals dying and in all stages of decomposition ... a fearful stench prevailed .... " It did not appear to have affected the buffalo, of which he came across a herd on the slopes of the Aberdares in March 1884. After this, on a cattle road in the forest, he passed numerous decomposing bodies of cattle in the midst of a Maasai cattle area (Angata bus), then meeting another herd of buffalo. In need of food (the Maasai objecting to the sound of a rifle), after rejecting a sick animal, he bought a bullock from the Maasai which was apparently healthy, but on opening it up "we all shrank back as a horrid stench assailed our nostrils ... Both sides of the ribs and the back-bone had all the appearance of being rotten, and many of the bones were so far decayed that a push with a stick was sufficient to make them crumble up." There were more encounters with buffalo on Laikipia, but no report of finding any dead. "Here also [Ngare Sugaroi] the epidemic was raging with great fury...." There were many more references to buffalo: "Game, especially buffalo, was in amazing abundance" (Nandi); "...we spied great numbers of buffalo" (north of Baringo); "...enormous herds of buffalo" (Nanyuki). The following year on the return journey, at the north end of Lake Naivasha on March 4, no food was obtainable "Owing to the cattle disease." On March 12: "The Masai of the surrounding district were in despair through the almost utter loss of their cattle." A drought was now contributing to their loss, but at Ulu, there were cattle in great numbers (Thomson, 1885). Thomson and his party were blamed for the deaths, and he was obliged to indulge in a show of magic designed to keep the disease at bay. Bishop Hannington, who was shown no hostility, noted on September 14, 1885 near Lake Elmenteita that the Maasai seemed to have immense herds of cattle but refused to sell, saying that they had but few, many having died of the plague (Dawson, 1888). When Captain Smith encountered the dying Maasai cattle at Naivasha in 1891, he noted, "The disease does not appear to be, and the Masai declare it not to be, the same as that found by Thomson. In no case in which we made a *post-mortem* examination was there any disease of the lungs..." (Thomas, 1959).

Von Höhnel, writing of Maasailand in July 1887, recorded:

> ...and of late so much cattle has been lost through disease, that oxen are becoming very difficult to get. In some parts, indeed, the natives are already suffering from famine on this account and are beginning seriously to devote their attention to the breeding of sheep...The Masai are of the opinion that the murrain from which oxen have been suffering for the past ten or twelve years, was introduced by an ox which had been stolen from Samburuland. The disease, which seems to be rapidly spreading and in some Masai districts is universal, threatens the very existence of the people. (Von Höhnel, 1894)

This epizootic is generally attributed to pleuropneumonia, which the Maasai reported as beginning about 1884, the evidence being that pleuropneumonia was known to have been present in Africa before rinderpest was. But Merker (1904) was assured by all his informants that rinderpest, which the Maasai called *ol odua*, was there first, and pleuropneumonia only killed the survivors, which is perhaps more reliable than the statement of a Maasai more than 50 years later that "Rinderpest was the first catastrophe and it started like this. First of all there was an eclipse of the sun and it took place at about five o'clock in the afternoon [December 22, 1889]...It was then that the rinderpest attacked the cattle. The epizootic finished the Maasai cattle..." (Hanley, 1971). However, pleuropneumonia does not appear to have spread from the north, and game seems to have been unaffected. The Fulani, who vaccinated their cattle against pleuropneumonia but found that the method did not work for rinderpest, were clearly conversant with a difference between the two diseases. That the disease Von Höhnel encountered had been present for 10–12 years suggests that it may have been the same disease present in Ethiopia in 1875–7. In 1888 at Lake Baringo (Njemps), the expedition's oxen suffered from what appears to have been pleuropneumonia, but that is not to say that this was the same disease they had encountered in Maasailand a year before. However, from Von Höhnel's descriptions of the quantities of game, it does not appear to have affected wild animals to any great extent, and therefore seems most likely to have been pleuropneumonia.

4. DUTCH CATTLE IMPORTS TO SOUTH AFRICA

Dutch settlement at the Cape began in 1652, but the date of first importation of Dutch cattle is not known. Mentzel, who stayed at the Cape from 1733 to 1741, wrote: "The company [Dutch East India Company] also brought over some horned cattle with much trouble." Abbott (1952) dismisses this as being an event which would have been 50 years before Mentzel was writing and therefore hearsay, but Mentzel does not say that the Company brought the cattle over at the first settlement, and he may have personally seen the different cattle. In 1773, Thunberg, a Swedish naturalist who visited the Cape between 1770 and 1773, wrote: "The horned cattle, of which the peasant's herds in this country consist, are of Dutch extraction it is true, but are greatly degenerated. A dutch cow that is brought hither ... gives more milk than three others, but its offspring degenerates and the third or fourth is exactly like the rest ...." Abbot suggests that this indicates that the "Fatherland" blood, as it was known before 1800, must have been widely disseminated by then, predating importation by Governor Van Plettenberg (1771–85), who is popularly credited with the first importation of pure bred Dutch-black cattle (a bull and a cow), but which must have been between 1772 and 1778. Prior to the mid-1760s, Dutch cattle would have been of the red type, which is what Thunberg would have been referring to when he stated that the cattle were of Dutch extraction. Rinderpest, prevalent in Holland from 1730 to 1765, continued to rage again from 1768 to 1770 and was active in 1774, continuing to 1786, with the loss of 300,000 head in 1781 alone. Dying down after 1786, it then reappeared in 1796. Thus, the disease was present in Holland during the whole era of importation of Dutch cattle into South Africa. Although the journey took 4 months, some cattle may have been carriers, and they were often imported with sheep at the same time. And what did Mentzel's remark "with much trouble" mean? That many died at sea? The presence of rinderpest in Holland may be why, in 1789, 50 black cattle[1] described as bastard European bulls and calves, including 35 cows, were imported from Britain by Colonel Gordon, together with a number of Spanish merino rams and ewes. In 1792, three cows for improving the local breed were imported from London on an American ship "The Favourite." In 1795, Britain seized control of the Cape, and in 1800, one Devon bull and two heifers were bought from the Duke of Bedford, sailing from London on May 15 and arriving at the Cape on September 11. Other importations from Britain were in 1815, 1825, and 1836. In 1830, a number of Cape Colony farmers petitioned the Governor to allow them to import Dutch cows from Holland, all Dutch vessels being prohibited from bringing any article into the Colony. Rinderpest was present in Austria and eastern Europe at this time but had not reached Holland. In 1854, imported cattle allegedly introduced pleuropneumonia via Mossel Bay (Henning, 1943).

If rinderpest could be imported to South America from Europe, as it was in 1921, then it could just as easily have been imported into South Africa, especially considering that Holland, from which the majority of the South African imports came, was a center of infection from the middle of the 18th century to the middle of the 19th century. Infected cattle which were imported and died would have simply been written off as being unable to acclimatize to their new environment, and until the 19th century, the Cape Dutch had little knowledge of what went on north of the Cape, so any possible introduced epizootics would have passed unnoticed.

5. EARLY PLAGUES IN EASTERN AFRICA

African oral tradition refers to several major cattle plagues reaching back much earlier than the period of Dutch settlement. Webster (1979) has a number of such references for western

[1]"Black cattle" was a term employed into the 19th century to describe all cattle regardless of color to distinguish them from horses.

Uganda. Among the Luo, one plague occurred about 1328–55 at the end of a lengthy period of low Nile levels, suggestive of a prolonged drought. At the same time, Ankole pastoralists' tradition relates that a whole generation grew to manhood without ever seeing a cow but reports only three cattle epizootics since 1400: about 1520, 1600–25, and 1890. Traditions also indicate that the growing power of one tribe of pastoralists in the north-west (Jo Oma) was interrupted by a cattle epizootic. Bunyoro tradition relates to a cattle epizootic at the end of 20 years of drought about 1625–52 which caused the king, Cwa, to attack Karagwe to replace his royal herds. A tribe on the west shores of Lake Victoria has a tradition relating to plague during 1741–68. For these references to persist in oral tradition for the centuries they have means they must have been significant events, and the majority predates the apparent importation of pleuropneumonia.

In view of the association of the spread of rinderpest with armies in Europe, it is perhaps not surprising that the 1890 panzootic was attributed to either the Nile Expedition of 1884–5 or the Italian landings at Massawa in 1887, but this overlooks the number of earlier references. It can now be fairly confidently asserted that a lack of enzooticity in game does not mean that rinderpest could not have been present formerly. There are, furthermore, no reports of rinderpest becoming enzootic in deer either in Britain or in Europe, or in wild ruminants in India, despite known centuries of contact and a high level of susceptibility. Some African antelopes show varying degrees of resistance to infection, hartebeest, for example, succumbing in some outbreaks and not apparently in others. Wildebeest were seemingly the last to be affected in the panzootic. Perhaps this varying response does indicate some degree of inherent resistance acquired from long contact with the disease. It is an aspect yet to be studied. Cattle would have been affected only on a limited scale if enzootic rinderpest had existed previously, for, despite popular belief, cattle stocks in Africa were probably relatively low at the end of the 19th century and herds reasonably widely separated, such that except in droughts, game–cattle contact was more limited. Excessive drought in the 1880s to 1890s may have disturbed a relative equilibrium.

In 1928, Tarantino reported that rinderpest was known as *gulhai* in Eritrea and *furuc* in Somaliland, the application of specific names for it perhaps suggesting a long association; but on the negative side, Curson (1919) reports that in the 1917–18 outbreak in south west Tanganyika, although it was at least the second visitation of the disease, the Wakonde people had no name for it, but they could recognize East Coast Fever, which was termed *matussi*. In southern Africa, Lerotholi, King of the Basotho, after being informed of how quickly animals died, wrote to the Resident Commissioner on December 25, 1896: "There has never been anything like it before."

6. ORIGIN OF THE AFRICAN VIRUS

Some believe that the current African viruses may be derived from strains introduced into Egypt and North Africa from Europe or the Middle East in the early part of the 19th century before the panzootic (Chamberlain *et al.*, 1993). The Kabete "O" strain maintained in vaccine form since about 1911 by passage in cattle is close to the strain of that of the great panzootic, but it is not possible to determine from its genetic sequence whether it was definitely of Asian origin as no contemporary Asian isolates exist. The phylogenetic evidence suggests that the virus evolved separately in Asia and Africa, but for how long cannot be determined, and there has been no later introduction from Asia. In the words of Barrett (personal communication, 1999), the phylogenetic evidence supports this view of separate evolution but is not conclusive evidence. However, in historical times, the greatest concentration of cattle in the world appears to have been in the Asiatic steppes, and it is suggested that the true origin may have been there.

We have seen how, at the end of 1999, a final push was required to eliminate persistence of the disease in Southern Sudan, Somali, and Pakistan; and the signs are that this has now succeeded in the first and last areas. But the method of elimination, by which cattle are vaccinated until the disease disappears whereupon vaccination is then halted, means that a vast susceptible cattle population has been created in Asia and Africa, making it imperative that the final focus is eliminated before a breakout occurs, which could cause losses on a scale never before experienced in the history of rinderpest. It is claimed that, if global eradication is successful, it will be the first animal disease to be eradicated from the world and one of the greatest achievements ever in veterinary science (FAO, 1999).

Appendix I

REGULATIONS NOW IN FORCE IN THE GERMAN EMPIRE IN REGARD TO MEASURES AGAINST RINDERPEST [1870]

1. ON THE OUTBREAK IN DISTANT REGIONS

ART. 1, Sec. 1. If the rinderpest appears in distant foreign parts, which are so connected with the home country by railways or navigation that cattle may be transported in a comparatively short time, then the importation of neat cattle, sheep, goats, and other ruminating animals is to be altogether prohibited from the infected regions.

ART. 2. The prohibition of importation is moreover to extend to all articles in a fresh state derived from ruminating animals (except butter, cheese, and milk). On the other hand, the trade in perfectly dry or salted hides and entrails, wool, hair, and bristles, melted tallow in casks and tubs, as well as perfectly air-dried bones, horns, and hoofs, free from soft animal matter, is not to be restricted.

ART. 3. The importation of ruminating animals from uninfected parts of the country in question may be restricted to certain stations, and made dependent upon the following conditions:

(a) That it be shown by official certificate that the animals in question have been in an uninfected place for at least thirty days immediately before their departure, and that the disease has not prevailed for thirty kilometres around them.
(b) That the transport has been through uninfected places.
(c) That the animals in question have been examined and found healthy by an official veterinary surgeon on their passing the frontier.

Easier regulations may, however, be made for the introduction of cattle for slaughter into such towns as have public slaughter-houses connected by branch rails with the railway which brings the cattle.

ART. 4. Further restrictions may be ordered in regard to the introduction of animals, animal products, and articles liable to infection, from countries whence, on account of extensive temporary or permanent infection, there may be great danger of introducing the rinderpest.

ART. 5. What is said about the introduction also applies to the transit.

2. ON THE OUTBREAK IN NEIGHBOURING PORTS

ART. 6. If the disease breaks out in regions of a neighbouring country which are not more than eighty kilometres distant from the frontier, then, for the frontier district, which is to be defined according to circumstances, the prohibition of importation is to be extended to all kinds of cattle, except horses, mules, and asses; to all animal substances derived from ruminating animals, in a fresh or dry state (excepting butter, cheese, and milk); to manure, fodder, straw, and other litter materials, used stable furniture, utensils, and leather work; to raw (or not thoroughly cleansed) wool, hair, and bristles, and to used apparel for trade, and rags.

Persons whose occupation brings them in contact with cattle, for example, butchers, cattle-dealers, and their servants, must pass the frontier only at stated places, and must undergo disinfection there.

Exceptions may he made with the special sanction of the authority, and with such precautionary measures as the special circumstances may require, in regard to the animal products mentioned in Article 2, paragraph 2, as well as in regard to rags packed in sacks, in so far as the importation takes place in closed railway carriages, with an official pass showing that the articles in question come from places entirely free from disease.

Hay and straw, only used for packing, are not liable to the prohibition of importation, but must be destroyed at the place of destination.

ART. 7. If the disease reaches the frontier regions, or continues to extend along the frontier at a distance within reach of the ordinary frontier traffic, then a complete cessation of traffic is to be enforced by a military cordon; but in the neighbouring parts of the home country the regulations of Section 2 come into force.

The transit of railway trains, mails, &c., is to be allowed, even during the cessation of traffic, under the restrictions and with the precautionary measures required by the circumstances.

ART. 8. If the prescribed isolation should be broken through in the cases mentioned in Articles 6 and 7, then the animals liable to the prohibition are to be immediately killed and buried, and all articles subject to infection are to be destroyed or disinfected.

Other articles as well as persons, in case of a violation of the cessation of traffic, directed by Article 7, if it appear that they cannot be disinfected, must be taken back over the frontier by the shortest way, without passing through inhabited places, if possible.

ART. 9. The following measures of supervision are to be introduced in the frontier districts for all localities which are within fifteen kilometres distance from the frontier:

A cattle-inspector is to be appointed in each place, who is to draw up an exact register of the existing stock of cattle, and to make a special record daily of the deductions from and additions to the stock, as well as every alteration therein.

The cattle-register is to be examined at least once a week by the superior authorities.

Notice is to be immediately given in case of disease or death among the cattle.

ART. 10. The directions contained in the present section are also to be applied, with the alterations required by the circumstances, if there is danger of the introduction of the disease by water.

3. MEASURES ON THE OUTBREAK OF RINDERPEST IN GERMANY

ART. 11. As soon as any case of disease or death suspected to be from rinderpest occurs among the cattle at a place in our country, the owner must give notice thereof to the local police authorities.

ART. 12. In that case the owner must not slaughter or kill the diseased animals, nor bury or otherwise dispose of any dead animals, until the nature of the disease is ascertained. Till then, the animal must be so kept that men and animals shall not have access to them.

ART. 13. On receiving notice the local police authorities are to send for the competent veterinary surgeon, so that he may verify the disease on the spot. If there be no carcass, an animal is to be killed for the dissection requisite for the purpose. The result of the examination is to be drawn up in a report.

ART. 14. If the disease is ascertained to be rinderpest, the inquiry is to be extended to find out how it was introduced. Further notice is then to be given to the superior authorities, and public announcement is to be made, in which the duty of notification is to be specially pointed out for the neighbouring districts as well.

ART. 15. If it be found that there is only a strong suspicion of the rinderpest, a preliminary closing of the farm-yard is to be directed until such time as the disease is undoubtedly proved by further attacks and the requisite dissection or the suspicion is shown to be unfounded. In doubtful cases a superior veterinary surgeon is to be called in.

If the suspicion arises in larger cattle-yards, which are under the regular control of the veterinary police, the preliminary closing may be restricted to a separate part of the cattle-yard in question, with the necessary precautionary means.

4. IF THE SUSPICION OF RINDERPEST EXISTS IN REGARD TO HERDS IN COURSE OF TRANSPORT, THE NECESSARY PRECAUTIONARY MEASURES ARE TO BE TAKEN ACCORDING TO THE CIRCUMSTANCES

ART. 16. The application, sale, and recommendation of preventive and curative means in the rinderpest are to be forbidden under penalty. Means of disinfection are not to be reckoned among the preventive means.

ART. 17. After the outbreak of the rinderpest, the holding of cattle markets, if necessary also other markets, and other large assemblages of men and animals, are to be forbidden within a circuit to be specially fixed, according to the circumstances, but which, as a rule, shall be laid down at not less than twenty kilometers from the infected place. The trade in cattle and the transport thereof, as well as of manure, fodder straw, and other litter materials are also to be forbidden unless under special permit. The neccessary cattle for meat consumption must only be bought under the superintendence of the authorities intrusted with the veterinary police.

In the threatened localities the measures of control mentioned in Article 9, paragraphs 2–4, are moreover to be introduced.

For towns of considerable traffic, and for the environs of such towns, special regulations differing from those of this article may be made.

ART. 18. In an infected place, the slaughtering of cattle must be restricted to actual local demand, and take place by direction of the police and under the superintendence of professional persons.

ART. 19. In an infected place the duty of notification extends to every case of ailment in cattle and other ruminants, with the exception of mere external injuries.

ART. 20. The farm-yard in which the rinderpest has broken out is to be, first of all, isolated by watchmen, who must neither enter the farm-yard nor hold communication with its inhabitants, and must not allow the ingress or egress of any persons, except those legally authorized, nor of living or dead animals, or articles of any kind. Only adult male persons are to be employed as watchmen, and they must be provided with an easily distinguishable mark.

Authority to enter the farm-yard can only be given to the persons who are engaged in extirpating the disease, and to clergymen, law officers, doctors, or midwives, for the exercise of their profession; and care must be taken that they are duly identified. At the entrance and round about the farm placards are to be fixed, with the inscription, "Rinderpest."

ART. 21. A partial local isolation is to be imposed on the whole of the circuit to which the infected farms belong, such isolation to consist in the following:

The inhabitants may communicate with each other, but must not leave the locality without special permission, which, as a rule, is only to be granted to such persons as having nothing to do with cattle.

All domestic animals, excepting horses, mules, and asses, must be kept in the stable or shut up. If they are found running about, they are to be seized and slaughtered; dogs and cats to be killed and buried. Conveyances are to be made only with horses, mules, and asses.

The bringing in, sending away, and passing through of all cattle, hay, straw, and other infectable things, are to be forbidden.

At all the entrances to and outlets from the locality, placards are to be fixed, with the inscription, "Rinderpest," and watchmen are to be stationed to enforce the observation of the foregoing regulations.

ART. 22. A local commissioner is to be appointed for each considerable circuit or for several neighbouring smaller circuits together, for the time of the continuance of the disease, and he is to have such special inspectors as may be necessary. The notifications prescribed in Article 19 are to be addressed to the local commissioner, who is to see that requisite measures are carried out.

When the outbreak of the disease in any locality is ascertained, the local commissioner is to take measures for ascertaining any new cases of ailment (Article 13).

ART. 23. If the disease prevails in the majority of the farms in the circuit, the superior authorities may order a complete local isolation.

The locality is then to be entirely surrounded by sentinels (in this case military) and isolated from every kind of intercourse, excepting that of authorized persons, and for the indispensable necessities of the inhabitants of the place, under special precautionary regulations.

The intercourse of the inhabitants is also to be restricted to what is unavoidable. Divine service, schools, and other assemblages are to be suspended. (See Article 17.) The beer shops and taverns are to be closed.

The roads leading through the locality are to be barred for the time. If the place is near a railway, no train must stop there, even though it may be a station, unless the station be so situated that it can be entirely isolated and the traffic of the railway station with other places can be carried on without touching the infected place.

ART. 24. According to the extent and the nature of the buildings of any locality visited by the rinderpest, the partial or the complete local isolation may be restricted to separate parts of the locality, and, on the other hand, separate houses and farms in neighbouring localities are to be included in the isolation, if necessary.

ART. 25. All cattle sick with the rinderpest, or suspected of being so, are to be immediately killed.

Cattle are to be considered as suspected whenever they have been in the same stable with diseased animals, have had the same attendants, fodder, utensils, or drink, or have otherwise come into direct or indirect contact with diseased animals.

Under what premises other ruminating animals are to be considered as suspected is to be decided according to the special circumstances of each case.

If by killing the suspected animals the stock of cattle on a farm is reduced to a proportionately small remainder, then that also is to be killed.

By authority of the superior power, healthy cattle may be killed for the more rapid extirpation of the disease, though the above conditions may not exist; and this measure may be extended to farms not as yet shown to be infected. (See Article 36, paragraph 1.)

In the larger towns and in slaughter-houses under regular supervision of the veterinary police, the disposal of the flesh and hides of animals which, upon examination, in a living or slaughtered state have been found healthy, may be allowed. But the slaughtering must be done in proper places, under supervision of the veterinary police, and the flesh and the inner parts removed only after cooling, and the hides taken away only after they are thoroughly dry or have lain three days in lime-water (1, 69).

ART. 26. The slaughtered animals, in reference to which the provision in the last paragraph of Article 25 does not apply, are to be buried. For this purpose suitable places are to be used, as far as possible from roads and farms, at such spots as cattle do not frequent. So far as possible waste spots are to be chosen, which are not cultivated at all, or but slightly. The burying-places are also, as a rule, to be hedged in, and to be planted with such plants as grow fast and strike deep roots.

The pits must be dug so deep that the earth may cover the carcasses for at least two metres in height.

ART. 27. The killing and burying is to be done, so far as possible, by the inhabitants of the infected farms, or by such persons of the locality as have no cattle themselves and do not usually come into contact with cattle.

Persons from other localities, especially butchers, are to be employed only when no suitable inhabitants of the locality can be found. To prevent the spreading of the rinderpest by such persons, the proper measures must be adopted. (See Article 42.)

ART. 28. The place where the cattle are to be killed is to be appointed by the local commissioner in consultation with the veterinary officer, taking care to avoid all danger of spreading the disease.

Excrement discharged by the animal in the transport must be removed and buried.

Carcasses must be transported to the pit only by horses or men, on carts, drags, or sleighs, and no parts must touch the ground. The means of transport must be carefully kept separate so long as further transports are expected, and are afterward to be destroyed.

ART. 29. The skinning of carcasses, in reference to which the provision in the last paragraph of Article 25 does not apply, is to be strictly forbidden. Before the burial the hide must be cut in several places and rendered useless. All kinds of refuse, blood, and earth soaked with blood are to be thrown into the pit together with the carcasses. So far as possible, the carcass must be covered with lime before the pit is filled up.

In filling up the pit, layers of stones or brush-wood, if possible, are to be introduced. Watchmen are to be placed at the pit until the isolation is over, or at least for three weeks.

ART. 30. If a stable in which diseased or suspected cattle have stood is emptied by killing off the stock, then, if the regular disinfection (Article 40 *et seq.*) cannot be undertaken immediately after the removal of the cattle, any remaining dung is to be burnt or to have disinfected liquid poured over it; the stable, after having all its openings made air-tight, is to be strongly fumigated with chlorine, and then the door of the stable is to be closed and sealed up until the commencement of the regular disinfection. All stable utensils and anything else used for the animals are to remain in the stable.

ART. 31. The foregoing regulations for the isolation of farms and localities may undergo the absolutely necessary modifications for the interests of business, if the disease appears at a time when agricultural labours and pasturage are going on. These modifications are to be decided upon by the superior authorities. The following points (Articles 32 and 33) are to be attended to in such cases.

ART. 32. The isolation of farms cannot, even then, be dispensed with or mitigated; but exertions are to be made to clean the farm as soon as possible. (See Article 25.)

Agricultural works which cannot be delayed are either to be performed by external aid or by the people belonging to the farm, with the necessary precautionary measures.

ART. 33. If the conditions for the isolation of the locality exist, then the isolation of the *Feldmark* (rural district) takes its place; that is, the isolating measures directed in Articles 21 and 23 *et seq*. are to be transferred to the boundary of the *Feldmark*. The roads leading through the *Feldmark* are to be cut off. The passage and transport of cattle, fodder, &c., on the roads leading along the boundary are to be forbidden.

All inhabitants of the locality who still have farms free from disease and unisolated may carry on their agricultural labours with their own people and have their own teams. Cattle-teams in such cases are to be kept as far as possible away from the neighbouring estate boundaries, and from the forbidden roads.

ART. 34. In case of necessity, pasturage is likewise to be forbidden in the environs of the infected locality, and the necessary restrictions of intercourse are to be ordered for the immediately adjoining lands, as well as precautionary measures for the management of fields.

ART. 35. In cases of complete isolation, care is to be taken to provide for the most pressing necessities of the inhabitants, victuals, fuel, &c., with the requisite precautionary measures.

ART. 36. In residential and commercial towns, as well as in other towns with brisk traffic, the partial and the complete isolation are not applicable. Other exceptions, required by circumstances, form the provisions of Article 18 *et seq*., are also allowable. But care must always be taken for the speedy extirpation of the disease by promptly killing all the cattle in infected yards, as well as by proper isolation of the infected localities, and immediate disinfection.

If the rinderpest is found to exist in a public slaughter-house, or in a separately established cattle-market of a large town, then the locality is to be immediately isolated to prevent the disposal of the ruminants and swine which are there. In this case if the disease has not spread so far as to require the immediate slaughter and destruction of the whole stock of ruminants, the animals not yet diseased may be killed off for the purpose of disposal. But the slaughter, which is also to extend to the swine, must take place in the same locality, and within three days, at most, under the supervision and direction of veterinary surgeons. For the disposal of the flesh and internal parts, as well as the hides of the slaughtered animals, the provisions of Article 25, paragraph 6, are to be observed.

5. MEASURES AFTER THE EXTINCTION OF THE DISEASE

ART. 37, Sec. 3. The disease is to be considered as extinct on a farm or in a locality if either all the cattle are dead or have been killed, or if three weeks have elapsed since the last case of disease or death, and if the disinfection has been performed according to the following rules:

ART. 38. The disinfection must be commenced, according to circumstances, as soon as a stable on a farm is cleared of cattle.

It is also to be performed if the stock of cattle has been killed, though no outbreak of rinderpest has been proved. (Article 25, paragraph 5.)

ART. 39. The disinfection must be done on official direction and under professional supervision.

ART. 40. When the stable has been shut up (Article 31), the disinfection begins with its reopening, which is to take place, when possible, within twenty four hours; care must be taken for sufficient ventilation during the disinfecting operations.

The dung must be taken away and burnt or buried deep in places where no cattle can come within the next three weeks. The liquid manure, collected in trenches, is to be properly disinfected by the application of sulphuric acid and chloride of lime, and to be conducted to pits sufficiently deep.

All walls are to be scraped (the joints cleaned), and then to be well whitewashed; the woodwork is also to be cleaned, washed with strong, hot lye, and after some days to have a coating of chloride of lime solution.

Earthen, gravel, and loam floors are to be dug up, the earth excavated a foot deep, and all treated like the dung. Paved floors of the ordinary kind, that is, with stones laid in sand or earth, are likewise to be taken up, the earth is to be excavated a foot deep, and treated like the dung. The stones may be cleaned, treated with a solution of chloride of lime, and when they have been for four weeks in the air, may be used again.

Wood floors, according to their nature, are either to be burnt or properly disinfected. If the floors are to he taken up, the earth must be excavated and treated as above. Firm impervious paving of asphalt, cement, or paving laid in cement, is to be cleaned and disinfected.

Instead of chloride of lime, other disinfecting substances known to be effective, such as carbolic acid, boiling water, &c., may be used.

All movable woodwork (racks, mangers, vessels, and other utensils, also the partitions if possible) is to be burnt; ironwork is to be thoroughly heated.

Receptacles for liquid manure and stable-drains are to be treated in the same way as stable-floors, or, if in brickwork, like the walls.

When the disinfection is completed, the stable is to be ventilated for fourteen days.

ART. 41. Only people belonging to infected farms must be employed in the disinfection, or persons who have no cattle themselves; these persons must remain on the farm until the cleaning is over. Only horse-teams must be used for conveyance.

The transport of dung and earth is to be effected according to Articles 28 and 29; the things used for transport, instead of being burnt, may be carefully disinfected, as directed for woodwork.

ART. 42. The clothing of people engaged with dead or diseased animals, and in the cleansing and disinfection, is either to be burnt, or, if it can be washed, is to be soaked in hot lye for twelve to twenty-four hours, then thoroughly washed with soap, and dried in the air; if it cannot be washed, it must be fumigated with chlorine or hung in a dry heat for twelve to twenty-four hours, and then be aired for fourteen days.

Shoes and boots, and other leather articles, must be carefully cleaned, washed with lye or a solution of chloride of lime, and freshly greased, then fumigated with chlorine, and aired for fourteen days.

The people themselves must change their clothes and thoroughly wash their bodies.

ART. 43. All fodder which, from its position, may appear liable to have become infected, is to be destroyed by fire at the very beginning of the disinfection.

ART. 44. Dung on the dung-hills, which has been put there during the prevalence of the disease or within ten days before its appearance, is to be treated like the stable dung (Article 40). The other manure on the dung-hills is to be taken to the fields by horse-conveyance and, if possible, to be ploughed in after three or four weeks.

Until this is done and for four weeks afterward, no cattle must go into those fields.

If all the dung cannot be taken away immediately, the uppermost layer must be saturated with a disinfecting fluid. But the removal, according to the above directions, must take place as soon as possible.

ART. 45. Even after the complete disinfection of a farm or a locality, and the cessation of the isolation, no fresh purchase or sale of cattle must take place until after an interval of time to be fixed by the competent authorities, which interval must not be less than three weeks from the time when the place was declared free from disease.

Pasture-grounds which have been used by diseased cattle, or cattle suspected of disease, must not be used again before the lapse of at least two months.

The time when the burying-places may be used again will be fixed by the superior authority, according to the local circumstances in each case.

ART. 46. The holding of cattle-markets is not to be allowed until the lapse of three weeks after the last place in the infected district has been declared free from disease.

When the rinderpest has broken out in residential and commercial towns, or in other towns with a brisk traffic, or in the neighbourhood thereof, special regulations may be made, differing from provisions of Article 45, paragraph 1, and Article 46, paragraph 1.

Appendix II

INSTRUCTIONS AND DECREES ISSUED BY CARL, DUKE OF BRAUNSCHWEIG AND LÜNEBURG, 1745–82

1. 1745

Mar 15. Issued at Wolfenbüttel. Orders that the cattle be watched carefully. If one suspects plague, or something in the neighbourhood; or hears of a good, tried method, against it; to report to the authorities.

Mar. 20. Wolfenbüttel. No raw skins to be imported unless accompanied by a certificate showing they are from an area free of all infection for the past 6 months. Any brought in clandestinely without such certificate to be seized and burned.

Aug. 7. Salzthal. All herdsmen are to swear an oath to observe to their utmost their duties in preventing the cattle plague.

Aug. 7. Salzthal. Decrees should be delivered speedily to villages by horsemen. Officials will be held responsible if they are not delivered within 1 hour.

Aug. 23. Braunschweig. Officials, particularly at border posts, are to ensure that cattle with passes do not come from infected areas, or areas supected to be infected.

Aug. 28. Salzthal. Orders the use of setons where still not employed and a heavy penalty for those not doing so.

Sep. Wolfenbüttel. Advertisement of a good method to prevent the cattle plague (a mixture of 2 parts sulphur plus 1 of kitchen salt on a piece of bread).

Sep. 25. Salzthal. Decree forbidding the importation of foreign livestock, and meat; fresh, salted or smoked. Penalty for contravention corporal punishment; the meat to be seized and buried, the livestock slaughtered and buried.

Oct. 2. Salzthal. Because the disease can be carried in the air, if plague breaks out, cattle should not be herded near sick bays, and when going to pasture should pass on the windward side.

Oct. 9. Salzthal. The use of powder or the root of Wrang or "black sneezing root" [black hellebore] is proposed as a useful remedy in an earlier decree, but in the event of plague one should use only hair-ropes as setons and the other as a powder. All officials are to see that this is properly observed. If an outbreak occurs, infected stalls are to be cleaned and the healthy stock remain within. Stalls where infected stock have been are to be fumigated with gunpowder and amber; and mangers, etc. smeared with strong-smelling substances such as stag's-horn, or sword oil, tar or turpentine; so that healthy stock can go back. If animals become sick in the autumn or winter, they are not to change stalls again before the spring.

Oct. 12. Salzthal. "We have learnt…" that bulls and calves are not being rowelled because of a common belief that it makes bulls sterile and calves are not strong enough to withstand it. But medical reports state it is in no way damaging, so this very good method should not have been omitted. However, because of the difficulty in applying to some bulls, these will be allowed to remain without setons if they are held in special stalls, and if there is any suspicious illness, all communication with others must be cut off and they must be treated with other methods. It is also a false belief that camphor in the recommended pouch [hung around the neck] strengthens the bulls, but this can be replaced with sulphur. The stalls are to be smeared with stinking stag-horn, wine-stone, sword oil, or good tar. It is false that calves are not strong enough for rowelling and calves which are kept through the winter should have setons.

Oct. 16. Salzthal. In addition to the law requiring livestock dealers to produce an attestation for imported stock, the stock must remain at the border and watched for 8 days. Dealers must prove there has been no disease where they have come from in the past 3 months. Also, no animal has died in the places they have passed through, and there has been no sign for 3 months of any infectious disease. If anything suspicious is seen the stock must be sent back and the authorities notified immediately.

Oct. 16. Salzthal. No livestock should be sold except in the immediate neighbouring lands, and then only if certified that it will be slaughtered and used in the immediate neighbourhood and not taken further.

Oct. 18. Saltzhal. A decree of 28 protective measures mainly repeating and consolidating previous decrees issued as a precaution, the plague being in neighbouring provinces. Imported livestock showing the least unusual sign is to be slaughtered by the knacker in the presence of witnesses, and buried 5 feet deep, covered with lime or sand. The borders are to be watched by guards composed of villagers, dragoons, and soldiers. No dealer is to be allowed across the border without a pass certifying he has not been in an infected area. Two persons are to make twice-weekly examinations of all livestock and report to the authorities anything suspicious, immediately isolating suspicious animals. If any animals should be struck, the first 3 only are to be slaughtered and buried. The rest are to be placed in sick bays constructed near each village, where they are to be smoked. All unaffected cattle are to be rowelled. Persons in contact with the sick animals are to wash their clothes afterwards with soap and water. All dung and milk is to be buried, and all stalls and equipment washed with soap and water after use. The stalls are not to be used again for 2 months. The village is to be isolated. All senior officials, court authorities and magistrates, are responsible for seeing that the orders are properly executed [Repeated in all decrees].

Oct. 25. Wolfenbüttel. "We hear with great displeasure…" that the repeated order on rowelling is not being properly applied everywhere. To further strengthen the decree, for each animal not with a seton there should be a (Mfl.) penalty for each death; and if through insubordination or several offences infection breaks out, an added 24 hours' imprisonment with bread and water. Fines are to be set aside and a decree will follow for their use at the end of the year. For each contravention the offender shall discharge (Mfl.), 10 (Mfl) for 10 deaths.

Oct. 29. Wolfenbüttel. For the Martinmas Market at Helmstedt dealers may drive stock only through Wolfenbüttel and Schöning Districts with border passes, with the route to be specified and not passing near any local livestock, and are only allowed to proceed if the pass is correct. Stock at the annual Helmstedt market must have a horn stamped if in possession of a pass. Stock must first remain for 8 days at the border before being admitted.

(Nov. 2)

Nov. 13. Wolfenbüttel. Refers to the decree of the 18th of October concerning guarding of sick animals in sick bays. A little house should be built nearby and all food and drink and other necessities brought to within a gunshot distance where they can be collected. Even if sick animals recover they must still not be brought back amongst the healthy, but kept for a long period under supervision. Food and water for the stock, both sick and healthy, must also be left at a distance and collected. The sick bay attendant must burn all used straw. The pit for burial must be of sufficient depth in accordance with the decree and the carcase covered with lime if available. Straw and dung from the sick bay must be buried. The burial place must be fenced in so that no livestock can come near it for a long time. With regard to fumigation, one must be careful to avoid starting fires.

Nov. 20. Wolfenbüttel. Because fogs at this time of year could be damaging to cattle going into the open fields, all cattle are to be kept at home and not driven out until early next year. Rowelling should be maintained throughout winter.

Dec. 1. Wolfenbüttel. Refers to the decree of October 18. The pit in which the knacker buries a slaughtered animal must be 5 feet deep whether plague is present or not. Slaughtered animals are only to be buried if suffering from an illness and not otherwise. Each village is to have 2, and each city 3 or 4, pits. Any animals slaughtered must be under the supervision of the landlords, in cities by a person of the court, who observe at a distance. The carcase should be dragged to the already-made pit on a sleigh. Anything suspicious must be reported.

Dec. 20. Wolfenbüttel. Because of the impending Braunschweig Candlemas Fair and arrival of foreign buyers and drovers staying in Braunschweig, Wolfenbüttel, and other cities; people with livestock are to keep it shut up so that foreigners or dogs can not come in contact. No foreign horses are to be stabled in cow stalls, and the stalls are to be smeared with strong-smelling oils such as stag's horn oil, sword-black and stinking wine. Inn-keepers are instructed how to feed and house travelers' horses and are not to keep livestock in the same stalls between the time the waggons come through in January 1746 until their exit in February; and clean them thoroughly before using again. They are not to sell, under heavy penalty, any straw and hay which the visiting waggons brought.

n.d. Wolfenbüttel. A "Proposal of some good remedies against the cattle plague ..." A lengthy document of some 33 sections describing the symptoms of the disease and listing various preventative cures. The symptoms were as described by the Royal Prussian Medical College, 1716; Mr. Rath Hoffman; Mr. D. Christian Wolff, who postmortemed 20 animals in 1716; and Mr. D. Arens of Plön. On finding a sick animal it should be separated immediately and placed in a shed. The smoke of sheep dung had a very penetrating effect as a fumigant on large livestock, causing them to sweat. As soon as an animal became sick one should bleed, first on both sides of the neck, removing half an Oessel, or 4 quarts; taking the remaining 1/2; quart from the tail vein. Three or four spoonfuls should also be taken from the veins under the tongue. If the animal already has a seton, Wrang root, hellebore root, or black sneezing root, should be put in if the seton is moved to a different position. Some place a piece of Wrang root in the loose skin behind the legs. Next give the animal a good deep warm bedding and administer 8 parts of best theriac, or cattle-theriac, in lukewarm beer. Then give something warm, with 3 or 4 handfuls of stone flour, bruised grain, or wheat bran. One could add oil cake but if the animal is weak then add a pound of white clay or dried forecastle-stove glue. Some books commenting on the 1682 outbreak claim a certain method was lime-tree resin, or a spoonful of powder in the drink; but this was not proven effective. One could without danger, and maybe with advantage, add some lime-tree rot, or flour, bruised grain or bran. If the animal won't drink, then it should be poured down the throat. Give the following in warm beer: saltpetre, antimony, Mary thistle or Cardobenedictene seeds, and unheated stag's horn oil. After the first 24 hours add, every 6 hours, a handful of Wulfferlen or Great Lucian's herb. If after 24 hours there is no aggravation, the animal is rowelling well and sweating, give the powder every 6 or 8 hours. If it becomes worse after 24 hours give 2 good handfuls of soot from the stove with 2 good spoonfuls of elder-berry jam stirred in warm beer and vinegar. If still no improvement give the saltpetre etc. powder with the addition of cascarilla bark.

The price of Venice treacle [theriac] with camphor is 2 Groschen 2 Pfennig; and of cattle theriac with camphor, 2 Groschen 4 Pfennig.

n.d. Wolfenbüttel. Certificate concerning the prevention and cure of the evil-raging cattle plague. Written in the first person, presumably referring to the Duke of Braunschweig and Lüneburg, it stresses the most effective means against the introduction and spread of cattle plague are the administrative measures and refers to Ramazzini quoting that infection was carried into Italy in 1711 by a single ox from Dalmatia. Kanold also wrote that there was no better method than administrative measures for preventing infection. Methods of cure in the public gazettes were so manifold and almost contradictory to one another, and it was surprising that an effective cure had not yet been found. A foreign doctor had stated there was no method of cure if an animal was already infected. To better cut off and impede infection the methods quoted from the best writers are wanted. Infection can take place from:

(a) driving in foreign stock which are already infected. There is the authentic case of the single ox in Italy which resulted in 26,252 head of cattle dying in the Papal State alone. It is not even necessary to smuggle in infected cattle, for healthy cattle passing over pastures on which infected cattle have been can contract the infection if the weather conditions are right (Ramazzini).
(b) Through the blood from slaughtered animals and poured-out washing-up water (Kanold).
(c) Through not burying carcases deeply enough (Ramazzini, Lancisi).
(d) Through imported raw skins (Lancisi).
(e) Through dogs eating infected carrion or having been in contact with infected cattle (Lancisi). Kanold quotes the case of a tame raven which fed on a dog and spread infection.
(f) Through persons tending and managing sick stock or circulating among them picking up infection on their clothes and thus transferring it to healthy cattle (Kanold, Ramazzini, Lancisi).

Lancisi quotes the strange case repeated in the King of Prussia's decree of 1716 of a knacker's farmhand who went into an infected stall and afterwards slept on some hay in a barn, after which cattle ate the hay and became infected.

The only way to prevent these methods of infection is through administrative precautions. Regarding infection carried in the air, if many animals die the air becomes charged with the smell and a contagious haze. Stalls should be fumigated with laurel, juniper, sulfur, gunpowder, stag-horn oil, etc.; but the smoke is present only for a short time so it is advisable to tie a bag of herbs under each animal's neck. The contents and method of making the pouch are described. This keeps the smell around the animal resisting the poisonous haze and provides a tenuous antidote for the body. If a careful house-keeper wants to do more, stag-horn oil can be spread now and then on the walls of the stalls.

From experience, nothing of an infectious nature attacks the body unless it is in a certain disposition, therefore one should ensure:

1. Good feeding.
2. Good management.
3. Use of precautionary methods.

1. Impure food should not be given, but feed sparingly for fat livestock suffers more than lean. Give clean and pure water, also add coarse flour, bruised barley or oil cakes.

2. Good management means that the stalls should be kept as clean as possible and animals not left standing on rotting dung. Fresh straw should be strewn on the floor. It is very good to keep stock clean. Grooming is also good and the mouth should be washed with vinegar and salt.
3. (a) If the cattle have already sickened, besides that given in the Hamburg newspaper (No. 49, 1745), blood-letting at the tail is recommended, also from the brown veins under the tongue.
   (b) The hair-seton which in all plagues has done good service, cannot be recommended enough. The same is proposed in the decree of 1716. The manner of use is well known but will be repeated. Twisted or woven horse hair the thickness of a fair piece of twine is smeared with olive-oil and inserted through the dewlap with a "pack-needle" and the ends tied together. In the Harlemmer newspaper (No. 11, 1745) a piece of hellebore root instead of hair is proposed, but it is better to use a hair-rope as the other is not easy to keep in place. One can sprinkle Christian or black Sneezing-root powder on a portion of the hair-rope if it moves, and pull it through the wound.
   (c) The laxative method. There is no previous experience to go on but Lancisi quoted a very good recipe for those already sick. Fig broth followed by 2–3 pounds of cooked plums with 3–4 handfuls of salt in a pail of water, half-cook then add 1/2 pound of senna-pods. Strain, and pour a quart of the liquid down the throat. Instead of senna-pods, 5 or 6 pieces of tobacco can be used.

Whatever laxative methods are used the cattle should be given plenty of salt. At many places each cow when going onto the pasture in the Spring has a pickled herring turned in tar put down the throat, but it is the salt alone which is of benefit.

Tar-water now has a good reputation for the preservation of livestock. [Instructions for its preparation and use are given.]

A powder of brimstone and antimony on buttered bread given from time to time is a healthy method.

The powder first given in the Hamburg newspaper of March 1714 and afterwards in the decree of 1716, still stands:

Smearwort, or Hollow-root leaves; speedwell leaves; lungwort leaves; hyssop leaves; viper's grass leaves; gentian root; angelica root; butterbur root; tormentil root; bay; juniper berries. [This is the Dutch recipe tried by Bates in 1714, lacking carline thistle root.]

Common Theriac (*Theriaca diatessaron*) or cattle Theriac (*Theriaca propecoribus*) in water will also be useful.

The high bailiff of Dornick states the remedy is well known and is incorporated in the 27th issue of the Braunschweig *Anseigungs* paper. Offered in the public interest it is not to be doubted that it will be very useful for preservation and cure.

2. 1746

Jan. 5. Wolfenbüttel. A decree that only two mill visits [for grinding corn] per week allowed. The reason should be explained, and if visitors find a sick animal they should not proceed further but notify the authorities and cleanse themselves. Visits to the mill should also cease and restrictions of the decrees in force applied. Each housekeeper, to avoid a heavy penalty, must report each morning where his livestock is and anything suspicious.

Jan. 8. Wolfenbüttel. Because of the cattle market on the 1st. of February at Königslutter, referring to the decrees of the 29th Oct. and 2nd Nov., incoming dealers must be questioned, passes examined, cattle horns marked, and the routes they may follow specified.

Apr. 16. Wolfenbüttel. "We hear with surprise…" that a few herdsmen in the local lands perform blood-letting and one recently was a disaster because when an animal sickened a competent person could not be found. Officials should see that someone in each office understands it properly and supervises it. Rowelling is also to be used.

Apr. 21. Wolfenbüttel. Until further decree no foreign cattle should be fattened or grazed. If grazing is insufficient for one's own cattle, land leaseholders should allow grazing and butchers and livestock dealers should take as many animals as they can.

Apr. 30. Wolfenbüttel. When the cattle are driven out the pastures should be divided up into marked sectors and the herds separated into groups which must each remain within their sector. Officials are to see that the areas are agreed upon and then organize them, marking the boundaries perhaps with tufts of straw. There should be areas between them where no one must drive cattle. If any place does not do this we are to be informed.

May 28. Wolfenbüttel. If a herd outside is moved and an animal sickens or dies, the herd must be left on the pasture until the sick or dead animal has been investigated and further actions prescribed. There will be a heavy penalty if anyone brings the herd into a city or village.

Jun. 18. Wolfenbüttel. The measured contribution for 2 to live on, if the milk finishes, could become distributed in several more smaller amounts. In order to manage and not to spare the attendants' wages, more money is to be supplied at the offices. Extra wages must be found from the congregations, but the less well-off should not have to pay. It is specified that for an infected place up to a mile distant must provide. Authorities are responsible for the distribution of the Braunschweig information leaflets which shall be post free. Money for producing them shall be pro rata and charged to the annual tax.

Jul. 29. Wolfenbüttel. Frequent importations of stock contrary to the decree are taking place, and if an animal belches it is not notified. Where the decree is violated punishment should be with the big cart. This decree is to be read aloud from the church pulpit on Sundays as well as being posted.

Sep. 10. Braunschweig. "We have with the greatest displeasure noted…" that methods against the cattle plague are not being properly continued. In some places not at all and in others incorrectly. The danger is still present and it is of no use waiting until the disease appears as it is then more difficult to control. It is therefore necessary to sharpen the decrees concerning precautionary methods. It is ordered that fresh pouches [of herbs] are hung on the cattle, or the horns smeared with strong-smelling substances; and the use of setons is not to cease.

Sep. 26. Braunschweig. In the event of cattle plague appearing in the winter. At this time when the cattle are collected and brought into stalls for the winter they could contract infection and bring it with them. Therefore, to decrease the possibility of disaster, all possible caution should be applied when bringing them to the stalls. If disease is present in a yard all contact with the yard is to be avoided, proper care taken, and all precautionary methods applied. Instructions follow for reporting outbreaks, and if it occurs in 1 to 4 yards and there is no danger of spread, the village is left open. But if a public or army road passes through, drovers are not to be allowed to pass. Any animal with a suspicious illness is to be slaughtered and buried. Those with it in the stalls are to be slaughtered also, but the owner can salt or smoke the meat although it is better to bury it. Stalls must then be cleansed and also the people who dealt with the sick animals. All infected yards must remain closed for at least 14 days after the outbreak ceases. If it enters a village and spreads then, the village must be isolated. A remote building, such as a sheep shelter, with a place for an attendant, must be set aside as a sick bay. Each local authority must decide beforehand which buildings to use.

The inhabitants of an infected place must slaughter and bury the animals. If they do not want to slaughter healthy stock, then these can be kept in stalls, but the infected must be brought to the sick bays. The attendants must remain at the sick bays constantly and no other persons are to be allowed there.

Oct. 11. Braunschweig. Sick bays must still be maintained. Because of their isolation one must guard against thieves stealing the materials. If a theft from a sick bay is committed the penalty should be heavier than normal with corporal punishment and the discretion to fine.

Oct. 22. Wolfenbüttel. All officials are to ensure that any suspicious animals are correctly slaughtered. Where people disobey the order and don't report slaughter, severe corporal punishment should be considered. The butcher, in the presence of witnesses, using proper precautions, is to carry out the slaughter.

Oct. 26. Wolfenbüttel. When the herds are taken into the stalls, they are not to be let out again before early next year if the plague appears.

3. 1750

Sep. 26. Braunschweig. Now and then in the neighbourhood and at several border posts, the cattle plague has broken out again. The 1745–1746 precautions are referred to, and how it is spread is described. Everyone is exhorted to help in its prevention:

1. Until further notice when the disease is present, otherwise for the next 6 months, no raw skins or cattle to be imported. Anyone knowingly doing so will be punished.
2. If cattle are imported contrary to this order, at such place where they are found, the knacker, in the presence of sworn witnesses, whether the animals are sick or not, will slaughter them and bury them in a pit 5 feet deep, covering them with sand or unslaked lime.
3. Confiscated skins are to be buried or burnt.
4. The prevention of contagion at such places is then necessary.
5. No cattle are to be imported from infected regions, but those from disease-free regions must have a certificate.
6. No cattle dealer or his belongings to be admitted without a pass.
7. Details of what is to appear on the pass are given.
8. The same pass is required if animals are being brought from uninfected places.
9. Those who want to sell stock must do so from stalls, and not in the open. Also stock brought in with passes must be kept separate from one's own stock.
10. An owner is free to sell healthy stock fattened locally, with a pass allowing the sale, but the buyer must be issued with a free certificate as well.
11. No unmelted fat, meat, and wrapping cloths, are to be imported without a free pass. If so the fat and meat is to be seized and buried, the cloths burnt.
12. Stock must not be driven through previously infected places for several months under threat of severe penalty.
13. Markets must follow orders under threat of severe penalty until the danger is wholly ceased, with separate orders for the Braunschweig markets.
14. Anything the least suspicious must be notified to the authorities, suspect animals isolated and slaughtered, and cleansing rules followed carefully. No sick stock are to be driven on the common pasture and there must be no grazing where a sick animal has been unless and until the cause is otherwise established.
15. Dogs are to be tied up under risk of severe penalty.

16. If a really infectious illness breaks out with 1 or a few animals, the entire village is to be isolated and a runner sent to report. Meanwhile all rules must be emphatically observed.
17. Stalls must be fumigated with bay, juniper, sulphur and gunpowder. The walls smeared with stag-oil, tar, or bruised garlic.
18. Stock must not be driven out too early in the morning. If there is fog they are to be kept in until 2 hours after it has lifted. If whole cloudy days occur, or rain, the stock are to be kept in.
19. Instructions are given for feeding and watering.
20. Stalls are to be kept clean and free of dung, and the animals kept clean. Frequent grooming is recommended as it opens the sweat pores. The mouths should be washed with vinegar and salt, and parts of the head smeared with tar so that the animal retains its smell.
21. In the autumn, and 3 months into the winter, then a third time 3 months after in next spring, the cattle should be bled on both sides of the neck, taking 2–3 pints from an adult.
22. Laxative powders should be given an hour before winter feeding: English salt, Jalap root and saltpetre in tepid water.
23. For the following 14 days give 2 or 3 drachms of powdered antimony wetted with the feed, or on buttered bread.
24. Instead for 14 days pour a quart of tar-water down the throat each morning. Instructions are given for preparing it.

4. 1752

Dec. 16. Wolfenbüttel. Because of the presence of cattle plague in Wernigerode and the forthcoming Braunschweig Candlemas Fair, no cattle, sheep, or pigs, person or cart, is allowed to enter on the route out of or through Wernigerode into the princedom of Braunschweig, even with a pass.

5. 1754

Nov. 28. Braunschweig. Warning against the possibility of the cattle plague entering from neighboring lands, ways are described in which it can be spread: by dogs, cats, infected stalls, clothes and vessels. Laws are in force from 1745, 1746 and 1750. The sections of the 1750 decree are repeated with the addition of:

3. Skins imported in contravention to be thrown in pits and covered with lime and water to destroy them as opposed to burning as formerly.
6. An exception is made concerning importation of foreign stock for foreign army cattle introduced in addition to market stock. These are to be placed in separated yards and barns overnight where there are no other cattle and not in ordinary livestock stalls. They must halt at the border before the land-defence guards until the pass has been sent in for examination and approved. The cattle are allowed before the city gate and what the butcher wants to buy brought into his stall but separated from other cattle, and slaughtered within 8 days or less before 2 witnesses of the commissaries.
8. No dealers or their stock are to be allowed in without being believed to be in good health. Jews wandering on foot, even if with passes, should not be allowed in as

long as the danger is present, also riding Jews. During the Braunschweig fairs, 8 days before and 8 days after, riding Jews may come in if with correct passes, but must convince the border authorities they do not deal in livestock or raw skins. As so few, Jews on foot may be admitted to the Braunschweig fairs.

11. Cattle for sale are not to be stalled but left in the open. When purchased they are not to be put into stalls with other cattle but kept apart for a while.
12. Locally fed and fattened cattle can be traded, but those from neighbouring lands must be slaughtered and used on the spot and not taken further.
14. If the cattle plague breaks out:
    - (i) If only in 1, 2 or 3 yards, there is hope of saving the rest of the stock, but if it spreads
    - (ii) sick animals must be slaughtered and buried at a distant place from the village and the coats of others washed with salt water.
    - (iii) A cheap cottage must be provided for a watchman.
    - (iv) The cottage must be constantly fumigated.
    - (v) No one who is not involved is to be allowed into the village.

    Instructions are given concerning the tanning of skins.
23. Where cattle plague breaks out, the stalling and separation of healthy cattle is shown to be the best method of prevention.
24. The fumigation of stalls is necessary and instructions are given. Inspectors are to inspect the equipment for fumigation and report if it is unsatisfactory. Walls should be smeared with stag-horn oil or bruised garlic and inspected that this is done.

Authorities must see that orders are obeyed precisely and not modified.

6. 1757

Aug. 22. Braunschweig. Outbreaks of cattle plague are reported in Oldenburg, Butjadinger and mainly East Friesland. The import of all stock without a pass is forbidden. Any imported in contravention will be seized, slaughtered and buried. The pass must show whence bought, that the area was healthy and there was no fear of infection.

Sep. 18. A Braunschweig military order in German and French by Germain Dubois, knight, Master of Crancé, Baltham and other places, First War Commissar of the army of Master Marshall Duke of Richelieu.

On the part of representations which have been made by the Prince of Brunswick, that an epizootic illness of cattle reigns. He published a decree for all slaughterers and butchers of the army who would bring in beasts, to submit without difficulty to the inspection of the official to this effect. In order to avoid the communication of all foreign cattle with those grazing, we renew the preceding regulations published on this topic. We order, besides, that all butchers or persons dealing in cattle, are held, under penalty of confiscation, to submit to the arrangements for grazing, especially those holding their cattle far away, of those of the inhabitants, and of satisfying themselves of the emplacements and pastures which will be assigned by the Prince.

7. 1761

Sep. 8. Braunschweig. The cattle plague having broken out in some places it is ordered that all cattle both in cities and in the country, bought within the past 6 months, be immediately placed in separate pastures and other livestock not allowed into the stalls up until winter.

Nov. 17. Braunschweig. The plague continues to spread in local and adjoining lands.

1. No movement of stock is allowed without a pass, whether being imported or within the country.
2. Up to the end of the year only army cattle are allowed and all trade in breeding, milk cattle, young bulls and breeding calves and others is forbidden. Until further notice breeding cattle will only be allowed to be imported from a few foreign places. No passes may be issued for any other cattle except for butchers.
3. Passes are granted for army cattle for oxen, cows, beef or calves; which are to be slaughtered within 24 hours.
4. All local people and butchers in other cities in the country are not to buy any army stock under severe penalty.

8. 1770

Jul. 5. Braunschweig. It is observed that sick or dead cattle lie unburied at the carrion pits and near the army routes. It is therefore ordered:

1. Carrion pits near cities and villages, at army routes and gardens, are to be covered.
2. At a remote place, 2, 3 or more 6 feet deep pits to be dug in which the knacker throws the carrion and covers with earth. When a pit is full another must be dug. If not immediately buried but the carrion left to rot, the penalty is 3 Thaler.
3. Cattle are no longer to be slaughtered at the inspector's place but at the carrion pit to which the owner must bring them, where they are put in, slaughtered, and immediately buried.
4. If horses or other draught animals pulling waggons die on the route, they must be brought to the carrion pit for burial and the village through which they have passed informed. Persons will be liable to a penalty for covering-up.

9. 1774

Oct. 5. Regulation Braunschweig. Recent outbreaks of cattle plague have occurred in some foreign areas, so particularly at the local city from the 18th to the 31st of October the following is to be observed to prevent the evil:

1. No different cattle to be admitted, unless with a certificate of health from places wholly free of cattle plague.
2. Owners and salesman are also to swear that there is nothing suspicious, none have sickened on the journey and all are healthy and free of all suspicion.
3. Cattle are allowed entry only if with a proper pass.
6. In the grazing areas groups are to be far enough from each other that no sector touches another.
7. In the unexpected event of disease breaking out in a cattle drive, the owner is entirely liable to prevent its spread.
8. Cattle are driven in at the Petri, Weden and August gates.
9. As few handlers as possible should accompany the cattle to the stalls.

All incoming foreign livestock dealers must pay attention to the orders.

10. 1775

Nov. 10. Braunschweig. Because of cattle plague in neighbouring countries officials must report immediately any cattle dying and the existing regulations and remedies are to be applied.

Nov. 13. An Advertisement issued by the Police Department at Braunschweig.

Because of the widespread danger of cattle plague, after consultation with experts, the following methods of prevention are recommended: keep animals in the stalls and keep them clean; fumigate the stalls 2 or more times daily using sharp vingear; don't give cold but lukewarm drinks, also give some with vinegar or urine added. Rub the tongue and inside of the mouth with salt. Give a few times each week 6 Lot of kitchen salt to clean the blood and a portion of wheat bran and flax oil in warm water for loosening of the intestines. Also for about 3 days 4 Lot of rue spirit. Also *Borstorffer* and *Rainetten* apples or sour apple disks.

Nov. 27. Braunschweig. Renewal of the decree concerning burial of animals dead of the cattle plague. All possible caution must be taken to ensure the plague does not spread further through the air. As the decree of the 1st December 1745 is no longer available to all it is repeated herewith and its literal content renewed.

Dec. 11. Braunschweig. For the conservation of stock and breeding stock no cow-calves may be slaughtered but must be sold.

Dec. 27. Braunschweig. Concerns the production of passes because cattle plague has broken out in different places. To prevent its spread, border passes must state:

1. The cattle have come from healthy, infection-free places.
2. They have not passed through any infected places or army posts or routes.
3. They have not lingered with waggons or merchandise on the route.
4. The pass must show a specified route, signed from place to place.
5. Raw and untanned skins, wool, hemp, smoked tallow, horns and meat, must not be loaded in cities and other places if cattle plague is raging, and not for 6 months afterwards.

11. 1776

(Jan. 28)

Mar. 18. A decree to prevent the further spread of cattle plague by the separation of stock on common grazing grounds. The isolation of infected houses, yards and whole villages, has prevented the widespread occurrence of the plague. Pastures can be infected by other animals and when healthy animals pass over them they themselves become infected. Therefore, before the cattle are driven out, the pastures must be divided up and each owner graze his cattle only in the sector assigned and not drive them over others. This is to apply throughout the year.

Jun. 27. Braunschweig. Precautions to be observed at cattle pastures this summer as the cattle plague is still raging in different parts of the princedom.

1. Every owner must provide a list of all stock and show any decrease. Anyone dealing for money against this decree is liable to corporal punishment.
2. It is unhealthy for stock to be let out to pasture too early or left out too late, so hours of pasturing for each month from July to November are given. Flexibility is allowed

for distant pastures and for bad weather. In some districts stock is left outside from Spring to Autumn and should, where possible, be sheltered at night.

3. A strong smell of tar is beneficial so tar should be spread on the head and nose and on stalls.
4. If there is an eclipse of the sun, cloudy days, cold weather or prolonged rain, cattle should not be driven out. Herdsmen will be liable if driving out other than at specified hours, or within 2 hours of fog lifting or if a heavy dew.
5. Clean water is essential. Wet places are to be drained and water courses cleaned to ensure a fresh supply and not stagnant or fouled with damaging insects.
6. At least 1 morning each week 6 Lot of kitchen salt should be given and the mouth frequently washed with vinegar and salt; and rue given to smell. In big yards salt licks should be provided.
7. Large neighbouring estates should work together to prevent the danger. Herdsmen should be instructed to observe the wind direction when grazing near the border and not to come near another herd when it changes due to evaporation.
8. The decree of the 18th March gives instructions which are again repeated concerning distribution and preservation of grazing, with heavy penalties for hersdsmen infringing the decree.
9. It remains prohibited to take sick cattle, sheep or horses onto the common pasture under the utmost severe corporal punishment; and to remove livestock from places which have not been without cattle plague for a full 6 months.
10. Regulation 28 [1752 28th Nov. ?] prescribed that where cattle dead of the plague are buried the place is not to be used for 2 years and herdsmen are to give it a wide berth with cattle and dogs. Sheep are also to avoid it as although they are not infected the poisonous haze clings to their wool.
11. If an animal sickens on the pasture, the event must be speedily notified and the animal separated. Until a thorough investigation shows that it is not cattle plague, cattle should not be collected from the pasture.
12. In the villages and elsewhere where it could happen on cattle routes, the fatstock, milk-cattle, bulls, and calves, should not be mixed or collected together.
13. If where cattle plague rages part of the stock is sick and another different part separated in the stalls is healthy, it should remain protected for 8 weeks in case of an attack.
14. If plague breaks out on a pasture the herd must remain there night and day and all communication severed so that it does not contaminate people and dogs.
15. Refers to the decree of the 13th November 1775 specifying certain actions to be taken when livestock are retained in stalls and infection breaks out on the pasture. A sick bay must be made, burial pits dug, dogs tied up. Slaughtered stock must be buried with the skin intact. A post must be set up, etc.
16. Milk from sick animals must be poured into a deep pit at the sick bay. Milk from healthy animals may only be used for one's own consumption and is not to be sold.
17. To prevent cattle plague entering cities, those having stock in the city and on the fields before the gate, must not increase the number. Butchers can increase army cattle and herd them on the common.
18. Refers to the refund of expenditure for inspections, reports, etc. Some duties are to be *ex officio* and performed gratis.

Jul. 4. Braunschweig. Renewal of decree of the 27th Dec 1775 with reference to passes for foreign imports. Article 5 is amended reducing the 6 month period to 3 months for the absence of disease from places where raw skins, wool, rough tallow, horns and meat are loaded.

(July 15th) "Circular-Rescript".

Jul. 30. Braunschweig. A decree relating to the preservation of paddock grazing of sheep, horses and pigs when cattle plague breaks out. Previous orders did not cover these animals and it is resolved that they can be brought to allotted pastures if there is no suspicious illness; but if cattle sicken then all other stock are to avoid the pasture until the danger has passed. The order is to apply regardless of commons rights for all sheep, horses and pigs.

Sep. 9. Braunschweig. Due to the Royal Prussian war decrees already enacted, concerning main army routes that requisite cattle health passes are immediately conferrable without requiring waiting at the border, it has been learnt that such passes to introduce army stock are being charged high fees. Such passes are issued without cost. The "Circular-Rescript" of the 15th July concerning breeding stock is extended.

12. 1777

Dec. 8. Braunschweig. How to allow the purchase of breeding stock in view of the restrictions. Inhabitants of places where cattle plague has raged, if they have followed the decree of the 28th Jan 1776 and there has been no outbreak for 6 months, can obtain breeding stock again. If special circumstances allow a shorter time with security they can do so as this is permitted under the decree. When however one wants to sell stock out of a cattle plague area, the plague must not have been present for a full year, and with an attestation of the judiciary that it has been absent a year and a day the stock can be brought into the home of the buyer and admitted into cities, but will be slaughtered if in contravention of the decree.

13. 1778

Sep. 3. Advertisement Braunschweig.

It is now sometime since cattle plague was present and the so-called Wolfenbüttel incident of the 20th October Gaul-livestock notices do not hold. No cattle are admitted and one must buy none from foreign public markets as one does not know where they are from. Horses, sheep and pigs, for the usual annual markets, can be brought in with a pass showing the route and that they have passed no infected places. The butcher and his staff can, with passes, bring healthy fat livestock into cities from local land, or such foreign provinces where cattle plague has not raged this spring and summer, and are brought on clean routes.

14. 1782

Feb. 25. Braunschweig. Renewal of decree of 27th November 1775 concerning burial of cattle which have died of the plague. Slaughterers are not observing proper caution and while there is still danger sick animal carcases should be buried. If a livestock illness exists in the next neighbourhood and an animal dies, it is to be opened at a laboratory in the presence of intelligent people and if found not to have the cattle plague also to be buried in a moderately deep place which is to be avoided. The decree of 28th November 1754 specified the precautions to be taken if the disease spreads.

No date, no place of issue. A description of the so-called "fire-sickness" of cattle and a recipe against the same. A description of the disease attacking cattle especially in 1765 and 1766 which still rages here and there. Rapid death without signs of sickness distinguishes it from cattle plague, the illness being an inward inflammation and stagnation of the blood. It can occur at 2, 3 or 4-year intervals, caused by very cold alternating spring and summers. Instructions are given for bleeding and a herb recipe.

Appendix III

STATUTES, ORDERS IN COUNCIL, AND PROCLAMATIONS CONCERNING RINDERPEST IN BRITAIN AND IRELAND 1746–1878

1. STATUTES

1.1. 1746

Feb. 13. An Act to enable His Majesty to make Rules, Orders, and Regulations, more effectually to prevent the spreading of the distemper which now rages amongst the horned Cattle in this Kingdom. 19 Geo. II.

1.2. 1747

Jan. 10. An Act to continue, explain, and amend an Act made in the last session of Parliament, intituled, *An Act to enable His Majesty to make Rules, Orders, and Regulations, more effectually to prevent the spreading of the distemper which now rages amongst the horned Cattle in this Kingdom*. 20 Geo. II.

Nov. 10. An Act ... to continue Two Acts, the One passed in the Nineteenth Year, the Other in the Twentieth Year of His present Majesty's reign, to prevent the spreading of the Distemper amongst the Horned Cattle. 21 Geo. II.

1.3. 1748

Mar. 22. A New Order of His Majesty in Council, dated the Twenty second of March, 1747, relating to the Distemper'd Cattle, For continuing the present Orders, which will expire on the Twenty fourth Instant, to the eighth of April next, when they are to cease; and for establishing New Rules and Regulations to be observed thenceforth, in lieu thereof. 22 Geo. II.

1.4. 1749

An Act to contuinue several Laws for preventing the Spreading of the Distemper which now rages amongst the Horned Cattle, and for impowering His Majesty to prohibit the Killing of Cow Calves. 23 Geo. II.

1.5. 1750

Jan. 17. An Act for explaining, continuing, and enforcing several Laws therein mentioned more effectually to prevent the spreading of the Distemper which now rages amongst the Horned Cattle in this Kingdom. 24 Geo. II.

1.6. 1751

An Act to continue, explain, and amend, several Laws more effectually to prevent the Spreading of the Distemper which now rages amongst the Horned cattle in this Kingdom. 25 Geo. II.

1.7. 1753

An Act to explain, and amend, and continue several Laws more effectually to prevent the spreading of the Distemper which now rages amongst the Horned cattle in this Kingdom. 26 Geo. II.

1.8. 1754

An Act to continue several laws relating to the Distemper now raging among the Horned cattle in this Kingdom. 27 Geo. II.

1.9. 1755

An Act…for continuing an Act made in the last session of Parliament, to continue several Laws relating to the Distemper now raging among the Horned Cattle in this Kingdom. 29 Geo. II.

1.10. 1756

An act more effectually to prevent the Spreading of the Distemper now raging amongst the Horned Cattle in this Kingdom. 30 Geo. II.

1.11. 1770

An Act for indemnifying all Persons, with respect to advising, or carrying into Execution His Majesty's Orders of Council made for preventing the Spreading of a contagious Distemper amongst the Horned Cattle, and for rendering the same valid and effectual, and for preventing Suits in consequence thereof; and to authorize the continuing, extending, and executing the same, for a further Time. 10 Geo. III.

An Act to prevent the further Spreading of the contagious Disorder among the Horned Cattle in Great Britain. 10 Geo. III.

1.12. 1772

An Act for continuing an Act, made in the Tenth Year of the Reign of His present Majesty, to prevent the further spreading of the contagious Disorder among the Horned Cattle in Great Britain. 12 Geo. III.

1.13. 1866

Feb. 20. The Cattle Diseases Prevention Act. 29 Vict.

Mar. 6. An Act to amend the Law relating to Contagious Diseases amongst Cattle and other Animals in Ireland. 29 Vict.

Apr. 23. An Act to amend the Act of the Eleventh and Twelfth Years of Her present Majesty, Chapter 107, to prevent the spreading of contagious or infectious Disorders among Sheep, Cattle, and other animals.

Jun. 11. An Act to give further Facilities for the Establishment of Societies for the Assurance of Cattle and other Animals.

Aug. 10. An Act to amend The Cattle Diseases Prevention Act. 29 & 30 Vict.

1.14. 1867

Aug. 20. The Contagious Diseases (Animals) Act. 30 & 31 Vict.

1.15. 1869

Aug. 9. The Contagious Diseases (Animals) Act. 32 & 33 Vict.

1.16. 1875

Aug. 11. An Act to amend the Contagious Diseases (Animals) Act, 1869. 38 & 39 Vict.

1.17. 1878

Contagious Diseases (Animals) Act. 41 & 42 Vict.

2. ORDERS IN COUNCIL

2.1. 1714

Dec. 1.

2.2. 1746

Mar. 12. For preventing the spread of the contagious disease among horned cattle.

Nov. 25. Appointment of Justices of the Peace in Middlesex.

Dec. 17. Re-issue of Order of March 12th, to be read in all Parish Churches after Prayers.

2.3. 1747

Jan. 15. Modifying the regulations for preventing the spread of the contagious disease among horned cattle.

Mar. 12. Further regulations for preventing the spread of the contagious disease amongst horned cattle.

Apr. 16. Further regulations for preventing the spread of the contagious disease amongst the horned cattle, by stopping the holding of fairs, etc.

Jun. 18. Further regulations for preventing the spread of the contagious disease amongst the horned cattle.

Sep. 8. Further regulations for preventing the spreading of the contagious disease amongst the horned cattle.

Oct. 27. Against abuses in claiming compensation for cattle slaughtered on account of the contagious disease.

Dec. 11. Renewing the orders for the preventing the spread of the contagious disease amongst horned cattle for a further period, with variations.

2.4. 1748

Jan. 23. Renewing the Order of the 11th December as to the contagious disease amongst horned cattle.

Feb. 25. Renewing the Order of the 11th December as to the contagious disease amongst horned cattle for fourteen days.

Feb. 28. North Riding restrictions.

Mar. 8. Renewing the Order of December 11 as to the contagious disease amongst horned cattle for fourteen days.

Mar. 22. Further Orders for preventing the spread of the contagious disease amongst horned cattle.

Apr. 19. North Riding of County of York Local Authority Order.

Apr. 29. North Riding of County of York Local Authority Order.

May 17. North Riding of County of York Local Authority Order.

May 27. North Riding of County of York Local Authority Order.

Jun. 9. North Riding of County of York Local Authority Order.

Jun. 27. North Riding of County of York Local Authority Order.

Oct. 4. North Riding of County of York Local Authority Order.

Nov. 9. North Riding of County of York Local Authority Order.

Nov. 21. North Riding of County of York Local Authority Order.

Dec. 5. North Riding of County of York Local Authority Order.

Dec. 8. North Riding of County of York Local Authority Order.

Dec. 9. North Riding of County of York Local Authority Order.

Dec. 12. North Riding of County of York Local Authority Order.

Dec. 13. North Riding of County of York Local Authority Order.

2.5. 1749

Feb. 20. Guisbrough Local Authority Order.

Dec. 14. Prohibiting the removal of horned cattle after January 14, except for an immediate slaughter or under licence.

2.6. 1750

Jan. 12. Repeating Order of the 14th December last, and renewing that of 14th March 1748 (8th March?), as to the disease amongst horned cattle.

Apr. 4. Shropshire restrictions.

2.7. 1751

Apr. 4. Shropshire restrictions.
Apr. 16. Cheshire restrictions.
Aug. 8. Prohibition on holding markets at Barnes and Harlow Bush.
Aug. 9. Against the holding of fairs likely to spread the distemper among horned cattle.

2.8. 1757

Oct. 4. Prohibition of fairs and markets in Essex.

2.9. 1763

Feb. 14. Against the importation of hides of horned cattle, from ports of Denmark and Germany infected with the distemper amongst horned cattle.

2.10. 1770

Apr. 5. For preventing the spread of the contagious disease amongst horned cattle in Scotland.

Oct. 5. Prohibiting the importation of cattle, hides, horns, etc., from Ports on the Coast of France in Picardy, Normandy, Brittany, etc.

Oct. 12. Prohibiting the importation of cattle, hides, horns, fat, straw, etc., from parts infected with the cattle disease.

2.11. 1774

Oct. 28. Continuing the prohibition to import cattle, hides, etc., from certain parts of France, imposed by Order of the 5 October 1770.

2.12. 1857

Apr. 22.

2.13. 1863

Nov. 17. Regulations as to the importation of cattle, for preventing the introduction and spread of contagious and infectious diseases.

2.14. 1865

Jul. 24. For preventing the spread of a contagious and infectious disease prevailing amongst the cattle of the Metropolis.

Jul. 28. (control of imports).

Aug. 11. Further regulations for preventing the spread of the cattle plague in London, and other parts adjacent.

Aug. 11. For extending the provisions of the Order in Council of 24th July to all parts of England and Wales.

Aug. 18. For extending the provisions of the Order in Council of the 11th August, with regard to the cattle plague to all parts of Scotland.

Aug. 25. Prohibiting the removal of cattle from Great Britain to Ireland.

Aug. 26. Revising Orders of the 24th July, 11th August and 18th August, as to the cattle plague in England and Scotland.

Sep. 7. Prohibiting the removal of hides, skins or horns from Great Britain into Ireland.

Sep. 13. For exempting from the operation of the Order in Council of the 7th September 1865, all hides, skins and horns brought from India, Australia, etc., into Great Britain, and thence transported to Ireland.

Sep. 22. Consolidating and amending several Orders for preventing the spread of the cattle plague in Great Britain.

Sep. 29. Prohibiting the removal of any lambs, sheep, skins, etc., from Great Britain to Ireland.

Sep. 29. Prohibiting the removal of any lambs, sheep, skins, etc., to the Island of Lewis, county Ross.

Oct. 31. Further regulations with a view to check the spread of the cattle plague.

Nov. 3. Revoking the Order in Council of 29th September 1865, and issuing other regulations for preventing the spread of the cattle plague in certain parts of Scotland.

Nov. 23. Consolidating and amending certain Orders, with a view to checking the spread of the cattle plague.

Nov. 27. Horns imported from India, South America, etc., to be excepted from the operation of the Order in Council of the 7th September 1865.

Dec. 16. Revising the Order in Council of the 23rd November 1865, as to the cattle plague.

2.15. 1866

Jan. 20. Revising certain regulations for checking the spread of the cattle plague in Great Britain.

Jan. 20. Extending certain provisions for checking the spread of the cattle plague to Ireland.

Feb. 1. Exempting sheep and lamb pelts that have been thoroughly limed, from the operation of the Order of September 29th 1865.

Feb. 6. Revising the regulations regarding the cattle plague, so far as they refer to the Metropolis.

Mar. 17. Regulating the carriage of cattle for hire in any part of Great Britain with a view to check the spread of the cattle plague.

Mar. 24. Prohibiting the removal of cattle on any railway before the 16th April.

Mar. 24. Prohibiting the importation of cattle from the Netherlands.

Mar. 24. Consolidating and amending divers Orders relating to the cattle plague.

Apr. 7. For establishing in Ireland a registry of drovers and dealers of cattle, with a view to check the spread of the cattle plague.

Apr. 14. Further regulations as to the cattle plague to be observed in South Wales, Monmouth, etc.

Apr. 16. Regulations as to the carriage of cattle per rail or boat, and for disinfecting all carriages, trucks, etc.

Apr. 17. For establishing registry offices for granting certificates for the removal of cattle from Ireland during the prevalence of the cattle plague in Great Britain.

Apr. 27. For regulating the removal of any cattle from any place on the River Thames or Medway to the Metropolitan cattle market, for immediate slaughter.

Apr. 27. Special provisions with regard to the cattle plague in Westmorland.

Apr. 27. Excepting wool imported from India, South America and Australia from the operation of the Order in Council of 29th September 1865.

May 5. Cattle removed from the Liverpool cattle market at Stanley, to be slaughtered within 48 hours of their arrival at Birkenhead.

May 9. Continues provisions of 29 Vic. cap 2 *Cattle Plague Act* until 30 June sections 12–16 concerning slaughter.

May 15. Prohibiting the holding of cattle fairs, markets, etc., in the baronies of Lower Iveagh, Kinclarty, and Upper Castlereagh, Co. Down, until after the 1st June 1866.

May 18. For preventing the spread of the cattle plague in the Isle of Man.

May 19. Prohibiting the holding of cattle fairs, markets, etc., in the counties of Antrim, Armagh and Down in Ireland, except as directed.

May 22. For preventing the spread of the cattle plague in Ireland.

May 26. Continuing certain regulations, and making others as to the landing of foreign cattle, the movement of cattle in England and Wales, etc.

May 26. Regulating the removal of cattle from the Channel Islands and Man into the United Kingdom, and varying the provisions of the Order in Council, of the 24th March 1866.

May 26. Rescinding the Order of the 27th April as to the cattle plague in Westmorland, and making similar provisions in respect of Northumberland, Westmorland and Montgomery.

May 31. Extending the provisions of the Order of the 5th May last as regards the cattle market at Stanley.

Jun. 1. As to disinfecting cattle trucks, boats, etc., used for the transport of cattle.

Jun. 7. Regulations with regard to the removal of cattle to be observed in Wolverhampton Cattle Market.

Jun. 7. Regulations with regard to the removal of cattle to be observed in Salford Cattle Market.

Jun. 7. Extending the limits of the North Western district of Scotland so as to include Inverness, Elgin and Nairn for purposes of the *Cattle Plague Act*.

Jun. 8. Prohibiting the removal of any cattle from Ireland to Great Britain without licence.

Jun. 11. Continues provisions of 29 Vic. cap 2 *Cattle Plague Act* concerning slaughter of diseased animals.

Jun. 19. Providing for the removal of cattle from the Newcastle-on-Tyne cattle market to Gateshead, for slaughter.

Jun. 22. Extending the permission to remove cattle for immediate slaughter from the cattle market at Stanley, to the towns specified.

Jun. 22. Empowering the local authorities of certain counties to make their own regulations as to admission and removal of cattle.

Jun. 25. For disinfecting all persons coming into Ireland who have been connected with cattle in England.

Jun. 27. Authorising the importation of cattle from Harlingen, Friesland, and Delfzyl, Gröningen, subject to certain regulations.

Jul. 3. Permitting the importation of cattle from Leinster, with form of licence, etc.

Jul. 17. Prohibiting the importation of offal from Great Britain into Ireland.

Jul. 26. Continuing the provisions of sections 12 to 16 of the *Cattle Diseases Act* until the 1st August.

Jul. 31. As to the removal of cattle, for immediate slaughter in the counties of Warwick and Worcester.

Jul. 31. Authorising the local authorities for the county of Denbigh to make regulations to prevent the spread of the cattle plague as required.

Jul. 31. As to publication in the local newspapers of orders and regulations made by the Local Authorities.

Aug. 3. For putting the provisions of the *Cattle Plague Act* into force throughout the whole of Ireland.

Aug. 8. Revoking Orders of 11th June and 3rd July 1866, as to removal of cattle.

Aug. 8. Altering the forms of licence set forth in the Order of 24th March last as to the removal of cattle.

Aug. 9. For a Special Prayer in England and Wales (12th August), for relief from the cattle plague and protection from the cholera.

Aug. 9. For a Special Prayer in Scotland (12th August), for relief from the cattle plague and protection from the cholera.

Aug. 11. Declaring Scotland to be free from the cattle plague, and suspending certain orders.

Aug. 31. Prohibiting the introduction of cattle, etc., from England and Ireland into Scotland.

Aug. 31. The provisions of the Order of the 8th August last as to licences for removal of cattle, to be extended to the county of Oxford.

Sep. 5. Prescribing the mode for carrying into effect the provisions of the Order of August 31 as to Oxford, Headington and Abingdon.

Sep. 18. Regulating the holding of markets in Scotland, at which foreign imported cattle may be exposed for sale.

Sep. 25. Continuing the provisions of the Order of the 7th June last, as to movement of cattle in Wolverhampton until 30th September.

Sep. 25. Continuing the provisions of the Order of the 19th June last, as to movement of cattle in Newcastle-on-Tyne until 31st December.

Oct. 4. Regulating the removal of cattle sold in the market at Newcastle-on-Tyne to South Shields, Tynemouth or Jarrow, for slaughter within four days.

Oct. 4. Permitting the importation of cattle from Ireland and Scotland into certain parts of England, under licence.

Oct. 4. Regulating the removal of cattle in Salford market for immediate slaughter.

Oct. 12. Prohibiting the importation into Ireland of cattle from foreign parts, excepting Spain and Portugal.

Oct. 12. Allowing the importation of offal from Scotland, under certain regulations.

Oct. 12. Regulating the exportation of cattle from Ireland to Great Britain.

Oct. 15. Regulating the removal and slaughter of cattle imported from Belgium.

Oct. 15. Cattle imported from Friesland and Groningen to be slaughtered within twentyfour hours, under the direction of the Customs.

Oct. 24. Authorising the Local Authority of Anglesey to grant licences for the holding of fairs.

Oct. 30. Authorising the Local Authority of Glamorgan to grant licences for the holding of cattle fairs, markets, etc.

Nov. 7. Providing for the licensing of cattle fairs, markets, etc., in places not affected by the cattle plague.

Nov. 10. For a Special Prayer and Thanksgiving throughout England and Wales (18th November), for relief from the cattle plague, and protection from cholera.

Nov. 10. For a Special Prayer and Thanksgiving throughout Scotland (November 18), for relief from the cattle plague, and protection from cholera.

Nov. 10. Regulations under the *Cattle Plague Act* for subjecting cattle to quarantine.

Nov. 14. For a Prayer and Thanksgiving to Almighty God, for relief from the plague amongst cattle.

Nov. 22. Regulations relating to cattle imported into Great Britain from foreign countries.

Dec. 3. Regulating the movement of cattle between sunset and sunrise.

Dec. 5. Making provisions to meet cases where licences for the removal of cattle may have been issued by inadvertence without the alterations contained in the Order of the 3rd December.

Dec. 7. Revoking the Order of the 7th April 1866, as to registration of cattle dealers, drovers, etc.

Dec. 21. Restricting the powers of local authorities to licence the removal of cattle.

Dec. 27. Continuing the Order of the 7th June last in force until 31st March, as far as relates to the Salford Cattle Market.

Dec. 27. Continuing the Order of the 7th June last in force until 31st March, as far as relates to the Wolverhampton Cattle Market.

Dec. 28. That the Order of the 10th November as to importation of foreign cattle does not apply to the Channel Islands.

2.16. 1867

Jan. 8. Permitting the importation of cow hair from Great Britain into Ireland under certain conditions.

Jan. 16. Making provision for the sale of cattle referred to in the Orders of the 24th March and 8th August 1866.

Jan. 23. Licensing the port of Southampton for the landing of foreign cattle.

Jan. 23. Licensing certain pens and sheds adjoining the Southampton Company's Docks for the reception of foreign cattle.

Jan. 29. As to the removal of cattle from Southampton to the Islington cattle market for immediate slaughter.

Feb. 1. Altering the boundary limits with regard to animals moved within the Metropolis.

Feb. 4. Regulating the sale and removal of cattle for immediate slaughter in and from the Birmingham cattle market.

Feb. 4. Authorizing removal of cattle from Westmorland into Scotland.

Feb. 14. Rescinding the Order of the 4th February as to removal of cattle from Birmingham market.

Feb. 14. Authorizing removal of cattle to certain places from Birmingham market, for slaughter within four days of exposure in market.

Feb. 15. Defining the area within which cattle brought by sea to Dee Quay, Chichester, shall be slaughtered.

Feb. 19. Authorising the removal of cattle from Scotland to licensed cattle markets in England and Wales with a store stock licence.

Feb. 19. Defining the limits within which cattle may be landed at the Port of Liverpool.

Feb. 19. Regulating the removal of cattle sold in Wolverhampton market.

Feb. 19. The Order of 5th May 1866 as to removal of cattle to be determined and no longer in force after the 22nd February.

Feb. 26. Continuing the compulsory slaughter of diseased animals until the 31st May next.

Mar. 5. Revoking the Orders of the 14th and 27th April, 26th May, 22nd June, and the 31st July 1866, prohibiting the removal of cattle in certain districts, and the Orders of the 26th May, 24th and 30th October 1866, relating to the licensing of fairs.

Mar. 5. As to the granting of licences for the sale of animals upon application of the local authority as defined by the Order of 24th March 1866.

Apr. 1. Permitting the removal of cattle sold in Salford market to a distance of twelve miles from the market for immediate slaughter until 1st August.

Apr. 1. Permitting the removal of cattle sold in the market at Newcastle-on-Tyne to any slaughterhouse in Sunderland until 1st August.

Apr. 1. Permitting the removal of cattle sold in Wolverhampton market to various places for immediate slaughter until 1st August.

Apr. 11. Permitting cattle brought on the Great Eastern Railway to Tottenham Station to be driven to the Metropolitan cattle market after sunset on Saturday and Sunday.

Apr. 11. Permitting cattle from foreign ports landed at Harwich to be removed by the Great Eastern Railway to the Metropolitan cattle market.

Apr. 12. Revoking the Order of 1st April relating to the Wolverhampton cattle market.

Apr. 12. For regulating the sale and movement of cattle at Wolverhampton cattle market.

Apr. 13. As to the licensing of markets for the sale of cattle otherwise than for immediate slaughter, providing that no foreign cattle shall be removed from such place unless to the nearest slaughterhouse.

Apr. 25. Revoking the Orders of the 17th March, 14th April, 26th May, 22nd June and 31st July 1866, as to the use of railway trucks, etc., for the conveyance of cattle.

Apr. 25. Permitting as may be expedient, the revocation of renewal of Orders made by the local authorities relating to commonable animals.

May 3. Requiring all animals imported from Spain to be slaughtered within twelve days, instead of within four days, as required by the Order of 12th October 1866.

May 6. For regulating the sale and movement of cattle at the Birmingham Smithfield market.

May 9. Providing for the holding of exhibitions of animals, with form of exhibition pass and other regulations.

May 10. Permitting removal of cattle sold in the Sheffield cattle market to certain other places for slaughter within four days, until 31st July.

May 15. Defining the area at the port of Littlehampton within which foreign cattle may be landed.

May 17. For regulating the removal of cattle to and from the Metropolitan cattle market.

May 17. For continuing the compulsory slaughter of diseased animals until 31st August.

May 22. Declaring the Butchers' Arms Inn, Islington, to be an infected place under the *Cattle Plague Act*.

May 22. Declaring the premises in the occupation of Mark Mills, Islington, to be an infected place under the *Cattle Plague Act*.

May 27. For regulating the removal of foreign cattle imported from ports outside the United Kingdom (except the Channel Islands and Man) to the Metropolitan cattle market.

Jun. 6. For regulating the removal of cattle from the Liverpool new cattle market.

Jun. 7. Extending the time to six days within which cattle are to be slaughtered after exposure in a licensed market.

Jun. 13. For regulating the removal of cattle sold in the cattle fair ground, Wakefield.

Jun. 13. For regulating the removal of cattle sold in the Rotherham cattle market.

Jun. 17. The provisions of the Order of the 27th May last, as to the removal of foreign cattle, to come into operation on the 25th inst.

Jun. 28. Permitting sales of animals intended for immediate slaughter to take place within the limits of the ports defined for the landing of foreign cattle.

Jun. 28. Revoking the Order of the 19th February 1867, with regard to the landing of cattle at Liverpool.

Jun. 13. Defining the limits of the port of Liverpool within which foreign cattle may be landed.

Jul. 4. The Butchers' Arms Inn, Islington, declared to be "free from disease."

Jul. 4. The premises of Mark Mills, cowkeeper, of Islington, declared to be "free from disease".

Jul. 12. For regulating the movement of foreign cattle to the Metropolitan cattle market.

Jul. 12. Defining the limits of the Port of Bristol within which foreign cattle may be landed.

Jul. 12. Defining the limits of the Port of Sunderland within which foreign cattle may be landed.

Jul. 16. For regulating tthe removal of infected dung and litter out of the Metropolis City of London.

Jul. 18. Making provision for the re-shipment of cattle from the Port of Sunderland.

Jul. 19. Further regulations as to the issuing and signing of store stock licences.

Jul. 23. As to the marking of foreign cattle imported into Great Britain.

Jul. 26. Defining the limits of the Port of Penzance within which foreign cattle may be landed.

Jul. 29. Defining the limits of the Port of Dartmouth within which foreign cattle may be landed.

Jul. 30. As to the landing of cattle expected to arrive at Thames Haven from St. Petersburg for the Metropolitan cattle market.

Jul. 31. For continuing the provisions of the Order of the 1st April as to removal of cattle from Salford market.

Jul. 31. For continuing the provisions of the Order of the 1st April as to removal of cattle from the market at Newcastle-on-Tyne.

Aug. 1. Prohibiting the removal of cattle arrived at the Port of London from Cronstadt, in the vessel "St. Petersburg," unless under the Inspector's certificate.

Aug. 3. Revoking so much of the Order of the 24th March 1866, as would apply to cattle passing through Limburgh, Holland.

Aug. 3. For regulating the importation of horns, hoofs and hides from certain parts of Holland.

Aug. 19. Revoking the Order of the 24th March 1866, relating to the removal of cattle from Harwich to London.

Aug. 20. Consolidating and amending the provisions now in force under the *Contagious Diseases (Animals) Acts*, to be known as the *Consolidated Cattle Plague Order of August 1867.*

Aug. 20. Amending the regulations now in force for preventing the spread of the cattle plague in the Metropolis, to be known as the *Metropolitan Cattle Plague Order of August 1867.*

Sep. 2. Revoking the Order of 1st June 1866, relating to the disinfection of cattle pens, trucks, carriages, etc.

Sep. 11. Consolidating and amending the regulations under the *Contagious Diseases (Animals) Act*, which are to remain in force in Scotland, to be known as the *Consolidated Cattle Plague Order (Scotland) of September 1867.*

Sep. 16. Confirming certain Orders as to the importation of cattle into Ireland, etc.

Sep. 26. That the provisions of the Orders of the 20th August and 11th September as to the landing of cattle shall come into operation on the 15 October.

Oct. 11. Defining limits at the Port of London within which foreign cattle may be landed.

Oct. 11. Revoking the licences and Order in Council of the 23rd January 1867, respecting pens and sheds belonging to the Southampton Docks Company, at Southampton.

Oct. 11. Sixteen separate Orders defining limits within which foreign cattle may be landed at the Ports of Southampton, Cowes, Dover, Falmouth, Glasgow, Granton, Hartlepool, Harwich, Hull, Leith, Middlesborough, Newcastle-on-Tyne, North Shields, Plymouth, Portsmouth and Shoreham.

Oct. 22. Defining limits at the Port of Grimsby within which foreign cattle may be landed.

Oct. 23. Revising certain articles of the *Consolidated Cattle Plague Order of August 1867.*

Oct. 28. Declaring a certain field on Hackney Marsh to be "free from cattle plague."

Nov. 4. Revoking certain Orders for quarantine, removal and slaughter of foreign cattle, etc., which are no longer necessary.

Nov. 7. Defining limits at the Port of London (Poplar, etc.), within which foreign cattle may be imported.

Nov. 7. Providing for the re-shipment of foreign cattle from Hartlepool to certain other ports.

Nov. 12. Declaring a meadow, etc., belonging to Lieut.-Col. Custance, at Weston, to be "free of the cattle plague."

Nov. 18. Revoking the Order of the 24th March 1866, excepting such part as relates to cattle imported from South Holland.

Nov. 18. Revoking the Order of the 27th June 1866, excepting such part as relates to hides, horns, etc., imported from South Holland.

Nov. 29. Defining limits at the Port of Littlehampton within which foreign cattle may be landed.

Nov. 29. Defining limits at the Port of Dartmouth within which foreign cattle may be landed.

Nov. 29. Defining limits at the Port of Sunderland within which foreign cattle may be landed.

Nov. 29. Revoking certain Orders as to landing of cattle in pursuance of the *Consolidated Cattle Plague Order of August 1867.*

Nov. 29. Revoking the Order of the 28th June 1867, as to the limits of the Port of Liverpool.

Nov. 29. Defining limits at the Port of Liverpool within which foreign cattle may be landed.

Nov. 29. Revoking what remains in force of the Order of the 24th March 1866, as to cattle from Holland.

Nov. 29. Revoking what remains in force of the Order of the 27th June 1866, as to hides, etc. from Holland.

2.17. 1868

Jan. 21. That the Orders now in force respecting the landing of foreign cattle, be extended to sheep and goats imported from Belgium.

Feb. 4. That foreign cattle may after the 11th February be kept alive for ten days after landing.

Feb. 12. Revoking the Order of the 4th October 1866, as to introduction of cattle into England and Wales from Ireland and Scotland.

Jun. 30. Revoking certain Orders as to the importation of cow-hair, offal, hides, etc. into Ireland.

Jul. 24. Extending the time within which cattle exposed in the Metropolitan cattle market must be slaughtered, to ten days.

Jul. 24. Providing for the movement out of the Metropolis of live cattle.

Jul. 24. Providing for the movement through the Metropolis of cattle arriving by the London and North Western Railway.

Jul. 24. Providing for the importation of cattle from Spain, Portugal, Normandy or Brittany into any port of Great Britain.

Jul. 29. Providing for the movement through the Metropolis of live cattle arriving by the Great Northern Railway.

Jul. 29. As to the importation of cattle from Tonning or Husum, Schleswig-Holstein, or from Hamburg.

Jul. 31. Revoking the Order of the 24th July last, as to the movement of cattle arriving in the Metropolis by the London and North Western Railway.

Jul. 31. Providing for the movement of cattle carried by the London and North Western Railway in and out of London.

Aug. 20. Permitting the removal through the Metropolis of cattle arriving by the Midland Railway.

Aug. 20. Permitting the removal through the Metropolis of cattle arriving by the Great Western Railway.

Sep. 16. Re-defining the limits of the Port of Shoreham for the landing of foreign cattle.

2.18. 1869

Apr. 9. Permitting the re-shipment of foreign cattle landed at Hull to the Port of Goole.

Aug. 10. Revoking all previous Orders relating to contagious and infectious diseases among cattle.

Aug. 10. General Order relating to the contagious and infectious diseases among animals, to be cited as the *Contagious Diseases (Animals) Order of August 1869*.

Aug. 10. Order for the Metropolis relating to the contagious and infectious diseases among cattle, to be cited as the *Metropolitan Contagious Diseases (Animals) Order of August 1869*.

Aug. 10. Defining the limits of the Port of London within which foreign cattle may be landed for slaughter.

Aug. 10. Defining the limits of the Port of London within which foreign cattle may be landed for slaughter, amended.

Aug. 10. 25 separate Orders defining the limits within which foreign cattle may be landed for slaughter of the Ports of Chichester, Bristol, Cowes, Dover, Hartlepool, Harwich, Hull, Newcastle-upon-Tyne, Plymouth, North Shields, North Shields (amended), Shoreham, Southampton, Granton, Leith, Glasgow, Middlesborough, Portsmouth, Portsmouth (amended), Grimsby, Dartmouth, Littlehampton, Sunderland, Goole and Liverpool.

Sep. 1. General Order with regard to the importation of cattle.

Dec. 16. Defining parts of Port of Hartlepool where foreign cattle may be landed for slaughter.

2.19. 1870

May 12. As to the transit of animals, to be known as the Order of May 1970.

May 12. Directing the various railway companies to provide water for cattle carried on their lines.

Jun. 17. Defining part of Port of Grimsby where foreign cattle may be landed for slaughter.

Sep. 9. That the provisions of the 4th Schedule of the *Contagious Diseases (Animals) Act of 1869*, shall apply to cattle brought from any part of France.

Sep. 20. For regulating the importation of cattle, sheep and goats from the German Confederation, or France.

Sep. 28. As to importation of cattle, sheep and goats from any port in Belgium.

1871.

Nov. 23. Defining part of Port of Sunderland where foreign cattle may be landed for slaughter.

Dec. 20. Defining part of Hull where foreign cattle may be landed for slaughter.

2.20. 1875

The Animals Order of 1875.

2.21. 1877

Jan. 17. Local Authority order declaring part of Foreign Cattle Market at Deptford an infected place.

Jan. 27. Prohibition on landing of stock and products from Belgium or the German Empire.

Jan. 31. Metropolitan Cattle Plague Order.

Feb. 2. Metroplitan Cattle Plague Order.

Feb. 8. (Scheduled ports).

Feb. 9. Revoking Local Authority Order of 17th January declaring part of Foreign Cattle Market at Deptford an infected place.

Feb. 13. Cattle Plague Order of 1877.

Feb. 20. East Riding Order.

Mar. 6. Lincoln Order.

Mar. 8. Cattle-Plague Record to be maintained.

Mar. 12. Amendment to Order of March 6th.

Apr. 12. City of London and Metropolitan Police District Order.

Apr. 27. Amendment to Order of April 12th. Middlesex Order.

Apr. 27. Amendment to Order of April 12th. Metropolis Order.

May 1. Declaration of infected place. Ledbury Mews.

May 1. Declaration of infected place. Chamberlayne's Farm.

May 3. Declaration of infected place. Devonshire-place.

May 4. Exclusion of certain areas from The Animals Order of 1875.

May 9. Slaughter of animals in the City of London and Metropolitan Police District.

May 18. Declaration of infected place. White's Row.

May 29. Revocation of Orders of March 6th and 12th.

May 31. Movement of animals out of Middlesex prohibited except into the Metropolis.

Jun. 14. Revocation of the Cattle Plague Order of 1877 and any regulations made by a Local Authority under the Order.

Jun. 14. Revocation of the Order of January 27th as concerns fresh meat only.

Jun. 26. Revocation of infected places orders.

Jul. 16. Revocation of an infected place order.

Jul. 16. Restriction on movement of cattle in the Metropolis north of the Thames.

Jul. 18. Revocation of Order of July 16th controlling movement of cattle in the Metropolis north of the Thames.

Jul. 18. Declaration of part of the Metropolis as an infected area.

Jul. 18. Prohibition on moving cattle in the Metropolis north of the Thames to the south of the Thames.

Aug. 14. Revocation of an infected place Order.

Aug. 14. Revocation of Order of July 18th concerning movement of cattle to the south of the Thames.

Aug. 14. Concerning movement of stock in and out of the Metropolis by railway.

Aug. 14. Concerning the port of Sunderland.

Oct. 5. Revocation of part of the Order of January 27th.

Nov. 21. Revocation of The Metropolitan Cattle-Plague Order 1877 of 31st. January; regulating the holding of markets and sales within the Metropolis of February 2nd.; the Metropolis (Movement) Order of August 14th.

Nov. 26. Revocation of Orders of December 16th 1869; June 17th; November 23rd, concerning ports.

Nov. 26. Defining area within Port of Hartlepool where foreign cattle may be landed for slaughter.

Nov. 26. Defining area within Port of Grimsby where foreign cattle may be landed for slaughter.

Nov. 26. Defining area within Port of Sunderland where foreign cattle may be landed for slaughter.

Dec. 4. Revocation of Order of December 20th 1871.

2.22. 1878

Aug. 16. The Revocation Order of 1878. Revokes outstanding Orders concerning cattle-plague restrictions.

REFERENCES

"A. B." (1750). Herefordshire. Feb. 21st. *Gent. Mag.* **20**, 106–107.

Abbott, C. W. (1952). Who was the first importer of cattle? *Farm. South Africa* March, 225–226.

Abildgaard, P. C. (1795). Abhandlung über die allgemeine Rindviehseuche in vorzüglicher Hinsicht auf ihre Einfälle in Dänemark und die daselbst angestellten Impfversuche. In Viborg, E. (ed.) *Sammlung von Abhandlungen für Thierärtze und Oekonomen*, vol. I. Prost, Copenhagen.

Adam, Th. (1862). *Die Veterinär-Polizei mit Berücksichtigung der neuesten Gesetzgebungen*. Joh. Palm, Munich.

Adami, P. (1781). *Beiträge zu Gesch. des Viehseuches in den k. k. Erbl.* Vienna.

Adams, A. H. (1919). *Report on a small outbreak of rinderpest in British Somaliland*. MS. Colonial Office.

Adlakha, S. C. (1982). *Rinderpest*. Indian Council of Agricultural Research, New Delhi.

"Agricola" (1750). The immediate killing diseased Cattle justify'd. *Gent. Mag.* **20**, 105–106.

Ainsworth, J. (1905). *Reports relating to the Administration of the East Africa Protectorate. Cd.2740*. HMSO, London.

Alexis, M. G. (1894). *Alexis Vrithoff compagnon des Capitaines Jacques et Joubert au Lac Tanganyika (Afrique centrale)*. Société de Saint-Augustin.

Ali, B. el H. (1973). A natural outbreak of rinderpest involving sheep, goats and cattle in Sudan. *Bull. Epizoot. Dis. Afr.* **21**, 421–423.

Allen, W. E. D. (1970). *Russian Embassies to the Georgian Kings (1589–1605)*, 2 vols. Cambridge University Press, Cambridge.

Alta, E. (1765). *Verhandeling over de natuurlijke oorzaken der ziekte onder het rundvee*. W. Wigeri, Leeuwarden.

Alta, E. (1769). *Nodige raadgevingen aan overheden en ingezetenen dewelke in het bijzonder voor den boer in deeze akelige omstandigheden, waar in het vaderland zig door de ziekte van het rundvee thans bevind, van zeer veel nut kunnen zijn*. H. A. de Chalmot, Leeuwarden.

Alta, E. (1780). Bericht van de inënting der kalveren. *Verhand. Maatsch. ter bevordering van den Landbouw. Amsterdam* **2**(1), 125–164.

Anderson, D. M. and Johnson, D. H. (1988). Introduction. In Johnson, D. H. and Anderson, D. M. (eds.) *The Ecology of Survival Case Studies from Northeast African History*. Lester Crook Academic Publishing, London, pp. 1–24.

Anderson, R. M. and May, R. M. (1979). Population biology of infectious diseases: Part II. *Nature* **280**, 455–461.

Angeloff, St. (1917). Auftreten und Bekämpfung der Rinderpest im Königreich Bulgarien während des Balkankrieges 1912–1913. *Arch. Wiss. Prakt. Tierheilk.* **43**, 383–420.

Anon. (1682). *Bericht der an etlichen Orten eingerissenen Vieh-Seuche Betreffend Sambt underschiedlichen Recepten für dieselbe*. J. H. Kramer, Hildesheim.

Anon. (1711). *Giornale de letterati d'Italia* **X**, 3: 141.

Anon. (1716a). *Gründliche Anweisung, woran man die leyder gar zu sehr eingerissene Krankheit des Viehes erkennen; das Gesunde davor praeserviren, und das Krancke curiren könne*. Hildesheim.

Anon. (1716b). *Kurtzer Unterricht, wie man bei der jetzo unter dem Horn-Vieh grassirenden Seuche sich zu verhalten*. Wolffenbüttel.

Anon. (n.d.) [1719]. *Entwurf einiger Hülfs-Mittel So bei hin und wieder sich äußernden Umbfal des Horn-Viehs zubeobachten und zugebrauchen, die von verständigen hauß = und Land = Wirthen gut befunden worden*.

Anon. (1720a). *Considerations on the Nature, Causes, Cure and Prevention of Pestilences. The Free-Thinker.* CCXCII. W. Watkins, London.

Anon. (1720b). *A Form of Prayer to be Used in all Churches and Chapels Throughout England and Wales, and the Town of Berwick upon Tweed; on Friday the Sixteenth Day of December, being the Day appointed by His Majesty for a General FAST and Humiliation, to be Observed in a most Solemn and Devout manner*. John Baskett, London.

Anon. (1732). *Memoire sur les Maladies dont les Bestiaux ont été attaquez dans differens Endroits de la Province de Languedoc, & les Remedes qu'on doit employer, tant pour prevenir cettle Maladie que pour la guerison des Bestiaux attaquez*. François Rochard, Montpellier.

Anon. (1740). *Heilsame Vorschlage mit der grassirenden Seuche unter dem Horn = Vieh vorzubauen und was vor Mittel dazu dienlich.* Collegii Sanitatis zu Halle, Memingen.

Anon. (1744). Observations on the contagious distemper among the oxen and cows in Frenche-comte. *Gent. Mag.* **14**, 585–588.

Anon. (1745a). Remedies for the disease among the black cattle. *Lond. Mag.* **14**, 598–600.

Anon. (1745b). *Bericht Von einem Hülfsmittel, welches aus Middelburg in Seeland von guter Hand übersand worden, mit glaubwürdiger Versicherung, daß dadurch mehr als Tausend Stück Hornvieh erhalten sind.* Amsterdam.

Anon. (1745c). Remedies for the distemper among the black cattle. *Lond. Mag.* **14**, 557–559.

Anon. (1745d). Extract of a letter from a physician at Paris to a physician of Provence, on the distemper among the black cattle. From the French. *Gent. Mag.* **15**, 600–601.

Anon. (1746a). Of the cause and method of cure of the distemper among the horned cattle. *Gent. Mag.* **16**, 79–80.

Anon. (1746b). For the fever in the horned cattle. *Gent. Mag.* **16**, 698.

Anon. (1746c). Receipt for the distemper of the cows; taken from the treatise on tar-water, by T. Prior, Esq; of Dublin, who collected an authentic account of several hundred surprising cures performed by it. *Gent. Mag.* **16**, 495–496.

Anon. (1747a). To the author. *Gloucester J.* October 6.

Anon. (1747b). Supplement to an essay on the disease among the Cattle. *Gent. Mag.* **17**, 599–600

Anon. (1747c). Extract of a letter from Norwich, Feb 12. *Gent. Mag.* **17**, 74–75.

Anon. (1747d). A pastoral. *Gent. Mag.* **17**, 443.

Anon. (1747e). Experimental queries in electricity. *Lond. Mag.* **16**, 135–136.

Anon. (1747f). For the distemper amongst cattle. *Gent. Mag.* **17**, 22.

Anon. (1748a). The monthly chronologer. *Lond. Mag.* **17**, 41–42.

Anon. (1748b). *An Essay, Occasioned by the Contagious Distemper Still Raging Among the Cattle in Many Parts of this Kingdom, with a Note on What Has Been Publish'd Concerning the Supposed Efficacy of Jesuit's Bark in the Case, Both by Way of Cure and Prevention.* J. Rivington, London.

Anon. (1748c). Of the distemper among the cattle etc. *Lond. Mag.* **17**, 152.

Anon. (1748d). A letter from Cheshire. *Gent. Mag.* **18**, 167.

Anon. (1748e). The monthly chronologer. *Lond. Mag.* **17**, 233.

Anon. (1749a). Recipe for the diseased cattle. *Gent. Mag.* **19**, 389.

Anon. (1749b). Newcastle, April 20. *Gent. Mag.* **19**, 151.

Anon. (n.d.) [1750a]. *An Essay, Occasioned by The Distemper of the Cattle Shewing, Probably the Cause of its Continuance in this Kingdom; and the Means By Which Alone, Humanly Speaking, it Can Be Exterminated*, 2nd edn. J. Rivington, London.

Anon. (1750b). The case of killing infected beasts consider'd. *Gent. Mag.* **20**, 489–492.

Anon. (1750c). Observations on the regulations which have been made for preventing the spreading of the distemper among the cattle, presented to the Council, Jan. 9. *Gent. Mag.* **20**, 22–25.

Anon. (1750d). Extract of a letter from Norwich to a gentleman in London. *Gent. Mag.* **20**, 525.

Anon. (n.d.) [c.1750]. *Preservatifs Contre la Maladie des Bestiaux*. Crevier, Trevoux.

Anon. (1751a). Charles-Town, S. Carolina, Nov. 10. *Gent. Mag.* **21**, 39–40.

Anon. (1751b). *Gent. Mag.* **21**, 476.

Anon. (1751c). Historical chronicle. *Gent. Mag.* **21**, 235.

Anon. (1751d). Epping. Jan. 18. *Gent. Mag.* **21**, 41.

Anon. (1753a). *Britain's Alarm, from the Continuance of the Contagion among the Cattle, and Other More Afflicting and Important Evils Threatening and Befalling Us*, 2nd edn.

Anon. (1753b). *A Probable scheme for Putting a Final Stop to the Distemper Among the Horned Cattle, and Preventing the Ruin of Farmers While it Continues; in a Letter to a Member of Parliament.* M. Cooper, at the Globe in Pater-noster-row; W. Owen, at Homer's Head, Temple-bar; and G. Woodfall, at Charing-Cross, London.

Anon. (1753c). *Einige von den Medicis approbirte und theils probirte Praeservativ- und Curativ-Recepta, Welche Bey der in dem Churfürstl. Pfleg-Gericht Schwaben, unter dem Huff=, Horn- und Klohn = Vieh eingerissenen Vieh-Sucht: den gelben Schölm genannt.* Vötter, Munich.

Anon. (1754a). Cattle inoculated. *Gent. Mag.* **24**, 493.

Anon. (1754b). *Praeservativ-Mittel Gegen die ansteckende Seuche unter dem Horn-Viehe. Nachdem die in Franckreich enstandene ansteckende Seuchen Unter Pferde, Rind und Schaaf-Viehe, sich leider bis in unser Nachbarschafft erstreckt.*

Anon. (1754c). *Unterricht wie der LandMann bei einer Hornviehseuche sowohl zur Verhütung als auch zur Dämpfung derselben sich zu verhalten habe.* Nuvemburg.

Anon. (1755a). The following directions for treating distemper'd cattle we have received from an honest farmer, who has tried it with success, and we shall therefore give it in his own words. *Gent. Mag.* **25**, 56.

Anon. (1755b). An almost infallible method of curing the contagious disease of horned cattle, by Dr Fischer, Physician to the late Empress Anne of Russia, now in general use all over hivonia. *Gent. Mag.* **25**, 464.

Anon. (1755c). Amsterdam, April 1755. *Gent. Mag.* **25**, 160–161.

Anon. (1761). Annual register for the year 1761. *Ann. Reg.* **4**, 122.

Anon. (1763). *Versuch einer nähern Erklärung der Hornviehseuche, nebst einigen Wahrnehmungen über die Einpropfung derselben*. Braunschweig.

Anon. (1767). *Beschreibung der sogenannten Feuer-Krankheit unter dem Horn-Vieh und Recept wider selbige*. O. O. & J., Hannover.

Anon. (1769). A list, shewing the state of the distemper among the horned cattle, in the districts of south and north Holland, in the month of September, 1769. *Ann. Reg.* **12**, 166.

Anon. (1775). *Reglement wie ein jeder sich bey der Viehseuche zu verhalten*. Hildesheim.

Anon. (1776). *Beitrag zur Gesichte der Rindviehseuche im Hannöverischen. Nebst zween Briefen über die Lentinischen Pulver*. Richterischen Buchhandlung, Altenburg.

Anon. [Uden, K. F.] (1777). *Beiträge zur Geschichte der Hornviehseuche in einigen Kreisen der Altmark, und deren Erkenntniß und Heilung*. Erste Sammlung. Joh. Christ. Dieterich, Stendal.

Anon. [Reuss, C. F.] (1784). *Rindvieh Arzneybuch sowohl vor die gewöhnlichste Viehkrankheiten, als auch vor die sonstige Viehseuchen*. Jac. Fried. Heerbrandt, Tübingen.

Anon. [Reuss, A. C.] (1797). *Entwurf eines wirtembergischen Arztes die gegenwärtig unter dem Rindvieh herrschende Blattern = Seuche zu behandeln*. Franz Christian Löflund, Stuttgart.

Anon. (1813). *Verordnung die in der Grafschaft Schaumburg sich ausbreitende Rind-Vieh-Seuche betreffend*.

Anon. (1818). The courier. *The Bombay Courier* 28 (1353). August 1.

Anon. (1838). *Trans. Med. Phys. Soc. Bombay* **1**, 70.

Anon. (1859). *The Veterinarian* **32**, 415.

Anon. (1863). Cattle-disease in town dairies, and its prevention: the question of legislation. *Br. Med. J.* **(ii)**, 474–476.

Anon. (1865a). *Report on the Inoculation of Cattle Plague. Experiments Performed Under the Direction of the Imperial Commission for the Advancement of Veterinary Art and the Investigation of Measures for Preventing Epizootics in the Empire*. In Russian.

Anon. (1865b). The cattle plague, *Lancet* (2) (2190). 212–213. August 19.

Anon. (1865c). Prayers against the cholera, *Pall Mall Gazette* October 9, 1.

Anon. (1865d). The competing prayer against the plague. *Spectator* **XXXVIII** October 14, 1139–1140.

Anon. (1865e). The cattle-plague. *Br. Med. J.* **(ii)**, 372–374.

Anon. (1865f). The cattle-plague. *Br. Med. J.* **(ii)**, 611–613.

Anon. (1865g). Fatality of epizootic diseases in Egypt. *The Veterinarian* **38**, 92–93.

Anon. (1866a). Observations on the cattle plague in Ireland and the cattle disease in the Isle of Man. *The Veterinarian* **39**, 708–713.

Anon. (1866b). The Appendix to the Third Report of the Cattle-Plague Commission. *Br. Med. J.* **(i)**, 207.

Anon. (1866c). The gazelles and the cattle plague. *Br. Med. J.* **(i)**, 48–49.

Anon. (1866d). What Lanzoni says of the cattle plague. *Br. Med. J.* **(i)**, 74–75.

Anon. (1866e). Decline of the cattle plague. *Br. Med. J.* **(i)**, 444.

Anon. (1866f). *Br. Med. J.* **(i)**, 23.

Anon. (1866g). The appendix to the Third Report of the Cattle Plague Commission. *Br. Med. J.* **(ii)**, 42.

Anon. (1866h). *The Times* February 14, 9.

Anon. (1866i). The failures of the Infallible. *Br. Med. J.* **(i)**, 284–285.

Anon. (1866j). *Br. Med. J.* **(i)**, 316.

Anon. (1866k). *The Times* March 21, 5.

Anon. (1866l). *Prayer and the Cattle Plague*. London.

Anon. (1866m). The position of the government and their party. *Blackwood's Edinb. Mag.* **99**, 389–397.

Anon. (1869). *Circular Letter. Government of Bengal, No. 1503. August 11.*

Anon. (1871). In Hallen, J. H. B., Charles, J. G., Jan, M. H., Kerr, H. C., and McLeod, K. (eds.) *Report of the Commissioners Appointed to Inquire into the Origin, Nature, Etc., of Indian Cattle Plagues. With Appendices*. Government Printer, Calcutta.

Anon. (1873). *Report from the Select Committee on Contagious Diseases (Animals); Together with the Proceedings of the Committee, Minutes of Evidence, and Appendix*. House of Commons, London.

Anon. (n.d.) [1874]. *Junius: Including Letters by the Same Writer, Under Other Signatures (Now First Collected). To Which are Added, His Confidential Correspondence with Mr. Wilkes, and His Private Letters Addressed to Mr. H. S. Woodfall*. George Routledge & Sons, London.

Anon. (n.d.) [1875]. *The Life and Eccentricities of Lionel Scott Pilkington, Alias Jack Hawley, of Hatfield, Near Doncaster*. Dale, Doncaster.

Anon. (1877). XI. The outbreak of cattle-plague. *J. R. Agricult. Soc. Engl.* Second Ser. **13**, 211–241.

Anon. (1880). *Contagious Diseases of Domesticated Animals. Continuation of Investigation by Department of Agriculture. Department of Agriculture. Special Report No. 22.* Government Printing Office, Washington, DC.

Anon. (1882). *Rinderpest in Aberdeenshire: The Outbreak of 1865, and How it was Stamped Out.* The Free Press Office, Aberdeen.

Anon. (1893). News from British Central Africa. *Gaz. Zanzibar East Afr.* **2**, 2.

Anon. (1896a). *Correspondence Relating to the Outbreak of Rinderpest in South Africa in March 1896. Presented to both Houses of Parliament by Command of Her Majesty. July 1896. C.8141.* HMSO, London.

Anon. (1896b). *Natal Witness* October 7, 2.

Anon. (1897a). The Rinderpest, *Cape Times* October 14, 5.

Anon. (1897b). *Times of Swaziland* (14). August 28.

Anon. (1897c). The Famine in the Transvaal, *Christian Express* May 1, 70, 71.

Anon. (1899). *Cape of Good Hope. Blue Book on Native Affairs* **71**, 81.

Anon. (1923). La sensibilité des chameaux à la peste bovine. *Rev. Vét. Constantinople* **1(i)**, quoted by Bey, 1926.

Anon. (1975). *Cheetal* **16(4)**, 17.

Anon. (1983a). The resurgence of rinderpest. *World Anim. Rev.* Special Issue, 5–9.

Anon. (1983b). *Pan-African Rinderpest Campaign. Project Operation and Funding Document.* OAU-IBAR, Nairobi.

Anon. (1997). The blueprint for the global eradication of rinderpest by the year 2010. In *Prevention and Control of Transboundary Diseases. FAO Animal Production and Health Paper 133.* FAO, Rome.

Anon. (1998). Rinderpest in Russia. *The Empres Transboundary Animal Disease Bulletin.* (7), 2.

Anon. (2000). *Empres Transboundary Animal Diseases Bulletin No. 13—Rinderpest.* FAO, Rome.

Anon. (2002a). *Empres Transboundary Animal Diseases Bulletin No. 21—Rinderpest.* FAO, Rome.

Anon. (2002b). *Global Rinderpest Now Reduced to One Area, But Risks Breakout FAO Warns.* EMPRES Publications, FAO, Rome.

Ansorge, W. J. (1899). *Under the African Sun. A Description of Native Races in Uganda, Sporting Adventures and Other Experiences.* William Heinemann, London.

Anton, P. 1712. *Conghietture del Doctor Pietre Anton. Michelotti, Filosofo, e Medico d'Arco, sopra la Natura, Cagione e Rimedi dell' infermita regnanti ne' Animali Bovini di molte Citta, &c. Nell' Autunno del' Anno cadente, 1711.* Venice.

Antonelli, P. (1882). *Scioa e i Scioani.* Rome.

Antonelli, P. (1890). *Camera dei deputati, Documenti diplomatichi presentati al parlamento italiano, Missione Antonelli in Ethiopia. Rome.* Quoted by Pankhurst, 1966.

Arkell-Hardwick, A. (1903). *An Ivory Trader in Northern Kenya. The Record of an Expedition Through Kikuyu to Galla-Land in East Equatorial Africa with an Account of the Rendili and Burkeneji Tribes.* Longmans, Green, & Co., London.

Arloing, S. (1905). La peste bovine en Égypte. *J. Méd. Vét. Zootéchnique* 321.

Armatage, G. (1894). *Cattle: Their Varieties and Management in Health and Disease.* Frederick Warne & Co., London.

Armstrong, T. (1975). *Yermak's Campaign in Siberia. A Selection of Documents Translated from the Russian by Tatiana Minorsky and David Wileman, and Edited With Introduction and Notes.* The Hakluyt Society, London.

Asano, K., Tsukiyama, K., Shibata, S., Yamaguchi, K., Momoki, T., Maruyama T., Kohara, M., Miki, K., Sugimoto, M., Yoshikawa, Y., Nagata, T., and Yamanouchi, K. (1991). Immunological and virological characterisation of improved construction of recombinant vaccinia virus expressing rinderpest virus haemagglutinin. *Arch. Virol.* **116**, 81–90.

Ashburnham, J. A. (1898). *Northern Protectorate: Annual Report. Colonial Reports—Annual. No. 226. Bechuanaland Protectorate. Annual Reports for 1896–7.* HMSO, London.

Ashe, R. P. (1894). *Chronicles of Uganda.* Hodder & Stoughton, London.

Askar, M. Bey. (1924). *Experiments to Determine the Susceptibility of Weaned Calves to Cattle Plague.* Government Press, Cairo.

Atkinson, J. C. (1891). *Forty Years in a Moorland Parish. Reminiscences and Researches in Danby in Cleveland.* Macmillan & Co., London.

Atkinson, T. W. (1860). *Oriental and Western Siberia: A Narrative of Seven Years' Explorations and Adventures in Siberia, Mongolia, the Kirghis Steppes, Chinese Tartary, and Part of Central Asia.* Hurst & Blackett, London.

Austin, H. H. (1902). *Among Swamps and Giants in Equatorial Aftrica. An Account of Surveys and Adventures in the Southern Sudan and British East Africa.* C. Arthur Pearson, London.

Austin, H. H. (1903). *With Macdonald in Uganda.* Edward Arnold, London.

Bache, E. (1934). *The Youngest Lion. Early Farming Days in Kenya.* Hutchinson & Co. Ltd., London.

Baden-Powell, R. S. S. (1897). *The Matabele Campaign. Being a Narrative of the Campaign in Suppressing the Native Rising in Matabeleland and Mashonaland 1896.* Methuen & Co., London

Bagehot, W. (1872). *The English Constitution*, 2nd edn. C. Kegan Paul & Co., London.

Baier, S. (1980). *An Economic History of Central Niger*. Clarendon Press, Oxford.

Baird, J. L. (1900). *Foreign Office 1/39, 20 May 1900.*

Baker, S. W. (1867). *The Nile Tributaries of Abyssinia and the Sword Hunters of the Hamran Arabs*, 2 vols. Macmillan & Co., London.

Baker, S. W. (1890). *Wild Beasts and Their Ways. Reminiscences of Europe, Asia, Africa and America*. Macmillan & Co., London.

Baldwin, T. (1866). In *Sequel to the Report of the Committee Convened by the Lord Lieutenant of Ireland, to Consider the Measures to be Adopted for Arresting the Progress of the Cattle Plague, in Case of its appearance in Ireland*. HMSO, Dublin.

Ballard, C. (1986). The repercussions of rinderpest: cattle plague and peasant decline in colonial Natal. *Int. J. Afr. Hist. Stud.* **19**, 421–450.

Baradat, R. (1929). *La Prophylaxie de la Peste Bovine au Cambodge*. Thèse Doctorat, École Nationale Véterinaire de Toulouse, Toulouse.

Barber, J. (1968). *Imperial Frontier. A Study of Relations Between the British and the Pastoral Tribes of North East Uganda*. East African Publishing House, Nairobi.

Barberet, D. (1766). *Mémoire sur les maladies épidemiques des bestiaux*. Veuve d'Houry, Paris.

Barber-Lomax, J. W. (1968). Animal disease and public health. In Poynter, F. N. L. (ed.) *Medicine and Science in the 1860s*. Wellcome Institute of the History of Medicine, London, pp. 243–252.

Barfield, T. J. (1989). *The Perilous Frontier. Nomadic Empires and China*. Basil Blackwell, Oxford.

Barker, J. [Anon.] (1745). *An Account of the Present Epidemical Distemper amongst the Black Cattle Together with a Method of Cure Deduced from a Knowledge of the Disease, and Confirmed by Experience. And also Some Arguments to Prove that it is not Contagious*. By a Member of the College of Physicians. A. Millar, London.

Barker, J. [Anon.] (1746). Of the cause and method of cure of the distemper among the horned cattle. *Gent. Mag.* **16**, 79–80.

Barnes, R. G. (1911). *Babes in the African Wood*. Longmans, Green & Co., London.

Barnes, T. A. (1922). *The Wonderland of the Eastern Congo*. G. P. Putnam's Sons, London.

Baron, M. D. and Barrett, T. (1995). The sequence of the N and L genes of rinderpest virus, and the 5[1] and 3[1] extra-genic sequences: the completion of the genome sequence of the virus. *Vet. Microbiol.* **44**, 175–185.

Baron, M. D. and Barrett, T. (1997). Rescue of rinderpest virus from cloned cDNA. *J. Virol.* **71**, 1265–1271.

Baronius, C. (1739). *Annales Ecclesiastici*. (Pagius, P. A., ed.), 38 vols. Leonardi Venturini, Lucae.

Barrau, P. B. (1805). *Projet d'Assurances Réciproques contre la Mortalité des Bestiaux*. Bénichet, Toulouse.

Barrett, T. (1996). Morbilliviruses into the twenty first century. In *Proceedings of the FAO Technical Consultation on the Global Rinderpest Eradication Programme, Rome, Italy, 22–24 July 1996*. FAO, Rome, pp. 25–38.

Barrett, T. (2001). *SGM Symposium*, Edinburgh 26–30 March 2001.

Barrett, T., Forsyth, M. A., Inui, K., Wamwayi, H. M., Kock, R., Wambua, J., Mwanzia, J., and Rossiter, P. B. (1998). Rediscovery of the second African lineage of rinderpest virus: its epidemiological significance. *Vet. Rec.* **142**, 669–671.

Barron, H. P. T. (1866). *Report on the Cattle Plague in Belgium*. Houses of Parliament, London.

Bary, M. De. (1910). *Grand Gibier et Terres Inconnues. Autour des Grands Lacs de l'Afrique Centrale. Le Mont Elgon*. Librairie Plon, Paris.

Bates, T. (1714). *Manuscript Letter. September 24, 1714. Herts County Archives. D/EP/F125.*

Bates, T. (1718). A brief account of the contagious disease which raged among the Milch cows near London in the year 1714, and of the methods that were taken for suppressing it. *Phil. Trans. R. Soc.* **30**, 872–885.

Baumann, O. (1894). *Durch Massailand zur Nilquelle*. Dietrich Reimer, Berlin.

Beale, L. S. (1866). In Bernard, M. *Second Report of the Commissioners Appointed to Inquire into the Origin and Nature &c. of the Cattle Plague*. HMSO, London.

Beale, L. S. (1872). *Disease Germs: Their Nature and Origin*. J. & A. Churchill, London.

Béhic, A. (1865). The cattle disease. *The Times* September 8, 11.

Bekenov, A. B., Blank, D. A., Grachev, Yu. A., and Plakhov, K. N. (2001). In Mallon, D. P. and Kingswood, S. C. (eds.) *Antelopes Part 4: North Africa, the Middle East, and Asia*. IUCN, Gland.

Bell, J. (1883). *The Cattle Plague in Westmorland and Cumberland 1745–54.*

Bergius, P. J. (1772). *Bedenklichkeiten bei der Einimpfung der Hornviehseuche Göttingen.*

Berk, J. T. (1986). Rinderpest in 19th Century Russia: an account by two Danish Veterinarians. *Hist. Med. Vet.* **11**, 104–110.

Berkeley, G. (1747). *Lond. Mag.* **16**, 127–128.

Bernard, C. E. (1870). In Hallen, J. H. B., Charles, J. G., Jan, M. H., Kerr, H. C., and McLeod, K. (eds.) 1871. *Report of the Commissioners Appointed to Inquire into the Origin, Nature, Etc., of Indian Cattle Plagues. With Appendices*. Government Printer, Calcutta.

Bernard, M. (1865). *First Report of Her Majesty's Commissioners Appointed to Inquire into the Origin and Nature &c. of the Cattle Plague*. HMSO, London.
Bernard, M. (1866a). *Second Report of the Commissioners Appointed to Inquire into the Origin and Nature &c. of the Cattle Plague*. HMSO, London.
Bernard, M. (1866b). *Third Report of the Commissioners Appointed to Inquire into the Origin and Nature &c. of the Cattle Plague*. HMSO, London.
Bettembourg (1888). *Les Missions Catholiques* **20**, 557.
Bevan, W. H. R. (1897). In *Annual Report 1897(B)*. Society for the Propagation of the Gospel, London.
Beveridge, W. (1939). *Prices and Wages in England*, vol. I. Longmans Green, London.
"Beverlac" (1865). The murrain. *Notes And Queries* **8** (3rd S), 335–336.
Bey, I. E. (1926). *La Peste Bovine en Turquie*. Thèse Doctorat, École Nationale Véterinaire d'Alfort, Paris.
Bidjeh, K., Ouagal, M., Diallo, A., and Bornarel, P. (1997). Transmission of rinderpest virus strains of different virulence to goats in Chad. *Ann. Méd. Vet.* **141**, 65–69.
Binger, L. G. (1892). *Du Niger au Golfe de Guinée par le Pays de Kong et le Mossi*, 2 vols. Librairie Hachette et Cie., Paris.
Bland-Sutton, J. (1911). *Man and Beast in Eastern Ethiopia*. Macmillan & Co., Ltd, London.
Blondet, M. (1748). *Dissertation sur la maladie épidemique des bestiaux*. Paris.
Blondet, M. (1751). *A Dissertation on the Epidemical Distemper Among the Horned Cattle Wherein, After Giving a Short History of the Origin, Progress, and Nature of that Distemper, The Most Proper Manner of Treating it is Laid Down, and According to the Principles of Physick and Natural Philosophy*. W. Owen, London.
Bliss, E. (1922). Rinderpest and its prevention. *Chin. Med. J.* **36**, 153–170.
Bloom, B. R. (1989). Vaccines for the Third World. *Nature* **342**, 115–120.
Blystone, M. (2001). Rinderpest: one virus's impact on veterinary history. *Vet. Herit.* **24**, 8–12.
Boahen, A. Adu. (1964). *Britain, the Sahara, and the Western Sudan 1788–1861*. Clarendon Press, Oxford.
Boddam, H. L. (1870). In Hallen, J. H. B., Charles, J. G., Jan, M. H., Kerr, H. C., and McLeod, K. (eds.) 1871. *Report of the Commissioners Appointed to Inquire into the Origin, Nature, Etc., of Indian Cattle Plagues. With Appendices*. Government Printer, Calcutta.
Bojanus, L. H. (1820). Anleitung zur kenntnis und behandlung der wichtigsten seuchen unter den hausthieren. Fr. Fleischer, Leipzig.
Bökönyi, S. (1974). *History of Domestic Mammals in Central and Eastern Europe*. Akadémiai Kiadó, Budapest.
Boldrini, G. (1954). Un episodio di peste bovina su una nave del Lloyd Triestino. *Vet. Ital.* **5**, 1182.
Bòttego, V. (1895). *Il Giuba Esplorato*. Loescher, Rome.
Bouley, H. (1871). La peste bovine et la consommation des betes atteintes. *Rev. Sci.* **1**, 38–39.
Bouley, H. (1872). Sanitary police and the cattle-plague. *C. R. Hebd. Seances Acad. Sci.* **74**, 1154–1165.
Bourgignon, H. (1865). *On the Cattle Plague: or, Contagious Typhus in Horned Cattle. Its History, Origin, Description, and Treatment*. John Churchill & Sons, London.
Bourke, Earl of Mayo (1876). *Sport in Abyssinia; or, The Mareb and Tackazzee*. John Murray, London.
Bowen, J. (1752). *The Observations and Experiments of John Bowen, of Bishop's-Castle, in the County of Salop*. T. Aris, Birmingham.
Boyer, N. (1745). A memoir on the disease infesting cattle. *Phil. Trans. R. Soc.* **43**, 477.
Boyes, J. (1911). *John Boyes King of the Wa-Kikuyu. A True Story of Travel and Adventure in Africa. Written by Himself.* Methuen, London.
Boynton, W. H. (1916). Rinderpest in swine, with experiments upon the transmission from cattle and carabaos to swine and vice versa. *Philippine Agricult Rev.* **9**, 288–336.
Boynton, W. H. (1917a). Preliminary report on the virulence of certain body organs in rinderpest. *Philippine Agricult. Rev.* **10**, 410–433.
Boynton, W. H. (1917b). Experiments on the treatment of rinderpest with various drugs. *Philippine Agricult. Rev.* **10**, 272–298.
Braddon, W. L. (1902). *Report on Researches on the Subject of Methods of Protection of Buffaloes and Cattle Against Rinderpest*. Kuala Lumpur.
Bradley, R. (1721). *The Plague at Marseilles Consider'd*, 3rd edn. W. Mears, London.
Branagan, D. and Hammond, J. A. (1965). Rinderpest in Tanganyika: a review. *Bull. Epizoot. Dis. Afr.* **13**, 225–246.
Brander, A. A. D. (1923). *Wild Animals in Central India*. Edward Arnold & Co., London.
Brauell, F. (1862). *Neue Untersuchungen betreffend die Pathologische Anatomie der Rinderpest*. E. J. Karow, Dorpat.
Bristowe, J. S. (1866). In Bernard, M. (ed.) *Third Report of the Commissioners Appointed to Inquire into the Origin and Nature &c. of the Cattle Plague*. HMSO, London.
Broad, J. (1983). Cattle Plague in eighteenth-century England. *Agricult. Hist. Rev.* **31**, 104–115.
Brocklesby, R. (1746a). *An Essay Concerning the Mortality, Now Prevailing Among the Horned Cattle, in Several Parts of Europe, and Chiefly About London*. Brindley, London.

Brocklesby, R. (1746b). Abstract of Doctor Brocklesby's essay on the Mortality among the horned cattle. *Gent. Mag.* **16**, 121–122.

Brompton, J. (1652). In Twysden, R. (ed.) *Historiæ Anglicanæ Scriptores X*. Cornelii Bee, London.

Bronson, E. B. (1910). *In Closed Territory*. A. C. McClurg & Co., Chicago.

Brown, A. M. (1884). *The Origin of Contagia in Relation to the Prevailing Theoretic Views and Scientific Applications*. John Bale & Sons, London.

Brown, D. [1842]. Report to S. Ludlow, esq. Superintending Surgeon, Barrackpur. No. XXVI. 1842. *Trans. Med. Phys. Soc. Calcutta* **8**, Appendix: xcvii–cvi.

Brown, G. T. (1877). In *Report from the Select Committee on Cattle Plague and Importation of Live Stock*. House of Commons, London.

Brown, G. T. (1878). *Annual Report of the Veterinary Department of the Privy Council Office for the Year 1877, with an Appendix*. HMSO, London.

Brown, H. T. (1929). Rinderpest. *Peking Soc. Nat. Hist. Bull.* **4**, 87–94.

Brown, T. (1751). *The Tryal of Thomas Colley, at the Assizes at Hertford, on Tuesday the 30th of July, 1751*. T. Brown, London.

Brown, W. H. (1899). *On the South African Frontier. The Adventures and Observations of an American in Mashonaland and Matabeleland*. Sampson Low, Marston, & Co. Ltd, London.

Browne, W. G. (1799). *Travels in Africa, Egypt; and Syria*. T. Cadell & W. Davies, London.

Bruce, C. P. (1870). In Hallen, J. H. B., Charles, J. G., Jan, M. H., Kerr, H. C., and McLeod, K. 1871. *Report of the Commissioners Appointed to Inquire into the Origin, Nature, Etc., of Indian Cattle Plagues. With Appendices*. Government Printer, Calcutta.

Bruce, J. (1790). *Travels to Discover the Source of the Nile, In the Years 1768, 1769, 1770, 1771, 1772, and 1773*, vol. IV. G. G. J. and J. Robinson, Edinburgh.

Bruins, L. H. (1951). *Leven en werken van Geert Reinders, de grondlegger van de immunologie*. Assen.

Bucard-Mauchart (1745). *Med. de lue vaccarum Tubingensi*. Tubingen.

"Bucolicus" (1746). *Gent. Mag.* **16**, 87–88.

Budd, W. (1863). Variola ovina, sheep's small-pox; or the laws of contagious epidemics illustrated by an experimental type. *Br. Med. J.* **ii**, 141–150.

Budd, W. (1864). Investigation of epidemic and epizootic diseases. *Br. Med. J.* **ii**, 354–357.

Budd, W. (1865a). The Siberian Cattle Plague; or, the typhoid fever of the ox. *Br. Med. J.* **ii**, 168–179.

Budd, W. (1865b). The Siberian Cattle Plague; or the typhoid fever of the ox. *The Veterinarian* **38**, 695–710.

Bulpett, C. W. L. (1907). *A Picnic Party In Wildest Africa*. Edward Arnold, London.

Buniva, M. F. (1808). *Recueil de mémoires et observations-pratiques sur l'épizootie*. Reymann, Lyons.

Burdett-Coutts, A. G. (1865). Miss Burdett Coutts on the cattle disease. *The Farmer's Magazine* **52**, 267–268.

Burdick, L. D. (1905). *Magic and Husbandry. The Folk-Lore of Agriculture*. The Otseningo Publishing Co., New York.

Burman, J. (1971). *Disaster Struck South Africa*. C. Struik (Pty) Ltd, Cape Town.

Burn, W. L. (1964). *The Age of Equipoise: A Study in the Mid-Victorian Generation*. George Allen & Unwin Ltd, London.

Burooah, N. C. (1870). In Hallen, J. H. B., Charles, J. G., Jan, M. H., Kerr, H. C., and McLeod, K. (eds.) 1871. *Report of the Commissioners Appointed to Inquire into the Origin, Nature, Etc., of Indian Cattle Plagues. With Appendices*. Government Printer, Calcutta.

Butler, W. F. (1887). *The Campaign of the Cataracts Being a Personal Narrative of the Great Nile Expedition of 1884–5*. Sampson Low, Marston, Searle, & Rivington, London.

Buxton, E. N. (1902). *Two African Trips with Notes and Suggestions on Big Game Preservation in Africa*. Edward Stanford, London.

Cammerarius, K. J. (1713). *Dissertation de lue vaccarum*. Tubingen.

Campbell, F. J. G. (1870). In Hallen, J. H. B., Charles, J. G., Jan, M. H., Kerr, H. C., and McLeod, K. (eds.) 1871. *Report of the Commissioners Appointed to Inquire into the Origin, Nature, Etc., of Indian Cattle Plagues. With Appendices*. Government Printer, Calcutta.

Campbell, J. D. (1870). In Hallen, J. H. B., Charles, J. G., Jan, M. H., Kerr, H. C., and McLeod, K. (eds.) 1871. *Report of the Commissioners Appointed to Inquire into the Origin, Nature, Etc., of Indian Cattle Plagues. With Appendices*. Government Printer, Calcutta.

Camper, P. (1769). *Lessen over de thans zweevende veesterfte openlyk gehouden ... in het Theatrum Anatomicum te Groningen, den 10, 11, 13, en 14 van Sprokkelmand, in den jaare 1769*. H. A. de Chalmot, Leeuwarden.

Camper, P. (1769). *Lessenover the thans zweevende Veesterfte te Leuwarden Kopenhague*.

Camper, P. (1777). Mémoire sur l'épizootie de la Hollande. *Soc. R. Méd.* August 12.

Camper, P. (1779). Ueber die wahre und eigentliche Ursache der Krankheiten, die unter dem grossen und kleinen Viehe als ansteckende seuchen, wüthen. Beschäft. d. Berlinischen Gesellsch. *Naturforsch. Freunde* **4**, 95–166.

Camper, P. (1783). *Ueber das Anstecken der Viehseuche, die wahre und eigenthümliche Ursache derselben, und die eigentlichen Vorbauubngsmittel darwider. Zwei von der Berlinischen Gesellschaft Nautrforschender Freunde gekrönte Preisschriften. Von Herrn Petrus Camper, der Arzeneigelahrtheit Doctor und Professor, und der mehrsten gelehrten Akademien und Gesellschaften Mitgliede, und Herrn Doctor Weiß, ausübenden Arzt zu Leer in Ostfriesland, und Mitglied der Gesellsch. Naturf. Freunde in Berlin. Mit einigen Zusätzen*. Anton Ferdinand Röse, Greiswald.

Camper, P. (1803). Leçons sur l'Epizoötie qui regna dans la Province de Gröningen en 1769. In *Oeuvres de Pierre Camper qui ont pour objet l'histoire naturelle, la physiologie et l'anatomie comparée*, vol. 3. H. J. Jansen, Paris.

Camper, P. and Munniks, W. (1769). *Voorloper van waarneemingen omtrent den uitslag van de inëntinge der besmettelijke veeziekte*. H. A. de Chalmont, Leeuwarden.

Canevese, P. F. (1770). *Compendio Di varie Ricette per Medicar Cavalli Esperimentate sì nella Scuderia dell'Illustriss Si. Co: Giuseppe Maria Arconati, uno de'Signori Sessanta Decurioni della Città di Milano, e Regio Feudatorio di Lomazzo, e sue pertinenze ec., come in quelle de'suoi Antenati. Raccolte, e date in luce da me Pietro Francesco Canevese Mastro di Stalla del medisimo Si. Conte. S'aggiunge inoltre alcune altre Ricette esposte con un nuovo Metodo esperimentato per curare il male epidemico delle Bestie Bovine. Opera di grande giovamento, ed utile a chi si diletta di numerosa comitiva di Cavalli*. Francesco Locatelli, Venice.

Canterbury, Archbishop of (1865). *Guardian* **20**, 1003.

Carlier, M. (1920). La Lutte contre la Peste bovine au Ruanda. *Bull. Agricult. Congo Belge* **11**, 191–206.

Carmichael, J. (1933). The virus of rinderpest and its relation to *Glossina morsitans*, Westw. *Bull. Ent. Res.* **24**, 337–342.

Carmichael, J. (1938). Rinderpest in African game. *J. Comp. Pathol.* **51**, 264–268.

Carmichael, J. (1973). Uganda 1909–1949. In Rollinson, D. H. L. and Callear, J. F. F. (eds.) *A History of the Overseas Veterinary Services*. Part Two. British Veterinary Association, London, pp. 341–362.

Carnegie, V. M. (1930). *A Kenyan Farm Diary*. William Blackwood & Sons Ltd, Edinburgh.

Carré, H. and Fraimbault. (1898). Note sur la contagiosité de la peste chez le porc. *Ann. Inst. Pasteur* **1898**, 848–856.

Carruthers, J. (1995). *The Kruger National Park. A Social and Political History*. University of Natal Press, Pietermaritzburg.

Cary, E. (1986). *The Roman Antiquities of Dionysius of Halicarnassus*, vol. IV. William Heinemann Ltd, London.

Caspar, J. (1777). *Unterricht von Criminalfällen, und wie sich ein Artz in Abgebung seines Gutachten hierüber zu verhalten habe. Nebst einem Anhang von der Hornviehseuche*. Wolfgang Schwarzkopf, Nuremburg.

Cathcart, Lord. (1866a). *J. R. Agricult. Soc. Engl.* **27**, 498.

Cathcart, Lord. (1866b). Abtract Report of Agricultural Discussions. *J. R. Agricult. Soc. Engl.* **27**, 282–284.

Cave, W. C. (1930). *Hieronymus Fracastorius. Contagion, Contagious Diseases and their Treatment. History of Medicine Series II*. G. P. Putnam's Sons, New York.

Chalmers (1823). In Scot (ed.) *Report of the Madras Medical Board*. Quoted in Orton, R. 1831. *An Essay on the Epidemic Cholera of India. Second Edition with a Supplement*. Burgess & Hill, London.

Chamberlain, R. W., Wamwayi, H. M., Hockley, E., Shaila, M. S., Goatley, L., Knowles, N. J., and Barrett, T. (1993). Evidence for different lineages of rinderpest virus reflecting their geographic isolation. *J. Genet. Virol.* **74**, 2775–2780.

Chanler, W. A. (1896). *Through Jungle and Desert. Travels in Eastern Africa*. Macmillan & Co., New York.

Chantrey, J., Meyer, H., Baxby, D., Begon, M., Bown, K. J., Hazel, S. M., Jones, T., Montgomery, W. I., and Bennett, M. (1999). Cowpox: reservoir hosts and geographic range. *Epidemiol. Infect.* **122**, 455–460.

Chapman, A. (1908). *On Safari Big-Game Hunting in British East Africa with Studies in Bird-Life*. Edward Arnold, London.

Charles, J. G. (1871). In Hallen, J. H. B., Charles, J. G., Jan, M. H., Kerr, H. C., and McLeod, K. (eds.) 1871. *Report of the Commissioners Appointed to Inquire into the Origin, Nature, Etc., of Indian Cattle Plagues. With Appendices*. Government Printer, Calcutta.

Chauveau, M. (1871). La science et la legislation, dans leurs rapports avec la police sanitaire du typhus épizootique ou peste bovine, en France. *Rev. Sci.* **1**, 77–87.

Chernyak, V. Z. (1959). Veterinary pathology as a part of comparative pathology (in Russian). *Arkh. Patologii.* **21**, 82–89.

Chesney, F. R. (1854). *The Russo-Turkish Campaigns of 1828 and 1829: With a View of the Present State of Affairs in the East*. Smith, Elder & Co., London.

Chevalier, A. (1908). *Mission Chari-Lac Tchad 1902–1904. L'Afrique Centrale Française. Récit du Voyage de la Mission*. Augustin Challamel, Paris.

Chomel, J. B. L. (1745). *Lettre d'un Médecin de Paris à un Médecin de Province, sur la maladie des Bestiaux*. Paris.

Christopher, S. A. (1916). *Big Game Shooting in Lower Burma*. Burma Pictorial Press, Rangoon.

Churchill, W. S. (1908). *My African Journey*. Hodder & Stoughton, London.

Cilli, V. (1951). Atteggiamenti biologici del virus K. A. G. (Kabete Attenuated Goat) della peste Bovina sulle capre Eritree. *Riv. Biol.* **43**, 255–284.
Cilli, V., Mazzaracchio, V., and Roetti, C. (1951). L'episodio di peste bovina al Giardino Zoologico di Roma. *Archo. Ital. Sci. Med. Trop. Parasit.* **32**, 83–94.
Cipolla, A. (1927). *Pagine africane di un esploratore*. Alpes, Milan.
Clark, J. (1806). *First Lines of Veterinary Physiology and Pathology*, vol. I. Printed for the Author, Edinburgh.
Claverie, J. (1931). La lutte contre la peste bovine en Guinée française. *Rec. Vét. exot.* **1931**, 121.
Cockayne, O. (1864–1866). *Leechdoms, Wortcunning, and Starcraft of Early England. Being a Collection of Documents, for the Most Part Never Before Printed, Illustrating the History of Science in this Country Before the Norman Conquest*, 3 vols. Longman, Green, Longman, Roberts & Green, London.
Cogrossi, C. F. (1714). *Nuova idea del male contagioso de' buoi'*. Marc' Antonio Pandolfo, Milan.
Coillard, F. (1897). *On the Threshold of Central Africa. A Record of Twenty Years' Pioneering Among the Barotsi of the Upper Zambesi*. Hodder & Stoughton, London.
Collin, L. (1747). *Gent. Mag.* **17**, 408–409.
Columella, L. J. M. (1745). *Of Husbandry in Twelve Books: and His Book Concerning Trees*. A. Millar, London.
Colville, H. (1895). *The Land of The Nile Springs*. Edward Arnold, London.
Comyn, D. C. E. ff. (1911). *Service & Sport in the Sudan. A Record of Administration in the Anglo-Egyptian Sudan. With some Intervals of Sport and Travel*. John Lane, The Bodley Head, London.
Conger, A. B. (1867). *First and Second Reports of the Special Committee Appointed by the Executive Board of the N.Y. State Agricultural Society, on the Statistics, Pathology and Treatment of the Epizoötic Disease Known as the Rinderpest*. Weed, Parsons & Co., Albany, NY.
Constable, H. S. (1865). *The Cattle Plague*, 2nd edn. Robert Sunter, York.
Constable, H. S. (1866a). Letter to the Editor. *The Times* March 31.
Constable, H. S. (1866b). *Observations Suggested by The Cattle Plague, about Witchcraft, Credulity, Superstition, Parliamentary Reform, and Other Matters*. Dalton & Lucy, London.
Constable, H. S. (1866c). *The Cattle Plague; with Remarks upon the Drainage of Farm Buildings and Stables*, 3rd edn. Robert Sunter, York.
Conti, G. (1913). La peste bovine chez le chameau. *Mod. Zooiat.* 1913, 25.
Cook, A. R. (1945). *Uganda Memories (1897–1940)*. The Uganda Society, Kampala.
Corfield, F. D. (1960). *The Origins and Growth of Mau Mau*. Government Printer, Nairobi.
Cornarium, J. (Trans.) (1549). *Ætii Medicii Graeci Contractae Ex Veteribus Medicinae Tetrabiblos, hoc est, quaternio, sive libri universales quatuor, singuli quatuor sermones complectentes, ut sint in summa quatuor sermomum quaterniones, id est.* G. and M. Bering, Lyons.
Cornell, R. L. (1934). Rinderpest in wildebeest. *Rep. Dept. Vet. Sci. Anim. Husb., Tanganyika 1933*, 37–42.
Cosserat, A. W. (1869). In Hallen, J. H. B., Charles, J. G., Jan, M. H., Kerr, H. C., and McLeod, K. (eds.) 1871. *Report of the Commissioners Appointed to Inquire into the Origin, Nature, Etc., of Indian Cattle Plagues. With Appendices*. Government Printer, Calcutta.
Cothenius, C. A. (1768). *Memoire to the Berlin Academy of Sciences*.
Cottrell, S. (1770). On the Cattle-distemper. *Scots Mag.* **32**, 169–174.
Courbet, P. (1927). *Assurance Contre la Mortalité du Bétail (Équidés–Bovidés). Influence des Maladies contagieuses sur l'Évolution et la Forme de cette Assurance*. Thèse doctorat, Ecole Nationale Vétérinaire d'Alfort. P. Macron, Paris.
Courtivron, Marqis de (1748). Observations sur la maladie du gros bétail. *Mém. de l'Academie des Sciences*. Paris.
Cranworth, Lord (1919). *Profit and Sport in British East Africa*. Macmillan & Co. Ltd, London.
Crellin, J. K. (1968). The dawn of the germ theory: particles, infection and biology. In Poynter, F. N. L. (ed.) *Medicine and Science in the 1860s*. Wellcome Institute of the History of Medicine, London, pp. 57–76.
Crisp, E. (1865). Abstract Report of Agricultural Discussions. *J. R. Agricult. Soc. Engl.* **2**, 280.
Crisp, E. (1866a). 2. Preparations exhibiting the various stages of cattle-plague. *Trans. Pathol. Soc. Lond.* **17**, 447–449.
Crisp, E. (1866b). 3. Specimens of the "cattle-plague," so called, in sheep. *Trans. Pathol. Soc. Lond.* **17**, 449.
Crisp, E. (1866c). *Farmer's Mag.* **29**, 53.
Crookes, W. (1866). The application of disinfectants in arresting the spread of the Cattle Plague. In Bernard, M. 1866b. *Third Report of the Commissioners appointed to inquire into the Origin and Nature &c. of the Cattle Plague*. HMSO, London.
Crosbie, R. (1896). In Faure, P. H. and Roos, J. De V. (eds.) 1896. *Minutes of the Conference of Delegates from the Cape Colony, Natal, Basutoland, the Bechuanaland Protectorate, South African Republic, Orange Free State and German West Africa, for the Purpose of Discussing Measures to Put a Stop to the Disease of Rinderpest*. G82-'96. Cape of Good Hope.
Cross, H. E. (1917). *The Camel and its Diseases*. Ballière Tindall & Co., London.

Cross, H. E. (1918). Are camels susceptible to blackquarter, haemorrhagic septicaemia and rinderpest? *Agricult. Res. Inst., Pusa, Bull.* **80**, 1–17.

Crouzet, J. (1889a). *Les Missions Catholiques* **21**, 62–63.

Crouzet, J. (1889b). *Les Missions Catholiques* **21**, 169–172.

Croveri, P. (1921). *La peste bovine*. A. Guidi, Buenos Aires.

Crowther, J. R. (1997). Rinderpest: at war with the disease of war. *Science Progress* **80**, 21–43.

Cullen, W. (1770). Memorial concerning the contagious disease affecting the Horned Cattle. *Scots Mag.* **32**, 172–174.

Curasson, G. (1921). Notes sur la peste bovine en A. O. F. et en Pologne. *Rev. Gén. Vét.* October, 569.

Curasson, G. (1932). *La Peste Bovine*. Vigot Frères, Paris.

Curasson, S. and Delpy, L. (1926). Sur l'immunisation contre la peste bovine par le virus formolé. *Rec. Méd. Vét.* **102**, 297.

Curson, H. H. (1919). Rinderpest in the New Langenberg Province of German East Africa during 1917–18. *J. Comp. Path. Ther.* **32**, 197–210.

Curasson, G. (1936). Theiler and the rinderpest epizootic of 1896–1903. *J. S.A.V.M.A.* **7**, 187–203.

Dalton, O. M. (1927). *The History of the Franks by Gregory of Tours*. The Clarendon Press, Oxford.

Dalyell, J. G. (1834). *The Darker Superstitions of Scotland, Illustrated from History and Practice*. Waugh & Innes, Edinburgh.

Danysz, M. J., Bordet, J., and Theiler, A. (1898). Report on Rinderpest to the Transvaal Government. *Vet. Rec.* **10**, 497–501.

d'Arboval, L. M. J. H. (1816). *Dictionnaire de medécine et de chirurgie vétérinaires*, 2nd edn. Ballière, Paris.

d'Arboval, L. M. J. H. (1826). *Dictionnaire de medécine et de chirurgie vétérinaires*, 3rd edn. Ballière, Paris.

Darwin, C. (1871). *The Descent of Man, and Selection in Relation to Sex*. John Murray, London.

Dash, Ya. and Sokolov, M. N. (1986). The reduction of stocks of Mongolian gazelles in Mongolia, its possible causes and recommendations for their conservation. In *Natural Conditions and Biological Resources of the Mongolian People's Republic. Proceedings of International Conference, Moscow* In Russian.

Daubney, R. (1928). Observations on Rinderpest. *J. Comp. Pathol. Ther.* **41**, 228–248; 263–297.

Daubney, R. (1943). *Report of the Veterinary Department Kenya 1942*. Government Printer, Nairobi.

Daubney, R. (1953). Rinderpest vaccination. *Bull. Epiz. Dis. Afr.* **1**, 12–15.

Davies, J. N. P. (1979). *Pestilence and Disease in the History of Africa*. Witwatersrand University Press, Johannesburg.

Dawson, E. C. (1888). *The Last Journals of Bishop Hannington*. Seeley & Co., London.

Dawson, H. (1865a). In Hallen, J. H. B., Charles, J. G., Jan, M. H., Kerr, H. C., and McLeod, K. (eds.) 1871. *Report of the Commissioners Appointed to Inquire into the Origin, Nature, Etc., of Indian Cattle Plagues. With Appendices*. Government Printer, Calcutta.

Dawson, H. (1865b). In Hallen, J. H. B., Charles, J. G., Jan, M. H., Kerr, H. C., and McLeod, K. (eds.) 1871. *Report of the Commissioners Appointed to Inquire into the Origin, Nature, Etc., of Indian Cattle Plagues. With Appendices*. Government Printer, Calcutta.

Dawson, W. H. (1882). *History of Skipton (W.R. Yorks)*. Simpkin, Marshall & Co., London.

Daye, P. (1918). *Avec les Vainqueurs de Tabora*. Perrin et Cie., Paris.

De Berg. (1778). *Mem. R. Soc. Paris*.

De Bozas, R. du Bourg. (1906). *De la Mer Rouge à l'Atlantique à Travers l'Afrique Tropicale (Octobre 1900-Mai 1903). Carnets de Route*. F. R. De Rudeval, Paris.

Decle, L. (1893). Scarcity of game in Africa. North of Lake Nyasa and east of Lake Tanganyika. *The Field* **2138**, 927.

Decle, L. (1896). *Three Years in Savage Africa*. Methuen & Co., London.

Deem, S. L., Karesh, W. B., Linn, M. J., Schaller, G., Lhagvasuren, B., Hyamsuren, and Olson, K. (2001). Health evaluation of Mongolian gazelles *Procapra gutturosa* on the Eastern Steppes. *Gnusletter* **20**, 18–20.

Defoe, D. (1722a). *Due Preparations for the Plague, as well as for Soul as Body*. London.

Defoe, D. (1722b). *A Journal of the Plague Year*. London.

De Leo, G., Dobson, A., and Goodman, A. (2002). Wildlife Perspectives on the Evolution of Virulence. In Dieckmann, U., Metz, J. A. J., Sabelis, M. W., and Sigmund, K. (eds.) *Adaptive Dynamics of Infectious Diseases. In Pursuit of Virulence Management*. Cambridge University Press, Cambridge, pp. 26–38.

Delpy, L.-P. (1931). *Contribution à l'étude de la Peste bovine et des Trypanosomiases animales en Afrique Occidentale Française*. Thèse Doctorat, Ecole Nationale Vétérinaire d'Alfort.

de Vries, J. (1994). De bestrijding van de runderpest in Friesland gedurende de 18 de eeuw. *Argos* **10**, 315–323.

Dewar, J. (1866). *Sulphurous Acid as a Fumigant Generally and with Especial Reference to "Rinderpest"*. John Hughes, Edinburgh.

Dhillon, S. S. (1959). Incidence of rinderpest in camels in Hissar district. *Indian Vet. J.* **36**, 603–607.

Dick, W. (1858). Non-contagious nature of epizootic diseases. *Trans. Highland & Agricult. Soc.*, 265–268.

Dickinson, F. A. (1908). *Big Game Shooting on the Equator*. John Lane The Bodley Head, London.

Di Cosmo, N. (1994). Ancient inner Asian nomads: their economic basis and its significance in Chinese history. *J. Asian Stud.* **53**, 1092–1126.

Didier (1722). *Traité de la Peste*. Marseilles.

Dieckerhoff, W. (1890). *Geschichte der Rinderpest und ihrer Literatur*. Enslin, Berlin.

Dion Lardner, E. G. (1912). *Soldiering & Sport in Uganda 1909–1910*. The Walter Scott Publishing Co. Ltd, London and Felling-on-Tyne.

Donaldson-Smith, A. (1897). *Through Unknown African Countries. The First Expedition from Somaliland to Lake Lamu*. Edward Arnold, London.

Dorwart, R. A. (1959). Cattle disease (rinderpest?)—prevention and cure in Brandenburg, 1665–1732. *Agricult. Hist.* **32**, 79–85.

Dossie, R. (1771). *Memoirs of Agriculture, and other Oeconomical Arts*, vol. II. J. Nourse, London.

Doutresoulle (1924). Rapport sur la recherche d'une méthode d'immunisation active des veaux non sevrés, provenant de mères réfractaires à la peste bovine. *Rec. Méd. vét.* **100**, 464–468.

Downing, J. (1797). *A Treatise on the Disorders Incident to Horned Cattle, Comprising a Description of their Symptoms, and the Most Rational Methods of Cure, Founded on Long Experience; to Which are Added Receipts, for Curing The Gripes, Staggers, and Worms in Horses, and an Appendix, Contraining Instructions for the Extracting of Calves*. J. Rollason, Stourbridge.

Dracopoli, I. N. (1914). *Through Jubaland to the Lorian Swamp. An Adventurous Journey of Exploration & Sport in the Unknown African Forests & Deserts of Jubaland to the Unexplored Lorian Swamp*. Seeley, Service & Co. Ltd, London.

Dubois, F. (1897). *Timbuctoo the Mysterious*. William Heinemann, London.

Dufot (1773). *Mémoire pour préserver les bêtes à cornes de la maladie épizootique*. Soissons.

Dugmore, A. R. (1910). *Camera Adventures in the African Wilds Being an Account of a Four Months' Expedition in British East Africa, for the Purpose of Securing Photographs from Life of Game*. William Heinemann, London.

Duke, H. L. (1919). An enquiry into the relations of *Glossina morsitans* and ungulate game, with special reference to rinderpest. *Bull. Ent. Res.* **10**, 7–20.

Dunlop, R. H. and Williams, D. J. (1996). *Veterinary Medicine an Illustrated History*. Mosby-Year Book Inc., St. Louis.

Du Plessis, D. (1896). In Faure, P. H. and Roos, J. De V. (eds.) 1896. *Minutes of the Conference of Delegates from the Cape Colony, Natal, Basutoland, the Bechuanaland Protectorate, South African Republic, Orange Free State and German West Africa, for the Purpose of Discussing Measures to Put a Stop to the Disease of Rinderpest*. G82-'96. Cape of Good Hope.

East, R. (ed.) (1988). *Antelopes Global Survey and Regional Action Plans Part 1. East and Northeast Africa*. IUCN, Gland.

East, R. (ed.) (1997). *Antelope Survey Update* **5**, 6–7. IUCN, Gland.

Eden, A. (1869). In Hallen, J. H. B., Charles, J. G., Jan, M. H., Kerr, H. C., and McLeod, K. (eds.) 1871. *Report of the Commissioners Appointed to Inquire into the Origin, Nature, Etc., of Indian Cattle Plagues. With Appendices*. Government Printer, Calcutta.

Edington, A. (1896). Rinderpest. *Agricult. J.* **9**, 253–254.

Edington, A. (1899). A Retrospect of the Rinderpest Campaign in South Africa. *The Lancet* **I**, 357–359.

Edwards, J. T. (1928). Rinderpest: active immunization by means of the serum simultaneous method; goat virus. *Agricult. J. India* **23**, 185–189.

Edwards, J. T. (1930). The problem of rinderpest in India. In *Bull. 199. Imperial Institute of Agricultural Research, Pusa*. Government Printer, Calcutta.

Edwards, J. T. (1949). The uses and limitations of the caprinized virus in the control of rinderpest among British and Near-Eastern cattle. *Br. Vet. J.* **105**, 209–253.

Egan, J. (1865). In Bernard, M. (ed.) 1866. *Second Report of the Commissioners Appointed to Inquire into the Origin and Nature, Etc. of the Cattle Plague*. HMSO, London.

Eland, G. (ed.) (1931). *Purefoy Letters, 1735–1753*. Sidgwick & Jackson, Ltd, London.

Elliott, G. (1973). Field systems of northwest England. In Baker, A. R. H. and Butlin, R. A. (eds.) *Studies of Field Systems in the British Isles*. Cambridge University Press, Cambridge.

Elsworthy, F. T. (1895). *The Evil Eye*. Murray, London.

Elton, C. (1927). *Animal Ecology*. Sidgwick & Jackson Ltd, London.

Engelman, J. (1763). Nader verhandeling over de rundveesterfte betreklijk tot de waarneemingen vervat in't VI de deel, II e stuk'. *Verhand. Holl. Maatsch. Wetenschap.* **7**, 247–318.

Epstein, H. (1969). *Domestic Animals of China*. Commonwealth Agricultural Bureaux, Farnham Royal.

Erman, A. (1848). *Travels in Siberia: Including Excursions Northwards, Down the Obi, to the Polar Circle, and Southwards, to the Chinese Frontier*, 2 vols. Longman, Brown, Green, & Longmans, London.

Ernes, W. (1865). In Bernard, M. (ed.) 1865. *First Report of Her Majesty's Commissioners Appointed to Inquire into the Origin and Nature &c. of the Cattle Plague*. HMSO, London.

Erxleben, J. C. P. (1771). *Praktischer Unterricht in der Vieharzneykunst*. Johann Ch. Dieterich, Göttingen.

Evans, E. P. (1895). Recent Recrudescence of superstition. *Pop. Sci. Monthly* **48**, 73–92.

Evans, G. H. (1910). *Elephants and their Diseases*. Superintendent, Government Printing, Rangoon.

Evans-Pritchard, E. E. (1940). *The Nuer. A Description of the Modes of Livelihood and Political Institutions of a Nilotic People*. Oxford University Press, Oxford.

Evans-Pritchard, E. E. (1956). *Nuer Religion*. Oxford University Press, London.

Ewen, C. L'E. (1929). *Witch Hunting and Witch Trials*. Kegan Paul, Trench, Trubner & Co. Ltd, London.

Faber, J. A. (1962). Cattle-plague in the Netherlands during the eighteenth century. *Meded. Landbouwhogeschool, Wageningen*. **62**, 1–7.

FAO (1984). *Pan-African Rinderpest Campaign Summary of Project Proposals*. FAO, Rome. Cyclostyled.

FAO (1999). *FAO Press Release 99/33. The EMPRES Transboundary Animal Disease Bulletin No.* ***10****—June 1999*: 2–3. FAO, Rome.

Falcone, G. (1612). *La nuova, vaga, et dilettevole villa di Guiseppe Falcone Piacentino, Opera d'agricoltura, pia che necessaria, per chi desidera d'accrescere l'entartet, de suoi poderi, utile a tutti quelli, che fano professione d'agricoltura per plantare, allevare, incalamare arbore, coltivar giradini, eminar camp, secundo la qualita de terreni & paesi, edificar palaggi, case, & edificii pertinenti all villa, con rimedii per varie infermita de buoi, cavalli, ed altri animal. Estratta da tutti gli autori, greci, latini, & italiana, che sin'hora hanno scritto di tal materia. Et di nuovo data in luce*. Lutz, Venice.

Farr, W. (1866). General cattle mutual insurance fund. *J. R. Agricult. Soc. Engl.* **2**, 455–471; 534–549.

Farrell, H. (1870). In Hallen, J. H. B., Charles, J. G., Jan, M. H., Kerr, H. C., and McLeod, K. (eds.) 1871. *Report of the Commissioners Appointed to Inquire into the Origin, Nature, Etc., of Indian Cattle Plagues. With Appendices*. Government Printer, Calcutta.

Farrell, H. (1871). In Hallen, J. H. B., Charles, J. G., Jan, M. H., Kerr, H. C., and McLeod, K. (eds.) 1871. *Report of the Commissioners Appointed to Inquire into the Origin, Nature, Etc., of Indian Cattle Plagues. With Appendices*. Government Printer, Calcutta.

"A Farmer" (1748). *Gent. Mag.* **18**, 166–167.

"A Farmer's Friend" (1864). *Shall We Prevent Cattle Diseases, or Continue to Eat Carrion?* Simpkin, Marshall, & Co., London.

Faust, B. C. (1797). *Über die Rindvieh = Pest (Viehseuche, Löserdürre,) die einzig und allein durch Ansteckung entsteht, und die durch Unvorsichtigkeit von einem Orte zum andern, von einem Lande zum andern, vorzüglich vermittelst der Viehmärkte und des Viehhandels, verbreitet wird: und über die Verhütung und Ausrottung derselben*. Paul Gotthelf Kummer, Leipzig.

Faversham, Lord (1866). Abstract report of agricultural discussions. *J. R. Agricult. Soc. Engl.* **27**, 278–279.

Fay, L. D. (1972). *Report to the Government of Kenya on Wildlife Disease Research*. FAO, Rome.

Ferguson, H. (1866). *Professor Ferguson's Report to His Excellency John Baron Wodehouse, Lord Lieutentant-General and General Governor of Ireland*. HMSO, Dublin.

Ferrandi, U. (1903). *Seconde Spedizione Bòttego. Lugh, emporio commerciale sul Giuba*. Socièta geografica italiana, Florence.

Ferraro, G. (1917). Perché persiste la peste bovine in Eritrea. *Moderno Zooiatro* **5**, 215–216.

Ficino, M. (1556). *De causis pestilentiae urbem Venetam opprimentis anno MDLVI*. G. Jolitus, Venice.

Filopatride, A. (1800). *Lettera Sopra La Facilita' Con Cui Si Può Communicare Il Contagio Nelle Epizootíe E Il Vero Mezzo Di Garantirsene*. Giacomo Marsoner, Rimino.

Findlay, F. R. N. (1903). *Big Game Shooting and Travel in South-East Africa. An Account of Shooting Trips in the Cheringoma and Gorongoza Divisions of Portuguese South-East Africa and in Zululuand*. T. Fisher Unwin, London.

Fischer, G. A. (1895). Am Ostufer des Victoria-Njansa. *Petermanns Mitteilungen* **41**, 66–72.

Fisher, A. G. B. and Fisher, H. J. (1980). *Sahara and Sudan Volume II Kawar, Bornu, Kanem, Borku, Ennedi by Gustav Nachtigal*. C. Hurst & Co., London.

Fisher, A. G. B. and Fisher, H. J. (1987). *Sahara and Sudan Volume III The Chad Basin and Bagirmi*. C. Hurst & Co., London.

Fisher, A. G. B., Fisher, H. J., and O'Fahey. R. S. (1971). *Sahara and Sudan. Volume Four: Wadai and Darfur By Gustav Nachtigal*. University of California Press, Berkeley, CA.

Fisher, J. R. (1980). The economic effects of cattle disease in Britain and its containment, 1850–1900. *Agricult. Hist.* **54**, 278–294.

Fisher, J. R. (1993). British physicians, medical science, and the Cattle Plague, 1865–66. *Bull. Hist. Med.* **67**, 651–669.

Fitzgerald, W. W. A. (1898). *Travels in the Coastlands of British East Africa and the Islands of Zanzibar and Pemba. Their Agricultural Resources and General Characteristics*. Chapman & Hall, London.

Fleischman, G. (1800). *Historia pestis bovillae*. Dissertatio inaugralis quam pro cappessendo gradu Doctoris Medicinae. Kunstmanni, Erlangae.

Fleming, G. (1864). Letters. *North British Agriculturist* March 16.

Fleming, G. (1866). *The Veterinary Record* **39**, 578.

Fleming, G. (1871). *Animal Plagues: Their History, Nature, and Prevention*. Chapman & Hall, London.

Fleming, G. (1875). *A Manual of Veterinary Sanitary Science and Police*, 2 vols. Chapman & Hall, London.

Flemming, M. (1755). *A Proposal in Order to Diminish the Progress of the Distemper Amongst the Horned Cattle, Supported by Facts*. London.

Flemynge, M. (1755). *Uni. Mag.* **17**, 153–154.

Foa, E. (1899). *After Big Game in Central Africa*. Adam & Charles Black, London.

Fonck, H. (1910). *Deutsch-Ost-Afrika*. Vossische Buchhandlung, Berlin.

Ford, J. (1971). *The Role of the Trypanosomiases in African Ecology*. Clarendon Press, Oxford.

Forsten, R. (1797). *De vee-pest welke zich in het Quartier van Nymegen geopenbaart heeft… waar bij gevoegd is voorbehoed en geneeskundig advys over dezelve vee-pest, uitgebragt door de Geneeskundige Faculteit te Harderwyk*. van Goor, Arnhem.

Foster, J. B. and Kearney, D. (1967). Nairobi National Park game census, 1966. *E. Afr. Wildl. J.* **5**, 112–120.

Foureau, F. (1902). *Mission Saharienne Foureau-Lamy. D'Alger au Congo par le Tchad*. Masson et Cie, Paris.

Fox, F. D. and Varian, E. F. (1911). Big game in Angola. *The Field Newspaper* **118**, 854–855.

Fracastoro, G. (1530). *Syphilis sive Morbus gallicus*. Stefano Nicolini da Sabbio, Verona.

Fracastoro, G. (1546). *De contagione et contagiosis morbis et eorum curatione. Libri III*. Lucaeantonii Iuntae Florentini, Venetiis. Trans. Wright, W. C. 1930. G. P. Putnam's Sons, New York.

Francis, J. (1963). John Gamgee's work on Cattle Plague. *Br. Vet. J.* **119**, 315–324.

Frank, G. R. (1802). *Ueber die Rinderpest und die Mittel sie zu Heilen und Ausrotten*. Heinrich August Rottmann, Berlin.

Frazer, J. G. (1922). *The Golden Bough. A Study in Magic and Religion*. Macmillan & Co. Ltd., London.

Friedberger, F. and Fröhner, F. (1886–1887). *Lehrbuch der speciellen Pathologie und Therapie der Haustiere*, 2 vols. F. Enke, Stuttgart.

Friend, E. (1900). With the C.I.V.'s at the Front. *Black and White Budget* **3**(27), 62.

Froehner, R. (1930). Tieropfer zur Abwendung der Rinderpest. *Veterinärhistorische Mitteilungen*. **10**(7): 65–66.

Fuchs, P. (1907). Wirtschaftliche Eisenbahn-Erkundungen im mittleren und nördlichen Deutsch-Ostafrika. *Beiheft zum Tropenpflanzer* **11**(8).

Fuller, C. (1923). *Tsetse in the Transvaal and Surrounding Territories An Historical Review. Department of Agriculture Entomology Memoirs No. 1*. Government Printer, Pretoria.

Fusella, L. (1963). Il Dãgmãwi Menilek di Afawårq Gabra Iyasus. *Rassegna di Studi Etiopici* **19**, 121.

Galleski, J. G. (1765). *Angestellte Untersuchung über die Rindviehseuche worin deren wahre und eigentliche Ursachen bemerket und die Mittel sie abzuwenden fürgeschlagen werden*. Hoff = und Akad. Hartungschen Buchdruckeren, Königsberg.

Gamgee, J. (1857a). The cattle plague. *North Br. Agricult.* April 9.

Gamgee, J. (1857b). Letter

Gamgee, J. (1857c). Letter.

Gamgee, J. (1861). *Dairy Stock Its Selection, Diseases, and Produce with a Description of the Brittany Breed*. Thomas C. Jack, Edinburgh.

Gamgee, J. (1863a). Letter. *The Times*.

Gamgee, J. (1863b). *Edinb. Vet. Rev.* **5**, 261.

Gamgee, J. (1865). In Bernard, M. (ed.) 1865. *First Report of Her Majesty's Commissioners Appointed to Inquire into the Origin and Nature &c. of the Cattle Plague*. HMSO, London.

Gamgee, J. (1866). *The Cattle Plague with Official Reports of The International Veterinary Congresses, Held in Hamburg, 1863, and in Vienna, 1865*. Robert Hardwicke, London.

Gamgee, J. (1867). Epizootics. *The Veterinarian* **40**, 537.

Gamgee, J. (1877). In *Report from the Select Committee on Cattle Plague and Importation of Live Stock*. House of Commons, London.

Garmonsway, G. N. (1953). *The Anglo-Saxon Chronicle*. J. M. Dent & Sons Ltd, London.

Garrison, F. H. (1910). *Science* April 1.

Gasparin, A. E. P. (1821). *Des Maladies Contagieuses des Bêtes à Laine*. Huzard, Paris.

Gayau, J. (1682). *Remède et Benédictions contre les Maladies et Mortalité du Bétail*. Agen.

Gedge, E. (1891). Diary entries for March 24–27. Quoted in Waller (1988).

Geil, W. E. (1905). *A Yankee in Pigmy Land*. Hodder & Stoughton, London.

Gerbezii. (1713). *Chronologia Medico-practica*. Frankfurt.

Gerlach, A. C. (1865). In Bernard, M. (ed.) 1865. *First Report of Her Majesty's Commissioners Appointed to Inquire into the Origin and Nature &c. of the Cattle Plague*. HMSO, London.

Gerlach, A. C. (1867). *Die Rinderpest*. Schmorl & von Seefeld, Hanover.

Gessi, F. (1892). *Seven Years in the Soudan Being a Record of Explorations, Adventures, and Campaigns Against the Arab Slave Hunters by Romolo Gessi Pasha*. Sampson Low, Marston & Co., London.

Ghika, N. D. (1898). *Cinq Mois au Pays des Somalis*. Georg & Co., Bale.

Giarola, A., Cantoni, M., and Magnone, E. (1967). La dottrina esogena, delle infezioni dall'antichità ai giorni nostri—X[e]—Un grande tropicalista ed epidemiologo del "700", Eusebio Valli. *Riv. Ital. med. Igiene della Scuola*. **XIII**(II), 191–199.

Giavedoni, L., Jones, L., Mebus, C., and Yilma, T. (1991). A vaccinia virus double recombinant expressing the F and H genes of rinderpest virus protects cattle against rinderpest and causes no pock lesions. *Proc. Natl. Acad. Sci. USA* **88**, 8011–8015.

Gibbons, A. St. H. (1898). *Exploration and Hunting in Central Africa 1895–96*. Methuen & Co., London.

Gibbs, C. S. (1925). Observations and experiments on rinderpest. *China J. Sci. Arts* **3**, 399–409.

Giblett, J. (1877). In *Report from the Select Committee on Cattle Plague and Importation of Live Stock*. House of Commons, London.

Giblin, J. (1996). The precolonial politics of disease control in the lowlands of northeastern Tanzania. In Maddox, G., Giblin, J., and Kimambo, I. (eds.) *Custodians of the Land*. James Currey, London, pp. 127–151.

Gielgud, V. (1901). Letter to the Secretary of the British South Africa Company. In Austen, E. A. (ed.) 1903. *A Monograph of the Tsetse-Flies*. British Museum (Natural History), London.

Gilchrist, W. (1848). *A Practical Treatise on the Diseases of the Elephant, Camel, and Horned Cattle, with Instructions for Preserving their Efficiency*. Military Orphan Press, Calcutta.

Girard, J. and Dupuy, A. C. (1816). Notice sur l'épizootie qui règne sur le gros bétail. Egron, Paris.

Glossop, B. R. M. (1906). *Sporting Trips of a Subaltern*. Harper & Brothers, London.

Gmelin, J. G. (1751–1752). *Reise durch Sibirien von dem Jahr 1733 bis 1743*. A. Vandenhoeck, Göttingen.

Godley, R. S. (1935). *Khaki and Blue*. Lovat Dickson & Thompson Ltd., London.

Goodenough, W. H. (1896). In *Correspondence Relating to the Outbreak of Rinderpest in South Africa in March 1896. Presented to both Houses of Parliament by Command of Her Majesty*. July 1896. C.8141. HMSO, London.

Gowlett, J. A. J. (1988). Human adaptation and long-term climatic change in northeast Africa: an archaeological perspective. In Johnson, D. H. and Anderson, D. M. (eds.) *The Ecology of Survival Case Studies from Northeast African History*. Lester Crook Academic Publishing, London, pp. 27–45.

Graf, J. (1796). *Abhandlung uber die gegenwartige Hornviehseuche, ihren Ursprung, Natur, Vorbeugungs = und Heilungsmittel*. Johann Baptist Strobel, Munchen.

Grant, J. A. (1864). *A Walk Across Africa*. William Blackwood & Sons, Edinburgh.

Gray, C. E. (1917). Rinderpest in East Africa. *MSS Reports. May 25 & July 26*, pp. 25. Colonial Office, London.

Greenhow, E. H. (1857). *Report on Murrain in Horned Cattle, the Public Sale of Diseased Animals, and the Effects of the Consumption of their Flesh on Human Health; Addressed to the Right Honourable the President of the General Board of Health*. HMSO, London.

Gregory, T. D. (1866). "Rinderpest." *The Veterinarian* **XXXIX**, 346–358.

Greig, T. (n.d.) [1866]. *Address on the Cattle Plague*.

Grimwood, I. R. (1961). *Report of the Game Department Kenya 1960*. Government Printer, Nairobi.

Grindle, E. S. (1866). *A Sermon Preached in the Parish Church of S. Lawrence, Sigglesthorne, on Wednesday, March 7, 1866, Being the Day Set Apart by Authority as a Day of Humiliation Throughout the Diocese of York, on Account of the Cattle Plague*. Joseph W. Leng, Kingston-upon-Hull.

Grogan, E. S. and Sharp, A. H. (1900). *From the Cape to Cairo. The First Traverse of Africa from South to North*. Hurst & Blackett, Ltd., London.

Gulland, F. M. D. (1995). The impact of infectious diseases on wild animal populations—a review. In Grenfell, B. T. and Dobson, A. P. (eds.) *Ecology of Infectious Diseases in Natural Populations*. Cambridge University Press, Cambridge, pp. 20–51.

Günther jun (1858). Kersting, über die Einimpfung der Rinderpest. *Mag. Ges. Thierheilk* **24**, 1–42.

Gupta, K. C. S. and Verma, N. S. (1949). Rinderpest in wild ruminants. *Indian J. Vet Sci* **XIX**, 219–224.

Gutsche, T. (1979). *There was a Man: The Life and Times of Sir Arnold Theiler K. C. M. G. of Onderstepoort*. Howard Timmins, Cape Town.

Haining, P. (1974). *The Witchcraft Papers. Contemporary Records of the Witchcraft Hysteria in Essex 1560–1700*. Robert Hale & Co., London.

Haji, C. S. J. (1932). Rinderpest in camels. *Indian Vet. J.* **9**, 13–14.

Hall, S. A. (1962). The cattle plague of 1865. *Med. Hist.* **6**, 45–58.

Hallen, J. H. B. (1871). *Manual of the More Deadly Forms of Cattle Disease in India*. Government Printer, Calcutta.

Hallen, J. H. B. (1903). *Manual of the More Deadly Forms of Cattle Disease in India*, 3rd edn. Government Printer, Calcutta.

Hallen, J. H. B., Charles, J. G., Jan, M. H., Kerr, H. C., and McLeod, K. (1871). *Report of the Commissioners Appointed to Inquire into the Origin, Nature, Etc., of Indian Cattle Plagues. With Appendices*. Government Printer, Calcutta.

Haller, A. von. (1773). *Abhandlung von der Viehseuche*. Bern.

Halpin, B. (1975). *Patterns of Animal Disease*. Ballière Tindall, London.

Haltermann, J. N. (1773). Geschichte der Hornviehseuche in den Herzogthümern Bremen und Verden, aus Regierungsacten von Ao. 1715 bis 1772. In Pratje, J. H. (ed.) *Altes und Neues aus den Herzogthümern Bremen und Verden*, Band **6**, Buchdruckerey Stade.

Hamers-Casterman, C., Atarhouch, T., Muyldermans, S., Robinson, G., *et al.* (1993). Naturally occurring antibodies devoid of light chains. *Nature* **363**, 446–448.

Hamilton, A. (1877). In *Report from the Select Committee on Cattle Plague and Importation of Live Stock*. House of Commons, London.

Hanley, G. (1971). *Warriors and Strangers*. Hamish Hamilton, London.

Hannæo, G. (1746). *Historische Beschreibung der Viehseuche welche seit dem Jahre 1745 in den hiesigen Gegenden grassiret*. Conrad König, E. Hochedl. & Hochw. Raths Buchdrucker, Hamburg.

Hannah, W. C. (1871). In Hallen, J. H. B., Charles, J. G., Jan, M. H., Kerr, H. C., and McLeod, K. (eds.) 1871. *Report of the Commissioners Appointed to Inquire into the Origin, Nature, Etc., of Indian Cattle Plagues. With Appendices*. Government Printer, Calcutta.

Harding, C. (1905). *In Remotest Barotseland Being an Account of a Journey of over 8,000 Miles Through the Wildest and Remotest Parts of Lewanika's Empire*. Hurst & Blackett Ltd., London.

Hardinge, A. H. (1897). *FO 107/81*. Public Record Office, London.

Hardynge, J. (1543). *The Chronicle of J. Hardynge from the firste begynnyng of Englande*. R. Graftoni, London.

Harrod, H. (1866). Some details of a murrain of the Fourteenth Century; from the court rolls of a Norfolk Manor. *Archaeologia* **41**, 1–14.

Harward, M. (1673). *The Herdsman's Mate*. Dublin.

Hattersley, C. W. (1908). *The Baganda at Home*. The Religious Tract Society, London.

Haubner. (1869). *Handbuch der Veterinär-Polizei*. Dresden.

Haughton, J. C. (1870). In Hallen, J. H. B., Charles, J. G., Jan, M. H., Kerr, H. C., and McLeod, K. (eds.) 1871. *Report of the Commissioners Appointed to Inquire into the Origin, Nature, Etc., of Indian Cattle Plagues. With Appendices*. Government Printer, Calcutta.

Havas, I. (1839). *Havas Ignatz Rendszerénti Föorvosnak Hivatalos Jelentése A'marhapestist enyhitö banasmodrol. (Official Communiqué of Ignatz Havas, chief physician of County Tolna, Concerning the Healing Treatment of Cattle Plague)*. Szekszard.

Havemann, A. C. (1814). *Ueber Viehpest oder Löserdürre*. Hannover.

Head, A. S. (1906). Cattle plague in the Anglo-Egyptian Sudan. *J. Comp. Pathol. Ther.* **18**, 12–18.

Hecker, J. F. C. (1844). *The Epidemics of the Middle Ages*. Trans. B. G. Babington. The Sydenham Society, London.

Heller, C., Stem, C., Wamwayi, H., and Grieve, A. (1998). Development of a filter paper-based ELISAfor rinderpest antibodies. *Vet. Rec.* **142**, 729.

Henderson, E. (1873). Cattle Disease in China—Rinderpest. *The Veterinarian* **46**, 35–51.

Henle, J. (1840). Von den Miasmen und Contagien und von den miasmatisch-contagiösen Krankheiten. In A. Hirschwald (ed.) *Pathologische Untersuchungen*. Berlin.

Henning, M. W. (1943). *Animal Diseases in South Africa*. Central News Agency Ltd, Cape Town.

Hepburn, A. B. (1913). *The Story of an Outing*. Harper & Brothers, New York.

Herdt, R. De. (1970). *Bijdrage tot de geschiedenis van de veeteelt in Vlaanderen, inzonderheid tot de geschiedenis van de rundveepest 1769–1785*. Belgisch Centrum voor de Landelijke Geschiedenis, Leuven.

Herskovits, M. J. (1926). The cattle-complex in East Africa. *Am. Anth.* **28**, 230–272; 361–380; 494–528; 633–664.

Hervey, J. (1747). *Meditations and Contemplations*. John Rivington jun., London.

Hibbert, C. (1980). *The French Revolution*. Allen Lane, London.

Higgs, H. E. (1869). In Hallen, J. H. B., Charles, J. G., Jan, M. H., Kerr, H. C., and McLeod, K. (eds.) 1871. *Report of the Commissioners Appointed to Inquire into the Origin, Nature, Etc., of Indian Cattle Plagues. With Appendices*. Government Printer, Calcutta.

Higgs, H. E. (1870). In Hallen, J. H. B., Charles, J. G., Jan, M. H., Kerr, H. C., and McLeod, K. (eds.) 1871. *Report of the Commissioners Appointed to Inquire into the Origin, Nature, Etc., of Indian Cattle Plagues. With Appendices*. Government Printer, Calcutta.

Hinde, S. L. and Hinde, H. (1901). *The Last of the Masai*. William Heinemann, London.

Hindlip, Lord (1906). *Sport and Travel Abyssinia and British East Africa*. T. Fisher Unwin, London.

Hird, W. (1753). *Remarks upon Pestilence and Pestilential Diseases, Interspersed with Some Observations on the Mortality Amongst the Horned Cattle, Wherein this Heavy Calamity, Both from Reason and Analogy, is Set in a New and Clear Light, and Certain Easy Means of Preservation and Cure are Recommended upon Rational and Experimental Principles*. W. Innys & J. Richardson, London.

Hobley, C. W. (1929). *Kenya From Chartered Company to Crown Colony*. H. & F. Witherby, London.

Hodges, N. (1720). *Loimologia: or, an account of the Plague in London in 1665: with Precautionary Directions Against the Like Contagion … to which is Added, an Essay On the Different Causes of Pestilential Diseases, and How They Become Contagious: with Remarks on the Infection now in France*, 2nd edn. E. Bell, London.

Hodgson (n.d.) History of Allertonshire and Birdforth, quoted by "W.D." (1865). *Notes And Queries* **8**(3rd S), 223–224.

Hoffman, F. (1721). *Medicinisches Gutachten über die in denen Jahren 1715 und 1716 graßirende Horn = Vieh = Seuche, wie dieselbe sowohl vorzubeugen als zu curiren*. (Ej. Cons. Med. Vol. I.)

Holberg, L. (1746). *Kurzes Bedencken über die jetziger Zeit regierende ViehSeuche, Mit einigen Oeconomischen Anmerckungen; Aus dem Dänischen übersetzt*. Johann Andreas Berger, Wismar.

Hole, H. M. (1926). *The Making of Rhodesia*. Macmillan & Co. Ltd, London.

Holland, H. (1839). *Medical Notes and Reflections*. Longman, London.

Holland, T. J. and Hozier, H. M. (1870). *Record of the Expedition to Abyssinia Compiled by Order of the Secretary of State for War*, 3 vols. HMSO, London.

Hollis, A. C. (1905). *The Masai. Their Language and Folklore*. Clarendon Press, Oxford.

Holmes, J. D. E. (1913). Researches into the susceptibility to rinderpest in several districts of India and a perfected method of preparation of anti-serum. *Indian Civil Dep. Mem. 1910–1911*, 98.

Holmes, J. D. E. (1914). A note on the effect of heat on the rinderpest immune bodies. *Bull. 43. Agricult. Res. Inst., Pusa*. Government Printer, Calcutta.

Holt, P. M. (1958). *The Mahdist State in the Sudan 1881–1898. A Study of its Origins, Development and Overthrow*. The Clarendon Press, Oxford.

Home, F. (1759). *Medical Facts and Experiments*. A. Miller, London.

Hopen, C. E. (1958). *The Pastoral Fulbe Family in Gwandu*. International African Institute, Oxford.

Hornby, H. E. (1926). Studies in rinderpest immunity. (2) Methods of infection. *Vet. J.* **82**, 348–355.

Hornby, H. E. (1928). The distribution of rinderpest virus in infected blood. *J. Comp. Pathol. Ther.* **41**, 17–24.

Hornby, H. E. (1931). *Tanganyika Territory. Annual Report of the Department of Veterinary Science and Animal Husbandry 1930*. Crown Agents, London.

Hornby, H. E. (1939). *Tanganyika Territory. Annual Report of the Department of Veterinary Science and Animal Husbandry 1938*. Crown Agents, London.

Howard, C. (1865). Central Farmers' Club. The Cattle Plague. *Farmer's Mag.* **58**, 483–498.

Howard, C. W. (1911). The distribution of Tstese-flies in the Province of Mozambique, East Africa. *Bull. Entemd. Res.* **2**, 39–43.

Huc, E. R. (1850). *Souvenirs d'un Voyage dans la Tartarie, le Thibet, et la Chine pendant les années 1844–1846*. Adrien le Clere, Paris.

Hudson, P. J., Rizzoli, A., Grenfell, B. T., Heesterbeek, and Dobson, A. P. (eds.) (2002). *The Ecology of Wildlife Diseases*. Oxford University Press, Oxford.

Hüpsch, J. W. C. A. baron von. 1776. *Patriotische Vorschläge die Ausbreitung der jetz herschenden Landesverderblichen Hornviehseuche auf eine leichte und wohlfeite Art zu verhindern, nebst einigen neuen und eigenen Beobachtungen über die besten Bewahrungs = und Heilungs = Mittel*. Heinrich Ludwig Brönner, Frankfurt am Main.

Hurel, E. (1911). Réligion et Vie domestique des Bakerewe. *Anthropos* **6**, 62–94.

Hussel, L. (1960). *Die Rinderpest*. S. Hirzel, Leipzig.

Hutcheon, D. (1896a). In *Correspondence Relating to the Outbreak of Rinderpest in South Africa in March 1896. Presented to Both Houses of Parliament by Command of Her Majesty. July 1896. C.8141*. HMSO, London.

Hutcheon, D. (1896b). In Faure, P. H. and Roos, J. Dev (eds.) *Minutes of the Conference Delegates from the Cape Colony, Natal, Basutoland, the Bechuanaland Protectorate, South African Republic, Orange Free State and German West Africa, for the Purpose of Discussing Measures to Put a Stop to the Spread of the Disease of Rinderpest. G.82-'96*. Cape of Good Hope.

Hutcheon, D. (1897). *Special Report on Rinderpest in South Africa, by the Colonial Veterinary Surgeon, from March, 1896, to February, 1897. G33-'97*. Cape of Good Hope.

Hutcheon, D. (1902). Rinderpest in South Africa. *J. Comp. Pathol. Ther.* **XV**, 301–324.

Hutcheon, D. (1903). *Report of the Colonial Veterinary Surgeon and the Assistant Veterinary Surgeons for the Year 1902*. G.33-1903. Department of Agriculture, Cape of Good Hope.

Hutcheon, D. (1904). *Report of the Colonial Veterinary Surgeon and the Assistant Veterinary Surgeons for the Year 1903*. G.41-1904. Department of Agriculture, Cape of Good Hope.

Hutchinson, J. (1834). In Hallen, J. H. B., Charles, J. G., Jan, M. H., Kerr, H. C., and McLeod, K. (eds.) 1871. *Report of the Commissioners Appointed to Inquire into the Origin, Nature, Etc., of Indian Cattle Plagues. With Appendices*. Government Printer, Calcutta.

Hutchinson, J. (1836). In Hallen, J. H. B., Charles, J. G., Jan, M. H., Kerr, H. C., and McLeod, K. (eds.) 1871. *Report of the Commissioners Appointed to Inquire into the Origin, Nature, Etc., of Indian Cattle Plagues. With Appendices*. Government Printer, Calcutta.

Hutyra, F. and Marek, J. (1916). *Die Orientalische Rinderpest*. Gustav Fischer, Jena.

Huxley, E. (1935). *White Man's Country. Lord Delamere and the Making of Kenya*, 2 vols. Macmillan & Co., Ltd, London.

Huygelen, C. (1997). The immunization of cattle against rinderpest in eighteenth-century Europe. *Med. Hist.* **41**, 182–196.

"Iater" [J. Davis] (1747). *An Essay Concerning Pestilential Contagion*. John Whiston, London.

Inoue, T., Harada, S., and Shimizu, T. (1929). Note préliminaire sur l'infection expérimentale avec le virus de la peste bovine de *Citellus mongolicus ramosus*. *J. Jap. Soc. Vet. Sci.* **1929**, 161.

Jack, E. M. (1914). *On the Congo Frontier*. T. Fisher Unwin, London.

Jack, R. W. (1923). Tsetse Fly. A four years' experiment in game elimination. *Rhod. Agricult. J. Bull.*, 460.

Jackson, F. J. (1894). In Phillipps-Wolley, C. (ed.) *Big Game Shooting*, 2 vols. Longmans, Green, & Co., London.

Jackson, F. J. (1930). *Early Days in East Africa*. E. Arnold & Co., London.

Jackson, J. G. (1820). *An Account of Timbuctoo and Housa Territories in the Interior of Africa by El Hage Abd Salam Shabeeny with Notes, Critical and Explanatory*. Longman, Hurst, Rees, Orme, & Brown, London.

Jackson, R. and Cabot, D. H. E. (1930). La résistance du virus de la peste bovine. *Bull. Off. Int. Epiz.* **1930**, 775.

Jacotot, H. (1927). Transmission expérimentale de la peste bovine à *Cervus aristotelis*. *C. R. Soc. Biol.* **96**, 1134–1135.

Jahn, U. (1884). Die Deutschen Opfergebräuche bei Ackerbau und Viehzucht. Ein Beitrag zur Deutschen Mythologie und Alterthumskunde. In Weinhold, K. (ed.) *Germanistische Abhandlung*. Wilhelm Koebner, Breslau.

Jameson, J. (1820). *Report on the Epidemic Cholera Morbus, Drawn Up under the Superintendence of the Medical Board of Bengal*. Calcutta.

Jargow. (1814). *Beilage zu den Mecklenburgischen Anzeigen*. Mittwochen.

Jassim, S. A. A. and Naji, M. A. (2001). The desert ship heritage and science. *Biologist* **48**, 268–272.

Jauffret, R. (1928). *Prophylaxie de la Peste Bovine en Cochinchine*. Thèse Doctorat, Ecole Nationale Vétérinaire de Lyon, Lyon.

Jessen, P. (1852). *Ueber die gänzlich Ausrottung der Rinderpest*. H. Laakmann, Dorpat.

Jessen, P. (1865). Sendschreiben an die österreichischen Veterinairkundigen. *Mag. Ges. Thierheilk.* **31**, 116–128.

Jewell, P. A. (1972). Social organisation and movements of topi (*Damaliscus korrigum*) during the rut, at Ishasha, Queen Elizabeth Park, Uganda. *Zool. Afr.*, **7**, 233–255.

Johnson, H. (1902). *The Uganda Protectorate*, 2 vols. Hutchinson & Co., London.

Johnson, R. H. (1958). *Vet. Rec.* **70**, 457.

Johnston, C. (1844). *Travels in Southern Abyssinia*, 2 vols. J. Madden & Co., London.

Jones, B. V. (1997). A nice address: Plague Cottages. *Cricklade Hist. Soc. Bull.* **2**, 7–9.

Jones, E. L. (1978). Disaster management and resource saving in Europe, 1400–1800. *Proc. 7th Int. Econ. Hist. Conf. Edinb.*, 21–28.

Jones, T. (1952). *Brut y Tywysogyon or The Chronicle of the Princes. Peniarth MS. 20 Version*. University of Wales Press, Cardiff.

Jones, T. (1955). *Brut y Tywysogyon or The Chronicle of the Princes. Red Book of Hengest Version*. University of Wales Press, Cardiff.

Jones, T. (1971). *Brenhinedd y Saesson or The Kings of the Saxons*. University of Wales Press, Cardiff.

Jones, W. H. S. (1923). *Hippocrates*, vol. I. William Heinemann Ltd., London.

Jordan, W. C. (1996). *The Great Famine. Northern Europe in the Early Fourteenth Century*. Princeton University Press, Princeton, NJ.

Joubert, L. (1567). *De peste liber unus. Accesserunt duo tractatus unus de quartana febre, alter de paralysi*. Joannem Frellonium, Lugduni.

"J. S." (1746). Success of Dr. Barker's Method of treating the Distemper among cows. *Gent. Mag.* **16**, 686–687.

"J. S." (1747a). Method of treating sick cattle. *Gent. Mag.* **17**, 16–18.

"J. S." (1747b). Another, by J. S. who says that it has often been successful. *Gent. Mag.* **17**, 409.

Kakizaki, C. (1918). Study on the glycerinated rinderpest vaccine. *Kitasato Arch. Exp. Med* **2**, 549–566.

Kakizaki, C. (1924). Experimental studies on the prophylactic inoculation against rinderpest. *Kitasato Arch. Exp. Med.* **6**, 139–197.

Kakizaki, C., Nakaniski, S.-H., and Nakamura (1928). Etudes éxperimentales sur la vaccination économique contre la peste bovine. *Révue générale*, 497.

Kakizaki, C., Nakaniski, S.-H., and Ozumi, T. (1926). Experimental researches on preventive inoculation against rinderpest. *J. Jap. Soc. Vet. Sci.* **5**, 221.

Kanold, J. (1713). *Historische Relation von der Pestilentz des Horn-Viehes, welche anno 1711 und 1712 in Schlesien, wie nicht weniger diese, und das vorhergegangene 1710. Jahr in Moscau, Polen, Ungarn, Pesterreich, Siebenbürgen, Italien, und andern Ländern starck grassiret/ausgefertiget von Johann Kanold.* Fellgiebel, Breslau.

Kanold, J. (1721). Jahreshistorie von den Seuchen des Viehes, von 1701 bis 1717. *Sammlung von Natur und Medicin* **4**, Budissin.

Karsten, F. C. L. (1814). *Wie sichern wir uns gegen die wiederkehrende Rindvieh-Pest? Ein Versuch zur Beantwortung dieser Frage zur Prüfung dargelegt von Franz Christian Lorenz Karsten. Hannibal ante portas!* Stiller, Rostock & Schwerin.

Kassner, T. (1911). *My Journey from Rhodesia to Egypt Including an Ascent of Ruwenzori and a Short Account of the Route from Cape Town to Broken Hill and Lado to Alexandria*. Hutchinson & Co., London.

Kausch, J. J. (1793). *Kameralprinzipien über Rindviehsterben*. Heinrich August Rottmann, Berlin.

Kennan, G. (1891). My last days in Siberia. *Century Mag.* **42**, 803–820.

Kennedy, J. T. (1929). Glossina morsitans *in Relation to Rinderpest Epizootic, West Nile District*. Entebbe.

Kershaw, I. (1973). The great famine and agrarian crisis in England 1315–1322. *Past & Present* **59**, 3–50.

Keylock, H. E. (1909). Cattle-Plague in China. *J. Comp. Pathol. Ther.* **22**, 192–213.

Khera, S. S. (1979). Rinderpest eradication campaign in India. *Proc. 2nd Int. Symp. Vet. Epidemiol. Econ.*, 581–586.

Kieckhefer, R. (1976). *European Witch Trials*. Routledge & Kegan Paul, London.

Kilian, H. (1934). *Die Bekämpfung der Rinderpest in Mecklenberg-Strelitz (1769–1780).* Inaugural Diss. Vet. med. Rudolph Pfau, Berlin.

King, S. (1750). *The Hand of the Lord upon the Cattle considered and improved, in a Sermon Preached at Northampton, December 7, 1749. On Occasion of the Present Mortality Amongst the Cattle*. J. Buckland, London.

Kirilyuk, V. and Laurie, A. (2001). Winter kill of Mongolian gazelles. *Gnusletter* **20**, 26–27.

Kirk, J. (1896). African rinderpest. *Nature* **55**, 53–54.

Kirkland, C. (1908). *Some African Highways. A Journey of Two American Women to Uganda and the Transvaal.* Duckworth & Co., London.

Kirkland, T. (1783). *An Account of the Distemper Among the Horned Cattle at Caulk in Derbyshire in 1783. And of the Remedies Recommended, for the Cure, with Observations*. J. Beadsmore, Ashby de la-Zouch.

Kishi, H. (1976). A historical study on outbreaks of rinderpest during the Yedo era in Japan. *Yamaguchi J. Vet. Med.* **3**, 33–40.

Kishi, H. (1982). A study on restoration after the outbreak of rinderpest in 1672–1673 in the western part of Japan. *Jap. J. Vet. Hist.* **16**, 84–85.

Kittredge, G. L. (1929). *Witchcraft in Old and New England.* Harvard University Press, Cambridge, MA.

Kjekshus, H. (1977). *Ecology Control and Economic Development in East African History. The Case of Tanganyika 1850–1950.* University of California Press, Berkeley, CA.

Knyghton, H. (1652). In Twysden, R. (ed.) *Historiæ Anglicanæ Scriptores X.* Cornelii Bee, London.

Koch, A. (1891). *Encyklopädie der gesammten Thierheilkunde und Thierzucht. 8*. Moritz Perles, Vienna.

Koch, R. (1897a). Special Report to the "British Medical Journal" by Professor R. Koch on his Researches into the Cause of Cattle Plague. *Br. Med. J.* **(i)**, 145–146.

Koch, R. (1897b). I. Prof. Robert Koch's Berichte über seine in Kimberley ausgeführten Experimentalstudien zur Bekämpfung der Rinderpest. *Deutsche Medicinische Wochenschrift* **15**, 225–227.

Koch, R. (1897c). I. Prof. Robert Koch's Berichte über seine in Kimberley ausgeführten Experimentalstudien zur Bekämpfung der Rinderpest. *Deutsche Medicinische Wochenschrift* **16**, 241–243.

Kock, R. A. (2000). Working to eradicate cattle plague from Africa. *Lifewatch* Summer, 18–20.

Kock, R. A., Wambua, J. M., Mwanzia, J., Wamwayi, H., Ndungu, E. K., Barrett, T., Kock, N. D., and Rossiter, P. B. (1999). Rinderpest epidemic in wild ruminants in Kenya 1993–97. *Vet. Rec.* **145**, 275–283.

Koczians, A. (1769). *Prufung der Ursachen von der Hornviehseuche, mit dem Vorschlage auf was für Art derselben am sichersten Einhalt gethan werden könne*. Joseph Kurzböcken, Vienna.

Koferle, J. C. von. (1796). *Anleitung uber die gegenwartige Hornviehseuche*. Ulm, Stettin.

Kolb, T. (1828). *Aphtharum Pecorinarum et Claudicationis Nuperrime Grassatarum Historia Succincta*. Diss. Inauguralis Med. Ludovici Friderici Fues, Tubingae.

Kolle, W. (1901). *Report on Cattle Plague in the Sudan*. National Records Office, Khartoum, [NRO] Cairint 3/11/204.

Kolle, W. and Turner, G. (1897). Uber den Fortgang der Rinderpestforschungen in Koch's Versuchestation in Kimberley. Mit spezieller Berucksichtigung einer Immunisirungsmethode durch gleichzeitige Einspritzung der Thiere mit virulentem Infectionsstoff (Blut kranker Thiere) und Serum hochimmunisirter Thiere. *Deutsche Medicinische Wochenschrift* **23**, 793–795.

Kolle, W. and Turner, G. (1898). Ueber Schutzimpfungen und Heilserum bei Rinderpest. *Zeitschirift Hyg. Infkr.* **29**, 309–376.

Kollman, P. (1899). *The Victoria Nyanza. The Land, the Races and their Customs, with Specimens of some of the Dialects*. Swan Sonnenschein & Co. Ltd, London.

Königshofen, J. von. (1698). *Die Alteste Teutsche so wol Allgemeine Als insonderheit Elsassiche und Strassburgische Chronicke*. Josias Städel, Strasburg.

Koponen, J. (1988). *People and Production in Late Precolonial Tanzania History and Structures*. Finnish Society for Development Studies, Jyväskylä.

Koponen, J. (1994). *Development for Exploitation. German Colonial Policies in Mainland Tanzania, 1884–1914*. Finnish Historical Society, Helsinki.

Kraiewsky, A. (1885). Désinfection et utilisation des peaux provenant d'animaux atteints de la peste bovine. *Arch. Méd. Vét. Saint-Pétersbourg*.

Krogh, T. J. (1896). In Faure, P. H. and Roos, J. De V. (eds.) 1896. *Minutes of the Conference of Delegates from the Cape Colony, Natal, Basutoland, the Bechuanaland Protectorate, South African Republic, Orange Free State and German West Africa, for the Purpose of Discussing Measures to Put a Stop to the Disease of Rinderpest. G82-'96*. Cape of Good Hope.

Kumm, H. K. W. (1910). *From Hausaland to Egypt, Through the Sudan*. Constable & Co. Ltd, London.

Kuper, R. (1978). *Sahara: 10,000 Jahre zwischen Weide und Wuste*. Museen der Stadt, Cologne.

Kurze, F. (1895). *Annales Regni Francorum Inde Ab A. 741 usque Ad A. 829. Qui Dicuntur Annales Laurissenses Maiores et Einhardi*. Bibliopolii Hahniani, Hanover.

Kylasamier, K. (1931). Studies on Madras fowl-pest. *Indian Vet. J.* **7**, 340–346.

Labrousse, C. E. (1943). *La crise de l'economie française á la fin de l'Ancien Régime et au début de la Révolution*. Presses Universitaires de France, Paris.

Lamb (1836). Letter to the Superintending Surgeon, Dacca. In Hallen, J. H. B., Charles, J. G., Jan, M. H., Kerr, H. C., and McLeod, K. (eds.) 1871. *Report of the Commissioners Appointed to Inquire into the Origin, Nature, Etc., of Indian Cattle Plagues. With Appendices*. Government Printer, Calcutta.

Lamb (1842). Copy of Mr. Lamb's letter, dated Dacca, May 9, 1836. *Trans. Med. Phys. Soc. Calcutta* **8**, Appendix: xciv–xcvii.

Lamond, E. (1890). *Walter of Henley's Husbandry Together with an Anonymous Husbandry, Seneschaucie and Robert Grosseteste's Rules*. Longmans, Green, & Co., London.

Lamphear, J. (1976). *The Traditional History of the Jie*. The Clarendon Press, Oxford.

Lancisi, G. M. (1715). *Dissertatio Historica de Bovilla Peste ex Campaniae finibus, anno 1713 Latio importata*. Joannis Mariæ Salvioni, Rome.

Landor, A. H. S. (1907). *Across Widest Africa. An Account of the Country and People of Eastern, Central and Western Africa as Seen During a Twelve Months' Journey from Djibuti to Cape Verde*, 2 vols. Hurst & Blackett Ltd, London.

Lange, D. (1972). L'intérieur de l'Afrique Occidentale d'après Giovanni Lorenzo Anania (XVIe siècle). *J. World Hist.* **14**, 299–351.

Lappenberg, J. M. (1845). *A History of England Under the Anglo-Saxon Kings*. Trans. Thorpe, B. John Murray, London.

Lawrence, D. A. (1970). The history of veterinary services in Rhodesia. 1. Earliest veterinary developments 1890 to 1899. *Rhod. Vet. J.* **1**, 29–30.

Layard, D. P. (1757). *An Essay on the Nature, Causes, and Cure of the Contagious Distemper Among the Horned Cattle of these Kingdoms*. John Rivington, London.

Layard, D. P. (1758). A Discourse on the Usefulness of Inoculation of the horned Cattle to prevent the contagious Distemper among them. *Phil. Trans. R. Soc.* **50**, 528–538.

Layard, D. P. (1780). A Letter to Joseph Banks, Esq … relative to the Distemper among the horned Cattle. *Phil. Trans. R. Soc.* **70**, 536–545.

Leclainche, E. (1955). *Histoire Illustrée de la Médecine Vétérinaire*. 2 vols. Albin Michel, Monaco.

Leese, A.-S. (1927). *The One-humped Camel*. Haynes & Son, Stamford.

Lefantiri, A. 1897. Letter. *Leselinyana [Morija]*. June 15.

Lefebvre, G. (1951). *La Révolution française*. Presses Universitaires de France, Paris.

Leibold, A. A. (1917). Trans. *Miessner, Epizoötics and their Control During War*. American Veterinary Publishing Co., Chicago.

Lepissier, H. E. (1971). *General Technical Report on OAU/STRC Joint Campaign Against Rinderpest in Central and West Africa (1961–69)*. OAU/STRC, Lagos.

Lerotholi. (1896). *Lesotho National Archives, S3/1/5/9*. December 25.

Le Roux, H. (n.d.) [1905] *Chasses et Gens d'Abyssinie*. Calmann-Lévy, Paris.

Le Roy, A. (1914). *Au Kilima-Ndjaro*. La Congrégation du Saint-Esprit, Paris.

Leutwein, T. (1906). *Elf Jahre Gouverneur in Deutsch=Südwestafrika*. Mittler und Sohn, Berlin.

Lewis, B. A. (1972). *The Murle. Red Chiefs and Black Commoners*. The Clarendon Press, Oxford.

Lewis, J. (1866). On agricultural statistics and the cattle census. *J. R. Agricult. Soc. Engl.* **2**, 393–429.

Lhagvasuren, B. and Milner-Gulland, E. J. (1997). The status and management of the Mongolian gazelle *Procapra gutturosa* population. *Oryx* **31**, 127–134.

Liebert. (1869). In Hallen, J. H. B., Charles, J. G., Jan, M. H., Kerr, H. C., and McLeod, K. (eds.) 1871. *Report of the Commissioners Appointed to Inquire into the Origin, Nature, Etc., of Indian Cattle Plagues. With Appendices*. Government Printer, Calcutta.

Liebert, E. von. (1898). *Neunzig Tage im Zelt: meine Reise nach Uhehe*. Mittler & Sohn, Berlin.

Liebig, J. (1843). *Chemistry in its Applications to Agriculture and Physiology*, 3rd edn. Taylor & Walton, London.

Lingard, A. (1905a). *Report on the Preparation of Rinderpest Protective Serum with Fifty Tabular Statements. Revised*. Government Printer, Calcutta.

Lingard, A. (1905b). Resistance against rinderpest and other infectious diseases conferred by the subcutaneous injection of certain bile products and also by the injection of substances prepared from animal testes and the seeds of plants. *Centralblatt Bakt. I Abt. Originale Bd.* **37**, Heft 2.

Littlejohn, H. D. (n.d.) [1866]. *Suggestions for the Prevention and Treatment of the Cattle Plague*. Edinburgh.

Littlewood, W. (1905). Cattle Plague in Egypt in 1903-04-05. *J. Comp. Pathol. Ther.* **17**, 312–321.

Litton, E. (1750). *The Theory of the Distemper Amongst the Horned Cattle, and Electrical Medicines Proposed*. W. Owen, London.

Livingstone, D. (1857). *Missionary Travels and Researches in South Africa*. John Murray, London.

Livy, T. (1922–1938). *Ab Urbe Condita*. Books IV xxv; xxx. V xiii; Trans. Foster, B. O., XLI xxi; Trans. Sage, E. T. and Schlesinger, A. C. William Heinemann, London.

Lloyd, A. B. (1906). *Uganda to Khartoum. Life and Adventure on the Upper Nile*. T. Fisher Unwin, London.

Lobb, T. (1745). *Letters Relating to the Plague, and other Contagious Distempers*. Buckland, London.

Lorimer, N. (1917). *By the Waters of Africa*. Robert Scott, London.

Lorinser, C. J. (1831). *Untersuchung über die Hornviehseuche*. Schüppel, Berlin.

Lowe, H. J. (1942). Rinderpest in Tanganyika Territory. *Empire J. Exp. Agricult.* **10**, 189.

Ludlow, S. (1842). To Dr. James Ranken, Officiating Secretary, Medical Board, Fort William. 23rd. May, 1836. *Trans. Med. Phys. Soc. Calcutta* **8**, Appendix: xciii–xcvii.

Ludolphus, J. (1682). *A New History of Ethiopia. Being a Full and Accurate Description of the Kingdom of Abessinia. Vulgarly, though Erroneously, called The Empire of Prester John*. Samuel Smith, London.

Lugard, F. D. (1892). Travels from the East Coast to Uganda, Lake Albert Edward, and Lake Albert. *Proc. R. Geog. Soc.* **12**, 817–841.

Lugard, F. D. (1893). *The Rise of Our East African Empire. Early Efforts in Nyasaland and Uganda*, 2 vols. William Blackwood & Sons, London.

Lugg, H. C. (1970). *A Natal Family Looks Back*. T. W. Griggs & Co. (Pty) Ltd, Durban.

Lund, J. (n.d.) *England and America's Cattle Doctor; or a Reformed Treatise of Medicines and Means for the Cure of Diseases in Oxen, Cows, Sheep, Swine and Dogs; with a Great Variety of Original Recipes. Directions for the Treatment of Rinderpest. And Other Valuable Information Whereby Every Man May Be His Own Cattle Doctor*. W. Nicholson & Sons, London.

Lushchekina, A. A., Nerenov, V. M., and Yamnikova, S. S. (1988). On the epizootic of influenza among Mongolian gazelles. In *The Natural Conditions, Vegetation Cover and Animal World of Mongolia*. Puschino, Moscow. In Russian.

Lydekker, R. (1924). *The Game Animals of India, Burma, Malaya and Tibet*, 2nd edn. revised by J. G. Dollman. Rowland Ward Ltd, London.

Lyons, M. (1989). In E. Crawford (ed.) *Famine: The Irish Experience 900–1900: Subsistence Crises and Famines in Ireland*. John Donald, Edinburgh, pp. 31–74.

Maberly, J. (1898). The rinderpest in South Africa. *Lancet*, **11**, 1189–1192.

Macaulay, J. W. (1973). Kenya. In *A History of the Overseas Veterinary Services. Part Two*. pp. 141–161. British Veterinary Association, London.

McCaul, J. B. (1866). *Darkness that May be Felt: A Warning to England at the Present Crisis*. London.

McCullough, K. C., Sheshberadran, H., Norrby, E, Obi, T. U., and Crowther, J. R. (1986). Monoclonal antibodies against morbilliviruses. *Rev. Sci. Tech. Off. Int. Epizoot.* **5**, 411–427.

Macdonald, J. R. L. (1897). *Soldiering and Surveying in British East Africa 1891–1894*. Edward Arnold, London.

McDougall, A. (n.d.) [1866?] *The Cattle Plague: its Prevention and Cure*. Simpkin, Marshall & Co., London.

Mack, R. (1970). The great African cattle plague epidemic of the 1890s. *Trop. Anim. Health. Prod.* **2**, 210–219.

Mackenzie, M. (1765). An Account of the Plague at Constantinople: In a Letter from Mordach Mackenzie, M.D. To Sir James Porter, His Majesty's Envoy Plenipotentiary at Brussels, and F. R. S. *Phil Trans. R. Soc.* **43**, 69–82.

M'Clean, J. R. (1865). In Bernard, M. (ed.) 1865. *First Report of Her Majesty's Commissioners Appointed to Inquire into the Origin and Nature &c. of the Cattle Plague*. HMSO, London.

McLeod, K. (1869). In Hallen, J. H. B., Charles, J. G., Jan, M. H., Kerr, H. C., and McLeod, K. (eds.) 1871. *Report of the Commissioners Appointed to Inquire into the Origin, Nature, Etc., of Indian Cattle Plagues. With Appendices*. Government Printer, Calcutta.

Macnab, F. (1897). *On Veldt and Farm. In Bechuanaland—Cape Colony—The Transvaal—and Natal*. Edward Arnold, London.

Macpherson. (1842). Copy of Mr. Macpherson's letter, dated Murshidabad, May 1, 1836. *Trans. Med. Phys. Soc. Calcutta* **8**, Appendix: xciv.

Madeira, P. C. (1909). *Hunting in British East Africa*. J. B. Lippincott Co., Philadelphia, PA.

Mageoghagan, C. (1896). *The Annals of Clonmacnoise being Annals of Ireland from the Earliest period to A.D. 1408*. Ed. Murphy, D. University Press, Dublin.

Mahamooth, T. M. Z. (1943). Rinderpest. *Trop. Agricult.* **99**, 20.

Mähle, G. (1813). *Mittel gegen die Verbreitung der Viehseuche*. Rostock.

Maistre, C. (1895). *A Travers L'Afrique Centrale du Congo au Niger 1892–1893*. Librairie Hachette et Cie, Paris.

Mallet, M. (1923). *A White Woman Among the Masai*. E. P. Dutton & Co., New York.

Mansfield, W. R. (1857). In *Papers Relating to the Cattle Plague in Poland*. House of Lords, London.

Mantovani, A. and Zanetti, R. (1993). Giovanni Maria Lancisi De Bovilla Peste and stamping out. *Hist. Med. Vet.* **18**, 97–110.

Marchal, J. Y. (1980). *Chronique d'un cercle de l'AOF*. O.R.S.T.O.M., Paris.

Marensky, A. (1896). Letter to Colonial Government. In Hutcheon, D. (ed.) 1897. *Special Report on Rinderpest in South Africa, by the Colonial Veterinary Surgeon, from March, 1896, to February, 1897*. G33-'97. Cape of Good Hope.

Mares, R. G. (1990). A Faraway Look Herbert Watkins-Pitchford and Rinderpest 1896–1912. *Vet. Hist.* **6**, 89–102.

Mares, R. G. (1995). A closer look. Herbert Watkins-Pitchford, CMG, FRCVS, FRSE. *Hist. Med. Vet.* **20**, 81–93.

Maresch, J. (1863). On the infection of sheep by the contagium of the Rinderpest. *Austria Quart. Vet. J.* **19**, 34.

Mariner, J. C., House, A., Sollod, A. E., Stem, C., Van Den Ende, M., and Mebus, C. A. (1990). Comparison of the effect of various chemical stabilizers and lyophilization cycles on the thermostability of a Vero cell-adapted rinderpest vaccine. *Vet. Microb.* **21**, 195–209.

Martini, S. (1880). Terzo viaggio del Cap. S. Martini allo Scioa. *Cosmos* **6**, 65–74.

Masch (1865). The cattle plague. *The Farmer's Mag.* **52**, 377–379.

Masius, G. H. (1807). *Ueber die Viehseuche oder Löserdürre, und die wirksamsten Vorbauungsmittel gegen dieselbe*. Adlers Erben, Rostock.

Matossian, M. K. (1989). *Poisons of the Past. Molds, Epidemics, and History*. Yale University Press, New Haven, CT.

Maurer, F. D., Jones, T. C., Easterday, B., and DeTray, D. (1955). The pathology of rinderpest. *Proc. 92nd Annual Meeting of the AVMA*, 201–211.

May, R. M. and Anderson, R. M. (1979). Population biology of infectious diseases: Part I. *Nature* **280**, 361–367.

Maydon, H. C. (1925). *Simen, its Heights and Abysses. A Record of Travel and Sport in Abyssinia, with some Account of the Sacred City of Aksum and the Ruins of Gondar*. H. F. & G. Witherby, London.

Mead, R. (1720). *A Short Discourse Concerning Pestilential Contagion, and the Methods Used to Prevent it*. Sam Buckley, London.

Mecklenberg, A. F. Duke of. (1910). *In the Heart of Africa*. Cassell & Co., Ltd, London.

Mecklenberg, A. F. Duke of. (1913). *From the Congo to the Niger and the Nile. An Account of the German Central African Expedition of 1910–1911*, 2 vols. Duckworth & Co., London.

Meikle, R. S. and Meikle, M. E. (n.d.) [1914]. *After Big Game. The Story of an African Holiday*. T. Werner Laurie Ltd, London.

Meinertzhagen, R. (1957). *Kenya Diary 1902–1906*. Oliver & Boyd, Edinburgh.

Melland, F. H. and Cholmeley, E. H. (1912). *Through the Heart of Africa. Being an Account of a Journey on Bicycles and on Foot from Northern Rhodesia, Past the Great Lakes, to Egypt, Undertaken when Proceeding Home on Leave in 1910*. Constable & Co. Ltd, London.

Memmo, G., Martoglio, and Adani, C. (1904). Peste bovine. *Annali d'Igiene Sperimentale* **14**, 256.

Mendes, A. M. (1998). Notas Historicas—A Peste Bovina em Angola e Sao Tome. *Rev. Port. Cien. Vet.* **93**, 104–114.

Mentzel, O. F. (1921). *A Geographical and Topographical Description of the Cape of Good Hope*. Trans. H. J. Mandelbrote. Pt. I. Van Riebeeck Society, Cape Town.

Merker, M. (1904). *Die Masai*. Die Reimer, Berlin.

Mettam, R. W. M. (1937). A short history of rinderpest with special reference to Africa. *Uganda J.* **5**, 22–26.

Michelet, J. (1847–1853). *Histoire de la Révolution française*. Le Vasseur, Paris.

Michelotti, P. A. (1720). Account of a Contagion among the Cattle in the Venetian Territories. *Phil. Trans. R. Soc.* **31**, 83–86.

Middleton, J. (1971). Some effects of colonial rule among the Lugbara. In Turner, V. (ed.) *Colonialism in Africa 1870–1960*, vol. 3. Cambridge University Press, Cambridge, pp. 6–48.
Miki, S. (1962). History of rinderpest in Korea. *Chosen Gakuho* **34**, 1–27 (in Japanese, quoted in Kishi, 1976).
Milner, J. (1746). *Phil. Trans. R. Soc.* **44**, 226.
Minett, F. C. (1954). Dissemination of animal disease in India. Role of man and carrion feeders. *Br. Vet. J.* **110**, 19–24.
Mingay, G. E. (1956). The agricultural depression, 1730–1750. *Econ. Hist. Rev. 2nd. Ser.* **8**, 323–338.
Molema, S. M. (1966). *Montshiwa 1815–1896 Barolong Chief and Patriot.* C. Struik (Pty.) Ltd., Cape Town.
Molinié, J.-P. (1930). *Considérations Générales sur la Pest Bovine dans la Région du Bas-Laos. Contamination a l'Espèce Porcine.* Imprimerie Toulousaine, Toulouse.
Monteil, P.-L. (1895). *De Saint-Louis à Tripoli par le Lac Tchad. Voyage au Travers du Soudan et du Sahara accompli pendant les Années 1890-91-92.* Félix Alcan, Paris.
Monter, E. W. (1976). *Witchcraft in France and Switzerland.* Cornell University Press, Ithaca, NY.
Montgomery, H. (1748). Letter from Hugh Montgomery, Esq; of North Cave, in the East Riding of Yorkshire, concerning distemper'd Cattle. *Gent. Mag.* **18**, 283.
Montigny (1775a). *Instructions et avis aux habitants des provinces méridionales de la France, sur la maladie putride et pestilentielle qui détruit le bétail.* Paris.
Montigny (1775b). *Second Mémoire instructif sur l'execution du plan etc. publié en Nov. 1775.* Paris.
Montigny (1776). *Of the Rots and Pestilential Illness of the Livestock.* Berlin (see Faust, 1797).
Moor, F. R. (1896). *Secretary for Native Affairs to Colonial Secretary, 24 April 1896. SNA 1/1/219, 516/1896.* South African Archives, Pietermaritzburg.
Moore, D. (1866). *Prayer and Providence: Two Sermons Designated to Vindicate the Use of Prayer in Relation to Plague and Pestilence.* London.
Moore, J. E. S. (1901). *To the Mountains of the Moon. Being an Account of the Modern Aspect of Central Africa, and of some Little Known Regions Traversed by the Tanganyika Expedition, in 1899 and 1900.* Hurst & Blackett Ltd, London.
Morcos, Z. (1953). Concise history of veterinary science in Egypt. *Br. Vet. J.* **109**, 345–346.
Morgan, M. (1946). *The English Lands of the Abbey of Bec.* Oxford University Press, London.
Morière, J. (1872). Typhus Contagieux de l'Espèce Bovine. *Annuaire des Cinq Départements de la Normandie.* Trente-Huitième Année 1872. F. LeBlanc-Hardel, Caen.
Mortimer, C. (1745a). Some account of the distemper raging among the cow-kind in the neighbourhood of London, together with some remedies proposed for their Recovery. *Phil. Trans. R. Soc.* **43**, 532–537.
Mortimer, C. (1745b). Further observations on the distemper now raging among the cow-kind. *Phil. Trans. R. Soc.* **43**, 549–555.
Mortimer, C. (1746). A third account of the distemper among the cows. *Phil. Trans. R. Soc.* **43**, 4–10.
Morton, W. J. T. and Simonds, J. B. (1865). Cattle plague. *The Veterinarian* **38**, 621–640.
"Mulo-Medicus" (1748). *Lond. Mag.* **17**, 516.
Muller (1794). *Avis aux Habitans de la Campagne, Pour la Santé de leur Bétail; Contennant une connaissance exacte de plusieurs maladies, & notamment celles qui sont les plus fréquentes dans ce pays, comme le Louet, ou Louat, la Blanche & la Jaune, & le Quartier, en français l'Avant-coeur, en allemand Huntsch. Avec un choix des meilleurs remedes faciles à faire & à se procurer.* Author, Lausanne.
Muller, C. (1877). In *Report from the Select Committee on Cattle Plague and Importation of Live Stock.* House of Commons, London.
Mullett, C. F. (1946). The cattle distemper in mid-eighteenth-century England. *Agricult. Hist.* **20**, 144–165.
Muratori, L. V. (1714). *Del Governo Della Peste, E delle maniere di guardarsene.* Niccolo Gavelli, Pessaro.
Murchison, C. (1865a). On the anatomical lesions of the cattle plague now prevalent in London. *Lancet* August 26.
Murchison, C. (1865b). Report to the Pathological Society on twenty-seven dissections of cattle that died of rinderpest. *Lancet* October 28.
Murchison, C. (1865c). On the points of resemblance between cattle plague and sheep pox. *Lancet* December 30.
Murchison, C. (1866). On the pathology of the Cattle Plague, and on its relation to human diseases. In Bernard, M. (ed.) *Third Report of the Commissioners Appointed to Inquire into the Origin and Nature &c. of the Cattle Plague.* HMSO, London.
Murchison, C. (1873). *A Treatise on the Continued Fevers of Great Britain*, 2nd edn. Longmans, Green & Co., London.
Murray, A. J. (1865). *The Cattle Plague: Its History, Treatment, and Prevention.* Taylor & Francis, London.
Murray, R. (1866). In Bernard, M. (ed.) 1866. *Second Report of the Commissioners Appointed to Inquire into the Origin and Nature &c. of the Cattle Plague.* HMSO, London.
Nakamura, J., Kishi, S., Matsuzawa, H., Kiuchi, J., and Miyamoto, T. (1957). Inoculation experiments with attenuated strains of rinderpest virus in goats, pigs and hamsters. *Bull. Nippon Inst. Bio. Sci. Tokyo* **2**, 1–11.
Neale, A. (1831). *Researches to Establish the Truth of the Linnaean Doctrine of Animate Contagions.* Longman, London.

Needham (1772). Mémoire sur la maladie des bêtes à cornes. *Journal de Physique* **2**, 120.

Nencki, M. (1900). Vaccination contre la peste bovine. *Gazeta-Ldkarska* **1900**, 711–934, 1150.

Nencki, M., Sieber, N., and Wyznikiewicz, W. (1898). Untersuchung über die Rinderpest. *Centralblatt Bakteriol. Parasitenkunde Infektionskrankheiten* **23**, 529–538.

Nencki, M., Sieber, N., and Wyznikiewicz, W. (1899). Recherches sur la peste bovine. *Arch. Sc. Biol. Saint-Petersburg* 303.

Neumann, A. H. (1898). *Elephant-hunting in East Equatorial Africa. Being an Account of Three Years' Ivory-hunting under Mount Kenya and among the Ndorobo Savages of the Lorogi Mountains, Including a Trip to the North End of Lake Rudolph*. Rowland Ward Ltd, London.

Newby, T. (1828). *Beta Depicta, or, Remarks on Mangel Wurzel*. Simpkin, Marshall & Co., London.

Newland, R. N. (1908). Review of the cattle trade in B.E.A. 1904–1908. *Agricult J. Br East Afr*. **1**, 264–258.

Newton, F. J. (1896). In *Correspondence Relating to the Outbreak of Rinderpest in South Africa in March 1896. Presented to Both Houses of Parliament by Command of Her Majesty. July 1896. C.8141*. HMSO, London.

Newton, F. J. (1898). *Colonial Reports—Annual. No. 226. Bechuanaland Protectorate. Annual Reports for 1896–7*. HMSO, London.

Nicolle, M. and Adil-Bey (1899). Etudes sur la peste bovine. *Ann. Inst. Past.* **13**, 319–336.

Nocard, E. and Leclainche, E. (1896). *Les maladies microbiennes des animaux*, 2 vols. Masson et Cie, Paris.

Nozeman, C., Kool, A. R., and Tak, J. (1755). *Eerste proefneeming over de uitwerkingen van de inentinge der besmettende ziekte in het rundvee, Gedaan in de Beverwijk*. K. van der Sys & K. de Veer, Amsterdam.

Nugent, T. (1768). *Travels through Germany*, 2 vols. Dilly, London.

Nugent, T. (1781). *Reisen durch Deutschland und vorzüglich durch Mecklenburg (ausl dem Englischen übersetzt)*. Berlin u. Stettin.

Ofcansky, T. P. (1981). The 1889–97 rinderpest epidemic and the rise of British and German colonialism in eastern and southern Africa. *J. Afr. Stud.* **8**, 31–38.

Oliphant, H. L. (1870). In Hallen, J. H. B., Charles, J. G., Jan, M. H., Kerr, H. C., and McLeod, K. (eds.) 1871. *Report of the Commissioners Appointed to Inquire into the Origin, Nature, Etc., of Indian Cattle Plagues. With Appendices*. Government Printer, Calcutta.

Ono, S. and Kondo, S. (1923). Studies on rinderpest in the deer (*Cervus sika*) and changes in the blood of infected animals. *J. Japan. Soc. Vet. Sci.* **2**, 158–161.

"O. O." (1755). Success of inoculating cattle in Holland. *Gent. Mag.* **25**, 160–161.

Organ, G. E. and Fosbrooke, H. A. (1963). *Ngorongoro's First Visitor. Being an Annotated and Illustrated Translation from Dr. O. Baumann's Durch Masailand zur Nilquelle*. East African Literature Bureau, Dar es Salaam.

Osmer, W. (1759). *A Treatise on the Diseases and Lameness of Horses*. T. Waller, London.

Ostertag, von R. (1916). Ueber Rinderpest. Ein beitrag zum Stande und zur Bekämpfung der Tierseuchen in Deutsch-Ostafrika. *Zeitschrift. Infektionsk. Parasit. Krankh. Hyg. Haustiere* **18**, 1–48.

Owaga, M. (1985). Observations on the efficacy of buffalo urine as a potent olfactory attractant for *Glossina pallidipes* Austen. *Insect Sci. Appl.* **6**, 561–566.

Oxley, T. (1841). *Trans. Med. Phys. Soc. Calcutta* **8**, 328.

Palmer, C. (1865). In Hallen, J. H. B., Charles, J. G., Jan, M. H., Kerr, H. C., and McLeod, K. (eds.) 1871. *Report of the Commissioners Appointed to Inquire into the Origin, Nature, Etc., of Indian Cattle Plagues. With Appendices*. Government Printer, Calcutta.

Pankhurst, R. (1966). The great Ethiopian famine of 1888–1892: a new assessment. *J. Hist. Med. Allied Sci.* **21**, 95–124, 271–294.

Pankhurst, R. (1968). *Economic History of Ethiopia 1800–1935*. Haile Selassie I University Press, Addis Ababa.

Pankhurst, R. (n.d.) [1985]. *The History of Famine and Epidemics in Ethiopia Prior to the Twentieth Century*. Relief and Rehabilitation Commission, Addis Ababa.

Pankhurst, R. and Johnson, D. H. (1988). The great drought and famine of 1888–92 in northeast Africa. In Johnson, D. H. and Anderson, D. M. (eds.) *The Ecology of Survival Case Studies from Northeast African History*. Lester Crook Academic Publishing, London, pp. 47–70.

Parke, T. H. (1891). *My Personal Experiences in Equatorial Africa*. Sampson Low, Marston & Co., London.

Parkin, J. (1865). The cattle plague and its origin. *Farmer's Mag.* **52**, 374–376.

Parsons, J. (1746). A letter from James Parsons, M. D. F.R.S. to the President; serving to introduce a Remark from John Milner, Esq; concerning the burying of the Cows, dead of the present reigning Distemper, in Lime or not. *Phil. Trans. R. Soc.* **44**, 224–225.

Parsons, S. (1866). What is the cattle plague? *Farmer's Mag.* **29**, 77.

Patil, P. K. Mondal, B., Goel, A. C., and Sharma, B. (1997). Potency testing of tissue culture rinderpest vaccine in rabbits. *Vaccine* **15**, 608–611.

Patterson, J. H. (1909). *In the Grip of the Nyika. Further Adventures in British East Africa*. Macmillan & Co. Ltd, London.

Pattison, I. (1990). *A Great British Veterinarian Forgotten: James Beart Simonds 1810–1904*. J. A. Allen & Co., London.

Paulet, M. (1775). *Recherches Historiques et Physiques sur les Maladies Epizootiques, Avec les Moyens d'y Remédier, dans tous les cas*, 2 vols. Ruault, Paris.

Peacock, E. H. (1933). *A Game-Book for Burma & Adjoining Territories*. H. F. & G. Witherby, London.

Pearsall, W. H. (1954). Biology and land-use in East Africa. *New Biol.* **17**, 9–26.

Pease, H. T. (1894). *Rinderpest-Cattle Plague. Civil Veterinary Department Ledger Series No. 1*. Office of the Superintendent of Government Printing, Calcutta.

Pécaud, G. (1924). Contribution à l'étude de la pathologie vétérinaire de la Colonie du Tchad. *Bull. Soc. Pathol. Exot.* **17**, 196–207.

Pelling, M. (1978). *Cholera, Fever, and English Medicine, 1825–1865*. Oxford University Press, Oxford.

Penning, C. A. (1894). Runderpest epizoötisch heerschende onder der varkens. *Veeartsenijkundige Bladen voor Nederlansche Indië*.

Percival, A. B. (1918). Game and disease. *J. East Afr. Uganda Nat. Hist. Soc.* **13**, 302–315.

Percival, A. B. (1924). *A Game Ranger's Note Book*. Nisbet & Co. Ltd, London.

Percival, A. B. (1928). *A Game Ranger on Safari*. Nisbet & Co. Ltd, London.

Perham, M. and Bull, M. (1959). *The Diaries of Lord Lugard*, 3 vols. Faber & Faber, London.

Perkin, S. (1995). Multiple land use in the Serengeti region: The Ngorongoro Conservation Area. In Sinclair, A. R. E. and Arcese, P. (eds.) *Serengeti II. Dynamics, Management, and Conservation of an Ecosystem*. University of Chicago Press, Chicago, pp. 571–587.

Perks, W. (1869). In Hallen, J. H. B., Charles, J. G., Jan, M. H., Kerr, H. C., and McLeod, K. (eds.) 1871. *Report of the Commissioners Appointed to Inquire into the Origin, Nature, Etc., of Indian Cattle Plagues. With Appendices*. Government Printer, Calcutta.

Pessina, I. J. (1808). *Anleitung zur Heilung der Viehpest mit der eisenhaltigen Salzsäure*.

P. H. F. (1866). In Simonds, J. B. (ed.) The Cattle-Plague. *J. R. Agricult. Soc. Engl.* **2**, 278.

"A Physician" [Dr. Noah] (n.d.) [1747]. *An Essay in order to fix the Nature and Cure of the Distemper now raging among the Horned Cattle*. J. Rose, Lincoln, NE.

"A Physician" (1749). *Considerations Concerning the Distemper Which Still Spreads Itself Among the Horned Cattle in This Kingdom*. J. & P. Knapton, London.

"A Physician" (1751). *A Treatise on the Plague and Pestilential Fevers, with Some Useful Hints for the Better Prevention and Cure, Together with Some Observations on the Pestilential Fever Now Raging Among the Horned Cattle*. Robinson, Edinburgh.

"Philo-Boas" (1750). The case of killing infected beasts consider'd. *Gent. Mag.* **20**, 489–492.

Phoofolo, P. (1993). Epidemics and revolutions: the rinderpest epidemic in late nineteenth-century southern Africa. *Past and Present* **138**, 112–143.

Piddington, H. (1836). An account of the *bosonto* which attacks the cattle of Bengal. *J. Agricult. Hort. Soc.* **3**, 126.

Pinto, S. (1881). *How I Crossed Africa: from the Atlantic to the Indian Ocean, Through Unknown Countries; Discovery of the Great Zambesi Affluents, &c*, 2 vols. Sampson Low, Marston, Searle, & Rivington, London.

Piot-Bey, M. (1922). A propos des vaccinations antipestiques par le procédé Gordziakowski. *Rec. Méd. Vét.* **98**, 194–197.

Pitman, C. R. S. (1931). *A Game Warden Among His Charges*. Nisbet & Co. Ltd, London.

Pitman, C. R. S. (1942). *A Game Warden Takes Stock*. James Nisbet & Co. Ltd, London.

Playfair, L. (1865). The cattle plague. *North Br. Rev.* **43**, 518.

Playfair, L. (1866). *The Cattle Plague in its Relation to Past Epidemics and to the Present Attack*. Edmonston & Douglas, Edinburgh.

Plenciz, M. A. (1762). *Opera medico physica*, 2 vols. J. I. Trattner, Vienna.

Plowright, W. (1963a). The role of game animals in the epizootiology of rinderpest and malignant catarrhal fever in East Africa. *Bull. Epiz. Dis. Afr.* **11**, 149–162.

Plowright, W. (1963b). Some properties of strains of rinderpest virus recently isolated in E. Africa. *Res. Vet. Sci.* **4**, 96–108.

Plowright, W. (1968). Rinderpest virus. *Virol. Monogr.* **3**, 25–110.

Plowright, W. (1982). The effects of rinderpest and rinderpest control on wildlife in Africa. *Symp. Zool. Soc. Lond.* **50**, 1–28.

Plowright, W. (1985). La peste bovine aujord'hui dans le monde Contrôle et possibilité d'éradication par la vaccination. *Ann. Méd. Vét.* **129**, 9–32.

Plowright, W., Laws, R. M., and Rampton, C. S. (1964). Serological evidence for the susceptibility of the hippopotamus (*Hippopotamus amphibius* Linnaeus) to natural infection with rinderpest virus. *J. Hyg. Camb.* **62**, 329–336.

Plowright, W. and McCulloch, B. (1967). Investigations on the incidence of rinderpest virus infection in game animals of N. Tanganyika and S. Kenya 1960/63. *J. Hyg. Camb.* **65**, 343–358.

Poel, J. M. G. (1949). *Heren en Boreen*. Veenman, Wageningen.

Polding, J. B., Simpson, R. M., and Scott, G. R. (1959). Links between Canine Distemper and Rinderpest. *Vet. Rec.* **71**, 643–644.

Pollock, A. J. O. (1894). *Sporting Days in Southern India*. Horace Cox, London.

Pollok, F. T. and Thom, W. S. (1900). *Wild Sports of Burma and Assam*. Hurst & Blackett, Ltd, London.

Portal, G. (1894). *The British Mission to Uganda in 1893*. Edward Arnold, London.

Postlethwaite, J. R. P. (1947). *I Look Back*. T.V. Boardman & Co. Ltd, London.

Poulton, W. F. (1927). *Game Preservation and Economic Development*. Government Printer, Kampala.

Powell-Cotton, P. H. G. (1902). *A Sporting Trip Through Abyssinia*. Rowland Ward, Ltd, London.

Powell-Cotton, P. H. G. (1904). *In Unknown Africa. A Narrative of Twenty Months' Travel and Sport in Unknown Lands and Among New Tribes*. Hurst & Blackett Ltd, London.

Pringle, J. (1722). *A Rational Enquiry into the Nature of the Plague: Drawn from Historical Remarks on Those That Have Already Happen'd*. J. Peele, London.

Prior, T. (1746). *An Authentick Narrative of the Success of Tar-Water in Curing a Great Number and Variety of Distempers; with Remarks and Occasional Papers Relative to the Subject*. W. Innys, C. Hitch, and M. Cooper, in Paternoster-Row; and C. Davis, in Holborn, London.

Pritchard, R. (1872). In Williams, A. (ed.) 1873. *Report of the Veterinary Department for the Year 1872 with an Appendix*, xii–xiii. HMSO, London.

Prosser, S. A. (1872). *The Days of the Cattle Plague*. The Religious Tract Society, London.

Provost, A. (1980). Queries about rinderpest in African wild animals. In Karstad, L., Nestel, B., and Graham, M. (eds.) *Wildlife Disease Research and Economic Development Proceedings of a Workshop Held in Kabete, Kenya, September 8 and 9, 1980* IDRC, Ottawa, pp. 19–20.

Provost, A., Villemot, J. M., and Queval, R. (1959). Recherches immunologiques sur la péripneumonie VII. Immunisation au moyen d'une souche avianisée de *Mycoplasma mycoides* var *mycoides* inoculée par la voie du mufle. *Rev. Elev. Méd. Vét. Pays Trop.* **12**, 381–403.

Prüsek, J. (1966). The steppe zone in the period of early Nomads and China of the 9th–7th Centuries B.C. *Diogenes* **54**, 23–46.

Purves, H. R. (1868). In Hallen, J. H. B., Charles, J. G., Jan, M. H., Kerr, H. C., and McLeod, K. (eds.) 1871. *Report of the Commissioners Appointed to Inquire into the Origin, Nature, Etc., of Indian Cattle Plagues. With Appendices*. Government Printer, Calcutta.

Puxley, F. L. (1929). *In African Game Tracks. Wanderings with a Rifle Through Eastern Africa*. H. F. & G. Witherby, London.

Quain, R. (1866). 4. Inoculation by cattle-plague poison producing a vaccine-like vesicle. *Trans. Pathol. Soc. Lond.* **17**, 449–452.

Rad-Pezechk, A. (1935). *L'état actuel de nos connaissances sur la peste bovine avec considérations particulières sur la peste bovine en Iran*. Thése Doctorat, Ecole Nationale Vétérinaire d'Alfort, Paris.

Rainey, H. J. (1869a). Letter. *Englishman* May.

Rainey, H. J. (1869b) Letter. *Englishman* September.

Rainey, H. J. (1870). In Hallen, J. H. B., Charles, J. G., Jan, M. H., Kerr, H. C., and McLeod, K. (eds.) 1871. *Report of the Commissioners Appointed to Inquire into the Origin, Nature, Etc., of Indian Cattle Plagues. With Appendices*. Government Printer, Calcutta.

Rainsford, W. S. (1909). *The Land of the Lion*. William Heinemann, London.

Rajgopalan, V. R. (1937). Is there a relationship between viruses of rinderpest and Doyle's disease. *Indian Vet. J.* **7**, 59–64.

Ramani, K., Charles, Y. S., and Ramachandran, S. (1974). Further studies on rinderpest virus of sheep origin. *Indian Vet. J.* **51**, 129–138.

Ramazzini, B. (1711). *De Contagiosa Epidemia Quæ in Patavino Agro, & tota ferè Veneta Ditione in Boves irrepsit*. G. B. Conzatti, Padua.

Ranger, T. O. (1967). *Revolt in Southern Rhodesia 1896–7. A Study in African Resistance*. Heinemann, London.

Ravitsch, J. (1863). Neue Untersuchungen über die pathologische Anatomie der Rinderpest. *Mag. Ges. Thierheilk.* **29**, 313–356.

Ravitsch, J. (1864). *Neue Untersuchungen über die pathologische Anatomie der Rinderpest*. August Hirschwald, Berlin.

Rea, G. (1877). In *Report from the Select Committee on Cattle Plague and Importation of Live Stock*. House of Commons, London.

Read, C. S. (1877). In *Report from the Select Committee on Cattle Plague and Importation of Live Stock*. House of Commons, London.

Redi, F. (1687). *Opusculorum pars prior, sive experimenta circa generationem insectorum ad illustrem virum Carolum Dati*. H. Wetstenium, Amstelaedami.

Reed, H. (1866). The cattle plague. *J. R. Agricult. Soc. Engl. 2nd Ser* **2**, 230–269.

Reeygasse (1889). *Les Missions Catholiques* **21**, 100.

Réfik-Bey and Réfik-Bey. (1899). La peste bovine en Turquie. *Ann. Inst. Past.* July, 596–611.

Reich, G. C. (1797). *Richtige und gewissenhafte Belehrung für den Landmann über die Rindviehseuche und die Inoculation derselben*. Raspesche Buchhandlung, Nürnburg.

Reid, A. G. (1899). *The Annals of Auchterarder and Memorials of Strathearn*. David Philips, Crieff.

Reid, E. T. M. (1954). Observations on feeding habits of adult *Acryophora. Proc. R. Entomol. Soc. Lond.* **23**, 200–204.

Reid, N. R. (1949). *In Report of the Department of Veterinary Science and Animal Husbandry Tanganyika 1948*. Government Printer, Dar es Salaam.

Reinders, G. (1774). *Brief aan den wel Edelen Hooggeleerden Heere Petrus Camper en aan… Wyn. Munnicks weegens eene voorname ontdekkinge door inentinge op het rundvee gedaan*. L. Huisingh, Gröningen.

Reinders, G. (1776). *Waarneemingen en proeven meest door inëntinge op het rundvee gedaan, dienende ten bewijze, dat wij onze kalvers van gebeterde koejen geboren, door inëntinge tegen de veepest kunnen beveiligen*. L. Huisingh, Gröningen.

Reinders, G. (1777). *Bericht uit Holland, Friesland en de provintie van Stad en Lande, wegens de inenting der kalveren van gebeterde koejen voor den jaare 1776*. L. Huisingh, Gröningen.

Reinders, G. (1778). *Conditien op welke de Inëntings Societeit, te garnwert opgeregt, presenteert, kalveren welke van de besmettelijke veeziekte zijn gebeterd, te leveren, tegens ongebeterde kalvers*. June 16. Garnwert.

Renault, E. (1856). *Typhus Contagieux du Gros Bétail*. Labe, Paris.

Renault, E. (1860). *Typhus Contagieux des Bêtes Bovines. Examen au point de vue de la Police Sanitaire Internationale*. Labe, Paris.

Renshaw, G. 1907. *Final Natural History Essays*. Sherratt & Hughes, London.

Reuss, J. J. (1814–1818). *Wesen der Exantheme*, 2 vols. Riegel & Wiessner, Nuremburg.

Reynal, J. (1873). *Traité de Police Sanitaire des Animaux Domestiques*. Asselin, Paris.

Reynier, J. F. (1762). *Le Louvet, Maladie du Bétail; Ses Causes, Ses Remedes: et les moyens de le prévenir*. Marc Chapuis, Lausanne.

Ribbe, J. C. (1814). *Anleitung zur richtigen Erkenntniß der Rinderpest, so wie zur Tilgung dieser verheerenden Seuche*. Andreas Füchsel, Zerbst.

Richardson, U. F. and Carmichael, J. (1926). Epizootic of rinderpest dealt with by local serum production in an outlying district of the Uganda Protectorate. *J. Comp. Pathol. Ther.* **39**, 311–317.

Richter, F. (1900). Einige weitere ethnographische Notizen über den Bezirk Bukoba. *Mitt. Deutschen Schutzgebieten* **13**, 61–75.

Rieck, W. (1930). Zwei Braunschweiger Erlasse gegen die Rinderpest. Anno 1801. *Veterinärhistorische Mitteilungen* **10**(1), 6.

Rieck, W. (1936). Die Rinderpest im Reiche Karls d. Gr. Anno 810. *Veterinärhistorische Mitteilungen* **16**(11), 97–100.

Rivers, J. P. W., Holt, J. F. J, Seaman, J. A., and Bowden, M. R. (1976). Lessons for epidemiology from the Ethiopian famines. *Ann. Soc. Belge. Med. Trop.* **56**, 345–357.

Roberts, G. A. (1921). Rinderpest (peste bovina) in Brazil. *J. Am. Vet. Med. Assoc.* **13**, 177–185.

Robinet, J. (1789). *Manuel du Bouvier, ou Traité de la Médecine Pratique des Bêtes à Cornes. Contenant l'âge de ces animaux, leur choix, avec la manière de les dresser pour le travail, de les conduire & gouverner, &c. &c. &c.*, 2 vols. Paris, Guillot.

Robinson, H. (1896). In *Correspondence Relating to the Outbreak of Rinderpest in South Africa in March 1896. Presented to Both Houses of Parliament by Command of Her Majesty. July 1896*. C.8141. HMSO, London.

Robson, J., Arnold, R. M., Plowright, W., and Scott, G. R. (1959). The isolation from an eland of a strain of rinderpest virus attenuated for cattle. *Bull. Epiz. Dis. Afr.* **7**, 97–102.

Roby, M. (1911). *My Adventures in the Congo*. Edward Arnold, London.

Roeder, P. (2002a). *Managing the Risk of Rinderpest to Afghanistan and Pakistan*. EMPRES Publications, FAO, Rome.

Roeder, P. (2002b). *Cleaning Out Rinderpest's Last Reserves*. EMPRES Publications, FAO, Rome.

Roeder, P. and Reichard, R. (1999). *Report of an OIE Mission to the Russian Federation Concerning the Outbreak of Rinderpest in Amur Region in 1998 Conducted in Collaboration with FAO. Dates of Mission: June 21 to July 2, 1999*, 14 pp. FAO, Rome.

Rogers, L. (1900). *Report on an Experimental Investigation of the Methods of Inoculation Against Rinderpest from the Imperial Bacteriological Laboratory, Muktesar, India, 1900*. Government Printer, Calcutta.

Röll, M. F. (1856). *Lehrbuch der Pathologie und Therapie der Hausthiere*. Wilhelm Braumüller, Vienna.

Romano, T. M. (1997). The cattle plague of 1865 and the reception of "The Germ Theory" in mid-Victorian Britain. *J. Hist. Med.* **52**, 51–80.

Roosevelt, T. (1910). *African Game Trails. An Account of the African Wanderings of an American Hunter-Naturalist*. John Murray, London.

Roscoe, J. (1922). *The Soul of Central Africa. A General Account of the Mackie Ethnologiocal Expedition*. Cassell & Co. Ltd, London.

Rosen, G. (1938). Jacob Henle "On miasmata and contagia" [1840]. *Bull. Hist. Med.* **6**, 907–983.

Ross, E. D. (1933). *Sir Anthony Sherley and his Persian Adventure*. George Routledge & Sons Ltd, London.

Rossiter, P. (1996). Epidemiological and clinical features of rinderpest in the 1990s. In *Proceedings of the FAO Technical Consultation on the Global Rinderpest Eradication Programme, Rome, Italy, July 22–24, 1996*, pp. 67–69.

Rossiter, P. B., Taylor, W. P., Bwangamoi, B., Ngereza, A. R. H., Moorhouse, P. D. S., Haresnape, J. M., Wafula, J. S., Nyange, J. F. C., and Gumm, I. D. (1987). Continuing presence of rinderpest virus as a threat in East Africa, 1983–1985. *Vet. Rec.* **120**, 59–62.

Rowlin, J. (1792). *The Complete Cow-Doctor, or Farmer's Companion*. Robinson, Edinburgh.

Rudkin, T. (1877). In *Report from the Select Committee on Cattle Plague and Importation of Live Stock*. House of Commons, London.

Ruediger, E. H. (1909). Observations on the cattle plague. *The Phillipine J. Sci.* **4**, Ser. B.5, 384–386.

"Rusticus" (1745). On the distemper among the cattle. *Gent. Mag.* **15**, 630–631.

Rweyamamu, M. (1996). The global rinderpest status in 1996. In *Proceedings of the FAO Technical Consultation on the Global Rinderpest Eradication Programme, Rome, Italy, 22–24 July 1996*. FAO, Rome, pp. 41–47.

Said, A. (1931). *La Peste Bovine L'Epidémie de 1925–1927 en Syrie*. Thèse Doctorat, Ecole Nationale Véterinaire d'Alfort, Paris.

Saini, R. K. (1986). Antennal responses of *Glossina morsitans morsitans* to buffalo urine, a potent olfactory attractant of tsetse. *Insect Sci. Applic.* **7**, 771–775.

Saint-Hilaire, M. A. G. (1865). Note sur le typhus contagieux au Jardin d'Acclimatation. *Bull. Soc. Imp. Zool. Acclim., Paris* **2**, 685.

Saker, H. and Aldridge, J. (1971). The origins of the Langeberg rebellion. *J. Afr. Hist.* **12**, 299–317.

Salchow, U. C. (1755). *Untersuchet die Rindviehseuche aus Gründen der Natur, Arzeneilehre, Chemie und Erfahrung, nach allen ihren Umständen, Ursachen und Verhältnissen: und eröfnet solchergestalt derselben gründliche Cur und übrige Verwahrungsmittel*. Haude und Spener, Berlin.

Salchow, U. C. (1780). *Durchseuchungscur*. Georg Ludewig Förster, Bremen.

Salt, H. S. (1921). *Seventy Years Among Savages*. George Allen & Unwin Ltd., London.

Sander, L. (1893). *Deutsches Kolonialblatt* **4**, 543–544.

Sanderson, G. P. (1878). *Thirteen Years Among the Wild Beasts of India*. John Grant, Edinburgh.

Sanderson, J. (1866). *The Cattle Plague. A Letter to the Right Hon. Sir George Grey, Bart., M.P.* J. B. Nichols & Sons, Westminster.

Sanderson, J. B. (1865). In Bernard, M. (ed.) 1865. *First Report of Her Majesty's Commissioners Appointed to Inquire into the Origin and Nature & c. of the Cattle Plague*. HMSO, London.

Sanderson, J. B. (1866a). Nature, propagation, progress and symptoms of the disease. In Bernard, M. (ed.) 1866b. *Third Report of the Commissioners Appointed to Inquire into the Origin and Nature & c. of the Cattle Plague*. HMSO, London.

Sanderson, J. B. (1866b). On the cattle plague in its epidemiological aspects. *Trans. Epid. Soc.* **2**, 463–470.

Sanderson, J. B. (1877). The Cattle-Plague. *Br. Med. J.* (**I**), 491.

Saunders, J. J. (1971). *The History of the Mongol Conquests*. Routledge & Kegan Paul, London.

Saunders, L. Z. (1980). *Veterinary Pathology in Russia, 1860–1930*. Cornell University Press, Ithaca, NY.

Sauvages, M. de. (1746). *Memoire sur la Maladie des Boeufs du Vivarais*. Jean Martel, Montpellier.

Scarfoglio, E. (1936). *Abissinie*. Edizione Roma, Rome.

Schafer, E. H. (1950). The Camel in China down to the Mongol Dynasty. *Sinologiya* **2**, 165–194.

Sharpe, A. H. (1893a). Disease among the wild game of Central Africa. *The Field* April 1, **2101**, 471.

Sharpe, A. H. (1893b). A journey from the Shire River to Lake Mweru and the Upper Luapula. *Geogr. J.* **1**, 524–533.

Schein, H. (1922). Note sur la peste bovine. *Rec. Méd. Vét.* **98**, 621–632.

Schein, H. and Jacotot, H. (1925). Expériences de conservation du virus pestique dans les peaux vertes. *Ann. Inst. Pasteur* May.

Schele, F. F. von. (1894). Ueber den werth und die Entwickelungsfähigkeit der Kolonie Deutsch-Ostafrika. *Deutsches Kolonialblatt* 230–232.

Schern, K., Mavrides, N., and Major, A. (1919). Ueber Rinderpest. (Dritte Mitteilung). Ueber Resistenz und Inkubation bei der experimentellen Rinderpest anatolischer und podolischer Rinder. *Zeitschrift Infektionskr. Haustiere* **20**, 117–121.

Schillings, C. G. (1906). *With Flashlight and Rifle. A Record of Hunting Adventures and of Studies in Wild Life in Equatorial East Africa*. Hutchinson & Co., London.

Schroëckius (1713). In Sydenham, T. (ed.) *Constitutio Epidemica Hassiaca. Ephemera Naturae Curiosa*, Appendix, 2 vols. Geneva.

Schulze (1763). Schreiben an einen Freund, wegen der seit einiger Zeit unter dem Hornviehe beobachteten Seuche. *Dresdnisches Magazin, oder Ausarbeitungen und Nachrichten, zum Behuf der Naturlehre, der Arzneikunst, der Sitten und der schönen Wissenschaften.* **2**(6), 337–346.

Schumacher, W. (1779). *Die sichersten Mittel wider die Gefahr beym Eintritte der Rindviehseuche aus erfahrungen und Urkunden bestätiget 1777*. Oemigke, Berlin.

Schumacher, W. (1793). *Die sichersten Mittel wider die Gefahr beym Eintritte der Rindviehseuche aus erfahrungen und Urkunden bestätiget 1777*, 2nd edn. Oemigke, Berlin.

Schutt, G. J., Römer, E. J., Lanckom, P. van, and Greeve, G. (1770). *Brief aan den Hoog Geleerden Heere Petrus Camper behelzende eenige proeven omtrent de inëntinge der besmettelijke veeziekte, genomen in de Vrijheid van Utrecht*. G. B. Timon van Paddenburg, Utrecht.

Schwabe, C. W. (1978). *Cattle, Priests, and Progress in Medicine*. University of Minnesota Press, Minneapolis, MN.

Schwenke, M. W. (1755). Letter on the subject of inoculating horned cattle. *Gent. Mag.* **25**, 464.

Schwenke, M. W. (1757). Proeven op het vee door inenting. *Bremisches Mag. Ausbreitung Wissensch.* **1**(47), 406.

Scott, G. R. (1955). Life expectancy of rinderpest virus. *Bull. Epiz. Dis. Afr.* **3**, 19.

Scott, G. R. (1970). In Davis, J. W., Karstad, L. H., and Trainer, D. O. (eds.) *Infectious Diseases of Wild Mammals*. The Iowa State University Press, Iowa, pp. 20–35.

Scott, G. R. (1993). Pan-African Rinderpest Campaigns. *Joint Conference on African Boundaries and Borderlands*. Edinburgh, pp. 1–12.

Scott, G. R. (1996). The history of rinderpest in Britain: 809–1799. *State Vet. J.* **6**, 8–10.

Scott, G. R. and Brown, R. D. (1958). A neutralization test for the detection of rinderpest antibodies. *J. Comp. Pathol. Ther.* **68**, 308–314.

Scott, G. R. and Brown, R. D. (1961). Rinderpest diagnosis with special reference to the agar gel double diffusion test. *Bull. Epiz. Dis. Afr.* **9**, 83–125.

Scott, G. R., Cowan, K. M., and Elliott, R. T. (1960). Rinderpest in Impala. *Vet. Rec.* **72**, 787.

Scott, G. R. and Macdonald, J. (1962). Kenya camels and rinderpest. *Bull. Epiz. Dis. Afr.* **10**, 495–497.

Scott, G. R., Taylor, W. P., and Rossiter, P. B. (1986). *Manual on the Diagnosis of Rinderpest*. FAO, Rome.

Scott, G. R. and Witcomb, M. A. (1958). Rinderpest virus in laboratory animals. *Report E. Afr. Vet. Res. Org. Cong. 1956–57*. Nairobi, pp. 15–17.

Scott-Elliot, G. F. (1896). *A Naturalist in Mid-Africa. Being an Account of a Journey to the Mountains of the Moon and Tanganyika*. A. D. Inness & Co., London.

Scully, W. C. (1899). The Prayer of the Cattle Smitten with Rinderpest. *The Pall Mall Magazine*. January–April, 433–435.

Scully, W. C. (1913). *Further Reminiscences of a South African Pioneer*. Fisher Unwin, London.

Seaman, I. (1866). Payment of cattle inspectors. *The Veterinarian* **34**, 1000–1002.

Sée, H. (1927). *Economic and Social Conditions in France During the Eighteenth Century*. Trans. E. H. Zeydel. F. S. Crofts & Co. Inc., New York.

Seifman, (1865). Memorandum upon the Cattle Disease. In *Further Papers Relating to the Cattle Plague in Poland 1857–65*. HMSO, London.

Seligman, C. G. and Seligman, B. Z. (1932). *Pagan Tribes of the Nilotic Sudan*. George Routledge & Sons Ltd, London.

Selbourne, Earl of (1897). *Public Record Office. C.O., 417/224, Minutes, February 26, 1897, on Rosemead to Chamberlain, January 28, 1897.*

Selous, F. C. (1896). *Sunshine and Storm in Rhodesia. Being a Narrative of Events in Matabeleland Both Before and During the Recent Native Insurrection up to the Date of the Disbandment of the Bulawayo Field Force*. Rowland Ward & Co. Ltd., London.

Selous, F. C. (1908). The tsetse fly in the Eastern Transvaal. Letters. *The Field* Ltd., **3**, 22.

Semmer, E. (1875). *Uber die Pathologische Anatomie der Rinderpest*. C. Mattiesen, Dorpat.

Semmer, E. (1893). Uber das Rinderpestcontagium und uber Immuniserung und Schutzimrfung gegen Rinderpest. *Berliner Tierärztliche Wochenschr*. **23**, 590–591.

Semmer, E. and Archangelski, A. (1883). Ueber das Rinderpestcontagium und dessen Mitigation. *Centralb. Med. Wissensch*. **21**, 306–307.

Serjeant, E. M. (2001). Your Times. March 9, 16. *The Oxford Times*.

Seshadri, B. (1969). *The Twilight of India's Wild Life*. John Baker, London.

Settegast, A. F. (1774). *Epitome Historica Febrium Putridarum Anno 1754 Nascentium in 1773. Preseverantium, Geneses, Mutationes, Causas, Dia et Pro Gnoses Nec Non Causarum Indoli, Vi, Moribus Oppositas Prophylaxis et Therapiae Leges Exhibens Ex ipsis causarum visceribus, mutationibus, duratione, successione, compliactione, multitudine, magnitudine, numero, tendentiis, & conspirationibus fideliter congesta, & sincere digesta*. Ferdinandi Rommerskirchen, Bonn.

"Several Hands" (1747). *A General System, or Summary Method, for Curing the Epidemical Distemper Which Rages Amongst the Horned Cattle. Calculated For all the Different Stages and Degrees of it etc*. M. Cooper in Pater-noster-Row, London.

Shaila, M. S., Shamaki, D., Forsyth, M., Diallo, A., Goatley, L. Kitching, P., and Barrett, T. (1996). Geographic distribution and epidemiology of *peste des petits ruminants* viruses. *Virus Res.* **43**, 149–153.

Sharpe, A. (1893a). A journey from the Shire River to Lake Mweru and the Upper Luapula. *Geog. J.* **1**, 524–533.

Sharpe, A. (1893b) Disease among the wild game of Central Africa. Letters. *The Field* **2101**, 471.

Sharpe, G. C. H. (1916). Rinderpest observation in German East Africa. *Report to Chief Veterinary Surgeon, Salisbury, Southern Rhodesia, from Neu Langenburg. July 25.* MS. Colonial Office.

Shillington, K. (1985). *The Colonisation of the Southern Tswana 1870–1900.* Ravan Press, Johannesburg.

Shilston, A. W. (1917). Vitalité du virus pesteux en dehors du corps de l'animal dans les conditions naturelles, Calcutta. *An. Rec. Vét. Alfort.* Nov. **1918**, 264.

Sick (1812). In Hering, K. L. (ed.) *Ueber die Rinderpest und deren Tilgung, besonders in Anwendung auf das Viehsterben zu Pyritz in Pommern im Jahre 1808, und die darüber erschienene Abhandlung des Dr. Roserus. Nach Grundsätzen des Professors Sick.* Kommission bei Saalfeldt, Berlin.

Sigl, A. (1892). Bericht über den Handelsverkehr von Tabora. *Deutsches Kolonialblatt* **3**, 164–166.

Simmons, R. J. (1957). In retrospect. *Bull. Epiz. Dis. Afr.* **5**, 359–364.

Simon, G. (1885). *L'Éthiopie, ses Moeurs, ses Traditions, le Négouss Iohannès, les Églises Monolithes de Lalibéla.* Callamel Ainé, Paris.

Simonds, J. B. (1857). Report on steppe murrain or rinderpest. *J. R. Agricult. Soc. Engl.* **18**, 197–270.

Simonds, J. B. (1865). In Bernard, M. (ed.) (1865). *First Report of Her Majesty's Commissioners Appointed to Inquire into the Origin and Nature & c. of the Cattle Plague.* HMSO, London.

Simonds, J. B. (1866). The cattle-plague. *J. R. Agricult. Soc. Engl.* **2**, 270–286.

Simonds, J. B. (1877). In *Report from the Select Committee on Cattle Plague and Importation of Live Stock.* House of Commons, London.

Simonsen, B. T., Siegismund, H. R., and Arctander, P. (1998). Population structure of African buffalo inferred from mtDNA sequences and microsatellite loci: high variation but low differentiation. *Mol. Ecol.* **7**, 225–237.

Sinclair, A. R. E. (1979). In Sinclair, A. R. E. and Norton-Griffiths, M. (eds.) *Serengeti. Dynamics of an Ecosystem.* The University of Chicago Press, Chicago, pp. 82–103.

Sinclair, A. R. E. (1995). Serengeti past and present. In Sinclair, A. R. E. and Arcese, P. (eds.) *Serengeti II. Dynamics, Management, and Conservation of an Ecosystem.* University of Chicago Press, Chicago, pp. 3–30.

Singh, K. V. and Ata, F. (1967). Experimental rinderpest in camels. A preliminary report. *Bull. Epizoot. Dis. Afr.* **15**, 19–23.

Sinor, D. (1990). *The Cambridge History of Early Inner Asia.* Cambridge University Press, Cambridge.

Skinner, R. P. (1906). *Abyssinia of To-Day An Account of the First Mission Sent by the American Government to the Court of the King of Kings (1903–1904).* Edward Arnold, London.

Slare, F. (1682). An Abstract of a Letter from Dr. Wincler chief Physitian of the Prince Palatine, Dat. Dec. 22. 1682, to Dr. Fred. Slare Fellow of the Royal Society, containing an account of a Murren in Switzerland, and the Method of its cure. *Phil. Trans. R. Soc.* **5**, 93–94.

Slare, F. (1683). A further confirmation of the above mentioned Contagion, of its nature, and manner of spreading, by way of Postscript from the ingenious Fred. Slare. M.D. and F.R.S. Dat. March 27. 1683. *Phil. Trans. R. Soc.* **5**, 94–95.

Slatin, R. C. (1895). *Fire and Sword in the Sudan. A Personal Narrative of Fighting and Serving the Dervishes. 1879–1895.* Edward Arnold, London.

Smith, A. (1866). In Bernard, M. (ed.) 1866b. *Third Report of the Commissioners Appointed to Inquire into the Origin and Nature &c. of the Cattle Plague.* HMSO, London.

Smith, A. B. (1992). *Pastoralism in Africa Origins and Development Ecology.* Hurst & Co., London.

Smith, B. F. (1866). *Prayer and the Cattle Plague: Two Sermons. London and Cambridge.*

Smith, F. (1933). *The Early History of Veterinary Literature.* Baillière Tindall & Cox, London. Reprinted 1976 J. A. Allen & Co., London.

Smith, J. (1941). *Report by Mr. J. Smith, O.B.E, M.R.C.V.S., D.V.H., Adviser on Animal Health to the Secretary of State for the Colonies on his Visit to East Africa April–July, 1940. C.A.C. 558.* Colonial Office, London.

Smith, W. (1868). The cattle plague in Norfolk. *The Veterinarian* **41**, 776–780.

Smith, W. D. (1994). *Hippocrates Volume VII.* Harvard University Press, London.

So, J. F. and Bunker, E. C. (1995). *Traders and Raiders on China's Northern Frontier.* University of Washington Press, Seattle, WA.

Sobania, N. (1988). Pastoralist migration and colonial policy: a case study from northern Kenya. In Johnson, D. H. and Anderson, D. M. (eds.) *The Ecology of Survival Case Studies from Northeast African History.* Lester Crook Academic Publishing, London, pp. 219–239.

Soga, A. K. (1896). In *Rinderpest—Report of visit of Native Representatives to inquire into the Disease and its Indications.1897. G.78-'97.* Ministerial Department of the Prime Minister, Cape of Good Hope.

Sokolov, V. and Luschekina, A. (1997). *Procapra gutturosa. Mamm. Species* **571**, 1–5.

Solotowsky, A. L. (1856). *The Significance of Rinderpest Inoculations in the Practice of Veterinary Medicine.* In Russian.

Somerville, M. (1848). *Physical Geography*. Chapman & Hall, London.

Speke, J. H. (1863). *Journal of the Discovery of the Source of the Nile*. William Blackwood & Sons, Edinburgh.

Spinage, C. A. (1992). The decline of the Kalahari wildebeest. *Oryx* **26**, 147–150.

Spinage, C. A., Guinness, F., Eltringham, S. K., and Woodford, M. H. (1972). Estimation of large mammal numbers in the Akagera National Park and Mutara Hunting Reserve, Rwanda. *La Terre et la Vie* **26**, 561–570.

Spinage, C. A. and Matlhare, J. M. (1992). Is the Kalahari cornucopia fact or fiction? A predictive model. *J. Appl. Ecol.* **29**, 605–610.

Spinola, W. T. J. (1846). *Mittheilungen über die Rinderpest*. Hirschwald, Berlin.

Spruit, J. (1769). Brief van den Heer Joh. Spruit, Heelmeester te Nichtwegt, aan den Schrijver, behelzende eenige waarneemingen over de inenting van de ziekte der runderen. In Sandifort E. (ed.) *Natuur-en Geneeskundige Bibliotheek* **6**, 899. P. Van Cleef, s'Gravenhage.

Stähle, R. (1981). *Verlauf und Verbreitung der Rinderpest im Herzogtum Bremen und Verden von 1715 bis 1781*. Inaugural-Dissertation Doctor Medicinae Veterinariae. Tierärztliche Hochschule Hannover, Hannover.

Stanley, E. (1870). The cattle murrain of India. *The Veterinarian* **43**, 14–28.

Stanley, H. M. (1878). *Through the Dark Continent or The Sources of the Nile Around the Great Lakes of Equatorial Africa and Down the Livingstone River to the Atlantic Ocean*, 2 vols. Sampson Low, Marston, Searle & Rivington Ltd, London.

Stanley, H. M. (1890). *In Darkest Africa or the Quest Rescue and Retreat of Emin Governor of Equatoria*, 2 vols Sampson Low, Marston, Searle and Rivington Ltd, London.

Stasiw, W. (1947). *Nouvelles données sur la vaccination contre la peste bovine*. Thèse doctorat, Ecole Nationale Véterinaire d'Alfort, Paris.

St. Croix, F. W. de. (1945). *The Fulani of Northern Nigeria*. The Government Printer, Lagos.

Stebbing, E. P. (1911). *Jungle By-Ways in India*. John Lane The Bodley Head, London.

Stenning, D. J. (1959). *Savannah Nomads*. Oxford University Press, London.

Stevenson, J. (1853). *The Chronicle of Florence of Worcester*. Seeleys, London.

Stevenson-Hamilton, J. (1907). The tsetse fly in the Eastern Transvaal. Letters. *The Field*. **110**, 1122.

Stevenson-Hamilton, J. (1911). The relation between game and tsetse-flies. *Bull. Ent. Res.* **2**, 113–118.

Stevenson-Hamilton, J. (1929). *The Low-Veld: Its Wild Life and its People*. Cassell & Co., London.

Stevenson-Hamilton, J. (1957). Tsetse fly and the rinderpest epidemic of 1896. *South Afr. J. Sci.* **58**, 216–218.

Stewart, D. (1844). *Report on Small-pox in Calcutta*. Calcutta.

Stigand, C. H. (1910). *To Abyssinia Through an Unknown Land. An Account of a Journey Through Unexplored Regions of British East Africa by Lake Rudolf to the Kingdom of Menelek*. Seeley & Co. Ltd, London.

Still, F. (Trans.) (1866). Regulations concerning cattle disease in Russia and in Bavaria. I. Russia. In Bernard, M. (ed.) 1866b. *Second Report of the Commissioners Appointed to Inquire into the Origin and Nature, etc. of the Cattle Plague*. HMSO, London.

Stolte, J. H. (1777). Schreiben wegen Einimpfung des jungen Rindviehs. Nov. 1. In Schumacher, W. 1777.

Stordy, R. J. (1901). Report on veterinary work in British East Africa and Uganda protectorates for the years 1898–1900. *Diplomatic and Consular Reports. Africa. No. 551 Miscellaneous Series*. HMSO, London.

Stordy, R. J. (1918). *British East Africa. Department of Agriculture. Annual Report of the Veterinary Department for the Year Ending March 31, 1918*. Government Printer, Nairobi, pp. 87–94.

Stuhlmann, Fr. (1892). Rückreise vom Victoria-Nyansa nach Bagamoyo. *Mittheilungen aus dem Deutschen Schutzgebieten* **5**, 185–188.

Stuhlmann, Fr. (1894). *Deutsch-Ostafrika; mit Emin Pascha ins herz von Afrika*. D. Reimer, Berlin.

Summers, M. (ed.) (1928). *Malleus Maleficarum by Jacobus Sprenger and Heinrich Kramer*. John Rodker, London.

Swayne, H. G. C. (1895). *Seventeen Trips Through Somaliland. A Record of Exploration and Big Game Shooting, 1885 to 1893*. Rowland Ward & Co., Ltd, London.

Swayne, R. G. (1866). *The Secret of the Cattle Plague*. London.

Swencke, Dr. (1755). Letter on the subject of inoculating horned cattle. *Gent. Mag.* **25**, 464.

Swynnerton, C. F. M. (1936). The tsetse flies of East Africa. A first study of their ecology, with a view to their control. *Trans. R. Entomol. Soc. Lond.* **84**, 1–579.

Swynnerton, G. M. (1958). Tanganyika. *Annual Report of the Game Department 1957*. Government Printer, Dar es Salaam.

Sykes, C. A. (1903). *Service and Sport on the Tropical Nile. Some Records of the Duties and Diversions of an Officer Among Natives and Big Game During the Re-occupation of the Nilotic Province*. John Murray, London.

Taine, H. A. (1875). *L'Ancien Régime*. Hachette, Paris.

Talbot, L. M. and Talbot, M. H. (1961). Preliminary observations on the population dynamics of wildebeest in Narok District, Kenya. *E. Afr. Agricult. For. J.* **27**, 108–116.

Tarantino, G. B. (1928). Peste bovina. *La Clin. Vet.* **51**, 739–745.

Tartakovskii, M. G. (1899). Experimental findings on the question of the susceptibility of camels to rinderpest. *Arkh. Vet. Nauk.* **29**, 228–254. In Russian.

Taylor, W. P., Plowright, W., Pillenger, R., Rampton, C. S., and Staple, R. F. (1965). Studies on the pathogenesis of rinderpest in experimental cattle. IV. Proliferation of the virus following contact infection. *J. Hyg. Camb.* **63**, 497–506.

Taylor, W. P. and Watson, R. M. (1967). Studies on the epizootiology of rinderpest in blue wildebeest and other game species of Northern Tanzania and Southern Kenya, 1965–67. *J. Hyg. Camb.* **65**, 537–545.

Teare, S. P. (1935). *Report of the Game Preservation Department Tanganyika 1934*. Government Printer, Dar es Salaam.

Teare, S. P. (1936). *Report of the Game Preservation Department Tanganyika 1935*. Government Printer, Dar es Salaam.

Ternan, T. (1930). *Some Experiences of an Old Bromsgrovian: Soldiering in Afghanistan, Egypt and Uganda*. Cornish Brothers Ltd., Birmingham.

Textor, J. N. (1739). *Vernunst und erfahrungsmäßige Versuche, wie die giftig ansteckenden Viehseuchen, unter dem Hornviehe und Pferden wohl erkennet, präserviret und curiret werden können. Nebst einer kurzen Abhandlung von Pferde und andern Viehkrankheiten*. Andreas Jacob Maschenbauer, Karlsrühe.

Thal, J. A. (1972). *Les maladies similaires à la peste bovine, études et lutte, Ndélé, République Centrafricaine. Enquête sur la peste bovine et les maladies similaires*. FAO, Rome.

Theiler, A. (1897a). Experimentaluntersuchungen über Rinderpest. *Schw. Arch. Tierheilk.* **39**, 193–213.

Theiler, A. (1897b). Rinderpest in Süd-Afrika. *Schw. Arch. Tierheilk.* **39**, 49–62.

Theiler, A. (1898). Blutserum immuner Thiere in Kampfe gegen die Rinderpest. *Deutsche Tierarz. Wochensch.* **6**, 205–208.

Theves, G. (1994). De la "maladie des bêtes à cornes," au Duché de Luxembourg pendant le XVIIIe siècle Traitement et prophylaxie. *Ann. Méd. Vét.* **138**, 81–88.

Thirsk, J. (ed.) (1985). *The Agrarian History of England and Wales 1640–1750. Vol. 5 (2)*. Cambridge University Press, Cambridge.

Thomas, A. D. and Reid, N. R. (1944). Rinderpest in game. A description of an outbreak and an attempt at limiting its spread by means of a bush fence. *Onderstepoort J. Vet. Res.* **20**, 7–23.

Thomas, H. B. (1959). Capt. Eric Smith's expedition to Lake Victoria in 1891. *Uganda J.* **23**, 135.

Thomé, M. (1964). *Rapport annuel 1964, Ministère de l'Agriculture et de la Production Animale, République du Tchad.* Direction de l'Elévage, Fasc. **7**, 5–17. Typescript.

Thompson (1866). *Farmer's Mag.* **29**, 55.

Thompson, F. (1896). *The Times*. 9 November.

Thompson, R. (1871). In Hallen, J. H. B., Charles, J. G., Jan, M. H., Kerr, H. C., and McLeod, K. (eds.) 1871. *Report of the Commissioners Appointed to Inquire into the Origin, Nature, Etc., of Indian Cattle Plagues. With Appendices*. Government Printer, Calcutta.

Thompson, R. D'Arcy (1974). *The Remarkable Gamgees. A Story of Achievement*. The Ramsay Head Press, Edinburgh.

Thomson, J. (1827). *The Works of William Cullen, M. D.*, 2 vols. William Blackwood, Edinburgh.

Thomson, J. (1859). *Account of the Life, Lectures, and Writings of William Cullen, M. D.*, 2 vols. William Blackwood & Sons, Edinburgh.

Thomson, J. (1885). *Through Masailand. A Journey of Exploration Among the Snowclad Volcanic Mountains and Strange Tribes of Eastern Equatorial Africa Being a Narrative of the Royal Geographic Society's Expedition to Mount Kenia and Lake Victoria Nyanza, 1883–1884*. Sampson Low, Marston, Searle, & Rivington, London.

Thorley, J. (1878). *Thorley's Illustrated Farmers' Almanack for 1878 and Diary*. Joseph Thorley, London.

Thorpe, L. (1969). *Charlemagne by Notker the Stammerer*. Penguin Books, London.

Thruston, A. B. (1900). *African Incidents. Personal Experiences in Egypt and Unyoro*. John Murray, London.

Thudichum, J. L. W. (1865). *Memorandum on the Principles and Practice of Disinfection, as Applicable to the Present Epidemic of Cattle Disease*. HMSO, London.

Thunberg, C. P. (1795). *Travels in Europe, Africa and Asia, Made Between the Years 1770 and 1779*, 4 vols. F. & C. Rivington, London.

Timoni, E. (1714). An account, or history, of the procuring the smallpox by incision, or inoculation, as it has for some time been practised at Constantinople. *Phil. Trans. R. Soc.* **29**, 72–82.

Tjader, R. (1911). *The Big Game of Africa*. D. Appleton & Co., New York.

Todd, C. (1912). *Description of the Serum Institute (Abbassia) and of the Preparation of Cattle Plague Serum*. Government Printer, Cairo.

Tode, J. C. (1775). *Geschichte der Einimpfungen der Hornviehseuche, welche in den Jahren 1770, 71 und 72 in Dännemark auf königliche Kosten angestellet worden*. Johann Gottlob Rothen, Copenhagen.

Tolberg, J. W. (1808). *Ueber das einzige Mittel bei der sich jetzt nahenden Viehseuche das Rindviehe zu retten*. Wilhelm Heinrichshofen, Magdeburg.

Torday, E. (1913). *Camp and Tramp in the African wilds. A Record of Adventure, Impressions, and Experiences During Many Years Spent Among the Savage Tribes Round Lake Tanganyika and in Central Africa, with a Description of Native Life, Character, and Customs*. Seeley, Service & Co. Ltd., London.

Treves, F. (1910). *Uganda for a Holiday*. Smith, Elder & Co., London.

Trier, Dukedom of (1719). *Verordnung von Anno 1719*. Ehrenbreitstein.

Trier, Dukedom of (1746). *Churfürstlich = Trierische Wasen = Verordnung De 4 Octobris 1746*. Ehrenbreitstein.

T. S. (1754). Cattle inoculated. *Gent. Mag.* **24**, 493.

Tucker, J. (1866). In *Sequel to the Report of the Committee, Convened by the Lord Lieutenenat of Ireland, to Consider the Measures to be Adopted for Arresting the Progress of the Cattle Plague, in Case of its Appearance in Ireland*. Alexander Thom, Dublin.

Tucker, A. R. (1908). *Eighteen Years in Uganda and East Africa*. Edward Arnold, London.

Turner, F. M. (1974). Rainfall, plagues, and the Prince of Wales: A chapter in the conflict of religion and science. *J. Br. Stud.* **13**, 46–65.

Turner, G. and Kolle, W. (1897). *Report to the Agricultural Department, Cape Colony*. Cape of Good Hope, Department of Agriculture, Cape Town.

Turner, G. and Kolle, W. (1898). *Report on the Cure and Prevention of Rinderpest*. Cape of Good Hope, Department of Agriculture, Capetown.

Tyndall, J. (1865). Letters. *Guardian* **20**, 1192.

Tyndall, J. (1866). Letter to A. Stanley, March 13th. In *Typescript of the Journal of John Tyndall, 1855–1872* **3**, 1318. Archives of the Royal Institution, London.

Unterberger, F. (1864). *Einige Worte uber die Heimath der Rinderpest*. St. Petersburg.

Urban, S. (1745). On the distemper among the cattle. *Gent. Mag.* **15**, 630–631.

Vageler, P. (1912). Ugogo. *Beihefte zum Tropenpflanzer* **13**, 1–127.

Vallisnieri, A. (1715). In Lancisi, G. M. (ed.) 1715. *Dissertatio Historica de Bovilla Peste ex Campaniae finibus, anno 1713 Latio importata*. Joannis Mariæ Salvioni, Rome.

Vandeleur, S. (1898). *Campaigning on the Upper Nile and Niger*. Methuen & Co., London.

van Doeveren, W. (1769). Raadevingen om de inentinge der ziekte van't rundvee, op eene gemakkelijke, onkostbare en voordeelige manier te doen. In Sandifort, E. (ed.) *Natuur-en Geneeskundige Bibliotheek* **6**, 436–442. P. Van Cleef, s'Gravenhage.

Vannutelli, L. and Citerni, C. (1899). *L'Omo: Viaggi de l'Esplorazione nell'Africa Orientale*. Socièta geografica italiana, Milan.

Van Onselen, C. (1972). Reactions to Rinderpest in Southern Africa 1896–97. *J. Afr. Hist.* **13**, 473–488.

Van Oordt, H. F. (1897). *Unpublished Report*. Transvaal Archives SS2031 R8009/89.

Van Oordt, H. F. (1898a). *Unpublished Report*. Transvaal Archives SS7606 R16403/98.

Van Oordt, H. F. (1898b). *Unpublished Report*. Transvaal Archives SS2031 R8009/89.

van Saceghem, R. (1923). La peau, voie de pénétration pour le virus de la peste bovine. *Bull. Agricult. Congo Belge* **14**, 56–57.

Van Zyl, D. J. (1897). *Petition on the Subject of the Shooting of Cattle by Rinderpest Quarantine Guards*. Cape of Good Hope, Cape Town.

Varian, H. F. (1953). *Some African Milestones*. George Ronald, Oxford.

Varnell, G. (1866). In Bernard, M. (ed.) 1866a. *Third Report of the Commissioners Appointed to Inquire into the Origin and Nature & c. of the Cattle Plague*. HMSO, London.

Vaughan Kirby, F. (1896). *In Haunts of Wild Game. A Hunter-Naturalist's Wanderings from Kahlamba to Libombo*. William Blackwood & Sons, Edinburgh.

Vegetius, R. 1528. *Artis Veterinariae sive Mulomedicinae*. J. F. Emmeus, Basle.

Verdernikoff. (1893). Les maladies du chameau. *Arch. Russes médec. Vét.* **1893**, 143.

Verney, F. A. (1898). The Rinderpest in South Africa. *J. Comp. Pathol. Ther.* **11**, 95–103.

Vicq d'Azyr, F. (1775). *Instruction sur la manière de désinfecter les cuirs des bestiaux morts de l'Epizootie et de les rendre propres à être travaillés dans les Tanneries sans y porter la Contagion*. Peterink-Cramé, Lille.

Vicq d'Azyr, F. (1776). *Exposé des Moyens Curatifs & Preservatifs qui peuvent être employés contre les maladies pestilentielles des bêtes à cornes*. Merigot l'aine, Paris.

Vignéras, S. (1897). *Une Mission Française en Abyssinie*. Armand Colin et Cie, Paris.

Villalba, D. J. de. (1803). *Epidemiologia Española ó Historia Cronólogica de las Pestes, Contagios, Epidemias y Epizootias que han acaecido en España desde la Venida de los Cartagineses Hasta el ano 1801, etc.* D. Fermin Villalpando, Madrid.

Villelume, E. de. (1909). *Au Coeur de l'Afrique*. Gabriel Beauchesne et Cie., Paris.

Vittoz, R. (1954). Considérations pratiques sur le role des animaux sauvages dans la transmission des maladies contagieuse et la prophylaxie de celles-ci dans le Sud-Est Asiatique. *Bull. Off. Int. Epiz.* **42**, 206–212.

Vogel, S. W. and Heyne, H. 1996. Rinderpest in South Africa—100 years ago. *J. S. Afr. Vet. Ass.* **67**, 164–170.

Von Höhnel, L. (1894). *Discovery of Lakes Rudolf and Stefanie. A Narrative of Count Samuel Teleki's Exploring and Hunting Expedition to Eastern Equatorial Africa in 1887 & 1888*, 2 vols. Longmans, Green, & Co., London.

von Oertzen, C. D. (1779). *Oeffentliche Bekanntmachung der nunmehr sattsam erprobten und in Mecklenburg allgemein gewordenen Inoculation der Rindviehseuche.* C. W. Meyn, Hamburg.

von Ostertag, R. (1921). La récente importation de la peste Bovine en Belgique. *Zeitschrift Fleish Milk Hygiène. An. Rev. Gén. Méd. Vét.* March 15, 135.

Waldegrave, S. (1866). *The Cattle Plague: A Warning Voice to Britain from the King of Nations*. London.

Walker, G. K. (1912). Treatment of rinderpest and haemorraghic septicaemia with potassium permanganate. *J. Comp. Pathol. Ther.* **25**, 185.

Wall, J. (1751). Thoughts upon the Distemper amongst the Horned Cattle, with a Method of Cure Proposed. *Gent. Mag.* **21**, 71–72.

Wallace, R. (1888). *India in 1887*. A. & C. Black, London.

Wallace, R. (1896). *Farming Industries of Cape Colony*. P. S. King & Son, London.

Waller, R. (1976). The Maasai and the British 1895–1905 the origins of an alliance. *J. Afr. Hist.* **17**, 529–553.

Waller, R. (1988). Emutai: crisis and response in Maasailand 1883–1902. In Johnson, D. H. and Anderson, D. M. (eds.) *The Ecology of Survival Case Studies from Northeast African History*. Lester Crook Academic Publishing, London, pp. 73–112.

Walley, T. (1879). *The Four Bovine Scourges: Pleuro-pneumonia, Foot-and-Mouth Disease, Cattle Plague, Tubercle (Scrofula). With an Appendix on the Inspection of Live Animals and Meat.* Maclachlan & Stewart, Edinburgh.

Walsh, J. (1987). War on cattle divides the troops. *Science* **237**, 1289–1291.

Walz, G. H. (1803). *Untersuchungen über die Natur und Behandlungs Weise der Rinder = Pest, mit einer kurzen Geschichte des Verlaufs derselben in Würtemberg in den Jahren 1795–1801*. J. H. Steinkopf, Stuttgart.

Wamwayi, H. M., Fleming, M., and Barrett, T. (1995). Characterisation of African isolates of rinderpest virus. *Vet. Microbiol.* **44**, 151–163.

Wamwayi, H. M., Kariuki, D. P., Wafula, J. S., Rossiter, P. B., Mbuthia, P. G., and Macharia, S. (1992). Observations on rinderpest in Kenya, 1986–1989. *Rev. Sci. Tech. l'O. I. E.* **11**, 769–784.

Warburton, R. E. E. (1866). *Documents and Letters Relating to the Cattle Plague in the Years 1747–9 from the Papers at Arley Hall, Cheshire*. Charles Sims & Co., Manchester.

Waterfield, G. (1969). *Letters from Egypt (1862–1869) by Lady Duff Gordon*. Routledge & Kegan Paul, London.

Watkins-Pitchford, H. (1897). Report on Rinderpest Investigation in the Transvaal. Reprinted in: *The Veterinarian* **70**, 530–535.

"W. D." (1865). The cattle disease, 1765, 1865. *Notes And Queries* **8** (3rd Ser.), 223–224.

Webster, J. B. (1979). *Chronology, Migration and Drought in Interlacustrine Africa*. Longman & Dalhousie University Press, London.

Webster, N. (1799). *A Brief History of Epidemic and Pestilential Diseases; with the Principal Phenomena of the Physical World, which Precede and Accompany them, and Observations Deduced from the Facts Stated*, 2 vols. Hudson & Goodwin, Hartford.

Weiß, Dr (1783). In *Ueber das Anstecken der Viehseuche, die wahre und eigenthümliche Ursache derselben, und die eigentlichen Vorbauubngsmittel darwider. Zwei von der Berlinischen Gesellschaft Nautrforschender Freunde gekrönte Preisschriften. Von Herrn Petrus Camper, der Arzeneigelahrtheit Doctor und Professor, und der mehrsten gelehrten Akademien und Gesellschaften Mitgliede, und Herrn Doctor Weiß, ausübenden Arzt zu Leer in Ostfriesland, und Mitglied der Gesellsch. Naturf. Freunde in Berlin. Mit einigen Zusätzen*. Anton Ferdinand Röse, Greiswald.

Wellby, M. S. (1901). *'Twixt Sirdar and Menelik. An Account of a Year's Expedition from Zeila to Cairo Through Unknown Abyssinia*. Harper & Brothers Publishers, London.

Wens, H. M. (1987). Beitrag zu der Inoculation der Rindeprest nach den Answeisungen von J.A. Kersting in den Jahren 1779 bis 1781. *Dtsch. Tierärtztl. Wschr.* **94**, 559–562.

White, J. (1870). In Hallen, J. H. B., Charles, J. G., Jan, M. H., Kerr, H. C., and McLeod, K. (eds.) 1871. *Report of the Commissioners Appointed to Inquire into the Origin, Nature, Etc., of Indian Cattle Plagues. With Appendices.* Government Printer, Calcutta.

White, J. B. (1869). In Hallen, J. H. B., Charles, J. G., Jan, M. H., Kerr, H. C., and McLeod, K. (eds.) 1871. *Report of the Commissioners Appointed to Inquire into the Origin, Nature, Etc., of Indian Cattle Plagues. With Appendices*. Government Printer, Calcutta.

White, J. B. (1870). In Hallen, J. H. B., Charles, J. G., Jan, M. H., Kerr, H. C., and McLeod, K. (eds.) *Report of the Commissioners Appointed to Inquire into the Origin, Nature, Etc., of Indian Cattle Plagues. With Appendices*. Government Printer, Calcutta.

Whitelock, D. (1955). *English Historical Documents c.500–1042*. Eyre & Spottiswoode, London.

Whitworth, R. (1750). *The Case of Killing Infected Beasts Briefly Considered*. Manchester.

Wilde, J. K. H. (1953). The game animal factor in the control of rinderpest in tropical Africa. *XVth International Veterinary Congress, Stockholm* **1**, 283–287.

Wilkes (1747). *A letter to the Gentlemen, Farmers, and Graziers in the County of Stafford, concerning the Mortality or Plague amongst the horned Cattle, &c.*

Wilkinson, L. (1984). Rinderpest and mainstream infectious disease concepts in the eighteenth century. *Med. Hist.* **28**, 129–150.

Wilkinson, L. (1992). *Animals and Disease. An Introduction to the History of Comparative Medicine*. Cambridge University Press, Cambridge.

Willems, L. (1900). *Cinquante années d'inoculation préventive de la péripneumonie contagieuse des bovidés (1850–1900)*. Hayez, Brussels.

William, F. (1776). *Von Gottes Gnaden Wir Friderich Wilhelm Bischof zu Hildesheim & c*. Hildesheim.

Williams, A. (1866). *Report on the Origin, Propagation, Nature, and Treatment of the Cattle Plague, from Information Received at the Veterinary Department of the Privy Council Office, from June, 1865, up to March 20, 1866; with Returns and Maps Showing the Progress of the Disease, Corrected to December 30, 1865, from Information Received up to January 31, 1866*. HMSO, London.

Williams, A. (1872). *Report of the Veterinary Department for the Year 1871 with an Appendix*. HMSO, London.

Williams, A. (1875). *Annual Report of the Veterinary Department of the Privy Council Office for the Year 1874 with an Appendix*. HMSO, London.

Williams, G. (1904). *Report of the Proceedings of the Conference on Disease Amongst Cattle and other Animals in South Africa, Held on the December 3, 4 and 5, 1903*. Argus Printing and Publishing Co. Ltd., Bloemfontein.

Willoughby, W. C. (1896). *Willoughby to London Missionary Society 21.8.1896*.

Willoughby, W. C. (1897). *Willoughby to Schofield 30.11.1896. Union Street Magazine*. February, Brighton.

Wilson, H. A. (1913). *A British Borderland. Service and Sport in Equatoria*. John Murray, London.

Wingate, F. R. (1892). *Ten Years' Captivity in the Mahdi's Camp 1882–1892. From the Original Manuscripts of Father Joeseph Ohrwalder Late Priest of the Austrian Mission Station at Delen, in Kordofan*. Sampson Low, Marston & Co., London.

Wise, T. (1836). In Hallen, J. H. B., Charles, J. G., Jan, M. H., Kerr, H. C., and McLeod, K. (eds.) 1871. *Report of the Commissioners Appointed to Inquire into the Origin, Nature, Etc., of Indian Cattle Plagues. With Appendices*. Government Printer, Calcutta.

Wollaston, A. F. R. (1908). *From Ruwenzori to the Congo. A Naturalist's Journey Across Africa*. John Murray, London.

Wolstein, J. G. (1782). *Ammerkungen über die Viehseuchen in Oesterrich nebst einer Abhandlung gegen das Umbringen der Thiere in Seuchen*. J. E. von Kurzbeck, Vienna.

Wolstein, J. G. (1799). *Belehrung der Landbleute über die Zeichen und Ursachen der Hornvieh = Seuchen*. Hamburg.

Wolverton, Lord (1894). *Five Months' Sport in Somali Land*. Chapman & Hall, Ltd., London.

Woodford, M. H. (1984). *Rinderpest in Wildlife in Sub-Sahelian Africa. Technical Analysis Report*. FAO, Rome.

Worseck, M. (1977). Die Bekämpfung der Rinderpest vor 200 Jahren in Mecklenberg. *Mh. Vet. Med.* **32**, 182–184.

Worboys, M. (1991). Germ theories of disease and British veterinary medicine, 1860–1890. *Med. Hist.* **35**, 308–327.

Wright, T. (1839). *The Political Songs of England*. Camden Soc., London.

Wurtz, R. (1898). Hygiène publique et privée en Abyssinie. *La Semaine Médicale* **8**, 489–494.

Wylde, A. B. (1901). *Modern Abyssinia* Methuen & Co., London.

Yamanouchi, K., Barrett, T., and Kai, C. (1998). New approaches to the development of virus vaccines for veterinary use. *Rev. Sci. Tech. Off. Int. Epiz.* **17**, 641–653.

Yilma, T., Hsu, D., Jones, L., Owens, S., Grubman, M., Medbus, C., Yamanaka, M., and Dale, B. (1988). Protection of cattle against rinderpest with vaccinia virus recombinants expressing the HA or F gene. *Science* **242**, 1058–1061.

Youatt, W. (1834). *Cattle; Their Breeds, Management, and Diseases*. Simpkin, Marshall, & Co., London.

Zimmerman (1814). *Ein Paar Worte über Rindviehseuche oder Rinderpest*. Stillerschen Buchhandlung, Rostock & Schwerin.

Zonchello, A. (1917). Il gulhai nel Sahel. *Clinica Vet.* **30**, 113–132.

INDEX